Status and Trends of the Nation's Biological Resources

Volume 2

Status and Trends
of the
Nation's Biological Resources

Volume 2

Project Director
Michael J. Mac
U.S. Geological Survey
Biological Resources Division

Science Editor
Paul A. Opler
U.S. Geological Survey
Biological Resources Division

Technical Editor
Catherine E. Puckett Haecker
U.S. Geological Survey
Biological Resources Division

Graphics Editor
Peter D. Doran
Bureau of Land Management
National Applied Resource Sciences Center

U.S. Geological Survey's National Wetlands Research Center provided editorial support and project administration.
The Bureau of Land Management's National Applied Resource Sciences Center provided graphic design and production.

U.S. Department of the Interior
U.S. Geological Survey
1998

Library of Congress Cataloging-in-Publication Data

Status and trends of the nation's biological resources / project director, Michael J. Mac; science editor, Paul A. Opler; technical editor, Catherine E. Puckett Haecker; graphics editor, Peter D. Doran. 2v. (xi, 964 p.): ill.; 28 cm. Includes bibliographical references and index.
ISBN 016053285X

1. Biological diversity—United States. 2. Biological diversity conservation—United States.
3. Species diversity—United States. 4. Indicators (Biology)—United States.
I. Mac, Michael J. II. Opler, Paul A. III. Haecker, Catherine E. Puckett. IV. Doran, Peter D.
V. Geological Survey (U.S.)
QH104.S74 1998
333.95'11'0973—dc21

This publication may be cited as follows:

Mac, M. J., P. A. Opler, C. E. Puckett Haecker, and P. D. Doran. 1998. Status and trends of the nation's biological resources. 2 vols. U.S. Department of the Interior, U.S. Geological Survey, Reston, Va.

The volumes may be cited individually as follows:

Mac, M. J., P. A. Opler, C. E. Puckett Haecker, and P. D. Doran. 1998. Status and trends of the nation's biological resources. Vol. 1. U.S. Department of the Interior, U.S. Geological Survey, Reston, Va. 1-436 pp.

Mac, M. J., P. A. Opler, C. E. Puckett Haecker, and P. D. Doran. 1998. Status and trends of the nation's biological resources. Vol. 2. U.S. Department of the Interior, U.S. Geological Survey, Reston, Va. 437-964 pp.

Volume 1 Contents

Part 2—Regional Trends of Biological Resources

Volume 2 Contents

Grasslands

"The prairie, in all its expressions, is a massive, subtle place, with a long history of contradiction and misunderstanding. But it is worth the effort at comprehension. It is, after all, at the center of our national identity."

William Least Heat Moon (1991)

Grasslands rank among the most biologically productive of all communities (Williams and Diebel 1996). Their high productivity stems from high retention of nutrients, efficient biological recycling, and a structure that provides for a vast array of animal and plant life (Estes et al. 1982). Grasses have contributed the hereditary material for the principal human food crops—rice, wheat, corn, and other grains. Worldwide production of such grain crops exceeds all other food crops combined. Grasslands also contribute immense value to watersheds and provide forage and habitat for large numbers of domestic and wild animals. Nevertheless, current levels of erosion in North America exceed the prairie soil's capacity to tolerate sediment and nutrient loss, thus threatening a resource essential to sustain future generations (Sampson 1981). Added to this threat is the potential for overgrazing by livestock and for other human activities to reduce the social and aesthetic values of grasslands and to restrict the commodities that grasslands can produce (National Research Council 1994), and the likelihood that severe degradation may be irreversible.

In North America, the prairie communities (about 1.5 million square kilometers) contain the majority of the continent's native grasslands (Fig. 1). Environmental features that describe the native North American grasslands embody similarity in vegetation, an abiotic environment to which the vegetation and structure respond, and the nature of the animal communities.

North American grasslands are similar in the general uniformity of their vegetation, dominance of grasses and grasslike plants, lack of shrubs, and absence of trees (Weaver 1968). The Great Plains grassland evolved in the rain shadow of the Rocky Mountains, where seasonal precipitation occurs mostly in spring and summer. From the Rocky Mountains east to the Mississippi River, the amount of precipitation increases and the frequency of droughts decreases (Simms 1988). Along a north-south gradient from central Texas to south-central Canada, the growing season becomes shorter, the average temperature decreases, and a greater proportion of annual precipitation occurs as snow. These broad-scale environmental gradients significantly influenced the evolutionary composition and distribution of prairie communities (Steinauer and Collins 1996; Weaver et al. 1996).

Many small and large grazing animals evolved on the North American prairie (Van Valkenburgh and Janis 1993), each with life-history and behavioral traits well adapted to the open character of prairie. For example, prairie dogs markedly affect the nutrient cycling, soil formation, and composition of grassland animal and plant communities (Miller et al. 1994). Moreover, it is important to realize that grasslands and their associated wildlife reflect events of the distant past (Knopf and Samson 1997). The incursion of animals—bison, elk, and others—into North America across land-bridges that once connected the Asian and North American continents is but one example of the role of past events. Thus, understanding how events of the distant past influenced both the isolation and interchange of plants and animals that interact with the more recent

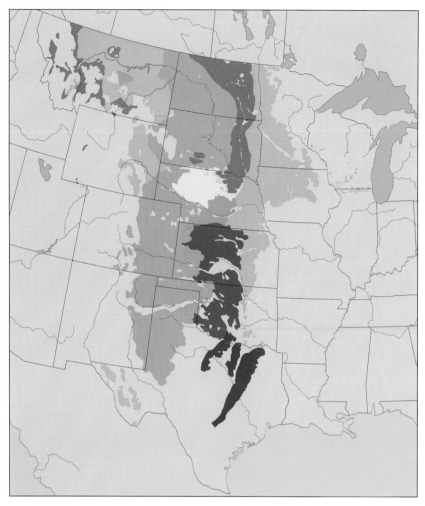

Küchler's potential vegetation

■ Bluestem–grama prairie

▨ Bluestem prairie

▨ Blue grama–buffalo grass

□ Nebraska Sandhills prairie

▨ Wheatgrass–blue stem–needlegrass

▨ Wheatgrass–blue grama–buffalo grass

▨ Blue grama–tobosa prairie

▨ Blue grama–needlegrass–wheatgrass

■ Blackland prairie

▨ Foothills prairie

▨ Wheatgrass–needlegrass

Fig. 1. The central grasslands, delineated by Küchler's potential vegetation types (Küchler 1964).

landscape is an essential step in interpreting any particular ecological community, including grasslands.

Understanding the biological resources of the Great Plains is difficult. The exact size of remaining grasslands in North America is unknown and difficult to estimate, and the community has undergone significant change since it was first described by early explorers and surveyors (Samson and Knopf 1994). Agriculture, urbanization, and mineral exploration have had both local and regional effects on biological resources. Invasions of nonindigenous plant species after fire suppression in the eastern, central, and southern prairies, as well as water developments in the western plains, have drastically altered grassland landscapes. Establishment of woodlots, shelterbelts, and tree-lined river and stream corridors within the prairie has contributed to a significant and ongoing loss of genetic diversity in North American grasslands (Knopf 1986).

This chapter highlights the status and trends of the main bodies of North American grasslands: the tall-grass prairie, the mixed-grass prairie, and the short-grass prairie. We feature

the animals and plants dependent on native grassland and attempt to provide insight into the relationship between remaining native grassland and biological resources by reviewing available, current information and by describing threats.

Prairie Past and Present

In the past, grassland dominated central North America (Fig. 2) and, during the warm, dry interglacial times, reached—as the prairie peninsula—into parts of Wisconsin, Illinois, Indiana, and eastern Ohio (Bazzaz and Parrish 1982). The main bodies of native grassland, now vastly altered, are the tall-grass prairie extending from Canada (Manitoba) and Minnesota south to Texas; the mixed-grass prairie from Canada, Montana, and North Dakota south to Texas; and the short-grass prairie extending from eastern Wyoming south to western Texas and eastern New Mexico. In the north, the natural grasslands are bordered on the west by coniferous forests of the Rocky Mountains and on the east by oak savannah (Anderson 1983) and aspen parkland in Manitoba and northwest Minnesota, with the transition from prairie to forest often abrupt (Great Plains Flora Association 1986). Across the Great Plains, coniferous and deciduous forest types meet only in the valley of the Niobrara River in north-central Nebraska, and isolated stands of both forest types occur in the Black Hills of South Dakota.

Tall-grass Prairie

Tall-grass prairie (Fig. 3) is the wettest of the grassland provinces and is predominantly composed of sod-forming bunch grasses. Like other grasslands, the tall-grass prairie has species originally from different geographical sources (Simms 1988). Grassland groupings of the tall-grass prairie are the bluestem prairie from southern Manitoba through eastern North Dakota and western Minnesota south to eastern Oklahoma, and the wheatgrass, bluestem, and needlegrass area from south-central Canada through east-central North Dakota and South Dakota to southern Nebraska.

Three additional areas are associated with tall-grass prairie: the Crosstimbers, a band of grassland and oak savanna at the southern edge of the bluestem prairie in Kansas to the Trinity River in Texas (Küchler 1964), the Blackland Prairie south of the Crosstimbers (Gould 1962), and the rice prairies. The rice prairies are former coastal prairies that have been converted to rice production (Hobaugh et al. 1989). The original vegetation in rice prairies was mainly tall grass

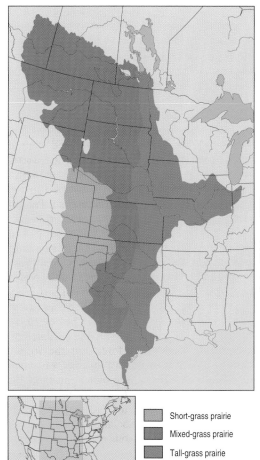

Courtesy J. T. Lokemoen, USGS

	Short-grass prairie
	Mixed-grass prairie
	Tall-grass prairie

Fig. 2. Extent of historical (pre-European) tall-grass, mixed-grass, and short-grass prairies on the North American Great Plains.

and extended across 9,000 square kilometers, largely along the Texas coast and inland as much as 125 kilometers, and into Louisiana. Little coastal prairie remains; Attwater's Prairie Chicken National Wildlife Refuge in Texas is the single major remnant.

Since 1830 declines in the area of tall-grass prairie within specific states and provinces are estimated to be 82.6 to 99.9% (Table 1) and exceed those reported for any other major ecological community in North America (Samson and Knopf 1994). Iowa, for example, has barely 12,140 hectares remaining of its original 12 million hectares of tall-grass prairie. Less than 1% of the presettlement tall-grass prairie remains in Manitoba, Illinois, Indiana, and North Dakota. Minnesota and Missouri, states active in prairie conservation, work with less than 9% of the presettlement tall-grass prairie. Tall-grass prairie remains important to ranching in the Osage and Flint Hills of Kansas and in tracts in South Dakota, Oklahoma, and Texas.

Fig. 3. Tall-grass prairie.

Mixed-grass Prairie

One can envision the short-grass and tall-grass prairies intergrading just east of an irregular line that runs from northern Texas through Oklahoma, Kansas, and Nebraska, northwestward into west-central North Dakota and South Dakota (Figs. 1 and 2). The perimeter is not well defined because of the array of

Table 1. Summary of the estimated past area, current area, and percent decline of tall-grass, mixed-grass, and short-grass prairies.

Prairie type Location	Past area (hectares)[a]	Current area (hectares)[a]	Decline (percent)
Tall-grass			
Manitoba	600,000	300	99.9
Illinois	8,500,000	930	99.9
Indiana	2,800,000	404	99.9
Iowa	12,000,000	12,140	99.9
Kansas	6,900,000	1,200,000	82.6
Minnesota	7,300,000	30,000–60,000	99.2–99.6
Missouri	6,000,000	32,000	99.5
Nebraska	6,100,000	123,000	98.0
North Dakota	130,000	120	99.9
Oklahoma	5,200,000	N/A[b]	N/A[b]
South Dakota	2,600,000	20,000	99.2
Texas	7,200,000	720,000	90.0
Wisconsin	2,400,000	1,000	99.9
Mixed-grass			
Alberta	8,700,000	3,400,000	60.9
Manitoba	600,000	300	99.9
Saskatchewan	13,400,000	2,500,000	81.3
Nebraska	7,700,000	1,900,000	75.3
North Dakota	14,200,000	4,500,000	68.3
Oklahoma	2,500,000	N/A[b]	N/A[b]
South Dakota	1,600,000	480,000	70.0
Texas	14,100,000	9,800,000	30.5
Short-grass			
Saskatchewan	5,900,000	840,000	85.8
Oklahoma	1,300,000	N/A[b]	N/A[b]
New Mexico	N/A[b]	1,255,200	N/A[b]
South Dakota	179,000	116,350	35.0
Texas	7,800,000	1,600,000	79.5
Wyoming	3,000,000	2,400,000	20.0

[a] Estimates of past and current area based on information from The Nature Conservancy's Natural Heritage Data Center Network; Provinces of Alberta, Manitoba, and Saskatchewan; universities; and state conservation organizations.
[b] N/A means information is not available.

short-stature, intermediate, and tall-grass species that make up an ecotone between the short-grass and tall-grass prairies (Bragg and Steuter 1996). In general, the mixed-grass prairie is characterized by the warm-season grasses of the short-grass prairie to the west and the cool- and warm-season grasses, which grow much taller, to the east (Fig. 4). Because of this ecotonal mixing, the number of plant species found in mixed-grass prairies exceeds that in other prairie types. Estimated declines in area of native mixed-grass prairie, although less than those of the tall-grass, range from 30.5% in Texas to over 99.9% in Manitoba (Table 1).

Courtesy F. L. Knopf, USGS

Fig. 4. Mixed-grass prairie in Nebraska Sandhills.

Short-grass Prairie

The short-grass prairie extends east from the Rocky Mountains and south from Montana through the Nebraska panhandle and southeastern Wyoming into the high plains of Oklahoma, New Mexico, and Texas (Figs. 1 and 2.). The short-grass prairie landscape (Fig. 5) was one of relatively treeless stream bottoms and uplands dominated by blue grama and buffalo grass, two warm-season grasses that flourish under intensive grazing (Weaver et al. 1996). Buffalo grass reproduces both sexually and by tillering sprouts from the base of grass clumps. Unlike the more eastern species, short-grass prairie species remain digestible and retain their protein content when dormant.

Declines in short-grass prairie have generally been much less than those of tall-grass and mixed-grass prairies (Table 1). However, perhaps in no other system than short-grass

prairie are historical and evolutionary impacts of grazing so apparent (Knopf 1996). Clearly, birds endemic to the short-grass prairie express life-history characteristics and habitat use in response to grazing (Fig. 6). The mountain plover responds to highly disturbed sites, the chestnut-collared longspur to moderately grazed areas, and the Baird's sparrow to sites with taller grasses. In the mid-1800's the numbers of individuals of native mammal species—bison, prairie dogs, pronghorn, elk, grizzly bears, and gray wolves—rivaled or exceeded those now in the African Serengeti (Howe 1994). Major antigrazing structures evolved in plants: thorns and spikes; thick or hard tissues difficult to bite, chew, or digest; and secondary compounds difficult to digest. These structures have arisen through the long coevolutionary association between plants and animals with grazing on grasslands.

At present, extensive areas of short-grass prairie are dominated by invasive perennial and annual species, whose presence is attributed to overgrazing by domestic livestock and dryland farming (Weaver et al. 1996). To the south, specifically the Texas high plains, much of the short-grass prairie is now farmland or shrubland invaded by prickly pear cacti and oaks. Only the short-grass prairie and, to a lesser extent, mixed-grass prairie remain in public ownership. These areas are largely on the national grasslands managed by the U.S. Forest Service.

Prairie Wetlands

The northern prairie contains numerous wetlands (Fig. 7), including the glaciated prairie pothole region (Fig. 8), the Nebraska Sandhills, and the Rainwater Basin (Kresl et al. 1996; Mack 1996). The 770,000-square-kilometer prairie pothole region extends from Alberta, Saskatchewan, and Manitoba across northeastern Montana, then southeast through North Dakota and eastern South Dakota into western Minnesota and northwestern and central Iowa. This landscape is pockmarked with numerous small, shallow depressions that capture snowmelt and rainwater or are within reach of subsurface waters. Estimated losses of prairie pothole wetland range from 35% in South Dakota to 99% in Iowa with loss rates upwards of 1,300 hectares per year (Tiner 1984).

The Nebraska Sandhills is the largest dune area (5,260 square kilometers) in North America (Mack 1996). About 526,100 hectares of wetlands are scattered throughout the area. The rapid movement of groundwater creates a continuum among lakes, wetlands, and streams, thus an alteration in one area may easily affect vegetation and wetlands over a larger landscape. Wetlands in the sandhills range from shallow, extremely alkaline basins to deeper, freshwater lakes to spring-fed streams. They are economically valuable, particularly as a source of irrigation water.

Another wetland area dependent on capturing water runoff is the Rainwater Basin of south-central Nebraska (6,720 square kilometers). Throughout the basin, rainwater is caught by scattered wind-excavated depressions underlain by an impermeable clay pan. Since the late 1800's efforts have been under way to drain Rainwater Basin; today, fewer than 400 depressions remain of an estimated 4,000, and they account for 22% of the former area. Other large, alkaline wetlands in Kansas include the Jamestown marsh, Talmo marsh, Lincoln salt marsh, Cheyenne Bottoms, Quivira National Wildlife Refuge (including Big and Little Salt marshes), and Slate Creek salt marsh. A similar alkaline lake is the Great Salt Plains Reservoir in Alfalfa County, Oklahoma.

Effects of collective water loss on the northern prairie range from significant declines in waterfowl breeding populations (Bethke and Nudds 1995) to elimination of the flood storage value of natural wetlands. About half of the continental waterfowl production comes from the prairie pothole region. Nebraska's Rainwater Basin is the major spring staging area for the buff-breasted sandpiper and the greater white-fronted goose, and it provides migratory habitat for endangered species such as the whooping crane and interior least tern. In addition, about 45% of North America's shorebird population east of the Rocky Mountains may stop at Cheyenne Bottoms during spring migration, including 90% of the North American populations of the white-rumped sandpiper, stilt sandpiper, Baird's sandpiper, long-billed dowitcher, and Wilson's phalarope, and over half of all pectoral sandpipers, marbled godwits, and Hudsonian godwits.

Interior wetland from the edge of the prairie pothole region across the central Great Plains is associated with major river systems. The area has few natural lakes, the largest of which is Inman Lake (78 hectares) in central Kansas (Carlander et al. 1986). Climate and past events account for the interior wetland's habitat characteristics (Cross and Moss 1987). On the central Great Plains, such areas have been transformed from wide, unvegetated channels (Fig. 9) to the current extensive cottonwood–willow woodlands lining narrow channels (Johnson 1994). This transformation is a result of human alteration of natural flow regimes, cessation of prairie fires, and elimination of the bison (Currier 1982). Overall, the presettlement near-river mosaic of meadow, marsh, and drier upland grassland is now an open- to

Courtesy F. L. Knopf, USGS

Fig. 5. Short-grass prairie in Laramie Plain, Wyoming.

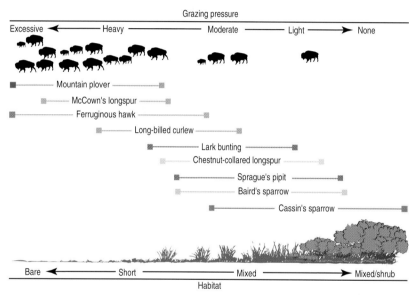

Fig. 6. Importance of coevolution between grazing and native prairie bird distributions and abundances (after Knopf 1996).

Courtesy USGS

Fig. 7. Prairie wetlands, showing the zonation of wetland plant communities.

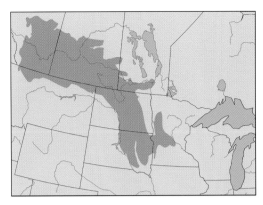

a.

Fig. 8. The prairie pothole region (in green) of the northern Great Plains.

b.

Fig. 9. a) Upper California Crossing, Oregon Trail, Fort Sedgwick (Ovid), Colorado, around 1900; b) Same locale, town of Ovid, in 1990.

Courtesy F. L. Knopf, USGS

Courtesy F. L. Knopf, USGS

closed-canopy woodland. Major changes in the habitat mosaic are expected in the future because secondary succession of woody vegetation will lead to a climax forest of regenerating stands of the nonindigenous Russian-olive along the Platte River and elsewhere across the entire central and western United States (Olson and Knopf 1986).

To the south are the Playa lakes. Upwards of 26,000 of these shallow, small wetlands dot the short-grass prairie in Texas and adjoining states, reaching northward to Colorado (Bolen et al. 1989). Because of evaporation, such lakes often are rimmed in a salty crust. Playa lakes have abundant populations of small, aquatic invertebrates (Carlander et al. 1986). At times, these lakes serve as vital waterfowl nesting and resting areas (U.S. Fish and Wildlife Service 1986) and as winter habitat for most North American sandhill cranes.

Aquifers and Waterways

The High Plains aquifer (formerly called the Ogallala aquifer) consists of one or more geological units connected belowground under the central Great Plains. The aquifer is essential to agricultural, urban, and environmental resources, containing about 20% of the irrigated farmland in the High Plains and about 30% of the water used for irrigation (Huntzinger 1996). Precipitation is the principal source of natural groundwater recharge, but recharge can also result from seepage loss from streams and lakes. Natural discharge occurs as evaporation from plants and soils where the water table is near the surface or as seepage to springs. Over the long term, natural recharge should compensate discharge.

The development of the High Plains aquifer for irrigation (1940–1980) is evident in an average area-weighted, water-level decline of 3 meters (0.07 meters per year; Dugan et al. 1994). Declines vary with locale, exceeding 30 meters in some parts of the central and southern High Plains; 6 meters in southwestern Kansas, east-central New Mexico, and the Oklahoma and Texas panhandles; and 3 to 6 meters in northeastern Colorado, northwestern Kansas, and southwestern Nebraska. Since 1980, water levels in such areas have continued to decline but at a slower rate, the result of greater than normal precipitation (10 to 15 centimeters annually), water conservation practices (particularly minimum tillage), and reduction in irrigated land area (about 540,000 hectares, 1979 to 1991). The extreme southern plains of Texas and eastern and east-central Nebraska have experienced water-level rises of 1 to 3 meters (Dugan et al. 1994).

Agricultural use of nitrogen fertilizers is the largest source of nitrates in near-surface aquifers in the midcontinent (Koplin et al. 1994). Over 100,000 metric tons of pesticides (herbicides, insecticides, and fungicides) were applied in the midcontinent in 1991, often to control nonindigenous plants and animals. In spring and summer 1991, concentrations of several herbicides exceeded U.S. Environmental Protection Agency standards in about half of the streams sampled in the upper Missouri River basin (Huntzinger 1996). Effects of these pollutants on the quality of human life and on the integrity of the ecological community are largely unknown. The U.S. Environmental Protection Agency has initiated an effort to develop stressor information to help recognize areas where urban development, agricultural nonpoint pollution (pesticides, toxic chemicals, nutrient pollution), and agricultural development may exacerbate ecological decline.

The rivers of the Great Plains flow from west to east; extreme turbidity is the key characteristic of the larger rivers. Whereas most water in these rivers originates from western mountains, sediments originate from thunderstorm runoff on the Great Plains. In small channels, fine particles held in suspension produced quicksands, which inhibited crossings by early travelers and caused extreme turbidity during low flows. In summer, open-river water temperatures often exceed 30°C, the salinity level is high because of salt- and gypsum-laden groundwaters, rates of evaporation are high, and the flow velocity is moderate.

The larger Great Plains rivers have been subjected to dewatering for irrigation, other consumptive uses, and reservoir construction (Cross and Moss 1987). In virtually all these river systems, such dewatering has altered the timing and extent of flows, downstream temperatures, levels of dissolved nutrients, sediment transport and deposition, and the structure of plant and animal communities. Few major interior rivers still exhibit the conditions evident before agricultural development and water management had occurred.

On the Missouri River, once a free-flowing river of more than 3,700 kilometers (from Three Forks, Montana, to St. Louis, Missouri), reservoir construction in Montana, North Dakota, and South Dakota has virtually stopped sediment transport below major reservoirs, as it also has on the Platte and Kansas rivers (Huntzinger 1996). Sediment deposition is part of reservoir design but remains a maintenance concern. Because of the value of surface waters, there is increasing interstate interest in surface waters such as those of the Missouri River. Such surface waters are important to wildlife, fish, and recreation, as well as to navigation and water-supply interests that rely on reservoir resources.

Ecologically, the effect of water management reaches far to the south because sediment deposits are needed to sustain the Mississippi Delta. Unchannelized reaches of the Missouri River are near Bismarck, North Dakota and in southeastern South Dakota and northeastern Nebraska. Such reaches host a number of pearlymussel, fish, and bird species that are federally listed or are candidates for listing under the Endangered Species Act of 1973.

The basic features of most Great Plains streams, such as flow and substrate, are unknown (Matthews 1988). In general, streams in the south are characterized by irregular flow, small particle size in substrates, and a distinct wet–dry cycle. Drought may be a more extreme disturbance than either the wet–dry cycle or floods. Streams in the northern prairies are more consistent in flow and tend to have cobble substrates, and winter precipitation (snow) is released with spring thaw. Great Plains streams fall into three categories: the shallow stream with shifting sand beds; clear brooks, ponds, and marshes supported by seeps and springs; and residual pools of intermittent streams (Cross and Moss 1987). All three stream types are affected by dewatering of channels and stabilized flows.

Soils

Grassland soils are fundamentally different from those found under a forest canopy (Simms 1988). Many factors—parent material, climate, time, and human intervention—influence grassland soil type and condition (Peterson and Cole 1996). For example, soil organic carbon, a significant influence on soil productivity, is greatest in the Northeast (cooler and wet) and lower in the Southwest (warmer and dry). Where evaporation is low, water is more likely to remain in the soil, increasing the rate of mineral weathering and allowing large amounts of nitrogen, phosphorus, and sulfur to accumulate in conjunction with carbon.

Managed agriculture began on the easternmost grasslands in the 1850's (Peterson and Cole 1996). Surface cover is reduced by agriculture, and soil structure is destabilized by reducing aggregate size—the result of mixing and grinding action by farm implements. Organic carbon loss is accelerated by agriculture, and cultivated crops (particularly in dryland areas) return little carbon to the soil. Few agricultural practices of the early settlers captured or retained moisture. The black-dust storms of the 1930's resulted from exposing vast areas of cultivated prairie soil to wind action and drought (Sampson 1981).

Several straightforward and well-documented relationships exist between plant productivity and soil organic material after native sod is cultivated. For example, soil productivity (indexed by corn grain yield) declined 71% and soil nitrogen 49% during a 28-year interval after cultivation began (Williams and Wolman 1986). Retention of organic matter—and thus the level of productivity—in grassland soils is only possible if the correct proportions of carbon, nitrogen, and phosphorus are present (Peterson and Cole 1996). Nitrogen fertilizers are used extensively to restore soil nitrogen levels, and more than 6.4 million metric tons of nitrogen fertilizers were applied to cropland in the Mississippi River basin in 1991 (Goolsby et al. 1993). In addition, removal of phosphorus in harvested plants and loss of organic phosphorus due to cultivation require fertilizer supplements to maintain productivity. Elevated concentrations of phosphorus may affect aquatic plant growth and reduce oxygen content in streams.

Soil formation is a slow, continuous process. About 2.5 centimeters of new topsoil is formed every 100 to 1,000 years, depending on climate, vegetation and other living organisms, topography, and the nature of the soil's parent material (Sampson 1981). Some soils, particularly where moisture is a limiting factor and growing seasons are short, may take 10,000 years to produce 2.5 centimeters of soil. On average, annual loss of topsoil in the United States is nearly three times greater than that being formed. Exact measurements of soil losses are difficult to obtain because soil eroded in one area is eventually deposited at another site (Peterson and Cole 1996). Smaller and lighter nutrient-rich organic soils are the most transportable, creating additional threat to the future productivity of prairie soils.

Prairie Processes

Climate and fire (Fig. 10) are thought to be most important to the spread and maintenance of grasslands (Anderson 1990). The air mass originating in the Gulf of Mexico spreads high humidity and precipitation as it moves north. As it moves from west to east, the Pacific air mass passes over several mountain ranges, giving up most of its moisture and, as it meets the gulf air mass, creating a climate gradient across the Great Plains. The north–south gradient in temperature and moisture, largely influenced by snow cover to the north, is an effect of the polar air mass. Long-term response of vegetation to climate, particularly water availability, is illustrated by regional differences in species composition and height of native grasses: arid western short-grass prairie, central mixed-grass prairie, and eastern tall-grass prairie. As documented in past droughts, grassland distribution is controlled by extremes of climate variability. These forces are evident in the changing eastern edge of the prairie peninsula and, to a great extent, the annual productivity of grasslands.

Grassland plants have growing points protected from fire beneath the soil surface.

Frequent fire is essential to maintaining native species diversity, and it affects other components, including nutrient cycling and productivity (Collins and Wallace 1990). On tall-grass prairie, the relationship between total plant species richness and the number of times a site is burned is important and positive (Collins 1991), at least on a small scale. Small animals create gaps and edges that influence tall-grass plant community structure and composition (Reichman et al. 1993). Recently burned tall-grass prairie has also provided stopover habitat for many long-distance migrant birds such as the lesser golden-plover and the now-endangered and possibly extinct Eskimo curlew. In the past, grazing was localized in these burned areas because of their greater productivity and the nutritive value of their forage (Risser 1990). Thus, the movement and impact of grazing animals on tall-grass prairie grasslands were bound to the spatial distribution of burned patches. For mixed-grass prairie, discussions of community composition and individual species should be set in a similar context of species patterns of both past grazing animals (ranging from bison to ants) and current grazing animals (domestic livestock) in relation to disturbance events (Umbanhowar 1992).

Grazing has direct and indirect effects at landscape and regional scales, which, in turn, interact with other small-scale and large-scale factors to heighten temporal and spatial diversity in grasslands (Gibson and Hulbert 1987; Risser 1990). A recent comparison of grazing over a global range of environments, however, suggests grazing is a factor in the conversion of grasslands to less desirable shrublands (Milchunas and Lauenroth 1993). Moreover, primary production on grasslands, largely the production of plant material, does not necessarily change when plant species composition changes. Current species-based management criteria by land management agencies, therefore, may lead to erroneous conclusions about the ability and future of grasslands to sustain productivity. Adequate assessment of the effects of grazing on grasslands, as with the effects of climate and fire, must be multiscaled and match management inferences and applications (Steinauer and Collins 1996).

Interactions among other factors, aside from climate, grazing, and fire, also influence grasslands (Burke et al. 1991). In the east, nitrogen normally restricts the annual production and composition of grasslands. In the semiarid west, the availability of nitrogen and water is important to composition and production. Long-term vegetative production on short-grass prairie is closely tied to precipitation (Lauenroth and Sala 1992). The most productive years are those when small precipitation events first stimulate

Fig. 10. Fire plays a major role in prairie dynamics.

Courtesy F. L. Knopf, USGS

nutrient availability, followed by large precipitation events that stimulate plant growth. Semiarid areas are thought of as especially variable in environmental conditions, particularly in precipitation, and the short-grass prairie is no exception. Effective grassland management requires understanding the effects of both the spatial and temporal patterns of precipitation on short-grass prairie.

In prairie wetlands, disruption of natural processes such as fire has led to domination by robust, emergent plants, particularly in the prairie pothole region. Cattail, once rare on the Great Plains, has spread across thousands of prairie wetlands, as has purple loosestrife, a species native to Europe which is now threatening waterways across the United States (U.S. Congress, Office of Technology Assessment 1993; Malecki and Blossey 1994). In the past, climate, fire, and grazing controlled the diversity and abundance of vegetation in northern prairie wetlands. As environmental conditions changed, some plant populations have declined and others have increased. Belowground seed reserves favor those species with seeds that germinate under a wide range of conditions, such as cattail, purple loosestrife, and other nonindigenous species.

More is known about the effects of grazing than fire. Nodal rooting, or underground branching, and unpalatability are evident evolutionary responses of wetland plants to grazing. Under certain conditions grazing can increase species diversity and the development of intricate patterns and sharp boundaries among prairie wetland plant communities (Bakker and Ruyter 1981).

Plant Assemblages

The Nature Conservancy, in a preliminary survey, has identified rare plant assemblages across the United States (Grossman et al. 1994). Of the 633 assemblages in the Great Plains, 107 (17%) are considered rare (Chaplin et al. 1996).

The 16 rare Great Plains forest assemblages are largely cottonwood and oak floodplain forests on the eastern and western edges of the plains (Grossman et al. 1994). The 20 rare canyon and mountain plant assemblages tend to be open pine, fir, and oak. The eight rare sparse woodland forests are primarily oak savannas on the eastern plains. The 19 rare shrubland assemblages include many sagebrush, hawthorn, and willow species.

Among 45 rare grassland assemblages on the Great Plains, 18 are found in tall-grass prairie, 13 in mixed-grass prairie, 7 in short-grass prairie, and 7 primarily in wetlands. Big bluestem is dominant in 9 of 18 rare tall-grass prairie communities, little bluestem in 3, and drop-seed species in 2. Similarly, little bluestem is common to 6 of 13 rare mixed-grass prairie communities, and sedges are important in 3. Buffalo grass, in part, distinguishes 5 of 8 rare short-grass prairie communities, and sedges are important to 2. The 7 remaining rare communities are dominated by forbs and embrace wetland plants.

Invertebrates

A wide diversity of terrestrial insects exists on grasslands. For example, in two years of sampling on a 1,400-hectare area of tall-grass prairie in northeastern Oklahoma, 16 orders, 131 families, and more than 3,000 insect species were noted (Risser et al. 1981). More than 1,600 insect species are known from a short-grass prairie in Colorado (Kumar et al. 1976), and this list is incomplete. Inventories are rarely representative; some taxa are present in hot, dry years, others in wet years, and no single sampling method is adequate.

Other terrestrial invertebrates are also abundant. Smolik (1974) found 2 to 6 million soil nematodes per square meter to the depth of 60 centimeters in South Dakota mixed-grass prairie soil. Terrestrial invertebrates are important to the prairie community: they feed on plant tissue, pollen, nectar, and seeds; regulate numbers of other insects and plants; and recycle energy and nutrients (Risser et al. 1981). Underground, earthworms accelerate the decomposition and mineralization of soil organic matter and affect soil structure through burrowing and casting. The soil formation activities of native and nonindigenous earthworms vary considerably; the latter have a negative effect on soil turnover, at least in tall-grass prairie soils (James 1991). Nevertheless, in short-grass prairie soils, 90% of invertebrate energy cycling occurs belowground, less in tall-grass and mixed-grass prairies.

Preliminary lists of Lepidoptera (butterflies and moths) are available for Montana, North Dakota, Wyoming, Colorado, and New Mexico and are in progress in Texas (Powell 1995). Species numbers in a few selected Lepidoptera families vary from 181 in North Dakota to 520 in Texas, and numbers in Nebraska (254) and Oklahoma (228) rank high (Opler 1995). Several prairie butterflies—the Dakota skipper (Fig. 11), regal fritillary, tawny crescent, and

© R. Dana, St. Paul, Minnesota

Fig. 11. The Dakota skipper, a rare prairie butterfly.

Tall-grass Prairie Butterflies and Birds

The destruction and degradation of North American prairie habitats, especially in the eastern tall-grass region, have considerably reduced the abundance of associated animals. The species most affected are those restricted to remaining prairie fragments. Prairie birds and butterflies often have specific habitat requirements, and reduction of the tall-grass prairie has caused serious declines in their populations. Because some species of prairie sparrows and butterflies have similar habitat requirements, trends in their populations are sometimes correlated.

Prairie Sparrows

In eastern and central North America, birds nesting in grasslands and ground-nesting birds have declined more than birds of any other North American behavioral or ecological guild (Knopf 1994). Data from the U.S. Geological Survey Breeding Bird Survey illustrate that, between 1966 and 1993, Henslow's sparrows have declined 91% rangewide, grasshopper sparrows have been reduced by 66%, and dickcissels by 39% (Fig. 1). Henslow's sparrows and grasshopper sparrows are among the fastest declining North American songbirds (Peterjohn et al. 1994). The Henslow's sparrow has more specialized habitat and management needs (Kahl et al. 1985; Zimmerman 1988; Smith and Smith 1992) than the other two species. The grasshopper sparrow, which is rarer and declining more rapidly than the dickcissel, seems to be the more sensitive of the two species to habitat changes and management methods (Skinner et al. 1984; Smith and Smith 1992; Zimmerman 1992).

Records indicate that these sparrows had declined considerably even before the Breeding Bird Survey began. Nineteenth-century sources summarized by Herkert (1994) reveal that Henslow's sparrow was at that time one of the most abundant birds in Illinois. Much of the Henslow's sparrow population losses actually occurred decades before Breeding Bird Survey monitoring began (Graber and Graber 1963; Herkert 1994), with a decline of 90% occurring between 1958 and 1979 in Illinois (Illinois Natural History Survey 1983). These three sparrow species have all undergone long-term range reductions in the eastern parts of their breeding ranges (Fretwell 1973; Smith and Smith 1992). The range of Henslow's sparrow is also contracting from the north (McNicholl 1988), and numerous local populations have disappeared in recent decades (Hands et al. 1989; Illinois Natural History Survey 1983).

Prairie Butterflies

No long-term monitoring program comparable to the Breeding Bird Survey is available for butterflies, and population dynamics are not as well documented for prairie butterflies as for birds. However, butterflies requiring prairie habitat have clearly experienced long-term declines, both along the fringes of their core ranges in the central United States and within the prairie province. The extinction wave of the regal fritillary (Fig. 2) from east to west and the species' increasingly localized occurrence within the prairie region are well documented (Swengel 1993). The Dakota skipper and the Poweshiek skipperling (Fig. 3) have also become more localized and restricted to prairie fragments (Opler and Krizek 1984; Johnson 1986).

Although declining grassland birds and prairie-specialist butterflies share the

Fig. 3. Poweshiek skipperling.

same habitats, their abundances are not necessarily correlated, in part due to differences in the geographic scales of their habitat use. For example, Henslow's sparrows and grasshopper sparrows are short-distance migrants, wintering primarily in the southern United States, while dickcissels winter mostly in northern South America (Fretwell 1973). Prairie-specialist butterflies are year-round residents on particular prairie patches, with relatively little dispersal among patches (Opler 1981; Opler and Krizek 1984; Moffat and McPhillips 1993). Although birds depend upon suitable habitat and resources being available at seasonally appropriate times in widely distributed regions, butterflies need resources and conditions to be consistently available within a particular habitat patch.

Key habitat features and required resources differ between birds and butterflies. Both butterflies and birds show preferences for either wet lowlands or dry upland habitats; however, vegetational structure is particularly important only to birds (Hopkins 1991). Henslow's sparrow prefers large grassland expanses with consistent patches of dense cover provided by dead vegetation, whereas grasshopper sparrows favor shorter, more open vegetation with sparse cover (Kahl et al. 1985). Butterflies often have specific associations with plant species, especially in the larval life stage (for example, larval food plants), and as adults they show some degree of preference for certain nectar flowers (Opler 1981; Opler and Krizek 1984).

It may be difficult to census butterfly and bird populations in the same survey because of their differing daily and seasonal activity patterns. Songbirds are active early and late in the day, a pattern that is weak or lacking in most grassland birds (Kantrud 1981).

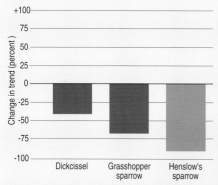

Fig. 1. Percent decline of three grassland sparrows in the U.S. Geological Survey Breeding Bird Survey between 1966 and 1993 (Peterjohn et al. 1994).

Fig. 2. Regal fritillary on a purple coneflower.

Butterflies tend to be more active in the warmer and sunnier parts of the day (Opler and Krizek 1984). Songbirds are more detectable during the breeding season in late spring to early summer, when they vocalize more. Butterflies are easier to see during their adult life stage, the timing of which varies considerably by species (Opler and Krizek 1984).

Population Trends

Despite the difficulties associated with simultaneous surveys of grassland birds and butterflies, some covariances are apparent between populations of the two groups (Fig. 4). In our study, prairie-specialist butterfly species were more strongly correlated with grassland sparrows than with butterflies less restricted in habitat. The regal fritillary, the most widely occurring prairie-specialist butterfly, showed the most consistent co-occurrence with grassland sparrows.

Prairie birds and butterflies present both conservation concerns and opportunities for preservation. Though their populations have been much reduced through habitat loss, no known species of prairie-specialist butterfly or North American grassland sparrow has yet become extinct. Conservation activities

that are effective at maintaining one prairie species may confer benefits to others. For example, in southwestern Missouri, prairie conservation management rotates midsummer haying (annually to triennially, though usually biennially) to benefit the greater prairie-chicken (Solecki and Toney 1986). This rotated haying also supports large populations of prairie-specialist butterflies (Swengel 1996) and grassland sparrows (Skinner et al. 1984). Grassland birds and butterflies also benefit from nonintensive grazing and hay-cutting regimes (Skinner 1975; Kantrud 1981; Smith and Smith 1992; Swengel 1996), and birds particularly benefit from idling of croplands, as in the United States Conservation Reserve Program, which rewards farmers for tilling a smaller percentage of their land (Johnson and Schwartz 1993).

Acknowledgments

We thank the funding providers of the surveys used in Fig. 4: Lois Almon Small Grants Research Program, The Nature Conservancy Minnesota Chapter, Wisconsin Department of Natural Resources, U.S. Fish and Wildlife Service, and W. and E. Boyce.

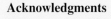
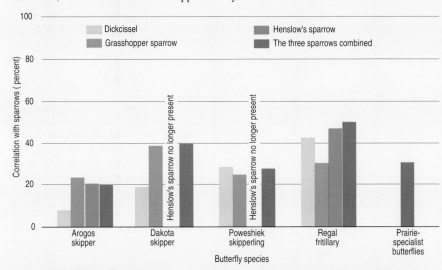

See end of chapter for references

Fig. 4. Percent correlation in abundance between species pairs. A 100% correlation indicates a total correspondence in abundance patterns; 0% correlation indicates a complete noncorrespondence in abundance patterns. Percentages are based on Pearson's product-moment correlations of log-transformed observations of each species per hour per survey, in surveys from mid-June to 31 July 1988–1995 at 104 prairie sites in Illinois, Iowa, Minnesota, Missouri, and Wisconsin (methods described in Swengel 1996). Although Henslow's sparrows historically occurred at sites in the range of the Dakota skipper and the Poweshiek skipperling, they do not occur at these sites today.

Authors

Ann B. Swengel
Scott R. Swengel
909 Birch Street
Baraboo, Wisconsin 53913

maculated manfreda skipper—and two moths—the rattlesnake-master borer moth and phlox moth—are federal species of concern. An additional six insects—the persius duskywing, poweshiek skipperling, ottoe skipper, byssus skipper, silver-bordered fritillary, and Ozark emerald—are considered species of concern by The Nature Conservancy, and another six are endemic to the prairie (Royer 1992).

Adequate inventory and distribution information is unavailable for predicting status and trends for most invertebrates. Ranges of a number of grasshopper species seem focused specifically on short-grass, a combination of short-grass and mixed-grass, and tall-grass prairies (Otte 1981). Ranges of many other grasshopper species center on grasslands but extend into adjacent forested and scrub habitats. The prairie mole cricket, a strikingly large insect (Fig. 12), and the superb spharagemon grasshopper are

federal species of concern. Additional species of concern are the Ozark snaketail dragonfly, a true bug, a fly, six amphipods, nine cave spiders, two cave beetles, a cave amphipod, a cave shrimp, and a number of beetles, including the widely distributed sixbanded longhorn beetle.

Leafhoppers are among the most diverse and well-studied terrestrial insects on the grasslands (Whitcomb et al. 1994). The ranges of some species span considerable distances across the

Fig. 12. The prairie mole cricket, an unusual and rare prairie species.

prairie, whereas other species are more restricted in distribution. Most are highly specialized, often endemic, and good indicators of grassland condition. On the short-grass prairie, 20 leafhopper species may colonize a single host, such as blue grama, which indicates a long record of coevolution. There is, however, a rather spectacular partitioning of resources: no more than 10 leafhopper species occur on the blue grama host at a given site. The "invisible wall" partitioning resources is climate, with each taxon showing singular limits to humidity, cold, and heat. Understanding any single assemblage of leafhoppers (and perhaps most other insects) requires knowledge of the communities in which they and their relatives reside and of the past structure and conditions of these communities.

Around 1889 abundant mussel populations of the Great Plains were recognized as an economic resource, particularly as the materials for button production. One example of the abundancy of this resource was a single mussel bed that covered an area of 2.4 kilometers by 288 meters in the Mississippi River near New Boston, Illinois (Carlander et al. 1986). The mussel bed was depleted by 1898, and large-scale propagation to restore the resource failed. No federal regulations restrict the harvest of mussels unless they are federally listed as endangered or threatened. Many states (Illinois, Iowa, Kansas, Minnesota, Missouri, Montana, Nebraska, North Dakota, Oklahoma, South Dakota, and Texas), however, have instituted harvest regulations.

Along with siltation and contamination, dams, with their altered flow regimes and accompanying reservoirs, are believed to have caused declines of mussels and other aquatic organisms. Four pearlymussels—the elktoe, spectacle-case, snuffbox, and scaleshell—are federal species of concern. The American Fisheries Society has identified 213 of 297 (71.7%) mussels found in the United States and Canada as threatened, endangered, or of special concern (Williams et al. 1993). Many of these species are endemic; for example, all three endangered and six threatened species, and about one-third (two of seven) of the mussels of special concern recently added by the American Fisheries Society list for Texas are species with restricted ranges. Several nonindigenous species, particularly the Asian clam and the zebra mussel, pose potential threats to the native mussel fauna on the Great Plains.

Across the grasslands, a number of snails, including both land and aquatic species, are federal species of concern: 10 are found in Texas, at least 8 in the upper Midwest (Illinois, Iowa, Minnesota), 5 in the central plains (Missouri and Oklahoma), and 3 in the West (Wyoming and the Dakotas). One snail, the Iowa Pleistocene snail, is listed as endangered. The same factors negatively affecting other species—habitat loss, drainage, water pollution, stream desiccation accompanying the lowering of water tables for agriculture and municipalities, and competition with and predation by introduced species—are thought to threaten freshwater native snails, crayfishes, and other aquatic invertebrates.

Fishes

In general, drainages east of the Rocky Mountains have a richer fish fauna than those to the west (Brooks and McLennan 1993). The western Mississippi basin occupies much of the grasslands, including the basins of the Missouri, Arkansas, and Red rivers. Geological and pre-settlement history and current information depict changing fish distributions and abundances across this vast region (Cross et al. 1986). Geological information provides insights into how glacial advances fragmented fish populations into smaller, subregional distributions and transported species to distant and distinct drainages. Government surveys seeking routes for human immigrants to the Southwest and Pacific coast provide important historical information. Recent concern for prairie (and other) fishes has encouraged the continuation of many ongoing surveys.

The western Mississippi basin contains at least 266 species of fishes from the United States, which is slightly more than one-third of the fish species in the United States and Canada (Cross et al. 1986). Thirty-one species were introduced to North America or now exist beyond their past ranges. Two species are diadromous (and thus reside at some time in their life cycles in marine waters), and 34 are endemic to one or more drainages in the vast basin.

Cross et al. (1986) described two physiographic regions that encompass the habitat of most fishes on the grasslands: the Great Plains, from eastern Montana south to western Texas, and the Central Lowlands, from North Dakota south to eastern Texas. Thirteen fishes are endemic to the two physiographic regions: 2 to the Central Lowlands, and 11 to the Great Plains. The Great Plains province has 77 fish species (86 including introduced species), that, with the exception of 5 species, are a subset of the Central Lowlands fish assemblage. The Central Lowlands has 139 native species with over 24 species introduced into the region, most intentionally as sport fish or forage for sport fish. Eleven grassland fishes are large-river species that have centers of origin in one or two adjacent physiographic regions but are shared among the Central Lowlands and Great Plains physiographic regions. Overall, the species inhabiting small streams and rivers outnumber those in large rivers about ten to one.

Decades of intensive agricultural development and modified flow regimes are held responsible for declines in the fishes endemic to the small streams and turbid rivers of the Great Plains (Cross and Moss 1987). Nevertheless, the first declines noted in regional endemic fishes were in those species in small, clear, spring-fed streams, particularly streams that were home to the Topeka shiner and Arkansas darter. Although agriculture and altered flow regimes may explain declines, patterns in decline differ among species, with colonization of suitable habitat of restricted stream and river reaches exceptionally rare. Several fish species of shifting-sand bottom streams in the Great Plains are considered federal species of concern, including the sicklefin chub, the sturgeon chub, the Arkansas River speckled chub (a subspecies of the speckled chub), and the Arkansas River shiner. The Topeka shiner, the plains topminnow, and the Arkansas darter, which occur in clear streams fed by springs and seeps, and the lake sturgeon are also candidates for species of concern (Echelle et al. 1995).

Overall, 9 of 10 species and subspecies of broad, shallow, and sandy-bottom streams seem to be in serious decline. Two additional species, the plains minnow and flathead chub, seem to be drastically declining (V. M. Tabor, U.S. Fish and Wildlife Service, Manhattan, Kansas, unpublished manuscript) and are candidates for federal listing. Marked increases are known in some fishes in the natural prairie community. Clear-water fishes, particularly sunfish, perch, and introduced species such as carps, replace species native to turbid conditions and tend to be increasing. Within the Missouri River drainage, the upper Missouri and Yellowstone rivers provide the best remaining habitat for the large-river natives. About 20% of the native large-river fish species are declining.

Reptiles and Amphibians

Most grassland reptiles and amphibians are widely distributed. Half of the species (6 of 12) found in Alberta also occur several hundred kilometers to the south in northern Mexico (N. J. Scott, Jr., U.S. Geological Survey, San Simeon, California, unpublished manuscript). More than 40 reptiles and amphibians are characteristic of prairie habitat (Table 2). The number of local species is influenced by the presence of water (which amphibians need to complete their life cycle), complex habitats, and sandy or loose soils needed for concealment by some species.

Habitat	Species
Grassland	Ornate box turtle
	Great Plains skink
	Prairie skink
	Western slender glass lizard
	Racer
	Western rattlesnake
	Night snake
	Prairie kingsnake
	Texas slender blind snake
	Massasauga (rattlesnake)
	Lined snake
Temporary water	Wyoming toad
	Great Plains narrow-mouthed red toad
	Spotted chorus frog
	Plains leopard frog
	Yellow mud turtle
	Plains garter snake
	Common garter snake
	Checkered garter snake
Bare ground	Lesser earless lizard
	Texas horned lizard
	Coachwhip
	Six-lined racerunner
Bare ground and water	Great Plains toad
	Green toad
	Plains spadefoot toad
	Black-spotted newt
Sandy soils	Glossy snake
	Western hog-nosed snake
Sandy soils and water	Woodhouse's toad
Trees or rocks	Reticulate collared lizard
	Spot-tailed earless lizard
	Collared lizard
	Eastern fence lizard
	Great Plains rat snake
	Plains black-headed snake
	Milk snake
Rocky canyons and water	Red-spotted toad
	Spiny softshell
	Plain-bellied water snake
	Diamondback water snake
	Slider

Table 2. Reptiles and amphibians of the North American prairie by habitat association (N. J. Scott, Jr., U.S. Geological Survey, San Simeon, California, unpublished manuscript).

Loss of small water areas, nonindigenous terrestrial and aquatic predators, grazing, exotic plantings, and prairie dog control are believed to contribute to reptile and amphibian declines. Most reptiles and amphibians rely on temporary ponds rather than streams or rivers. Permanent water provides habitat for the especially predatory bullfrog, catfish, and sunfish. Woody vegetation near permanent water favors mammalian predators such as the Virginia opossum, raccoon, and skunk (Schwalbe and Rosen 1989). Moderate grazing increases habitat structure and patchiness important to reptile and amphibian abundance, but overgrazing reduces needed habitats, as does planting of nonindigenous species such as buffel grass (Scott 1997). Prairie dog burrows provide winter retreats and summer nesting sites for reptiles and amphibians, thus their destruction may cause local reptile and amphibian declines.

Most grassland reptiles and amphibians seem widespread and secure. Several, however, particularly those with very restricted ranges, are thought to be declining. Habitat for the

Amphibians of the Northern Grasslands

What Do We Know?

No cry of alarm has been sounded over the fate of amphibian populations in the northern grasslands of North America, yet huge percentages of prairie wetland habitat have been lost, and the destruction continues. Scarcely 30% of the original mixed-grass prairie remains in Nebraska, South Dakota, and North Dakota (See Table 1 in this chapter). If amphibian populations haven't declined, why haven't they? Or, have we simply failed to notice?

Amphibians in the northern grasslands evolved in a boom-or-bust environment: species that were unable to survive droughts lasting for years died out long before humans were around to count them. Species we find today are expert at seizing the rare, wet moment to rebuild their populations in preparation for the next dry season. When numbers can change so rapidly, who can say if a species is rare or common? A lot depends on when you look.

Some changes brought upon the northern Great Plains by human enterprise mimic this climatically induced variability. Frogs, toads, and salamanders that find themselves in a rare remaining wetland will thrive. Progeny will issue forth from the wetland in comfortably large numbers, and those that return to breed the next year will be rewarded with a reasonable likelihood of success. Those that strike out for new breeding territory, as some percentage must, will likely be less fortunate but also less conspicuous: who searches wheat fields for the frog that didn't make it?

Other changes have no precedent. Aquaculture—which involves modifying the water regime in a previously semipermanent wetland so that it can support stocked fish populations—brings vulnerable larval amphibians into contact with predators against which they have no defense. In addition to supporting stocked fish, these modified wetlands provide new habitat for predatory bullfrog larvae that require two years of stable water to mature. As predatory species take hold, native amphibians die out (Hayes and Jennings 1986).

One way to decide if amphibian populations are changing is to conduct surveys in areas that were studied long ago. Often past data were collected as an offshoot of a detailed study of a particular species. The danger here is that the past data were probably collected at the very best sites investigators could find; populations at these sites are far more likely to decline than to increase,

simply as a result of natural fluctuations (Johnson and Larson 1994). Nonetheless, it is possible to find past surveys (as opposed to ancillary species lists), and comparisons between current and former occurrences are, at the least, instructive.

Another method of assessing the stability of amphibian populations is to enlist volunteers in broad-scale monitoring programs, similar to the well-known U.S. Geological Survey Breeding Bird Survey. The North American Amphibian Monitoring Program, administered through the U.S. Geological Survey Biological Resources Division's Inventory and Monitoring Program, aims to do just that. Volunteers are being mobilized to conduct surveys of calling frogs and toads on carefully selected routes that will stand up to statistical scrutiny. Such a program could prove especially valuable for the northern Great Plains, where relatively few species are silent (tiger salamanders, skinks, and mudpuppies) and thus uncountable. After a reasonable number of years, such surveys will yield valuable information to help us distinguish between natural year-to-year fluctuation and real changes in population sizes.

Two Case Studies

Amphibians at the Cottonwood Lake Study Area

The Cottonwood Lake Study Area is a 49-hectare complex of 17 wetlands in the mixed-grass prairie of central North Dakota. We have monitored adult and larval amphibians there by using drift fences, pitfall traps, and aquatic funnel traps since the spring of 1992. Several noteworthy changes have occurred.

We have observed only four amphibian species at Cottonwood Lake: the gray tiger salamander, the striped chorus frog, the wood frog, and the northern leopard frog. Only tiger salamanders and chorus frogs have been common. Toads have been noticeably absent in our study. Although four species of toads are known to occur in the region (the American toad, the Canadian toad, the Great Plains toad, and Woodhouse's toad), we have not captured or observed a single individual of any toad species since monitoring began in 1992.

The dynamic nature of the Great Plains is well illustrated at Cottonwood Lake. In 1992, water levels were the lowest since record-keeping began in 1967, then record

precipitation in 1993 and 1994 resulted in the highest water levels ever recorded at the study area. Tiger salamanders have responded accordingly: in 1992, we captured only 32 individuals; with the progressive return of water to wetlands, the number of captures jumped to 567 in 1993, 1,270 in 1994, and 2,862 in 1995. As water levels in wetlands have become more stable, tiger salamander larvae have begun to successfully overwinter. Throughout all of 1992, 1993, and 1994, only two overwintered salamander larvae were captured during our spring sampling (both in 1994); in 1995, we captured 50.

Striped chorus frog numbers also increased at Cottonwood Lake, but not until 1994. We captured only eight chorus frogs in our funnel traps in 1993. Captures increased to more than 300 in both 1994 and 1995.

Although leopard frogs are reported as abundant throughout the prairie pothole region, we have captured only 13 in our funnel traps (3 in 1993 and 10 in 1995). All 13 were adults, and all were captured in late summer, suggesting that they were newly metamorphosed individuals dispersing into the study area from surrounding wetlands. To date, we have observed no evidence of the reestablishment of a resident leopard frog population at Cottonwood Lake.

Wood frogs have been, and continue to be, captured in very small numbers at Cottonwood Lake. We captured three individuals in our funnel traps in 1992, two in 1993, three in 1994, and only one in 1995. The absence of an increase in wood frog numbers during this period of increasing water levels suggests that drought is not a key factor limiting wood frog populations.

Prairie Pothole Amphibians: Changes Since the 1920's

In the summer of 1920, Frank Blanchard visited the Iowa Lakeside Laboratory in Dickinson County in northwestern Iowa, where he conducted what may have been the earliest study of prairie pothole amphibian populations. His expressed purpose was to provide baseline data for future herpetological surveys: "It is highly important that faunistic studies be undertaken here, and throughout our country, at as early a date as possible if we are to have any record of the composition and distribution of our native fauna, and if we are to deal intelligently with its preservation" (Blanchard 1923).

We repeated Blanchard's survey (Lannoo et al. 1994) and recorded the current amphibian diversity and relative abundance in Dickinson County. In addition to Blanchard's results, we relied on two locally written natural history accounts to determine the changes in amphibian populations. The first account (Anonymous 1907) estimated the number of northern leopard frogs taken from the region by commercial hunters. The second account (Barrett 1964) is a reminiscence about this early 1900's "frogging" industry in Dickinson County. These two accounts, while anecdotal, provide independent observations of the same hunting events and corroborate each other.

Five species reported by Blanchard persist: eastern tiger salamander, American toad, striped chorus frog, gray treefrog, and northern leopard frog. Two species reported by Blanchard were not found: mudpuppy and Blanchard's cricket frog. We collected two species not found by Blanchard: Great Plains toad and bullfrog. Great Plains toads may have migrated into Dickinson County from the west. Bailey and Bailey (1941) found Great Plains toads only west of Dickinson County; by 1984, Reeves (1984) found them east of Dickinson County.

Together, these results suggest that between the early 1940's and the early 1980's, the Great Plains toad may have expanded its range eastward and entered Dickinson County. The bullfrog was introduced by state fisheries biologists.

Several changes have occurred in the relative abundances of amphibian species since 1920. American toads and striped chorus frogs now rank higher in relative abundance; tiger salamanders now rank lower. Blanchard (1923) stated that nearly every wetland he sampled had tiger salamander larvae. Today, only 13 out of 32 wetlands contain tiger salamanders. One cause for concern over this decline is that Dickinson County tiger salamanders, unlike any other known population of the eastern tiger salamander, have larvae that are polymorphic (that is, they exist in different forms), exhibiting typical, cannibal, and intermediate physical forms.

From descriptions of the commercial frogging industry in Dickinson County at the turn of the century, we estimate that the number of leopard frogs has declined by at least two—and probably three—orders of magnitude. This decline may be due more to the loss of wetland habitat than to past hunting pressure. In our opinion, the most immediate threat to the existing populations of native amphibians in northwestern Iowa comes from the introduced bullfrog, although aquacultural practices have been important in eliminating and isolating amphibian habitat in other portions of the eastern prairie pothole region.

See end of chapter for references

Authors

Diane L. Larson
Ned Euliss
U.S. Geological Survey
Biological Resources Division
Northern Prairie Science Center
Jamestown, North Dakota 58401-7317

Michael J. Lannoo
Muncie Center for Medical Education
Ball State University
Muncie, Indiana 47306-0230

David M. Mushet
U.S. Geological Survey
Biological Resources Division
Northern Prairie Science Center
Jamestown, North Dakota 58401-7317

---

reticulate collared lizard and spot-tailed earless lizard, species restricted to south Texas, is threatened by exotic buffelgrass (Scott 1997). Agriculture in the Lower Rio Grande valley in Texas has reduced the number of temporary ponds, which are needed by the black-spotted newt. Pesticides may negatively affect the Wyoming toad, an endangered species, but conclusive evidence is lacking. Across its range, the yellow mud turtle, a federal species of concern, is restricted to a few widely distributed ponds (Dodd 1983).

Birds

Of the 435 bird species that breed in the United States, 330 breed on the Great Plains (Knopf and Samson 1995). Nevertheless, few North American bird species are believed to have evolved within the Great Plains. Mengel (1970) suggested that only 12 bird species are endemic to the grasslands. An additional 25 species are believed to have evolved on the grassland, though they range widely into adjoining vegetation provinces. Five of these 25 species are specifically associates of sagebrush landscapes of the Great Basin (Knopf and Samson 1995).

As a group, the endemic grassland bird species have shown more consistent, widespread, and steeper declines (Table 3) than any other guild of North American bird species

Species	Rate of change
Endemic species	
Ferruginous hawk	+1.6%
Mountain plover	-3.7%
Long-billed curlew	-1.7%
Marbled godwit	+0.8%
Wilson's phalarope	-0.1%
Franklin's gull	-0.9%
Sprague's pipit	-3.6%
Lark bunting	-2.1%
Baird's sparrow	-1.8%
Cassin's sparrow	-2.5%
McCown's longspur	+7.3%
Chestnut-collared longspur	+0.4%
Secondary species	
Mississippi kite	+0.9%
Swainson's hawk	+1.4%
Northern harrier	-0.4%
Prairie falcon	+0.3%
Greater prairie-chicken	-6.9%
Lesser prairie-chicken	N/A[a]
Sharp-tailed grouse	+1.1%
Upland sandpiper	+2.7%
Burrowing owl	-0.2%
Short-eared owl	-0.6%
Horned lark	-0.7%
Eastern meadowlark	-2.3%
Western meadowlark	-0.5%
Dickcissel	-1.6%
Savannah sparrow	-0.5%
Grasshopper sparrow	-4.1%
Henslow's sparrow	-5.0%
Vesper sparrow	-0.3%
Lark sparrow	-3.5%
Clay-colored sparrow	-1.2%

[a] N/A means sampling effort was inadequate.

Table 3. Birds of the North American grassland with annual rates of change in populations. U.S. Geological Survey (Breeding Bird Survey data 1966–1993; Knopf 1986).

Courtesy F. L. Knopf, USGS

Fig. 13. A mountain plover, an endemic bird species of the short-grass prairie that evolved with intensive grazing pressure from bison, pronghorn, and prairie dogs in Colorado.

(Knopf 1992, 1996). Individually, populations of the mountain plover (Figs. 13 and 14a), Cassin's sparrow (Fig. 14b), and clay-colored sparrow (Fig. 14c) are declining throughout their breeding ranges. Breeding habitats are disappearing locally for the Franklin's gull (Fig. 14d), the dickcissel (Fig. 14e), the Henslow's sparrow (Fig. 14f), the grasshopper sparrow (Fig. 14g), and the western meadowlark (Fig. 14h). Breeding ranges are shifting for the ferruginous hawk (Fig. 14i), the Mississippi kite (Fig. 14j), the upland sandpiper (Fig. 14k), the horned lark, the vesper sparrow, the savannah

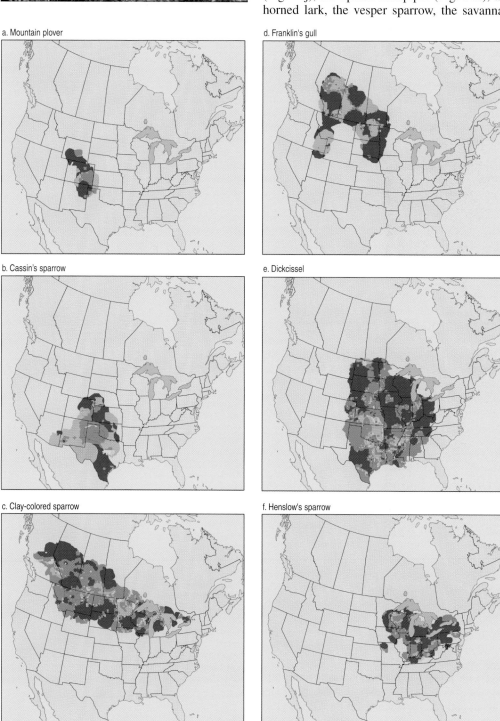

a. Mountain plover

d. Franklin's gull

b. Cassin's sparrow

e. Dickcissel

c. Clay-colored sparrow

f. Henslow's sparrow

Change per year (percent)

- ■ < -1.5
- ■ -1.5 – -0.26
- □ -0.25 – 0.25
- ■ 0.25 – 1.5
- ■ > 1.5

Fig. 14. Distribution and trends of selected grassland bird species in the United States and Canada.

g. Grasshopper sparrow

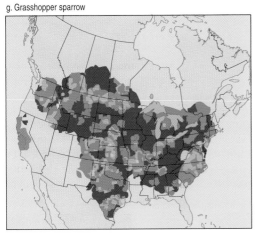

j. Mississippi kite

h. Western meadowlark

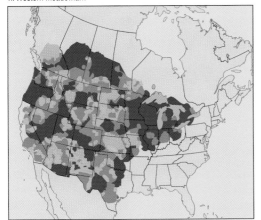

k. Upland sandpiper

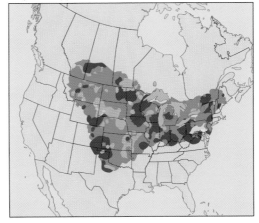

i. Ferruginous hawk

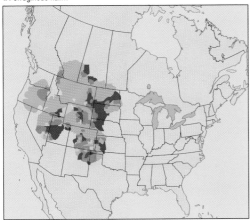

l. McCown's longspur

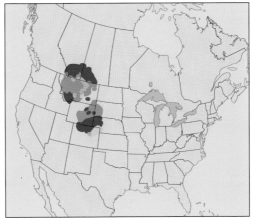

Fig. 14. continued.

sparrow, and the Henslow's sparrow. Populations of the wetland-associated marbled godwit and Wilson's phalarope seem stable, and populations of the upland sandpiper and McCown's longspur (Fig. 14l) have increased markedly.

Most of the endemic bird species of the short-grass and mixed-grass prairies are associated with large grazing animals (Fig. 15); others, such as the ferruginous hawk, prairie falcon, and burrowing owl, are either somewhat or strongly associated with prairie dog colonies.

Courtesy F. L. Knopf, USGS

Fig. 15. A bison herd in southwestern South Dakota.

Wetland Birds in the Northern Great Plains

When the Wisconsin glacier retreated about 10,000 years ago, it left innumerable depressions scattered throughout the northern Great Plains. These depressional wetlands, called *prairie potholes*, contain water for various lengths of time in most years (Kantrud et al. 1989). Their size, permanence, hydrology, water chemistry, plant associations, and invertebrate communities vary widely among wetlands and, within a basin, through time (Cowardin et al. 1979).

These diverse wetlands support a breeding avifauna as rich and varied as the wetlands themselves. Johnsgard (1979) listed 72 breeding bird species associated with freshwater pond environments in the Great Plains. Other species, such as the northern harrier, marbled godwit, Le Conte's sparrow, and Nelson's sharp-tailed sparrow, are associated with grasslands but extensively use these prairie wetlands. Stewart (1975) identified 63 breeding bird species as wetland associates in North Dakota alone. Since 1975, several species could be added to Stewart's list (Faanes and Stewart 1982), including the reintroduced Canada goose (Lee et al. 1989) and several herons, egrets, and ibises that have expanded their breeding range into the state (Lokemoen 1979). Most wetland birds are short-distance migrants, wintering primarily north of the United States–Mexico border (Igl and Johnson 1995).

Table 1. Densities (breeding pairs per square kilometer) of breeding birds by wetland class in North Dakota (Kantrud and Stewart 1984).

Table 2. Breeding bird populations in North Dakota: numbers in 128 randomly selected quarter-sections and statewide population estimates, 1967 and 1992–1993.

Species	Number of breeding pairs			Population estimate	
	1967	1992	1993	1967	1992–1993
Pied-billed grebe	11	4	7	24,000	12,000
Horned grebe	2	1	0	4,000	1,000
Red-necked grebe	0	1	1	0	2,000
Eared grebe	40	48	22	90,000	78,000
Western grebe	0	2	1	0	3,000
American white pelican	0	0	2	0	2,000
Double-crested cormorant	0	1	10	0	12,000
American bittern	9	2	8	19,000	10,000
Great blue heron	2	1	3	4,000	4,000
Black-crowned night-heron	17	5	5	37,000	11,000
Northern harrier	15	21	34	33,000	60,000
Yellow rail	0	1	0	0	1,000
Virginia rail	3	5	2	7,000	8,000
Sora	32	41	78	68,000	128,000
American coot	348	76	124	761,000	220,000
Piping plover	5	2	1	11,000	3,000
Killdeer	105	112	142	227,000	280,000
American avocet	14	6	13	31,000	21,000
Willet	18	16	27	39,000	48,000
Spotted sandpiper	12	12	9	26,000	22,000
Marbled godwit	17	8	14	37,000	24,000
Common snipe	0	2	7	0	10,000
Wilson's phalarope	73	30	36	157,000	72,000
Franklin's gull	22	79	56	48,000	148,000
Ring-billed gull	1	49	11	2,000	65,000
California gull	0	0	2	0	2,000
Forster's tern	3	6	4	6,000	11,000
Common tern	6	6	3	13,000	10,000
Black tern	118	39	39	254,000	84,000
Belted kingfisher	0	1	1	0	2,000
Sedge wren	10	20	37	22,000	62,000
Marsh wren	51	113	153	112,000	293,000
Common yellowthroat	134	91	175	285,000	286,000
Le Conte's sparrow	6	2	14	12,000	16,000
Nelson's sharp-tailed sparrow	3	3	13	7,000	34,000
Red-winged blackbird	945	597	710	2,038,000	1,421,000
Yellow-headed blackbird	89	155	175	193,000	356,000

Species	Wetland class					
	Temporary	Seasonal	Semipermanent	Permanent	Alkali	Fen
Pied-billed grebe		5.4	11.9	1.3		12.2
Horned grebe		1.4	0.6	0.3		
Eared grebe		3.9	1.9	1.6		
Western grebe			0.2	2.8		
American bittern	5.8	3.3	5.8	0.1		8.4
Black-crowned night-heron		0.4	4.2	1.0		17.2
Northern harrier		0.6	1.5			7.5
Virginia rail		0.4	1.8			7.5
Sora	10.1	12.9	12.6		0.2	25.1
American coot	25.2	73.8	180.5	8.9		52.8
Piping plover			0.2		2.2	
Killdeer	20.6	7.2	5.3	1.2	2.0	
American avocet		3.2	3.2		19.5	
Willet	10.1	12.3	7.0	1.0	1.3	2.5
Marbled godwit	10.1	6.9	3.6	0.1	2.7	5.0
Wilson's phalarope	45.4	28.9	11.5	0.1	4.3	5.0
Black tern	5.8	19.0	44.9	3.3		17.2
Marsh wren		4.9	43.8			52.6
Common yellowthroat		7.9	7.8			55.4
Savannah sparrow	188.9	21.5	15.4		5.0	8.6
Red-winged blackbird	300.0	99.8	106.8	16.9	14.9	125.5
Yellow-headed blackbird	11.1	18.1	253.3			271.0
Total	633.1	331.8	723.8	38.6	52.1	673.5

Wetland birds are not easily monitored by standard census techniques (Bibby et al. 1992). Dense vegetation reduces the visibility of some species. Many species lack territorial songs or rarely call; others that make diagnostic sounds do so primarily or only at night. Some species, such as rails, are notoriously elusive, even within a few meters of an observer (Burt 1994). Others are colonial, resulting in tremendous spatial variability in their numbers. Thus, no single technique works well for censusing all wetland species. Accurate censusing of wetland birds requires a variety of techniques, including nocturnal surveys, nest counts, intensive efforts involving walking or canoeing through marshes, and the use of recorded calls to elicit responses (Weller 1986). Recently, an informal group that monitors marsh birds has formed to address such issues.

Kantrud and Stewart (1984) surveyed breeding populations of wetland bird species other than waterfowl on 1,321

Table 3. Trends from the U.S. Geological Survey Breeding Bird Survey for the central region, 1966–1994, 1966–1979, and 1980–1994. Also given is average number recorded per route (R.A.) for the entire period.

Species	1966–1994 R.A.	1966–1994 Trend[a]	1966–1979 Trend[a]	1980–1994 Trend[a]
Pied-billed grebe	0.39	0.5	1.2	5.8
Eared grebe	0.67	5.2	24.0 ↑	−16.2 ↓
American white pelican	1.22	3.5 ↑	0	1.1
Double-crested cormorant	0.56	26.6 ↑	6.7 ↑	11.1 ↑
American bittern	0.50	−3.1	−4.9	0.1
Great blue heron	0.89	3.0 ↑	6.4 ↑	0.6
Black-crowned night-heron	0.37	3.3	−10.3 ↓	2.2
Yellow-crowned night-heron	0.53	0.7	19.6	−2.9
Great egret	2.84	3.8	4.2	4.4 ↑
Snowy egret	1.66	27.5	87.2	15.7 ↑
Little blue heron	2.98	−1.8	−0.9	−3.9 ↓
Tricolored heron	1.20	10.4	88.0	−4.1
Cattle egret	21.67	2.2	5.2	−2.8 ↓
Green heron	1.00	0.6	1.2	−3.4 ↓
White ibis	6.10	22.3 ↑	80.0	17.4 ↑
White-faced ibis	5.90	8.5	9.1	−9.9
Northern harrier	0.61	−2.1 ↓	−1.9	−0.3
King rail	0.82	−1.4	5.0	−2.0
Sora	0.94	−2.6	−8.5 ↓	11.0 ↑
Common moorhen	1.79	6.1	22.4 ↑	0.6
American coot	2.14	−0.8	−1.5	3.6
Killdeer	8.88	−0.3	3.0 ↑	−2.0 ↓
American avocet	0.63	−0.2	11.2 ↑	−1.8
Willet	1.03	−1.8	4.7 ↑	−0.4
Spotted sandpiper	0.08	2.1	9.4	−2.2
Marbled godwit	1.36	0.7	7.9 ↑	—
Common snipe	1.22	0.7	6.7 ↑	−1.0
Wilson's phalarope	1.24	−3.2	−5.7 ↓	6.7
Franklin's gull	7.55	−7.6 ↓	−17.0 ↓	42.3 ↑
Ring-billed gull	2.47	6.4 ↑	−5.6 ↓	10.3 ↑
California gull	0.95	17.6 ↑	−9.3	11.3 ↑
Laughing gull	12.95	5.4 ↑	7.5	−3.2
Forster's tern	0.54	0.7	12.1 ↑	−0.9
Black tern	2.70	−5.0 ↓	−13.0 ↓	2.7
Belted kingfisher	0.16	−1.6	−1.4	0.8
Sedge wren	1.25	1.3	−4.1 ↓	5.7 ↑
Marsh wren	1.34	3.6	−4.9 ↓	6.7 ↑
Common yellowthroat	6.87	−0.9 ↓	1.8 ↑	−2.1 ↓
Le Conte's sparrow	0.84	0.7	6.5	7.3
Nelson's sharp-tailed sparrow	0.15	5.2	0	17.7 ↑
Swamp sparrow	0.25	1.3	6.3 ↑	2.5
Red-winged blackbird	85.09	−0.5 ↓	1.1 ↑	−1.3 ↓
Yellow-headed blackbird	15.77	0.5	3.3	−2.1 ↓

[a]Average percentage annual change between 1967 and 1993: ↓ indicates statistically significant population decline; ↑ indicates statistically significant population increase.

wetland basins in the prairie pothole region of North Dakota. Densities of each species were reported for six different classes of wetlands (as defined by Stewart and Kantrud 1971; Table 1). Four of the wetland classes (permanent, semipermanent, seasonal, and temporary) were distinguished by water permanence as indicated by the vegetative zone occupying the deepest part of the basin. Alkali wetlands were recognized by the occurrence of hypersaline surface water, and fens were identified by a characteristic zone of fen vegetation that develops where groundwater seeps saturate the soil. Most wetland species were found on semipermanent and seasonal wetlands, reflecting the variety of habitats within these wetland classes (Table 1). Although some species were found in all wetland classes, most

species showed a preference for one or two classes.

Population estimates and trends of wetland bird species, exclusive of waterfowl, are limited. In 1967 Stewart and Kantrud (1972) conducted an extensive survey of breeding bird populations throughout North Dakota to obtain baseline estimates of statewide breeding bird abundance and frequency of occurrence. In 1992 and 1993 Igl and Johnson (1997) repeated the survey by using the same sample units and methods as the 1967 survey to examine changes in breeding bird populations. These data offer both overall population estimates and some indication of population changes from 1967 to now (Table 2). According to habitat affinities, wetland species composed the largest proportion (32%) of species and 22% of the

observed breeding pairs during the three years covered in the two Igl and Johnson surveys. The species that declined in North Dakota were mostly grassland and wetland species, whereas increasing species were predominantly resident species and species associated with human structures and woody vegetation (Igl and Johnson 1995). Similarly, results from the U.S. Geological Survey Breeding Bird Survey for North Dakota showed that 23 of the 28 (82%) observed species with statistically significant decreasing trends in the state were associated with wetland or grassland habitats.

We obtained trends for abundance of 43 wetland birds from the Breeding Bird Survey (Robbins et al. 1986) for the central region (from the Rocky Mountains to the Mississippi River) during the entire survey period (1966–1994) and for two subperiods: early (1966–1979) and recent (1980–1994) (Table 3). Population increases outnumbered decreases for each of the three time intervals. The percentage of species with increasing trends was 67% during the early subperiod (1966–1979) and 55% during the recent subperiod (1980–1994). For the entire survey period (1966–1994), seven species increased significantly; these included mostly colonial-nesting species (American white pelican, double-crested cormorant, and three gull species). Five species decreased significantly; all of these frequently nest in emergent wetland vegetation (northern harrier, Franklin's gull, black tern, common yellowthroat, and red-winged blackbird).

These data are consistent with earlier reports showing that breeding bird populations in wetland ecosystems are as dynamic as they are rich. The divergent patterns observed among the species and studies reflect the species' disparate habitat requirements, geographic ranges, and unique responses to natural and anthropogenic changes in their environments. Determining the status and trends of wetland bird populations is a necessary first step toward the more daunting challenge of understanding the mechanisms that drive population changes.

See end of chapter for references

Authors

Lawrence D. Igl
Douglas H. Johnson
U.S. Geological Survey
Biological Resources Division
Northern Prairie Science Center
Jamestown, North Dakota 58401-7317

Waterfowl in the Prairie Pothole Region

The prairie pothole region of the northern Great Plains is one of the most important areas for duck reproduction in North America. The region produces, on average, 50% of the primary species of game ducks on the continent (Smith 1995), yet accounts for only 10% of the waterfowl breeding habitat in North America (Smith et al. 1964). Twelve of the 34 species of North American ducks are common breeders in the region. For seven species—mallard, gadwall, blue-winged teal, northern shoveler, northern pintail, redhead, and canvasback—the prairie pothole region accounts for more than 60% of the breeding population (Smith 1995). The region is also a major migration corridor during fall and spring for other ducks, geese, and other water birds.

Annual variations in the number and distribution of ducks are strongly influenced by dynamic water conditions in the prairie pothole region (Batt et al. 1989). Breeding duck population sizes and reproduction are positively related to the number of wetland basins holding water in May and July (Reynolds 1987; Johnson and Grier 1988; Batt et al. 1989). During periods of widespread drought in the grassland portion of the region, many ducks move into the parkland; when both regions are dry, ducks may be displaced to the boreal forest or tundra regions (Johnson and Grier 1988). Species such as pintail and blue-winged teal tend to be more affected by drought conditions because of their preference for temporary and seasonal wetlands, whereas canvasbacks and lesser scaup, which use more stable semipermanent and permanent wetlands, are less likely to be displaced unless drought is severe. For all species, however, productivity is generally reduced during drought conditions because of poorer nesting effort and success and low survival rates of young.

The U.S. Fish and Wildlife Service and Canadian Wildlife Service have conducted annual surveys of breeding waterfowl populations in the principal breeding areas of North America since 1955. The surveys provide data on habitat conditions (numbers of ponds), breeding population sizes, and production. The Waterfowl Breeding Population and Habitat Survey, conducted each May, uses east–west aerial transects spaced within 50 habitat strata (Smith 1995). Data from aerial counts in the region are adjusted for visibility based on results from selected ground survey areas to provide population estimates in each stratum. This annual survey is among the most extensive and comprehensive animal surveys conducted in the world, and it provides an important long-term data base for population and habitat trends and population management. I used these data (Smith 1995) for southern Alberta, southern Saskatchewan, southern Manitoba, eastern Montana, North Dakota, and South Dakota to assess the status and trends of 10 common duck species (mallard, gadwall, American wigeon, green-winged teal, blue-winged teal, northern shoveler, northern pintail, redhead, canvasback, and lesser scaup) over the last decade. Information on Canada geese breeding in the prairie pothole region is derived from May surveys and midwinter counts from the U.S. Fish and Wildlife Service. For information on trends in waterfowl populations before 1986, see Batt et al. (1989).

Trends in Water Conditions Affect Duck Populations

During the last decade, spring water conditions in the prairie pothole region, as measured by the number of wetland basins holding water (ponds), ranged from good in 1986 (more than 5 million) to very poor during the 1988–1993 drought (3.6 million or less) to excellent by 1994–1995 (Table 1). In 1986 pond numbers were above the long-term averages (1974–1995) for both the U.S. and Canadian portions of the region. Conditions in both areas degraded markedly over the next 3 to 5 years with the onset of drought conditions. Severe drought conditions continued in the United States through 1992, then rebounded dramatically in 1994–1995. Water conditions across much of the region in early 1996 remained good to excellent.

Trends in Breeding Duck Numbers

Over the past 10 years, the number of breeding ducks in the prairie pothole region averaged 15,195,000 (Table 1), 16% lower than the long-term average (1955–1995) of 18,166,000. Duck numbers were lowest during 1988–1993, when drought conditions were widespread. With the return of heavy precipitation and excellent water conditions in most areas in 1994–1995, duck numbers responded rapidly and exceeded the long-term average by 5% and 20%, respectively.

Numbers of mallards and blue-winged teal were relatively low during the drought but rebounded in 1994–1995 (Table 2). Although the gadwall is more closely associated with the prairie pothole region than other ducks (more than 90% of the continental surveyed gadwall population occurs in the region), gadwall numbers dropped only slightly below their long-term average during 1987–1988; by 1995 gadwall numbers were 111% above the long-term average (Table 2). Numbers of pintails in the region reached record lows during the drought as large portions of the population were displaced to northern areas, and their reproduction rate was low (Hestbeck 1995). Wigeon numbers in the region changed little over the past decade and remained below their long-term average despite some increase in 1994–1995.

The breeding population of northern shovelers in the prairie pothole region (Table 2) followed a pattern of decline and recovery similar to mallards and blue-winged teal. Green-winged teal populations did not decline markedly during drought years but did increase substantially in 1994–1995. Of diving ducks, lesser scaup showed the greatest response to the drought and the return of good water conditions. Numbers of redheads and canvasbacks were slightly depressed during drought years but, like scaup, responded to the return of good water conditions in 1994–1995. By 1995, redhead and canvasback numbers exceeded long-term averages.

Status and Trends of Canada Geese

Three populations of migratory Canada geese occur in the Great Plains. The Highline population breeds in southeastern Alberta, southwestern Saskatchewan, eastern Montana, eastern Wyoming, and north-central Colorado. January surveys in Colorado and New Mexico indicated that the Highline population has grown an average of 10% per year over the past 10 years, from about 75,000 geese in 1985 to a record 174,400 geese in 1995 (U.S. Fish and Wildlife Service, unpublished data). The Western Prairie population nests in eastern Saskatchewan and western Manitoba, and the Great Plains population is a restored population breeding in Saskatchewan, North Dakota, Nebraska, Kansas, Oklahoma, and

Table 1. Breeding population estimates for all ducks in the prairie pothole region and estimates of numbers of ponds during May, 1986–1995.

Year	Number of breeding ducks	Number of ponds		
		Total	United States	Canada
1986	18,429,000	5,760,000	1,735,000	4,025,000
1987	16,521,000	3,872,000	1,348,000	2,524,000
1988	13,515,000	2,901,000	791,000	2,110,000
1989	12,725,000	2,983,000	1,290,000	1,693,000
1990	13,399,000	3,508,000	691,000	2,817,000
1991	11,944,000	3,200,000	706,000	2,494,000
1992	14,256,000	3,609,000	825,000	2,784,000
1993	12,180,000	3,611,000	1,350,000	2,261,000
1994	18,997,000	5,985,000	2,216,000	3,769,000
1995	21,892,000	6,336,000	2,443,000	3,893,000

seldom synchronized for more than 2 or 3 years, and such high duck production is likely in only 2 or 3 years out of 10 where habitats are altered by agriculture (Lynch et al. 1963). High production of waterfowl, especially of early-nesting species such as mallards and pintails, has become more difficult to achieve during years of moderate water conditions (Lynch 1984; Batt et al. 1989).

The ability of duck populations to recover from naturally occurring droughts has been reduced by continued loss of nesting habitat to agricultural activities, primarily grain cropping and intensified grazing. These agricultural effects are likely to be greatest in the grassland portion of the

Table 2. Breeding populations (in thousands) of 10 duck species in the prairie pothole region in 1986–1995 (Smith 1995). TYA = 10-year average; LTA = long-term average (1955–1995).

Species	Year										TYA	LTA
	1986	1987	1988	1989	1990	1991	1992	1993	1994	1995		
Mallard	3,900	3,678	2,726	2,957	2,800	2,863	3,326	3,188	4,516	5,352	3,531	4,678
Blue-winged teal	3,892	2,800	2,761	2,438	2,318	3,113	3,572	2,409	4,199	4,847	3,235	3,594
Gadwall	1,463	1,244	1,237	1,301	1,458	1,443	1,916	1,636	2,201	2,734	1,663	1,293
Northern pintail	1,655	1,398	674	1,002	966	524	905	1,075	2,066	1,805	1,207	3,112
American wigeon	544	440	440	398	508	510	685	504	763	852	564	1,021
Northern shoveler	1,609	1,349	930	930	1,080	1,078	1,195	1,290	2,187	2,177	1,382	1,444
Lesser scaup	1,311	856	1,023	621	741	822	919	738	1,020	1,253	930	1,107
Redhead	509	479	398	458	416	349	498	403	581	855	495	512
Green-winged teal	297	307	345	309	356	331	403	281	574	686	389	530
Canvasback	285	309	240	201	238	247	215	210	293	491	273	329

Texas. Because of the way the survey design relates to the breeding range boundaries, separate estimates for these two populations are not available. Estimates from May surveys for ducks suggest that the Western Prairie and Great Plains populations combined in the region have increased from 108,000 in 1986 to 228,000 in 1995 (U.S. Fish and Wildlife Service, unpublished data).

Factors Affecting Recent Waterfowl Populations in the Prairie Pothole Region

The dynamics of water conditions and duck populations observed over the past decade are characteristic of the prairie pothole region (Lynch 1984). Widespread drought during 1988–1993 reduced wetland habitat available to waterfowl, causing a marked reduction in waterfowl production. Displacement of ducks, particularly pintails and mallards, to northern regions also reduced populations during the drought. Pintail populations also may have been depressed by intensification of agricultural activities (drainage, cropping, and grazing)

in key breeding areas (Hestbeck 1995). The dramatic recovery of most duck species in the region in 1994–1995 resulted primarily from heavy precipitation patterns that began in late 1993 and replenished many wetlands, providing abundant food and habitat for breeding ducks.

Two other factors probably were significant in contributing to the large and rapid recovery by most species. First, changes in the predator community have altered predation pressures on nesting waterfowl in many areas (Greenwood and Sovada 1996). After the drought, red fox population sizes were low, favoring high nest success, and mink numbers were also low, enhancing duckling survival. Second, about 1 million hectares of cropland in the U. S. portion of the prairie pothole region were restored to perennial grassland through the Conservation Reserve Program. Fields in the Conservation Reserve Program provide attractive and often highly productive nesting cover for upland-nesting ducks (Kantrud 1993). The combination of good to excellent water conditions, reduced predator pressure, and improved availability of nesting habitat in the United States provided conditions for a dramatic rebound in duck numbers after the drought. However, these conditions are

prairie pothole region, which experiences the greatest variability in water conditions and has had the greatest expansion of agricultural activities (Bethke and Nudds 1995). Whereas parkland and boreal areas can sustain duck populations over time, it is the grasslands' capacity for high duck production during wet periods that is critical to the growth of waterfowl populations (Lynch 1984).

See end of chapter for references

Author

Jane E. Austin
U.S. Geological Survey
Biological Resources Division
Northern Prairie Science Center
Jamestown, North Dakota 58401-7317

Duck Plague: Emergence of a New Cause of Waterfowl Mortality

Disease-causing organisms are natural components of biological systems. Too often, however, the occurrence of disease in wildlife is directly associated with changes due to human actions. Duck plague is an example of an infectious disease of domestic waterfowl that has begun to infect migratory waterfowl populations. Our growing concern is that duck plague will soon become a major cause of death for North American waterfowl and that the threat may be heightened because actions necessary to combat duck plague appear to have become embroiled in controversy. The ensuing debate has affected both the response to outbreaks and disease prevention efforts, and it may be aiding the geographic spread of this disease.

Duck plague is an acute, contagious, and often fatal herpesvirus infection of ducks, geese, and swans (Leibovitz 1991). However, not all species of waterfowl are equally susceptible. For example, northern pintails are highly resistant, blue-winged teals are highly susceptible, and mallards are moderately susceptible. The amount of duck plague virus required to cause disease in these different species differs by orders of magnitude (Spieker 1977). Occurrence of disease is also complicated because different strains of duck plague virus vary greatly in their ability to cause disease and death (Jansen 1961; Spieker 1977).

An outbreak of an acute hemorrhagic disease of domestic ducks in the early 1920's in the Netherlands is generally believed to be the first scientific documentation of this disease (Baudet 1923). However, identification of duck plague as a distinct disease of domestic waterfowl did not occur until 1942 (Bos 1942), despite numerous outbreaks in the Netherlands from 1923 to 1942 (Jansen 1964). The name "duck plague" was proposed by Jansen and Kunst (1949) and accepted as official terminology for this disease. "Duck virus enteritis" is the name based on the principal pathological features of the disease and is used to distinguish it from fowl plague in the United States (Leibovitz 1991).

Reports of duck plague in Europe and Asia have essentially been confined to domestic waterfowl, and outbreaks with severe economic effects have occurred in those countries. Therefore, the 1967 entry of duck plague into the Pekin duck industry of Long Island, New York, (Leibovitz and Hwang 1968) quickly resulted in aggressive actions by the U.S. Department of Agriculture to eradicate this foreign animal

disease. Outbreaks also occurred in upstate New York during 1967. Duck plague reached Pennsylvania and Maryland in 1968, but in 1970 the U.S. Department of Agriculture declared success in its efforts to eradicate this disease from the United States. Secretary of Agriculture Clifford M. Harden issued certificates honoring New York State agencies and the Long Island Pekin duck industry for their part in the eradication effort. "The certificates officially proclaimed the eradication of duck virus enteritis (DVE) or duck plague from commercial waterfowl in the United States" (U.S. Department of Agriculture 1970). Unfortunately, this was not the end of duck plague.

No one knows how duck plague entered the western United States. The disease, however, appeared in San Francisco in 1972 after additional outbreaks in New York, Pennsylvania, and Maryland during 1970 and 1971. A city pond at the Palace of Fine Arts was the affected site in San Francisco (Snyder et al. 1973). The following year duck plague appeared in the Midwest and by 1975 had reached the Gulf of Mexico coast (Fig. 1) and Canada (Bernier and Filion 1975). Not only has there been a continual geographic expansion of duck plague (Fig. 2), but in some regions the number of outbreaks has also increased during each

decade. Increased numbers of duck plague outbreaks and deaths have occurred within the eastern United States since 1970, but outbreaks in other regions have declined during the past decade (Friend 1995). The greatest number of outbreaks has occurred in Maryland (Fig. 1); New York, Virginia, and California also have relatively frequent occurrences of this disease (National Wildlife Health Center, Madison, Wisconsin, unpublished records).

In addition to domestic ducks, small numbers of wild waterfowl and wild species being maintained in avicultural collections or free as feral birds became infected during the 1967 Long Island duck plague outbreak. Collectively, more feral, captive, and wild waterfowl have become infected and

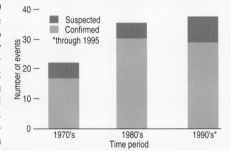

Fig. 2. Number of duck plague outbreaks in the United States by decade since 1970.

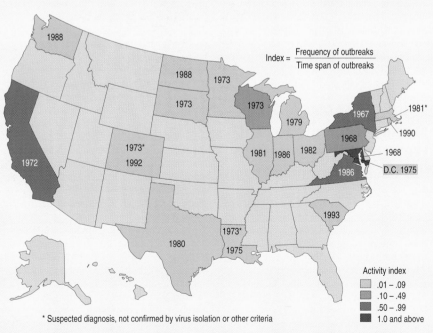

Fig. 1. Geographic distribution, spread, and relative activity of duck plague since its 1967 introduction into the United States. Years represent introduction into individual states.

survived subsequent outbreaks than have domestic ducks. The biological significance of those infections lies in the observation by Wobeser (1981) that birds surviving primary infection become carriers and potentially serve as future sources of additional duck plague outbreaks. To prevent the disease from spreading to other waterfowl populations, wildlife conservation agencies generally seek to destroy infected flocks. Infected waterfowl cannot be cured of the disease nor can disease carriers be readily detected by conventional technology.

Public opposition to the destruction of infected flocks is fueled by the fact that nearly all duck plague outbreaks occur in nonmigratory waterfowl in city parks and other public areas, or in captive flocks. Inherent human feelings for protecting animals, the beliefs that wild birds are the source for infection of urban and suburban waterfowl populations and that disease control is ineffective, and personal economic considerations are viewpoints often expressed by those opposing destruction of infected flocks. The objective of those opposing flock destruction is often to protect the individual birds surviving the outbreak. This is unfortunate since, as noted by Leibovitz (1991), the disease can best be controlled by eliminating entire infected populations.

The devastating 1973 duck plague outbreak in southeastern South Dakota at the Lake Andes National Wildlife Refuge demonstrated the negative effect that this disease can have on wild waterfowl (Fig. 3). Approximately 40,000 mallards and small numbers of other waterfowl species died. This was the first duck plague event involving a large-scale loss of wild waterfowl (Friend and Pearson 1973). Concern among the wildlife agencies responsible for managing the die-off resulted in the most aggressive actions ever taken to combat a disease outbreak in wild birds. Destruction of the large number of waterfowl that did not die from infection was considered but rejected because of arguments that methods were not available to prevent accidental dispersal of actively infected birds. It was feared that dispersed birds would initiate additional outbreaks as they relocated within other populations. Instead, the control effort focused on moving the birds from the confined area of Owens Bay at the refuge to the more open areas of the nearby Missouri River and then decontaminating the environment the birds had been removed from to kill any residual virus in the areas the birds had used. In a 6-month effort that involved approximately 90 people, researchers and refuge personnel decontaminated the environment by applying 7,000 pounds of chlorine gas, 2,000 pounds of calcium

Courtesy M. Friend, USGS

Fig. 3. During the 1973 outbreak of duck plague at Lake Andes National Wildlife Refuge in South Dakota, more than 40,000 mallards died.

hypochlorite, and 64,000 pounds of soda ash. In addition, all dead birds that could be found were picked up daily and incinerated. Owens Bay was drained as much as possible and kept dry throughout the following summer months. The vegetation around the bay was burned, and all areas where disease-control activities took place on the refuge and other areas, including road surfaces and ice surfaces, were decontaminated. A sample of mallards was tested for exposure to duck plague and then color-marked before the birds dispersed from Owens Bay. The color marking was used to provide visible evidence of the location of birds that had dispersed so that researchers could survey their new sites for additional mortality and, if such mortality were detected, disease-control actions could begin quickly. Additionally, all workers in the area had to follow strict decontamination procedures to prevent disease spread. These unprecedented efforts are credited with minimizing the

potential for additional outbreaks—there have been no occurrences of duck plague at Lake Andes since the 1973 event.

To date, the 1973 Lake Andes outbreak and a February 1994 outbreak that killed about 1,200 wild waterfowl in the Finger Lakes region of upstate New York (National Wildlife Health Center, unpublished records) are the only major outbreaks in wild waterfowl, although the pattern of duck plague outbreaks within the United States is one of an emerging disease (Figs. 1 and 2). The interface that often occurs between migratory and nonmigratory waterfowl provides opportunity for the transmission of disease between different populations (Tables 1 and 2). The international mobility of migratory waterfowl provides a means for widely disseminating duck plague; avian cholera has provided a precedent to support this concern. An important disease of domestic poultry within the United States since at least 1867 (Rhoades and Rimler 1991), avian cholera first appeared in wild waterfowl within the United States in 1944

Table 1. Classification of waterfowl involved in duck plague.

Commercial	Birds raised for consumptive markets, such as white Pekin ducks
Captive collections	Zoological and other collections of birds for display and research
Game farms	Birds raised for release for sporting programs, such as mallards
Feral	Nonmigratory, nonconfined waterfowl of various species
Nonmigratory	Resident populations of native wild species such as mallards and Canada geese
Migratory	North American waterfowl that breed in one geographic area and winter in another before returning to breeding grounds

Table 2. Status of duck plague in the United States.

Waterfowl classification	Occurrence of disease	
	Mortality events	Trends, 1967–1995
Commercial	Infrequent	Decreasing
Captive collections	Occasional	None; sporadic
Game farms	Occasional	None; sporadic
Feral	Common	Increasing
Nonmigratory	Occasional	None; sporadic
Migratory	Rare	None; rare

(Quortrup et al. 1946; Rosen and Bischoff 1949). It wasn't until the mid-1970's, however, that avian cholera emerged as a major cause of death for migratory waterfowl across the nation (Friend 1995).

The recent emergence of new diseases and reemergence of old diseases have become a focus for increasing concerns involving human and domestic animal health. The subject of emerging diseases has captured the attention of filmmakers, as depicted by the movie *Outbreak,* and is the basis of popular books such as *The Coming Plague* (Garrett 1994). The subject has also earned extensive media coverage and has been the focus for scientific symposia, meetings, and publications. Duck plague is an emerging disease in the United States and

must be addressed as such. The occurrence of this disease varies with different types of waterfowl populations. Migratory waterfowl have rarely been the primary species involved in outbreaks (Tables 1 and 2). It would be folly, when the opportunity exists to prevent it, to allow duck plague to join the ranks of avian botulism, avian cholera, and lead poisoning as another major cause of mortality for wild waterfowl.

Advances in molecular biology offer new potential for the detection of duck plague carriers, allow researchers to determine the origin of individual outbreaks, and aid in the search for answers to other important questions that will help resolve public and scientific debate involving this disease. Research is needed to develop required

technology and to implement methodical epizootiological investigations. The resulting information will provide wildlife management agencies with the information necessary to develop strategies to make duck plague a matter of historical record rather than a major cause of death for migratory waterfowl.

See end of chapter for references

Author

Milton Friend
U.S. Geological Survey
Biological Resources Division
National Wildlife Health Center
6006 Schroeder Road
Madison, Wisconsin 53711

Locally, drought tolerance seems to be the principal ecological process influencing the composition of grassland bird communities (Wiens 1974), with grazing (Hobbs and Huenneke 1992) and wildfire (Zimmerman 1992) playing secondary roles. Unlike most forest birds that winter in the Neotropics, virtually all endemic grassland birds winter on the continent; thus influences on their status and trends occur within North America.

The prairie pothole region is a key breeding area for species such as the mallard, blue-winged teal, and northern pintail. Researchers believe that preserving native grassland and wetlands is essential for slowing declines in duck numbers, including the mallard, American widgeon, and northern pintail (Canadian Wildlife Service and U.S. Fish and Wildlife Service 1994). Predation on eggs and hatchlings by red foxes, striped skunks, raccoons, and other species substantially reduces the abundances of ducks (Ball et al. 1994).

Mammals

The grasslands have a fertile history of plant- and seed-eating mammals (Hall and Kelson 1959). A rich array of large herbivores, including camels, rhinoceroses, mammoths, mastodons, and bison, evolved on the North American grasslands only to disappear during the last Ice Age (Van Valkenburgh and Janis 1993). The surviving large grassland herbivore, the plains bison, evokes a mystique not shared by any other North American mammal and which is largely derived from Native American and frontier heritages (Meagher 1978). In the past, bison (Fig. 15) numbered from about 60 to 70 million and roamed in large herds that, in the 1860's, often required horseback riders several days to successfully penetrate and cross.

As many as 5 billion prairie dogs may have been in North America before European settlement. An estimated 98% decline in prairie dog numbers has occurred since European settlement (Summers and Linder 1978). The black-tailed prairie dog may occupy less than 0.5% of its original range, the short-grass and mixed-grass prairies. As a result, a variety of species closely associated with the prairie dog are either federally listed as endangered or are being considered for listing as threatened or endangered. These include the black-footed ferret, the swift fox, and the mountain plover.

The ranges of more than 100 native mammals extend into the prairie; nearly half of these occur in the forest–grassland ecotone and others in diverse habitat types (Risser et al. 1981). Nevertheless, surveys of eastern and western species whose ranges stop short of the Great Plains support the effectiveness of the grassland as an evolutionary barrier to dispersal (Hall and Kelson 1959). Estimates of Great Plains–restricted mammals range from as few as 10 (Risser et al. 1981) to as many as 18 (Jones et al. 1985). These species include one lagomorph (the white-tailed jack rabbit), eight rodents (thirteen-lined ground squirrel, Franklin's ground squirrel, black-tailed prairie dog, plains pocket gopher, olive-backed pocket mouse, plains pocket mouse, plains harvest mouse, and prairie vole), and two carnivores (swift fox and black-footed ferret).

One recent extinction, the Audubon bighorn sheep, was a subspecies found along the upper Missouri River, including North Dakota, South Dakota, Wyoming, and Nebraska. Populations of the caribou, a species once common across northern North Dakota but which is now extirpated, are still common in Canada. Similarly, the gray wolf and elk, once common on the grasslands, and the less common mountain lion and wolverine are found elsewhere.

Population Trends for Prairie Pothole Carnivores

Since settlement of the prairie pothole region of the northern Great Plains by Europeans in the late 1800's, carnivore populations have changed considerably—mostly due to habitat alteration and human-inflicted mortality. At least 19 species of carnivorous mammals once occurred in the prairie pothole region (Jones et al. 1983). Presently, only eight are common throughout the region—coyote, red fox, raccoon, American badger, striped skunk, mink, ermine, and long-tailed weasel (Sargeant et al. 1993). Other species that occur locally or intermittently are mountain lion, lynx, bobcat, gray wolf, gray fox, swift fox, spotted skunk, and least weasel. Grizzly bears, wolverines, and river otters once occurred in the region but are now extirpated.

Competition among species affects the distribution of coyotes, wolves, and foxes (Carbyn 1982; Rudzinski et al. 1982; Sargeant et al. 1987; Bailey 1992). These larger canids are keystone species that suppress the distribution of smaller canids (Johnson and Sargeant 1977; Dekker 1989; Johnson et al. 1989).

Gray Wolf

Human influences on canid communities began in the late 1800's, first with the killing of wolves and later, coyotes, because the predators killed livestock (Johnson and Sargeant 1977; Andelt 1987). Treatment of wolves was especially harsh; consequently, they were essentially eliminated from the prairie pothole region.

Coyote

Coyotes are most abundant in the northwestern prairie pothole region (Sargeant et al. 1993; Fig. 1). In spring, populations are composed of family groups, each generally a mated pair and one or more associated adults (Voigt and Berg 1987), which occupy relatively exclusive territories (Sargeant et al. 1987). In spring, recorded home ranges of family groups averaged 61 square kilometers in North Dakota (Allen et al. 1987) but only 12 square kilometers in Alberta, where coyote densities were higher (Roy and Dorrance 1985). Home ranges are smaller where coyotes are more abundant (Allen et al. 1987).

Coyotes increased in the prairie pothole region after European settlement and the extirpation of the gray wolf (Bailey 1926; Criddle 1929). Coyotes were numerous in

Fig. 1. Coyote hunting in prairie grassland.
© H. Umber, North Dakota Game and Fish Department

Fig. 2. Red fox in prairie grassland.
© H. Umber, North Dakota Game and Fish Department

the prairie pothole region in the early 1900's, but populations were noticeably lower by the 1950's and remained low through the 1970's. Population reduction by killing was greatest in farmed areas of the southeastern portion, especially after the 1940's, and some populations were completely eliminated. Survival of coyotes increased after government bans in 1972 of certain predator control methods (Johnson and Sargeant 1977) and with reduced harvest pressure in response to low pelt values in the 1980's. In 1985 and subsequent years, millions of hectares of cropland in the prairie pothole region were enrolled in the Conservation Reserve Program of the U.S. Department of Agriculture (Young and Osborn 1990). Security provided by these extensive grasslands likely enhanced coyote survival. In the early 1990's, coyotes were present throughout most of the prairie pothole region (Sargeant et al. 1993; Sovada et al. 1995). However, an outbreak of sarcoptic mange is now causing declines in coyote populations in North Dakota (Allen 1996) as well as in adjacent states and Canadian provinces. This outbreak may lead to notable population changes in coming years.

Red Fox

Red foxes generally are more abundant in the central and southeastern portions of the prairie pothole region (Sargeant et al. 1993; Fig. 2). In spring, red fox populations are composed of family groups, usually mated pairs (Sargeant 1972). In North Dakota, documented family groups occupied exclusive territories that ranged from 3 to 21 square kilometers but that generally were smaller than 12 square kilometers (Sargeant 1972). Home ranges are smaller when red foxes are more abundant and are

affected by nearness of resident coyotes (Sargeant et al. 1987).

In the prairie pothole region, red fox population trends are generally opposite those of coyotes (Fig. 3). Red fox populations expanded greatly in the prairie pothole region between the 1890's and 1930's, especially in the southern portion; expansion began in the 1950's in the northern portion, especially in the northeast (Bird 1961; Johnson and Sargeant 1977). Red foxes were abundant throughout the eastern part of the prairie pothole region between the 1940's and 1970's. In the late 1970's to mid-1980's, red fox populations declined in the prairie pothole region during periods of high fur prices (Sargeant 1982). More recently, red fox populations have declined in the southern portion, apparently in response to competition from increased coyote populations (Sovada et al. 1995). Mange is expected to affect red fox populations in coming years (Allen 1996).

Striped Skunk

Striped skunk populations fluctuate erratically throughout the prairie pothole region (Rosatte 1987; Sargeant et al. 1993). Striped skunks are solitary, except in winter when they may occupy communal dens (Verts 1967). Recorded densities of adults in North Dakota in spring ranged from 1.2 to 1.5 animals per square kilometer, but densities were difficult to estimate because of the species' movements (Greenwood et al. 1985). During April in North Dakota, the minimum home range size averaged 242 hectares for females and 308 hectares for males (Greenwood et al. 1985). Striped skunks were less abundant in the presence of badgers and coyotes, both of which are predators (Sargeant et al. 1982; Johnson et

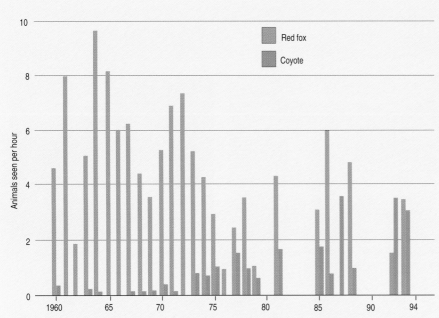

Fig. 3. Number of red foxes and coyotes seen per hour during winter aerial deer censuses in the Missouri Coteau of North Dakota (information from *North Dakota Outdoors*).

al. 1989). Striped skunks in the prairie pothole region die primarily because of diseases such as rabies, but they are also killed by humans, and some starve to death in the winter (Bjorge et al. 1981; Sargeant et al. 1982; Wade-Smith and Verts 1982). High reproduction (Greenwood and Sargeant 1994) and excellent dispersal (Sargeant et al. 1982) allow striped skunks to rapidly repopulate areas where populations are decimated.

American Badger

Badgers are at the northern limit of their range in the prairie pothole region (Hall 1981). The presence of badgers is tied to extensive grasslands (Messick et al. 1981; Sargeant et al. 1993), and they avoid cultivated areas (Messick and Hornocker 1981). Burrowing rodents found in grasslands are important prey of badgers (Messick 1987). Badgers are solitary, and little is known of their home range size or density in the prairie pothole region. Their populations in this region seem to have declined significantly since European settlement and have remained relatively low, probably because of the large areas of grassland converted to cropland. Human-inflicted mortality (for example, trapping and shooting) also influences badger populations (Messick et al. 1981). Recently, badgers have extended their range to the east and north in the prairie pothole region, probably as a result of clearing of trees in parkland, draining of wetlands, and increases in rodent populations (Nugent and Choate 1970; Lintack and

Voigt 1983; Long and Killingley 1983). Based on numbers of captures in the last 5–10 years, badger populations appear to have increased slightly in the region (U.S. Department of Agriculture, Animal Damage Control, Bismarck, North Dakota, unpublished data). The population increase may be related to extensive grassland habitat provided by the Conservation Reserve Program and to reduced hunting caused by low pelt value.

Mink

Mink population sizes are erratic in the prairie pothole region, especially where shallow basin wetlands, their preferred habitat, predominate (Arnold and Fritzell 1987a; Eagle and Whitman 1987; Sargeant et al. 1993). In spring, populations are composed of territorial males that occupy large areas and females that occupy smaller areas (Eagle and Whitman 1987). In the prairie pothole region of Manitoba during May–July, home ranges of male mink averaged 646 hectares (Arnold and Fritzell 1987b) but were larger during the breeding season. Drought reduces mink reproduction (Eberhardt 1974) and has catastrophic effects on their populations (Sargeant et al. 1993). Widespread wetland drainage probably affects mink populations in ways similar to drought.

During the droughts of the 1980's, mink were undetected in many areas of the prairie pothole region (Sargeant et al. 1993). Population lows also probably occurred during the droughts of the 1910's, 1930's, and

1960's. During droughts, rivers and lakes harbor mink populations. The high reproductive potential of mink (Eagle and Whitman 1987) allows the rapid recovery of their populations when prairie wetlands recover from drought. Mink populations likely were extremely low in much of the prairie pothole region of North Dakota in the early 1990's, after the drought of the 1980's. Since 1993, however, populations have begun to recover with improved wetland conditions (M. A. Sovada, U.S. Geological Survey, Jamestown, North Dakota, personal observation).

Weasels

Populations of weasels fluctuate greatly in response to availability of prey, primarily small mammals (Gamble 1981; Jones et al. 1983). Information on distribution and population trends of weasels is scant. Long-tailed weasel and ermine populations are believed to be fragmented and small throughout the prairie pothole region (Fagerstone 1987; Sargeant et al. 1993). Local occurrence is influenced by prey diversity and abundance, availability of water and den sites, and harvest pressure (Simms 1979; Jones et al. 1983; Fagerstone 1987). Extensive conversion of native uplands and wetlands to cropland probably has reduced the distribution and abundance of ermines and long-tailed weasels throughout the prairie pothole region.

Long-tailed weasel populations appear relatively stable, in contrast to ermine populations, which may fluctuate markedly. This stability reflects the long-tailed weasel's wide range of foods, which differs from the specialist ermine's narrow diet (Simms 1979; Gamble 1981; Fagerstone 1987). From the mid-1980's to early 1990's, the long-tailed weasel was considered threatened in the prairie provinces of Canada as a result of habitat loss and increased use of pesticides (Gamble 1982). However, based on a survey conducted in 1991, Proulx and Drescher (1993) reported the species was present throughout central and southern Alberta. Sargeant et al. (1993) observed large weasels (no distinction was made between long-tailed or ermines) in 20 of 33 study areas in Alberta, Manitoba, Saskatchewan, North Dakota, and South Dakota. The Conservation Reserve Program likely has enabled weasel populations to increase in the southern part of the prairie pothole region.

Raccoon

Raccoons are most common in the southeast part of the region (Sargeant et al.

1993). In spring, density of adult raccoons in North Dakota was estimated at about one or less per square kilometer (Fritzell 1978). Raccoon populations in the prairie pothole region consist of territorial males with large home ranges; females, often attended by young from the previous year, have smaller home ranges (Fritzell 1978). In North Dakota, home ranges of males averaged 2,560 hectares, and those of pregnant females or females with young averaged 806 hectares. Females did not maintain exclusive home ranges.

Raccoons are semiaquatic omnivorous carnivores that have benefited from European settlement of the prairie pothole region. Agricultural crops (cereal grain, corn, and sunflowers) provide a stable food base that replaces mast (fallen nuts on the forest floor) consumed by raccoons in forested areas (Greenwood 1982). Raccoons probably occurred only in riparian and wooded areas in the southeastern portion of the region before European settlement (Bailey 1926). Raccoons were absent in Canada, except possibly in southern Manitoba (Houston and Houston 1973). After European settlement, raccoons became more widely distributed in the southern portion of the prairie pothole region, although populations were low. In the 1940's, raccoons were abundant throughout much of the southeastern portion of the prairie pothole region. In the 1950's, populations in Canada expanded (Lynch 1971; Houston and Houston 1973), and by the 1960's, raccoons were considered a major predator of nesting canvasbacks in southwestern Manitoba (Stoudt 1982). Principal causes of raccoon death in North Dakota are related to human activities (shooting, vehicle impact, and so forth). There is evidence of a negative relationship between coyotes and raccoons (Johnson et al. 1989), suggesting that coyotes may suppress raccoon populations. Suppression of coyotes in the 1950's may have contributed to the range extension of raccoons.

Acknowledgments

We thank H. Umber, North Dakota Department of Fish and Game, Bismarck, for providing his photographs, and *North Dakota Outdoors* for permission to reprint the information in Figure 3.

See end of chapter for references

Authors

Raymond J. Greenwood
Marsha A. Sovada
U.S. Geological Survey
Biological Resources Division
Northern Prairie Science Center
Jamestown, North Dakota 58401-7317

Prairie Integrity and Legacies

Intercommunity Management: Prairie Integrity

Integrity here means maintaining species and ecological processes characteristic to a particular landscape (Samson and Knopf 1993); this is an emerging goal in resource conservation (Angermeier and Karr 1994). The human-caused breakdown of barriers to dispersal that has permitted invasion of nonindigenous species has caused the extinction of more grassland species than any factor except habitat loss (D'Antinio and Vitousek 1992). Nonindigenous species include exotics, which are transported beyond their natural range, and aliens, those that colonize an altered landscape. Introducing nonindigenous species may increase the number of local species, but it reduces integrity, above- and belowground, and also the number of native species, both aquatic and terrestrial.

A more subtle threat to integrity is loss of genetic diversity. Species hybridize along forested corridors that now fragment the Great Plains (Knopf 1986). Human activity, either accidental or deliberate, moves species from one place to another at ever-increasing rates (Knopf 1992). As a result, species that evolved in isolation from one another are forced into contact. In terms of conservation of biological diversity, the loss of six bird subspecies due to the hybridization arising from these forested stepping stones and artificial corridors (Knopf 1986) rivals the loss of three species attributed to fragmentation of the eastern deciduous forest.

These recent forest patches and woody corridors that border rivers on the Great Plains also favor movement of reptiles and mammals from east to west, which contributes to the degradation of the biological diversity and integrity of the Great Plains (Knopf and Scott 1990). In 1842, in eastern Colorado, the explorer John C. Frémont observed that "antelope were tolerably abundant, wolves were seen in great numbers, and buffalo absolutely covered the plains on both sides of the (South Platte) river" but reported no deer (*in* Nevins 1956). In recent years, deer abundance has increased markedly, particularly that of the eastern white-tailed deer, which may replace the mule deer, known to have occurred on the western plains since before European settlement (Kufeld and Bowden 1995). Hybridization between the two deer is known to occur (Stubblefield et al. 1986).

Nonindigenous species now account for 13% to 30% of prairie species (U.S. Congress, Office of Technology Assessment 1993; C. Freeman, Kansas Natural Heritage Program, Lawrence, personal communication). Increases in distribution and abundance are inevitable without action to prevent them, as evident in the naturalization of Russian-olive trees in the western United States (Olson and Knopf 1986). Russian-olives, which were introduced from Europe in colonial times, range across the Great Plains into the West. In agricultural areas, this species interferes with farming operations; it also hinders management activities on national wildlife refuges, increases degradation of river channels, contributes to declines in river levels, and supplants native riparian tree species. Its adaptability and resistance to control measures, two life-history traits shared with other non-indigenous species, will continue to add to the management concerns associated with the ever-growing number of other nonindigenous plant and animal species.

Intracommunity Management: Prairie Legacies

The approximate action to take after inter-community management is to identify and retain a set of species and natural processes that sustain communities characteristic of a particular landscape (Chaplin et al. 1996). The Nature Conservancy has identified significant concentrations—legacies—of prairie species that are rare or of declining abundance (Figs. 16–23). Principally these are species that are federally listed or are species of concern that occur within certain communities and ranges of environmental features, including those from prairie wetlands to cottonwood savannahs.

Recommending the restoration of ecological processes in conservation is not new (Leopold 1933). Understanding scale, spatial and temporal, in management is new (Gibson et al. 1993).

☐ Great Plains

■ Landscapes of biological significance

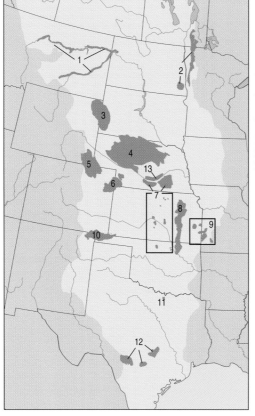

Fig. 17. Confluence of turbid waters of the Yellowstone and clear waters of the Missouri as a result of the Ft. Peck Reservoir just upstream.

Fig. 16. Priority landscapes of biological significance in the Great Plains: 1) Upper Missouri and Yellowstone rivers and watersheds in Montana and North Dakota; this area is an example of a free-flowing Great Plains river and watershed (see Fig. 17) ; 2) Glacial Lake Agassiz Interbeach Area in North Dakota and Minnesota; this area has a number of large, intact expanses of tall-grass prairie; 3) Black Hills and grasslands in South Dakota and Wyoming; these are two of the largest publicly owned examples of short-grass and mixed-grass prairies (see Fig. 18); 4) Sandhills in Nebraska and South Dakota; this is the largest dune system and one of the largest expanses of native grassland left in North America; 5) Western high plains grassland in Colorado, Nebraska, and Wyoming, which includes the Pawnee National Grassland and other adjacent short-grass prairie habitat; 6) Arikaree Sandsage Prairie in Colorado, Kansas, and Nebraska, which includes sandsage and an example of the rare cottonwood–switchgrass savanna (see Fig. 19); 7) Central Plains Wetlands in Nebraska, Kansas, and Oklahoma (areas in box plus areas indicated above box); this area is a string of significant wetlands (see Fig. 20); 8) Flint Hills in Kansas and Oklahoma, which is the largest remaining area of native tall-grass prairie; 9) Osage Cuestas Tallgrass in Kansas (in box), which has rolling to level tall-grass prairie with examples of wet savanna and bottomland forest; 10) Upper Cimarron Mesas in Colorado, Kansas, New Mexico, and Oklahoma; within this area are extensive grasslands and mesas, along with the head-waters of the Cimarron River (see Fig. 21); 11) Fort Worth Prairie in Oklahoma and Texas, which is an unbroken tall-grass prairie; 12) Texas Hill Country in Texas, which has four differing land-scapes, each significant to a unique assemblage of natural and rare communities (see Fig. 22); and 13) Central Platte River in Kansas; this is a shallow, braided river of immense wetland importance (see Fig. 23).

Fig. 18. Black Hills: Cathedral Spires, South Dakota.

Fig. 19. Arikaree River, Colorado.

Fig. 20. Rainwater Basin, Nebraska.
© Nebraskaland Magazine/Nebraska Game and Parks Commission

Fig. 22. Texas Hill Country: San Marcos River, Texas. The San Marcos runs through the Hill Country and is important to a large number of endemic salamanders.

Fig. 21. Upper Cimarron Mesas: Lasa, Mesa de Maya, Colorado, Cobert Mesa.

Resource and Research Needs

Solutions to the deterioration of grassland resources appear to revolve around a single emerging concept—*sustainability*—and conform to a proposed strategy. It is important to increase our understanding of the long-term sustainability both of populations of species and of the overall ecosystem. The strategy arises from the following conceptual principles:

• Support public, private, and governmental prairie conservation initiatives as a step toward the long-term goal of grassland sustainability.

The Western Governors' Association's Great Plains Program, the first broad-scale ecosystem management effort in the United States, seeks to demonstrate that economic and environmental interests are served by preventing declines in the numbers of prairie species and their host ecosystems (Clark 1996). The program is built on broad-based science (Johnson and Bouzaher 1996).

Fig. 23. Platte River with sandhill cranes.

© Nebraskaland Magazine/Nebraska Game and Parks Commission

- Develop information that recognizes the biological and ecological significance of prairie communities.

The International Institute for Sustainable Development's Sustainable Development for the Great Plains Policy Analysis (Tyrchniewicz and Wilson 1994) links, at an ecosystem level, the well-being of grassland biological resources, particularly soil and water, and society in general. Increasing evidence suggests that dry ecosystems, whether in South America (Mares 1992) or North America (Samson and Knopf 1994), are unusually diverse compared with wet or rain-forest ecosystems. In the face of global warming, the health of planet Earth may depend on grasslands because they are superior carbon sinks compared to forests with similar environmental characteristics (Seastedt and Knapp 1993).

- Identify, inventory, and conserve prairie-limited animals and plants, particularly the large number of plant and invertebrate species.

Sustainability depends on biological diversity to keep all ecological systems, aquatic and terrestrial, functioning and healthy (Lubchenco et al. 1991). Sustainability and biodiversity are two sides of the same coin (Raven 1991). Information and appropriate actions are required to minimize any negative effects on prairie genetic stock and thus on biodiversity, because diverse natural ecosystems help maintain hydrological cycles, regulate climate, absorb and break down pollutants, and contribute to the process of soil formation (Tyrchniewicz and Wilson 1994).

- Evaluate the status of species of concern and other sensitive species, and encourage measures to reverse downward trends in population numbers of prairie species and rare communities.

Putting the maintenance of diversity as a top priority builds ecological knowledge accessible to the public and environmental decision makers (public and private) and provides opportunities to cooperate in conservation of rare species and communities (Chaplin et al. 1996). Sites for tourism and recreation are often identified in the process as well. The net payoff of understanding and displaying diversity is identification of areas of endemism as an essential aid in planning for the conservation of the nation's biodiversity. An urgent need exists to further develop and refine this process on a national and international basis (International Council for Bird Protection 1992).

- Prairie management should mimic the natural disturbance regime to take advantage of preselected traits of prairie species.

The importance of disturbance in shaping grassland communities (Figs. 3 and 10) and ecosystem dynamics is recognized, yet significant questions remain to be answered on the relationship between disturbance and species persistence (Bragg and Steuter 1996; Steinauer and Collins 1996; Weaver et al. 1996). The purpose of conservation is not to conserve species per se but to conserve interactions among species and processes that maintain the health and productivity of communities and ecosystems (Odum 1992).

The Conservation Reserve Program is one of the most popular and successful conservation programs ever implemented by the U.S. Department of Agriculture. By establishing needed grassland, the Conservation Reserve Program has improved game species habitat, prevented loss of topsoil, improved water quality by reducing pesticide and fertilizer runoff, and provided billions of dollars in environmental benefits over the life of the program. The act, however, needs refocusing to be of major conservation benefit to endemic grassland species (Allen 1993).

- Education programs should play an integral role in ensuring conservation of grasslands.

The premise in prairie conservation is that attitudes of individual landowners and the community as a whole play decisive roles in determining the eventual fate of grasslands (Mack 1996). Programs must foster a climate favorable to grassland conservation as an integral component of agricultural land management (Dyson 1996), to the development of covenant agreements to protect remnants (World Wildlife Fund Canada 1988), to cooperative conservation programs between private and neighboring government-managed lands (Bueseler 1996), and to the importance of science-based management (Johnson and Bouzaher 1996).

Almost a half-century has passed since Weaver (1954) noted that the disappearance of a major unit of vegetation—the North American prairie—is an event worth considering. Fortunately, to a growing segment of our society, prairie "looms as large as the universe, as intimate as a village" (Least Heat Moon 1991).

Authors

Fred B. Samson
U.S. Forest Service
Northern Region
200 East Broadway
Missoula, Montana 59807

Fritz L. Knopf
U.S. Geological Survey
Biological Resources Division
Midcontinent Ecological Science
Center
4512 McMurry Avenue
Fort Collins, Colorado 80525

Wayne R. Ostlie
The Nature Conservancy
Great Plains Program
1313 Fifth Street, Suite 323
Minneapolis, Minnesota 55414

Cited References

Allen, A. W. 1993. Wildlife habitat criteria in relation to future use of CRP lands. Pages 41–88 *in* Proceedings of the Great Plains Agricultural Council, Rapid City, S. Dak.

Anderson, R. C. 1983. The eastern prairie–forest transition—an overview. Pages 8–92 *in* R. Brewer, editor. Proceedings of the eighth North American prairie conference, Western Michigan University, Kalamazoo.

Anderson, S. L. 1990. The historic role of fire in the North American grassland. Pages 8–18 *in* S. L. Collins and L. L. Wallace, editors. Fire in North American tall-grass prairies. University of Oklahoma Press, Norman.

Angermeier, P. L., and J. R. Karr. 1994. Biological integrity versus biological diversity as policy directives. BioScience 44:690–697.

Bakker, J. P., and J. C. Ruyter. 1981. Effects of five years of grazing on a salt-water marsh. Vegetatio 44:81–100.

Ball, I. J., T. E. Martin, and J. K. Ringleman. 1994. Conservation of non-game birds and waterfowl: conflict or compliment. Transactions of the North American Wildlife and Natural Resources Conference 59:337–347.

Bazzaz, F. A., and J. A. D. Parrish. 1982. Organization of grassland communities. Pages 233–254 *in* J. R. Estes, R. J. Tyrl, and J. N. Brunken, editors. Grasses and grasslands. Systematics and ecology. University of Oklahoma Press, Norman.

Bethke, R. W., and T. D. Nudds. 1995. Effects of climate change and land use on duck abundance in Canadian prairie–parklands. Ecological Applications 5:588–600.

Bolen, E. G., G. A. Baldassarre, and F. S. Guthrey. 1989. Playa lakes. Pages 341–366 *in* L. M. Smith, R. L. Pederson, and R. M. Kaminski, editors. Habitat management for migrating and wintering waterfowl. Texas Tech University Press, Lubbock.

Bragg, T. B., and A. A. Steuter. 1996. Mixed-grass prairies of the North American Great Plains. Pages 53–66 *in* F. B. Samson and F. L. Knopf, editors. Prairie conservation: preserving North America's most endangered ecosystem. Island Press, Covelo, Calif.

Brooks, D. R., and D. A. McLennan. 1993. Historical ecology: examining phylogenetic components of community evolution. Pages 267–280 *in* R. E. Ricklefs and D. Schluter, editors. Species diversity in ecological communities. University of Chicago Press, Ill.

Bueseler, P. 1996. Sustainable grassland utilization and conservation in prairie agricultural landscapes. Pages 221–230 *in* F. B. Samson and F. L. Knopf, editors. Prairie conservation: preserving North America's most endangered ecosystem. Island Press, Covelo, Calif.

Burke, I. C., T. G. F. Kittel, W. R. Laurenroth, P. Snook, C. M. Yonker, and W. J. Parton. 1991. Regional analysis of the central Great Plains. BioScience 41:685–692.

Canadian Wildlife Service and U.S. Fish and Wildlife Service. 1994. Waterfowl population status, 1994. U.S. Fish and Wildlife Service, Washington D.C. 39 pp.

Carlander, K. D, R. S. Campbell, and W. H. Irwin. 1986. Midcontinent states. Pages 317–348 *in* D. G. Frey, editor. Limnology of North America. University of Wisconsin Press, Madison.

Chaplin, S. J., W. R. Ostlie, R. E. Schneider, and J. S. Kenney. 1996. Conservation planning on the Great Plains. Pages 187–202 *in* F. B. Samson and F. L. Knopf, editors. Prairie conservation: preserving North America's most endangered ecosystem. Island Press, Covelo, Calif.

Clark, J. 1996. The Great Plains initiative. Pages 169–174 *in* F. B. Samson and F. L. Knopf, editors. Prairie conservation: preserving North America's most endangered ecosystem. Island Press, Covelo, Calif.

Collins, S. L. 1991. Fire frequency and community structure in tall-grass prairie vegetation. Ecology 73:2001–2006.

Collins, S. L., and L. L. Wallace, editors. 1990. Fire in the North American tall-grass prairies. University of Oklahoma Press, Norman. 175 pp.

Cross, F. B., R. L. Mayden, and J. D. Stewart. 1986. Fishes in the western Mississippi drainage. Pages 363–412 *in* C. H. Hocutt and E. O. Wiley, editors. The zoogeography of North American freshwater fishes. John Wiley & Sons, New York.

Cross, F. B., and R. E. Moss. 1987. Historic changes in fish communities and aquatic habitats in plains streams of Kansas. Pages 155–165 *in* W. J. Matthews and D. C. Heins, editors. Community and evolutionary ecology of North American stream fishes. University of Oklahoma Press, Norman.

Currier, P. J. 1982. The floodplain vegetation of the Platte River: phytosociology, forest development and seedling establishment. Ph.D. dissertation, Iowa State University, Ames. 212 pp.

D'Antinio, C. M., and P. M. Vitousek. 1992. Biological invasions by exotic grasses, the grass/fire cycle, and global change. Annual Review of Ecology and Systematics 23:63–87.

Dodd, C. K., Jr. 1983. A review of the status of the Illinois mud turtle *Kinosternon flavescens spooneri* Smith. Biological Conservation 27:141–156.

Dugan, J. T., T. McGrath, and R. B. Zelt. 1994. Water-level changes in the High Plains aquifer predevelopment to 1992. U.S. Geological Survey Water-Resources Investigations Report 94-4027. 34 pp.

Dyson, I. W. 1996. Canada's prairie conservation action plan: new directions for the millennium. Pages 175–186 *in* F. B.

Samson and F. L. Knopf, editors. Prairie conservation: preserving North America's most endangered ecosystem. Island Press, Covelo, Calif.

Echelle, A. A., G. R. Luttrell, R. D. Larson, A. V. Zale, W. L. Fisher, and D. M. Leslie, Jr. 1995. Decline of native prairie fishes. Pages 303–305 *in* E. T. LaRoe, G. S. Farris, C. E. Puckett, P. D. Doran, and M. J. Mac, editors. Our living resources: a report to the nation on the distribution, abundance, and health of U.S. plants, animals, and ecosystems. U.S. Department of the Interior, National Biological Service, Washington, D.C.

Estes, J. R., R. J. Tyrl, and J. N. Brunken, editors. 1982. Grasses and grasslands. Systematics and ecology. University of Oklahoma Press, Norman. 312 pp.

Gibson, D. J., and L. C. Hulbert. 1987. Effects of fire, topography and year-to-year climatic variation on species composition in tall-grass prairie. Vegetatio 72:175–182.

Gibson, D. J., T. R. Seastedt, and J. Briggs. 1993. Management practices in tall-grass prairie: large- and small-scale experiments on species conservation. Journal of Applied Ecology 30:247–255.

Goolsby, D. A., W. A. Battaglin, and E. M. Thurman. 1993. Occurrence and transport of agricultural chemicals in the Mississippi River Basin, July through August 1993. U.S. Geological Survey Circular 1120-C. 22 pp.

Gould, F. W. 1962. Texas plants—a checklist and ecological summary. Texas Agricultural Experiment Station Bulletin MP-585. 17 pp.

Great Plains Flora Association. 1986. Flora of the Great Plains. University of Kansas Press, Lawrence. 1392 pp.

Grossman, D. H., K. L. Goodin, and C. L. Ruess. 1994. Rare plant communities of the conterminous United States. The Nature Conservancy, Arlington, Va. 347 pp.

Hall, E. R., and K. R. Kelson. 1959. Mammals of North America. Volumes 1 and 2. The Ronald Press, New York. 1083 pp.

Heat Moon, W. L. 1991. Prairy-Erth. Houghton Mifflin Company, Boston. 624 pp.

Hobaugh, W. C., C. D. Stutzenbaker, and E. L. Flickinger. 1989. The rice prairies. Pages 367–384 *in* L. M. Smith, R. L. Pederson, and R. M. Kaminski, editors. Habitat management for migrating and wintering waterfowl in North America. Texas Tech University Press, Lubbock.

Hobbs, R. J., and L. F. Huenneke. 1992. Disturbance, diversity, and invasion: implications for conservation. Conservation Biology 6:324–337.

Howe, H. F. 1994. Managing species diversity in tall-grass prairie: assumptions and implications. Conservation Biology 8:705–712.

Huntzinger, T. L. 1996. Surface water: a critical Great Plains resource. *In*

S. Johnson and A. Bouzaher, editors. Conservation of Great Plains ecosystems: current science, future options. Kluwer Academic Press, Dordrecht, The Netherlands.

International Council for Bird Protection. 1992. Putting biodiversity on the map: priority areas for conservation. International Council for Bird Protection, Cambridge, England. 90 pp.

James, S. W. 1991. Soil, nitrogen, phosphorus, and organic matter processing by earthworms in tall-grass prairie. Ecology 72:2101–2109.

Johnson, S., and A. Bouzaher, editors. 1996. Conservation of Great Plains ecosystems: current science, future options. Kluwer Academic Press, Dordrecht, The Netherlands.

Johnson, W. C. 1994. Woodland expansion in the Platte River, Nebraska: patterns and causes. Ecological Monographs 64:45–84.

Jones, J. K., D. M. Armstrong, and J. R. Choate. 1985. Guide to mammals of the plains states. University of Nebraska Press, Lincoln. 371 pp.

Knopf, F. L. 1986. Changing landscapes and the cosmopolitism of eastern Colorado avifauna. Wildlife Society Bulletin 14:132–142.

Knopf, F. L. 1992. Faunal mixing, faunal integrity, and the biopolitical template for diversity conservation. Transactions of the North American Wildlife and Natural Resources Conference 57:330–342.

Knopf, F. L. 1996. Prairie legacies—birds. Pages 135–148 in F. B. Samson and F. L. Knopf, editors. Prairie conservation: preserving North America's most endangered ecosystem. Island Press, Covelo, Calif.

Knopf, F. L., and F. B. Samson. 1995. Conserving the biotic integrity of the Great Plains. Pages 121–133 in S. Johnson and A. Bouzaher, editors. Conservation of Great Plains ecosystems: current science, future options. Kluwer Academic Press, Dordrecht, The Netherlands.

Knopf, F. L., and F. B. Samson. 1997. Ecology and conservation of Great Plains vertebrates. Ecological Studies 125. Springer, New York. 320 pp.

Knopf, F. L., and M. L. Scott. 1990. Altered flows and created landscapes in the Platte River headwaters, 1840–1990. Pages 47–70 in J. M. Sweeney, editor. Management of dynamic ecosystems. North Central Section of the Wildlife Society, West Lafayette, Ind.

Koplin, D. W., M. R. Burkart, and E. M. Thurman. 1994. Herbicides and nitrate in near-surface aquifers in the midcontinental United States, 1991. U.S. Geological Survey Water-Supply Paper 2413. 34 pp.

Kresl, S., J. Leach, and C. A. Lively. 1996. Working partnerships for conserving the nation's prairie pothole ecosystem. Pages 203–210 in F. B. Samson and F. L. Knopf, editors. Prairie conservation: preserving North America's most endangered ecosystem. Island Press, Covelo, Calif.

Küchler, A. W. 1964. Potential natural vegetation of the conterminous United States [map]. American Geographical Society, New York.

Kufeld, R. C., and D. C. Bowden. 1995. Mule deer and white-tailed deer inhabiting eastern Colorado plains river bottoms. Colorado Division of Wildlife Technical Publication 41. Denver. 58 pp.

Kumar, R., R. J. Lavigne, J. E. Lloyd, and R. E. Pfadt. 1976. Insects of the Central Great Plains Experimental Range, Pawnee National Grasslands. Agricultural Experiment Station Scientific Report 32. University of Wyoming, Laramie. 74 pp.

Lauenroth, W. K., and O. E. Sala. 1992. Long-term forage production of North American steppe. Ecological Applications 2:397–403.

Leopold, A. 1933. Game management. Charles Scribner, New York. 438 pp.

Lubchenco, J., A. M. Olson, L. B. Brubacker, S. R. Carpenter, M. M. Holland, S. P. Hubbell, S. A. Levin, J. A. MacMahon, P. A. Matson, J. M. Melillo, H. A. Mooney, H. R. Pulliam, L. A. Real, P. J. Regal, and P. G. Risser. 1991. The sustainable biosphere initiative: an ecological research agenda. Ecology 72:371–412.

Mack, G. 1996. The Sandhill management plan: a partnership initiative. Pages 241–248 in F. B. Samson and F. L. Knopf, editors. Prairie conservation: preserving North America's most endangered ecosystem. Island Press, Covelo, Calif.

Malecki, R., and B. Blossey. 1994. Insect biological weed control: an important and underutilized management tool for maintaining native plant communities threatened by exotic plant introductions. Transactions of the North American Wildlife and Natural Resources Conference 59:400–404.

Mares, M. A. 1992. Neotropical mammals and the myth of Amazonian biodiversity. Science 255:976–979.

Matthews, W. J. 1988. North American prairie streams as systems for ecological study. Journal of the North American Benthological Society 7:387–409.

Meagher, M. M. 1978. Bison. Pages 123–134 in J. L. Schmidt and D. L. Gilbert, editors. Big game of North America. Stackpole Books, Harrisburg, Pa.

Mengel, R. M. 1970. The North American central plains as an isolating agent in bird speciation. Pages 280–340 in W. Dort and J. K. Jones, editors. Pleistocene and Recent environments of the central Great Plains. University of Kansas Press, Lawrence.

Milchunas, D. G., and W. K. Lauenroth. 1993. Quantitative effects of grazing on vegetation and soils over a global range of environments. Ecological Monographs 63:327–366.

Miller, B., G. Ceballas, and R. Reading. 1994. The prairie dog and biotic diversity. Conservation Biology 8:677–681.

National Research Council. 1994. Rangeland health. New methods to classify, inventory, and monitor rangelands. National Academy Press, Washington, D.C. 180 pp.

Nevins, A., editor. 1956. Narratives of exploration and adventure by John Charles Fremont. Longmans, Green, and Company, New York. 532 pp.

Odum, E. P. 1992. Great ideas in ecology for the 1990's. BioScience 42:542–545.

Olson, T. E., and F. L. Knopf. 1986. Naturalization of Russian-olive in the western United States. Journal of Applied Forestry 1:65–69.

Opler, P. A. 1995. Species richness and trends of western butterflies and moths. Pages 172–174 in E. T. LaRoe, G. S. Farris, C. E. Puckett, P. D. Doran, and M. J. Mac, editors. Our living resources: a report to the nation on the distribution, abundance, and health of U.S. plants, animals, and ecosystems. U.S. Department of the Interior, National Biological Service, Washington, D.C.

Otte, D. 1981. The North American grasshoppers 1, Acrididae. Harvard University Press, Cambridge, Mass. 275 pp.

Peterson, G. A., and C. V. Cole. 1996. Productivity of Great Plains soils: past, present and future. Pages 325–342 in S. Johnson and A. Bouzaher, editors. Conservation of Great Plains ecosystems: current science, future options. Kluwer Academic Press, Dordrecht, The Netherlands.

Powell, J. A. 1995. Lepidoptera inventories in the continental United States. Pages 168–171 in E. T. LaRoe, G. S. Farris, C. E. Puckett, P. D. Doran, and M. J. Mac, editors. Our living resources: a report to the nation on the distribution, abundance, and health of U.S. plants, animals, and ecosystems. U.S. Department of the Interior, National Biological Service, Washington, D.C.

Raven, P. H. 1991. The politics of preserving biodiversity. BioScience 40:769–774.

Reichman, O. J., J. H. Benedix, Jr., and T. R. Seastadt. 1993. Distinct animal-generated edge effects in a tall-grass prairie community. Ecology 74:1281–1285.

Risser, P. G. 1990. Landscape processes and the vegetation of the North American grassland. Pages 133–146 in S. L. Collins and L. L. Wallace, editors. Fire in North American tall-grass prairies. University of Oklahoma Press, Norman.

Risser, P. G., E. C. Birney, H. D. Blocker, S. W. May, W. J. Parton, and J. A. Wiens. 1981. The true prairie ecosystem. Hutchinson Ross Publishing Company, Stroudsburg, Pa. 557 pp.

Royer, R. A. 1992. Butterflies of North Dakota: an atlas and guide. Minot State University, Minot, N. Dak. 192 pp.

Sampson, R. N. 1981. Farmland or wasteland: a time to choose: overcoming the threat to America's farm and food future. Rodale Press, Emmaus, Pa. 481 pp.

Samson, F. B., and F. L. Knopf. 1993. Managing biological diversity. Wildlife Society Bulletin 21:509–514.

Samson, F. B., and F. L. Knopf. 1994. Prairie conservation in North America. BioScience 44:418–421.

Schwalbe, C. R., and P. C. Rosen. 1989. Preliminary report on effect of bullfrogs on wetland herptofaunas in southeastern Arizona. Pages 166–179 *in* R. C. Szaro, K. E. Severson, and D. R. Patton, editors. Management of amphibians, reptiles, and small mammals in North America. U.S. Forest Service, Rocky Mountain Forest and Range Experiment Station, General Technical Report RM-166.

Scott, N. J. 1997. Evolution and management of the North American grassland herpetofauna. Conservation Biology. In press.

Seastedt, T. R., and A. K. Knapp. 1993. Consequences of nonequilibrium resource availability across multiple time scales: the transient maxima hypothesis. American Naturalist 141:621–633.

Simms, P. L. 1988. Grasslands. Pages 265–286 *in* M. G. Barbour and W. D. Billings, editors. North American terrestrial vegetation. Cambridge University Press, New York.

Smolik, J. D. 1974. Nematode studies at the Cottonwood site. United States International Biological Program Technical Report 251. Colorado State University, Fort Collins. 80 pp.

Steinauer, E. M., and S. L. Collins. 1996. Disturbance as a management component in tall-grass prairie. Pages 39–52 *in* F. B. Samson and F. L. Knopf, editors. Prairie conservation: preserving North America's most endangered ecosystem. Island Press, Covelo, Calif.

Stubblefield, S. S., R. J. Warren, and B. R. Murphy. 1986. Hybridization of free-ranging white-tailed and mule deer in Texas. Journal of Wildlife Management 50:688–690.

Summers, C. A., and R. L. Linder. 1978. Food habits of the black-tailed prairie dog in western South Dakota. Journal of Range Management 31:134–136.

Tiner, R. W. 1984. Wetlands of the United States: current status and recent trends. U.S. Fish and Wildlife Service, Washington, D.C. 43 pp.

Tyrchniewicz, A., and A. Wilson. 1994. Sustainable development for the Great Plains policy analysis. International Institute for Sustainable Development, Winnipeg, Manitoba, Canada. 35 pp.

Umbanhowar, C. E. 1992. Abundance, vegetation, and environment of four patch types in a northern mixed prairie. Canadian Journal of Botany 70:277–284.

U.S. Congress, Office of Technology Assessment. 1993. Harmful nonindigenous species in the United States. OTA-F-565. U.S. Congress, Office of Technology Assessment, Washington, D.C. 391 pp.

U.S. Fish and Wildlife Service. 1986. North American Waterfowl Management Plan. U.S. Department of the Interior, Washington, D.C. 38 pp.

Van Valkenburgh, B., and C. M. Janis. 1993. Historical diversity patterns in North American large herbivores and carnivores. Pages 330–340 *in* R. E. Ricklefs and D. Schluter, editors. Species diversity in ecological communities. University of Chicago Press, Ill.

Weaver, J. E. 1954. North American prairie. Johnson Publishing Company, Lincoln, Nebr. 318 pp.

Weaver, J. E. 1968. Prairie plants and their environment. A fifty-year study in the Midwest. University of Nebraska Press, Lincoln. 276 pp.

Weaver, T., E. M. Payson, and D. L. Gustafson 1996. Adequacy for conservation for short-grass prairie. Pages 67–76 *in* F. B. Samson and K. L. Knopf, editors. Prairie conservation: preserving North America's most endangered ecosystem. Island Press, Covelo, Calif.

Whitcomb, R. F., A. L. Hicks, H. D. Blocker, and D. E. Lynn. 1994. Biogeography of leafhoppers: specialists of the short-grass prairie: evidence for the roles of phenology and phylogeny in determination of biological diversity. American Entomologist 40:19–35.

Wiens, J. A. 1974. Climatic instability and the "ecological saturation" of bird communities in North American grasslands. Condor 76:385–400.

Williams, J., and P. Diebel. 1996. The economic value of prairie. Pages 19–35 *in* F. B. Samson and F. L. Knopf, editors. Prairie conservation: preserving North America's most endangered ecosystem. Island Press, Covelo, Calif.

Williams, J. D., M. L. Warren, Jr., K. S. Cummings, J. L. Harris, and R. J. Neves. 1993. Conservation status of freshwater mussels of the United States and Canada. Fisheries 18:6–22.

Williams, W. G., and M. G. Wolman. 1986. Effects of dams and reservoirs on surface-water hydrology changes in rivers downstream from dams. Pages 83–88 *in* D. W. Moody, E. G. Chase, and D. R. Aroson, editors. National water summary 1985—hydrologic events and surface water reservoirs. U.S. Geological Survey Water-Supply Paper 2300.

World Wildlife Fund Canada. 1988. Prairie conservation action plan. World Wildlife Fund Canada, Toronto. 38 pp.

Zimmerman, J. L. 1992. Density-dependent factors affecting the avian diversity of the tall-grass prairie. Wilson Bulletin 104:85–94.

Tall-grass Prairie Butterflies and Birds

Fretwell, S. 1973. The regulation of bird populations on Konza Prairie: the effects of events off the prairie. Pages 71–76 *in* L. C. Hulbert, editor. Third Midwest prairie conference proceedings. Kansas State University, Manhattan.

Graber, R. R., and R. W. Graber. 1963. A comparative study of bird populations in Illinois, 1906–1909 to 1956–58. Illinois Natural History Bulletin 28:383–519.

Hands, H. M., R. A. Drobney, and M. R. Ryan. 1989. Status of the Henslow's sparrow in the North Central United States. U.S. Fish and Wildlife Service, Twin Cities, Minn. 12 pp.

Herkert, J. R. 1994. Status and habitat selection of the Henslow's sparrow in Illinois. Wilson Bulletin 106:35–45.

Hopkins, J. J. 1991. Management of semi-natural lowland dry grasslands. Pages 119–124 *in* P. D. Goriup, L. A. Batten, and J. A. Norton, editors. The conservation of lowland dry grasslands in Europe. Joint Nature Conservation Council, Peterborough, England.

Illinois Natural History Survey. 1983. The declining grassland birds. Illinois Natural History Survey Notes 227:1–2.

Johnson, D. H., and M. D. Schwartz. 1993. The Conservation Reserve Program: habitat for grassland birds. Great Plains Research 3:273–295.

Johnson, K. 1986. Prairie and plains disclimax and disappearing butterflies in the central United States. Atala 10–12:20–30.

Kahl, R. B., T. S. Baskett, J. A. Ellis, and J. N. Burroughs. 1985. Characteristics of summer habitats of selected nongame birds in Missouri. Missouri Agricultural Experiment Station Research Bulletin 1056. Columbia. 155 pp.

Kantrud, H. A. 1981. Grazing intensity effects on the breeding avifauna of North Dakota native grasslands. Canadian Field-Naturalist 94:404–417.

Knopf, F. L. 1994. Avian assemblages on altered grasslands. Pages 247–257 *in* J. R. Jehl, Jr., and N. K. Johnson, editors. A century of avifaunal change in western North America. Cooper Ornithological Society, Studies in Avian Biology 15.

McNicholl, M. K. 1988. Ecological and human influences on Canadian populations of grassland birds. Pages 1–25 *in* P. D. Goriup, editor. Ecology and conservation of grassland birds. Technical Publication 7. International Council for Bird Preservation, Cambridge, England.

Moffat, M., and N. McPhillips. 1993. Management for butterflies in the northern Great Plains: a literature review and guidebook for land managers. U.S. Fish and Wildlife Service, Pierre, S. Dak. 19 pp.

Opler, P. A. 1981. Management of prairie habitat for insect conservation. Journal of the Natural Areas Association 1(4):3–6.

Opler, P. A., and G. O. Krizek. 1984. Butterflies east of the Great Plains. The Johns Hopkins University Press, Baltimore, Md. xvii + 294 pp.

Peterjohn, B. G., J. R. Sauer, and W. A. Link. 1994. The 1992 and 1993 summary of the North American Breeding Bird Survey. Bird Populations 2:46–61.

Skinner, R. M. 1975. Grassland use patterns and prairie bird populations in Missouri. Pages 171–180 *in* M. K. Wall, editor. Prairie: a multiple view. University of North Dakota, Grand Forks.

Skinner, R. M., T. S. Baskett, and M. D. Blenden. 1984. Bird habitat on Missouri prairies. Terrestrial Series 14. Missouri Department of Conservation, Jefferson City. 37 pp.

Smith, D. J., and C. R. Smith. 1992. Henslow's sparrow and grasshopper sparrow: a comparison of habitat use in Finger Lakes National Forest, New York. Bird Observer 20:187–194.

Solecki, M. K., and T. Toney. 1986. Characteristics and management of Missouri's public prairies. Pages 168–171 *in* G. K. Clambey and R. H. Pemble, editors. Proceedings of the ninth North American prairie conference. Tricollege University Center for Environmental Studies, Fargo, N. Dak.

Swengel, A. 1993. Regal fritillary: prairie royalty. American Butterflies 1(1):4–9.

Swengel, A. B. 1996. Effects of fire and hay management on abundance of prairie butterflies. Biological Conservation 76:73–85.

Zimmerman, J. L. 1988. Breeding season habitat selection by the Henslow's sparrow (*Ammodramus henslowii*). Wilson Bulletin 100:17–24.

Zimmerman, J. L. 1992. Density-independent factors affecting the avian diversity of the tall-grass prairie community. Wilson Bulletin 104:85–94.

Amphibians of the Northern Grasslands

Anonymous. 1907. Frogs. Okoboji Protective Association Bulletin 3:5.

Bailey, R. M., and M. K. Bailey. 1941. The distribution of Iowa toads. Iowa State College Journal of Science 15:169–177.

Barrett, W. 1964. Frogging in Iowa. Annals of Iowa, 3rd Series 37:362–365.

Blanchard, F. N. 1923. The amphibians and reptiles of Dickinson County, Iowa. University of Iowa Studies in Natural History, Lakeside Laboratory Studies 10:19–26.

Hayes, M. P., and M. R. Jennings. 1986. Decline of ranid frog species in western North America: are bullfrogs (*Rana catesbeiana*) responsible? Journal of Herpetology 20:490–504.

Johnson, D. H., and D. L. Larson. 1994. Declines in amphibian populations: real or perceived? Proceedings of the North Dakota Academy of Science 48:15.

Lannoo, M. J., K. Lang, T. Waltz, and G. Phillips. 1994. An altered amphibian assemblage: Dickinson County, Iowa, 70 years after Frank Blanchard's survey. American Midland Naturalist 131:311–319.

Reeves, D. A. 1984. Iowa frog and toad survey. Department of Natural Resources Report, Iowa Department of Natural Resources, Boone. 19 pp.

Wetland Birds in the Northern Great Plains

Bibby, C. J., N. D. Burgess, and D. A. Hill. 1992. Bird census techniques. Academic Press, London. 257 pp.

Burt, W. 1994. Shadowbirds: a quest for rails. Lyons & Burford, New York. 172 pp.

Cowardin, L. M., V. Carter, F. C. Golet, and E. T. LaRoe. 1979. Classification of wetlands and deepwater habitats of the United States. U.S. Fish and Wildlife Service FWS/OBS–79/31. 131 pp.

Faanes, C. A., and R. E. Stewart. 1982. Revised checklist of North Dakota birds. Prairie Naturalist 14:81–92.

Igl, L. D., and D. H. Johnson. 1995. Migratory bird population changes in North Dakota. Pages 298–300 *in* E. T. LaRoe, G. S. Farris, C. E. Puckett, P. D. Doran, and M. J. Mac, editors. Our living resources: a report to the nation on the distribution, abundance, and health of U.S. plants, animals, and ecosystems. U.S. Department of the Interior, National Biological Service, Washington, D.C.

Igl, L. D., and D. H. Johnson. 1997. Changes in breeding bird populations in North Dakota: 1967 to 1992–93. Auk. In press.

Johnsgard, P. A. 1979. Birds of the Great Plains: breeding species and their distribution. University of Nebraska Press, Lincoln. 539 pp.

Kantrud, H. A., G. L. Krapu, and G. A. Swanson. 1989. Prairie basin wetlands of the Dakotas: a community profile. U.S. Fish and Wildlife Service Biological Report 85(7.280). 111 pp.

Kantrud, H. A., and R. E. Stewart. 1984. Ecological distribution and crude density of breeding birds on prairie wetlands. Journal of Wildlife Management 48:426–437.

Lee, F. B., C. H. Schroeder, T. L. Kuck, L. J. Schoonover, M. A. Johnson, H. K. Nelson, and C. A. Beauduy. 1989. Rearing and restoring giant Canada geese in the Dakotas. North Dakota Game and Fish Department, Bismarck. 79 pp.

Lokemoen, J. T. 1979. The status of herons, egrets and ibises in North Dakota. Prairie Naturalist 11:97–110.

Robbins, C. S., D. Bystrak, and P. H. Geissler. 1986. The Breeding Bird Survey: its first fifteen years, 1965–1979. Resource Publication 157. U.S. Fish and Wildlife Service, 196 pp.

Stewart, R. E. 1975. Breeding birds of North Dakota. Tri-College University Center for Environmental Studies, Fargo, N. Dak. 295 pp.

Stewart, R. E., and H. A. Kantrud. 1971. Classification of natural ponds and lakes in the glaciated prairie region. U.S. Fish and Wildlife Service. Resource Publication 92. 57 pp.

Stewart, R. E., and H. A. Kantrud. 1972. Population estimates of breeding birds in North Dakota. Auk 89:766–788.

Weller, M. W. 1986. Marshes. Pages 201–224 *in* A. Y. Cooperrider, R. J. Boyd, and H. R. Stuart, editors. Inventory and monitoring of wildlife habitat. U.S. Bureau of Land Management Service Center, Denver, Colo.

Waterfowl in the Prairie Pothole Region

Batt, B. D., M. G. Anderson, C. D. Anderson, and F. D. Caswell. 1989. The use of prairie potholes by North American ducks. Pages 204– 227 *in* A. van der Valk, editor. Northern prairie wetlands. Iowa State University Press, Ames.

Bethke, R. W., and T. D. Nudds. 1995. Effects of climate change and land use on duck abundance in Canadian prairie-parklands. Ecological Applications 5:588–600.

Greenwood, R. J., and M. A. Sovada. 1996. Prairie duck populations and predation management. *In* Transactions of the North American Wildlife Natural Resource Conference 61:31–42.

Hestbeck, J. B. 1995. Decline of northern pintails. Pages 38–39 *in* E. T. LaRoe, G. S. Farris, C. E. Puckett, P. D. Doran, and M. J. Mac, editors. Our living resources: a report to the nation on the distribution, abundance, and health of U.S. plants, animals, and ecosystems. U.S. Department of the Interior, National Biological Service, Washington, D.C.

Kantrud, H. A. 1993. Duck nest success on Conservation Reserve Program land in the prairie pothole region. Journal of Soil and Water Conservation 48:238–242.

Johnson, D. H., and J. W. Grier. 1988. Determinants of breeding distributions of ducks. Wildlife Monographs 100. 37 pp.

Lynch, J. 1984. Escape from mediocrity: a new approach to American waterfowl hunting regulations. Wildfowl 35:5–13.

Lynch, J., C. D. Evans, and V. C. Conover. 1963. Inventory of waterfowl environments of Prairie Canada. Transactions of the North American Wildlife Natural Resources Conference 28:93–109.

Reynolds, R. E. 1987. Breeding duck population, production, and habitat surveys, 1979–1985. Transactions of the North American Wildlife and Natural Resource Conference 52:186–205.

Smith, G. W. 1995. A critical review of aerial and ground surveys of breeding waterfowl in North America. National Biological Service Biological Science Report 5. 252 pp.

Smith, R., H. F. Dufresne, and H. A. Hanson. 1964. Northern watersheds and deltas. Pages 51–66 *in* J. P. Linduska, editor. Waterfowl tomorrow. U.S. Fish and Wildlife Service, Washington, D.C.

Duck Plague: Emergence of a New Cause of Waterfowl Mortality

Baudet, A. E. R. F. 1923. Mortality in ducks in the Netherlands caused by a filterable virus; fowl plague. Tijdschrift voor Diergeneeskunde 50:455–459.

Bernier, G., and R. Filion. 1975. Enterite a virus du canard. Observation au Quebec d'une condition ressemblant a la peste du canard. Canadian Veterinary Journal 16:215–217.

Bos, A. 1942. Some new cases of duck plague. Tijdschrift voor Diergeneeskunde 69:372–381.

Friend, M. 1995. Increased avian diseases with habitat change. Pages 401–405 *in* E. T. LaRoe, G. S. Farris, C. E. Puckett, P. D. Doran, and M. J. Mac, editors. Our living resources: a report to the nation on the distribution, abundance, and health of U.S. plants, animals, and ecosystems. U.S. Department of the Interior, National Biological Service, Washington, D.C. 530 pp.

Friend, M., and G. L. Pearson. 1973. Duck plague: the present situation. Proceedings of the Western Association of State Game and Fish Commissioners 53:315–325.

Garrett, L. 1994. The coming plague: newly emerging diseases in a world out of balance. Farrar, Straus, and Giroux, New York. 750 pp.

Jansen, J. 1961. Duck plague. British Veterinary Journal 117:349–356.

Jansen, J. 1964. Duck plague (a concise survey). Indian Veterinary Journal 41:309–316.

Jansen, J., and H. Kunst. 1949. Is duck plague related to Newcastle disease or to fowl plague? Proceedings of the 14th International Veterinary Congress 2:363–365.

Leibovitz, L. 1991. Duck virus enteritis (duck plague). Pages 609–618 *in* B. W. Calnek, H. J. Barnes, C. W. Beard, W. M. Reid, and H. W. Yoder, Jr., editors. Diseases of poultry, 9th edition. Iowa State University Press, Ames.

Leibovitz, L., and J. Hwang. 1968. Duck plague on the American continent. Avian Diseases 12:361–378.

Quortrup, E. R., F. B. Queen, and L. Mervoka. 1946. An outbreak of pasteurellosis in wild ducks. Journal of the American Veterinary Medical Association 108:94–100.

Rhoades, K. R., and R. B. Rimler. 1991. Fowl cholera. Pages 145–162 *in* B. W. Calnek, H. J. Barnes, C. W. Beard, W. M. Reid, and H. W. Yoder, Jr., editors. Diseases of poultry, 9th edition. Iowa State University Press, Ames.

Rosen, M. N., and A. I. Bischoff. 1949. The 1948–1949 outbreak of fowl cholera in birds in the San Francisco Bay area and surrounding counties. California Fish and Game 35:185–192.

Snyder, S. B., J. G. Fox, L. H. Campbell, K. F. Tam, and A. O. Soave. 1973. An epornitic of duck virus enteritis (duck plague) in California. Journal of the American Veterinary Medical Association 163:647–652.

Spieker, J. O. 1977. Virulence assay and other studies of six North American strains of duck plague virus tested in wild and domestic waterfowl. Ph.D. dissertation, University of Wisconsin, Madison. 110 pp.

U.S. Department of Agriculture. 1970. USDA honors cooperators in duck disease eradication program. News Release, USDA 1079-70, Washington, D.C. 2 pp.

Wobeser, G. A. 1981. Diseases of wild waterfowl. Plenum Press, New York. xii+300 pp.

Population Trends for Prairie Pothole Carnivores

Allen, S. H. 1996. Coyotes on the move. North Dakota Outdoors 58:6–11.

Allen, S. H., J. O. Hastings, and S. C. Kohn. 1987. Composition and stability of coyote families and territories in North Dakota. Prairie Naturalist 19:107–114.

Andelt, W. F. 1987. Coyote predation. Pages 128–140 *in* M. Novak, J. A. Baker, M. E. Obbard, and B. Malloch, editors. Wild furbearer management and conservation in North America. Ontario Ministry of Natural Resources, Ottawa.

Arnold, T. W., and E. K. Fritzell. 1987a. Food habits of prairie mink during the waterfowl breeding season. Canadian Journal of Zoology 65:2322–2324.

Arnold, T. W., and E. K. Fritzell. 1987b. Activity patterns, movements, and home ranges of prairie mink. Prairie Naturalist 19:25–32.

Bailey, E. P. 1992. Red foxes, *Vulpes vulpes*, as biological control agents for introduced arctic foxes, *Alopex lagopus*, on Alaskan islands. Canadian Field-Naturalist 106:200–205.

Bailey, V. 1926. A biological survey of North Dakota. North American Fauna 49. 229 pp.

Bird, R. D. 1961. Ecology of the aspen parkland of western Canada in relation to land use. Canadian Department of Agriculture Publication 1066. 155 pp.

Bjorge, R. R., J. R. Gunson, and W. M. Samuel. 1981. Population characteristics and movements of striped skunks (*Mephitis mephitis*) in central Alberta. Canadian Field-Naturalist 95:149–155.

Carbyn, L. N. 1982. Coyote population fluctuations and spatial distribution in relation to wolf territories in Riding Mountain National Park, Manitoba. Canadian Field-Naturalist 96:176–183.

Criddle, N. 1929. Memoirs of the eighties. Canadian Field-Naturalist 43:176–181.

Dekker, D. 1989. Population fluctuations and spatial relationships among wolves, *Canis lupus*, coyotes, *Canis latrans*, and red foxes, *Vulpes vulpes*, in Jasper National Park, Alberta. Canadian Field-Naturalist 103:261–264.

Eagle, T. C., and J. S. Whitman. 1987. Mink. Pages 615–624 *in* M. Novak, J. A. Baker, M. E. Obbard, and B. Malloch, editors. Wild furbearer management and conservation in North America. Ontario Ministry of Natural Resources, Ottawa.

Eberhardt, L. E. 1974. Food habits of prairie mink (*Mustela vison*) during the waterfowl breeding season. M.S. thesis, University of Minnesota, Minneapolis. 49 pp.

Fagerstone, K. A. 1987. Black-footed ferret, long-tailed weasel, short-tailed weasel, and least weasel. Pages 549–573 *in* M. Novak, J. A. Baker, M. E. Obbard, and B. Malloch, editors. Wild furbearer management and conservation in North America. Ontario Ministry of Natural Resources, Ottawa.

Fritzell, E. K. 1978. Aspects of raccoon (*Procyon lotor*) social organization. Canadian Journal of Zoology 56:260–271.

Gamble, R. L. 1981. Distribution in Manitoba of *Mustela frenata longicauda* Bonaparte, the long-tailed weasel, and the interrelation of distribution and habitat selection in Manitoba, Saskatchewan, and Alberta. Canadian Journal of Zoology 59:1036–1039.

Gamble, R. L. 1982. Status report of the prairie long-tailed weasel *Mustela frenata longicauda*. COSEWIC Report, Ottawa, Ontario. 23 pp.

Greenwood, R. J. 1982. Nocturnal activity and foraging of prairie raccoons (*Procyon lotor*) in North Dakota. American Midland Naturalist 107:238–243.

Greenwood, R. J., and A. B. Sargeant. 1994. Age-related reproduction in striped skunks (*Mephitis mephitis*) in the upper Midwest. Journal of Mammalogy 75:657–662.

Greenwood, R. J., A. B. Sargeant, and D. H. Johnson. 1985. Evaluation of mark-recapture for estimating striped skunk abundance. Journal of Wildlife Management 49:332–340.

Hall, E. R. 1981. Mammals of North America, Volume 2. John Wiley and Sons, Inc., New York. 1181 pp.

Houston, S., and M. I. Houston. 1973. A history of raccoons in Saskatchewan. Blue Jay 31:103–104.

Johnson, D. H., and A. B. Sargeant. 1977. Impact of red fox predation on the sex ratio of prairie mallards. U.S. Fish and Wildlife Service, Wildlife Research Report 6. 56 pp.

Johnson, D. H., A. B. Sargeant, and R. J. Greenwood. 1989. Importance of individual species of predators on nesting success of ducks in the Canadian prairie pothole region. Canadian Journal of Zoology 67:291–297.

Jones, J. K., Jr., D. M. Armstrong, R. S. Hoffmann, and C. Jones. 1983. Mammals of the northern Great Plains. University of Nebraska Press, Lincoln. 379 pp.

Lintack, W. M., and D. R. Voigt. 1983. Distribution of the badger, *Taxidea taxus*, in southwestern Ontario. Canadian Field-Naturalist 97:107–109.

Long, C. A., and C. A. Killingley. 1983. The badgers of the world. Charles C. Thomas Publishers, Springfield, Ill. 404 pp.

Lynch, G. M. 1971. Raccoons increasing in Manitoba. Journal of Mammalogy 52:621–622.

Messick, J. P. 1987. North American badger. Pages 587–597 *in* M. Novak, J. A. Baker, M. E. Obbard, and B. Malloch, editors. Wild furbearer management and conservation in North America. Ontario Ministry of Natural Resources, Ottawa.

Messick, J. P., and M. G. Hornocker. 1981. Ecology of the badger in southwestern Idaho. Wildlife Monographs 76. 53 pp.

Messick, J. P., M. C. Todd, and M. G. Hornocker. 1981. Comparative ecology of two badger populations. Pages 1290–1304 *in* J. A. Chapman and D. Pursley, editors. Proceedings of the 1980 Worldwide Furbearer Conference, Frostburg, Md.

Nugent, R. F., and J. R. Choate. 1970. Eastward dispersal of the badger *Taxidea taxus* into the northeastern United States. Journal of Mammalogy 51:626–627.

Proulx, G., and R. K. Drescher. 1993. Distribution of the long-tailed weasel *Mustela frenata longicauda* in Alberta as determined by questionnaires and interviews. Canadian Field-Naturalist 107:186–191.

Rosatte, R. C. 1987. Striped, spotted, hooded, and hog-nosed skunks. Pages 599–613 *in* M. Novak, J. A. Baker, M. E. Obbard, and B. Malloch, editors. Wild furbearer management and conservation in North America. Ontario Ministry of Natural Resources, Ottawa.

Roy, L. D., and M. J. Dorrance. 1985. Coyote movements, habitat use, and vulnerability in central Alberta. Journal of Wildlife Management 49:307–313.

Rudzinski, D. R., H. B. Graves, A. B. Sargeant, and G. L. Storm. 1982. Behavioral interactions of penned red and arctic foxes. Journal of Wildlife Management 46:877–884.

Sargeant, A. B. 1972. Red fox spatial characteristics in relation to waterfowl predation. Journal of Wildlife Management 36:225–236.

Sargeant, A. B. 1982. A case history of a dynamic resource—the red fox. Pages 121–137 *in* G. C. Sanderson, editor. Midwest furbearer management, proceedings, 1981 symposium, Midwest Fish and Wildlife Conference, Wichita, Kansas. North Central Section, Central Mountains and Plains Section, and Kansas Chapter, The Wildlife Society.

Sargeant, A. B., S. H. Allen, and J. O. Hastings. 1987. Spatial relations between sympatric coyotes and red foxes in North Dakota. Journal of Wildlife Management 51:285–293.

Sargeant, A. B., R. J. Greenwood, J. L. Piehl, and W. B. Bicknell. 1982. Recurrence, mortality, and dispersal of prairie striped skunks, *Mephitis mephitis*, and implications to rabies epizootiology. Canadian Field-Naturalist 96:312–316.

Sargeant, A. B., R. J. Greenwood, M. A. Sovada, and T. L. Shaffer. 1993. Distribution and abundance of predators that affect duck production—prairie pothole region. U.S. Fish and Wildlife Service Resource Publication 194. 96 pp.

Simms, D. A. 1979. North American weasels: resource utilization and distribution. Canadian Journal of Zoology 57:504–520.

Sovada, M. A., A. B. Sargeant, and J. W. Grier. 1995. Differential effects of coyotes and red foxes on duck nest success. Journal of Wildlife Management 59:1–9.

Stoudt, J. H. 1982. Habitat use and productivity of canvasbacks in southwestern Manitoba, 1961–72. U.S. Fish and Wildlife Service Special Science Report 248. 31 pp.

Verts, B. J. 1967. The biology of the striped skunk. University of Illinois Press, Urbana. 218 pp.

Voigt, D. and W. E. Berg. 1987. Coyote. Pages 345–357 *in* M. Novak, J. A. Baker, M. E. Obbard, and B. Malloch, editors. Wild furbearer management and conservation in North America. Ontario Ministry of Natural Resources, Ottawa.

Wade-Smith, J., and B. J. Verts. 1982. *Mephitis mephitis*. Mammalian Species 173. American Society of Mammalogists, New York. 7 pp.

Young, C. E., and C. T. Osborn. 1990. Costs and benefits of the Conservation Reserve Program. Journal of Soil and Water Conservation 45:370–373.

Rocky Mountains

"A climb up the Rockies will develop a love for nature, strengthen one's appreciation of the beautiful world outdoors, and put one in touch with the Infinite."

Enos A. Mills (1924)

The Rocky Mountains, the great backbone of North America, extend 5,000 kilometers from New Mexico to Canada. The elevations range from about 1,500 meters along the plains to 4,399 meters, and the widths range from 120 to 650 kilometers (Lavender 1975). The Rocky Mountains are composed of many mountain ranges with unique ecological features. For example, 20 ranges make up the Rocky Mountains in and adjacent to Wyoming (Knight 1994). The natural beauty, abundant wildlife, and fresh water have attracted human inhabitants for the last 10,000–12,000 years (Fig. 1).

Geology and Hydrology

The younger ranges of the Rocky Mountains uplifted during the late Cretaceous period (140 million–65 million years ago), although some portions of the southern mountains date from uplifts during the Precambrian (3,980 million–600 million years ago). The mountains' geology is a complex of igneous and metamorphic rock; younger sedimentary rock occurs along the margins of the southern Rocky Mountains, and volcanic rock from the Tertiary (65 million–1.8 million years ago) occurs in the San Juan Mountains and in other areas. Millennia of severe erosion in the Wyoming Basin transformed intermountain basins into a relatively flat terrain. The Tetons and other north-central ranges are magnificent granitic intrusions of folded and faulted rocks of Paleozoic and Mesozoic age (Peterson 1986; Knight 1994).

Periods of glaciation occurred from the Pleistocene Epoch (1.8 million–70,000 years ago) to the Holocene Epoch (fewer than 11,000 years ago). Recent episodes included the Bull Lake Glaciation that began about 150,000 years ago and the Pinedale Glaciation that probably remained at full glaciation until 15,000–20,000 years ago (Pierce 1979). Ninety percent of Yellowstone National Park was covered by ice during the Pinedale Glaciation (Knight 1994). The "little ice age" was a period of glacial advance that lasted a few centuries from about 1550 to 1860. For example, the Agassiz and Jackson glaciers in Glacier National Park reached their most forward positions about 1860 during the little ice age (Grove 1990).

Water in its many forms sculpted the present Rocky Mountain landscape (Athearn 1960). Runoff and snowmelt from the peaks feed Rocky Mountain rivers and lakes with the water supply for one-quarter of the United States. East of the Continental Divide, the Arkansas, Missouri, North and South Platte, and Yellowstone rivers flow to the Gulf of Mexico. The Colorado, Columbia, Green, Salmon, San Juan, and Snake rivers flow westward to the Pacific Ocean. Water, the "transparent gold" of the West, supports agriculture, municipal supplies, recreation, and hydroelectric power generation and is the lifeblood of all plants and animals. As Emperor Yu of China realized in 1600 B.C., "To protect your rivers, protect your mountains."

Courtesy B. R. McClelland, National Park Service (retired)

Fig. 1. The Rocky Mountains.

Paleoecology

Paleoecological data from the Holocene in the central and northern Rocky Mountains are limited because the interpretation of pollen records is difficult—high winds distribute pollen locally and regionally—and because packrat middens, usually a good source of data, are rare and restricted to lower elevations. Evidence of the climate and vegetation changes since the last ice age is sketchy (Nichols 1982; Whitlock 1993). However, a scenario can be pieced together from paleoecological studies at treeline in Colorado (Benedict 1981; Nichols 1982), from current distributions of forest species in the northern Rocky Mountains (Alexander 1985), and from extensive work in and near Yellowstone National Park (Whitlock et al. 1991; Whitlock 1993), the Great Basin (Wells 1983), the Colorado Plateau (Betancourt et al. 1990), and the Colorado River (Cole 1990).

Paleoclimatic research at treeline in the central Rocky Mountains revealed that the last glacial age waned about 12,000 years ago. The climate changed from cool–moist to warm–moist about 10,000 years ago and then to a warm–dry climate about 4,000 years ago. The climate began to cool again about 2,500 years ago (Nichols 1982). During the Pinedale Glaciation (22,400–12,200 years ago), treeline averaged 500 meters lower than at present. Upper treeline advanced to as high as 300 meters above its present position during the warmest period. Cooling during the past 2,500 years has forced the treeline down to its present location; its decline was perhaps more rapid during the little ice age, between 1550 and 1860 (Nichols 1982).

No one has determined exactly how lower and midelevation tree species migrated, adapted, or changed in response to climate change. Some inferences may be made from recent research in the Grand Teton and Yellowstone national parks (Whitlock 1993) and from studies in the highest mountains in the Southwest, Colorado Plateau, and Great Basin. The vegetation diversity and complex elevation-moisture gradients of these areas are similar to those of the Rocky Mountains (Nichols 1982; Wells 1983; Betancourt et al. 1990; Cole 1990). One probable scenario is that toward the end of the last Ice Age (12,000 years ago), the central Rocky Mountains had relicts of alpine tundra at higher elevations. The lower elevations had subalpine forests dominated by limber pine and Engelmann spruce with an understory of juniper in wet areas and sagebrush outcrops in dry areas. Quaking aspens filled forest openings, and lush meadows lined the streams. During the Holocene transition (12,000–8,000 years ago), subalpine forests migrated upslope, and montane forests of lodgepole pines and Douglas-firs began to expand in the lower elevations. The Ice Age ended gradually; the southern Rocky Mountains and south-facing slopes warmed much earlier than north-facing slopes and the northern Rocky Mountains. As the climate warmed between 9,000 and 4,500 years ago, ponderosa pines, a species absent from the fossil record in the central and northern Rocky Mountains and Great Basin, became abundant in the lower elevations. An accompanying northern expansion of the summer monsoon may have extended the range of ponderosa pines and Douglas-firs. Drier and cooler conditions maintained the ponderosa pines and allowed further expansion of lower-elevation sagebrush-grass communities well up into the Rocky Mountains. Cooler conditions in the past 2,500 years decreased the midelevation fire frequencies that favor spruce-fir and lodgepole pine forests over limber pines; the frequency of fire increased at lower elevations because of the highly flammable structure of ponderosa pine–grassland fuels. The little ice age from 1550 to 1860 forced a descent of the upper treeline and perhaps lengthened intervals between fires. Tree rings from the region record a warming trend in recent decades (Weisberg and Baker 1995). The present rates of ecological change can be gauged against these background levels.

Current Vegetation Patterns

Vegetation patterns in the Rocky Mountains can be explained by elevation, aspect, and precipitation. Merriam (1890) recognized that two-dimensional diagrams of elevation and aspect described plant community distribution in the southern Rocky Mountains. Other ecologists generally embraced this two-dimensional view until the complexities of environmental gradients such as temperature, precipitation, solar radiation, wind, soils, and hydrology could be described and modeled (Peet 1981). Several authors (Alexander 1985; Peet 1988; Allen et al. 1991; Cooper et al. 1991; Knight 1994) described the vegetation patterns in different areas of the Rocky Mountains. Peet (1988) provided the most complete description of 11 major forest community types, which are summarized here. Two nonforested vegetation types, plains and alpine tundra, described by Sims (1988) and Billings (1988), were added to emphasize that nonforested communities play a large role in wildlife conservation in the Rocky Mountains (Knight 1994). The status and trends of many plant and animal populations in montane communities are inseparably linked to the adjacent and interwoven communities of the tundra and the plains (Fig. 2). Because of the variations in latitude and precipitation along this huge mountain range, the elevations presented here are gross generalizations.

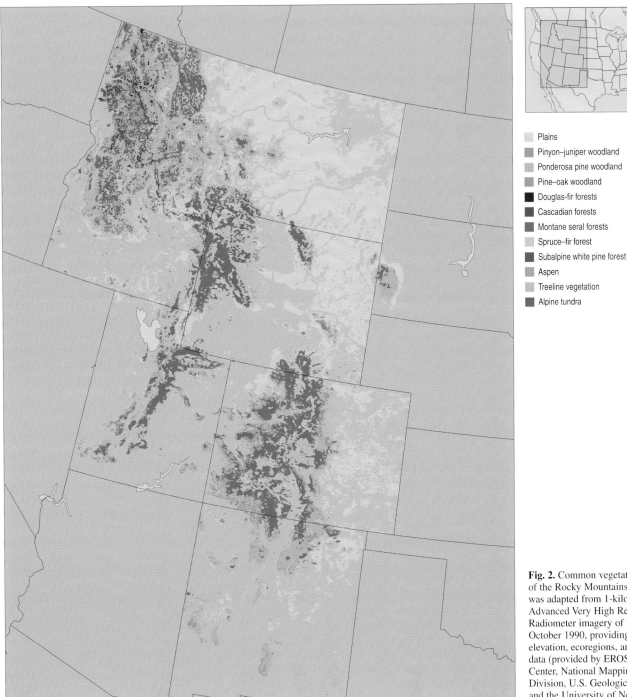

Plains
Pinyon–juniper woodland
Ponderosa pine woodland
Pine–oak woodland
Douglas-fir forests
Cascadian forests
Montane seral forests
Spruce–fir forest
Subalpine white pine forest
Aspen
Treeline vegetation
Alpine tundra

Fig. 2. Common vegetation types of the Rocky Mountains. This map was adapted from 1-kilometer Advanced Very High Resolution Radiometer imagery of March–October 1990, providing digital elevation, ecoregions, and climate data (provided by EROS Data Center, National Mapping Division, U.S. Geological Survey, and the University of Nebraska–Lincoln).

Plains

The eastern side of the Rocky Mountains is bordered by mixed-grass prairie to the north and by short-grass prairie to the south (see Plains; Fig. 2); prairie generally extends to elevations of 1,800 meters. Dominant plants of the mixed grass prairie include little bluestem, needle-grasses, wheatgrasses, sand-reeds, and gramas, with dropseeds and cottonwoods in riparian zones. Short-grass prairie species include little bluestem, buffalo grass, western wheatgrass, sand dropseed, ringgrass, needle-and-thread, Junegrass, and galleta (Sims 1988). Extensions of these vegetation types reach well into the Rocky Mountains along the valleys and on dry slopes. Plant species composition varies locally with changes in soil characteristics and topographic position—that is, from hilltops to valley bottoms (Knight 1994).

Riparian and Canyon Forests

Broad-leaved deciduous cottonwoods, alders, and willows line streamsides and canyons. The herbaceous layer in riparian communities is often more diverse than upslope areas and adjacent forests (Peet 1978; Baker 1990). Riparian and canyon vegetation types are generally too thin or too small to be displayed on regional vegetation maps, but the habitat is extremely important in the arid West.

Pinyon–Juniper Woodland

In the southern Rocky Mountains, a transition occurs between about 1,800 and 2,500 meters, where plains communities are accompanied by pinyon pines (Fig. 2). Mexican pinyons and singleleaf pinyons are found in western Utah, alligator junipers and Rocky Mountain junipers grow to the south, and Utah junipers grow to the north. Many shrubs and grasses of the plains occupy the gaps between tree outcrops. Heavy livestock grazing is associated with the spread of junipers (by reducing competition from grasses), and fire suppression is partly responsible for their continued dominance (West et al. 1975).

Ponderosa Pine Woodland

The appearance of ponderosa pine woodlands varies from scattered individuals in low-elevation or rocky areas to dense forests at higher elevations or on deeper soils (Peet 1981, 1988; Knight 1994). Although ponderosa pines dominate the biomass of this community, other tree species such as Douglas-fir and Rocky Mountain juniper, shrubs (for example, raspberries, big sagebrush, gooseberries, currants, bitterbrush), and herb layers (such as mountain muhly, sedges, and sagebrushes) can develop. Typical intervals between natural fires are less than 40 years in most ponderosa pine forests (Laven et al. 1980; Keane et al. 1990).

Pine–Oak Woodland

In the southern Rocky Mountains, lower slopes of ponderosa pine communities can be accompanied by Gambel oaks, other oak species (for example, Emory oaks, silverleaf oaks, netleaf oaks), and shrubs (such as sumacs, buckbrushes, and mountain-mahoganies). In the absence of fire, the oak stands may be invaded by pines (Peet 1988; Knight 1994).

Douglas-Fir Forest

Douglas-firs grow in a broad range from Mexico to British Columbia, generally from near lower treeline upward in elevation to spruce–fir forests (Fig. 2). In Colorado, the species ranges from about 1,650 to 2,700 meters and is often found in mixed stands with ponderosa pine, blue spruce, or lodgepole pine. Like ponderosa pine, Douglas-fir is tolerant of frequent, low-intensity surface fires. Fire intervals in Douglas-fir forests in Wyoming average 50–100 years (Loope and Gruell 1973).

Cascadian Forest

Several tree species commonly associated with the Cascade Mountains grow on the rain-swept western slopes of the northern Rocky Mountains. These include western hemlock, western redcedar, grand fir, mountain hemlock, and larches. These forests are subject to infrequent, high-intensity fires.

Montane Seral Forest

Lodgepole pine forests interspersed with stands of quaking aspens are fire-resilient forests that dominate the central and north-central Rocky Mountains (Fig. 2). Usually found between 2,500 and 3,200 meters in Colorado, lodgepole pines and aspens grow rapidly after fire in mostly even-aged stands.

Intervals between fires typically range from 100 to 300 years (Romme and Knight 1981). As evidenced by the fires in the Yellowstone National Park in 1988, lodgepole pine forests are rejuvenated by crown fires that replace tree stands. Although aspen stands generally cover less than 1% of the landscape (for example, Rocky Mountain National Park, Grand Teton National Park), they are keystone communities for hundreds of birds and mammals and are especially important forage for deer and elk (Mueggler 1993).

Spruce–Fir Forest

The subalpine forests of the Rocky Mountains are characterized by spruces and firs and are floristically and structurally similar to the boreal conifer forests to the north. Dominant tree species in the Colorado Rocky Mountains subalpine forests include Engelmann spruce and subalpine fir. In the Black Hills of South Dakota, white spruce replaces Engelmann spruce. Stand-replacing fires typically occur at 200- to 400-year intervals. Widespread insect outbreaks in spruce–fir forests occur more frequently (Veblen et al. 1991).

Subalpine White Pine Forest

On exposed, dry slopes at high elevations, subalpine white pine forests replace spruce–fir forests. Common species of the white pine forests include whitebark pine in the northern Rocky Mountains, limber pine in the central and north-central Rocky Mountains, and bristlecone pine in the southern Rocky Mountains. Typical intervals between fires range from 50 to 300 years (Kendall 1995). The white pines are tolerant of extreme environmental conditions and can be important postfire successional species.

Treeline Vegetation

Treeline is the elevation above which trees cannot grow. It is controlled by a complex of environmental conditions, primarily soil temperatures and the length of the growing season—which becomes shorter with higher elevations. The elevation of treeline rises steadily at the rate of 100 meters per degree of latitude from the northern to the southern Rocky Mountains. Dominant treeline species, including spruces, firs, and white pines, often have a shrublike form in response to the extreme conditions at the elevational limits of their physiological tolerance; such dwarfed trees are called *krummholz*. Krummholz islands may actually move about 2 centimeters per year in response to the wind; they reproduce by vegetative layering on their lee sides, while dying back from wind damage on their windward sides. Under favorable climatic conditions, krummholz can assume an upright treelike form or can increase their cone crops and seedling establishment.

Alpine Tundra

Alpine tundra is a complex of high-elevation meadows, fell (barren) fields, and talus (rock) slopes above treeline (above 3,400–4,000 meters). Grasses and sedges dominate the meadow communities, and fens (a type of wet meadow) and willows exist in wet soils. Vegetation in the alpine zone is similar to that in the Arctic: 47% of the plant species in the alpine zone of the Beartooth Mountains in Wyoming and Montana are also found in the Arctic (Billings 1988). This high-diversity area includes alpine sagebrush, tufted hairgrass, clovers, pussytoes, and succulents, and hundreds of grasses and wildflower species (Billings 1988; Popovich et al. 1993).

Extensive investigations have been made of the forests of the Rocky Mountains (Peet 1981; Allen et al. 1991; Veblen et al. 1991). Weber (1976:4–5) cautioned that the vegetation zones "overlap and telescope into each other considerably" in a landscape that is "always full of surprises." The resulting patchwork mosaic of vegetation types and disturbance regimes leads to a complex of side-by-side communities, wildlife habitats, and species distributions.

Wildlife

The charismatic megafauna of the Rocky Mountains includes elk, moose, mule and white-tailed deer, pronghorns, mountain goats, bighorn sheep, black bears, grizzly bears (Fig. 3), coyotes, lynxes, and wolverines. Equally important contributors to the region's biological diversity include small mammals, fishes, reptiles, amphibians, hundreds of bird species, and tens of thousands of species of terrestrial and aquatic invertebrates and soil organisms.

Courtesy R. Stottlemyer, USGS

Fig. 3. Fewer than 420 grizzly bears remain in the United States portion of the Rocky Mountains.

Human History and Cultural Development

Since the last great Ice Age, the Rocky Mountains were a sacred home first to Paleo-Indians and then to the Native American tribes of the Apache, Arapaho, Bannock, Blackfoot, Cheyenne, Crow, Flathead, Shoshoni, Sioux, Ute, and others (Johnson 1994). Paleo-Indians hunted the now-extinct mammoth and ancient bison (an animal 20% larger than modern bison) in the foothills and valleys of the mountains. Like the modern tribes that followed them, Paleo-Indians probably migrated to the plains in fall and winter for bison and to the mountains in spring and summer for fish, deer, elk, roots, and berries. In Colorado, along the crest of the Continental Divide, rock walls that Native Americans built for driving game date back 5,400–5,800 years (Benedict 1981; Buchholtz 1983). A growing body of scientific evidence indicates that Native Americans had significant effects on mammal populations by hunting and on vegetation patterns through deliberate burning (Kay 1994).

Recent human history of the Rocky Mountains is one of more rapid change (Lavender 1975; Knight 1994). The Spanish explorer Francisco Vásquez de Coronado—with a group of soldiers, missionaries, and African slaves—marched into the Rocky Mountain region from the south in 1540. The introduction of the horse, metal tools, rifles, new diseases, and different cultures profoundly changed the Native American cultures. Native American populations were extirpated from most of their historical ranges by disease, warfare, habitat loss (eradication of the bison), and continued assaults on their culture.

The Lewis and Clark expedition (1804–1806) was the first scientific reconnaissance of the Rocky Mountains. Specimens were collected for contemporary botanists, zoologists, and geologists (Jackson 1962). The expedition was said to have paved the way to (and through) the Rocky Mountains for European-Americans from the East, although Lewis and Clark met at least 11 European-American mountain men during their travels. Meriwether Lewis sent this description to Thomas Jefferson on 21 September 1806:

> The passage by land of 340 miles from the Missouri (River) to the Kooskooske (Clearwater River) is the formidable part of the tract proposed across the Continent: of this distance 200 miles is along a good road, and 140 over tremendous mountains which for 60 mls (miles) are covered with eternal snows. (Jackson 1962:320)

Mountain men, primarily French, Spanish, and American fur traders and explorers, roamed the Rocky Mountains from 1800 to 1850. The more famous of these include William Henry Ashley, Jim Bridger, Kit Carson, John Colter, Thomas Fitzpatrick, Andrew Henry, and Jedediah Smith. Beavers had been trapped to near-extinction by the 1840's.

The Mormons began to settle near the Great Salt Lake in 1847. In 1859 gold was discovered near Cripple Creek, Colorado, and the regional economy of the Rocky Mountains was changed forever. The transcontinental railroad was completed in 1869, and Yellowstone National Park was established in 1872. While settlers filled the valleys and mining towns, conservation and preservation ethics began to take hold. President Harrison established several forest reserves in the Rocky Mountains in 1891–1892. In 1905 President Theodore Roosevelt extended the Medicine Bow Forest Reserve to include the area now managed as Rocky Mountain National Park (Buchholtz 1983). Economic development began to center on mining, forestry, agriculture, and recreation, as well as on the service industries that support them (Lavender 1975). Tents and camps became ranches and farms, forts and train stations became towns, and some towns became cities.

Economic resources of the Rocky Mountains are varied and abundant. Minerals found in the Rocky Mountains include significant deposits of copper, gold, lead, molybdenum, silver, tungsten, and zinc. The Wyoming Basin and several smaller areas contain significant reserves of coal, natural gas, oil shale, and petroleum. Forestry is a major industry. Agriculture includes dryland and irrigated farming and livestock grazing. Livestock are frequently moved between high-elevation summer pastures and low-elevation winter pastures. Every year the scenic splendor and recreational opportunities of the Rocky Mountains draw millions of tourists. The National Park System units include Glacier, Grand Teton, Yellowstone, Rocky Mountain, and 16 others.

Abandoned mines with their wakes of mine tailings and toxic wastes dot the Rocky Mountain landscape. Eighty years of zinc mining profoundly polluted the river and bank near Eagle River in north-central Colorado. High concentrations of the metal carried by spring runoff harmed algae, moss, and trout populations. An economic analysis of mining effects at this site revealed declining property values, degraded water quality, and the loss of recreational opportunities. The analysis also revealed that cleanup of the river could yield $2.3 million in additional revenue from recreation. In 1983 the former owner of the zinc mine was sued by the Colorado Attorney General for the $4.8

million cleanup costs; 5 years later, ecological recovery was considerable (Brandt 1993).

Human Population Trends

The human population grew rapidly in the Rocky Mountain states between 1950 and 1990 (Fig. 4). The 40-year statewide increases in population range from 35% in Montana to about 150% in Utah and Colorado. The populations of several mountain towns and communities have doubled in the last 40 years. Jackson Hole, Wyoming, increased 260%, from 1,244 to 4,472 residents, in 40 years.

This rapid population growth increases demands for water, power, and natural resources. Ironically, the montane valley used for the filming of *The Unsinkable Molly Brown* (1964), which features the *Titanic* voyage, is now underwater, flooded by the Blue Mesa Reservoir (Stephanie Two Eagles, Location Specialist, Colorado Film Commission, personal communication).

Status and Trends of Ecosystems

Determining the status and trends of ecosystems requires fairly complete biotic inventories of major biological groups and detailed understanding of the behavior and interactions of plants and animals in complex environments. It also requires the systematic monitoring of key ecological processes (for example, disturbance, predation, competition, nutrient cycling, and energy flow). Although detailed information about some components and processes in certain Rocky Mountain ecological communities is available, information about entire communities is incomplete. Inferences about ecological trends usually are made from data describing selected components or processes in larger landscapes. Some of the most obvious changes that affect Rocky Mountain ecosystems are described in this chapter.

Bordering Prairie and Intermountain Ecosystems

Prairie dogs are keystone species in the plains, piedmont valleys, mesas, and foothills of the Rocky Mountains and throughout the plains of North America (Fig. 5). Prairie dog ecosystems support about 170 vertebrate species, including higher numbers of birds and mammals than the adjacent grasslands without prairie dogs (Miller et al. 1994). Prairie dog towns also have more plant species and more specialized insects and allies than adjacent areas (Knight 1994). Although exact causes of

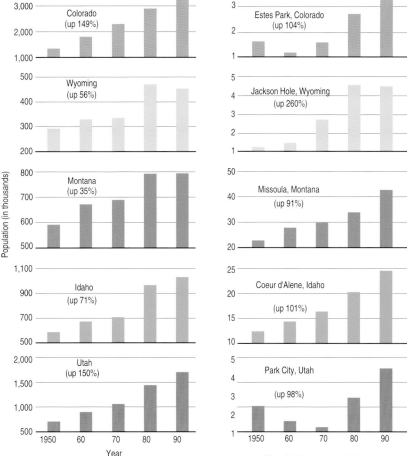

population changes in the Rocky Mountain foothills are unknown, the estimated 98% decline in prairie dog populations throughout North America is in part the result of federal- and state-sponsored prairie dog control (Miller et al. 1990). The U.S. Environmental Protection Agency and the Animal Plant Health Inspection Service estimated that 80,000 hectares of prairie dog habitat are eliminated annually (Captive Breeding Specialist Group 1992). Habitat loss and prairie dog control led to the endangered status (U.S. Endangered Species Act of 1973) of the black-footed ferret and the candidate listings of mountain plovers, ferruginous hawks, and swift foxes.

Because prairie vegetation and associated wildlife extend into the valleys and drier slopes of the Rocky Mountains, the food chain and ecological processes of adjacent ecosystems may be affected in subtle and poorly understood ways. For example, many of the 170 vertebrate species that are supported by prairie dog ecosystems have home ranges that extend into the montane forests of the Rocky Mountains. Reductions in the abundance of prairie dogs probably reduce the numbers of birds, insects (flower pollinators), snakes, and other species with important roles in adjacent ecosystems.

Fig. 4. Human population growth in the Rocky Mountain states and selected towns, 1950–1990, with rate of growth (percentage in 40 years).

Fig. 5. The black-tailed prairie dog, a species whose habitat in North America has suffered a 98% loss.

Livestock have played (and continue to play) an even more important role in changes to these ecosystems. The most widespread influence throughout most natural ecosystems in the Rocky Mountains (from the plains to alpine meadows) is that of livestock grazing (Wagner 1978; Crumpacker 1984); 70% of the western United States is grazed. Undisturbed herbaceous ecosystems across the western United States are rare. Native vegetative ecosystems have previously been directly converted to agricultural fields and to livestock grazing areas (Grossman and Goodin 1995). A precise determination of the ecological effects of grazing often is difficult to obtain because ungrazed land is extremely rare, exclosures are small, exact figures on grazing intensities are scarce, and approaches to evaluate the effects of grazing are not standardized (Fleischner 1994). Johnson (1987) used paired photographs from 1870 and from the 1980's to show that only minor changes had occurred in landscapes in southern Wyoming during the twentieth century. Still, a recent synthesis of field studies (Fleischner 1994) suggested that in many habitats, grazing may have profound ecological effects that include reduced species richness and reduced density and biomass of native grasses, altered ecosystem processes (nutrient cycling and succession), and altered ecosystem structure caused by changing vegetation patterns, which contribute to soil erosion and decreased availability of water to biotic ecosystems. Livestock grazing aids in the spread of weeds (D'Antonio and Vitousek 1992), prevents native plains cottonwood regeneration in riparian zones (Glinski 1977), and influences the conditions and widths of riparian zones (Knopf and Cannon 1982; Chaney et al. 1990).

Personnel of the U.S. Environmental Protection Agency (Chaney et al. 1990) concluded that riparian conditions throughout the West are now the worst in American history (Knight 1994); however, they also correctly caution that generalizations are not possible because of the complexity of the responses to grazing in different habitats over time. Removal of livestock lets riparian communities recover quickly from damage by grazing, except erosion. Trout habitat along Summit Creek in Idaho improved only 2 years after the elimination of grazing (Keller and Burnham 1982); 9 years after cattle exclusion, beavers and waterfowl returned to Camp Creek, Oregon (Winegar 1977). Thus, the potential for rehabilitation and restoration of excessively grazed riparian areas is good.

Ponderosa Pine Ecosystems

In geological time, ponderosa pine ecosystems are relatively new to the foothills of the central Rocky Mountains (Fig. 6). An even newer addition to the ecosystem, European–American settlers, devastated the ponderosa pine forests through logging for houses, fencing, firewood, mine timbers, and railroad ties, and with fire. The ponderosa pine forests were close to the developing population centers at the forest–prairie edge. The scale of the loss of ponderosa pine habitat is demonstrated best in several hundred paired photographs from the early 1900's and 1980's (Gruell 1983; Veblen and Lorenz 1991). However, nearly all the paired photographs also reveal that the most important feature of the ponderosa pine ecosystem is its resilience. Ponderosa pine seedlings establish quickly in disturbed sites. Research in the Front Range of Colorado shows a tenfold increase in ponderosa pine biomass since 1890 in many stands (M. Arbaugh, U.S. Forest Service, unpublished data). This regeneration has restored habitat for many wildlife species. More than 60 years of fire suppression, however, has created hazardous fuels in a forest ecosystem that naturally burned at 20- to 40-year intervals in many areas.

Old-Growth Forest Ecosystems

Old-growth forests were more common in the Rocky Mountains before European–Americans arrived than in recent times. However, presettlement old-growth lodgepole pine or Engelmann spruce–subalpine fir forests were rarer in the south-central Rocky Mountains south of Wyoming (Roovers and Robertus 1993) and more common in the northern Rocky Mountains (Romme and Knight 1981).

Crown fires that lead to the replacement of subalpine forests typically occur at 200- to

Fig. 6. Ponderosa pine.

© R. R. Bachand, Estes Park, Colorado

400-year intervals (Peet 1988). Extensive fires occurred in the mid-1700's (Robertus et al. 1991), and large-scale logging and the careless use of fire between 1850 and 1950 decimated forest resources (Gruell 1983; Veblen and Lorenz 1991). Insect outbreaks and windthrow can affect thousands of square kilometers of forests every few decades (Veblen et al. 1991; Knight 1994). In the south-central Rocky Mountains, the harm to spruce–fir forests from spruce beetle outbreaks can be as significant as that from fire (Baker and Veblen 1990; Veblen et al. 1991). Five major spruce beetle outbreaks, including a 290,000-hectare outbreak in the White River National Forest in Colorado in 1965, have affected thousands of square kilometers since the mid-1800's (Hinds et al. 1965). An additional natural disturbance, a windstorm, blew trees down in an area of 6 square kilometers in the Teton Wilderness Area in August 1987 (Knight 1994).

Because old-growth forest ecosystems were once more common in Rocky Mountain landscapes, it seems likely that species that depend on these ecosystems probably were also more common in presettlement times. Many threatened, endangered, and vulnerable wildlife species are largely dependent on intact old-growth ecosystems (Finch 1992). Caribou in northern Idaho, for example, feed on lichens that grow only in old-growth hemlock and cedar forests. Wolverines, martens, Abert's squirrels (Fig. 7), and fishers are associated almost exclusively with old-growth forests (Finch 1992). The southern red-backed vole is more abundant and has better body condition in old-growth spruce–fir forests (Nordyke and Buskirk 1991). Several bird species, such as the northern goshawk, flammulated owl, Mexican spotted owl, boreal owl, and olive-sided flycatcher, depend on old-growth conifer resources for either nesting or foraging (Finch 1992; Hayward and Verner 1994). The purple martin resides primarily in mature aspens, and like many other species that depend on old-growth forests, it is showing widespread declines of population sizes (Finch 1992). Given the historical threats faced by old-growth forests from fire, wind, and pathogens, one may ask how well these ecosystems have been protected in the past 40 years.

A recent U.S. Forest Service report on forest resources of the United States contains mixed news on the preservation of old-growth resources in the Rocky Mountains (Powell et al. 1993). Old-growth losses from wildfire (including the fires in the Yellowstone National Park in 1988) have been greatly reduced since the 1950's, but losses from the sawtimber industry have climbed. In the intermountain region (Arizona, New Mexico, Colorado, Wyoming, Utah, Idaho, and Montana), the total volume of available softwood sawtimber increased 3.7% between 1952 and 1992; however, available trees of the largest size class measured (those trees measuring more than 73 centimeters diameter at breast height) decreased 31.1% during the same period (Fig. 8). This represents a change in sawtimber volume from 151.9 million cubic meters to 104.6 million cubic meters. In 1992 only 45,000 cubic meters of these largest trees remained from the original 164,000 cubic meters in the nearby Great Plains (Kansas, Nebraska, North Dakota, and South Dakota), a 72.3% decline in 40 years.

The 35-year annual rate of decline in large tree volume (1952–1987) was 0.7% in the intermountain subregion. From 1987 to 1992, however, the annual rate of decline in large tree volume almost tripled to 2.0%. Old-growth forests are being rapidly replaced by younger, faster-growing forests (Fig. 9). Loss of habitat and increasing fragmentation of habitat may impede the protection of old-growth-dependent species for a long time.

Postdisturbance Ecosystem

Fire suppression has been effective, perhaps too effective for some species (see box on Fire Suppression in Land Use chapter). Nationwide losses of forestlands to wildfires decreased from about 20 million hectares per year in 1930 to fewer than 800,000 hectares per year in the late 1980's (MacCleery 1992). Even the 560,000-hectare fires in the Yellowstone National Park in 1988 that were reminiscent of fires in the 1700's burned less than 12% of the forestlands in Wyoming, Montana, and Idaho (Romme and Despain 1989; Schullery 1989). Thus, fire suppression may be profoundly affecting disturbance-dependent species. For example, the number of three-toed woodpeckers increases greatly for 3–5 years after a fire because the birds feed on larval spruce beetles found in

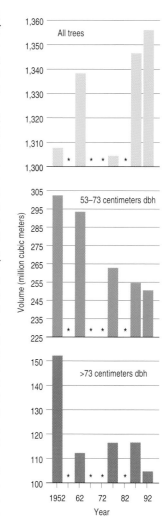

Fig. 8. Sawtimber volume change in the Rocky Mountain states (adapted from Powell et al. 1993); "dbh" = diameter of tree at breast height. "*" = data not available.

Fig. 7. Abert's squirrel, a species dependent on old-growth forests.

© L. D. Schell, Fort Collins, Colorado

Fig. 9. Large trees (such as the one shown here) and old-growth forests may be rapidly disappearing from Rocky Mountain forests.

burned stands. These woodpeckers may significantly affect nearby forest stands by reducing the severity of spruce beetle epidemics. The black-backed woodpecker was absent in the Yellowstone National Park area before the 1988 fire, then it appeared for a few years in the burned forests to feed on insects and larvae.

Herbaceous plant diversity increases abruptly and drastically after a burn. Seventeen years after the 390-hectare burn in the Ouzel Lake area in the Rocky Mountain National Park, twice as

many understory species had grown in the burn area than in adjacent stands of unburned forest (Stohlgren et al. 1997; Fig. 10). Species that depend on postdisturbance stands suffer habitat loss with each suppressed fire.

Air Pollution Effects on Many Ecosystems

Air pollution, primarily from the combustion of fossil fuels in the Denver–Boulder–Fort Collins metropolitan corridor, may dramatically harm montane forests in Colorado. Chemical analyses of the high-elevation Colorado snowpack are revealing high concentrations (about 15 microequivalents per liter) of sulfate and nitrate in areas northwest of Denver (Turk et al. 1992). Some areas in the Colorado Front Range may have experienced a ninefold increase in wet deposition of nitrogen in the past several decades. Remote areas of the world typically receive less than 0.5 kilograms per hectare per year of inorganic nitrogen, whereas the high-elevation sites in the Colorado Front Range now receive as many as 4.7 kilograms per hectare per year of inorganic nitrogen (M. Williams and colleagues, University of Colorado, Boulder, unpublished data). In February 1995 the Colorado Air Quality Commission increased the Denver metropolitan area's particulate pollution limit from the current 41.2 tons per day to 44 tons per day in the next 20 years. Terrestrial biota of high-elevation areas may not be greatly affected by this increased nitrogen loading (Nams et al. 1993), but we can expect direct and indirect effects on ecosystem functions in forested catchments (Baron et al. 1994).

Aquatic Ecosystems and Wetlands

Information on aquatic ecosystems in the Rocky Mountains is highly fragmented, but available information suggests that the region may be typical of the United States: 80% of the nation's flowing waters are characterized by poor quantity and quality of fish habitat and fish community composition (Flather and Hoekstra 1989). All rivers in the Rocky Mountain region have been altered by reservoirs or other water projects (transbasin canals, irrigation ditches, and small water impoundments; see chapter on Water Use). A major transbasin water import project in Colorado carries about 370 million cubic meters per year from the Colorado River (west of the Continental Divide) through a 7.8-kilometer tunnel to the Big Thompson River (east of the Continental Divide). Ten reservoirs were constructed to support this project. There are approximately 40 major reservoirs in the watersheds of the Arapaho–Roosevelt national

© L. D. Schell, Fort Collins, Colorado

Fig. 10. The Ouzel burn area of the Rocky Mountain National Park, Colorado. The burn area has five times as many species of plants, including twice as many understory species, as nearby forests that were not burned.

Whitebark Pine

Whitebark pine is a picturesque tree of the subalpine forest and treeline of the Rocky Mountains, Coast and Cascade ranges, and the Sierra Nevada (Fig.1). Slow growing and long-lived, it is typically more than 100 years old before it produces cones. Whitebark pine's growth form ranges from a krummholz mat to a moderately tall, upright tree, but it is often short and heavily branched, with multiple stems.

Whitebark pine typically grows with other high mountain conifers but can form nearly pure stands in relatively dry mountain ranges (Arno and Hoff 1989). Where associated trees are capable of forming closed stands, whitebark pine can be a long-lived dominant seral species if periodic disturbance, such as fire, removes its shade-tolerant competitors. On a broad range of dry, windy sites, however, whitebark pine is a climax tree because it is hardier and more durable than subalpine fir and other tree species (Arno and Hoff 1989). The sites where whitebark pine is seral tend to be moister and more productive than sites where the tree is climax (Arno 1986).

Until recently, little was known of whitebark pine status because it occurs in rugged terrain and has limited use as a commercial timber species. In the last decade, however, its role as a keystone species has been recognized. Whitebark pine seeds are a preferred food of the threatened grizzly bear and many other mammals and birds (Fig. 2).

Because whitebark pine can grow in cold, dry, and windy conditions tolerated by no other tree, and because it pioneers disturbed sites, it plays an important role in tree establishment in high-elevation open sites. Whitebark pine helps stabilize snow, soil, and rocks on steep terrain and has potential for use in high-elevation land reclamation projects (Arno and Hoff 1989). Because of their spreading crowns and penchant for establishing on windswept ridges (Fig. 3), whitebark pines accumulate and retain snow, extending the snowmelt period into the growing season, when water is needed. With the growing appreciation of these values has come more interest and mounting concern over dramatic declines in whitebark pine stands (Arno 1986; Kendall and Arno 1990; Keane and Arno 1993; Lanner 1993).

Causes of Decline

Sixty years of fire suppression have advanced forest succession at the expense of seral whitebark pine communities. The transport and caching of whitebark pine seed by Clark's nutcrackers and the hardiness of seedlings on exposed microsites give a competitive edge to whitebark pine over less hardy wind-dispersed conifers, such as spruce and fir, in reforesting large burns. However, without fire, whitebark pines are shaded out by other trees, and there are few

Fig. 2. Whitebark pine cones and large, wingless seeds.

Courtesy K. C. Kendall, USGS

Fig. 3. Whitebark pine occupying a windswept ridge in Glacier National Park, Montana.

Courtesy B. R. McClelland, National Park Service (retired)

open sites for whitebark pine regeneration. Before fire exclusion, the average whitebark pine stand burned every 50 to 300 years. Even with the prescribed natural fires that have been allowed to burn in wilderness and national parks in the past 25 years, fewer than 1% of seral whitebark pine stands have burned during that period—an average fire return interval of more than 3,000 years (Arno 1986; Keane 1995a). The Selway–Bitterroot Wilderness in Montana has one of the most extensive prescribed fire programs in the United States, yet between 1979 and 1990, burning in the whitebark pine zone was less than half the area burned each year in presettlement times (Brown et al. 1994; Arno 1995).

An exotic fungus, white pine blister rust, has killed many whitebark pine trees in the moister parts of its range. White pine blister rust, which was introduced from Europe to western North America around 1910, has spread to most whitebark pine forests. Although white pine blister rust can damage all North American white pine species, whitebark pine is the most vulnerable (Fig. 4); fewer than 1 in 10,000 trees is resistant to blister rust. Because whitebark pine cones form in the top third of the tree and blister rust tends to kill trees from the top down, a tree's ability to produce seed is eliminated by the rust long before the tree dies (Fig. 5).

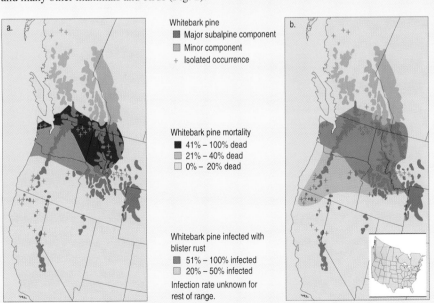

Whitebark pine
- ■ Major subalpine component
- ▨ Minor component
- + Isolated occurrence

Whitebark pine mortality
- ■ 41% – 100% dead
- ▨ 21% – 40% dead
- ☐ 0% – 20% dead

Whitebark pine infected with blister rust
- ■ 51% – 100% infected
- ☐ 20% – 50% infected

Infection rate unknown for rest of range.

Fig. 1. a) Natural distribution of whitebark pine with amount of mortality from all causes since presettlement; b) white pine blister rust infection rates in whitebark pine. In northern Canada and the southern United States, blister rust is present but infection rates are unknown (revised from original map in Kendall 1995).

Courtesy S. Gniadek, National Park Service

Fig. 4. Heavy mortality from blister rust in a whitebark pine stand in Glacier National Park, Montana.

Courtesy K. C. Kendall, USGS

Fig. 5. An ancient whitebark pine in the Mission Mountain Wilderness, Montana, dying from the top down from the introduced fungus, white pine blister rust.

rust and fire control, whitebark pine was an important component on about 10%–15% of the forested landscape in the Rocky Mountains of Montana, Idaho, and northwestern Wyoming (Arno 1986). On about 1.2 million hectares of this area, whitebark pine communities are seral.

Although there is not comprehensive information on whitebark pine throughout its range, recent studies have begun to piece together the current status of this species. An assessment of the interior Columbia River basin found that the area of whitebark pine cover types has declined 45% since the turn of the century (Keane 1995b). Most of this loss occurred in the more productive, seral whitebark pine communities; 98% of them have been lost. Practically all the remaining whitebark pine stands are old. In southwestern Montana, a project to reconstruct landscape patterns found that 14% of the sampled stands were dominated by whitebark pine around 1900, but none of them were by the early 1990's (Arno et al. 1993). Moreover, the extent of stands with significant cone-bearing whitebark pine trees had declined by half.

Nearly half of the whitebark pine trees in Glacier National Park and the Bob Marshall Wilderness Complex in northwestern Montana are dead (Fig. 6). Of the remaining live trees, more than 80% are infected with rust and more than a third of their cone-bearing crowns are dead (Keane et al. 1994; Kendall et al. 1996a). Much of this mortality has been recent; few whitebark pines had suffered significant damage from rust in the early 1970's (Keane and Arno 1993). Blister rust is now present throughout the range of whitebark pine in the Canadian Rockies,

Courtesy B. R. McClelland, National Park Service (retired)

Fig. 6. A ghost forest of whitebark pine in Glacier National Park, Montana.

Declines from Historical Levels

Natural whitebark pine abundance before the recent decline has been summarized by Arno and Hoff (1989). Near the northern end of its range in the British Columbia coastal mountains, whitebark pine is a minor component of treeline communities. In the Olympic Mountains and on the west slope of the Cascades, it grows primarily on exposed sites near treeline. East of the Cascade crest, it is abundant within both the subalpine forest and treeline zone.

Whitebark pine is a major component of high-elevation forests in the Cascades of southern Oregon and northern California. Near the northern end of their distribution in the Rockies of Alberta and British Columbia, whitebark pines are generally small, scattered, and confined to dry, exposed sites at treeline. Whitebark pine becomes increasingly abundant southward, especially in Montana and central Idaho. It is a major component of high-elevation forests and the treeline zone in western Montana. In western Wyoming, it is abundant between elevations of 2,440 meters to 3,200 meters. Before the advent of blister

with the highest rust infection rates and mortality within 125 kilometers of the United States border (Smith 1971; R. Hunt, Forestry Canada, unpublished data).

In southern Montana and Wyoming, whitebark pine health improves as the climate becomes drier (Fig. 7). In the Gallatin National Forest and Yellowstone and Grand Teton national parks, an average of 7% of the whitebark pines are dead and 5% of the live trees are infected with rust (Kendall et al. 1996a,b). The highest infection rates (up to 44%) are found in the Teton Range, where conditions are moister than in neighboring areas to the north. Whitebark pine is reported to be functionally extinct on the Mallard Larkins Pioneer area in the Idaho Panhandle National Forest (Zack 1995). Rust infection rates in the Sawtooth National Recreation Area in central Idaho are generally light, but low elevations may harbor some heavily infected sites (Smith 1995).

Fig. 7. A healthy whitebark pine in Yellowstone National Park, Wyoming.

There is less information about the status of whitebark pine west of Idaho. As a rule, blister rust is present and whitebark pine infection levels and mortality are high in the Cascade and Coast ranges. For a time, the dry conditions in the Sierra Nevada were believed to protect most white pine stands there, but in 1976 and 1983, unusually favorable weather produced heavy waves of rust infection in California white pines (Kinloch and Dulitz 1990). Although sugar pine has been the most affected and studied of these, rust is also present at low levels in some whitebark pine stands. In Kings

Canyon and Sequoia national parks, fewer than 1% of the whitebark pine sampled in 1995 was infected with rust (Duriscoe 1995).

In Washington state, northern Idaho, northwest Montana, and southern Alberta and British Columbia, 40%–100% of the whitebark pine is dead in most stands, and 50%–100% of the live trees are infected with rust (Fig. 1) and have lost most of their capacity to produce cones (Kendall and Arno 1990; Kendall 1994a,b; Kendall 1995). Mortality and rust infection levels decline in the drier areas to the south.

Future Trends

Successional replacement due to fire exclusion is a major cause of whitebark pine decline (Keane et al. 1994). Whitebark pine cannot maintain its functional role in mountain ecosystems unless areas suitable for its regeneration are available across the landscape. Modern fires are restricted in whitebark pine habitats because they normally burn only at the height of very active fire seasons and, under those conditions, managers choose to suppress new fires (Arno 1995). Options for providing sites for whitebark pine regeneration include allowing wildfires to burn near historical levels, having more management-ignited burns with slash cut to help carry the fire in moderate fire weather, and selectively removing whitebark pine's competitors.

It is clear that the blister rust epidemic in whitebark pine has not yet stabilized, even in regions with the longest history and highest infection levels of rust. The most likely prognosis for whitebark pine in sites already heavily infected with rust is that they will

continue to die until most trees are gone. In the southern Rockies and Sierra Nevada where there is currently little or no infection of whitebark pine, waves of infection are expected to occur within a few decades (Kinloch and Dulitz 1990; Kendall et al. 1996a). Eventually, whitebark pine in these areas is likely to suffer heavy losses.

Whitebark pine possesses some ability to defend itself from white pine blister rust (Arno and Hoff 1989), and there is evidence that natural selection has already started to enhance that ability. Forty percent more seedlings from stands with high blister rust mortality survived artificial inoculation with rust than seedlings from low mortality stands (Hoff 1994). In the future, whitebark pine trees will be all but absent in most areas, and small, isolated populations will be lost until rust-resistant types evolve. Without intervention, this is expected to require hundreds—if not thousands—of years, because whitebark pine matures slowly and most of the population soon will be lost (Fig. 8). Management strategies such as breeding whitebark pine for rust resistance and establishing natural selection stands will speed this evolution (Hoff et al. 1994).

See end of chapter for references

Author

Katherine C. Kendall
U.S. Geological Survey
Biological Resources Division
Glacier Field Station Science Center
West Glacier, Montana 59936-0128

Fig. 8. Winter comes to a whitebark pine stand in Yellowstone National Park, Wyoming.

Limber Pine

Limber pine is a five-needled pine widely distributed in the mountains and foothills of the Rocky Mountains in the western United States and southern Canada (Fig. 1). It is adapted to dry and windy conditions and can grow in some of the driest sites capable of producing trees (Pfister et al. 1977). Limber pine ranges from upper treeline and midelevation sites to lower treeline where the mountain forests give way to shrub steppes or prairie grasslands (Fig. 2). In most old stands, limber pines are widely spaced dominant trees with a short, bushy form; on moister sites, though, limber pines are moderately tall trees. The large wingless seeds of limber pines are a favored food of many animals. Limber pines are not usually commercially harvested because of their low productivity and poor form, thus, information about the species status is scarce. Recent observations of limber pine mortality have sparked increased interest in trends of limber pine communities (Kendall et al. 1996).

Limber pine, like the related whitebark pine, has been damaged extensively in some areas by white pine blister rust (Kendall 1995; Kendall et al. 1996). Blister rust is an exotic fungus for which limber pine has evolved few defenses, so the tree is extremely susceptible to this deadly disease. Limber pine is less affected by fire suppression than whitebark pine; in some areas, limber pine

Fig. 2. Typical limber pine savannah near the town of Gardiner in southwestern Montana.

Courtesy K. C. Kendall, USGS

has expanded its range by invading grasslands where it was previously excluded by fire.

Status

Limber pine has suffered extensive mortality and blister rust infection in northwest Montana and southern Alberta (Fig. 1). On average, more than a third of the limber pines are dead and 90% of the remaining live trees are infected with rust (Kendall et al. 1996; Fig. 3). The status of limber pine in the northern Canadian Rockies is not known (Smith 1971). To the south, limber pine rust infection is reduced. In southwestern Montana, northwestern Wyoming, and adjoining areas of Idaho, limber pine mortality and incidence of rust are low to moderate, with a few hot spots of heavy infection. Blister rust incidence on limber pine in the Bighorn Mountains of north-central Wyoming has increased dramatically in the past few years (Lundquist 1993). No rust has been found in Craters of the Moon National Monument in southern Idaho (Smith 1995; Kendall et al. 1996). Blister rust has not been reported in limber pine south of Wyoming (Hawksworth 1990; Duriscoe 1995), and little is known of its status in Utah, Nevada, and California.

Outlook

Because limber pine grows in very dry areas, biologists hoped that blister rust, with its higher moisture requirements, would not be able to make significant inroads in limber pine stands. Unfortunately, it is now apparent that for most sites it may be just a matter of time before the necessary climatic conditions combine to produce a large wave of infection, even in drier climates in the southern parts of the limber pine range (Kinloch and Dulitz 1990). Once infected, most trees will die.

White pine blister rust was recently discovered in southwestern white pines in southern New Mexico (Hawksworth 1990). The nearest known rust occurrence is 1,000 kilometers to the north on limber pines in southern Wyoming. It is not clear whether this outbreak is a result of infected cultivars brought to the region or a result of long-distance transport of rust spores, but it is likely that few limber pine stands are ultimately safe from rust.

Other North American white pines have a small degree of rust resistance that can be strengthened by natural selection or tree-breeding programs. It is likely that the same is true for limber pine, but this potential remains unexplored. Because it is also likely that some individual trees are naturally resistant to blister rust, limber pine probably is not threatened with extinction. Some isolated populations, however, will be lost, and limber pine will be functionally extinct in areas suffering from heavy mortality for the hundreds of years that will be required for rust-resistant types to emerge. Natural selection could be speeded by a breeding program and establishment of stands where more limber pine seedlings are available for natural rust-resistant type selection because all other competing species are removed (Hoff et al. 1994).

See end of chapter for references

Author

Katherine C. Kendall
U.S. Geological Survey
Biological Resources Division
Glacier Field Station Science Center
West Glacier, Montana 59936-0128

■ Low to moderate mortality and infection
■ High mortality and blister rust infection
■ Limber pine distribution

Fig. 1. Distribution of limber pine showing mortality zones and blister rust infection rates (adapted from Little 1971).

Fig. 3. Ghost limber pine forest on the Blackfoot Indian Reservation, with the mountains of Glacier National Park in the background.

Courtesy R. Keane, USDA, Intermountain Research Station

forests in Colorado alone (C. Chambers, U.S. Forest Service, personal communication). The relationship between reservoir building and human population increases is particularly obvious (Fig. 11). Domestic water use accounts for less than 6% and agriculture for about 90% of the total water use (U.S. Geological Survey 1990). Reservoir building may also correlate with increased irrigation or other uses indirectly related to population growth.

Water quality is a growing concern in the Rocky Mountains. The U.S. Geological Survey

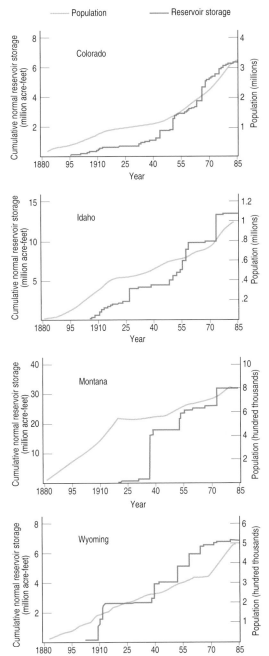

Fig. 11. Relationship between human population growth and cumulative storage of water in reservoirs (adapted from U.S. Geological Survey 1990).

(1993) stated that all of Colorado's major drainages are affected to some degree by pollution. Past mining operations still contribute toxic trace elements to more than 2,100 kilometers of rivers and streams in Colorado. More than 900 kilometers of Colorado's 50,300 kilometers of streams do not meet water-quality criteria for fishing. Other Rocky Mountain states report similar problems (U.S. Geological Survey 1993). Although water developments affect riparian zones upstream and downstream from dams (Mills 1991), regional information on the biotic effects of water projects and pollution is either extremely limited, fragmentary, or inaccessible.

Throughout the conterminous United States, the surface area of wetlands has decreased from 11% to 5% (Brady and Flather 1994). Most of the loss is attributed to agricultural land-use conversions and urban expansion. Although little specific information is available on the status of wetlands in the Rocky Mountains region, urban development and water-control projects on floodplains along the major rivers are clearly modifying critical wildlife habitat.

Beavers helped shape many riparian zones and wetlands in the Rocky Mountains for thousands of years (Knight 1994). Their debris dams influenced vegetation patterns, sedimentation rates, flood severity, and water quality (Knight 1994; Schlosser 1995). Trapping the beaver to near extinction reduced the abundance of willow and moist-grass communities and increased erosion downstream (Knight 1994).

Nearly all native fisheries in the Rocky Mountains have been compromised by introduced fishes (Trotter 1987; Behnke 1992). Of the 13 subspecies of native cutthroat trout once found in the interior West, 2 are extinct and 10 have suffered catastrophic declines (Behnke 1992).

The Greater Rocky Mountain National Park Ecosystem

The Greater Rocky Mountain National Park ecosystem is typical of many park and forest ecosystems in the Rocky Mountains. It serves as an example of how natural ecosystems are confronted by a multitude of simultaneous threats including encroachment from urbanization and development, habitat fragmentation, fire suppression, nonindigenous species invasion, and global climate change (Stohlgren et al. 1995a). The response of the forest to turn-of-the-century logging and wildfires was a fivefold increase in ponderosa pine bole (trunk) biomass (Fig. 12). This is good news for wildlife dependent on ponderosa pine, but fire suppression continues to create a growing wildfire hazard.

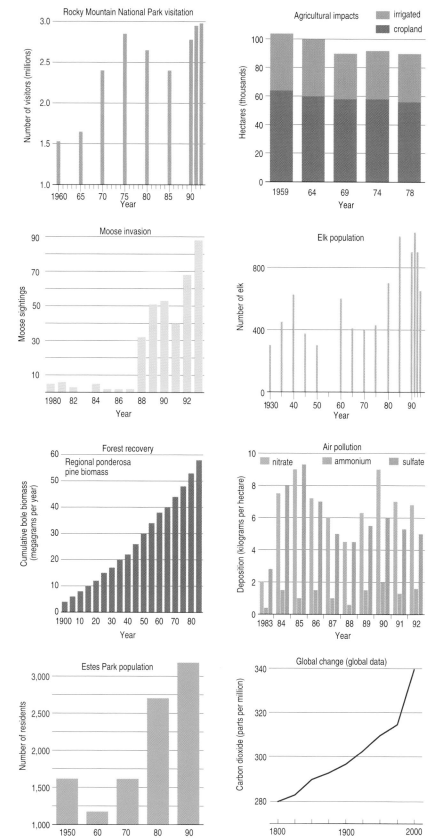

Visitor use of the Rocky Mountain National Park has almost doubled since 1960, and the human population in Estes Park has more than doubled (Fig. 12). Urban development throughout the Front Range of Colorado caused habitat loss and fragmentation and increased air pollution. Annual values of wet deposition of nitrate, sulfate, and ammonium in the Loch Vale watershed of the Rocky Mountain National Park are significantly greater than the average value of 0.5 kilograms per hectare found in remote areas of the world (Fig. 12).

Elk and moose populations continue to increase in the park (Fig. 12) because of a complex of causes including reduced hunting and predation (wolves have been extirpated) and diminished habitat and migratory corridors outside the park. Agricultural land use in Larimer County has declined slightly in recent years (Fig. 12), but landscape and ecosystem integrity are challenged by fire suppression, nonindigenous species invasions, weather modification (that is, cloud seeding; Stohlgren et al. 1995b), and global climate change (Stohlgren et al. 1993).

The Greater Yellowstone Ecosystem

Scientists who study the Greater Yellowstone National Park ecosystem have long acknowledged the ecosystem-scale issues surrounding wildlife management (Leopold et al. 1963). Migratory populations of elk, bison, and deer; a wild grizzly bear population; the Yellowstone cutthroat trout fishery; and world-renowned geothermal resources are challenged by increasing visitor use, introduced diseases and competitors, and global climate change

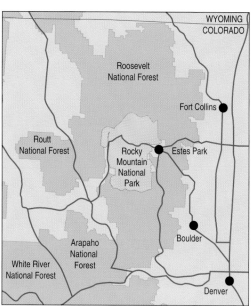

Fig. 12. Trends in Rocky Mountain National Park, Colorado (see inset map): number of visitors to the park, agricultural land use, elk populations, moose invasion, forest recovery, air pollution, human population of nearby Estes Park, and global change in carbon dioxide.

(Schullery 1995). Increasing numbers of elk (31,000 summer in the area) may be having adverse effects on the elks' winter range (Knight 1994). Livestock grazing is permitted on about 40% of grizzly bear habitat in the Greater Yellowstone ecosystem, and many bears have been killed in these areas (Keiter 1991). Evidence now shows that only 15%–20% of the park was subjected to the crown fires of 1988, and the burned areas are recovering rapidly (Knight 1994). The park's geologic features were unscathed by the fires, large mammal populations are thriving, and visitation (and tax revenues) are increasing (Knight 1991).

Pressures from outside the park include commercial development, logging—in some instances up to the border of the park—threatened exploitation of geothermal resources, and hunting of stray animals. Several population-size trends of plant and animal species are issues in the Greater Yellowstone ecosystem requiring consideration (Schullery 1995). Park personnel must manage grizzly bear persistence, wolf reintroduction, elk population increases, introduced diseases (brucellosis in bison, white pine blister rust), and introduced fishes, all while coping with more than 3.2 million visitors per year.

Status and Trends of Plant and Animal Populations

Because information on trends in population sizes of most species in the Rocky Mountains is incomplete, the status and trends of only a few species can be reported. Species such as the black bear and mountain lion, many small mammals, and common bird and plant species are described because, in most instances, the populations are persistent and not rapidly increasing or declining. Even basic regional information is not available on many nocturnal species (for example, bats, raccoons, and so forth); invertebrates; lichens, mosses, and fungi; and soil microorganisms. In essence, information for only the high-profile species is available and accessible. The status of each individual species, however, affects and is affected by the status and trends of other species in its ecosystem.

Invertebrates

Although most of the animals in the Rocky Mountains are invertebrates, little is known about this component of the region's fauna (Mason 1995). As one entomologist stated, "We do not know how many species of moths and butterflies live in any state, county, or locality in North America" (Powell 1995:170). In a few areas in the western United States, information is available on the species richness of moths and butterflies. Most of the Rocky Mountain states and the Front Range of Colorado in particular support high species richness of butterflies and moths (Opler 1995). In Colorado, the diverse habitats—from prairie to tundra—support about 2,000 species of butterflies, moths, and skippers; more than 1,000 species are in the Front Range (Opler 1995). Some species of grasshoppers are unique to individual mountaintops in Colorado, New Mexico, Arizona, Nevada, and Utah (Otte 1995). The Rocky Mountain locust, a common pest to farmers in the 1800's, is now extinct. Heavy grazing along river valleys in Montana and Idaho is thought to have irreparably destroyed locust breeding areas (Otte 1995).

Amphibians

Globally, populations of amphibians are declining in size as a result of habitat loss, predation by nonindigenous sport fishes, timber harvest, increased ultraviolet radiation, and disease (Bury et al. 1995). The widespread declines of amphibian populations throughout the Rocky Mountains mirror these global trends. Western toads, once common between altitudes of 2,300 and 4,200 meters throughout the central and northern Rocky Mountains, now occupy less than 20% of their previous range, from southern Wyoming to northern New Mexico (Bury et al. 1995). Eleven populations of western toads disappeared from the West Elk Mountains of Colorado between 1974 and 1982 because of a bacterial infection and, perhaps, multiple sublethal environmental causes (Carey 1993). In the past two decades, western toads disappeared from 83% of their historical range in Colorado and from 94% of Wyoming sites (Bury et al. 1995). Populations of northern leopard frogs are significantly declining throughout the Rocky Mountains (Corn and Fogelman 1984; Bury et al. 1995).

Fishes

Greenback Cutthroat Trout

Greenback cutthroat trout historically inhabited the cold-water streams in the mountains of Colorado (Fig. 13). The species was near extinction by the early 1900's because of broadscale stocking of these streams with nonindigenous brown trout from Europe and rainbow trout from the Pacific Coast, and because of land and water exploitation, mining, and logging. Native genetic diversity is now lost in greenback cutthroat trout because this species hybridizes with the introduced fishes. Interagency efforts began in 1959 to save the species. With viable populations now at 48 sites, the greenback cutthroat trout is one of the

Amphibians of Glacier National Park

Reports of amphibian declines worldwide have raised concerns about the status of the amphibians of Glacier National Park, a 410,360-hectare federal reserve in the northwest corner of Montana. The headwaters of three continental river systems flow from the park: the North Fork and the Middle Fork of the Flathead River drainages of the Columbia River basin; the upper Missouri River drainage; and the South Saskatchewan (Hudson Bay) River drainage. More than 700 lakes and 3,660 kilometers of streams create a mosaic of aquatic habitats ranging in elevation from 925 meters to more than 3,000 meters.

As many as a dozen species of amphibians have been reported to occur in the park (Manville 1957; Davis and Weeks 1963; Metter 1967; Black 1970a,b; Thompson 1982; R. B. Brunson, University of Montana [retired], unpublished manuscript). However, a recent study, which used geographic information system processing of species-sighting data, confirmed only five species as park residents. Field investigations were supplemented by examination of specimens and collection records from the park museum and 13 other museums that had specimens from Glacier National Park.

Columbia Spotted Frog

The Columbia spotted frog is the most common frog in Glacier National Park. It occurs in both green and brown color phases and prefers small shallow ponds, marshes, and bogs with mud bottoms and dense emergent nonwoody vegetation (Fig. 1).

Courtesy L. Marnell, National Park Service

Fig. 1. Natural camouflage of the adult Columbia spotted frog makes it difficult to see in its natural habitat.

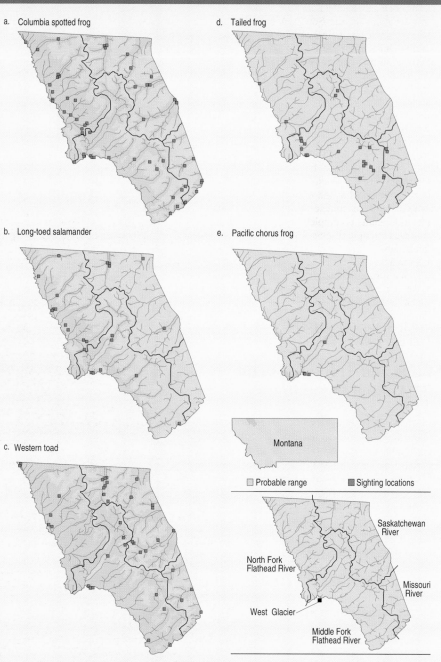

a. Columbia spotted frog

d. Tailed frog

b. Long-toed salamander

e. Pacific chorus frog

c. Western toad

Montana

☐ Probable range   ■ Sighting locations

Saskatchewan River

North Fork Flathead River

Missouri River

West Glacier

Middle Fork Flathead River

Fig. 2. Sighting locations and probable ranges in Glacier National Park of a) Columbia spotted frogs, b) long-toed salamanders, c) western toads, d) tailed frogs, and e) Pacific chorus frogs.

The species occurs on both sides of the Continental Divide at elevations between 1,050 meters and 1,845 meters (Fig. 2a). Columbia spotted frog egg clusters were observed by early May, with hatching from late May through mid-June. High egg mortality was observed at some locations. Many of the documented breeding locations were beaver ponds.

Long-Toed Salamander

The long-toed salamander is the only resident salamander species in Glacier National Park. Local forms display a prominent dorsal stripe variable from dull green to bright yellow, and the rest of the body is typically dusky black or charcoal with fine white or bluish-white speckles on the lower sides. This salamander is common through-out valleys and midelevation forests west of the Continental Divide at elevations from 1,055 meters to 1,515 meters (Fig. 2b). Long-toed salamanders are found at only a few locations in the South Saskatchewan River drainage and are common there only in the Waterton Valley. A large breeding population occurs near Waterton Townsite.

Adults are nocturnal and are active aboveground for only a short period during spring and early summer. After breeding, these salamanders disperse rapidly into woodland habitats. Adults can be found in the rotting bark of downed trees or under rocks and logs. Small clutches containing 5–20 salamander eggs have been seen by late April in most years. Hatching typically occurs by mid-May and is complete by late May at low to midelevations. Metamorphosis to adults is complete by early August followed by rapid dispersal to nearby woodlands. However, breeding periods and incubation times vary considerably. For example, in a spring-fed pool below Upper Kintla Lake at 1,397 meters elevation, salamander larvae were still relatively small by late August 1991. In late October the pool was mostly frozen over, and salamander larvae were still visible on the mud bottom beneath the ice.

Western Toad

Western toads in Glacier National Park vary from medium brown to black with a pale yellow or cream-colored dorsal stripe. Toads occur on both sides of the Continental Divide from low-elevation valleys upward to timberline (Fig. 2c). Western toads are the most wide-ranging amphibians in the park; sightings were recorded from 1,045 meters to 2,255 meters elevation. Among 41 sighting locations, 17 (41%) were above 1,650 meters, and 7 sites (17%) were higher than 1,980 meters. Manville (1957) reported a toad sighting near the Mount Brown fire lookout at 2,320 meters. A breeding site was located at 2,088 meters elevation at a glacial tarn near Logan Pass. Although western toads are distributed parkwide below timberline, they are not abundant at most locales.

Breeding times for western toads are strongly correlated with elevation and water temperatures. Egg strings typically appear at low elevations in early May in shallow ponds with mud or silt bottoms, often attached to the edges of algal mats in water 10 to 20 centimeters deep. About half of the 14 documented breeding sites for western toads were beaver ponds. Hatching occurred at most locations by the second or third week of June.

Tailed Frog

Tailed frogs seldom stray far from their preferred habitat of cold turbulent headwater streams with cobble substrates. Glacier National Park lies near the eastern limit of their range. The body color of adult tailed frogs in Glacier National Park is typically gray-brown to rust, fading to a lighter-colored belly (Fig. 3). Adult tailed frogs are nocturnal and are often difficult to locate. The species occurs throughout much of the Middle Fork of the Flathead River drainage but is intermittent and widely dispersed in the North Fork and upper Missouri River drainages of Glacier National Park (Fig. 2d). Tailed frogs are most often seen at elevations between 1,045 meters and 2,140 meters. The range map (Fig. 2d) probably underrepresents the distribution of tailed frogs in Glacier National Park because only a small proportion of suitable habitats was searched.

Courtesy L. Marnell, National Park Service

Fig. 3. Adult tailed frogs are nocturnal and are rarely seen during daylight.

Little is known about the breeding activities of tailed frogs in the streams of Glacier National Park. Researchers believe that development time from hatching to emergence is highly variable for this species (Metter 1967). Time to metamorphosis is up to 4 years in the Washington Cascades and an additional 5–6 years is required for the froglets to attain sexual maturity (Leonard et al. 1993). From 3 to 5 years may be required for metamorphosis in the higher-elevation streams of Glacier National Park.

Pacific Chorus Frog

Pacific chorus frogs have the most restricted distribution of any frog or toad in the park. Glacier National Park is near the eastern limit of this species. Most sightings have been made near the community of West Glacier at about 1,035 meters elevation (Fig. 2e). Breeding occurs in several small ponds

above the floodplain of the Middle Fork of the Flathead River, upstream from West Glacier. Chorus frogs appear to travel on warm summer nights and are capable of dispersing several kilometers from breeding sites.

Pacific chorus frogs begin breeding in Glacier National Park in late April and early May. Vocalizations usually peak during the first week of May. Egg deposition has not been observed during daylight, but egg clutches containing 10 to 30 eggs are present at several of the ponds by mid-May. Hatching occurs from late May through the first week of June, and by the end of June most adult chorus frogs have left the breeding ponds. Development is rapid, with emergence occurring by the first week in August. Juvenile chorus frogs are present in shallow waters among shoreline vegetation at the edges of several of the ponds by mid-August. Dispersal into nearby riparian zones occurs a few days after emergence. Most juvenile frogs have left the breeding ponds by late August.

Effects of Fish Introductions

The introduction of sport fish into a large number of formerly fishless lakes (Marnell et al. 1987; Marnell 1988) may have contributed to the loss or decline of several amphibians in portions of Glacier National Park. The presence of fish has been implicated in the decline of some amphibian species (Bradford 1989; Bradford et al. 1993; Corn et al. 1997). Long-toed salamanders were particularly vulnerable to predation by introduced fishes in portions of the Cascade Mountains in western Washington and Oregon (Leonard et al. 1993). Long-toed salamander larvae were not observed in any Glacier National Park water harboring fish, and this species existed close to fish at only 2 of 25 sites. Concerns about fish predation may be especially warranted in parts of the upper Missouri River drainage and in the Many Glacier Valley (South Saskatchewan River drainage). The extent of damage to native amphibians in Glacier National Park as a consequence of fish introductions may never be fully understood.

See end of chapter for references

Author

Leo Marnell
U.S. Geological Survey
Biological Resources Division
Glacier Field Station Science Center
West Glacier, Montana 59936–0128

Fig. 13. Restoration of the green-back cutthroat trout in Colorado has been highly successful.

few species that will be removed from the endangered species list (Colorado Division of Wildlife 1986; Henry and Henry 1991). Three of the four other native subspecies of the cutthroat trout are extinct (Greenback Cutthroat Trout Recovery Team 1983).

Westslope Cutthroat Trout

Most aquatic ecosystems in the Rocky Mountains are now influenced by nonindigenous brown trout and rainbow trout. One of the more subtle, yet devastating, trends in the Rocky Mountain fishery is loss of genetic diversity in native fishes resulting from introductions of nonindigenous fishes. For example, only 15 of 32 lakes in Glacier National Park, Montana, contain pure genetic strains of the native cutthroat trout. The others contain totally non-indigenous fishes or hybrids with the introduced Yellowstone cutthroat trout or rainbow trout. Introductions of nonindigenous fishes to improve the sport fishery have compromised about 84% of the historical range of the native cutthroat trout (Marnell 1995).

Yellowstone Cutthroat Trout

Yellowstone Lake in Yellowstone National Park, Wyoming, is the site of the most recent catastrophic species invasion. *The Washington Post* (2 October 1994) reported that the non-indigenous lake trout, a native of the Great Lakes, had been insidiously introduced into one of the nation's premier fisheries. The native Yellowstone cutthroat trout may not compete well with lake trout because lake trout eat cutthroat trout. The potential ecological repercussions are staggering. If populations of cutthroat trout decline, grizzly bears could lose an important posthibernation food because the native cutthroat trout spawn in the streams and are easy prey for the bears, whereas the nonindigenous lake trout spawn in deep water.

White Sturgeon

The largest freshwater fish in the Rocky Mountains (and North America) is also in trouble. The white sturgeon historically ranged from the mouth of the Columbia River to the Kootenai River upstream to Kootenai Falls, Montana. The Kootenai River population of the white sturgeon is unstable and declining in size (Miller et al. 1995); fewer than 1,000 remain, 80% are older than 20 years, and virtually no recruitment has occurred since 1974, soon after Libby Dam in Montana began regulating flows (Apperson and Anders 1990).

Birds

Bald Eagles

The coniferous and deciduous forests of North America have long been the home of bald eagles (Fig. 14). Bald eagle populations are now recovering after years of hunting, habitat destruction, and pesticide-induced deaths (Finch 1992). In the early 1970's, Colorado had just one breeding pair of bald eagles but by 1993 biologists counted 19 breeding pairs (Colorado Division of Wildlife 1993). In Wyoming nesting attempts increased from 20 in 1978 to 42 in 1988 (Finch 1992). The bald eagle is not yet fully recovered, however; pesticide residues continue to inhibit bald eagle reproduction, and habitat loss and lead poisoning remain serious threats (Henry and Anthony 1989).

Fig. 14. The bald eagle is recovering throughout the United States.

Peregrine Falcons

Peregrine falcons are cliff-dwelling raptors that once ranged through most of North America. Like the bald eagle, this species was driven to near extinction by pesticides. By 1965 fewer than 20 breeding pairs were known west of the Great Plains (Finch 1992). Even in the Greater Yellowstone ecosystem, federal spruce budworm control relied on DDT, which accumulates in the food chain, causing eggshell thinning and reduced reproductive success in raptors (Boyce 1991). Six breeding pairs of American peregrine falcons were found in Colorado in the early 1970's (Colorado Division of Wildlife 1986). By 1994, 53 pairs were breeding in Colorado. In Wyoming, Montana, and Idaho combined, 8 of 59 historical sites were used by falcons in 1987. Low breeding densities, reproductive isolation, habitat loss, and pesticide poisoning on wintering grounds remain threats to peregrine falcon recovery (Finch 1992).

White-Tailed Ptarmigans

White-tailed ptarmigans have been monitored in Rocky Mountain National Park, Colorado, since 1966 (Colorado Division of Wildlife 1994). Short-term population cycles are well documented in populations that are not hunted but not in populations outside the park, which are hunted. Although detailed population size data are available from more than 28 years of monitoring (Fig. 15), scant information is available on habitat change, predator populations, or other potential causes of change in ptarmigan populations. A 2-year study (Melcher 1992) revealed lower ptarmigan densities where elk use was greater, although characteristics of willow, which is ptarmigan habitat, did not significantly differ in the high- and low-use elk sites (Melcher 1992). Furthermore, a 2-year study of ptarmigan habitat cannot explain 28-year trends in population size. Habitat loss and other factors partly responsible for ptarmigan deaths—such as predation and competition—were not studied during the 28-year period.

Trumpeter Swans

Trumpeter swan populations were seriously threatened in the 1930's; fewer than 70 birds were thought to exist (Boyce 1991). Now protected from hunting, more than 1,500 swans winter in the Greater Yellowstone ecosystem, but the size of the breeding population has declined in recent years because of habitat loss (Boyce 1991).

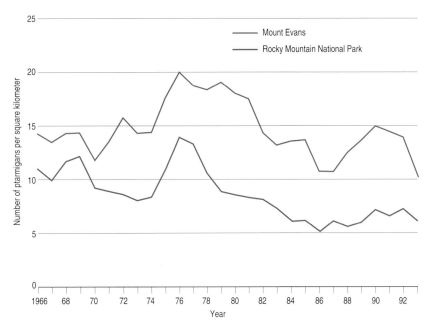

Fig. 15. White-tailed ptarmigan populations show cycles in hunted populations (Mount Evans) and in unhunted populations in the Rocky Mountain National Park, but the causes of these cycles remain unclear (Colorado Division of Wildlife 1994).

Neotropical Migrant Songbirds

Many forest-dwelling songbirds breed in the Rocky Mountains and winter in Central and South America. Wildlife biologists suspect that population size declines in the songbirds may be partly the result of increased predation and brood parasitism. Brood parasitism by brown-headed cowbirds, for example, increases as a result of nearby logging (Evans and Finch 1994). In conifer forests in west-central Idaho, common songbirds benefited from timber harvest, whereas the abundances declined of rare species that inhabit old-growth forests (hermit thrush, Swainson's thrush, and pileated woodpecker; Evans and Finch 1994).

Mammals

Grizzly Bears

Grizzly bears once roamed throughout the Rocky Mountains and the western Great Plains. They were hunted relentlessly by European settlers in the 1800's and early 1900's (Mattson et al. 1995). The last known grizzly bear in Colorado was killed in 1979. The decline of the bears to just 2% of their original range (Fig. 16) tells of the human-caused extirpation of large predators in the Rocky Mountain region. Only 700–900 grizzly bears may be alive today in the conterminous United States (Serveen 1990). During the last 20 years, about 88% of all grizzly bears studied in the northern Rocky Mountains were killed by humans (Mattson et al. 1995).

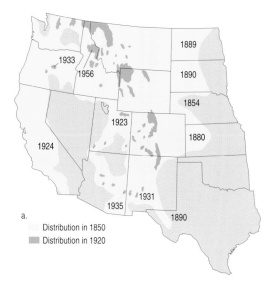

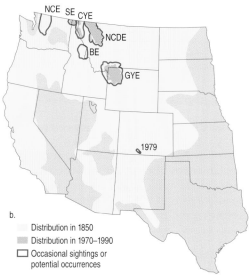

Fig. 16. a) Approximate distribution of grizzly bears in 1850 compared with 1920 and with b) 1970–1990. Local extinction dates appear in a). Populations identified in b) are NCE—North Cascades ecosystem, SE—Selkirk ecosystem, CYE—Cabinet–Yaak ecosystem, BE—Bitterroot ecosystem, NCDE—Northern Continental Divide ecosystem, and GYE—Greater Yellowstone ecosystem. As indicated in b), a grizzly bear was killed in the San Juan Mountains of Colorado in 1979 (Mattson et al. 1995).

Fig. 17. The gray wolf is slowly returning to a small portion of its former range.

Gray Wolves

Gray wolves once were common throughout the Rocky Mountains (Cook 1993). They were shot, poisoned, and trapped into local extinction by early settlers and federal agents (Brandt 1993; Cook 1993; Fig. 17). The last gray wolf in Colorado was killed in 1940, and the wolf was first listed as an endangered species in 1967 (Mech et al. 1995). Wolves from southeastern British Columbia recolonized northwestern Montana in 1986; by 1994 the population had grown to 7 packs and about 70–75 wolves (N. Bishop, Yellowstone National Park, personal communication). Wolves from Glacier National Park have dispersed naturally as far away as northeastern Idaho and just south of Yellowstone National Park. A wolf was shot near Yellowstone National Park in 1992. From January to March 1995, 15 adult wolves from 7 different packs in Canada were introduced into central Idaho wilderness areas. Several pairs have bred and produced the first litters of wolf pups born in Idaho in more than 50 years. Fourteen wolves (three family groups) were released in the Yellowstone National Park in late March 1995 (Bishop, personal communication). An adult female from each group may have bred. The status of gray wolves in the northern Rocky Mountains is improving, but the species is still persecuted and probably occupies less than 10% of its historical range in the contiguous United States. More than 70% of Colorado residents support a restoration of wolves in Colorado, but no such effort is under way (Manfredo et al. 1994).

The restoration of the gray wolf to the Yellowstone National Park not only restores an important ecosystem component (the wolf) and process (predation by wolves) to bring the park into better ecological balance, but it also is economically sound. After weighing the costs (including full reimbursement to ranchers for the loss of livestock) and benefits (increased revenues from hunting and tourism), economists estimated (before the actual restoration took place) a net $18 million return during the first year after the wolves were returned, and about $110 million in 20 years (Duffield 1992; Brandt 1993). More tourists are expected to visit the area of the Yellowstone National Park and to stay longer in hope of hearing or seeing wolves in the wild (Duffield 1992). Compensatory payments to ranchers for the loss of cattle and sheep to wolves averaged about $1,800 per year in northwestern Montana (Bishop, personal communication).

Caribou

Caribou were once common in the northern Rocky Mountains. In fact, the head of a caribou

Courtesy L. Rogers, International Wolf Center

killed by Theodore Roosevelt hangs in Jack's Bar in Bonners Ferry, Idaho (Speart 1994). Hunting and habitat loss (logging of old-growth Rocky Mountain juniper and hemlock forests) reduced the herd in Idaho's Selkirk Mountains to 100 animals by the 1950's. By the 1980's the few remaining caribou had crossed the border to the Canadian Rocky Mountains. A lone male reintroduced the species by wandering back into Idaho in 1984 (Speart 1994). The Endangered Species Act of 1973 helped protect some of the last remaining old-growth Rocky Mountain juniper and hemlock stands with the old-growth-dependent lichens that are the primary food of the caribou. A series of three caribou transplants from Canada has helped maintain the newly reestablished herd of about 30 animals.

North American Elk, Deer, Pronghorns, and Moose

Population trends in North American elk and deer (mule deer and white-tailed deer combined) may be heading in opposite directions. The number of elk has increased steadily in Colorado and Wyoming, whereas the abundances of deer are showing signs of decline (Fig. 18). Elk on U.S. Forest Service lands in the Rocky Mountains increased from 268,000 in 1965 to 372,000 in 1984 (Flather and Hoekstra 1989). Similarly, the number of elk on Bureau of Land Management lands rose from 35,000 in 1966 to 114,000 in 1985. Meanwhile, the number of deer on U.S. Forest Service lands declined from 1,742,000 in 1965 to 1,197,000 in 1984. Deer populations also declined on Bureau of Land Management lands. Thus, in some areas in the last 20 years, the abundances of elk have increased by about 40%, whereas deer have decreased by about 30% (Flather and Hoekstra 1989). Possible reasons for the increase in elk populations include mild winters, range extension into lowlands and highlands, increased adaptability to human-modified landscapes, and lack of predation in spite of increased hunting (F. Singer, U.S. Geological Survey, personal communication). The causes of the deer population declines remain unknown (Connolly 1981) but may include excessive harvest in the 1970's and habitat overlap with elk, intensifying competition for similar resources.

Pronghorn populations have fluctuated but generally have increased in the past 20 years in Colorado (Fig. 18) and Wyoming. Moose populations have increased 50% since 1980 in Wyoming and have been rapidly increasing since their introduction into Colorado in 1978 and 1979 (Fig. 18).

Bighorn Sheep

Populations of bighorn sheep are at only about 2% to 8% of their sizes at the time of European settlement (Singer 1995). Causes for the rapid decline from 1870 through 1950 included unregulated harvesting, excessive grazing of livestock on rangelands, and diseases transmitted by domestic sheep. In recent years, 115 translocations were made to restore bighorn sheep into the Rocky Mountains and into many national parks. Only 39% of the 115 bighorn sheep translocations are persisting in 6 Rocky Mountain states. Populations of 100 or more sheep now occur in 10 national park units, populations of 100–200 sheep in 5 units, and populations of more than 500 sheep in 5 units. Populations of fewer than 100 animals exist in 5 other park units (Singer 1995).

Beaver

Beavers once played important roles in shaping vegetation patterns in riparian and meadow communities in the Rocky Mountains (Knight 1994). Studies of beaver populations in one small area in Yellowstone National Park (Tower Junction area) in the early 1920's reported 232 beavers and extensive beaver dams. Repeated surveys in the same area in the early 1950's and in 1986 revealed no beavers or dams (Chadde and Kay 1991). Beavers need aspens or tall willows for food and building materials—resources that are made scarce by lack of both fires and floods and by herbivory by elk, moose, and domestic livestock. Beaver ponds are known to maintain fish and invertebrate populations (Schlosser 1995) and to create and maintain riparian zones that are critical to wildlife (Chadde and Kay 1991; Knight 1994), yet the beaver is virtually absent in many areas (Chadde and Kay 1991).

Introduced Diseases and Plant and Animal Species

Introduced Pathogens

In addition to the pathogen of the white pine blister rust, many other introduced pathogens are having profound effects on native species. When bison were restored to Yellowstone National Park in 1902, they may have transported unwanted guests. The origin and management implications of brucellosis, a disease caused by bacteria in domesticated animals (cattle, horses), bison, elk, and even rodents, are controversial. Although the wild bison may not transmit brucellosis directly to cattle (Meagher and Meyer 1994), some level of human intervention in and adjacent to Yellowstone National

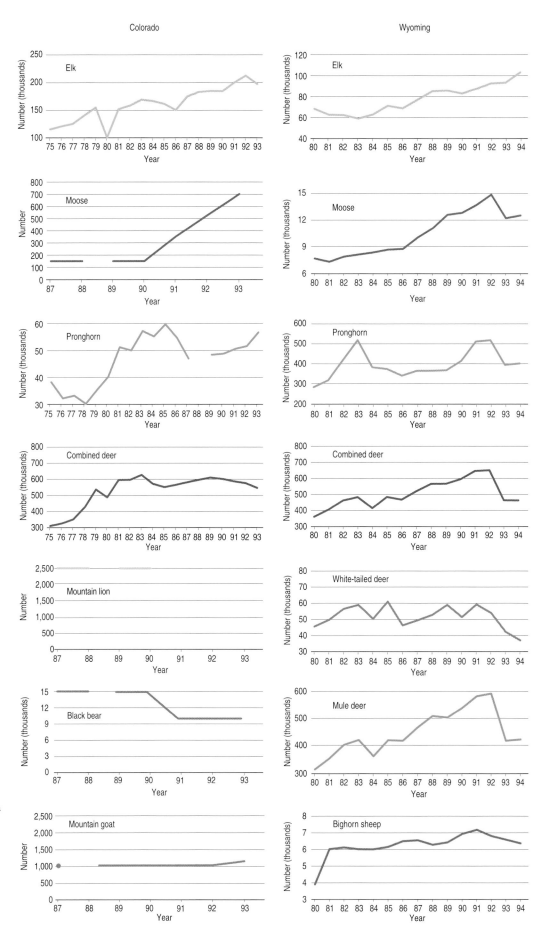

Fig. 18. Big game populations in Colorado from 1975 to 1993 (data from the Colorado Division of Wildlife, Department of Natural Resources Database, Fort Collins) and in Wyoming from 1980 to 1994 (Wyoming Game and Fish Department, Cheyenne, unpublished data). Missing data represent years when populations were not monitored.

Park may be needed to prevent spread of the disease (Aguirre and Starkey 1994).

The lungworm-pneumonia complex is a bacterial disease that causes spontaneous death in the lambs of bighorn sheep in summer. Although some strains of the disease complex are native to bighorn sheep and others are related to domestic sheep, disease exchange can be fatal to both groups (Aguirre and Starkey 1994). Proximity to domestic sheep is highly correlated with deaths in newly reestablished bighorn sheep populations (Singer 1995). In spite of the disease, however, bighorn sheep populations in national forests in the Rocky Mountain region increased from 11,533 individuals in 1965 to 17,658 in 1984 (Flather and Hoekstra 1989).

Whirling disease, introduced from Europe, is a parasitic infection that attacks recently hatched trout. It now affects wild trout populations in Rocky mountain regions. The disease was first thought to affect only hatchery fishes, but the native greenback cutthroat trout also may be susceptible.

Introduced Plants

Cheatgrass has invaded not only the Great Basin but also significant portions of the western pinyon–juniper woodlands and ponderosa pine and Douglas-fir ecosystems in the Rocky Mountains (Peters and Bunting 1994). Many native shrubs and perennial grasses cannot survive the increased competition from cheatgrass (as many as 3,000–10,000 plants per square meter; Barbour et al. 1987). Several rare plant species are being displaced by introduced plants in western rangelands (Rosentreter 1994). A dense cover of cheatgrass has increased the fire frequency in many of these areas. With each fire, the dominance of nonindigenous annual grasses is enhanced at the expense of native perennial grasses. The establishment of cheatgrass is also enhanced by livestock grazing (Fleischner 1994).

More than 1.4 million hectares of Oregon, Washington, Idaho, and Montana are covered with spotted knapweed (Langner and Flather 1994). How much of this coverage is in the Rocky Mountains cannot be estimated.

Purple loosestrife, another European weed, is beginning to invade Rocky Mountain wetlands and streamsides. Purple loosestrife spreads quickly and crowds out native plants that animals use for food and shelter. This invader has no natural enemies in the United States and therefore spreads unchecked (Thompson et al. 1987). The effects of these introduced plants on Rocky Mountain ecosystems are poorly understood.

The battle against introduced weeds has intensified with the use of biological control agents. Biological control usually entails the purposeful introduction of a pathogen or insect that inhibits the establishment, growth, or reproduction of the target species. The biological control agent must be highly specific to the target weed. Extensive laboratory and field tests must be completed before the release of a control agent. A large project is under way in Bozeman, Montana, and in Regina, Saskatchewan, where European insects are being released to control the poisonous European leafy spurge (DeLoach 1991).

Introduced Mammals

Two species of hooved mammals—the mountain goat and the moose—were deliberately introduced into Colorado. Although these species occasionally wandered into Colorado in presettlement times, breeding populations did not occur until after deliberate introduction. Accidentally introduced mammals in Colorado and Wyoming include the house mouse and the Norway rat (Armstrong 1993). The potential effects of these introduced mammals on Rocky Mountain ecosystems are poorly understood.

Challenge of Assessing Status and Trends of Biotic Resources

Assessing the status and trends of most biotic resources in the Rocky Mountains is difficult for many reasons, most importantly because of incomplete inventories, fragmented monitoring, poorly standardized inventory and monitoring, and inaccessible data and information. Future efforts to determine the status and trends of biotic resources in the Rocky Mountains would be helped greatly by improving the accessibility of information and by integrating multiscale research, inventory, and monitoring.

Existing Biotic Inventories Are Incomplete

The best-studied areas in the Rocky Mountains may be within the national park system, but even the biotic inventories in national parks are incomplete (Stohlgren et al. 1995c). Accurate vegetation maps are available for less than half of the national park units and detailed soils and geology maps for less than 25%. Species lists of vascular plants, mammals, reptiles, amphibians, and birds in most park units are generally only 50%–80% complete (Stohlgren et al. 1995c). Even less is known about nonvascular plants and invertebrates (Stohlgren and Quinn 1992; Opler 1995; Powell 1995).

Lists of threatened, endangered, and sensitive species are readily available from the U.S. Forest Service and U.S. Fish and Wildlife Service, but information about population-size trends in most species is not available. In Colorado the status of the river otter, lynx, and wolverine is unknown. The last known occurrence of the river otter was in 1906, of the lynx in 1980, and of the wolverine in 1890 (Colorado Division of Wildlife 1986).

Often, the best information on wildlife populations is available from state departments of fish and game, primarily in the form of brochures and unpublished reports. In general, a great deal of attention is given to the hunted megafauna and little to nongame species. Better information on the plants of the Rocky Mountains is forthcoming; a multiple-volume flora is being developed (R. Hartman and colleagues, University of Wyoming, Laramie). The three best sources of information on biotic resources of the Rocky Mountains have included a recent natural history and ecology text on Wyoming (Knight 1994), a synthesis about rare and threatened species (Finch 1992), and *Our Living Resources*—a relatively new type of publication of the National Biological Service (now the U.S. Geological Survey) that attempts to consolidate biotic status and trends information from diverse sources (LaRoe et al. 1995).

One indication of the incompleteness of biotic inventories is that many species are added to the species list of a park, refuge, or national forest each time an inventory is made (Stohlgren and Quinn 1992). Often, common but previously overlooked species are added, and species thought to be locally rare are found to be less rare. In Rocky Mountain National Park, for example, more than 100 vascular plant species (none of them threatened or endangered) were added to the park checklist in the past 6 years (L. Yeates, Denver Botanical Garden, personal communication). Rocky Mountain National Park has been well studied by botanists, but many more plant species will probably be found there, and even more in the surrounding, less well-studied areas of the Rocky Mountains.

Accurate, landscape-scale surveys are a necessary component of successful resource management. Total recorded amphibian and reptile diversity of the United States increased 12% since 1978 (McDiarmid 1995) as a result of biological surveys. As new subpopulations of declining species are found, hope for species persistence in the larger landscape increases. The hot springs in Yellowstone National Park are home for thousands of species of microbes that are potentially useful to humans for DNA fingerprinting and for the removal of oil spills and low-level radioactive wastes (Robbins 1994). Yet little is known of the taxonomy and diversity of these ecologically important species. Biological surveys may also lead to the removal of species from the endangered species category, as was done with the bald eagle, whose status was downlisted when its soaring numbers were documented.

Most Monitoring Efforts Are Fragmentary

Monitoring of biotic resources has been highly fragmentary. Often, researchers have monitored wildlife populations without also monitoring their habitats. Recording quantitative information about species abundance but not information about habitat quantity, quality, spatial extent, fragmentation, or connectivity is common in bird surveys (for example, the white-tailed ptarmigan; Colorado Division of Wildlife 1994). For studies of birds, a mixture of low- and high-intensity surveys and monitoring is needed to integrate information from local and regional scales. Short- and long-term monitoring are both needed to link changes of habitat with changes in populations (Butcher et al. 1993; Droege 1993).

What researchers have learned from monitoring amphibian populations has yet to be backed up with other needed information. Although amphibian population declines are well documented, amphibian metapopulation dynamics and the highly variable causes of decline are poorly understood (Corn 1994; Bury et al. 1995).

Additional monitoring of grazing effects is also needed (Vavra et al. 1994). Bock et al. (1993a: 304) stated that "virtually nothing is known about effects of grazing on birds of coniferous forests." Areas that are ungrazed are scarce and small, however. New, larger exclosures and sophisticated, statistically sound sampling designs are needed so that whole ecosystems can be monitored. Because the effects of grazing vary spatially (Knight 1994; Vavra et al. 1994), more detailed information is needed on regional range conditions and use. Bock et al. (1993b) proposed that some tracts of public rangelands be protected from livestock grazing in order to study the tracts in detail. Investigations of exclosures may be necessary to provide information for sound management of grasslands and forests.

Inventories of forests continue to focus on timber commodities and net volume increases (Powell et al. 1993) but are not designed to provide information about other resources such as habitat quality, habitat fragmentation or connectivity, nonconsumptive uses, or aesthetic values. Information on the effects of harvesting on old-growth habitat fragmentation and connectivity are perhaps more important to conservation biology and ecosystem management than information on net volume growth.

In short, nearly all of today's monitoring should be improved in four important ways. First, detailed population monitoring should be accompanied by detailed habitat monitoring. Second, observations should be augmented with well-designed experimental research to deduce the causes of population change. Third, intensive population and habitat monitoring at study sites should be coupled with extensive landscape-scale research to assess metapopulation dynamics to determine whether a local population is typical of other populations. Last, spatially explicit, predictive models should be used for a proactive approach to biological conservation. Without these consolidated efforts, the causes of the declines in abundance of many species in the Rocky Mountains—from amphibians to white-tailed deer—may never be determined. Even worse, future declines in abundance of many species may never be predicted or prevented.

Previous Inventories and Monitoring Were Not Standardized

Determining the status and trends of biotic resources in the Rocky Mountains is problematic because inventories and monitoring have been conducted without standardized procedures (Stohlgren and Quinn 1992; Stohlgren 1994; Stohlgren et al. 1995c). Recently, standardized field protocols were recommended for the study of amphibians (Heyer et al. 1994), but such protocols are lacking (and long overdue) for the studies of other biological groups. For example, personnel from no two national parks in the Rocky Mountains have consistently collected bird information with the same categories for habitat, nesting, abundance, or observed behavior. Surveys of vegetation in the same ecosystem by the National Park Service in Rocky Mountain National Park and by the U.S. Forest Service in the adjacent Arapaho–Roosevelt national forests used different remote sensing data, vegetation classification schemes, sampling designs, and field methods for validating the respective vegetation maps. Thus, the consolidation of data on vegetation in the region is a formidable task.

Accessibility of Data is Limited

Existing data are often inaccessible. Different agencies and nongovernment organizations tend to use different hardware, software, geographic information systems (GIS), data management programs, file formats, categories of information, field names, geographic and regional boundaries, classification schemes, and storage media. For example, lists of threatened and endangered plants and animals are accessible, but population-size estimates usually are not (Finch 1992). Information on forest size-class distributions is accessible, but not mapped locations of old-growth forests (Powell et al. 1993). Information on trends in water use and reservoir storage is accessible (U.S. Geological Survey 1990), but information on cumulative regional biotic effects of water projects is not readily available. Information is available on the potential effects of single proposed mining operations in environmental impact statements, but regional effects of mining are lacking.

Conclusion

Based on the previous rates of ecological change and extinctions documented in the paleoecology literature, the current rates of change in species and habitat losses in the Rocky Mountains have increased significantly during the last 200 years (Cole 1995). The human population in the Rocky Mountain region will probably double in the next 20–40 years and will proportionally increase demands for and pressures on natural resources. National parks, forests, and wilderness areas will probably become increasingly insular, and habitat fragmentation will increase in nature reserves and urban areas. Continued declines of species (for example, prairie dogs, amphibians, deer, and species dependent on old-growth forests), habitat loss (wetlands, riparian zones, and old-growth forests), increased air pollution and water developments, and introduced species and diseases will continue to affect many Rocky Mountain ecosystems.

Ecosystem science is not a panacea for declining abundances and degraded ecosystems, but, coupled with standardized biotic resource inventories, predictive models, and long-term monitoring, it is the logical approach to responsible stewardship. Humans have been and will forever be an integral component of Rocky Mountain ecosystems. The greatest challenge is not simply monitoring the status and trends of biotic resources in the Rocky Mountains but preventing future problems by using a proactive approach to conservation biology and ecosystem management. In the words of Chief Seattle, "We do not inherit the land from our ancestors, we borrow it from our children."

Author

Thomas J. Stohlgren
U.S. Geological Survey
Biological Resources Division
Rocky Mountain Field Station
Natural Resource Ecology Laboratory
Colorado State University
Fort Collins, Colorado 80523

Acknowledgments

K. Alper, C. Orth, L. Schell, and H. Fields assisted with library research, graphics, and clerical support. S. Haire developed the vegetation map. Rocky Mountain National Park staff and several friends generously provided photographs. R. Bachand, G. Chong, and C. Stohlgren took on additional tasks to free up my time. K. Alper, D. Binkley, G. Chong, H. Fields, and L. Schell provided constructive comments on an earlier draft of the manuscript. D. Knight and three anonymous reviewers provided many helpful suggestions. To all, I am grateful.

Cited References

Aguirre, A. A., and E. E. Starkey. 1994. Wildlife disease in U.S. national parks: historical and coevolutionary perspectives. Conservation Biology 8:654–661.

Alexander, R. R. 1985. Major habitat types, community types and plant communities in the Rocky Mountains. U.S. Forest Service General Technical Report RM-123. 4 pp.

Allen, R. B., R. K. Peet, and W. L. Baker. 1991. Gradient analysis of latitudinal variation in southern Rocky Mountain forests. Journal of Biogeography 18:123–139.

Apperson, K. A., and P. J. Anders. 1990. Kootenai River white sturgeon investigations and experimental culture. Annual Progress Report 1989. Division of Fish and Wildlife Project 88-65. Bonneville Power Administration, Portland, Oreg. 50 pp.

Armstrong, D. M. 1993. Lions, ferrets and bears. A guide to the mammals of Colorado. Colorado Division of Wildlife and University of Colorado Museum, Denver. 65 pp.

Athearn, R. G. 1960. High country empire: the High Plains and Rocky Mountains. McGraw-Hill, New York. 358 pp.

Baker, W. L. 1990. Species richness of Colorado riparian vegetation. Journal of Vegetation Science 1:119–124.

Baker, W. L., and T. T. Veblen. 1990. Spruce beetles and fires in the nineteenth century subalpine forests of western Colorado, U.S.A. Arctic and Alpine Research 22:65–80.

Barbour, M. G., J. H. Burk, and W. D. Pitts. 1987. Terrestrial plant ecology. 2nd edition. Benjamin/Cummings, Menlo Park, Calif. 604 pp.

Baron, J. S., D. S. Ojima, E. A. Holland, and W. J. Parton. 1994. Analysis of nitrogen saturation potential in Rocky Mountain tundra and forest: implications for aquatic systems. Biogeochemistry 27:61–82.

Behnke, R. J. 1992. Native trout of western North America. American Fisheries Society Monograph 6, Bethesda, Md. 275 pp.

Benedict, J. B. 1981. The Fourth of July Valley: glacial geology and archeology of the timberline ecotone. Research Report 2. Center for Mountain Archeology, Ward, Colo. 139 pp.

Betancourt, J. L., T. R. Van Devender, and P. S. Martin, editors. 1990. Late Quaternary biogeography of the Colorado Plateau. Pages 259–292 *in* Packrat middens: the last 40,000 years of biotic change. University of Arizona Press, Tucson.

Billings, W. D. 1988. Alpine vegetation. Pages 391–420 *in* M. G. Barbour and W. D. Billings, editors. North American terrestrial vegetation. Cambridge University Press, New York.

Bock, C. E., J. H. Bock, and H. M. Smith. 1993b. Proposal for a system of federal livestock exclosures on public rangelands in the western United States. Conservation Biology 7:731–732.

Bock, C. E., V. A. Saab, T. D. Rich, and D. S. Dobkin. 1993a. Pages 296–309 *in* Status and management of Neotropical migratory birds. U.S. Forest Service General Technical Report RM-229.

Boyce, M. S. 1991. Natural regulation or the control of nature. Pages 183–208 *in* R. B. Keiter and M. S. Boyce, editors. Greater Yellowstone ecosystem: redefining America's wilderness heritage. Yale University Press, New Haven, Conn.

Brady, S. J., and C. H. Flather. 1994. Changes in wetlands on nonfederal rural lands of the conterminous United States from 1982–1987. Environmental Management 18:693–705.

Brandt, E. 1993. How much is a gray wolf worth? National Wildlife 31:4–12.

Buchholtz, C. W. 1983. Rocky Mountain National Park: a history. Colorado Associated University Press, Boulder. 255 pp.

Bury, R. B., P. S. Corn, C. K. Dodd, Jr., R. W. McDiarmid, and N. J. Scott, Jr. 1995. Amphibians. Pages 124–126 *in* E. T. LaRoe, G. S. Farris, C. E. Puckett, P. D. Doran, and M. J. Mac, editors. Our living resources: a report to the nation on the distribution, abundance, and health of U.S. plants, animals, and ecosystems. U.S. Department of the Interior, National Biological Service, Washington, D.C.

Butcher, G. S., B. Peterjohn, and C. J. Ralph. 1993. Overview of national bird population monitoring programs and databases. Pages 192–203 *in* Status and management of Neotropical migratory birds. U.S. Forest Service General Technical Report RM-229.

Captive Breeding Specialist Group. 1992. Black-footed ferret recovery plan review, Washington D.C. Captive Breeding Specialist Group, Apple Valley, Minn. 15 pp.

Carey, C. 1993. Hypothesis concerning the causes of the disappearance of boreal toads from the mountains of Colorado. Conservation Biology 7:355–362.

Chadde, S. W., and C. E. Kay. 1991. Tall-willow communities on Yellowstone's northern range: a test of the natural regulation paradigm. Pages 231–262 *in* R. B. Keiter and M. S. Boyce, editors. Greater Yellowstone ecosystem: redefining America's wilderness heritage. Yale University Press, New Haven, Conn.

Chaney, E., W. Elmore, and W. S. Platts. 1990. Livestock grazing on western riparian areas. U.S. Environmental Protection Agency, Region 8, Denver, Colo. 45 pp.

Cole, K. L. 1990. Reconstruction of past desert vegetation along the Colorado River using packrat middens. Palaeogeography, Palaeoclimatology, Palaeoecology 76:349–366.

Cole, K. L. 1995. Vegetation change in national parks. Pages 224–228 *in* E. T. LaRoe, G. S. Farris, C. E. Puckett, P. D. Doran, and M. J. Mac, editors. Our living resources: a report to the nation on the distribution, abundance, and health of U.S. plants, animals, and ecosystems. U.S. Department of the Interior, National Biological Service, Washington, D.C.

Colorado Division of Wildlife. 1986. Wildlife in danger: the status of Colorado's threatened or endangered fish, amphibians, birds, and mammals. Fort Collins, Colo. 39 pp.

Colorado Division of Wildlife. 1993. Annual report. Fort Collins, Colo. 2 pp.

Colorado Division of Wildlife. 1994. Annual report. Fort Collins, Colo. 2 pp.

Connolly, G. E. 1981. Trends in populations and harvest. *In* O. C. Wallmo, editor. Mule and black-tailed deer of North America. University of Nebraska Press, Lincoln. 605 pp.

Cook, R. S., editor. 1993. Ecological issues on reintroducing wolves into Yellowstone National Park. National Park Service Scientific Monograph NPS/NRYELL/NRSM-93-22, Denver, Colo. 328 pp.

Cooper, S. V., K. E. Neiman, and D. W. Roberts. 1991. Forest habitat types of northern Idaho: a second approximation. U.S. Forest Service General Technical Report INT-236. 143 pp.

Corn, P. S. 1994. What we know and don't know about amphibian declines in the West. Pages 59–67 *in* W. W. Covington and L. F. DeBano, technical coordinators. Sustainable ecological systems: implementing an ecological approach to land management. U.S. Forest Service General Technical Report RM-247.

Corn, P. S., and J. C. Fogelman. 1984. Extinction of montane populations of the northern leopard frog (*Rana pipiens*) in Colorado. Journal of Herpetology 18:147–152.

Crumpacker, D. W. 1984. Regional riparian research and a multi-university approach to the special problems of livestock grazing in the Rocky Mountains and Great Plains. Pages 413–422 *in* R. E. Warner and K. Hendrix, editors. California riparian systems: ecology, conservation, and productive management. University of California Press, Berkeley and Los Angeles.

D'Antonio, C. M., and P. M. Vitousek. 1992. Biological invasions of exotic grasses, the grass/fire cycle, and global change. Annual Review of Ecology and Systematics 23:63–87.

DeLoach, C. J. 1991. Past successes and current prospects in biological control of weeds in the United States and Canada. Natural Areas Journal 11:129–142.

Droege, S. 1993. Monitoring Neotropical migrants on managed lands: when, where, why? Pages 189–191 *in* Status and management of Neotropical migratory birds. U.S. Forest Service General Technical Report RM-229.

Duffield, J. W. 1992. An economic analysis of wolf recovery in Yellowstone: park visitor attitudes and values. Pages 2–33 and 2–87 *in* J. D. Varley and W. G. Brewster, editors. Wolves for Yellowstone? A report to the United States Congress IV, Research and Analysis, National Park Service, Yellowstone National Park, Yellowstone, Wyo.

Evans, D. M., and D. M. Finch. 1994. Relationships between forest songbird populations and managed forests in Idaho. Pages 308–314 *in* W. W. Covington and L. F. DeBano, technical coordinators. Sustainable ecological systems: implementing an ecological approach to land management. U.S. Forest Service General Technical Report RM-247.

Finch, D. M. 1992. Threatened, endangered, and vulnerable species of terrestrial vertebrates in the Rocky Mountain region. U.S. Forest Service General Technical Report RM-215. 38 pp.

Flather, C. H., and T. W. Hoekstra. 1989. An analysis of the wildlife and fish situation in the United States 1989–2040. U.S. Forest Service General Technical Report RM-178. 146 pp.

Fleischner, T. L. 1994. Ecological costs of livestock grazing in western North America. Conservation Biology 8:629–644.

Glinski, R. L. 1977. Regeneration and distribution of sycamores and cottonwoods along Sonoita Creek, Santa Cruz County, Arizona. Pages 166–174 *in* R. R. Johnson and D. A. Jones, technical coordinators. Importance, preservation, and management of riparian habitat: a symposium. U.S. Forest Service General Technical Report RM-43.

Greenback Cutthroat Trout Recovery Team. 1983. Greenback cutthroat trout recovery plan revision. U.S. Fish and Wildlife Service and Colorado Division of Wildlife, Fort Collins. 21 pp.

Grossman, D. H., and K. L. Goodin. 1995. Rare terrestrial ecological communities of the United States. Pages 218–221 *in* E. T. LaRoe, G. S. Farris, C. E. Puckett, P. D. Doran, and M. J. Mac, editors. Our living resources: a report to the nation on the distribution, abundance, and health of U.S. plants, animals, and ecosystems. U.S.

Department of the Interior, National Biological Service, Washington, D.C.

Grove, J. M. 1990. The little ice age. Rutledge Press, New York. 498 pp.

Gruell, G. E. 1983. Fire and vegetative trends in the northern Rockies: interpretations from 1871–1982 photographs. U.S. Forest Service General Technical Report INT-158. 117 pp.

Hayward, D. G., and J. Verner, editors. 1994. Flammulated, boreal, and great gray owls in the United States: a technical conservation assessment. U.S. Forest Service General Technical Report RM-253. 213 pp.

Henry, C. J., and R. G. Anthony. 1989. Bald eagle and osprey. Pages 66–82 *in* Proceedings of the western raptor management symposium and workshop. National Wildlife Federation, Washington, D.C.

Henry, D., and M. Henry. 1991. Greenbacks are back. Colorado Outdoors 40:23–25.

Heyer, R. W., M. A. Donnelly, R. W. McDiarmid, L. C. Hayek, and M. S. Foster. 1994. Measuring and monitoring biological diversity: standard methods for amphibians. Smithsonian Institution Press, Washington, D.C. 364 pp.

Hinds, T. E., F. G. Hawksworth, and R. W. Davidson. 1965. Beetle-killed Engelmann spruce: its deterioration in Colorado. Journal of Forestry 63:536–542.

Jackson, D. 1962. Letters of the Lewis and Clark expedition with related documents 1783–1854. University of Illinois Press, Urbana. 728 pp.

Johnson, K. L. 1987. Rangeland through time: a photographic study of vegetation change in Wyoming (1870–1986). Miscellaneous Publication 50. Wyoming Agricultural Experiment Station, Laramie. 188 pp.

Johnson, M. G. 1994. The native tribes of North America: a concise encyclopedia. Macmillian Publishing, New York. 210 pp.

Kay, C. E. 1994. Aboriginal overkill. Human Nature 5:359–398.

Keane, R. E., S. F. Arno, and J. K. Brown. 1990. Simulating cumulative fire effects in ponderosa pine/Douglas-fir forests. Ecology 7:189–203.

Keiter, R. B. 1991. An introduction to the ecosystem management debate. Pages 3–18 *in* R. B. Keiter and M. S. Boyce, editors. Greater Yellowstone ecosystem: redefining America's wilderness heritage. Yale University Press, New Haven, Conn.

Keller, C. R., and K. P. Burnham. 1982. Riparian fencing, grazing, and trout habitat preference on Summit Creek, Idaho. North American Journal of Fisheries Management 2:53–59.

Kendall, K. C. 1995. Whitebark pine: ecosystem in peril. Pages 228–230 *in* E. T. LaRoe, G. S. Farris, C. E. Puckett, P. D. Doran, and M. J. Mac, editors. Our living resources: a report to the nation on the distribution, abundance, and health of U.S. plants, animals, and ecosystems. U.S. Department of the Interior, National Biological Service, Washington, D.C.

Knight, D. H. 1991. The Yellowstone fire controversy. Pages 87–103 *in* R. B. Keiter and M. S. Boyce, editors. Greater Yellowstone ecosystem: redefining

America's wilderness heritage. Yale University Press, New Haven, Conn.

Knight, D. H. 1994. Mountains and plains: the ecology of Wyoming landscapes. Yale University Press, New Haven, Conn. 338 pp.

Knopf, F. L., and R. W. Cannon. 1982. Structural resilience of a willow riparian community to changes in grazing practices. Pages 198–207 *in* L. Nelson, J. M. Peek, and P. D. Dalke, editors. Proceedings of the wildlife–livestock relationships symposium. Forest, Wildlife, and Range Experiment Station, University of Idaho, Moscow.

Langner, L. L., and C. H. Flather. 1994. Biological diversity: status and trends in the United States. U.S. Forest Service General Technical Report RM-244. 21 pp.

LaRoe, E. T., G. S. Farris, C. E. Puckett, P. D. Doran, and M. J. Mac, editors. 1995. Our living resources: a report to the nation on the distribution, abundance, and health of U.S. plants, animals, and ecosystems. U.S. Department of the Interior, National Biological Service, Washington, D.C. 530 pp.

Laven, R. D., P. N. Omi, J. G. Wyant, and A. S. Pinkerton. 1980. Interpretation of fire scar data from a ponderosa pine ecosystem in the central Rocky Mountains, Colorado. Pages 46–49 *in* M. A. Stokes and J. H. Dieterich, editors. Proceedings of the fire history workshop. U.S. Forest Service General Technical Report RM-81.

Lavender, D. 1975. The Rockies. Harper and Row, New York. 433 pp.

Leopold, A. S., S. A. Cain, C. M. Cottam, I. N. Gabrielson, and T. L. Kimball. 1963. Wildlife management in the national parks. Transactions of North American Wildlife and Natural Resources Conference 28:28–45.

Loope, L. L., and G. E. Gruell. 1973. The ecological role of fire in Jackson Hole, northwestern Wyoming. Quaternary Research 3:425–443.

MacCleery, D. W. 1992. American forests: a history of resilience and recovery. U.S. Forest Service FS-540. Forest History Society, Durham, N.C. 59 pp.

Manfredo, M. J., A. D. Bright, J. Pate, and G. Tischbein. 1994. Colorado residents' attitudes and perceptions toward reintroduction of the gray wolf (*Canis lupus*) into Colorado. Human Dimensions in Natural Resources Unit Project Report 21. Colorado State University, Fort Collins. 99 pp.

Marnell, L. F. 1995. Cutthroat trout in Glacier National Park, Montana. Pages 153–154 *in* E. T. LaRoe, G. S. Farris, C. E. Puckett, P. D. Doran, and M. J. Mac, editors. Our living resources: a report to the nation on the distribution, abundance, and health of U.S. plants, animals, and ecosystems. U.S. Department of the Interior, National Biological Service, Washington, D.C.

Mason, W. T., Jr. 1995. Invertebrates. Pages 159–160 *in* E. T. LaRoe, G. S. Farris, C. E. Puckett, P. D. Doran, and M. J. Mac, editors. Our living resources: a report to the nation on the distribution, abundance, and health of U.S. plants, animals, and ecosystems. U.S. Department of the Interior, National Biological Service, Washington, D.C.

Mattson, D. J., R. G. Wright, K. C. Kendall, and C. J. Martinka. 1995. Grizzly bears. Pages 103–105 *in* E. T. LaRoe, G. S. Farris, C. E. Puckett, P. D. Doran, and M. J. Mac, editors. Our living resources: a report to the nation on the distribution, abundance, and health of U.S. plants, animals, and ecosystems. U.S. Department of the Interior, National Biological Service, Washington, D.C.

McDiarmid, R. W. 1995. Reptiles and amphibians. Pages 117–118 *in* E. T. LaRoe, G. S. Farris, C. E. Puckett, P. D. Doran, and M. J. Mac, editors. Our living resources: a report to the nation on the distribution, abundance, and health of U.S. plants, animals, and ecosystems. U.S. Department of the Interior, National Biological Service, Washington, D.C.

Meagher, M., and M. E. Meyer. 1994. On the origin of brucellosis in bison of Yellowstone National Park: a review. Conservation Biology 8:645–653.

Mech, L. D., D. H. Pletscher, and C. J. Martinka. 1995. Gray wolves. Pages 98–100 *in* E. T. LaRoe, G. S. Farris, C. E. Puckett, P. D. Doran, and M. J. Mac, editors. Our living resources: a report to the nation on the distribution, abundance, and health of U.S. plants, animals, and ecosystems. U.S. Department of the Interior, National Biological Service, Washington, D.C.

Melcher, C. P. 1992. Avifauna responses to intensive browsing by elk in Rocky Mountain National Park. M.S. thesis, Colorado State University, Fort Collins. 86 pp.

Merriam, C. H. 1890. Results of a biological survey of the San Francisco Mountain region and desert of the Little Colorado, Arizona. North American Fauna 3. 208 pp.

Miller, A. I., T. D. Counihan, M. J. Parsley, and L. G. Beckman. 1995. Columbia River Basin white sturgeon. Pages 154–157 *in* E. T. LaRoe, G. S. Farris, C. E. Puckett, P. D. Doran, and M. J. Mac, editors. Our living resources: a report to the nation on the distribution, abundance, and health of U.S. plants, animals, and ecosystems. U.S. Department of the Interior, National Biological Service, Washington, D.C.

Miller, B., G. Ceballos, and R. Reading. 1994. The prairie dog and biotic diversity. Conservation Biology 8:677–681.

Miller, B., C. Wemmer, D. Biggins, and R. Reading. 1990. A proposal to conserve black-footed ferrets and the prairie dog ecosystem. Environmental Management 14:763–769.

Mills, E. A. 1924. Wild life on the Rockies. Houghton Mifflin Company, New York. 263 pp.

Mills, J. D. 1991. Wyoming's Jackson Hole dam, horizontal channel stability, and floodplain vegetation dynamics. M.S. thesis, University of Wyoming, Laramie. 54 pp.

Mueggler, W. F. 1993. Forage. Pages 129–134 *in* N. V. DeByle and R. P. Winokur, editors. Aspen: ecology and management in the

western United States. U.S. Forest Service General Technical Report RM-119.

Nams, V. O., N. F. G. Folkard, and J. N. M. Smith. 1993. Effects of nitrogen fertilization on several woody and nonwoody boreal forest species. Canadian Journal of Botany 71:93–97.

Nichols, H. 1982. Review of late Quaternary history of vegetation and climate in the mountains of Colorado. Pages 27–33 *in* J. C. Halfpenny, editor. Ecological studies in the Colorado alpine: a festschrift for John W. Marr. Institute of Arctic and Alpine Research, Occasional Paper 37. University of Colorado, Boulder.

Nordyke, K. A., and S. W. Buskirk. 1991. Southern red-backed vole, *Clethrionomys gapperi*, populations in relation to stand succession and old-growth character in the central Rocky Mountains. Canadian Field-Naturalist 105:330–334.

Opler, P. A. 1995. Species richness and trends of western butterflies and moths. Pages 172–174 *in* E. T. LaRoe, G. S. Farris, C. E. Puckett, P. D. Doran, and M. J. Mac, editors. Our living resources: a report to the nation on the distribution, abundance, and health of U.S. plants, animals, and ecosystems. U.S. Department of the Interior, National Biological Service, Washington, D.C.

Otte, D. 1995. Grasshoppers. Pages 163–166 *in* E. T. LaRoe, G. S. Farris, C. E. Puckett, P. D. Doran, and M. J. Mac, editors. Our living resources: a report to the nation on the distribution, abundance, and health of U.S. plants, animals, and ecosystems. U.S. Department of the Interior, National Biological Service, Washington, D.C.

Peet, R. K. 1978. Forest vegetation of the Colorado Front Range: patterns of species diversity. Vegetatio 37:65–78.

Peet, R. K. 1981. Forest vegetation of the Colorado Front Range. Vegetatio 45:3–75.

Peet, R. K. 1988. Forests of the Rocky Mountains. Pages 63–101 *in* M. G. Barbour and W. D. Billings, editors. North American terrestrial vegetation. Cambridge University Press, New York.

Peters, E. F., and S. C. Bunting. 1994. Fire conditions pre- and post-occurrence of annual grasses on the Snake River plain. Pages 31–36 *in* S. B. Monsen and S. G. Kitchen, editors. Proceedings—Ecology and Management of Annual Rangelands. U.S. Forest Service General Technical Report INT-GTR-313.

Peterson, J. A., editor. 1986. Paleotectonics and sedimentation in the Rocky Mountain Region, United States. Memoir 41, American Association of Petroleum Geologists. Tulsa, Okla. 693 pp.

Pierce, K. L. 1979. History and dynamics of glaciation in the northern Yellowstone National Park area. Professional Paper 729-F. U.S. Geological Survey, Washington, D.C. 90 pp.

Popovich, S. J., W. D. Shepperd, D. W. Reichert, and M. A. Cone. 1993. Flora of the Fraser Experimental Forest, Colorado. U.S. Forest Service General Technical Report RM-233. 62 pp.

Powell, D. S., J. L. Faulkner, D. R. Darr, Z. Zhu, and D. W. MacCleery. 1993. Forest

resources of the United States, 1992. U.S. Forest Service General Technical Report RM-234. 132 pp.

Powell, J. A. 1995. Lepidoptera inventories in North America. Pages 168–170 *in* E. T. LaRoe, G. S. Farris, C. E. Puckett, P. D. Doran, and M. J. Mac, editors. Our living resources: a report to the nation on the distribution, abundance, and health of U.S. plants, animals, and ecosystems. U.S. Department of the Interior, National Biological Service, Washington, D.C.

Robbins, J. 1994. The microbe miners. Audubon 96:90–95.

Robertus, A. J., B. R. Burns, and T. T. Veblen. 1991. Stand dynamics of *Pinus flexilis*-dominated subalpine forests in the Colorado Front Range. Journal of Vegetation Science 2:445–458.

Romme, W. H., and D. G. Despain. 1989. Historical perspective on the Yellowstone fires of 1988. BioScience 39:695–699.

Romme, W. H., and D. H. Knight. 1981. Fire frequency and subalpine forest succession along a topographic gradient in Wyoming. Ecology 62:319–326.

Roovers, L. M., and A. J. Robertus. 1993. Stand dynamics and conservation of an old-growth Engelmann spruce-subalpine fir forest in Colorado. Natural Areas Journal 13:256–267.

Rosentreter, R. 1994. Displacement of rare plants by exotic grasses. Pages 170–175 *in* S. B. Monsen and S. G. Kitchen, editors. Proceedings—Ecology and management of annual rangelands. U.S. Forest Service General Technical Report INT-GTR-313.

Schlosser, I. J. 1995. Dispersal, boundary processes, and trophic-level interactions in streams adjacent to beaver ponds. Ecology 76:908–925.

Schullery, P. 1989. The fire and fire policy. BioScience 39:686–694.

Schullery, P. 1995. The greater Yellowstone ecosystem. Pages 312–314 *in* E. T. LaRoe, G. S. Farris, C. E. Puckett, P. D. Doran, and M. J. Mac, editors. Our living resources: a report to the nation on the distribution, abundance, and health of U.S. plants, animals, and ecosystems. U.S. Department of the Interior, National Biological Service, Washington, D.C.

Servheen, C. 1990. The status and conservation of the bears of the world. Page 20 *in* 8th international conference on bear research and management. Monograph Series 2, Victoria, British Columbia.

Sims, P. L. 1988. Grasslands. Pages 265–286 *in* M. G. Barbour and W. D. Billings, editors. North American terrestrial vegetation. Cambridge University Press. New York.

Singer, F. 1995. Bighorn sheep in the Rocky Mountain National Park. Pages 332–333 *in* E. T. LaRoe, G. S. Farris, C. E. Puckett, P. D. Doran, and M. J. Mac, editors. Our living resources: a report to the nation on the distribution, abundance, and health of U.S. plants, animals, and ecosystems. U.S. Department of the Interior, National Biological Service, Washington, D.C.

Speart, J. 1994. The case of the wandering bull. National Wildlife 32:14–17.

Stohlgren, T. J. 1994. Planning long-term vegetation studies at landscape scales. Pages 209–241 *in* T. M. Powell and J. H. Steele, editors. Ecological time series. Chapman and Hall, New York. 491 pp.

Stohlgren, T. J., J. Baron, and T. Kittel. 1993. Understanding coupled climatic, hydrological, and ecosystem responses to global climate change in the Colorado Rockies biogeographical area. Pages 184–200 *in* W. E. Brown and S. D. Veirs, Jr., editors. Partners in stewardship: proceedings of the 7th conference on research and resource management in parks and on public lands. George Wright Society, Hancock, Mich.

Stohlgren, T. J., J. Baron, T. G. F. Kittel, and D. Binkley. 1995a. Ecosystem trends in the Colorado Rockies. Pages 310–312 *in* E. T. LaRoe, G. S. Farris, C. E. Puckett, P. D. Doran, and M. J. Mac, editors. Our living resources: a report to the nation on the distribution, abundance, and health of U.S. plants, animals, and ecosystems. U.S. Department of the Interior, National Biological Service, Washington, D.C.

Stohlgren, T. J., J. F. Quinn, M. Ruggiero, and G. Waggoner. 1995b. Status of biotic inventories in U.S. national parks. Biological Conservation 71:97–106.

Stohlgren, T. J., M. B. Falkner, and L. D. Schell. 1995c. A modified-Whittaker nested vegetation sampling method. Vegetatio 117:113–121.

Stohlgren, T. J., M. B. Coughenour, G. W. Chong, D. Binkley, M. A. Kalkhan, L. D. Schell, D. J. Buckley, and J. K. Berry. 1997. Landscape analysis of plant diversity. Landscape Ecology. In press.

Stohlgren, T. J., and J. F. Quinn. 1992. An assessment of biotic inventories in western U.S. national parks. Natural Areas Journal 12:145–154.

Thompson, D. Q., R. L. Stuckey, and E. B. Thompson. 1987. Spread, impact, and control of purple loosestrife (*Lythrum salicaria*) in North American wetlands. U.S. Fish and Wildlife Service, Fish and Wildlife Research 2. 55 pp.

Trotter, P. C. 1987. Cutthroat: native trout of the West. Colorado Associated University Press, Boulder. 219 pp.

Turk, J. T., D. H. Campbell, G. P. Ingersoll, and D. A. Clow. 1992. Initial findings of synoptic snowpack sampling in the Colorado Rocky Mountains. U.S. Geological Survey Open-File Report 92-645. Denver, Colo. 6 pp.

U.S. Geological Survey. 1990. National water summary 1987—hydrologic events and water supply and use. U.S. Geological Survey Water-Supply Paper 2350. Denver, Colo. 243 pp.

U.S. Geological Survey. 1993. National water summary 1990–91. Hydrologic events and stream water quality. U.S. Geological Survey Water-Supply Paper 2400. Washington, D.C. 243 pp.

Vavra, M., W. A. Laycock, and R. D. Pieper, editors. 1994. Ecological implications of livestock herbivory in the West. Society of Range Management, Denver, Colo. 297 pp.

Veblen, T. T., K. S. Hadley, M. S. Reid, and A. J. Rebertus. 1991. Stand response to spruce beetle outbreak in Colorado subalpine forests. Ecology 72:213–231.

Veblen, T. T., and D. C. Lorenz. 1991. The Colorado Front Range: a century of ecological change. University of Utah Press, Salt Lake City. 186 pp.

Wagner, F. H. 1978. Livestock grazing and the livestock industry. Pages 121–145 *in* H. P. Brokaw, editor. Wildlife and America. Council on Environmental Quality, Washington, D.C.

Weber, W. A. 1976. Rocky Mountain flora. Colorado Associated University Press, Boulder. 479 pp.

Weisberg, P. J., and W. L. Baker. 1995. Spatial variation in tree seedlings and krummholz growth in the forest-tundra ecotone of Rocky Mountain National Park, Colorado, U.S.A. Arctic and Alpine Research 27:116–129.

Wells, P. V. 1983. Paleobiogeography of montane islands in the Great Basin since the last glaciopluvial. Ecological Monograph 53:341–382.

West, N. E., K. H. Rea, and R. J. Tausch. 1975. Basic synecological relationships in juniper–pinyon woodlands. Pages 41–53 *in* G. F. Gifford and F. E. Busby, editors. The pinyon–juniper ecosystem: a symposium. Utah Agriculture Experiment Station, Utah State University, Logan.

Whitlock, C. 1993. Post-glacial vegetation and climate of Grand Teton and southern Yellowstone national parks. Ecological Monographs 63:173–198.

Whitlock, C., S. C. Fritz, and D. R. Engstrom. 1991. A prehistoric perspective on the northern range. Pages 289–305 *in* R. B. Keiter and M. S. Boyce, editors. Greater Yellowstone ecosystem: redefining America's wilderness heritage. Yale University Press, New Haven, Conn.

Winegar, H. H. 1977. Camp Creek channel fencing—plant, wildlife, soil, and water responses. Rangeman's Journal 4:10–12.

Whitebark Pine

Arno, S. F. 1986. Whitebark pine cone crops—a diminishing source of wildlife food? Western Journal of Applied Forestry 1(3):92–94.

Arno, S. 1995. Notes on fire in whitebark pine. U.S. Forest Service, Intermountain Research Station, Missoula, Mont. Nutcracker Notes 5:7.

Arno, S. F., and R. J. Hoff. 1989. Silvics of whitebark pine (*Pinus albicaulis*). U.S. Forest Service General Technical Report INT-253. Ogden, Utah. 11 pp.

Arno, S. F., E. D. Reinhardt, and J. H. Scott. 1993. Forest structure and landscape patterns in the subalpine lodgepole pine type: a procedure for quantifying past and present conditions. U.S. Forest Service General Technical Report INT-294. 17 pp.

Brown, J. K., S. F. Arno, S. W. Barrett, and J. P. Menakis. 1994. Comparing the prescribed natural fire program with presettlement fires in the Selway–Bitterroot Wilderness. International Journal of Wildland Fire 4(3):157–168.

Duriscoe, D. 1995. White pine blister rust in Kings Canyon and Sequoia national parks: preliminary results of an extensive survey. Sequoia and Kings Canyon national parks, U.S. National Park Service, Three Rivers, Calif. 12 pp.

Hoff, R. 1994. Artificial rust inoculation of whitebark pine seedlings—rust resistance across several populations. U.S. Forest Service Intermountain Research Station, Missoula, Mont. Nutcracker Notes 4:7–9.

Hoff, R. J., S. K. Hagle, and R. G. Krebill. 1994. Genetic consequences and research challenges of blister rust in whitebark pine forests. Pages 118–126 *in* W. Schmidt and F. K. Holtmeier, editors. Proceedings—International workshop on subalpine stone pines and their environment: the status of our knowledge. U.S. Forest Service General Technical Report INT-GTR-309.

Keane, R. E. 1995a. Whitebark pine ecosystem restoration and research: where are we and where are we going? U.S. Forest Service Intermountain Research Station, Missoula, Mont. Nutcracker Notes 6:2–3.

Keane, R. E. 1995b. A coarse scale assessment of whitebark pine in the interior Columbia River basin. U.S. Forest Service Intermountain Research Station, Missoula, Mont. Nutcracker Notes 5:5–6.

Keane, R. E., and S. F. Arno. 1993. Rapid decline of whitebark pine in western Montana: evidence from 20-year remeasurements. Western Journal of Applied Forestry 8:44–47.

Keane, R. E., P. Morgan, and J. P. Menakis. 1994. Landscape assessment of the decline of whitebark pine (*Pinus albicaulis*) in the Bob Marshall Wilderness Complex, Montana, USA. Northwest Science 68(3):213–229.

Kendall, K. C. 1994a. Whitebark pine monitoring network. Pages 110–118 *in* K. C. Kendall and B. Coen, editors. Workshop proceedings: research and management in whitebark pine ecosystems. U.S. National Biological Service, Glacier National Park, West Glacier, Mont.

Kendall, K. C. 1994b. Whitebark pine conservation in North American national parks. Pages 302–307 *in* W. Schmidt and F. K. Holtmeier, editors. Proceedings—International workshop on subalpine stone pines and their environment: the status of our knowledge. U.S. Forest Service General Technical Report INT-GTR-309.

Kendall, K. C. 1995. Whitebark pine: ecosystem in peril. Pages 228–230 *in* E. T. LaRoe, G. S. Farris, C. E. Puckett, P. D. Doran, and M. J. Mac, editors. Our living resources: a report to the nation on the distribution, abundance, and health of U.S. plants, animals, and ecosystems. U.S. Department of the Interior, National Biological Service, Washington, D.C.

Kendall, K. C., and S. F. Arno. 1990. Whitebark pine—an important but endangered wildlife resource. Pages 264–273 *in* Proceedings of symposium on whitebark pine ecosystems: ecology and management of a high-mountain resource. U.S. Forest Service General Technical Report INT-270.

Kendall, K. C., D. Schirokauer, E. Shanahan, R. Watt, D. Reinhart, R. Renkin, S. Cain, and G. Green. 1996a. Whitebark pine health in northern Rockies national park ecosystems: a preliminary report. U.S. Forest Service Intermountain Research Station, Missoula, Mont. Nutcracker Notes 7:16.

Kendall, K. C., D. Tyers, and D. Schirokauer. 1996b. Status of whitebark pine in the Gallatin National Forest, Montana. U.S. Forest Service Intermountain Research Station, Missoula, Mont. Nutcracker Notes 7:200.

Kinloch, B. B., Jr., and D. Dulitz. 1990. White pine blister rust at Mountain Home Demonstration State Forest: a case study of the epidemic and prospects for genetic control. U.S. Forest Service Research Paper PSW-204. 7 pp.

Lanner, R. M. 1993. Is it doomsday for whitebark pine? Western Journal of Applied Forestry 8(2):47,70.

Smith, G. J. 1971. Distribution of white pine blister rust in the Canadian Rocky Mountains. Department of the Environment, Ottawa, Canada, Bi-monthly Research Notes 27(6):43.

Smith, J. 1995. U.S. Forest Service Region 4 disease survey of whitebark pine. U.S. Forest Service, Intermountain Research Station, Missoula, Mont. Nutcracker Notes 6:8.

Zack, A. 1995. Whitebark pine and mountain hemlock in the Mallard Larkins Pioneer Area of Idaho. U.S. Forest Service Intermountain Research Station, Missoula, Mont. Nutcracker Notes 6:9–10.

Limber Pine

Duriscoe, D. 1995. White pine blister rust in Kings Canyon and Sequoia national parks: preliminary results of an extensive survey. U.S. National Park Service, Sequoia and Kings Canyon national parks, Three Rivers, Calif. 12 pp.

Hawksworth, F. G. 1990. White pine blister rust in southern New Mexico. Plant Disease 74(11):938.

Hoff, R, J., S. K. Hagle, and R. G. Krebill. 1994. Genetic consequences and research challenges of blister rust in whitebark pine forests. Pages 118–126 in W. Schmidt and F. K. Holtmeier, editors. Proceedings—International workshop on subalpine stone pines and their environment: the status of

our knowledge. U.S. Forest Service General Technical Report INT-GTR-309.

Kendall, K. C. 1995. Whitebark pine: ecosystem in peril. Pages 228–230 in E. T. LaRoe, G. S. Farris, C. E. Puckett, P. D. Doran, and M. J. Mac, editors. Our living resources: a report to the nation on the distribution, abundance, and health of plants, animals, and ecosystems. U.S. Department of the Interior, National Biological Service, Washington, D.C.

Kendall, K. C., D. Ayers, and D. Schirokauer. 1996. Limber pine status in Montana, Wyoming, and Idaho. U.S. Forest Service, Intermountain Research Station, Missoula, Mont. Nutcracker Notes 7:23–24.

Kinloch, B. B., Jr., and D. Dulitz. 1990. White pine blister rust at Mountain Home Demonstration State Forest: a case study of the epidemic and prospects for genetic control. U.S. Forest Service Research Paper PSW-204. 7 pp.

Little, E. L., Jr. 1971. Atlas of United States trees. volume 1: conifers and important hardwoods. U.S. Forest Service Miscellaneous Publication 1146, Washington, D.C.

Lundquist, J. E. 1993. Large scale spatial patterns of conifer diseases in the Bighorn Mountains, Wyoming. U.S. Forest Service Research Note RM-523. 8 pp.

Pfister, R. D., B. L. Kovalchik, S. F. Arno, and R. C. Presby. 1977. Forest habitat types of Montana. U.S. Forest Service General Technical Report INT-34.

Smith, G. J. 1971. Distribution of white pine blister rust in the Canadian Rocky Mountains. Department of the Environment Bi-Monthly Research Notes 27(6):43. Ottawa, Canada.

Smith, J. 1995. U.S. Forest Service Region 4 disease survey of whitebark pine. U.S. Forest Service Intermountain Research Station, Missoula, Mont. Nutcracker Notes 6:8.

Amphibians of Glacier National Park

Black, J. 1970a. Montana wildlife—amphibians. Montana Fish and Game Department, Publication–Education Division, Animals of Montana Series 1. 32 pp.

Black, J. 1970b. Checklist and distribution of the amphibians and reptiles of Montana. 12 pp.

Bradford, D. F. 1989. Allopatric distribution of native frogs and introduced fishes in high Sierra Nevada lakes of California: implication of the negative effect of fish introductions. Copeia 1989:775–778.

Bradford, D. F., F. Tabatabai, and D. M. Graber. 1993. Isolation of remaining populations of the native frog, R. mucosa by introduced fishes in Sequoia and Kings Canyon national parks, California. Conservation Biology 7(4):882–888.

Corn, P. S., M. L. Jennings, and E. Muths. 1997. Survey and assessment of amphibian populations in Rocky Mountain National Park. Northwest Naturalist 78:34-35.

Davis, C. V., and S. E. Weeks. 1963. Montana snakes. Montana Department of Fish and Game, Helena.

Leonard, W. P., H. A. Brown, L. C. Jones, K. R. McAllister, and R. M. Storm. 1993. Amphibians of Washington and Oregon. Seattle Audubon Society, Seattle, Wash. 168 pp.

Manville, R. H. 1957. Amphibians and reptiles of Glacier National Park, Montana. Copeia 4:308–309.

Marnell, L. F. 1988. Status of the westslope cutthroat trout in Glacier National Park, Montana. American Fisheries Society Symposium 4:61–70.

Marnell, L. F., R. J. Behnke, and F. W. Allendorf. 1987. Genetic identification of cutthroat trout Salmo clarki, in Glacier National Park, Montana. Canadian Journal of Fisheries and Aquatic Sciences 44:1830–1839.

Metter, D. E. 1967. Variation in the ribbed frog Ascaphus truei (Stejneger). Copeia 3:634–649.

Thompson, L. S. 1982. Distribution of Montana amphibians, reptiles, and mammals. Montana Audubon Council, Helena. 24 pp.

Great Basin–Mojave Desert Region

The Great Basin–Mojave Desert region is a land of striking contrasts. Forming an expansive wedge between the Sierra Nevada, the Transverse ranges, the Rocky Mountains, and the Columbia and Colorado plateaus, the region harbors great biological diversity. This high diversity is produced by a blending of the surrounding region's flora and fauna with the unique species of the Great Basin and Mojave Desert. The Great Basin–Mojave Desert region is home to the oldest living organisms on Earth, such as Great Basin bristlecone pines, which can live 4,900 years (Schmid and Schmid 1975), and the creosotebush clones in the Mojave Desert, which are an estimated 10,000–11,000 years old (Vasek et al. 1975).

Biologists easily distinguish the Mojave Desert and the Great Basin from other regions by their flora and fauna. However, the boundaries between them and the surrounding regions are often unclear. Various scientific disciplines define the Great Basin–Mojave Desert region somewhat differently (Grayson 1993). Anthropologists define the region by cultural attributes of the aboriginal inhabitants (d'Azavedo 1986), botanists by species composition of the vegetation (Billings 1951; Holmgren 1972; Vasek and Barbour 1990), and geologists by the structure of the land (Hunt 1967). The region includes nearly all of Nevada, much of eastern California, western Utah, southeastern Oregon, and portions of southern Idaho. Our descriptions include the hydrographic Great Basin, the floristic Mojave Desert, and the Muddy, Virgin, and White rivers, which are tributaries of the Colorado River with headwaters deep in the floristic Great Basin (Fig. 1).

Hydrographic Great Basin

The hydrographic Great Basin is the area of internal drainage between the Rocky Mountains and the Sierra Nevada. The waters of streams in this area never reach the ocean but are confined to closed basins. The hydrographic Great Basin includes most of the Mojave Desert and exceeds 500,000 square kilometers (Morrison 1991). When the explorer John C. Frémont realized that this region did not drain to the ocean, he coined the term Great Basin (Frémont 1845).

Physiographic Region

The landforms of the Great Basin and the Mojave Desert define the region as part of the Basin and Range Physiographic Province (Hunt 1967) that extends south to include the Sonoran and Chihuahuan deserts of Arizona and Mexico. The Basin and Range Physiographic Province is characterized by hundreds of long, narrow, and roughly parallel mountain ranges that are separated by deep valleys (Hunt 1967; King 1977; Fiero 1986). These features have evoked several colorful descriptions, including "an army of caterpillars crawling northward out of Mexico" (Dutton *in* King 1977:156) and "washboard topography" (Houghton 1978:vii).

The Great Basin mountains are geologically recent, and the landforms of the region are a product of the formation of the Rocky Mountains and the Sierra Nevada. The structures of the more than 400 mountain ranges in the region are similar, but their compositions are diverse. Many

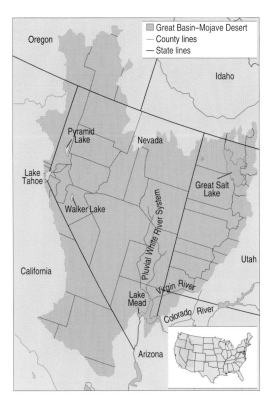

Fig. 1. The Great Basin–Mojave Desert region.

granite ranges occur in the west, basalt ranges occupy the northwest, rhyolite mountains form the center, and limestone mountains dominate the east and southwest. In addition to this diversity of substrates, there is great topographic relief throughout the region, the greatest of which is 4,494 meters in the 135 kilometers between Badwater in Death Valley and the summit of Mount Whitney, both in California. In a transect across the Great Basin at 39° north latitude, Grayson (1993) found that the difference in elevation between mountaintops and valley bottoms ranged from 1,158 to 2,316 meters, with an average relief of 1,768 meters.

The topographic relief in the region creates powerful elevation gradients to which all the organisms in the region respond. As elevation increases, air density decreases and solar radiation and precipitation increase. The interaction of these factors produces different temperature regimes at different elevations, which significantly affect the distribution of plants (Billings 1970) and the animals that depend on them (Hall 1946). This mountainous terrain thus provides many opportunities for a multitude of organisms with diverse life strategies.

Floristic Region

The Mojave Desert can be distinguished from the Great Basin by the presence and abundance of its different plant species. Again, the boundaries are imprecise. The principal distinguishing feature of the two floristic regions is the presence of creosotebush in the Mojave Desert and its absence from the Great Basin (Billings 1951; Holmgren 1972). Alternatively, big sagebrush dominates much of the Great Basin floristic region, but it is mostly absent from the Mojave Desert except at moderate to high elevations in the mountains. The separation of the Mojave from the Colorado and Sonoran deserts is less clear because creosotebush is also dominant in these deserts.

Climate

The climate of the Great Basin–Mojave Desert region is one of the most varied and extreme in the world (Hidy and Klieforth 1990). The Sierra Nevada is primarily responsible for creating this arid, continental climate by capturing moisture from Pacific storm fronts before the moisture reaches the desert (Houghton et al. 1975); similarly, the Rocky Mountains intercept storms from the Gulf of Mexico (Hidy and Klieforth 1990). Local weather patterns are complicated by the mountain ranges that uplift the dispersed moisture, creating mountain storms (Hidy and Klieforth 1990). Thus, precipitation increases with elevation (Billings 1951), and average annual precipitation can be highly variable over small distances.

The region can be separated climatically into hot and cold deserts. The lower, hot, Mojave Desert receives most of its precipitation as rain, and the higher, cold, Great Basin Desert receives most of its moisture as snow (MacMahon 1988a). This distinction is complicated by a strong east–west gradient that allows more summer rains from the Gulf of Mexico to reach the eastern regions of the Great Basin and Mojave than their western regions. This pattern changes around 40° north latitude, where seasonal precipitation regimes are nearly equal between the Warner Mountains in the west and the Jarbidge Mountains in the east (Charlet 1991).

Climatic History and Its Influence

The Pleistocene Epoch, in which several major glaciation and deglaciation events occurred, lasted from about 3 million to about 11,000 years before present (Flint 1971). During the Pleistocene Epoch, enormous lakes and marshes formed in many of the valleys (Mifflin and Wheat 1979; Williams 1982; Benson et al. 1990; Morrison 1991; Fig. 2), and glaciers spotted the high mountains (Blackwelder 1931, 1934; Flint 1971; Piegat 1980; Porter et al. 1983; Osborn 1989). The modern distribution and ecology (interactions of organisms with the environment) of the organisms of the Great Basin–Mojave Desert region largely reflect the effects of increased precipitation and cooler temperatures of the Pleistocene Epoch (Reveal 1979; Wharton et al. 1990).

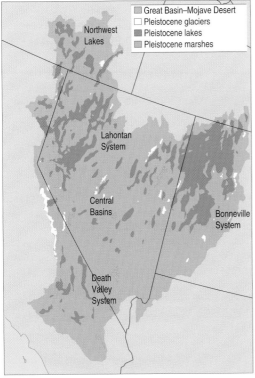

Fig. 2. Estimate of the maximum extent of marshes, pluvial lakes, and mountain glaciers during the Pleistocene Epoch in the Great Basin–Mojave Desert region. Data from Blackwelder (1931, 1934), Flint (1971), Piegat (1980), Williams (1982), Porter et al. (1983), and Osborn (1989).

Plants and animals of the region responded to the Pleistocene climate by changing their relative distributions and abundances (Brown 1971, 1978; Billings 1978; Harper et al. 1978; Wells 1983; Charlet 1991). Therefore, community composition changed as species moved in search of more congenial climates (C. L. Nowak 1991), adapted in place, or became either locally or globally extinct (Grayson 1993). Species were generally forced downslope and southward in response to Pleistocene conditions (Axelrod 1950, 1976, 1983; Van Devender and Spaulding 1979; Wells 1983; Mehringer and Wigand 1990; Wigand et al. 1994).

The Holocene Epoch of the last 11,000 years is characterized by the recession of mountain glaciers and a warmer and often drier climate than existed during the Pleistocene Epoch. Drought- and heat-tolerant species from southern deserts have moved northward into the Great Basin–Mojave Desert region during the Holocene Epoch (Reveal 1979), while drought- and heat-intolerant species have been forced farther north or higher into the cooler and wetter conditions of the mountains, or were extirpated (Billings 1978). This dry period has affected the animals (Brown 1978), the people (Bettinger 1991; Thomas 1997), and the plants (Billings 1978) of the region.

Gould (1991) maintains that the Holocene Epoch is simply another interglacial period before renewed glaciation. This view echoes that voiced earlier by Wright (1983:xi), "there is little doubt that another major glaciation will ensue." In the face of either a new glaciation period or human-induced global warming, how will the biota of the Great Basin–Mojave Desert region respond? To understand the answer, we must examine the spectacularly varied and numerous isolated communities of the region with history in mind.

Montane Islands in a Sea of Shrubs

The last 3 million years of climatic fluctuations left a biological legacy on the landscapes of the Great Basin–Mojave Desert region. Remnants of ancient pluvial lakes include Pyramid, Walker, Carson, Eagle, and Mono lakes; each has its unique plants and animals. The many mountaintops are refugia for species left behind from the Pleistocene Epoch in communities that are vastly different from the adjacent desert valleys. Within 10 kilometers, a single basin-range unit can host environments that range from treeless alpine bogs and rocky slopes to montane coniferous forests, diverse mountain shrublands, pygmy woodlands of pinyon pine or juniper, lower slopes of sagebrush and grasses, lake shores that support an entirely different array of shrubs and flowers, barren sand dunes, or playas (saline flats that sometimes form shallow lakes). Dozens of montane habitat islands in the region are now separated from each other by deserts.

Transition Zones

Contact between the flora and fauna of two regions creates transition zones (Meyer 1978; MacMahon 1988b). These zones have characteristics of both regions and constitute a unique type of habitat. Although transition zones may be important sites for plant evolution (G. L. Stebbins 1974), few researchers have examined the animals of these zones in detail (Pianka 1967, 1970; Parker and Pianka 1975; Robinson 1988; Wilson 1991).

The eastern boundary of the Great Basin–Mojave Desert region is sharply defined by the high elevations of the Wasatch Mountains and the Colorado Plateau. The western boundary along the Sierra Nevada is even sharper and has been called one of the world's sharpest boundaries between biological regions (Billings 1990). However, the northern and southern boundaries are subtler, and the transition zones there are much broader.

The most complex transition zone is about 115 kilometers wide and occurs between the Great Basin and the Mojave Desert (Beatley 1976; O'Farrell and Emery 1976; Fig. 3). This zone contains a mixture of smaller transition zones and sharp habitat edges (Billings 1949, 1951; Beatley 1974a,b, 1976; Meyer 1978; El-Ghonemy et al. 1980a,b; Turner 1982; Callison and Brotherson 1985; Young et al. 1986; MacMahon 1988b; West 1988; Tueller et al. 1991), which result from increasing elevation, decreasing temperatures, and a south–north shift in predominant precipitation from rain to snow (Billings 1951; MacMahon 1988a; Tueller et al. 1991).

Many terrestrial animals reach their northern or southern distribution limits in this zone (Hall 1946; Holmgren 1972; Stebbins 1985), and the boundaries of several subspecies occur here, such as the northern and southern desert horned lizards and the northern and desert side-blotched lizards (Stebbins 1985). The turnover of some bird and mammal species is also abrupt between the Great Basin and the Mojave Desert (Hall 1946; Behle 1978).

Another gradual transition zone occurs toward the northern edge of the region, where sagebrush declines in favor of perennial bunch-grasses. Pristine conditions in the north once included such a high grass component that these communities are commonly referred to as sage-brush–steppe or shrub–steppe. The gradation continues north until grasses dominate the steppe of the Columbia Plateau (West 1988).

Status of Major Communities

Aquatic Communities

Riparian Communities

Riparian communities occur along the major watercourses in most intermountain valleys of the Great Basin–Mojave Desert region and in association with isolated springs, seeps, and smaller streams. In the Great Basin, riparian communities are dominated by various mixtures of cottonwood, aspen, and willow species that are used by the native Humboldt beaver (Fig. 4). Thickets of chokecherry are common, and occasional patches of silver buffaloberry provide important overwintering sites for North American porcupines (Sweitzer 1990). Impressive groves of cottonwoods occur along the larger rivers (Fig. 5). In the low-lying Mojave Desert, these dominants are largely replaced by mesquite, cat-claw acacia, and velvet ash. Other riparian associations dominated by saltgrass and iodine bush occur on saline soils and along streams such as Salt Creek in Death Valley (Fig. 6).

Fig. 4. Humboldt beaver at the headwaters of the Humboldt River, Jarbidge Mountains, Nevada.

Although extremely small in total area, riparian communities in this region are critical centers of biodiversity. More than 75% of the species in the region are strongly associated with riparian vegetation (U.S. General Accounting Office 1993), including 80% of the birds (Dobkin 1998) and 70% of the butterflies (Brussard and Austin 1993). Several butterflies are completely restricted to these habitats (for

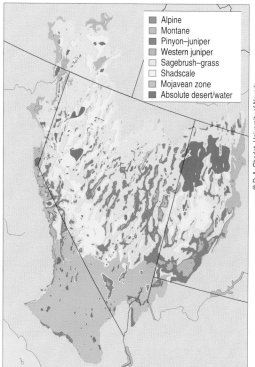

Fig. 3. Vegetation zones in the Great Basin–Mojave Desert region. Compiled from Shreve (1942), Billings (1951), Holmgren (1972), Ertter (1992), and D. A. Charlet (University of Nevada, unpublished data).

■ Alpine
■ Montane
■ Pinyon–juniper
■ Western juniper
□ Sagebrush–grass
□ Shadscale
■ Mojavean zone
■ Absolute desert/water

Fig. 5. Cottonwood groves along the Walker River in Antelope Valley, California.

Fig. 6. A Mojave Desert riparian system at Salt Creek, Death Valley National Park, California.

example, the Apache nokomis fritillary butterfly and the Weidemeyer's admiral butterfly). Riparian corridors also provide migration pathways for many species, and subalpine and montane conifers descend to much lower elevations in riparian situations than in upland stands (Charlet 1996, 1997). Much of this vegetation has been destroyed or degraded by water diversions, agricultural development, and livestock grazing (U.S. General Accounting Office 1993), which is a major cause of riparian habitat degradation in the Great Basin (Platts 1990; Chaney et al. 1993; National Research Council 1994). Effects of grazing include a reduction of natural vegetation, stream channel widening and aggradation, and lowered water tables (Kauffman and Krueger 1984; Armour et al. 1991; Chaney et al. 1993; Fleischner 1994). A structurally diverse flora in riparian areas that has not been grazed supports a broad assemblage of wildlife species, whereas reductions of shrub and herbaceous cover following heavy cattle grazing modify many bird and small-mammal communities (Schulz and Leininger 1991; Dobkin 1994a). The effect of livestock on riparian habitats in the Great Basin is often so severe that those habitats no longer represent natural vegetation (Ehrlich and Murphy 1987; Chaney et al. 1993), and the associated faunal communities support largely widespread,

ecological generalists that are adapted to such highly disturbed conditions (Dobkin 1998).

A striking example of generalist species in disturbed areas is found along the Virgin, Mojave (Lovich et al. 1994), Humboldt, and Walker rivers and in many other parts of the region, where riparian communities are now largely dominated by aggressive, nonindigenous tamarisks. These trees reduce the abundance of native riparian vegetation, and their rapid migration is profoundly altering species composition and community functions (Vitousek 1986). Tamarisks use water more efficiently than other trees do and tolerate saline conditions (Busch and Smith 1993). Tamarisks are fire-tolerant, thus increased fire frequency favors them over the native riparian vegetation (Anderson et al. 1977; Busch and Smith 1993). Tamarisks are also more tolerant of boron, a toxic element that concentrates in soil after fires (Busch and Smith 1993), than are native willow and cottonwood species. These postfire adaptations permit tamarisks to dominate riparian communities at the expense of the native trees (Anderson et al. 1977).

Stream flow also is altered by the encroachment of tamarisks into riparian communities. Tamarisks create narrower stream channels (Blackburn et al. 1982) and promote local flooding (Graf 1980). Although expensive (Graf 1980; Barrows 1993), the eradication of tamarisks has been shown to rejuvenate marshes, such as those along Eagle Borax Spring in Death Valley (Vitousek 1986).

Wet-Meadow Communities

Permanently wet meadows, like persistent streams, are scarce in this arid region, yet they contribute significantly to the region's plant diversity (Linsdale et al. 1952; Loope 1969; Charlet 1991). These habitats also are frequently destroyed by livestock grazing and water diversion (Hammond and McCorkle 1983). Several species at risk are characteristic of these communities. For example, many populations of the silver-bordered fritillary butterfly have been lost because springs were capped (Pyle et al. 1981).

Terminal-Wetland Communities

A distinguishing feature of most of the Great Basin–Mojave Desert region is the drainage of its waters to terminal basins rather than into the ocean. This feature creates isolated terminal lakes, marshes, and playas (Fig. 7), many of which support unique aquatic species. Although most are small and only seasonally filled with water, these wetlands are surprisingly numerous and critically important to the biological diversity and ecology of the region (Table 1).

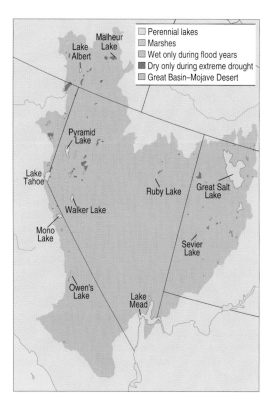

Fig. 7. Terminal wetland ecosystems in the Great Basin–Mojave Desert region.

Water levels in Great Basin–Mojave Desert wetlands are lowered by diversions for agriculture and urban growth; for example, Nevada lost 52% of its wetlands between 1780 and the mid-1980's (Dahl and Johnson 1991). Low water levels are accompanied by an increase in total dissolved solids (TDS) and increasing accumulations of heavy metals such as boron, selenium, arsenic, and mercury. Many of the region's wetlands, including some of the larger ones like Honey Lake, now go dry during low-water years.

All the wetlands of the region are important to waterbirds. Mono Lake and the Great Salt Lake already have high levels of TDS, and if TDS concentrations increase further, these lakes may lose their tremendous populations of brine shrimp and brine flies, which are key food sources for migrating waterbirds. Further increases in TDS levels at Walker Lake (associated with lower water levels) could prove fatal to fishes and eliminate them from the lake (Koch et al. 1979; C. A. Stockwell, Biological Resources Research Center, University of Nevada, Reno, unpublished manuscript).

Terrestrial Communities

Absolute Desert

Absolute desert communities (Table 2; Fig. 3) include playas, salt barrens, and sand dunes. When playas fill from rare rains or snow melt, an invertebrate fauna develops that consists of various crustaceans (for example, fairy shrimp) and insects, which in turn supports large populations of migratory waterbirds. This fauna has not been studied extensively and may include several undescribed species. The effects of off-highway vehicles on these organisms are unknown but are probably detrimental (R. C. Stebbins 1974).

Mojavean Zone

The Mojavean vegetation zone (Fig. 8) encompasses most of the geographic area of the Mojave Desert except the higher mountains. The flora and communities in this zone are characterized by great diversity, and only half of its 545 plant species also occur in the Sonoran Desert (Vasek and Barbour 1990). Several vegetation types are recognized in the Mojavean zone: creosotebush scrub, saltbush scrub, shadscale scrub, blackbrush scrub, Joshua-tree woodland, and annual vegetation (Vasek and Barbour 1990). The highest mountains within the boundaries of the Mojavean zone (the Sheep Range, and White, Spring, and Panamint mountains) exhibit most of the Great Basin vegetation zones and thus resemble high mountains in the Great Basin.

Las Vegas, the fastest growing city in the United States, lies in the northeastern part of the Mojave Desert. This growth and the lack of local persistent streams create an unprecedented demand on groundwater in the area. Exploitation of groundwater in the Pahranagat and Moapa valleys has significantly altered the hydrology of the area and has harmed riparian communities (Runeckles 1982; Franklin et al. 1987; Busch and Smith 1995). The fishes are restricted to springs and streams in the area. Tamarisk has invaded and now dominates many riparian areas, causing additional problems. Ironically, "Las Vegas" means "the meadows" in Spanish, but the inspiration for the name vanished early in the city's sprawling growth. This sprawl covers a large portion of the Mojavean zone vegetation with concrete and places at risk the desert tortoise, many other reptiles, and the California bearpaw poppy.

Shadscale Zone

The shadscale vegetation zone is named after its dominant plant species and is the third largest in total area in the region (Table 2; Fig. 3). In addition to shadscale, this area is populated by many other species of widely spaced, drought-tolerant shrubs that are usually thorny with small leaves. In the western Great Basin and in the Bonneville Basin, this zone occurs primarily on slightly saline soils on the periphery of ancient lake beds. In the Mojave Desert and in the mountains of southern Utah, it also occurs on rocky upland sites that are too dry to support sagebrush (Billings 1949). Greasewood is the

Table 1. Status of terminal wetlands in the Great Basin–Mojave Desert region.

Terminal wetland	Status
Nevada	
Artesia Lake	Dry during 1987–1994; currently refilling
Carson Lake	Selenium and boron are contaminants during low-water years; Public Law 101–618 should effect an increase of water levels and quality (U.S. Fish and Wildlife Service 1992)
Duck Flat	Episodically dries
Fallon National Wildlife Refuge	Dry during 1987–1994; currently refilling
Fernley Sink	Dry during 1987–1994; currently refilling
Humboldt Wildlife Management Area	Water levels were low from 1987 to 1994; total dissolved solids were high enough to kill duck chicks; high arsenic and selenium levels caused problems through bioaccumulation; current water levels are high
Mason Valley Wildlife Management Area	The hatchery outflow provides a reliable source of water; effects of a recent drought are being alleviated
Pyramid Lake	Very reduced lake levels; reduced inflows prevented spawning by the cui-ui and the Lahontan cutthrout trout until 1995
Ruby Lake and Ruby Marsh	Fed by 200 springs; water level in 1994 was 61 centimeters below targeted level; heavy metals were expected to become harmful
Soda Lake	Some effects from drought; high brine fly production; important to migrating red-necked phalaropes
Stillwater Wildlife Management Area	Water levels were low from 1987 to 1994; selenium, boron, and salinity levels were high; Public Law 101–618 should effect an increase of water levels and quality during drought years; water levels currently high
Washoe Lake	Dry during the 1930's and 1960's and during 1991–1994; currently high water level; historically very productive
Winnemucca Lake	Dry since 1938
Walker Lake	Rapidly decreasing water levels until 1994, when total dissolved solids approached levels that are fatal to fishes; water level currently stabilized
Southeastern Oregon	
Abert Lake	Lowest water level in past 18 years because of drought and water diversion; relatively clean; naturally high level of total dissolved solids; brine fly populations are low because of increased total dissolved solids
Alkali Lake	Chemical residue in water because of nearby disposal site
Alvord Hot Springs	Dry during recent drought
Bacus Lake	Water levels follow those of Malheur Lake
Malheur Lake	High precipitation in 1980–1986 increased the size of the lake; by fall 1992, lake size had been reduced to a historical low; in 1994, water levels continued to decline
Silver Lake	Dry since early 1980's from irrigation diversions
Stinking Lake	Dry during drought
Summer Lake	At 0.25 capacity as of fall 1994; relatively pesticide-free; wetlands are maintained regardless of lake level
Warner Wetlands	During 1983–1984, highest water level in recorded history; during 1985–1990, completely dry because of diversions and drought
Eastern California	
Clear Lake	Irrigation reservoir; turbid
Eagle Lake	Steady decline of water level from 1987 to 1994 because of drought; currently refilling
Goose Lake	No contamination; water diverted for farming
Honey Lake	Seasonally dry; possible chemical contamination; currently refilling
Lake Tahoe	1994 water level was below rim; in June 1995 water level was above rim; record high levels in January 1997
Modoc National Wildlife Refuge	Irrigation reservoir; no contamination
Mono Lake	High level of total dissolved solids; in accordance with legal settlements, water level will be raised
Owens Lake	Dry since 1957

common phreatophyte (plant requiring groundwater) in the zone (Holmgren 1972).

The rapid decline in the last 100 years of the important browse shrub, winterfat, and its replacement by nonindigenous annual species are clearly documented with photographs (Rogers 1982). This decline is due to heavy grazing by domestic livestock (Billings 1951; Holmgren 1972). An associated change of particular concern is the increase of the nonindigenous weed halogeton, which is often fatal to domestic stock (Holmgren 1972). Tamarisk is also rapidly invading and dominating riparian areas in this zone.

Sagebrush–Grass Zone

Although the principal shrub in this zone is big sagebrush, several other sagebrush species and subspecies also occur and can be locally dominant. Because precipitation is higher in the sagebrush–grass zone than in the shadscale zone (Billings 1949), this zone can

support a surprisingly high richness of shrubs, grasses, and perennial forbs (Fig. 9). The sagebrush–grass zone is the largest (Table 2) and most contiguous vegetation zone in the region (Fig. 3).

One of the most significant changes in the sagebrush–grass zone has been the invasion of introduced plant species such as cheatgrass, halogeton, and other annuals, at the expense of the native bunchgrasses and forbs (Young et al. 1972; Rogers 1982). The sagebrush–grass zone

Table 2. Total area and percent of region occupied by the eight vegetation zones in the Great Basin–Mojave Desert region.

Vegetation zone	Total area in region (square kilometers)	Percent of region
Absolute desert	29,510	5.54
Mojavean	98,068	18.40
Shadscale	91,317	17.13
Sagebrush–grass	206,071	38.66
Pinyon–juniper	60,556	11.36
Western juniper	7,187	1.35
Montane	39,037	7.32
Alpine	1,224	0.23
Total	532,970	99.9[a]

[a]Does not total 100% because of rounding.

Fig. 8. Mojavean vegetation zone, eastern base of the Spring Mountains, Nevada.

is contracting because of the downward expansion of the pinyon–juniper zone in the central portion of the region (Blackburn and Tueller 1970) and by western juniper woodlands in the north (Burkhardt and Tisdale 1976; Miller and Wigand 1994; Miller and Rose 1995). Sagebrush–grass communities are being fragmented by agriculture and human population increases, both of which increase the probability of invasion by nonindigenous plant species. This zonal contraction and fragmentation and the increase of nonindigenous plants are further aggravated by livestock grazing and fire suppression. The combination of these factors decreases grass density and increases shrub density (Rice and Westoby 1978; Young and Evans 1978; Anderson and Holte 1981). More than 99% of the sagebrush–grass zone has

been negatively affected by livestock, with serious effects in about 30% of the zone (Noss et al. 1995).

Pinyon–Juniper and Western Juniper Zones

The pinyon–juniper zone was defined by Billings (1951) as the lowest in elevation of the montane zones. Here we follow Holmgren (1972) and map the unit as a separate zone because of its large size and ecological significance. Billings (1954a) found that thermal belts and higher precipitation on the mountain slopes contributed to the formation and maintenance of the zone, which is characterized by woodlands of pinyon pine and several species of juniper in various combinations. Understories are composed of grasses, perennial forbs, and shrubs (principally sagebrush and bitterbrush), and several gooseberry species are also common. By far the most abundant pinyon pine is single-leaf pinyon, but Colorado pinyon pine occurs in the Utah portion of the region (Lanner 1984). Utah juniper is the dominant juniper in the zone, but western juniper and Rocky Mountain juniper may enter the zone at its higher reaches or along streams. This zone is extensive throughout the Great Basin south of the Humboldt and Truckee river drainages and is sparse in the northern Lahontan River basin; it disappears altogether in Oregon. The zone barely extends into Idaho, east of the Humboldt River drainage. Many higher mountains in the Mojave Desert (for example, Providence, Panamint, Sheep, and Spring mountains) also support pinyon–juniper woodlands.

Higher precipitation and colder climate distinguish western juniper woodlands from those with pinyon or Utah juniper (Wigand and Nowak 1992). This zone dominates much of southeastern and central Oregon, northeastern California, and extreme northwestern Nevada (Fig. 3).

The pinyon–juniper and western juniper zones are spreading throughout the Great Basin. Tree densities increased dramatically in the last 100–150 years (Tausch et al. 1981; Rogers 1982). The abundance of Utah juniper clearly has increased in the Bonneville Basin since the late 1800's (Rogers 1982), and Utah juniper and singleleaf pinyon are also invading black sagebrush communities in east-central Nevada west of the Bonneville Basin (Blackburn and Tueller 1970). Increased tree density is accompanied by a decline in understory density and diversity, and in severely degraded juniper stands virtually all herb species can be excluded (Tausch and Tueller 1990; Charlet 1996, 1997). Such a landscape has little value as livestock forage or wildlife habitat.

Singleleaf pinyon also is spreading wherever it occurs in the Great Basin, particularly in western Nevada (Billings 1951). During the

Fig. 9. Sagebrush–grass vegetation zone, Pueblo Valley, Oregon–Nevada border. The Pine Forest Range is in the distance.

last 250 years, the species advanced north across the Truckee River into Peavine Mountain, the Virginia Mountains, the Junction House Range, and the Pah Rah Range (Charlet 1996). Western juniper is increasing its range and abundance in the northern part of the region in southern and central Oregon (Miller and Wigand 1994; Miller and Rose 1995) and in southwestern Idaho (Burkhardt and Tisdale 1976).

This woodland expansion is largely a result of a combination of fire suppression and overgrazing. These factors lead to a decline of browse and grass species that competitively exclude juniper and provide the fuels to carry fires that restrict junipers to rocky sites (Burkhardt and Tisdale 1976). Extreme measures, including chaining (Tidwell 1986), burning (Bunting 1986), and poisoning (Johnson 1986) have been used in attempts to control this expansion.

Montane Zone

Billings (1951) divided the mountain vegetation above the sagebrush–grass zone into several montane zones in three geographic series. We, however, followed Holmgren (1972) by mapping the pinyon–juniper zone and the alpine zone separately from the remainder of the montane series (Fig. 3). Smaller species of sagebrush (for example, Vasey's sagebrush and little sagebrush) are dominant plants above the pinyon–juniper zone and are associated with a different suite of shrubs and grasses than those in sagebrush–grass communities below the pinyon–juniper zone. These are productive communities that provide important forage for wildlife and livestock.

A common feature of the montane zone is the widespread occurrence of quaking aspens (Fig. 10). In many northern mountain ranges, aspens form extensive pure forest stands. Because these ranges usually contain few or no conifer (cone-bearing) species, aspens provide important habitats for birds and other wildlife. Curlleaf mountain-mahogany also provides important shelter and nesting sites and usually occurs in rocky, ridgeline situations. A close relative, the Mexican cliffrose, replaces it in the Mojave Desert mountains in similar but hotter and drier situations.

Forests of western juniper occur in the montane zone on the northern slope of Steens Mountain, Oregon (Holmgren 1972), and the species also extends into the montane zone of a few western Great Basin mountain ranges (Vasek 1966; Charlet 1996). High-diversity coniferous forests are abundant in the Carson Range, a spur of the Sierra Nevada along the western boundary of the region. Only 2 of the 15 conifer species in the Carson Range do not

© D. A. Charlet, University of Nevada

also inhabit western Great Basin mountain ranges (Charlet 1996).

Along the eastern boundary of the Great Basin region, forest fragments and extensive woodlands characteristic of the Rocky Mountains occur in the Wasatch Range (Billings 1951, 1990; Holmgren 1972). A few Great Basin–Mojave Desert mountain ranges (for example, Snake Range, Spring Mountains) support montane coniferous forests, but these are usually small and have few conifer species. However, subalpine woodlands (Fig. 11), with various mixtures of whitebark, bristlecone, limber, and lodgepole pines, and other conifers, occur in most of the higher mountains of the region (Charlet 1996).

Several forest trees have been extirpated during this century. For example, Douglas-fir has been lost from two mountain ranges in northern

Fig. 10. Quaking aspens along Lye Creek in the Santa Rosa Range, Nevada. Granite Peak is in the distance.

© D. A. Charlet, University of Nevada

Fig. 11. Subalpine woodlands and meadows of the montane zone at Cougar and Prospect peaks, Jarbidge Mountains, Nevada.

Nevada since 1937 (Loope 1969; Charlet 1996), and California white fir probably was extirpated from Petersen Mountain, Nevada, by a fire in 1994 (Charlet 1996). Many conifer stands and their associates are susceptible to extirpation throughout the region, and fires may leave bald mountains (Billings and Mark 1957) in areas that can only marginally support trees. The loss of trees subsequently causes a rapid deterioration of conditions that are suitable to forest development (Billings and Mark 1957).

Increased success of subalpine conifer seedlings at higher elevations has been observed throughout the region. For example, whitebark pine and subalpine fir are moving toward the summits of the Jarbidge Mountains on the northern border of the region (Charlet 1996). Studies of tree rings revealed accelerated growth rates during the last 100 years in upper treeline trees such as limber pine in the Toquima Range and Great Basin bristlecone pine in the White Mountains (M. Rose, Desert Research Institute, Reno, Nevada, personal communication).

Alpine Zone

The alpine zone begins at the limit of the upper treeline. In the Great Basin–Mojave Desert region, the zone is tiny and restricted to only a few of the highest mountain ranges (Fig. 3). True alpine tundra is present in small local areas in the Ruby (Loope 1969), Sweetwater (Lavin 1981), White (Billings 1978), Toiyabe (Linsdale et al. 1952), and Toquima (Charlet 1997) mountain ranges. Alpinelike habitats also occur on many of the mountain ranges that do not have an upper treeline, such as the Spring Mountains and the Santa Rosa and Pine Forest ranges in Nevada and Steens Mountain in Oregon. In most of the mountains with an alpine zone, more alpine area is occupied by talus slopes, cliffs, rocky ridges, and peaks than by alpine tundra. Nevertheless, small patches of tundra, alpine bogs, and springs host many alpine plant species that occur nowhere else in that mountain range. The alpine flora of the Great Basin, with about 600 species, is as diverse as that of either the Rocky Mountains or the Sierra Nevada at equivalent latitudes (Charlet 1991). This richness of alpine species suggests that there was a greatly expanded alpine zone throughout much of the region during the Pleistocene Epoch.

Subnormal precipitation and higher temperatures during 1971–1994 (H. Klieforth, Desert Research Institute, University of Nevada System, Reno, personal communication) reduced snow cover and caused the demise of many perennial snowfields in the region. This allowed subalpine woodlands to encroach on the alpine zone, reducing its already tiny size.

Reduced snow cover is detrimental for many plant species in alpine communities such as Eschscholtz's buttercup, which occurs around snowbanks; white bog-orchid, swamp laurel, and sphagnum moss, which are associated with bogs; and louseworts, shooting stars, and ladies' tresses, which are characteristic of springs (Charlet 1991).

Biodiversity Hot Spots

The Nevada Natural Heritage Program (1992) recognizes 101 locations in Nevada as areas with many rare species—biodiversity hot spots—based on the total number of sensitive species of all groups. Eighteen of these sites were identified and ranked on the basis of worldwide endangered subspecies and varieties, protection urgency, and management urgency (Nevada Natural Heritage Program 1992). Detailed maps of sensitive species of all taxonomic groups have been prepared for the Spring Mountains of southern Nevada (The Nature Conservancy 1994). The Spring Mountains are of particular concern for conservation because they are one of only two ranges (the other is the White Mountains) that contain all the vegetation zones in the Great Basin–Mojave Desert region. As a consequence, the Spring Mountains and the White Mountains have a great many species. The Spring Mountains are also unique because about 20% of the endemic plant species of Nevada—those found only in that state—occur there. In the exploratory study of the Spring Mountains, 23 areas of concern were mapped and described as biodiversity hot spots (The Nature Conservancy 1994).

Recognition of areas with particularly high biological diversity, regardless of the status of the taxa, is only at its beginning stages in the Great Basin–Mojave Desert region. Most of this exploratory work has focused on lists of species for local areas. Checklists for vascular plants were prepared for a few highly diverse mountain ranges, and 988 species were identified in the White Mountains (Morefield et al. 1988), more than 625 species in the Ruby and Sweetwater mountains (Charlet 1991), and many more than 500 species in other high ranges such as the Jarbidge, Warner, Toiyabe (Charlet 1991), Toquima, Santa Rosa, and Independence mountains (Charlet, unpublished data). Future mapping efforts of the Nevada Biodiversity Initiative are expected to reveal other areas of high biological diversity.

Montane Islands

Ecological islands of montane and alpine vegetation occur throughout the Great Basin–Mojave Desert region. These isolated, generally small communities have been known to exist for many years (Sudworth 1898, 1913) and have been the subject of serious inquiry for decades (Billings 1950, 1951, 1954b; Vasek 1966; Critchfield and Allenbaugh 1969; Loope 1969; Brown 1971, 1978; Little 1971; Johnson 1975, 1978; Axelrod 1976; Harper et al. 1978; Reveal 1979; Thompson and Mead 1982; Wells 1983; Critchfield 1984a,b; Axelrod and Raven 1985; Wilcox et al. 1986; Grayson 1987; Charlet 1991; Morefield 1992). The wide distribution of these habitats and the large number of species they support add greatly to the biological diversity of the region and to the potential resilience of the region in response to climatic change (Wharton et al. 1990; Charlet 1991).

Contemporary populations of 16 montane mammal species are presently isolated on mountains, and probably have been since the Pleistocene Epoch (Brown 1971, 1978; Grayson 1987; Mead 1987). Palmer's chipmunk is the only full species of mammal that is endemic to the Great Basin, while 15 endemic subspecies of 8 other species are montane isolates (Hall 1981; Table 3). Fossils indicate that populations of montane mammals occurred throughout the region during the Pleistocene Epoch (Grayson 1987). Exploratory research on the extinction of the region's montane mammals has included comparisons of current species assemblages above 2,300 meters in individual mountain ranges with species assemblages in the Rocky Mountains and Sierra Nevada (Brown 1971, 1978; Patterson 1984; Patterson and Atmar 1986; Belovsky 1987). The studies revealed that warm, dry periods during the Holocene Epoch reduced population sizes of montane mammals in the region and led to the extinction of many of their populations. If extirpated, relict mammal populations that are isolated on montane islands probably could not recolonize under current climatic conditions. Such extirpations may eliminate genetically

unique populations; this threat has clear implications for the management of high-altitude environments in the Great Basin (Grayson 1987).

Lowland Aquatic and Riparian Endemism

Fishes are the best-known native aquatic species in the region. Aquatic invertebrates may be equally diverse (Hershler and Pratt 1990), but relatively little is known about them. Riparian butterflies, however, have been well studied and show rather striking endemism in the region, far more so than in montane areas (Austin 1985, 1992). In the northern Great Basin, 27 species of butterflies are closely associated with lowland riparian vegetation. Of these, 9 (33%) have differentiated into 23 endemic subspecies. The most extensive subspeciation has occurred in the common wood nymph butterfly; nine subspecies are endemic to the Great Basin (Austin 1992). These butterflies evidently evolved in response to Holocene droughts that resulted in the isolation of remnant waters and their associated riparian communities.

Sand Dunes

At least 52 unconsolidated sand dunes occur in the region, 32 in Nevada (Fig. 12). These dunes were formed during the Holocene Epoch (Morrison 1964; Smith 1982; Lancaster 1988a,b; 1989; Dohrenwend et al. 1991) and are unique habitats because they are rare, small, of recent origin, and spatially dynamic. The dunes

Table 3. Mammal taxa confined to montane habitats in the Great Basin–Mojave Desert region (Hall 1981).

Scientific name	Common name
Eutamias panamintinus panamintinus	Panamint chipmunk
E. panamintinus acrus	
E. umbrinus inyoensis	Uinta chipmunk
E. umbrinus nevadensis	
E. palmeri	Palmer's chipmunk
E. amoenus monoensis	Yellow-pine chipmunk
Spermophilus lateralis certus	Golden-mantled ground squirrel
S. lateralis bernardinus	
Neotoma cinerea lucida	Bushy-tailed woodrat
Marmota flaviventris parvula	Yellow-bellied marmot
M. flaviventris fortirostris	
Ochotona princeps nevadensis	Pika
O. princeps tutelata	
O. princeps sheltoni	
O. princeps albata	
Microtus longicaudus latus	Long-tailed vole
Zapus princeps curtatus	Western jumping mouse

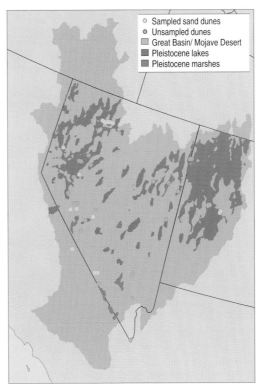

Fig. 12. Sand dunes of the Nevada portion of the Great Basin–Mojave Desert region.

support a diverse group of animals (Hall 1946; Brown 1973) and plants (Pavlik 1985, 1989) that have responded with special adaptations to the unique challenges of this harsh environment (Fig. 13).

Beetles are the best-studied group of sand dune invertebrates (Fig. 14). Most are either predators or are dependent as adults on fine organic debris, and the larvae of many species feed on plant roots. Most dune beetles depend on vegetation around the base of the dunes for adult or larval forage, mating sites, and protective cover. Ten Nevada sand dune beetle species are species of concern (U.S. Fish and Wildlife Service 1994a), including Giuliani's dune scarab from the Big Dune, the large aegialian scarab from the Big and Lava dunes, and several unnamed species of scarabs on the Big, Crescent, and Sand Mountain dunes. Additional sand dune insects that are species of concern from the region include the Sand Mountain pallid blue butterfly, Nelson's miloderes weevil,

Fig. 13. Eureka Dune in Eureka Valley, Death Valley National Park, California. Eureka Dune is the tallest sand dune in North America.

the Kelso giant sand treader cricket, and the Kelso Jerusalem cricket (U.S. Fish and Wildlife Service 1994a). Many of these species are highly endemic (confined to one or a few dunes). Small geographic distributions and habitat destruction by off-highway vehicles (Luckenbach and Bury 1983) are the greatest threats to their persistence.

Edaphic Communities

Communities that resemble the montane coniferous forest vegetation of the Sierra Nevada are scattered throughout much of the west-central Great Basin in a matrix of pinyon–juniper woodlands (DeLucia et al. 1988). More than 140 of these communities occur in edaphic sites (Billings 1950, 1990). These sites are well removed from the main body of Sierran forests and range from 1 to 15 hectares (DeLucia and Schlesinger 1990). The

Fig. 14. *Eusattus muricatus*, a darkling ground beetle at Sand Mountain, Nevada.

endemic altered buckwheat, a candidate for federal listing, occurs exclusively in these disjunct Sierran communities. The closest relative of this species, Lobb buckwheat, is restricted to high elevations in the Sierra Nevada. Seventy-four species of wild buckwheat and many more subspecies and varieties are native to Nevada (Reveal 1985, 1989; Kartesz 1988). One of the features of the buckwheat genus that contributes to its high diversity is its ability to occupy unusual soils. For example, Tiehm buckwheat is a candidate species that occurs in clays in the Silver Peak Range of western Nevada, Sulphur Springs buckwheat is from the Ruby Valley of northeastern Nevada, and the endangered Steamboat buckwheat grows at Steamboat Hot Springs in western Nevada.

In the Mojave Desert, gypsum forms an unusual soil that supports many narrowly endemic plant species especially adapted to these conditions (Meyer 1986), including the endangered California bearpaw poppy. The Bureau of Land Management is currently mapping these edaphic units in detail (S. D. Smith, University of Nevada, Las Vegas, personal communication). Endemics that are restricted to limestone are important in canyons east of Death Valley (Raven 1988).

Status of the Biota

Fungi

Fungi are essential for community health because they decompose dead organic material, making nutrients again available for plants. Unfortunately, the fungi of the Great Basin–Mojave Desert region are virtually unknown. A survey of stalked puffballs was started in 1975 to help fill this void. About 700 collections of this group and all other encountered macrofungi have been made, but vast areas of the region have not yet been covered (Fig. 15).

Plants

Although large numbers of sensitive plant species are found in local areas, sensitive plants display no distinctive geographic or ecological patterns in the Great Basin–Mojave Desert region. Sensitive species occur in all areas and vegetation zones (J. Morefield, Nevada Natural Heritage Program, Reno, personal communication), and rare and endemic plant species occur in many types of habitats, from alpine summits to sand dunes. Many sensitive plants are restricted to edaphic habitat islands such as alkaline, limestone, or gypsum soils in the southern part of the region and soils derived from volcanic ash in the north. Throughout the

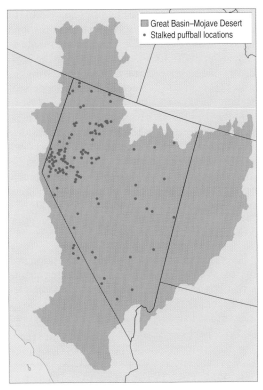

Fig. 15. Collection sites for stalked puffballs, a fungus, in Nevada and adjacent California.

region, hot springs often host narrowly restricted native species.

Populations of most sensitive plant species are presumably stable (Morefield, personal communication), but there are several exceptions. The California bearpaw poppy on gypsum soils in the Mojave Desert is seriously threatened by the expanding urbanization in the Las Vegas Valley. The persistence of the endangered Steamboat buckwheat, known only from ancient hot springs south of Reno, is of concern. This site is on private land that was recently developed for a geothermal power plant, and the species is therefore not protected by the Endangered Species Act of 1973. However, Nevada law (Nevada R.S. 527.260–527.300) prohibits the destruction of sensitive species listed by the state forester fire warden except by special permit. This statute was invoked to protect the species, and a cooperative venture was initiated between the developer, The Nature Conservancy, the State of Nevada Division of Forestry, and the Nevada Natural Heritage Program. As a result, the development proceeded, and a small portion of the Steamboat buckwheat population was transplanted to a nearby location. So far, the transplanted buckwheats are doing well, but a large portion of the plant's habitat is still on private lands that may be developed.

Assistance with the preservation of Grime's vetchling, a species of concern (U.S. Fish and Wildlife Service 1994a), was recently provided by a mining company in northeastern Nevada. This relative of the sweet pea was known only from a single population in the Independence Mountains (Barneby 1989). The mining company had proposed expansion of its exploration into the area occupied by Grime's vetchling and provided a helicopter to conduct aerial surveys of the species. Ground inspection of sites spotted from the air revealed 60 additional populations of the species (Morefield, personal communication).

Many plant species in alpine and montane habitat islands are uncommon in the region but are not candidates for listing because they are abundant outside of the Great Basin. Although the species are not globally threatened, the potential for dispersal and recolonization in the Great Basin and Mojave Desert in a different climatic regime is now small.

Invertebrates

Invertebrates make up 95% of the animals on Earth and are critical for ensuring community health (Mason 1995). Invertebrates generally experience the environment on much smaller spatial and shorter temporal scales than larger animals. As a result, invertebrates are extraordinarily susceptible to natural and human-caused habitat modifications (Mason 1995). Larvae and adults of the same species may rely on entirely different resources, including food plants, microhabitats, and solar insolation regimes (Weiss et al. 1988; New 1991). Consequently, many invertebrates are sensitive not only to habitat alterations that change vegetation structure but also to those that perturb microclimate (Thomas 1983, 1991; Dobkin 1985; Dobkin et al. 1987; Murphy et al. 1990). Therefore, land use patterns are important for the persistence of invertebrate populations at the regional scale and at the level of microhabitats (Hammond and McCorkle 1983; Dobkin et al. 1987; Weiss et al. 1988; Kremen 1992). Thomas (1984:337) noted that some invertebrates may be "extremely particular about the niches they occupy," and that "seemingly minor" modifications in land management or habitat structure may have major repercussions on the suitability of a given site for eggs and larvae. Land management is also critical to invertebrates that depend on a level of plant quality and abundance specific to a given successional stage (Scott 1986; Murphy et al. 1990; Thomas 1991).

The restricted mobility of many invertebrates that inhabit the diverse habitat islands in the Great Basin–Mojave Desert region renders them increasingly vulnerable to habitat disturbance. Many invertebrates have highly specialized adaptations, and the reestablishment of extirpated populations becomes less probable in degraded and fragmented habitats. Thus, invertebrates are of substantial conservation concern.

Terrestrial Invertebrates

Several groups of terrestrial invertebrates are well known in the region. The ants and butterflies are perhaps the most collected and studied groups of insects (Austin 1992; Wheeler and Wheeler 1986). The mosquitoes were studied by Chapman (1966), and the grasshoppers and crickets are well represented in collections of 179 known species and more than 3,000 identified specimens (J. B. Knight, Nevada Department of Agriculture, Reno, personal communication). However, most systematic entomologists consider collections of almost all other insect groups from the Great Basin to be insufficient, and most investigators report new, undescribed species (Rust 1986; Austin 1992) or geographic range extensions. Until the taxonomy of the region's insect fauna is better known, an assessment of species of concern is difficult.

Aquatic Invertebrates

The aquatic invertebrate fauna in the region shares species with the adjacent regions, especially the Rocky Mountains; species that also occur in the Sierra Nevada are not as widely distributed in the Great Basin (Nelson 1994). *Speciation* (the evolutionary formation of new species) in the region has probably occurred frequently because of recurring fragmentation and reconnection of the wetlands during the past million years. This pattern is now being documented for springsnails (Hershler and Sada 1987; Hershler 1989, 1994; Hershler and Pratt 1990). As with the terrestrial species, hundreds of native aquatic invertebrates probably exist in the Great Basin but have not been described (see Channel Islands and California Desert Snail Fauna box in California chapter).

General phenomena that affect aquatic invertebrates have been studied in other areas and probably also apply to the Great Basin. The composition of aquatic insect communities changes after the damming of waterways (Ward 1984; Williams and Feltmate 1992) and the release of mining waste (Ward 1984; Wiederholm 1984; Williams and Feltmate 1992). Furthermore, changes in water temperature after removal of vegetation, the presence of heavy metals in the water, and eutrophication of lakes are detrimental to aquatic insects (Ward 1984; Wiederholm 1984).

Kremen et al. (1993) suggested that terrestrial arthropods are appropriate indicators of the need for conservation, and aquatic insects are probably equally appropriate indicators of the need for conservation of aquatic communities.

Aquatic invertebrate communities are useful for evaluating water quality because they respond to low-level disturbances, function as monitors (Chandler 1970; Erman 1991), and may be indicators of conditions for terrestrial plant communities (Erman 1991). Because the abundances of aquatic organisms are declining much more rapidly than those of their terrestrial counterparts (Moyle and Yoshiyama 1994), studies are clearly needed. Although aquatic invertebrates in other areas of the country have been used for evaluating alterations to stream communities, few studies have been done in the Great Basin–Mojave Desert region.

Vulnerable Invertebrates

Specific information on the conservation status of almost all sensitive invertebrates in the Great Basin–Mojave Desert region is completely lacking. For example, only the Ash Meadows naucorid is listed as threatened, and 1 grasshopper, 13 beetles, 27 moths and butterflies, 16 snails, 1 mussel, 1 stonefly, and 1 wasp in Nevada are species of concern (U.S. Fish and Wildlife Service 1994a; Table 4). As studies of invertebrates progress, this number will probably increase. A previously undescribed riffle beetle was recently discovered in Ash Spring in the Pahranagat Valley and will probably be recommended for federal listing (Schmude and Brown 1991).

Two examples of vulnerable invertebrates include the Apache nokomis fritillary and the Giuliani's dune scarab. The Apache nokomis fritillary is a showy, monarch-sized butterfly that is confined to wet areas, including riparian corridors, seeps, and damp meadows across the Great Basin. The species of violet on which the larvae feed and the thistles from which the adults take nectar are confined to these moist habitats. Although the butterfly is physically capable of dispersing over large distances, observations and genetic studies have revealed that it rarely ventures far from its small, insular patches of habitat. Thus, each colony is a semi-isolated population that is rarely recolonized after extirpation. Several colonies were

Table 4. Sensitive aquatic insects in the Great Basin–Mojave Desert region.

Common name	Scientific name	Category listing[a]
Lake Tahoe benthic stonefly	*Capnia lacustra*	S
Ash Meadows naucorid	*Ambrysus amargosus*	T
Death Valley agabus diving beetle	*Agabus rumppi*	S
Devil's Hole warm spring riffle beetle	*Stenelmis calida calida*	S
Moapa warm spring riffle beetle	*S. calida moapa*	S
Travertine bandthigh diving beetle	*Hygrotus fontinalis*	S
Amargosa naucorid	*Pelocoris shoshone amargosus*	S

[a] S = species of concern (previously treated as Category 2 species; U.S. Fish and Wildlife Service 1994a); T = threatened

destroyed by agricultural development in recent years. A related although still unnamed subspecies of the Apache nokomis fritillary from the Carson Valley is a species of concern (U.S. Fish and Wildlife Service 1994a).

Giuliani's dune scarab occurs only at Big Dune and Lava Dune in the southern Great Basin of Nevada and, because of its restricted geographic range and rarity, it became a candidate for federal listing in May 1984, although it is now considered a species of concern (U.S. Fish and Wildlife Service 1994a). Surveys of the beetle in 1993 and 1994 revealed the presence of fewer than 10 individuals at each survey site. This unique species is at high risk of extinction because it is present at only a few isolated sites and is abundant nowhere. No one knows how many other invertebrate species are in similar situations across the region.

Fishes

Fifty-six species and 75 subspecies of fishes occurred historically in the Great Basin–Mojave Desert region. Of these 131 taxa, 10 (8%) are now extinct. Of the remaining 121, 75 (62%) are listed, are candidates for federal listing, or are species of concern (U.S. Fish and Wildlife Service 1994a). More than 40% of the species and 90% of the subspecies are endemics. The causes of nearly all, if not all, declines in desert fish populations can be traced directly to human activities. Deacon (1979) recognized that habitat modification (including damming, diverting, and channelizing waterways; overgrazing; and other forms of human disturbance) was the largest contributor to the endangerment of desert fishes. Biotic interactions (including hybridization, competition, and predation), primarily with introduced fish species, are the second largest contributor to the endangerment of desert fishes. Given the pervasiveness of water development and introductions of nonindigenous fishes, probably few fish populations in the West remain unaffected. Minckley and Douglas (1991:15) succinctly summarize problems faced by desert fishes: "All major streams in the western United States are dammed, controlled, and overallocated. . . . Groundwaters are pumped at rates greatly exceeding those at which aquifers can be recharged." Clearly, this does not bode well for the future of the region's fishes or for its human inhabitants.

The entire fish fauna of the Virgin River in the southeastern part of the region is imperiled. The native fauna includes two endangered species (the woundfin and the Virgin River chub), one subspecies proposed as threatened (the Virgin spinedace), and at least one other candidate species. Despite management, the long-term persistence of native fish populations in the Virgin River continues to be doubtful. The

recent use of rotenone to rid the system of nonindigenous red shiners severely harmed the native species and may have resulted in the loss of genetic variation in at least the Virgin River chub (Demarais et al. 1993).

The survival of native fishes in the Colorado River is also in doubt. The damming of this river drastically altered flow and temperature regimes, and many nonindigenous species have been introduced. Most of the larger native fishes such as the Colorado squawfish, the humpback and bonytail chubs, and the razorback sucker are endangered; it is unlikely that an effective preserve for these fishes can be established. Although dams could be regulated to maintain a more natural flow regime, this action is politically impossible on a river with overallocated water resources on which so many people depend.

The extensive isolation and close adaptation to specific circumstances by most desert fish species are illustrated by the six distinct subspecies of the Amargosa pupfish: the Amargosa, Tecopa, Ash Meadows, Saratoga Springs, Warm Springs, and Shoshone pupfishes. Each of these subspecies occurs in a single spring or small group of springs in the Amargosa Valley of southern Nevada and eastern California (Fig. 16). Two of the subspecies, the Ash Meadows and the Warm Springs pupfishes, are listed as endangered. Two more of the subspecies are species of concern (U.S. Fish and Wildlife Service 1994a). At least one subspecies, the Tecopa pupfish, is probably extinct (Sigler and Sigler 1994).

The problems of the Amargosa pupfish and all desert fishes can be summarized quite simply: water in the desert is scarce, humans want more than they require, and the fishes

© D. A. Charlet, University of Nevada

Fig. 16. The Amargosa Valley and Ash Meadows from Devils Hole Hills, Nevada. This area supports many endemic species.

must have it. Although desert fishes have a tenuous hold on survival under natural conditions, occurring only in the few permanent springs, rivers, and lakes of the region, their plight has been exacerbated greatly by human activities. The pumping of groundwater for agriculture nearly eliminated several Amargosa pupfish populations, including the Devils Hole pupfish and several other native endemic species (Deacon and Williams 1991). Grazing by cattle in riparian areas has increased sedimentation and siltation, reducing spawning of the threatened Lahontan cutthroat trout. Diversion of most of the flow of the Truckee River by the Newlands Project in 1906 to support agriculture in the Fallon area eliminated the largest known strain of cutthroat trout, the Pyramid Lake strain, and nearly wiped out another endemic fish, the endangered cui-ui. Anglers also pose a threat to desert fishes, not through direct take but by stocking waters with nonindigenous game fishes. Eighty-one species of nonindigenous fishes have been introduced in the region, and 28 species of nonindigenous game fishes have become naturalized or are currently released. Many of these have been implicated in the decline of native species.

One of the greatest concerns affecting survival of many fishes in the desert is the lack of available information on the status of these species. Few long-term studies or monitoring of any fishes other than those listed or proposed for federal listing have been made, and the distributions of many native fishes are poorly known. For example, many undescribed subspecies of the tui chub and speckled dace may exist, and the discovery and formal descriptions of other species or subspecies are possible.

Amphibians

Amphibians are one of the rarest animal groups in the Great Basin–Mojave Desert region because of their high water requirements. Only 22 native species and 2 introduced species occur, and the ranges of many of these barely extend into the region along the eastern Sierra–Cascades or along the western Colorado Plateau and the Wasatch Mountains in Utah. Four native species of frogs and toads are widely distributed throughout: the Great Basin spadefoot, western toad, Pacific chorus frog, and northern leopard frog. Three narrowly distributed natives also are residents: the relict leopard frog, Amargosa toad, and black toad. The introduced bullfrog occurs throughout the region. Of the 22 native amphibian species, 10 (45%) are species of concern or candidates for federal listing (U.S. Fish and Wildlife Service 1994a; Table 5), and several others seem to be declining. Surveys were conducted recently at

Table 5. Amphibians of the Great Basin–Mojave Desert region, compiled from Stebbins (1985) and Zeiner et al. (1988).

Common name	Status [a]
Long-toed salamander	N
Tiger salamander	N
Inyo Mountains slender salamander	S
Western toad	N
Yosemite toad	S
Great Plains toad	N
Black toad	S
Southwestern toad	S
Amargosa toad	C
Red-spotted toad	N
Woodhouse's toad	N
Mount Lyell salamander	S
Owens Valley web-toed salamander	S
Boreal chorus frog	N
Pacific chorus frog	N
Canyon treefrog	N
California red-legged frog	PE
Bullfrog	I
Mountain yellow-legged frog	S
Relict leopard frog	N
Northern leopard frog	N
Spotted frog	C
Great Basin spadefoot toad	N

[a] C = Category 1 species; species for which there is adequate information on which to base a proposal to list as threatened or endangered under the Endangered Species Act; S = species of concern (previously treated as Category 2 species; U.S. Fish and Wildlife Service 1994a); I = introduced to the region; PE = proposed endangered; N = not listed.

several sites in northern Nevada where leopard frogs, spotted frogs (Fig. 17), western toads, chorus frogs, and spadefoots had been abundant earlier in this century. All species except the nonindigenous bullfrog are now difficult or impossible to find (P. Hovingh, University of Utah, personal communication).

The Amargosa toad, which is endemic to the Mojave Desert, is probably the most imperiled amphibian in this region. Thirty years ago, the estimated population size of this species was several thousand, but now as few as 15 breeding pairs and perhaps fewer than 100 adults may exist (G. Clemmer, Nevada Natural Heritage Program, Carson City, personal communication; K. Hoff, Biology Department, University of Nevada, Las Vegas, personal communication). Habitat loss and degradation—from grazing, off-highway vehicles, development, and introduced predators—are implicated in the decline of the Amargosa toad. The species is currently a candidate for federal listing (Table 5), although conservation at the local level may make such a listing unnecessary (Clemmer, personal communication).

The shortage of quantitative data on amphibian populations in this region is serious. Management of these species is difficult or impossible without data on current distributions, population trends, and species-specific ecological requirements. Amphibian populations may well disappear from the region even before they have been discovered.

Fig. 17. Spotted frogs in the Great Basin.

Mojave Desert Reptiles

The Pet Trade and Mojave Desert Lizards

In 1993, 19 commercial pet collectors reported a harvest of 21,794 reptiles in Nevada at an estimated total value of $250,000 (S. Albert, Nevada Division of Wildlife, personal communication). Collectors took 308 chuckwallas (a species of concern; U.S. Fish and Wildlife Service 1994a) from Nevada in 1993 and may have collected more than 400 in 1994. This volume of collecting may harm chuckwalla populations and populations of other large, long-lived species such as desert iguanas and desert collared lizards. Commercially popular species such as California king snakes, Panamint rattlesnakes, long-nosed leopard lizards, Gila monsters, and Mojave Desert sidewinders are similarly at risk from pet-trade collecting.

States adjacent to Nevada eliminated the commercial take of reptiles for the pet trade through law enforcement. Because baseline data on reptile populations in Nevada are not available and the effect of unlimited harvesting of the state's reptiles has never been quantified, the Nevada Fish and Wildlife Commission has been reluctant to eliminate commercial harvesting. However, preliminary studies on several species of large-bodied lizards in southern Nevada revealed declines of lizard populations in areas that are affected by humans (U.S. Fish and Wildlife Service 1989). Leopard lizard abundances are much lower adjacent to roads and where off-highway vehicles are used than in areas that are removed from such effects.

Desert Tortoise

Declines of desert tortoise populations became a major concern of many biologists in the 1970's. The desert tortoise is a long-lived animal with a low reproduction rate, and population persistence depends on the long-term survival of adults (Fig. 18). High adult mortality seems to be the main cause of the population decline. Many factors are responsible for this mortality: direct take by humans (for example, collection for pets or food, shooting, killing and injuring with motor vehicles); habitat loss, degradation, and fragmentation (for example, from roads, agriculture, and residential development); trampling by livestock; predation by common ravens; diseases; and recent droughts (U.S. Fish and Wildlife Service 1994b).

The reversal of population declines in the desert tortoise is possible only if pressure from these factors is greatly alleviated. Protection of the desert tortoise began in 1980 when it was placed on the federal list of threatened species on the Beaver Dam Slope, Utah. In April 1990, the population in the Mojave Desert was listed as threatened (U.S. Fish and Wildlife Service 1990). Critical habitat was delineated in 1994 (U.S. Fish and Wildlife Service 1994c), following guidelines set by a desert tortoise recovery plan, which was released in the same year (U.S. Fish and Wildlife Service 1994b). The recovery plan recommended six recovery units based on genetic, ecological, and behavioral differences found in the species throughout the Mojave Desert. Several desert wildlife management areas on federal lands in these recovery units were proposed to provide protection for the tortoises.

Fig. 18. A desert tortoise in the Mojave Desert.

More than 150,000 square kilometers of desert lands recently gained some additional protection under the California Desert Protection Act (Public Law 103–433), which created the Mojave Desert National Preserve. Although the preserve includes parts of two major desert tortoise populations, many activities that contribute to declines of desert tortoise populations—such as residential development and agriculture (U.S. Fish and Wildlife Service 1994b)—are to continue in the preserve. On the positive side, however, the Death Valley and Joshua Tree national monuments were expanded and gained full national park status.

In addition, 69 wilderness study areas were designated wilderness areas by the Bureau of Land Management and were thereby afforded additional protection. The same preserves may protect many other desert species whose populations have declined.

Birds

Vulnerable Birds

Although only three of the Great Basin–Mojave Desert region's bird species are accorded official status as endangered or threatened (bald eagle, peregrine falcon, and southwestern willow flycatcher), an additional 20 species that regularly breed here are candidates for federal listing or are considered sensitive species by either the U.S. Fish and Wildlife Service (1994a), the Bureau of Land Management, or the region's state wildlife agencies. The Columbian sharp-tailed grouse historically occurred in the region but has been extirpated in recent decades, and the western yellow-billed cuckoo's population has been reduced to only a few individuals. Most species of concern are jeopardized by the loss and degradation of habitats which they use during breeding, wintering, or migration.

Almost all of the region's bird species depend on wetland and riparian habitats during at least some phase of their annual cycle. Among the 134 species of migratory land birds that breed regularly in the Great Basin, more than half are associated primarily with riparian habitats (Dobkin 1998). Throughout the arid and semiarid West, an extraordinary diversity of bird species depends on these habitats (Carothers et al. 1974; Knopf et al. 1988a,b; Dobkin 1994a), and degradation and destruction of riparian areas are widely viewed as the most important causes of the decline of land bird populations in western North America (Bock et al. 1993; DeSante and George 1994; Ohmart 1994). Restoration of riparian habitats must become a top priority for land management agencies, many of which are now aware of the importance of protecting and restoring these habitats for birds and other wildlife (Thomas et al. 1979; Warner and Hendrix 1984; Johnson et al. 1985). However, information about the response of bird populations to riparian habitat restoration is scarce (Krueper 1993; Ohmart 1994).

In conjunction with the cessation of livestock grazing on the Hart Mountain and Sheldon national wildlife refuges in the northwestern Great Basin, a long-term study was initiated in 1991 to examine the relation of bird populations to riparian habitat recovery (Dobkin 1994b; Dobkin et al. 1995b). Although many species used riparian habitats during the breeding season (96 at Hart Mountain, 86 at Sheldon), most were represented by only a few individuals. The ten most abundant species accounted for more than half of the total abundance of breeding birds in riparian habitats on both refuges (Dobkin 1994b). The overall composition of birds in riparian habitats was characterized by low species richness and disproportionate representation of a small number of abundant, widespread species. The riparian habitats of the Sheldon National Wildlife Refuge (elevation 1,335–1,855 meters) supported less diverse avifaunas than comparable habitats on Hart Mountain (elevation 1,750–2,250 meters) and were quintessential examples of degraded, lower-elevation Great Basin riparian communities that are numerically dominated by blackbirds. Long-term studies are needed to determine whether recovery of quality riparian bird communities will be as slow and unpredictable in degraded Great Basin habitats as suggested thus far (Dobkin 1994b, 1998).

Nonriparian, shrub-dominated habitats support nearly 20% of the migratory land birds that breed in the Great Basin, and the number of breeding bird species there is second only to that in riparian habitats (Dobkin 1998). In addition, resident species of conservation importance, such as the sage grouse (Dobkin 1995), are linked inextricably to sagebrush-dominated steppe. Although abundances of migratory shrub–steppe birds often vary greatly in time and space (Wiens and Rotenberry 1981; Wiens 1989), DeSante and George (1994) detected widespread population declines of grassland and shrub–steppe birds throughout the western United States and attributed these declines to destruction of grasslands and livestock overgrazing in shrub–steppe habitats (Fleischner 1994). As indicated by Bock et al. (1993), almost all shrub–steppe-nesting birds are probably harmed by livestock grazing. The extreme modification of vegetation structure and plant species composition from overgrazing creates communities that are poor in both plant and bird species. To understand the dynamics of shrub–steppe birds, restored examples are urgently needed of the shrub–steppe communities that existed before livestock were introduced into the region. Such communities are dominated by native perennial grasses and forbs and have only moderate shrub densities (Bock et al. 1993; Dobkin 1995).

Population declines of many bird species that nest in forests in eastern North America have been linked to fragmentation of forest and woodland habitats (Robbins et al. 1989; Terborgh 1989; Ehrlich et al. 1992; Bohning-Gaese et al. 1993; Rich et al. 1994), but few studies have been conducted on the effects of habitat fragmentation on western birds (Dobkin 1994a). Habitat fragmentation, degradation, and loss are threats to birds in the riparian and shrub–steppe communities in the Great Basin. The distribution patterns of many riparian species are area-dependent, and many such species are lost from smaller riparian fragments, such as the Toiyabe Mountains of Nevada (Dobkin and Wilcox 1986). In shrub–steppe habitats of the Snake River Plains in southern Idaho, species such as the sage thrasher, Brewer's sparrow, and sage sparrow select nesting territories based on landscape-scale patterns of size and distribution of shrub habitat patches (Knick and Rotenberry 1995). Fragmentation of native shrub–steppe by major wildfires that now let nonindigenous plant species invade may be an important factor in the decline of bird populations of the shrub–steppe and grasslands in the region. The potential influence of habitat fragmentation in coniferous woodlands and forests of the Great Basin has not been examined.

The shrinking and disturbance of suitable habitat also present problems for many birds during migration. On their annual migration, large numbers of birds of prey and many shorebird species move through the Great Basin. The larger saline and alkaline lakes of the region are critical stopover points for waterbirds such as the eared grebe, Wilson's phalarope, and red-necked phalarope (Jehl 1994). The large concentrations of these and other birds depend on these lakes for rest and food and would be seriously threatened by the shrinkage or human-caused disturbance of this habitat.

Neotropical Migrant Birds

Recent interest in the management of Neotropical migrant land birds (Finch and Stangel 1993; Dobkin 1994a) arose when the declining abundance of many of these species in the eastern United States was recognized. Neotropical migrants depend on habitat in their North American nesting areas and on habitat in their wintering areas in the Neotropics, which include the Caribbean, Central America, and South America. In the Great Basin–Mojave Desert region, Neotropical migrants are a heterogeneous group that includes more than half of all nesting bird species.

To assess changes in the abundances of North American birds, the U.S. Fish and Wildlife Service established the Breeding Bird Survey. Annual surveys have been conducted in the western United States since 1968. Data from the survey have been used to determine trends in the population sizes of some western birds (Robbins et al. 1986), but the reliability of the data on most species in the Great Basin–Mojave Desert region is questionable because of limited sampling. To overcome these

limitations, statistical analyses of Breeding Bird Survey data have become increasingly complex (Peterjohn and Sauer 1993; Peterjohn et al. 1994). Population sizes of some bird species in the Great Basin have fluctuated during a 23-year period, but no trend was statistically significant in the recent 10-year period (Fig. 19). However, population sizes and trends in population sizes have not been estimated for all species, and the reliability of the estimated trends for some species is not absolute.

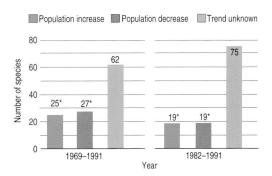

Fig. 19. Long-term population-size trends of 110 species of Neotropical migratory land birds in the Great Basin, based on rankings from population-trend indices developed with Breeding Bird Survey data (Dobkin 1998). Asterisks denote inclusion of species that exhibited upward and downward trends in two different portions of the region. For most species, data are insufficient to establish reliable trends.

Many species with downward trends in population size are associated primarily or exclusively with shrub–steppe or riparian habitats. In the shrub–steppe, such species include the northern harrier, mourning dove, horned lark, loggerhead shrike, green-tailed towhee, vesper sparrow, and sage sparrow. In pinyon–juniper woodlands with a sagebrush and bunchgrass understory, such species are the common nighthawk, northern flicker, gray flycatcher, northern mockingbird, chipping sparrow, and Scott's oriole. Trends in the population sizes of the rock wren, sage thrasher, Brewer's sparrow, black-throated sparrow, and western meadowlark were upward in one portion of the Great Basin and downward in another.

Species with downward population-size trends and associated most closely with riparian habitats include the killdeer, violet-green swallow, warbling vireo, yellow warbler, lazuli bunting, savannah sparrow, song sparrow, yellow-headed blackbird, and Brewer's blackbird. The population-size trend of the red-tailed hawk has been downward for 23 years.

Three riparian species with upward population-size trends are the broad-tailed hummingbird, red-winged blackbird, and Bullock's oriole. Other species with upward population-size trends are the American kestrel, house wren, mountain bluebird, American robin, and black-headed grosbeak—all of which frequently occur in riparian habitats and in pinyon–juniper or mountain-mahogany woodlands—and the northern rough-winged swallow, cliff swallow, and barn swallow, which are aerial insectivores and feed over riparian meadows and other open habitats. More species with upward population-size trends are the brown-headed cowbird, a nest parasite of many Neotropical migrants, and several wide-ranging birds of prey such as the turkey vulture, Swainson's hawk, golden eagle, and prairie falcon. The lark sparrow was the only songbird species with an upward population-size trend in the shrub–steppe.

Data have been largely insufficient to determine population-size trends of nearly any of the 80 species of Neotropical migrant land birds that occur in the Mojave Desert (Fig. 20). The population sizes of the mourning dove and the loggerhead shrike, whose abundances are declining widely in western North America (DeSante and George 1994), are decreasing in the Great Basin. Upward population-size trends were found for the red-tailed hawk, ash-throated flycatcher, northern mockingbird, and black-throated sparrow.

Fig. 20. Long-term population-size trends of 80 Neotropical migratory land birds in the Mojave Desert, based on rankings from population-size trend indices developed with Breeding Bird Survey data (Dobkin 1998). Because of inadequate data, reliable trends could be determined for only 6 of 80 species.

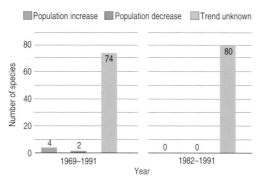

The preponderance of downward population-size trends of birds of the shrub–steppe indicates continuing problems with the health of these communities. As noted elsewhere (DeSante and George 1994), declining abundances of several of these species may also be linked to habitat degradation on wintering grounds in the southwestern United States and northern Mexico. The population-size trends were downward for several riparian specialists and upward for only two riparian specialists (broad-tailed hummingbird and Bullock's oriole). These trends are further evidence of the continuing deterioration of the riparian habitats of the Great Basin.

Although more than 90% of the region's Neotropical migrant land bird species were seen during Breeding Bird Surveys, inadequate sample sizes precluded the determination of population-size trends of almost all species, especially those in the Mojave Desert. Some species cannot be sampled adequately with the standard techniques of the Breeding Bird Survey, regardless of sampling effort. The monitoring of such species requires specialized techniques.

Waterbirds

Because of the tremendous past and continuing loss of wetlands, many waterbird species in the Great Basin–Mojave Desert region should be considered sensitive. Data on waterfowl are available from long-term surveys, but few data are available on other species.

Long-term surveys of waterfowl indicate that none seems to be in immediate danger of extirpation. The only waterfowl species that are considered sensitive in North America and occur in the Great Basin–Mojave Desert region are the trumpeter swan and the canvasback. Swans have a relatively stable, small population (about 25; Alcorn 1988), and canvasback populations seem to be stable.

Common loons and several species of grebes migrate through Nevada. Their primary stopover site, Walker Lake, is in danger of becoming too saline to support fishes, their primary food. The largest breeding colony of American white pelicans in the United States is on Anaho Island in Pyramid Lake, and the size of this population is believed to be decreasing. Survey records from the 1960's and 1970's showed that the population sizes of double-crested cormorants also were declining.

Population-size trends in wading birds are largely unknown. Population sizes of the great blue heron declined in the late 1980's in association with an extended drought, but white-faced ibis populations seem to be increasing. No information is available on population-size trends of two species of concern, the American bittern and the western least bittern.

Surveys of shorebirds in western North America are inadequate, and a determination of the importance of habitats in the Great Basin for many of these species is difficult. Most data are provided by surveys in staging areas, and little is known about species that migrate in small groups. Wetlands of the Great Basin–Mojave Desert provide critical stopover habitat during migration for great numbers of the Wilson's and red-necked phalaropes, long-billed dowitcher, and American avocet, and for smaller numbers of the least and western sandpipers. Significant numbers of the breeding snowy plover, long-billed curlew, American avocet, black-necked stilt, spotted sandpiper, and common snipe occur in the western Great Basin. Of these, the snowy plover is probably of greatest concern. The western snowy plover has been declining in abundance throughout its range, including Nevada (from 878 in 1980 to 349 in 1988 and 139 in 1991; see box on Western Snowy Plovers in California chapter) and in southeastern Oregon. The Franklin's gull and the black tern occur in the Great Basin, and their numbers are declining in many parts of their ranges. No data are available on which to base population-size trends of either species in the region, but both seem to warrant concern.

Mammals

One hundred fifty species of mammals occur in the Great Basin–Mojave Desert region (Zeveloff 1988). Thirty (20%) of these species are either species of concern, candidates for listing as endangered under the Endangered Species Act of 1973, or are already listed as endangered (U.S. Fish and Wildlife Service 1994a,d; Table 6). The Mojave ground squirrel, Tehachapi pocket mouse, and Owens Valley vole are small rodents whose entire geographic ranges are in the Mojave Desert. The Mojave ground squirrel occupies a restricted range in the northwestern Mojave Desert in California (Hafner and Yates 1983), and the Owens Valley vole is restricted to the Owens Valley of southeastern California. The flora and fauna of the northwestern portion of the Mojave Desert were isolated in a wet relict habitat during the last glaciation and consequently became an ecologically unique area (Hafner 1992). This corner of the Mojave Desert borders on the expanding Los Angeles metropolitan area, and encroaching urban development is believed responsible for the declining abundance of the Mojave ground squirrel. Little is known about the ecology, the current population size, or the distribution of the Owens Valley vole.

The Sierra Nevada red fox, Pacific fisher, California wolverine, and Sierra Nevada snowshoe hare are montane mammals whose distributions in the Great Basin are confined to the east slope of the Sierra Nevada. The declining abundances of the Pacific fisher and California wolverine are attributed to trapping in the late nineteenth and early twentieth centuries (Ingles 1947; Hall 1981; Jameson and Peeters 1988; R. M. Nowak 1991). Humans perceived the wolverine as a nuisance and eliminated the species (Ingles 1947; R. M. Nowak 1991).

The California bighorn sheep is a species of concern (U.S. Fish and Wildlife Service 1994a). The historical distribution of this subspecies extended from northeastern California into northern Nevada and eastern Oregon (Hall 1981), but by the early part of the twentieth century, sheep populations were declining because of hunting (Zeveloff 1988). Bighorn sheep were reintroduced into the Great Basin by Nevada and Oregon wildlife agencies in an attempt to reestablish them as a game species. California bighorn sheep from populations in British Columbia are thriving at the Hart Mountain National Wildlife Refuge where the current population exceeds 450 individuals. Forty-two

Table 6. Mammal species of concern, previously treated as Category 2 status of the U.S. Fish and Wildlife Service, in the Great Basin–Mojave Desert region. "States present" refers to the occurrence of the species in this region.

Species	States present
Amargosa vole	California
Pygmy rabbit	Nevada, Oregon, Idaho, Utah
Spotted bat	Nevada, Idaho, Utah, California, Arizona
Greater western mastiff-bat	Nevada, California, Arizona
Palmer's chipmunk	Nevada
Hidden Forest Uinta chipmunk	Nevada
North American lynx	Nevada, Oregon, California
North American wolverine	Nevada, Idaho, Utah, California
California wolverine	Nevada, California
Allen's big-eared bat	Nevada, Utah, California, Arizona
Sierra Nevada snowshoe hare	Nevada, California
Southwestern otter	Nevada, Utah, Arizona
California leaf-nosed bat	California?, Nevada?
Desert Valley kangaroo mouse	Nevada
Fletcher dark kangaroo mouse	Nevada
Pahranagat Valley montane vole	Nevada
Ash Meadows montane vole	Nevada
Small-footed myotis	Nevada?
Long-eared myotis	Nevada, California, Utah, Arizona
Fringed myotis	Nevada, Oregon, Idaho, California
Cave myotis	Nevada, Arizona
Long-legged myotis	Nevada, Oregon, Idaho, California, Utah, Arizona
Yuma myotis	Nevada, Oregon, Idaho, California, Utah, Arizona
Big free-tailed bat	Nevada, California, Utah, Arizona
Pale Townsend's big-eared bat	Nevada, Oregon, Idaho, Utah, Arizona
Preble's shrew	Nevada, Oregon, California
Fish Spring pocket gopher	Nevada
San Antonio pocket gopher	Nevada
Mojave ground squirrel	California
Owens Valley vole	California
Sierra Nevada red fox	California
Pacific fisher	California
California bighorn sheep	Nevada, Oregon, California
Tehachapi white-eared pocket mouse	California

hunting tags were issued in 1994 (R. Cole, Hart Mountain National Wildlife Refuge, Oregon, personal communication).

Gaps in Knowledge

Recent research in Nevada revealed significant extensions of the known distributions of 14 of 22 conifer species that are native to the state (Charlet 1996). For example, the range of the western juniper was extended to 43 additional mountain ranges (Fig. 21), and the range of white fir was extended to seven more mountain ranges (Fig. 22). Many of these stands in Nevada were unknown to scientists, although conifers are easy to see on the landscape even from long distances, and their visual prominence is mirrored by their ecological importance wherever they occur. If not even the locations of the trees are known, how much can be known about other floral and faunal distributions and community processes of the region? Conifer species represent only 0.07% of the more than 3,000 known plant species in Nevada (Kartesz 1988). Among the Nevada flora, 127 plants are listed, are candidates for federal listing, or are species of concern (U.S. Fish and Wildlife Service 1994a,d), but virtually nothing

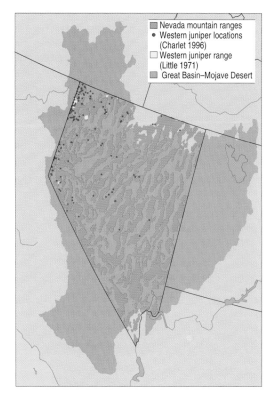

Fig. 21. Distribution of western juniper in Nevada. Red dots indicate known distributions based on collections (Charlet 1996). Yellow polygons are distributions as reported by Little (1971).

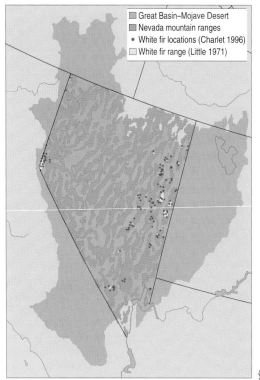

Fig. 22. Distribution of white fir in Nevada. Red dots indicate known distributions based on collections (Charlet 1996). Yellow polygons are distributions as reported by Little (1971).

such information would greatly assist managers of public lands with balancing mandates for the conservation of biological diversity and for multiple use by humans.

Habitats

Caves

Most of the 300 described caves in the Great Basin are in limestone mountains (McLane 1975; Fiero 1986). Anthropological information has been a major focus of cave explorations in the region because of the long history of cave use by Native Americans (Thomas 1985). Woodrat middens, in which many important fossils have been found, are an additional rich resource in cave habitats throughout the region (Betancourt et al. 1990; C. L. Nowak 1991; Wigand and Nowak 1992; Nowak et al. 1994).

The Lehman Caves in Nevada and the Timpanagos Cave National Monument in Utah were developed for tourism (Fiero 1986). The Lehman Caves receive 46,000 visitors yearly, and changes in the cave community resulting from such heavy use are a source of concern (Great Basin National Park 1988). Invertebrates in caves are rarely surveyed (Desert Research Institute 1968), so it is difficult to determine if any invertebrate groups are at risk. Large colonies of bats in caves were killed or extirpated when the caves were mined for guano; bats that hibernate or raise young in caves are especially susceptible to disturbance (Pierson and Brown 1992; Dobkin et al. 1995a; Fig. 23).

Mines

Tunnel mines that were excavated since the mid-1800's provide potential roosting sites for 19 of the 22 bat species in the region (Burt and Grossenheider 1976; Pierson et al. 1991). In Nevada, most of the more than 50,000 known inactive mines are unsuitable roosts, although a few may house significant colonies. Inactive mines in California provide most known colony sites for the cave myotis and Townsend's big-eared bat. This situation is probably mirrored in

is known about them except that they are probably rare. The basic information about plant distributions is skeletal at best, yet plants are the primary producers in communities, and all organisms ultimately depend on them. Information about the ecology and evolution of most organisms has not been gathered, although

© D. A. Charlet, University of Nevada

Fig. 23. Long-eared myotis, a species of concern, at Emerald Lake, Jarbidge Mountains, Nevada.

the Great Basin (Pierson et al. 1991), although these species make extensive use of natural caves in forested lava flows adjacent to the region in Oregon and Idaho (Genter 1986; Dobkin et al. 1995b). The loss of mines as colony sites for rearing young and for hibernation is magnified by the degradation of natural roosts in caves through human disturbances such as recreation or guano mining (Pierson and Brown 1992). In addition, cyanide ponds used for gold extraction at active gold mines have been implicated in the deaths of many bats who drink the contaminated water (Clark and Hothem 1991).

The Abandoned Mines Program of the Nevada Division of Minerals identifies and secures abandoned mines that pose a hazard to human safety. Approximately half of the 3,000 mines that were closed in Nevada since 1987 were not inspected for wildlife and became unusable to bats and other wildlife after they were filled (D. Driesner, Bureau of Abandoned Mine Lands, Nevada Division of Minerals, Reno, personal communication). Since October 1993, however, mines have instead been closed by fencing and signs where possible, to retain wildlife habitat. Education of land management agencies about the use of inactive mines by bats in Nevada increased the frequency of bat surveys and proactive management. The Bureau of Land Management entered a cooperative agreement with the Nevada Division of Mines about management of orphaned mine sites on Bureau of Land Management land (Bureau of Land Management 1994). Population and natural history data of the region's bats will become increasingly important as managers focus on the 11 bat species in the region that recently became species of concern (Burt and Grossenheider 1976; U.S. Fish and Wildlife Service 1994a).

Bogs

Although eagles were believed to require only the water they obtained from their prey (Brown and Amadon 1968), Charlet and Rust (1991) observed golden eagles drinking, bathing, and preening at high mountain bogs and springs in the northern Great Basin (Fig. 24). Five eagles at one time were seen on the ground on three occasions (Charlet and Rust 1991), and more than 45 landings at one bog occurred in an 8-hour period (Charlet, unpublished data). To date, golden eagles have been observed drinking at high mountain springs in 15 mountain ranges in the northern and central Great Basin (Charlet and Rust, unpublished data). Several of these bogs are in areas that attract many visitors or are claimed for mining. Trails pass near some of the bogs, and people camp at others to use the water. Golden eagles

© D. A. Charlet, University of Nevada

Fig. 24. Golden eagle leaving a bog after drinking and bathing at the headwaters of the Humboldt River, Jarbidge Mountains, Nevada.

will not come to the sites if they detect humans (Charlet and Rust 1991); however, it is unclear how these birds are affected when they are displaced by humans.

Sand Dunes

Despite the uniqueness of the flora and fauna of sand dunes across the Great Basin–Mojave Desert region, little is known about these island-like habitats or the organisms that inhabit them. The locations of some dunes in the region are not even portrayed on U.S. Geological Survey maps. Virtually nothing is known about the spatial dynamics of these dunes and how these dynamics affect the biological communities. Biological inventories exist only from the most heavily used dunes, often from only one survey (Rust 1994). Popular dunes in the region (for example, Sand Mountain and Nevada and Little Sahara dunes, Utah) may receive several thousand visitors during one holiday weekend, but the ecological effects of this heavy recreational use have been assessed in only a few cases. Hardy and Andrews (1979) and Luckenbach and Bury (1983) linked the effects of off-highway vehicles to the loss of vertebrate and invertebrate species richness, a reduction in vertebrate and invertebrate populations, and a disruption of mating behaviors in insects that depend on dune-margin vegetation for mating habitat. More information is needed on the effects of off-highway vehicles on the dune biota and on how these effects can be minimized or mitigated. Nearly all sand dunes in the region are on land that is administered by the Bureau of Land Management, and this agency should take the lead on conserving the unique biota.

A good start to conservation of sand dune biota was made with the Sand Mountain pallid blue butterfly. This species was originally discovered in a small patch of Kearney buckwheat—the plant on which its caterpillars feed—near the parking lot at the base of Sand Mountain, Nevada. Sand Mountain is a large dune used extensively for recreational use of off-highway vehicles and was officially designated by the Bureau of Land Management for that purpose. Because the distribution of the butterfly at Sand Mountain was not known and because trampling and traffic seemed to threaten the species, it became a species of concern (U.S. Fish and Wildlife Service 1994a).

When biologists visited the site in summer 1994, they found that much of the Kearney buckwheat at the parking lot site had been beaten down by vehicles. At this point the options seemed to be fencing off the remaining plants or closing part of this popular recreational area, and recreationists feared the whole dune would then be closed. However, a recreationist overheard the agency biologists and took on the task of finding additional food plants. And indeed he found them—the plants were widely distributed all around the base of the dune. Thus, considerable potential habitat for the butterfly existed in areas that are rarely visited by off-highway vehicles. The question remained whether the butterfly also existed in these areas. Their presence at these sites was verified later by biologists during the butterfly's flight season. The Bureau of Land Management developed a management plan that includes monitoring the butterfly and its food plant and directing off-highway vehicle traffic away from the vegetated dune fringes, which are rarely used for recreation anyway. As long as the off-highway-vehicle community cooperates, the butterfly and its food plant should be secure without a federal listing.

Potential Ecological Crises

Invasion by Nonindigenous Plants

Europeans brought to the Great Basin–Mojave Desert region their farming techniques, plants, and animals; these created different disturbance patterns than the native biota of the region had previously experienced (Mack 1986). The newly disturbed environments created excellent seedbeds for highly invasive species like the Russian thistle, tumble mustard, cheatgrass, halogeton (Holmgren 1972), filaree (Mack 1986), and tamarisk (Robinson 1965).

The total effect of nonindigenous plant species in the region is unknown, but in North America invasions of nonindigenous grasses are most severe in the arid western United States (D'Antonio and Vitousek 1992). In Nevada alone, at least 315 nonindigenous plant species

(about 10% of the state flora) are now established (J. Kartesz, Department of Biology, University of North Carolina at Chapel Hill, personal communication), and their biomass sometimes exceeds that of native species (Hunter 1990). Although most of the effect of nonindigenous plant species in the region occurs in sagebrush–grass, pinyon–juniper, and wetland zones, it is apparent to some extent in every community.

Several highly invasive nonindigenous species are recent introductions, and others, such as the yellow star-thistle, are poised on its edges. The medusahead is considered a particular threat because of its ability to invade undisturbed sites (R. J. Tausch, U.S. Forest Service, Intermountain Research Station, Reno, Nevada, and J. A. Young, Agricultural Research Service, University of Nevada, Reno, personal communication). Red brome continues to spread in the Mojave Desert (Beatley 1966; Hunter 1990, 1991). The barbwire thistle resembles and hybridizes with the Russian thistle, thus the barbwire thistle's introduction and dispersal during the past 20 years went virtually unnoticed until recently (J. A. Young, U.S. Forest Service Intermountain Research Station, personal communication).

The invasion of cheatgrass caused drastic changes in fire regimes and secondary plant succession (Piemeisel 1951) across at least a million hectares. Species composition (Piemeisel 1951) and competitive interactions in the affected plant communities are greatly modified by this invasion (Melgoza et al. 1990; D'Antonio and Vitousek 1992). Because of the rapid accumulation of flammable fuels from stalks, cheatgrass-infested areas experience more frequent fires (Pickford 1932; Whisenant 1990) and a longer fire season (Roberts 1990) than areas without grass or with sparse grass cover. The pattern of fire damage was altered: smaller, patchier burns that were typical before European settlement were replaced by fires that tend to destroy vegetation over large, more continuous stretches of land (Whisenant 1990; D'Antonio and Vitousek 1992). When fire-intolerant native plant species are burned, the resident fungi necessary for their establishment and growth also are lost, favoring the establishment of nonindigenous species such as the Russian thistle, tumble mustard (Goodwin 1992), and cheatgrass (Yensen 1981).

Although some areas that are dominated by cheatgrass can be reclaimed by native species if left undisturbed (Hironka and Tisdale 1963), the vulnerability of these sites to burning and susceptibility to midsummer erosion nearly ensures the continued persistence of cheatgrass (Yensen 1981). Thus, the naturalization of cheatgrass in sagebrush–grass communities is a

Human-Induced Changes in the Mojave and Colorado Desert Ecosystems: Recovery and Restoration Potential

Large parts of the Mojave and Colorado desert (see chapter on Southwest) ecosystems have been affected by humans and their activities, especially in the last 100 years. Urbanization, agriculture, off-highway vehicle use, construction of roads and utility corridors, livestock grazing, and military training activities have all created measurable changes in the structure and stability of the ecosystem. Air pollution, although difficult to measure accurately (Reible et al. 1982), has noticeably damaged plants in the Mojave and Colorado deserts (Fisher 1978). The effects of mining have been limited to specific sites in the region. Like all ecosystems, the Mojave and Colorado deserts are resilient to some stresses but not to others. Unfortunately, the weakly defined, thin soil layers that characterize the region are easily disturbed. This vulnerability, coupled with the extreme aridity and temperature ranges of the region, means that conditions necessary for germination and survival of many plants occur only infrequently, and recovery of plant populations to predisturbance conditions is very slow (Table).

Understanding the changes that any system has undergone requires baseline information on conditions that existed before disturbance. The creosotebush-dominated shrub–steppe ecosystem that many recognize as characteristic of the Mojave and Colorado deserts (Burk 1977; Vasek and Barbour 1977) may have existed only for the last 10,000 years or so (Grayson 1993). Before the arrival of Europeans, this vast area was characterized by communities of widely spaced, long-lived, fire-sensitive shrubs, including creosotebush, burro-weed, Joshua-trees, various cacti, and other less familiar associates. Since then, the combined effects of human activities have increased erosion, the destruction of soil-stabilizing microorganisms, soil compaction, fire frequency, habitat destruction and fragmentation, the number of nonindigenous species, and the number of threatened and endangered species.

Much of what is known about ecological succession in the Mojave and Colorado deserts comes from research on recovery from human disturbances in the region. The rate of change in plant communities over time, or succession, is a function of the type of disturbance, its magnitude and frequency, and, to a lesser extent, the type of plant community affected. Short-term, partial recovery of Mojave and Colorado Desert plant communities may take place in a span of 9 to 78 years (Vasek 1979; Prose et al. 1987; Webb et al. 1987), but 1,000 or more years may be required for disturbed sites to recover to predisturbance vegetation structure and species composition (Webb and Newman 1982), though cover of colonizing plants may equal predisturbance cover values within 20–50 years.

Construction of Utility Corridors

In the Mojave and Colorado deserts, the rates of increase and composition of colonizing plant species vary considerably following construction of utility corridors for power lines and pipelines, demonstrating how difficult it is to predict succession relative to adjacent undisturbed areas. Ground cover of short-lived perennial species actually increases in areas of severe disturbance, under the central wires, and along the edges of maintenance roads. After 33 years, there is a noticeable but incomplete recovery of vegetation (Vasek et al. 1975a). Natural revegetation (to 41% ground cover) by long-lived perennials was observed 12 years after construction of a pipeline by trenching, piling, and refilling (Vasek et al. 1975b). Disturbed and control areas appear to have similar cover, biomass, and densities of plants following partial recovery; however, long-lived species are poorly represented on disturbed sites (Lathrop and Archbold 1980a,b).

Impacts of Military Activities

Large areas of the Mojave and Colorado deserts have been, and continue to be, affected by military training activities. The recovery of such areas of the eastern Mojave Desert was studied almost 36 years after the region was first subjected to military activities (Lathrop 1983a). Disturbed areas included tent sites, roads, and tank tracks. All of these areas exhibited reduced plant density and cover relative to control areas. Reductions of cover and density were greatest in tank tracks and least in tent areas. Recovery to predisturbance levels of cover and density varied according to disturbance type. Tent areas showed the greatest recovery, and roadways showed the least, reflecting the intensity of disturbance. Recovery in tank tracks was intermediate. Diversity of

Table. Estimated natural recovery times (years) for California desert plant communities subjected to various human-induced disturbances.

Disturbance	Location	Recovery time (years)	Reference
Tank tracks (military)	Eastern Mojave	65,[a] 76[b]	Lathrop (1983a)
Tent areas (military)	Eastern Mojave	45,[a] 58[b]	Lathrop (1983a)
Dirt roadways (military)	Eastern Mojave	112,[a] 212[b]	Lathrop (1983a)
Tent sites (military)	Eastern Mojave	8–112[c]	Prose and Metzger (1985)
Tent roads (military)	Eastern Mojave	57–440[c]	Prose and Metzger (1985)
Parking lots (military)	Eastern Mojave	35–440[c]	Prose and Metzger (1985)
Main roads (military)	Eastern Mojave	100–infinity[c]	Prose and Metzger (1985)
Military	Eastern Mojave	1,500–3,000[d]	Prose and Metzger (1985)
Town sites	Northern Mojave	80–110,[e] 20–50,[b] 1,000+[f]	Webb and Newman (1982)
Pipeline	Southern Mojave	Centuries[g]	Vasek et al. (1975b)
Power line	Southern Mojave	33[h]	Vasek et al. (1975a)
Fire	Western Colorado Desert	5[b,i]	O'Leary and Minnich (1981)
Off-road vehicle use	Western Mojave	Probably centuries	Webb et al. (1983)
Pipeline (berm and trench)	Mojave Desert	100[j]	Lathrop and Archbold (1980a)
Pipeline (road edge)	Mojave Desert	98[j]	Lathrop and Archbold (1980a)
Power line pylons and road edges	Mojave Desert	100[j]	Lathrop and Archbold (1980a)
Under power line wires	Mojave Desert	20[j]	Lathrop and Archbold (1980a)

[a]Recovery time to reach control density.
[b]Recovery time to reach control cover.
[c]Estimated recovery time for creosotebush to reach control densities.
[d]Estimated recovery time (if at all) to reach original vegetative structure, assuming establishment of control densities.
[e]Compaction recovery time.
[f]Total estimated recovery time.
[g]30–40 years assuming linear rates of succession; 3,000 years until formation of large creosote bush clonal rings.
[h]Incomplete recovery time in areas of high impact.
[i]Time for appearance of perennial seedlings.
[j]Biomass recovery assuming that successional vegetative growth is approximated by a straight line. Recovery of long-lived species is estimated to take at least three times longer than indicated.

dominant perennials also varied between disturbed and nondisturbed areas, but results were clouded by low species richness at the study sites and few individuals of subdominant species. However, diversity in disturbed transects at the Camp Ibis study site was low relative to control sites. The more intense or frequent the use of the site, the less similar the species composition was to that of undisturbed control sites.

Overall, recovery of plant density is slow relative to increases in cover: the number of individuals present at a site changes little following recovery from disturbance, but surviving individuals cover larger areas. Lathrop (1983a,b) concluded that recovery of the disturbed eastern Mojave sites to some original level of community composition and stability may not occur in the foreseeable future. Similar observations and conclusions were reached by Prose and Metzger (1985) and Prose et al. (1987) at abandoned military camps in the eastern Mojave. Long-lived species such as creosotebush were dominant in all control areas, but their cover and density were reduced in disturbed areas. Dominant plants in disturbed areas included pioneer species such as burro-weed and burrobrush. Ground cover by pioneer species in disturbed areas was equal to or greater than their cover in control sites.

Differences in vegetative structure between control and disturbed plots were due to soil compaction, removal of the top layer of soil, and alteration of drainage channel density at the military sites (Prose et al. 1987). Penetrometer measurements show that the compaction caused by a single pass by a "medium" tank can increase average resistance values in the upper 20 centimeters of soil by 50% relative to adjacent untracked soil; values of up to 73% were recorded. Dirt roadways could not be penetrated below 5–10 centimeters because of extreme compaction. Physical modifications to the soil beneath tank tracks extended to a depth of 25 centimeters and outward from the track edge to 50 centimeters (Prose 1985).

Effects of Off-Highway Vehicles

Off-highway vehicles have also disturbed large areas of the Mojave and Colorado deserts. The effects of these vehicles have been well documented and include destruction of soil stabilizers, soil compaction, increased wind and water erosion, noise, decreased abundance of lizard populations (Busack and Bury 1974), and destruction of vegetation (Webb and Wilshire 1983). Susceptibility of soils to

damage is generally high in all areas except barren sand dunes (but see Bury and Luckenbach 1983) and the clay flats of playas (Dregne 1983). Soil damage caused by off-highway vehicles is environmentally significant because desert soils may take 10,000 years to develop. From this estimate, Dregne (1983) concluded that it was futile to speak of disturbed soil recovery in time periods related to human occupancy of the affected areas (Fig. 1).

A major effect of off-highway vehicles is the physical destruction of plants. Plants are destroyed when their stems and foliage, root systems, and germinating seeds are crushed. Lathrop (1983b) examined aerial photographs of nine disturbed and undisturbed areas in the Mojave Desert to assess the effects of off-highway vehicle use. Perennial plant density and cover were dramatically reduced in areas disturbed by vehicles, and total plant cover and density were less than 15% of that in three undisturbed control sites examined.

Weeds and Fire

Like the rest of the western United States, the Mojave and Colorado deserts have been hit hard in the last century by invasive nonindigenous plants. Nonindigenous annual grasses have become the dominant understory plants in areas formerly occupied by native perennial grasses (D'Antonio and Vitousek 1992). Large areas of the Mojave and Colorado deserts are infested with Mediterranean grasses, cheatgrass, and other exotics (Beatley 1969; Bowers 1987; Hunter 1991).

The proliferation of nonindigenous annual plants has dramatically increased the fuel load and frequency of fires in parts of the Colorado Desert in recent years (O'Leary and Minnich 1981; Brown and Minnich 1986). The frequency of fires in the Colorado Desert of California is further enhanced by the proximity of previously burned areas (Chou et al. 1990). Native perennial shrubs are poorly adapted to fire, as evident in their low rates of recovery. In the upper Coachella Valley on the east scarp of the San Jacinto Mountains near Palm Springs, California, burned creosotebush scrub is replaced by open stands of brittlebush, native ephemerals, and nonindigenous annual grasses (Brown and Minnich 1986).

Although fire had a role in the evolution of the desert plant community, it was probably minor, with limited effect and long intervals between fires. With the invasion of species that serve as fine fuels, like nonindigenous annual grasses, the fire cycle has been significantly shortened and fires are more likely to spread. The result has been the conversion of desert scrub landscapes to "weedscapes" dominated by nonindigenous plants.

Livestock Grazing

The effects of livestock grazing in the Mojave and Colorado deserts, while controversial, have been locally significant (General Accounting Office 1992). No published studies have yet fully documented the impact of grazing by livestock or estimated the time required by heavily grazed areas of the desert to recover to

Courtesy Bureau of Land Management

Fig. 1. Large areas of the Mojave and Colorado desert ecosystems have been affected by off-highway vehicles. The road scars shown in this photograph will be clearly visible for decades or longer.

pregrazing levels of plant diversity, density, and cover (Oldemeyer 1994).

Webb and Stielstra (1979) observed that, relative to ungrazed control areas, soils in the Mojave Desert exhibited greater compaction in areas where sheep bedded and grazed. Compaction was greatest in the upper 10 centimeters of soil but was also observed at lower depths. Surface soils trampled by grazing animals lose stabilizers composed of microorganisms, which increases erosion potential (Fig. 2).

The Role of Restoration

Establishment and growth of native plants are naturally slow processes under the extreme conditions of the desert, and disturbance makes these conditions even more severe. Natural recovery is thus extremely slow (Table) and does not necessarily result in communities that resemble predisturbance conditions. Revegetation and restoration can help mitigate many of these negative impacts and speed recovery.

Unfortunately, our ability to restore degraded habitats relies on current technologies that are sharply constrained by the harsh conditions imposed by the desert environment. Furthermore, the costs of large-scale restoration are prohibitive, and the chances of long-term success are low or unknown. Brum et al. (1983) estimated that power line corridors in the Mojave Desert could be restored for $9,221 per hectare by using seeding and irrigation. Estimates of costs involved in restoring degraded land at the Yucca Mountain site in Nevada were much higher, ranging from $73,969 to $115,754 per hectare, depending on the nature of the disturbance (Malone 1991). However, recent advances in desert restoration technology offer hope for the success of localized restoration efforts (Bainbridge and Virginia 1990; Bainbridge et al. 1995). Given the sensitivity of desert habitats and their slow rate of natural recovery, the best management option is to limit the extent and intensity of disturbances as much as possible.

See end of chapter for references

Author

Jeff Lovich
U.S. Geological Survey
Biological Resources Division
Department of Biology
University of California
Riverside, California 92521-0427

Courtesy J. Lovich, USGS

Fig. 2. Grazing can have locally significant effects in the Mojave Desert, particularly around watering tanks. Notice the almost complete absence of perennial plants in the immediate vicinity of the tank. Soil compaction in these areas is very high relative to undisturbed areas.

"catastrophic change to a new system" (Turner et al. 1993:223). The conversion of sagebrush–steppe to annual grasslands makes these rangelands virtually worthless (Yensen 1981; Morrow and Stahlman 1984; Roberts 1990), and the cost of rehabilitation is often higher than the value of the land on the open market (Roberts 1990).

Degradation of Springs

Isolation and small size render many spring communities in the Great Basin–Mojave Desert region particularly vulnerable to disturbance and loss. Groundwater pumping in Mojave Desert areas, such as Pahrump Valley, caused the drying of springs, complete loss of habitat, and extinction of subspecies of native fishes. Pumping in Ash Meadows nearly led to the loss of springs in the 1970's, but the first intervention of the Endangered Species Act of 1973 prevented these losses (Soltz and Naiman 1978). Continuing development of hot springs for electric power in the Great Basin also poses

questions about the persistence of spring habitats. Presently, pumping of groundwater from gold mines is one of the greatest threats to spring communities. In the north-central region of Nevada in particular, large open-pit gold mines are rapidly altering groundwater conditions in many areas, and many spring communities there are at risk. Although relatively few fish populations may be lost by these practices, the invertebrate faunas of the affected areas are poorly known, thus the effects on these organisms cannot be determined.

On a smaller scale, the continuing development of springs for livestock by ranchers and state and federal agencies also poses a threat to the continued existence of spring biota. These actions typically involve fencing of the area immediately adjacent to springs, placing a springbox over the water source, and piping most or all of the water off the site into livestock tanks. Although some of the riparian vegetation may be retained with such practices, the essential flowing character of the spring is lost, and often no exposed water remains on the surface. Populations of endemic species of springsnails have been lost under such circumstances (G. L. Vingard, University of Nevada, Reno, personal observation), and the consequences for other invertebrates must also be severe.

Livestock grazing continues to pose another serious threat to spring communities. Heavy trampling by livestock often reduces the substrate to mud, can completely eliminate riparian vegetation, and alters flow characteristics. Although the magnitude of these effects on the spring biota is largely unknown, it is probably great because of the complete alteration of the vegetation and substrate structure.

In springs throughout the region, introductions of nonindigenous organisms—particularly fishes, snails, crayfishes, and frogs—also have had adverse effects. Fish species have been lost in some springs, and changes in the invertebrate fauna have been substantial in others (Soltz and Naiman 1978). It is difficult to assess the magnitude of the effects from introduced nonindigenous species on spring biota because data on native organisms are lacking. Many populations and species will probably be destroyed before their presence has even been documented.

Development in the Las Vegas Valley

Future losses of species and decreased biological diversity attributable to water diversions seem inevitable in light of estimated human population growth throughout the Great Basin–Mojave Desert region and particularly in Clark County, southern Nevada. Water supplies in Clark County today include the Colorado River (300,000 acre-feet or 365.9 million cubic meters per year), groundwater in Las Vegas Valley (35,000 acre-feet or 42.7 million cubic meters per year net), and wastewater reuse. Forecasts indicate that at current rates of use, existing supplies will meet local needs until the year 2013; however, by the year 2020, the population of Clark County is expected to increase by 63%, from 919,388 in 1993 to an estimated 1,450,409 (Clark County 1994). To meet the water needs of this population, the Las Vegas Valley Water District filed applications to obtain surface water from the Virgin River and groundwater diversions from approximately 20 basins (Table 7); this includes all of the unappropriated perennial

yield that would otherwise be lost to evaporation (Las Vegas Valley Water District 1992, 1993). Although hydrologic models are used to predict steady-state groundwater flow and to ascertain the effects of groundwater withdrawals, no one knows the level of success reached by predictions of the magnitude of these effects over the long term and on a regional scale.

Developments on the Truckee and Carson Rivers

The Truckee River originates as overflow from Lake Tahoe (Fig. 25) in the Sierra Nevada, flows through the Truckee Meadows (now occupied by the Reno-Sparks metropolitan area), and terminates at Pyramid and Winnemucca lakes. In 1906, water was diverted from the Truckee River for the Newlands Project in the Fallon area, the first project in an effort to make the desert bloom in the United States. While an agricultural economy was created in Fallon, Pyramid Lake dropped 25 meters, the endemic cui-ui became endangered, the Pyramid Lake cutthroat trout went extinct, and Winnemucca Lake completely dried shortly after it was established as a national wildlife refuge for waterbirds. Because of the loss of Winnemucca Lake, the Stillwater National Wildlife Refuge near Fallon became one of the only resources for migrating shorebirds in the area. The Stillwater marshes were naturally fed by the Carson River, but most of those waters are also diverted for irrigation. Added to these complications are the demands on the Truckee River from the rapidly growing Reno–Sparks area. These demands conflict with those of the Pyramid Lake Paiute tribe, the Fallon farmers, and the Stillwater refuge.

Heroic water importation schemes to solve Reno's insatiable thirst have included draining groundwater from central and eastern Nevada mines, sending this water down the Humboldt River, and pumping it to Reno. Pumping groundwater from Honey Lake Valley in northeastern California for delivery to Reno was also proposed. Estimates on the yield of the aquifer were made by using the rate of water use by the greasewood in Honey Lake Valley (L. Crowe, Air Quality Management, County of Washoe, Nevada, personal communication). Water used by greasewood was considered wasted, but this water could support additional development in Reno. Yet, that water supports not only greasewood, but an entire natural community that includes at least six species of scale insects from four families (D. R. Miller, Agricultural Research Service, U.S. Department of Agriculture, Logan, Utah, personal communication) and several

Table 7. Sources and amounts of water currently consumed by Clark County and those sources in the permit application process with the Nevada Division of Water Resources (Las Vegas Valley Water District 1992; Clark County 1994; Nevada Division of Water Resources 1995, personal communication).

Water sources	Acre-feet per year	(Cubic meters per year)
Currently consumed		
Colorado River	300,000	(369,900,000)
Groundwater (from Las Vegas Valley basin)	35,000	(43,155,000)
Total consumed	335,000	(413,055,000)
In permit application		
Virgin River	125,000	(154,130,000)
Groundwater (from approximately 20 basins)	180,000	(221,940,400)
Total in permit application	305,000	(376,070,000)

desert rodents (W. Longland, Agricultural Research Service, Reno, Nevada, personal communication).

Walker River Basin and Walker Lake

The Walker River basin is a medium-sized drainage in eastern California and western Nevada. The eastern and western forks of the Walker River originate on the eastern slope of the Sierra Nevada in California, flow into Nevada via the Smith and Mason valleys, merge, and eventually terminate in Walker Lake (Fig. 26). During the last 100 years or so, upstream water diversions for irrigation created a vigorous agricultural economy in the Smith and Mason valleys, but these diversions also diminished flows into Walker Lake. The lake's level has dropped considerably, and the concentration of total dissolved solids has increased to the extent that the lake will not be able to support fishes much longer (Koch et al. 1979; Horne et al. 1995; Stockwell, unpublished manuscript).

One hundred twenty percent of the average flow in the Walker River is allocated to upstream uses, primarily for agriculture in Smith and Mason valleys (California Department of Water Resources 1992). Thus, a runoff of 120% of normal—or about 420,000 acre-feet (512.2 million cubic meters)—is necessary for the righted allocations of all upstream users. Under drought conditions, flows are about 40%–60% of average, and only negligible amounts of water reached Walker Lake from 1986 to 1994. Under these conditions, the total dissolved solids in the lake will soon reach a level that will shift a fish-dominated community to one dominated by invertebrates (Stockwell, unpublished manuscript). One effect of this shift will be the disappearance of the fish-eating birds that depend on the lake's resources during migration; another will be the loss of a major recreational fishery that is important to the economy of Hawthorne, a town at the southern end of the lake.

The only feasible way to forestall these changes is to increase flows into Walker Lake. If 80,000–90,000 acre-feet (97.6–109.8 million cubic meters) of water were to reach the lake annually, the total dissolved solids would fluctuate around the present level, which is marginal for fish life. An annual inflow of more than 109.8 million cubic meters would result in a gradual reduction of the total dissolved solids (Stockwell, unpublished manuscript). However, providing about another 100,000 acre-feet (122 million cubic meters) of water annually for Walker Lake would require annual flows of 520,000 acre-feet (634.2 million cubic meters), or 157% of average. Because this amount of

water probably will not become available in the basin unless it consistently receives much higher than average precipitation, various water redistribution schemes have been proposed by groups intent on preserving the current Walker Lake community. These schemes include the purchase of water rights from willing sellers, voluntary contributions to instream flow, or more draconian measures such as reallocation. However, the economy of the Smith and Mason valleys would probably suffer if substantial amounts of irrigation water were diverted into the lake.

© D. A. Charlet, University of Nevada

Fig. 25. Lake Tahoe, Nevada and California. Overflow from Lake Tahoe is the primary source of the Truckee River.

© D. A. Charlet, University of Nevada

Fig. 26. Walker Lake, Nevada, viewed from the summit of Mount Grant.

Authors

Peter F. Brussard
David A. Charlet*
Department of Biology
and Biological Resources
Research Center
University of Nevada
Reno, Nevada 89557

David S. Dobkin
High Desert Ecological Institute
15 SW Colorado, Suite 300
Bend, Oregon 97702

*current address:
Community College of
Southern Nevada
Department of Science S2B
3200 East Cheyenne Avenue
North Las Vegas, NV 89030

Contributing authors

Lianne C. Ball
Kathleen A. Bishop
Hugh B. Britten
Erica Fleishman
Scott A. Fleury
Tom Jenni*
Tom B. Kennedy
Christine O. Mullen
Mary M. Peacock
Don Prusso
Michael Reed
Lynn Riley
Richard W. Rust
Janice L. Simpkin
Gary Vinyard
Ulla G. Yandell
Biological Resources Research
Center
University of Nevada
Reno, Nevada 89557

Ron Marlow
Department of Biology
University of Nevada
Reno, Nevada 89557

*(currently employed by Nevada
Natural Heritage Program
Carson City, Nevada)

Furthermore, irrigation in these valleys may have created habitats that also support important elements of biological diversity. Upstream riparian and wetland habitats originally covered only a fraction of the land area in the Smith and Mason valleys, but they expanded considerably under irrigation. At present, modern irrigation practices (such as replacement of the original earthen ditches by concrete-lined ditches) and the recent drought have resulted in the degradation or loss of riparian segments. The extent to which these habitats support important elements of biological diversity must be quantified so that the overall effects of potential water redistribution can be predicted.

A second potential effect of water redistribution on upstream biological diversity may be the conversion of former grazing or agricultural lands into residential areas. If irrigation is no longer possible, the most profitable course for a landowner is to sell the land to a developer. Residential development creates habitat types that are not used by most native species, and pets (particularly house cats) have a substantial negative effect on bird, reptile, and small mammal populations. If the density of houses in a development is sufficiently high, most native animals disappear altogether.

The Walker River basin serves as an example for many others facing resource management dilemmas in the arid West. The land in the Walker River drainage basin has a variety of public and private ownership (Fig. 27), including two states with their respective fish and wildlife agencies, the U.S. Forest Service, the Bureau of Land Management, the Department of Defense, private landowners, and an Indian reservation. Resident organisms include a threatened species (the Lahontan cutthroat trout) and many migratory birds, and all vegetation zones of the Great Basin are represented in the Walker River basin. Will this area become the focus of yet another conflict over land and water use or an example of cooperative ecological restoration?

Conclusions

Biological diversity in the Great Basin and Mojave Desert region is concentrated in remnant waters, montane islands, and specialized habitats. Throughout the region, there is much endemism, mostly at the subspecies but also at the species level. However, the human population in Nevada is growing at one of the fastest rates in the nation (6.7% annually, which represents a doubling time of 10 years), and the portions of the region in adjacent states are growing nearly as fast. Most people who move into the region are from nondesert areas and do not understand the fragile ecology of the desert and the value of its biological heritage.

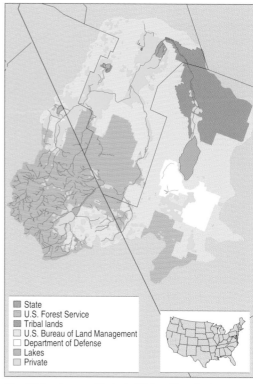

State
U.S. Forest Service
Tribal lands
U.S. Bureau of Land Management
Department of Defense
Lakes
Private

Fig. 27. Land ownership in the Walker River basin, California and Nevada.

In addition to being subjected to the new, severe effects brought on by population growth, the biota of the region already has been severely harmed by water development, mining, grazing, and the introduction of nonindigenous species. Many rare species are at risk in the Great Basin–Mojave Desert region, but even common species are now imperiled by human enterprises. All of these factors profoundly alter community structure, function, and integrity in many ways. In the face of these accelerating changes, it is difficult to be optimistic. Unless major changes are made in the interaction of people with natural communities in this region—one of the last large expanses of wild land in the nation—hope for the retention of the natural character and important ecological role of the Great Basin–Mojave Desert is small.

Acknowledgments

We gratefully acknowledge the assistance of the following individuals and their institutions: G. Clemmer, K. Cooper, and J. Morefield of the Nevada Natural Heritage Program, Carson City; R. Cole, Hart Mountain National Wildlife Refuge, Oregon; S. Bassett, T. Edwards, and C. Homer, U.S. Geological Survey, Utah Cooperative Fish and Wildlife Research Unit, Department of Fisheries and Wildlife, Utah State University, Logan; R. Hamlin, U.S. Fish and Wildlife Service, Nevada

State Office, Reno; J. Kartesz, Department of Biology, University of North Carolina, Chapel Hill; A. Launner and J. Reaser, Center for Conservation Biology, Stanford University, Stanford, California; Q. Ly and M. Rahn, Biological Resources Research Center, University of Nevada, Reno; T. Charlet, Department of Biochemistry, University of Nevada, Reno; K. Geluso, Department of Biology, University of Nevada, Reno; J. Nachlinger, The Nature Conservancy, Northern Nevada Project Office, Reno; P. Wigand and M. Rose, Quaternary Sciences, Desert Research Institute, Reno, Nevada; G. Stephens, Idaho Conservation Data Center, Idaho Department of Fish and Game, Boise; S. Smith, Department of Biology, University of Nevada, Las Vegas; and J. Young, U.S. Department of Agriculture, Agricultural Research Service, Reno, Nevada. The text was improved by the comments of four anonymous reviewers. Research at Hart Mountain and Sheldon National Wildlife Refuges was supported by the U.S. Fish and Wildlife Service and by the High Desert Ecological Research Institute of Bend, Oregon. Research in Nevada was supported by the Nevada Biodiversity Initiative and the Biological Resources Research Center and the Nevada Agricultural Experiment Station, University of Nevada, Reno.

Cited References

Alcorn, J. R. 1988. The birds of Nevada. Fairview West Publishing, Fallon, Nev. 418 pp.

Anderson, B. W., A. Higgins, and R. D. Ohmart. 1977. Avian use of saltcedar communities in the lower Colorado River Valley. Pages 128–136 *in* R. R. Johnson and D. A. Jones, editors. Importance, preservation and management of riparian habitat: a symposium. U.S. Forest Service General Technical Report RM-43.

Anderson, B. W., and K. E. Holte. 1981. Vegetation development over 25 years without grazing on sagebrush-dominated rangeland in southeastern Idaho. Journal of Range Management 34:25–29.

Armour, C. L., D. A. Duff, and W. Elmore. 1991. The effects of livestock grazing on riparian and stream ecosystems. Fisheries 16:7–11.

Austin, G. T. 1985. Lowland riparian butterflies of the Great Basin and associated areas. Journal of Research on the Lepidoptera 24:117–131.

Austin, G. T. 1992. *Cercyonis pegala* (Fabricius) (Nymphalidae: Satyrinae) in the Great Basin: new subspecies and biogeography. Bulletin of the Allyn Museum 135. 59 pp.

Axelrod, D. I. 1950. Evolution of desert vegetation in western North America. Contributions to Paleontology. Carnegie Institute of Washington Publication 590 VI. 306 pp.

Axelrod, D. I. 1976. History of the coniferous forests, California and Nevada. University of California Publications in Botany 70. 62 pp.

Axelrod, D. I. 1983. Paleobotanical history of the western deserts. Pages 113–129 *in* S. G. Wells and D. R. Haragan, editors. Origin and evolution of deserts. University of New Mexico Press, Albuquerque.

Axelrod, D. I., and P. H. Raven. 1985. Origins of the Cordilleran flora. Journal of Biogeography 12:21–47.

Barneby, R. C. 1989. Fabales. Volume 3, part B. *In* A. Cronquist, A. H. Holmgren, N. H. Holmgren, J. L. Reveal, and P. K. Holmgren, editors. Intermountain flora: vascular plants of the Intermountain West, U.S.A. New York Botanical Garden, New York.

Barrows, C. W. 1993. Tamarisk control II: a success study. Restoration and Management Notes 11:35–38.

Beatley, J. 1966. Ecological status of introduced brome grasses (*Bromus* spp.) in desert vegetation of southern Nevada. Ecology 47:548–554.

Beatley, J. C. 1974a. Effects of rainfall and temperature on distribution and behavior of *Larrea tridentata* (creosotebush) in the Mojave Desert of Nevada. Ecology 55:245–261.

Beatley, J. C. 1974b. Phenological events and their environmental triggers in Mojave Desert ecosystems. Ecology 55:856–863.

Beatley, J. C. 1976. Vascular plants of the Nevada Test Site and central-southern Nevada: ecologic and geographic distributions, TID-26881. Energy Research and Development Administration, National Technical Information Service, Springfield, Va.

Behle, W. H. 1978. Avian biogeography of the Great Basin and Intermountain Region. Intermountain biogeography: a symposium. Great Basin Naturalist Memoirs 2:55–80.

Belovsky, G. E. 1987. Extinction models and mammalian persistence. Pages 35–58 *in* M. E. Soulé, editor. Viable populations for conservation. Cambridge University Press, Cambridge, England.

Benson, L. V., D. R. Currey, R. I. Dorn, K. R. Lajoie, C. G. Oviatt, S. W. Robinson, G. I. Smith, and S. Stine. 1990. Chronology of expansion and contraction of four Great Basin lake systems during the past 35,000 years. Palaeogeography, Palaeoclimatology, Palaeoecology 78:241–286.

Betancourt, J. L., T. R. Van Devender, and P. S. Martin. 1990. Introduction. Pages 2–11 *in* J. L. Betancourt, T. R. Van Devender, and P. S. Martin, editors. Packrat middens: 40,000 years of biotic change. University of Arizona Press, Tucson.

Bettinger, R. L. 1991. Native land use: archaeology and anthropology. Pages 463–486 *in* C. A. Hall, Jr. Natural history of the White–Inyo Range, eastern California. University of California Press, Berkeley and Los Angeles.

Billings, W. D. 1949. The shadscale vegetation zone of Nevada and eastern California in relation to climate and soils. American Midland Naturalist 42:87–109.

Billings, W. D. 1950. Vegetation and plant growth as affected by chemically altered rocks in the western Great Basin. Ecology 67:62–74.

Billings, W. D. 1951. Vegetational zonation in the Great Basin of western North America. Les bases écologiques de la régénération de la végétation des zones arides. Pages 101–122 *in* International Union of Biological Sciences, Series B, No. 9.

Billings, W. D. 1954a. Temperature inversions in the pinyon–juniper zone of a Nevada mountain range. Butler University Botanical Studies 11:112–118.

Billings, W. D. 1954b. Nevada trees. University of Nevada, Reno, Agricultural Extension Service Bulletin 94. 125 pp.

Billings, W. D. 1970. Plants and the ecosystem. Wadsworth Publishing Company, Inc., Belmont, Calif. iv + 154 pp.

Billings, W. D. 1978. Alpine phytogeography across the Great Basin. Intermountain biogeography: a symposium. Great Basin Naturalist Memoirs 2:105–117.

Billings, W. D. 1990. The mountain forests of North America and their environments. Pages 47–86 *in* C. B. Osmond, L. F. Pitelka, and G. M. Hidy, editors. Plant biology of the Basin and Range. Springer-Verlag, New York.

Billings, W. D., and A. F. Mark. 1957. Factors involved in the persistence of montane treeless balds. Ecology 38:140–142.

Blackburn, W. H., R. W. Knight, and J. L. Schuster. 1982. Saltcedar influence on sedimentation in the Brazos River. Journal of Soil and Water Conservation 37:298–301.

Blackburn, W. H., and P. Tueller. 1970. Pinyon and juniper invasion in black sagebrush communities in east-central Nevada. Ecology 51:841–848.

Blackwelder, E. 1931. Pleistocene glaciation in the Sierra Nevada and basin ranges. Geological Society of America Bulletin 42:865–922.

Blackwelder, E. 1934. Supplementary notes on Pleistocene glaciation in the Great Basin. Journal of the Washington Academy of Sciences 24:212–222.

Bock, C. E., V. A. Saab, T. D. Rich, and D. S. Dobkin. 1993. Effects of livestock grazing on Neotropical migratory land birds in western North America. Pages 296–309 in D. M. Finch and P. W. Stangel, editors. Status and management of Neotropical migratory birds. U.S. Forest Service General Technical Report RM-229.

Bohning-Gaese, K., M. L. Taper, and J. H. Brown. 1993. Are declines in North American insectivorous songbirds due to causes on the breeding range? Conservation Biology 7:76–86.

Brown, J. H. 1971. Mammals on mountaintops: non-equilibrium insular biogeography. American Naturalist 105:467–478.

Brown, J. H. 1973. Species diversity of seed-eating rodents in sand dune habitats. Ecology 54:775–787.

Brown, J. H. 1978. The theory of insular biogeography and the distribution of boreal birds and mammals. Great Basin Naturalist Memoirs 2:209–227.

Brown, L. H., and D. Amadon. 1968. Eagles, hawks, and falcons of the world. McGraw-Hill, New York. 945 pp.

Brussard, P. F., and G. A. Austin. 1993. Nevada butterflies: check list and ecological distribution. Biological Resources Research Center, Reno. 5 pp.

Bunting, S. C. 1986. Use of prescribed burning in juniper and pinyon–juniper woodlands. Pages 141–144 in R. L. Everett, editor. Proceedings: pinyon–juniper conference. U.S. Forest Service General Technical Report INT-215.

Bureau of Land Management. 1994. Cooperative agreement with the Nevada Division of Minerals on Securing Mine Safety Standards. Instruction Memorandum NV-95-022. 14 pp.

Burkhardt, J. W., and E. W. Tisdale. 1976. Causes of juniper invasion in southwestern Idaho. Ecology 76:472–484.

Burt, W. H., and R. P. Grossenheider. 1976. Field guide to the mammals. Peterson field guide series 5. Houghton Mifflin Company, Boston, Mass. xvii + 289 pp.

Busch, D. E., and S. D. Smith. 1993. Effects of fire on water and salinity relations of riparian woody taxa. Oecologia 94:186–194.

Busch, D. E., and S. D. Smith. 1995. Mechanisms associated with the decline of woody species in riparian ecosystems

of the southwestern United States. Ecological Monographs 65(3):347–370.

California Department of Water Resources. 1992. Walker River atlas. State of California, The Resources Agency, Sacramento. ix + 99 pp.

Callison, J., and J. D. Brotherson. 1985. Habitat relationships of the blackbrush community (*Coleogyne ramosissima*) of southwestern Utah. Great Basin Naturalist 45:321–326.

Carothers, S. W., R. R. Johnson, and S. W. Aitchison. 1974. Population structure and social organization of southwestern riparian birds. American Zoologist 14:97–108.

Chandler, J. R. 1970. A biological approach to water quality management. Water Pollution Control 4:415–422.

Chaney, E., W. Elmore, and W. S. Platts. 1993. Livestock grazing on western riparian areas. U.S. Government Printing Office, Washington, D.C. 45 pp.

Chapman, H. C. 1966. The mosquitoes of Nevada. Entomology Research Division, Agriculture Research Service, U.S. Department of Agriculture and the Max C. Fleischmann College of Agriculture, University of Nevada, Reno. 41 pp.

Charlet, D. A. 1991. Relationships of the Great Basin alpine flora: a quantitative analysis. M.S. thesis, University of Nevada, Reno.

Charlet, D. A. 1996. Atlas of Nevada conifers: a phytogeographic reference. University of Nevada Press, Reno. xiv + 320 pp.

Charlet, D. A. 1997. Floristics of the Mt. Jefferson alpine flora. In D. H. Thomas, editor. The archaeology of Monitor Valley. 4. Alta Toquima and the Mount Jefferson Complex. American Museum of Natural History Anthropological Papers, New York. In press.

Charlet, D. A., and R. W. Rust. 1991. Visitation of high mountain bogs by golden eagles in the northern Great Basin. Journal of Field Ornithology 62:46–52.

Clark County. 1994. Clark County draft Desert Conservation Plan, August 1994, Clark County, Nevada. 123 pp. + appendixes.

Clark, D. R., Jr., and R. L. Hothem. 1991. Mammal mortality at Arizona, California, and Nevada gold mines using cyanide extraction. California Fish and Game 77:61–69.

Critchfield, W. B. 1984a. Crossability and relationships of Washoe pine. Madroño 31:144–170.

Critchfield, W. B. 1984b. Impact of the Pleistocene on the genetic structure of North American conifers. Pages 70–118 in R. L. Lanner, editor. Proceedings of the Eighth North American Forest Biology Workshop, 30 July–1 August 1984, Utah State University, Logan.

Critchfield, W. B., and G. L. Allenbaugh. 1969. The distribution of Pinaceae in and near northern Nevada. Madroño 20:12–26.

Dahl, T. E., and C. E. Johnson. 1991. Status and trends of wetlands in the conterminous United States, mid-1970's to mid-

1980's. U.S. Fish and Wildlife Service. 28 pp.

D'Antonio, C. M., and P. M. Vitousek. 1992. Biological invasions by exotic grasses, the grass/fire cycle, and global change. Annual Review of Ecology and Systematics 23:63–87.

d'Azevedo, W. 1986. Handbook of North American Indians, Volume 11: Great Basin. Smithsonian Institution, Washington, D.C. 852 pp.

Deacon, J. E. 1979. Endangered and threatened fishes of the West. Great Basin Naturalist Memoirs 3:41–64.

Deacon, J. E., and C. D. Williams. 1991. Ash Meadows and the legacy of the Devils Hole pupfish. Pages 69–87 in W. L. Minckley and J. E. Deacon, editors. Battle against extinction. The University of Arizona Press, Tucson.

DeLucia, E. H., and W. H. Schlesinger. 1990. Ecophysiology of Great Basin and Sierra Nevada vegetation on contrasting soils. Pages 143–178 in C. B. Osmond, L. F. Pitelka, and G. M. Hidy, editors. Plant biology of the Basin and Range. Springer-Verlag, Berlin, Germany.

DeLucia, E. H., W. H. Schlesinger, and W. D. Billings. 1988. Water relations and the maintenance of Sierran conifers on hydrothermally altered rock. Ecology 69:303–311.

Demarais, B. D., T. E. Dowling, and W. L. Minckley. 1993. Post-perturbation genetic changes in populations of endangered Virgin River chubs. Conservation Biology 7:334–341.

DeSante, D. F., and T. L. George. 1994. Population trends in the land birds of western North America. Pages 173–190 in J. R. Jehl, Jr., and N. K. Johnson, editors. A century of avifaunal change in western North America. Studies in Avian Biology, Volume 15.

Desert Research Institute. 1968. Final reports on the Lehman Caves studies to the Department of the Interior, National Park Service, Lehman Caves National Monument. The Laboratory of Desert Biology, Desert Research Institute, Reno, Nev. 57 pp.

Dobkin, D. S. 1985. Heterogeneity of tropical floral microclimates and the response of hummingbird flower mites. Ecology 66:536–543.

Dobkin, D. S. 1994a. Conservation and management of Neotropical migrant land birds in the northern Rockies and Great Plains. University of Idaho Press, Moscow. 220 pp.

Dobkin, D. S. 1994b. Community composition and habitat affinities of riparian birds on the Sheldon-Hart Mountain refuges, Nevada and Oregon, 1991–93. Final report. U.S. Fish and Wildlife Service, Lakeview, Oreg. 287 pp.

Dobkin, D. S. 1995. Management and conservation of sage grouse, denominative species for the ecological health of shrub–steppe ecosystems. Technical note. Bureau of Land Management, Portland, Oreg. 26 pp.

Dobkin, D. S. 1998. Conservation and management of Neotropical migrant land birds in the Great Basin. University of Idaho Press, Moscow. In press.

Dobkin, D. S., A. C. Rich, J. A. Pretare, and W. H. Pyle. 1995a. Nest-site relationships among cavity-nesting birds in riparian and snowpocket aspen woodlands in the northwestern Great Basin. Condor 97:694–707.

Dobkin, D. S., R. D. Gettinger, and M. G. Gerdes. 1995b. Springtime movements, roost use, and foraging activities of Townsend's big-eared bat (*Plecotus townsendii*) in central Oregon. Great Basin Naturalist 55:315–321.

Dobkin, D. S., I. Olivieri, and P. R. Ehrlich. 1987. Rainfall and the interaction of microclimate with larval resources in the population dynamics of checkerspot butterflies (*Euphydryas editha*) inhabiting serpentine grassland. Oecologia 71:161–166.

Dobkin, D. S., and B. A. Wilcox. 1986. Analysis of natural forest fragments: riparian birds in the Toiyabe Mountains, Nevada. Pages 293–299 *in* J. Verner, M. L. Morrison, and C. J. Ralph, editors. Wildlife 2000: modeling habitat relationships of terrestrial vertebrates. University of Wisconsin Press, Madison.

Dohrenwend, J. C., W. B. Bull, L. D. McFadden, G. I. Smith, R. S. U. Smith, and S. G. Wells. 1991. Quaternary geology of the Great Basin. Pages 321–352 *in* R. B. Morrison, editor. Quaternary nonglacial geology: conterminous U.S. Volume K-2. Geological Society of America, Boulder, Colo.

Ehrlich, P. R., D. S. Dobkin, and D. Wheye. 1992. Birds in jeopardy: the imperiled and extinct birds of the United States and Canada including Hawaii and Puerto Rico. Stanford University Press, Stanford, Calif. x + 259 pp.

Ehrlich, P. R., and D. D. Murphy. 1987. Monitoring populations on remnants of native vegetation. Pages 201–210 *in* D. A. Saunders, G. W. Arnold, A. A. Burbidge, and A. J. M. Hopkins, editors. Nature conservation: the role of remnants of native vegetation. Surrey Beatty and Sons Pty. Limited, Chipping Norton, New South Wales, Australia.

El-Ghonemy, A. A., A. Wallace, and E. M. Romney. 1980a. Multivariate analysis of the vegetation in a two-desert interface. Great Basin Naturalist Memoirs 4:42–58.

El-Ghonemy, A. A., A. Wallace, and E. M. Romney. 1980b. Socioecological and soil-plant studies of the natural vegetation in the northern Mojave Desert–Great Basin Desert interface. Great Basin Naturalist Memoirs 4:71–86.

Erman, N. A. 1991. Aquatic invertebrates as indicators of biodiversity. Proceedings of the symposium on biodiversity of northwestern California, 28–30 October 1991. Santa Rosa, Calif.

Ertter, B. 1992. Floristic regions of Idaho. Journal of the Idaho Academy of Science 28:57–70.

Fiero, B. 1986. Geology of the Great Basin. Max C. Fleischmann series in Great Basin natural history, University of Nevada Press, Reno. 197 pp.

Finch, D. M., and P. W. Stangel, editors. 1993. Status and management of Neotropical migratory birds. U.S. Forest Service General Technical Report RM-229. 422 pp.

Fleischner, T. L. 1994. Ecological costs of livestock grazing in western North America. Conservation Biology 8:629–644.

Flint, R. F. 1971. Glacial and quaternary geology. John Wiley & Sons, New York. 892 pp.

Franklin, J. F., H. H. Shugart, and M. E. Harmon. 1987. Tree death as an ecological process. BioScience 37:550–556.

Frémont, J. C. 1845. Report of the exploring expedition to the Rocky Mountains in the year 1842 and to Oregon and California in the years 1843–1844. Goles and Seaton, Washington, D.C. 693 pp.

Genter, D. L. 1986. Wintering bats of the upper Snake River plain: occurrence in lava-tube caves. Great Basin Naturalist 46:241–244.

Goodwin, J. 1992. The role of mycorrhizal fungi in competitive interactions among native bunchgrasses and alien weeds: a review and synthesis. Northwest Science 66:251–260.

Gould, S. J. 1991. Abolish the recent. Natural History 5(91):16–21.

Graf, W. L. 1980. Riparian management: a flood control perspective. Journal of Soil and Water Conservation 35:158–161.

Grayson, D. K. 1987. The biogeographic history of small mammals in the Great Basin: observation on the last 20,000 years. Journal of Mammalogy 68:359–375.

Grayson, D. K. 1993. The desert's past: a natural prehistory of the Great Basin. Smithsonian Institution Press, Washington, D.C. 356 pp.

Great Basin National Park. 1988. Resource baseline inventory of the Great Basin National Park. Cooperative National Park Resources Study Unit, University of Nevada, Las Vegas, and National Park Service. Great Basin National Park Special Publication 1, NPS/WRGR-BA/92-01. 56 pp.

Hafner, D. J. 1992. Speciation and persistence of a contact zone in Mojave Desert ground squirrel, subgenus *Xerospermophilus*. Journal of Mammalogy 73:770–778.

Hafner, D. J., and T. L. Yates. 1983. Systematic status of the Mojave ground squirrel, *Spermophilus mohavensis* (subgenus *Xerospermophilus*). Journal of Mammalogy 64:397–404.

Hall, E. R. 1946. Mammals of Nevada. University of California Press, Berkeley. xi + 710 pp.

Hall, E. R. 1981. Mammals of North America. John Wiley & Sons, New York. xv + 1,181 pp.

Hammond, P. C., and D. V. McCorkle. 1983. The decline and extinction of *Speyeria* populations resulting from human environmental disturbances (Nymphalidae: Argynninae). Journal of Research on the Lepidoptera 22:217–224.

Hardy, A. R., and F. G. Andrews. 1979. An inventory of selected Coleoptera from Algodones Dune. Bureau of Land Management contract report CA-060-CT-8-98:1–35.

Harper, K. T., D. L. Freeman, W. K. Ostler, and L. G. Klikoff. 1978. The flora of Great Basin mountain ranges: diversity, sources, and dispersal ecology. Pages 81–104 *in* K. T. Harper and J. L. Reveal, editors. Intermountain biogeography: a symposium. Great Basin Naturalist Memoirs 2.

Hershler, R. 1989. Springsnails (Gastropoda: Hydrobiidae) of Owens and Amargosa River (exclusive of Ash Meadows) drainages, Death Valley system, California–Nevada. Proceedings of the Biological Society of Washington 102:176–248.

Hershler, R. 1994. A review of the North American freshwater snail genus *Pyrgulopsis* (Hydrobiidae). Smithsonian Contribution to Zoology 554:1–115.

Hershler, R., and W. L. Pratt. 1990. A new *Pyrgulopsis* (Gastropoda: Hydrobiidae) from southeastern California, with a model for historical development of the Death Valley hydrographic system. Proceedings of the Biological Society of Washington 103:279–299.

Hershler, R., and D. Sada. 1987. Springsnails (Gastropoda: Hydrobiidae) of Ash Meadows, Amargosa basin, California–Nevada. Proceedings of the Biological Society of Washington 100:776–843.

Hidy, G. M., and H. E. Klieforth. 1990. Atmospheric processes affecting the climate of the Great Basin. Pages 17–45 *in* C. B. Osmond, L. F. Pitelka, and G. M. Hidy, editors. Plant biology of the Basin and Range. Springer-Verlag, New York.

Hironka, M., and E. W. Tisdale. 1963. Secondary succession in annual vegetation in southern Idaho. Ecology 44:810–812.

Holmgren, N. 1972. Plant geography in the Intermountain Region. Pages 77–161 *in* A. Cronquist, A. H. Holmgren, N. H. Holmgren, and J. L. Reveal, editors. Intermountain flora. Volume I. Hafner Publishing Company, New York.

Horne, A. J., J. C. Roth, and N. J. Barratt. 1995. Walker Lake, Nevada: state of the lake 1992–1994. Report to the Nevada Department of Environmental Protection. 83 pp. + appendixes.

Houghton, J. G. 1978. Foreword. Pages vii–viii *in* A. McLane. Silent cordilleras. Camp Nevada, Reno.

Houghton, J. G., C. M. Sakamoto, and R. O. Gifford. 1975. Nevada's weather and climate. Nevada Bureau of Mines and Geology, Special Publication 2, Reno. 78 pp.

Hunt, C. B. 1967. Physiography of the United States. W. H. Freeman, San Francisco, Calif. 480 pp.

Hunter, R. 1990. Recent increases in *Bromus* populations on the Nevada test site. Pages 22–25 *in* E. D. McArthur, E. M. Romney, S. D. Smith, and P. T. Tueller, editors. Proceedings of a symposium on cheatgrass invasion, shrub die-off, and other aspects of shrub biology and management. Las Vegas, Nevada, 5–7 April 1989. U.S. Forest Service Intermountain Research Station General Technical Report INT-276.

Ingles, L. G. 1947. Mammals of California. University of California Press, Berkeley. xix + 258 pp.

Jameson, E. W., Jr., and H. J. Peeters. 1988. California mammals. University of California Press, Berkeley. xi + 403.

Jehl, J. R., Jr. 1994. Changes in saline and alkaline lake avifaunas in western North America in the past 150 years. Pages 258–272 *in* J. R. Jehl, Jr., and N. K. Johnson, editors. A century of avifaunal change in western North America. Studies in Avian Biology, Volume 15.

Johnson, N. K. 1975. Controls of number of bird species on montane islands in the Great Basin. Evolution 29:545–567.

Johnson, N. K. 1978. Patterns of avian geography and speciation in the Great Basin. Pages 137–159 *in* K. T. Harper and J. L. Reveal, editors. Intermountain biogeography: a symposium. Great Basin Naturalist Memoirs 2.

Johnson, R. R., C. D. Ziebell, D. R. Patton, P. F. Ffolliott, and R. H. Hamre, technical editors. 1985. Riparian ecosystems and their management: reconciling conflicting uses. U.S. Forest Service General Technical Report RM-120. 523 pp.

Johnson, T. N. 1986. Using herbicides for pinyon–juniper control in the Southwest. Pages 330–334 *in* R. L. Everett, editor. Proceedings: pinyon–juniper conference. U.S. Forest Service General Technical Report INT-215.

Kartesz, J. T. 1988. A flora of Nevada. Ph.D. dissertation, University of Nevada, Reno. 1,729 pp.

Kauffman, J. B., and W. C. Krueger. 1984. Livestock impacts on riparian ecosystems and stream management implications: a review. Journal of Range Management 37:430–438.

King, P. B. 1977. The evolution of North America. Princeton University Press, Princeton, N. J. 197 pp.

Knick, S. T., and J. T. Rotenberry. 1995. Landscape characteristics of fragmented shrub–steppe habitats and breeding passerine birds. Conservation Biology. 9:1059–1071.

Knopf, F. L., R. R. Johnson, T. Rich, F. B. Samson, and R. C. Szaro. 1988a. Conservation of riparian ecosystems in the United States. Wilson Bulletin 100:272–284.

Knopf, F. L., J. A. Sedgwick, and R. W. Cannon. 1988b. Guild structure of a riparian avifauna relative to seasonal cattle grazing. Journal of Wildlife Management 52:280–290.

Koch, D. L., J. J. Cooper, E. L. Lider, R. L. Jacobson, and R. J. Spencer. 1979.

Investigations of Walker Lake, Nevada: dynamic ecological relationships. Bioresources Center, Desert Research Institute, University of Nevada System, Reno. 189 pp.

Kremen, C. 1992. Assessing the indicator properties of species assemblages for natural areas monitoring. Ecological Applications 2:203–217.

Kremen, C., R. K. Colwell, T. L. Erwin, D. D. Murphy, R. F. Noss, and M. A. Sanjayan. 1993. Terrestrial arthropod assemblages: their use in conservation planning. Conservation Biology 7:796–808.

Krueper, D. J. 1993. Effects of land use practices on western riparian ecosystems. Pages 321–330 *in* D. M. Finch and P. W. Stangel, editors. Status and management of Neotropical migratory birds. U.S. Forest Service General Technical Report RM-229.

Lancaster, N. 1988a. Controls of eolian dune size and spacing. Geology 16:972–975.

Lancaster, N. 1988b. On desert sand seas. Episodes 11:12–17.

Lancaster, N. 1989. The dynamics of star dunes: an example from Gran Desierto, Mexico. Sedimentology 36:273–289.

Lanner, R. M. 1984. Trees of the Great Basin. University of Nevada Press, Reno. 215 pp.

Las Vegas Valley Water District. 1992. Environmental report of the Virgin River Water Resource Development Project, Clark County, Nevada. Cooperative Water Project, Report 2, Hydrographic Basin 222. 130 pp.

Las Vegas Valley Water District. 1993. Addendum to environmental report of the Virgin River Water Resource Development Project, Clark County, Nevada. Cooperative Water Project, Report 2, Hydrographic Basin 222. 27 pp.

Lavin, M. T. 1981. The floristics of the headwaters of the Walker River, California and Nevada. M.S. thesis, University of Nevada, Reno. 141 pp.

Linsdale, M. A., J. T. Howell, and J. M. Linsdale. 1952. Plants of the Toiyabe Mountains area, Nevada. Wasmann Journal of Biology 10:129–200.

Little, E. L., Jr. 1971. Atlas of United States trees. Volume 1. Conifers and important hardwoods. U.S. Forest Service Miscellaneous Publication 1146. 203 pp.

Loope, L. L. 1969. Subalpine and alpine vegetation of northeastern Nevada. Ph.D. dissertation, Duke University, Durham, N.C. 292 pp.

Lovich, J. E., T. B. Egan, and R. C. de Gouvenain. 1994. Tamarisk control on public lands in the desert of southern California: two case studies. Pages 166–177 *in* 46th Annual California Weed Conference, California Weed Science Society.

Luckenbach, R. A., and R. B. Bury. 1983. Effects of off-road vehicles on the biota of Algodones Dunes, Imperial County, California. Journal of Applied Ecology 20:265–286.

Mack, R. N. 1986. Alien plant invasion into the Intermountain West: a case history. Pages 191–213 *in* H. A. Mooney and J. A. Drake, editors. Ecology of biological invasions in North America and Hawaii. Springer-Verlag, New York.

MacMahon, J. A. 1988a. Warm deserts. Pages 231–264 *in* M. G. Barbour and W. D. Billings, editors. North American terrestrial vegetation. Cambridge University Press, New York.

MacMahon, J. A. 1988b. North American deserts: their floral and faunal components. Pages 21–81 *in* M. G. Barbour and W. D. Billings, editors. North American terrestrial vegetation. Cambridge University Press, New York.

Mason, W. T., Jr. 1995. Invertebrates. Pages 159–160 *in* E. T. LaRoe, G. S. Farris, C. E. Puckett, P. D. Doran, and M. J. Mac, editors. Our living resources: a report to the nation on the distribution, abundance, and health of U.S. plants, animals, and ecosystems. U.S. Department of the Interior, National Biological Service, Washington, D.C.

McLane, A. R. 1975. A bibliography of Nevada's caves. Center for Water Resources Research, Desert Research Institute, Reno, Nev. 99 pp.

Mead, J. I. 1987. Quaternary records of pika, *Ochotona*, in North America. Boreas 16:165–171.

Mehringer, P. J., Jr., and P. E. Wigand. 1990. Comparison of late-Holocene environments from woodrat middens and pollen: Diamond Craters, Oregon. Pages 294–325 *in* J. L. Betancourt, T. R. Van Devender, and P. S. Martin, editors. Packrat middens—the last 40,000 years of biotic change. University of Arizona Press, Tucson.

Melgoza, G., R. S. Nowak, and R. J. Tausch. 1990. Soil water exploitation after fire: competition between *Bromus tectorum* (cheatgrass) and two native species. Oecologia 83:7–13.

Meyer, S. E. 1978. Some factors governing plant distributions in the Mojave–Intermountain transition zone. Intermountain biogeography: a symposium. Great Basin Naturalist Memoirs 2:197–207.

Meyer, S. E. 1986. The ecology of gypsum communities in the Mojave Desert. Ecology 67:1303–1313.

Mifflin, M. D., and M. M. Wheat. 1979. Pluvial lakes and estimated pluvial climates of Nevada. University of Nevada, Reno. Mackay School of Mines Bulletin 94. 57 pp.

Miller, R. F., and J. A. Rose. 1995. Historic expansion of *Juniperus occidentalis* (western juniper) in southeastern Oregon. Great Basin Naturalist 55:37–45.

Miller, R. F., and P. E. Wigand. 1994. Holocene changes in semiarid pinyon–juniper woodlands. BioScience 44:465–474.

Minckley, W. L., and M. E. Douglas. 1991. Discovery and extinction of western fishes: a blink of the eye in geologic time. Pages 7–17 *in* W. L. Minckley and J. E.

Deacon, editors. Battle against extinction. University of Arizona Press, Tucson.

Morefield, J. D. 1992. Spatial and ecologic segregation of phytogeographic elements in the White Mountains of California and Nevada. Journal of Biogeography 19:33–50.

Morefield, J. D., D. W. Taylor, and M. DeDecker. 1988. Vascular flora of the White Mountains of California and Nevada: an updated, synonymized working checklist. Appendix *in* C. A. Hall, Jr., and V. Doyle-Jones, editors. The Mary DeDecker Symposium: plant biology of eastern California. White Mountain Research Station, University of California, Los Angeles.

Morrison, R. B. 1964. Lake Lahontan: geology of southern Carson Desert, Nevada. U.S. Geological Survey Professional Paper 401. 156 pp.

Morrison, R. B. 1991. Quaternary stratigraphic, hydrologic, and climatic history of the Great Basin, with emphasis on Lakes Lahontan, Bonneville, and Tecopa. Pages 283–320 *in* R. B. Morrison, editor. Quaternary nonglacial geology: conterminous United States. The Geology of North America, Volume K-2.

Morrow, L. A., and P. W. Stahlman. 1984. The history and distribution of downy brome (*Bromus tectorum*) in North America. Weed Science 32, Supplement 1:2–7.

Moyle, P. B., and R. M. Yoshiyama. 1994. Protection of aquatic biodiversity in California: a five-tiered approach. Fisheries 19:6–18.

Murphy, D. D., K. E. Freas, and S. B. Weiss. 1990. An environment–metapopulation approach to population viability analysis for a threatened invertebrate. Conservation Biology 4:41–51.

National Research Council. 1994. Rangeland health: new methods to classify, inventory, and monitor rangelands. National Academy Press, Washington, D.C. xvi + 180 pp.

Nelson, C. R. 1994. Insects of the Great Basin and Colorado Plateau. Pages 211–238 *in* K. T. Harper, L. L. St. Clare, K. H. Thorne, and W. M. Hess, editors. Natural history of the Colorado Plateau and Great Basin. University Press of Colorado, Niwot.

Nevada Natural Heritage Program. 1992. Summary of elements that occur on the top priority conservation planning sites. Nevada Natural Heritage Program, Carson City.

New, T. R. 1991. Butterfly conservation. Oxford University Press, Melbourne, Australia. xi + 224 pp.

Noss, R. F., E. T. LaRoe, III, and J. M. Scott. 1995. Endangered ecosystems of the United States: a preliminary assessment of loss and degradation. National Biological Service Biological Report 28. 58 pp.

Nowak, C. L. 1991. Reconstruction of postglacial vegetation and climate history in western Nevada: evidence from plant macrofossils in *Neotoma* middens. M.S.

thesis, University of Nevada, Reno. viii + 69 pp.

Nowak, C. L., R. S. Nowak, R. J. Tausch, and P. E. Wigand. 1994. A 30,000 year record of vegetation dynamics at a semiarid locale in the Great Basin. Journal of Vegetation Science 5:579–590.

Nowak, R. M. 1991. Walker's mammals of the world. Fifth edition. Johns Hopkins University Press, Baltimore, Md. Volumes 1 and 2.

O'Farrell, T. P., and L. A. Emery. 1976. Ecology of the Nevada test site: a narrative summary and annotated bibliography. Report NVO-167. U.S. Department of Energy, National Technical Information Services, U.S. Department of Commerce, Springfield, Va.

Ohmart, R. D. 1994. The effects of human-induced changes on the avifauna of western riparian habitats. Pages 273–285 *in* J. R. Jehl, Jr., and N. K. Johnson, editors. A century of avifaunal change in western North America. Studies in Avian Biology, Volume 15.

Osborn, G. 1989. Glacial deposits and tephra in the Toiyabe Range, Nevada, U.S.A. Arctic and Alpine Research 21:256–267.

Parker, W. S., and E. R. Pianka. 1975. Comparative ecology of populations of the lizard *Uta stansburiana*. Copeia 4:615–632.

Patterson, B. D. 1984. Mammalian extinction and biogeography in the southern Rocky Mountains. Pages 247–293 *in* M. H. Nitecke, editor. Extinctions. University of Chicago Press, Ill.

Patterson, B. D., and W. Atmar. 1986. Nested subsets and the structure of insular mammalian faunas and archipelagos. Biological Journal of the Linnaean Society 28:65–82.

Pavlik, B. M. 1985. Sand dune flora of the Great Basin and Mojave deserts of California, Nevada, and Oregon. Madroño 32:197–213.

Pavlik, B. M. 1989. Phytogeography of sand dunes in the Great Basin and Mojave deserts. Journal of Biogeography 16:227–238.

Peterjohn, B. G., and J. R. Sauer. 1993. North American Breeding Bird Survey annual summary 1990–1991. Bird Populations 1:1–15.

Peterjohn, B. G., J. R. Sauer, and W. A. Link. 1994. The 1992–1993 summary of the North American Breeding Bird Survey. Bird Populations. 2:246–261.

Pianka, E. R. 1967. On lizard species diversity: North American flatland deserts. Ecology 50:1012–1030.

Pianka, E. R. 1970. Comparative autecology of the lizard (*Cnemidophorus tigris*) in different parts of its geographic range. Ecology 51:703–720.

Pickford, G. D. 1932. The influence of continued heavy grazing and of promiscuous burning on spring–fall ranges in Utah. Ecology 13:159–171.

Piegat, J. J. 1980. Glacial geology of central Nevada. M.S. thesis, Purdue University, West Lafayette, Ind.

Piemeisel, R. L. 1951. Causes affecting change and rate of change in a vegetation of annuals in Idaho. Ecology 32:53–72.

Pierson, E. D., and P. E. Brown. 1992. Saving old mines for bats. Bats 10:11–13.

Pierson, E. D., W. E. Rainey, and D. M. Koontz. 1991. Bats and mines: experimental mitigation for Townsend's big-eared bat at the McLaughlin mine in California. Pages 31–42 *in* Proceedings V: issues and technology in the management of impacted wildlife. Thorne Ecological Institute, Snowmass Resort, Calif.

Platts, W. S. 1990. Managing fisheries and wildlife on rangelands grazed by livestock. Nevada Department of Wildlife, Reno.

Porter, S. C., K. L. Pierce, and T. D. Hamilton. 1983. Late Wisconsin mountain glaciation in the western United States. Page 71–111 *in* S. C. Porter, editor. Late-Quaternary environments of the United States. Volume 1. University of Minnesota Press, Minneapolis.

Pyle, R., M. Bentzien, and P. Opler. 1981. Insect conservation. Annual Review of Entomology 26:233–258.

Raven, P. H. 1988. The California flora. Pages 109–137 *in* M. G. Barbour and J. Major, editors. Terrestrial vegetation of California. California Native Plant Society, Special Publication 9.

Reveal, J. L. 1979. Biogeography of the Intermountain Region: a speculative appraisal. Mentzelia 4:1–92.

Reveal, J. L. 1985. Annotated key to *Eriogonum* (Polygonaceae) of Nevada. Great Basin Naturalist 45:495–519.

Reveal, J. L. 1989. A checklist of the Eriogonoideae (Polygonaceae). Phytologia 66:266–294.

Rice, B., and M. Westoby. 1978. Vegetative responses of some Great Basin shrub communities protected against jackrabbits or domestic stock. Journal of Range Management 31:28–34.

Rich, A. C., D. S. Dobkin, and L. J. Niles. 1994. Defining forest fragmentation by corridor width: the influence of narrow forest-dividing corridors on forest-nesting birds in southern New Jersey. Conservation Biology 8:1109–1121.

Robbins, C. S., D. Bystrak, and P. H. Geissler. 1986. The Breeding Bird Survey: its first 15 years, 1965–1979. U.S. Fish and Wildlife Service Research Publication 157. iii + 196 pp.

Robbins, C. S., D. K. Dawson, and B. A. Dowell. 1989. Habitat area requirements of breeding forest birds of the Middle Atlantic states. Wildlife Monographs 103. 34 pp.

Roberts, T. C., Jr. 1990. Cheatgrass: management implications in the '90's. Pages 9–21 *in* E. D. McArthur, E. M. Romney, S. D. Smith, and P. T. Tueller, editors. Proceedings: symposium on cheatgrass invasion, shrub die-off, and other aspects of shrub biology and management. Las Vegas, Nevada, 5–7 April 1989. U.S. Forest Service Intermountain Research Station General Technical Report INT-276.

Robinson, S. K. 1988. Reappraisal of the costs and benefits of habitat heterogeneity for nongame wildlife. Pages 145–155 *in* Transactions of the 53rd North American Wildlife and Natural Resources Conference.

Robinson, T. W. 1965. Introduction, spread, and areal extent of saltcedar (*Tamarix*) in the western states. U.S. Geological Survey Professional Paper 491-A. 12 pp. + 1 plate.

Rogers, G. F. 1982. Then and now: a photographic history of vegetation change in the central Great Basin desert. University of Utah Press, Salt Lake City. 152 pp.

Runeckles, V. C. 1982. Relative death rate: a dynamic parameter describing plant response to stress. Journal of Applied Ecology 19:295–303.

Rust, R. W. 1986. New species of *Osmia* (Hymenoptera: Megachilidae) from the southwestern United States. Entomological News 97:147–155.

Rust, R. W. 1994. Survey and status of federal category 2 candidate beetle species from Big Dune and Lava Dune in the Amargosa Desert of Nevada. Bureau of Land Management Report for Proposal NV-0546631. 26 pp.

Schmid, R., and M. J. Schmid. 1975. Living links with the past. Natural History 84:38–45.

Schmude, K. L., and H. P. Brown. 1991. A new species of *Stenelmis* (Coleoptera: Elmidae) found west of the Mississippi River. Proceedings of the Entomological Society of Washington 93:51–61.

Schulz, T. T., and W. C. Leininger. 1991. Nongame wildlife communities in grazed and ungrazed montane riparian sites. Great Basin Naturalist 51:286–292.

Scott, J. A. 1986. The butterflies of North America. Stanford University Press, Stanford, Calif. xii + 583 pp.

Shreve, F. 1942. The desert vegetation of North America. Botanical Review 8:195–246.

Sigler, J. W., and W. F. Sigler. 1994. Fishes of the Great Basin and the Colorado Plateau: past and present forms. Pages 163–208 *in* K. T. Harper, L. L. St. Clair, K. H. Thorne, and W. M. Hess, editors. Natural history of the Colorado Plateau and Great Basin. University Press of Colorado, Niwot.

Smith, R. S. U. 1982. Sand dunes in the North American deserts. Page 481–524 *in* G. L. Bender, editor. Reference handbook of the deserts of North America. Greenwood Press, Westport, Conn.

Soltz, D. L., and R. J. Naiman. 1978. The natural history of native fishes in the Death Valley system. Natural History Museum of Los Angeles County in conjunction with the Death Valley Natural History Association, Science Series 30. Los Angeles, Calif. 76 pp.

Stebbins, G. L. 1974. Flowering plants. Belknap Press, Cambridge, Mass. 399 pp.

Stebbins, R. C. 1974. Off-road vehicles and the fragile desert. American Biology Teacher, National Association of Biology Teachers 36:203–208 and 294–304.

Stebbins, R. C. 1985. A field guide to western reptiles and amphibians. Petersen field guide series 16. Houghton Mifflin Company, Boston, Mass. xiv + 336 pp.

Sudworth, G. B. 1898. Check list of the forest trees of the United States, their names and ranges. U.S. Division of Forestry Bulletin 17. 144 pp.

Sudworth, G. B. 1913. Forest atlas. Geographic distribution of North American trees. Part I. Pines. U.S. Forest Service. 36 maps (folio).

Sweitzer, R. A. 1990. Winter ecology and predator avoidance in porcupines (*Erethizon dorsatum*) in the Great Basin Desert. M.S. thesis, University of Nevada, Reno. 64 pp.

Tausch, R. J., and P. T. Tueller. 1990. Foliage biomass and cover relationships between tree- and shrub-dominated communities in pinyon–juniper woodlands. Great Basin Naturalist 50:121–134.

Tausch, R. J., N. E. West, and A. A. Nabi. 1981. Tree age and dominance patterns in Great Basin pinyon–juniper woodlands. Journal of Range Management 34:259–264.

Terborgh, J. 1989. Where have all the birds gone? Princeton University Press, Princeton, N.J. xvi + 207 pp.

The Nature Conservancy. 1994. Spring Mountains National Recreation Area: biodiversity hot spots and management recommendations. Report to the Bureau of Land Management, U.S. Fish and Wildlife Service, and U.S. Forest Service. The Nature Conservancy, Reno, Nev. 52 pp. + 23 maps.

Thomas, D. H. 1985. The archaeology of Hidden Cave, Nevada. Anthropological papers of the American Museum of Natural History, New York. Volume 61, Part 1. 430 pp.

Thomas, D. H. 1997. The archaeology of Monitor Valley. 4. Alta Toquima and the Mount Jefferson Complex. American Museum of Natural History Anthropological Papers, New York. In press.

Thomas, J. A. 1983. A quick method for estimating butterfly numbers during surveys. Biological Conservation 27:195–211.

Thomas, J. A. 1984. The conservation of butterflies in temperate countries: past efforts and lessons for the future. Pages 333–353 *in* R. I. Vane-Wright and P. R. Ackery, editors. The biology of butterflies. Princeton University Press, Princeton, N.J.

Thomas, J. A. 1991. Rare species conservation: case studies of European butterflies. Pages 141–197 *in* I. F. Spellerberg, M. G. Morris, and F. B. Goldsmith, editors. The scientific management of temperate communities for conservation. 29th Symposium of the British Ecological Society, Blackwell Scientific Publications, Oxford, England.

Thomas, J. W., C. Maser, and J. E. Rodiek. 1979. Wildlife habitats in managed rangelands—the Great Basin of southeastern Oregon. Riparian zones. U.S. Forest Service General Technical Report PNW-80. 18 pp.

Thompson, R. S., and J. I. Mead. 1982. Late Quaternary environments and biogeography in the Great Basin. Quaternary Research 17:39–55.

Tidwell, D. P. 1986. Multi-resource management of pinyon–juniper woodlands: times have changed, but do we know it? Pages 5–8 *in* R. L. Everett, editor. Proceedings: pinyon–juniper conference. U.S. Forest Service General Technical Report INT-215.

Tueller, P. T., R. J. Tausch, and V. Bostick. 1991. Species and plant community distribution in a Mojave–Great Basin Desert transition. Vegetatio 92:133–150.

Turner, M. G., W. H. Romme, R. H. Gardner, R. V. O'Neill, and T. K. Kratz. 1993. A revised concept of landscape equilibrium: disturbance and stability on scaled landscapes. Landscape Ecology 8:213–227.

Turner, R. M. 1982. Cold-temperate desertlands. Pages 145–155 *in* D. E. Brown, editor. Biotic communities of the American Southwest—United States and Mexico. Volume 4: Desert plants.

U.S. Fish and Wildlife Service. 1989. Endangered and threatened wildlife and plants, emergency determination of endangered status for the Mojave population of the desert tortoise. Federal Register 54:32326.

U.S. Fish and Wildlife Service. 1990. Endangered and threatened wildlife and plants, determination of threatened status for the Mojave population of the desert tortoise. Federal Register 55:12178–12191.

U.S. Fish and Wildlife Service. 1992. Scoping report: proposed water acquisition program for Lahontan Valley wetlands under Public Law 101–618. September: i, 24, A1–A3, B1–B4.

U.S. Fish and Wildlife Service. 1994a. Endangered and threatened wildlife and plants; animal candidate review for listing as endangered or threatened species. Federal Register 59:58982–59028.

U.S. Fish and Wildlife Service. 1994b. Desert Tortoise (Mojave population) Recovery Plan. U.S. Fish and Wildlife Service, Portland, Oreg. 73 pp. + appendixes.

U.S. Fish and Wildlife Service. 1994c. Endangered and threatened wildlife and plants: determination of critical habitat for the Mojave population of the desert tortoise: final rule. Federal Register 59:5820–5846.

U.S. Fish and Wildlife Service. 1994d. Endangered and threatened wildlife and plants: 50 CFR 17.11 and 17.12. 380–789/20165. Washington, D.C. 42 pp.

U.S. General Accounting Office. 1993. Livestock grazing on western riparian areas. Gaithersburg, Md. 44 pp.

Van Devender, T. R., and W. G. Spaulding. 1979. Development of vegetation and climate in the southwestern United States. Science 204:701–710.

Vasek, F. C. 1966. The distribution and taxonomy of three western junipers. Brittonia 18:350–372.

Vasek, F. C., and M. G. Barbour. 1990. Mojave Desert scrub vegetation. Pages 835–867 *in* M. G. Barbour and J. Major, editors. Terrestrial vegetation of California. California Native Plant Society, Special Publication 9.

Vasek, F. C., H. B. Johnson, and D. H. Elsinger. 1975. Effects of pipeline construction on creosotebush scrub vegetation of the Mojave Desert. Madroño 23:1–13.

Vitousek, P. M. 1986. Biological invasions and ecosystem properties: can species make a difference? Pages 163–176 *in* H. A. Mooney and J. A. Drake, editors. Ecology of biological invasions of North America and Hawaii. Springer-Verlag, New York.

Ward, J. V. 1984. Ecological perspectives in the management of aquatic insect habitat. Pages 558–577 *in* V. H. Resh and D. M. Rosenberg, editors. The ecology of aquatic insects. Prager Publishers, Westport, Conn.

Warner, R. E., and K. M. Hendrix. 1984. California riparian systems: ecology, conservation, and productive management. University of California Press, Berkeley. xxix + 1,035 pp.

Weiss, S. B., D. D. Murphy, and R. R. White. 1988. Sun, slope, and butterflies: topographic determinants of habitat quality for *Euphydryas editha*. Ecology 69:1486–1496.

Wells, P. V. 1983. Paleobiogeography of montane islands in the Great Basin since the last glaciopluvial. Ecological Monographs 53:341–382.

West, N. E. 1988. Intermountain deserts, shrub steppes, and woodlands. Pages 209–230 *in* M. G. Barbour and W. D. Billings, editors. North American terrestrial vegetation. Cambridge University Press, New York.

Wharton, R. A., P. E. Wigand, M. R. Rose, R. L. Reinhardt, D. A. Mouat, H. E. Klieforth, N. L. Ingraham, J. O. Davis, C. A. Fox, and J. T. Ball. 1990. The North American Great Basin: a sensitive indicator of climatic change. Pages 323–359 *in* C. B. Osmond, L. F. Pitelka, and G. M. Hidy, editors. Plant biology of the Basin and Range. Springer-Verlag, New York.

Wheeler, G. C., and J. N. Wheeler. 1986. The ants of Nevada. Natural History Museum of Los Angeles County, Los Angeles, Calif. vii + 138 pp.

Whisenant, S. J. 1990. Changing fire frequencies on Idaho's Snake River plains: ecological and management implications. Pages 4–10 *in* E. D. McArthur, E. M. Romney, S. D. Smith, and P. T. Tueller, editors. Proceedings: symposium on cheatgrass invasion, shrub die-off, and other aspects of shrub biology and management, 5–7 April 1989, Las Vegas, Nevada. U.S. Forest Service Intermountain Research Station General Technical Report INT-276.

Wiederholm, T. 1984. Responses of aquatic insects to environmental pollution. Pages 508–557 *in* V. H. Resh and D. M. Rosenberg, editors. The ecology of aquatic insects. Prager Publishers, Westport, Conn.

Wiens, J. A. 1989. The ecology of bird communities. Volume 2. Processes and variations. Cambridge University Press, New York. xii + 316 pp.

Wiens, J. A., and J. T. Rotenberry. 1981. Habitat associations and community structure of shrub–steppe environments. Ecological Monographs 51:21–41.

Wigand, P. E., M. L. Hemphill, and S. M. Patra. 1994. Late Holocene climate derived from vegetation history and plant cellulose stable isotope records from the Great Basin of western North America. Pages 2574–2583 *in* Proceedings of the High-Level Radioactive Waste Management Conference and Exposition, 22–26 May 1994, Las Vegas, Nev.

Wigand, P. E., and C. L. Nowak. 1992. Dynamics of northwest Nevada plant communities during the last 30,000 years. Pages 40–61 *in* C. A. Hall, Jr., V. Doyle-Jones, and B. Widawski, editors. The history of water: eastern Sierra Nevada, Owens Valley, White-Inyo Mountains. White Mountain Research Station Symposium Volume 4.

Wilcox, B. A., D. D. Murphy, P. R. Ehrlich, and G. T. Austin. 1986. Insular biogeography of the montane butterfly faunas in the Great Basin: comparison with birds and mammals. Oecologia 69:188–194.

Williams, D. D., and B. W. Feltmate. 1992. Aquatic insects. Redwood Press Ltd., Melksham, England. xiii + 358 pp.

Williams, T. R. 1982. Late Pleistocene lake level maxima and shoreline deformation in the Basin and Range Province, western United States. M.S. thesis, Colorado State University, Fort Collins. 52 pp.

Wilson, B. S. 1991. Latitudinal variation in activity season mortality rates of the lizard *Uta stansburiana*. Ecological Monographs 61:393–414.

Wright, H. E., Jr. 1983. Introduction. Pages xi–xvii *in* H. E. Wright, Jr., editor. Late-Quaternary environments of the United States. Volume 2. The Holocene. University of Minnesota Press, Minneapolis.

Yensen, D. I. 1981. The 1900 invasion of alien plants into southern Idaho. Great Basin Naturalist 41:176–183.

Young, J. A., and R. A. Evans. 1978. Population dynamics after wildfires in sagebrush grasslands. Journal of Range Management 31:283–289.

Young, J. A., R. A. Evans, and J. Major. 1972. Alien plants in the Great Basin. Journal of Range Management 25:194–201.

Young, J. A., R. A. Evans, B. A. Roundy, and J. A. Brown. 1986. Dynamic landforms and plant communities in a pluvial lake basin. Great Basin Naturalist 46:1–21.

Zeiner, D. C., W. F. Laudenslayer, Jr., and K. E. Mayer, editors. 1988. California's wildlife. Volume I. Amphibians and reptiles. State of California Department of Fish and Game, Sacramento, Calif. ix + 272 pp.

Zeveloff, S. I. 1988. Mammals of the Intermountain West. University of Utah Press, Salt Lake City. xxiv + 365 pp.

Human-Induced Changes in the Mojave and Colorado Desert Ecosystems: Recovery and Restoration Potential

Bainbridge, D. A., M. Fidelibus, and R. MacAller. 1995. Techniques for plant establishment in arid ecosystems. Restoration and Management Notes 13:190–197.

Bainbridge, D. A., and R. A. Virginia. 1990. Restoration in the Sonoran Desert of California. Restoration and Management Notes 8:3–13.

Beatley, J. C. 1969. Biomass of desert winter annual plant populations in southern Nevada. Oikos 20:261–273.

Bowers, M. A. 1987. Precipitation and the relative abundances of desert winter annuals: a 6-year study in the northern Mojave Desert. Journal of Arid Environments 12:141–149.

Brown, D. E., and R. A. Minnich. 1986. Fire and changes in creosote bush scrub of the western Sonoran Desert, California. American Midland Naturalist 116:411–422.

Brum, G. D., R. S. Boyd, and S. M. Carter. 1983. Recovery rates and rehabilitation of powerline corridors. Pages 303–314 *in* R. H. Webb and H. G. Wilshire, editors. Environmental effects of off-road vehicles: impacts and management in arid regions. Springer-Verlag, New York.

Burk, J. H. 1977. Sonoran Desert. Pages 869–889 *in* M. G. Barbour and J. Major, editors. Terrestrial vegetation of California. John Wiley & Sons, New York.

Bury, R. B., and R. A. Luckenbach. 1983. Vehicular recreation in arid land dunes: biotic responses and management alternatives. Pages 207–221 *in* R. H. Webb and H. G. Wilshire, editors. Environmental effects of off-road vehicles: impacts and management in arid regions. Springer-Verlag, New York.

Busack, S. D., and R. B. Bury. 1974. Some effects of off-road vehicles and sheep grazing on lizard populations in the Mojave Desert. Biological Conservation 6:179–183.

Chou, Y. H., R. A. Minnich, L. A. Salazar, J. D. Power, and R. J. Dezzani. 1990. Spatial autocorrelation of wildfire distribution in the Idyllwild quadrangle, San Jacinto Mountains, California. Photogrammetric Engineering and Remote Sensing 56:1507–1513.

D'Antonio, C. M., and P. M. Vitousek. 1992. Biological invasions by exotic grasses, the grass/fire cycle, and global change. Annual Review of Ecology and Systematics 23:63–87.

Dregne, H. E. 1983. Soil and soil formation in arid regions. Pages 15–30 *in* R. H. Webb and H. G. Wilshire, editors. Environmental effects of off-road vehicles: impacts and management in arid regions. Springer-Verlag, New York.

Fisher, J. C., Jr. 1978. Studies relating to the accelerated mortality of *Atriplex hymenelytra* in Death Valley National Monument. M.S. thesis, University of California, Riverside. 81 pp.

General Accounting Office. 1992. Rangeland management: BLM's hot desert grazing program merits reconsideration. U.S. General Accounting Office. GAO/RCED-92-12. Washington, D.C.

Grayson, D. K. 1993. The desert's past: a natural prehistory of the Great Basin. Smithsonian Institution Press, Washington, D.C. 356 pp.

Hunter, R. 1991. *Bromus* invasions on the Nevada Test Site: present status of *B. rubens* and *B. tectorum* with notes on their relationships to disturbance and altitude. Great Basin Naturalist 51:176–182.

Lathrop, E. W. 1983a. Recovery of perennial vegetation in military maneuver areas. Pages 265–277 *in* R. H. Webb and H. G. Wilshire, editors. Environmental effects of off-road vehicles: impacts and management in arid regions. Springer-Verlag, New York.

Lathrop, E. W. 1983b. The effect of vehicle use on desert vegetation. Pages 154–166 *in* R. H. Webb and H. G. Wilshire, editors. Environmental effects of off-road vehicles: impacts and management in arid regions. Springer-Verlag, New York.

Lathrop, E. W., and E. F. Archbold. 1980a. Plant responses to utility right of way construction in the Mojave Desert. Environmental Management 4:215–226.

Lathrop, E. W., and E. F. Archbold. 1980b. Plant response to Los Angeles aqueduct construction in the Mojave Desert. Environmental Management 4:137–148.

Malone, C. R. 1991. The potential for ecological restoration at Yucca Mountain, Nevada. The Environmental Professional 13:216–224.

Oldemeyer, J. L. 1994. Livestock grazing and the desert tortoise in the Mojave Desert. Pages 95–103 *in* R. B. Bury and D. J. Germano, editors. Biology of North American tortoises. National Biological Service, Washington, D.C.

O'Leary, J. F., and R. A. Minnich. 1981. Postfire recovery of creosote bush scrub vegetation in the western Colorado Desert. Madroño 23:61–66.

Prose, D. V. 1985. Persisting effects of armored military maneuvers on some soils of the Mojave Desert. Environmental Geology and Water Science 7:163–170.

Prose, D. V., and S. K. Metzger. 1985. Recovery of soils and vegetation in World War II military base camps, Mojave Desert. U.S. Geological Survey Open-File Report 85-234. 114 pp.

Prose, D. V., S. K. Metzger, and H. G. Wilshire. 1987. Effects of substrate disturbance on secondary plant succession; Mojave Desert, California. Journal of Applied Ecology 24:305–313.

Reible, D. D., J. R. Ouimette, and F. H. Shair. 1982. Atmospheric transport of visibility degrading pollutants into California Mojave Desert. Atmospheric Environment 16(3): 599–613.

Vasek, F. C. 1979. Early successional stages in Mojave Desert scrub vegetation. Israel Journal of Botany 28:133–148.

Vasek, F. C., and M. G. Barbour. 1977. Mojave Desert scrub vegetation. Pages 835–867 *in* M. G. Barbour and J. Major, editors. Terrestrial vegetation of California. John Wiley & Sons, New York.

Vasek, F. C., H. B. Johnson, and G. D. Brum. 1975a. Effects of power transmission lines on vegetation of the Mojave Desert. Madroño 23:114–131.

Vasek, F. C., H. B. Johnson, and D. H. Eslinger. 1975b. Effects of pipeline construction on creosote bush scrub vegetation of the Mojave Desert. Madroño 23:1–13.

Webb, R. H., and E. B. Newman. 1982. Recovery of soil and vegetation in ghost-towns in the Mojave Desert, southwestern United States. Environmental Conservation 9:245–248.

Webb, R. H., J. W. Steiger, and R. M. Turner. 1987. Dynamics of Mojave Desert shrub assemblages in the Panamint Mountains, California. Ecology 68:478–490.

Webb, R. H., and S. S. Stielstra. 1979. Sheep grazing effects on Mojave Desert vegetation and soils. Environmental Management 3:517–529.

Webb, R. H., and H. G. Wilshire. 1983. Environmental effects of off-road vehicles: impacts and management in arid regions. Springer-Verlag, New York. 534 pp.

Webb, R. H., H. G. Wilshire, and M. A. Henry. 1983. Natural recovery of soils and vegetation following human disturbance. Pages 279–302 *in* R. H. Webb and H. G. Wilshire, editors. Environmental effects of off-road vehicles: impacts and management in arid regions. Springer-Verlag, New York.

Southwest

The southwestern region of the United States is a land of extremes and contrasts. Elevations vary from below sea level in the Imperial Valley of California to mountain peaks approaching 4,000 meters. Landscapes are striking and variable and include mountains, foothills, canyons, deserts, plains, and rivers. The area is arid or semiarid and, depending on the location, may have mild winters and summers, periods of bitter cold, or intervals of intense heat. Climate is inextricably tied to water and its availability. Historically, water varied from abundant to sparse over the span of a year, and adaptations of native plants and animals reflect those extremes. Annual precipitation, usually in the form of rain, varies from 30 to 40 millimeters in the low-elevation Sonoran Desert to more than 1,000 millimeters in the high mountains (Brown 1982a; Bahre and Shelton 1993). This variation in topography and climate has produced great floral and faunal diversity.

The Southwest, as discussed here, includes Texas west of the Pecos River, most of New Mexico, Arizona, the Colorado Desert of southeastern California, and the Colorado Plateau of western Colorado and eastern Utah (Fig. 1). Our review of the status and trends of the biota of this region was aided by several particularly useful sources, especially Brown (1982a) for Arizona, Dick-Peddie (1993) for New Mexico, Harper et al. (1994) for the Colorado Plateau, and MacMahon (1979, 1988a) and West (1988) for North American arid ecosystems in general.

Brown (1982a) discussed the biotic resources of the Southwest within a life-zone context, in which ecosystem expression is strongly tied to gradual changes in elevation and latitude. This general approach was first formulated from visits made by C. Hart Merriam (Merriam and Steineger 1890) to the San Francisco Peaks, just south of the Grand Canyon in Arizona. Merriam's studies allowed him to postulate a relationship among latitude, elevation, and resulting climatic gradients; his conclusions significantly influenced early biological thought about the American West. He used descriptors such as Lower Sonoran to refer to desert, Upper Sonoran to refer to grasslands and woodlands, Transition to refer to pine forests, and Canadian to refer to higher-elevation forests. Although later work has refined, if not supplanted, most of Merriam's work, his basic approach to life zones in the West is still useful in understanding biological diversity.

Plants and animals in the Southwest respond to a multitude of gradients, and the interactions of gradients such as aspect, elevation, and latitude increase the complexity of biological interactions as well as our ability to understand them. Merriam was especially interested in elevation gradients of vegetation and how such gradients reflected similar north–south latitudinal gradients. The presence of northern conifer forests dominated by fir and spruce on the mountaintops of the Southwest, well to the south of the main concentration of such forests, is an example of this phenomenon (Fig. 2).

In the Southwest, plant and animal distributions are affected by general climatic zones such as significant north–south gradients in temperature and precipitation. To the north, the Colorado Plateau region tends to have extremely cold winters, with winter precipitation predominantly occurring as snow. In contrast, summers in the Colorado Plateau can be quite hot and dry. To the south of the plateau in central and western Arizona, winters are milder and precipitation occurs as rain. Summers are

Courtesy C. D. Allen, USGS

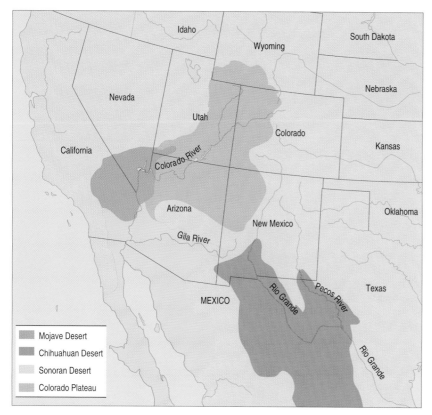

Fig. 1. The southwestern region of North America.

railroad (Hastings and Turner 1965; Bahre 1991; Bahre and Shelton 1993). Recent, rapid population growth (Fig. 3), continued economic transformation, and ongoing resource use in the Southwest have meant that humans have greatly influenced the region's natural systems—and still do.

Climatic Variability

Reconstructions of southwestern climate from studies of tree growth rings have been especially useful in understanding long-term environmental variation (Fritts and Swetnam 1989). Such tree-ring reconstructions reveal that high variability in precipitation occurred at annual, decadal, and centennial time scales extending back for the last 2,000 years (Grissino-Mayer 1995; Fig. 4). Variability in precipitation amounts and timing is related in part to El Niño events in the Pacific Ocean, which bring wetter winters and springs to portions of the Southwest about every 3 to 5 years (Ropelewski and Halpert 1986). Precipitation variability affects biotic productivity and diversity on local to regional scales (Pieper 1994) and also affects disturbance processes such as fires (Swetnam and Betancourt 1990), floods (Molles and Dahm 1990; Webb and Betancourt 1992), and insect population outbreaks (Betancourt et al. 1993; Swetnam and Lynch 1993).

still relatively hot and dry, particularly at the eastern edge of the Mojave Desert. Farther south and eastward, in the Sonoran and Chihuahuan deserts, winters are also less severe, and precipitation shifts, occurring predominantly in late summer as monsoonlike rains. Late spring and early summer are hot and dry.

Ecosystems are also shaped by changes in temperature and moisture along elevation gradients and aspect (that is, north- and south-facing slopes) of local mountain ranges. For example, as elevation increases in the mountains in southeastern Arizona, the vegetation shifts from grasslands, woodlands, and forests of Mexican Sierra Madrean affinity to forests and meadows of Rocky Mountain affinity (Brown 1982a; Muldavin et al. 1996).

People and Processes of the Southwest

Ecosystems in the Southwest have been extensively shaped by their histories of climatic variability and by disturbances such as fires and floods. Superimposed on these natural sources of variation are the prehistoric and recent activities of humans in the Southwest. Most human-induced change has occurred since about 1870, after the Southwest was settled primarily by Anglo-Americans and market forces were developed, as represented by the arrival of the

Fig. 2. Forests of Engelmann spruce, corkbark fir, aspen, and Douglas-fir in New Mexico.

Human Settlement

People have inhabited the Southwest for more than 10,000 years, from Clovis Paleo-Indians to sedentary farmers (for example, Anasazi, Mogollon, and Hohokam peoples) and nomadic peoples ancestral to the modern Navajos, Apaches, and Utes (Stewart and Gauthier 1986). Prehistoric peoples possibly contributed to the extinction of the Pleistocene megafauna (Martin 1967) and to localized deforestation of woodlands and depletion of wildlife and soils (Betancourt and Van Devender 1981; Samuels and Betancourt 1982; Kohler 1992). At the time of Spanish incursions in the 1500's, about 100,000 Native Americans lived in about 100 communities (pueblos) in northern and central New Mexico (Schroeder 1979), but colonial disruptions of their cultures and diseases introduced by Hispanic colonists drastically reduced their numbers. Today, the Southwest is home to more than 330,000 Native Americans, who live on millions of hectares of tribal lands. For example, in Arizona, about 8 million hectares (27%) are tribal lands, and in

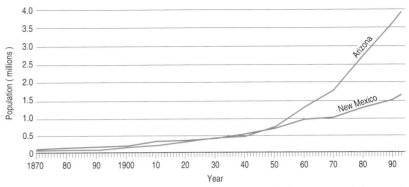

New Mexico, about 3 million hectares (9.4%) are tribal lands (Williams 1986; Utter 1993).

European settlers also introduced new plants, animals, technologies, and land-use patterns to the Southwest. Over the past century, these factors, along with the rapidly increasing human populations and their accompanying resource consumption, have resulted in unprecedented effects on southwestern ecosystems. Agents of change that have significantly affected the status of southwestern ecosystems and are largely responsible for ecosystem trends

Fig. 3. Human population growth in Arizona and New Mexico, 1870–1993 (data from U.S. Bureau of the Census 1975, 1994).

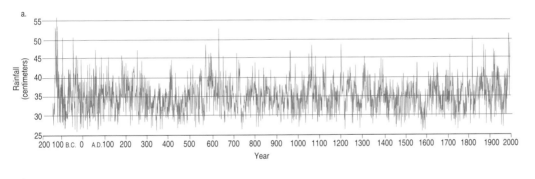

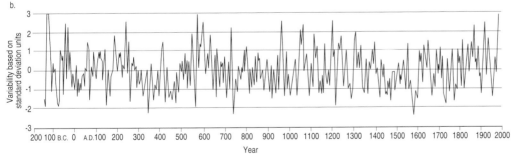

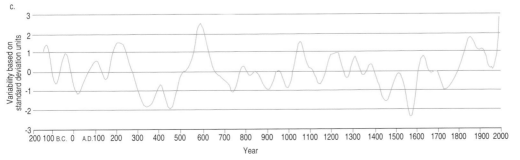

Fig. 4. a) Reconstruction of annual rainfall from El Malpais National Monument, New Mexico, based on tree-ring analysis; b) 10-year smoothing line of the reconstruction, using standard deviation units to highlight short-term (fewer than 25 years) climatic episodes; c) 100-year smoothing line showing long-term (more than 50 years) climatic trends (Grissino-Mayer 1995).

that we can identify today include grazing, fire suppression, logging, dams and water diversion, biocides, agriculture, fragmentation of wildlands by roads and other construction, introduction of nonindigenous plants and animals, and urbanization.

Grazing

Widespread grazing of domestic livestock has caused major cumulative effects on the ecology of the arid Southwest (see Fleischner 1994 for a comprehensive review). Even before this area became part of the United States, livestock production was an important activity (Denevan 1967). Sheep were the primary stock favored by the Spanish, with at least several hundred thousand sheep inhabiting New Mexico from about 1788 onward (Denevan 1967). After Native Americans were confined to reservations and railroads were linked to external markets, an immense boom in the livestock industry, particularly cattle, occurred in the early 1880's (Wooton 1908; Stewart 1936; Denevan 1967). In Arizona, livestock increased from about 142,000 in 1880 (Stewart 1936) to 650,000 in 1885 (Hastings and Turner 1965), and roughly 1,500,000 in 1891 (Cameron 1896; Morrisey 1950; Bahre 1991). In New Mexico, cattle numbered about 137,000 in 1880 and reached 1,380,000 by 1889 (Wooton 1908; Williams 1986), while numbers of sheep in New Mexico increased from 619,000 in 1870 (Denevan 1967) to roughly 2,000,000 in 1880 and 5,400,000 in 1884 (Wooton 1908). In general, livestock numbers peaked between the 1880's and 1920's and have declined since (Schickedanz 1980; Branson 1985).

Overall stocking rates have decreased primarily because of a transition from sheep grazing to cattle grazing. In 1906 in New Mexico there were a little more than a million cattle and almost 6 million sheep; by 1979, the number of cattle had increased to 1.5 million and the number of sheep had declined 90% to 604,000. Reductions in livestock numbers are associated with the end of open ranges, the Taylor Grazing Act of 1934, and some increasing control asserted by federal land management agencies over public lands (deBuys 1985; Bahre and Shelton 1993). In New Mexico, about 44% of total lands are privately held; the remainder belong to federal, state, and tribal governments (Williams 1986). Figures for private land in Arizona are similar. Most southwestern lands continue to be grazed, including even apparent reserves (for example, wilderness areas and some national and state parks). Even relatively secure areas (for example, Department of Defense lands) have experienced some level of grazing pressure in recent years that is mostly undocumented (E. Muldavin, New Mexico Natural Heritage Program, Albuquerque, personal observation).

The extremely high historical stocking rates and concomitant overgrazing and livestock preferences for certain more palatable plants (for example, grasses) led to significant alterations in the species composition of vegetation across the Southwest (Leopold 1924; Cottam and Stewart 1940; Cooper 1960; Buffington and Herbel 1965; Humphrey 1987; Grover and Musick 1990; Archer 1994; Fleischner 1994; Pieper 1994). Cool-season grasses and other preferred forage species declined (Bohrer 1975), while unpalatable and weedy species, such as broom snakeweed, and shrubs, such as creosotebush and mesquite, increased (Wooton 1908; Bahre and Shelton 1993). Cole (1995) has shown that over the last 5,450 years vegetation in one area of the southern Colorado Plateau remained fairly stable up to the last few hundred years. Since settlement, plants preferred by sheep and cattle, such as winterfat and ricegrass, have disappeared entirely from pollen profiles, whereas plants associated with overgrazing (such as whitebark rabbitbrush, snakeweed, and greasewood), which were not recorded in pollen profiles before settlement, are now present. Livestock also altered vegetation composition by serving as an agent for the spread of weedy and nonindigenous plant species such as Lehmann lovegrass (Warshall 1995). Concentrated livestock use of riparian zones has had particularly significant negative ecological effects (General Accounting Office 1988; Szaro 1989; Bahre and Shelton 1993; Fleischner 1994).

The year-round, high-intensity grazing of open ranges that occurred in the past also led to marked reductions in herbaceous plant and litter ground cover (Fleischner 1994). Because the productivity (and thus grazing capacity) of many southwestern rangelands varies markedly in response to annual variability in precipitation (Pieper 1994; Fig. 4), the tendency to stock ranges at the carrying capacity of the wetter years results in severe overgrazing effects during drought conditions (Cameron 1896) unless livestock numbers are rapidly reduced to track actual range conditions. Overgrazing is also widely considered a major trigger of soil erosion, flooding, and arroyo cutting in the Southwest (Wooton 1908; Leopold 1924; Cooperrider and Hendricks 1937; Cottam and Stewart 1940; Smith 1953; Cooke and Reeves 1976; Bahre and Bradbury 1978; Branson 1985; Bahre 1991), although climatic fluctuation has also been considered an important factor by other authors (Leopold 1951; Hastings and

Turner 1965; Denevan 1967; Graf 1986; Humphrey 1987; Webb and Betancourt 1992). Similarly, while Turner (1990) provided evidence of climatic influence on Sonoran Desert vegetation, Bahre and Shelton (1993) generally dismissed the effect of climate change as the primary agent of long-term directional vegetation change in the Southwest; rather, they attributed such change to livestock grazing and fire suppression. Fleischner (1994:637) said, "For now, the best historic evidence seems to support the idea that livestock grazing, interacting with fluctuations in climatic cycles, has been a primary factor in altering ecosystems of the Southwest." However, Brown and McDonald (1995), among others, have questioned the objectivity of Fleischner's presentation. We recognize that significant, ongoing efforts have been made to make livestock grazing more ecologically sensitive and sustainable (for example, rest and rotation systems, better spatial distribution, and immediate stocking reductions when droughts occur). Still, livestock grazing clearly remains a key process affecting southwestern ecosystems, and eliminating livestock entirely may be the only way to allow some systems to recover.

Fire

Fire is an integral process in many southwestern ecosystems, and fire scars on trees demonstrate that past fires were frequent and widespread in forested landscapes (Swetnam and Baisan 1996). Even desert grasslands apparently sustained significant fires (Bahre 1991). Fires affect ecosystems at scales ranging from site-level nutrient cycling (White 1994) to formation of landscape-scale vegetation patterns (Allen 1989).

The cessation of frequent natural surface fires all across the Southwest in the late 1800's apparently was due to reduced vegetation caused by intense grazing by livestock (Savage and Swetnam 1990; Swetnam 1990; Swetnam and Baisan 1996). Except in a few isolated localities protected from grazing, this region-wide reduction in fires began several decades before people began to actively initiate fire suppression (Madany and West 1983; Touchan et al. 1995) and testifies to the ubiquity and intensity of regional livestock grazing effects during the late 1800's. The initial suppression of natural surface fires by livestock grazing graded into the period of active suppression of all fires by land management agency personnel after about 1910 (Swetnam 1990). Fire suppression over the past century pervasively affected many southwestern ecosystems (Covington and Moore 1994).

Forest Logging and Forest Health Issues

Forestry practices on large land grants and public lands in the Southwest have gone through several phases in the past century. Relatively indiscriminate cutting practices characterized the late 1800's and early 1900's (deBuys 1985), selective logging the mid-1900's, and even-aged management the 1960's through 1980's; now we are back to more selective, uneven-aged silvicultural practices. Until recently, forest harvest practices tended to remove all old-growth trees to bring forest stands under "control," and extensive road networks proliferated to aid forestry activities (Allen 1989). More recently, new management plans support the maintenance of at least some old-growth forests and roadless areas as important for ecological and social reasons (see compilation in Kaufmann et al. 1992).

Fire suppression, commercial forestry practices, and overgrazing have pervasively altered the structure and species composition of most southwestern forests (U.S. Forest Service 1993; Covington and Moore 1994). Old-growth forests have been greatly reduced by high-grading and even-aged management practices that targeted the most valuable old trees, especially ponderosa pine. In addition, until 20 years ago, snags (dead trees) were systematically removed as fire and forest health hazards, while extensive road networks aided those who poached fuelwood (poachers often focused on snags). Hence, most managed forests now lack desired numbers of large-diameter snags, which serve important ecological roles such as cavity-nesting sites for many breeding birds (Thomas et al. 1979; Hejl 1994) and probably for many bats as well (H. Green, U.S. Forest Service, Arizona, personal communication; M. Bogan, U.S. Geological Survey, Albuquerque, New Mexico, unpublished data). Today's forests are characterized by unnaturally dense stands of young trees, a variety of forest health concerns, increasing potential for widespread insect population outbreaks (Swetnam and Lynch 1993), and unnatural crown fires (Covington and Moore 1994; Sackett et al. 1994; Samson et al. 1994).

Southwestern forests, while limited in areal distribution and regional importance for timber production, do provide a multitude of essential market and nonmarket benefits, ranging from human recreation to protection of watersheds and biodiversity. Increased forest densities lead to decreases in total streamflow, peak flow, and base flow (Troendle and Kaufmann 1987; Ffolliott et al. 1989), important concerns in the water-limited Southwest. Timber harvests from public forests have declined markedly in recent years because of the depletion of large, high-value trees and increased attention to environmental constraints (for example, sensitive species, water quality, old-growth set-asides). Forest health issues drive current efforts directed at thinning understory trees mechanically and with prescribed fires to restore more natural, sustainable forest structures and processes.

Biocides

Extensive applications of biocides such as DDT on forests and agricultural lands during the 1950's and 1960's have left persistent concentrations of these compounds in southwestern ecosystems (also see Environmental Contaminants chapter). Our knowledge of biocide use in the Southwest is largely anecdotal. For example, U.S. Forest Service operations against western spruce budworm included spraying 514,664 kilograms of DDT on 478,055 hectares of the Santa Fe National Forest and adjacent Carson National Forest (both in New Mexico) between 1955 and 1963 (Brown et al. 1986). Likewise, over the past decade, about 40,000 hectares of Bureau of Land Management lands in southeastern New Mexico have been aerially treated with tebuthiuron to enhance livestock range values by converting Havard's oak sandhills to grass-dominated systems (C. Painter, New Mexico Department of Game and Fish, Santa Fe, personal communication). Schmitt and Bunck (1995) and Glaser (1995) reviewed persistent environmental contaminants and their recent

effects on fish and wildlife. The magnitude and full effects of past and potential biocide use in the Southwest remain unassessed.

Dams, Water Diversions, and Roads

Most declines and extirpations of aquatic organisms in the Southwest can be traced to the construction of dams, either for water storage or flood control, and to other development on or near waterways, such as diversion structures and drainage of wetlands. Dam building and water diversions have significantly degraded most major river systems (Szaro 1989; Crawford et al. 1993), causing dire consequences for native fishes. In addition, such structures often contribute to habitat fragmentation (Crawford et al. 1993), with plant and animal distributions often reduced to increasingly small areas. Similarly, landscapes have been fragmented by proliferating road networks (Fig. 5), with numerous negative ecological implications (Allen 1989).

Southwestern Ecosystems

The biotic communities of the Southwest have responded in different ways to the prehistoric and recent effects of human activities on the natural southwestern environment. We focus on several important ecosystem types, which are defined primarily by dominant vegetation and are arranged along an elevation gradient. We consider the status and trends of each in the context of specific biotic composition, climate considerations, and human uses.

New Mexico

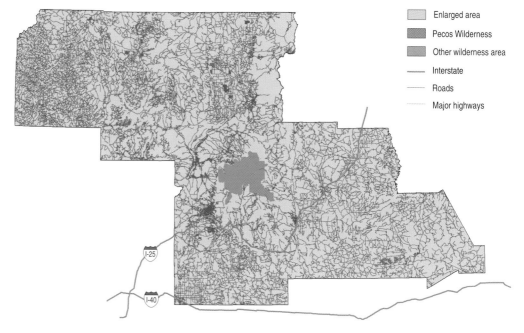

Fig. 5. Proliferating road networks in northern New Mexico reflecting human land uses in this relatively sparsely populated area. Note the absence of roads in the Pecos Wilderness. Map from Earth Data Analysis Center, University of New Mexico.

Alpine Tundra, Subalpine and Montane Grasslands, and Meadows

Alpine tundra occurs above treeline atop the highest mountain peaks and covers relatively little area in the Southwest; it occurs principally in northern Arizona, northern New Mexico, and western Utah (see summaries by Pase 1982; Baker 1983; Moir 1993). Alpine tundra is characterized by prostrate woody shrubs (commonly willows), herbs, lichens, and mosses growing in cool, usually wet, conditions within a relatively short growing season. The status and trends of tundra and its associated organisms in the Southwest are little known. Most southwestern tundra ecosystems occur within national forests, commonly within wilderness areas, but they are still subject to grazing by cattle, sheep, and horses during the growing season (Dick-Peddie 1993). Effects of grazing on alpine tundra, such as changes in plant species composition and soil compaction, need to be assessed. Researchers are also concerned about the effects of localized recreational use, warming climates, and acid deposition in the alpine tundras (Moir 1993). In the highlands of the Southwest, extensive grasslands (on uplands) and meadows (in drainages) occur above and below treeline (Brown 1982b). These grasslands are dominated by perennial bunchgrasses such as fescues, oatgrasses, and wheatgrasses; the meadows are characterized by herbaceous plants, grasses, sedges, and rushes.

By the late 1800's, the species composition and vegetation structure of these high-elevation herbaceous ecosystems were being altered by intense grazing by livestock, especially sheep. For example, moist meadows in the Jemez Mountains of New Mexico are often dominated now by nonindigenous plants such as Kentucky bluegrass, white clover, and dandelion (Allen 1989; Wolters 1996). Additionally, increased tree densities in surrounding forests may be transpiring water that previously kept many meadows moist. Channel incision, caused by elimination of beavers and continued livestock grazing, also has caused many meadow sites to become drier, allowing trees (often blue spruce) to invade their margins. Reduced competition from grasses, fire suppression, and climatic warming also have enabled widespread tree invasion of montane and subalpine grasslands (Allen 1989; Moir and Huckaby 1994; J. Elson, W. deBuys, W. Moir, and C. Allen, Santa Fe, New Mexico, unpublished data); parks and grasslands throughout the region are being diminished by tree invasion (Fig. 6). Better management practices and reduced livestock numbers, especially far fewer sheep, have allowed improvements in grassland conditions in many areas from their low point in the early

twentieth century (Branson 1985), for example in the Pecos Wilderness (J. Elson, W. deBuys, W. Moir, and C. Allen, unpublished data). Many moist meadows, though, are still heavily grazed by domestic livestock and burgeoning elk populations (Allen 1996a; Wolters 1996).

Fig. 6. Conifer tree invasion of a montane grassland in northern New Mexico.

Courtesy C. D. Allen, USGS

Subalpine Forests

Forests cover about 6.5% of Utah, 6.8% of Arizona, and 8.4% of New Mexico (Powell et al. 1993), although Van Hooser et al. (1993) reported that only 6.2% of New Mexico has more than 10% forest canopy cover. The highest-elevation forests in the Southwest are subalpine forests, characterized by stands of conifers such as Engelmann spruce and subalpine fir. These forests most often occur in relatively small, isolated mountaintop stands (Pase and Brown 1982a; Moir 1993). Spruce–fir forests grade into bristlecone pine stands on some treeline sites and into mixed conifer forests at lower elevations. Large stands of quaking aspen commonly occur in subalpine forests, particularly after fires. In closed-canopy conifer forests few herbaceous plants grow amidst the blankets of needle and wood litter, whereas aspen stands may have luxuriant herbaceous plant-dominated understories. Spruce–fir forests cover about 0.2% of Arizona and 0.5% of New Mexico (Alexander 1987).

Major natural disturbances in the subalpine forest are the uprooting and blowdown of trees by wind (windthrow), spruce beetle outbreaks, and fire. Windthrow of remnant trees is a problem because partial cutting of spruce–fir forest stands exposes remaining old trees to new wind stresses (Alexander 1987). Because spruce beetles prefer downed trees, major outbreaks usually originate in material from blowdowns or logging operations (Schmid and Frye 1977).

Fire histories of southwestern subalpine forests have been little studied, but recent work indicates that some spruce–fir stands experienced mixed fire regimes, with patchy crown fires occurring about every several hundred years and more frequent surface fires occurring every 15 to 30 years (Grissino-Mayer et al. 1995; Touchan et al. 1996). Because of these longer fire intervals, subalpine forests have probably been altered less by modern fire suppression than have other southwestern forest types. Also, because of the natural paucity of herbage in these forests, they have been spared much livestock grazing.

The greatest human effects on spruce–fir forests have resulted from silvicultural practices, especially clear-cuts of old-growth forests. Windthrow, spruce beetle infestation, and tree regeneration problems have occurred in the process (Alexander 1987). Projected climate changes might eliminate some subalpine forests from isolated mountain ranges in the Southwest (Gosz 1992).

Many aspen forests in the Southwest are now composed of trees more than 100 years old; these trees are subject to increased insect and disease problems as they decline in vigor. Modern fire suppression has prevented aspen regeneration, and conifer understories are now widely overtopping aspen stands; for example, between 1962 and 1986, the area of aspen stands declined by 46% in Arizona and New Mexico (U.S. Forest Service 1993). Elk herbivory on aspen sprouts now retards regeneration on small burns or clear-cuts (Moir 1993; Allen 1996a). Without major fires, aspen stands will continue to decline, although aspen clones are able to persist in a suppressed state in the understories of conifer forests for many years. The high probability of intense fires in southwestern conifer forests in the coming decades suggests that new aspen stands will develop again soon, changing their status from declining to increasing.

Mixed-Conifer Forests

Diverse forests of mixed-conifer species blanket many southwestern mountains. Dominant species include Douglas-fir, white fir, limber pine (in the north), southwestern white pine (in the south), ponderosa pine, and blue spruce (Pase and Brown 1982b; Moir 1993; Muldavin et al. 1996). Aspen, along with Gambel oak, is prominent in these forests following disturbances (DeByle and Winokur 1985; Moir 1993). Numerous herbaceous plant species may occur in these forests, and understory conditions vary widely, from dry, open-canopy forests with grassy undergrowth on open slopes and ridges to moist, closed-canopied stands dominated by herbaceous plants in the canyons and ravines.

Fire histories vary with forest composition and landscape characteristics, from frequent surface fires to infrequent, patchy crown fires (Grissino-Mayer et al. 1995; Swetnam and Baisan 1996; Touchan et al. 1996). Mean fire return intervals were about 10 years at many sampled sites until the late 1800's, when widespread fires stopped. As a result of fire cessation and suppression, southwestern mixed-conifer forests have undergone major changes in structure and species composition in the past century, as they have elsewhere in the interior West (Samson et al. 1994). Ponderosa pine was once codominant in many mixed-conifer forests with relatively open stand structures, but fire suppression has allowed the development of dense sapling understories, with regeneration dominated by the more fire-sensitive Douglas-fir and white fir. Forest stand inventory data from Arizona and New Mexico show an 81% increase in the area of mixed-conifer forests between 1962 and 1986 (U.S. Forest Service 1993). Because white fir and Douglas-fir are the preferred host species for spruce budworms, the increases in fir abundance and distribution have led to increasingly intense and synchronous spruce budworm outbreaks throughout mixed-conifer forests in the southern Rocky Mountains (Swetnam and Lynch 1993). Herbaceous understories have been reduced by denser canopies and needle litter, and nutrient cycles have been disrupted. Heavy surface fuels and a vertically continuous ladder of dead branches have developed, resulting in increased risks of crown fires (U.S. Forest Service 1993). When droughts such as the drought of the 1950's recur, such conditions could lead to catastrophic widespread fires, which could significantly reduce, or even eliminate, mixed-conifer forests from the smaller sky island mountains along the United States–Mexico border. As these forests become reduced or fragmented, local endemic plants could become threatened or endangered (Warshall 1995).

Ponderosa Pine Forests

Forests dominated by ponderosa pine cover extensive portions of the Southwest (Pase and Brown 1982b; Moir 1993; but see Betancourt 1990 for the Holocene nature of the phenomenon) and grade into mixed-conifer forests at higher elevations and into woodlands below. On rugged mountain slopes and in canyons, these forests can form closed-canopy stands with a complex undergrowth structure of shrubs—particularly oaks (for example, Gambel oak, gray oak, wavyleaf oak, Arizona white oak)—and a high diversity of herbs. On more gentle terrain, such as on plateaus and mesas or in wide valley bottoms, the forests tend to form more open parklike stands with grassy ground cover and scattered clumps of shrubs (Fig. 7). An array of herbaceous plant species occurs across the range of these forests, and understory conditions vary widely.

Before European settlement, ponderosa pine forests were generally open stands with well-developed herbaceous understories (Cooper 1960). Widespread surface fires that occurred every 2 to 15 years favored grasses and kept pine densities in check (Swetnam and Baisan 1996). Fire suppression since the late 1800's has had pervasive effects on ponderosa pine forests (Covington and Moore 1994) similar to those described for many mixed-conifer forests. The most obvious result has been great increases in the density of young pine trees, with associated buildups of thick blankets of needle litter; these buildups have markedly reduced the

Courtesy C. D. Allen, USGS

Fig. 7. Open-stand structure of ponderosa pine forest characteristic of presettlement conditions.

A Ponderosa Pine Natural Area Reveals Its Secrets

Monument Canyon Research Natural Area preserves an unlogged 259-hectare stand of old-growth ponderosa pine in the Jemez Mountains of New Mexico. This preserve, established in 1932, is the oldest research natural area in the state. This two-tiered forest displays an old-growth density of 100 stems per hectare (Muldavin et al. 1995), with an understory thicket of stagnant saplings and poles that raises the total stand density to an average of 5,954 stems per hectare, with concentrations as high as 21,617 stems per hectare (Fig. 1).

The old overstory trees in the research area are declining and dying, possibly from altered nutrient and water availabilities and competition with the dense understory saplings (Sackett et al. 1994). A thick layer of ponderosa pine needles blankets the forest floor, and herbaceous plant diversity is low. Although 34 herbaceous species have been found on the 259-hectare site (Deichmann 1980), as many as 60 might be expected in more open stands that have regularly experienced fire and which lack the dense sapling layer. Further, those herbs that do occur are usually found only as isolated individuals across the landscape.

Fire scars show that surface fires burned through this forest about every 6 years for at least 300 years until 1892, when the last significant fire occurred (Fig. 2). Understory saplings that have been dated reveal that tree establishment greatly increased in the early 1900's (Muldavin et al. 1995; Fig. 3) after grass competition was reduced because of

Courtesy C. D. Allen, USGS

Fig. 1. Dense doghair thicket of young ponderosa pine in the Monument Canyon Research Natural Area.

livestock grazing and after the fires that once inhibited pine regeneration were suppressed. As a result, a highly unnatural stand structure has developed, with the doghair thickets representing extreme density conditions, although similar conditions are found in many southwestern ponderosa pine stands (Covington and Moore 1994).

Analogous stand structures and high fuel loads prevailed in the nearby area burned by the 1977 La Mesa fire (Foxx and Potter 1984), when an intense crown fire converted 4,000 hectares of dense ponderosa pine forest into largely treeless grasslands. It is clear that fire is a keystone process in ponderosa pine forests—without it the whole ecosystem changes markedly. From this perspective the admonition emblazoned on the Monument Canyon Research Natural Area boundary sign acquires ironic overtones: "This area must be preserved in a natural state as near as possible."

A number of small archaeological sites from ancestral Pueblo people are found within this research natural area; these sites are typified by a two-room field house dated to between 1330 and 1630. This house was probably used while the people tended adjoining farm plots. The ponderosa pine thicket that shrouds these sites today would render even small-plot farming impossible without wholesale forest clearance. Fire suppression is the primary reason for the profound transformation this forest has experienced in the past century.

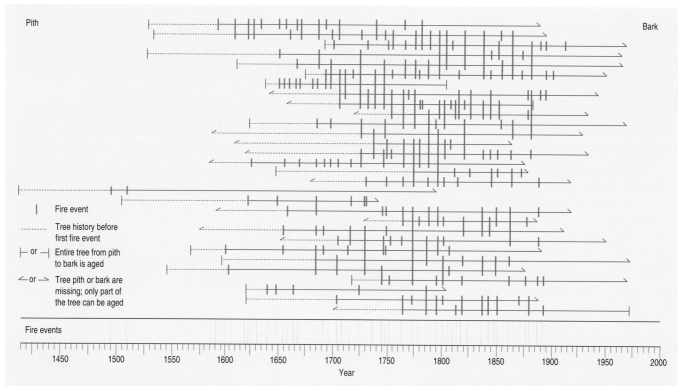

Fig. 2. Fire-scar chronology for Monument Canyon Research Natural Area. The horizontal lines represent the life spans of individual trees, while fire-scar events are shown by short vertical bars. The longer vertical lines at the bottom indicate the dates of fire events in which at least 10% of the sampled trees recorded a fire. Note the cessation of spreading fires after 1892. For additional information on the fire history of ponderosa pine forests and on the use of fire scars to date fires, see the U.S. Geological Survey Biological Resources Division LUHNA (Land Use History of North America) website at http://biology. usgs.gov/luhna/southwest/southwest.html.

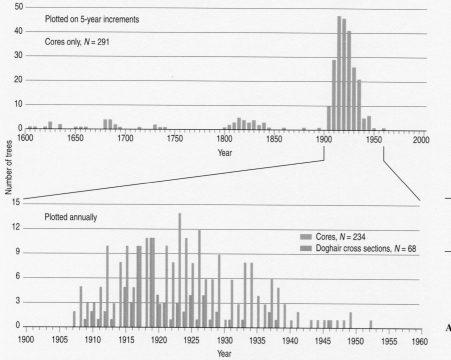

Fig. 3. Ponderosa pine recruitment dates from Monument Canyon Research Natural Area (Muldavin et al. 1995). Dates are estimated from increment cores taken from living trees over 10 cm diameter at breast height and from cross sections taken at root crown from poles in doghair thickets. Because increment cores were taken near ground level but often did not intersect the pith, dates from cores are considered less accurate than dates from cross sections.

See end of chapter for references

Author

Craig D. Allen
U.S. Geological Survey
Biological Resources Division
Midcontinent Ecological Science Center
Jemez Mountains Field Station
Bandelier National Monument
Los Alamos, New Mexico 87544

quantity and diversity of herbaceous understory plants. Fire suppression has also changed ponderosa pine forests in more subtle but important ways. For example, the buildup of flammable aromatic, monoterpenoid compounds in the needle litter caused by fire suppression inhibits nitrification in ponderosa pine ecosystems (Covington and Sackett 1992; White 1994), which in turn disrupts nitrogen cycling and contributes to poor tree growth.

Although crown fires in ponderosa are unknown from regional fire histories (Swetnam 1990), by the 1950's crown fires had begun to occur as ponderosa pine forest structures changed. Today, the potential for crown fire is widespread in southwestern pine forests. Thus, fire suppression has led us into a "catch-22" situation, where the more we suppress fires, the more hazardous the fuels become, leading to further suppression efforts. Yet, despite these efforts, more fires are occurring that cannot be suppressed, as crown fires develop during extreme weather conditions in the extreme fuel conditions that active fire suppression helped create. Increasing urbanization in and near forests further complicates the fire suppression picture.

The dramatic ecological results of unnatural crown fires on ponderosa pine forests are illustrated by the relatively well-studied 1977 La Mesa fire in the Jemez Mountains (Foxx 1984; Allen 1996b). By trying to "protect" ponderosa pine forests from fire we have actually fostered conditions that can destroy these forests (Moir and Dieterich 1988). Historically, these forests have been the focus of timber harvest in the Southwest. As a result, relatively little old-growth ponderosa pine forest remains. On marginal low-elevation sites the harvest of ponderosa pine can cause site conversion to pinyon–juniper woodland.

Almost all of these forests are grazed by livestock in the Southwest; there are few ponderosa pine forests left that are protected from grazing and from which we can assess the effects of grazing on biological diversity. Livestock grazing undoubtedly affects species composition by reducing or removing palatable species and replacing them with thorny, less palatable, or even poisonous species and nonindigenous species. We suspect a significant trend in the reduction of biodiversity in these forest ecosystems is a function of fire suppression and grazing, but further research is needed.

Noss et al. (1995) reported that old-growth ponderosa pine forests in the intermountain west, including Arizona and New Mexico, are endangered, having suffered 85%–98% areal declines due to destruction, conversion to other uses, and significant degradation in structure, function, or composition since settlement.

Madrean Forests

Along the border with Mexico in southeastern Arizona and southwestern New Mexico occur highly diverse forests dominated by species that are characteristic of the mountains of the Sierra Madre Occidental of northern Mexico (Whittaker and Niering 1965). Within the United States, these forest ecosystems are very restricted and occur among the various sky island mountain ranges that rise from the desert floor (Warshall 1995). The sky islands are the only areas in the United States where trees such as Apache pine, Chihuahua pine, Arizona pine, Mexican pinyon, and Arizona cypress dominate the forests (Brown 1982c; McPherson 1992; Moir 1993; Barton 1994; Muldavin et al. 1996). Associated with these conifers are several evergreen oaks that are extensions of populations in northern Mexico (for example, netleaf oak, silverleaf oak, Emory oak, Arizona white oak, Mexican blue oak, and gray oak). Plant and animal biodiversity levels are high in Madrean forests; in fact, some groups of animals, such as certain insects and hummingbirds, reach their highest diversity levels in these forests (Marshall 1957; Whittaker and Niering 1965; Niering and Lowe 1984; Opler 1995). Herbaceous ground cover is also highly diverse, with as many as 50 different species occurring within a given forest stand (Muldavin et al. 1996).

Like most forests of the Southwest, grazing is ubiquitous, with only a few areas currently protected (for example, Saguaro National Monument in the Rincon Mountains, Ramsey Canyon [The Nature Conservancy] and Garden Canyon [U.S. Army] in the Huachuca Mountains, and Chiricahua National Monument in the Chiricahua Mountains). Historically, these forests were logged for mining timbers (Bahre and Hutchinson 1985), but little commercial forestry is practiced now (U.S. Forest Service, Coronado National Forest Management Plan).

Historically, Madrean forests had high-frequency fire regimes similar to those of other southwestern forest types (Caprio and Zwolinski 1992; Swetnam and Baisan 1996); frequent fires still persist in some Mexican forests. Fire suppression increases potential for crown fires in these forests just as it does in other forest types. Although commercial forest exploitation occurred historically (Bahre 1991), Madrean forests in the United States are now used primarily for recreation, fuelwood, and livestock grazing (U.S. Forest Service, Coronado National Forest Management Plan).

Pinyon–Juniper Woodlands and Juniper Savannas

Pinyon–juniper woodlands cover about 15 million hectares in the Southwest, including about 11.4% of New Mexico (Van Hooser et al. 1993). These woodlands are diverse; in Arizona and New Mexico the U.S. Forest Service distinguishes 32 pinyon and 23 juniper plant community types (Bassett et al. 1987; Larson and Moir 1987). Dominant woody plants include several species of pinyon and juniper, along with many kinds of shrubs (for example, oaks, sagebrush, mountain-mahoganies, and rabbitbrush).

Pinyon–juniper woodlands harbor relatively few endemic vertebrate animals (for example, see Brown 1982d) but contain significant levels of biodiversity in less prominent organisms within the ecosystem, particularly herbaceous vegetation and soil organisms (Gottfried et al. 1995). Recent compilations of information on southwestern pinyon–juniper woodlands include Everett (1987), Evans (1988), Aldon and Shaw (1993), Gottfried et al. (1995), and Shaw et al. (1995).

At higher elevations the woodlands are dominated by pinyons and tend to form more closed-canopied stands that exhibit forest-like dynamics and species composition, commonly including a significant shrub component of oaks and alderleaf mountain-mahogany and limited grasses. In contrast, at lower elevations the woodlands are dominated by junipers in open savannas of scattered trees without a significant shrub component, except in areas where big sagebrush has become dominant in the grasslands. These savannas are a broad ecotone between true woodlands and grasslands or sagebrush shrublands (Dick-Peddie 1993).

The long history of livestock grazing has markedly diminished and altered herbaceous vegetation in most of these semiarid woodlands (Branson 1985; West and Van Pelt 1987; Miller and Wigand 1994), leading to widespread desertification of understory conditions (Gottfried et al. 1995). Major changes occurred in understory herbaceous species composition, density, vigor, and productivity (Wooton 1908; Bahre and Bradbury 1978), including decreases in cool-season grasses and increases in grazing-resistant plants such as snakeweed (Archer 1994). Researchers believe that across the landscape, year-round grazing suppressed former fire regimes in the late 1800's (Branson 1985), because surface fires could no longer spread through the bare interspaces between the trees; however, there is little firm documentation of woodland fire histories in the Southwest (Allen 1989; Despain and Mosley 1990; Gottfried et al. 1995). Fire suppression and reduced

herbaceous competition allowed tree densities to increase within these woodlands, and pinyon–juniper ecosystems expanded upslope into ponderosa pine forests and downslope into grass and shrub communities (Leopold 1924; Bradley et al. 1992; Dick–Peddie 1993; Miller 1994; Miller and Wigand 1994). Van Hooser et al. (1993) estimated that 2.6 billion trees occur on 3.6 million hectares of pinyon–juniper woodland in New Mexico, mostly in young age classes. Tree densities have increased to the point that larger proportions of pinyon–juniper woodland can now support crown fires.

Accelerated precipitation runoff and soil erosion (Fig. 8) commonly occur in the often barren, desertlike interspaces between woodland trees. Ongoing soil erosion is causing significant, permanent losses of site productivity in many southwestern woodlands; Van Hooser et al. (1993) reported evidence of soil erosion on 78% of 1,014 woodland plots in New Mexico. Sampling of more than 3 kilometers of line transects in Bandelier National Monument (New Mexico) woodlands revealed tree canopy coverage ranges of 12% to 45%, herbaceous plant coverage (basal intercept) of only 0.4% to 9%, and exposed soils covering between 38% and 75% of ground surfaces, with widespread sheet erosion evident (C. D. Allen, U.S. Geological Survey, unpublished data). High erosion rates can be expected wherever such large

Courtesy C. D. Allen, USGS

Fig. 8. An eroding pinyon–juniper site in northern New Mexico. Note the barren interspaces between the clumps of trees.

percentages of exposed soil occur in woodlands. At current erosion rates, one intensively studied watershed at Bandelier National Monument will lose its entire mantle of soil, averaging 35 centimeters, in about 100 years.

Extensive efforts to convert woodlands to grassland, largely to improve forage conditions for livestock, have occurred in the Southwest. For example, about 600,000 hectares of pinyon–juniper woodlands were mechanically treated in Arizona during the 1950's and early 1960's (Cotner 1963), as were about 223,000 hectares of woodland on U.S. National Forests alone between 1950 and 1985 (Dalen and Snyder 1987). Herbicide applications have also been used to control woodland trees in the Southwest (Johnsen 1987). Efforts to convert pinyon–juniper woodlands to open grasslands have been greatly reduced on public lands since the 1970's because of environmental concerns and the questionable economics of such treatments given that trees tend to reestablish dominance rather quickly (Dalen and Snyder 1987). Harvests of fuelwood apparently peaked in the 1970's, when harvest levels proved unsustainable in at least some areas (deBuys 1993). The U.S. Forest Service has attempted to focus more attention on active management of southwestern woodlands to restore less erosive watershed conditions (Shaw et al. 1995); these efforts include harvest of fuelwood and thinning projects (Edwards 1995). Given current concerns over woodland influences on watershed conditions, a thorough review should be conducted of the areal extent and ecological effects of the widespread woodland conversion efforts of the 1950's to 1980's. Gottfried et al. (1995) outlined additional ecological research needs for southwestern pinyon–juniper woodlands.

Encinal Oak Woodlands

In the United States, encinal oak woodlands are limited to southeastern Arizona and southwestern New Mexico. These woodlands form moderately closed to open stands (savannas) dominated by a high diversity of evergreen oaks (for example, gray oak, Mexican blue oak, Arizona white oak, Emory oak, silverleaf oak, and netleaf oak), along with border pinyon and alligator juniper. Ground cover is rich in herbaceous plants and grasses, creating one of the more diverse ecosystems in the United States (McLaughlin 1986, 1989).

Because this ecosystem is distributed mostly within forest reserves established between 1902 and 1910, it has not been subjected to extensive settlement or agricultural conversion, and its overall distribution remains comparatively stable when compared with grasslands and shrublands. Before 1902, these woodlands were extensively used for fuelwood for homes and in copper smelters (Bahre and Hutchinson 1985). After the turn of the century, use of fuelwood declined until the energy crisis of the 1970's (Bennett 1992).

Although grazing may have been more concentrated in adjacent lowlands, grazing in oak woodlands exceeded carrying capacity up until the 1960's (Allen 1992; McClaran et al. 1992) and had significant effects. Grazing continues today but at much lower stocking rates (McPherson 1992). As seen elsewhere in the Southwest, grazing by large livestock herds of the late 1800's reduced fuels, which subsequently lowered fire frequencies. Reduced fire frequency has increased the overall density of trees and may be leading to limited expansion of woodlands into semidesert grassland and to juniper encroachment into short-grass prairie. This expansion, though, may be counterbalanced by increased oak seedling mortality due to grazing (McPherson 1992). Grazing may also adversely affect several rare species through changes in species composition and vegetation structure (McClaran et al. 1992). Long-term trends are difficult to ascertain, and we need much more research on the effects of livestock use, lower fire frequency, and consumption of fuelwood on overall diversity and ecosystem dynamics of encinal woodlands.

Desert Shrublands and Semidesert Grasslands

In the Southwest, desert shrublands and semidesert grasslands form a highly diverse and complex mosaic of vegetation across arid landscapes. They extend from the "cold" desert of the elevated Colorado Plateau region of southern Utah, south into the "warm" Sonoran and Chihuahuan deserts of southeastern California, Arizona, New Mexico, and west Texas. We discuss desert shrublands and grasslands together because they occur near one another and are commonly related to one another through time by desertification processes (Schlesinger et al. 1990). The coming of the railroad and the cattle industry to these arid lands in the late 1800's extensively changed ecosystems that, unlike the grasslands of the Great Plains, previously lacked large grazing animals (except for pronghorn). The current mosaic of shrublands and grasslands in the Southwest is in large part a reflection of continuing desertification (Grover and Musick 1990) in concert with urbanization and conversion to agriculture.

In the cool-temperate Colorado Plateau region of southern Utah, northern Arizona, and northwestern New Mexico, most precipitation falls in the winter as snow. In response, much of the native vegetation was at one time a mosaic of deep-rooted shrubs and grassland dominated by cold-tolerant, cool-season bunchgrasses (for example, western wheatgrass, Indian ricegrass, galleta), which are most productive during spring and early summer. The extent of these grasslands has been greatly reduced, and what remains may represent one of the rarest ecosystems in the Southwest (Cottam 1947; Hull and Hull 1974). The decline can be attributed to several causes: native cool-season bunchgrasses sustain high mortality when grazed heavily in spring (Stoddard 1946) and are generally not tolerant of grazing (Branson 1985); fire suppression has favored shrub species; accelerated soil erosion has permanently altered site conditions; and bare ground is aggressively colonized by nonindigenous Eurasian annual grass species such as cheatgrass. The loss of perennial grasses occurred in only 10–15 years of overgrazing by cattle near the end of the last century (Hull 1976).

Much of what was grassland in the early nineteenth century has been converted to shrublands dominated by saltbush, greasewood, and big sagebrush, a hardy, cold-tolerant shrub that shapes the ecosystems it dominates. The expansion of big sagebrush on the Colorado Plateau has been remarkable (Gross and Dick-Peddie 1979; Dick-Peddie 1993); the shrubs are commonly widely spaced with herbaceous plants and grasses living beneath them and with intershrub spaces that are barren or contain microphytic crusts composed of lichens and algae. The shrubs form islands of fertility that concentrate water and nutrients and that are not easily altered (West 1983). There are sites where these shrublands have always existed as natural soil- and climate-determined communities, but only a few are protected enough to give us a sense of what presettlement conditions were like (Edwards 1995).

Grazing on these ranges continues today, and extensive amounts of land also are being converted to agricultural production, particularly in the Four Corners area, where Arizona, Colorado, New Mexico, and Utah meet. Conversion of dominant plant cover to annual nonindigenous species such as cheatgrass is expected to continue (Rogers 1982). Many of these nonindigenous plants are flammable (Fig. 9). Once these shrublands—with or without nonindigenous annuals—become established, there is only limited potential for conversion to native grasslands, either mechanically or by removal of grazing (Potter and Krenetsky 1967; Rice and Westoby 1978; West et al. 1984).

To the south of the Colorado Plateau, the Sonoran Desert covers much of southeastern California, south-central and southwestern Arizona, and northwestern Mexico; summers

there are very warm and winters are mild. In the northern and western parts of the desert, precipitation falls mostly in winter but as rain rather than as the snow that occurs in the northern plateau region. In the southern and eastern portions of the desert, the rainfall pattern shifts to a predominantly summer regime (McClaran 1995). The Sonoran Desert is characterized by great floral richness, including large columnar cacti, such as saguaro, cardon, and organpipe, as well as many shrub species and an undergrowth of grasses and herbaceous plants with subtropical affinities (Brown 1982a).

The western lowland basins (lower Colorado River valley of Brown 1982a) are dominated by creosotebush, white bursage, and saltbushes. In uplands and to the east, saguaro, yellow paloverde, and ocotillo become more prominent and form the classical Arizona desert. This area has an extremely high diversity of spiny shrubs and subtrees, cacti, yuccas, agaves, herbaceous plants, and grasses (MacMahon 1979), which make this ecosystem one of the most diverse in the Southwest or perhaps in the entire United States.

To the east across the Continental Divide lies the Chihuahuan Desert, which has colder winters and summer-dominated precipitation (monsoons). The Chihuahuan Desert extends from south-central New Mexico southward into western Texas and Mexico. Creosotebush and saltbushes are still common dominants but share

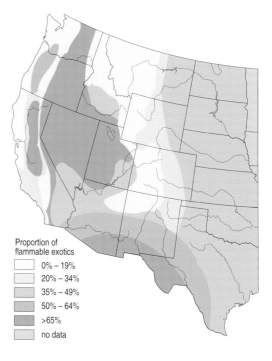

Proportion of
flammable exotics

☐	0% – 19%
☐	20% – 34%
☐	35% – 49%
☐	50% – 64%
☐	>65%
☐	no data

Fig. 9. Geographic distribution of flammable nonindigenous plants in the western United States (U.S. Environmental Protection Agency, Environmental Systems Monitoring Laboratory, Las Vegas, Nevada).

the landscape with honey mesquite, cat-claw acacia, tarbush, and a variety of noncolumnar cacti.

Semidesert grasslands are intermixed with the desert shrubs and extend farther northward or to higher elevations, where they meet encinal and pinyon–juniper woodlands or the short-grass prairie of the Great Plains (McClaran and Van Devender 1995). These desert grasslands are dominated by warm-season grasses, including black grama and other "southern" gramas, tobosa grass, and sacatons. These grasslands, with distinctive shrubby components of sotols, joint-firs or Mormon teas, and yuccas, are especially productive after the summer rains arrive in late summer.

At present, Sonoran and Chihuahuan desert shrubs are expanding their ranges at the expense of semidesert grasslands through a widespread desertification process brought on by livestock grazing in climatically marginal ecosystems of low production and resilience (Grover and Musick 1990; Bahre and Shelton 1993); these grasslands have been declining for more than a century. Grazing pressure was severe across the range in the last half of the nineteenth century; in Arizona alone livestock numbers rose from 5,000 in 1870 to 1,500,000 in 1891 (Hastings and Turner 1965). As a result, these grasslands were greatly altered by 1900 in the United States and by 1940 in Mexico (Brown 1982a); most areas were converted to degraded forms of Sonoran Desert shrubland (Humphrey 1987), and this trend is continuing. A recent satellite imagery study in southern Arizona (Fig. 10) revealed a 35% decrease in overall grassland cover, a 60% decrease in the size of grassland patches, and an increase in the number of patches over time. Only a few relict areas in southern Arizona (for example, the Muleshoe Preserve and around Sonoita) continue to support semidesert grasslands in the Sonoran Desert.

Similarly, in New Mexico and western (Trans-Pecos) Texas, researchers have documented extensive conversions of Chihuahuan semidesert grasslands to Chihuahuan Desert shrubland (Buffington and Herbel 1965; York and Dick-Peddie 1969; Branson 1985; Grover and Musick 1990). Even though livestock numbers in these areas have stabilized well below historical levels, these are marginal grazing habitats, and the combination of reduced fire frequency (Moir 1980) and continued topsoil erosion is likely sustaining an irreversible decline in which much of the remaining grassland is being converted to desert shrubland (Dick-Peddie 1993). The introduced species, honey mesquite, in particular is greatly expanding its range and now dominates extensive areas that once were grasslands. More than 28 million hectares are now dominated by mesquite,

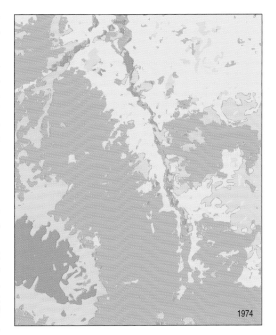

Fig. 10. Grassland cover (green areas) is decreasing in Southern Arizona (drawn from a recent satellite photo [Kepner 1994]).

an increase of 100% or more over a 100-year period (Parker and Martin 1952). Once this aggressive invader controls a site, significant changes in ecosystem functions occur: water and nutrients become nonuniformly distributed in space and time and are confined to zones beneath shrubs (Schlesinger et al. 1990). This creates a new, relatively heterogenous and stable soil environment when compared with the original grasslands. Such changes may be irreversible, although Robinett (1995) has documented the effective use of fire to reduce mesquite and increase grass cover in the Sonoran Desert.

Extant Chihuahuan semidesert grasslands in good condition in New Mexico and western

Texas are rare today. Some limited de facto protection occurs on military reserves such as White Sands Missile Range in New Mexico and Fort Bliss in Texas. Large stands of these grasslands have been protected as an incidental outcome of military activities that preclude livestock use. Increased military activities on these sites, coupled with the presence of nonindigenous grazers (such as gemsbok, feral horses, and cattle that have trespassed) may be significantly affecting the ecosystems and leading to degradation. The only known sites that are truly protected from grazing by nonindigenous species are within national parks and national wildlife refuges (for example, Big Bend National Park in Texas and Bosque del Apache National Wildlife Refuge in New Mexico) and a few small, scattered research natural areas and

private preserves. Outside of these sites the actual amount of semidesert grassland that has not been degraded is known imprecisely. To assess and analyze the trends of the remaining extent of functional semidesert grassland ecosystems in the Southwest, researchers need to use high-resolution satellite imagery (for example, Loveland and Hutcheson 1995).

Within the Sonoran Desert, there are significant indications of decline of biodiversity caused by agricultural conversion, nonindigenous species invasion, and urbanization. Much of the saltbush ecosystem has been converted to irrigated agriculture (Brown 1982a). In addition, nonindigenous European grasses and herbaceous plants such as sandburs, Lehmann lovegrass, and weedy mustards are displacing perennial native species (Waser and Price

Soils and Cryptobiotic Crusts in Arid Lands

Cryptobiotic crusts (Fig. 1) are important features of arid and semiarid ecosystems throughout the Southwest, including pinyon–juniper woodlands and deserts. More data on the ecological role played by these crusts in the Southwest are needed, given the widespread (but unsubstantiated) belief among many range managers that the breaking up of such crusts by livestock hoof action may be beneficial (Belnap 1990; Ladyman and Muldavin 1996).

Living soil crusts are found throughout the world, from the hottest deserts to the polar regions. In arid regions, these soil crusts are dominated by blue-green algae

and also include soil lichens, mosses, green algae, microfungi, and bacteria (Belnap 1990; Johansen 1993; Ladyman and Muldavin 1996). In the cold deserts of the Colorado Plateau region (parts of Arizona, Colorado, New Mexico, and Utah), these crusts are extraordinarily well developed, often representing more than 70% of the living ground cover (Belnap 1990).

Blue-green algae occur as single cells or filaments; the most common form found in desert soils is the filamentous type. The cells or filaments are surrounded by sheaths that are extremely persistent in these soils. When moistened, the blue-green algal filaments

become active, moving through the soils and leaving behind a trail of the sticky, mucilaginous sheath material, which sticks to surfaces such as rock or soil particles, forming an intricate webbing of fibers in the soil. In this way, loose soil particles are joined, and otherwise unstable and highly erosion-prone surfaces become resistant to wind and water erosion. The soil-binding action is not dependent on the presence of living filaments, however—layers of abandoned sheaths, built up over long periods, can still be found clinging tenaciously to soil particles at depths greater than 15 centimeters in sandy soils, thereby providing cohesion and stability in loose sandy soils (Belnap and Gardner 1993).

The crusts are important in the interception of rainfall. When moistened, the sheaths absorb up to 10 times their volume of water. The roughened surface of the crusts slows precipitation runoff and increases water infiltration into the soil, which is especially important in arid areas with sporadic, heavy rainfall. Vascular plants growing in crusted areas have higher levels of many essential nutrients than plants growing in areas without crusts. Electron micrographs of sheaths (Fig. 2) show that they are covered with fine clay particles upon which essential nutrients cling, thereby keeping the nutrients from being leached out of the upper soil horizons or from being bound in a form unavailable to plants. In addition to stabilizing surfaces and increasing water harvesting, crustal organisms also contribute nitrogen and organic matter to ecosystems, functions that are especially important in desert ecosystems where nitrogen levels are low and often limit productivity.

Fig. 1. Cryptobiotic crusts on the Colorado Plateau.

Courtesy J. Belnap, USGS

Courtesy J. Belnap, USGS

Fig. 2. Micrograph of filamentous cryptobiotic crust showing sheaths with attached clay particles.

Unfortunately, many human activities are incompatible with maintaining these blue-green algal crusts. The blue-green algal fibers that confer such tensile strength to these crusts are no match for the compressional stress placed on them by machinery or by being stepped on by cows or people, especially when the crusts are dry and brittle. Crushed crusts not only contribute less nitrogen and organic matter to the ecosystem, but the impacted soils are also highly susceptible to wind and water erosion. In addition, raindrop erosion increases and consequent overland water flows carry detached material away, a severe problem when the destruction has occurred in a continuous strip, as it does with vehicular or bicycle tracks. Such tracks are highly susceptible to water erosion and quickly form channels, especially on slopes. After such damage, wind blows away pieces of the pulverized crust and also blows around the underlying loose soil, covering nearby crusts. Since crustal organisms depend on photosynthesis, burial can mean death. When large sandy areas are impacted in dry periods, previously stable areas can become a series of moving sand dunes in only a few years.

Large areas that are disturbed may never recover. Under the best circumstances, a thin veneer may return in 5 to 7 years. When the crust is disturbed, nitrogen fixation stops and underlying sheath material is crushed. Damage done to the abandoned sheath material underneath the surface cannot be repaired because the living organisms occur only on the surface. Instead, sheaths must build up slowly after many years of blue-green algal growth.

See end of chapter for references

Author

Jayne Belnap
U.S. Geological Survey
Biological Resources Division
Midcontinent Ecological Science Center
Canyonlands National Park Field Station
2282 S. West Resource Boulevard
Moab, Utah 84532

1981; Chew 1982; P. Warren, The Nature Conservancy, Phoenix, Arizona, personal communication). Ongoing urban sprawl, particularly around Phoenix and Tucson, creates air and noise pollution, site conversion to asphalt with its subsequent associated localized increases in air temperatures, and accelerated recreational use. Consequently, desert wildernesses are changing, if not vanishing, even within national parks and monuments.

With respect to grazing, some researchers have suggested that Sonoran shrublands, particularly in lowland areas, naturally lack a significant grass component and historically were less extensively grazed (Hastings and Turner 1965; Branson 1985). In contrast, Bahre and Shelton (1993) presented convincing data that these areas have experienced significant changes, particularly from the increased invasion of mesquite. Niering et al. (1963) concluded that even limited grazing in marginal ecosystems dominated by saguaro has caused long-term, possibly irreversible, declines of that cactus species. Bryant et al. (1990) presented evidence that vegetation conditions deteriorate even more immediately across the United States–Mexico border, apparently because of more severe grazing effects. Assessments of range conditions and trends are badly needed in the border area.

Historically, in the Chihuahuan Desert, most of the range was grazed extensively; it continues to be grazed even though plant and livestock production is low. As a result, desertification resulting in conversion of desert grassland to desert shrubland since the 1850's has been well documented (Buffington and Herbel 1965; York and Dick-Peddie 1969); this process continues today where livestock are not well-managed and stocking levels have not been reduced (Holechek 1991; Dick-Peddie 1993). As with the Sonoran Desert, urbanization and agricultural conversion are increasing as human populations expand. More information is needed to assess the long-term effects of humans on overall biological diversity of desert scrub ecosystems and to assess the specific effects of human activities, including the removal of woody plants that serve as nurse plants to other species, poaching of cacti and other plants for landscaping, and the introduction of nonindigenous grasses as livestock forage.

Scattered across the Southwest are smaller-scale landforms that include sand dunes, warm springs, and seasonally flooded playa lakes. Some of these azonal (in the sense of MacMahon 1979) areas, such as washes or arroyos, are locally dominant and may exhibit considerable endemism, as in the White Sands of New Mexico. Playas are low, internally drained basins that, because of water inundation and evaporation, become saline over time; vegetation in such areas typically consists of salt-loving plants such as saltbush. These smaller-scale landforms add considerably to regional biological diversity of plants and invertebrates and some are important to larger vertebrate wildlife, at least seasonally. Desert springs, for example, are quite important to many endemic plants and fishes.

Riparian Ecosystems

Riparian landscapes occur along streams and rivers and represent an interface, of varying width, between dry and wet systems; these landscapes are defined by the plants that inhabit them, that is, plants occurring near streams or rivers and not in drier environments. Riparian plants are dependent on an intact hydrological regime where groundwater is maintained and natural surface flows occur. Vegetation dynamics are integrated with, and dependent on, annual cycles of flooding and minimal flows, as well as sediment availability (Crawford et al. 1993). At lower elevations in the Southwest, riparian zones are the only places where sufficient moisture exists for large, broad-leaved trees to germinate and grow. These zones typically consist of cottonwood gallery forests, usually with an admixture of willows and other water-dependent plants. Historically, disturbances in flood-prone channels in the Southwest created vegetational mosaics of wetland communities, with different combinations of species dominating different areas (Campbell and Green 1968). Along the Rio Grande in New Mexico, for example, the historical vegetation probably consisted of patches of cottonwood–willow forest (bosque) and other wetlands, including wet meadows, marshes, and newly deposited alluvium (Crawford et al. 1993; Durkin et al. 1995). Riparian systems are now increasingly characterized by nonindigenous plants such as saltcedar, Russian-olive, and Siberian elm (Crawford et al. 1993).

In the Southwest, riparian landscapes are invaluable. Although they represent less than 1% of the region's area (Knopf et al. 1988), a large proportion (75%–80%; Gillis 1991) of vertebrate wildlife species depends on riparian areas for food, water, cover, and migration routes. Riparian zones also improve water quality because they filter sediments and nutrients; accumulated sediments in riparian zones store large amounts of water, which helps sustain streamflow during drier times. Abundant riparian vegetation, however, sustains high levels of photosynthesis and transpiration; 99% of water entering a plant through the root system is lost by transpiration from the leaves (Knight 1994). This water transpired by plants is lost to downstream users. Most land managers accept such losses in exchange for benefits of shade (which lowers losses by evaporation), increased forage, sediment accumulation, biological diversity, wildlife habitat, and longer sustained flows in late summer.

In many areas, especially at higher elevations, beaver probably played an important role in creating and maintaining riparian areas (Knopf and Scott 1990). Tree cutting and dam building by beavers trap alluvial sediments, provide opportunities for new plant growth, and increase the diversity of wildlife habitats. Removal of beavers or their dams, together with livestock grazing, has contributed to arroyo cutting and gullying of the landscape (see Denevan 1967 for a review on relative roles of grazing and climate on arroyo-cutting episodes). As the channel cuts deeper and the gradient increases, the water table is lowered and surface sediments begin to dry out; gradually, the vegetation becomes composed of plants tolerant of drier conditions.

Many of the conditions that make riparian zones relatively rare and valuable, particularly in a semiarid landscape, also make them fairly sensitive to disturbance and change. When free-flowing water is impounded or diverted from the main channel (by dams, diversions, irrigation, or channelization), the nature of the riparian landscape changes (see box on Impounded River Systems in Water Use chapter). Construction of impoundments, whether for flood control or water storage, has largely decreased or eliminated the shifting of river channels that historically created mosaics of riparian vegetation, especially cottonwood and willow habitat (Crawford et al. 1993). With less flooding, there is less channel shifting and less suitable habitat for establishment of cottonwood seedlings, which are dependent on recently inundated sediments to become established. Where inundation does not occur, seedlings will not establish, thereby leaving the ground available for other species (Durkin et al. 1995). Modification of historical disturbance regimes results in a decline in diversity of native species because when the frequency or intensity of a natural disturbance is decreased, competitively superior nonindigenous plants may invade (Hobbs and Huenneke 1992). As existing riparian forests age without replacement, they become a monoculture of maturing trees that eventually senesce, die, become victims of fires (caused by vandals or carelessness), and disappear from the landscape (Howe and Knopf 1991).

Overgrazing has been a major factor in the alteration and degradation of riparian areas (Cooperrider and Hendricks 1937; Armour et al. 1991). Livestock grazing typically results in reduction of plant species diversity and density, especially of palatable species such as willows and cottonwood saplings (Rickard and Cushing 1982; Cannon and Knopf 1984; General Accounting Office 1988; Schultz and Leininger 1990). Changes in species composition, relative abundance of species, and plant density cause overall plant community structure to change.

With heavy grazing, whether by big game or livestock, stabilizing vegetation deteriorates, banks are eroded, water storage capacity declines, water quality declines, streambeds become wider and stream depths shallower, water temperatures increase, and fish and aquatic invertebrate habitat quality declines (Crawford et al. 1993). Cottonwoods do not tolerate water stress well and may decline as groundwater becomes less available.

Additional causes of bank erosion include road building, construction, and other developments. Stream degradation also is caused by nutrients and fertilizers entering the water and resulting in increased eutrophication and by interactions of factors such as logging, fires, and overgrazing.

We probably know more about the responses of southwestern bird communities to riparian landscape changes due to grazing than we do about the responses of other animal groups (Rea 1983; U.S. Fish and Wildlife Service 1995). Reduction, modification, or removal of cattle grazing led to increases in abundance for several bird species associated with cottonwood–willow habitat on the San Pedro River (Krueper 1993). At other sites, 40% of the riparian bird species were negatively affected by livestock grazing (Bock et al. 1993), and a negative correlation between recent cattle grazing and abundance of several riparian birds was found (Taylor 1986). Farley et al. (1994) showed the importance of even young cottonwood riparian habitat to migrant birds in the middle Rio Grande ecosystem.

About 90% of total water consumption in western states is accounted for by agriculture and evaporation from reservoirs (Crawford et al. 1993). Irrigation by flooding has some negative effects, including nutrient leaching, which leads to fertilizer losses and to eutrophication of streams and groundwater. In addition, salts that have accumulated in soils may be transported downstream, creating saline water supplies. Irrigation water that is high in dissolved salts also favors nonindigenous saltcedar, which is more tolerant of high salt levels, at the expense of native cottonwood and willow (Kerpez and Smith 1987; Busch and Smith 1993).

Increased channelization of the lower Colorado River appears to have led to floodplain groundwater declines and has isolated riparian vegetation from its moisture source (Busch et al. 1992). Cottonwoods are now rare along the lower Colorado River and are represented mostly by old senescent individuals with little or no regeneration. Willows are somewhat more common but are increasingly challenged by nonindigenous species, particularly saltcedar (Busch et al. 1992). These changing water regimes give saltcedar a competitive edge, as it does not need floods to establish. Also, cattle won't eat saltcedar but do graze on shoots and seedlings of cottonwood and willow. The deep root system of saltcedar allows it to survive when water tables are lowered and surface flows are no longer present. Furthermore, establishment of saltcedar results in a regime of episodic fires, which researchers believe are uncommon in most native riparian woodlands (Busch and Smith 1993).

Noss et al. (1995) ranked riparian forests in Arizona and New Mexico as endangered, with 85%–98% declines due to destruction, conversion to other uses, or significant degradation in structure, function, or composition since settlement by Europeans. Large-scale losses of southwestern wetlands have occurred; the cottonwood–willow plant community has declined the most with modern river management (Rosenberg et al. 1991; U.S. Fish and Wildlife Service 1995). Exact amounts of loss vary from site to site, but in some places, such as the lower Colorado, lower Gila, lower Salt, and Rio Grande rivers, loss or modification of riparian habitat may be close to 100%. Overall, a 90% loss of presettlement riparian ecosystems has occurred in Arizona and New Mexico (Arizona State Parks 1988). Of the riparian areas under Bureau of Land Management control, an estimated 83% is in unsatisfactory condition (Almand and Krohn 1979). All major watercourses in southern Arizona suffered entrenchment and became more ephemeral in flow by about 1890, and riparian habitats were significantly altered by 1900 (Hastings and Turner 1965; Bahre 1991).

Riparian areas are not the only wetland losses in the Southwest. Dahl (1990) estimated losses of wetlands between 1780 and 1980 at 36% in Arizona, 33% in New Mexico, and 30% in Utah. *Cienegas*, a unique southwestern form of wetland upon which many animals depend, have suffered a 70% loss in Arizona since settlement, which places them in the threatened category (70%–84% decline) of Noss et al. (1995).

Even before settlement by Europeans, people and animals congregated along riparian strips in the Southwest. Following settlement by Europeans, however, livestock congregated there too. Metropolitan centers usually occur in riparian areas, and land ownership is overwhelmingly private. Urbanization, as exemplified by growth in large southwestern cities such as Albuquerque, El Paso, Phoenix, and Tucson, is changing riparian landscapes dramatically— in some cities the rivers have ceased flowing (for example, the Gila and Salt rivers in Phoenix and the Santa Cruz River in Tucson). At present, only relatively small patches of riparian habitat remain that are managed for their inherent values by local municipalities, Native American tribes, and the U.S. Fish and Wildlife Service.

Changing Landscapes of the Middle Rio Grande

Before the fourteenth century, the Rio Grande between Cochiti and San Marcial, New Mexico, was a perennially flowing, sinuous, and braided river (Crawford et al. 1993). The river migrated freely over the floodplain, limited only by valley terraces and bedrock outcroppings; this shifting of the river created ephemeral mosaics of riparian vegetation (forests and shrublands) and wetlands (ponds, marshes, wet meadows) (Durkin et al. 1995). Water diversion for irrigated agriculture by Native Americans and later by European immigrants may have somewhat diminished river flows during growing seasons before 1900. Increased sediment loading, the result of climatic variations and agriculture, caused the river's channel to become broader and shallower, which increased the river's tendency to flood.

In the late 1800's, groundwater levels in the Rio Grande floodplain rose dramatically because of a rising riverbed, irrigation, and poor return of irrigation water. Salts, leached upward by the rising groundwater, created salinity problems. Levees, built in the 1920's and 1930's to cope with floods, tended to constrain the floodway and channel, thereby reducing the river's tendency to meander, which is critical for establishment of native bosque (cottonwood–willow) vegetation. In addition, the riverbed aggraded inside the levees so that by the 1950's it was higher than adjacent downtown Albuquerque. Upstream dams were built largely for flood and sediment control, as well as water storage, and drainage systems were established to lower water tables in the floodplain. These actions, combined with water diversion channels and increased groundwater pumping in Albuquerque, disrupted the connection between the river water and groundwater in the floodplain; thus, hydrological conditions in the riparian zones were no longer linked in a natural historical way (Crawford et al. 1993).

Cottonwood–willow forests have also been reduced by land clearing, tree harvesting, water diversion, and agricultural uses. About 90% of the Rio Grande's water is used for agriculture in the middle Rio Grande valley (Crawford et al. 1993). Livestock graze back new riparian vegetation (young cottonwood and willow), which contributes to watershed erosion and leads to increased sediment loading in the river. Groundwater drainage and the absence of periodic flooding caused most of the valley's wetlands to dry up. Plant and animal species dependent on such areas have disappeared or are confined to restricted habitats. Cottonwood and willow have been widely replaced by species that are not as reliant on spring flooding and inundation to reproduce—saltcedar in southern reaches and Russian-olive in northern ones (Figure).

Roelle and Hagenbuck (1995), who documented surface cover changes in the Rio Grande floodplain from 1935 to 1989, found that five of eight wetland cover types declined by 17,000 hectares (45%) in that period; largest gains during the period were in urban and agricultural cover types. Only

Courtesy J. N. Stuart, USGS

Courtesy J. N. Stuart, USGS

Figure. Riparian vegetation. Mature cottonwood site (top) at Bosque del Apache National Wildlife Refuge, Socorro County, New Mexico, showing the relatively open nature of such stands, and a stand of nonindigenous saltcedar (bottom) on the Rio Hondo, Chaves County, New Mexico, showing the almost impenetrable nature of invading stands.

three wetland or riparian cover types increased: lake, wetland forest, and dead forest or scrub–shrub. The lake increase, though, was due to higher water levels in a large impoundment (Elephant Butte Reservoir), and wetland forest increase was primarily due to increasing forest cover between levees and the river channel, which has become narrower and straighter because of channel stabilization. Only 27% of the area forested in 1935 still supports forests. The flow regime of the river has been altered significantly, with lower peak flows, which means that cottonwood regeneration rarely occurs. Under current hydrological conditions, Russian-olive and saltcedar are likely to continue to replace cottonwood. Even though the middle Rio Grande valley in New Mexico supports the most extensive cottonwood gallery forest remaining in the entire Southwest (W. Howe, U.S. Fish and Wildlife Service, Albuquerque, New Mexico, personal communication), human-induced changes in hydrology and land use are rapidly shrinking remaining forests.

See end of chapter for references

Author

Michael A. Bogan
U.S. Geological Survey
Biological Resources Division
Midcontinent Ecological Science Center
Department of Biology
University of New Mexico
Albuquerque, New Mexico 87131

Because riparian ecosystems are critically dependent on disturbance caused by occasional high water flows, if these flows have been diminished or halted, natural riparian zones are being lost, along with their associated faunas. Some way must be found to produce or preserve flows that aid in the establishment, growth, and survival of riparian-dependent species.

Status and Trends of Plants

The diversity of vascular plant species is high throughout the Southwest; New Mexico and Arizona, especially, may be the most floristically rich areas in the United States (Morin 1995). New Mexico is estimated to have about 3,900 taxa of vascular plants (Martin and Hutchins 1980, 1981), and Arizona is similarly diverse with about 3,370 species of flowering plants and ferns (Kearney and Peebles 1960). Statewide, Utah has about 2,500 species of indigenous vascular plants (MacMahon 1988b) and 580 introduced species (Welsh et al. 1987). Despite the richness of regional species, endemism among southwestern vascular plants is not striking. About 150 taxa of vascular plants are endemic to New Mexico (Dick-Peddie 1993)—less than 4% of the total flora. Of these endemics, 24 are cacti, 10 are annual plants, 9 are woody shrubs, 3 are biennials, and 2 are indeterminate, having been collected only once, more than a century ago. About 5% of Arizona plants are endemic with 46 species confined to northern Arizona, 28 to central Arizona, 74 to southern Arizona, and 16 others widely distributed through the state (Kearney and Peebles 1960). Researchers believe that endemism is high among plants on the Colorado Plateau of Utah (Welsh et al. 1987).

Regionwide, the U.S. list of endangered and threatened plants shows that about 40 southwestern species are jeopardized. Although vertebrates dominated these lists during the early years of the U.S. Endangered Species Act, invertebrates (13%) and plants (48%) now represent the largest proportion of listed taxa (Flather et al. 1994). Arizona, New Mexico, and Utah each list from 50 to 140 taxa of plants as sensitive (formerly candidates for listing under provisions of the U.S. Endangered Species Act of 1973; memorandum from the director of the U.S. Fish and Wildlife Service, 19 July 1995). Sensitive plants in the region include cacti, wild buckwheats, prickly poppies, milk-vetches, paintbrushes, penstemons, sagebrushes, saltbushes, and others. The sunflower, pea, cactus, and figwort families account for more than half of the species of special concern in New Mexico (Dick-Peddie 1993).

Habitat alteration and incompatible land use are major threats to the region's rare plants. Some species, primarily cacti and those with showy flowers, are also threatened by overcollecting for the horticultural trade (Morse et al. 1995). Other species (for example, cacti, yuccas, agaves, penstemons) also are subject to poaching and commercial exploitation (Dick-Peddie 1993).

In general, most southwestern life zones have sensitive species of plants, and most of these zones have been subjected to perturbations such as grazing, fire suppression, road construction, and urbanization, which account for many of the threats to sensitive plants.

Arizona, California, New Mexico, Texas, and Utah have some of the highest proportions of globally rare native plants in the country (Morse et al. 1995), with proportions ranging from 12% to 17%. Still, less than 1% of the native flora may have been extirpated in these states (Morse et al. 1995). From a community perspective, 56% of rare plant communities occur in the western United States, including the Southwest (Grossman and Goodin 1995).

Many southwestern plants are poorly known, especially the small, unobtrusive species. For example, the moss flora of the

Southwest is one of the least known in the United States, and our knowledge of liverworts and hornworts in New Mexico, Arizona, and surrounding regions is the poorest in the country (Whittemore and Allen 1995). Some of the most poorly known areas for vascular plants are also in the Southwest, including parts of Arizona north of the Colorado River, portions of New Mexico, and parts of the Colorado Plateau (Morin 1995). Thus, in much of the Southwest, a considerable amount of basic floristic research needs to be done.

Status and Trends of Animals

Invertebrates

Invertebrates, like plants, are relative latecomers to state and federal lists of endangered and threatened species. At present, invertebrates represent about 13% of all federally listed threatened and endangered species (Flather et al. 1994), although worldwide they represent 90% of all animals (Mason 1995). As of 31 August 1992, there were 91 taxa of invertebrates on the U.S. list of endangered and threatened wildlife. Invertebrate species listed as threatened or endangered in the Southwest include one ambersnail in Arizona and Utah, and two springsnails and one isopod in New Mexico—about 4% of the total number of federally protected invertebrates.

Increasing numbers of invertebrates are showing up on more recent listings of sensitive species (U.S. Fish and Wildlife Service 1994a). In the last few years, listed sensitive invertebrates included a millipede from New Mexico, a pseudoscorpion from Arizona, two

Rare Aquatic Snails

State Natural Heritage Programs in the Southwest have identified 30 species of rare aquatic snails (Figure a), most in the mollusk family Hydrobiidae. The Hydrobiidae are at risk throughout North America, with 16 endangered, threatened, or U.S. Fish and Wildlife Service Category 1 candidate species, and 90 species of concern (Mehlhop and Vaughn 1994).

Several physiological and ecological aspects of rare southwestern snails render them vulnerable to extirpation. All are gill breathers and thus are intolerant of drying or anaerobic conditions. Individual snails tend to live about one year, making annual reproduction essential. Most snail species are geographically restricted to natural springs and nearby wetlands, with 83% of the species having a total range of less than 10 square kilometers. These mostly isolated habitats inhibit migration—of 30 snail species, most occur at only a single spring, and most of the others are found at only two or three springs (Figure b). Water-use activities that have altered the quantity or quality of many spring waters also threaten the snails (Figure c). Of 26 species for which threats have been assessed, only two were found to have no substantial identified threats.

Although the status of most of these rare aquatic snails is vulnerable (Mehlhop and Vaughn 1994), there are reasons for optimism. More than half (53%) of the snail habitats are in springs that are managed fully or in part by federal or state agencies or by private conservation organizations (Figure d). Also, although water-use activities appear to pose significant threats to the

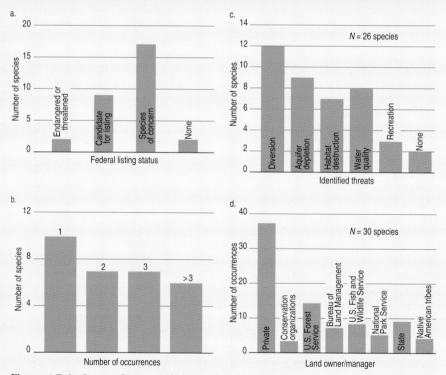

Figure. a) Federal status of rare or declining snails in the Southwest; b) number of known occurrences per species of rare aquatic snails; c) reported threats to rare aquatic snails in the Southwest; d) landowner or management agency of sites where aquatic snails in this study occur.

long-term viability of these species, these threats have existed for most of these species for decades, suggesting that such activities and snails may be able to coexist.

See end of chapter for references

Author

Patricia Mehlhop
New Mexico Natural Heritage Program
Department of Biology
University of New Mexico
Albuquerque, New Mexico 87131

amphipods—one from Arizona and another from New Mexico—and several mollusks, mostly springsnails, talussnails, and physas. Almost all sensitive invertebrates are from aquatic habitats and probably are subject to the same type of environmental threats that also affect southwestern fishes. Beyond that, individual state lists of sensitive species seem to represent local and existing expertise and concerns.

Sensitive insect species include 18 taxa from Arizona, 3 from New Mexico, and perhaps 4 from southern Utah. Lepidoptera (moths and butterflies) represent about 13% of North American insect species and attract attention from laypersons and professionals. Nonetheless, much remains to be learned of the basic taxonomy, distribution, and status of butterflies and moths (Powell 1995). A survey for large Lepidoptera in a 10-square-kilometer area in Arizona has taken 13 years, is 95% complete, and has identified over 900 species (Powell 1995). The highest species richness of butterflies and selected moth families in the western United States is found in Arizona (437 species), New Mexico (410 species), Utah (281 species), and western Texas (Big Bend, 199 species) (Opler 1995). Species richness of western Lepidoptera is usually highest in areas that adjoin the Mexican border. Each of five subregions, mostly in the Southwest, has many endemic species, of which 20 or more are potential candidates for listing as endangered species (Opler 1995). Threats to moths and butterflies include overgrazing, urbanization, and excessive modification or recreational use of specialized habitats such as wetlands or dunes (Opler 1995).

About 90,000 species of insects have been described from North America (north of Mexico) alone, with an estimated 72,500 species yet to be described (Hodges 1995). Given the diversity of invertebrates in the Southwest, we suspect that they are underrepresented as listed or sensitive species. Clearly, much basic collection and taxonomic work remains before we can meaningfully interpret status and trends of many invertebrates. Nonetheless, many invertebrates are threatened by the same factors that threaten other animal species—changes in water regimes, habitat modification and alteration, pollutants, urbanization, and nonindigenous species. For example, Africanized honey bees recently invaded the Southwest. (The ecological implications of this invasion on the biota of the region were discussed by Kunzmann et al. 1995; additional studies are needed.) Status studies of southwestern invertebrates and delineation of specific threats are clearly and urgently needed. Notably missing from most lists of sensitive

invertebrates are desert arthropods, especially those of specialized habitats such as dunes, playas, and oases; their status needs to be assessed.

Fishes

Freshwater fishes are the most imperiled vertebrate group in the United States (Williams et al. 1989; Minckley and Deacon 1991; Warren and Burr 1994). In the United States, about 20% of fishes are extinct or imperiled, as compared with 7% of the country's mammals and birds (Master 1990). Almost 30% of the surface land area in the conterminous United States occurs west of the Continental Divide, but only about 21% of the roughly 800 freshwater fishes native to the United States are found there (Page and Burr 1991). Aquatic ecosystems in western North America, however, particularly in the Southwest, are endowed with some of the highest rates of endemism on the continent. In the Colorado River basin, for example, 35% of all native genera and 64% of the 36 fish species are endemic (Carlson and Muth 1989). The other southwestern watershed that demonstrates a high degree of fish endemism (30%) is the Rio Grande in New Mexico. Endemism in western fishes generally reaches its highest level in small systems such as isolated lakes and desert springs.

The level of threats to western fishes is also high and is probably best reflected in the number of imperiled species found in each state. Southwestern states have some of the highest percentages of threatened fish fauna: Arizona, 85%; California, 72%; New Mexico, 30%; and Utah, 42% (Warren and Burr 1994; Fig. 11).

Fig. 11. Numbers of native freshwater fish species in the lower 48 states (purple number) and number of fishes recognized by fisheries professionals as endangered, threatened, or of special concern (red number; based on Warren and Burr 1994).

Regionally, more than 48% of the fishes in the Southwest have been identified as jeopardized, compared with 19% in the Northwest and 10% in the Southeast (Warren and Burr 1994).

Because we cannot present all possible situations affecting southwestern fishes, we emphasize factors relevant to the decline and demise of these fishes by using shared ecological traits to illustrate the problems faced by these fishes. Fish communities of the main-stem Colorado River and associated major tributaries, mountain headwater systems in the Gila River drainage (see box on Gila Trout), the main-stem Rio Grande, and desert spring systems are discussed here (Fig. 12). The fishes in these communities range from long-lived, large-bodied fishes found in large, highly variable rivers to small, specialized fishes that have been isolated for thousands of years in relatively stable environments.

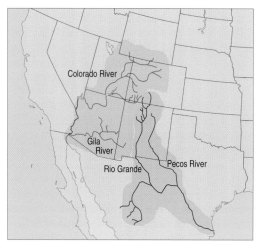

Fig. 12. Major river drainages in the Southwest.

Colorado River Fishes

Perhaps no other group of fishes better exemplifies the problems confronting aquatic ecosystems in the Southwest than the fishes of the Colorado River basin. More studies have been conducted and reports written on this group of fishes than any other assemblage of nongame fishes in the region. The Colorado River basin is the largest watershed in the Southwest, draining portions of seven western states, including about one-twelfth of the land area in the contiguous United States. The threats to this ecosystem are numerous and synergistic. Numerous dams on the main-stem river represent the most significant environmental perturbation facing these fishes. Main-stem impoundments have drastically changed water temperature, converted the river from sediment-laden to relatively clear, altered historical patterns of spring floods and the general water-flow regime, and blocked migratory pathways for fishes. Consequent modification or loss of habitat for native species and the creation of

suitable habitat for nonindigenous fishes have irreversibly altered the Colorado River aquatic ecosystem.

Although only 36 native freshwater fish species formerly lived in the Colorado River basin—a low number relative to basins east of the Continental Divide—species-level endemism is high (64%) (Carlson and Muth 1989). The number of native species in the major Colorado River basin drainages ranges from 5 (Bill Williams River) to 18 (Gila River). Although many of these endemic species are restricted to specific river systems in the Colorado River (for example, Virgin spinedace—Virgin River; spikedace and loach minnow—Gila River; woundfin—Virgin, Salt, and Gila rivers), several endemic taxa once were found generally distributed in main-stem habitats throughout the basin.

Fishes that inhabit and evolved in large rivers in the Colorado River basin are members of the chub complex (roundtail chub, humpback chub, and bonytail), Colorado squawfish (Fig. 13), and the razorback sucker. Of this group, four species are listed as endangered and one (roundtail chub) is a sensitive species (formerly Category 2 candidate; U.S. Fish and Wildlife Service 1994a). Most recent fisheries investigations have been directed at the three endangered Colorado River cyprinids—Colorado squawfish, humpback chub, and bonytail.

Courtesy S. P. Platania, University of New Mexico, Albuquerque

Fig. 13. Colorado squawfish, held by D. L. Propst, New Mexico Department of Game and Fish. This female fish was captured in the San Juan River, near its confluence with the Mancos River in New Mexico. The fish, 780 millimeters in total length and an estimated 8 kilograms in weight, was released unharmed. Anecdotal accounts refer to squawfish 2 meters long and weighing 45 kilograms. The maximum weights recorded in historical collections were 30 to 40 kilograms.

Studies on the Colorado squawfish demonstrated that it is a highly migratory species that formerly occurred throughout the basin but is now reduced to about one-third of its original range. Natural populations of this large-river fish have been eliminated from the lower basin, and the species is rare in the upper basin.

Humpback chubs inhabit relatively inaccessible reaches of the Colorado River system. Although this endemic Colorado River fish probably was found historically in most large-river habitats of the Colorado River, it now exists in only five canyon reaches in the upper basin—in the Grand Canyon in Arizona and near the confluence of the Colorado and Little Colorado rivers. Researchers believe that this last locality harbors one of the largest remaining humpback chub populations and that it is the spawning locality of Grand Canyon populations.

Bonytails (also called bonytail chubs) are the rarest of the endemic big-river fishes of the Colorado River. This fish species experienced the most abrupt decline of any of the long-lived fishes native to the main-stems of the Colorado River system and, because no young individuals have been found in recent years, has been called functionally extinct (Carlson and Muth 1989). Bonytails were one of the first fish species to reflect the changes that occurred in the Colorado River basin after the construction of Hoover Dam; the fish was extirpated from the lower basin between 1926 and 1950. In reference to the rapid demise of bonytails, Behnke and Benson (1980:20) said, "If it were not for the stark example provided by the passenger pigeon, such rapid disappearance of a species once so abundant would be almost beyond belief." Bonytails were also extirpated from several upper basin rivers (Green, Gunnison, and Yampa rivers) where they were once common. Populations in free-flowing waters now apparently survive only in the Colorado River in Colorado and Utah. The largest population of bonytails occurs in Lake Mohave (Mueller and Marsh 1995), but this population consists only of old individuals, and there is no evidence of reproduction.

Only recently have investigations begun to focus on the razorback sucker—one of the most threatened big-river fishes in the Colorado River basin (Mueller and Marsh 1995). This fish was formerly so abundant throughout the main-stem and major tributaries of the Colorado River basin that in the early 1900's it was one of the principal fishes taken by a commercial fishery in southern Arizona (Hubbs and Miller 1953). Additional anecdotal accounts of the abundance of razorback suckers occur in historical reports from the 1880's through the 1940's. This unique fish now inhabits only 1,208 (river) kilometers in the upper basin, while the only substantial population in the lower basin occurs in Lake Mohave (McAda and Wydoski 1980; Marsh and Minckley 1989). The most serious problem for the razorback sucker is the lack of any significant reproduction in recent years.

The dilemma of big-river fishes in the Colorado River basin is not limited to these species, which are accorded some level of federal protection. For example, the flannelmouth sucker, one of the most common suckers in many portions of the upper basin, has been eliminated from the Gila River drainage.

Concurrent with the decline in native fish species has been an increase in the species richness and abundance of nonindigenous fishes. Maddux et al. (1993) reported the introduction of at least 72 fish species, twice the number of native fishes, into the Colorado River basin. Many of these introduced fishes have established successful populations in parts of the Colorado River system and now are serious predators of young suckers, chubs, and squawfish.

Rio Grande Fishes

Fish communities of the Rio Grande consist of plains fishes that belong to the westernmost drainages of the Mississippi River basin. Fishes in the New Mexico portion of the Rio Grande exhibit a high degree of endemism and special biological characteristics that reflect their evolution in this variable system. Researchers believe that the native fish community of these middle reaches of the Rio Grande has between 16 and 27 species (Hatch 1985; Smith and Miller 1986; Propst et al. 1987; Sublette et al. 1990). Three of the six Rio Grande endemics—Rio Grande shiner, phantom shiner, and Rio Grande bluntnose shiner—no longer occur in the New Mexico portion of the Rio Grande (Bestgen and Platania 1990). In addition, the endemic Rio Grande chub and Rio Grande silvery minnow are reduced in range and abundance, with Rio Grande chubs now generally found only in tributary streams and Rio Grande silvery minnows only in main-stem habitats.

The history of the decline and extirpation of these formerly widespread species is relatively well documented. The Rio Grande shiner was last found in the middle Rio Grande in 1949, whereas phantom shiners and Rio Grande bluntnose shiners were last collected in the upper Rio Grande in 1939 and 1964, respectively (Chernoff et al. 1982; Bestgen and Platania 1990). The speckled chub, a nonendemic main-stem cyprinid, was last found in the middle Rio Grande in 1964. The Rio Grande silvery minnow is the only endemic main-stem Rio Grande cyprinid that survives in New Mexico (Bestgen and Platania 1990), although the species had formerly been abundant, occurring from the confluence of the Chama River and Rio Grande to the Gulf of Mexico (Fig. 14). The 95% reduction in the Rio Grande silvery minnow's range resulted in its listing as an endangered species (U.S. Fish and Wildlife Service 1994b).

The decline and demise of the middle Rio Grande fish fauna may be partly explained because habitats that formerly supported small isolated outlier populations of Rio Grande fishes may have relied on emigrants from upstream or downstream reaches to supplement isolated populations. When dams eliminated these dispersal avenues, populations of these short-lived fishes dwindled and eventually were extirpated. Life-history characteristics of these fishes sup-

port this assumption. For example, these species take advantage of changes in natural water flow for reproduction because their eggs and larvae occur in times of peak water flow; a generalized hydrograph for plains streams indicates that spawning takes place during spring runoff or summer rainstorms. Many fishes that are found only in these streams produce semibuoyant eggs that are carried along with these spates and hatch 24 to 48 hours after being spawned. The larval fish develop quickly in the warm waters characteristic of summer flow, and after about 3 days, they move into the highly productive waters typical of slow-velocity habitats, where they grow quickly (S. P. Platania and C. S. Altenbach, University of New Mexico, Albuquerque, personal observation).

Perils Facing the Gila Trout

The plight of Gila trout provides a case study of problems confronting fish inhabitants of mountain headwater streams in the Southwest. This endemic salmonid once occurred in much of the Gila River drainage—a Colorado River basin tributary in New Mexico and Arizona—but by the time the species was first described in 1950, its range had been reduced to a few headwater streams in New Mexico (Propst et al. 1992). The first attempts to conserve and recover populations of this fish occurred in 1923, when the New Mexico Department of Game and Fish established hatchery stocks of Gila trout and prohibited stocking of nonindigenous trout species in stream reaches where the Gila trout occurred. The U.S. Endangered Species Preservation Act of 1966, the U.S. Endangered Species Act of 1973, and the New Mexico Wildlife Conservation Act of 1974 legally protected the remaining populations.

The Gila trout recovery plans of 1979 and 1983 (U.S. Fish and Wildlife Service 1987) identified the need to replicate and safeguard the remaining populations as genetically distinct evolutionary units. These plans provided the guidelines to recovery, requiring the selection of stream reaches suitable for reestablishment, mechanical and chemical removal of nonindigenous trout, and transplanting Gila trout to the renovated systems. Streams selected as donor sites for Gila trout populations were generally headwater reaches within the historical range of the fish. By 1987 recovery efforts had resulted in the establishment of Gila trout populations at nine localities, eight in New Mexico and one in Arizona. After meeting benchmark

requirements of the recovery plans, the U.S. Fish and Wildlife Service proposed downlisting the Gila trout from endangered to threatened (U.S. Fish and Wildlife Service 1987); concurrently, the New Mexico Department of Game and Fish delisted the McKnight Creek population and downlisted all other New Mexico populations from endangered to threatened (Propst et al. 1992). No one anticipated the natural events and their repercussions that would transpire over the next 6 months.

Before the end of the comment period for the proposed downlisting, a flood eliminated more than 80% of the Gila trout population from McKnight Creek, a forest fire and subsequent flooding eliminated the Main Diamond Creek population (the type locality), and drought and forest fire eliminated more than 90% of the South Diamond Creek population. The occurrence of these chance events concurrent with attempts to downlist this fish demonstrates the fragility of small populations and dramatically shows the potentially serious problems in recovery strategies implemented during the 20-year recovery period.

Threats to Gila trout were comparable to those identified for other interior North American headwater stream trout—competition, predation, and hybridization by introduced trout such as the rainbow trout and brown trout, habitat loss and degradation, and decreased water quality and quantity. In addition, fragmentation of range and recovery attempts relying on establishment of isolated populations are principal problems for Gila trout. Additional threats identified after the catastrophes of the late 1980's were associated with the preferred habitat of this

species or with changes within that habitat (for example, fire suppression and riparian habitat degradation).

The attempt to propagate Gila trout by adhering to the historical approach of multiple but small populations had several flaws. Translocated fish were isolated from other populations in habitats that could support only relatively small populations. The limited area for reintroduction and the small population sizes made Gila trout at these sites extremely susceptible to loss through natural events (fire and flooding) or excessive human-related activities (livestock grazing, timber harvest, and mining). This propagation strategy—where more populations were equated with higher levels of protection—caused a false sense of security. Another, and more difficult, conclusion reached by Gila trout biologists concerned the potential need to place the survival of the species as a higher priority than the maintenance of genetically distinguishable populations. They concluded that there was a need to investigate transplanting populations to longer and larger streams outside of the current range of the species and to introduce more individuals more frequently (Propst et al. 1992).

See end of chapter for references

Author

Steven P. Platania
Division of Fishes
Museum of Southwestern Biology
University of New Mexico
Albuquerque, New Mexico 87131

Disturbances in natural river-flow regimes undoubtedly have adversely affected reproduction in these fishes.

Desert Spring Fishes

Desert springs and associated isolated pools are characteristic of the Southwest. These habitats are often marginal for fishes and are typically geographically isolated, have small surface volumes and relatively constant physicochemical features, and discharge warm, oxygen-poor water. Although each spring or pool is species-poor, most aquatic inhabitants of each pool are endemic. Most fishes in these habitats are short-lived (1–2 years) and are native to only a single locality. The fish communities of these habitats are generally composed of pupfishes, poolfishes and springfishes, killifishes, and livebearers.

Minckley et al. (1991) summarized the conservation status of 76 taxa of killifishes from the southwestern United States and northern Mexico. Of this group, 38% are classified as threatened or endangered, 30% as rare or vulnerable, and 9% as extinct. Minkley et al. (1991) reported that 6 poolfishes, springfishes, and pupfishes are extinct, and only 2 of the remaining 46 fish species are considered free from jeopardy. All 10 species and subspecies of poolfishes and springfishes are considered imperiled (Minckley et al. 1991).

Although there are numerous problems confronting desert spring fishes, perhaps the best known are those facing the Devils Hole pupfish. The life-history traits of this fish and the geologic nature of its habitat are similar to those of other southwestern pupfish. Devils Hole pupfishes became isolated at the mouth of a water-filled limestone cave in Ash Meadows of southern Nevada about 20,000 years ago when waters retreated from the last Ice Age. These fish congregate around a 6- by 3-meter limestone shelf, the smallest known habitat of any vertebrate animal. The algal mat that grows on this shelf is the base of a food chain composed of protozoa, diatoms, amphipods, and pupfish. The Devils Hole pupfish population numbers are greatest in summer (about 700 individuals), when primary productivity is also highest. In winter, when less sunlight reaches the shelf, the algal mat dies and the number of pupfish may be as low as 200 (Ono et al. 1983).

The greatest threat to this species occurred when farmers in the Ash Meadows Valley drilled numerous wells to pump groundwater from the aquifer. As predicted, these pumping activities lowered the water table, thereby threatening the pupfish with the exposure and subsequent drying of their shelf habitat. In addition, the lowering of the water table also imperiled the other endemic animals in several other

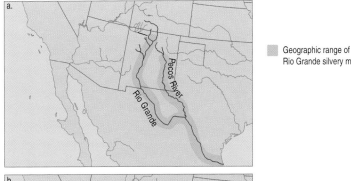

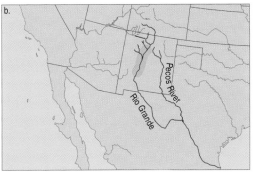

Geographic range of Rio Grande silvery minnow

Fig. 14. a) Historical and b) present distribution of the Rio Grande silvery minnow.

springs in the valley. When groundwater-pumping activities stopped in the mid-1970's, temporary relief occurred for this species and the other inhabitants of isolated spring pools throughout Ash Meadows Valley.

Recovery efforts involving fishes of these isolated habitats seem relatively simple compared with those required to recover long-lived fishes in large, free-flowing rivers. Such efforts require that the spring, its associated spring pool, and the aquifer be secured. Although the first two conditions may be met, securing an aquifer is a daunting task. Other recovery options include measures such as those developed for the Pahrump poolfish, a species that was formerly restricted to Manse Spring in Pahrump Valley, Nevada. Overpumping of the water table and the introduction of nonindigenous fishes severely imperiled this species. In the 1970's and 1980's, stockings of this species occurred in state and federal refuge complexes, and the species was successfully established outside of its native range, providing it with additional security for its continued survival (Minckley et al. 1991).

Causes of Imperilment

Little debate occurs among fisheries professionals about the causes of imperilment and extinctions of southwestern fishes. Most frequently mentioned causes are construction of dams (Fig. 15), loss of physical habitat (Fig. 16), habitat degradation, chemical pollution, overexploitation, and introduction of nonindigenous species. Dam construction and regulation probably had the greatest adverse effect on native fishes of southwestern rivers, while

Fig. 15. Downstream side of a diversion dam on the Rio Grande at San Acacia, New Mexico, in autumn 1986.

Fig. 16. A Rio Grande pool drying near Socorro, New Mexico, in autumn 1987.

perceived problems and life-history strategies. In general, researchers know what is required to recover most threatened and endangered fishes, but some solutions (for example, removing dams) may be unrealistic and controversial. The most frequently cited solution is habitat preservation, which may be relatively simple for fishes with restricted distributions and small population sizes. Recovery strategies for long-lived, wide-ranging species, such as the Colorado squawfish, are more complex and require a long-term commitment. The unique ecological requirements of the various life-history stages of long-lived species dictate the need for protection of extensive river reaches and perhaps changes in water use. Realistic possibilities for recovery of the native fish fauna of the Southwest are decreasing as human populations increase and formerly uninhabited lands become developed, causing native fish populations to decline further.

Amphibians and Reptiles

Diversity of Southwestern Herpetofauna

More than 150 species of amphibians and reptiles occur in the Southwest; this is partly due to the region's diverse habitats. Compared with the moister habitats of eastern North America, the southwestern amphibian fauna is relatively poor, with only 3 salamander species and about 30 frog and toad species. About 25% of the amphibians, though, are found nowhere else in the United States. About 50 lizard and 56 snake species account for more than 70% of the reptiles and amphibians inhabiting this region; most of these live in the arid and semiarid desert grasslands and shrublands that make up most of the Southwest. Although aquatic turtles are relatively scarce in the Southwest (11 species), most are important components of the fragile riparian ecosystems in this semiarid region. In the Colorado Plateau, 18 amphibian and 62 reptile species have been described (Drost and Deshler 1995). Current status and trends for amphibians and reptiles are not as well known as those for fishes and birds. Additionally, these organisms face some threats not faced by other vertebrate animals (for example, rattlesnake roundups, and commercial collection for the pet trade).

Montane Forest Herpetofauna

Conservation concerns relative to southwestern amphibians and reptiles can be effectively summarized by elevation, as many species in similar habitats are often threatened by the same factors. Amphibians and reptiles in the Southwest are common in the warm, lowland deserts and grasslands but are relatively uncommon at cool, high elevations. Still, the presence

the effects of excessive groundwater pumping have imperiled many spring systems and their associated fauna. The number of nonindigenous fish species in the Southwest is considerable: Arizona has 71 species; New Mexico, 75 species; Utah, 55 species; and Texas, 96 species (Boydstun et al. 1995).

As a whole, fishes in the western United States are clearly more imperiled than those in the eastern United States. More than half of the fishes listed as endangered or threatened by the U.S. Fish and Wildlife Service, or being considered for such listing, occur west of the Continental Divide. The commonly observed pattern is the disappearance of the most sensitive fishes, followed by the collapse of whole fish faunas in major western river basins. If current efforts directed at recovery of native western fishes are not continued and successful, we could witness the disappearance of most of the region's endemic fish fauna.

Recovery strategies for aquatic organisms in the Southwest vary depending on their

of several species found only in these high-elevation habitats makes certain coniferous forests important for herpetological diversity. For example, two species restricted to montane sky islands in New Mexico—the Jemez Mountains and Sacramento Mountain salamanders (Fig. 17)—are the only salamanders endemic to New Mexico. Both are federally listed as sensitive species (U.S. Fish and Wildlife Service 1994a) and are considered endangered by the state of New Mexico. Studies suggest that these salamanders are adversely affected by intensive logging operations, although to what extent remains unclear (Ramotnik and Scott 1989; Scott and Ramotnik 1992). The U.S. Forest Service has made significant efforts to protect known populations, including developing an agreement with the U.S. Fish and Wildlife Service and the New Mexico Game and Fish Department to protect all habitat occupied by the Jemez Mountains salamander while a management plan is produced.

Fig. 17. Sacramento Mountains salamander, one of two endemic salamanders in New Mexico.

A few endemic reptiles also occur in montane habitats, and some are prone to being over-collected because of their novelty to reptile enthusiasts. Commercial collection of snakes such as the twin-spotted rattlesnake, the New Mexican ridge-nosed rattlesnake, and the rock rattlesnake (Fig. 18) for the pet trade has been a concern in recent years because of the vulnerability of these species to extirpation (Ernst 1992). Mining and logging development may also moderately threaten some of these species (Ernst 1992).

Pinyon–Juniper Woodland Herpetofauna

Lower-elevation coniferous woodland is widely distributed in the Southwest and supports a variety of amphibians and reptiles, although there are few endemic species because few are specialists in using these habitats. Studies are needed on the potential effects of the increasing urban development that is

occurring in pinyon–juniper woodland and savanna habitats, especially in north-central New Mexico and northern Arizona. At present, this habitat type is sufficiently common that current effects on reptiles and amphibians are probably not significant.

Encinal Woodland Herpetofauna

Encinal forests in southwestern New Mexico and southeastern Arizona support a distinctive assemblage of species whose range occurs primarily in Mexico—most are found in the Sierra Madre Occidental and occur in the United States at the northern limits of their ranges. These species include the Tarahumara frog, canyon spotted whiptail, and Mexican garter snake—all sensitive species (U.S. Fish and Wildlife Service 1994a), and the endangered New Mexico ridge-nosed rattlesnake. The Tarahumara frog, once known from 30 localities in southern Arizona and northern Mexico, appears to be extirpated in the United States but is not threatened with extinction throughout its range (Hale et al. 1995). Disappearance of U.S. populations of the Tarahumara frog may be tied to acid rain, heavy metals, air particulates, or solar radiation (Hale et al. 1995), whereas habitat modification and collection for the pet trade (Bender 1981) threaten the other species mentioned.

Sonoran Desert Herpetofauna

In the Sonoran Desert, reptiles such as the chuckwalla, rosy boa, and desert tortoise have

Fig. 18. Banded rock rattlesnake in the Magdalena Mountains, New Mexico (top), and ridge-nosed rattlesnake in the Animas Mountains, New Mexico.

Courtesy C. W. Painter, New Mexico Department of Game and Fish, Santa Fe

apparently declined (U.S. Fish and Wildlife Service 1994a), in part because of overcollection for the pet trade, although habitat modification may also be a factor. The effects of grazing on herpetofaunal communities in the Sonoran Desert are not well studied, although trampling of young desert tortoises and damage to burrows and vegetation have been attributed to livestock (Berry 1978; Campbell 1988). Likewise, Szaro et al. (1985) found the wandering garter snake to be five times more abundant in ungrazed sites than in grazed sites in the White Mountains. Other communities appear to show few effects from grazing. For example, Jones (1981) found that lizard abundance and diversity in Arizona's Sonoran Desert were not significantly affected by heavy cattle grazing if overall vegetation structure was unaltered. However, lizard abundance and species diversity were greater on lightly grazed sites than on heavily grazed sites (Jones 1981). More intrusive activities, such as grazing by sheep, off-road vehicle use, and urban development, significantly alter vegetation structure (Busack and Bury 1974) and affect reptiles more. Declines of reptile species in the eastern Mohave Desert have been associated with increased disturbance (Stewart 1994); such changes are predictable as human populations and accompanying urbanization increase in the adjacent Sonoran Desert.

Chihuahuan Desert Herpetofauna

Relatively few studies have addressed the status and trends of Chihuahuan Desert herpetofauna. Conant (1977) and Scudday (1977) discussed human effects on this fauna, mainly in terms of aquatic amphibians and reptiles. Degenhardt (1977) suggested that some lizard species in Big Bend National Park had declined as plant cover increased in response to cessation of livestock grazing. Conversion of desert grassland to Chihuahuan Desert scrub over a 30-year period was implicated in the decline of several snake species and an increase in others along a road transect on the Arizona–New Mexico border (Mendelson and Jennings 1992). Bock et al. (1990) reported significantly reduced numbers of the bunch grass lizard and other lizard species in grazed grassland in southeastern Arizona. Long-term data, though, are generally lacking for most grassland herpetofaunal communities in the Southwest, thus permitting only tentative conclusions about local extirpations and distributional changes.

Grassland Herpetofauna

A number of amphibian and reptile species in the Southwest are restricted to plains and mesa grasslands, desert grasslands, or both. For many of these species we have little information

on status and trends. The status of spadefoots, for example, is poorly known (Corn 1995). The advancement of juniper savanna into disturbed grasslands in New Mexico and Arizona (Dick-Peddie 1993) is expected to affect some grassland reptiles; for example, the grassland-dwelling little striped whiptail has been replaced by the woodland-adapted plateau striped whiptail at some localities in northern New Mexico (J. Stuart, U.S. Geological Survey, Albuquerque, New Mexico, unpublished data).

Aquatic Herpetofauna

Many frogs and toads and almost all aquatic turtles in the Southwest are dependent on rivers or surface waters associated with river floodplains. Among the species that have suffered significant losses in the Southwest are the Chiricahua and lowland leopard frogs, the Mexican and narrow-headed garter snakes, and the southwestern toad—all sensitive species (U.S. Fish and Wildlife Service 1994a). The status of the Big Bend slider, a little-known turtle endemic to the Rio Grande basin, is uncertain in New Mexico (Degenhardt et al. 1996).

Documented declines of amphibians in the Southwest include some populations of the Huachuca tiger salamander, northern leopard frog, and western toad (see summary in Corn 1995). Most documented declines for southwestern amphibians have occurred in the leopard frog complex (Clarkson and Rorabaugh 1989; Degenhardt et al. 1996; Fig. 19). The Vegas Valley leopard frog is now extinct (Jennings 1988), and the relict leopard frog in southern Nevada and southwestern Utah was thought extinct until three small populations were recently rediscovered (Jennings 1993). In Arizona, biologists found northern leopard frogs were absent from 46% and Chiricahua

Fig. 19. Northern leopard frog.

Arizona Leopard Frogs: Balanced on the Brink?

The low humidity, high summer temperatures, and other natural forces that helped shape the saguaro, spectacular canyonlands, and other familiar features of the Southwest would seem to be inhospitable to the highly aquatic frogs of the genus *Rana,* known as ranids. In spite of these conditions, leopard frogs were, until recently, common inhabitants of Arizona's wetland and riparian ecosystems; in fact, the Southwest boasts the greatest diversity of leopard frog species in the United States. The recent dramatic declines of some members of this species complex need to be examined and addressed immediately.

Establishing a Historical Baseline

In the Southwest, native members of the true frog family (Ranidae) include the leopard frogs and the Tarahumara frog. Arizona's leopard frog fauna, among the largest in North America, is surprisingly diverse and includes at least five (possibly six) native leopard frogs and one introduced species (Platz et al. 1990; Table). While it is unclear exactly when declines of these frogs began, they were first noticed during the 1970's, when the status of the Tarahumara frog was of concern. Population studies of the Tarahumara frog in the mid-1970's indicated that the future existence of populations of this species in Southeast Arizona and the northern Sonora could be in jeopardy (Hale and May 1983; Hale and Jarchow 1988; Hale 1992). During these studies, researchers suspected that native leopard frogs were also declining, but there were no baseline data to support this suspicion. One of the first studies to investigate the status of Southwest leopard frogs on a large landscape was that of Clarkson and Rorabaugh (1989), who surveyed 56 historical and 7 new localities between 1983 and 1987 for four species of native leopard frogs.

Although this seminal investigation concluded that all species of leopard frogs examined were declining, several points relevant to status determination and conservation planning were not addressed by the authors. First, by focusing on historical localities, the data set gathered by Clarkson and Rorabaugh (1989) was biased toward concluding that declines have occurred (P. Geissler, U.S. Geological Survey, Laurel, Maryland, unpublished report). Second, because the study was not specifically designed to identify new localities or systematically evaluate threats, this information, which is important to status assessment and recovery planning, was not discussed.

In an effort to gather such data, the Arizona Game and Fish Department began assessing the status and current distribution of all native Arizona ranids by conducting statewide visual encounter surveys (in the sense of Crump and Scott 1994) of historical and high potential habitats, and by recording detailed habitat data and herpetofauna observations. Because of the confused taxonomic history of this group (Hillis 1988) and the likelihood that specimens of this complex are misidentified even in museum collections (Jennings 1994), biologists at the Arizona Game and Fish Department gathered historical locality data from selected sources in the published literature or local museums whose collections reflect the current taxonomy. Our data base presently contains over 4,000 herpetofaunal observations, collected from more than 1,500 localities.

I present a preliminary analysis of our survey data for all species of Arizona ranids except the Tarahumara frog, whose status has been recently reviewed elsewhere (Hale et al. 1995). Presence or absence of frogs at historical localities has been determined since 1990, the beginning of our statewide survey efforts. Because large temporal differences exist in activity of many amphibians, determining their presence or absence from a locality is difficult and requires multiple visits. Corn and Fogleman (1984), studying northern leopard frogs in Colorado, suggested that if frogs or reproduction are not observed at a locality over a three-year period, a population can be considered extirpated. In most, but not all instances, I have determined the presence or absence of leopard frogs at a site by examining survey results from three or more visits to that site during times of peak activity (April through October).

Preliminary Status of Arizona Leopard Frogs

Rio Grande Leopard Frog

Researchers believe that this species was inadvertently released into the Colorado River near Yuma, Arizona, in the 1960's during sport-fish stockings (Platz et al. 1990). Although all native leopard frogs have declined in at least some part of their Arizona ranges, it is ironic that this nonindigenous leopard frog continues to expand its Arizona range and is now known from the Colorado River near Yuma, the Gila River up to its confluence with the Salt River, and the Gila, Salt, Agua Fria, and Hassayampa rivers and adjacent agricultural areas near Phoenix (M. J. Sredl, Arizona Game and Fish Department, unpublished data; J. C. Rorabaugh, U.S. Fish and Wildlife Service, Phoenix, Arizona, personal communication).

Plains Leopard Frog

Most of the range of the plains leopard frog occurs in the central and southern Great Plains, where it is an inhabitant of aquatic habitats in prairie and desert grassland ecosystems (Stebbins 1985). In Arizona, this species is restricted to the Sulphur Springs

Common name	Number of localities/number of surveys	Historical absent	Historical present	Historical unsurveyed	New sites
Rio Grande leopard frog[a]	—	0	1	1	9
Plains leopard frog	85/98	14	1	1	2
Chiricahua leopard frog	679/871	80	18	35	45
Relict leopard frog[b]	—				
Northern leopard frog	477/566	24	2	21	29
Ramsey Canyon leopard frog	—		1	0	5
Yavapai leopard frog	648/797	35	28	35	167

[a] Introduced to Arizona.
[b] No verifed Arizona records.

Table. For each species of Arizona leopard frog, numbers of localities visited, numbers of surveys conducted within the range of that species, and frequency within each status category are listed. The status of each locality was evaluated relative to pre- and post-1990 surveys.

Valley in southeastern Arizona (Frost and Bagnara 1977), an area separated from the main portion of the range of this taxon by about 350 kilometers (Mecham et al. 1973). Specimens have also been collected from Ashurst Lake, southeast of Flagstaff in northern Arizona (J. E. Platz, Creighton University, Omaha, Nebraska, personal communication), but there have been no recent leopard frog records to verify this species at that site. Within the Arizona range of the plains leopard frog, Arizona Game and Fish biologists have conducted 98 surveys at 85 localities and found the species absent from nearly 93% of the localities surveyed historically (Table). Recent surveyors have only found this species at three localities, two of which were found after 1990.

Chiricahua Leopard Frog

Of all of Arizona's leopard frogs, the Chiricahua leopard frog has undergone perhaps the largest, most dramatic decline (Sredl and Waters 1995). The range of this species includes the southern edge of the Colorado Plateau in Arizona and New Mexico, southeastern Arizona, southwestern New Mexico, and the Sierra Madre Occidental in Mexico (Platz and Mecham 1979). The Arizona range of this frog consists of a northern part, which extends from the White Mountains and Mogollon Rim (the southern edge of the Colorado Plateau) in central Arizona, and a southern part in drainages associated with the Madrean oak woodlands and semidesert grasslands of the Madrean Archipelago (Sredl and Howland 1995). To evaluate the status of this frog, Arizona Game and Fish biologists have conducted 871 surveys at 679 localities, 98 of them historical. To date, the Chiricahua leopard frog appears absent from 82% of the historical localities surveyed since 1990 (Table). While surveyors found the Chiricahua leopard frog at 45 new sites, many of these observations consisted of as few as one or two frogs observed at localities adjacent to extant populations, making it likely that these individuals had dispersed from nearby populations.

Relict Leopard Frog

The relict leopard frog has the dubious distinction of being the first North American amphibian thought to have become extinct (Platz 1984) only to be rediscovered (Jennings 1993). No Arizona records of this species exist, but further clarification of relationships of southwestern leopard frogs may result in populations now classified as other species being included in this species.

Northern Leopard Frog

Until the late 1960's, all currently recognized leopard frogs, including those found throughout the Southwest, were classified as the species *Rana pipiens*, now known by the common name of northern leopard frog. During the 1960's, though, taxonomists realized that what was recognized as *R. pipiens* was really a multispecies complex (Hillis 1988). In Arizona, *R. pipiens* has been found in the lakes, earthen tanks, springs, creeks, and rivers of the Colorado Plateau in the northeast portion of the state. During these surveys, Arizona Game and Fish biologists visited 477 localities and conducted 566 surveys within the Arizona range of this frog, and found it at 8% of the historical localities visited. Although Arizona Game and Fish biologists found 29 new sites of occurrence, few of these populations are large. Within Arizona, there are probably fewer than five metapopulations, many of which are small.

Ramsey Canyon Leopard Frog

The Ramsey Canyon leopard frog is the most recently described of Arizona's leopard frog species (Platz 1993). This frog is known only from the Huachuca Mountains in southeastern Arizona, where six populations have been found in three drainages. A conservation team composed of agency and academic biologists as well as biologists from private institutions is developing a conservation agreement that is expected to be implemented soon. The continued viability of this species will depend on swift conservation actions by this group.

Yavapai Leopard Frog

Within the Southwest, the Yavapai leopard frog occurs in aquatic systems from desert scrub to pinyon–juniper habitats (Platz and Frost 1984). While the continued existence of this species in some parts of its range appears uncertain (Jennings 1995; Jennings and Hayes 1995), the status of this species in central Arizona seems good. Our intensive surveys of the range of this species have revealed that the species was present in 44% of the historical localities we visited. This percentage of occupancy of historical localities is by far the highest of any species of Arizona leopard frog. In addition, this frog has been found at 167 new localities, mostly in central Arizona (Table).

Mitigating Threats and Searching for Solutions

Analysis of historical and recent information reveals that the status of Arizona's six native ranids falls along a spectrum of endangerment. Although none is federally listed as threatened or endangered, two are candidates or species of special concern, and if declines continue, more species will be added. A general approach to stabilizing these declines or initiating recovery must incorporate traditional wildlife management techniques and techniques from endangered species recovery efforts. To have the greatest chance of success we need to begin formulating these plans now. Conservation and management activities should be implemented long before they become actions of last resort (Griffith et al. 1989).

See end of chapter for references

Author

Michael J. Sredl
Arizona Game and Fish Department
2221 West Greenway Road
Phoenix, Arizona 85023-4399

leopard frogs from 94% of known localities previously studied by Clarkson and Rorabaugh (1989). Yavapai leopard frogs have been extirpated from the Imperial Valley in California and the lower Colorado River of California and Arizona.

Causes for declines of semiaquatic amphibians and reptiles remain elusive but include habitat destruction and alteration, introduced predators, pollutants, disease, and climate change (Corn 1995). Most declines are human-induced. For example, urban and agricultural water demands can result in drainage of isolated surface waters (such as springs in arid areas) through lowering of water tables; increasing human populations in riparian areas have led to reduced surface waters. Urbanization also may cause more pollutants to be introduced into surface water and groundwaters; pollutants may be further concentrated by reduced water supplies.

In addition, drainage of shallow wetlands and playas can eliminate breeding habitat for many amphibians. Riparian corridors and their associated amphibian and reptile populations may become fragmented by dam construction, agricultural and urban development in floodplains, and dewatering of streams and rivers (Knopf 1989). Such fragmented, reduced populations of reptiles and amphibians may have lowered genetic variability and may be more vulnerable to environmental perturbations (Brode and Bury 1984; Corn and Fogleman 1984; Bradford et al. 1993).

Effects of Introduced Species

Although the effects of introduced species on native fishes in southwestern aquatic systems have been well documented (Miller 1961; Sublette et al. 1990), such effects have been less documented for amphibians and reptiles. The nonindigenous bullfrog has been implicated in the decline of native leopard frogs in California (Hayes and Jennings 1986) and of native leopard frogs and garter snakes in southern Arizona (Schwalbe and Rosen 1988). Rosen and Schwalbe (1995) provided additional evidence of the effect of bullfrogs on native herpetofauna and of the possibility of recovery of affected populations once bullfrogs are removed. The establishment of nonindigenous aquatic turtles in several southwestern river systems has been verified (Hulse 1980; Degenhardt et al. 1996), but the effect of these introductions on the native turtle fauna is largely unknown. Extirpation of native frogs by introduced warmwater fishes has occurred in some California drainages (Hayes and Jennings 1986; Bradford et al. 1993) and may also be a factor in the decline of some southwestern amphibian populations. Studies are urgently needed to determine the effects of such introduced predators on vulnerable populations of amphibians and reptiles.

Birds

Baseline Surveys

Recent survey efforts focusing on Neotropical migrant birds have highlighted the ecological problems these birds face. Overall, moderate amounts of data on status and trends exist for birds. Although our knowledge of population trends in breeding birds is augmented considerably by surveys such as the U.S. Geological Survey Breeding Bird Survey (Peterjohn 1994), reliance on such indexes is not always possible in the topographically diverse Southwest, where higher-elevation habitats may not be routinely surveyed for breeding birds. Lowland areas, however, including some important habitats, are usually well covered by such surveys.

Riparian Birds

Riparian areas of the Southwest are the most productive habitats in terms of abundance and diversity of birds (Knopf et al. 1988). An estimated 78 of the 166 (47%) bird species that breed in the complex deciduous vegetation associated with watercourses breed only in this vegetation type (Knopf and Samson 1994). The presence of foliage of various height classes, the richness of plant species and forms, the heterogeneous mix of open and densely vegetated areas, and the relatively high frequency of nesting cavities form a complex association that can support a large variety of birds. These corridors of woody vegetation also appear important for migrant land birds, including species that overwinter in the Neotropics and short-distance migrants that usually winter in the southern United States and northern Mexico (Farley et al. 1994).

In New Mexico and Arizona, 11 of 40 (27.5%) land bird species known to have declined in numbers over the last 100 years may have done so because of degradation and destruction of riparian habitats (DeSante and George 1994). Maintenance of historical overbank flooding regimes, if combined with cessation of cattle grazing or at least if allowing grazing only at times other than nesting season, results in more successful recruitment of native vegetation in riparian areas and consequent increased use of these habitats by breeding and migratory birds (Farley et al. 1994). Many of the birds in riparian habitats migrate long distances to the Neotropics during the nonbreeding season, and 25% of those that breed in New Mexico are considered high-priority conservation risks (Carter and Barker 1993; Mehlman and Williams 1995).

Data from the Breeding Bird Survey for 1966–1993 show significant declines for a number of birds associated with southwestern riparian ecosystems. For example, vermilion flycatchers declined an average of 3.1% annually, and Lucy's warblers declined 0.5% annually (Peterjohn et al. 1994). Statistically significant trends require repeated sampling over large areas for lengthy periods, and the restricted distribution and limited sampling of this habitat (Peterjohn et al. 1994) reduced the number of statistically testable declines.

Woodland and Forest Birds

Fluctuating bird populations also have been recorded in pinyon–juniper and ponderosa pine woodlands. In these habitats, the processes associated with the decline of birds may be less clear, but researchers believe that overgrazing and timber and snag removal are at least partially responsible (DeSante and George 1994;

Hejl 1994). Although there is no significant overall decline among all woodland-breeding birds in the western United States (Peterjohn and Sauer 1994), a number of species have experienced quantifiable reductions in their populations. For example, the ladder-backed woodpecker, Bendire's thrasher, and gray vireo, which all breed primarily in pinyon–juniper habitat, have exhibited declines in numbers based on Breeding Bird Survey sampling from 1966 to 1993 (2.4%, 3.7%, and 0.8% per year, respectively).

The buff-breasted flycatcher, a regular breeder in ponderosa pine woodlands, has also experienced significant reductions in range and population size. Trends using Breeding Bird Survey data are unavailable for the buff-breasted flycatcher because this bird is observed irregularly on census routes.

Many primarily Mexican highland bird species range into parts of Arizona, New Mexico, and Texas. Although several of these species are listed as endangered or threatened in the southwestern states (for example, thick-billed parrot, ferruginous pygmy-owl, violet-crowned hummingbird, elegant trogon, and varied bunting; Flather et al. 1994), accurate estimates of their population trends are frequently lacking. The presence of a few relatively small, isolated populations at the periphery of the ranges of these species makes field biology difficult and effective management challenging. Moreover, complementary or more extensive data from the relatively densely populated central portions of the ranges of these species in Mexico are mostly unavailable.

Recovery of Peregrine Falcons

The American peregrine falcon is a southwestern bird that is experiencing a notable positive trend even though the species had experienced significant declines 40 years ago. Concentrated reintroduction and recovery efforts led by the Peregrine Fund are now coming to fruition in many areas. For example, the canyon country of northern Arizona and southern Utah supports the highest densities of nesting peregrines south of Canada (Brown et al. 1992).

Demographic Studies Needed

A valuable form of data collection for many of the Neotropical migrant birds in this region could involve determining their demographic variables at the population level. For example, scientists do not know the levels of reproductive success, survivorship, interyear site fidelity, or population persistence for the southwestern willow flycatcher, a riparian

obligate recently listed as federally endangered (U.S. Fish and Wildlife Service 1995). Although population location and habitat associations are essential for the short-term maintenance of this species, its effective conservation will require additional detailed information.

The MAPS (Monitoring Avian Productivity and Survivorship) program (DeSante et al. 1993) is a sample protocol that defines standardized methods for determining and monitoring the species-specific demographic patterns of birds over regional and continental scales. To better understand the causes of population declines of southwestern birds, additional studies of individual bird species in each major ecosystem are necessary.

Mammals

Diversity of Southwestern Mammals

Cole et al. (1994) estimated that the mammalian fauna of the New World may be less threatened than that of other areas of the world. Of the estimated 1,750 New World mammal species, fewer than 5% are endangered, and about 15% are vulnerable or potentially vulnerable. Within temperate North America there are 37 families with 643 species of mammals; these figures include mammals in the Southwest. Cole et al. (1994) estimated that in North America about 65% of terrestrial mammal species were stable, 4% endangered, and 10% vulnerable; no assessment was possible for 20%.

The Southwest contributes impressively to continentwide diversity of mammals; native mammal species in southwestern states number about 120 in Texas, 138 in Arizona, 139 in New Mexico, and 163 in California (Findley et al. 1975). No other region in the country has so many mammal species—and many of these species and their named subspecies are endemic to the Southwest.

Effects of Human Settlement

It was this fundamental diversity, mammalian and otherwise, that helped attract early settlers to the region, although their concerns about the barren and arid nature of the region were clear. After European settlement, especially in the late 1800's and early 1900's, the first wave of changes in mammalian diversity was noted, with the first native species affected those that humans least tolerated—species originally characterized by large geographic ranges and home territories (Hall 1981). With the determined efforts and encouragement of employees of the Bureau of Biological Survey, large numbers of "predatory" animals, such as grizzly bears, gray wolves, and mountain lions (one subspecies of which, the Yuma puma, is a federal candidate species), were trapped, and some were eventually extirpated from the Southwest (Brown 1983). These large native carnivores came into conflict with human activities, especially the raising of domestic livestock. With the decline of native ungulate populations (Mackie et al. 1982) during and following settlement, domestic livestock may have provided replacement protein for native carnivores before they too were extirpated.

Other human activities, especially agricultural conversion of natural habitats, resulted in the demise or decline of at least two other species whose life histories are intertwined, the black-tailed prairie dog and the black-footed ferret. The prairie dog's own activities, grazing and construction of burrows and mounds, put it on a collision course with farmers and ranchers. Eradication campaigns reduced prairie dog distribution from 40,000,000 hectares to 600,000 hectares by 1960—about a 98% decline in the original geographic distribution of the species (Miller et al. 1990). These eradication campaigns, primarily using poison, also doomed the black-footed ferret.

Other mostly Neotropical carnivores, such as the jaguar, jaguarundi, and ocelot, are now uncommon or absent from the northern portions of their ranges, which include southern Arizona, New Mexico, and Texas (Hall 1981; Hoffmeister 1986). The retraction of their ranges is probably due to their intolerance of humans and to some level of predator control in the Southwest. These carnivore species, though, were never as common as the grizzly bear, the gray wolf, or the black-footed ferret, although some information indicates that jaguars were more common than is often believed (Nowak 1994).

Increases of Deer and Elk

After the extirpation or decline of these carnivores and the black-tailed prairie dog, humans and other mammals coexisted for much of the twentieth century, albeit with some continued level of "damage control" on rodents (mostly prairie dogs) and carnivores (for example, coyotes). Continued trapping of carnivores (including mountain lions) in the Southwest, population management, and more restrictive hunting regulations have resulted in recovery of ungulates from turn-of-the-century population lows (Mackie et al. 1982). Elk numbers are now at an all-time high, with an estimated 782,000 elk occupying more suitable habitat than at any time in this century (Peek 1995; see box on Elk Reintroductions). In much of the

Elk Reintroductions

Rocky Mountain elk are native to north-central New Mexico, including the Jemez Mountains, whereas a different subspecies, Merriam's elk, inhabited southern New Mexico, east-central Arizona, and the Mexican border region (Hall 1981). Merriam's elk went extinct around 1900 in New Mexico, and native Rocky Mountain elk were extirpated by 1909 (Findley et al. 1975). Although elk were known to early inhabitants of the Jemez Mountains (Fig. 1), elk remains are seldom found in archaeological sites there. Indeed, two of three known elk remains from the Jemez Mountains (Table) came from archaeological sites dating to the late 1880's, while the third is represented by a single bone tool dated at A.D. 1390 to 1520. This scarcity of elk in archaeological remains suggests that only small, local elk populations were present between A.D. 1150 and A.D. 1600. Elk numbers may have been suppressed by the many ancestral Pueblo people who inhabited the area, as suggested for nearby Arroyo Hondo by Lang and Harris (1984) and for the intermountain West by Kay (1994). The gray wolf, the most important natural predator of elk in the Jemez Mountains, was extirpated

Fig. 1. A drawing of elk from a rock art site in the Jemez Mountains, New Mexico. The elk was probably painted in the late 1800's.

Courtesy C. D. Allen, USGS

Table. Ungulate remains (minimum number of individual animals) recovered from archaeological sites in and near the Jemez Mountains.

Locality	Deer	Elk	Bighorn	Antelope	Bison	Ungulate
Jemez Mts. (45 sites)	154	3	30	24	7	58
Arroyo Hondo (Santa Fe)	157	6	-	56	7	213
Total	311	9	30	80	14	271

from the area by the 1940's (Findley et al. 1975). Hunting has reduced local populations of another elk predator, the mountain lion (Allen 1989).

Although Merriam's elk was driven to extinction, the Rocky Mountain subspecies survived farther north at places such as Yellowstone National Park; these northern animals were introduced widely in New Mexico (Findley et al. 1975). In 1948, for example, the New Mexico Department of Game and Fish released 21 cows and calves and 7 bulls captured at Yellowstone into the Jemez Mountains (S. Keefe, 25 September 1948, report on file at Bandelier National Monument, New Mexico). In 1964–1965, another 58 elk from Jackson Hole,

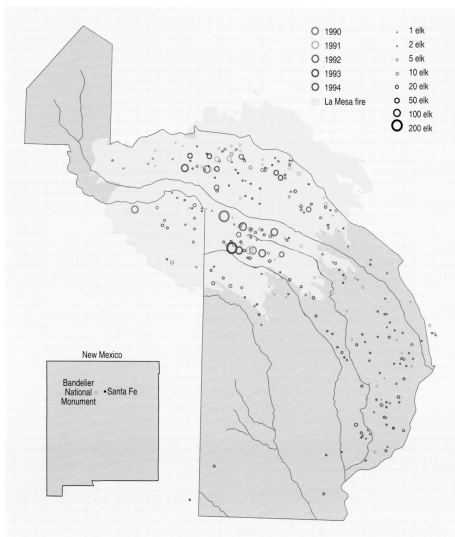

○ 1990   · 1 elk
○ 1991   · 2 elk
○ 1992   ∘ 5 elk
○ 1993   ∘ 10 elk
○ 1994   ∘ 20 elk
▢ La Mesa fire   ○ 50 elk
                 ○ 100 elk
                 ○ 200 elk

New Mexico

Bandelier
National • Santa Fe
Monument

Fig. 2. Distribution of elk at Bandelier National Monument and vicinity, Jemez Mountains, New Mexico.

as throughout the West (Peek 1995). Elk were reintroduced into New Mexico beginning in 1911, with estimated statewide population levels of 3,500 animals by 1934 and 10,000–12,000 elk by 1976; the 1992 population was estimated at 40,000 elk (D. Weybright, New Mexico Department of Game and Fish, Santa Fe, personal communication). If we assume 10,000 elk in 1976, this translates into a 9% annual population growth rate and an 8-year doubling time for 1976–1992 (for 1976 = 12,000 elk; the result is a 7.8% annual increment and a 9.2-year doubling time). Because these large ungulates compete with domestic livestock for herbaceous forage, conflicts with ranchers have emerged in some areas.

Existing data are inadequate to determine whether rapid population growth continues today in the Jemez Mountains, although observations over the past several years clearly reveal that local elk populations are now colonizing lower-elevation sites in ever-increasing numbers, which is indicative of range expansions, if not continued overall population growth. These large elk populations are affecting resources ranging from plant communities to soils and even archaeological sites throughout the Jemez Mountains, especially in the Bandelier National Monument area. Given the uncertainties associated with current data, further population increases should be discouraged until the effects of large elk numbers on the area's resources can be quantified, desirable population levels that are based on resource-carrying capacities identified, and appropriate management strategies determined and implemented.

Wyoming, were released into the mountains of Los Alamos County adjacent to Bandelier National Monument (White 1981).

Elk populations have exhibited exponential population growth since at least the 1970's in Bandelier National Monument and the surrounding Jemez Mountains. The size of the Bandelier National Monument area elk herd has increased dramatically since the 1977 La Mesa fire (Allen 1996a; Fig. 2), which created about 6,000 hectares of grassy winter range. Estimated size of the wintering elk population in Bandelier National Monument has increased from less than 100 in 1977–1978 to 296 in 1978–1979 (Conley et al. 1979); 200–400 elk occurred in the La Mesa fire area in 1979–1980 (Rowland et al. 1983). By 1989, though, 1,000–2,000 elk wintered on Bandelier National Monument and adjacent Los

Alamos National Laboratories and U.S. Forest Service lands (R. Isler, New Mexico Department of Game and Fish, Santa Fe, personal communication). Aerial surveys over Bandelier National Monument counted 907 elk in 1991, 867 in 1992, and 939 in 1994 (Allen 1996b). If we accept that 100 elk wintered in Bandelier National Monument in 1978 and 1,500 elk in 1992, the annual population growth rate was 21.3% with a 3.6-year population doubling time. Some of this population increase reflects concentration of animals into favorable wintering habitat from surrounding areas of the Jemez Mountains, where New Mexico Department of Game and Fish officials estimate that overall elk populations are between 3,500 and 8,000 animals.

Similar rapid population growth has also occurred in New Mexico as a whole, as well

See end of chapter for references

Author

Craig D. Allen
U.S. Geological Survey
Biological Resources Division
Midcontinent Ecological Science Center
Jemez Mountains Field Station
Bandelier National Monument
Los Alamos, New Mexico 87544

Southwest, though, recovered elk populations are composed of nonindigenous elk introduced by state game departments (and other game agencies) from areas to the north, such as Yellowstone National Park (Findley et al. 1975; Hoffmeister 1986). In many areas, mule deer and white-tailed deer also have made remarkable recoveries from earlier lows, although in some areas nonindigenous deer have been transplanted (Hoffmeister 1986).

Other ungulate populations have remained at low levels. Despite considerable transplant efforts, bighorn sheep may only make up 2%–8% of their population levels at the time of European settlement; these animals have never recovered from unregulated harvesting, habitat destruction, overgrazing of rangelands, and diseases contracted from domestic livestock (Singer 1995). Desert bighorn sheep numbers are low, currently estimated at about 19,000, although the population trends have been upward since the 1960's (McCutchen 1995). Numbers of Sonoran pronghorn, listed as endangered by the U.S. Fish and Wildlife Service, are also low, and overall pronghorn numbers in the Southwest are probably below historical highs noted in the 1800's (Hoffmeister 1986).

Concern for Smaller Mammals

Only recently has concern been focused on other less conspicuous mammals that differ in habits from those that were in conflict with humans during settlement. The last list of candidate and sensitive species (U.S. Fish and Wildlife Service 1994a) includes a broad spectrum of mammals that are of concern. The list no longer emphasizes large carnivores, or many carnivores at all, but instead emphasizes smaller species such as insectivores (shrews), a wide array of bats, small relatives of rabbits known as pikas, cottontails and hares, pocket gophers, tree squirrels, and a variety of mice and rats. In general, we know little of the status and trends of these species.

Several unifying trends exist among mammals on these lists, excluding bats (U.S. Fish and Wildlife Service 1994a). Many of the mammals, including shrews, pikas, gophers, tree squirrels, mice, and rats, have small or restricted ranges, often on single mountaintops or other functional islands as well as on actual islands—such as those in the Great Salt Lake. Many mammals on the lists are isolated subspecies of wide-ranging species and often occur on different mountain ranges, a phenomenon particularly true of pikas and gophers (Hall 1981). Also, and perhaps more importantly from an ecosystem perspective, many of the species or subspecies live in wet or moist habitats, often montane. This is clearly the case for the shrews,

voles, meadow jumping mice—a unique family of mice found in North America and China—and some other mice. A smaller number of mammals adapted to arid habitats (for example, kangaroo rats, pocket mice, and cotton rats), primarily in Arizona and Utah, are also on the sensitive species list (U.S. Fish and Wildlife Service 1994a).

Lists of species are sometimes idiosyncratic in nature, in that they reflect the activity, expertise, and concern of people, as well as their access to sources of data about particular species. Some mammals probably have been listed solely because of their restricted range and the absence of data on population trends. Although little information exists about the status or trends of mammals on mountaintops, it is unclear what might be threatening isolated populations of pikas on mountaintops that are not subject to extensive development or visitation. Although global climate change may result in different conditions at high elevations, such effects have not been proven. Likewise, the proliferation of pocket gophers on the sensitive species list (more than 20 southwestern species or subspecies; U.S. Fish and Wildlife Service 1994a) is somewhat puzzling but not yet addressed by any concerted information-gathering effort. Pocket gophers exhibit astounding diversity, with apparently distinct subspecies (or varieties) occurring within miles or even yards of one another (Thaeler 1968; Hall 1981; Patton and Smith 1990). Their subterranean existence promotes such diversity, but it is unknown whether the diversity in turn promotes concern because these different pocket gophers represent perceived dwindling endemism of small areas. Possible environmental threats to pocket gophers include habitat change through overgrazing, lowering of water tables, and poisoning campaigns directed against them or other rodents, such as prairie dogs. Gophers, like prairie dogs, perform ecologically important roles through their habit of turning over and aerating soils, thereby providing for percolation of water and creation of new substrates for vegetative succession.

Some additional species of concern in the Southwest include white-tailed deer in the Sonoran Desert, white-sided jackrabbits in southern New Mexico, and Mexican voles in Arizona. Baker (1977) cited other species for the Chihuahuan Desert, but for most taxa there are no data on status or trends. Species in mountainous areas have been affected by habitat alteration due to water diversions, forestry practices, fire suppression, and livestock grazing. Species in more arid areas are jeopardized by urbanization, agriculture, water diversion, off-road activities, and perhaps climate change.

Endemic Mammals of the Henry Mountains, Utah

Some species of small mammals may appear on lists of sensitive species (for example, U.S. Fish and Wildlife Service 1994) because they are known from only one or a few localities, have not been the subject of any studies, and are genetically distinct. The three endemic subspecies of mammals from the Henry Mountains of Utah illustrate some possibilities and problems faced by management agencies when such species are listed as sensitive species or as candidates for protection as endangered or threatened species. The Henry Mountains in south-central Utah were the last western mountain range to be discovered (Hunt et al. 1953; Figure), after being seen during one of John Wesley Powell's trips down the Colorado River and subsequently explored by a party sent out by Powell. The mountains provide habitat to three endemic subspecies of mammals: Mt. Ellen chipmunk, Mt. Ellen pocket gopher, and Mt. Ellen long-tailed vole (Hall 1981). Recent research in the Henry Mountains showed that the chipmunk is abundant with no apparent threats to its widespread habitat (Mollhagen and Bogan 1995). The vole, though less abundant, is commonly found in good habitat. Voles, however, cycle in abundance (Taitt and Krebs 1985) and surveys for them should be conducted over several seasons or years. Habitat for the vole could be jeopardized by overgrazing or drying of wetlands, but no studies in the Henry Mountains have documented such threats. These two species, the chipmunk and vole, are probably not in real danger from any perceivable threat. The pocket gopher,

Courtesy M. Bogan, USGS

Figure. East face of the Henry Mountains in southeastern Utah.

though, may represent a distinctly different case. Almost 40 years ago, Durrant (1958) commented that the gopher might be extinct, though the basis for his comment is unknown. More recent studies also found no sign of pocket gophers in the Henry Mountains (Mollhagen and Bogan 1995), though surveys are incomplete, and considerable habitat remains that must be searched. Additional surveys are needed to clearly determine the status of the Mt. Ellen pocket gopher so that management decisions can be made.

See end of chapter for references

Author

Michael A. Bogan
U.S. Geological Survey
Biological Resources Division
Midcontinent Ecological Science Center
Department of Biology
University of New Mexico
Albuquerque, New Mexico 87131

Introduced and Feral Mammals

Many kinds of introduced and feral mammals are now naturalized in the Southwest: Barbary sheep, gemsbok, ibex (Findley et al. 1975), and nonindigenous deer introduced for sport hunting; wild burros, horses, cattle, pigs, goats, sheep, dogs, and cats whose ancestors escaped from captivity; and mice and rats living commensally with humans. In general, little is known about the effect of these species on native vertebrate wildlife and their habitats. Drost and Fellers (1995) provided information on distribution and effects of some of these species on public lands, and Pogacnik (1995) discussed the

status of wild horses and burros on public lands. Arizona is estimated to have about 4,000 wild horses, asses, and mules; New Mexico around 500 (most on White Sands Missile Range), and Utah about 2,000 (Pogacnik 1995). In New Mexico, gemsbok appear to be expanding their range northward from White Sands Missile Range, where they were introduced by the New Mexico Department of Game and Fish; they now routinely encroach on the San Andres National Wildlife Refuge and the White Sands National Monument (M. Bogan, unpublished data). In some areas, gemsbok seem to cause damage by trampling, but no studies have documented their effects on public lands.

Southwestern Bats

Concern for bats in the Southwest (Figure) is a relatively recent phenomenon, although two southwestern bats are listed as endangered by the U.S. Fish and Wildlife Service. Before November 1994, seven species or subspecies of bats in the Southwest were listed as candidates for eventual listing under provisions of the U.S. Endangered Species Act of 1973. The last list (U.S. Fish and Wildlife Service 1994) contained 15 bats, including 10 full species. The total bat fauna of the Southwest is about 30 species, depending on locality, so one-third to one-half of the region's bat species are now considered sensitive. No other group of mammals is a target for such recent concern. In general, we have no long-term population data on status and trends of these bats, mostly because bats, long-lived and with a low reproductive potential, are difficult to study, and their numbers are even more difficult to quantify. No nationwide survey of bat populations exists, and no attempt has been made to quantify existing data so that trends in bat populations can be discerned, although the U.S. Geological Survey has initiated such a study (T. J. O'Shea, U.S. Geological Survey, Fort Collins, Colorado, personal communication). Many of the species added as candidates—now considered species of concern—in November 1994 are widespread in the West and are thought to be relatively secure.

Although many bat biologists and state and national groups (for example, Bat Conservation International and Colorado Bat Society) have expressed concern about threats to bats for many years, most land managers have become aware of the potential for declining bat populations only in the last few years. In the West in general, a preeminent concern for bat welfare comes from the widespread closure of abandoned mine entrances. These mines, a historical and common feature of the American West, represent an extreme safety hazard to humans. With funding from the federal government, many western states have set up programs to close such mines, and thousands have been closed. Closures usually occur with no thought or concern that bats might be using these mines as roosting sites, even though many mines have become havens for western bats since the mines were abandoned (K. Navo, Colorado Division of Wildlife, Denver, and J. S. Altenbach, University of New Mexico, Albuquerque, personal communications). Bats have moved into abandoned mines after being driven or excluded from other roosting sites, such as vandalized caverns.

Because topographic diversity, specifically the amount and nature of available roost sites, determines the species diversity of bats (Humphrey 1975), protection of roosts is essential. Thus, many individuals and groups are focusing their efforts on surveying the most likely mines that bats may be inhabiting before these mines are closed. In turn, many states are incorporating results of such surveys into their planning and, when necessary, erecting bat-friendly gates at the mine entrance instead of absolute closures (K. Navo, Colorado Division of Wildlife, Denver, personal communication). The National Biological Service (now the U.S. Geological Survey Biological Resources Division) recently helped fund a cooperative program (Bat Conservation International, Bureau of Mines Management and U.S. Geological Survey) aimed at assessing the magnitude of the threat that mine closures represent to bats. More data, especially long-term data, are needed to assess the status and trends of bats in the Southwest.

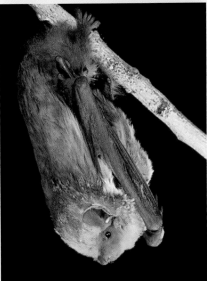

Figure. Endangered lesser long-nosed bat (top), western red bat (center), and Townsend's big-eared bat (bottom).

© J. S. Altenbach, University of New Mexico

See end of chapter for references

Author

Michael A. Bogan
U.S. Geological Survey
Biological Resources Division
Midcontinent Ecological Science Center
Department of Biology
University of New Mexico
Albuquerque, New Mexico 87131

Summary and Recommendations

Information Gaps

A striking lack of information exists about the status and trends of southwestern biota and ecosystems. To remedy this, researchers need to systematically collect and summarize the rigorously collected data on status and trends of plant and animal communities in the Southwest. For most groups we still lack information on population size and variability, recruitment, trends, effects of change due to human activity, interactions with nonindigenous species, and known threats to populations. However, for selected species of fishes and birds we have sufficient information not only to reveal the problems but also to start to understand the answers as well.

One wonders how far the current conservation effort for birds, Partners in Flight, would have gotten without the availability of detailed trend information gathered over the last 32 years by the U.S. Geological Survey Breeding Bird Survey. With well-designed efforts over the next 30 years, we eventually could have equivalent surveys for butterflies, snails, or other lesser known invertebrates or plants. We think caution is required in applying conservation strategies developed for well-understood groups to little-understood groups because such strategies may not provide workable answers in all cases. Some efforts, such as those directed at bats, are now under way (see box on Southwestern Bats), but additional conservation efforts need planning and implementation now.

The effect of habitat fragmentation—the division of solid habitat blocks by roads, right-of-ways, or development—on ecosystem integrity and on the species of the area affected is still incompletely understood and is a pervasive issue in the Southwest. For many areas our understanding of even the extent of habitat fragmentation is poor; available technology (such as geographic information systems and satellite imagery) should be used to assess status and trends of fragmentation. In addition, field studies are required to better understand the effects of fragmentation on populations and communities of southwestern plants and animals. Particularly important is acquiring knowledge of how organisms respond to shrinking habitats and whether linking habitats by corridors will compensate for some degree of absolute habitat loss (Simberloff et al. 1992).

Given that many human-related disturbances, such as livestock grazing and timbering, were more severe 90 or more years ago, and that some ecosystems are in better condition now

than at the turn of the century, research should address the responses of ecosystems and their components to such recovery. Not only have many deer, elk, and related grazer populations recovered from population lows, but some are now posing real management concerns (see box on Elk Reintroductions). Issues such as the control of ungulate (or carnivore) populations on federal lands, especially on national parks, can best be resolved with rigorous studies that assess current effects of expanding animal populations on localized environments. Studies that are firmly grounded in knowledge of the history of southwestern ecosystems seem to us to provide the best opportunity to resolve such issues.

Large information gaps persist regarding the effects on ecosystems and species of the three most important types of human landscape management in the Southwest. In spite of the plethora of studies cited in Fleischner's (1994) overview, the effects of grazing on natural ecosystems and species continue to be controversial. We believe that more studies on these effects are needed and that they must be carefully designed so that it is possible to "tease out" effects due to grazing from those due to climate, soils, or other factors. Furthermore, it is likely that results from some grazing studies are not widely applicable beyond the local environment; this implies a need for more site-specific studies. Likewise, although we have learned much of the history and effects of fire suppression, a great deal of additional site-specific information is needed to understand the ecological effects of fire on southwestern ecosystems and biological diversity. Finally, although the pervasive effects of water diversion and impoundment structures on southwestern aquatic habitats and their biota are becoming clear (see chapter on Water Use), little knowledge exists of how to mitigate or reverse such trends. A key approach to addressing these information gaps is through additional experimental studies (e.g., alternative grazing regimes, spring versus fall burning, altering reservoir water levels and release rates), which would rigorously assess the response of biota and ecosystem processes to alternative management actions at specific sites. Once enough site-specific information is available, it may be possible to develop regional summaries to allow improved extrapolation of results to unstudied areas.

In many portions of the Southwest, human populations can be expected to continue to grow rapidly, yet there appears to be little research that directly addresses the effect of increasing human numbers on natural communities and their constituent species. Although some national park research programs attempt to address the issue of visitor effects on park ecosystems,

Authors

Michael A. Bogan
U.S. Geological Survey
Biological Resources Division
Midcontinent Ecological
Science Center
Department of Biology
University of New Mexico
Albuquerque, New Mexico 87131

Craig D. Allen
U.S. Geological Survey
Biological Resources Division
Midcontinent Ecological
Science Center
Jemez Mountains Field Station
Bandelier National Monument
Los Alamos, New Mexico 87544

more rigorous studies are needed. For nature reserves (for example, wildlife refuges, national parks, wilderness areas) we particularly need continued but expanded baseline inventories to assure that we fully understand what ecosystem components are present. Once baseline information is adequate, monitoring of critical populations needs to begin. In addition, more data are needed on the effects of activities such as mining, timber harvest, road building, application of biocides, pollution, and conversion of land to agricultural or urban uses. With new and better data, we will better understand exactly how human population growth will affect southwestern ecosystems and their biota.

Fundamentally, many conservation biology concerns are related to habitat management issues. If sufficient habitat is maintained, then most biotic components of that ecosystem will likewise be maintained. At present, however, we have an inadequate understanding of how biotic components interact in the mosaic of habitats that characterize the Southwest. New studies should include attempts to better understand the history of southwestern ecosystems; the historical picture can then provide a context for understanding current patterns and predicting future trends.

An important avenue of research that we have not mentioned previously is the need for studies on sociological aspects of status and trends of southwestern biota. To what extent are citizens committed to maintaining some semblance of natural conditions in the Southwest? Studies that assess the nonmarket values of natural resources will be important in helping to set priorities for future research on southwestern

ecosystems. Fully successful conservation results will require a greater degree of social consensus than has been achieved recently. We believe that the lack of adequate unbiased data on both natural and cultural systems in the Southwest is a major impediment to achieving social agreement on needed actions.

Conclusion

Finally, we recognize that this overview sketch of the condition of southwestern biota necessarily shares the flaws of incompleteness and uncertainty common to this genre of summary reviews, reflecting the backgrounds of the few authors involved. Much additional, unsummarized information currently exists on the status and trends of southwestern biota, but this information is in widely dispersed and disparate forms, and much remains unpublished and generally inaccessible. One of several possible ways to begin to fill in known information gaps is to begin a series of actual or "virtual" regional scale workshops that bring together available experts to present and collate existing knowledge on the status and trends of southwestern biota, including much currently unavailable information. The initial workshops might be organized by regional ecosystem (for example, the Colorado Plateau or the Sonoran Desert) or by state, building to a regional scale synthesis. Such workshops or similar information-gathering efforts could play a key role in the development of the National Biological Information Infrastructure and would allow research needs to be better prioritized.

Esteban H. Muldavin
New Mexico Natural
Heritage Program
Department of Biology
University of New Mexico
Albuquerque, New Mexico 87131

Steven P. Platania
Division of Fishes
Museum of Southwestern Biology
University of New Mexico
Albuquerque, New Mexico 87131

James N. Stuart
U.S. Geological Survey
Biological Resources Division
Midcontinent Ecological
Science Center
Department of Biology
University of New Mexico
Albuquerque, New Mexico 87131

Greg H. Farley
Division of Biological Sciences
Fort Hays State University
Hays, Kansas 67613

Patricia Mehlhop
New Mexico Natural
Heritage Program
Department of Biology
University of New Mexico
Albuquerque, New Mexico 87131

Jayne Belnap
U.S. Geological Survey
Biological Resources Division
Midcontinent Ecological
Science Center
Canyonlands National Park
Field Station
2282 S. West Resource Boulevard
Moab, Utah 84532

Cited References

Aldon, E. F., and D. W. Shaw, technical coordinators. 1993. Managing piñon–juniper ecosystems for sustainability and social needs. U.S. Forest Service General Technical Report RM-236. 169 pp.

Alexander, R. R. 1987. Ecology, silviculture, and management of the Engelmann spruce–subalpine fir type in the central and southern Rocky Mountains. U.S. Department of Agriculture, Agriculture Handbook 659, Washington, D.C. 144 pp.

Allen, C. D. 1989. Changes in the landscape of the Jemez Mountains, New Mexico. Ph.D. dissertation, University of California, Berkeley. 346 pp.

Allen, C. D., technical editor. 1996a. Fire effects in southwestern forests: proceedings of the second La Mesa fire symposium. U.S. Forest Service General Technical Report RM-GTR-286. 216 pp.

Allen, C. D. 1996b. Response of elk populations to the La Mesa fire. Pages 179–195 *in* C. D. Allen, technical editor. Fire

effects in the southwestern forests: proceedings of the second La Mesa fire symposium. U.S. Forest Service General Technical Report RM-GTR-286. 216 pp.

Allen, L. S. 1992. Livestock–wildlife coordination in the encinal oak woodlands: Coronado National Forest. Pages 109–110 *in* P. F. Ffolliott, G. J. Gottfried, D. A. Bennett, J. M. Hernandez, C. A. Ortega, and R. H. Hamre, technical coordinators. Ecology and management of oak and associated woodlands: perspectives in the southwestern United States and northern Mexico. U.S. Forest Service General Technical Report RM-218. 224 pp.

Almand, J., and W. Krohn. 1979. The position of the Bureau of Land Management on the protection and management of riparian ecosystems. Pages 259–361 *in* R. Johnson and F. McCormick, technical coordinators. Strategies for protection and management of floodplain wetlands and other riparian ecosystems. Proceedings of the symposium, 11–13 December 1978. U.S. Forest Service General Technical Report WO-12, Washington, D.C.

Archer, S. 1994. Woody plant encroachment into southwestern grasslands and savannas: rates, patterns and proximate causes. Pages 13–68 *in* M. Vavra, W. A. Laycock, and R. D. Pieper, editors. Ecological implications of livestock herbivory in the West. Society for Range Management, Denver, Colo.

Arizona State Parks. 1988. Arizona wetlands priority plan. Arizona State Parks, Phoenix.

Armour, C. L., D. F. Duff, and W. Elmore. 1991. The effects of livestock grazing on riparian and stream ecosystems. Fisheries 16:7–11.

Bahre, C. J. 1991. A legacy of change: historic human impact on vegetation of the Arizona borderlands. University of Arizona Press, Tucson. 231 pp.

Bahre, C. J., and D. E. Bradbury. 1978. Vegetation change along the Arizona–Sonora border. Annals of the Association of American Geography 68:145–165.

Bahre, C. J., and C. F. Hutchinson. 1985. The impact of historical fuelwood cutting on semidesert woodlands of southeastern

Arizona. Journal of Forest History 29:175–186.

Bahre, C. J., and M. L. Shelton. 1993. Historic vegetation change, mesquite increases, and climate in southeastern Arizona. Journal of Biogeography 20:489–504.

Baker, R. H. 1977. Mammals of the Chihuahuan Desert region—future prospects. Pages 221–225 *in* R. H. Wauer and D. H. Riskind, editors. Transactions of the symposium on the biological resources of the Chihuahuan Desert region, United States and Mexico. National Park Service Transactions and Proceedings Series 3.

Baker, W. L. 1983. Alpine vegetation of Wheeler Peak, New Mexico, USA: gradient analysis, classification, and biogeography. Arctic and Alpine Research 15:223–240.

Barton, A. M. 1994. Gradient analysis of relationships among fire, environment, and vegetation in a southwestern USA mountain range. Bulletin of the Torrey Botanical Club 121:251–265.

Bassett, R., M. Larson, and W. Moir. 1987. Forest and woodland habitat types (plant associations) of Arizona south of the Mogollon Rim and southwestern New Mexico. U.S. Forest Service, Southwest Region, Albuquerque, N. Mex. 91 pp.

Behnke, R. J., and D. E. Benson. 1980. Endangered and threatened fishes of the upper Colorado River basin. Colorado State University Cooperative Extension Service Bulletin 503A:1–34.

Bender, M. 1981. "Sting" operation reveals massive illegal trade. Endangered Species Technical Bulletin 6:1,4.

Bennett, D. A. 1992. Fuelwood extraction in southeast Arizona. Pages 96–97 *in* P. F. Ffolliott, G. J. Gottfried, D. A. Bennett, J. M. Hernandez, C. A. Ortega, and R. H. Hamre, technical coordinators. Ecology and management of oak and associated woodlands: perspectives in the southwestern United States and northern Mexico. U.S. Forest Service General Technical Report RM-218.

Berry, K. H. 1978. Livestock grazing and the desert tortoise. Transactions of the North American Wildlife and Natural Resources Conference 43:505–519.

Bestgen, K. R., and S. P. Platania. 1990. Extirpation and notes on the life history of *Notropis simus simus* and *Notropis orca* (Cypriniformes: Cyprinidae) from the Rio Grande, New Mexico. Occasional Papers of the Museum of Southwestern Biology 6:1–8.

Betancourt, J. L. 1990. Late Quaternary biogeography of the Colorado Plateau. Pages 259–292 *in* J. L. Betancourt, T. R. Van Devender, and P. S. Martin, editors. Packrat middens: the last 40,000 years of biotic change. University of Arizona Press, Tucson. 468 pp.

Betancourt, J. L., and T. R. Van Devender. 1981. Holocene vegetation in Chaco Canyon, New Mexico. Science 214:656–658.

Betancourt, J. L., E. A. Pierson, K. A. Rylander, J. A. Fairchild-Parks, and J. S. Dean. 1993. Influence of history and climate on New Mexico piñon–juniper woodlands. Pages 42–62 *in* E. F. Aldon and D. W. Shaw, technical coordinators. Managing piñon–juniper ecosystems for sustainability and social needs. U. S. Forest Service General Technical Report RM-236.

Bock, C. E., H. M. Smith, and J. H. Bock. 1990. The effect of livestock grazing upon abundance of the lizard, *Sceloporus scalaris*, in southeastern Arizona. Journal of Herpetology 24:445–446.

Bock, C. E., V. A. Saab, T. D. Rich, and D. S. Dobkin. 1993. Effects of livestock grazing on Neotropical migratory landbirds in western North America. Pages 296–309 *in* D. M. Finch and P. W. Stangel, editors. Status and management of Neotropical migratory birds. U.S. Forest Service General Technical Report RM-229.

Bohrer, V. L. 1975. The prehistoric and historic role of the cool-season grasses in the Southwest. Ethnobotany 29:199–207.

Boydstun, C., P. Fuller, and J. D. Williams. 1995. Nonindigenous fish. Pages 431–433 *in* E. T. LaRoe, G. S. Farris, C. E. Puckett, P. D. Doran, and M. J. Mac, editors. Our living resources: a report to the nation on the distribution, abundance, and health of U.S. plants, animals, and ecosystems. U.S. Department of the Interior, National Biological Service, Washington, D.C.

Bradford, D. F., F. Tabatabai, and D. M. Graber. 1993. Isolation of remaining populations of the native frog, *Rana muscosa*, by introduced fishes in Sequoia and Kings Canyon national parks, California. Conservation Biology 7:882–888.

Bradley, A. F., N. V. Noste, and W. C. Fischer. 1992. Fire ecology of forests and woodlands in Utah. U.S. Forest Service General Technical Report INT-287. 128 pp.

Branson, F. A. 1985. Vegetation changes on western rangelands. Range Monograph 2, Society for Range Management, Denver, Colo. 76 pp.

Brode, J. M., and R. B. Bury. 1984. The importance of riparian systems to amphibians and reptiles. Pages 30–36 *in* R. E. Warner and K. M. Hendrix, editors. California riparian systems: ecology, conservation, and productive management. University California Press, Berkeley and Los Angeles. 1035 pp.

Brown, B. T., G. S. Mills, R. L. Glinski, and S. W. Hoffman. 1992. Density of nesting peregrine falcons in Grand Canyon National Park, Arizona. Southwestern Naturalist 37:188–193.

Brown, D., S. M. Hitt, and W. H. Moir, editors. 1986. The path from here: integrated forest protection for the future. Integrated pest management working group, U.S. Forest Service, Southwest Regional Office, Albuquerque, N. Mex. 1986-0-676-098/20078. 194 pp. + exhibits a–o.

Brown, D. E., editor. 1982a. Biotic communities of the American Southwest—United States and Mexico. Desert Plants 4:1–341.

Brown, D. E. 1982b. Alpine and subalpine grasslands, montane meadow grassland. Pages 108–114 *in* D. E. Brown, editor. Biotic communities of the American Southwest—United States and Mexico. Desert Plants 4:1–341.

Brown, D. E. 1982c. Madrean evergreen woodland. Pages 59–65 *in* D. E. Brown, editor. Biotic communities of the American Southwest—United States and Mexico. Desert Plants 4:1–341.

Brown, D. E. 1982d. Great Basin conifer woodland. Pages 52–57 *in* D. E. Brown, editor. Biotic communities of the American Southwest—United States and Mexico. Desert Plants 4:1–341.

Brown, D. E. 1983. The wolf in the Southwest: the making of an endangered species. University of Arizona Press, Tucson. 195 pp.

Brown, J. H., and W. McDonald. 1995. Livestock grazing and conservation on southwestern rangelands. Conservation Biology 9:1644–1647.

Bryant, N. A., L. F. Johnson, A. J. Brazel, R. C. Balling, C. F. Hutchinson, and L. R. Beck. 1990. Measuring the effect of overgrazing in the Sonoran Desert. Climatic Change 17:243–264.

Buffington, L. C., and C. H. Herbel. 1965. Vegetation changes on a semi-desert grassland range. Ecological Monographs 35:139–164.

Busack, S. D., and R. B. Bury. 1974. Some effects of off-road vehicles and sheep grazing on lizard populations in the Mohave Desert. Biological Conservation 6:179–183.

Busch, D. E., and S. D. Smith. 1993. Effects of fire on water and salinity relations of riparian woody taxa. Oecologia 94:186–194.

Busch, D. E., N. L. Ingraham, and S. D. Smith. 1992. Water uptake in woody riparian phreatophytes of the southwestern United States: a stable isotope study. Ecological Applications 2:450–459.

Cameron, C. 1896. [The Arizona livestock industry]. Pages 220–231 *in* Report of the Governor of Arizona to the Secretary of Interior. Report INT 96, Volume 3, 15, Washington, D.C.

Campbell, C. J., and W. Green. 1968. Perpetual succession of stream-channel vegetation in a semiarid region. Journal of the Arizona Academy of Science 5:86–98.

Campbell, F. T. 1988. The desert tortoise. Pages 567–581 *in* W. J. Chandler, editor. Audubon Wildlife Report 1988/1989. Academic Press, San Diego, Calif.

Cannon, R. W., and F. L. Knopf. 1984. Species composition of a willow community relative to seasonal grazing histories in Colorado. Southwestern Naturalist 29:234–237.

Caprio, A. C., and M. J. Zwolinski. 1992. Fire effects on Emory and Mexican blue oak in southeastern Arizona. Pages

150–154 *in* P. F. Ffolliott, G. J. Gottfried, D. A. Bennett, J. M. Hernandez, C. A. Ortega, and R. H. Hamre, technical coordinators. Ecology and management of oak and associated woodlands: perspectives in the southwestern United States and northern Mexico. U.S. Forest Service General Technical Report RM-218.

Carlson, C. A., and R. T. Muth. 1989. The Colorado River: lifeline of the American Southwest. Pages 220–239 *in* D. P. Dodge, editor. Proceedings of the international large river symposium. Canadian Special Publication of Fisheries and Aquatic Sciences 106.

Carter, M. F., and K. Barker. 1993. An interactive database for setting conservation priorities for western Neotropical migrants. Pages 120–144 *in* D. M. Finch and P. W. Stangel, editors. Status and management of Neotropical migratory birds. U.S. Forest Service General Technical Report RM-229.

Chernoff, B., R. R. Miller, and C. R. Gilbert. 1982. *Notropis orca* and *Notropis simus*, cyprinid fishes from the American Southwest, with description of a new subspecies. Occasional Papers of the Museum of Zoology, University of Michigan 698:1–49.

Chew, R. M. 1982. Changes in herbaceous and suffrutescent perennials in grazed and ungrazed desertified grassland in southeastern Arizona, 1958–1978. American Midland Naturalist 108:159–169.

Clarkson, R. W., and J. C. Rorabaugh. 1989. Status of leopard frogs (*Rana pipiens* complex: Ranidae) in Arizona and southeastern California. Southwestern Naturalist 34:531–538.

Cole, F. R., D. M. Reeder, and D. E. Wilson. 1994. A synopsis of distribution patterns and the conservation of mammal species. Journal of Mammalogy 75:266–276.

Cole, K. 1995. Vegetation change in national parks. Pages 224–227 *in* E. T. LaRoe, G. S. Farris, C. E. Puckett, P. D. Doran, and M. J. Mac, editors. Our living resources: a report to the nation on the distribution, abundance, and health of U.S. plants, animals, and ecosystems. U.S. Department of the Interior, National Biological Service, Washington, D.C.

Conant, R. 1977. Semiaquatic reptiles and amphibians of the Chihuahuan Desert and their relationships to drainage patterns of the region. Pages 455–491 *in* R. H. Wauer and D. H. Riskind, editors. Transactions of the symposium on biological resources of the Chihuahuan Desert region: United States and Mexico. National Park Service Transactions and Proceedings Series 3.

Cooke, R. R., and R. W. Reeves. 1976. Arroyos and environmental change in the American Southwest. Clarendon Press, Oxford. 213 pp.

Cooper, C. F. 1960. Changes in vegetation, structure, and growth of southwestern pine forests since white settlement. Ecological Monographs 30:129–164.

Cooperrider, C. K., and B. A. Hendricks. 1937. Soil erosion and stream flow on range and forest lands of the upper Rio Grande watershed in relation to land resources and human welfare. U.S. Department of Agriculture Technical Bulletin 567, Washington, D.C. 88 pp.

Corn, P. S. 1995. What we know and don't know about amphibian declines in the West. Pages 59–67 *in* W. W. Covington and L. F. DeBano, technical coordinators. Sustainable ecological systems: implementing an ecological approach to land management. U.S. Forest Service General Technical Report GTR-RM-247.

Corn, P. S., and J. C. Fogleman. 1984. Extinction of montane populations of the northern leopard frog (*Rana pipiens*) in Colorado. Journal of Herpetology 18:147–152.

Cotner, M. L. 1963. Controlling pinyon–juniper on southwestern rangelands. Arizona Agricultural Experiment Station Report 210. 28 pp.

Cottam, W. P. 1947. Is Utah Sahara bound? University of Utah Bulletin 37. 40 pp.

Cottam, W. P., and G. Stewart. 1940. Plant succession as a result of grazing and of meadow desiccation by erosion since settlement in 1862. Journal of Forestry 38:613–626.

Covington, W. W., and M. M. Moore. 1994. Southwestern ponderosa pine forest structure: changes since Euro–American settlement. Journal of Forestry 92:39–47.

Covington, W. W., and S. S. Sackett. 1992. Soil mineral nitrogen changes following prescribed burning in ponderosa pine. Forest Ecology and Management 54:175–191.

Crawford, C. S., A. C. Cully, R. Leutheuser, M. S. Sifuentes, L. H. White, and J. P. Wilber. 1993. Middle Rio Grande ecosystem: bosque biological management plan. Middle Rio Grande Biological Interagency Team, U.S. Fish and Wildlife Service, Albuquerque, N. Mex. 291 pp. + maps.

Dahl, T. E. 1990. Wetland losses in the United States 1780's to mid-1980's. U.S. Fish and Wildlife Service, Washington, D.C. 13 pp.

Dalen, R. S., and W. R. Snyder. 1987. Economic and social aspects of pinyon–juniper treatment—then and now. Pages 343–350 *in* R. L. Everett, compiler. Proceedings—Pinyon–juniper conference. U.S. Forest Service General Technical Report INT-215.

deBuys, W. 1985. Enchantment and exploitation: the life and hard times of a New Mexico mountain range. University of New Mexico Press, Albuquerque. 394 pp.

deBuys, W. 1993. The sociological and ecological consequences of managing piñon woodlands. Pages 82–84 *in* E. F. Aldon and D. W. Shaw, technical coordinators. Managing piñon–juniper ecosystems for sustainability and social needs. U.S. Forest Service General Technical Report RM-236.

DeByle, N. V., and R. P. Winokur, editors. 1985. Aspen: ecology and management in the western United States. U.S. Forest Service General Technical Report RM-119. 283 pp.

Degenhardt, W. G. 1977. A changing environment: documentation of lizards and plants over a decade. Pages 533–555 *in* R. H. Wauer and D. H. Riskind, editors. Transactions of the symposium on biological resources of the Chihuahuan Desert region: United States and Mexico. National Park Service Transactions and Proceedings Series 3.

Degenhardt, W. G., C. W. Painter, and A. H. Price. 1996. The amphibians and reptiles of New Mexico. University of New Mexico Press, Albuquerque. 431 pp.

Denevan, W. M. 1967. Livestock numbers in nineteenth-century New Mexico and the problem of gullying in the Southwest. Annals of the Association of American Geography 57:691-703.

DeSante, D. F., and T. L. George. 1994. Population trends in the landbirds of western North America. Pages 173–190 *in* J. R. Jehl, Jr. and N. K. Johnson, editors. A century of avifaunal change in western North America. Studies in Avian Biology 15.

DeSante, D. F., O. E. Williams, and K. M. Burton. 1993. The monitoring avian productivity and survivorship (MAPS) program: overview and progress. Pages 208–222 *in* D. M. Finch and P. W. Stangel, editors. Status and management of Neotropical migratory birds. U.S. Forest Service General Technical Report RM-229.

Despain, D. W., and J. C. Mosley. 1990. Fire history and stand structure of a pinyon–juniper woodland at Walnut Canyon National Monument, Arizona. National Park Service Technical Report 34, Cooperative Park Studies Unit, Flagstaff, Ariz. 27 pp.

Dick-Peddie, W. A. 1993. New Mexico vegetation: past, present, and future. University of New Mexico Press, Albuquerque. 244 pp.

Drost, C. A., and E. Deshler. 1995. Amphibian and reptile diversity on the Colorado Plateau. Pages 326–328 *in* E. T. LaRoe, G. S. Farris, C. E. Puckett, P. D. Doran, and M. J. Mac, editors. Our living resources: a report to the nation on the distribution, abundance, and health of U.S. plants, animals, and ecosystems. U.S. Department of the Interior, National Biological Service, Washington, D.C.

Drost, C. A., and G. M. Fellers. 1995. Non-native animals on public lands. Pages 440–442 *in* E. T. LaRoe, G. S. Farris, C. E. Puckett, P. D. Doran, and M. J. Mac, editors. Our living resources: a report to the nation on the distribution, abundance, and health of U.S. plants, animals, and ecosystems. U.S. Department of the Interior, National Biological Service, Washington, D.C.

Durkin, P., E. Muldavin, M. Bradley, S. E. Carr, A. Metcalf, R. A. Smartt, S. P. Platania, C. Black, and P. Mehlhop. 1995. The biodiversity of riparian ecosystems of the Ladder Ranch. New Mexico Natural Heritage Program, Albuquerque. Final report to the Turner Foundation. 118 pp. + appendix.

Edwards, T. C. 1995. Protection status of vegetation cover types in Utah. Pages 463–464 *in* E. T. LaRoe, G. S. Farris, C. E. Puckett, P. D. Doran, and M. J. Mac, editors. Our living resources: a report to the nation on the distribution, abundance, and health of U.S. plants, animals, and ecosystems. U.S. Department of the Interior, National Biological Service, Washington, D.C.

Ernst, C. H. 1992. Venomous reptiles of North America. Smithsonian Institution Press, Washington, D.C. 236 pp.

Evans, R. A. 1988. Management of pinyon–juniper woodlands. U.S. Forest Service General Technical Report INT-249. 34 pp.

Everett, R. L., compiler. 1987. Proceedings—Pinyon–Juniper conference. U.S. Forest Service General Technical Report INT-215, Intermountain Research Station, Ogden, Utah. 581 pp.

Farley, G. H., L. M. Ellis, J. N. Stuart, and N. J. Scott, Jr. 1994. Avian species richness in different-aged stands of riparian forest along the middle Rio Grande, New Mexico. Conservation Biology 8:1098–1108.

Ffolliott, P. F., G. S. Gottfried, and M. B. Baker, Jr. 1989. Water yield from forest snowpack management: research findings in Arizona and New Mexico. Water Resources Research 25:1999–2007.

Findley, J. S., A. H. Harris, D. E. Wilson, and C. Jones. 1975. Mammals of New Mexico. University of New Mexico Press, Albuquerque. 360 pp.

Flather, C. H., L. A. Joyce, and C. A. Bloomgarden. 1994. Species endangerment patterns in the United States. U.S. Forest Service General Technical Report RM-241. 42 pp.

Fleischner, T. L. 1994. Ecological costs of livestock grazing in western North America. Conservation Biology 8:629–644.

Foxx, T. S., compiler. 1984. La Mesa fire symposium. LA-9236-NERP. Los Alamos National Laboratory, Los Alamos, N. Mex. 172 pp.

Fritts, H. C., and T. W. Swetnam. 1989. Dendroecology: a tool for evaluating changes in past and present forest environments. Advances in Ecological Research 19:111–189.

General Accounting Office. 1988. Public rangelands: some riparian areas restored but widespread improvement will be slow. General Accounting Office, Washington, D.C. 85 pp.

Gillis, A. M. 1991. Should cows chew cheatgrass on common lands? BioScience 41:668–675.

Gilpin, M. E., and M. E. Soulé. 1986. Minimum viable populations: processes of species extinction. Pages 19–34 *in* M. E. Soulé, editor. Conservation Biology. Sinauer, Sunderland, Mass.

Glaser, L. C. 1995. Wildlife mortality attributed to organophosphorus and carbamate pesticides. Pages 416–418 *in* E. T. LaRoe, G. S. Farris, C. E. Puckett, P. D. Doran, and M. J. Mac, editors. Our living resources: a report to the nation on the distribution, abundance, and health of U.S. plants, animals, and ecosystems. U.S. Department of the Interior, National Biological Service, Washington, D.C.

Gosz, J. R. 1992. Gradient analysis of ecological change in time and space: implications for forest management. Ecological Applications 2:248–261.

Gottfried, G. J., T. W. Swetnam, C. D. Allen, J. L. Betancourt, and A. Chung-MacCoutbrey. 1995. Pinyon–juniper woodlands of the middle Rio Grande basin, New Mexico. Pages 95–132 *in* D. M. Finch and J. A. Tainter, technical editors. Ecology, diversity, and sustainability of the middle Rio Grande basin. U.S. Forest Service General Technical Report RM-268.

Graf, W. L. 1986. Fluvial erosion and federal public policy in the Navajo Nation. Physical Geography 7:97–115.

Grissino-Mayer, H. D. 1995. Tree-ring reconstructions of fire and climate history at El Malpais National Monument, New Mexico. Ph.D. dissertation, University of Arizona, Tucson.

Grissino-Mayer, H. D., C. H. Baisan, and T. W. Swetnam. 1995. Fire history in the Pinaleño Mountains of Southeastern Arizona: effects of human-related disturbances. Pages 399–407 *in* L. DeBano, P. F. Ffolliott, and J. Gottfried, technical coordinators. Biodiversity and management of the Madrean Archipelago: the sky islands of the southwestern United States and northwestern Mexico, September 19–23, 1994, Tucson, Arizona. U.S. Forest Service General Technical Report RM-264.

Gross, F. A., and W. A. Dick-Peddie. 1979. A map of primeval vegetation in New Mexico. Southwestern Naturalist 24:115–122.

Grossman, D. H., and K. L. Goodin. 1995. Rare terrestrial ecological communities of the United States. Pages 218–221 *in* E. T. LaRoe, G. S. Farris, C. E. Puckett, P. D. Doran, and M. J. Mac, editors. Our living resources: a report to the nation on the distribution, abundance, and health of U.S. plants, animals, and ecosystems. U.S. Department of the Interior, National Biological Service, Washington, D.C.

Grover, H. D., and H. B. Musick. 1990. Shrubland encroachment in southern New Mexico, U.S.A.: an analysis of desertification processes in the American Southwest. Climatic Change 17:305–330.

Hale, S. F., C. R. Schwalbe, J. L. Jarchow, C. J. May, C. H. Lowe, and T. B. Johnson. 1995. Disappearance of the Tarahumara frog. Pages 138–140 *in* E. T. LaRoe, G. S. Farris, C. E. Puckett, P. D. Doran, and M. J. Mac, editors. Our living resources: a report to the nation on the distribution, abundance, and health of U.S. plants,

animals, and ecosystems. U.S. Department of the Interior, National Biological Service, Washington, D.C.

Hall, E. R. 1981. The mammals of North America. John Wiley & Sons, New York. 1181 pp.

Harper, K. T., K. L. St. Clair, K. H. Thorne, and W. M. Hess, editors. 1994. Natural history of the Colorado Plateau and Great Basin. University Press of Colorado, Niwot. 294 pp.

Hastings, J. R., and R. M. Turner. 1965. The changing mile: an ecological study of vegetation change with time in the lower mile of an arid and semiarid region. University of Arizona Press, Tucson. 317 pp.

Hatch, M. L. 1985. Native fishes of the major drainages east of the Continental Divide, New Mexico. M.S. thesis, Eastern New Mexico University, Portales. 85 pp.

Hayes, M. P., and M. R. Jennings. 1986. Decline of ranid frog species in western North America: are bullfrogs (*Rana catesbeiana*) responsible? Journal of Herpetology 20:490–509.

Hejl, S. J. 1994. Human-induced changes in bird populations in coniferous forests in western North America during the past 100 years. Studies in Avian Biology 15:232–246.

Hobbs, R. J., and L. F. Huenneke. 1992. Disturbance, diversity, and invasion: implications for conservation. Conservation Biology 6:324–337.

Hodges, R. W. 1995. Diversity and abundance of insects. Pages 161–163 *in* E. T. LaRoe, G. S. Farris, C. E. Puckett, P. D. Doran, and M. J. Mac, editors. Our living resources: a report to the nation on the distribution, abundance, and health of U.S. plants, animals, and ecosystems. U.S. Department of the Interior, National Biological Service, Washington, D.C.

Hoffmeister, D. F. 1986. Mammals of Arizona. University of Arizona Press and Arizona Game and Fish Department, Tucson. 602 pp.

Holechek, J. L. 1991. Chihuahuan Desert rangeland, livestock grazing, and sustainability. Rangeland 13:115–120.

Howe, W. H., and F. L. Knopf. 1991. On the imminent decline of Rio Grande cottonwoods in central New Mexico. Southwestern Naturalist 36:218–224.

Hubbs, C. L., and R. R. Miller. 1953. Hybridization in nature between the fish genera *Catostomus* and *Xyrauchen*. Papers of the Michigan Academy of Science, Arts, and Letters 38:207–233.

Hull, A. C., Jr. 1976. Rangeland use and management in the Mormon West. Symposium on agriculture, food and man—a century of progress. Brigham Young University, Provo, Utah.

Hull, A. C., Jr., and M. K. Hull. 1974. Presettlement vegetation of Cache Valley, Utah and Idaho. Journal of Range Management 27:27–29.

Hulse, A. C. 1980. Notes on the occurrence of introduced turtles in Arizona. Herpetological Review 11:16–17.

Humphrey, R. R. 1987. 90 Years and 535 miles: vegetation changes along the Mexican border. University of New Mexico Press, Albuquerque. 448 pp.

Jennings, M. R. 1988. *Rana onca.* Catalogue of American Amphibians and Reptiles. 417:1–2.

Jennings, R. D. 1993. Rediscovery of *Rana onca* in southern Nevada and a taxonomic reevaluation of some southwestern leopard frogs. Annual meeting of the American Society of Ichthyologists and Herpetologists and Herpetologists' League. Austin, Texas, 27 May–5 June.

Johnsen, T. N. 1987. Using herbicides for pinyon–juniper control in the Southwest. Pages 330–334 *in* R. L. Everett, compiler. Proceedings—Pinyon–juniper conference. U. S. Forest Service General Technical Report INT-215.

Jones, K. B. 1981. Effects of grazing on lizard abundance and diversity in western Arizona. Southwestern Naturalist 26: 107–115.

Kaufmann, M. R., W. H. Moir, and R. L. Bassett. 1992. Old-growth forests in the Southwest and Rocky Mountain regions: proceedings of a workshop. U.S. Forest Service General Technical Report RM-213. 201 pp.

Kearney, T. H., and R. H. Peebles. 1960. Arizona flora. 2nd edition with supplement. University of California Press, Berkeley and Los Angeles. 1085 pp.

Kepner, W. G. 1994. EMAP, a landscape approach to environmental assessment. National Fish and Wildlife Foundation, Partners in Flight 4(2):18–19.

Kerpez, T. A., and N. S. Smith. 1987. Saltcedar control for wildlife habitat improvement in the southwestern United States. U.S. Fish and Wildlife Service Resource Publication 169. 17 pp.

Knight, D. H. 1994. Mountain and plains: the ecology of Wyoming landscapes. Yale University Press, New Haven, Conn. 338 pp.

Knopf, F. L. 1989. Riparian wildlife habitats: more, worth less, and under invasion. Pages 20–22 *in* K. Mutz, D. Cooper, M. Scott, and L. Miller, technical coordinators. Restoration, creation, and management of wetland and riparian ecosystems in the American West. Society of Wetland Scientists, Rocky Mountain Chapter, Boulder, Colo.

Knopf, F. L., R. R. Johnson, T. Rich, F. B. Samson, and R. C. Szaro. 1988. Conservation of riparian ecosystems in the United States. Wilson Bulletin 100:272–284.

Knopf, F. L., and F. B. Samson. 1994. Scale perspectives on avian diversity in western riparian ecosystems. Conservation Biology 8:669–676.

Knopf, F. L., and M. L. Scott. 1990. Altered flows and created landscapes in the Platte River headwaters, 1840–1990. Pages 47–70 *in* J. M. Sweeney, editor. Management of dynamic ecosystems. North-central section, The Wildlife Society, West Lafayette, Ind.

Kohler, T. A. 1992. Prehistoric human impact on the environment of the upland North American Southwest. Population and Environment: A Journal of Interdisciplinary Studies 13:255–268.

Krueper, D. J. 1993. Effects of land use practices on western riparian ecosystems. Pages 321–330 *in* D. M. Finch and P. W. Stangel, editors. Status and management of Neotropical migratory birds. U.S. Forest Service General Technical Report RM-229.

Kunzmann, M. R., S. L. Buchmann, J. F. Edwards, S. C. Thoenes, and E. H. Erickson. 1995. Africanized honeybees in North America. Pages 448–451 *in* E. T. LaRoe, G. S. Farris, C. E. Puckett, P. D. Doran, and M. J. Mac, editors. Our living resources: a report to the nation on the distribution, abundance, and health of U.S. plants, animals, and ecosystems. U.S. Department of the Interior, National Biological Service, Washington, D.C.

Larson, M., and W. H. Moir. 1987. Forest and woodland habitat types (plant associations) of northern New Mexico and northern Arizona. U.S. Forest Service, Southwest Region, Albuquerque, N. Mex. 160 pp.

Leopold, A. S. 1924. Grass, brush, timber, and fire in southern Arizona. Journal of Forestry 12:1–10.

Leopold, L. B. 1951. Vegetation of southwestern watersheds in the nineteenth century. Geographical Review 41:295–316.

Loveland, T. R., and H. L. Hutcheson. 1995. Monitoring changes in landscapes from satellite imagery. Pages 468–473 *in* E. T. LaRoe, G. S. Farris, C. E. Puckett, P. D. Doran, and M. J. Mac, editors. Our living resources: a report to the nation on the distribution, abundance, and health of U.S. plants, animals, and ecosystems. U.S. Department of the Interior, National Biological Service, Washington, D.C.

Mackie, R. J., K. L. Hamlin, and D. V. Pac. 1982. Mule deer. Pages 862–877 *in* J. A. Chapman and G. A. Feldhamer, editors. Wild mammals of North America: biology, management, economics. Johns Hopkins University Press, Baltimore, Md.

MacMahon, J. A. 1979. North American deserts: their floral and faunal components. Pages 21–82 *in* D. W. Goodal and R. A. Perry, editors. Arid-land ecosystems: structure, functioning, and management. Volume 1. Cambridge University Press, New York.

MacMahon, J. A. 1988a. Warm deserts. Pages 231–264 *in* M. G. Barbour and W. D. Billings, editors. North American terrestrial vegetation. Cambridge University Press, Cambridge, England.

MacMahon. J. A. 1988b. Introduction. Pages xiii–xvi *in* B. J. Albee, L. M. Shultz, and S. Goodrich, editors. Atlas of the vascular plants of Utah. Utah Museum of Natural History, Occasional Publication 7, Salt Lake City, Utah.

Madany, M. H., and N. E. West. 1983. Livestock grazing—fire regime interactions within montane forests of Zion National Park, Utah. Ecology 64:661–667.

Maddux, H. R., L. A. Fitzpatrick, and W. R. Noonan. 1993. Colorado River endangered fishes critical habitat draft biological support document. U.S. Fish and Wildlife Service, Salt Lake City, Utah. 225 pp.

Marsh, P. C., and W. L. Minckley. 1989. Observations on recruitment and ecology of razorback sucker: lower Colorado River, Arizona–California–Nevada. Great Basin Naturalist 49:71–78.

Marshall, J. T., Jr. 1957. Birds of pine–oak woodland in southern Arizona and adjacent Mexico. Cooper Ornithological Society, Pacific Coast Avifauna 3:1–125.

Martin, P. S. 1967. Prehistoric overkill. Pages 75–120 *in* P. S. Martin and H. E. Wright, Jr., editors. Pleistocene extinctions: the search for a cause. Yale University Press, New Haven, Conn.

Martin, W. C., and C. R. Hutchins. 1980. A flora of New Mexico. Volume 1. J. Cramer, Hirschberg, Germany. 1276 pp.

Martin, W. C., and C. R. Hutchins. 1981. A flora of New Mexico. Volume 2. J. Cramer, Hirschberg, Germany. 1315 pp.

Mason, W. T., Jr. 1995. Invertebrates: overview. Pages 159–160 *in* E. T. LaRoe, G. S. Farris, C. E. Puckett, P. D. Doran, and M. J. Mac, editors. Our living resources: a report to the nation on the distribution, abundance, and health of U.S. plants, animals, and ecosystems. U.S. Department of the Interior, National Biological Service, Washington, D.C.

Master, L. 1990. The imperiled status of North American aquatic animals. Biodiversity Network News 3:1–2, 7–8.

McAda, C. W., and R. S. Wydoski. 1980. The razorback sucker, *Xyrauchen texanus,* in the upper Colorado River basin, 1974–76. U.S. Fish and Wildlife Service Technical Paper 99:1–15.

McClaran, M. P., L. S. Allen, and G. B. Ruyle. 1992. Livestock production and grazing in the encinal oak woodlands of Arizona. Pages 57–64 *in* P. F. Ffolliott, G. J. Gottfried, D. A. Bennett, J. M. Hernandez, C. A. Ortega, and R. H. Hamre, technical coordinators. Ecology and management of oak and associated woodlands: perspectives in the southwestern United States and northern Mexico. U.S. Forest Service General Technical Report RM–218.

McClaran, M. P. 1995. Desert grasslands and grasses. Pages 1–30 *in* M. P. McClaran and T. R. Van Devender, editors. The desert grassland. University of Arizona Press, Tucson.

McClaran, M. P., and T. R. Van Devender, editors. 1995. The desert grassland. University of Arizona Press, Tucson. 346 pp.

McCutchen, H. E. 1995. Desert bighorn sheep. Pages 333–336 *in* E. T. LaRoe, G. S. Farris, C. E. Puckett, P. D. Doran, and M. J. Mac, editors. Our living resources: a report to the nation on the distribution, abundance, and health of U.S. plants, animals, and ecosystems. U.S. Department of the Interior, National Biological Service, Washington, D.C.

McLaughlin, S. P. 1986. Floristic analysis of the southwestern United States. Great Basin Naturalist 46:46–65.

McLaughlin, S. P. 1989. Natural floristic areas of the western United States. Journal of Biogeography 16:239–248.

McPherson, G. R. 1992. Ecology of oak woodlands in Arizona. Pages 24–33 *in* P. F. Ffolliott, G. J. Gottfried, D. A. Bennett, J. M. Hernandez, C. A. Ortega, and R. H. Hamre, technical coordinators. Ecology and management of oak and associated woodlands: perspectives in the southwestern United States and northern Mexico. U.S. Forest Service General Technical Report RM-218.

Mehlman, D. W., and S. O. Williams III. 1995. Priority Neotropical migrants in New Mexico. New Mexico Ornithological Society Bulletin 23:3–8.

Mendelson, J. R., III, and W. B. Jennings. 1992. Shifts in the relative abundance of snakes in a desert grassland. Journal of Herpetology 26:38–45.

Merriam, C. H., and L. Steineger. 1890. Results of a biological survey of the San Francisco Mountain region and desert of the Little Colorado, Arizona. North American Fauna 3. 208 pp.

Miller, B., C. Wemmer, D. Biggins, and R. Reading. 1990. A proposal to conserve black-footed ferrets and the prairie dog ecosystem. Environmental Management 14:763–769.

Miller, M. E. 1994. Historic vegetation change in the Negrito Creek watershed, New Mexico. M.S. thesis, New Mexico State University, Las Cruces. 144 pp.

Miller, R. F., and P. E. Wigand. 1994. Holocene changes in semiarid pinyon–juniper woodlands. BioScience 44:465–474.

Miller, R. R. 1961. Man and the changing fish fauna of the American Southwest. Papers of the Michigan Academy of Science, Arts, and Letters 46:365–404.

Minckley, W. L., and J. E. Deacon, editors. 1991. Battle against extinction: native fish management in the American West. University of Arizona Press, Tucson. 517 pp.

Minckley, W. L., G. K. Meffe, and D. L. Soltz. 1991. Conservation and management of short-lived fishes: the cyprinodontoids. Pages 247–282 *in* W. L. Minckley and J. E. Deacon, editors. Battle against extinction: native fish management in the American West. University of Arizona Press, Tucson.

Moir, W. H. 1980. A fire history of the high Chisos, Big Bend National Park, Texas. Southwestern Naturalist 27:87–98.

Moir, W. H. 1993. Alpine tundra and coniferous forest. Pages 47–84 *in* W. A. Dick-Peddie. New Mexico vegetation: past, present, and future. University of New Mexico Press, Albuquerque. 244 pp.

Moir, W. H., and J. H. Dieterich. 1988. Old-growth ponderosa pine from succession in pine–bunchgrass forests in Arizona and New Mexico. Natural Areas Journal 8:17–24.

Moir, W. H., and L. S. Huckaby. 1994. Displacement ecology of trees near upper timberline. International Conference for Bear Research and Management 9(1):35–42.

Molles, M. C., Jr., and C. N. Dahm. 1990. A perspective on El Niño and La Niña: global implications for stream ecology. Journal of the North American Benthological Society 9:68–76.

Morin, N. 1995. Vascular plants of the United States. Pages 200–205 *in* E. T. LaRoe, G. S. Farris, C. E. Puckett, P. D. Doran, and M. J. Mac, editors. Our living resources: a report to the nation on the distribution, abundance, and health of U.S. plants, animals, and ecosystems. U.S. Department of the Interior, National Biological Service, Washington, D.C.

Morrisey, R. J. 1950. The early range cattle industry in Arizona. Agricultural History 24:151–156.

Morse, L. E., J. T. Kartesz, and L. S. Kutner. 1995. Native vascular plants. Pages 205–209 *in* E. T. LaRoe, G. S. Farris, C. E. Puckett, P. D. Doran, and M. J. Mac, editors. Our living resources: a report to the nation on the distribution, abundance, and health of U.S. plants, animals, and ecosystems. U.S. Department of the Interior, National Biological Service, Washington, D.C.

Mueller, G., and P. Marsh. 1995. Bonytail and razorback sucker in the Colorado River basin. Pages 324–326 *in* E. T. LaRoe, G. S. Farris, C. E. Puckett, P. D. Doran, and M. J. Mac, editors. Our living resources: a report to the nation on the distribution, abundance, and health of U.S. plants, animals, and ecosystems. U.S. Department of the Interior, National Biological Service, Washington, D.C.

Muldavin, E. H., R. DeVelice, and F. Ronco, Jr. 1996. A classification of forest habitat types of southern Arizona and portions of the Colorado Plateau. U.S. Forest Service General Technical Report GTR-287.

Niering, W. A., R. H. Whittaker, and C. H. Lowe. 1963. The saguaro: a population in relation to environment. Science 142(3588):15–23.

Niering, W. A., and C. H. Lowe. 1984. Vegetation of the Santa Catalina Mountains: community types and dynamics. Vegetatio 58:3–28.

Noss, R. F., E. T. LaRoe III, and J. M. Scott. 1995. Endangered ecosystems of the United States: a preliminary assessment of loss and degradation. National Biological Service Biological Report 28. 58 pp.

Nowak, R. 1994. Jaguars in the United States. Endangered Species Technical Bulletin 19:6.

Ono, R. D., J. D. Williams, and A. Wagner. 1983. Vanishing fishes of North America. Stone Wall Press, Washington, D.C. 257 pp.

Opler, P. A. 1995. Species richness and trends in western butterflies and moths. Pages 172–174 *in* E. T. LaRoe, G. S. Farris, C. E. Puckett, P. D. Doran, and M. J. Mac, editors. Our living resources: a report to the nation on the distribution, abundance, and health of U.S. plants, animals, and ecosystems. U.S. Department of the Interior, National Biological Service, Washington, D.C.

Page, L. M., and B. M. Burr. 1991. A field guide to freshwater fishes of North America north of Mexico. Peterson field guide 42. Houghton Mifflin Company, Boston, Mass. 432 pp.

Parker, K. W., and S. C. Martin. 1952. The mesquite problem on southern Arizona ranges. U.S. Department of Agriculture Circular 908. 69 pp.

Pase, C. P. 1982. Arctic and alpine tundras. Pages 24–33 *in* D. E. Brown, editor. Biotic communities of the American Southwest—United States and Mexico. Desert Plants 4:1–341.

Pase, C. P., and D. E. Brown. 1982a. Rocky Mountain (Petran) subalpine conifer forest. Pages 37–39 *in* D. E. Brown, editor. Biotic communities of the American Southwest—United States and Mexico. Desert Plants 4:1–341.

Pase, C. P., and D. E. Brown. 1982b. Rocky Mountain (Petran) and Madrean montane conifer forests. Pages 43–48 *in* D. E. Brown, editor. Biotic communities of the American Southwest—United States and Mexico. Desert Plants 4:1–341.

Patton, J. L., and M. F. Smith. 1990. The evolutionary dynamics of the pocket gopher *Thomomys bottae*, with emphasis on California populations. University of California Publications in Zoology 123:1–161.

Peek, J. M. 1995. North American elk. Pages 115–116 *in* E. T. LaRoe, G. S. Farris, C. E. Puckett, P. D. Doran, and M. J. Mac, editors. Our living resources: a report to the nation on the distribution, abundance, and health of U.S. plants, animals, and ecosystems. U.S. Department of the Interior, National Biological Service, Washington, D.C.

Peterjohn, B. G. 1994. The North American Breeding Bird Survey. Birding 26:386–398.

Peterjohn, B. G., and J. R. Sauer. 1994. Population trends of woodland birds from the North American Breeding Bird Survey. Wildlife Society Bulletin 22:155–164.

Peterjohn, B. G., J. R. Sauer, and W. A. Link. 1994. The 1992 and 1993 summary of the North American Breeding Bird Survey. Bird Populations 2:46–61.

Pieper, R. D. 1994. Ecological implications of livestock grazing. Pages 177–211 *in* M. Vavra, W. A. Laycock, and R. D. Pieper, editors. Ecological implications of livestock herbivory in the West. Society for Range Management, Denver, Colo.

Pogacnik, T. 1995. Wild horses and burros on public lands. Pages 456–458 *in* E. T. LaRoe, G. S. Farris, C. E. Puckett, P. D. Doran, and M. J. Mac, editors. Our living resources: a report to the nation on the distribution, abundance, and health of U.S. plants, animals, and ecosystems. U.S. Department of the Interior, National Biological Service, Washington, D.C.

Potter, L. D., and J. C. Krenetsky. 1967. Plant succession with released grazing on New Mexico rangelands. Journal of Range Management 20:145–151.

Powell, D. S., J. L. Faulkner, D. R. Darr, Z. Zhu, and D. W. MacCleery. 1993. Forest resources of the United States, 1992. U.S. Forest Service General Technical Report RM-234. 132 pp.

Powell, J. A. 1995. Lepidoptera inventories in the continental United States. Pages 168–170 *in* E. T. LaRoe, G. S. Farris, C. E. Puckett, P. D. Doran, and M. J. Mac, editors. Our living resources: a report to the nation on the distribution, abundance, and health of U.S. plants, animals, and ecosystems. U.S. Department of the Interior, National Biological Service, Washington, D.C.

Propst, D. L., G. L. Burton, and B. H. Pridgeon. 1987. Fishes of the Rio Grande between Elephant Butte and Caballo reservoirs, New Mexico. Southwestern Naturalist 32:408–411.

Ramotnik, C. A., and N. J. Scott, Jr. 1989. Habitat requirements of New Mexico's endangered salamanders. Pages 54–63 *in* R. C. Szaro, K. E. Severson, and D. R. Patton, technical coordinators. Management of amphibians, reptiles, and small mammals in North America. U.S. Forest Service General Technical Report RM-166.

Rea, A. M. 1983. Once a river: bird life and habitat changes on the middle Gila. University of Arizona Press, Tucson. 212 pp.

Rice, B., and M. Westoby. 1978. Vegetation responses of some Great Basin shrub communities protected from jackrabbits or domestic stock. Journal of Range Management 31:28–33.

Rickard, W. H., and C. E. Cushing. 1982. Recovery of streamside woody vegetation after exclusion of livestock grazing. Journal of Range Management 35:360–361.

Robinett, D. 1995. Prescribed burning on upper Sonoran rangelands. Pages 361–363 *in* B. A. Roundy, E. D. McArthur, J. S. Haley, and D. D. Mann, compilers. Proceedings: wildland shrub and arid land restoration symposium, 1993, October 19–21. U.S. Forest Service General Technical Report INT-315.

Rogers, G. F. 1982. Then and now: a photographic history of vegetation change in the central Great Basin desert. University of Utah Press, Salt Lake City. 152 pp.

Ropelewski, C. F., and M. S. Halpert. 1986. North American precipitation and temperature patterns associated with the El Niño/Southern Oscillation (ENSO). Monthly Weather Review 114:2352–2362.

Rosen, P. C., and C. R. Schwalbe. 1995. Bullfrogs: introduced predators in southwestern wetlands. Pages 452–454 *in* E. T. LaRoe, G. S. Farris, C. E. Puckett, P. D. Doran, and M. J. Mac, editors. Our living resources: a report to the nation on the distribution, abundance, and health of U.S. plants, animals, and ecosystems. U.S. Department of the Interior, National Biological Service, Washington, D.C.

Rosenberg, K. V., R. D. Ohmart, W. C. Hunter, and B. W. Anderson. 1991. Birds of the lower Colorado River valley. University of Arizona Press, Tucson. 416 pp.

Sackett, S. S., S. Haase, and M. G. Harrington. 1994. Restoration of southwestern ponderosa pine ecosystems with fire. Pages 115–121 *in* W. W. Covington and L. F. DeBano, technical coordinators. Sustainable ecological systems: implementing an ecological approach to land management. U.S. Forest Service General Technical Report RM-247.

Samson, F. B., E. L. Adams, S. Hamilton, S. P. Nealey, R. Steele, and D. Van De Graaff. 1994. Assessing forest ecosystem health in the inland West. Forest Policy Center, Washington, D.C. 19 pp.

Samuels, M. L., and J. L. Betancourt. 1982. Modeling the long-term effects of fuelwood harvests on piñon–juniper woodlands. Environmental Management 6:505–515.

Savage, M., and T. W. Swetnam. 1990. Early and persistent fire decline in a Navajo ponderosa pine forest. Ecology 70:2374–2378.

Schickedanz, J. G. 1980. History of grazing in the Southwest. Pages 1–9 *in* K. McDaniel and C. Allison, editors. Proceedings: grazing management systems for Southwest rangelands, April 1980, Albuquerque, New Mexico. Range Improvement Task Force, New Mexico State University, Las Cruces.

Schlesinger, W. H., J. F. Reynolds, G. L. Cunningham, L. F. Huenneke, W. M. Jarrel, R. A. Virginia, and W. G. Whitford. 1990. Biological feedbacks in global desertification. Science 257:1043–1048.

Schmid, J. M., and R. H. Frye. 1977. Spruce beetle in the Rockies. U.S. Forest Service General Technical Report RM-49. 38 pp.

Schmitt, C. J., and C. M. Bunck. 1995. Persistent environmental contaminants in fish and wildlife. Pages 413–416 *in* E. T. LaRoe, G. S. Farris, C. E. Puckett, P. D. Doran, and M. J. Mac, editors. Our living resources: a report to the nation on the distribution, abundance, and health of U.S. plants, animals, and ecosystems. U.S. Department of the Interior, National Biological Service, Washington, D.C.

Schroeder, A. H. 1979. Pueblos abandoned in recent times. Pages 236–254 *in* A. Ortiz, editor. Handbook of North American Indians: Southwest. Volume 9. Smithsonian Institution, Washington, D.C.

Schultz, T. T., and W. C. Leininger. 1990. Differences in riparian vegetation structure between grazed areas and exclosures. Journal of Range Management 43:295–299.

Schwalbe, C. R., and P. C. Rosen. 1988. Preliminary report on effects of bullfrogs on wetland herpetofauna in southeastern Arizona. Pages 166–173 *in* R. C. Szaro, K. E. Severson, and D. R. Patton, editors. Management of amphibians, reptiles, and small mammals in North America. U.S. Forest Service General Technical Report RM-166.

Scott, N. J., Jr., and C. A. Ramotnik. 1992. Does the Sacramento Mountain salamander require old-growth forests? Pages 170–178 *in* M. R. Kaufman, W. H. Moir, and R. L. Bassett, editors. Old-growth forests in the Southwest and Rocky Mountain regions. Proceedings of a workshop. U.S. Forest Service General Technical Report RM-213.

Scudday, J. F. 1977. Some recent changes in the herpetofauna of the northern Chihuahuan Desert. Pages 513–522 *in* R. H. Wauer and D. H. Riskind, editors. Transactions of the symposium on biological resources of the Chihuahuan Desert region: United States and Mexico. National Park Service Transactions and Proceedings Series 3.

Shaw, D. W., E. F. Aldon, and C. LoSapio. 1995. Desired future conditions for piñon–juniper ecosystems. U.S. Forest Service General Technical Report RM-258. 226 pp.

Simberloff, D., J. A. Farr, J. Cox, and D. W. Mehlman. 1992. Movement corridors: conservation bargains or poor investments? Conservation Biology 6:493–504.

Singer, F. 1995. Bighorn sheep in the Rocky Mountain national parks. Pages 332–333 *in* E. T. LaRoe, G. S. Farris, C. E. Puckett, P. D. Doran, and M. J. Mac, editors. Our living resources: a report to the nation on the distribution, abundance, and health of U.S. plants, animals, and ecosystems. U.S. Department of the Interior, National Biological Service, Washington, D.C.

Smith, E. R. 1953. History of grazing industry and range conservation developments in the Rio Grande Basin. Journal of Range Management 6:405–409.

Smith, M. L., and R. R. Miller. 1986. The evolution of the Rio Grande basin as inferred from its fish fauna. Pages 457–485 *in* C. H. Hocutt and E. O. Wiley, editors. Zoogeography of North American freshwater fishes. John Wiley & Sons, New York.

Stewart, D. E., and R. Gauthier. 1986. Prehistory: the riverine period. Pages 89–91 *in* J. L. Williams, editor. New Mexico in maps. University of New Mexico Press, Albuquerque.

Stewart, G. 1936. History of range use. Pages 119–133 *in* The Western Range. U.S. Senate Document 199. Government Printing Office, Washington, D.C.

Stewart, G. R. 1994. An overview of the Mohave Desert and its herpetofauna. Pages 55–69 *in* P. R. Brown and J. W. Wright, editors. Herpetology of the North American deserts. Southwestern Herpetological Society Special Publication 5.

Stoddard, L. A. 1946. Some physical and chemical responses of *Agropyron spicatum* to herbage removal at various seasons. Utah State Agricultural College, Agricultural Experiment Station Bulletin 324. 24 pp.

Sublette, J. E., M. D. Hatch, and M. Sublette. 1990. The fishes of New Mexico. University of New Mexico Press, Albuquerque. 393 pp.

Swetnam, T. W. 1990. Fire history and climate in the southwestern United States. Pages 6–17 *in* J. S. Krammes, technical coordinator. Effects of fire management of southwestern natural resources: proceedings of the symposium, November 15–17, 1988, Tucson, Arizona. U.S. Forest Service General Technical Report RM-191. 293 pp.

Swetnam, T. W., and C. H. Baisan. 1996. Historical fire regime patterns in southwestern United States since A.D. 1700. Pages 11–32 *in* C. D. Allen, technical editor. Fire effects in southwestern forests: proceedings of the second La Mesa fire symposium. U.S. Forest Service General Technical Report RM-GTR-286.

Swetnam, T. W., and J. L. Betancourt. 1990. Fire–southern oscillation relations in the southwestern United States. Science 249:1017–1020.

Swetnam, T. W., and A. M. Lynch. 1993. Multi-century, regional-scale patterns of western spruce budworm history. Ecological Monographs 63:399–424.

Szaro, R. C. 1989. Riparian and forest and scrubland community types of Arizona and New Mexico. Desert Plants 9:70–135.

Szaro, R. C., S. C. Belfit, J. K. Aitkin, and J. N. Rinne. 1985. Impact of grazing on a riparian garter snake. Pages 359–363 *in* R. R. Johnson, C. D. Ziebell, D. R. Patton, P. F. Ffolliott, and F. G. Hamre, technical coordinators. Riparian ecosystems and their management: reconciling conflicting uses. U.S. Forest Service General Technical Report RM-120.

Taylor, D. M. 1986. Effects of cattle grazing on passerine birds nesting in riparian habitat. Journal of Range Management 39:254–258.

Thaeler, C. S., Jr. 1968. An analysis of the distribution of pocket gopher species in northeastern California (genus *Thomomys*). University of California Publications in Zoology 86:1–46.

Thomas, J. W., R. G. Anderson, C. Maser, and E. L. Bull. 1979. Snags. Pages 60–77 *in* J. W. Thomas, editor. Wildlife habitats in managed forests; the Blue Mountains of Oregon and Washington. U.S. Department of Agriculture, Agriculture Handbook 553, Washington, D.C.

Touchan, R. T., C. D. Allen, and T. W. Swetnam. 1996. Fire history and climatic patterns in ponderosa pine and mixed-conifer forests of the Jemez Mountains, northern New Mexico. Pages 33–46 *in* C. D. Allen, technical editor. Fire effects in southwestern forests: proceedings of the second La Mesa fire symposium. U.S. Forest Service General Technical Report RM-GTR-286.

Touchan, R., T. W. Swetnam, and H. Grissino-Mayer. 1995. Effects of livestock grazing on pre-settlement fire regimes in New Mexico. Pages 268–272 *in* J. K. Brown, R. W. Mutch, C. W. Spoon, and R. H. Wakimoto, technical coordinators. Symposium on fire in wilderness and park management. Missoula, Montana, 30 March to 1 April, 1993. U.S. Forest Service General Technical Report INT-320.

Troendle, C. A., and M. R. Kaufmann. 1987. Influence of forests on the hydrology of the subalpine forest. Pages 68–78 *in* C. A. Troendle, M. R. Kaufman, R. H. Hamre, and R. P. Winokur, technical coordinators. Management of subalpine forests: building on 50 years of research. U.S. Forest Service General Technical Report RM-149.

Turner, R. M. 1990. Long-term vegetation change at a fully protected Sonoran desert site. Ecology 71:464–477.

U.S. Bureau of the Census. 1975. Historical statistics of the United States. Colonial times to 1970. Part 2. Bicentennial edition. Washington, D.C. 609 pp.

U.S. Bureau of the Census. 1994. Statistical abstract of the United States: 1994. 114th edition. Washington, D.C. 1011 pp.

U.S. Fish and Wildlife Service. 1994a. Endangered and threatened wildlife and plants; animal candidate review for listing as endangered or threatened species; proposed rule. Federal Register 59(219):58982–59028.

U.S. Fish and Wildlife Service. 1994b. Endangered and threatened wildlife and plants; final rule to list the Rio Grande silvery minnow as an endangered species. Federal Register 59(138):36988–36995.

U.S. Fish and Wildlife Service. 1995. Endangered and threatened species: southwestern willow flycatcher; final rule. Federal Register 60:10694–10715.

U.S. Forest Service. 1993. Changing conditions in southwestern forests and implications on land stewardship. U.S. Forest Service, Southwest Region, Albuquerque, N. Mex. 8 pp.

Utter, J. 1993. American Indians: answers to today's questions. National Woodlands Publishing Company, Lake Ann, Mich. 331 pp.

Van Hooser, D. D., R. A. O'Brien, and D. C. Collins. 1993. New Mexico's forest resources. U.S. Forest Service Resource Bulletin INT-79. 110 pp.

Warren, M. L., Jr., and B. M. Burr. 1994. Status of freshwater fishes of the United States: overview of an imperiled fauna. Fisheries 19:6–18.

Warshall, P. 1995. Southwestern sky island ecosystems. Pages 318–322 *in* E. T. LaRoe, G. S. Farris, C. E. Puckett, P. D. Doran, and M. J. Mac, editors. Our living resources: a report to the nation on the distribution, abundance, and health of U.S. plants, animals, and ecosystems. U.S. Department of the Interior, National Biological Service, Washington, D.C.

Waser, N. M., and M. V. Price. 1981. Effects of grazing on diversity of annual plants in the Sonoran Desert. Oecologia 50:407–411.

Webb, R. H., and J. L. Betancourt. 1992. Climatic variability and flood frequency of the Santa Cruz River, Pima County, Arizona. U.S. Geological Survey Water-Supply Paper 2379. 40 pp.

Welsh, S. L., N. D. Atwood, S. Goodrich, and L. C. Higgins, editors. 1987. A Utah flora. Great Basin Naturalist Memoirs 9: 1–894.

West, N. E. 1983. Great Basin–Colorado Plateau sagebrush semi-desert. Pages 331–349 *in* N. E. West, editor. Temperate deserts and semi-deserts. Volume 5. Ecosystems of the World. Elsevier Scientific Publishing, Amsterdam and New York.

West, N. E. 1988. Intermountain deserts, shrub steppes, and woodlands. Pages 209–230 *in* M. G. Barbour and W. D. Billings. North American terrestrial vegetation. Cambridge University Press, Cambridge, England. 434 pp.

West, N. E., and N. S. Van Pelt. 1987. Successional patterns in pinyon–juniper woodlands. Pages 43–52 *in* R. L. Everett, compiler. Proceedings—Pinyon–Juniper conference. U.S. Forest Service General Technical Report INT-215.

West, N. E., F. D. Provenza, P. S. Johnson, and M. K. Owens. 1984. Vegetation change after 13 years of livestock grazing exclusion on sagebrush semidesert in west central Utah. Journal of Range Management 37:262–264.

White, C. S. 1994. Monoterpenes: their effects on ecosystem nutrient cycling. Journal of Chemical Ecology 20:1381–1406.

Whittaker, R. H., and W. A. Niering. 1965. Vegetation of the Santa Catalina Mountains, Arizona: a gradient analysis of the south slope. Ecology 46:429–452.

Whittemore, A., and B. Allen. 1995. Floristic inventories of U.S. bryophytes. Pages 198–200 *in* E. T. LaRoe, G. S. Farris, C. E. Puckett, P. D. Doran, and M. J. Mac, editors. Our living resources: a report to the nation on the distribution, abundance, and health of U.S. plants, animals, and ecosystems. U.S. Department of the Interior, National Biological Service, Washington, D.C.

Williams, J. D., J. E. Johnson, D. A. Hendrickson, S. Contreras-Balderas, J. D. Williams, M. Navarro-Mendoza, D. E. McAllister, and J. E. Deacon. 1989. Fishes of North America: endangered, threatened, or of special concern: 1989. Fisheries 14:2–20.

Williams, J. L. 1986. New Mexico in maps. University of New Mexico Press, Albuquerque. 409 pp.

Wolters, G. L. 1996. Elk effects on Bandelier National Monument meadows and grasslands. *In* C. D. Allen, technical editor. Fire effects in southwestern forests: proceedings of the second La Mesa fire symposium. U.S. Forest Service General Technical Report RM-GTR-286.

Wooton, E. O. 1908. The range problem in New Mexico. New Mexico College of Agriculture and Mechanic Arts, Agriculture Experiment Station Bulletin 66, Las Cruces. 46 pp.

York, J. C., and W. A. Dick-Peddie.1969. Vegetation changes in southern New Mexico during the past hundred years. Pages 157–199 *in* W. G. McGinnies and B. J. Goldman, editors. Arid lands in perspective. University of Arizona Press, Tucson.

A Ponderosa Pine Natural Area Reveals its Secrets

Covington, W. W., and M. M. Moore. 1994. Southwestern ponderosa pine forest structure: changes since Euro-American settlement. Journal of Forestry 92:39–47.

Deichmann, J. W. 1980. Botanical survey of the Monument Canyon Research Natural Area. U.S. Forest Service, Final Report on Contract 53-32-FT-8-22. 26 pp.

Foxx, T. S., and L. D. Potter. 1984. Fire ecology at Bandelier National Monument. Pages 11–38 *in* T. S. Foxx, compiler. 1984. La Mesa fire symposium. LA-9236-NERP. Los Alamos National Laboratory, Los Alamos, N. Mex.

Muldavin, E. H., T. W. Swetnam, and M. Stuever. 1995. Age structure and density of an old-growth ponderosa pine forest in relation to changes in fire regime and climate fluctuations. Supplement to Bulletin of the Ecological Society of America 76(3):368. Abstract.

Sackett, S. S., S. Haase, and M. G. Harrington. 1994. Restoration of southwestern ponderosa pine ecosystems with fire. Pages 115–121 *in* W. W. Covington and L. F. DeBano, technical coordinators. Sustainable ecological systems: implementing an ecological approach to land management. U.S. Forest Service General Technical Report RM-247.

Soils and Cryptobiotic Crusts in Arid Lands

Belnap, J. 1990. Microbiotic crusts: their role in past and present ecosystems. Park Science 10:3–4.

Belnap, J., and J. S. Gardner. 1993. Soil microstructure in soils of the Colorado Plateau: the role of the cyanobacterium *Microcoleus vaginatus*. Great Basin Naturalist 53:40–47.

Johansen, J. 1993. Cryptogamic crusts of semi-arid and arid lands of North America. Journal of Phycology 29:146–147.

Ladyman, J. A., and E. H. Muldavin. 1996. Terrestrial cryptogams of pinyon–juniper woodlands in southwestern United States: a review. U.S. Forest Service General Technical Report GTR-280. 33 pp.

Changing Landscapes of the Middle Rio Grande

Crawford, C. S., A. C. Cully, R. Leutheuser, M. S. Sifuentes, L. H. White, and J. P. Wilber. 1993. Middle Rio Grande ecosystem: bosque biological management plan. Middle Rio Grande Biological Interagency Team, U.S. Fish and Wildlife Service, Albuquerque, N. Mex. 291 pp. + maps.

Durkin, P., E. Muldavin, M. Bradley, S. E. Carr, A. Metcalf, R. A. Smartt, S. P. Platania, C. Black, and P. Mehlhop. 1995. The biodiversity of riparian ecosystems of the Ladder Ranch. New Mexico Natural Heritage Program, Albuquerque, N. Mex. Final report to the Turner Foundation. 118 pp. + 8 page appendix.

Roelle, J. E., and W.W. Hagenbuck. 1995. Surface cover changes in the Rio Grande floodplain, 1935–89. Pages 290–292 *in* E. T. LaRoe, G. S. Farris, C. E. Puckett, P. D. Doran, and M. J. Mac, editors. Our living resources: a report to the nation on the distribution, abundance, and health of U.S. plants, animals, and ecosystems. U.S. Department of the Interior, National Biological Service, Washington, D.C.

Rare Aquatic Snails

Mehlhop, P., and C. A. Vaughn. 1994. Threats to and sustainability of ecosystems for freshwater mollusks. Pages 69–77 *in* W. W. Covington and L. F. DeBano, technical coordinators. Sustainable ecological systems: implementing an ecological approach to land management. U.S. Forest Service General Technical Report RM-247.

Perils Facing the Gila Trout

Propst, D. L., J. A. Stefferud, and P. R. Turner. 1992. Conservation and status of Gila trout, *Oncorhynchus gilae*. Southwestern Naturalist 37:117–125.

U.S. Fish and Wildlife Service. 1987. Endangered and threatened wildlife; proposed reclassification of the Gila trout (*Salmo gilae*) from endangered to threatened. Federal Register 52:37424–37427.

Arizona Leopard Frogs: Balanced on the Brink?

Clarkson, R. W., and J. C. Rorabaugh. 1989. Status of leopard frogs (*Rana pipiens* complex: Ranidae) in Arizona and southeastern California. Southwestern Naturalist 34(4):531–538.

Corn, P. S., and J. C. Fogleman. 1984. Extinction of montane populations of the northern leopard frog (*Rana pipiens*) in Colorado. Journal of Herpetology 18:147–152.

Crump, M. L., and N. J. Scott, Jr. 1994. Standard techniques for inventory and monitoring—visual encounter surveys. Pages 84–92 *in* W. R. Heyer, M. A. Donnelly, R. W. McDiarmid, L. C. Hayek, and M. S. Foster, editors. Measuring and monitoring biological diversity: standard methods for amphibians. Smithsonian Institution Press, Washington, D.C.

Frost, J. S., and J. T. Bagnara. 1977. Sympatry between *Rana blairi* and the southern form of leopard frog in southeastern Arizona (Anura: Ranidae). Southwestern Naturalist 22(4):443–453.

Griffith, B., J. M. Scott, J. W. Carpenter, and C. Reed 1989. Translocation as a species conservation tool: status and strategy. Science 245:477–480.

Hale, S. F. 1992. A survey of historical and potential habitat for the Tarahumara frog (*Rana tarahumarae*) in Arizona. J. M. Howland and M. J. Sredl, editors. Report to the Arizona Game and Fish Department, Phoenix, and U.S. Forest Service, Tucson, Ariz. 42 pp.

Hale, S. F., C. R. Schwalbe, J. L. Jarchow, C. J. May, C. H. Lowe, and T. B. Johnson. 1995. Disappearance of the Tarahumara frog. Pages 138–140 *in* E. T. LaRoe, G. S. Farris, C. E. Puckett, P. D. Doran, and M. J. Mac, editors. Our living resources: a report to the nation on the distribution, abundance, and health of U.S. plants, animals, and ecosystems. U.S. Department of the Interior, National Biological Service, Washington, D.C.

Hale, S. F., and J. L. Jarchow. 1988. The status of the Tarahumara frog (*Rana tarahumarae*) in the United States and Mexico: Part 2. *In* C. R. Schwalbe and T. B. Johnson, editors. Report to the Arizona Game and Fish Department, Phoenix, and U.S. Fish and Wildlife Service, Albuquerque, N. Mex. 101 pp.

Hale, S. F., and C. J. May. 1983. Status report for *Rana tarahumarae* Boulenger. U.S. Fish and Wildlife Service Report, Albuquerque, N. Mex. 99 pp.

Hillis, D. M. 1988. Systematics of the *Rana pipiens* complex: puzzle and paradigm. Annual Review of Ecology and Systematics 19:39–63.

Jennings, M. R. 1994. Use of unverified museum databases for land management decisions: the case of native California frogs. Page 108 *in* American Society of Icthyologists and Herpetologists Annual Meeting, June 2–8. Abstract.

Jennings, M. R., and M. P. Hayes. 1995. Amphibian and reptiles of special concern in California. California Department of Fish and Game, Rancho Cordova. 255 pp.

Jennings, R. D. 1993. Rediscovery of *Rana onca* in southern Nevada and taxonomic reevaluation of some Southwest leopard frogs. Page 178 *in* Annual Meeting of American Society of Ichthyologists and

Herpetologists, Austin, Texas 27 May–2 June. Abstract.

Jennings, R. D. 1995. Investigations of recently viable leopard frog populations in New Mexico: *Rana chiricahuensis* and *Rana yavapaiensis*. New Mexico Department of Game and Fish, Santa Fe. 36 pp.

Mecham, J. S., M. J. Littlejohn, R. S. Oldham, L. E. Brown, and J. R. Brown. 1973. A new species of leopard frog (*Rana pipiens* complex) from the plains of the central United States. Occasional Papers of the Museum, Texas Tech University 18:1–18.

Platz, J. E. 1984. Status report for *Rana onca* Cope. U.S. Fish and Wildlife Service Biological Report. 27 pp.

Platz, J. E., R. W. Clarkson, J. C. Rorabaugh, and D. M. Hillis. 1990. *Rana berlandieri*: recently introduced populations in Arizona and southeastern California. Copeia 1990:324–333.

Platz, J. E. 1993. *Rana subaquavocalis*, a remarkable new species of leopard frog (*Rana pipiens* complex) from southeastern Arizona that calls under water. Journal of Herpetology 27:154–162.

Platz, J. E., and J. S. Frost. 1984. *Rana yavapaiensis*, a new species of leopard frog (*Rana pipiens* complex). Copeia 1984:940–948.

Platz, J. E., and J. S. Mecham. 1979. *Rana chiricahuensis*, a new species of leopard frog (*Rana pipiens* complex) from Arizona. Copeia 1979:383–390.

Sredl, M. J., and J. M. Howland. 1995. Conservation and management of Madrean populations of the Chiricahua leopard frog. Pages 379–385 *in* L. F. DeBano, G. J. Gottfried, R. H. Hamre, C. B. Edminster, P. F. Ffoliot, and A. Ortega-Rubio, editors. Biodiversity and management of the Madrean Archipelago: the Sky Islands of the southwestern United States and northwestern Mexico. U.S. Forest Service General Technical Report RM-GTR-264.

Sredl, M. J., and D. L. Waters. 1995. Status of (most of the) leopard frogs in Arizona. Declining Amphibian Populations Task Force, Southwestern United States Working Group Meeting, Phoenix, Ariz. 5–6 January. Abstract.

Stebbins, R. C. 1985. A field guide to western reptiles and amphibians. 2nd edition. Houghton Mifflin Company, Boston, Mass. 336 pp.

Elk Reintroductions

Allen, C. D. 1989. Changes in the ecology of the Jemez Mountains, New Mexico. Ph.D. dissertation, University of California, Berkeley. 346 pp.

Allen, C. D. 1996a. Response of elk populations to the La Mesa Fire. Pages 179–195 *in* C. D. Allen, technical editor. Fire effects in southwestern forests: Proceedings of the second La Mesa fire symposium. U.S. Forest Service General Technical Report RM-GTR-286.

Allen, C. D., technical editor. 1996b. Fire effects in southwestern forests: proceedings of the second La Mesa fire symposium. U.S. Forest Service General Technical Report RM-GTR-286. 216 pp.

Conley, W., R. Sivinski, and G. C. White. 1979. Responses of elk (*Cervus elaphus*) and mule deer (*Odocoileus hemionus*) to wildfire: changes in utilization and migration patterns. Final report to the National Park Service, Contract CX7029-7-0057. 88 pp.

Findley, J. S., A. H. Harris, D. E. Wilson, and C. Jones. 1975. Mammals of New Mexico. University of New Mexico Press, Albuquerque. 360 pp.

Hall, E. R. 1981. The mammals of North America. John Wiley & Sons, New York. 1181 pp.

Kay, C. 1994. Aboriginal overkill: the role of North Americans in structuring western ecosystems. Human Nature 5:359–398.

Lang, R., and A. H. Harris. 1984. The faunal remains from Arroyo Hondo Pueblo, New Mexico: a study in short-term subsistence change. School of American Research Press, Santa Fe, N. Mex. 316 pp.

Peek, J. M. 1995. North American elk. Pages 115–116 *in* E. T. LaRoe, G. S. Farris, C. E. Puckett, P. D. Doran, and M. J. Mac, editors. Our living resources: a report to the nation on the distribution, abundance, and health of U.S. plants, animals, and ecosystems. U.S. Department of the Interior, National Biological Service, Washington, D.C.

Rowland, M. M., A. W. Alldredge, J. E. Ellis, B. J. Weber, and G. C. White. 1983. Comparative winter diets of elk in New Mexico. Journal of Wildlife Management 47:924–932.

White, G. C. 1981. Biotelemetry studies on elk. LA-8529-NERP. Los Alamos National Laboratory, Los Alamos, N. Mex. 24 pp.

Endemic Mammals of the Henry Mountains, Utah

Durrant, S. D. 1958. Annotated checklist of the mammals. Pages 209–219 *in* R. Anderson, editor. Preliminary report on biological resources of the Glen Canyon Reservoir. University of Utah Anthropological Papers 31.

Hall, E. R. 1981. The mammals of North America. John Wiley & Sons, New York. 1181 pp.

Hunt, C. B., P. Everitt, and R. L. Miller. 1953. Geology and geography of the Henry Mountains region, Utah. U.S. Geological Survey Professional Paper 228. 234 pp.

Mollhagen, T. R., and M. A. Bogan. 1995. Baseline surveys for mammals in the Henry Mountains, Utah. Annual Report submitted to Bureau of Land Management, Hanksville, Utah. 5 pp. + appendixes.

Taitt, M. J., and C. J. Krebs. 1985. Population dynamics and cycles. Pages 567–620 *in* R. H. Tamarin, editor. Biology of New World *Microtus*. Special Publication 8, American Society of Mammalogists.

U.S. Fish and Wildlife Service. 1994. Endangered and threatened wildlife and plants; animal candidate review for listing as endangered or threatened species; proposed rule. Federal Register 59(219):58982–59028.

Southwestern Bats

Humphrey, S. R. 1975. Nursery roosts and community diversity of Nearctic bats. Journal of Mammalogy 56:321–346.

U.S. Fish and Wildlife Service. 1994. Endangered and threatened wildlife and plants; animal candidate review for listing as endangered or threatened species; proposed rule. Federal Register 59(219):58982–59028.

California

This chapter presents information on that part of California to the west of the crests of the Peninsular Ranges, Transverse Ranges, Sierra Nevada, and Cascades. This portion of California, which makes up 70% of the state, is referred to as cismontane California or, more simply, westside California. Portions of California excluded from this chapter include the northern steppe to the east of the Cascades and Sierra Nevada, as well as the Mojave Desert—these areas are discussed in the Great Basin chapter. The Colorado Desert, to the east of the Transverse and Peninsular ranges of southern California, is discussed in the Southwest chapter. For some considerations of animals or plants, we also discuss species found in eastside California or discuss the diversity of the state as a whole.

The state of California encompasses 411,015 square kilometers and is 1,326 kilometers long from corner to corner (Donley et al. 1979; Kreissman 1991); westside California covers 287,560 square kilometers of this area (Raven 1977). Westside California is an ecological island: its complex geology and topography, biogeographic history, and Mediterranean climate combine to make the state's animals and plants distinctive (Bakker 1972). Many groups of westside California's native animals and plants exhibit both a high level of uniqueness or endemism and a high total species richness, yet some groups are relatively poor in numbers of species compared with other regions of the United States. Like true island species, many of California's endemic animals and plants do not fare well against the competition of invading nonindigenous species (see chapter on Nonindigenous Species; Bury and Luckenbach 1976; Moyle 1976a; Bradford et al. 1993).

From east to west, westside California stretches from the crests of the Cascades, Sierra Nevada, Transverse Ranges, and Peninsular Ranges to the Pacific Ocean. The Cascades are not a continuous range in California but are variously elevated volcanic flows and cones punctuated by two impressive volcanic peaks, Mt. Shasta (4,305 meters) and Mt. Lassen (3,187 meters), which is still active. The Sierra Nevada is a more impressive, continuous barrier, with boreal conditions extending its full length from southern Plumas County south to Tulare County. The Transverse and Peninsular ranges of southern California are not as high or as continuous as the northern California mountains, but there is still a clear dividing line between the coastal Mediterranean climate and the more rigorous interior continental climate.

Between the dividing mountain crests lies the Central Valley (the combined valleys of the Sacramento and San Joaquin rivers), and beyond it the Klamath Mountains in the north and the complex Coast Ranges to the south. Immediately west of the Coast Ranges lies a relatively narrow Coastal Plain and the Pacific Ocean. In southern California, the Transverse Ranges interrupt the Coast Ranges and form the southern limit of the Central Valley. Only a narrow strip of westside California occurs south of Los Angeles. The often spectacular California coastline is alternately rocky (Fig. 1) and sandy—mostly rocky to the north and more sandy to the south. Where rivers and smaller drainages reach the coast, there may be protected bays, salt marshes, and coastal dunes. Off the coast of southern California lie the Channel Islands, most of which are now a national park.

Fig. 1. Rocky coastline, Carmel, California.

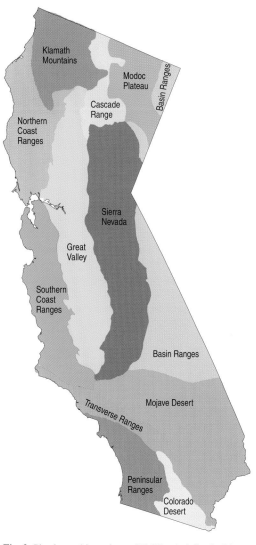

Fig. 2. Physiographic regions of California (after Jenkins 1952).

Environments and History

With 12 physiographic regions (Jenkins 1952; Fig. 2) from high mountains, foothill woodland, chaparral, moist forests, and an alternating rocky and sandy coast, California has high topographic diversity, including the highest land in the conterminous 48 states (Mt. Whitney—4,406 meters). Huge differences in daily and annual temperatures, precipitation, and evaporation have led to strongly differing vegetation patterns (Major 1977) and centers of plant endemism (Stebbins and Major 1965).

Climate

Westside California has a Mediterranean climate, typified by winter and spring precipitation and summer drought (Sprague 1941; Bailey 1966; Gilliam 1966; Major 1977; Mallette 1981). In the Sierra Nevada below about 1,000 meters elevation, precipitation usually falls as rain, whereas above that elevation it falls as snow in winter. Annual precipitation is greatest along the north coast and at middle elevations (2,000 to 2,500 meters) on the west slope of the Sierra Nevada (as much as 1,500 to 2,800 millimeters per year). The least amount of precipitation occurs at low elevations in southern California and in rain shadows along the eastern slopes of the southern Coast Ranges (as little as 140 to 275 millimeters per year). Along the coast, dense fog is common during spring and summer. Moisture-laden offshore breezes emanate from the Pacific High, an offshore high pressure zone, and then move across coldwater upwelling zones to form these fogs (Gilliam 1966), which pour through gaps in the Coast Ranges and bring as much as 200 millimeters of additional moisture in the form of *fog drip* (Azevado and Morgan 1974) along the coast. Coast redwood is limited to this fog belt along the coast from Point Sur north to extreme southern Oregon (Griffin and Critchfield 1972). Most of westside California is dry from May through August, but summer thunderstorms produce small amounts of precipitation at higher elevations in the Sierra Nevada.

Topographic Features

Most of westside California's topographic relief may be explained by the juxtaposition of the Farallon and North American tectonic plates along the western edge of the state (Howard 1979). The crustal movements of these plates give rise to earthquake fault zones (including the well-known San Andreas Fault), volcanoes, and mountain building through faulting or crustal folding (Howard 1979).

Westside California's topographic relief can be divided into nine subregions: the Sierra Nevada, Cascade province, Klamath Mountains, Central Valley, San Francisco Bay and delta, Coast Ranges, California Channel Islands, Transverse Ranges, and Peninsular Ranges. The subregions do not necessarily coincide with vegetation provinces or ecoregions.

The most notable topographic feature of westside California is the Sierra Nevada, a mountain range that extends roughly 650 kilometers from southern Plumas County to north-central Kern County (Storer and Usinger 1964; Howard 1979). The Sierra Nevada is a large granite block that faulted and tilted up along its eastern edge, primarily during the Pliocene and Pleistocene, beginning about 5 million years ago. From the west, the range rises gradually (2%–3% slopes) over its 70- to 90-kilometer east–west breadth but rises dramatically and precipitously along its eastern escarpment (Storer and Usinger 1964; Howard 1979).

Running north from the Sierra Nevada toward the Oregon Cascades (see chapter on Pacific Northwest) is a portion of the Cascade province that consists largely of dissected lava flows of early Tertiary origin and is punctuated by Mt. Lassen and Mt. Shasta, two volcanic cones that rise dramatically above the surrounding landscape (Howard 1979).

The Central Valley, drained by the San Joaquin River in the south and the Sacramento River to the north, lies between the Sierra Nevada and the Coast Ranges. It is a mostly flat plain, 60–120 kilometers wide, which extends about 680 kilometers from Shasta County south to Kern County (Howard 1979). Its elevation ranges from about sea level to a few hundred meters. The soils of the Central Valley consist largely of alluvial sediments derived from various Sierra Nevada uplifts and subsequent erosions (Howard 1979). Where the San Joaquin and Sacramento rivers join and flow westward toward the Golden Gate and the Pacific Ocean, an alluvial delta with alluvial, sandy, or peat soils has formed, braided with sloughs that form numerous islands.

In the northwest corner of California, extending into extreme southwest Oregon, is the Klamath Mountain block, which includes several subsidiary ranges—the Marble, Salmon, Scott, Scott Bar, and Siskiyou mountains; the Trinity Alps; and Mt. Eddy. This block consists mainly of ancient metamorphic and plutonic rocks and is the oldest geomorphic feature in westside California (Hickman 1993). The Klamath Mountain block extends about 200 kilometers south of the Oregon boundary and abuts against the largely sedimentary northern Coast Ranges near the upper headwaters of the South Fork of the Trinity River.

The Coast Ranges are a series of large mountain ranges oriented north to south and extending from the southern limit of the Klamath Mountain block for roughly 720 kilometers south to the Cuyama River, which forms the San Luis Obispo County–Santa Barbara County boundary. These ranges are complex, primarily folded fault blocks composed largely of sedimentary formations with some volcanic intrusions such as Mt. Konocti (1,310 meters) and Mt. Diablo (1,173 meters). Lying between these ranges are north–south-oriented valleys filled with recent alluvial soils. The Coast Ranges, with a few exceptions, lie below 2,000 meters.

The San Francisco Bay and delta include the Sacramento River–San Joaquin River delta, several water bodies and associated wetlands (including San Francisco Bay, San Pablo Bay, and Suisun Bay) that lie between the delta and the Pacific Ocean, and the adjacent low hills. All of the water bodies are tidally influenced well into the delta, and there is a gradual gradation from the seawater that occurs at the entrance of the Golden Gate upstream to the brackish waters of the delta and the fresh water to the east. The hills surrounding the bay are primarily of recent alluvial origin, but other formations are present, including sandstones and serpentine outcrops (Howard 1979).

California has relatively few offshore islands. The Channel Islands occur off the coast of southern California and comprise four northern islands (Anacapa, San Miguel, Santa Cruz, and Santa Rosa) off Santa Barbara and Ventura counties, and four southern islands (San Clemente, San Nicolas, Santa Barbara, and Santa Catalina) off Los Angeles and San Diego counties. The islands are low and generally rolling, with maximum elevations of no more than a few hundred meters, although some islands have steep cliffs. Most of the islands are now included in Channel Islands National Park.

The Transverse Ranges are mainly oriented east to west. These granitic ranges lie between the southern terminus of the Coast Ranges and the Los Angeles basin and trough (Jaeger and Smith 1966), from near the coast inland to the western and southern edges of the Mojave Desert. They include the San Gabriel Mountains, San Bernardino Mountains, Mt. Pinos, and the Santa Ynez, Topatopa, Santa Susanna, Santa Monica, and Liebre mountains. The mountains of the Transverse Ranges are geologically complex, and some reach elevations of more than 3,000 meters, such as Mt. San Gorgonio (3,506 meters) and San Antonio Mountain, which is also called Mt. Baldy (3,068 meters).

The Peninsular Ranges include several more or less north–south-oriented, largely granitic ranges that lead toward the Sierra Juarez of Baja California Norte, Mexico, and that separate the narrow strip of westside California between Los Angeles and the Mexican boundary from the Colorado Desert to the east (see chapter on Southwest). The Peninsular Ranges include the Laguna, San Jacinto, Santa Ana, Santa Rosa, and Vallecito mountains (Jaeger and Smith 1966). Exceptionally high Peninsular Range peaks include Mt. San Jacinto (3,286 meters) and Santa Rosa Mountain (2,452 meters). The low-elevation gaps and passes that occur between the various Peninsular Ranges have allowed some desert animals and plants to extend their ranges almost to the relatively arid southern coast.

Faunal and Floral History

During the Tertiary (65 million to 1.6 million years ago) the area that is now California slowly changed from wet tropical conditions to a more arid environment (Axelrod 1977; Wilken 1993). At that time, westside California had an impressive vertebrate fauna that included such species as sabertooth cats, mammoths, camels, giant bison, rhinoceroses, ground sloths, and ancestors of modern horses and pronghorns (Stirton 1951).

Fossil plant remains tell us that 75 million years ago (near the beginning of the Tertiary), the westside California flora was composed of three major floral elements—the Neotropical-Tertiary Geoflora, the Arcto-Tertiary Geoflora, and the Madro-Tertiary Geoflora (Raven 1977; Raven and Axelrod 1978). As conditions became more arid during the Tertiary, elements of the Neotropical-Tertiary Geoflora disappeared from California; today their relatives are found in the humid tropics of Mexico and Central America. Elements of the other two geofloras evolved in California and became mixed in various ecosystems and habitats with Arcto-Tertiary elements and their derivatives, which tend to be found at higher elevations and at locations with higher precipitation. Madro-Tertiary elements and derivatives tend to be found at lower elevations and at locations with lower precipitation (Axelrod 1977; Raven 1977; Raven and Axelrod 1978; Wilken 1993).

The uplift of the Sierra Nevada was a major geologic event, separating much of westside California from the Great Basin steppe to the east (Hinds 1952). This uplift, which began in the Pliocene (5 million years ago) and accelerated during the Pleistocene (1 million years ago), formed a significant topographic barrier and made considerable changes in the climate on both sides of the mountains (Gilliam 1966;

Howard 1979). Species adapted to arid conditions were largely isolated to the east in the range's rain shadow, whereas species adapted to Mediterranean climates were confined to the west side.

The ancestors of California's present plants and animals were largely in place before the Pleistocene ice ages (up to 1 million years ago), when most of westside California did not experience ice sheet advances, and only the high Sierra Nevada experienced valley glaciers, as in Yosemite Valley (Hinds 1952).

Stebbins and Major (1965) analyzed the genetic variability and origins of modern California's endemic plants and divided them into paleoendemics—those species of ancient origins—and neoendemics—those species that have evolved relatively recently in California or in nearby regions. Predictably, paleoendemics are concentrated in the ancient Klamath Mountain block of northwestern California and the low deserts of southern California, and neoendemics are concentrated in relatively young geological formations in the central portion of westside California.

Human History

Human occupation of California has increasingly altered the state's natural resources since the first human occupation of the land 11,000–12,000 years ago (Eargle 1986). The number of Native Americans at the time of European or European-American contact is estimated at 300,000. Today the human population of the nation's most populous state is nearing 32 million and is likely to continue increasing.

Before European contact, more than 100 Native American tribes inhabited California (Rawls 1984). They modified local landscapes by burning vegetation and by hunting and gathering. Tribes in the northwest part of the state were culturally similar to those of the Pacific Northwest. The northeast part of the state was thinly populated, and life there was difficult because of the harsh climate. Peoples of the Central Valley lived a relatively sedentary, peaceful life. Their staple food was meal made from acorns of the valley oaks. Southern California was the most populous part of the state, especially along the coast, where people survived primarily on marine resources.

Although the first Spanish explorers reached California in 1542, and Sir Francis Drake landed near San Francisco Bay in 1579, European colonization did not begin until the Spanish Franciscan missionaries arrived in 1769 (Rawls 1984). Over the next few decades, the Franciscans built 21 missions along the coast from San Diego to San Francisco. These missions served religious and secular purposes: to protect Spanish interests in the area, as well as to convert the natives to Christianity and make them "useful" citizens of the Spanish empire. Native Americans were relocated near the missions and forced to work. Nearly two-thirds of the native population died as the result of introduced diseases (Rawls 1984).

The mission period ended in 1821, when Mexico became independent from Spain. In the years that followed, the Mexican government made many private land grants in California, and Mexican landowners emigrated to Alta (upper) California. Many of the remaining native people worked as bond laborers on Mexican haciendas (Eargle 1986). Spanish and Mexican exploration and colonization in the eighteenth and nineteenth centuries brought ranching and agriculture and the first of a stream of nonindigenous plants and animals to California.

In the early 1800's American trappers arrived in the state, attracted by the wealth to be had from selling sea otter and beaver pelts. Both animals were trapped to near extinction. By 1841 settlers from the eastern United States were arriving in California via the Oregon Trail. Mexico ceded title to California to the United States in 1848; in the same year, gold was discovered at Sutter's Mill, resulting in a flood of prospectors, miners, farmers, merchants, and their families. The European–American population of California grew from 15,000 in 1848 to 380,000 by 1860 (Eargle 1986).

The remaining Native American population suffered greatly from this influx of settlement and development. Whole tribes were exterminated or displaced by settlers who considered them an impediment to development, and during the mid-nineteenth century, many were sold into slavery. Reservations were set aside as early as 1853, but only in the latter part of the nineteenth century were tribal lands secured by the federal government and aid made available for the remaining native peoples in California (Eargle 1986).

Experience with hydraulic gold-mining techniques, combined with the region's rainless summer climate pattern, quickly encouraged industrious settlers to control and divert the rivers for agricultural uses. By 1867 miners had constructed 6,780 kilometers of water ditches and canals (Kahrl 1978). A little more than a hundred years later, almost all of the Central Valley, the large coastal valleys, and the Imperial-Coachella Valley were under irrigation (Donley et al. 1979). In addition, most of the larger rivers and streams—except the Smith River, which drains to the Pacific near the well-watered California–Oregon border—were dammed and regulated.

Through the last half of the nineteenth century and until the 1930's, native populations of many game animals were greatly reduced by subsistence and market hunting to support the growing human population. Populations of native species—from elk, deer, ducks, and geese to frogs and fishes—were shot, gigged, and netted, and sold fresh, dried, or canned to the growing urban populations. Populations of many game species declined until fish and game laws, habitat protection, and game management practices were introduced to protect fragments of the once amazing array of native wildlife. As trade and commerce grew in California, cities also grew, and forests were exploited to supply sawmills with the timber necessary for local construction and also for timber export around the Pacific Rim.

Today, almost all of California's rivers are dammed and fed into federal and state water distribution systems, providing water for the state's intensive agriculture and extensive urbanization. The Central Valley rivers are diked to provide flood control for the farms and cities that now occupy land that was once seasonally flooded. Twentieth-century urban development has claimed much of the southern coast and the area around San Francisco Bay. Millions of hectares of native grasslands, marshes, and seasonal wetlands in the Central Valley and delta—habitats formerly home to native herbivores and large migratory waterfowl populations—have been converted to agriculture. Grizzly bears no longer occur in the valleys and the foothills of the region, and anadromous fisheries of the watersheds draining the Sierra Nevada have much-reduced runs of salmon and other native fishes. As water supplies were acquired by large development interests with the political and financial ability to move water to the semideserts of southern California, the growth of cities and agriculture greatly accelerated, resulting in the loss of the incredible richness of the Central Valley.

As the population increased, so did the harvest of Sierra Nevada and Coast Range forests. Early logging for homesteads and mining activities expanded to accommodate construction of railroads and cities. By the late 1800's, however, some people began to fear that some of the world's most majestic forests and scenic places would be completely destroyed and therefore advocated the establishment of parks on the forested slopes and glaciated valleys of the Sierra Nevada, beginning with Yosemite in 1890 (Fig. 3), and in the coast redwood forests, including California Redwood Park (Big Basin Redwood State Park) in 1902 and Muir Woods National Monument in 1908. Although early timber harvest was extensive, improved logging equipment, better roads, tractors, and trucks

© P. A. Opler

Fig. 3. El Capitan and Half Dome in Yosemite National Park.

Society's Inventory of Rare and Endangered Vascular Plants of California is the most comprehensive source of available information and is regularly updated (Skinner and Pavlick 1994). The society's new publication on California vegetation (Sawyer and Keeler-Wolf 1995) begins to address the relative status of plant communities in California and provides systematic methods for their description.

The most human-altered ecosystems in California are those that have been affected by grazing and agriculture, followed by those susceptible to urbanization and timber harvest. Noss and Peters (1995) report significant reductions in the native vegetation of several westside California plant communities and formations (Table 1). Endangered plant communities in southern California include grasslands, coastal sage–scrub, riparian woodlands, and estuarine wetlands (Schoenherr 1990). In most instances, except coastal redwood forest, native vegetation has been replaced by nonindigenous vegetation, croplands, or development.

Table 1. Human-caused reductions in westside California plant communities and formations (after Noss and Peters 1995).

Community/formation	Vegetation reduced (percent)
Native grasslands	99
Needlegrass steppe	99.9
Southern San Joaquin Valley alkali sink scrub	99
Southern California coastal sage–scrub	70–90
Vernal pools	91
Wetlands	91
Riparian woodlands	89
Coast redwood forest	85

combined to feed the more demanding needs of the building boom after World War II. From 1950 to 1975, approximately 5.3 billion board feet of lumber were produced annually from California forests (Donley et al. 1979).

Status of Ecosystems

Many ways have been proposed to divide westside California into ecological subgroupings or ecosystems (Van Dyke 1919; Dice 1943; Munz and Keck 1949, 1950, 1959, 1968; Küchler 1977; Bailey 1995). The complex patterns of vegetation have been variously classified; Munz and Keck (1959, 1968) identified 29 plant communities in 11 vegetation types. Küchler (1977) classified California's natural vegetation into 54 communities with nine formations. More recently, Bailey (1995) divided westside California into five provinces.

We follow Hickman's (1993) divisions, which refer to all of terrestrial westside California as the Californian Floristic Province and divide the region into six smaller subregions that combine the major topographic features discussed previously. Each of the regions is divided at least once, and several have a further hierarchical geographic division so that, in all, 27 subdivisions are recognized. Even within each subdivision, though, there are hundreds of distinctive communities or formations with characteristic plants and animals, each based on local climate, soils, elevation, and exposure (Munz and Keck 1959, 1968; Barbour and Major 1977). The California Native Plant

Awareness of the relationship among habitat conversion and fragmentation, and the endangerment of habitat-limited, narrowly distributed, or ecologically specialized plants and animals has led to an increased effort by public entities and private developers to deal with land-use issues on bioregional and ecosystem bases. For example, the interagency California Natural Areas Coordinating Committee divided the state into 11 bioregional planning areas in 1990 (Fig. 4), and in 1991 California adopted a bioregional strategy for resource conservation of biodiversity; this strategy involves both California and U.S. agencies (California Governor Pete Wilson, press release, Sacramento, 1991).

Much of the high-elevation forests in the Sierra Nevada, Cascade Range, and Klamath Mountains, and in the drier southern California mountain ranges, were only lightly developed and were eventually included in the national forest system, which is now composed of some 9,161,910 hectares, about one-fifth of California. Much of the land considered to be of

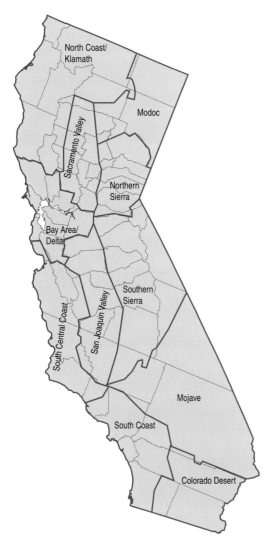

Fig. 4. Bioregional planning areas in California.

lesser economic value in the last century has now become the relatively unaltered wilderness and natural areas conserved within national parks, national forests, and Bureau of Land Management wilderness areas.

Ecological Provinces

Northwestern California

Northwestern California has the wettest, most consistent climate in the state. It is composed mainly of the coastline and several metamorphic mountain ranges, including the Klamath Mountains and the north Coast Ranges. The coastal region, from the Oregon border south to Bodega Bay, is dominated by areas of coastal prairie (Fig. 5), some coastal marsh, closed-cone pine and cypress forests on poor soils, and grand fir–Sitka spruce forests on better soils (Hickman 1993). In California, serpentine soils are common in isolated patches

where intrusions of molten material forced their way into the Earth's crust. In these areas, the reddish or greenish rock usually weathers into infertile soils that are high in magnesium and poor in phosphates, potassium, and calcium and are often too thin to hold moisture. Only the hardiest plants can live on these serpentine soils, where they benefit from the absence of competing plants and the absence of many harmful soil organisms (Bakker 1972; Fig. 6).

The closed-cone pines and cypresses form unique communities scattered along the coast, mountains, and islands (Fig. 7). Many of the closed-cone pines and cypresses are endemic (8 of the 10 species of cypress), including Santa Cruz, Modoc, Tecate, Monterey, and Sargent cypresses, as well as knobcone pine, bishop pine, and Monterey pine, a species that is widely planted in reforestation projects all over the world. All of these species are dependent on fire for regeneration and are relics of a previously more widespread ecosystem that retreated when the climate changed in the Tertiary (Vogl et al. 1977).

Fig. 5. Coastal prairie, Point Reyes National Seashore.

Fig. 6. Serpentine grassland, Santa Clara County.

Fig. 7. Monterey cypress in Point Lobos State Preserve, Monterey County.

© P. A. Opler

Most of the cypress and closed-cone pine forests in California were part of Spanish and early American ranches and have been grazed by domestic livestock. Because the trees are generally dwarfed and gnarled in form, they have not been logged except occasionally for fenceposts and firewood. The most significant human effects on these forests have come from suppression and alteration of fire (Vogl et al. 1977; see box on Torrey Pine).

Many of the cypress groves are associated with chaparral, rock outcrops, or serpentine soils. Cypress seeds are sealed inside their cones with resin; they do not germinate in litter but require bare mineral soils. Cypresses are intolerant of shade and do not reproduce well in the absence of fire. Although cypress trees are usually killed by fire, their cones open when heat from the fire melts the resin. During the following months, the seeds drop, creating dense stands of seedlings. Recently though, some cypress groves have become smaller, possibly because these cypress species could be continuing the natural decline they have experienced since the Tertiary by becoming more restricted in their range following every unfavorable event (for example, fire suppression and ecological changes caused by grazing). In addition, fire suppression over the last century has altered fire frequencies and intensities in the groves, creating less favorable conditions for cypress reestablishment (Vogl et al. 1977).

The closed-cone pines are generally small in stature and, like the cypresses, are associated with chaparral, fire, and shallow, acidic, nutrient-poor soils, often serpentine or

sandstone. These pines are short-lived (50–100 years), and their seeds can only germinate on bare mineral soils. Like the cypresses, the closed-cone pines require fire for successful reproduction. Their cones remain attached to the parent tree for years and are sealed shut with resin, rarely opening unless the resin is melted by fire (Vogl et al. 1977). Knobcone pine is the most widespread of the closed-cone pines, ranging nearly the length of the state.

The Klamath Mountains are geologically old and support mixed evergreen forests of Douglas-fir, ponderosa pine, and sugar pine, with mountain hemlock, white fir, and chinquapin found at higher elevations. Serpentine soils are common in the Klamath Mountains. On the west side, Douglas-fir–hardwood forests grow at low elevations, giving way at higher elevation to white fir–Douglas-fir forests, white fir–California red fir forests, and finally to mountain hemlock–California red fir at the highest elevations. East and south of the highest ridges, the climate is drier and more continental. At low elevations, forests are dominated by ponderosa pine, which is replaced by white fir–pine forests at higher elevations, then red fir–white fir forests, and finally mountain hemlock–red fir, with whitebark pine occurring at the highest elevations. The Klamath Mountains have a high floristic diversity, in part because they have acted as refugia supporting many endemics and relict species, including Pacific silver fir, subalpine fir, Alaska-cedar, Brewer spruce, Engelmann spruce, and foxtail pine. The complex vegetation patterns in the Klamath Mountains seem based primarily on differences in soils and secondarily on elevation and soil moisture (Sawyer and Thornburgh 1977).

The northern Coast Ranges occur immediately south of the Klamath Mountains. Coast Range forests do not include hemlock and have noble or red fir replacing grand fir, with rhododendron replacing chinquapin in the understory. Hardwoods increase in frequency on the drier slopes inland. The outer northern Coast Ranges, those farthest to the west, receive a great deal of rain (Hickman 1993).

Riparian areas and north-facing slopes of the Coast Range fog belt support redwood forests (Fig. 8), which thrive where coastal fog is frequent. Redwood is a California endemic and is the tallest (112 meters) and fastest-growing tree in the world (Zinke 1977); one of these trees may live more than 2,000 years (Bakker 1972). Although redwoods were common in the Tertiary over much of North America, they are now restricted to the fog belt of maritime central and northern California. Proximity to the sea moderates temperatures, and fog helps prevent evapotranspiration (moisture loss from leaves). Fog drip contributes considerable

Courtesy M. Madej, USGS

Fig. 8. Coast redwood forest.

Courtesy M. Madej, USGS

Fig. 9. Coast redwood forest with rhododendron understory.

moisture to the soil during the otherwise dry summer season (18–30 centimeters per year; Zinke 1977). The continuous moisture enables redwood forests to be home to a number of amphibians, including ensatinas, ocelot-spotted giant salamanders, tailed frogs, and seep salamanders, as well as the more common banana slugs (Bakker 1972).

Redwoods have thick, nonresinous bark, and mature trees are able to withstand fires. Redwood seedlings are tolerant of shade and germinate on logs and root wads rather than in the duff that accumulates under mature trees. Redwoods also sprout from basal buds if the main trunk is damaged or killed by fire (Veirs 1982). Redwoods are also very resistant to feeding by herbivorous insects. Douglas-fir is often a codominant in redwood forests, becoming established after fires, and tanoak, California bay, madrone, and western hemlock are common understory trees where enough light penetrates the canopy (Zinke 1977; Fig. 9).

Redwood is a valuable timber tree because of its size and because of the wood's unique resistance to rot. More than 85% of the old-growth coast redwood forests has been logged, but much of the original distribution of about 810,000 hectares remains in second-growth redwood forests of varying ages. Second-growth redwood forests support most of the same native vascular plants as old-growth forests, but habitat for species that depend on old-growth forests—such as spotted owls, marbled murrelets, some arthropods, mollusks, and canopy lichens—has been greatly reduced (U.S. Fish and Wildlife Service 1995a). Logging of redwood continues, although most old-growth stands are now protected in state parks and in Redwood National Park.

Drier slopes of the Coast Ranges support mixed-evergreen and mixed-hardwood forests, whereas montane forests of subalpine fir and pines are found at higher elevations. Vegetation on the highest peaks is similar to that found at high elevations in the Sierra Nevada; peaks above 1,500 meters are treeless and experience heavy winter snows. Summers are hot and rainfall is low in the inner northern Coast Ranges, especially on eastern slopes in the rain shadow of the peaks. Serpentine soils are common, and dry eastern slopes support chaparral and pine–oak woodland. (Hickman 1993).

Cascade Ranges

The Cascades are volcanic mountains to the east of the Northwest province. The foothills, ranging in elevation from 100 to 500 meters, support a mosaic of chaparral and blue oak–foothill pine woodland. The southern foothills form an ecotone with the grasslands on the north and east end of the Central Valley. The high Cascades, above 500 meters, support forests of ponderosa pine. At higher elevations the vegetation is montane in character, with fir–pine and lodgepole pine forests. True mixed-conifer forests predominate on slopes between 500 and 2,000 meters, with sugar pine and white fir more common on moist sites and ponderosa pine and incense-cedar common on drier slopes. Alpine vegetation occurs at the summits of Mt. Shasta and Lassen Peak. Mudflows and avalanches are regular natural disturbances on these peaks, periodically destroying large areas of forest (Rundel et al. 1977).

Sierra Nevada

The Sierra Nevada province includes the long, north–south-oriented mountain range that borders the Central Valley on the east and forms a formidable topographic barrier between the Great Basin and westside California. The foothills of the Sierra Nevada, between 500 and 800 meters in elevation, support blue oak, foothill digger pine woodlands (Fig. 10), and chaparral along the entire length of the range. Above 500 meters, high Sierra Nevada forests vary with elevation. In the lower montane zone, the drier slopes are forested with ponderosa pine, whereas moister slopes support white fir and giant sequoia. Slightly higher, in the upper montane zone, forests of red fir, Jeffrey pine (Fig. 11), and lodgepole pine dominate (Rundel et al. 1977). In the montane zones of the Sierra Nevada, natural fire regimes have been altered by fire control and suppression, and species composition and vegetation structure have changed significantly, as demonstrated in forests protected from timber harvest in the Sierra Nevada (see box on Fire and Fuel in the Sierra Nevada).

Most trees of the Sierra Nevada are common in mountain ranges across the West, but giant sequoia is a California endemic. Although coast redwoods are taller, giant sequoias are the most massive plants on earth, and they may live more

than 3,000 years (Bakker 1972). Closely related to coast redwood, giant sequoia is scattered in some 75 disjunct groves throughout the middle elevations of the central Sierra Nevada. The unusual distribution of this species is believed to be due to changing climates in the Holocene, and giant sequoias are probably more abundant now than they were 10,000 years ago (Anderson 1994). At the time of European settlement, giant sequoia populations seem to have been increasing or at least stable (Stephenson 1994).

Seedlings are intolerant of shade and need bare mineral soil to establish. Individual treefalls provide such conditions, but not enough to sustain populations. Giant sequoias are quite resistant to fire and insects; in fact, successful regeneration of the species seems to require locally intense fires that kill the forest canopy, followed by one or more wet summers to aid seedling establishment. Most living sequoias occur in even-aged groups that probably correspond to localized *hot spots* in past fires. Successful management of giant sequoias requires reintroducing fires in which local patches burn intensely within a mosaic of frequent, gentle surface fires (Stephenson 1994).

Giant sequoias are usually dominants in mixed-conifer stands; associated tree species include California white fir, sugar pine, and incense-cedar. Red fir is an important associate at higher elevations, whereas ponderosa pine and California black oak are common associates at lower elevations. Some giant sequoia groves were logged in the late nineteenth century, but most are now protected in national parks (Bakker 1972).

The subalpine zone of the Sierra Nevada is forested by mountain hemlock and whitebark pine, and many of the peaks are high enough to have treeless alpine zones; Mt. Whitney, for example, is more than 4,400 meters high (Fig. 12). The Tehachapi Mountains are a small region of foothill and montane vegetation below 2,000 meters in elevation, at the south end of the Sierra Nevada. These mountains support a unique mixture of vegetation types found in the neighboring provinces, including pinyon–juniper woodland, prairie, and mixed woodlands (Hickman 1993).

Led by the U.S. Forest Service, with the cooperation of other federal and state agencies and the University of California, the Sierra Nevada Ecosystem Project is nearing its conclusion. This bioregional program, which is driven by federal legislation, attempts to describe resource conditions in the Sierra Nevada ecosystem by using existing information; it also attempts to anticipate the direction of future change and the effect of potential resource use on the natural resources and human communities in the area.

Fig. 10. Digger pine woodland, Contra Costa County.

© P. A. Opler

Fig. 11. Jeffrey pines, Mono County.

© P. A. Opler

Fig. 12. Subalpine vegetation, Sierra Nevada.

Central Valley

The Central Valley extends 640 kilometers from north to south through the central region of California. Bounded on the east by the Sierra Nevada foothills and on the west by the Coast Ranges, the valley floor averages 64 kilometers in width and encompasses nearly 42,000 square kilometers. The Klamath Mountains form the valley's northern end and the Tehachapi Mountains form its southern end. The valley is divided into three major regions: the Sacramento Valley, which drains southward; the San Joaquin Valley, which drains northward; and the delta and Suisun Marsh areas where the Sacramento and San Joaquin river systems meet. Much of the Central Valley is now agricultural land, but fragmented native ecosystems still exist. The Central Valley once supported extensive prairie (Fig. 13), marshes, riparian woodlands, and valley oak savannas.

Native grasslands in the Central Valley were composed mainly of drought-resistant perennial bunchgrasses, such as needlegrasses, blue wild rye, various bromes, melicgrass, and deergrass (Bakker 1972). However, the native California prairie has almost disappeared, either converted to agriculture or to nonindigenous annual grasslands. Around the middle of the nineteenth century, heavy grazing by cattle and sheep caused native perennials to be replaced by fast-growing

Fig. 13. California poppies blooming at Shotgun Pass, Merced County, in native prairie in the Central Valley.

annual grasses, which are able to take advantage of spring rains and produce seeds before the dry heat of summer. The native perennial grasses, which are more palatable to livestock than the annuals, were damaged by grazing and trampling during the Mexican Rancho period. Native annuals such as six-weeks fescue, three awns, and lovegrass, whose ranges had been restricted by the dominant perennials, became more prominent and were soon followed by nonindigenous, weedy species such as wild oats, soft chess, and goatgrasses. The disappearance of perennial grasses caused both the extirpation of pronghorns and the near-disappearance of Tule elk from the Central Valley (Bakker 1972). Tule elk, a small, endemic California subspecies, now inhabits California only in managed (usually transplanted) herds in parks and preserves. Pronghorns, however, are still common elsewhere in the West and the Great Plains.

The Sacramento Valley is composed of five drainages: the Butte, Colusa, Sutter, Yolo, and American river basins. The San Joaquin Valley consists of the San Joaquin basin in the north and the Tulare basin, which forms a closed drainage system at the southern end of the valley. The delta consists of a network of sloughs and islands at the confluence of the Sacramento and San Joaquin rivers. The islands, diked and reclaimed in the 1800's, contain rich peat soils that support a thriving agricultural industry. The Suisun Marsh is an estuarine wetland system that forms a transition zone between the fresh water of the delta and the marine environment of the San Francisco Bay estuary. Within the last 50 years, public works projects designed to respond to the water demands of agriculture and large metropolitan areas have produced a great network of artificial lakes and rivers interconnected by a system of Central Valley aqueducts. The federally administered Central Valley Project and the associated State Water Project are the most important of these systems. Their primary function is to transport water from major sources in northern California to arid regions in the south. This reliable water source, the rich soils, and the ideal climate have made California the nation's leading agricultural state for the last 50 years.

Wetlands in California historically have hosted one of the largest concentrations of wintering waterfowl and other migratory birds in the world. In the mid-1800's, an estimated 2 million hectares of wetlands were present in California, and early explorers reported vast concentrations of waterfowl and other marsh and shorebirds (California Department of Fish and Game 1983). About 1.6 million of these hectares were in the Central Valley. Since then, more than 95% of the historical wetlands in

California have been destroyed or modified (Gilmer et al. 1982). Of about 116,000 hectares that remain in the Central Valley (Heitmeyer et al. 1989), two-thirds are privately owned and managed for duck hunting; the remaining one-third is divided between state and federal ownership and is managed by the California Department of Fish and Game as Wildlife Management Areas (15,282 hectares) or by the U.S. Fish and Wildlife Service as national wildlife refuges (20,337 hectares). Most of these wetlands are seasonal and intensively managed.

Closely associated with the historical wetlands were extensive riparian forests that flourished among wetlands and along waterways (Warner and Hendrix 1984). Recent estimates indicate that only about 11% of the original riparian forest remains in the Central Valley (Katibah 1984). Destruction of riparian forests throughout the Central Valley has had a detrimental effect on species such as the yellow-billed cuckoo and Bell's vireo, which depend on these forests for breeding or wintering habitats (Warner and Hendrix 1984).

About every three to five winters, approximately 38,500 hectares of additional wetland habitat are created, when bypasses (diked agricultural lands) channel Sacramento River overflows to the delta (Kahrl 1978). Major losses of natural wetlands occurred when these flooded areas were converted to cultivation of rice, which has been an important crop in California since 1912. Because rice is grown in flooded fields, marsh soils are ideal for its production. If rice fields are flooded for fall waterfowl hunting they provide considerable benefits to waterbirds, because of the increased availability of waste grain and invertebrates. The total area of harvested rice increased through the 1980's as world markets expanded. For example, in the Sacramento Valley, where rice is primarily grown, land used for this crop has recently increased from about 81,000 to nearly 243,000 hectares. However, if farmers switched from rice cultivation to other agricultural crops, the carrying capacity of the Central Valley for wintering waterbirds could greatly decrease.

The San Joaquin Valley has experienced major wetland losses caused by the conversion of natural ecosystems to the cultivation of cotton and a variety of row crops. A notable exception to this trend has been the Grasslands, a 26,325-hectare area of private duck clubs that have preserved much of the property's high-quality wetlands. The Grasslands represents the largest tract of waterfowl habitat in the San Joaquin Valley.

Vernal pools, called *hog wallows* in the Central Valley, are unique wetlands found mostly on protected areas in the San Joaquin Valley (Bakker 1972). Vernal pools are small depressions with hardpan floors that fill with water during winter rains and evaporate through the spring. Most pools are between 1 to 2 square meters in area, though they vary considerably in size. These pools were more widespread before the advent of agriculture and urbanization, but they still are scattered throughout the Central Valley grasslands, mostly on old terrace soils (Holland and Jain 1977). Vernal pools support specialized invertebrate and plant communities and are veritable islands of unique vegetation. Of the 101 plants found in vernal pools in one study (Holland and Jain 1977), 70% were native annuals, and only 7% were introduced (non-indigenous) annuals; 55 plants were endemic to California, and another 14 were near-endemics. High species diversity occurs between pools, but less diversity occurs within a single pool. No trees, shrubs, or stem succulents are associated with vernal pools.

Because California vernal pools are a relatively recent phenomenon in geologic time, most of their endemic plants are relatively young species (Stone 1990), many with close ecological and evolutionary relations with highly specialized insects, particularly bees (Thorp 1990). The soil and water of vernal pools are very alkaline, and the pH of the water increases as it evaporates in the spring. As the pools dry, annual plants bloom in concentric rings determined by their proximity to the standing water (Figs. 14 and 15). Within the standing water, one might find the plants hogwallow starfish, Howell's quillwort, marsilea ferns, pygmy epilobium, monkeyflower, or mouse-tail, whereas around the pool margins, foxtail, common stickyseed, annual hairgrass, hedge-hyssop, toad rush, leafybract dwarf rush, and smooth tidytips can be found. Sharp ecotones exist between the pool flora and the annual grassland on the mounds between them. On the mounds one might find brome grasses, fescues, wild oats, and occasionally the native perennial purple needlegrass. Because the soil of these pools is replete with salts and other solutes (alkali), and because of their ephemeral nature, vernal pools have successfully resisted invasion by nonindigenous species, unlike nearby marsh habitats.

Water of sufficient quantity and quality is a major limiting factor for wetlands and wildlife populations in the Central Valley. Historically, legislation governing the allocation of surface water by the Central Valley Project and the State Water Project assigned higher priority to agricultural and municipal needs than to fish and wildlife requirements. About 87% of the water provided by these systems is used for irrigation.

Fig. 14. Vernal pool vegetation: California goldfields, Butte County.

© P. A. Opler

Fig. 15. Vernal pool vegetation: meadow foam, Butte County.

Migratory birds of special interest that breed in the valley include Swainson's hawk, yellow-billed cuckoo, and tricolored blackbird. Wetlands in the Central Valley are critical habitat for the threatened giant garter snake, a species of limited distribution that is being studied.

Central Western California

The central western coast is divided into three regions: the central coast, the San Francisco Bay area, and the south coast. The central coast, right along the ocean, includes true coastal vegetation, coastal sage–scrub in the south, and salt marsh and coastal prairie around San Francisco Bay. Inland from the immediate coast, the San Francisco Bay area supports wet redwood forest in riparian areas and seaward slopes, and dry oak–pine woodlands and chaparral farther inland and on dry slopes. The southern Coast Ranges, which extend along the coast from Santa Clara County to Santa Barbara County, have many small serpentine outcrops. The northern ends of the ranges are forested with redwood and mixed hardwoods near the coast. Farther south, southern oak and blue oak–foothill pine woodland dominate along the coast, with chaparral occurring farther inland.

Oak woodlands are one of the most common and characteristic vegetation types in California, covering 3 million hectares of rolling foothill topography (Huntsinger et al. 1991; Fig. 16). The term *oak woodland* encompasses a variety of environmental conditions and vegetation, representing a group of variable communities geographically located between grassland or scrub and montane forests, with their boundaries often obscured by a zone of chaparral. Open stands of deciduous white oaks (that is, valley oak, blue oak, and Oregon oak) are typical of huge areas, but

evergreen black (or red) oaks (that is, coast live oak and interior live oak) are often present and sometimes dominant (Griffin 1977). Oregon oak is important in northern oak woodlands, blue oak and valley oak in central woodlands (Fig. 17), and Engelmann oak in southern woodlands. In coastal foothills, coast live oak is prominent, whereas interior live oak is more common inland. The scrub oaks of chaparral, such as leather oak, which is limited to serpentine soils, and Nuttall's scrub oak, are true shrubs. Nine of California's oak species are endemic, including all the species just mentioned, and many oaks hybridize with each other, making classification difficult in some cases. Tanoak is actually a separate genus and is not a true oak, though it occurs with oaks in woodlands and chaparral. Chaparral often forms a mosaic with grassland and woodland communities on poorer, shallower soils (Griffin 1977).

Oak woodlands often have an understory of grasses and forbs, which now include many nonindigenous species, and have been used for livestock grazing since the time of the Mexican Ranchos. They still provide about one-third of the total rangeland forage supply for California's livestock industry. Oaks are not usually logged, except for firewood. The acorns

Fig. 16. Blue oak woodland, Lake County.

Fig. 17. Valley oaks in fog, Santa Clara County.

they produce, though, were the staple food for Native Americans, who may have actively managed them by setting grass fires. They are also an important food for wildlife. Most (82%) of California's oak woodlands are privately owned (Huntsinger et al. 1991) and are still used for livestock grazing and recreational hunting (Morrison et al. 1991). Since the mid-nineteenth century, oaks have increased in both density and extent at low elevations, probably partly in response to climatic warming, changes in grazing pressure and woodcutting practices, and cessation of regular burning by Native Americans (Byrne et al. 1991).

Southwestern California

This portion of California is made up of several geographic units and a multitude of complex ecological systems and communities (Jaeger and Smith 1966). The area includes the Channel Islands, the Transverse Ranges, the Peninsular Ranges, intervening valleys, and coastal lowlands. The coast is highly urbanized, and most natural vegetation and habitat have been lost. The many endemic plants and animals of this area are threatened or extirpated by urban development. However, some native vegetation remains, consisting mostly of coastal sage–scrub and chaparral.

The Transverse Ranges are unusual in that they are oriented east to west. They include the San Bernardino, San Jacinto, Santa Monica, Santa Ynez, Topatopa, Santa Susanna, and the Liebre and Sierra Pelona ranges. The mountains become hotter and drier to the east, where they border the Mojave Desert. Chaparral dominates the vegetation at lower elevations, grading into southern oak forest, then dry montane forest with white fir, Jeffrey pine, sugar pine, and lodgepole pine at the highest elevations. The Peninsular Ranges are farther south and include the Santa Ana, Cuyamaca, Santa Rosa, Laguna, Jacumba, and San Jacinto ranges (Hickman 1993). The coastal sage–scrub, which is restricted to coastal plateaus and lower slopes of the coastal ranges, changes in character from north to south; evergreen species are more common in the north, whereas many species in the south are drought-deciduous or succulent. Most of the plants grow actively only during the cool, wet winters.

Coastal sage–scrub occupies drier, usually lower elevation sites than the chaparral and tends to be shorter; on sites that can support chaparral, sage–scrub eventually succeeds to chaparral following disturbances. Sage–scrub habitat also adjoins the annual grasslands, and islands of sage may be included within the grasslands. Usually, though, a bare zone nearly a meter wide occurs between scrub and grassland wherever they meet; scientists suspect that

this is due to inhibitory compounds produced by the sagebrush plants (Mooney 1977). Dominant species in most sage–scrub sites are California sagebrush, white sage, black sage, purple sage, California buckwheat (also see Fig. 18), and mahogany sumac. Farther south, where conditions are drier, such distinctive species as California buckeye (Fig. 19), California adolphia, golden-spined cereus, and Shaw's agave join the sages (Mooney 1977).

Chaparral is a word of Spanish derivation that originally referred to a thicket of shrubby evergreen oaks but is now applied to dense brushland in general. It is the most extensive vegetation type in California, covering about 5% of the state (Hanes 1977), but it reaches its maximum development in southern California, where it ranges in elevation from 300 to 3,000 meters. Chaparral develops on alluvial fans and

Fig. 18. Wild buckwheat, Granite Creek, Madera County.

Fig. 19. California buckeye, Mariposa County.

washes adjacent to coastal sage–scrub and riparian woodland. In northern California, chaparral is scattered, ringing the Central Valley and generally occurring on drier, south-facing slopes of the inner ranges and coast ranges (Fig. 20). Chaparral consists of dense shrubs with distinctive sclerophyll leaves, which are small, stiff, thick, and usually evergreen. Chaparral occurs in areas with cool, wet winters and hot summers with prolonged drought.

Fire has shaped chaparral communities for more than 2 million years. Most fires in California natural areas are in these communities, occurring every 10 to 40 years (Hanes 1977). Chaparral shrubs have adapted to fire by producing seeds at an early age, by producing fire-resistant seeds that can live in the soil for decades, or by sprouting from a lignotuber or root-crown burl after the main-stem is destroyed. As California's human population has grown, this fire-adapted vegetation is now interspersed with housing developments in many areas, and fires in this urban–wildland interface are dangerous to human life and costly in terms of property loss.

Chamise chaparral, the most common chaparral type in California, is so named because it is dominated by chamise. It occurs on hot, dry, infertile south- and west-facing slopes and is 1 to 2 meters tall, with little or no understory and few associated species, such as manzanitas, different kinds of ceanothus, giant wild-rye, California buckwheat, Nuttall's scrub oak, sugar bush, white sage, and Our Lord's candle. Shrubs recover slowly after fire, and ephemeral annuals may occupy spaces between shrubs in the first few seasons following a fire. Because the shrubs hold the soil on steep slopes, landslides and erosion are common following fires.

Ceanothus chaparral is successional to other communities in southern California, but it is common as a climax community in northern California. It grows on moister sites than chamise, rarely above 1,200 meters. Several species in ceanothus chaparral—including hoaryleaf ceanothus, hairy ceanothus, and blue blossom (also see Fig. 21)—may be codominant with manzanitas. Ceanothus chaparral may also form the understory for deciduous oak woodland or ponderosa pine forest. The crowns are not as dense as those of chamise and reach a height of 1–3 meters at maturity. Some species are short-lived and between fires may die out of a stand. Associated species are usually sprouters, including chamise, Nuttall's scrub oak, toyon, and sugar bush, and elements of coastal sage–scrub also occur in ceanothus chaparral, such as California buckwheat and California sagebrush (Hanes 1977).

Scrub oak chaparral occupies wetter north-facing slopes below 900 meters but is more

Fig. 20. Coastal chaparral, Contra Costa County.

common above 900 meters in elevation in southern California. In the north, it occurs above chamise chaparral where soil is sufficiently deep. The dominant species is Nuttall's scrub oak. The canopy is so short (2–4 meters tall) that the trees' lowest branches nearly touch the ground. Scrub oak chaparral has the greatest species richness of all the chaparral communities; woody vines and ferns are common on its deep litter. Scrub oak chaparral develops rapidly after fire because it usually occurs on more mesic sites and because many of the species sprout then (Hanes 1977).

Manzanita chaparral occurs on deeper soils and at higher elevations than chamise or ceanothus, thus it obtains most of its moisture from fog drip, freezing precipitation, and snow. It forms very dense stands from 1 to 2 meters tall. Only about half of the manzanita species are sprouters, so many of them must seed after fire. Manzanitas are long-lived, and several species may reach tree size (Hanes 1977).

Montane chaparral occurs at higher elevations in scattered thickets in the Cascades and Sierra Nevada. It is often seral or an understory to coniferous forest and rarely exceeds 2 meters in height; usually it is less than 1 meter tall. Associated shrub species include manzanitas, huckleberry oak, bush chinquapin, and many species of ceanothus.

Red shank chaparral occurs in only four locations in southern California and Baja California: the San Jacinto and Santa Rosa mountains, and in the interior valleys of Riverside and San Diego counties. It is dominated by red shank and grows between 600 and 1,800 meters in elevation on granitic soils. Individual shrubs are 2–3 meters tall and grow in relatively open sites (Hanes 1977).

Serpentine chaparral is associated with serpentine soils from San Luis Obispo County northward and is characterized by a dwarfed stature due to the poor soils on which it grows. Chamise and toyon are the dominant shrubs; whiteleaf manzanita and musk brush are endemic members of this community. Sargent cypress, interior silk-tassel, and leather oak are also characteristic of serpentine chaparral, which can be associated with foothill

Fig. 21. Blooming deer brush, a ceanothus species, Butte County.

woodlands or montane coniferous forests. The shrubs are 0.5 to 2 meters in height, compact, and close to the ground. Their leaves are often reduced, curled, or thickened (Hanes 1977; Fig. 22).

Desert chaparral is associated with desert scrub communities in the inner Coast Ranges and Transverse Ranges. It is more open than most chaparral communities and does not burn as often as these other forms. Woodland chaparral has two phases, one of large woody shrubs associated with tree communities and the other a live oak chaparral dominated by Nuttall's scrub oak, interior live oak, and canyon live oak.

Fig. 22. Serpentine chaparral, Lake County.

© P. A. Opler

Fire has shaped nearly all of California's vegetation communities. From the closed-cone pines and cypresses on the northwest coast to the chaparral-covered hills above Los Angeles, fire has been a constant in the seasonally dry climate. Before European settlement, most fires were probably started by lightning, though Native Americans also deliberately used fire to create both habitat for animals and travel routes, and to select for useful plants (Martin and Sapsis 1995). In presettlement times, fire burned 5.5%–13% of California's total area each year (Martin and Sapsis 1995).

Fires that occurred during the settlement period (1800's) tended to be large and destructive in terms of life and lost resources because they were indiscriminately set and could not be effectively suppressed. These fires helped convince followers of the early conservationist movement in the late 1800's that fire was the enemy; thus, most forestry practices in the early

1900's excluded fire (Martin and Sapsis 1995). From about 1910 until the 1960's, fire in California was aggressively suppressed with increasing effectiveness. But although humans were able to suppress the less intense fires, we have not reduced the number or magnitude of large and disastrous wildland fires; if anything, these have increased in frequency since fire suppression began (Martin and Sapsis 1995).

The state of California now has the best wildland fire suppression capability in the world, with 20,000 firefighters and support staff available within 72 hours, aircraft, and an incident command system (Irwin 1995). But fires still occur, and though only 3% of fires do excessive damage, all are costly. The Tunnel fire near Oakland in October 1991 killed 25 people, injured 150, destroyed 3,810 dwellings, and burned 1,500 acres at an estimated cost of $1 billion (Sapsis et al. 1995). In recent decades, more people have moved out of California's large cities and into the urban fringes, where human housing is mixed with native vegetation that is adapted to periodic fire. Often such communities are not designed with the likelihood of fire in mind, exacerbating the problem (Irwin 1995). Thus, fires in this wildland–urban interface are difficult to suppress and tend to be costly.

Since the 1960's, fire itself has been used as a tool to control fuel buildups and to mitigate the effects of fire. Prescribed fire techniques were pioneered in California (Green 1981), particularly in chaparral. Prescribed fires are deliberately set by land management professionals under a predetermined set of weather and fuel conditions that are predicted as closely as possible so that the fire can be controlled to produce the desired effect. The usual goal is a slow-burning surface fire, which reduces fuels within a prearranged area. Because the dead fuels carry the fire, chaparral younger than 25 years old is difficult to burn. Topography, temperature, relative humidity, condition of the fuels, timing, and especially wind are key factors in the success of a prescribed burn. Most of these factors can be predicted and many are related to one another; topography helps determine wind direction, and temperature and relative humidity help determine the flammability of fuels (Green 1981).

Prescribed fire is not used as widely as it could be, in part because of concerns for air quality (Hurley 1995). Though a prescribed fire generally produces less smoke than a wildfire, air quality restrictions and complaints from the public have curtailed some prescribed burning in wildland–urban interface areas. Also, prescribed burning is inherently risky; though it is possible to predict many conditions of fuels and weather, it is always possible, however unlikely, that a prescribed fire could escape control or at least have different effects than those expected. Because suppression is generally funded first and is extremely expensive, fuel mitigation is often underfunded (Irwin 1995). Prescribed fire, though, is important in order to restore California ecosystems to a healthier, more natural state and to protect property in the wildland–urban interface.

Channel Islands

The California Channel Islands are a group of eight islands lying from 19 to 97 kilometers off the southern California coast, in the nearshore section of the Pacific Ocean known as the southern California bight. These islands are generally segregated into the four adjacent northern islands of Anacapa, Santa Cruz, Santa Rosa, and San Miguel, and the southern four, more widely scattered islands of Santa Barbara, San Nicolas, Santa Catalina, and San Clemente. Four additional islands range south to Punta Eugenia, Baja California, and are known as the Mexican Channel Islands. The islands are owned and managed by a variety of federal, Mexican, state, and private agencies.

Fire and Fuel in a Sierra Nevada Ecosystem

Early travelers and photographers in the mid-1800's recorded the forests of the Sierra Nevada as parklike, with little undergrowth and wide expanses of meadows. The forests were a mixture of conifers dominated by ponderosa pine, with some incense-cedar and California black oak at lower elevations and increasing numbers of sugar pine and white fir at the higher elevations. Beneath the larger trees, the forest floor was carpeted with needles, forbs, and grasses. The understory, where present, consisted of young trees and some chaparral shrubs. These open conditions were attributed to low-intensity surface fires set by lightning and augmented by Native Americans.

Lightning fires have unique spatial and temporal distribution patterns in relation to topography and vegetation. The ecological role of fire is a manifestation of those patterns. The simultaneous occurrence of a lightning strike, flammable fuel, and conducive weather determines the frequency, size, and intensity of a fire. The prevalence of lightning strikes and fires shows conclusively that fire is an integral and pervasive part of Sierra Nevada ecosystems rather than an external disturbance (van Wagtendonk 1994).

Nearly a century of fire control in the Sierra Nevada has led to conditions that now threaten the very forests they were designed to protect. Suppression of naturally occurring surface fires has allowed the forest floor to become a tangle of understory vegetation and accumulated debris. Open forests and meadows have been invaded by trees and chaparral. Thickets of shade-tolerant incense-cedar and white fir have increased and have deflected succession away from the less shade-tolerant ponderosa and sugar pines.

As undergrowth has increased, fuel volumes have expanded, and a continuous ladder of fuel extends from the ground to the forest canopy. The understory is now so thick with dead trees, branches, needles, and other debris that inevitable wildfires will soon reach catastrophic proportions. Today, millions of hectares of forest and grasslands in the Sierra Nevada face abnormally high risks of wildfire because of these altered conditions. This situation is worsened by the increasing numbers of vacation homes and other human developments within the forest and foothill vegetation.

The cycle can be illustrated by examining simulated fuel accumulation rates during periods before and after the initiation of fire suppression, by using graphs generated by the FYRCYCL model based on data collected in Yosemite National Park (van Wagtendonk 1985; Figure). As long as fires are suppressed, this new, longer cycle of extremely intense wildfires will continue.

If natural conditions and processes are to be restored and perpetuated in the Sierra Nevada, fire must be reintroduced. In large wilderness areas and parks, naturally occurring lightning fires should be allowed to burn under prescribed conditions. Where this is not possible because the area is too small or because other human factors (such as the presence of human dwellings, timber harvest areas, and so forth) preclude the implementation of a program to monitor wildland fires, surrogates for fire must be found. Prescribed burns, mechanical manipulation, and artificial cutting are possible options. In any case, it is important that naturally managed ecosystems not be denied ecologically significant processes such as fire.

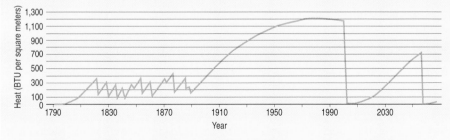

Figure. Simulated fire cycle from Yosemite National Park as generated by the model FYRCYCL. Starting in 1790 with open ground, as might be expected following a stand-replacing fire, a forest began to grow and fuels accumulated. By 1820 enough fuel had accumulated to carry a fire, and lightning strikes occurred when weather conditions were conducive to burning. The resulting fire reduced fuels. According to the model, during the next 70 years, lightning-caused surface fires occurred in a cycle averaging one every 7 years. In 1890 a policy of total fire suppression was implemented, and all subsequent fires were extinguished. Without the frequent surface fires, fuels will have accumulated to such a volume by the years 2000 and 2055 that when fires do occur, they will be intense crown fires that exceed suppression capability. The amount of fuel on the ground is measured in units of heat per square meter.

See end of chapter for references

Author

Jan W. van Wagtendonk
U.S. Geological Survey
Biological Resources Division
Western Ecological Research Center
Yosemite Field Station
El Portal, California 95318

The eight U.S. California Channel Islands range in size from 249 square kilometers (Santa Cruz Island) to 2.6 square kilometers (Santa Barbara Island). In addition to Santa Cruz Island, three other islands exceed 100 square kilometers in size—Santa Rosa (217 square kilometers), Santa Catalina (194 square kilometers), and San Clemente (145 square kilometers; Philbrick 1967). Because the islands were never connected to the mainland in their geologic history, they are rich in endemic plants and animals. Furthermore, the islands are near an area of cold-water upwelling where the Pacific Current meets the warm waters of the bight. As a consequence, they have an extremely rich marine flora and fauna.

The four northern islands are most likely a geological extension of the Santa Monica Mountains, separated from the mainland by a deep ocean trench. During the Pliocene they were joined into one large island called Anacapia. The four southern islands, also known as Catalinia (Weaver and Doerner 1967), are more scattered and include islands farther from the mainland. They seem geologically related to the Palos Verdes Peninsula, which was itself an island at various times in the Cenozoic, at the edge of the Greater Los Angeles basin (Weaver and Doerner 1967). San Clemente Island is the largest island that is far from the mainland, and it is perhaps not surprising that it has the largest number of endemic plants—40 species, subspecies, and varieties (Raven 1967). All the Channel Islands have varied in size during sea-level shifts that have occurred since their formation (Valentine and Lipps 1967; Weaver and Doerner 1967). Today, their terrestrial fauna and flora most closely resemble those of adjacent southern California, although some elements are more related to today's biota of central coastal California (Raven 1967).

The Channel Islands are rich in endemic species, subspecies, and varieties of animals and plants. Most of the species did not evolve on the islands alone—they are relict populations of species that once occurred on the mainland during the Tertiary and Pleistocene but which became extinct there because of harsher climatic conditions or competition (Axelrod 1967; Powell 1985). Approximately half of the endemic species occur on more than one island. Genetic divergence of island populations continues, however, and there are distinct subspecies and varieties of many taxa endemic to single islands (Raven and Axelrod 1978).

The islands were used for ranching, farming, and military activities beginning in the mid-1800's. Grazing by sheep, cattle, goats, donkeys, and pigs has limited certain species to tiny inaccessible pockets on canyon walls and coastal bluffs (Raven 1963). The accompanying soil compaction and erosion have changed community dynamics, and many habitats are now dominated by nonindigenous Mediterranean annual grasses and herbs.

Conservation is now a major land-use priority on the Channel Islands. Anacapa, Santa Rosa, San Miguel, and Santa Barbara constitute the Channel Islands National Park, and Santa Cruz is owned by The Nature Conservancy. There are large areas in conservation reserves on San Nicolas and San Clemente islands, which are owned by the U.S. Navy, and also on Santa Catalina, a privately owned island managed largely by the Catalina Island Conservancy.

The Channel Islands are rich in endemic plants; Raven (1967) cites 76 endemic plant species, subspecies, and varieties. One tree genus in the rose family, Catalina ironwood (Fig. 23), is entirely endemic to the Channel Islands and is found on the four largest islands. More than 40 species of plants, and a number of additional subspecies and varieties, are endemic to the islands, including island live oak, wild lilacs, wild cherries, native sages, and several woody shrubs such as manzanitas.

Five endemic Channel Island plants are listed as endangered on the U.S. list of endangered and threatened plants and wildlife (U.S. Fish and Wildlife Service 1994a): San Clemente Island broom, San Clemente Island bush-mallow, San Clemente Island Indian paintbrush, San Clemente Island larkspur, and Santa Barbara Island liveforever. Three additional plants from the southern Channel Islands and 16 plants from the northern Channel Islands have been proposed as endangered (U.S. Fish and Wildlife Service 1995b,c). The proposed listings summarize the status and population trends of each plant and detail the detrimental effects on each caused by intentionally introduced grazing mammals and rooting pigs (U.S. Fish and Wildlife Service 1995b,c). Thus, about a third of the Channel Islands' endemic flora is listed as endangered or is likely to be listed soon.

Notable invertebrates of the Channel Islands include several genera of snails (see box on Channel Islands and California Desert Snail Fauna), including island snails, cactus snails, and a shelled slug, as well as the Avalon hairstreak, a species of butterfly found only on Santa Catalina Island. The Avalon hairstreak has one of the smallest ranges of any North American butterfly (Gall 1985). The status of Channel Islands insects has been the topic of a symposium (Menke and Miller 1985), and Miller (1985) lists the Channel Islands' endemic arthropods. Powell (1985) described the patterns of apparent endemism in the

Torrey Pine

Torrey pines occur naturally only in two limited coastal California sites and are the rarest species of North American pines. The first population was discovered on the coast at the mouth of the Soledad River near San Diego. This population, now within the Torrey Pine State Reserve, includes an estimated 4,000–5,000 trees, mostly larger, reproductive individuals, some of which were planted by concerned caretakers. In 1988, following a damaging wind storm and several seasons of drought, this mainland stand was severely attacked by bark beetles, and hundreds of trees were killed. The beetles were controlled by trapping using a chemical attractant, which probably saved the bulk of the mainland trees (Berson 1992).

The second population occurs on Santa Rosa Island (now a part of Channel Islands National Park) and was described (Brandegee 1888) as a stand of about 100 trees with plenty of vigorous young trees (Fig. 1). This stand was probably limited by the species' sensitivity to the occasional fires that burned through the island vegetation (S. Veirs, U.S. Geological Survey, Davis, California, unpublished fire history; Fig. 2). Sheep were introduced to the island by Europeans about 1840, and they controlled fires by consuming most of the island's vegetation. The sheep were removed early in the twentieth century and were replaced by cattle. In the continued absence of fire, the stand has increased to 4,200 individuals, including many seedlings and young trees (Veirs, unpublished data; Fig. 3).

Fig. 1. Torrey pines on Santa Rosa Island.

Fig. 2. Scars from surface fires at the base of a Torrey pine on Santa Rosa Island. The tree is about 275 years old and its scars record fires that occurred in 1756, 1773, 1803, 1814, 1837, and 1840; there have been no surface fires recorded in the last 150 years, since the introduction of sheep to the island.

The native populations of Torrey pines are at considerable risk because of their small numbers and limited natural distribution. Natural influences, including drought, insects, and fire, pose serious hazards and

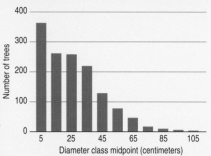

Fig. 3. Numbers of Torrey pines on Santa Rosa Island by size class; note that there are many more small trees than large ones (seedlings omitted.

probably account for the present limited distribution of this species. Torrey pines are relics of the Pleistocene flora of California; this relict flora has apparently been pushed to the brink of extinction by natural processes—probably climatic changes—over the past 10,000 years (Vogl et al. 1988). Careful monitoring, understanding, and management are essential to perpetuating the native populations of Torrey pines.

See end of chapter for references

Author

Stephen D. Veirs, Jr.
U.S. Geological Survey
Biological Resources Division
Western Ecological Research Center
University of California
Davis, California 95616

islands' butterfly and moth faunas and analyzed the species richness. Generally, an island's fauna is proportional to its size, except that islands closer to the mainland have a richer fauna than those more distant (Powell 1985).

The endemic vertebrate fauna includes the Channel Islands gray fox, a species protected by the state of California and which is found on the six largest islands, each with its own subspecies (von Bloecker 1967); an island subspecies of spotted skunk found on Santa Rosa and Santa Cruz; and the island night lizard, a genus and species endemic to San Clemente, San Nicholas, and Santa Barbara islands (Savage 1967). The Santa Cruz Island gopher snake, a candidate for federal protection, occurs on Santa Rosa and Santa Cruz; each of these islands also has a distinct subspecies of deer

Fig. 23. Channel Island scrub-jay on Catalina ironwood; both species are Channel Island endemics.

mouse. The San Clemente sage sparrow and the San Clemente loggerhead shrike are both on the U.S. list of threatened and endangered wildlife (U.S. Fish and Wildlife Service 1994b).

Six species of pinnipeds feed in the waters surrounding the islands: California sea lions, northern elephant seals, northern sea lions, harbor seals, northern fur seals, and Guadalupe fur seals. These species have breeding grounds and rookeries on the sand beaches of the islands, most notably on San Miguel, San Nicolas, and San Clemente. Populations of these species are recovering from the near extinctions at the turn of the century that resulted from hunting and from pesticide poisoning of their food base.

Some of the richest marine communities in the world are found in the waters of the Channel Islands. Intertidal areas are dominated by algae and invertebrates. Nearshore waters support forests of giant kelp, one of the fastest-growing algae in the world and home to more than 1,000 species of fishes, other algae, plants, and a wide variety of invertebrates. The outer waters are traversed by migrating gray whales, humpback whales, blue whales, and other whales, as well as dolphins and large fishes; all these species compete for traveling space with commercial freighters, private fishing vessels, and recreational boat traffic.

The marine communities of the Channel Islands have been heavily exploited for their fishes, shellfish, marine invertebrates, and kelp. These organisms are largely unprotected by federal and state laws, and serial depletion of one species after another has been a long-term trend evident in commercial harvest records since World War II (see box on California Abalone). Conservation efforts now focus on the establishment of marine reserves in and around the islands managed by the National Park Service.

Intertidal, Beach, and Dune Communities

Intertidal, beach, and dune communities may be divided into northern and southern biotic units at Point Conception, Santa Barbara County. Major habitat types are rocky intertidal, sandy beach intertidal, dune (including coastal dunes), and salt marsh.

The California outer coast is 1,326 kilometers long, with an additional 365–370 square kilometers of salt marshes and somewhat extensive areas of brackish marshes. Major California estuaries, from north to south, are the adjacent Humboldt and Arcata bays (Barnhart et al. 1992), Bodega Bay, Tomales Bay, Drakes Bay, San Francisco Bay (together with adjacent San Pablo and Suisun bays; Josselyn 1983), Morro Bay, and San Diego Bay (Macdonald 1977; Zedler 1982). San Francisco Bay accounts for about 90% of California's remaining salt marshes (Macdonald 1977).

The habitats and biota of the outer coast have been dealt with by researchers emphasizing the central coast (Ricketts et al. 1952), Bodega Head in Sonoma County (Barbour et al. 1973), southern California estuarine species (MacGinitie and MacGinitie 1949), and the ecology and systematics of marine invertebrates of the central California coast (Light et al. 1961). Seashore plants, including both marine and strand species of northern and southern California, respectively, are summarized by Dawson (1966a,b) and Barbour and Johnson (1977).

The status and trends of intertidal species of the outer coast's rocky and sandy shores are influenced by several natural and human-related factors. Some species, such as abalones (see box), clams, and sea urchins (see chapter on Marine Resources), are harvested for human consumption, whereas several species in the lower intertidal zones are preyed on by southern sea otters, an endangered species (Riedman and Estes 1990; Estes et al. 1995; see boxes on Sea Otters in the Pacific Northwest chapter and in this chapter). Episodic changes in sea temperatures and the related occurrence of coastal fogs triggered by El Niño events may cause periodic die-offs of intertidal species, but most recover after cooler sea temperatures return (see chapter on Marine Resources).

The trends of most species associated with estuarine habitats in California have been declining disastrously (Macdonald 1977; Harvey et al. 1992). There has been a tremendous loss of habitat due to the filling of much of California's estuarine habitats, especially in San Francisco Bay (Harvey et al. 1992; Monroe et al. 1992), in the vicinity of Los Angeles, and in San Diego Bay (Zedler 1982); estuarine wetlands have been almost completely replaced by landfills, marinas, docks, and other development (Zedler 1982; Monroe et al. 1992). As a result, several endangered species—including the salt marsh harvest mouse, California clapper rail, light-footed clapper rail, and salt marsh bird's-beak—are protected by the U.S. Fish and Wildlife Service (1994a; Fig. 24). The Belding's savannah sparrow is listed as endangered under California law. Coastal wetland-dependent species listed as sensitive or of special concern by the state or federal government include 1 plant, 7 insects, 2 reptiles, and 14 birds. Each of these species depends on different habitat types or combinations within salt-marsh ecosystems.

The Humboldt Bay estuarine ecosystem and its biota have been summarized by Barnhart et al. (1992), and the biotic resources of San Francisco Bay and southern California coastal marshes have been summarized by Josselyn (1983) and Zedler (1982), respectively.

Emerging Diseases in Southern Sea Otters

The southern sea otter is a large mustelid that spends its life in the nearshore marine community along the California coast. Prized for its fur, this subspecies was thought extirpated until a remnant population of approximately 50 animals was discovered near Big Sur in the early 1900's. The slowly recovering population was listed as threatened by the U.S. Fish and Wildlife Service in 1977. Although the sea otter's range has expanded to cover more than 320 kilometers of the central California coast and the population is now about 2,500 animals, the rate of recovery has been slower than biologists expected (Riedman and Estes 1990; Riedman et al. 1994). Concern that excess mortality was hindering recovery prompted the U.S. Fish and Wildlife Service to ask the U.S. Geological Survey's National Wildlife Health Center to perform an intensive necropsy survey of wild southern sea otters beginning in 1992.

The necropsies we performed from 1992 through 1995 yielded unexpected results (Figure). In particular, we found that the frequency and variety of infectious diseases were unusual for wildlife species. Forty percent of the sea otters examined died from parasitic, fungal, or bacterial infections. Traumatic injuries are generally common in wildlife, and injuries such as shark attack or shooting were also common (21%) in southern sea otters. Eleven percent of the otters were emaciated at death with no specific cause identified for this debility. In 10% of the sea otters, we diagnosed a variety of other problems, such as gastrointestinal or urinary tract obstructions or tumors. For 18%, we could not ascertain the cause of death.

Peritonitis induced by acanthocephalan parasites was the most frequent (15%) cause of death by infectious disease we identified in the otters. Peritonitis occurs when acanthocephalan parasites that inhabit the intestine migrate aberrantly through the intestinal

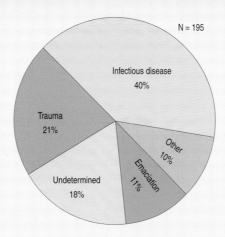

Figure. Causes of mortality in southern sea otters from 1992 through 1995.

wall, perforate the intestine, and allow bacteria to enter the abdominal cavity. Although this parasite (*Polymorphus* sp.) has been present in the sea otter population for many years (Hennessy and Morejohn 1977), the infection and mortality rates observed in the 1990's appear unprecedented. The invertebrate intermediate hosts that transmit this parasite are largely unconfirmed, and the roles of other hosts, such as birds, are unexplained.

Another parasitic disease, protozoan encephalitis, was newly identified in this survey, indicating that it may be an emerging disease. We are investigating the identity of the causative parasite; the ubiquitous organism *Toxoplasma gondii* was isolated from several otters, but the results are confounded by evidence that another protozoan may also be involved. Fatal toxoplasmosis is more common in animals and humans that have impaired or immature immune systems than it is in healthy organisms.

Before our study, coccidioidomycosis, or San Joaquin Valley fever, had been described in a sea otter only once, in 1976

(Cornell et al. 1979), but one or more cases were diagnosed during each year of our survey (Thomas et al. 1996). This fungal disease affects humans and animals in the deserts of the lower Sonoran life zone of the southwestern United States, so its prevalence in the marine environment is puzzling. The fungus *Coccidioides immitis* thrives in arid and semiarid soil. Inhalation of the airborne fungal spores may produce respiratory disease, or the spores may disseminate to many organs in susceptible individuals. The cases of San Joaquin Valley Fever in sea otters coincide with a human epidemic of the fever that began in California in the fall of 1991. The human epidemic was tentatively attributed to unusual weather and environmental conditions rather than to human-related factors, a hypothesis supported by the coincident occurrence in sea otters.

Researchers continue to monitor southern sea otter abundance and distribution to assess the population's progress toward recovery goals. The emergence of several diseases as important causes of death raises concern about the otters' immune status and resistance. By elucidating factors in the individual disease cycles we are trying to determine both an explanation for the emergence of these diseases and a means of controlling them. The challenge remains to identify not only the overt but also the underlying factors that may have more far-reaching effects in the marine environment.

See end of chapter for references

Authors

Nancy J. Thomas
Lynn H. Creekmore
Rebecca A. Cole
Carol U. Meteyer
National Wildlife Health Center
6006 Schroeder Road
Madison, Wisconsin 53711

Beaches and dunes, which are separate but usually connected sandy coastal habitats (Barbour and Johnson 1977), are limited to 305 kilometers of California's coastline, whereas the more limited occurrence of major dunes along the California coast is described by Cooper (1967), who does not include many smaller dunes in his discussion. The Antioch Dunes National Wildlife Refuge is a unique dune system found along the tidal Sacramento–San Joaquin River (Cooper 1967).

Beaches, as defined by Barbour and Johnson (1977), are limited to the salt spray-saturated zone below the highest high-tide line, typically seaward of the foredune crest. Beaches have a very limited flora of salt-tolerant plants that includes European beachgrass, maritime sea-rocket, silverweed cinquefoil, coastal sand verbena, and iceplant as dominants, and pink sand verbena and beach saltbush as subdominants (Barbour and Johnson 1977).

California Abalone

The recent demise of abalone in southern California evokes memories of the fate of bison in the United States in the nineteenth century. Once so abundant their disappearance was unimaginable, the abalone that supported huge commercial and sport fisheries 30 years ago are now on the brink of extinction.

Abalone are marine snails, with some 70 modern species occurring globally. Fossil abalone first appear in Cretaceous rocks dating from about 70 million years ago, in what is now California (Lindberg 1992). Humans have exploited abalone for food, tools, and jewelry for millennia (Shepherd et al. 1992). Five of the eight eastern Pacific abalones were abundant enough to support multimillion-dollar fisheries through most of the twentieth century. However, abalone populations in southern California recently collapsed under a flawed management approach (Davis et al. 1992; Richards and Davis 1993; State of California 1995). This situation has caused worldwide concern because abalone management practices in Australia, New Zealand, South Africa, and elsewhere were based on the same assumptions and strategies as in California (Shepherd et al. 1992). In the mid-1990's, California fisheries for four abalone species were closed to prevent harvest-induced extinctions, but even that drastic action may have been too late to save the white abalone (Davis et al. 1996). The story of these populations, the fisheries they supported, and efforts to restore them is a harbinger for coastal marine fisheries. The lessons learned in efforts to restore and sustain California abalone can be applied profitably to coastal resources worldwide.

Biology

Abalone cling to rocks, from wave-swept intertidal ledges down into the twilight zone of deep reefs at 65 meters, wherever they can catch drifting fronds of kelp and other algae. In California, species separate themselves roughly by depth and latitude (Haaker et al. 1986). Black abalone live in tidal pools from Oregon to the southern tip of Baja California. Green abalone, pink abalone, and white abalone prefer southern climes, with each species occupying increasingly deeper waters, respectively, from Point Conception into Baja California. Red abalone, the largest species, occupies the broadest range, from tidal pools in Oregon to deep reefs as far south as Bahia Tortugas, Baja California. Flat abalone, pinto abalone, and threaded abalone are

relatively rare and are only incidentally involved in California fisheries.

Abalone have separate sexes. To reproduce, they broadcast sperm and eggs into the sea, relying on high gamete densities for successful fertilization, a reproductive strategy requiring densely aggregated adults for success. The larvae are free-swimming for only a few days before settling to the bottom as juveniles. California abalones mature between 3 and 7 years of age and may live for 35 to 54 years, commonly reaching sizes of 15–25 centimeters in length and 1–2 kilograms. Fecundity increases exponentially with size. Newly mature females produce only a few hundred thousand eggs each year, whereas older individuals produce 10–15 million eggs (Hahn 1989).

Management Strategies

California's management strategy to sustain exploited abalone populations was based on a surplus yield model. It was implemented through minimum harvest sizes, based on growth rates and size at maturity. Under this scheme, abalone were permitted to reproduce for the first few years of maturity and then were harvested. Since no relationship between spawning stock and recruitment had been defined, fishery managers assumed cohorts of young abalone could sustain harvests with no other constraints (Tegner et al. 1989). Closed seasons protected spawning aggregations. Other regulations, such as bag limits on sport fishers and limited entry to the commercial fishery, attempted to allocate limited resources equitably.

In northern California, an additional management measure protects more brood stock of red abalone, thus assuring a sustained fishery. Only sport breath-hold diving or shore-picking in the intertidal zone is allowed. Zoning use in this way separates commercial and sport fishing and protects a large spawning stock of big abalone. The inherent depth limit imposed by breath-hold diving creates a refuge at greater depths, thereby protecting sufficient brood stock to replenish the harvest in adjacent shallow waters. This refugia-based red abalone fishery is the only sustained abalone fishery in California today.

Serial Depletion

Truly sustainable fisheries are based on sustained populations of target species. The frontier approach to fishery management

practiced in many areas today produces a pseudo-sustained fishery. This approach unrealistically assumes the availability of an endless supply of new, unexploited populations, but it has the virtue of economic expediency.

Following a fishing hiatus during World War II, southern California abalone fisheries grew rapidly. Soon, readily available and well-known abalone populations along mainland shorelines were exhausted. Then lightweight, mobile, inexpensive diving gear, fast boats, modern navigational aids, and improved knowledge of abalone and the coastal environment made available virtually all of the pristine abalone habitat on offshore reefs. After 25 years of apparently sustained fisheries, abalone landings began declining in the 1970's. A careful examination of this harvest shows it was not truly sustained but rather the result of serial depletion (Dugan and Davis 1993; State of California 1995). Fishery landings and fleet income were sustained at the expense of a series of abalone populations in different areas. Only after landings of pink abalone and red abalone declined in the 1970's did the fishery shift to green and white abalone in shallow and deep water, respectively. When white and green abalone populations collapsed, harvest shifted to intertidal black abalone. Then, as the remnants of the black abalone population succumbed to disease, the diving fishery shifted to red sea urchins. Income to the diving fleet during this period remained relatively stable. The fishery had been sustained, but the productivity of the exploited populations was destroyed in the process.

The success of serial depletion in sustaining fishery income obscured the need to restore severely depleted stocks and to protect more reproductive capacity of abalone populations. Denial that abalone populations were imperiled obstructed efforts to improve management. The virtual absence of fishery-independent information made it difficult to assess population status and gave fishery landings data more credibility than they warranted. These gaps in accessible information delayed remedial actions, making restoration more costly and perhaps impossible for some species.

Current Population Status

Few fishery-independent data exist for abalone in California. Since 1982 the National Park Service and the California Department of Fish and Game have jointly monitored population dynamics of nearly

100 marine taxa, including the five common abalones, in Channel Islands National Park (Davis et al. 1994). Other population trends must be inferred from fishery landings, with considerable uncertainty and ambiguity.

Black abalone populations in California survived with harvests of 2,000 metric tons per year for the last third of the nineteenth century. Large populations accumulated after harvest ceased in 1900, with densities often exceeding 125 abalone per square meter on the Channel Islands (Richards and Davis 1993; Figure). When harvest resumed in 1968, annual landings quickly rose to 870 metric tons. By the mid-1980's black abalone were found primarily on offshore islands and inaccessible sections of the coast north of Santa Barbara. A withering foot disease caused massive deaths in these remnant populations, beginning in 1985 on the Channel Islands and spreading to the mainland (Haaker et al. 1992). Relict populations of apparently disease-resistant individuals survive on the islands at less than 1% of their former abundance.

White abalone occurred at average densities of 10,000 per hectare in the early 1970's (Tutschulte 1976). Ten years later, densities at the historical center of their abundance were 10 per hectare. By the early 1990's densities were only 1 per hectare, and the species was in danger of extinction (Davis et al. 1996).

In Channel Islands National Park, exploited pink abalone population densities fell from 250 per hectare to less than 14 per hectare in the 1980's, while a population protected in an ecological reserve remained relatively stable at about 400 per hectare (Davis et al. 1992). Red abalone population densities in the park dropped from more

than 1,000 per hectare to less than 10 per hectare in the 1980's. Fishery landings reflected these population trends. Pink abalone landings fell from 48 metric tons to 7 metric tons, and red abalone landings dropped from 235 metric tons in 1980 to 120 metric tons in 1986. There are no recent fishery-independent data for green abalone populations.

Fishery Status

California abalone fisheries landed more than 4,000 metric tons per year during the 1950's and 1960's, split roughly equally among sport and commercial interests (State of California 1995). After harvest started in 1968, black abalone landings peaked quickly at 870 metric tons in 1973 and stabilized at 225 metric tons in the early 1980's. Then, reflecting the rapid population collapse, landings dropped to less than one metric ton in 1993, when the fishery was closed to protect disease-resistant stock for use in restoration efforts. Landings of black, white, pink and green abalones fell in the 1990's to less than 4% of the early 1970's landings, from 882 metric tons to 32 metric tons. The California Fish and Game Commission closed the pink, green, and white abalone fisheries in 1996 to prevent extinction of reproductive stocks (State of California 1995).

For more than 20 years, sport and commercial abalone fisheries generated $15–$20 million per year for the state's economy. The abalone population trends of the 1980's, caused largely by a flawed harvest scheme, must be reversed if the productivity of these fisheries is to be restored.

Only red abalone still support fisheries in California. The sport-only red abalone fishery in northern California annually produces about 1,000 metric tons (Tegner et al. 1992). Although generally stable, continued productivity of that fishery is now threatened by poaching, induced by the extremely high value of individual abalone ($32 per pink abalone to $100 per white abalone). Abalone fisheries in southern California are now concentrated on a small population of red abalone around San Miguel Island, in western Channel Islands National Park and National Marine Sanctuary. This population is located at the edge of the southern-flowing California current, and recruitment to this population may be provided from unharvested populations to the north. Annual commercial red abalone landings have apparently stabilized at about 170 metric tons, less than 15% of the 1,200 ton landings of 30 years ago.

Restoration Plans

Restoring populations of slow-growing, long-lived abalones to levels that can sustain productive fisheries will take decades and will require active intervention. Closing the Orange County shoreline to abalone harvest in 1977 and waiting 15 years for populations to recover spontaneously was ineffective (Tegner 1992). Abalone are not unusual in this respect. Recent analysis of 128 marine fish stocks revealed that only 3 species might be able to recover spontaneously from severe harvest-induced reductions (Meyers et al. 1995). Active brood-stock husbandry now seems to offer the only promising abalone restoration approach (Tegner 1992, 1993).

Formal comprehensive plans have not yet been made to restore the productivity of California abalone populations. Limited research on recruitment dynamics, larval and juvenile stocking feasibility, and brood-stock husbandry are under way. White abalone stocks are so low that extremely expensive, large-scale surveys of deep reefs are needed just to find enough individuals for captive breeding and rearing programs and to test new strategies, such as aggregating adults in refugia (Davis and Haaker 1995).

See end of chapter for references

Author

Gary E. Davis
National Park Service
Channel Islands National Park
1901 Spinnaker Drive
Ventura, California 93001-4354

Figure. Intertidal black abalone on Santa Rosa Island, California, before the catastrophic population collapse in the late 1980's.

Courtesy G. E. Davis, USGS

© P. A. Opler

Fig. 24. A tidal salt marsh in southern San Francisco Bay provides habitat for the endangered California clapper rail.

Human foot traffic has eliminated the strip of vegetation (primarily the nonindigenous American sea-rocket and silver bur ragweed) found just above the highest high-tide line from most California beaches. As a result, the globose dune beetle, a restricted denizen of that narrow habitat, has been eliminated from most California beaches, although it may be common on only a few relatively undisturbed Channel Island beaches (Doyen 1976; J. Doyen, University of California [retired], El Cerrito, California, and P. A. Opler, U.S. Geological Survey, Fort Collins, Colorado, unpublished data).

Beach and dune habitats in California are negatively affected by human foot traffic and by off-road vehicles, but they are also severely affected by the presence of nonindigenous plants, many of which were planted intentionally for "beach stabilization." Additionally, development, primarily residential, has eliminated huge areas of formerly extensive sand dune systems; the San Francisco, Monterey, and El Segundo dunes have been almost entirely replaced by development. The native animals and plants of other extensive dune systems, such as the dunes at Point Reyes National Seashore (Fig. 25), have been almost completely replaced by introduced European beachgrass

and iceplant (Slobodchikoff and Doyen 1977; Doyen, personal communication). At the Oso Flaco dunes south of San Luis Obispo, almost all life forms have disappeared under the onslaught of off-road vehicles (Powell 1981).

Above the high-tide line are dunes and deflation plains that form the habitat of several nesting birds, numerous native plants, and many insects (Powell 1981; Doyen, personal communication) and other invertebrates. Many of these species are restricted to such beach dune habitats. The endangered California least tern and the threatened western snowy plover (Page and Stenzel 1981; Page et al. 1991; Page and Gill 1994) are two federally listed birds that nest in this habitat (U.S. Fish and Wildlife Service 1994a; also see box).

Status of Species

Plants

The pioneering botanical work of Willis Linn Jepson (Jepson 1925) was the first comprehensive treatment of the higher plants of California. The revised Jepson flora of California plants (Hickman 1993) includes 5,800 species in 173 families and 1,222 genera; about 24% are endemic. Much of the California flora is specifically adapted to the state's Mediterranean climate, and many of the dominants in the plant communities are endemic.

Completely new plant species may yet be identified in more remote and undeveloped parts of California. For example, in 1993 a new genus, Shasta snow-wreath, was discovered near Redding in northern California (Taylor 1994), and two plants that had been believed to be extinct were rediscovered (Corbin et al. 1993).

With development and exploitation of grasslands and other vegetation easily converted to human use, native vegetation has been disappearing. Indeed, most of the plant communities with the largest numbers of rare plant occurrences in California are those that are poorly represented in parks and preserves (Skinner and Pavlick 1994), largely because of their locations and their relatively high value in the 1800's and early 1900's. Today, of some 6,300 native vascular plant taxa (including subspecies and varieties), some 857 (13.6%) are considered rare or endangered in California and elsewhere, and 34 (0.05%) are considered extinct (Skinner and Pavlick 1994).

Invertebrates

California has a rich terrestrial and aquatic invertebrate fauna that features a high level of endemism and a number of species and groups

© P. A. Opler

Fig. 25. Coastal dunes, Point Reyes National Seashore.

with specialized life histories or behaviors. A conservative estimate of the state of California's insect fauna is 27,000 to 28,000 species (Powell and Hogue 1979). This does not include other arthropods such as mites, spiders, harvestmen, and crustaceans, nor does it include mollusks or other invertebrate phyla. The total count of invertebrates for California is probably in excess of 50,000 species. The exact distribution and population status are not known for most described species. At least several hundreds of undescribed invertebrates also inhabit California.

Unlike much of the eastern United States, where between-habitat diversity of invertebrates is relatively low, California's between-habitat diversity is high. Even though the number of insects found in a single habitat might be relatively low, after visiting several habitats one would soon appreciate that California has a vast array of invertebrates (Powell and Hogue 1979). For example, in the Sierra Nevada, one might find no more than 30 species of butterflies at a single locality on a very good day but would find more than 60 species at three or four sites at different elevations or in different plant communities, and more than 100 different species of butterflies at these localities over the course of a year (Swengel 1995; Opler, unpublished data). In lowland and coastal areas, one would have to make visits almost year-round to find most of the moth species at a single locality (for example, Opler and Buckett 1971; Powell 1995).

Land Snails

California's native land snail fauna is large and diverse, reflecting the geological and ecological diversity of the state. The state's known fauna is composed of about 200 species, and new species continue to be discovered and named (for example, see Roth and Hochberg 1988, 1992). Several genera are endemic or nearly endemic to California or have undergone their greatest diversification in the state.

Land snails are abundant in most terrestrial ecosystems, although they are often inconspicuous because of secretive habitats, seasonal inactivity, or—for some species—their very small size. Most feed on fungi, decaying plant or animal material, or green plants, but a few are predators of snails and other invertebrates. Land snails are fed on by a variety of amphibians, reptiles, birds, and mammals. California does have nonindigenous garden pests, such as the brown gardensnail and several nonindigenous slug species, but the native fauna is essentially harmless.

California includes portions of four molluscan faunal provinces (Pilsbry 1948): the Oregonian province includes humid coastal areas of northwestern California through Oregon and Washington to Alaska; the Southwestern province as defined by Bequaert and Miller (1973) includes the Mojave and Colorado deserts of southeastern California and extends to Arizona, New Mexico, and adjacent parts of Mexico; the Rocky Mountain province is along the eastern border of California north of the Mojave Desert but contributes relatively little to the faunal diversity of the state; and the Californian province occupies the remainder of the state. These provinces are not precisely or consistently defined and serve here only to broadly describe the diversity and affinities of the molluscan fauna.

Because land snails are highly dependent on ambient moisture for their activities, their richness in California, where most regions are at least seasonally dry, may seem surprising. Land snails have adapted to these conditions by finding shelter and becoming dormant within deep plant litter, rockslides, or fissures in soil and rock, emerging only when rains supply sufficient moisture. In very arid areas, conditions for such shelter may be in short supply and widely scattered. These small, scattered refuges, together with the poor dispersal ability of snails, have promoted speciation in many groups and are partly responsible for the high species richness of California's snail fauna.

The shoulderband genus of snails, with 52 species recognized, constitutes perhaps the most remarkable and distinctly Californian genus of snails (Turgeon et al. 1988). This group extends into northern Baja California Norte and southwestern Oregon but is otherwise restricted to California. Shoulderbands are found in a range of habitats, including moist areas on the northwest coast and dry hills in the Mojave Desert.

Many populations of native land snails have been lost because of urban and agricultural development. The species most vulnerable to extinction are the highly localized endemics, but this localization can also simplify their protection. Many species occur on public lands where, if land managers are aware of their presence and habitat needs, protective measures can ensure their persistence (Fig. 26). In many instances, overall ecosystem management will enhance native land snail populations. For example, restoration of native plant communities in the Channel Islands National Park should also restore native land snails. Releases of nonindigenous predatory snails—such as the decollate snail—to control brown garden snails, are an additional threat to the native fauna.

Fig. 26. The Coachella desert snail (Deep Canyon, Riverside County, California).

Patterns of Insect Diversity

As elsewhere in the United States, the richest habitats for insects in California occur in areas with at least moderate topographic relief

Channel Islands and California Desert Snail Fauna

The land snail faunas of the Channel Islands and the California deserts include many helminthoglyptid species (desert snails, shoulderbands, island snails, and allies). These are relatively well known, and their classifications are well documented and easy to describe. These regions provide only two examples of the diversity of California's land snail fauna. Snails and slugs found elsewhere in California are equally deserving of attention and conservation.

Channel Islands Snails

Southern California's Channel Islands are sometimes separated into two groups. The northern group (including San Miguel, Santa Rosa, Santa Cruz, and Anacapa islands) has been considered a westward extension of the Santa Monica Mountains on the mainland, although existence of a Pleistocene or Holocene connection to the mainland is now doubted (Junger and Johnson 1980). The southern group (Santa Barbara, San Nicolas, Santa Catalina, and San Clemente) is generally more remote from the mainland than the northern group.

The San Miguel shoulderband is endemic to the northern group of islands. This species has its closest known affinities with two coastal species from north of Point Conception in Santa Barbara and San Luis Obispo counties: the surf shoulderband and the endangered Morro shoulderband (Roth 1973; U.S. Fish and Wildlife 1994). Even without an earlier land bridge connection to the mainland, the ancestors of these snails could have reached these islands through over-water dispersal. Many land snails of arid or semiarid areas have features (reviewed by Chambers 1991) that permit them successful island colonization by dispersal on floating rafts of woody debris.

The southern group of islands is occupied by a much more diverse fauna that includes island snails and cactus snails. The island snails are single-island endemics on these islands and on Guadalupe Island, Baja California Norte, except the San Nicolas island snail, which also occurs as fossils on San Clemente Island (Roth 1975a,b). The cactus snails (formerly included with island snails by Pilsbry 1939) are made up of four living species. The plain cactus snail and the wreathed cactus snail are extant, single-island endemics that occur on San Clemente Island. The Catalina cactus snail has also been reported by Pilsbry (1939) from the Palos Verdes Peninsula, Los Angeles County, which is the only report of the genus outside of the southern group of Channel Islands. The co-occurrence of species of these two groups on each of these islands is another notable feature of their land snail fauna.

The Santa Barbara shelled slug has been found living only on Santa Barbara Island, although fossil shells have been found on San Nicolas Island (Roth 1975a; Hochberg 1979). The only other occurrences of this small group are the Guadalupe shelled slug of Guadalupe Island, Baja California Norte, and forms of limited known occurrence on the mainland of Baja California Norte (Roth 1975b).

An extremely curious element of the snail fauna of these islands is the Catalina mountain snail, which is endemic to Santa Catalina Island. This population occurs several hundred miles away from the nearest living related populations in southeastern Arizona and Baja California Sur. Because the original collector (Henry Hemphill in 1905) was not known for his care in handling specimens, and because subsequent workers failed to locate the species on Santa Catalina Island, this seemingly anomalous record was doubted for many years. Hochberg et al. (1987) provide a fascinating account of their rediscovery and verification of the identity of the species, along with documentation of its likely status as a relict of the group's formerly wider distribution.

The high incidence of endemism of land snails in the Channel Islands has probably resulted from the low frequency of inter-island dispersal and the age of the islands. Formerly larger island areas during periods of lower sea levels (Roth 1975b) may have contributed to the high diversity seen today in the southern group of islands.

California Desert Snails

A diverse helminthoglyptid fauna has evolved in the Mojave and Colorado deserts of southeastern California (Pilsbry 1939; Bequaert and Miller 1973). Most distinctive are two desert snail groups limited to these California deserts and extending only to western Arizona and northern Mexico. Species of these groups mainly occupy rockslides in the bordering mountain ranges and the smaller, isolated ranges that are scattered throughout these deserts. Some of these ranges support endemic desert snails.

Two additional groups endemic to the northern Mojave Desert are each represented by a single species. The El Paso shoulderband is endemic to the El Paso Mountains, and the Argus desert snail is endemic to the Argus and Slate mountains (Pilsbry 1939).

The land snails of this desert fauna have probably evolved from ancestors that were more widespread during prolonged periods of moister climate (Wells and Berger 1967). They have probably survived in this most inhospitable (for snails) dry climate by remaining inactive deep within rockslides between rains. Isolated because of poor dispersal capabilities and lack of refuge from desiccation between rockslide areas, they have differentiated into the relict fauna observed today.

The shoulderband group is also represented by several localized species in the Mojave Desert. Endemic species occur in the El Paso (mimic shoulderband) and Panamint mountains (Panamint shoulderband). Two additional species occur near the Mojave River, which receives flow from snowmelt from the San Bernardino Mountains. Other desert shoulderbands occur at the edges of the deserts, such as near the San Gabriel (Soledad shoulderband) and Tehachapi mountains (Mojave shoulderband).

See end of chapter for references

Author

Steven M. Chambers
Division of Endangered Species
U.S. Fish and Wildlife Service
P.O. Box 1306
Albuquerque, New Mexico 87103

and with the richest array of native woody vines, trees, and shrubs. Such localities are often found along or adjacent to streambeds or valleys at low to moderate elevations in mountainous areas.

Notable invertebrate habitats in California that are either unique or limited elsewhere in the United States include coast redwood forests, serpentine grasslands, coastal and riverine dunes, chaparral, evergreen oak woodland, and coastal sage–scrub. The uniqueness and local nature of many of California's insects have resulted in either the extinction or endangerment of many species and subspecies. Fifteen of 27 U.S. insects listed as endangered or threatened under the U.S. Endangered Species Act are from California (U.S. Fish and Wildlife Service 1994b; Table 2).

Table 2. California invertebrates listed as endangered or threatened on the U.S. list of endangered and threatened wildlife (U.S. Fish and Wildlife Service 1994a).

Species	Status
Mollusks	
Morro shoulderband snail	Endangered
Insects	
Ash Meadows naucorid	Threatened
Bay checkerspot butterfly	Threatened
Delhi Sands flower-loving fly	Endangered
Delta green ground beetle	Threatened
El Segundo blue butterfly	Endangered
Kern primrose sphinx moth	Threatened
Lange's metalmark butterfly	Endangered
Lotis blue butterfly	Endangered
Mission blue butterfly	Endangered
Myrtle's silverspot butterfly	Endangered
Oregon silverspot butterfly	Threatened
Palos Verdes blue butterfly	Endangered
San Bruno elfin butterfly	Endangered
Smith's blue butterfly	Endangered
Valley elderberry longhorn beetle	Threatened
Crustaceans	
Shasta crayfish	Endangered
California freshwater shrimp	Endangered

Most listed California insects are subspecies of butterflies that occur in extremely localized habitats within 80 kilometers of the coast (Arnold 1983; Opler 1991; New 1993). The population structure and detailed status of six listed California butterflies were described by Arnold (1983), and the population size of the Lange's metalmark butterfly at the Antioch Dunes National Wildlife Refuge 80 kilometers east of San Francisco has been estimated during most years for more than 12 years. Management actions taken to improve the number and density of butterfly host plants at the refuge have been rewarded by fairly dramatic increases in the population size of the Lange's metalmark (Opler and Robinson 1986).

One of the longest continuously monitored populations of any animal is that of the Bay checkerspot butterfly, another listed species, on Jasper Ridge above the campus of Stanford University in central California. More than 120 insects and many other westside California invertebrates are species of special concern that are being monitored by The Nature Conservancy on behalf of the U.S. Fish and Wildlife Service (U.S. Fish and Wildlife Service 1994b).

There have been no specific studies of invertebrates of old-growth forests in westside California, but a literature survey, review of museum material, and interviews with expert entomologists indicate that many arthropods are found primarily in north coast Douglas-fir and coast redwood old-growth forests (U.S. Fish and Wildlife Service 1995a; Opler, U.S. Geological Survey, Fort Collins, Colorado, and J. Lattin, Oregon State University, Corvallis, unpublished manuscript).

Some California habitats dominated by nonindigenous plants are virtually devoid of native insects, and presumably of other native invertebrates as well. Examples are urban environments dominated by plantings of nonindigenous ornamental trees, shrubs, and grasses; Coast Range, Sierra Nevada, or Transverse Range foothills dominated by Mediterranean grasses such as wild oats; or coastal strand and dunes taken over by European beachgrass (Slobodchikoff and Doyen 1977).

Fishes

The fish fauna of westside California is composed of 100 species (Moyle 1976a). The losses of fishes and fisheries represent national and global declines as well, because 31 of the 52 (60%) native westside California fish species are endemic (Table 3), and most of the remainder are confined to the Pacific coast. A report on the status of the California fish fauna (Moyle et al. 1995:1) addresses this regrettable situation:

> In the event these fishes are lost from California, they will be globally extinct; there are no populations in some distant or remote location that can be used to resurrect the local populations. These fishes represent millions of years of evolutionary response to the fluctuating and often harsh aquatic environments of the state. As a result, there is an extraordinary diversity of form and function among the native fishes. They are found in habitats ranging from tiny desert springs, to rivers that have huge fluctuations in flow, to high mountain streams, to shallow alkaline lakes, to salty estuaries. Although the native fishes are admirably suited for surviving the vagaries of nature, they have done poorly when forced to compete with humans for the waters which are their homes.

Table 3. California fishes of conservation concern, including endemic species (modified from Moyle 1976b; U.S. Fish and Wildlife Service 1994a,b). Status codes are C = full species endemic to California, to California and adjacent state, or to Baja California; E = presumed extinct in California; FE = federally endangered; and FT = federally threatened.

Species	Status
Arroyo chub	C
Blue chub	C
California killifish	C
California roach	C
Clear Lake splittail	C, E
Delta smelt	C, FE
Green sturgeon	C
Hardhead	C
Hitch	C
Kern brook lamprey	C, E
Klamath largescale sucker	C, E
Klamath River lamprey	C
Klamath smallscale sucker	C
Little Kern golden trout[a]	FT
Longjaw mudsucker	C
Lost River sucker	C, E
Marbled sculpin	C
Modoc sucker	C, E
Pit sculpin	C
Pitt–Klamath brook lamprey	C
Riffle sculpin	C
Rough sculpin	C
Sacramento blackfish	C
Sacramento perch	C
Sacramento splittail	C
Sacramento squawfish	C
Sacramento sucker	C
Santa Ana sucker	C
Shortnose sucker	C, E
Thicktail chub	C, E
Tidewater goby	C, FE
Tule perch	C
Unarmored three-spined stickleback[a]	FE

[a] Subspecies.

Although water diversion and use have seriously affected waterfowl, the native freshwater fishes and fisheries in California have been more seriously affected. Unique local populations and species are being extirpated or becoming extinct on a regular basis, and major fisheries have greatly declined (Moyle 1976a; Moyle and Williams 1990). The disappearance of native fishes and fisheries reflects the degree to which California's streams have been dammed and diverted, the poor condition of many watersheds that support the streams, and the detrimental effects of the many nonindigenous fishes and invertebrates that have been introduced into California's waters.

At present, 9 of the 116 fish taxa (species and subspecies) are extinct in California, 16 are formally listed as threatened or endangered, 25 probably qualify for listing as threatened or endangered, and 27 others require special management to prevent further declines that might put them in jeopardy (Moyle and Yoshiyama 1994; Moyle et al. 1995). In short, 67 (57%) of the taxa are either extinct or on the road to extinction if present trends continue. Not surprisingly, valuable fisheries for native fishes are also in decline, as salmon, steelhead, and other fish populations dwindle. Many of these fisheries are increasingly dependent on hatchery production, but the added fish production has not halted the declines, it has only slowed them down and may make recovery more difficult. The problems with fishes and fisheries can be seen by examining the status of anadromous fishes, Sacramento–San Joaquin estuarine fishes, Sierra Nevada fishes, and southern California fishes.

Anadromous Fishes

California supports the southernmost populations of 13 species of native anadromous (sea-run) fishes, plus 2 nonindigenous species (striped bass and American shad). All these species are in decline (Moyle 1994), which has resulted in major economic losses related to fisheries. The most visible declines are those of coho salmon, chinook salmon, and steelhead. Coho salmon were once abundant spawners in most coastal streams north of Monterey Bay; they have suffered a 90% decline in abundance over the past 50 years, and their populations have disappeared from about half the streams in which they once spawned (Brown et al. 1994). The principal causes of coho salmon decline seem to be loss of spawning habitat and juvenile rearing habitat in streams as the result of logging, road building, urbanization, and other factors.

Chinook salmon are the most abundant salmon in California and show remarkable adaptations to local conditions for spawning. In rivers of the Central Valley, for example, four distinct runs exist, maximizing the ability of the salmon to take advantage of unique conditions. In the nineteenth century, the annual runs of these fishes were 2–3 million per year in Central Valley rivers alone. Today, the Central Valley chinook runs are around 130,000 fishes per year (and declining), about 90% of which are fall-run chinook, a run supported mainly by fish hatcheries (Fisher 1994). The winter run is formally listed as endangered, and the spring and late fall runs qualify as such (Moyle et al. 1995). The single biggest cause of chinook salmon declines is that dams have cut off access to most of their historical spawning grounds. Their continuing decline in recent years is due to many factors, including habitat degradation and diversion of water from the Sacramento–San Joaquin estuary.

These same factors have contributed to the continuing decline of steelhead in the Central Valley, which now number only about 35,000 fishes annually, 90% of them of hatchery origin (California Department of Fish and Game 1990; Mills et al. 1996). In coastal streams, steelhead are still widely distributed as far south as Malibu Creek in Los Angeles County, but their numbers are greatly reduced, especially in southern California, where the populations are genetically distinct from more northern populations (Moyle et al. 1995). Steelhead co-occur in many streams with coho salmon and have declined for similar reasons. A number of Pacific salmon are on the threatened and endangered species list (see chapter on Marine Resources for current listings).

Sacramento–San Joaquin Estuary Fishes

The Sacramento–San Joaquin estuary is one of the most modified estuaries in North America, and its fishes and fisheries have suffered as a consequence (Herbold et al. 1992). Declines in striped bass, chinook salmon, steelhead, and American shad have been noted for the past 30 years, but the declines of other species, such as the endemic delta smelt and the longfin smelt, have only been noted more recently (Moyle et al. 1992). The decline in the fish populations has largely been linked to ever-increasing freshwater diversions from the estuary, but also to recently introduced nonindigenous species, pesticides, and other factors. The crisis in fish population declines, including two endangered species, led to the development of a 1994 Delta Native Fishes Recovery Plan by the U.S. Fish and Wildlife Service and an agreement (December 1994) on new estuarine standards among California and federal agencies, urban and agricultural water users, and environmental groups.

Sierra Nevada Fishes

The Sierra Nevada is the backbone of California and is the source of much of its agricultural and urban water. Sierra Nevada lakes and streams support 40 kinds of native fishes but are best known for 7 native trouts (California golden, Kern rainbow, Little Kern golden, Eagle Lake rainbow, Paiute cutthroat, steelhead, and Lahontan cutthroat; Moyle et al. 1995). Five of these forms would probably be extinct today (three are still listed as threatened species; U.S. Fish and Wildlife Service 1994a) if drastic action had not been taken by fisheries agencies to protect and improve their habitats, move fishes to new locations, and remove competing nonindigenous trouts from their streams.

The western Sierra Nevada also was once a major spawning region for anadromous fishes, primarily chinook salmon, steelhead, and Pacific lamprey, but these species are now relatively unimportant there. Overall, 22 of the 40 native fishes are threatened, declining rapidly, or otherwise in need of special protection (Moyle et al. 1995). As a consequence, the sport fisheries of the Sierra Nevada consist primarily of introduced nonindigenous fishes in reservoirs and streams.

Southern California Fishes

Nowhere in California is the fish fauna in more trouble than in the urbanized coastal areas of southern California (Swift et al. 1993; Moyle et al. 1995). Although only three of the native fish species (unarmored threespine stickleback, tidewater goby, southern steelhead) are listed as endangered, strong cases can be made for listing all of the native forms as endangered, including arroyo chub, Santa Ana sucker, and Santa Ana speckled dace. Indeed, a strong case can be made for designating the native aquatic ecosystems as endangered, if such a delineation were possible. The declines are all linked to the region's expanding human population; specific causes of the declines include dams and diversions, watershed urbanization, channelization of streams for flood control, pollution, and heavy recreational use of stream corridors.

Accounts similar to the decline of fishes, fisheries, and suitable fish habitat in southern California could be written for other regions of the state (for example, the Sacramento, Klamath, and Pit rivers). Widespread decline of native fishes (including economically and culturally important species) is associated with major alterations of aquatic habitats (Moyle 1995) and has occurred in the entire state. There are many indications that such trends can be halted and perhaps reversed when private landowners and public land users form cooperative alliances with public agencies and environmental groups to find solutions to the problems. To function well, such alliances require strong leadership, economic incentives, and good scientific information. In the long run, however, conservation of the fish fauna of California has to be part of a statewide strategy for the conservation of aquatic biodiversity (Moyle and Yoshiyama 1995).

Amphibians

For the past 5 years, considerable attention has been focused on the apparent decline of many amphibian species in North America and in other parts of the world (see reviews in Blaustein 1994). Although, as with the fishes, much of the overall decline of amphibians can be directly attributed to extensive habitat alteration and loss, a small component of amphibian population extinctions in seemingly pristine areas is currently unexplained (Blaustein and Wake 1995). California is one of the few places where such unexplained declines have been observed and are being studied (for example, Fellers and Drost 1993; Bradford et al. 1994; Jennings 1995). We describe here the present status of California's declining amphibian fauna and the species most at risk of extinction.

California currently has 77 amphibian species, which include all known species, subspecies, and undescribed forms. At least 43 (56%) are endemic or nearly so (Stebbins 1985; Jennings 1995; Fig. 27; Table 4). New species continue to be discovered and described (Wake 1994). Historically, only a handful of species were commercially exploited to the point of localized extinction (for example, see Bury and Stewart 1973; Jennings and Hayes 1985). However, a recent survey of the status of the state's native amphibian fauna revealed that 18 of 45 (40%) salamanders and 15 of 28 (54%) frogs are already protected or are in need of protection at the state or federal level (Jennings and Hayes 1994a; Table 4). Four salamanders and frogs are introduced forms that became established in the state during the twentieth century. The current status of native true toads and frogs is of particular concern, because during the past 25 years, 4 of 10 (40%) of the toad species and 7 of 8 (88%) of the frog species have disappeared from 45% or more of their historical California ranges (Table 5). The decline is especially significant in southern California, where all of the native frogs are either extirpated or on the verge of extinction (Jennings and Hayes 1994b; Jennings 1995).

The reasons for the decline and loss of certain native amphibians in California are varied. For example, all of the salamanders (Fig. 28) and some of the frogs (such as the California red-legged frog, foothill yellow-legged frog, and Yavapai leopard frog) have declined because of outright habitat loss from agriculture, livestock grazing, urbanization, placer mining, road construction, and large water-development projects (Jennings and Hayes 1994a). Some of the aquatic habitats relied on by these organisms have also been extensively altered by human-induced hydrological alterations and by nonindigenous vegetation (such as water hyacinth and saltcedar), so that they

Fig. 28. Adult male Santa Cruz long-toed salamander, Santa Cruz County.

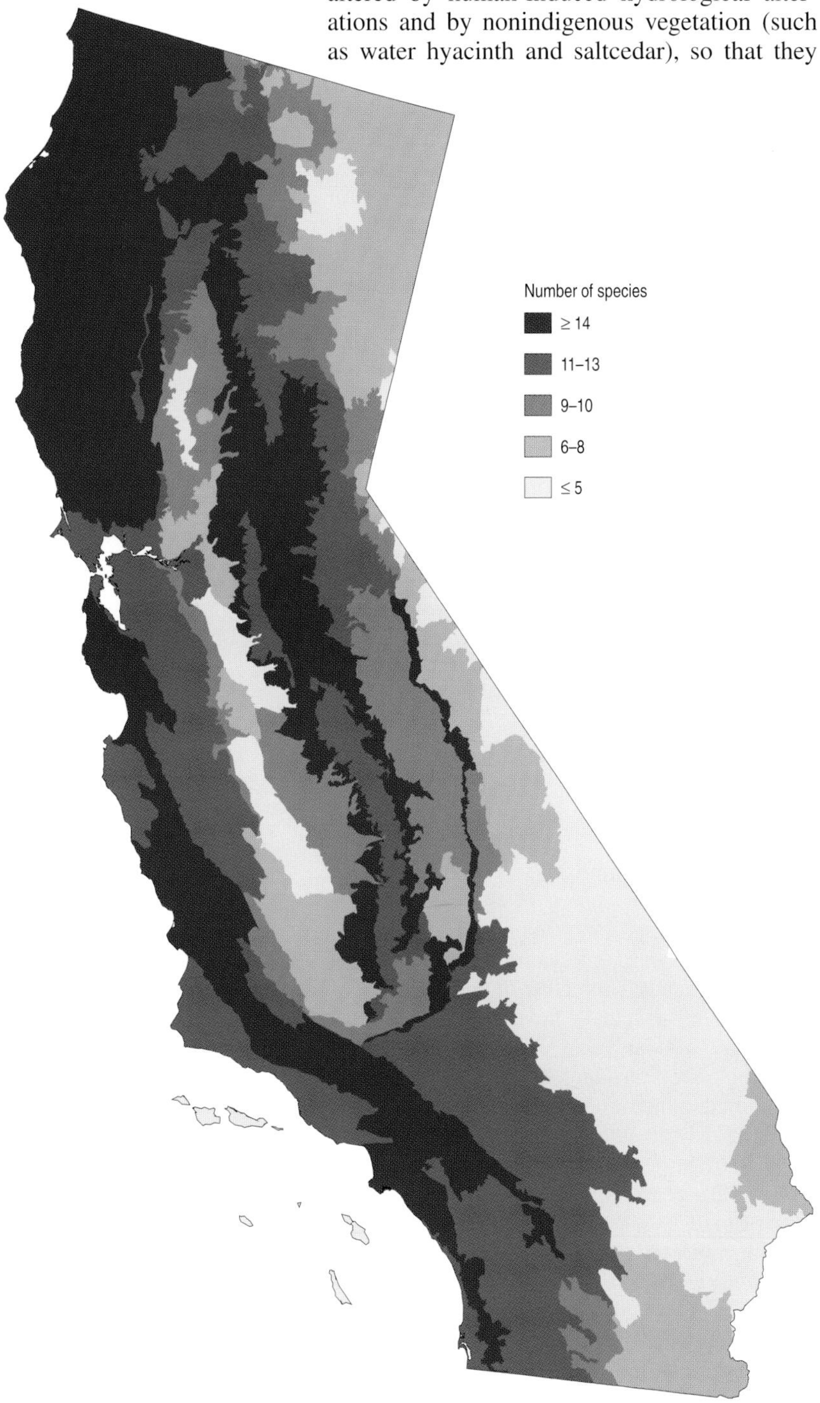

now support largely nonindigenous aquatic faunas. Virtually all of the major river systems and native grasslands of California have been changed by human activities to the point that they can no longer support several imperiled amphibians, such as Colorado River toads in the Colorado River Valley and California tiger salamanders in the Central Valley (Jennings and Hayes 1994a).

Additionally, the introduction of a wide variety of nonindigenous predatory fishes, bullfrogs, and crayfishes into natural and artificial waterways has resulted in the extirpation of much of the local native amphibian fauna through predation on vulnerable eggs, larvae, and juvenile life stages (Hayes and Jennings 1986; Bradford 1989; Bradford et al. 1993; Jennings and Hayes 1994b; Rosen and Schwalbe 1995). Predation on amphibians has been exacerbated by the extensive transplantation of trout into the originally fishless lakes of the Sierra Nevada, as well as the interbasin transfer of water that has occurred throughout California since the 1930's. The interbasin transfer of water has resulted in the establishment of many nonindigenous species in locations far from the area where they were originally introduced (Moyle 1976a).

Besides the human-related factors just listed, some amphibian declines and extinctions can be attributed to natural disasters such as flooding, drought, and fires (both natural and human-caused). For example, a pair of 500-year floods (that is, floods of a magnitude that usually occur only once every 500 years) in 1968–1969 effectively eliminated many native frog populations in southern California (Jennings and Hayes 1994b). In the Sierra Nevada foothills, 5 years of drought from 1988 to 1992 resulted in the extirpation of most of the remaining California red-legged frog populations (Jennings and Hayes 1994a; Jennings 1996; Fig. 29) and negatively affected many high-elevation populations of Yosemite toads as well (Sherman and Morton 1993). Sweet (1991) documented the

Number of species

- ≥ 14
- 11–13
- 9–10
- 6–8
- ≤ 5

Fig. 27. Potential species richness of amphibians (total of 46 species) in California; 5-quantile classes by U.S. Department of Agriculture ecological subsection. Each class contains an equal number of data records. Note the concentrations of amphibians along the coast and in the mountains. Maps provided courtesy of the California Wildlife Habitat Relationships Program, Wildlife Management Division, California Department of Fish and Game, Sacramento.

Table 4. California amphibians, including endemic species and those of conservation concern (modified from Stebbins 1985; Jennings 1987; and Jennings and Hayes 1994a). Status codes are C = full species endemic to California, to California and a small part of an adjacent state, or to Baja California; E = presumed extinct in California; I = introduced; FE = federally endangered; FT = federally threatened; PE = proposed endangered; PT = proposed threatened; N = none; ST = state threatened; and SC = state species of special concern.

Species	Status
Salamanders	
Arboreal salamander	C
Black salamander	C
Black-bellied slender salamander	C
Breckenridge Mountain slender salamander	C, E, PE
California giant salamander	C
California slender salamander	C
California tiger salamander	C, PT[a]
Channel Islands slender salamander	C
Clouded salamander	N
Coast Range newt[b]	C, SC[c]
Del Norte salamander	SC
Desert slender salamander	C, FE
Dunn's salamander	N
Fairview slender salamander	C
Gabilan slender salamander	C
Garden slender salamander	C
Guadalupe slender salamander	C
Hell Hollow slender salamander	C
Inyo Mountains slender salamander	C, PT
Kern Canyon slender salamander	C, ST
Kern Plateau slender salamander	C
Large-blotched ensatina[b]	C, SC
Limestone salamander	C, ST
Monterey ensatina[b]	C
Mount Lyell salamander	C, SC
Northern rough-skinned newt[b]	N
Northwestern salamander	N
Oregon ensatina[b]	N
Owens Valley web-toed salamander	C, SC
Pacific giant salamander	N
Painted ensatina[b]	N
Red-bellied newt	C
Relictual slender salamander	C, SC
San Gabriel slender salamander	C
Santa Cruz long-toed salamander[b]	C, FE
Santa Lucia slender salamander	C
Shasta salamander	C, ST
Sierra Nevada ensatina[b]	C
Sierra newt[b]	C
Siskiyou Mountains salamander	ST
Southern long-toed salamander[b]	N
Southern seep salamander	C, PT[d]
Tehachapi slender salamander	C, ST
Tiger salamander	I?
Yellow-blotched ensatina[b]	C, SC
Yellow-eyed ensatina[b]	C
Frogs and toads	
African clawed frog	I
Arizona toad[b]	N
Arroyo toad[b]	FE
Black toad	C, ST
Bullfrog	I
California red-legged frog[b]	C, FT[e]
California toad[b]	C
California treefrog	N
Cascades chorus frog[b]	N
Cascades frog	SC, PE
Coast chorus frog[b]	N
Couch's spadefoot	SC
Foothill yellow-legged frog	C, SC, PT, PE
Great Basin spadefoot	N
Great Plains toad	N
Mountain yellow-legged frog	C, PT, PE[f]
Northern leopard frog	PE[g]

Species	Status
Northern red-legged frog[b]	SC
Oregon spotted frog	PE[h]
Pacific chorus frog[b]	N
Red-spotted toad	N
Rio Grande leopard frog	I
Sierra chorus frog[b]	C
Sonoran Desert toad	E, PE
Southwestern chorus frog[b]	N
Tailed frog	SC, PT
Western spadefoot	C, PT
Western toad[b]	N
Woodhouse's toad[b]	N
Yavapai leopard frog	E, PE
Yosemite toad	C, PT

[a] This salamander was found to be warranted but precluded from listing by the most recent ruling of the U.S. Fish and Wildlife Service (1994c).

[b] Subspecies.

[c] Southern California populations only (Jennings and Hayes 1994a).

[d] This salamander was recently petitioned to the U.S. Fish and Wildlife Service for listing under the Endangered Species Act (H. H. Welsh, U.S. Forest Service, Redwood Sciences Laboratory, Arcata, California, personal communication).

[e] A final rule for listing this frog as threatened was recently published by the U.S. Fish and Wildlife Service (1996).

[f] Southern California populations of this frog have been petitioned to the U.S. Fish and Wildlife Service for listing under the Endangered Species Act (M. C. Long, Eaton Canyon Nature Center, Pasadena, California, personal communication).

[g] Some populations in the Lake Tahoe basin are known to have been introduced (Bryant 1917; Jennings and Hayes 1994b).

[h] This frog was petitioned for listing under the Endangered Species Act by the Board of Directors of the Utah Nature Society (U.S. Fish and Wildlife Service 1989). Although a ruling has been made on this petition, the status of peripheral populations (such as those in California) is still under review (M. P. Hayes, Portland State University, Portland, Oregon, personal communication).

Species	Reduction in range (percent)
Arroyo toad	76
Breckenridge Mountain slender salamander	100
California red-legged frog[b]	75
California tiger salamander	55
Cascades frog	50
Foothill yellow-legged frog	45
Mountain yellow-legged frog	50/99[a]
Northern leopard frog	95
Northern red-legged frog[b]	15
Oregon spotted frog	99
Sonoran Desert toad	100
Western spadefoot	30/80[a]
Yavapai leopard frog	100
Yosemite toad	50

Table 5. Percentage of reduction of historical range of selected native California amphibians (data from Jennings and Hayes 1994a; Jennings 1995).

[a] Values are for central/southern California populations, respectively.

[b] Subspecies.

Fig. 29. Adult male California red-legged frog, Pescadero Natural Area, San Mateo County.

Fig. 30. Adult female arroyo toad, San Diego County.

loss of an arroyo toad population in southern California because of a wildfire burning the riparian zone it inhabited (Fig. 30).

In recent years, several kinds of native true toads and frogs have disappeared from seemingly pristine areas such as wilderness areas and national parks (Bradford 1991; Fellers and Drost 1993; Drost and Fellers 1994; Jennings and Hayes 1994a). Such losses are especially troubling since they indicate that even the traditional conservation practice of setting aside large protected areas may be insufficient to keep these amphibians from becoming extinct (Jennings 1995). A number of hypotheses have been put forth in an attempt to explain these declines and localized losses, including acid precipitation (Bradford et al. 1992, 1994), air pollution (T. Cahill, University of California, Davis, personal communication), increased ultraviolet radiation levels (Blaustein et al. 1994a), introduced pathogens (Blaustein et al. 1994b), and pesticides (Stebbins and Cohen 1995). None of these hypotheses, though, can convincingly explain the widespread amphibian declines, and there is great debate among herpetologists over their validity (for example, Pechmann et al. 1991; Blaustein 1994; Pechmann and Wilbur 1994).

In California and over much of the American West, unexplained amphibian declines seem to have the following in common: only true frogs and toads appear to be affected (for example, lists in Hayes and Jennings 1986; Scott 1993), death occurs in the postmetamorphic stages (that is, juvenile and adult frogs and toads), populations are able to successfully reproduce until all adults are extirpated, and die-offs are most pronounced at higher elevations (Carey 1993; Scott 1993). The most likely causes of this kind of mortality are natural or human-induced stressors (Scott 1993), such as increased UV-B levels or air pollution, which weaken the immune systems of host organisms. Disease organisms could either be natural or introduced. Whatever is occurring, it is apparent that mass die-offs of native frogs and toads have been observed in natural areas of California over the past 25 years (Bradford 1991; Scott

1993; M. R. Jennings, U.S. Geological Survey, unpublished data). Clearly, more study is needed to determine the exact causes of these amphibian losses.

Reptiles

California has a diverse nonmarine reptile fauna, including 5 freshwater turtles, 1 tortoise, 38 lizards, and 37 snakes (Stebbins 1985; Jennings 1987; Laudenslayer et al. 1991; Fig. 31). Three turtle species and one gecko are nonindigenous species. The California population of one native reptile, the Sonoran mud turtle, has probably been extirpated (Jennings 1987). Many of California's reptiles are common in much of western North America, but there are 14 endemic species (15%) with restricted ranges that include only some part of California or California and a portion of an adjacent state and Baja California (Stebbins 1985; Table 6). In addition, many species have one or more subspecies with limited ranges that include a portion of California.

Nomenclature of reptiles in California is in flux, and taxonomic relationships are in the process of being revised for a number of groups. Unlike the amphibians, which are threatened by factors that often appear to be systemic in nature, most terrestrial reptiles are threatened only by habitat conversion. Reptile species richness increases from north to south in California, along with an increase in average temperature and aridity (Fig. 31). Only a few species are found in the cool, moist northwestern corner of the state, whereas the southern tier of counties hosts a wide array of species (Stebbins 1985). Eleven taxa of special concern are listed in California's Natural History Diversity Database (California Department of Fish and Game 1996). Although some taxa, like the San Francisco garter snake, have very restricted ranges, limited range per se is not the primary threat to continued existence. In general, habitat destruction is the main cause of reptile population declines in California. This is evident because the distribution of species identified by either the state or federal governments as being at risk occurs primarily in areas where the greatest habitat manipulation has occurred in California: coastal urban development, Central Valley agriculture, and desert livestock and recreational habitat alteration.

Giant Garter Snake

The giant garter snake, which is federally listed as a threatened species in the Central Valley of California, demonstrates the effects of habitat changes on population size and species viability (Fig. 32). Historically, the giant garter snake ranged throughout the San Joaquin and

Sacramento River basins, inhabiting wetland habitats along streams and rivers, and perhaps parts of the major delta marshes as well (Brode 1988). Much of the Sacramento Valley wetland was probably suitable habitat for the giant garter snake when Europeans first arrived in California (Bryan 1923; Hinds 1952). By 1971, though, much of the Central Valley had been drained and converted to dryland agriculture, eliminating almost all the natural habitat of this snake (Brode 1988). In the Sacramento River basin, where rice cultivation has maintained some seasonal wetland habitat, the giant garter snake has occupied the rice fields and associated water delivery systems. Relatively healthy populations can still be found in the Sacramento basin.

The San Joaquin River basin has been more intensively manipulated than the Sacramento River basin. Most of the area was converted to dryland farms such as orchards or cotton, which are unsuitable habitats for giant garter snakes (Hansen and Brode 1980; Brode 1988), and it is not known whether any giant garter snakes remain in the San Joaquin basin. Any populations there must be very small and isolated. In the San Joaquin Valley, no individuals were captured during extensive surveys in 1985–1986, nor were they captured in additional surveys in 1995 (Brode 1988; J. Brode, California Department of Fish and Game, Sacramento, personal communication). Although there is

Table 6. California reptiles of conservation concern, including endemic species (modified from Stebbins 1985; Jennings 1987; Jennings and Hayes 1994a). Status codes are C = full species endemic to California, to California and an adjacent state, or to Baja California; E = presumed extinct in California; CE = California endangered; FE = federally endangered; FT = federally threatened; and ST = state threatened.

Species	Status
Tortoises and turtles	
Desert tortoise	FT, ST
Western pond turtle	C
Lizards, skinks, and geckos	
Blunt-nosed leopard lizard	C, FE, CE
California legless lizard	C
Coachella Valley fringe-toed lizard	C, CE, FT
Gilbert's skink	C
Granite night lizard	C
Island night lizard	C, FT
Mojave fringe-toed lizard	C
Panamint alligator lizard	C
Switak's banded gecko[a]	ST
Snakes	
Alameda striped racer[a]	ST
California black-headed snake	C
California mountain kingsnake	C
Giant garter snake[a]	ST
San Francisco garter snake[a]	FE, CE
Sharp-tailed snake	C
Sierra garter snake	C
Southern rubber boa[a]	ST
Striped racer	C

[a] Subspecies.

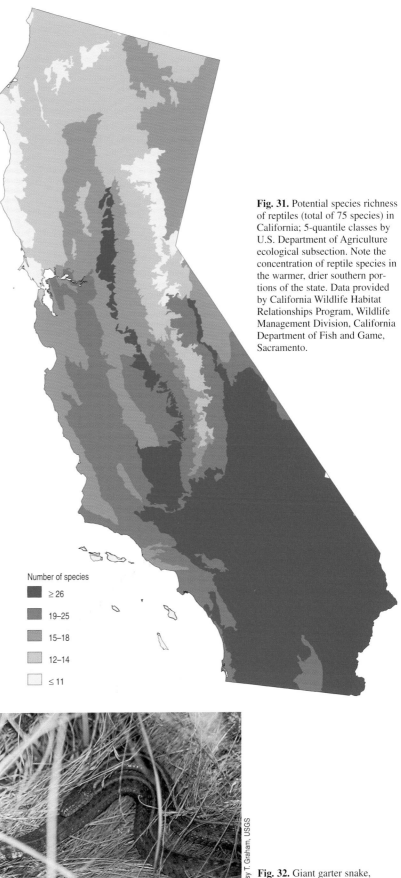

Fig. 31. Potential species richness of reptiles (total of 75 species) in California; 5-quantile classes by U.S. Department of Agriculture ecological subsection. Note the concentration of reptile species in the warmer, drier southern portions of the state. Data provided by California Wildlife Habitat Relationships Program, Wildlife Management Division, California Department of Fish and Game, Sacramento.

Number of species

■ ≥ 26
■ 19–25
■ 15–18
□ 12–14
□ ≤ 11

Courtesy T. Graham, USGS

Fig. 32. Giant garter snake, Gilsizer Slough, Sutter County, 1996.

some concern that agricultural pesticides may have played a role in the giant garter snake's decline, habitat alteration has unquestionably been the major factor.

Birds

California is rich in bird species, with 581 recorded as of 1991 (Zeiner et al. 1990a; Laudenslayer et al. 1991). This number includes migrants and species that breed or winter in California, as well as accidental and vagrant species. California birds range in size from the largest flying land bird in North America—the California condor (Pattee and Mesta 1995)—to the smallest bird in the United States, the tiny calliope hummingbird (Small 1994). Nearshore and pelagic marine birds are not considered here; they are discussed in the chapter on Marine Resources and are treated in detail elsewhere (Stallcup 1990). California is an important region for breeding, migrant, and wintering species, each of which we consider separately.

Breeding Birds

California's breeding avifauna consists of 293 species (Fig. 33), of which about 26 are exclusively nearshore and pelagic breeders. Twenty-three of these species breed only in eastside California, further reducing the nonmarine westside California avifauna to about 244 species. Sixteen bird species are either relatively narrow Pacific coast or Californian endemics, whereas only five (2%; California condor, Pacific-slope flycatcher, yellow-billed magpie, coastal California gnatcatcher, and tricolored blackbird) are either entirely limited to California (condor and magpie) or extend only a short distance north or south into Oregon or Baja California, respectively. Two of the five endemics are either ecologically extinct (California condor) or are in need of conservation intervention and management to assure their long-term survival (Table 7).

Unique species of westside California's bird fauna are those which breed primarily in the Central Valley (yellow-billed magpie and tricolored blackbird), in various forms of chaparral or brushland, and in foothill oak woodland. In the montane portions of westside California as well as in moist coastal forests, bird species tend to be widespread in at least western North America, although certain species (mountain quail, white-headed woodpecker, hermit warbler, and gray-crowned rosy-finch) are relatively restricted to the Pacific coast and possibly to small areas away from the coastal states.

Wintering Birds

While there are many more birds that winter in California than breed, there are similar numbers of species (Fig. 33). In addition to the breeding birds, many of which are year-round residents, massive numbers of birds that breed in Alaska, western Canada, and other Pacific Northwest states winter in California because of its relatively mild winter weather and rich food resources. Including birds that winter in the nearshore and pelagic marine zone, California supports 289 wintering species.

Wintering birds include huge numbers of waterfowl in freshwater, brackish, and saltwater environments. California, principally its coastal areas, harbors significant numbers of wintering shorebirds (Gill et al. 1995), gulls, and wading birds, although there are probably more shorebirds that pass through the state as spring and fall migrants. Some shorebird species, such as surfbirds, black turnstones, and black oystercatchers, winter only on the rocky coastline, but most species are found in California's restricted and shrinking estuarine mudflat habitats.

Waterbirds

California has a rich diversity of breeding, migratory, and wintering waterbirds (Cogswell 1977). In westside California, there are at least 64 breeding waterbird species. Sixteen of these species breed in nearshore or pelagic marine environments and are discussed in the chapter

Table 7. California birds of conservation concern (U.S. Fish and Wildlife Service 1994a; California Department of Fish and Game 1995). Status codes are FE = federally endangered; FT = federally threatened; CE = California endangered; and CT = California threatened.

Species	Status
Aleutian Canada goose[a]	FT
American peregrine falcon[a]	FE, CE
Arizona Bell's vireo[a]	CE
Bald eagle	FT, CE
Bank swallow	CT
Belding's savannah sparrow[a]	CE
California black rail[a]	CT
California clapper rail[a]	FE, CE
California condor	FE, CE
California least tern[a]	FE, CE
Coastal California gnatcatcher	FT
Elf owl	CE
Gila woodpecker	CE
Gilded flicker	CE
Great gray owl	CE
Greater sandhill crane[a]	CT
Inyo California towhee[a]	FT, CE
Least Bell's vireo[a]	FE, CE
Light-footed clapper rail[a]	FE, CE
Marbled murrelet	FE, CE
Northern spotted owl[a]	FT
San Clemente loggerhead shrike[a]	FE
San Clemente sage sparrow[a]	FT
Southwestern willow flycatcher[a]	FE
Swainson's hawk[a]	CT
Western snowy plover[a]	FT
Western yellow-billed cuckoo[a]	CE
Willow flycatcher	CE
Yuma clapper rail[a]	FE, CT

[a] Subspecies.

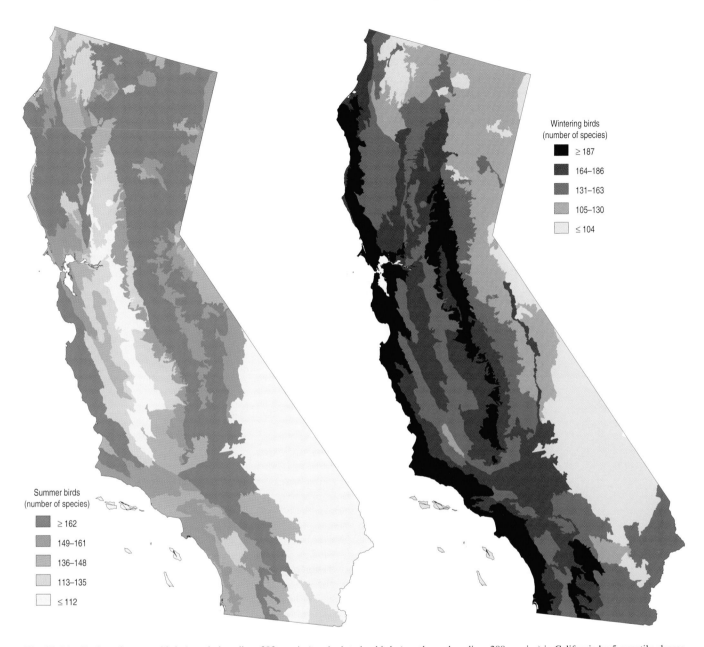

Fig. 33. Distribution of summer birds (mostly breeding; 293 species) and wintering birds (mostly nonbreeding; 289 species) in California by 5-quantile classes by quadrangle. Summer and winter range maps for California Wildlife Habitat Relationships System bird species were intersected with the boundaries of 2,831 1:24,000 scale topographic quadrangles. Each covers approximately 153 square kilometers and is labeled with the number of species that potentially occupy it. Data from California Wildlife Habitat Relationships Program, Wildlife Management Division, California Department of Fish and Game, Sacramento.

on Marine Resources, and 48 are coastal or inland breeders (Cogswell 1977; Stallcup 1990; Table 8). A number of additional waterbirds breed in northeastern or central eastern California (Mono and Topaz lakes), and the Salton Sea east of the Peninsular Ranges (Cogswell 1977). American white pelicans and fulvous whistling-ducks formerly bred in the San Joaquin Valley but no longer breed in westside California.

Westside California is rich in wintering, migratory, and postbreeding waterbirds, with more than 200 recorded species, including some rare visitors or vagrants (Cogswell 1977; Stallcup 1990). Although some species occur in both marine and more inland settings, an arbitrary division was made between marine and coastal or inland waterbirds, with 81 species considered entirely or primarily marine and 120 species being primarily coastal or inland in their occurrence (Table 8). Some terrestrial members of usual "waterbird" groups, such as mountain plovers and long-billed curlews, are arbitrarily grouped as waterbirds.

Several of the westside California waterbirds are on the U.S. list of threatened and endangered wildlife (U.S. Fish and Wildlife Service 1994a). These include the California and light-footed clapper rails, western snowy plover, California least tern, and marbled murrelet.

Table 8. Waterbirds of westside California.

Family or group	Breeding		Nonbreeding[a]	
	Marine	Nonmarine	Marine	Nonmarine
Loons	0	0	4	0
Grebes	0	3	2	4
Albatrosses	0	0	5	0
Shearwaters and petrels	0	0	10	0
Storm-petrels	3	0	8	0
Tropicbirds	0	0	3	0
Pelicans	1	1[b]	1	1
Boobies	0	0	3	0
Cormorants	2	1	2	1
Frigatebirds	0	0	1	0
Herons and allies	0	8	0	12
Storks	0	0	0	1
Ibises	0	1	0	3
Waterfowl	0	14[b]	11	27
Cranes	0	0	0	1
Rails and allies	0	7	0	6
Oystercatchers	1	0	2	0
Avocets and stilts	0	2	0	2
Plovers	0	2	0	9
Probing shorebirds	0	2	4	30
Phalaropes	0	1	1	2
Jaegers and skuas	0	0	5	0
Gulls and terns	3	3	7	18
Skimmers	0	0	0	1
Murres and allies	6	1	12	0
Kingfishers	0	1	0	1
Dippers	0	1	0	1
Total	16	48	81	120

[a] Includes wintering, migrant, or vagrant species.

[b] Includes one species that formerly bred in the San Joaquin Valley.

Waterfowl

As recently as the 1970's, an estimated 10–12 million waterfowl wintered in or migrated through California (U.S. Fish and Wildlife Service 1978). While California's upland birds depend on a variety of vegetation and habitat types, the great majority of the Pacific Flyway's migratory and overwintering waterfowl depend upon habitat in California's Central Valley.

Migratory waterfowl concentrations in the Central Valley are greatest during the fall and winter when migrants join local breeding birds. In recent years, approximately 2.5 million waterfowl wintered in the Central Valley (U.S. Fish and Wildlife Service 1987). This represents approximately 60% of all waterfowl wintering in the Pacific Flyway and about 20% of those wintering in the entire United States (Heitmeyer et al. 1989). Most of the continental populations of northern pintails, white-fronted geese, Ross's geese, cackling Canada geese, and Great Basin Canada geese and the entire populations of Aleutian Canada geese and Tule white-fronted geese winter in the Central Valley (Heitmeyer et al. 1989).

Land Birds

Although many insect-eating land birds leave California in winter, many fruit-eating, seed-eating, and bird of prey species winter in California. Examples of species that are absent or rare in summer but common to abundant in winter include the varied thrush and golden-crowned sparrow. Additionally, birds that breed at high elevations in summer may winter at lower elevations.

The California Fish and Game Commission lists 15 land birds as endangered or threatened under California law (California Department of Fish and Game 1995; Table 7), and there are 5 fully protected bird species in California (California Department of Fish and Game 1988).

Species of concern, also known as species at risk, are species monitored by The Nature Conservancy and are under consideration for possible addition to the list of federal endangered and threatened wildlife (U.S. Fish and Wildlife Service 1994a,b). The species of management concern list, developed and maintained by the Office of Migratory Bird Management of the U.S. Fish and Wildlife Service, identifies "32 species, subspecies, and populations of all migratory nongame birds that, without additional conservation action, are likely to become candidates for listing under the Endangered Species Act of 1973" in California (U.S. Fish and Wildlife Service 1995d).

Similar to the two federal species of concern programs, the California species of special concern is an informal designation used by the California Department of Fish and Game to identify declining and vulnerable species in the state (Remsen 1978). Presently, there are 45 land birds on the species of special concern list in California (California Department of Fish and Game 1992). Many species of special concern also occur on other California or federal species protection or management lists. The California species of special concern list is being revised; 20 land bird species or subspecies that are not mentioned on preceding lists are proposed for addition (S. Laymon, Kern River Research Center, Weldon, California, personal communication). Many species on California's species of special concern list are widespread elsewhere in the United States and are not included in Table 7.

Of the 342 species of land birds that occur in California, 73 (21.3%) are listed as California or federally threatened and endangered species (U.S. Fish and Wildlife Service 1994a; California Department of Fish and Game 1995), California fully protected species (California Department of Fish and Game 1988), federally designated species of concern (U.S. Fish and Wildlife Service 1994b) and species of management concern (U.S. Fish and Wildlife Service 1995d), or California species of concern (California Department of Fish and Game 1992). An additional 19 species show significant population declines (U.S. Geological

Western Snowy Plovers and California Least Terns

Western Snowy Plover

Western snowy plovers are small shorebirds that breed along the Pacific coast of the United States and northern Mexico as well as at interior sites in several western states (U.S. Fish and Wildlife Service 1993). The Pacific coast population was recently listed as threatened (U.S. Fish and Wildlife Service 1992, 1993). This population nests in Washington, Oregon, California, and Baja California, Mexico, and is associated with coastal wetlands and coastal dune habitat (Palacios and Alfaro 1994; Page et al. 1995). As much as half of the Pacific coast population may breed in Mexico (Palacios and Alfaro 1994). This population winters along the coasts of southern Oregon, California, and Baja California, Mexico (Page et al. 1995). Some snowy plovers that nest along the coast of California do not migrate in winter but remain on their breeding grounds (Stenzel et al. 1994; Powell et al. 1995; Fig. 1).

Fig. 1. Western snowy plover nest, with two newly hatched chicks.

The decline and loss of western snowy plover populations along the Pacific coast have been attributed to habitat loss and disturbance caused by urbanization. At northern sites, the invasion of nonindigenous beach grasses has reduced available breeding habitat, including dunes with scant vegetation, dredge-spoil islands, natural salt panne, and salt evaporation pond levees. The greatest loss of plover habitat has occurred along the southern California coast (U.S. Fish and Wildlife Service 1993). In southern California, many of the plover's nesting sites are associated with breeding colonies of California least terns.

Causes of low reproductive success in western snowy plovers include loss and degradation of breeding habitat, inclement weather, human disturbance, and increased numbers of predators associated with urban areas, including domestic and feral dogs, feral cats, red foxes, American kestrels, common ravens, American crows, striped skunks, Virginia opossums, and raccoons. Predators may take adults, chicks, or eggs. Plovers are highly susceptible to human disturbance and, if disturbed sufficiently, may abandon their nests. In addition, eggs have been lost from being trampled and run over by vehicles. At one site in coastal California, humans were directly responsible for the loss of at least 14% of nests over a 6-year period (Warriner et al. 1986). Chicks that become separated from adults through human disturbance or predators may die of exposure. Annual reproductive success for coastal snowy plovers in California has ranged from 0.8–0.9 fledglings per female near Monterey Bay to 0.8–1.1 fledglings per female in San Diego County (Warriner et al. 1986; Powell et al. 1995). Predation rates and levels of human disturbance are probably higher in central and northern California because plovers nesting in southern California benefit from site protection and predator management associated with California least tern colonies.

Western snowy plovers have high breeding-site fidelity, but some movement occurs between sites within and between years (Stenzel et al. 1994; Page et al. 1995; Powell et al. 1995). In addition, there is site fidelity associated with wintering areas (Page et al. 1995; A. Powell, U.S. Geological Survey, San Diego, California, unpublished data). Although some plovers return to their natal site to breed, there are few data on natal site fidelity. Little is known about the genetic makeup of snowy plover populations; however, banding studies indicate little mixing occurs among breeding sites in southern California (Powell et al. 1995).

Regular, standardized monitoring of western snowy plovers along the Pacific coast has not been conducted on an annual basis. However, a 20% reduction in population size was reported from surveys between the late 1970's and late 1980's, and winter numbers obtained from Christmas Bird Counts along the California coast declined significantly between 1962 and 1984 (Page et al. 1995). Other evidence of population decline has come from the documentation of the loss of breeding sites. Before 1970

snowy plovers nested at 53 sites along the California coast; the number of sites available has since been reduced by 62%. Currently, about 78% of the California breeding population is supported at only eight sites (U. S. Fish and Wildlife Service 1993). The breeding range along California's coast has been significantly interrupted by the loss of all historical breeding sites in Los Angeles County and most of Orange County. Loss of habitat in these areas has been attributed to high levels of recreational beach use and the raking of beach sand (for removal of debris) on a regular basis. Only one site in Orange County has supported a few nesting pairs in recent years (Powell et al. 1995).

Breeding populations of western snowy plovers in California continue to decline despite relatively high reproductive success at selected sites (Fig. 2). Numbers of snowy plovers surveyed in California during the middle of the breeding season in 1989, 1991, and 1995 were 1,139, 1,180, and 967, respectively (G. Page, Point Reyes Bird Observatory, Stinson Beach, California, unpublished data). Western snowy plovers have only been afforded protection by the Endangered Species Act for a short time, and populations appear to be steady or in decline.

Management plans for snowy plovers need to include designation of critical habitat, protection of nesting areas from recreational use during the species' breeding season, increased monitoring of populations and their reproductive success, predator management, and education. There is some evidence of higher reproductive success for snowy plovers nesting in areas protected as California least tern breeding habitat, probably because of the limited recreational use and the predator management in these locations. However, many of the largest snowy plover breeding areas in California do not overlap with least tern colonies.

California Least Tern

California least terns (Fig. 3) are migratory and spend the breeding season, from April through August, along the central and southern California coast, as well as along northern Baja California, Mexico. Historically, the breeding range stretched from Monterey County, California, to Cabo San Lucas, Baja California Sur, Mexico (Atwood and Minsky 1983). California least terns nest in colonies on sandy beaches that are usually associated with river mouths or

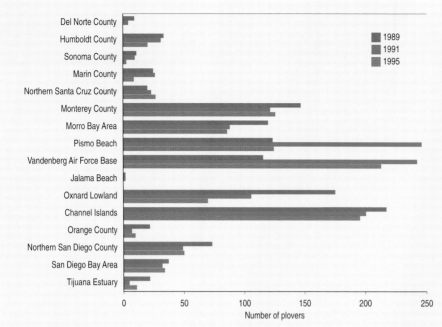

Fig. 2. Populations of breeding western snowy plovers during statewide surveys in California in 1989, 1991, and 1995. Sites are listed from north (top) to south (bottom). Data compiled by Gary Page, Point Reyes Bird Observatory, Stinson Beach, California.

estuaries. Nesting habitat has been degraded by high levels of human disturbance in sandy dune areas as well as by the effects of urbanization, including industrial, recreational, and residential development of the shoreline. Least terns, however, have successfully used created sites for nesting, including areas on dredge-spoil islands, open areas adjacent to airport runways, and industrial ports. Like snowy plovers, least terns are ground-nesting birds. They feed themselves and their chicks with small fish such as anchovies and topsmelts, captured from nearshore waters, estuaries, river mouths, and bays (Massey 1974; Atwood and Minsky 1983).

Low rates of reproductive success of California least terns have been linked to several factors. El Niño events, which cause nearshore water temperatures to rise, have depressed food availability for terns, which may in turn reduce tern productivity. The lowest annual production ever recorded for

California least terns occurred after the 1982–1983 El Niño event, when fish populations off the shores of southern California plummeted (Massey et al. 1992). In addition to their vulnerability to catastrophic events, least tern colony sites in California have become restricted to fewer and smaller areas that are often surrounded by highly developed settings, leaving tern colonies susceptible to human disturbance as well as to intense predation. Predators associated with urban landscapes, such as domestic and feral dogs and cats, red foxes, American kestrels, American crows, common ravens, coyotes, raccoons, striped skunks, and Virginia opossums, eat least tern adults, chicks, and eggs.

Contaminants bioaccumulated in fish eaten by least terns may be another

contributing factor to least terns' low reproductive success. Preliminary research on contaminants shows elevated levels of PCB's in California least tern eggs collected from sites around San Francisco Bay (Hothem and Zador 1995).

Although California least terns can and do nest again after losing eggs or chicks, some adults may abandon further breeding attempts that season (Fancher 1992). Least terns are fairly faithful to breeding sites and return year after year regardless of past nesting success. In addition, there is some evidence that least terns tend to return to their natal sites to breed (Atwood and Massey 1988). This may have major conservation implications because the average expected breeding life of California least terns is estimated at more than 9 years (Massey et al. 1992). Least terns breed after their second year, and first-time breeders are more likely to nest later in the breeding season (Massey and Atwood 1981).

Between 1978 and 1994, approximately 50 sites in California supported nesting least terns (Fancher 1992; Caffrey 1995). Fewer sites have been used in recent years; for example, only 36 sites were used in 1994 (Caffrey 1995). Furthermore, most California least terns nest at only a few select sites. In 1994, 76% of the population nested at nine sites, all in southernmost coastal California. Four of the nine sites (in Los Angeles, Orange, and San Diego counties) supported 48% of the breeding pairs (Caffrey 1995). In 1970, when California least terns were listed as endangered by the federal government and California, their population in California was estimated at 600 breeding pairs (Fancher 1992).

By 1994 the population had increased to an estimated 2,792 pairs (Fig. 4), which represents more than a fourfold increase (Caffrey 1995). Although the increase in the breeding population has not been consistent from year to year, long-term trends have

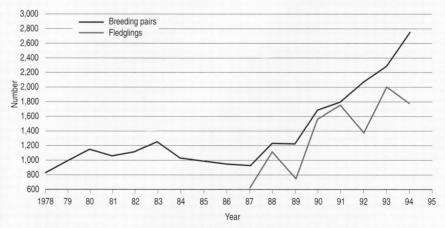

Fig. 4. Populations of breeding California least terns in California. Data compiled by the U.S. Fish and Wildlife Service and the California Department of Fish and Game.

Courtesy A. Powell, USGS

Fig. 3. California least tern and chick.

shown steady population growth (Fig. 4). Tern population growth has been sustained even though ratios of fledglings to adults have fluctuated between colony sites and years (Massey et al. 1992; Caffrey 1995). Population growth rates have increased, especially since the mid-1980's, when active management for least terns was initiated.

Management of California least tern colonies has included intensive monitoring of nesting colonies, site preparation to reduce vegetative cover, protection of sites by means of reduced access to humans, and predator management. Although individual nesting sites may not be used every year, and reproductive success varies among sites and years, the population of least terns in California continues to grow. Historical breeding sites should be preserved and managed for least terns because their adaptability to new or different sites depends on past reproductive success, predation pressure, and food supplies.

Author

Abby Powell
U.S. Geological Survey
Biological Resources Division
Northern Prairie Wildlife Research Center
Department of Biological Sciences
University of Arkansas
Fayetteville, AR 72701

See end of chapter for references

Survey Breeding Bird Survey, Laurel, Maryland, unpublished data) but have no protection or management designation. However, this additional number of declining or vulnerable land bird species in California is likely an underestimate because it does not consider species that are underreported or unreported by the Breeding Bird Survey, nor does it account for all of the species that depend upon vulnerable or restricted habitats (for example, riparian obligate species). Finally, this estimate does not include all the species that are proposed for addition to protection or management lists (for example, species of special concern) but have not yet been officially approved.

For a majority of California's land bird species, population trends are unknown. According to the recently analyzed (1966–1994) Breeding Bird Survey data covering 145 of the 342 land bird species in California, for 110 of the 145 species, the population trend is either insignificant (no real change exhibited) or the number of Breeding Bird Survey routes is too small to provide a reliable estimate of the species' statewide population trend (B. Peterjohn, U.S. Geological Survey Breeding Bird Survey, Laurel, Maryland, personal communication). Thus, of California's 342 land bird species, there are only 35 species for which present Breeding Bird Survey population monitoring efforts in California provide a reliable estimate of the species' statewide population trend.

Further analysis of the land bird monitoring situation in California reveals that—excluding the 83 land bird species that are rare visitors to California (and are therefore difficult to monitor), the 20 state or federally listed land birds whose populations are already being monitored, and the 35 species with reliable Breeding Bird Survey population trend estimates—there are roughly 204 land bird species in California that require additional or alternative population monitoring efforts. More conventional monitoring (that is, additional Breeding Bird Survey routes) as well as alternative land bird popula-tion monitoring programs (for example, habitat-specific monitoring, night surveys, soaring bird surveys, winter surveys) are sorely needed to accurately assess the status and trends of land bird species in California.

The population trends of many of California's land bird species are estimated from data collected annually by the U.S. Geological Survey's Breeding Bird Survey. To furnish a reliable (statistically sound) analysis of land bird population trends in California, we consider only species that were observed on 33 or more Breeding Bird Survey routes (Peterjohn, personal communication) and whose 29-year (1966–1994) population trend is significantly increasing, decreasing, or stable.

According to the 29 years of Breeding Bird Survey data that are available between 1966 and 1994, 13 California land bird species show significant population increases: red-shouldered hawk, red-tailed hawk, nonindigenous rock dove, white-headed woodpecker, black phoebe, scrub-jay, American crow, common raven, white-breasted nuthatch, northern mockingbird, phainopepla, solitary vireo, and common yellowthroat (U.S. Geological Survey Breeding Bird Survey, unpublished data).

California land bird species that show significant population declines between 1966 and 1994 are American kestrel, band-tailed pigeon, mourning dove, belted kingfisher, olive-sided flycatcher, western wood-pewee, horned lark, northern rough-winged swallow, barn swallow, mountain chickadee, chestnut-backed chickadee, plain titmouse, Bewick's wren, American robin, Wilson's warbler, chipping sparrow, black-chinned sparrow, dark-eyed junco, western meadowlark, Brewer's blackbird, pine siskin, and American goldfinch (U.S. Geological Survey Breeding Bird Survey, unpublished data).

Five land bird species are subject to sport or commercial harvest under regulations of the California Fish and Game Commission: band-tailed pigeon, spotted dove, white-winged dove, mourning dove, and American crow

(Laudenslayer et al. 1991). Populations of both the band-tailed pigeon and the mourning dove have declined significantly from 1966 to 1994, whereas the American crow shows a significant population increase. No Breeding Bird Survey population trend estimates are reported for the spotted dove and the white-winged dove (U.S. Geological Survey Breeding Bird Survey, unpublished data).

Four species of feral introduced land birds are established and successfully breeding in California: rock dove, spotted dove, European starling, and house sparrow (Laudenslayer et al. 1991; Small 1994). The rock dove appears to be thriving in California and shows a significant increasing population trend (U.S. Geological Survey Breeding Bird Survey, unpublished data). Also, two native species, the ruddy ground-dove and the plumbeous solitary vireo, appear to be expanding their ranges into California (Small 1994). In addition, there are 83 land bird species that are occasionally seen in California and are considered rare visitors to the state (Laudenslayer et al. 1991; Small 1994).

A discussion of land bird population trends would not be complete without also mentioning the impact of the brown-headed cowbird on many species of land birds. Cowbird parasitism is the main reason that the least Bell's vireo and the willow flycatcher are endangered in California. The effect of the cowbird has increased dramatically as a result of agricultural rangeland expansion, which creates more foraging habitat for cowbirds as well as habitat fragmentation, both of which enable cowbirds to find a larger proportion of host nests (Palazzo 1994; Laymon, personal communication). Although efforts by California Fish and Game and U.S. Fish and Wildlife Service personnel to trap and remove cowbirds continue in areas throughout the state (Franzreb et al. 1994), the brown-headed cowbird does not yet show a significant population decline in California (U.S. Geological Survey Breeding Bird Survey, unpublished data).

Riparian habitats, mostly composed of willow, alder, cottonwood, and dense undergrowth bordering streams and lakes, are the richest terrestrial habitats for breeding and wintering land birds in California. Riparian habitats, however, have been converted to other uses in California at a faster rate than any other habitat, with the possible exception of perennial grasslands. Only 5%–10% of the original riparian habitat in California exists today, and much of what remains continues to be developed for flood control or is degraded by grazing, logging, water diversions, and the introduction of invasive nonindigenous plants. Riparian habitats support more endangered land bird species than any other habitat type in California. Birds that use or nest in riparian habitat are limited not only by the loss and degradation of this habitat but also by cowbird brood parasitism that is aided by riparian habitat fragmentation. Of the 36 land bird species that rely heavily or depend exclusively upon riparian habitat, 21 have undergone substantial population declines and are either legally protected by state or federal governments or appear on species of concern lists. Nine additional riparian obligate species show signs of population decline (U.S. Geological Survey Breeding Bird Survey, unpublished data) or are suggested for addition to the California list of species of special concern (Laymon, personal communication).

Mammals

There are 181 species of terrestrial and flying mammals that regularly occur in California (Fig. 34), 29 of which are endemics (16%) whose entire ranges are limited to California or to California and portions of another state and Baja California (Burt and Grossenheider 1976; Table 9). Fourteen mammal species are non-indigenous introductions (Ingles 1957; Zeiner et al. 1990b).

Five mammals that occurred in California at the time of European settlement are now extirpated from the state but can still be found elsewhere. The last California grizzly bear was killed in 1922 in Sequoia National Forest (Storer and Tevis 1955). The other four mammals—gray wolf, jaguar, bison, and white-tailed deer—were restricted in range and abundance, and our information about them is sparse and often contradictory (Steinhart 1990). Wolves were recorded with some regularity in the Modoc Plateau region of northeastern California, where they were apparently an extension of Oregon and Nevada populations; the last record for that area is 1922 (Jameson and Peeters 1988). The Modoc Plateau country was also where bison and white-tailed deer populations were found in the early 1800's. Jaguars may have ranged north from Mexico as far as the Monterey Bay area in the early nineteenth century, but they were probably always rare. The last recorded jaguar in California was killed in Palm Springs in 1860 (Jameson and Peeters 1988).

Mammals at Risk

As large portions of many ecosystems have been destroyed, fragmented, or altered by development and subdivision, many populations of native mammals have decreased in direct proportion to the loss in their habitats. Fourteen nonmarine California mammals are on the U.S. list of endangered and threatened wildlife (U.S. Fish and Wildlife Service 1994a) or are listed as endangered or threatened under California law. The federally listed species include the San Joaquin kit fox, Fresno kangaroo rat, giant kangaroo rat, Morro Bay kangaroo rat, Stephens' kangaroo rat, Tipton kangaroo rat, salt marsh harvest mouse, and Amargosa vole; California lists the Mojave ground squirrel, Sierra Nevada red fox, Channel Islands gray fox, wolverine, California bighorn sheep, and peninsular bighorn sheep.

As a further indication of declining or at-risk California mammal populations, 49 additional species or subspecies are listed by the state as mammals of special concern, and the state is reconsidering a further 38 taxa for that status (including those already on the state list; P. Brylski, Santa Barbara Museum of Natural History, California, personal communication). Most of these are federal species of special concern.

In earlier times, mammals most at risk were large carnivores such as the grizzly bear, jaguar, and mountain lion, as well as furbearing species including the fisher, marten, red fox, river otter, and wolverine. These species declined as a result of being hunted or trapped; many appear to have recovered significantly under the protection afforded them since trapping days (Schempf and White 1977).

Table 9. California mammals of conservation concern, including endemic species (modified from Burt and Grossenheider 1976; U.S. Fish and Wildlife Service 1994a). Status codes are C = full species endemic to California, to California and adjacent state, or to Baja California; E = presumed extinct in California; CE = California endangered; FE = federally endangered; and ST = state threatened.

Species	Status
Agile kangaroo rat	C
Amargosa vole[a]	FE, CE
Broad-footed mole	C
Brush rabbit	C
California bighorn sheep[a]	ST
California ground squirrel	C
California mouse	C
California pocket mouse	C
California vole	C
Channel Islands gray fox	ST
Dusky-footed woodrat	C
Fresno kangaroo rat	C, FE, CE
Giant kangaroo rat	C, FE, CE
Heermann's kangaroo rat	C
Lodgepole chipmunk	C
Long-eared chipmunk	C
Merriam's chipmunk	C
Mohave ground squirrel	C, ST
Morro Bay kangaroo rat[a]	FE, CE
Mountain pocket gopher	C
Mt. Lyell shrew	C
Narrow-faced kangaroo rat	C
Nelson's antelope squirrel	C, ST
Ornate shrew	C
Pacific shrew	C
Panamint kangaroo rat	C
Peninsular bighorn sheep[a]	ST
Red tree vole	C
Salt marsh harvest mouse	C, FE, CE
San Diego pocket mouse	C
San Joaquin kit fox[a]	FE, ST
San Joaquin pocket mouse	C
Sierra Nevada red fox[a]	ST
Sonoma chipmunk	C
Stephens' kangaroo rat	C, FE, ST
Tipton kangaroo rat[a]	FE, CE
Wolverine	ST
Yellow-eared pocket mouse	C

[a] Subspecies.

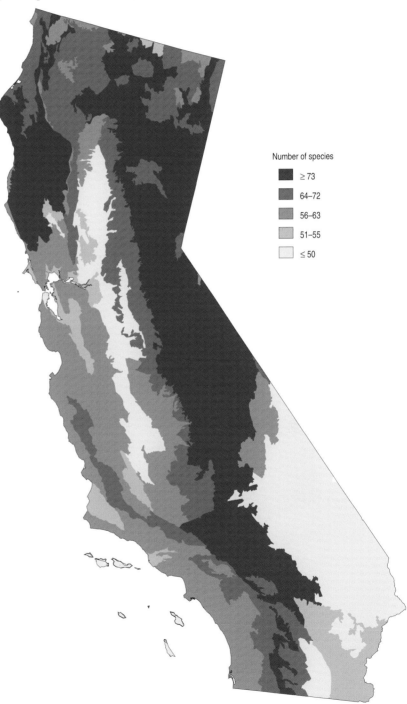

Fig. 34. Potential species richness of mammals (181 species) in California; 5-quantile classes by U.S. Department of Agriculture ecological subsection. Data provided by California Wildlife Habitat Relationships Program, Wildlife Management Division, California Department of Fish and Game, Sacramento.

Number of species

≥ 73

64–72

56–63

51–55

≤ 50

On the other hand, taxa on contemporary lists are mostly rodents, bats, and other insectivores. These creatures are listed largely because of habitat loss or modification (see chapter on Land Use), and in the case of bats, perhaps because of pesticide poisoning as well (see box on Southwestern Bats in Southwest chapter). Many of the small rodents, especially kangaroo rats, are moderately to highly localized subspecies occupying habitats that have been almost completely converted to agriculture or to housing developments. Most species at risk are not native to montane or desert habitats but occupy the valleys and coastal plains where settlement and agriculture are intense. The most endangered biological communities in California are those of the Central Valley grasslands and the coastal scrub and chaparral communities (Fig. 35).

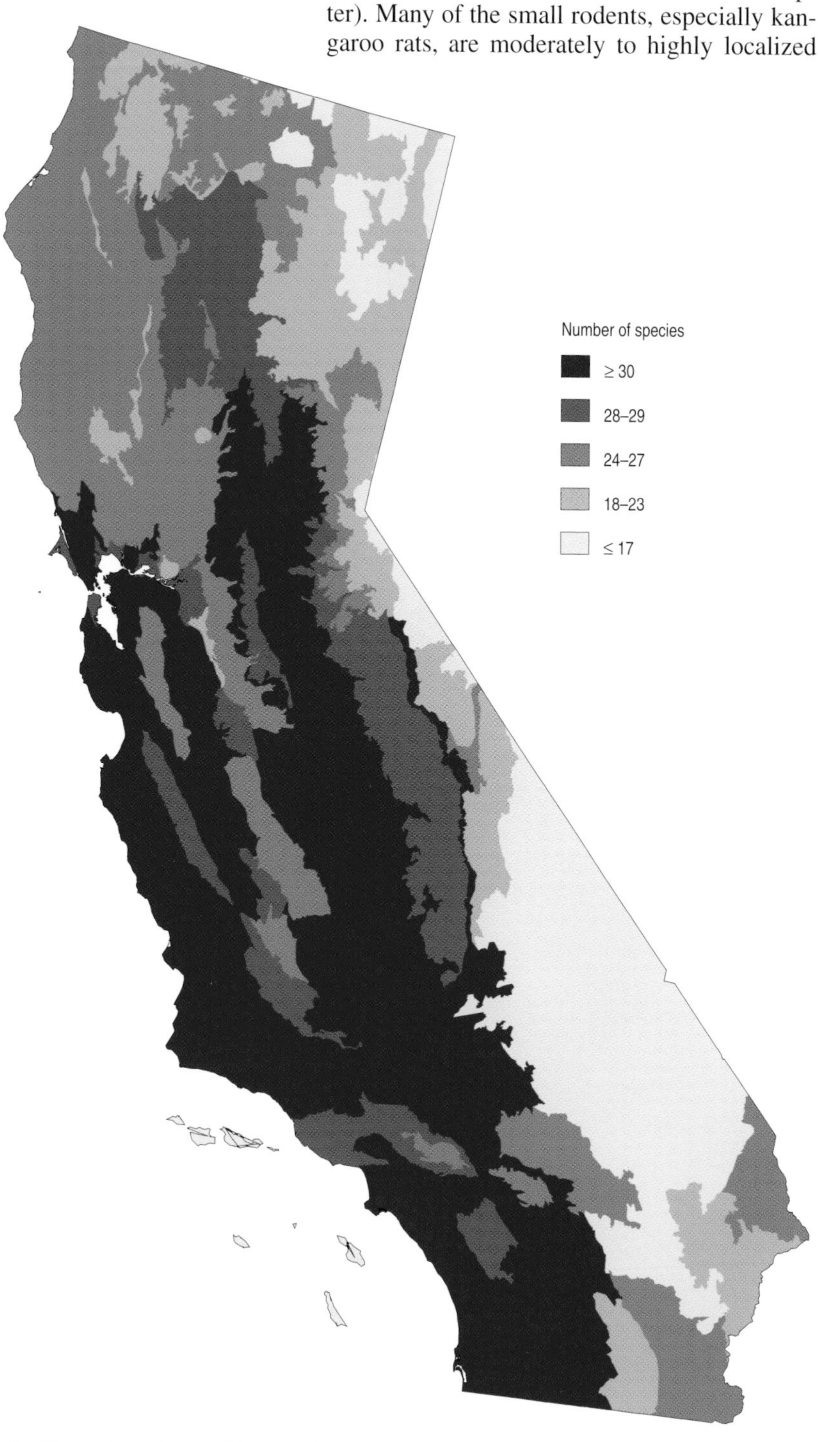

It is important to note that for most California mammals, abundance and distribution records are poor, localized, and often quite old. Many of these species and their distributions were described in the first half of the century, during the "golden age" of trapping.

Nonindigenous Mammals

The nonindigenous mammals of California include the classic human symbionts—house mouse, Norway rat, and black rat; these are virtually restricted to regions of moderate to dense settlement. Several mammals were introduced for fur trapping: Virginia opossums in Palm Springs in 1910, which are now widespread in all but deserts and mountains above about 1,500 meters; muskrats, introduced to the Central Valley and now widespread in waterways there; and nonindigenous subspecies of the red fox, which escaped from fur farms and now occur in small populations over much of western California. All three of these species can be considered pests to varying degrees. Opossums overturn garbage cans and consume the eggs of native birds. Muskrats have compromised the integrity of the artificial water systems of California by their burrowing. Introduced red foxes pose a potential threat to the genetic integrity of the native Sierra Nevada red fox and ravage the nests of endangered coastal shorebirds.

Wild pigs, introduced throughout westside California as game animals (as well as released or escaped domestic forms), are now widely recognized as significant ecological and agricultural pests of western parks and farmlands. Fallow deer, sambar, Axis deer, and Barbary sheep are all introduced game species now highly localized in California. The population of feral goats, also localized, has probably arisen from escaped and released individuals.

Mammals of the Sierra Nevada

In the Sierra Nevada, 84 terrestrial vertebrate species are considered dependent on riparian habitat (including wet meadow or lakeshore) to sustain viable populations; 24% of these are considered at risk. Eighteen species are similarly dependent upon late successional forests; 28% of these are considered at risk. There are

Number of species

- ≥ 30
- 28–29
- 24–27
- 18–23
- ≤ 17

Fig. 35. Species richness of threatened, endangered, and candidate terrestrial vertebrates (94 species) in California by 10-quantile classes; range maps for species were intersected with 1:24,000 scale topographic quadrangles of approximately 153 square kilometers. Each map is labeled with the number of threatened, endangered, or candidate species that potentially occupy the region. Note the concentration of species in urbanized coastal areas and the Central Valley. Data provided by California Department of Fish and Game.

85 species that require west-slope foothill savannah, woodland, chaparral, or riparian habitats (some species are also counted as riparian-dependent) for Sierran population viability; 16% of these species are listed as at risk. This latter number is misleadingly low because many of these species are more widely distributed elsewhere, such as in the Coast Ranges (California Department of Fish and Game 1994; D. Graber, National Park Service, Three Rivers, California, unpublished data).

Seventeen bat species are believed to inhabit the Sierra Nevada. Of these, seven have been nominated for listing under the Endangered Species Act. Three of those and one additional species have been listed as sensitive or of special concern. Concerns began about many bat species when populations using known historical roosts were noticeably smaller or had disappeared entirely. An obvious potential culprit in these declines has been pesticides, because bats are insectivorous. But habitat requirements of most bat species have been based on studies at a very small number of sites. More recent work by E. Pierson (University of California, Berkeley) and others in California suggests that the large, old trees and snags associated with late successional forests may be quite important to long-eared myotis and long-legged myotis, as healthy populations have been found only in late successional forests. The large trees and snags of conifers are riddled with cavities and crevices that provide thermal protection for these bats. The presence of spotted bats, Brazilian free-tailed bats, and western mastiff-bats is correlated with meadows, whereas many, if not most, Sierran bats forage over water, especially in riparian corridors. As bats use lower elevations for part of the year, loss of high-quality riparian habitat there may be a factor in the apparent decline of many species. Relatively high densities of spotted and mastiff-bats have been found only near the substantial cliffs in large river drainages such as the Kings, Kaweah, Merced, and Tuolumne rivers (E. Pierson and W. Rainey, W. Rainey, and E. Pierson and P. Heady, University of California, Berkeley, unpublished reports; Pierson, personal communication).

As in other places in the West, bighorn sheep populations in the Sierra Nevada were decimated following the arrival of Europeans in the mid-nineteenth century (Buechner 1960). Sheep populations in the Sierra Nevada were originally scattered along the crest and east slope from Sonora Pass south, and along the Great Western Divide of what is now Sequoia National Park; there was also a population in the Truckee River drainage (Jones 1950; Wehausen 1988). Likely causes for the precipitous population decline include market hunting, severe overgrazing by domestic stock, and, probably most importantly, the transmission from domestic sheep of respiratory bacteria fatal to mountain sheep.

Bighorn sheep had been extirpated from the Yosemite region before the turn of the century (Grinnell and Storer 1924). By the 1970's, only two populations remained: one near Mt. Baxter (about 220 sheep) and the other on Mt. Williamson (about 30 sheep), west of Independence. The Mt. Baxter herd increased during the 1970's (Wehausen 1980), and from 1979 until 1988 this population was used by the California Department of Fish and Game, in cooperation with the U.S. Forest Service and the National Park Service, to successfully reestablish herds near Wheeler Ridge, Mt. Langley, and Lee Vining Canyon. Some mountain lions were removed from the Lee Vining Canyon area to reduce significant losses while that sheep herd was becoming established. By 1990 the three introduced herds were all increasing, and the overall Sierra bighorn population was at least 300 (Bleich et al. 1990).

Between 1977 and 1987, reports of mountain lion depredation of mountain sheep in Inyo and Mono counties, as well as in California as a whole, increased dramatically (Foley et al. 1996). During that period, 50 mountain sheep from the Mt. Baxter herd were lost to predation on the herd's escarpment base winter range. Losses from mountain lion predation were detected in the other herds as well. During the extended drought of the late 1980's and early 1990's, the herds gradually abandoned their low-elevation winter ranges for much higher elevation sites that, although inferior from the standpoint of forage and protection from cold, were relatively snow-free during the drought and afforded protection from predation. This profound behavior change is attributed to heavy mountain lion predation pressure in the traditional low-elevation ranges (J. D. Wehausen, University of California, Bishop, personal communication). Concurrent with this behavior change has been a steady decline in the bighorn sheep populations. The Mt. Baxter population included 108 ewes in 1978; no more than 20 were counted in 1995. Twelve sheep died in a single avalanche on Wheeler Ridge in 1995, and only 10 ewes remain as a reproductive base. The Lee Vining Canyon population declined from 36 ewes in 1993 to 14 in 1995. Whether because of accidents or an inferior energetic balance, the new situation is distinctly pessimistic. As of summer 1996, the rangewide Sierra Nevada bighorn sheep population had been reduced to about 150 individuals—well below the 250 recorded in the original reintroduction in 1979.

There is no reason to assume that mountain lion populations were smaller before settlement, although they may well have fluctuated significantly over time. But whereas mountain sheep were widespread in the Sierra Nevada at settlement, they only persist in scattered small pockets of high-elevation habitat where snow depths are tolerable and mountain lions are absent.

Information Gaps

In California, the status and trends of vegetation are better known than those of the fauna. Researchers are proceeding to describe regional overviews of the distribution and conservation status of major terrestrial plant species and communities (Davis et al. 1995). In addition, a framework for characterizing communities statewide exists in the new *Manual of California Vegetation* (Sawyer and Keeler-Wolf 1995). California botanists are working together to define future needs in the study of California floristics (Mishler et al. 1995). Rare plant taxa and state and federally listed species are fairly well described and managed. Bioregional conservation and development planning that focuses on plant communities, as demonstrated by the California coastal sage–scrub Natural Community Conservation Planning process, is increasingly recognized as a tool for protecting natural diversity in an increasingly urbanized landscape (State of California 1993; McCaull 1994; Fig. 36). The threats to biodiversity presented by invasion of nonindigenous weeds and altered fire regimes are increasingly well understood, but implementation of effective programs to systematically monitor or manage these threats lags severely (van Wagtendonk 1985; Keeley 1995; Schierenbeck 1995). Land management agencies have begun to inventory and monitor the vegetation under their stewardship. Although much of the systematic monitoring is in relation to consumptive resource uses, such as timber harvest and grazing, some monitoring, including the model broad-spectrum program of the National Park Service at the Channel Islands National Park (Davis and Halvorson 1988), encompass natural communities not now subjected to consumptive uses. Rare plants are tracked by the Rare Plant Program of the California Native Plant Society; Skinner and Pavlick (1994) identify research needed for their perpetuation.

Many of the programs we have mentioned are pilot programs or are still being developed. Actual practices lag far behind. In 1991 only 11% of Bureau of Land Management lands in California had been adequately inventoried for special status plants, and only 6% of those were being monitored (Willoughby 1995). Although the U.S. Fish and Wildlife Service is responsible for enforcing the Endangered Species Act, complete botanical inventories for the agency's refuge lands in California do not exist (Knight 1995). The National Park Service is assembling a national flora of parklands (NPFLORA data base), but the species lists from most parklands in California are old and usually incomplete. The California Department of Fish and Game administers more than 300,000 hectares of land, including 3,665 hectares in 14 ecological reserves managed specifically for native or rare plant populations, but most of these lands have not been thoroughly inventoried, and only a few priority plant populations are monitored (Morey 1995). Lower plant taxa are much more poorly known and less well monitored. Lichens are extremely sensitive indicators of air quality changes, but they are poorly monitored and not widely known.

The status and trends of most of California's more than 50,000 invertebrate species are either unknown or poorly known. In fact, the estimated number probably includes many hundreds of undescribed species. On the other hand, some of California's invertebrates are among the best-monitored animal populations on the continent, and it is not appropriate to assume that the status of any particular species or population is unknown without having sought the information. For example, the Jasper Ridge population of the threatened bay checkerspot butterfly above Stanford University has been monitored in great detail for more than 30 consecutive years (P. R. Ehrlich et al., Stanford University, Stanford, California, unpublished data). To a lesser degree, populations of several other listed butterflies have been closely monitored, some sporadically, for 20 years or more. Populations of other insects, such as the listed California elderberry longhorn beetle, are relatively well surveyed. The status of other insect groups, insects of certain localities or habitats, and some economically important species is well known, but much of the information is scattered in the literature or is unpublished.

Still there is a need for a much greater effort at surveying California's invertebrates that are or may be of conservation concern. The fact that parks or reserves have been established for vertebrates or plants implies neither that the state's invertebrates are equally protected nor that appropriate management actions are being taken on behalf of invertebrates on existing tracts of protected land.

Data for the status of most vertebrate taxa in California are poor and are largely based on surveys conducted in the 1930's. Data for population trends often represent short time periods or exist for only a few locations. Existing status and trend data are largely inadequate for the

Fig. 36. Urbanized landscape, looking across San Francisco Bay to San Francisco and the Bay Bridge.

basic policy and land management decisions essential for protection, listing, delisting, or habitat management. Better surveys and monitoring of density, distribution, and habitat relations—such as those being conducted in the coastal sage–scrub community for amphibians, reptiles, small mammals, and birds, and in the Channel Islands National Park terrestrial and nearshore marine communities—are much needed in many other California habitat types.

Time-series and geographic data for status and trends exist for only a few groups of vertebrates, including the long-term transects of the North American Breeding Bird Survey in California and long-term monitoring of important game species, such as mule deer and waterfowl. Observations of birds, often made by volunteers who contribute to data bases such as the annual Christmas Bird Counts, provide a large, if limited, data set. For other species, range and habitat type maps have been developed by drawing upon the location data from museum specimens and extrapolating this data onto maps (U.S. Geological Survey, Western Ecological Research Center, Davis, unpublished data). These voucher specimens were collected over a period as great as a hundred years, and many or most locations are represented by specimens collected in the early or middle part of this century. Few surveys have been repeated. Recent repeat surveys, such as those for the amphibians in the national parks of California (Drost and Fellers 1994, 1996) and in California in general (Jennings and Hayes 1994a), have been launched as a result of recognition by individual herpetologists of broad species losses and not as a result of systematic assessments of status and trends by government agencies.

Files of national forest, national park, and Department of Fish and Game biologists, some county agencies, and many private land managers and landowners, often contain records of observations of rare or unusual wildlife, many with behavioral or habitat-use information attached. Although this information is of far less value than systematic scientific surveys, long-term studies, or investigations of species–habitat relations, it can be valuable for improving the resolution of distribution information and for making inferences about habitat preferences and other ecological characteristics of the species. At present, there is no efficient way to locate these data.

The California Natural Diversity Database, managed by the California Department of Fish and Game, keeps site records of agency-listed plant and animal species in the state. It has the potential to serve as a clearinghouse and manager of data on all species throughout California but to be effective would require a budget many times its present one. This would be an invaluable service to land managers, landowners, and government agencies throughout California. The California Environmental Resources Evaluation System is a pilot effort to make natural resource information available to the public, agencies, and businesses via the Internet; however, the value of the information delivery system is limited by the quality and quantity of information available.

A promising synthesis of the Breeding Bird Survey and the California Wildlife Habitat Relationships System, entitled Avesbase, has been produced by the U.S. Forest Service, Pacific Southwest Region (Davidson and Manley 1993). This computer database and analytical program combines information about population trends and habitat distributions for Neotropical migrant birds as an aid in assessing their risk in California.

Principal Authors

Stephen D. Veirs, Jr.
U.S. Geological Survey
Biological Resources Division
Western Ecological Research Center
University of California
Davis, California 95616

Paul A. Opler (USGS retired)*
Midcontinent Ecological Science Center
4512 McMurry Avenue
Fort Collins, Colorado 80525

*Current address:
Department of Bioagricultural Sciences
Colorado State University
Fort Collins, Co. 80523

Contributing Authors

David S. Gilmer
U.S. Geological Survey
Biological Resources Division
Western Ecological Research Center
6924 Tremont Road
Dixon, California 95620

David M. Graber
Sequoia and Kings Canyon National Parks
Three Rivers, California 93271

Tim Graham
U.S. Geological Survey
Biological Resources Division
Forest and Rangeland Ecosystem Science Center
82 Dogwood Avenue
Moab, Utah 84532

Laurie S. Huckaby
Remtech Services Inc.
4512 McMurry Avenue
Fort Collins, Colorado 80525

Mark R. Jennings
U.S. Geological Survey
Biological Resources Division
Western Ecological Research
Center
P.O. Box 70
San Simeon, California 93452

Kathryn McEachern
U.S. Geological Survey
Biological Resources Division
Channel Islands National Park
Field Station
1901 Spinnaker Drive
Ventura, California 93001–4354

Peter B. Moyle
Department of Wildlife, Fish,
and Conservation Biology
University of California
Davis, California 95616

Rosemary A. Stefani
Division of Environmental Studies
University of California
Davis, California 95616

The status of fishes in the waters of the state as reported here is subject to rapid change as a result of changes in land use and continuing accidental and intentional introductions. Some threatened species and populations are currently monitored, but statewide trends are not systematically tracked. Invertebrates are poorly monitored, except for a few taxa of special concern. Because the distribution and abundance of many invertebrates are extremely variable in time, meaningful monitoring is difficult, and our knowledge of lower forms of all taxa is extremely limited. We still have much work ahead of us to understand the effects of human-created processes upon other organisms, and conversely, to understand the significance to humans of changes in the micro- and macrobiota of California's valuable ecosystems.

Acknowledgments

L. Comrack, S. Laymon, B. Meese, and B. Peterjohn, U.S. Geological Survey, Breeding Bird Survey, Laurel, Maryland; J. Sauer, U.S. Geological Survey, Patuxent Wildlife Science Center, Laurel, Maryland; and R. Sayers for their assistance in providing information used in the preparation of the terrestrial bird section of the manuscript. C. Rutherford, U.S. Fish and Wildlife Service, Ventura, California, provided extensive information on the species and ecosystems of the Channel Islands, and personnel at the San Francisco Bay National Wildlife refuge complex, Newark, California, provided information on the status and trends of animals and plants found on the refuge's components. T. Lupo, Wildlife Management Division, California Department of Fish and Game, Sacramento, California, provided the GIS maps on vertebrate diversity. R. Hothem, U.S. Geological Survey, Davis, California, provided several references that were useful in preparing the chapter.

Cited References

Anderson, R. S. 1994. Paleohistory of a giant sequoia grove: the record from Log Meadow, Sequoia National Park. Pages 49–55 *in* P. S. Aune, technical coordinator. Proceedings of the symposium on giant sequoias: their place in the ecosystem and society. 23–25 June 1992, Visalia, California. U.S. Forest Service Pacific Southwest Research Station General Technical Report PSW-GTR-151.

Arnold, R. A. 1983. Ecological studies of six endangered butterflies (Lepidoptera: Lycaenidae): island biogeography, patch dynamics, and the design of habitat preserves. University of California Publications in Entomology 99:1–161.

Axelrod, D. I. 1967. Geologic history of the Californian insular flora. Pages 267–316 *in* R. N. Philbrick, editor. Proceedings of the symposium on the biology of the California Islands. Santa Barbara Botanic Garden, Santa Barbara, Calif.

Axelrod, D. I. 1977. Outline history of California vegetation. Pages 139–193 *in* M. G. Barbour and J. Major, editors. Terrestrial vegetation of California. John Wiley & Sons, New York.

Azevado, J., and D. L. Morgan. 1974. Fog precipitation in coastal California forests. Ecology 55:1135–1141.

Bailey, H. P. 1966. The climate of southern California. California Natural Guide 17. University of California Press, Berkeley and Los Angeles. 87 pp.

Bailey, R. G. 1995. Description of the ecoregions of the United States. Second edition. U.S. Forest Service Miscellaneous Publication 1391. 108 pp.

Bakker, E. 1972. An island called California. University of California Press, Berkeley and Los Angeles. 357 pp.

Barbour, M. G., R. B. Craig, F. R. Drysdale, and M. T. Ghiselin. 1973. Coastal ecology. University of California Press, Berkeley and Los Angeles. 338 pages.

Barbour, M. G., and A. F. Johnson. 1977. Beach and dune. Pages 223–261 *in* M. G. Barbour and J. Major, editors. Terrestrial vegetation of California. John Wiley & Sons, New York.

Barbour, M. G., and J. Major, editors. 1977. Terrestrial vegetation of California. John Wiley & Sons, New York.

Barnhart, R. A., M. J. Boyd, and J. A. Pequegnat. 1992. The ecology of Humboldt Bay, California: an estuarine profile. U.S. Fish and Wildlife Service Biological Report 1. 121 pp.

Bequaert, J. C., and W. B. Miller. 1973. The mollusks of the arid Southwest with an Arizona check list. University of Arizona Press, Tucson. 271 pp.

Blaustein, A. R. 1994. Chicken Little or Nero's fiddle? A perspective on declining amphibian populations. Herpetologica 50:85–97.

Blaustein, A. R., P. D. Hoffman, D. G. Hokit, J. M. Kiesecker, S. C. Walls, and H. B. Hays. 1994a. UV repair and resistance to solar UV-B in amphibian eggs: a link to population declines? Proceedings of the National Academy of Sciences 91:1791–1795.

Blaustein, A. R., D. G. Hokit, R. K. O'Hara, and R. A. Holt. 1994b. Pathogenic fungus contributes to amphibian losses in the Pacific Northwest. Biological Conservation 67:251–254.

Blaustein, A. R., and D. B. Wake. 1995. The puzzle of declining amphibian populations. Scientific American 272:52–57.

Bleich V. C., J. D. Wehausen, K. R. Jones, and R. A. Weaver. 1990. Status of bighorn sheep in California, 1989, and translocations from 1971 through 1989. Desert Bighorn Council Transactions 34:24–26.

Bradford, D. F. 1989. Allotropic distribution of native frogs and introduced fishes in the high Sierra Nevada lakes of California: implications of the negative effects of fish introductions. Copeia 1989:775–778.

Bradford, D. F. 1991. Mass mortality and extinction in a high elevation population of *Rana muscosa*. Journal of Herpetology 25:174–177.

Bradford, D. F., M. S. Gordon, D. F. Johnson, R. D. Andrews, and W. B. Jennings. 1994. Acidic deposition as an unlikely cause for amphibian population declines in the Sierra Nevada, California. Biological Conservation 69:155–161.

Bradford, D. F., D. M. Graber, and F. Tabatabai. 1993. Isolation of remaining populations of the native frog, *Rana muscosa*, by introduced fishes in Sequoia and Kings Canyon national parks, California. Conservation Biology 7:882–888.

Bradford, D. F., C. Swanson, and M. S. Gordon. 1992. Effects of low pH and aluminum on two declining species of amphibians in the Sierra Nevada, California. Journal of Herpetology 26:369–377.

Brode, J. M. 1988. Natural history of the giant garter snake (*Thamnophis couchii gigas*). *In* H. F. De Lisle, editor. Proceedings of the conference on California herpetology. Southwestern Herpetologists Society, Special Publication 4, Van Nuys, Calif.

Brown, R., P. B. Moyle, and R. M. Yoshiyama. 1994. Historical decline and current status of coho salmon in California. North American Journal of Fisheries Management 14:237–261.

Bryan, K. 1923. Geology and ground-water resources of Sacramento Valley, California.

U.S. Geological Survey Water-Supply Paper 495. 285 pp.

Bryant, H. C. 1917. The leopard frog in California. California Fish and Game 3(2):90.

Buechner, H. K. 1960. The bighorn sheep in the United States, its past, present and future. Wildlife Monograph 4:6–174.

Burt, W. H., and R. P. Grossenheider. 1976. A field guide to the mammals. Third edition. Peterson Field Guide 5. Houghton Mifflin Co., Boston. 289 pp.

Bury, R. B., and R. A. Luckenbach. 1976. Introduced amphibians and reptiles in California. Biological Conservation 10:1–14.

Bury, R. B., and G. R. Stewart. 1973. California protects its herpetofauna. HISS News-Journal 1(2):43–48.

Byrne, R., E. Edlund, and S. Mensing. 1991. Holocene changes in the distribution and abundance of oaks in California. Pages 182–188 *in* R. B. Standiford, technical coordinator. Proceedings of the symposium on oak woodlands and hardwood rangeland management. 31 October–2 November 1990, Davis, California. U.S. Forest Service Pacific Southwest Research Station General Technical Report PSW-GTR-126.

California Department of Fish and Game. 1983. A plan for protecting, enhancing, and increasing California's wetlands for waterfowl. California Department of Fish and Game, Sacramento. 59 pp.

California Department of Fish and Game. 1988. California's fully protected birds, mammals, reptiles, amphibians and fish—March 1988. California Department of Fish and Game, Sacramento. 4 pp.

California Department of Fish and Game. 1990. Central Valley salmon and steelhead restoration and enhancement plan. California Department of Fish and Game, Sacramento. 115 pp.

California Department of Fish and Game. 1992. Bird species of special concern (July 1992). California Department of Fish and Game, Wildlife Management Division, Nongame Bird and Mammal Section, Sacramento. 1 p.

California Department of Fish and Game. 1994. California wildlife habitat relationships database system. V. 5.0. Computer database and program. California Department of Fish and Game, Sacramento.

California Department of Fish and Game. 1995. Endangered and threatened animals of California. List generated from California Code of Regulations (Title 14, Section 670.5). California Department of Fish and Game, Natural Heritage Division, Natural Diversity Database, Sacramento. 13 pp.

California Department of Fish and Game. 1996. Natural diversity database: endangered and threatened animals of California. The Resources Agency, State of California.

Carey, C. 1993. Hypothesis concerning the causes of the disappearance of boreal toads from the mountains of Colorado. Conservation Biology 7:355–362.

Cogswell, H. L. 1977. Water birds of California. California Natural History Guide 40. University of California Press, Berkeley and Los Angeles. 399 pp.

Cooper, W. S. 1967. Coastal dunes of California. Geological Society of America Memoir 104. 131 pp.

Corbin, B. 1993. Discovery of the lost *Scheuchzeria*. Fremontia 204:19–20.

Davidson, C., and P. Manley, 1993. Avesbase: a conservation database for California birds, user's guide and documentation. U.S. Forest Service Pacific Southwest Region, San Francisco. 104 pp.

Davis, F. G., P. A. Stine, D. M. Stoms, M. I. Borchert, and A. D. Hollander. 1995. Gap analysis of the actual vegetation of California: 1. The southwestern region. Madroño 42(1):40–78.

Davis, G. E., and W. L. Halvorson. 1988. Inventory and monitoring of natural resources in Channel Islands National Park, California National Park Service, Channel Islands National Park, Ventura, Calif. 31 pp.

Dawson, E. Y. 1966a. Seashore plants of northern California. California natural history guide 20. University of California Press, Berkeley and Los Angeles. 103 pp.

Dawson, E. Y. 1966b. Seashore plants of southern California. California natural history guide 19. University of California Press, Berkeley and Los Angeles. 101 pp.

Dice, L. R. 1943. The biotic provinces of North America. University of Michigan Press, Ann Arbor. 73 pp.

Donley, M. W., S. Allen, P. Caro, and C. Patton. 1979. Atlas of California. The Pacific Book Center, Culver City, Calif. 191 pp.

Doyen, J. T. 1976. Systematics and biology of the genus *Coelus* (Coleoptera: Tentyriidae). Journal of the Kansas Entomological Society 49:595–624.

Drost, C. A., and G. M. Fellers. 1994. Decline of frog species in the Yosemite section of the Sierra Nevada. California National Park Resources Studies Unit, University of California, Davis, Technical Report NPS/WRUC/NRTR-94-02. 56 pp.

Drost, C. A., and G. M. Fellers. 1996. Collapse of a regional frog fauna in the Yosemite area of the California Sierra Nevada. Conservation Biology 10:(2):414–425.

Eargle, D. H., Jr. 1986. The Earth is our mother. A guide to the Indians of California, their locales and historic sites. Tree Company Press, San Francisco. 193 pp.

Estes, J. A., R. J. Jameson, J. L. Bodkin, and D. R. Carlson. 1995. California sea otters. Pages 110–112 *in* E. T. LaRoe, G. S. Farris, C. E. Puckett, P. D. Doran, and M. J. Mac, editors. Our living resources: a report to the nation on the distribution, abundance, and health of U.S. plants, animals, and ecosystems. U.S. Department of the Interior, National Biological Service, Washington, D.C.

Fellers, G. M., and C. A. Drost. 1993. Disappearance of the Cascades frog *Rana cascadae* at the southern end of its range, California, USA. Biological Conservation 65:177–181.

Fisher, F. 1994. Past and present status of Central Valley chinook salmon. Conservation Biology 8:870–873.

Foley, F., P. Foley, and S. Torres. 1996. Mountain lion depredation in California 1972–1992. California Department of Fish and Game, Sacramento. Unpublished report.

Franzreb, K., J. Greaves, and R. McKernan. 1994. Least Bell's vireo. Pages 216–217 *in* C. G. Thelander and M. Crabtree, editors. Life on the edge: a guide to California's endangered natural resources: wildlife. BioSystems Books, Santa Cruz, Calif.

Gall, L. F. 1985. Santa Catalina Island's endemic Lepidoptera. Pages 95–104 *in* A. S. Menke and D. R. Miller, editors. The Avalon hairstreak, *Strymon avalona*, and its interaction with the recently introduced gray hairstreak, *Strymon melinus* (Lycaenidae). Volume 2. Entomology of the California Channel Islands: proceedings of the first symposium. Santa Barbara Museum of Natural History, Calif.

Gill, R. E., Jr., C.M. Handel, and G. W. Page. 1995. Western North American shorebirds. Pages 60–65 *in* E. T. LaRoe, G. S. Farris, C. E. Puckett, P. D. Doran, and M. J. Mac, editors. Our living resources: a report to the nation on the distribution, abundance, and health of U.S. plants, animals, and ecosystems. U.S. Department of the Interior, National Biological Service, Washington, D.C.

Gilliam, H. 1966. Weather of the San Francisco Bay region. California natural history guide 6. University of California Press, Berkeley and Los Angeles. 72 pp.

Gilmer, D. S., M. R. Miller, R. D. Bauer, and J. R. LeDonne. 1982. California's Central Valley wintering waterfowl: concerns and challenges. Transactions of the North American Wildlife and Natural Resources Conference 47:441–452.

Green, L. R. 1981. Burning by prescription in chaparral. U.S. Forest Service Pacific Southwest Research Station General Technical Report PSW-GTR-51. 36 pp.

Griffin, J. R. 1977. Oak woodland. Pages 383–415 *in* M. G. Barbour and J. Major, editors. Terrestrial vegetation of California. John Wiley & Sons, New York.

Griffin, J. R., and W. B. Critchfield. 1972. The distribution of the forest trees of California. U.S. Forest Service Research Paper PSW-82/1972. Berkeley, Calif. 114 pp.

Grinnell, J., and T. I. Storer. 1924. Animal life in the Yosemite. University of California Press, Berkeley and Los Angeles. 752 pp.

Hanes, T. L. 1977. California chaparral. Pages 417–469 *in* M. G. Barbour and J. Major, editors. Terrestrial vegetation of California. John Wiley & Sons, New York.

Hansen, G. E., and J. M. Brode. 1980. Status of the giant garter snake *Thamnophis couchii gigas* (Fitch). California Department of Fish and Game, Inland Fisheries Endangered Species Program Special Publication 80-5, Sacramento. 14 pp.

Harvey, T. E., K. J. Miller, R. L. Hothem, M. J. Rauzon, G. W. Page, and R. A. Keck. 1992. Status and trends report on wildlife of the San Francisco Bay estuary. San Francisco Estuary Project. U.S. Fish and Wildlife Service, Sacramento, Calif. 283 pp.

Hayes, M. P., and M. R. Jennings. 1986. Decline of native ranid frog species in western North America: are bullfrogs (*Rana catesbeiana*) responsible? Journal of Herpetology 20:490–509.

Heitmeyer, M. E., D. P. Connelly, and R. L. Pederson. 1989. The Central, Imperial, and Coachella valleys of California. Pages 475–505 *in* L. M. Smith, R. L. Pederson, and R. M. Kaminski, editors. Habitat management for migrating and wintering waterfowl in North America. Texas Tech University Press, Lubbock.

Herbold, B., A. D. Jassby, and P. B. Moyle. 1992. Status and trends report on aquatic resources in the San Francisco estuary. San Francisco Estuary Project, Oakland, Calif. 257 pp.

Hickman, J. C., editor. 1993. The Jepson manual: higher plants of California. University of California Press, Berkeley and Los Angeles. 1400 pp.

Hinds, N. E. A. 1952. Evolution of the California landscape. California Division of Mines Bulletin 158. 240 pp.

Holland, R. H., and S. Jain. 1977. Vernal pools. Pages 515–533 *in* M. G. Barbour and J. Major, editors. Terrestrial vegetation of California. John Wiley & Sons, New York.

Howard, R. D. 1979. Geologic history of middle California. California natural history guide 43. University of California Press, Berkeley and Los Angeles. 113 pp.

Huntsinger, L., J. W. Bartolome, and P. F. Starrs. 1991. A comparison of management strategies in the oak woodlands of Spain and California. Pages 300–306 *in* R. B. Standiford, technical coordinator. Proceedings of the symposium on oak woodlands and hardwood rangeland management. 31 October–2 November 1990, Davis, California. U.S. Forest Service Pacific Southwest Research Station General Technical Report PSW-GTR-126.

Hurley, J. 1995. Prescribed burning in the 21st century. Pages 67–69 *in* D. R. Weise and R. E. Martin, technical coordinators. The Biswell symposium: fire issues and solutions in urban interface and wildland ecosystems. 15–17 February 1994, Walnut Creek, California. U.S. Forest Service Pacific Southwest Research Station General Technical Report PSW-GTR-158.

Ingles, L. G. 1957. Mammals of California and its coastal waters. Stanford University Press, Stanford, Calif. 396 pp.

Irwin, R. L. 1995. "What do we do now, Ollie?" Pages 17–21 *in* D. R. Weise and R. E. Martin, technical coordinators. The Biswell symposium: fire issues and solutions in urban interface and wildland ecosystems. 15–17 February 1994, Walnut Creek, California. U.S. Forest Service Pacific Southwest Research Station General Technical Report PSW-GTR-158.

Jaeger, E. C., and A. C. Smith. 1966. Introduction to the natural history of southern California. California natural history guide 13. University of California Press, Berkeley and Los Angeles. 104 pp.

Jameson, E. W., Jr., and H. J. Peeters. 1988. California mammals. University of California Press, Berkeley and Los Angeles. 403 pp.

Jenkins, O. P. 1952. Map of California showing natural provinces. *In* N. E. A. Hinds, editor. Evolution of the California landscape. San Francisco: California Division of Mines Bulletin 158.

Jennings, M. R. 1987. Annotated checklist of the amphibians and reptiles of California. Second revised edition. Southwestern Herpetologists Society Special Publication 3:1–48.

Jennings, M. R. 1995. Native ranid frogs in California. Pages 131–134 *in* E. T. LaRoe, G. S. Farris, C. E. Puckett, P. D. Doran, and M. J. Mac, editors. Our living resources: a report to the nation on the distribution, abundance, and health of U.S. plants, animals, and ecosystems. U.S. Department of the Interior, National Biological Service, Washington, D.C.

Jennings, M. R. 1996. Status of amphibians. Pages 31-1 to 31-24 in Sierra Nevada ecosystem project: final report to Congress. Volume 2. Assessments and scientific basis for management options. Centers for Water and Wetland Resources, University of California, Davis.

Jennings, M. R., and M. P. Hayes. 1985. Pre-1900 overharvest of the California red-legged frog (*Rana aurora draytonii*): the inducement for bullfrog (*Rana catesbeiana*) introduction. Herpetologica 41:94–103.

Jennings, M. R., and M. P. Hayes. 1994a. Amphibian and reptile species of special concern in California. California Department of Fish and Game, Inland Fisheries Division, Rancho Cordova. 255 pp.

Jennings, M. R., and M. P. Hayes. 1994b. The decline of native ranid frogs in the desert southwest. Pages 183–211 *in* P. R. Brown and J. W. Wright, editors. Herpetology of the North American deserts: proceedings of a symposium. Southwestern Herpetologists Society Special Publication 5. 300 pp.

Jepson, W. L. 1925. A manual of the flowering plants of California. University of California Press, Berkeley. 1238 pp.

Jones, F. L. 1950. A survey of the Sierra Nevada bighorn. Sierra Club Bulletin 35:29–76.

Josselyn, M. 1983. The ecology of San Francisco Bay tidal marshes: a community profile. U.S. Fish and Wildlife Service Biological Report FWS/OBS-82/23. 102 pp.

Kahrl, W. L. 1978. The California water atlas. California Department of Water Resources, Sacramento. 118 pp.

Katibah, E. F. 1984. A brief history of the riparian forests in the Central Valley of California. Pages 23–29 *in* R. E. Warner, editor. Proceedings of the California riparian systems conference. 17–19 September 1981, University of California, Davis. California Water Resources Center Report 55. University of California, Davis.

Keeley, J. E. 1995. Future of California floristics and systematics: wildfire threats to the California flora. Madroño 42(2):175–179.

Knight, J. 1995. The future of California's floristic heritage on public lands: overview of the Fish and Wildlife Service. Madroño 42:245–247.

Kreissman, B. 1991. California: an environmental atlas and guide. Bear Klaw Press, Davis, Calif. 255 pp.

Küchler, A. W. 1977. The map of the natural vegetation of California. *In* M. G. Barbour and J. Major, editors. Terrestrial vegetation of California. John Wiley & Sons, New York. Reported by the California Native Plant Society. 1988. Sacramento, Calif. 1000 pp. and supplement.

Laudenslayer, W. F., Jr., W. E. Grenfell, Jr., and D. C. Zeiner. 1991. A checklist of the amphibians, reptiles, birds and mammals of California. California Department of Fish and Game 77:109–141.

Light, S. F., R. I. Smith, F. A. Pitelka, D. P. Abbott, and F. M. Weesner. 1961. Intertidal invertebrates of the central California coast. University of California Press, Berkeley and Los Angeles. 446 pp.

Macdonald, K. B. 1977. Coastal salt marsh. Pages 263–294 *in* M. G. Barbour and J. Major, editors. Terrestrial vegetation of California. John Wiley & Sons, New York.

MacGinitie, G. E., and N. MacGinitie. 1949. Natural history of marine animals. McGraw-Hill, New York. 473 pp.

Major, J. 1977. California climate in relation to vegetation. Pages 11–74 *in* M. G. Barbour and J. Major, editors. Terrestrial vegetation of California. John Wiley & Sons, New York.

Mallette, E. E. 1981. California climate zone descriptions. California Energy Commission, Conservation Division, Sacramento. 39 pp.

Martin, R. E., and D. B. Sapsis. 1995. A synopsis of large or disastrous wildland fires. Pages 35–38 *in* D. R. Weise and R. E. Martin, technical coordinators. The Biswell symposium: fire issues and solutions in urban interface and wildland ecosystems. 15–17 February 1994, Walnut Creek, California. U.S. Forest Service Pacific Southwest Research Station General Technical Report PSW-GTR-158.

McCaull, J. 1994. The natural community conservation planning program and the coastal sage scrub ecosystem of southern California. Pages 281–292 *in* R. E. Grumbine, editor. Environmental Policy and Biodiversity. Island Press, Washington, D.C.

Meng, L., and P. B. Moyle. 1995. Status of the splittail in the Sacramento–San Joaquin estuary. Transactions of the American Fisheries Society 124:538–549.

Menke, A. S., and D. R. Miller, editors. 1985. Entomology of the California Channel Islands: proceedings of the first symposium. Santa Barbara Museum of Natural History, Santa Barbara, Calif. 178 pp.

Miller, S. E. 1985. The California Channel Islands—past, present, and future: an entomological perspective. Pages 3–28 *in* A. S. Menke and D. R. Miller, editors. Entomology of the California Channel Islands: proceedings of the first symposium. Santa Barbara Museum of Natural History, Santa Barbara, Calif.

Mills, T. J., D. R. McEwan, and M. R. Jennings. 1996. California salmon and steelhead: beyond the crossroads. Pages 91–111 *in* D. J. Stouder, P. A. Bisson, and

R. J. Naiman, editors. Pacific salmon and their ecosystems: status and future options. Chapman and Hall, New York.

Mishler, B. D., B. G. Baldwin, and S. D'Alcamo. 1995. Introduction, special issue. The future of California floristics and systematics: research, education and conservation. Madroño 42:93–95.

Monroe, M. W., J. Kelly, and N. Lisowski. 1992. State of the estuary. San Francisco Estuary Project. Association of Bay Area Governments, Oakland, Calif. 269 pp.

Mooney, H. A. 1977. Southern coastal scrub. Pages 471–489 *in* M. G. Barbour and J. Major, editors. Terrestrial vegetation of California. John Wiley & Sons, New York.

Morey, S. C. 1995. The future of California's floristic heritage on public lands: overview of the California Department of Fish and Game. Madroño 42:255–257.

Morrison, M. L., W. M. Block, and J. Verner. 1991. Wildlife–habitat relationships in California's oak woodlands: where do we go from here? Pages 105–109 *in* R. B. Standiford, technical coordinator. Proceedings of the symposium on oak woodlands and hardwood rangeland management. 31 October–2 November 1990. Davis, California. U.S. Forest Service Pacific Southwest Research Station General Technical Report PSW-GTR-126.

Moyle, P. B. 1976a. Fish introductions in California: history and impact on native fishes. Biological Conservation 2:101–118.

Moyle, P. B. 1976b. Inland fishes of California. University of California Press, Berkeley and Los Angeles. 405 pp.

Moyle, P. B. 1994. The decline of anadromous fishes in California. Conservation Biology 6:869–870.

Moyle, P. B. 1995. Conservation of native freshwater fishes in the Mediterranean-type climate of California, U.S.A.: a review. Biological Conservation 72:271–279.

Moyle, P. B., B. Herbold, D. E. Stevens, and L. W. Miller. 1992. Life history and status of delta smelt in the Sacramento–San Joaquin estuary. Transactions of the American Fisheries Society 121:67–77.

Moyle, P. B., and J. E. Williams. 1990. Biodiversity loss in the temperate zone: decline of the native fish fauna of California. Conservation Biology 4:275–284.

Moyle, P. B., and R. M. Yoshiyama. 1994. Protection of aquatic biodiversity in California: a five-tiered approach. Fisheries 19:6–18.

Moyle, P. B., R. M. Yoshiyama, J. E. Williams, and E. Wikramanayake. 1995. Fish species of special concern in California. California Department of Fish and Game, Sacramento. 222 pp.

Munz, P. A., and D. D. Keck. 1949. California plant communities. El Aliso 2:87–105.

Munz, P. A., and D. D. Keck. 1950. California plant communities. El Aliso 2:199–202.

Munz, P. A., and D. D. Keck. 1959. A California flora. University of California Press, Berkeley and Los Angeles. 1681 pp.

Munz, P. A., and D. D. Keck. 1968. A California flora. University of California Press, Berkeley and Los Angeles. 1905 pp.

New, T. R., editor. 1993. Conservation biology of Lycaenidae (butterflies). The IUCN Species Survival Commission. IUCN—The World Conservation Union, Gland, Switzerland. 173 pp.

Noss, R. F., and R. L. Peters. 1995. Endangered ecosystems: a status report on America's vanishing habitat and wildlife. Defenders of Wildlife, Washington, D.C.

Opler, P. A. 1991. North American problems and perspectives in insect conservation. Pages 9–31 *in* N. M. Collins and J. A. Thomas, editors. The conservation of insects and their habitats. Academic Press, London.

Opler, P. A., and S. B. Buckett. 1971. Seasonal distribution of macrolepidoptera in Santa Clara County, California. Journal of the Research on the Lepidoptera 9:75–88.

Opler, P. A., and L. Robinson. 1986. Lange's metalmark butterfly. Pages 910–916 *in* A. S. Eno, R. L. Di Silvestro, and W. J. Chandler, editors. Audubon wildlife report 1986. The National Audubon Society, New York.

Page, G. W., and R. E. Gill, Jr. 1994. Shorebirds in western North America: late 1800's to late 1900's. Studies in Avian Biology 15:147–160.

Page, G. W., and L. E. Stenzel, editors. 1981. The breeding status of the snowy plover in California. Western Birds 12:1–40.

Page, G. W., L. E. Stenzel, W. D. Shuford, and C. R. Bruce. 1991. Distribution and abundance of the snowy plover on its western North American breeding grounds. Journal of Field Ornithology 62:245–255.

Palazzo, T. L. 1994. Nest invaders. Pages 224–225 *in* C. G. Thelander and M. Crabtree, editors. Life on the edge: a guide to California's endangered natural resources: wildlife. BioSystems Books, Santa Cruz, Calif.

Pattee, O. H., and R. Mesta. 1995. California condors. Pages 80–81 *in* E. T. LaRoe, G. S. Farris, C. E. Puckett, P. D. Doran, and M. J. Mac, editors. Our living resources: a report to the nation on the distribution, abundance, and health of U.S. plants, animals, and ecosystems. U.S. Department of the Interior, National Biological Service, Washington, D.C.

Pechmann, J. H. K., D. E. Scott, R. D. Semlitsch, J. P. Caldwell, L. J. Vitt, and J. W. Gibbons. 1991. Declining amphibian populations: the problem of separating human impacts from natural fluctuations. Science 253(5022):892–895.

Pechmann, J. H. K., and H. M. Wilbur. 1994. Putting declining amphibian populations in perspective: natural fluctuations and human impacts. Herpetologica 50:65–84.

Philbrick, R. N. 1967. Introduction. Pages 3–8 *in* R. N. Philbrick, editor. Proceedings of the symposium on the biology of the California Islands. Santa Barbara Botanic Garden, Santa Barbara, Calif.

Pilsbry, H. A. 1948. Land Molluska of North America (north of Mexico). The Academy of Natural Sciences of Philadelphia Monograph 3.

Powell, J. A. 1981. Endangered habitats for insects: California coastal sand dunes. Atala 6:41–55.

Powell, J. A. 1985. Faunal affinities of the Channel Islands Lepidoptera: a preliminary overview. Pages 69–94 *in* A. S. Menke and D. R. Miller, editors. Entomology of the California Channel Islands: proceedings of the first symposium. Santa Barbara Museum of Natural History, Santa Barbara, Calif.

Powell, J. A. 1995. Lepidoptera inventories in the continental United States. Pages 168–170 *in* E. T. LaRoe, G. S. Farris, C. E. Puckett, P. D. Doran, and M. J. Mac, editors. Our living resources: a report to the nation on the distribution, abundance, and health of U.S. plants, animals, and ecosystems. U.S. Department of the Interior, National Biological Service, Washington, D.C.

Powell, J. A., and C. L. Hogue. 1979. California insects. California natural history guide 44. University of California Press, Berkeley and Los Angeles. 388 pages.

Raven, P. H. 1963. A flora of San Clemente Island, California. Aliso 5:289–347.

Raven, P. H. 1967. The floristics of the California Islands. Pages 57–67 *in* R. N. Philbrick, editor. Proceedings of the symposium on the biology of the California Islands. Santa Barbara Botanic Garden, Santa Barbara, Calif.

Raven, P. H. 1977. The California flora. Pages 109–137 *in* M. G. Barbour and J. Major, editors. Terrestrial vegetation of California. John Wiley & Sons, New York.

Raven, P. H., and D. I. Axelrod. 1978. Origins and relationships of the California flora. University of California Publication in Botany 72:1–134.

Rawls, J. J. 1984. Indians of California. The changing image. University of Oklahoma Press, Norman. 293 pp.

Remsen, J. V., Jr. 1978. Bird species of special concern in California: an annotated list of declining or vulnerable bird species. California Department of Fish and Game, Nongame Wildlife Investigations, Sacramento. Wildlife Management Branch Administrative Report 78-1. 54 pp.

Ricketts, E. F., J. Calvin, and J. W. Hedgpeth. 1952. Between Pacific tides. Third edition. Stanford University Press, Stanford, Calif. 502 pp.

Riedman, M. L., and J. A. Estes. 1990. The sea otter (*Enhydra lutris*): behavior, ecology, and natural history. U.S. Fish and Wildlife Service Biological Report 90(14). 126 pp.

Rosen, P. C., and C. R. Schwalbe. 1995. Bullfrogs: introduced predators in southwestern wetlands. Pages 452–454 *in* E. T. LaRoe, G. S. Farris, C. E. Puckett, P. D. Doran, and M. J. Mac, editors. Our living resources: a report to the nation on the distribution, abundance, and health of U.S. plants, animals, and ecosystems. U.S. Department of the Interior, National Biological Service, Washington, D.C.

Roth, B., and F. G. Hochberg. 1988. A new species of *Helminthoglypta* (Coyote) (Gastropoda: Pulmonata) from the

Tehachapi Mountains, California. The Veliger 31:258–261.

Roth, B., and F. G. Hochberg. 1992. Two new species of *Helminthoglypta* (Gastropoda: Pulmonata) from southern California, with comments on the subgenus *Charodotes* Pilsbry. The Veliger 35:338–346.

Rundel, P. W., D. J. Parsons, and D. T. Gordon. 1977. Montane and subalpine vegetation of the Sierra Nevada and Cascade ranges. Pages 559–599 *in* M. G. Barbour and J. Major, editors. Terrestrial vegetation of California. John Wiley & Sons, New York.

Sapsis, D. B., D. V. Pearman, and R. E. Martin. 1995. Progression of the Oakland/Berkeley Hills "Tunnel Fire." Pages 187–188 *in* D. R. Weise and R. E. Martin, technical coordinators. The Biswell symposium: fire issues and solutions in urban interface and wildland ecosystems. 15–17 Feb. 1994, Walnut Creek, Calif. U.S. Forest Service Pacific Southwest Research Station General Technical Report PSW-GTR-158.

Savage, J. M. 1967. Evolution of the insular herpetofaunas. Pages 219–228 *in* R. N. Philbrick, editor. Proceedings of the symposium on the biology of the California Islands. Santa Barbara Botanic Garden, Santa Barbara, Calif.

Sawyer, J. O., Jr., and T. Keeler-Wolf. 1995. A manual of California vegetation. California Native Plant Society, Sacramento. 471 pp.

Sawyer, J. O., Jr., and D. A. Thornburgh. 1977. Montane and subalpine vegetation of the Klamath mountains. Pages 699–732 *in* M. G. Barbour and J. Major, editors. Terrestrial vegetation of California. John Wiley & Sons, New York.

Schempf, P. F., and M. White. 1977. Status of six furbearer populations in the mountains of northern California. U.S. Forest Service, San Francisco, Calif. 51 pp.

Schierenbeck, K. A. 1995. The threat to the California flora from invasive species, problems and possible solutions. Madroño 42(2):168–174.

Schoenherr, A. A., editor. 1990. Endangered plant communities of southern California. Proceedings of the 15th annual symposium. Southern California Botanists, Rancho Santa Ana Botanical Gardens, Claremont. 114 pp.

Scott, N. J., Jr. 1993. Postmetamorphic death syndrome. Froglog (7):1–2.

Sherman, C. K., and M. L. Morton. 1993. Population declines of Yosemite toads in the eastern Sierra Nevada of California. Journal of Herpetology 27:186–198.

Skinner, M. W., and B. M. Pavlik, editors. 1994. Inventory of rare and endangered vascular plants of California. Special Publication No. 1, fifth edition. California Native Plant Society, Sacramento.

Slobodchikoff, C. N., and J. T. Doyen. 1977. Effects of *Ammophila arenaria* on sand dune arthropod communities. Ecology 58:1171–1175.

Small, A. 1994. California birds: their status and distribution. Ibis Publishing, Vista, Calif. 342 pp.

Sprague, M. 1941. Climate of California. Pages 783–797 *in* Climate and man. U.S.

Department of Agriculture, Washington, D.C.

Stallcup, R. 1990. Ocean birds of the nearshore Pacific. Point Reyes Bird Observatory, Stinson Beach, Calif. 214 pp.

State of California. 1993. Department of Fish and Game. Southern California coastal sage scrub natural community conservation planning process guidelines (as amended). Sacramento, Calif. 39 pp.

Stebbins, G. L., and J. Major. 1965. Endemism and speciation in the California flora. Ecological Monographs 35:1–35.

Stebbins, R. C. 1985. A field guide to western reptiles and amphibians. 2nd edition. Houghton Mifflin Co., Boston. 336 pp.

Stebbins, R. C., and N. W. Cohen. 1995. Natural history of the amphibians. Princeton University Press, Princeton, New Jersey. 304 pp.

Steinhart, P. 1990. California's wild heritage. Threatened and endangered animals in the Golden State. California Department of Fish and Game, California Academy of Sciences, Sierra Club Books, San Francisco.

Stephenson, N. L. 1994. Long-term dynamics of giant sequoia populations: implications for managing a pioneer species. Pages 56–63 *in* P. S. Aune, technical coordinator. Proceedings of the symposium on giant sequoias: their place in the ecosystem and society. 23–25 June 1992, Visalia, Calif. U.S. Forest Service Pacific Southwest Research Station, General Technical Report PSW-GTR-151.

Stirton, R. A. 1951. Prehistoric land animals of the San Francisco Bay region. Pages 177–192 *in* O. P. Perkins, editor. Geologic handbook of the San Francisco Bay counties. California Department of Natural Resources, Division of Mines, Bulletin 154.

Stoms, D. M., and F. Davis. 1995. Biodiversity in the southwestern Californian region. Pages 465–466 *in* E. T. LaRoe, G. S. Farris, C. E. Puckett, P. D. Doran, and M. J. Mac, editors. Our living resources: a report to the nation on the distribution, abundance, and health of U.S. plants, animals, and ecosystems. U.S. Department of the Interior, National Biological Service, Washington, D.C.

Stone, R. D. 1990. California's endemic vernal pool plants: some factors influencing their rarity and endangerment. Pages 89–107 *in* D. H. Ikeda, R. A. Schlising, F. J. Fuller, L. P. Janeway, and P. Woods, editors. Vernal pool plants. Their habitat and biology. Studies from the Herbarium, California State University, Chico. Number 8. 178 pp.

Storer, T. I., and L. P. Tevis. 1955. California grizzly. University of California Press, Berkeley and Los Angeles. 335 pp.

Storer, T. I., and R. L. Usinger. 1964. Sierra Nevada natural history. University of California Press, Berkeley and Los Angeles. 374 pp.

Sweet, S. S. 1991. Ecology and status of the arroyo toad (*Bufo microscaphus californicus*) on the Los Padres National Forest of southern California, with management recommendations. Report to the U.S. Forest Service, Los Padres National Forest, Goleta, California, under contract. 198 pp.

Swengel, A. B. 1995. Fourth of July butterfly count. Pages 171–172 *in* E. T. LaRoe, G. S. Farris, C. E. Puckett, P. D. Doran, and M. J. Mac, editors. Our living resources: a report to the nation on the distribution, abundance, and health of U.S. plants, animals, and ecosystems. U.S. Department of the Interior, National Biological Service, Washington, D.C.

Swift, C. C., T. Haglund, M. Ruiz, and R. Fisher. 1993. Status and distribution of freshwater fishes of southern California. Bulletin of the Southern California Academy of Sciences 92:101–167.

Taylor, D. 1994. Shasta snow-wreath; a new genus in California. Fremontia 22(3):3–4.

Thorp, R. 1990. Vernal pool flowers and host-specific bees. Pages 109–122 *in* D. H. Ikeda, R. A. Schlising, F. J. Fuller, L. P. Janeway, and P. Woods, editors. Vernal pool plants. Their habitat and biology. Studies from the Herbarium, California State University, Chico. Number 8.

Turgeon, D. D., A. E. Bogan, E. V. Coan, W. K. Emerson, W. G. Lyons, W. L. Pratt, C. F. E. Roper, A. Scheltema, F. G. Thompson, and J. D. Williams. 1988. Common and scientific names of aquatic invertebrates from the United States and Canada: mollusks. American Fisheries Society Special Publication 16. American Fisheries Society, Bethesda, Md. 277 pp.

U.S. Fish and Wildlife Service. 1978. Concept plan for waterfowl wintering habitat preservation—Central Valley, California. U.S. Fish and Wildlife Service, Portland, Oreg. 116 pp.

U.S. Fish and Wildlife Service. 1987. Concept plan for waterfowl wintering habitat preservation, an update—Central Valley. U.S. Fish and Wildlife Service, Portland, Oreg. 14 pp.

U.S. Fish and Wildlife Service. 1989. Endangered and threatened wildlife and plants; finding on petition to list the spotted frog. Federal Register 54(199):42529.

U.S. Fish and Wildlife Service. 1994a. Endangered and threatened wildlife and plants, 50 CFR 17.11 & 17.12. U.S. Fish and Wildlife Service, Washington, D.C. 42 pp.

U.S. Fish and Wildlife Service. 1994b. Endangered and threatened wildlife and plants; animal candidate review for listing as endangered or threatened species; proposed rule. Federal Register 59(219):58982–59028.

U.S. Fish and Wildlife Service. 1994c. Endangered and threatened wildlife and plants; 12-month petition finding for the California tiger salamander. Federal Register 59(74):18353–18354.

U.S. Fish and Wildlife Service. 1995a. Draft environmental alternatives analysis for a 4(d) rule for the conservation of the northern spotted owl on nonfederal lands. U.S. Fish and Wildlife Service, Portland, Oreg. 468 pp.

U.S. Fish and Wildlife Service. 1995b. Migratory nongame birds of management concern in the United States: the 1995 list. U.S. Fish and Wildlife Service, Office of Migratory Bird Management, Washington, D.C. 25 pp.

U.S. Fish and Wildlife Service. 1995c. Endangered and threatened wildlife and plants; proposed rule to list three plants from the Channel Islands of southern California as endangered. Federal Register 60:37987–37993.

U.S. Fish and Wildlife Service. 1995d. Endangered and threatened wildlife and plants; proposed rule for 16 plant taxa from the northern Channel Islands, California. Federal Register 60:37993–38011.

U.S. Fish and Wildlife Service. 1996. Endangered and threatened wildlife and plants; determination of threatened status for the California red-legged frog. Federal Register 61 (101): 25813–25833.

Valentine, J. W., and J. H. Lipps. 1967. Late Cenozoic history of the southern California islands. Pages 21–36 *in* R. N. Philbrick, editor. Proceedings of the symposium on the biology of the California Islands. Santa Barbara Botanic Garden, Santa Barbara, Calif.

van Wagtendonk, J. W. 1985. Fire suppression effects on fuels and succession in short fire interval wilderness ecosystems. Pages 119–126 *in* J. E. Lotan, B. M. Kilgore, W. C. Fischer, and R. W. Mutch, technical coordinators. Proceedings of a symposium and workshop on wilderness fire. U.S. Forest Service General Technical Report INT-182.

Van Dyke, E. C. 1919. The distribution of insects in western North America. Annals of the Entomological Society of America 12:1–12.

Veirs, S. D. 1982. Coast redwood forest: stand dynamics, successional status and the role of fire. Pages 119–141 *in* J. E. Means, editor. Forest succession and stand development research in the Northwest. Forest Research Laboratory, Oregon State University, Corvallis.

Vogl, R. J., W. P. Armstrong, K. L. White, and K. L. Cole. 1977. The closed-cone pine and cypress. Pages 295–358 *in* M. G. Barbour and J. Major, editors. Terrestrial vegetation of California. John Wiley & Sons, New York.

von Bloecker, J. C., Jr. 1967. The land mammals of the southern California islands. Pages 245–264 *in* R. N. Philbrick, editor. Proceedings of the symposium on the biology of the California Islands. Santa Barbara Botanic Garden, Santa Barbara, Calif.

Wake, D. B. 1994. New species of slender salamanders, genus *Batrachoseps*, from California [abstract]. Program and abstracts of the 74th annual meeting of the American Society of Ichthyologists and Herpetologists, 10th annual meeting of the American Elasmobranch Society, and the 6th annual meeting of the Neotropical Ichthyological Association; hosted by the University of Southern California and the Natural History Museum of Los Angeles County, Los Angeles 2–8 June 1994:171.

Warner, R. E., and K. M. Hendrix, editors. 1984. California riparian systems. Ecology, conservation, and productive management. University of California Press, Berkeley and Los Angeles. 1035 pp.

Weaver, D. W., and D. P. Doerner. 1967. Western Anacapia—a summary of the Cenozoic history of the northern Channel Islands. Pages 13–20 *in* R. N. Philbrick, editor. Proceedings of the symposium on the biology of the California Islands. Santa Barbara Botanic Garden, Santa Barbara, Calif.

Wehausen, J. D. 1980. Sierra Nevada bighorn sheep: history and population ecology. Ph.D. dissertation, University of Michigan, Ann Arbor. 240 pp.

Wehausen, J. D. 1988. The historical distribution of mountain sheep in the Owens Valley region. Pages 97–105 *in* Mountains to desert, selected Inyo readings. Friends of the Eastern California Museum, Independence.

Wilken, D. H. 1993. California's changing climates and flora. Pages 55–58 *in* J. C. Hickman, editor. The Jepson manual: higher plants of California. University of California Press, Berkeley and Los Angeles.

Willoughby, J. 1995. The future of California's floristic heritage on public lands: overview of the Bureau of Land Management. Madroño 42(2):242–244.

Zedler, J. B. 1982. The ecology of southern California coastal salt marshes: a community profile. U. S. Fish and Wildlife Service Biological Report FWS/OBS-81/54. 110 pp.

Zeiner, D. C., W. F. Laudenslayer, Jr., K. E. Mayer, and M. White, editors. 1990a. California's wildlife. Vol. 2. Birds. California Department of Fish and Game, Sacramento. 732 pp.

Zeiner, D. C., W. F. Laudenslayer, Jr., K. E. Mayer, and M. White, editors. 1990b. California's wildlife. Vol. 3. Mammals. California Department of Fish and Game, Sacramento. 407 pp.

Zinke, P. J. 1977. The redwood forest and associated north coast forest. Pages 679–698 *in* M. G. Barbour and J. Major, editors. Terrestrial vegetation of California. John Wiley & Sons, New York.

Fire and Fuel in a Sierra Nevada Ecosystem

van Wagtendonk, J. W. 1985. Fire suppression effects on fuels and succession in short fire interval wilderness ecosystems. Pages 119–126 *in* J. E. Lotan, B. M. Kilgore, W. C. Fischer, and R. W. Mutch, technical coordinators. Proceedings of a symposium and workshop on wilderness fire. U.S. Forest Service General Technical Report INT-182.

van Wagtendonk, J. W. 1994. Spatial patterns of lightning strikes and fires in Yosemite National Park. Proceedings of the 12th conference on fire and forest meteorology 12:223–231.

Torrey Pine

Berson, D. R. 1992. Pheromones help save Torrey pines. Forestry West Research, Sept.:17–20.

Brandegee, T.S. 1888. *Pinus torreyana* on the Island of Santa Rosa [Letter from the Discoverer]. Pages 111–112 *in* Second Biennial Report of the California State Board of Forestry for the years 1887–1888 to Governor R. W. Waterman, Sacramento.

Vogl, R. J., W. P. Armstrong, K. L. White, and K. L. Cole. 1988. The closed cone pines and cypresses. Pages 295–358 *in* M. G. Barbour and J. Major, editors. Terrestrial vegetation of California. California Native Plant Society Special Publication 9, Davis.

Emerging Diseases in Southern Sea Otters

Cornell, L. H., K. G. Osborn, and J. E. Antrim, Jr. 1979. Coccidioidomycosis in a California sea otter (*Enhydra lutris*). Journal of Wildlife Diseases 15:373–378.

Hennessy, S. L., and G. V. Morejohn. 1977. Acanthocephalan parasites of the sea otter, *Enhydra lutris*, off coastal California. California Fish and Game 63:268–272.

Riedman, M. L., and J. A. Estes. 1990. The sea otter (*Enhydra lutris*): behavior, ecology, and natural history. U.S. Fish and Wildlife Service Biological Report 90(14).

Riedman, M. L., J. A. Estes, M. M. Staedler, A. A. Giles, and D. R. Carlson. 1994. Breeding patterns and reproductive success of California sea otters. Journal of Wildlife Management 58:391–399.

Thomas, N. J., D. Pappagianis, L. H. Creekmore, and R. M. Duncan. 1996. Coccidioidomycosis in southern sea otters. Pages 163–173 *in* H. E. Einstein and A. Catanzaro, editors. Coccidioidomycosis: proceedings of the 5th international conference on coccidioidomycosis. National Foundation for Infectious Diseases, Washington, D.C.

California Abalone

Davis, G. E., D. V. Richards, P. L. Haaker, and D. O. Parker. 1992. Abalone population declines and fishery management in southern California. Pages 237–249 *in* S. A. Shepherd, M. J. Tegner, and S. A. Guzmán del Próo, editors. Abalone of the world. Blackwell Scientific Publications, Oxford, England.

Davis, G. E., K. R. Faulkner, and W. L. Halvorson. 1994. Ecological monitoring in Channel Islands National Park, California. Pages 465–482 *in* W. L. Halvorson, editor. The fourth California islands symposium: update on the status of resources. Santa Barbara Museum of Natural History, Santa Barbara, Calif.

Davis, G. E., and P. L. Haaker. 1995. A strategy for restoration of white abalone, *Haliotis sorenseni*. Journal of Shellfish Research 14:263.

Davis, G. E., P. L. Haaker, and D. V. Richards. 1996. Status and trends of white abalone at the California Channel Islands. Transactions of the American Fisheries Society 125:42–48.

Dugan, J. E., and G. E. Davis. 1993. Applications of marine refugia to coastal fisheries management. Canadian Journal of Fisheries and Aquatic Sciences 50:2029–2042.

Haaker, P. L., K. C. Henderson, and D. O. Parker. 1986. California abalone. State of California, The Resources Agency, Department of Fish and Game, Long Beach. Marine Resources Leaflet No. 11. 16 pp.

Haaker, P. L., D. V. Richards, C. S. Friedman, G. E. Davis, D. O. Parker, and H. A. Togstad. 1992. Mass mortality and withering syndrome in black abalone, *Haliotis cracherodii*, in California. Pages 214–224 *in* S. A. Shepherd, M. J. Tegner, and S. A. Guzmán del Próo, editors. Abalone of the world. Blackwell Scientific Publications, Oxford, England.

Hahn, K. O., editor. 1989. Handbook of culture of abalone and other marine gastropods. CRC Press, Boca Raton, Fla. 348 pp.

Lindberg, D. R. 1992. Evolution, distribution and systematics of Haliotidae. Pages 3–18 *in* S. A. Shepherd, M. J. Tegner, and S. A. Guzmán del Próo, editors. Abalone of the world. Blackwell Scientific Publications, Oxford, England.

Meyers, R. A., N. J. Barrowman, J. A. Hutchings, and A. A. Rosenberg. 1995. Population dynamics of exploited fish stocks at low population levels. Science 269:1106–1108.

Richards, D. V., and G. E. Davis. 1993. Early warnings of modern population collapse of black abalone *Haliotis cracherodii* (Leach 1814) on the California Channel Islands. Journal of Shellfish Research 12:189–194.

Shepherd, S. A., M. J. Tegner, and S. A. Guzmán del Próo, editors. 1992. Abalone of the world. Blackwell Scientific Publications, Oxford, England. 608 pp.

State of California. 1995. Pink, green, and white abalone fishery closure. Draft Environmental Document. The Resources Agency, Department of Fish and Game, Sacramento, Calif. 192 pp.

Tegner, M. J. 1992. Brood stock transplants as an approach to abalone stock enhancement. Pages 461–473 *in* S. A. Shepherd, M. J. Tegner, and S. A. Guzmán del Próo, editors. Abalone of the world. Blackwell Scientific Publications, Oxford, England.

Tegner, M. J. 1993. Southern California abalones: can stocks be rebuilt using marine harvest refugia? Canadian Journal of Fisheries and Aquatic Sciences. 50:2010–2018.

Tegner, M. J., P. A. Breen, and C. E. Lennart. 1989. Population biology of red abalone, *Haliotis rufescens*, in southern California and management of the red and pink, *H. corrugata*, abalone fisheries. U.S. Fishery Bulletin 87:313–339.

Tegner, M. J., J. D. Demartini, and K. A. Karpov. 1992. The California red abalone fishery: a case study in complexity. Pages 370–383 *in* S. A. Shepherd, M. J. Tegner, and S. A. Guzmán del Próo, editors. Abalone of the world. Blackwell Scientific Publications, Oxford, England.

Tutschulte, T. C. 1976. The comparative ecology of three sympatric abalones. Ph.D. dissertation, University of California, San Diego. 335 pp.

Channel Islands and California Desert Snail Fauna

Bequaert, J. C., and W. B. Miller. 1973. The mollusks of the arid Southwest with an Arizona checklist. University of Arizona Press, Tucson. 271 pp.

Chambers, S. M. 1991. Biogeography of Galapagos land snails. Pages 307–325 *in* M. J. James, editor. Galapagos marine invertebrates: taxonomy, biogeography, and evolution in Darwin's islands. Topics in Geobiology. Volume 8. Plenum Press, New York.

Hochberg, F. G. 1979. VI. Invertebrate zoology: land mollusks. Part 6 *in* D. M. Power, editor. Natural resources study of the Channel Islands National Monument, California. Volume 1. Prepared for the National Park Service by the Santa Barbara Museum of Natural History.

Hochberg, F. G., Jr., B. Roth, and W. B. Miller. 1987. Rediscovery of *Radiocentrum avalonense* (Hemphill in Pilsbry, 1905) (Gastropoda; Pulmonata). Bulletin of the Southern California Academy of Sciences 86:1–12.

Junger, A., and D. L. Johnson. 1980. Was there a Quaternary land bridge to the northern Channel Islands? Pages 33–39 *in* D. M. Power, editor. The California Islands: proceedings of a multidisciplinary symposium. Santa Barbara Museum of Natural History, Santa Barbara, California.

Pilsbry, H. A. 1939–1948. Land Molluska of North America (north of Mexico). The Academy of Natural Sciences of Philadelphia Monograph 3.

Roth, B. 1973. *Helminthoglypta traskii fieldi* Pilsbry, 1930 (Gastropoda: Stylommatophora). Bulletin of the Southern California Academy of Sciences 72:148–155.

Roth, B. 1975a. Description of a new terrestrial snail from San Nicolas Island, California (Gastropoda: Stylommatophora). Bulletin of the Southern California Academy of Sciences 74:94–96.

Roth, B. 1975b. Investigations of the survival status of certain terrestrial mollusks of the California Channel Islands and in northern California. Report prepared for the Office of Endangered Species and International Activities, U.S. Fish and Wildlife Service, Washington D.C.

U.S. Fish and Wildlife Service. 1994. Endangered or threatened status for five plants and the Morro shoulderband snail from western San Luis Obispo County, California. Federal Register 59:64613–64623.

Wells, P. V., and R. Berger. 1967. Late Pleistocene history of coniferous woodland in the Mojave Desert. Science 155:1640–1647.

Western Snowy Plovers and California Least Terns

Atwood, J. L., and B. W. Massey. 1988. Site fidelity of least terns in California. Condor 90:389–394.

Atwood, J., and D. E. Minsky. 1983. Least tern foraging ecology at three major California breeding colonies. Western Birds 14:57–72.

Caffrey, C. 1995. California least tern breeding season: 1994 season. Bird and Mammal Conservation Program Report 95-3. California Department of Fish and Game, Sacramento. 49 pp.

Fancher, J. M. 1992. Population status and trends of the California least tern. Transactions of the Western Section of the Wildlife Society 28:59–66.

Hothem, R. L., and S. G. Zador. 1995. Environmental contaminants in eggs of California least terns (*Sterna antillarum browni*). Bulletin of Environmental Contamination and Toxicology 55:658–665.

Massey, B. W. 1974. Breeding biology of the California least tern. Proceedings of the Linnean Society of New York 72:1–24.

Massey, B. W., and J. L. Atwood. 1981. Second wave nesting of the California least tern: age composition and reproductive success. Auk 98:596–605.

Massey, B. W., D. W. Bradley, and J. L. Atwood. 1992. Demography of a California least tern colony including the effects of the 1982–1983 El Niño. The Condor 94:976–983.

Page, G. W., J. C. Warriner, and P. W. C. Paton. 1995. Snowy plover (*Charadrius alexandrinus*). *In* A. Poole and F. Gill, editors. The birds of North America, No. 154. Academy of Natural Sciences, Philadelphia, Pa., and The American Ornithologists' Union, Washington, D.C.

Palacios, E., and L. Alfaro. 1994. Distribution and abundance of breeding snowy plovers on the Pacific coast of Baja California. Journal of Field Ornithololology 65:490–497.

Powell, A. N., C. L. Collier, and B. Peterson. 1995. Status of western snowy plovers (*Charadrius alexandrinus nivosus*) in San Diego County, 1995. Report to U.S. Fish and Wildlife Service, Portland, Oregon, and the California Department of Fish and Game, Sacramento. 31 pp.

Stenzel, L. E., J. C. Warriner, J. S. Warriner, K. S. Wilson, F. C. Bidstrup, and G. W. Page. 1994. Long-distance breeding dispersal of snowy plovers in western North America. Journal of Animal Ecology 63:887–902.

U. S. Fish and Wildlife Service. 1992. Western snowy plover (*Charadrius alexandrinus nivosus*). Endangered Species Technical Bulletin 17:1–6.

U.S. Fish and Wildlife Service. 1993. Determination of threatened status for the Pacific coast population of the western snowy plover. Federal Register 58(42):12864–12874.

Warriner, J. S., J. C. Warriner, G. W. Page, and L. E. Stenzel. 1986. Mating system and reproductive success of a small population of polygamous snowy plovers. Wilson Bulletin 98:15–37.

Pacific Northwest

The Pacific Northwest has a tremendous wealth of natural resources. Spectacular forests, expansive rangelands, and plentiful salmon have drawn people from throughout the country for more than a century. Many biodiversity issues are debated in this region, but four primary concerns are currently emphasized and are the main focus of this chapter: status of mature and old-growth coniferous forests west of the Cascade Range (commonly called the westside); status and health of forests east of the Cascade Range (commonly called the eastside); status and health of eastside rangelands; and status and health of riparian and aquatic ecosystems and anadromous fish populations.

Physiographic Overview

To help with regional assessments, scientists divide the North American continent into a series of ecoregions (biogeographical regions) that represent distinct landscape units characterized by unique combinations of vegetation, geology, topography, hydrology, climate, and ecological history (Omernik 1987, 1995). The Pacific Northwest (Fig. 1a) includes either all or portions of 12 ecoregions in the United States (Fig. 1b).

Forests of large, coniferous trees (red cedars, Douglas-firs, firs, pines, spruces, and others) dominate most of the Pacific Northwest landscape. Coniferous forests cover about 80% of western Oregon and Washington (Tesch 1995). The number of conifer species in this region is unequaled, and several are the tallest and most massive trees in the world (Franklin et al. 1981). Cool, wet winters and warm, dry summers (largely a result of oceanic influence) favor evergreen species, whereas mild temperatures and rich soils promote fast and prolonged growth (Waring and Franklin 1979).

Although conifers predominate in many areas, the Pacific Northwest landscape is highly complex (Franklin and Dyrness 1973; Barbour and Billings 1988; Tesch 1995). The ocean and mountains are the principal factors responsible for the complexity. At lower elevations west of the Olympic Mountains and Coast Range (Fig. 1b), temperatures remain consistently mild and summer fog reduces moisture stress during an otherwise dry season. Aridity and temperatures progressively increase farther inland, especially east of the Cascade Range, because of rainshadow effects caused by mountains. The warmest and driest habitats in this region occur at low elevations in the Snake River Basin–High Desert ecoregion (Fig. 1b).

Regional variation in climate, soils, and topography contributes to differences in vegetation that distinguish each ecoregion. Dense, moist forests of primarily western hemlock and Douglas-fir predominate west of the Cascades (Cascades, western North Cascades, and Coast Range ecoregions). However, the Klamath Mountains ecoregion supports a diverse mixture of drought-resistant conifers and hardwoods, a result of lower precipitation and a complex geological and ecological history. In addition, the lowland river valleys of western Oregon and Washington support extensive oak woodlands, grasslands, and wetlands composed of herbaceous plants (Willamette Valley and Puget Lowland ecoregions). On the eastside, increased aridity and frequent fires promote open, parklike stands of ponderosa pine, lodgepole pine, and western larch in montane areas (eastern North Cascades, Eastern Cascades Slopes and Foothills,

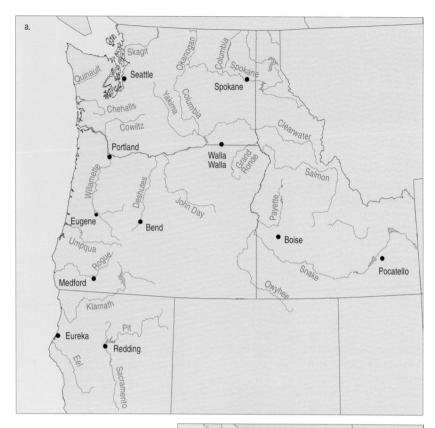

- Coast Range
- Puget Lowlands
- Willamette Valley
- Cascades
- Eastern Cascades Slopes and Foothills
- Columbia Basin
- Blue Mountains
- Snake River Basin–High Desert
- Okanogan Highlands subregion
- North Cascades
- Klamath Mountains
- Olympic Mountains

Fig. 1. a) The Pacific Northwest (rivers in blue); b) Pacific Northwest ecoregions (after Omernik 1987 © Blackwell Scientific Publishers; current version provided by S. Azevedo, U.S. Environmental Protection Agency, Corvallis, Oregon).

(Johannessen et al. 1971; Agee 1991, 1993). Scientists generally agree, however, that most presettlement forests included diverse mixtures of conifers of different age and size classes, with each class occurring in large patches (Ripple et al. 1991; Ripple 1994; Spies et al. 1994).

Fifty years of even-age timber management by clear-cut logging and replanting greatly altered the structure of forests in the region (Franklin and Forman 1987; Spies and Franklin 1988, 1991; Morrison and Swanson 1990; Ripple 1994; Spies et al. 1994). Today's forests are composed primarily of intermixed patches of stands less than 50 years old with small trees (average diameter at breast height less than 51 centimeters) and older stands with much larger trees (average diameter greater than 102 centimeters, height greater than 70 meters). Forests classified as old-growth (Fig. 2) are generally at least 200–250 years old (Old-Growth Definition Task Force 1986).

Scientists estimate that about 17% of the old-growth Douglas-fir forests that existed in the early 1800's remained in 1988 (Spies and Franklin 1988; Marcot et al. 1991) and that 96% of the original coastal rain forests in Oregon and 75% in Washington had been logged by 1988 (Kellogg 1992). About 51% (2,383,624 hectares) of the federal lands in western Oregon

and Blue Mountains ecoregions, and the Okanogan Highlands subregion) and juniper woodlands, sagebrush–steppe, and grasslands at lower elevations (Columbia Basin and Snake River Basin–High Desert ecoregions).

Status and Trends of Major Ecosystems

Westside Forests

A complex history of disturbances from natural and Native American-induced fires complicates assessments of the typical pre-European settlement in the westside landscape

Courtesy T. B. Thomas, U.S. Forest Service

Fig. 2. Old-growth Douglas-fir forest in the Bull Run watershed, Mount Hood National Forest, Oregon.

and Washington currently supports older forests (average tree diameter greater than 51 centimeters; Forest Ecosystem Management Assessment Team 1993), and about 56% (1,333,049 hectares) of these older stands (28% of total forest cover on federal lands) is classified as old-growth (Bolsinger and Waddell 1993).

Forest fragmentation (Fig. 3) is an important factor to consider when estimating remaining old-growth. Morrison (1988) estimated that 37% of the old-growth in the national forests of western Oregon and Washington occurs in fragmented landscapes (that is, stands of fewer than 162 hectares surrounded by clear cuts, young plantations, and nonforest habitats). The distribution and viability of some wildlife and plant species may be adversely affected by forest fragmentation (Harris 1984; Franklin and Forman 1987; Lehmkuhl and Ruggiero 1991). For species that depend on habitat characteristics that are unique to old-growth forests, a patchwork of small, disjunct stands of old-growth may not be equivalent to one solid patch of equal area.

Eastside Forests

In the past, wildfire, pathogens, and periodic droughts shaped the eastside forest landscape, which was much different from today's landscape (Mutch et al. 1993; Agee 1994; Arno and

Ottmar 1994; Lehmkuhl et al. 1994). Open, parklike, old-growth stands of ponderosa pine (Fig. 4), western larch, and lodgepole pine with grassy understories once dominated eastside forests at low and moderate elevations. Frequent, low-intensity fires maintained these forests. The typical landscape pattern was a fine-scale mosaic of stands of varying ages and stages of development, with young stands a result of infrequent, stand-replacing fires. A long history of selective harvesting of mature pines and larches, intensive grazing, and fire suppression greatly altered this forest landscape, however.

Sixty years ago, 74% of the commercial forests in the region was classified as ponderosa pine timber, much of which was old-growth (Cowlin et al. 1942). Recent estimates suggest that 92%–98% of the old-growth ponderosa pine that once existed in the Deschutes, Fremont, and Winema national forests has been logged or lost (Henjum et al. 1994). Stands classified as mixed-conifer timber, frequently overstocked with second-growth pines and firs tolerant of crowding, now cover most of the eastside forest landscape (Mason and Wickman 1994).

The current focus on assessments of forest health and options for restoring old-growth stands reflects concern that more than a century of human disturbance now threatens the stability and sustainability of eastside forest ecosystems (Everett et al. 1994; Tesch 1995). Large areas of dead trees exist now because of past harvest practices and fire suppression that led to overcrowding, epidemic outbreaks of insects, and unusually intense and extensive wildfires (Fig. 5).

■ Unharvested forest	▨ Nonforest
▨ Harvested forest 1972-1991	▨ Lakes and reservoirs

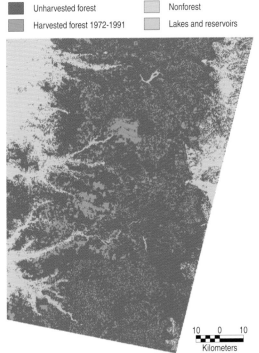

10　0　10
Kilometers

Fig. 3. Forest fragmentation in the central Oregon Cascades (image processing by W. B. Cohen, U.S. Forest Service, Corvallis, Oregon).

© S. Garrett and the McKay Collection, Bend, Oregon

Fig. 4. Classic old-growth ponderosa pine forest with open, grassy understory near Bend, Oregon.

Courtesy D. W. Scott, U.S. Forest Service

Fig. 5. Examples of unhealthy forests in the Blue Mountains ecoregion: a) overstocked firs (brown, dead, and dying trees) are susceptible to defoliation by insects such as the western spruce budworm; these insects do not feed on ponderosa pine and western larch (green trees); b) close-up of understory with overstocked grand firs that were killed by western spruce budworm—healthy, sparsely stocked, mature western larch remain in the overstory; c) combined mortality caused by attacks of budworms and bark beetles leaves a dead, fire-susceptible forest.

Suppression of low-intensity fires and selective harvesting of large, old trees have generally homogenized the eastside landscape, especially in mixed-conifer forests at middle elevations (Lehmkuhl et al. 1994). Unnaturally large, adjoining areas of densely stocked and stressed trees provide an increased food base for defoliating insects (Mason and Wickman 1988; Gast et al. 1991; Hessburg et al. 1994) and are more favorable to the growth of parasitic plants (Zimmerman and Laven 1984; Gast et al. 1991) and fungal pathogens (Filip and Schmitt 1990). The western spruce budworm is native to North America, but extensive, prolonged outbreaks of the intensity illustrated in Figure 6 were not recorded before the late 1940's (Gast et al. 1991). Historically, defoliating insects (for example, western spruce budworm, bark beetles, and mountain pine beetle) affected only small patches of forest, but such insects now tend to occupy large areas at periodic outbreak levels (Hessburg et al. 1994). From 1986 to 1991, for example, bark beetles killed an estimated 1.5 million cubic meters of timber in northeastern Oregon. Similarly, in 1991 western spruce budworms defoliated about 1,618,760 hectares of forest in the same region (Oester et al. 1992).

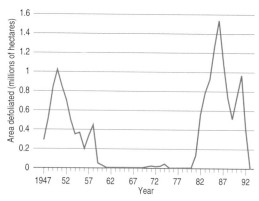

Fig. 6. Defoliation by western spruce budworms in the Blue Mountains ecoregion (data provided by D. Scott, U.S. Forest Service, LaGrande, Oregon).

Fire suppression is a principal cause of current problems with forest health in the region. The threat of extreme fire is now a serious problem because widespread tree mortality due to insect feeding and prolonged drought resulted in large accumulations of fuel. Wildfires are now more often of stand-replacing intensity and burn larger areas than they did before European settlement (Arno and Ottmar 1994). Thus, although the return of a more natural disturbance regime including frequent, low-intensity fires is the key to restoring forest health in the region, the reintroduction of fire to produce the desired results is difficult (Mutch et al. 1993; Tanaka et al. 1995).

Eastside Rangelands

Human activities during the past 100–150 years significantly altered most lowland, eastside ecosystems. More than 99% of the fertile Palouse Prairie grasslands of southeastern Washington and adjacent areas of Oregon and Idaho was converted to cropland, primarily for annual wheat (Noss et al. 1995). More than 99% of basin big sagebrush communities on the Snake River plain of Idaho and a total of 10% of the sagebrush–steppe (about 45 million hectares; Fig. 7) in the Intermountain West (that part of the United States between the Rocky Mountains and the coastal mountains), were converted to agriculture (Hironaka et al. 1983; Noss et al. 1995; West 1996). Livestock altered nearly all of the remaining sagebrush–steppe; about 30% was heavily grazed (West 1996). Because vegetation in this region did not coevolve with large herbivores, it is sensitive to grazing by domestic livestock (Daubenmire 1970). Considerable expanses (2–2.5 million hectares) of grassland and sagebrush–steppe have been replaced by invasive, nonindigenous annual vegetation, primarily because of overgrazing (Pellant 1990; Whisenant 1990; Pyke and Borman 1993; Noss et al. 1995).

Livestock grazing pressure and rangeland condition vary depending on who owns or manages the land. About 70% of nonfederal rangelands and 65% of Bureau of Land Management rangelands in the Pacific Northwest were in fair to poor condition in the mid-1980's (Joyce 1989). A similar regional rating for U.S. Forest Service lands was not available; however, a national assessment showed that 80% of these lands was in satisfactory condition for livestock in 1986 (Joyce 1989). From 1953 to 1986, grazing increased about 25% on U.S. Forest Service lands but remained at constant levels on Bureau of Land Management lands in the Pacific Northwest (Joyce 1989). These trends coincided with a 15%–20% decrease in total pasture and rangeland area in the three Pacific Coast states (Joyce 1989). Broader, integrated assessments of the health of public and private rangeland have been attempted (Society for Range Management 1989) but suffer from a lack of standardized surveys and common assessment criteria (National Research Council 1994).

Successional relations in grassland and sagebrush–steppe communities are more complex than formerly thought (Laycock 1991). The mere removal of disturbances such as grazing does not ensure the return of original plant communities (Anderson and Holte 1981; Anderson 1986). About 9% (2.8 million hectares) of the public land administered by the Bureau of Land Management in the region is sufficiently

© T. N. Kaye, Oregon Department of Agriculture

degraded so that changes in land management alone will not significantly improve rangeland quality (Pyke and Borman 1993).

Invasion and expansion of nonindigenous plants are major concerns in rangeland ecosystems of the Pacific Northwest (Pyke and Borman 1993). The introduction and initial spread of nonindigenous annual grasses (primarily cheatgrass [Fig. 8] and medusahead) occurred from the late 1800's to the mid-1900's during a period of severe overgrazing (Mack 1986). The competitive abilities of these nonindigenous grasses allow them to invade even areas that are not grazed, especially after wildfires (West and Hassan 1985; Whisenant 1990; Kindschy 1994). From 1981 to 1987, 202 fires burned 88,322 hectares in the Snake River Birds of Prey National Conservation Area in Idaho and converted a considerable area of a native shrub–steppe into sites dominated by nonindigenous grasses, primarily cheatgrass (Pellant 1990; Zarriello et al. 1994; Fig. 9). Cheatgrass is dominant on about 6.8 million hectares of sagebrush–steppe in the Intermountain West and could eventually invade an additional 25 million hectares (Pellant and Hall 1994).

The intentional introduction of nonindigenous perennial grasses such as crested wheatgrass is another problem. This species is widely

Fig. 7. Sagebrush–steppe near Fish Fin Rim, Lake County, Oregon.

Courtesy D.A. Pyke, USGS

Fig. 8. Nonindigenous cheatgrass (short, bright-green grass in foreground) invading native sagebrush–steppe near Orchard, Idaho.

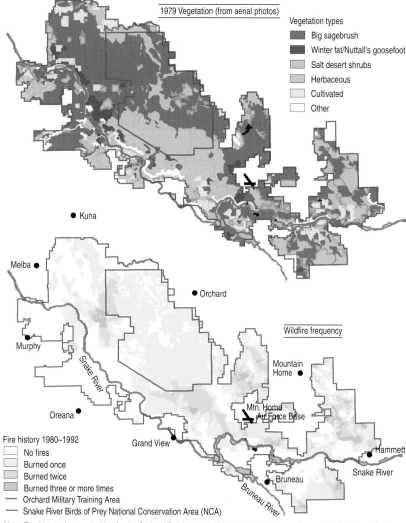

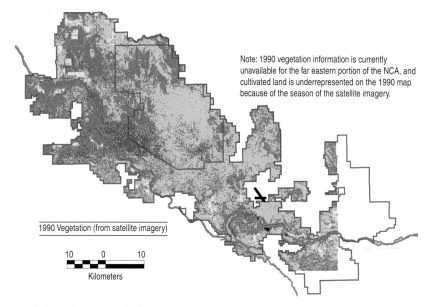

Fig. 9. Shifts from native shrub–steppe (big sagebrush and salt desert shrub) to primarily non-indigenous grasslands (herbaceous cover) between 1979 and 1990 in areas burned by wildfire in the Snake River Birds of Prey National Conservation Area (NCA), Idaho (maps prepared with the help of T. Zarriello, USGS, Forest and Rangeland Ecosystem Science Center, Corvallis, Oregon).

planted during revegetation projects because it is desirable as livestock forage and grows faster than most native species (Pellant and Monsen 1993). Even when planted as a mixture with native species, crested wheatgrass often dominates the site and thereby further reduces habitat diversity (DePuit and Coenenberg 1979; Schuman et al. 1982; Redente et al. 1984).

Juniper woodlands are also rapidly expanding, to the detriment of preferred rangeland vegetation (Miller and Wigand 1994; Fig. 10). The range of western juniper has expanded and contracted several times during the past 30,000 years. These shifts can largely be explained by changes in the global climate (Miller and Wigand 1994); however, the recent, accelerated increase in juniper coverage is a result of a combination of favorable climate, livestock grazing, and a lack of fire (Eddleman 1987; Miller and Rose 1995).

Riparian and Aquatic Ecosystems

The Pacific Northwest includes a diverse network of rivers and streams (Fig. 1a). The Columbia River is one of the largest rivers in North America, and several of its tributaries (for example, the Snake, Salmon, and Willamette rivers) are among the largest rivers in the conterminous United States. These river systems are major sources of natural resources; however, competing developmental interests, associated disturbances, and unsustainable resource extraction have led to widespread degradation of most of the river and stream networks in the Pacific Northwest.

In 1988, 95% (23,174 kilometers) of streams surveyed throughout Oregon had been moderately or severely degraded because of excessive sedimentation, high water temperatures, bank instability, or other problems with water quality related primarily to logging and removal of large woody debris from stream channels (Forest Ecosystem Management Assessment Team 1993). In 1992, 4,828 kilometers (54%) of rivers surveyed in Washington were classified as not fully supporting designated beneficial uses because of various types of pollution and degradation (Forest Ecosystem Management Assessment Team 1993). In the north coast and Klamath River basin regions of California, 51% of surveyed streams failed to meet requirements for compliance with the Clean Water Act (Forest Ecosystem Management Assessment Team 1993). About 83% of the riparian habitat on lands administered by the Bureau of Land Management was in need of extensive restoration in the late 1970's (Almand and Krohn 1979); a more recent assessment showed that this was still true in Idaho in 1991 (U.S. General Accounting

Office 1988). Degradation of these streams was mostly due to grazing-related damage to riparian vegetation, bank stability, and water quality (Kauffman and Krueger 1984; Platts 1991; Wissmar et al. 1994). Thus, throughout the Pacific Northwest, most riparian ecosystems are at least moderately if not severely degraded because of logging, grazing, mining, water diversions, hydroelectric dams, and other human activities.

The landscape along the largest river in Oregon, the Willamette River, has been drastically altered by a long history of agricultural development, channelization, and diking to control flooding (Sedell and Froggatt 1984). The southern half of the river was once a braided system with a diverse network of oxbows, small side channels, ponds, and sloughs that supported extensive marshlands and riparian gallery forests, but the river landscape is much simpler today. Moreover, large deposits of organic wastes from agricultural and urban operations greatly reduced water quality in the river until the 1950's, when secondary sewage treatment began (Hughes and Gammon 1987). These physical and biological changes have profoundly altered the associated riparian and aquatic communities along the river (Dimick and Merryfield 1945; Sedell and Froggatt 1984; Hughes and Gammon 1987; Farr and Ward 1993).

Widespread removal of large woody debris from streams (Fig. 11), lack of recruitment of new woody debris, and increased sedimentation caused by logging and other land uses have reduced the structural diversity of instream habitats for fishes and other aquatic organisms in many of the region's streams. The abundance of large, deep pools—essential components of high-quality fish habitat that frequently form behind large logs—declined in streams throughout the region during the past 50–60 years, in some areas by as much as 50%–60% (Forest Ecosystem Management Assessment Team 1993; McIntosh et al. 1994).

In the past, the status of small headwater springs, seeps, and creeks received little attention in the timber, management, and scientific communities, but this is not as true today (Johnson et al. 1991; Forest Ecosystem Management Assessment Team 1993; Thomas et al. 1993; Henjum et al. 1994; U.S. Forest Service and Bureau of Land Management 1994a). Although small and often ephemeral headwater streams typically do not support fishes, the integrity of these watershed areas is a critical determinant of water and habitat quality in downstream areas where fishes do occur (Vannote et al. 1980; Swanson et al. 1982; Naiman et al. 1992). Logging, mining, grazing, or road building in headwater areas may alter

© Bowman Museum, Crook County Historical Society, Prineville, Oregon

© J. B. Kauffman, Oregon State University

Fig. 10. Change in coverage of juniper woodland on hillsides above Keystone Ranch near Prineville, Oregon, between 1910 (top) and 1991.

downstream flow regimes, increase sediment loads, and reduce recruitment of woody debris and nutrients to fish-bearing streams. Moreover, small headwater seeps and creeks are habitats for many amphibians (Bury 1988) and may be important water sources for other forest wildlife (Brown 1985).

Many eastside river systems (for example, the John Day, Grand Ronde, Yakima, Wenatchee, Entiat, and Methow rivers) were recently studied to assess the effects of logging, grazing, and mining on riparian habitat and aquatic communities (Mullan et al. 1992; Li et al. 1994; McIntosh et al. 1994; Wissmar et al. 1994). Substantial portions of many of these forested river systems were badly degraded by a long history of logging and grazing (Henjum et al. 1994; U.S. Forest Service and Bureau of Land Management 1994b). Degradation of the rivers, creeks, and lakes in the Goose Lake–Pit River and Klamath basins of southern Oregon and northern California is also a problem because these systems harbor many endemic species and subspecies of fishes (Puchy and Marshall 1993), including two suckers (Lost River sucker and shortnose sucker) on the U.S. List of Endangered and Threatened Wildlife (U.S. Fish and Wildlife Service 1995).

Courtesy G. H. Reeves, U.S. Forest Service

Fig. 11. A healthy stream ecosystem in the Siskiyou National Forest, Oregon, with ample large woody debris that enhances habitat diversity for fishes and other aquatic organisms.

Eastside lowland ecoregions include the primary stretches of the extensive Columbia and Snake river systems. The main-stem and major tributary sections of these two rivers contain several large hydroelectric dams. Large dams create barriers to salmon and white sturgeon movements and alter river flow rates and patterns to the detriment of many fish populations (Northwest Power Planning Council 1986, 1994; Beamesderfer and Nigro 1993). In addition, most riparian cottonwood forests in Idaho are no longer self-sustaining because dams have eliminated the spring flooding that exposed the mineral soil needed for seed germination (Noss et al. 1995).

The Pacific Northwest contains more than 8,000 lakes, ponds, and reservoirs. However, aside from recent intensive research in North Cascades National Park Service Complex, Mt. Rainier National Park, and Crater Lake National Park (Liss et al. 1995; other unpublished work by W. J. Liss, Oregon State University, and G. L. Larson, U.S. Geological Survey, Corvallis, and their colleagues) and some work in Lake Lenore, Washington (Luecke 1990), these water bodies have not been well studied (Bahls 1992).

Status and Trends of Fungi, Plants, and Animals

Fungi

The wet, temperate, locally variable forest environment and high richness of plant species in the western Pacific Northwest are highly conducive for growth of a wide variety of fungi. Forest fungi are ecologically important to the region, because of their important role in forest food webs (Forest Ecosystem Management Assessment Team 1993); some species are eaten by small mammals such as western redbacked voles and northern flying squirrels (Maser et al. 1978, 1985; Maser and Maser 1988; Fig. 12), which, in turn, are primary prey species for northern spotted owls (Forsman et al. 1984). Moreover, 234 rare and endemic species of fungi are identified in the Northwest Forest Plan (Forest Ecosystem Management Assessment Team 1993; U.S. Forest Service and Bureau of Land Management 1994a) as associates of mature and old-growth forests. These concerns challenge resource managers to integrate fungal productivity, diversity, and conservation into strategies for ecosystem management in the region. The status and biology of most fungal species are poorly known, however.

Mycorrhizal and Old-Growth Forest Fungi

Many commercially harvested mushrooms (aboveground) and the truffles (belowground)

favored by small mammals are the reproductive structures of mycorrhizal fungi (Fig. 13). These fungi are particularly important components of forests in the region because they form beneficial, mutualistic symbioses with roots of most conifers (Trappe and Luoma 1992). Recent research has revealed that the diversity and productivity of mycorrhizal fungi are higher in mature stands of Douglas-fir than in young or clear-cut stands (Amaranthus et al. 1994; R. Molina, J. E. Smith, and M. A. Castellano, U.S. Forest Service, Corvallis, Oregon, unpublished data).

Scientists have identified 527 species of fungi that are closely associated with mature and old-growth forests in the Pacific Northwest (Forest Ecosystem Management Assessment Team 1993). About 150 of these species are mycorrhizal fungi that were found in the H. J. Andrews Experimental Forest in the Oregon Cascades, and 109 species are regional endemics that may be sensitive to alteration of the forest environment. One endemic species that may be an important indicator of old-growth forest conditions in the Cascade Range is *Piloderma fallax* (Smith, Molina, and Castellano, unpublished data; Figs. 14 and 15), an unusual fungus that does not produce fleshy reproductive structures (mushrooms or truffles).

Fig. 14. The golden-yellow threads running through a patch of forest soil are hyphae of the fungus *Piloderma fallax,* which is strongly associated with old-growth forests in the Oregon Cascades.

Scientists have conducted considerable research focused on inoculating tree seedlings used in forest regeneration with spores of mycorrhizal fungi to improve tree growth (Castellano and Molina 1989). Natural mycorrhizal inoculum potential declines when live trees remain absent from harvested sites for several years (Perry et al. 1987). Retention of some live host trees and abundant woody debris on harvested sites promotes forest tree and fungal regeneration (Amaranthus et al. 1994). Maintenance of mycorrhizal fungi on harvested sites also may depend on recovery of nontimber, pioneer trees, and shrubs that are normally

Fig. 12. Truffles, the belowground reproductive structures of many mycorrhizal fungi, are favored foods of small mammals such as this northern flying squirrel.

© W. Colgan III, Oregon State University

Fig. 13. This delicate coral fungus is a mycorrhizal species associated with Douglas-fir and other conifers in the Oregon Cascades.

Courtesy J. E. Smith, U.S. Forest Service

eliminated by forest management procedures (Perry et al. 1987; Amaranthus and Perry 1989).

Pathogenic Fungi

The biological diversity of microfungi (molds, mildews, rusts, and smuts) is not well known; however, some fungal pathogens have been extensively studied because of the economic consequences of timber loss. For instance, Douglas-fir is highly susceptible to the fungal infection called laminated root rot (Gast et al. 1991; Thies and Sturrock 1995).

The distribution of pathogenic fungi in eastside forests expanded as forest density and fire suppression increased. In 1986 and 1987, a minimum of 2%–20% of the trees (variable depending on the species; highest incidence among Douglas-fir) on managed stands in the Wallowa–Whitman National Forest was infected with some form of fungal pathogen (Schmitt et al. 1991). Trees weakened by fungal pathogens may be more susceptible to insect attack, and vice versa. Gast et al. (1991) discussed the biology, incidence, and effects of 11 different fungal diseases that are now widespread in the Blue Mountains ecoregion.

Fungal pathogens and forest insects reduce the growth and survival of trees; however, weakened live trees and standing or downed dead trees and snags are important habitats for many invertebrate and vertebrate species (Thomas 1979; Brown 1985). In addition, the patchy pattern of tree death and subsequent distribution of canopy gaps caused by root fungi could be a key landscape feature that enhances species and habitat diversity in mature forests (Gast et al. 1991; J. C. Tappeiner III, U.S. Geological Survey, Corvallis, Oregon, personal communication). Landscape-scale studies of the ecological roles and importance of fungi have only recently become a focus of attention, however (Amaranthus and Perry 1994).

Plants

Bryophytes and Lichens

The moist, low-elevation forests of the Pacific Northwest support a variety of plants, but they are particularly rich in bryophytes (mosses, liverworts, and hornworts) and lichens (Brinkley and Graham 1981; McCune 1993). Lobaria lichens (Fig. 16) are some of the few species of plants strongly associated with mature forests in the Cascades and Coast Range of Oregon (Spies 1991). The abundance of many mosses and liverworts is closely correlated with the abundance of coarse woody debris (Lesica et al. 1991), a primary distinguishing characteristic of mature forests. Thus, although systematic population surveys are not common,

the regional abundances of lichens and bryophytes probably declined as old-growth forests were extensively clear-cut (Forest Ecosystem Management Assessment Team 1993; McCune 1993).

Relatively few species of bryophytes and lichens occur in eastside rangelands, but these plants are major components of many arid ecosystems. Mixtures of bryophytes, lichens, fungi, algae, bacteria, and cyanobacteria, collectively called cryptobiotic crusts, fill the spaces between shrubs and bunchgrasses in many steppe ecosystems (see box on Soils and Cryptobiotic Crusts in Southwest Chapter). Crusts may improve soil fertility and stability, influence water infiltration, interfere with establishment of nonindigenous plants such as cheatgrass, and improve seed germination for various native plant species (St. Clair and Johansen 1993). In some areas, crusts may constitute 70%–80% of the living cover (Belnap 1990).

The area covered by cryptobiotic crusts in the United States is much smaller now than in the past (McCune and Rosentreter 1992). The survival of cryptobiotic crusts in the Pacific Northwest is threatened by three major factors: invasions of nonindigenous annual grasses and subsequent increases in fire frequency, the conversion of rangelands to agriculture and suburban developments, and livestock trampling (Belnap 1990; West 1990; Johansen et al. 1993; St. Clair and Johansen 1993; St. Clair et al. 1993). One cryptobiotic crust species, the woven-spored lichen, is particularly uncommon; it is listed by the Bureau of Land Management as a sensitive species in Idaho (McCune and Rosentreter 1992; Conservation Data Center 1994).

Vascular Plants

Accurate, comprehensive information on the status and population trends of most species of nontimber trees, shrubs, and herbs is lacking. The status of most species can only be inferred from studies of associations with broad forest types. One such study revealed that the relative abundance of most (80%) understory plant species does not differ among young, mature,

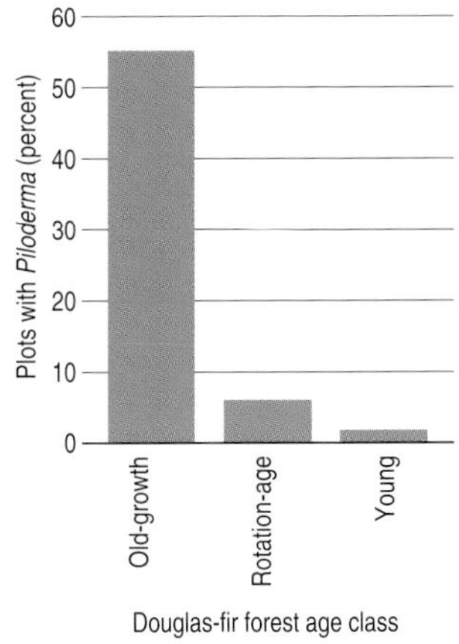

Fig. 15. Frequency distribution of the fungus *Piloderma fallax* in three age classes of Douglas-fir forest in the western Oregon Cascades.

© B. McCune, Oregon State University

Fig. 16. Lobaria lichens are characteristically most abundant in cool, moist, old-growth forests in the Pacific Northwest.

and old-growth Douglas-fir forests in the Cascades and Coast Range (Spies 1991).

Nine species of vascular plants in this region are on the U.S. List of Endangered and Threatened Plants (Table 1); however, many more species are candidates for federal listing (for example, 26 species in Oregon and 15 species in Washington; U.S. Fish and Wildlife Service 1993a) or are assigned special status in states where they occur (Fig. 17). In addition, state listings are increasing (Table 2). Most of these sensitive plant species are local or regional endemics with populations that are small, scattered, or highly restricted in distribution. The most commonly identified cause of declining populations is habitat loss caused by urban or agricultural development; however, the status of many of these plants is poorly known.

As of 1992, the Bureau of Land Management listed 144 plants in Oregon and Washington as special status species, but only 8% of these species had been monitored regularly (Willoughby et al. 1992). Among candidates for federal listing that occur in Oregon and Washington, 12 species have declining populations, 16 species have stable populations, and 11 species are listed with unknown status because of insufficient information (Category 1 species in U.S. Fish and Wildlife Service 1993a).

MacFarlane's four-o-clock (Fig. 18) is one example of an endangered plant whose status is well known and for which intensive management has proven successful. This plant occupies open rangeland and canyon habitats near the Snake River drainage in the Blue Mountains

Table 1. Vascular plants in the Pacific Northwest on the U.S. List of Endangered and Threatened Plants (U.S. Fish and Wildlife Service 1995).

Species	Federal or state status[a]					Distribution	Population trend	References[b]
	Federal	California	Oregon	Washington	Idaho			
Applegate's milk-vetch	E	—	E	—	—	Klamath Basin	Downward	
Bradshaw's desert-parsley	E	—	E	E	—	Willamette Valley	Stable	
MacFarlane's four-o-clock	E	—	E	—	S1	Blue Mountains, Western Rockies	Stable/upward	U.S. Fish and Wildlife Service 1992, 1993b
Malheur wire-lettuce	E	—	E	—	—	Snake River Basin–High Desert	Downward or stable but near extinction	
Marsh sandwort	E	E	—	ext	—	Puget Lowlands, Southern California	Nearly extinct	Washington Natural Heritage Program 1995
McDonald's rock-cress	E	E	*	—	—	Klamath Mountains	Stable or downward in California, unknown in Oregon	U.S. Fish and Wildlife Service 1992
Nelson's checker-mallow	T	—	T	E	—	Klamath Mountains, Coast Range, Willamette Valley, Puget Lowlands	Stable or upward in Oregon, upward? in Washington	
Water howellia	T	ext	ext	E	S1	Puget Lowlands, Columbia Basin, Snake River Basin–High Desert	Critically imperiled	Washington Natural Heritage Program 1995
Western lily	E	*	E	—	—	Klamath Mountains, Coast Range	Stable in California, downward? in Oregon	

[a] — = does not occur in state, * = no special status, E = endangered, T = threatened, ext = extirpated, S1 = critically imperiled. References: Oregon Natural Heritage Program 1993; Conservation Data Center 1994; Washington Department of Natural Resources 1994; California Department of Fish and Game 1995a; U.S. Fish and Wildlife Service 1995.

[b] Other comprehensive references: California Department of Fish and Game 1992; Skinner and Pavlik 1994; Oregon Natural Heritage Program 1995; Washington Natural Heritage Program 1995.

Fig. 17. Numbers of vascular plant species in the Pacific Northwest on state lists of extinct or extirpated, endangered (including Idaho Native Plant Society Category 1 species), threatened (including Idaho Native Plant Society Category 2 species), and sensitive (designated rare in California) or candidate species (Oregon Natural Heritage Program 1993; Conservation Data Center 1994; Skinner and Pavlik 1994; Washington Department of Natural Resources 1994; tallies for California and Idaho are based on county occurrences).

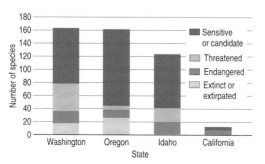

ecoregion. Since the plant was listed as an endangered species in 1979, the number of known individuals of MacFarlane's four-o-clock has increased from 25–30 individuals to around 8,600 (U.S. Fish and Wildlife Service 1993b). Efforts by government agencies and private landowners to restrict grazing during critical growing seasons are the keys to expansion of this species (U.S. Fish and Wildlife Service 1992).

About 25% of all the rare and sensitive vascular plant species in Oregon and 34% in Idaho occur in arid lowlands (Oregon Natural Heritage Program 1993; Conservation Data Center 1994). Many sensitive, lowland plant

Fig. 18. MacFarlane's four-o-clock, an endangered plant that occupies open rangeland and canyon habitats near the Snake River, has benefited from restriction of grazing during critical growing seasons.

Table 2. Status changes from 1990 to 1994 of vascular plant species listed as endangered, threatened, or sensitive in Washington (Washington Department of Natural Resources 1994).

Status category	Number of species in 1994	Number of species whose status was elevated since 1990	Number of species added since 1990	Number of species whose status was downgraded since 1990
Possibly extinct or extirpated	18	1	0	0
Endangered	19	7	1	0
Threatened	41	11	3	1
Sensitive	84	1	13	14
Total	162	20	17	15

species that occur in Idaho (13% of all such species) and Oregon (23%) are milk-vetches. Several such species are the focus of ongoing research by the U.S. Geological Survey on maintaining and restoring native plant communities on eastside rangelands (Pyke and Borman 1993).

Terrestrial Invertebrates

Mollusks

More than 150 species of terrestrial snails and slugs have been identified in moist forests of the Pacific Northwest (Forest Ecosystem Management Assessment Team 1993), and scientists are still discovering new species (Frest and Johannes 1993; Roth 1993). Of the 45 species and subspecies closely associated with mature westside forests, 17 snails and 7 slugs are given special consideration in the Northwest Forest Plan. Most mollusks have limited geographic ranges and narrow ecological requirements because they are poor dispersers. For these reasons, many species are sensitive to forest fragmentation, other aspects of forest management, and disturbances such as fire and grazing (Forest Ecosystem Management Assessment Team 1993).

Arthropods

The diverse mountains of the Pacific Northwest may support as many as 50,000–70,000 species of arthropods (insects, spiders, and their allies; P. A. Opler, U.S. Geological Survey, Fort Collins, Colorado, personal communication). Nevertheless, scientists have only recently begun to conduct comprehensive surveys of the arthropod fauna to document habitat affinities and community composition and structure. With few exceptions, it is impossible to discuss with certainty the status and trends of specific arthropod species.

Many arthropods in the region complete their long life cycles inside slowly decomposing, coarse woody debris in mature forests (Lattin 1993). The deep, moist, complex litter layers in mature westside forests support an even greater variety of arthropods. Scientists commonly recover 250 species of beetles, flies, bugs, mites, millipedes, centipedes, springtails, and sow bugs from a square meter of undisturbed forest floor (Lattin 1990; Moldenke 1990; Fig. 19). Temperate forest soil and litter layers may contain the greatest number of species and densities of arthropods in the world (Anderson 1975).

The wide variety of soil invertebrates is of scientific interest with regard to quantifying biological diversity. Perhaps more importantly, these tiny organisms play key roles in processing and recycling organic matter into nutrients that become available to trees and other forest vegetation (Maser and Trappe 1984; Moldenke and Lattin 1990; Lattin 1993). More than 200 known species of arthropods are closely associated with mature westside forests in the Pacific Northwest (Forest Ecosystem Management Assessment Team 1993; Thomas et al. 1993; P. A. Opler, U.S. Geological Survey, Fort Collins, Colorado and J. D. Lattin, U.S. Geological Survey, Oregon State University, Corvallis, unpublished manuscript). Accordingly, reduction of arthropod diversity and consequent impairment of forest health caused by extensive logging and other land uses could be significant. Landscape-scale information for the assessment of the extent of this potential problem is scarce, however.

The high species richness of arthropods in several westside areas is particularly interesting. Scientists have identified more than 3,400 species of terrestrial and aquatic arthropods in the H. J. Andrews Experimental Forest in the central Oregon Cascades (Parsons et al. 1991). This tally includes 76 species of butterflies or just under half the total known to occur in Oregon (Miller, unpublished data). The fauna of the Klamath Mountains ecoregion is also particularly diverse; it includes some primitive insects (Wygodzinsky and Stys 1970) and 44 species of butterflies in one township near Mt. Ashland (Miller, unpublished data). Mountain localities in the Coast Range and Olympic

Fig. 19. A typical collection of mites extracted from about 25 square centimeters of forest soil (field of view is about 1,500 microns wide, magnified here approximately 27 times).

© V. Behan-Pelletier, Agriculture Canada

Mountains also harbor rich and unique arthropod assemblages (Forest Ecosystem Management Assessment Team 1993), including many glacial-relict species (Lattin, unpublished data) and a diverse butterfly community around Mary's Peak near Corvallis, Oregon (60–62 species; Miller, unpublished data).

Butterflies are the best-known arthropods. Standardized, annual Fourth of July surveys to monitor populations are conducted in many areas. Fourteen counts are conducted in the Pacific Northwest (one in northern California, four in Oregon, eight in Washington, one in Idaho); however, most of the counts have been conducted for only 2–4 years (the longest for 8 years), which precludes accurate trend analyses (Swengel 1995).

The Oregon silverspot butterfly (Fig. 20), which occurs along the immediate coast, is the only terrestrial invertebrate in the Pacific Northwest on the U.S. List of Endangered and Threatened Wildlife (U.S. Fish and Wildlife Service 1995). This butterfly inhabits coastal, salt-spray meadow and open-field habitats at the fringes of coastal forests from southwestern Washington to central Oregon, with an additional disjunct population occurring in northern California (Pickering et al. 1992; Washington Department of Wildlife 1993a). The species now occurs in only 7 locations (20 in the past) along the northern Pacific Coast (1 in California, 6 in Oregon). Habitat has been lost to development and is degraded by livestock grazing, recreational disturbance, fire suppression, and invasion of nonindigenous plants (McCorkle et al. 1980; Washington Department of Wildlife 1993a). The species is listed as endangered in Washington, but aggressive management and habitat restoration have been successful in Oregon, where some populations are now stable (Hammond 1993).

The arthropod fauna seems less rich in eastside forests than in westside forests where the climate is milder and plant richness is higher (Horning and Barr 1970; Parsons et al. 1991). Few species occur on both sides of the Cascades; the abundances of those that do occur on both sides may drastically differ from one side to the other (Perry and Pitman 1983). The most current information on eastside arthropods relates to only a few species, primarily butterflies and forest insects of economic concern; thousands of other species are poorly known (Horning and Barr 1970; Cobb et al. 1981).

As previously discussed, the current focus of attention on arthropods in eastside forests is with epidemic outbreaks of defoliating insects and bark beetles rather than with declining populations (Furniss and Carolin 1977; Gast et al. 1991; Hessburg et al. 1994). Some species that now thrive in eastside forests are native to the region (for example, western spruce budworm, Douglas-fir tussock moth, mountain pine beetle), but others were introduced (for example, larch casebearer). Gast et al. (1991) and Hessburg et al. (1994) provide detailed information on the biology, ecology, history, and current status of ecologically and commercially important species.

Eastside lowlands have received limited attention from entomologists. Two examples of efforts to quantify insect diversity in the region are noteworthy. Horning and Barr (1970) catalogued 2,064 species of insects representing 248 families from Craters of the Moon National Monument in south-central Idaho, and Cobb et al. (1981) catalogued 1,055 insect species from the Alvord Desert sand dunes of southeastern Oregon.

The widespread conversion of natural grasslands and sagebrush–steppe to agriculture and pastures of nonindigenous crested wheatgrass greatly altered some components of the arthropod fauna in eastside lowlands. As in eastside forests, some native insects responded favorably to the introduction of new food sources. For example, populations of native black grass bugs expanded to epidemic proportions in habitats densely planted with crested wheatgrass (Lattin et al. 1994). Increasing the number of species in grass plantings, thereby mimicking the original state of the grasslands, helps reduce forage losses to these insects (Araya and Haws 1991).

Grasshoppers also concern agricultural and rangeland managers in the region because of the threat of epidemic outbreaks and widespread damage to crops and range grasses. Grasshopper outbreaks occur every 5–10 years and seem correlated with rainfall amounts (Fielding and Brusven 1996). Heavily disturbed sites that are dominated by nonindigenous annual vegetation typically support higher densities but fewer species of grasshoppers than sites with native shrubs and grasses (Fielding and Brusven 1993, 1994). The abundances of some native grasshoppers declined because of habitat conversion to nonindigenous vegetation; for example, the Idaho pointheaded grasshopper is one of only three species of terrestrial insects in Idaho that is listed as sensitive by the Bureau of Land Management (Otte 1981; Conservation Data Center 1994).

The conservation of arthropods has generally received little attention, but ensuring diverse and healthy insect populations could ultimately prove essential for maintaining the integrity of forest and lowland ecosystems (Moldenke and Lattin 1990; Lattin 1993). Successful conservation and restoration depend on accounting for the location of remaining healthy populations. For example, in western Oregon and

© P. Opler

Fig. 20. The federally threatened Oregon silverspot butterfly inhabits open fields and salt-spray meadows along the northern Pacific Coast.

Washington, valley landscapes are so highly altered and overrun with nonindigenous insects (Asquith and Lattin 1991; Lattin 1994) that the greatest richness and abundance of native insect species now seem concentrated at higher elevations in adjacent foothill and montane regions (Lattin, unpublished data).

Aquatic Invertebrates

A variety of environmental factors influence the abundance, species richness, and distribution of stream invertebrates. The distribution of benthic (bottom-dwelling) species may depend on substrate characteristics (Minshall 1984), particularly fine sediment concentration (Richards and Bacon 1994), which is often high in logged or grazed areas. Disturbance of riparian vegetation by logging or grazing affects the distribution of functional feeding groups by changing light levels, algal growth rates, and deposition rates of organic debris from riparian vegetation (Vannote et al. 1980; Hawkins and Sedell 1981; Murphy and Hall 1981; Murphy et al. 1981; Hawkins et al. 1982; Gregory et al. 1991). Removal of woody debris from streams during logging reduces the supply of food and stable substrates for benthic organisms and reduces retention of the detritus that other invertebrates eat (Gregory et al. 1991). Nutrient or toxic pollution in urban and agricultural areas can also lead to drastic shifts in community composition (Wiederholm 1984; Li and Gregory 1993).

Information on the status of aquatic invertebrates is generally limited to assessments of community composition rather than species-specific population trends. Because of the previously described sensitivities, inventories of aquatic invertebrates, primarily insects, are used as indicators of stream health (Anderson 1992; Plotnikoff 1992, 1994). Primary indicators of healthy cold-water streams with good water quality and abundant resources from riparian vegetation include high overall invertebrate richness with a high proportion of mayflies, stoneflies, and caddisflies (Fig. 21). Species

from these groups are intolerant of disturbances that yield low oxygen concentrations, high silt loads, and high temperatures. The Oregon Natural Heritage Program (1993) lists the Wahkeena Falls flightless stonefly as threatened with extinction and lists 24 caddisflies as probable sensitive species for which additional information is needed—more evidence that the integrity of many stream ecosystems in the region has been compromised.

Aquatic Invertebrates in Westside Rivers and Streams

Undisturbed streams in the Cascades ecoregion support a diverse aquatic invertebrate fauna, including many mayflies, stoneflies, and caddisflies. In Oregon, the diversity of aquatic invertebrates is generally highest in streams in the Cascades (Whittier et al. 1988); these streams also exhibit other faunal characteristics that indicate good-quality habitat (J. Li and A. Herlihy, Oregon State University, Corvallis, unpublished data; Table 3). During one spring–summer period on three streams in the H. J. Andrews Experimental Forest, Anderson (1992) collected more than 256,000 emerging-adult aquatic insects representing more than 250 taxa and 5 orders. In this study, a stream flowing through a 450-year-old conifer forest supported a more diverse insect fauna than streams flowing through a recent clear-cut forest and through a 40-year-old deciduous forest.

The diversity of aquatic insects is naturally low in the glacier-fed streams of the Olympic Mountains (Table 3); however, diversity is reduced further by disturbance. The number of taxa averaged 19 in undisturbed Hoh River tributaries and decreased to 9 where debris torrents occurred (McHenry 1991). Similarly, the number of taxa averaged 18 in undisturbed sections of the Elwha River and decreased to 12 below small dams (Li 1991). Sensitive aquatic insect taxa dominate these systems (Table 3).

Table 3. Characteristics of aquatic macroinvertebrate assemblages in different regions of the Pacific Northwest (averages across several streams).

Fig. 21. Mayfly nymphs like this one and other similar stream insects are considered indicators of healthy cold-water streams in the Pacific Northwest because they do not tolerate excessive disturbance.

© C. Dewberry, Florence, Oregon

Region	Density[a]	Total taxa per stream[b]	Intolerant taxa per stream[c] (percent)	Ratio of intolerant taxa to tolerant[d]	Proportion of organisms in most common taxon (percent)
Willamette Valley		29	3		
Puget Lowlands		16			
Mainstem Snake River	49	6	33		70
Western Cascades		60	51		21
Olympic Mountains		9–19	93		
Eastern Cascades	6,707	37	49	8.5	39
Columbia Basin	3,729	27	48		36
Grande Ronde River basin	3,550	42	57	10.5	31
Ochoco Mountains/John Day River basin	3,707	40	60	5.5	28
Oregon high desert	5,936	26	38	5.5	43

[a] Average total number of organisms per square meter.

[b] Typically genera.

[c] Taxa consist of intolerant (to habitat disturbance) mayflies, stoneflies, and caddisflies.

[d] Numbers of intolerant (to habitat disturbance) mayflies, stoneflies, and caddisflies to tolerant chironomid midges.

Streams in the Willamette Valley support fewer aquatic invertebrates (Table 3) and higher proportions of non-insect invertebrate species than streams in adjacent foothills and mountains (Whittier et al. 1988; Li and Herlihy, unpublished data). Similarly, the diversity of aquatic insects is generally lower in streams of the Puget Lowlands than in the Cascades (Plotnikoff 1992, 1994; Table 3). The diversity of freshwater mollusks is also usually highest in montane, spring-fed streams and pools (Forest Ecosystem Management Assessment Team 1993). Thus, decreased habitat diversity and low hydraulic variability in lowland streams seem to support fewer aquatic invertebrate species. However, part of the difference could be that human activities have affected the lowland streams for much longer periods. The number of species increased slowly in the low-gradient Tualatin River as caddisflies, riffle beetles, and mayflies returned after installation of wastewater treatment plants (Li and Gregory 1993).

Aquatic Invertebrates in Eastside Rivers and Streams

Based on regional summaries of insect community characteristics, the health of eastside streams averages fair to good (Li et al. 1995; Table 3). Most subregions, though, contain healthy (mostly at high elevations) and degraded streams (Whittier et al. 1988; Mangum 1992; Wisseman 1992a,b; U.S. Environmental Protection Agency 1994).

Comparative studies in the Ochoco National Forest and John Day River basin revealed that overgrazing, dewatering for irrigation, and loss of riparian vegetation during logging produced high water temperatures and excessive siltation that harmed aquatic invertebrate communities in many streams (Li et al. 1994; Tait et al. 1994). Still, the potential for restoration of degraded streams remains high because some mostly undisturbed streams at higher elevations have well-developed riparian canopies and still support diverse assemblages dominated by sensitive taxa.

On Catherine Creek in the Grand Ronde River basin, 10 years of extensive restoration with grazing exclosures returned the number of aquatic insect taxa (61) and the mayfly–caddisfly–stonefly to chironomid ratio (29:1; midge larvae are considered tolerant species) to high levels (McIntosh 1992). The causes of degradation are not always clear, however; the total number of taxa and the number of mayflies, caddisflies, and stoneflies are low (14 and 7 taxa, respectively), and dominance of the most common taxon is high (56%; lack of numerical or biomass dominance by one or a few taxa is another indicator of healthy communities) at Thirty-mile Creek

despite an absence of logging and light grazing by livestock (Carlson et al. 1990).

Benthic insect communities from the lower reaches of the Snake River are particularly depauperate (average 6 taxa) and are dominated by tolerant species (Dorband 1980; Table 3). The reduced fauna in low-elevation rivers and springs of the Columbia and Snake River basins is probably a result of greater human disturbance. Several species of aquatic snails from the Snake River basin (Snake River physa snail, Bliss Rapids snail, Utah valvata snail, Banbury Springs limpet, and Bruneau Hot Springs snail) are on the U.S. List of Threatened and Endangered Wildlife (U.S. Fish and Wildlife Service 1995) because of habitat degradation. In contrast, uniformly low diversity in the high desert region of southeastern Oregon (Robinson and Minshall 1991; U.S. Environmental Protection Agency 1994; M. Vinson, Bureau of Land Management, Logan, Utah, unpublished data) is probably related to the constancy of aridity and high temperatures.

The piedmont-fringe foothills of the Columbia basin ecoregion include spring-fed and seasonally intermittent streams that generally support more diverse aquatic insect communities than systems at lower elevations (Gaines 1987; Whittier et al. 1988; Gaines et al. 1992; Plotnikoff 1992; Wenatchee National Forest 1993; U.S. Environmental Protection Agency 1994). Strong representation of sensitive mayflies, stoneflies, and caddisflies further suggests that these foothill streams are relatively undisturbed natural systems that may provide source populations for restorations at lower elevations.

Aquatic Invertebrates in Lakes

Information on invertebrate communities in lakes is limited, but ongoing studies in North Cascades National Park Service Complex and in Mt. Rainier National Park are beginning to fill the void for montane lakes (Liss et al. 1995; Liss and Larson, unpublished data). The distribution of many benthic and planktonic invertebrates is naturally restricted because of sensitivities to temperature, water chemistry, and substrate type. The distribution of some species, however, is further restricted by the presence of introduced predators such as trout and sockeye salmon (see box on Complex Interactions of Introduced Trout and Native Biota in High-Elevation Lakes; Liss et al. 1995). Most lakes in the Cascades were without fishes before the onset of trout stocking; given that stocking is now widespread (Bahls 1992), the effect on invertebrate communities may be extensive (Anderson 1980; Luecke 1990; Liss et al. 1995).

Complex Interactions of Introduced Trout and Native Biota in High-Elevation Lakes

Mountain lakes in the West often occur in remote and pristine environments (Fig. 1); nevertheless, they are subject to human-induced disturbances, including global climate change, acid precipitation, ultraviolet radiation, and nonindigenous species. Most mountain lakes in the Pacific Northwest were formed by glacial activity thousands of years ago and do not naturally contain fish. Many, however, are now stocked with trout to provide opportunities for recreational fishing (Bahls 1992). Introduced trout and other nonindigenous fishes may selectively feed upon larger, more visible and active prey (Zaret 1990), and thereby adversely affect the stability of native prey populations. Negative effects include reduced densities, extirpated populations, reduced long-term regional persistence of species, and shifts in community structure toward smaller species. Such effects have been documented for amphibians (Taylor 1983; Bradford 1989; Bradford et al. 1993; Liss et al. 1995) and crustacean zooplankton (Anderson 1971, 1972, 1980; Stoddard 1987; Liss et al. 1995). Zooplankton, in particular, are important ecological components of the pelagic zone of mountain lakes. Recent research in the North Cascades National Park Service Complex in Washington suggests that the effects of trout on native biota may not be uniform on different prey species or in different lakes (Liss et al. 1995).

Salamander larvae are the top vertebrate predator in many North Cascades National Park Service Complex lakes without fishes (Fig. 2). The long-toed salamander is the most common amphibian in high-elevation North Cascade National Park Service Complex lakes. Larvae of the long-toed salamander, which are most abundant in smaller lakes and ponds, seem highly vulnerable to predation by cutthroat trout and rainbow trout (Liss et al. 1995). Snorkeling surveys during the day and at night revealed very few long-toed salamander larvae in lakes with reproducing trout populations. In contrast, larvae of the northwestern salamander (Fig. 3) seem less vulnerable to trout predation; embryo masses and larvae are often abundant in lakes with fish. The northwestern salamander has apparently developed behavioral and morphological adaptations, which may include skin secretions noxious to predators as well as reclusive and nocturnal behavior, that render it more capable of avoiding predation by trout.

Fig. 1. Tapto Lakes, typical high-elevation lakes in the Cascade Mountains of Washington.

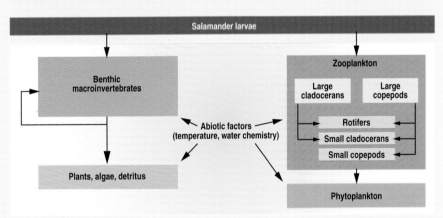

Fig. 2. Diagram of a typical food web in high-elevation lakes without fishes in the Cascade Mountains.

As in most nutrient-poor mountain lakes, North Cascades National Park Service Complex lakes generally contain few species of crustacean zooplankton. Dominance of communities by a single species is common; often one species, usually a large *Diaptomus* copepod (Fig. 4), makes up more than 90% of crustacean zooplankton abundance. Large copepods such as *Diaptomus kenai*, which is the most common species of crustacean zooplankton in this area, are apparently able to feed on a variety of foods including phytoplankton and other zooplankton. Large diaptomid

Fig. 3. Northwestern salamander adult.
© K. R. McAllister, Washington Department of Fish and Wildlife

copepods are probably the dominant pelagic, invertebrate predator when fish are absent, because other species of predaceous zooplankton (for example, phantom midges and cyclopoid copepods) are not common in North Cascades National Park Service Complex mountain lakes.

Another common species of zooplankton is the small, herbivorous copepod, *Diaptomus tyrrelli*. Whereas *D. kenai* occurs in lakes with a wide range of physical and chemical characteristics, *D. tyrrelli* seems to prefer smaller, shallower (maximum depth 10 meters) lakes with slightly higher nutrient levels (Liss et al. 1995). Apparently, predation by *D. kenai* adults eliminates *D. tyrrelli* from many lakes. Large copepods either do not occur or are uncommon in shallow lakes where the density of reproducing trout is high, whereas small copepods are abundant in these lakes. In contrast, large copepods are often abundant but small copepods are uncommon in lakes where the density of fishes is low.

The effects of introduced trout in North Cascades National Park Service Complex lakes may depend on the size distribution and density of trout, the native prey species present, lake nutrient levels, and the physical characteristics of the lake basin (Liss et al. 1995). The greatest effect of introduced trout may occur when reproducing trout occur at high densities in small, shallow lakes. Many of these lakes are productive for long-toed salamander larvae, and trout seem able to reduce or eliminate larval salamanders and large diaptomid copepods in such lakes. In lakes with higher levels of

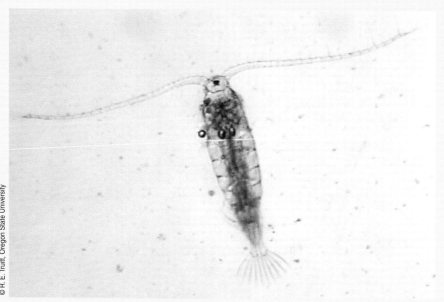

© R. E. Truitt, Oregon State University

Fig. 4. Large copepods (genus *Diaptomus* [this specimen is about 1.4 millimeters long]) are the dominant pelagic predator in many high Cascade lakes that lack fishes.

nutrients, reduced densities of large copepods brought about by fish predation may allow the abundance of smaller copepods to increase. Effects of trout on native biota may be more difficult to detect in larger, deeper lakes with low densities of fish.

See end of chapter for references

Authors

William J. Liss
Oregon State University
Department of Fish and Wildlife
Nash Hall
Corvallis, Oregon 97331

Gary L. Larson
U.S. Geological Survey
Biological Resources Division
Forest and Rangeland and
Ecosystem Science Center
3200 S.W. Jefferson Way
Corvallis, Oregon 97331

Vertebrate Animals

Fishes

Other fish species may play important roles in aquatic ecosystems of the region, but the fish fauna in the Pacific Northwest is most easily distinguished by an unequaled richness of salmon and trout. With few exceptions, scientists know relatively little about nonsalmonid species in the Pacific Northwest. Thus, our discussion is limited to the status of salmon and trout, except that we also provide a table of current status information (based mostly on literature review) for all species on federal and state sensitive species lists (Table 4).

Salmon and trout predominate in the cold, fast-flowing streams of montane forests. In the past, anadromous salmon and steelhead also filled the waterways at lower elevations as they made their way from ocean habitats to freshwater streams to breed. However, most stocks (genetically distinct, locally adapted, and

reproductively isolated populations) of salmon and trout in the Pacific Northwest are now greatly reduced or extirpated (Nehlsen et al. 1991; Wilderness Society 1993; Fig. 22).

The only salmon and trout on the U.S. List of Endangered and Threatened Wildlife (U.S. Fish and Wildlife Service 1995, 1996a,b, 1997) that occur in our region of interest are spring–summer and fall runs of chinook salmon from the Snake River in Idaho, and one stock of sockeye salmon from the Snake River, the Southern Oregon/Northern California Coast stock of coho salmon, and the Umpqua River cutthroat trout. All other species of salmon and trout are given some level of special status in one or more states where they occur, and several stocks of steelhead trout have been proposed for federal listing (Table 4).

In a landmark publication, Nehlsen et al. (1991) identified more than 100 stocks of extinct anadromous salmon and trout, 102 stocks now at high risk of extinction, 58 stocks at moderate risk of extinction, and 54 stocks of

Table 4. Status of fishes on federal and state sensitive species lists.

Species	Federal or state status[a]					Endemic	Anadromous	Range	Population trend	References[b]
	Federal	California	Oregon	Washington	Idaho					
Alvord chub	*	—	S–v	—	—	Yes		Snake River Basin–High Desert	Unknown, locally common	Williams and Bond 1983
Alvord cutthroat trout	*	—	Extinct	—	—	Yes		Snake River Basin–High Desert	Extinct	Miller et al. 1989; Williams and Bond 1983
Bigeye marbled sculpin	*	SC	—	—	—	Yes		Eastern Cascades Slopes and Foothills	Stable or downward	
Borax Lake chub	E	—	E	—	—	Yes		Snake River Basin–High Desert	Stable	U.S. Fish and Wildlife Service 1992; Oregon Natural Heritage Program 1995
Bull trout	C	E	S–c	*	SC–a			All forest ecoregions	Extirpated in California, downward elsewhere	Howell and Buchanan 1992; Washington Department of Wildlife 1992; Mongillo 1993; Rieman and McIntyre 1993
California (Pit) roach	*	SC	S–p	—	—			Snake River Basin–High Desert	Downward in California, stable in Oregon	
Catlow Valley redband trout	*	—	S–v	—	—	Yes		Snake River Basin–High Desert	Stable or downward	
Catlow Valley tui chub	*	—	S–v	—	—	Yes		Snake River Basin–High Desert	Unknown, locally common	
Chum salmon	C	*	S–c	*			Yes	All westside, extirpated in eastside, most of California	Stable or upward only in NW Washington	Frissell 1993; Moyle 1994; Kostow 1997; Mills et al. 1997
Coastal cutthroat trout	E/C[c]	SC	S–c	*	—		Yes	Klamath Mountains, Coast Range, Willamette Valley	Mostly downward	Moyle 1994; Mills et al. 1997; U.S. Fish and Wildlife Service 1996a
Coho salmon	T/C[d]	*	S–c	*			Yes	All westside, extirpated eastside	Stable or upward only in NW Washington	Frissell 1993; Brown et al. 1994; Kostow 1997; Mills et al. 1997; U.S. Fish and Wildlife Service 1996b, 1997
Cowhead Lake tui chub	C	SC	—	—	—			Snake River Basin–High Desert	Downward	
Foskett speckled dace	T	—	T	—	—	Yes		Snake River Basin–High Desert	Upward	Oregon Natural Heritage Program 1995
Goose Lake lamprey	*	SC	S–c	—	—	Yes		Eastern Cascades Slopes and Foothills, Snake River Basin–High Desert	Downward	
Goose Lake redband trout	*	SC	S–v	—	—	Yes		Eastern Cascades Slopes and Foothills, Snake River Basin–High Desert	Downward	
Goose Lake sucker	*	SC	S–c	—	—	Yes		Eastern Cascades Slopes and Foothills, Snake River Basin–High Desert	Stable or downward	
Goose Lake tui chub	*	SC	S–c	—	—	Yes		Eastern Cascades Slopes and Foothills, Snake River Basin–High Desert	Locally common but possibly downward	
Hutton Spring tui chub	T	—	T	—	—	Yes		Snake River Basin–High Desert	Downward	Oregon Natural Heritage Program 1995
Interior redband trout	*	—	S–v	—	SC–a			All eastside	Downward native, upward stocked	
Jenny Creek sucker	*	—	S–p	—	—	Yes		Eastern Cascades Slopes and Foothills	Unknown, highly restricted distribution	
Pitt-Klamath brook lamprey	*	SC	*	—	—	Yes		Eastern Cascades Slopes and Foothills	Downward?	

— continued —

Table 4. Continued.

Species	Federal or state status[a]					Endemic	Anadromous	Range	Population trend	References[b]
	Federal	California	Oregon	Washington	Idaho					
Klamath largescale sucker	*	SC	*	—	—	Yes		Eastern Cascades Slopes and Foothills	Downward	
Lahontan cutthroat trout	T	*	T	—	—	Yes		Eastern Cascades Slopes and Foothills, Snake River Basin–High Desert	Stable or downward	Wood et al. 1993; Oregon Natural Heritage Program 1995 Database
Lahontan shiner	*	*	S–p	—	—	Yes		Snake River Basin–High Desert	Stable or upward	
Lake chub	*	—	—	M	*			Eastern Cascades Slopes and Foothills, Columbia Basin, Okanogan Highlands, Blue Mountains	Unknown	
Leatherside chub	*	—	—	—	SC–c			Snake River Basin–High Desert	Unknown	
Lost River sucker	E	E	E	—	—	Yes		Eastern Cascades Slopes and Foothills	Downward	Coleman et al. 1987; Stubbs and White 1993; Oregon Natural Heritage Program 1995
Lower Columbia River fall chinook salmon	C	—	S–c	*	—	Yes		Cascades, Willamette Valley	Downward	Kostow 1997
Malheur mottled sculpin	*	—	S–c	—	—	Yes		Snake River Basin–High Desert	Unknown	
Margined sculpin	*	—	S–v	C	—	Yes		Columbia Basin, Blue Mountains	Unknown but highly restricted distribution	Lonzarich 1993
McCloud River redband trout	C	SC	—	—	—			Eastern Cascades Slopes and Foothills	Stable or downward	
Miller Lake lamprey	*	—	Extinct	—	—	Yes		Eastern Cascades Slopes and Foothills	Extinct	Miller et al. 1989
Millicoma dace	*	—	S-u	—	—			Coast Range	Unknown, locally common	
Mountain sucker	*	SC	*	M	*			Willamette Valley, Cascades, Eastern Cascades Slopes and Foothills, Columbia Basin, Snake River Basin–High Desert	Downward? in California, unknown otherwise	
Nooksack dace	*	—	—	M	—			Coast Range, Puget Lowlands	Downward in some areas, common in others	McPhail 1994; P. Mongillo, Washington Department of Fish and Wildlife, personal communication
Olympic mudminnow	*	—	—	C	—	Yes		Coast Range, Puget Lowlands	Unknown	P. Mongillo, Washington Department of Fish and Wildlife, personal communication
Oregon chub	E	—	S–c	—	—	Yes		Willamette Valley	Severely reduced range, possibly upward	Markle et al. 1989; Scheerer et al. 1994; Oregon Natural Heritage Program 1995
Oregon Lakes tui chub	*	—	S–v	—	—	Yes		Eastern Cascades Slopes and Foothills, Snake River Basin–High Desert	Unknown, locally common	
Pacific lamprey	*	—	S–v	*	E		Yes	All ecoregions except Snake River Basin–High Desert	Downward	Moyle 1994; Close et al. 1995
Paiute sculpin	*	*	*	M	*			Willamette Valley, Cascades, all eastside	Stable or upward in California, common in Washington, unknown otherwise	P. Mongillo, Washington Department of Fish and Wildlife, personal communication

— continued —

Table 4. Continued.

Species	Federal or state status[a]					Endemic	Anadromous	Range	Population trend	References[b]
	Federal	California	Oregon	Washington	Idaho					
Pink salmon	*	SC	*	*	—		Yes	All westside, extirpated in California	Downward	Moyle 1994; Mills et al. 1997
Pit sculpin	*	*	S–p	—	—	Yes		Eastern Cascades Slopes and Foothills	Stable or upward in California, downward in Oregon	
Pygmy whitefish	*	—	*	C	*			Coast Range, Cascades, North Cascades, Okanogan Highlands	Unknown	P. Mongillo, Washington Department of Fish and Wildlife, personal communication
Reticulate sculpin	*	SC	*	M	—			All westside	Downward in California, common in Washington, unknown otherwise	
Riffle sculpin	*	*	*	M	—			Coast Range, Willamette Valley, Puget Lowlands, Olympic Mountains	Stable or upward in California, unknown otherwise	
River lamprey	*	SC	*	*	—		Yes	Coast Range, Klamath Mountains, Willamette Valley, Puget Lowlands, Olympic Mountains	Stable or downward	Moyle 1994
Rough sculpin	*	T	—	—	—			Eastern Cascades Slopes and Foothills	Downward?	
Salish sucker	*	—	—	M	—	Yes		Puget Lowlands	Downward?	J. D. McPhail, University of British Columbia, personal communication
Sand roller	*	—	*	M	SC–c			Coast Range, Willamette Valley, Puget Lowlands, Eastern Cascades Slopes and Foothills, Columbia Basin	Stable in Washington, unknown otherwise	P. Mongillo, Washington Department of Fish and Wildlife, personal communication
Sheldon tui chub	*	—	S–c	—	—	Yes		Snake River Basin–High Desert	Unknown, highly restricted distribution	
Shortnose sucker	E	E	E	—	—	Yes		Eastern Cascades Slopes and Foothills	Downward	Coleman et al. 1987; Stubbs and White 1993; Oregon Natural Heritage Program 1995
Shoshone sculpin	*	—	—	—	SC–a	Yes		Snake River Basin–High Desert	Stable or upward	Wallace et al. 1982, 1984; Griffith and Daley 1984; Kuda et al. 1992; Kuda and Griffith 1993
Slimy sculpin	*	—	—	M	*			Columbia Basin, Okanogan Highlands	Common in Washington, unknown otherwise	P. Mongillo, Washington Department of Fish and Wildlife, personal communication
Snake River chinook salmon	T	—	S–c	*	E/T	Yes	Yes	Columbia Basin, Snake River Basin–High Desert	Downward	Hassemer 1993; Hassemer et al. 1997; Kostow 1997
Snake River fine-spotted cutthroat trout	*	—	—	—	SC–a	Yes		Snake River Basin–High Desert	Downward native, upward stocked	
Snake River sockeye salmon	E	—	*	*	E	Yes	Yes	Columbia Basin, Snake River Basin–High Desert	Downward	Hassemer 1993; Hassemer et al. 1997
South coast fall chinook salmon	C	*	S–c	—	—	Yes		Cascades, Klamath Mountains	Downward	Kostow 1997; Mills et al. 1997
Steelhead	PE/PT/C[e]	*	*	*	SC–a		Yes	All ecoregions	Stable or downward	National Marine Fisheries Service 1996

— continued —

Table 4. Continued.

| Species | Federal or state status[a] | | | | | Endemic | Anadromous | Range | Population trend | References[b] |
	Federal	California	Oregon	Washington	Idaho					
Summer Basin tui chub	*	—	S–c	—	—	Yes		Snake River Basin–High Desert	Unknown, probably downward	
Tahoe sucker	*	*	S–p	—	—			Snake River Basin–High Desert	Stable or upward in California, unknown in Oregon	
Tidewater goby	E	SC	—	—	—			Klamath Mountains	Downward	Frissell 1993
Umpqua chub	*	—	S–v	—	—	Yes		Coast Range, Klamath Mountains	Downward	
Warner Basin tui chub	*	—	S–c	—	—	Yes		Snake River Basin–High Desert	Downward?	
Warner sucker	T	—	E	—	—	Yes		Snake River Basin–High Desert	Downward	Williams et al. 1990; Oregon Natural Heritage Program 1995
Warner Valley redband trout	*	*	S–v	—	—	Yes		Snake River Basin–High Desert	Downward	
Westslope cutthroat trout	*	—	S–v	*	SC–a			Eastern Cascades Slopes and Foothills, Columbia Basin, Snake River Basin–High Desert	Downward native, upward stocked	
Wood River sculpin	*	—	—	—	SC–a	Yes		Snake River Basin–High Desert	Stable	

[a] — = does not occur in state, * = no special status, E = endangered, T = threatened, PT = proposed for federal threatened species listing, C = federal candidates for threatened or endangered species listing; California—CT = candidate for threatened status, SC = species of special concern; Oregon—S–c,v,p,u = sensitive species with critical, vulnerable, peripheral, or unknown status; Washington—S = sensitive species, M = monitor species; Idaho—SC–a,b,c = species of special concern category a (priority), b (peripheral), or c (undetermined status). References: Oregon Natural Heritage Program 1993; California Department of Fish and Game 1994, 1995b; Conservation Data Center 1994; U.S. Fish and Wildlife Service 1994, 1995. Washington Department of Wildlife 1994.

[b] Comprehensive references: general — Williams et al. 1989; California — McGinnis 1984; Moyle et al. 1989; Moyle and Williams 1990; California Department of Fish and Game 1992; Oregon — Puchy and Marshall 1993; Marshall et al. 1996; Washington — Wydoski and Whitney 1979; Rodrick and Milner 1991; Idaho — Simpson and Wallace 1982; salmon and trout — Konkel and McIntyre 1987; Nehlsen et al. 1991; Behnke 1992; Nickelson et al. 1992; Washington Department of Fisheries et al. 1993; Huntington et al. 1994; U.S. Fish and Wildlife Service 1994; National Marine Fisheries Service 1997.

[c] Umpqua River cutthroat trout is endangered; all other sea-run stocks are candidates.

[d] The Southern Oregon/Northern California Coast stock is threatened; the Oregon Coast stock is a candidate subject to additional review in 2000.

[e] The Upper Columbia River stock is proposed for endangered status; Snake River Basin, Lower Columbia River, Oregon Coast, Klamath Mountains Province, and Northern California stocks are proposed for threatened status; the Middle Columbia River stock is a candidate.

special concern because of low numbers or restricted distributions. Since then, researchers from Washington (Washington Department of Fisheries et al. 1993), Oregon (Nickelson et al. 1992; Kostow 1997), California (Higgins et al.

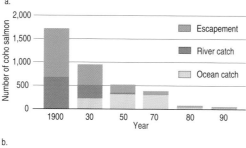

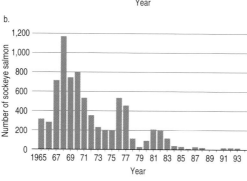

Fig. 22. Examples of declining populations of anadromous salmon: a) estimated abundance of coho salmon on the north coast of Oregon (adapted from Kostow 1997) and b) counts of sockeye salmon at upper dams on the lower Snake River in Idaho (adapted from Hassemer et al. 1997).

1992; Moyle 1994; Mills et al. 1997), and Idaho (Hassemer 1993; Hassemer et al. 1997) have conducted their own intensive assessments of stocks in their states. These more recent assessments resulted in subdivisions of some stocks and other additions to the list initiated by Nehlsen et al. (1991). In the Forest Ecosystem Management Assessment Team (1993) report, 314 out of a total of about 400 stocks that occur in the range of the northern spotted owl are identified as at risk.

In another effort, the Wilderness Society (1993) compared the past ranges with the current ranges of the 10 species and major races (for example, spring, summer, and winter runs of chinook salmon) of anadromous salmon and trout. The society determined that all species and races except the pink salmon are extinct or at risk of extinction over most of their former ranges (also see McIntosh et al. 1994; Fig. 23).

Most recently, Huntington et al. (1994) released a report on healthy native stocks in the Pacific Northwest and California. They classified 121 stocks as healthy and another 59 as marginally healthy (disputed health status); however, they cautioned that 95% of these stocks are threatened by some form of habitat degradation and that without increased conservation efforts these stocks may not remain healthy.

Remaining healthy stocks of anadromous salmon and trout are most concentrated in the upper tributaries of the Puget Sound drainage in Washington and in coastal drainages of the middle and northern Oregon Coast Range (Wilderness Society 1993; Huntington et al. 1994; Fig. 23). Few anadromous stocks from Idaho and California are healthy (Huntington et al. 1994); however, many northern coast, winter-run populations of California steelhead are still widely distributed (Mills et al. 1997), and six stocks of summer-run steelhead in Idaho are at least marginally healthy (Hassemer et al. 1997). Most anadromous stocks in the mid-Columbia basin are also relatively stable (Mullan et al. 1992; Wilderness Society 1993); however, spring runs of chinook salmon in the Grand Ronde and John Day river basins are rapidly dwindling in size (Li, personal communication), and dams eliminated all anadromous stocks from large sections of the upper Columbia and Snake river drainages (Fig. 23). Stocks that inhabit the western slope of the Cascade Range and interior drainages are at greater risk because of greater logging and grazing, both past and current (Wilderness Society 1993; McIntosh et al. 1994; Wissmar et al. 1994). Principal human-caused factors that harm salmon and trout are listed in Table 5.

Documenting changes in the status of anadromous salmon and trout is complicated because obtaining accurate indexes of abundance is difficult and because population sizes may naturally fluctuate from year to year depending on oceanic and freshwater environmental conditions (Platts 1988; Lawson 1993; Washington Department of Fisheries et al. 1993; Botkin et al. 1994). For instance, the El Niño phenomenon was linked to changes in ocean productivity (Mysak 1986) that probably affected the growth and productivity of anadromous species for multiyear periods (Nickelson and Lichatowich 1983; Nickelson 1986; Lawson 1993). Factors such as these mean that managers must have long-term (10–20 years or more) information on which to base status assessments. Some long-term data sets on commercial harvest and other indirect measures of abundance are available, but the accuracy of these indexes has been questioned (Botkin et al. 1994), and data on many watersheds and stocks do not exist.

The northern squawfish is not a member of the salmon family but is of interest with regard to management of anadromous salmon and trout. Unlike sturgeon (Beamesderfer and Nigro 1993) and anadromous salmon, the northern

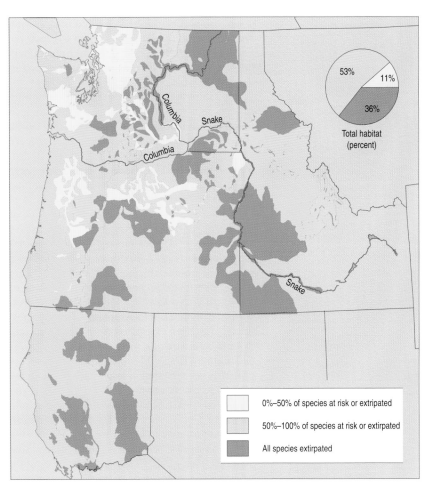

Fig. 23. Relative levels of endangerment of anadromous salmon and trout in the Pacific Northwest (adapted from Wilderness Society 1993).

Table 5. Human-caused adverse effects on salmon and trout in the Pacific Northwest.

Factor	Effects	References[a]
Hydroelectric dams	Impede migratory passage; aid predation by squawfish and sea lions	Northwest Power Planning Council 1986, 1994; Willis et al. 1994
Changes in water flow from hydroengineering, agriculture, and forest management	Alter habitat structure and water quality	Hicks et al. 1991
Siltation from logging, grazing, mining, and other land uses	Clogs spawning gravels; reduces habitat diversity	Everest et al. 1987; Bjornn and Reiser 1991; McIntosh et al. 1994
Removal of large woody debris and lack of accumulation of new debris after logging	Reduces habitat diversity, detritus and nutrient retention, and substrates for invertebrate prey	Anderson et al. 1978; Everest et al. 1985; Harmon et al. 1986; Bisson et al. 1987; Sedell et al. 1988; Bilby and Ward 1991; Reeves et al. 1993; McIntosh et al. 1994
Removal of riparian vegetation by logging or grazing	Increases water temperatures; reduces nutrient additions and sediment filtration; destabilizes stream banks	Kauffman and Krueger 1984; Beschta et al. 1987; Platts and Nelson 1989; Beschta 1991; Platts 1991; Li et al. 1994
Hybridization and competition with hatchery fish and nonindigenous species	Eliminate and reduce vigor of native stocks	Rieman and McIntyre 1993; Reisenbichler 1997
Excessive fishing	Reduces recruitment	Pacific Fishery Management Council 1992

[a] Other comprehensive references: Salo and Cundy 1987; Nehlsen et al. 1991; Meehan 1991; Bisson et al. 1992; Higgins et al. 1992; Naiman et al. 1992; Frissell 1993; Kaczynski and Palmisano 1993; Palmisano et al. 1993; Washington Department of Fisheries et al. 1993; Wilderness Society 1993; Botkin et al. 1994; Moyle 1994.

squawfish seems to have benefited from habitat changes caused by large Columbia River dams. Fast-flowing stretches of stream immediately downstream of the dams, commonly referred to as tailrace sections, are favored habitats for young salmon. Squawfish eat these young salmon and seem to benefit by congregating in tailrace areas (Rieman and Beamesderfer 1990). Because of this effect, squawfish are a concern for managers of salmon fisheries and hydroelectric facilities and are the focus of extensive predator control (Northwest Power Planning Council 1986, 1994; Willis et al. 1994).

All native trout species have populations or subspecies on federal or state lists of sensitive species (Table 4), but among resident species the bull trout (Fig. 24) is most recognized as a species in trouble. In the past, bull trout were widely distributed in cool mountain streams of the Pacific Northwest, southeastern Alaska, British Columbia, and Montana. This species, though, has been extirpated from California and from several streams in the Pacific Northwest and northern Rocky Mountains, and now occurs primarily as scattered local populations, many of which are declining in size (Howell and Buchanan 1992; Mongillo 1993; Rieman and McIntyre 1993). The U.S. Forest Service considers bull trout an indicator species because the fish is unusually sensitive to disturbances that degrade the quality of forested mountain streams. The integrity of many bull trout populations is further compromised by hybridization with nonindigenous brook trout and by competition with expanding and introduced populations of rainbow trout, brown trout, and lake trout (Rieman and McIntyre 1993). Accordingly, the bull trout is a species of concern throughout its range in the United States (U.S. Fish and Wildlife Service 1994).

Fig. 24. Bull trout are considered indicators of healthy cold-water streams in the mountains of the Pacific Northwest.

© T. H. Johnson, Washington Department of Fish and Wildlife

Amphibians

Amphibians flourish in the cool, wet forests of the western Pacific Northwest and are dominant vertebrates in many stream and upland habitats (Murphy and Hall 1981; Bury 1988; Welsh and Lind 1988; Bury et al. 1991a). Of the 33 species of amphibians that occur in the region, 15 salamanders and 2 frogs are endemic to the Pacific Northwest (Nussbaum et al. 1983; Leonard et al. 1993; Bury 1994; Table 6). Amphibians in this region have more unique and geographically isolated species than any other vertebrate group in the Pacific Northwest (Bury 1994).

An apparent global decline in the abundance of amphibians is generating considerable concern among scientists (Hayes and Jennings 1986; Blaustein and Wake 1990; Welsh 1990; Wyman 1990; Griffiths and Beebee 1992; Vial and Saylor 1993) and attention often focuses on western North America (Corn 1994). Only one Pacific Northwest amphibian is a candidate for federal listing: the Oregon spotted frog (Table 6). Many more species of salamanders and frogs are listed as sensitive species in states where they occur (Table 6). Widespread declines are of concern because amphibians are functionally significant components of the region's forest and aquatic ecosystems (Walls et al. 1992; Forest Ecosystem Management Assessment Team 1993).

Amphibians in Westside Forests

Seventeen species of salamanders, including 14 endemic species, and the tailed frog (Fig. 25) are closely associated with mature and old-growth westside forests (Forest Ecosystem Management Assessment Team 1993). Tailed frogs and 12 species of salamanders are obligate riparian species, whereas the other species are largely terrestrial. Four of the riparian species (Van Dyke's salamander, Cascade torrent salamander, southern torrent salamander, and tailed frog) and four of the terrestrial species (Del Norte salamander, Siskiyou Mountains salamander, Larch Mountain salamander, and Oregon slender salamander) are listed as sensitive, candidate, or monitor species in the states where they occur (Table 6). Several studies revealed changes in the sizes of local populations in the wake of logging and other disturbances, and repetitive sampling revealed

Fig. 25. The tailed frog is a primitive amphibian that is closely associated with cool mountain streams in the Pacific Northwest.

© D. Veseley, Oregon State University

Table 6. Status and habitat associations of amphibians in the Pacific Northwest. Ranges in parentheses indicate that the species' occurrence in the province is highly restricted.

Species	Endemic	Range	Federal or state status[a]					Population trend[b]	Habitat associations[c]
			Federal	California	Oregon	Washington	Idaho		
Black salamander		Klamath Mountains	*	*	S–p	—	—	Unknown	Large woody debris in forest; talus
California slender salamander		Klamath Mountains	*	*	S–p	—	—	Unknown	Woody debris and moist forest litter
Cascade torrent salamander	Yes	Cascades	*	—	S–v	M	—	Unknown	Seeps, springs, creeks in mature conifer forest
Cascades frog	Yes	Cascades, North Cascades, Eastern Cascades Slopes and Foothills	*	SC	S–v	*	—	Downward	Alpine and subalpine pools near streams and wet sphagnum moss
Clouded salamander		Coast Range, Klamath Mountains, Cascades	*	*	S–u	—	—	Unknown	Large Douglas-fir logs with intact bark in moist forest; talus
Columbia torrent salamander	Yes	Coast Range	*	—	S–v	M	—	Unknown	Seeps, springs, creeks in mature conifer forest
Cope's giant salamander	Yes	Coast Range, Cascades	*	—	S–u	M	—	Unknown	Riparian habitats in mature conifer forest
Del Norte salamander	Yes	Klamath Mountains	*	SC	S–v	—	—	Unknown	Talus and adjoining areas in mature conifer forest
Dunn's salamander	Yes	Coast Range, Klamath Mountains, Cascades	*	*	*	C	—	Stable?	Talus and rubble in riparian areas in mature conifer forest
Ensatina		All westside	*	*	*	*	—	Unknown	Coarse woody debris in forests
Foothill yellow-legged frog		Coast Range, Klamath Mountains, Cascades, Willamette Valley	*	SC	S–v	—	—	Downward	Warm streams and rivers
Great Basin spadefoot		Columbia Basin, Snake River Basin–High Desert	*	*	*	*	*	Downward in Idaho	Open, arid grass and shrub–steppe and pine forest
Larch Mountain salamander	Yes	Cascades	*	—	S–v	S	—	Unknown	Talus and woody debris in mature conifer forest
Long-toed salamander		All ecoregions	*	*	*	*	*	Downward in Idaho and high Cascades	Streams, ponds, lakes
Northern leopard frog		Okanogan Highlands, Columbia Basin, Snake River Basin–High Desert	*	SC	S–v	*	SC–a	Downward	Ponds, lakes, sluggish streams
Northern red-legged frog		All westside	*	SC	S–u	*	—	Downward	Ponds and other shallow, quiet waters
Northwestern salamander	Yes	All westside	*	*	*	*	—	Unknown	Streams, ponds, lakes
Olympic torrent salamander	Yes	Olympic Mountains	*	—	—	M	—	Unknown	Seeps, springs, creeks in mature conifer forest
Oregon slender salamander	Yes	Cascades, Eastern Cascades Slopes and Foothills	*	—	S–v	—	—	Unknown	Large woody debris in moist forest
Pacific chorus frog		All ecoregions	*	*	*	*	*	Downward in Idaho	Shrubby habitats
Pacific giant salamander	Yes	All westside, (Columbia Basin)	*	*	*	*	—	Stable?	Moist litter, riparian areas, and lakes in mature conifer forest
Rough-skinned newt		All westside, Eastern Cascades Slopes and Foothills, (Columbia Basin)	*	*	*	*	—	Unknown	Moist forest
Shasta salamander	Yes	Klamath Mountains	*	T	—	—	—	Stable or downward	Moist limestone formations
Siskiyou Mountains salamander	Yes	Klamath Mountains	*	T	S–v	—	—	Unknown	Talus and adjoining areas in mature conifer forest
Southern torrent salamander	Yes	Coast Range, Klamath Mountains, Cascades	*	CT	S–v	—	—	Downward ?	Seeps, springs, creeks in mature conifer forest
Spotted frog[d]		All ecoregions except Klamath Mountains	C	SC	S–c; S–u	*	SC–a	Downward	Marshy fringes of sluggish streams, lakes, and ponds
Striped chorus frog		Snake River Basin–High Desert	*	—	—	—	*	Downward	Marshy riparian, damp shrub, and woodland habitats
Tailed frog	Yes	Coast Range, Klamath Mountains, Cascades, North Cascades	*	SC	S–v	M	*	Downward	Cool, rocky streams in mature conifer forest
Tiger salamander		Eastern Cascades Slopes and Foothills, Okanogan Highlands, Columbia Basin, Snake River Basin–High Desert	*	*	S–u	M	*	Stable or upward	Lakes, ponds, reservoirs in grass and shrub steppe
Van Dyke's salamander	Yes	Coast Range, Cascades	*	—	—	C	—	Unknown	Moist litter, talus, riparian areas in mature conifer forest
Western toad		All ecoregions	*	*	S–v	*	SC–b	Locally variable	Marshes, ponds, humid terrestrial with moderate vegetative cover
Western red-backed salamander	Yes	All westside	*	—	*	*	—	Unknown	Coarse woody debris in young forest
Woodhouse's toad		Columbia Basin, Snake River Basin–High Desert	*	*	*	M	*	Stable or upward	Grass and shrub–steppe, riparian areas

[a] — = does not occur in state, * = no special status, C = candidate for federal threatened or endangered species listing; California — CT = candidate for threatened status, SC = species of special concern, T = state-threatened; Oregon — S–c,v,p,u = sensitive species with critical, vulnerable, peripheral, or unknown status; Washington — S = sensitive species, C = candidate for sensitive species status, M = monitor species; Idaho — SC–a,b = species of special concern category a (priority) or b (peripheral). References: Oregon Natural Heritage Program 1993; California Department of Fish and Game 1994, 1995b; Conservation Data Center 1994; U.S. Fish and Wildlife Service 1994; Washington Department of Wildlife 1994.

[b] Primary comprehensive references: Rodrick and Milner 1991; Groves and Peterson 1992; Marshall et al. 1996; see text for others.

[c] Primary comprehensive references: Nussbaum et al. 1983; Leonard et al. 1993; see text for others.

[d] Recently split into two species, Oregon and Columbia spotted frogs; both species show evidence of decline (see box on spotted frogs in the Western Pacific Northwest).

some large-scale extirpations; however, little quantitative data exist to accurately document regional population trends (Corn 1994).

Undisturbed streambeds in Pacific Northwest forests include mostly sediment-free gravel and large woody debris that provide vital cover and breeding substrates for stream amphibians (Bury and Corn 1988; Bury et al. 1991a). Woody debris in streams also provides substrate and food for invertebrate prey (Anderson et al. 1978). Land uses that remove riparian vegetation and woody debris and increase water temperatures and sedimentation in streams are detrimental to many amphibians (Murphy and Hall 1981; Welsh and Lind 1988; Corn and Bury 1989; Welsh 1990; Bull 1994). The number of species, the density, and the biomass of amphibians range from two to seven times higher in streams flowing through natural forests than in streams flowing through logged areas in the Coast and Cascade ranges of Oregon (Corn and Bury 1989).

The negative effect on amphibians of large-scale habitat modification and chronic sedimentation in streams may persist for decades (Corn and Bury 1989). Logging, though, is not always detrimental to stream productivity. In high-gradient streams where sedimentation is less of a problem, larvae of the Pacific giant salamander are often more abundant in streams that flow through clear-cuts than in streams that flow through intact forests. The salamanders respond to the greater primary productivity and abundance of invertebrate prey that result from increased insolation after removal of riparian vegetation (Murphy and Hall 1981; Murphy et al. 1981; Hawkins et al. 1983).

Forest fragmentation from clear-cutting, excessive thinning of stands, and removal of woody debris desiccate forest floors and harm amphibians. Southern torrent salamanders and tailed frogs often occur as isolated populations because they are unable to disperse across dry forests (Bury and Corn 1988; Bury et al. 1991a). Ensatinas, five endemic species of woodland salamanders, Oregon slender salamanders, clouded salamanders, and black salamanders are closely associated with woody debris and moist microhabitats in the riparian and forest-floor areas they prefer (Herrington and Larsen 1985; Aubry et al. 1988; Bury and Corn 1988; Raphael 1988; Welsh 1990; Bury et al. 1991b; Corn and Bury 1991; Gilbert and Allwine 1991a; Washington Department of Wildlife 1993b). In the Klamath Mountains ecoregion, in particular, the richness and abundance of terrestrial salamanders are generally higher in older and more mesic forests than in younger and drier forests (Welsh and Lind 1988).

Widespread Declines of Ranid Frogs

Much of the concern over global amphibian declines is focused on frogs, which may be particularly sensitive to a variety of modern environmental disturbances. Evidence suggests that every species of ranid frog in the western United States has experienced local or regional population declines (Hayes and Jennings 1986). The reasons for the declines vary depending on the species and locale; however, the reasons also often remain obscure (Corn 1994). Widespread loss and degradation of permanent and ephemeral waterways are common causes of amphibian declines. One probable threat to aquatic frogs may be a worldwide problem—increasing ultraviolet-B radiation caused by the thinning of the ozone layer could be causing widespread mortality of frog eggs (Blaustein et al. 1994a).

Large populations of foothill yellow-legged frogs and northern red-legged frogs still occur in parts of the Klamath Mountains ecoregion; however, most populations farther south are extirpated or declining (Blaustein and Wake 1990; Fellers and Drost 1993; Corn 1994; Jennings and Hayes 1995; Marshall et al. 1996), and northern red-legged frogs have disappeared from much of the Willamette Valley (Blaustein et al. 1994b). Population sizes of Cascades frogs are declining throughout the Cascades (Blaustein and Wake 1990; Marshall et al. 1996), and 24 of 30 populations monitored since the 1970's in Lassen Volcanic National Park in California are almost extirpated (Blaustein and Wake 1990; Fellers and Drost 1993). The spotted frog (now reclassified as two species, the Oregon spotted frog and Columbia spotted frog) was once widespread in a variety of low and high-elevation habitats on both sides of the Cascades. Now this species is almost extirpated west of the Cascades (see box on Spotted Frogs in the Western Pacific Northwest), and populations are declining in many other areas (Nussbaum et al. 1983; Hayes and Jennings 1986; McAllister et al. 1993; Hayes 1994).

Surveys in the late 1950's, 1970's, and early 1980's revealed robust populations of northern leopard frogs in several areas of eastern Washington (Metter 1960; Leonard and McAllister 1996). More recently, however, biologists from the Washington Department of Fish and Wildlife and Department of Natural Resources failed to find leopard frogs in most of these historical localities (Leonard and McAllister 1996). Introduced bullfrogs were usually abundant in areas without northern leopard frogs and absent in areas with leopard frogs (also see Bury and Whelan 1984).

Spotted Frogs in the Western Pacific Northwest

Before the turn of the century, most zoological collections were concentrated near forts and the early precursors of today's major cities. The spotted frog (Fig. 1) was first described from specimens collected by scientists based at Fort Steilacoom near present-day Tacoma, Washington (Baird and Girard 1853). Early naturalists also wrote about spotted frogs near Seattle and Portland (Dickerson 1906; Jewett 1936). Between 1850 and 1940, specimens were deposited in herpetological collections across the country. These specimens provide the primary evidence of the species's formerly broad distribution in western Washington and Oregon.

Fig. 1. Oregon spotted frog.

After 1940 the decline of spotted frog populations in western Oregon and Washington was probably well under way. During the 1950's and 1960's, researchers in western Washington and western Oregon had difficulty finding animals for their studies (Dumas 1966; Storm 1966). In the last several years, scientists expanded their searches for spotted frogs in the western halves of both states, but with few positive results. A single population near Olympia, Washington (discovered in 1990), is the last known population in the heavily developed lowlands of the Puget Lowland and Willamette Valley ecoregions (McAllister et al. 1993; Hayes 1994; Fig. 2).

Recent genetic investigations demonstrated significant differences between the spotted frogs of western Oregon and Washington and those from other parts of the species' range, which includes much of eastern Washington and Oregon and parts of Idaho (Leonard et al. 1993). These results warranted recent reclassification of the Oregon spotted frog and the Columbia spotted frog as distinct species (Green et al. 1996, 1997; Fig. 2). The Oregon spotted frog is now rare. In addition to the population near Olympia, these frogs are still found in northwestern Klickitat County in Washington, in several lakes and marshes in the Oregon Cascades (Hayes 1994), and at one site in southern British Columbia.

Nonindigenous bullfrogs are probably a primary cause of declining populations of spotted frogs (Storm 1966; Nussbaum et al. 1983; McAllister et al. 1993). Bullfrogs were brought into Washington and other western states to be farmed for culinary purposes. By the mid-1900's, breeding populations of bullfrogs were well established in the Puget Lowland and Willamette Valley ecoregions (Slater 1939; Nussbaum et al. 1983; Bury and Whelan 1984). Bullfrogs, though, are probably not the only cause of declining populations of spotted frogs. Introduced fishes, particularly warm-water species such as largemouth bass, sunfishes, perch, and bullhead catfishes, could also be involved because these species prey on both spotted frog tadpoles and adults (Hayes and Jennings 1986). In addition, human developments have altered or eliminated wetlands and introduced a wide array of contaminants to many aquatic systems.

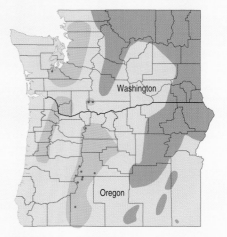

- Historical range of Columbia spotted frog.
- Historical range of Oregon spotted frog.
- Currently known populations of Oregon spotted frog.

Fig. 2. Historical and current range of the Oregon spotted frog, and historical range of the Columbia spotted frog in the United States.

The means for conservation of remaining populations of spotted frogs are not readily apparent. If nonindigenous species are the primary problem, some spotted frog populations remain vulnerable. Scientists continue to discover bullfrogs in new localities in Washington and Oregon. Habitat alterations from human activities also continue in most areas. Detailed knowledge of the effects of various habitat alterations and contaminants on spotted frogs is lacking.

See end of chapter for references

Author

Kelly R. McAllister
Washington Department of Fish and Wildlife
600 Capital Way N
Olympia, Washington 98501

Other Eastside Amphibians

A recent survey of experts in Idaho provides information about the possible status of some eastside amphibians (Groves and Peterson 1992; Table 7), including additional evidence of declines among spotted frogs and northern leopard frogs. Other sources suggest that some populations of Great Basin spadefoots (Fig. 26) benefited from increased irrigation (Nussbaum et al. 1983), whereas other populations suffered when breeding habitats in sagebrush were

Species	Number of responses	Percent stable or upward	Percent downward
Great Basin spadefoot toad	4	0	100
Long-toed salamander	7	43	57
Northern leopard frog	9	22	78
Pacific chorus frog	12	33	67
Spotted frog	4	25	75
Striped chorus frog	5	20	80
Tailed frog	4	75	25
Tiger salamander	4	75	25
Western toad	14	50	50
Woodhouse's toad	1	100	0

Table 7. Population trends of amphibians in Idaho based on a 1990 mail questionnaire sent to regional experts (Groves and Peterson 1992).

Fig. 26. Great Basin spadefoot.
© D. Quinney, State of Idaho, Military Division

replaced with grainfields and altered by hydro-electric development (Leonard et al. 1993). Otherwise, accurate status information and quantitative data on populations of eastside species are lacking (Groves and Peterson 1992).

Reptiles

The greatest richness of reptiles occurs in drier, warmer habitats on the eastside and in the mixed evergreen forests and oak woodlands of the Klamath Mountains ecoregion. Northwestern garter snakes and Oregon garter snakes are the only two species of reptiles endemic to the cool, moist forests of the Pacific Northwest (Nussbaum et al. 1983; Stebbins 1985). Populations of garter snakes (including those of three other species) are secure throughout the region (Marshall et al. 1996).

Sharp-tailed snakes (a sensitive species in Oregon), ring-necked snakes, western rattlesnakes, and common kingsnakes are largely extirpated from their former ranges in the southern Willamette Valley. Common causes for these declines include urban and agricultural development, removal of woody debris during logging, and loss or degradation of oak woodland habitats (Puchy and Marshall 1993; Marshall et al. 1996). Similarly, the California mountain kingsnake is listed as sensitive in Oregon and is a candidate for sensitive species status in Washington because of possible associations with coarse woody debris and sensitivities to disturbances from logging (Washington Department of Wildlife 1994; Marshall et al. 1996).

Many other reptiles that occur in the region, especially lizards, seem to respond positively to the opening of forests by logging, at least until the canopy closes in 10–30 years (Bury and Corn 1988; Raphael 1988). This is consistent with the trend of increasing diversity in warmer, drier habitats. No lizards that occur in the western Pacific Northwest are on state or federal lists of endangered, threatened, or sensitive species, but the actual status of these species is poorly known.

The western pond turtle (Fig. 27) is the only nonmarine reptile listed as endangered or threatened in Washington (Washington Department of Wildlife 1994), and it is one of only two reptiles listed as sensitive with critical status in Oregon (Oregon Natural Heritage Program 1993). Western pond turtles are abundant in parts of the Klamath Mountains ecoregion, but they have been extirpated or are rare from Seattle south through the Willamette Valley where threats from agriculture, pollution, and urbanization are greater (Holland 1991; Washington Department of Wildlife 1993c; Marshall et al. 1996; Holland and Bury 1997). Because these turtles do not reproduce until they are 8–11 years old and then lay small clutches of eggs, depleted populations rebound slowly, if at all (Holland and Bury 1997).

Fig. 27. Western pond turtle.
© D. Holland, Camp Pendleton Amphibian and Reptile Survey, Fallbrook, California

Eastside lowlands support the highest richness of reptiles in the Pacific Northwest, including several species that reach their northernmost range limits in southern Oregon and Idaho. The status of these species is poorly known, but reptiles are functionally important in arid steppe ecosystems. Striped whipsnakes, long-nosed snakes, ground snakes, desert collared lizards, and desert horned lizards are afforded some level of special status in Oregon, Washington, or Idaho. All of these species are widespread in the Great Basin (Stebbins 1985); the listings primarily reflect concern for local, peripheral populations that may be particularly susceptible to habitat degradation or other disturbances. However, desert collared lizards and desert horned lizards may be threatened by excessive collecting for the pet trade (Idaho Department of Fish and Game 1994; Marshall et al. 1996). In addition, large-scale changes in the structure and distribution of shrub–steppe vegetation and related habitats may adversely affect reptiles (Werschkul 1982); however, conservation is greatly hampered by the general lack of information on the status of reptilian populations.

Birds

The assessment of population trends of birds has been more successful than that for other vertebrate groups because the North American Breeding Bird Survey conducted by the U.S. Geological Survey provides long-term (1968–97 in the Pacific Northwest) abundance data for many species (Robbins et al. 1986; Peterjohn and Sauer 1993; Peterjohn 1994; Peterjohn et al. 1995). Breeding Bird Survey methods, though, do not adequately account for sparsely distributed, secretive, or nocturnal species. Consequently, Breeding Bird Survey data exist for only about 64% of the bird species that occur in westside forests (a total of about 140 species), and similar or more severe limitations apply to other regions. Moreover, the reliability of the data available for assessing trends in particular regions is highly dependent on the number of routes (25-kilometer segments along roads) surveyed each year. For the purpose of this report, we consider only species with sample sizes of at least 20 routes, and consider data for species with samples sizes between 20 and 40 routes only marginally reliable. In most eastside regions, in particular, many bird species are sparsely distributed and the survey effort was limited until just recently. In some cases, other survey data and information help fill the gaps in knowledge and increase the reliability of the indicated trends.

The proportion of species for which raw or modified Breeding Bird Survey data (modified data after Carter and Barker 1993; discussed further in eastside sections below) indicate significant population trends varies across regions of the Pacific Northwest (Fig. 28). Most important, the proportion of species with downward trends in population sizes is greatest in low- to moderate-elevation westside forests (in the South Pacific Rainforest region). Probable causes for this trend include widespread loss and degradation of mature riparian woodlands and extensive replacement of structurally diverse coniferous forests with early-seral vegetation because of clear-cutting (Mannan and Meslow 1984; Rosenberg and Raphael 1986; Raphael et al. 1988; Lehmkuhl et al. 1991; McGarigal and McComb 1993; Hejl 1994). This regional comparison may be misleading, however, because the proportion of species for which no data are available (many of which may have declining population sizes) is high in all regions.

Birds in Westside Ecosystems

The only birds on the U.S. List of Endangered and Threatened Wildlife (U.S. Fish and Wildlife Service 1995) that nest and regularly occur in westside forests are bald eagles, northern spotted owls, and marbled murrelets.

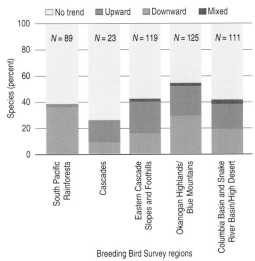

Fig. 28. Regional variation in proportions of bird species for which raw (westside forests) and modified (eastside habitats; after Carter and Barker 1993) Breeding Bird Survey data (U.S. Geological Survey, Washington, D.C.) indicate significant population trends (mixed means that the trends reversed during the past decade or so) since the late 1960's. (*N* = sample size. The South Pacific Rainforests region includes the Klamath Mountains, Coast Range, and low- to moderate-elevation western Cascades and North Cascades ecoregions. Cascades region includes the higher-elevation Cascades).

Each of these threatened species depends on habitat resources that occur mostly in mature conifer or riparian forests (Brown 1985; Thomas et al. 1993; Ralph et al. 1995; see box on Northern Spotted Owl). Several other species that state governments list as sensitive (for example, northern goshawk, harlequin duck, black-backed woodpecker, pileated woodpecker [Fig. 29], and Vaux's swift) also usually find optimal habitat in such mature forests (Brown 1985; Thomas et al. 1993). Breeding Bird Survey data indicate no significant population trends for pileated woodpeckers and Vaux's swifts in westside forests since 1968, but plots of annual abundance indexes show slight nonsignificant declines (Fig. 30). Breeding Bird Survey data on the other species are not available.

Other survey data show that the number of breeding attempts by bald eagles in Oregon steadily increased and that productivity (young per successful territory) remained stable from 1979 to 1994 (Fig. 31). Still, the productivity of some populations along the Columbia River may yet be compromised by waterborne toxins (C. J. Henny, U.S. Geological Survey, Corvallis, Oregon, personal communication).

Bird populations of westside forests are often affected by direct physical changes to their habitat. Widespread logging of marbled

Fig. 29. Pileated woodpecker at its nest cavity.

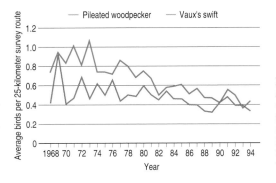

Fig. 30. Trends in the annual abundance indexes of pileated woodpeckers and Vaux's swifts in the western Pacific Northwest (Breeding Bird Survey data for the South Pacific Rainforest region, U.S. Geological Survey, Washington, D.C.).

Northern Spotted Owl

The northern spotted owl (Fig. 1) is an inconspicuous, medium-sized, dark brown owl that inhabits forests of the Pacific Coast region from southwestern British Columbia to central California (Fig. 2). It has been the centerpiece of debate regarding forest management on federal lands in the Pacific Northwest because of its apparent preference for large tracts of old-growth forest (Thomas et al. 1990, 1993; Forest Ecosystem Management Assessment Team 1993). In 1990 the species was federally listed as threatened.

Spotted owls are primarily nocturnal and normally spend their days perched in a protected roost. They nest in cavities or on platforms in large trees in nests built by other species (Forsman et al. 1984). Established pairs normally remain in the same territories from year to year; annual foraging areas may exceed 1,000 hectares (Forsman et al. 1984; Thomas et al. 1990; Carey et al. 1992a). Spotted owls eat a broad range of

Courtesy Bureau of Land Management

Fig. 1. Northern spotted owl adult and nestlings on nest in old-growth snag.

Fig. 2. Current demographic study sites for northern spotted owls and the distribution of all owl pairs located in California, Oregon, and Washington between 1986 and 1994 (map prepared by the U.S. Forest Service, Olympia Forestry Sciences Laboratory, Washington).

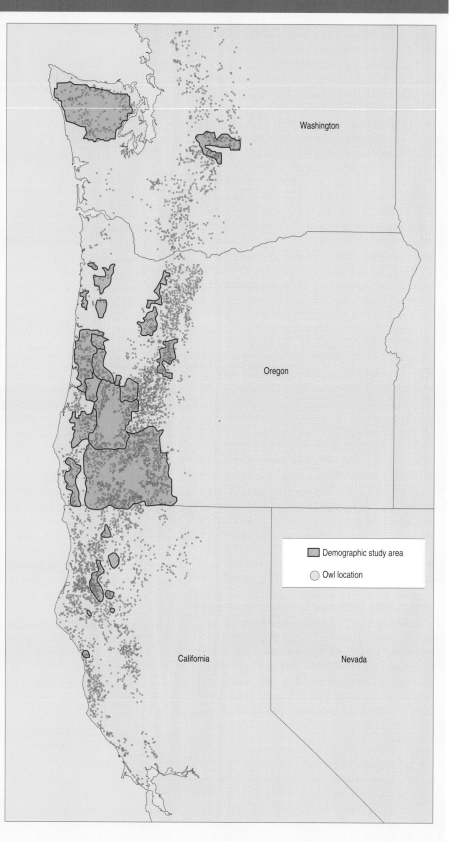

Washington

Oregon

Demographic study area

Owl location

California

Nevada

mammals, birds, insects, amphibians, and reptiles, but northern flying squirrels, voles, mice, and woodrats are their primary prey (Forsman et al. 1984; Thomas et al. 1990; Carey et al. 1992b). Predators of spotted owls include great-horned owls and northern goshawks.

Spotted owls occur in many types and age-classes of forests, but most occur in older forests, and most scientists believe that young forests are marginal habitat for spotted owls. Studies of habitat use (Forsman et al. 1984; Carey et al. 1992b) and of landscape features around nest sites of spotted owls (Ripple et al. 1991; Lehmkuhl and Raphael 1993) confirm selection of older forests for nesting and foraging. Nevertheless, most landscapes occupied by spotted owls include diverse mixtures of old and young forest patches that are created by natural disturbances and timber harvest.

An intensive and extensive survey by many federal and state agencies, consulting firms, and private landowners revealed that moderately large populations of northern spotted owls still exist (Thomas et al. 1993). The number of known or suspected pairs is approximately 30 in British Columbia, 860 in Washington, 2,900 in Oregon, and 2,300 in northern California (E. D. Forsman, U.S. Forest Service, Corvallis, Oregon, unpublished data). Studies of banded birds, however, suggested that adult survival has declined in recent years and has caused the population size of territorial owls to dwindle at an accelerating rate (Burnham et al. 1994). Population assessments based on studies of banded birds are controversial (Thomas et al. 1993), but the reliability of such indexes should increase as more years of data are included.

Population assessments are further complicated by the fact that responses by spotted owls to forest management seem to vary from region to region. For example, in some portions of northwestern California, spotted owls are relatively common in forests aged 60–100 years (L. V. Diller, Simpson Timber Co., Korbel, California, personal communication), whereas few owls occur in such forests in the central Oregon Coast Range (Forsman, unpublished data). Differential use of young forests by spotted owls probably depends on regional differences in prey populations, forest structure, and climate.

Because spotted owls use a wide range of forest types, managers have had difficulty developing a simple description of owl habitat that can be applied to all areas. This has led to considerable debate over how much habitat is still available for spotted owls. More is known about the distribution and abundance of the northern spotted owl than about any other owl in the world, but the status of the species is still hotly debated.

The productivity and occurrence of spotted owls also are affected by expanding populations of barred owls. The range of barred owls has been expanding from the eastern United States since the early 1900's and now includes western Canada, the Pacific Northwest, and northern California (Taylor and Forsman 1976; Hamer et al. 1994). Barred owls have invaded many forest areas that were previously occupied by spotted owls. In some cases, barred owls seemed to displace resident spotted owls. In other cases, individuals of the two species hybridized. The long-term effects of the barred owl invasion on spotted owl populations will probably remain unclear for many decades.

Current studies of spotted owls are many and diverse, including studies of population dynamics, diet, habitat, prey, dispersal, behavior, physiology, and genetics. Several large-scale demographic studies (Fig. 2) designed to monitor the survival and rates of reproduction of spotted owls are the most controversial (Burnham et al. 1994). These demographic studies cover a large portion of the owl's range and are the source of most of the current information on population trends. Despite this large investment in research and monitoring, spotted owl population trends are still not fully understood, especially in relation to changing habitat conditions.

Because spotted owls are a focus of debate about forest management practices in the Pacific Northwest, surveying and monitoring these owls will probably remain a high priority on federal forest lands. Although most current monitoring involves long-term studies of banded birds, other less costly methods of population assessment (for example, transect surveys of calling birds or habitat-based monitoring) are needed (U.S. Fish and Wildlife Service 1992). The ultimate objective of regional monitoring of a species such as the spotted owl is to learn if implementation of proposed management plans maintains viable populations. A meaningful effort will require extensive tracking of owl and prey populations and habitat changes for several decades.

See end of chapter for references

Author

Eric D. Forsman
U.S. Forest Service
Pacific Northwest Research Station
3200 S.W. Jefferson Way
Corvallis, Oregon 97331

---

murrelet nesting habitat in coastal old-growth forests probably has depleted this species' populations and reduced its range in the Pacific Northwest (Ralph 1994; Ralph et al. 1995). The U.S. Forest Service considers northern goshawks an indicator of old-growth forest conditions, and goshawks seem sensitive to forest fragmentation (Reynolds 1989; Marshall et al. 1996). Harlequin ducks nest in riparian areas in mature forests where disturbance is minimal, where abundant woody debris provides loafing areas and shelter for nests, and where healthy streams provide adequate macroinvertebrate prey (Rodrick and Milner 1991; Marshall et al. 1996). Black-backed woodpeckers and other woodpeckers may be adversely affected by current efforts to eliminate insect-occupied, diseased, and cavity-bearing trees and snags from

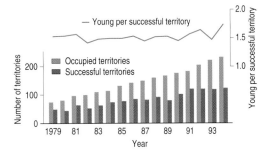

Fig. 31. Trends in the population size and productivity of nesting bald eagles in Oregon (adapted from Isaacs and Anthony 1994).

timber stands (Bull et al. 1986; Goggans et al. 1989; Marshall et al. 1996). Despite knowledge of these sensitivities, quantitative population data for these species are scarce.

Thomas et al. (1993) listed 38 species of birds closely associated with mature and old-growth westside forests. Breeding Bird Survey data for 17 of these species (including Vaux's swifts and pileated woodpeckers) are at least marginally reliable. Three of the 17 species show significant long-term or recent downward population trends (winter wren, golden-crowned kinglet, and Wilson's warbler; Table 8; Fig. 32); none show significant upward trends. Nine other species that often forage or nest in mature conifer or riparian forests, but which may also find suitable habitat in younger forests, also show downward trends (Table 8). For instance, rufous hummingbirds occur in a variety of shrub and forest habitats but are often most abundant around older forests (Gilbert and Allwine 1991b; Ralph et al. 1991). These downward trends are consistent with evidence of large-scale shifts in habitat availability from old-growth to young forests.

About one-third of the species that occur in westside forests use cavities in primarily large trees and snags for nesting and roosting (Brown 1985). Reliable Breeding Bird Survey data are available for 19 of these species. Winter wrens, which nest in a variety of cavities, crevices, and brush piles in mature forests, and four species that nest in residual, old-forest snags in open

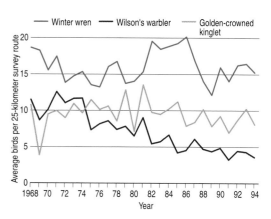

Fig. 32. Examples of downward trends in the annual abundance indexes of bird species that occur in mature and old-growth forests in the western Pacific Northwest (Breeding Bird Survey data for the South Pacific Rainforest region, U.S. Geological Survey, Washington, D.C.).

young-forest, early-seral, or riparian habitats (Bewick's wren, black-capped chickadee, western bluebird, American kestrel) show long-term or recent downward population trends (Table 8). Breeding Bird Survey data indicate no significant trends for five species of woodpeckers—

Table 8. Population trends (Breeding Bird Survey data, U.S. Geological Survey, Washington, D.C.) and habitat associations of bird species with significant long-term (1968–1994) or recent (1980–1994) downward abundance trends in westside forests.

Species	Trends by region[a,b] South Pacific Rainforests (long-term/recent)	Cascade Mountains (long-term/recent)	Old-growth forest	General forest	Shrub/ early seral forest	Meadow/ grassland	Riparian woodland	Herbaceous wetland	Cavity-nesting
American kestrel	D*/NS*		P		P		P		P
Band-tailed pigeon	D/D		P	P			P		
Barn swallow	D/D					P		P	
Bewick's wren	D/NS			S	P		P		S
Black-capped chickadee	NS*/D*		S	P			P		P
Brown-headed cowbird	D/NS	NS*/ND		P	P		P		
Bullock's oriole	D*/NS*						P		
Bushtit	D/D*			S	P				
Chipping sparrow	D/D*	NS*/ND	P	P	P		P		
Common nighthawk	D*/ND				P	P			
Dark-eyed junco	D/NS	NS*/NS*	P	P	P				
Golden-crowned kinglet	D/NS	NS*/D*	P	P					
Killdeer	D/D					P		P	
Lazuli bunting	D*/NS*			S	P		P		
Lesser goldfinch	D*/NS*			S	P		P		
MacGillivray's warbler	D/NS	NS*/NS*					P		
Mourning dove	D/D			P			P		
Olive-sided flycatcher	D/D	D*/NS*	P	S					
Orange-crowned warbler	D/D			P	S		P		
Pine siskin	D/D	NS*/NS*	P	S			P		
Rufous hummingbird	D/D*	NS*/ND	P	P	P		P		
Song sparrow	D/D	NS*/ND		S	P		P		
Western bluebird	D*/D*				P		S		P
Western meadowlark	D/D*					P		S	
Western tanager	D/NS	NS*/NS*	P	S					
Western wood-pewee	D/NS		P	S			P		
White-crowned sparrow	D/D			S	P	P	P		
Wilson's warbler	NS/D		P	S			P		
Winter wren	NS/D		P				P		S
Wrentit	NS*/D*				P				
Yellow-rumped warbler	NS*/D*	NS*/ND	P	P			P		
Total downward	31	2	14	21	15	5	21	3	5

[a] Breeding Bird Survey regions: South Pacific Rainforests equals the Coast Range, Klamath Mountains, low- to moderate-elevation western Cascades and North Cascades ecoregions; Cascade Mountains equals high-elevation Cascades.

[b] D = downward, NS = no significant trend, ND = insufficient data, and * = a problem with sample size (number of routes = 20–39).

[c] P = primary, S = secondary. References: Brown 1985; Ruggiero et al. 1991a; Puchy and Marshall 1993; Andelman and Stock 1994a,b.

primary cavity excavators—that occur in west-side forests and woodlands (northern flicker, acorn woodpecker, downy woodpecker, hairy woodpecker, and pileated woodpecker). Nest-site availability, though, could become a more important limiting factor for secondary and primary cavity-nesting species unless management aids in recruitment of large snags (Mannan and Meslow 1984).

The reasons for apparent downward trends in westside populations of species such as orange-crowned warblers, MacGillivray's warblers, yellow-rumped warblers, and bushtits (Table 8) are not obvious. These species typically inhabit forest–shrub edge habitats (Brown 1985). Clear-cutting often creates a version of this habitat type; however, natural gaps in older forests and riparian corridors in undisturbed, mature forests—both habitats that have been degraded in many areas—may provide suitable edge habitat without increased risk of exposure to predation by nonforest species or nest parasitism by brown-headed cowbirds (Yahner 1989; Robinson et al. 1992). In addition, all of these species except the bushtit are Neotropical migrants that may be experiencing problems during migration or on wintering grounds (Finch and Stangel 1992).

Only Steller's jays and common yellowthroats show long-term or recent upward population trends in westside forests (Fig. 33). These are widely distributed species with general habitat requirements.

Loss of large, riparian trees and snags along the Willamette River may once have reduced nesting by ospreys, but their local abundance has steadily increased since 1976 when the birds began nesting on utility poles (Henny and Kaiser 1996; Fig. 34). The Breeding Bird Survey does not cover the central Willamette Valley, and data for westside forests reveal no significant trend in osprey population sizes.

The Canada goose is another species that is common in the Willamette Valley, especially during winter. Its populations have increased in abundance during the past two decades (see box on Wintering Canada Geese in the Willamette Valley).

Birds in Eastside Forests

Breeding Bird Survey data for birds that occur in eastside habitats are limited because of the relative rarity and limited distributions of the associated species. Sample sizes for Breeding Bird Survey data on birds in the eastern Cascades (Breeding Bird Survey Pitt–Klamath Plateau region) are all below 40 routes, indicating poor or marginal reliability. More extensive and consistent monitoring of bird populations in eastside forests is needed. However, the reliability of trend interpretation

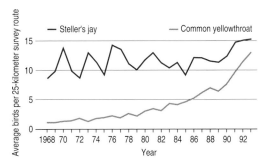

Fig. 33. Upward trends in the annual abundance indexes of Steller's jays and common yellowthroats in the western Pacific Northwest (Breeding Bird Survey data for the South Pacific Rainforest region, U.S. Geological Survey, Washington, D.C.).

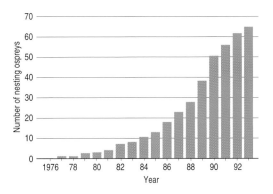

Fig. 34. Osprey abundance along the Willamette River, Oregon (adapted from Henny and Kaiser 1996).

is improved by indexes that incorporate information about the estimated magnitude of trends over time and the accuracy of the data. Carter and Barker (1993) developed such a modified index from Breeding Bird Survey data for Neotropical migrant land birds of the western United States. The trend data given here for birds of the Intermountain West, calculated for long-term (1969–1991) and recent (1982–1991) Breeding Bird Survey periods, were derived in a similar way.

In the Eastern Cascades Slopes and Foothills ecoregion, 29 species show significant upward population trends, whereas only 19 species show significant downward trends (Fig. 28). In contrast, slightly more species show downward trends than upward trends in the combined Okanogan Highlands–Blue Mountains region (Fig. 28). The strongest contrast is evident in recent data; many more species show recent downward trends and fewer species show recent undetermined trends in the Okanogan Highlands–Blue Mountains region than in the eastern Cascades region. A greater proportion of species show significant trends in the Okanogan Highlands–Blue Mountains region than in the eastern Cascades region, but still no trends can be discerned for 46% of the species that occur there (Fig. 28).

Ten species that are primarily associated with old-growth forests and four other species that often use old-growth forests show long-term or recent downward population trends in the eastern Cascades or Okanogan Highlands–Blue Mountains regions (Table 9). Only three

Wintering Canada Geese in the Willamette Valley

Only about 2,500 Canada geese wintered in the southern Willamette Valley between 1938 and 1948 (Gullion 1951), but during the subsequent 20 years, the size of the valley population greatly increased (Fig. 1). The U.S. Fish and Wildlife Service (Region 1, Portland, Oregon) began systematic surveys of wintering Canada geese in 1953 when the peak count was fewer than 10,000 birds. The typical winter population increased to about 20,000 birds by 1967. Throughout this period most of the geese counted were dusky Canada geese (Fig. 2). The southern Willamette Valley was the major harvest area for dusky Canada geese in the early 1950's (Chapman et al. 1969). In the mid-1960's, three national wildlife

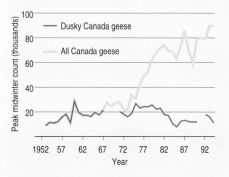

Fig. 1. Trends in winter population sizes of Canada geese in the Willamette Valley, Oregon (data provided by U.S. Fish and Wildlife Service, Portland, Oregon).

refuges were established to provide the geese with some refuge from hunting.

The winter abundance of dusky Canada geese in the Willamette Valley began to decline in the 1980's because of long-term ecological changes that occurred on the birds' nesting grounds in Alaska. In 1964 an earthquake on the Copper River Delta lifted the nesting grounds 0.5–2 meters (Shepherd 1965) and caused the habitat to change from tidal wetlands to uplands. Predation of nests by grizzly bears, coyotes, bald eagles, and mew gulls subsequently increased. By the 1980's nest productivity had significantly declined, leading to a population decline that continues today (Campbell 1992).

In the late 1960's, Taverner's Canada goose, a smaller subspecies that nests in arctic Alaska and Canada, began wintering in the Willamette Valley and greatly complicated management of geese in the valley. As the population of dusky Canada geese began to decline, the winter population size of Taverner's Canada geese increased exponentially during the mid-1970's and early 1980's (Jarvis and Cornely 1985). Managers were faced with the dilemma of protecting a declining subspecies threatened by changes on their breeding grounds, while the size of another subspecies population was rapidly expanding to the point of providing a considerable surplus for harvest. Moreover, certain behavioral tendencies of dusky Canada geese make them particularly vulnerable to hunters (Simpson and Jarvis 1979).

Low numbers of a third subspecies, the cackling Canada goose, wintered at the northern end of the Willamette Valley in the mid-1960's (Chapman et al. 1969), but most individuals wintered in the Central Valley of California. The population size of this subspecies subsequently declined precipitously, resulting in a flyway-wide hunting closure from 1984 to 1993. During the mid-1980's, the cackling Canada goose remained an insignificant part of the winter population of Canada geese in the Willamette Valley. The population size of this subspecies greatly increased during the last decade, however; the current winter population is about 60,000 birds. Four other subspecies also contribute to the winter population in the Willamette Valley: Vancouver Canada goose, the federally threatened Aleutian Canada goose, western Canada goose, and lesser Canada goose. Thus, 7 of the 11 recognized subspecies of Canada goose in North America (Delacour 1954) winter in the Willamette Valley. Moreover, the western Canada goose began nesting in the valley about 20 years ago and its nesting population continues to increase.

All subspecies of Canada goose that winter in the Willamette Valley graze on green vegetation. Numerous types of grass and clover are planted by valley farmers. The area planted in grasses for seed in the Willamette Valley changed from limited amounts in the 1950's to about 160,000 hectares in 1993 (Fig. 3). Estimates of grass seed production steadily increased from 13,900 to 243,300 metric tons from 1940 to 1990 (Oregon State University Extension Service, Corvallis, unpublished data). Modern farm practices steadily improved yields and this, in turn, probably resulted in increasing forage potential for wintering

Fig. 2. Dusky Canada geese.

Courtesy C. Stahr, W. L. Finlay, National Wildlife Refuge, Corvallis, Oregon

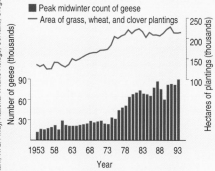

Fig. 3. Changes in the area planted for production of grass seed, wheat, and clover in the Willamette Valley, Oregon, since the 1950's (data provided by Oregon State University Extension Service, Corvallis).

geese. The cultivated grasses grow during the winter and provide fresh green vegetation, whereas most native plants are dormant in winter and do not provide forage for grazing geese.

The fact that Canada geese now winter farther north than before is not unique to the Willamette Valley; similar situations occur in northern portions of the Mississippi and Atlantic flyways. As a consequence, the Willamette Valley now supports about ten times more geese in winter than it did 40 years ago.

Authors

Charles J. Henny
U.S. Geological Survey
Biological Resources Division
Forest and Rangeland Ecosystem Science Center
Willamette Field Station
3080 S.E. Clearwater Drive
Corvallis, Oregon 97333

Maura B. Naughton
U.S. Fish and Wildlife Service
William L. Finley
National Wildlife Refuge
26208 Finley Refuge Road
Corvallis, Oregon 97333

See end of chapter for references

such species (Hammond's flycatcher, hermit thrush, and golden-crowned kinglet) show upward trends in the eastern Cascades region; no such species show upward trends in the Blue Mountains–Okanogan Highlands region. These trends are consistent with evidence of widespread loss and degradation of old-growth forests. The data, however, are insufficient to establish trends for most of the species that are closely associated with old-growth eastside forests. Specialized monitoring that targets specific habitats is needed for these species.

Four species that are closely associated with old-growth eastside forests have been extensively studied: pileated woodpecker (Bull 1987; Bull et al. 1992; Bull and Holthausen 1993), flammulated owl (Goggans 1985; Bull et al. 1990), great gray owl (Bryan and Forsman 1987; Bull et al. 1988; Bull and Henjum 1990; Fig. 35), and Vaux's swift (Bull 1991; Bull and Cooper 1991; Bull and Hohmann 1993). Although these studies were not designed to provide information on population trends, they showed how each species depends on habitat characteristics of old-growth forests (for example, cavities in large trees and snags for nesting and roosting). Because of such requirements, many scientists believe that losses of old-growth roost and nest trees and conversion of structurally complex old-growth pine forest to mostly homogeneous young and fir-dominated forests reduced populations of these species (Bryan and Forsman 1987; Marshall et al. 1996). Vaux's swifts show a significant downward population trend in the Okanogan Highlands–Blue Mountains and Columbia River basin–High Desert regions (Table 9), but the data are not sufficient to confirm trends for Vaux's swifts in the eastern Cascades ecoregion or for the other species in any eastside region.

Six species (house wren, mountain bluebird, tree swallow, violet-green swallow [Fig. 36], northern flicker, and western bluebird) that nest in cavities in the eastern Cascades region show long-term upward population trends. All of the secondary cavity nesters (that is, all of

Courtesy E. L. Bull, U.S. Forest Service

Fig. 35. Adult female and juvenile great gray owls in Oregon.

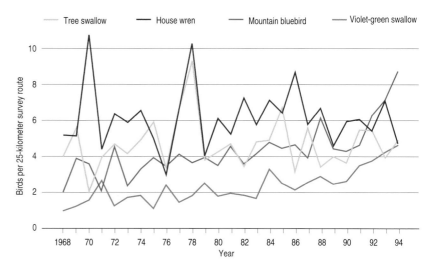

Fig. 36. Examples of upward trends in the annual abundance indexes of bird species that nest in tree cavities in the eastern Cascades (Breeding Bird Survey data for the Pitt–Klamath Plateau region, U.S. Geological Survey, Washington, D.C.).

Table 9. Population trends (after Carter and Barker 1993) and primary habitat associations of selected land birds with significant long-term (1969–1991) or recent (1982–1991) downward abundance trends in one or more eastside regions.

Species	Eastern Cascades Slopes and Foothills (long-term/recent)	Okanogan Highlands and Blue Mountains (long-term/recent)	Columbia Basin and Snake River Basin–High Desert (long-term/recent)	Old-growth forest	General forest	Early seral forest	Riparian woodland	Montane or riparian meadow	Grass or shrub–steppe	Juniper woodland	Cavity-nesting
American goldfinch	D/NS	D/D	NS/NS					X			
American kestrel	NS/D	NS/D	NS/NS				X			X	X
American robin	NS/NS	NS/D	U/U	X	X		X			X	
Bank swallow	U/NS	D/D	U/U					X			
Barn swallow	D/D	NS/D	NS/NS					X			
Brewer's blackbird	U/NS	D/D	NS/D				X	X	X	X	
Brewer's sparrow	NS/NS	D/D	D/D						X	X	
Brown creeper	D/NS	NS/NS	NS/NS	X							
Calliope hummingbird	D/NS	NS/NS	NS/NS					X			
Cassin's finch	NS/NS	NS/D	NS/NS	X						X	
Cedar waxwing	NS/NS	U/D	NS/NS				X				
Chipping sparrow	D/D	D/U	D/D	X	X	X				X	
Common nighthawk	U/U	U/U	U/D							X	
Common yellowthroat	NS/NS	D/D	NS/NS				X	X			
Dark-eyed junco	U/U	D/D	NS/NS			X		X			
Dusky flycatcher	D/NS	U/U	NS/NS				X	X			
Golden-crowned kinglet	U/NS	D/NS	NS/NS	X							
Grasshopper sparrow	NS/NS	D/D	D/D						X		
Gray catbird	NS/NS	NS/D	NS/NS				X				
Green-tailed towhee	U/D	NS/NS	NS/U						X	X	
Hammond's flycatcher	U/NS	D/D	NS/NS	X							
Hermit thrush	U/U	U/D	NS/NS	X							
Horned lark	D/NS	D/D	D/D						X		
Killdeer	D/D	D/D	D/D						X	X	
Lark sparrow	D/D	NS/NS	D/D					X	X	X	
Lewis' woodpecker	D/NS	D/NS	NS/NS	X			X				X
Loggerhead shrike	NS/NS	NS/NS	D/NS						X	X	
Long-billed curlew	NS/NS		NS/D						X		
MacGillivray's warbler	U/U	D/D	NS/NS			X	X				
Mountain bluebird	U/D	U/U	U/U				X			X	X
Mourning dove	D/NS	D/U	D/NS						X	X	
Northern flicker	U/U	D/D	U/U	X			X			X	X
Northern harrier	NS/NS	D/NS	D/NS					X	X		
Northern rough-winged swallow	D/NS	NS/U	U/U					X			
Olive-sided flycatcher	D/NS	D/D	NS/NS	X							
Pine siskin	NS/D	D/NS	NS/NS	X	X						
Prairie falcon	NS/NS	U/NS	D/NS						X	X	
Red-eyed vireo	NS/NS	D/NS					X				
Red-winged blackbird	NS/D	NS/NS	D/U					X			
Rock wren	U/NS	NS/D	D/D		X		X		X	X	
Sage sparrow	NS/NS	NS/NS	D/D						X		
Savannah sparrow	D/NS	NS/D	U/U					X			
Say's phoebe	NS/NS	U/D	NS/D					X			
Short-eared owl	NS/NS	NS/NS	D/D					X	X		
Song sparrow	U/U	D/D	U/U			X	X				
Spotted towhee	U/U	D/D	U/NS			X	X			X	
Swainson's thrush	NS/NS	NS/D	NS/NS	X			X				
Townsend's solitaire	U/D	U/D	NS/NS		X					X	
Townsend's warbler	NS/NS	NS/D		X							
Turkey vulture	NS/NS	D/D	D/D						X	X	
Vaux's swift	NS/NS	D/D	NS/D	X							X
Veery	NS/NS	D/D	NS/NS				X				
Violet-green swallow	U/U	U/U	D/U				X	X			X
Western kingbird	D/D	NS/NS	NS/D				X				
Western meadowlark	D/NS	NS/NS	NS/D					X	X		
White-crowned sparrow	NS/NS	D/NS	NS/NS					X		X	
Willow flycatcher	NS/NS	U/U	NS/D				X	X			
Wilson's warbler	U/U	NS/D	NS/NS				X	X			
Yellow-headed blackbird	NS/U	U/U	D/D					X			
Total downward	22	39	25	14	5	5	21	21	17	20	6

[a] D = downward, U = upward; NS = no significant trend.

[b] No trend given indicates species does not occur in that region or is not at sufficient level to appear in Breeding Bird Survey records.

[c] X = found in that habitat association; references: Puchy and Marshall 1993; Andelman and Stock 1994a,b; Dobkin 1994a.

the just-listed species except northern flicker) also show upward trends in the Okanogan Highlands–Blue Mountains region. These population responses may reflect higher densities of small- and medium-sized standing dead trees that resulted from outbreaks of disease and insects.

Eleven species (American kestrel, Brewer's blackbird, cedar waxwing, common yellowthroat, gray catbird, Lewis' woodpecker, MacGillivray's warbler, red-eyed vireo, spotted towhee, song sparrow, and Wilson's warbler) that regularly nest in woody riparian habitats in the Okanogan Highlands–Blue Mountains region show long-term or recent downward population trends (Table 9). Four other species (American robin, northern flicker, Swainson's thrush, and veery) that nest in woody riparian or mature upland forests in the Okanogan Highlands–Blue Mountains region also show downward population trends. These data are generally consistent with evidence that much of the riparian woodland and old-growth habitat in the region has been lost or degraded (Dobkin 1994a).

Wilson's warblers and spotted towhees in the eastern Cascades region, all of the secondary cavity nesters mentioned previously, and 11 more species (cedar waxwing, black-headed grosbeak, dusky flycatcher, lazuli bunting, Bullock's oriole, orange-crowned warbler, warbling vireo, western wood-pewee, willow flycatcher, yellow-breasted chat, and yellow warbler) in the Okanogan Highlands–Blue Mountains region also often nest in woody riparian habitats but show long-term or recent upward population trends (Table 9). Many of these species also will nest in shrub-dominated, early-seral habitats created by clear-cutting. Other such species with increasing populations in the eastern Cascades region include dark-eyed junco, MacGillivray's warbler, Nashville warbler, and song sparrow. In addition, it is probably not simply coincidental that the brown-headed cowbird—an obligate brood parasite of many of these species but a species that avoids large expanses of forest—also shows long-term or recent upward population trends in both eastside forest regions.

Nine species in the eastern Cascades region (barn swallow, calliope hummingbird, dusky flycatcher, killdeer, lark sparrow, northern rough-winged swallow, red-winged blackbird, savannah sparrow, and western meadowlark) and 12 species in the Okanogan Highlands–Blue Mountains region (American goldfinch, bank swallow, barn swallow, belted kingfisher, Brewer's blackbird, common yellowthroat, dark-eyed junco, killdeer, northern harrier, savannah sparrow, white-crowned sparrow, and Wilson's warbler) that nest in montane

or riparian meadow habitats show long-term or recent downward population trends (Table 9).

A few species that usually occur in upper-elevation juniper woodlands show long-term or recent downward population trends (Table 9). However, although the trends are not always consistent across the two regions, many more such species in the eastern Cascades (for example, Brewer's blackbird, common nighthawk, gray flycatcher, northern flicker, red-tailed hawk, rock wren, spotted towhee, solitary vireo, and vesper sparrow) and the Okanogan Highlands–Blue Mountains (for example, common nighthawk, golden eagle, mountain bluebird, prairie falcon, red-tailed hawk, ruby-crowned kinglet, solitary vireo, and Swainson's hawk) regions show long-term or recent upward population trends, with no indication of declines in the respective regions. This pattern is consistent with regional expansion of juniper woodlands.

Birds in Eastside Rangelands

As in other eastside habitats, raw Breeding Bird Survey data are not sufficient to confirm population trends for most species that occur in eastside rangelands (Columbia Basin and Snake River–High Desert). Among species for which significant population trends can be confirmed by using the modified index (Carter and Barker 1993), about equal numbers show upward and downward trends (Fig. 28).

Nearly two-thirds of the 25 species that show downward population trends in eastside rangelands are primarily or exclusively associated with grassland and shrub–steppe habitats (Fig. 37; Table 9). Two additional species that show downward population trends (rock wren and chipping sparrow) occur primarily in open

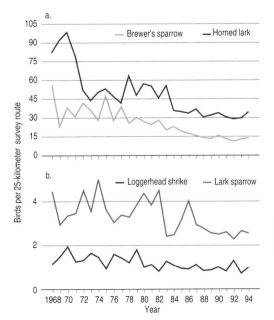

Fig. 37. Examples of downward trends in the annual abundance indexes of bird species that occur in eastside grassland and shrub–steppe habitats (Breeding Bird Survey data for the Columbia Plateau region; U.S. Geological Survey, Washington, D.C.).

juniper woodlands with a sagebrush and bunch-grass understory. Other than some birds of prey, the sage thrasher is the only grassland or shrub–steppe species for which adequate data are available that shows a significant upward population trend. The Columbian sharp-tailed grouse is also intimately tied to native grasslands but was extirpated from Oregon and California by the late 1960's and is declining

throughout the remainder of its range (Puchy and Marshall 1993; Tirhi 1995). Similarly, the sage grouse is closely associated with native sagebrush habitats, but its populations have greatly declined in Oregon (see box on Sage Grouse in Oregon). Widespread population declines of grassland and shrub–steppe birds in the western United States have been attributed to conversion of habitat to agriculture and to

Sage Grouse in Oregon

Sage grouse (Fig. 1) were once common in shrub–steppe habitats in central and eastern Oregon (Gabrielson and Jewett 1940), but concern about the status of sage grouse began nearly a century ago. Between 1900 and 1940 the range of sage grouse in Oregon was reduced by approximately 50%, and since 1940 the abundance of birds within the remaining range has declined by at least 60% (Crawford and Lutz 1985; Fig. 2). A reduction in the ability of the bird to nest successfully and to recruit chicks into the fall population seems to be the cause of the decline.

Fig. 1. Male sage grouse in breeding display.

The key factor that caused the initial decline of sage grouse in Oregon was probably the conversion of sagebrush-dominated lands for agricultural purposes, pastures, or plantings of nonindigenous grasses. These changes, though, do not account for declines in the vast areas of Oregon where sagebrush is still the dominant cover type. Research during the past 10 years revealed much information about the answer to this apparent dilemma. Sagebrush is important to sage

grouse; the grouse use it for food during fall and winter and for cover during most of the year (Patterson 1952). Still, other components of the sagebrush ecosystem, such as herbaceous grasses and forbs, may play the key role in determining sage grouse reproductive success (Gregg 1992).

Recent research in Oregon (Barnett 1993; Barnett and Crawford 1994) revealed that hen sage grouse readily consume herbaceous plants, such as clovers, desert parsleys, mountain-dandelions, hawksbeard, and phloxes, immediately before beginning to nest, and that the amount of these highly nutritious foods could be related to reproductive success. Hens also ate sagebrush, but the herbaceous plants contained 2–3 times more protein and much higher levels of other nutrients (Barnett and Crawford 1994).

The abundance of herbaceous plants influenced patterns of habitat use by hens with broods and by broodless hens (Gregg 1992; Gregg et al. 1993; Drut et al. 1994a; Ramsey et al. 1994). During summer, herbaceous plants are a very important source of nutrition for hens and their chicks. Without adequate amounts of certain forbs, chicks forage largely on nutrient-poor sagebrush and consequently are less likely to survive (Drut et al. 1994b). Important chick foods during the critical period of summer growth

include herbaceous plants such as hawksbeard, clovers, mountain-dandelions, milk-vetches, and microsteris, and insects such as ants and beetles, which are closely associated with plant stands that are rich in herbaceous plants (Drut et al. 1994b).

Other research in Oregon revealed the importance of grass cover, in particular residual grass cover with a height of about 20 centimeters or more (Gregg 1992; DeLong 1994; Gregg et al. 1994; Ramsey et al. 1994; DeLong et al. 1995). The amount of tall residual grass cover and medium-height sagebrush were the key factors that determined whether a sage grouse nest was successful. Sagebrush provides overhead concealment from predators and protection from environmental forces, whereas residual grass cover provides ground-level concealment for the hen. Sage grouse nest early in the year (March and April) when new grasses are just beginning to grow. The birds must, therefore, rely on residual cover from the previous year's growth for the protection they need.

In combination, these factors tremendously influence the nesting success of sage grouse and, therefore, sage grouse abundance. Maintenance of diverse herbaceous communities in existing sagebrush habitats and restoration of former sage grouse habitat will be keys to ensuring the continued existence of sage grouse in Oregon and elsewhere. Priorities for sage grouse research, management, and monitoring are outlined in a report derived from a recent conference on comprehensive approaches to sage grouse management and conservation (Dobkin 1995).

See end of chapter for references

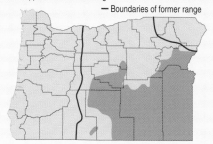

■ Approximate current range
— Boundaries of former range

Fig. 2. Comparison of past range with the approximate present range of sage grouse in Oregon (adapted from Crawford and Lutz 1985).

Author

John A. Crawford
Oregon State University
Department of Fisheries and Wildlife
Nash Hall
Corvallis, Oregon 97331

overgrazing by livestock (DeSante and George 1994; Fleischner 1994; Dobkin 1995; Tirhi 1995). Grazing affects fire patterns and aids in invasions of nonindigenous plant species, which in turn alter the structural characteristics of the landscape and reduces the diversity and abundance of native plants to the detriment of these bird species.

As in eastside forest regions, several species that often occur in juniper woodlands show upward population trends in eastside rangelands (American robin, gray flycatcher, green-tailed towhee, house wren, mountain bluebird, northern flicker, and red-tailed hawk). Again, this is consistent with regional expansion of juniper woodlands. The common nighthawk also occurs in juniper woodlands and shows a long-term upward population trend, but the trend seems to have recently reversed (Table 9).

Across the arid lands of western North America, many species of birds occur in riparian habitats (Knopf et al. 1988; Dobkin 1994a). Many bird biologists believe that degradation and elimination of riparian habitats are the most important causes of declining landbird populations in the region (DeSante and George 1994; Ohmart 1994). However, only two exclusively riparian species (willow flycatcher and yellow-headed blackbird) show long-term and recent downward population trends in eastside rangelands. Two other riparian species (violet-green swallow and red-winged blackbird) have long-term downward population trends, but these trends recently reversed. Fourteen riparian species show consistent upward population trends in eastside rangelands (bank swallow, belted kingfisher, black-headed grosbeak, Bullock's oriole, cliff swallow, lazuli bunting, northern rough-winged swallow, savannah sparrow, song sparrow, tree swallow, warbling vireo, western wood-pewee, yellow-breasted chat, and yellow warbler). Several of the species associated with juniper woodlands that show upward population trends also often nest in woody riparian habitats (American robin, house wren, mountain bluebird, northern flicker, and red-tailed hawk). In addition, the abundance of brown-headed cowbirds increased in eastside rangelands as the population sizes of potential host species grew. These upward trends may signal the beginning of riparian habitat recovery in some parts of the region; however, the magnitude of the increases was very small for nearly all of these species.

An ongoing study at Hart Mountain National Wildlife Refuge in southeastern Oregon is assessing the recovery rates of riparian vegetation and bird communities after removal of livestock (Dobkin 1994b; Dobkin et al. 1995a). The riparian avifauna is now characterized by a low number of species and a disproportionate representation of a few abundant and widespread species (principally house wren, American robin, and red-winged blackbird). It is encouraging, though, that two species which usually indicate mature riparian habitat, MacGillivray's warbler and lazuli bunting, began nesting during the third year of recovery.

Eastside rangelands harbor many birds of prey that are often closely associated with riparian habitats in the region (Knight 1988). Several species that occur throughout the arid rangelands of the Pacific Northwest (burrowing owl, red-tailed hawk, Swainson's hawk [Fig. 38], ferruginous hawk, and golden eagle) show long-term or recent upward population trends. This is a particularly positive sign for burrowing owls because their populations are declining throughout much of their western range (DeSante and George 1994).

The Snake River Birds of Prey National Conservation Area in southwestern Idaho and neighboring river canyons support one of the highest nesting densities of noncolonial birds of prey anywhere in the world (Knight 1988; Lehman et al. 1994). Prairie falcons and golden eagles (Fig. 39) have been closely monitored since the 1970's at the Snake River site. The number of nesting territories occupied by prairie falcons has fluctuated around a consistent mean since 1976, whereas the number of nests occupied by golden eagles has declined slightly since 1971; the productivity of both species has been variable (Fig. 40;

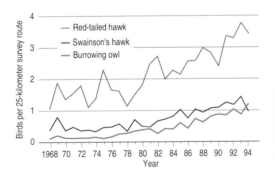

Fig. 38. Examples of upward trends in the annual abundance indexes of birds of prey in eastside lowlands (Breeding Bird Survey data for the Columbia Plateau region, U.S. Geological Survey, Washington, D.C.).

Courtesy M.W. Collopy, USGS

Fig. 39. Golden eagle with nestling in Idaho.

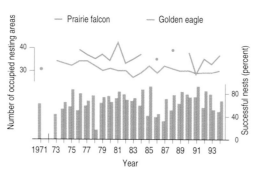

Fig. 40. Trends in the nesting populations and nesting success of prairie falcons (population trends in three regularly monitored 10-kilometer survey units) and golden eagles in the Snake River Birds of Prey National Conservation Area, Idaho (adapted from Lehman et al. 1994).

Fig. 41. The marten is one of the most threatened species closely associated with undisturbed old-growth forests in the western Pacific Northwest.

Lehman et al. 1994). Golden eagles vacated nesting territories in areas that were converted to agriculture or that were taken over by non-indigenous grasses.

Mammals

Carnivores

Several medium to large carnivores are on state lists of sensitive species and have regional populations that are either precariously small and probably nonviable or declining (Table 10). All except the kit fox are primarily forest species, and all have suffered from habitat loss and degradation, excessive hunting, or persecution. Martens (Fig. 41) and fishers are of particular interest because they are two of the most threatened terrestrial mammals closely associated with old-growth forests in the Pacific Northwest (Forest Ecosystem Management Assessment Team 1993; Thomas et al. 1993).

The geographic distribution of martens in the three Pacific Coast states markedly decreased during this century primarily because of habitat loss from logging of mature forests (Maser et al. 1981; Buskirk and Ruggiero 1994;

Table 10. Status of terrestrial, carnivorous mammals in the Pacific Northwest that are on federal or state lists of sensitive species.

Species	Federal or state status[a]					Population trend	References
	Federal	California	Oregon	Washington	Idaho		
Fisher	*	*	S-c	C	SC–a	Downward	Maser et al. 1981; Rodrick and Milner 1991; Aubry and Houston 1992; Marshall et al. 1996; Thomas et al. 1993; Powell and Zielinski 1994
Gray wolf	E	ext	E	E	E	Upward? in Washington, unknown otherwise	Laufer and Jenkins 1989; Washington Department of Wildlife 1991; U.S. Fish and Wildlife Service 1992
Grizzly bear	E	ext	ext	E	T	Stable or downward	Almack et al. 1993
Kit fox	*	—	T	—	SC–b	Downward	Keister and Immell 1994
Marten	*	*	S-c	C	*	Downward	Maser et al. 1981; Rodrick and Milner 1991; Marshall et al. 1996; Thomas et al. 1993; Buskirk and Ruggiero 1994; Oregon Department of Fish and Wildlife trapping records
North American lynx	*	—	*	T	SC–b	Downward	Washington Department of Wildlife 1993d; Koehler and Aubry 1994
Wolverine	*	T	T	M	SC–a	Upward? in Idaho, unknown otherwise	Groves 1987; California Department of Fish and Game 1992; Banci 1994

[a] — = does not occur in state, * = no special status, E = endangered, T = threatened, ext = extirpated; Oregon—S–c = sensitive species with critical status; Washington—C = candidate for sensitive species status, M = monitor species; Idaho—SC–a,b = species of special concern category a (priority) or b (peripheral). References: Oregon Natural Heritage Program 1993; California Department of Fish and Game 1994, 1995b; Conservation Data Center 1994; Washington Department of Wildlife 1994; U.S. Fish and Wildlife Service 1995.

Marshall et al. 1996). Accurate, quantitative population data are difficult to obtain, but the number of martens trapped in western Oregon steadily declined during the past 50 years (Fig. 42). The range of fishers in Washington remained constant during the past 40 years, but a scarcity of recent sighting records and incidental trappings suggests that the density of fishers is precariously low in Washington (Aubry and Houston 1992; Powell and Zielinski 1994). Fisher populations are probably also depleted in Oregon, but accurate assessments are unavailable (Maser et al. 1981; Powell and Zielinski 1994; Marshall et al. 1996). Fishers are sensitive to forest fragmentation in northern California (Rosenberg and Raphael 1986), and it is widely believed that martens are also sensitive to alteration of the mature forest environment they prefer (Buskirk and Ruggiero 1994). Current assessments suggest that both species will remain at risk.

In contrast, coyotes, red foxes, and black bears are widespread, and their populations are stable or increasing (Maser et al. 1981; Aubry 1984; MacCracken and Hansen 1987; Vaughan and Pelton 1995). These species easily adapt to the presence of humans and have prospered despite widespread persecution and trapping in many regions because they are highly fecund and readily exploit many habitats and food sources.

The status of large predators is relatively well known because they have concerned ranchers and trappers for decades and will

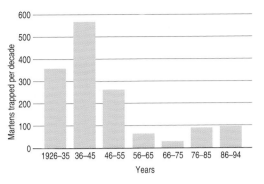

Fig. 42. Average number of martens trapped per decade in western Oregon (based on Oregon Department of Fish and Wildlife fur-trapper records).

probably always remain concerns of outdoor enthusiasts and inhabitants of rural communities (Laufer and Jenkins 1989; Almack et al. 1993; Washington Department of Wildlife 1993d). However, efforts to forecast population trends and manage even these species suffer from a lack of reliable data on past abundances and population demographics (Washington Department of Wildlife 1993d; North Cascades Grizzly Bear Steering Committee 1994).

Small Mammals in Westside Forests

Accurate population information about rodents, squirrels, and other small mammals is scarce. Eleven species of small mammals are endemic to conifer forests of the Pacific Northwest and northern California, but the status of most is unknown (Table 11).

The endemic red tree vole (Fig. 43) is another threatened species that is closely associated with old-growth forests in the western Pacific Northwest (Forest Ecosystem Management

Species	State status[a]			Distribution	Habitat association	Population trend
	California	Oregon	Washington			
Baird's shrew	*	*	*	Willamette Valley, Klamath Mountains, Coast Range/ Cascades—Oregon	Generalist	Unknown
Fog shrew	*	*	*	Klamath Mountains, Willamette Valley, Coast Range/Eastern Cascades Slopes and Foothills—Oregon	Generalist	Unknown
Mountain beaver	*	*	*	All westside, Eastern Cascades Slopes and Foothills	Early seral	Upward
Creeping vole	*	*	*	All westside, Eastern Cascades Slopes and Foothills—Washington, Okanogan Highlands	Early seral	Upward
Pacific shrew	*	*	*	Klamath Mountains, Coast Range—Oregon	Riparian	Unknown
Pacific marsh shrew	*	*	M	All westside, Eastern Cascades Slopes and Foothills	Riparian	Unknown
Red tree vole	SC	*	*	Klamath Mountains, Coast Range/Cascades—Oregon	Old-growth	Decreasing?
Shrew–mole	*	*	*	All westside, Eastern Cascades Slopes and Foothills	Riparian	Unknown
Southern red-backed vole	*	*	*	Blue Mountains, Cascades/North Cascades/Eastern Cascades Slopes and Foothills—Washington, Okanogan Highlands	Late seral	Unknown
Western red-backed vole	*	*	*	Klamath Mountains, Coast Range, Cascades, Eastern Cascades Slopes and Foothills	Late seral	Unknown
White-footed vole	SC	S–u	*	Klamath Mountains, Coast Range/Cascades—Oregon	Riparian	Unknown

[a] Status codes: SC = California state species of special concern; S–u = Oregon state sensitive species with unknown status; M = Washington state monitor species. References: Oregon Natural Heritage Program 1993; Washington Department of Wildlife 1994; California Department of Fish and Game 1994; * = no special status.

Table 11. Status, habitat associations, and distributions of small mammals that are endemic to Douglas-fir forests of the Pacific Northwest (based on Ingles 1965; Brown 1985; Puchy and Marshall 1993; Thomas et al. 1993).

Translocated Sea Otter Populations off the Oregon and Washington Coasts

The historical distribution of sea otters extended from the northern islands of Japan north and east across the Aleutian chain to the mainland of North America then south along the west coast to central Baja California, Mexico (Riedman and Estes 1990). By the beginning of the twentieth century, after 150 years of being intensively hunted for their valuable fur, sea otters had been extirpated from most of their range (Kenyon 1969). In 1911 sea otters were protected by the passage of the International Fur Seal Treaty. Unfortunately, only 13 remnant populations survived the fur-hunting period, and two of those, British Columbia and Mexico, would also ultimately disappear, leaving only a small group of sea otters south of Alaska, along the rugged Big Sur coast of California (Kenyon 1969).

The earliest attempts to reestablish sea otters to unoccupied habitat were begun in the early 1950's by R. D. (Sea Otter) Jones, then manager of the Aleutian National Wildlife Refuge (Kenyon 1969). These early efforts were experimental, and all failed to establish populations. However, the knowledge gained from Jones's efforts and the seminal work of Kenyon (1969) and others during the 1950's and early 1960's ultimately led to the successful efforts to come.

During the mid-1960's the Alaska Department of Fish and Game began translocating sea otters to sites where the species had occurred before the fur-trade period. The first translocations were restricted to Alaska, but beginning in 1969 and continuing through 1972, the effort expanded beyond Alaska. During this period, 241 sea otters were translocated to sites in British Columbia, Washington, and Oregon (Jameson et al. 1982). The work was done cooperatively between state and provincial conservation agencies, with much of the financial support for the Oregon and Washington efforts coming from the Atomic Energy Commission (now ERDA). Follow-up studies of the Oregon population began in 1971 and continued through 1975. After 1975, surveys in Oregon occurred infrequently. In Washington no follow-up surveys were conducted until 1977, although the population has been monitored closely since then (Jameson et al. 1982, 1986; Jeffries and Jameson 1995).

Oregon

Sea otters were extirpated by fur-trade hunters in Oregon by the early twentieth century. Most of Oregon's sea otter habitat occurs in the southern half of the state, where the only extensive nearshore rocky reef systems are found. Ninety-three translocated sea otters were liberated here: 29 near Port Orford in 1970 and 24 near Port Orford and 40 near Cape Arago in 1971. Counts never reached anywhere near the number of otters released, but from 1972 to 1974 they ranged from 20 to 23 otters. In 1975, however, the population began to decline, and in 1981 only one sea otter could be found (Jameson et al. 1982). By then the population was clearly no longer viable, and no subsequent sightings were made until the summer of 1992, when a single sea otter was observed at Cape Arago. No sea otters have been seen since then. Sea otters are once again extirpated in Oregon, and the translocation should be classed as a failure.

Washington

As in Oregon, the Washington sea otter population had also been extirpated by fur-trade hunting by the early twentieth century. Fifty-nine sea otters were released off the west coast of the Olympic Peninsula of Washington during the summers of 1969 and 1970 (Jameson et al. 1982); all had been translocated from Amchitka Island, Alaska. In 1969, 29 sea otters were released directly to the open ocean near Point Grenville (Fig. 1), with no time to acclimate or to recondition their fur. Sixteen of those 29 translocated sea otters were found dead on beaches near the release site within 2 weeks after translocation. No doubt some carcasses went undiscovered.

In 1970, release procedures and the release site were changed. The release location was changed to La Push (Fig. 1), located within the boundaries of Olympic National Park and near the middle of the best sea otter habitat in Washington. In midsummer, 30 sea otters were flown to La Push and released into holding pens anchored in a protected cove just beyond the La Push harbor entrance. The 30 otters were fed and allowed to acclimate for several days in the pens before release. All were liberated in excellent condition, and known mortality after release was low. Thus, the initial nuclear population in Washington could never have been larger than 43 otters and may have dropped to fewer than 10 individuals by the early 1970's (Jameson et al. 1982). No follow-up surveys of the Washington population were done until 1977 (Jameson et al. 1982, 1986; Table).

All sites within the survey area are located off the west coast of Washington's Olympic Peninsula between Destruction Island and Neah Bay (Fig. 2). From 1977 to 1984, surveys were conducted by U.S. Fish and Wildlife Service biologists (Table). Since 1985 surveys have been conducted cooperatively by the U.S. Fish and Wildlife Service's research division (now the Biological Resources Division of the USGS) and the Washington Department of Fish and Wildlife biologists (Table).

Population Growth

Growth of the population has continued at a finite rate of about 12% per year since 1989, when the current survey method began. From 1977 to 1988 the rate was higher, at 21% per year (R. Jameson, U.S. Geological Survey, Corvallis, Oregon, unpublished data; Fig. 3). Whether the difference between the rates indicates a slowdown of population growth or simply a difference in survey techniques (the method was modified in 1989) is still open to question. Pups were only noted separately from independent otters at ground count locations; thus the number of pups noted in the Table is probably low. However, pup counts at ground stations from 1993 to 1995 averaged 24 pups for every 100 independent otters, which suggests pup production has remained good.

The majority of sea otters in Washington occur between Makah Bay and Destruction Island (Fig. 2). Several significant changes in distribution have occurred recently, however. At the southern end of their range, sea otters now regularly occur inshore from Perkins Reef. As many as 20 sea otters have been counted in this area recently, although no more than one had been seen there before. At the northern end of the range, scattered individuals were regularly seen near Cape Flattery and between there and Neah Bay. In 1995, however, more than 100 otters moved into this area. This appears to be a seasonal phenomenon, occurring during the late winter and early spring. In late 1995, a small group of females rounded Cape Flattery and took up residence near Slant Rock. This area was previously inhabited almost entirely by male sea otters.

In 1988 and again in 1991, the outer coast of the Olympic Peninsula was hit by spills of bunker fuel oil, both from shipping accidents. The 1988 spill occurred in December, the 1991 spill in late July. In both cases about 230,000 gallons were spilled.

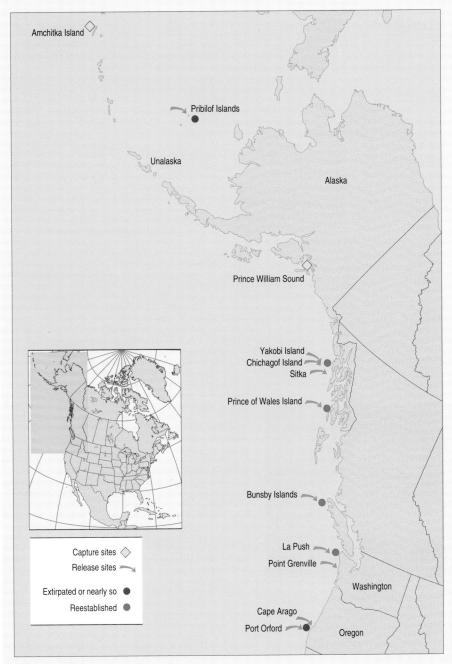

Fig. 1. General locations of capture and release sites and the status of translocated sea otter populations in Oregon and Washington (modified from Jameson et al. 1982).

Table. Results of surveys of the sea otter population in Washington, 1977–1995.

Year	Number of independents (adults and subadults)	Pups	Total
1977	15	4	19
1978[a]	12	0	12
1979	No survey		
1980	No survey		
1981	35	1	36
1982	No survey		
1983	48	4	52
1984	No survey		
1985	60	5	65
1986	No survey		
1987	89	5	94
1988	No survey		
1989	198	10	208
1990	197	15	212
1991 (July)	259	17	276
1991 (October)	242	20	262
1992	283	30	313
1993	283	24	307
1994	325	35	360
1995	341	54	395

[a] The 1978 results are probably not indicative of the actual number of sea otters in the population because inclement weather conditions precluded a thorough survey of the southern portion of the range.

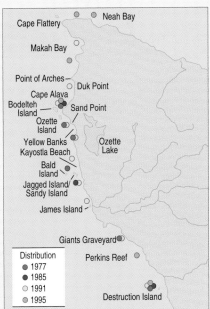

Fig. 2. Distribution of sea otters on the Olympic Peninsula coast, 1977–1995 (area blown up for detail indicated above).

The sea otter population was relatively unaffected by both spills, although thousands of seabirds died in each. No oiled sea otters were found in 1988, and only one was found in 1991. This animal did, however, die of complications caused by oiling (N. J. Thomas, National Wildlife Health Research Center, Madison, Wisconsin, necropsy report).

When we began our surveys in 1977 (Jameson et al. 1982), the sea otter population was distributed between Destruction Island and Cape Alava, a distance of about 60 kilometers. In 1992 the population was distributed between Destruction Island and Makah Bay, a distance of about 80 kilometers. By 1996 the range had increased to more than 110 kilometers, Destruction Island to Neah Bay (Fig. 2). Before 1991 the distribution had changed little from what it was in 1977. Until then all the population growth had taken place within the 1977 boundaries. In 1991 a large group broke away from the main population and established itself in Makah Bay about 15 kilometers north of where they were the previous

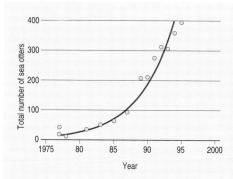

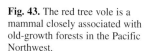

Fig. 3. Growth of the Washington translocated sea otter population, 1977–1995. Circles represent actual survey counts.

year. The distribution in 1992 was similar to 1991. Females with pups now occur from near Cape Flattery to Destruction Island.

Important Surveys

The Washington sea otter population is important because it is the only translocated population having the dual distinction of becoming successfully established and of being intensively monitored. Other translocations have been successful, but few data are available on their patterns of growth. Others that have been intensively monitored, in Oregon (Fig. 1) and San Nicolas Island, California, have failed, or appear to be heading toward failure (Jameson et al. 1982; Rathbun et al. 1990). The Washington sea otter population will continue to be monitored, and in 1994 a project was initiated to collect data on female reproductive rates, pup survival rates, foraging behavior, and activity and time budgets. This information, coupled with the population growth data, will provide a basis for comparison among populations that are either stable, growing at expected rates, or growing at rates below what is expected for populations reoccupying historical habitat.

The southern sea otter (Wilson et al. 1991) population in California, listed as threatened under the Endangered Species Act in 1977, is such a population. Since 1982 this population has grown at about 5% per year (Riedman and Estes 1990), considerably more slowly than the Washington population and slower than most growing sea otter populations (Estes 1990). Contrasting the reproductive and pup survival rates of the Washington and California populations will hopefully provide insight into why the growth rates are so different. Once that point is reached, researchers can attempt to uncover the cause or causes of the differences. Recent information from California and Kodiak Island, Alaska, suggests that preweaning survival of pups may be a quite significant factor in determining rates of sea otter population growth (Riedman et al. 1994; Monson and DeGange 1995).

See end of chapter for references

Author

Ronald J. Jameson
U.S. Geological Survey
Biological Resources Division
California Science Center
Sea Otter Project
200 S.W. 35th Street
Corvallis, Oregon 97333

Fig. 43. The red tree vole is a mammal closely associated with old-growth forests in the Pacific Northwest.

© Oregon State University, Forest Science Media Center

Assessment Team 1993; Thomas et al. 1993). The number of species of small mammals may not differ between old-growth and young, managed forests in Oregon and Washington, but old-growth stands with diverse understory vegetation and abundant snags and coarse woody debris typically support 1.5 times more biomass and individuals than do young forests (Carey 1995; Carey and Johnson 1995). Thus, the abundances of these and other sensitive species (for example, Douglas's squirrel, shrew-mole, northern flying squirrel, and western red-backed vole) probably declined in proportion to the loss of old-growth forests during the past 50 years (Thomas et al. 1993). In contrast, mountain beavers and creeping voles are two species endemic to westside forests that are increasing in abundance and seem to thrive in the early-seral habitats that result from logging (Black 1992).

Ungulates

Deer and elk are ecologically and economically important. Elk thrive in undisturbed old-growth forest in Olympic National Park and significantly influence the structure of forest vegetation (see box on Roosevelt Elk and Forest Structure in Olympic National Park). Other long-term ecological studies of elk are ongoing at the Starkey Experimental Forest near LaGrande, Oregon (Skovlin 1991), and on the Arid Land Ecology Reserve administered by the U.S. Department of Energy in south-central Washington (McCorquodale et al. 1988).

The endangered Columbian white-tailed deer has suffered from habitat loss and degradation. These deer were once widespread but now occur only in a small area of island and floodplain habitat near the mouth of the Columbia River and in the upper Umpqua Valley of Oregon (U.S. Fish and Wildlife Service 1983; Rodrick and Milner 1991). Their preferred habitat is a mosaic of woodlands that provide cover and small- to moderate-sized patches (less than 500 meters wide) of brushland or open pasture that provide food (Suring 1975). The causes of decline are urban and agricultural development, competition with Columbian black-tailed deer (stable populations in western Oregon during the past 20 years; Fig. 44) in remaining habitat, and avoidance of pastures occupied by cattle (Rodrick and Milner 1991). However, habitat enhancement and protection allowed populations to rebound from 300 to 400 individuals in 1976 to 6,000–7,000 individuals in 1992. The U.S. Fish and Wildlife Service (1992) is considering a reclassification from endangered to threatened, and the Oregon Department of Fish and Wildlife recently removed the species from the state's list of endangered species because of these successes.

Small Mammals in Eastside Rangelands

Small mammals can be very abundant in eastside shrub–steppe and grassland ecosystems; total biomass ranges from 650 grams per hectare in bunchgrass stands of south-central Washington (French et al. 1976) to as high as 5,000 grams per hectare in southeastern Idaho

Roosevelt Elk and Forest Structure in Olympic National Park

Roosevelt elk occur along the Pacific Coast of North America from Vancouver Island to northern California. Large populations occupy Washington's Olympic Peninsula (1.38 million hectares); the 5,000 or so elk in Olympic National Park (370,000 hectares) represent the last large population in mostly undisturbed natural habitat, which includes old-growth forests (Houston et al. 1990; Fig. 1).

The elk of Olympic National Park include year-round residents and populations that migrate to high elevations during summer. From 3,000 to 4,000 elk reside entirely in the park along drainages on the north and west sides, including the Elwha, Hoh, Queets, and Quinault rivers. Winter densities are around 6–7 elk per square kilometer. Censuses revealed that subpopulations on the west side of the park remained stable during the 1980's (Houston et al. 1990). Old-growth forests of massive western hemlock, Sitka spruce, western red-cedar, and Douglas-fir provide much of the habitat used by these subpopulations.

Scientists study elk–vegetation relations in the park, primarily in forests on the west side, to increase understanding of the long-term effects of native ungulates in forest communities. Elk consume ferns, shrub foilage, and coniferous foilage during fall and winter, and grasses and herbaceous plants during spring and summer. Seasonally important dietary items include western hemlock, sword fern, red alder, and oxalis. Digestible energy in these foods is usually low, indicating that elk densities may be limited by the quality and quantity of winter forage (Leslie et al. 1984).

Twenty-five ungulate exclosures were established in Olympic National Park (23) and the surrounding Olympic National Forest (2) in the 1930's and 1950's. Woodward et al. (1994) recently (1987–1990) resampled vegetation inside and adjacent to the original exclosures. When past vegetation was compared with present vegetation on either side of the exclosures, scientists learned that ungulates, mainly elk, are a powerful force shaping plant communities (Fig. 2). Ungulate

Courtesy National Park Service, Olympic National Park

Fig. 1. A bull Roosevelt elk in characteristic old-growth habitat in Olympic National Park, Washington.

Courtesy National Park Service, Olympic National Park

Fig. 2. Comparison of vegetation inside an ungulate exclosure (background) with vegetation outside (foreground) the exclosure near old-growth forest in Olympic National Park.

herbivory influenced the species composition, morphology, and standing crop of forest vegetation at all structural layers (herbaceous understory, shrubs, lower tree canopy, and overstory canopy). In communities on valley floors initially dominated by grasses, exclusion of ungulates resulted in decreased cover of grasses and usually forbs, decreased species richness of forbs, and sometimes increased height and abundance of ferns. Shrub size and density also increased in the absence of herbivory, and ungulates influenced the recruitment and morphology of vine maple in the lower forest canopy. Ungulates had variable

effects on the establishment of overstory species; browsing seemed to affect recruitment of Pacific silver fir and western redcedar on Olympic National Forest after clear-cut logging, but effects on other tree species outside and inside the park were unclear.

The intensity of ungulate herbivory varies in time and space, and the effects on vegetation are complex. As in other recent studies, the Olympic National Park studies show that mammalian herbivores strongly interact with vegetation and are not just passive components of the ecosystems they occupy.

See end of chapter for references

Authors

Douglas B. Houston
Edward S. Schreiner
Andrea Woodward
U.S. Geological Survey
Biological Resources Division
Forest and Rangeland Ecosystem Science Center
Olympic Field Station
600 East Park Avenue
Port Angeles, Washington 98362

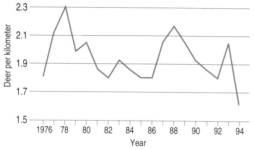

Fig. 44. Average number of Columbian black-tailed deer spotlighted per kilometer of road in western Oregon from 1976 to 1994 (data provided by the Oregon Department of Fish and Wildlife, Corvallis).

(Groves and Keller 1983). Livestock grazing and widespread seeding of ranges with non-indigenous crested wheatgrass have drastically altered small mammal habitats throughout the region, however (Boula and Sharp 1985). Livestock grazing may reduce the number of species but increase the total biomass of small mammals in rangeland habitats (Medin and Clary 1990). In contrast, small mammal biomass typically decreases after replacement of native shrubs with crested wheatgrass (Boula and Sharp 1985; Koehler and Anderson 1991). Small mammals also may be adversely affected by changes in fire frequency and intensity that drastically alter habitats (Groves and Steenhof 1988).

As elsewhere, the regional population status of most eastside small mammals is unknown. Still, Bureau of Land Management and U.S. Geological Survey biologists have been studying populations of Townsend's ground

squirrels (Fig. 45) in the Snake River Birds of Prey National Conservation Area in Idaho for more than a decade as part of a landscape-scale study of golden eagles and other birds of prey (Smith and Johnson 1985; Yensen et al. 1992). The abundance and size of squirrels are generally greater on sites with native shrubs than on recently burned or grassy sites, and squirrels on undisturbed shrubby sites are more likely to survive drought (Groves and Steenhof 1988; Yensen et al. 1992; Van Horne et al. 1993a). Burned sites, however, may provide a rich supply of succulent forage during wet years (Van Horne et al. 1993a,b), and the open habitat may aid in the ability of small mammals to be vigilant against birds of prey (Sharpe et al. 1994). Obtaining accurate population estimates has been problematic because of seasonal and annual variation in the squirrels' use of complex burrow networks that lead to various habitats (Knick 1993). Unraveling these complexities requires long-term data sets and careful, comprehensive analyses. The same applies to other mammals, including black-tailed jack rabbits, whose population sizes dramatically fluctuate on cycles of 10–12 years (Fig. 46).

Fig. 45. Townsend's ground squirrels have been a focus of long-term studies in the Snake River Birds of Prey National Conservation Area in Idaho.

© J. Weaver, State of Idaho, Military Division

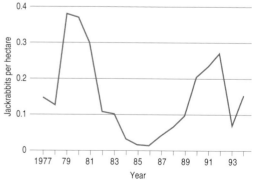

Fig. 46. Estimated density of black-tailed jack rabbits during spring in the Snake River Birds of Prey National Conservation Area, Idaho (adapted from Watts and Knick 1994).

Townsend's ground squirrels and black-tailed jack rabbits are common and thriving, but Washington ground squirrels, Idaho ground squirrels, and pygmy rabbits are not faring as well. The distribution of the Washington ground squirrel, once a widespread inhabitant of native grasslands in the Columbia River basin, is now greatly reduced (Betts 1990). Incompatible farming practices, livestock grazing, and limited distribution are factors responsible for its decline (Marshall et al. 1996) and those of the Idaho ground squirrel, which inhabits sagebrush–steppe along the middle portion of the Snake River. Because of similar factors, the pygmy rabbit has been eliminated from most of its former range in Washington (Washington Department of Wildlife 1993e); the decline of this species is coincident with estimates of a 60% reduction in shrub–steppe habitat (Dobler 1992).

Bats

Bats make up a unique group of small mammals because of their ability to fly, their nocturnal habits, and their insectivorous diet. Bats naturally contribute to regulation of insect populations and may contribute to nutrient cycling, especially in riparian–upland ecotones (Cross 1988; Christy and West 1993). Twelve species of bats occur in Douglas-fir forests (Christy and West 1993), and the diversity of bats is even higher in some eastside regions. Life-history details and the status of most species are poorly known, however.

Assessment of bat population trends is difficult because of variation in life-history strategies and a consequent lack of standardized monitoring. One large, summer-resident colony of Brazilian free-tailed bats (200,000 or more, the largest known congregation of bats on the Pacific Coast) at the Lava Beds National Monument in northern California has been monitored since 1988 (Cross 1991; B. Stoffel,

Lava Beds National Monument, Tule Lake, California, personal communication), and populations of several species have been monitored at the Oregon Caves National Monument in the Klamath Mountains ecoregion since the mid-1970's (Cross 1977; Cross and Schoen 1989). Populations at both sites have remained stable during the periods of record.

In contrast, the abundance of Townsend's big-eared bats has declined by 58% west of the Cascades and 16% east of the Cascades since 1975–1985 because of habitat alteration and human disturbance (Perkins 1990). Populations of other species associated with forests also seem to be declining.

Nine forest-dwelling bats, including Townsend's big-eared bat, are assigned some level of special status in states where they occur (Table 12). Most of these species also were targeted for special consideration in the Northwest Forest Plan. Most common bat species in the Washington Cascades and Oregon Coast Range are more abundant (2.5–9.8 times) in old-growth forests than in younger forests (Thomas and West 1991). Several species are closely associated with old-growth forests because they roost in large trees and snags with deeply furrowed bark and cavities (Perkins and Cross 1988; Thomas 1988; Thomas and West 1991; Cross 1993; Cross and Waldien 1994, 1995; Perkins 1994). Definitive studies are scarce, but timber harvest substantially reduces activity levels of the little brown bat (Lunde and Harestad 1986) and probably those of other forest species of bats (Thomas and West 1991).

In eastside rangelands, bats typically roost in rock crevices or caves and therefore are not limited by the availability of trees and snags. Most of the remaining 3,000 or so Townsend's big-eared bats roost in lava tubes in central and southeastern Oregon (Perkins and Levesque 1987; Dobkin et al. 1995b). Restricted water availability may be as important a limiting

Table 12. Status and habitat associations of bats in the Pacific Northwest that are on state sensitive species lists.

Species	State status[a]				Habitat associations[b]	Population trend[c]
	California	Oregon	Washington	Idaho		
Fringed myotis	*	S–v	M	SC–c	Forest, clearings	Downward? in Oregon
Keen's myotis	—	—	M	—	Forest	Unknown
Long-eared myotis	*	S–u	M	*	Forest, clearings, steppe	Unknown
Long-legged myotis	*	S–u	M	*	Forest, clearings, meadows, steppe	Unknown
Pallid bat	SC	S–v	M	*	Forest, steppe	Downward? in Oregon
Red bat	*	*	M	—	Forest	Unknown
Silver-haired bat	*	S–u	*	*	Forest	Unknown
Spotted bat	SC	*	—	SC–c	Steppe	Unknown
Townsend's big-eared bat	SC	S–c	C	SC–c	Forest, steppe	Downward
Western pipistrelle	*	*	M	SC–c	Steppe	Unknown
Western small-footed myotis	*	S–u	M	*	Steppe	Unknown
Yuma myotis	*	S–u	*	*	Forest, steppe	Unknown

[a] — = does not occur in state, * = no special status; California—SC = species of special concern; Oregon—S–c,u,v = sensitive species with critical, vulnerable or unknown status; Washington—C = candidate for sensitive species status, M = monitor species; Idaho—SC–c = species of special concern category c (undetermined status). References: Oregon Natural Heritage Program 1993; California Department of Fish and Game 1994; Conservation Data Center 1994; Washington Department of Wildlife 1994; Marshall et al. 1996.
[b] References: Brown 1985; Christy and West 1993; Puchy and Marshall 1993; Thomas et al. 1993.
[c] References: Perkins and Levesque 1987; Rodrick and Milner 1991; Marshall et al. 1996; U.S. Fish and Wildlife Service 1994.

factor as roost availability in this region. Maintenance of healthy open-water habitats is important because all Pacific Northwest bats need water for drinking, and many species concentrate their feeding in such areas because insects are plentiful there (Whitaker et al. 1977; Christy and West 1993).

Knowledge Gaps, Current Efforts, and Prospects for the Future

Westside Forests

One of the most highly coordinated wildlife research and monitoring efforts in the country exists in the western Pacific Northwest. The Northwest Forest Plan and the Old-Growth Forest Wildlife Habitat Research and Development Program (Ruggiero et al. 1991b) are virtually unprecedented. In addition, long-term research on the northern spotted owl, although highly controversial, has enhanced the understanding of forest structure and stimulated interest in species that otherwise would have been largely ignored (for example, fungi and small mammals).

Forest management to aid in the creation of old-growth characteristics requires an understanding of functional processes in existing old-growth forests (Forest Ecosystem Management Assessment Team 1993). Managers, therefore, require information on a diverse array of species and functional relations rather than on only a few endangered species, which have been the focus in the past. The Survey and Manage category of species listed in the Northwest Forest Plan is a response to this gap in knowledge. Most of these species are fungi, lichens, bryophytes, mollusks, and arthropods that researchers believe are functionally significant associates of old-growth forests, but for which little or no information exists.

Eastside Forests

The extensive loss of parklike old-growth forests and unprecedented increases in insect and disease epidemics in eastside forests spurred scientists to develop strategies for restoring past ranges of variability in species composition and structure (Gast et al. 1991; Mutch et al. 1993; U.S. Forest Service 1993; Tanaka et al. 1995; Tesch 1995). Reducing stand density to minimize moisture and nutrient stress for individual trees and reintroducing fire—the natural thinning agent—are primary objectives, but both these actions are controversial (Agee 1994; Arno and Ottmar 1994). Scientists emphasize that restoration efforts must be focused at the landscape level to ensure the restoration of a mosaic of forest types that will, in turn, reduce the continuity of food sources for defoliating insects (Mason and Wickman 1994) and provide natural fuel breaks (Arno and Ottmar 1994). Without a dynamic, landscape-level effort to restore a balanced mosaic of healthy trees, insect and disease outbreaks will continue to increase in frequency, severity, and length (Lehmkuhl et al. 1994), and wildfires will continue to be large and difficult to control (Arno and Ottmar 1994).

Eastside Rangelands

Restoration of diverse mixtures of native plants on sites dominated by nonindigenous species is a great challenge facing rangeland managers (Pyke and Borman 1993). As in eastside forests, reestablishing a natural fire regime is often emphasized, but additional research is needed to determine how fire can be effectively reintroduced without exacerbating problems with nonindigenous species. Natural, often unpredictable weather cycles may determine the short-term success of restoration projects; large-scale annual and seasonal variation in the size and activity patterns of wildlife populations are common. Therefore, unraveling the complex dynamics of these arid ecosystems and generating accurate assessments of the success of restoration efforts will require heretofore uncommon intensive and long-term research and monitoring.

Riparian and Aquatic Ecosystems

Several major studies were recently initiated in the Pacific Northwest to expand our understanding of riparian ecosystems and the ecology of salmon and trout. The congressionally appointed Northwest Power Planning Council (1986, 1994) developed an early comprehensive management plan for the Columbia River and associated hydroelectric dams. The development of a regionwide strategy for managing watersheds and fishery resources in the Pacific Northwest began with the Gang-of-Four Report (Johnson et al. 1991), which formed the basis for the Aquatic Conservation Strategy in the Northwest Forest Plan (also see Reeves and Sedell 1992). The Aquatic Conservation Strategy identifies key watersheds with good water quality, good fish habitat, and populations of threatened or endangered stocks of anadromous salmon and trout, and prescribes undisturbed forest buffer zones along streams to protect stream integrity (that is, to ensure low sediment loads, adequate shade, continued recruitment of woody debris and nutrients, and so on). Other scientists suggested revised criteria for

Principal Authors

Jeff P. Smith*
Oregon State University
Department of Forest Science and
U.S. Geological Survey
Forest and Rangeland Ecosystem
Science Center
3200 S.W. Jefferson Way
Corvallis, Oregon 97331

*Current address:
Hawkwatch International
P.O. Box 660
Salt Lake City, Utah 84110

Michael W. Collopy
U.S. Geological Survey
Forest and Rangeland Ecosystem
Science Center
3200 S.W. Jefferson Way
Corvallis, Oregon 97331

Contributing Authors

R. Bruce Bury
U.S. Geological Survey
Forest and Rangeland Ecosystem
Science Center
Willamette Field Station
3080 S.E. Clearwater Drive
Corvallis, Oregon 97333

selecting watersheds for conservation (Nickelson et al. 1992; American Fisheries Society 1993; Pacific Rivers Council 1994), but the essential common theme is that to ensure protection of multiple scales of biological diversity and landscape-level ecological processes, the minimum scale of attention must be entire watersheds rather than single stream reaches (Naiman et al. 1992; Reeves and Sedell 1992).

The U.S. Forest Service and Bureau of Land Management (1994a) are also conducting a joint study commonly called PACFISH. The goal of this study is to develop "an ecosystem-based, aquatic habitat and riparian-area management strategy" for lands the agencies administer outside the range of the northern spotted owl in the Pacific Northwest and California. The PACFISH effort is an analog of the Aquatic Conservation Strategy in the Northwest Forest Plan.

Two major efforts were recently commissioned by the Oregon state legislature to synthesize information and to develop a comprehensive plan for coordinating restoration efforts in the state. The first was designed to assess the effects of forestry practices on anadromous fishes in western Oregon and northern California. It resulted in recommendations about how forest management could help restore anadromous fish populations (Botkin et al. 1994; Cummins et al. 1994). The intent of the second study, commonly known as the Bradbury Commission and recently transformed into a cooperative venture called For the Sake of the Salmon (B. Bradbury, executive director, Portland, Oregon), is to develop a management framework to "protect and restore native fishes by focusing on strategies that provide the greatest ecological benefits for native fishes and ecosystems (with priority given to anadromous salmonids)" (W. Nehlsen, Pacific Rivers Council, Portland, Oregon, unpublished report). Two key principles guide this study: protection of relatively intact ecosystems must be the first priority because such areas provide source populations for recovery of nearby degraded systems (Pacific Rivers Council 1994), and conservation must accommodate biological diversity at several spatial scales (landscape, river basins, and individual watersheds).

Specific Biota

Information about most species, particularly at the level of landscape ecology and population dynamics, is incomplete. Scientists are beginning to achieve this level of understanding for closely monitored vertebrate species such as northern spotted owls and some forest trees such as Douglas-fir and ponderosa pine. Many other species, though, are important links in critical ecosystem processes but are poorly known beyond information from small-scale studies.

Key examples of species for which there are obvious gaps in knowledge include fungi, arthropods, small mammals, and nongame fishes. Fungi are biologically and ecologically diverse and functionally significant to food webs, nutrient cycles, and plant productivity in forest and lowland ecosystems (Trappe and Luoma 1992). The number of species of arthropods is far greater than the number of all other species combined, and scientists know that arthropods are functionally important in many aquatic (Anderson et al. 1978; Anderson and Sedell 1979) and terrestrial (Mattson 1977; Moldenke and Lattin 1990; Miller 1993) ecosystems. Moreover, diverse fungal and arthropod communities could be important hallmarks of old-growth forest structure (Forest Ecosystem Management Assessment Team 1993). Nevertheless, little is known about the ecology of most fungal and arthropod species. Similarly, small mammals and fishes are important links in terrestrial and aquatic food webs that support many of the larger vertebrates that so often garner public attention, yet many species are poorly known. If the goal is to maintain healthy, functioning ecosystems, the well-being of these less obvious and perhaps ultimately more important elements of the systems must be ensured.

Many different species have shown signs of decline from human-caused disturbances such as logging; however, the apparent widespread decline of amphibians is especially noteworthy, and the Pacific Northwest is increasingly a focus of attention in this regard. Recent studies have revealed localized declines of aquatic and terrestrial amphibians because of logging and riparian habitat alterations. Similarly, extensive and intensive surveys left little doubt that the ranges of some species (for example, spotted frog and northern leopard frog) are greatly reduced. Nevertheless, regional and landscape-level information about most species is unavailable. The status of entire species or subspecies cannot be adequately assessed from responses of a few local populations to management (Corn 1994).

These examples emphasize particularly obvious gaps of knowledge that should receive attention. However, our intent is not to suggest that these subjects should necessarily become top-priority foci of concern at the expense of current emphases. In fact, many scientists may argue that a taxonomic focus should be avoided in favor of a process-oriented focus on, for example, the landscape-level dynamics and

Michael A. Castellano
U.S. Forest Service
Pacific Northwest Research
Station
3200 S.W. Jefferson Way
Corvallis, Oregon 97331

Stephen P. Cross
Southern Oregon State College
Department of Biology
Ashland, Oregon 97520

David S. Dobkin
High Desert Ecological Research
Institute
15 S.W. Colorado Avenue
Suite 300
Bend, Oregon 97702

Joan Hagar
Oregon State University
Department of Forest Science
3200 S.W. Jefferson Way
Corvallis, Oregon 97331

John D. Lattin
Oregon State University
Department of Entomology
Cordley Hall
Corvallis, Oregon 97331

Judith Li
Oregon State University
Department of Fisheries and
Wildlife, Nash Hall
Corvallis, Oregon 97331

William C. McComb
University of Massachusetts
Department of Forestry and
Wildlife Management
Holdsworth Natural Resources
Center
Amherst, Massachusetts 01003

Karl J. Martin
Oregon State University,
Department of Forest Science
3200 S.W. Jefferson Way
Corvallis, Oregon 97331

Jeffrey C. Miller
Oregon State University
Department of Entomology
Cordley Hall
Corvallis, Oregon 97331

Randy Molina
U.S. Forest Service,
Pacific Northwest Research
Station
3200 S.W. Jefferson Way
Corvallis, Oregon 97331

J. Mark Perkins
2217 E. Emerson
Salt Lake City, Utah 84108

David A. Pyke
U.S. Geological Survey
Forest and Rangeland Ecosystem
Science Center
3200 S.W. Jefferson Way
Corvallis, Oregon 97331

Roger Rosentreter
Bureau of Land Management
1387 S. Vinnell Way
Boise, Idaho 83709

Jane E. Smith
U.S. Forest Service
Pacific Northwest Research
Station
3200 S.W. Jefferson Way
Corvallis, Oregon 97331

Edward G. Starkey
U.S. Geological Survey
Forest and Rangeland Ecosystem
Science Center
3200 S.W. Jefferson Way
Corvallis, Oregon 97331

Steven D. Tesch
Oregon State University
Department of Forest Engineering
Peavy Hall
Corvallis, Oregon 97331

effects of natural disturbances (Perry 1988; Agee 1993; Swanson et al. 1994).

The accounts presented in this regional report provide only a glimpse of the depth and breadth of research that has taken place in the Pacific Northwest. For decades, federal and state agencies and private industry have supported considerable research on a wide array of organisms, communities, and ecosystems. Today, some of the best long-term, ecological data sets come from this region. Nevertheless, our knowledge remains grossly incomplete for numerous species, and we have only begun to unravel the ecological complexities of most regional ecosystems.

Acknowledgments

Numerous people from the following state agencies provided valuable assistance with identifying key contacts, issues, and research programs, and with securing data and graphics: Idaho Department of Fish and Game; California Department of Fish and Game; Washington Department of Natural Resources; Washington Department of Fish and Wildlife; Oregon Department of Fish and Wildlife; Oregon Department of Agriculture; Oregon, Washington, and California natural heritage programs; Idaho Conservation Data Center; and the California Native Plant Society. Many other people contributed to the development of the chapter; the following were particularly instrumental in this regard: R. Anthony, H. Li, and J. Tappeiner III, U.S. Geological Survey, Corvallis, Oregon; B. Peterjohn, U.S. Geological Survey, Laurel, Maryland; S. Knick and T. Zarriello, U.S. Geological Survey, Boise, Idaho; G. Lienkaemper, D. Olson, G. Reeves, and T. Spies, U.S. Forest Service, Corvallis, Oregon; K. Aubry and M. Raphael, U.S. Forest Service, Olympia, Washington; L. Jurs, Bureau of Land Management, Spokane, Washington; J. Lint, Bureau of Land Management, Roseburg, Oregon; M. Borman, S. Gregory, B. McIntosh, and A. Moldenke, Oregon State University, Corvallis; S. Balakoff, The Wilderness Society, Seattle, Washington; B. Hughes, U. S. Environmental Protection Agency, Corvallis, Oregon; and J. E. Smith, Pacific Watershed Institute, Seattle, Washington. Constructive reviews of earlier drafts by E. Multer, U.S. Geological Survey, Corvallis, Oregon; J. E. Williams, Bureau of Land Management, Boise, Idaho; and two anonymous reviewers improved the quality of the manuscript.

Cited References

Agee, J. K. 1991. Fire history of Douglas-fir forests in the Pacific Northwest. Pages 25–34 *in* L. F. Ruggiero, K. W. Aubry, A. B. Carey, and M. H. Huff, editors. Wildlife and vegetation in unmanaged Douglas-fir forests. U.S. Forest Service General Technical Report PNW-GTR-285. Pacific Northwest Research Station, Portland, Oreg.

Agee, J. K. 1993. Fire ecology of Pacific Northwest forests. Island Press, Washington, D.C. 493 pp.

Agee, J. K. 1994. Fire and weather disturbances in terrestrial ecosystems of the eastern Cascades. U.S. Forest Service General Technical Report PNW-GTR-320. Pacific Northwest Research Station, Portland, Oreg. 52 pp.

Almack, J. A., W. L. Gaines, R. H. Naney, P. H. Morrison, J. R. Eby, G. F. Wooten, M. C. Snyder, S. H. Fitkin, and E. R. Garcia. 1993. North Cascades grizzly bear ecosystem evaluation: final report. Interagency Grizzly Bear Committee, Denver, Colo. 156 pp.

Almand, J., and W. Krohn. 1979. The position of the Bureau of Land Management on the protection and management of riparian ecosystems. Pages 259–361 *in* R. Johnson and F. McCormick, technical coordinators. Strategies for protection and management of floodplain wetlands and other riparian ecosystems. Proceedings of the symposium, 11–13 December 1978, Callaway Gardens, Ga. U.S. Forest Service General Technical Report WO-GTR-12. Washington, D.C.

Amaranthus, M. P., and D. A. Perry. 1989. Interaction effects of vegetation type and Pacific madrone soil inocula on survival, growth, and mycorrhiza formation of Douglas-fir. Canadian Journal of Forestry Research 19:550–556.

Amaranthus, M. P., and D. A. Perry. 1994. The functioning of ectomycorrhizal fungi in the field: linkages in space and time. Plant and Soil 159:133–140.

Amaranthus, M., J. M. Trappe, L. Bednar, and D. Arthur. 1994. Hypogeous fungal production in mature Douglas-fir forest fragments and surrounding plantations and its relation to coarse woody debris and animal mycophagy. Journal of Forest Research 24:2157–2165.

American Fisheries Society. 1993. Oregon critical watersheds database. American Fisheries Society, Oregon Chapter, Corvallis.

Andelman, S. J., and A. Stock. 1994a. Management, research and monitoring priorities for the conservation of Neotropical migrant landbirds that breed in Oregon. Washington Natural Heritage Program, Washington Department of Natural Resources, Olympia. 24 pp. + appendixes.

Andelman, S. J., and A. Stock. 1994b. Management, research and monitoring priorities for the conservation of Neotropical migrant landbirds that breed in Washington State. Washington Natural Heritage Program, Washington Department of Natural Resources, Olympia. 25 pp. + appendixes.

Anderson, J. E. 1986. Development and structure of sagebrush–steppe plant communities. Pages 10–12 *in* P. J. Joss, P. W. Lynch, and O. B. Williams, editors. Rangelands: a resource under siege. Proceedings of the Second International Rangeland Congress, Australian Academy of Science, Canberra.

Anderson, J. E., and K. E. Holte. 1981. Vegetation development over 25 years without grazing on sagebrush-dominated rangeland in southeastern Idaho. Journal of Range Management 34:25–29.

Anderson, J. M. 1975. The enigma of soil animal species diversity. Pages 51–88 *in* J. Vanek, editor. Progress in soil zoology: proceedings of the 5th international colloquium on soil zoology, 17–22 September, 1973. Academia, Publishing House of the Czechoslovak Academy of Sciences, Prague.

Anderson, N. H. 1992. Influence of disturbance on insect communities in Pacific Northwest streams. Hydrobiologia 248:79–92.

Anderson, N. H., and J. R. Sedell. 1979. Detritus processing by macroinvertebrates in stream ecosystems. Annual Review of Entomology 24:351–377.

Anderson, N. H., J. R. Sedell, L. M. Roberts, and F. J. Triska. 1978. The role of aquatic invertebrates in processing woody debris from coniferous forest streams. American Midland Naturalist 100:64–82.

Anderson, R. S. 1980. Relationships between trout and invertebrate species as predators and the structure of the crustacean and rotiferan plankton in mountain lakes. Pages 635–641 *in* W. C. Kerfoot, editor. Ecology and evolution of zooplankton communities. Special symposium volume 3, American Society of Limnology and Oceanography. University Press of New England, Hanover, N.H.

Araya, J. E., and B. A. Haws. 1991. Arthropod populations associated with a grassland infested by black grass bugs, *Labops hesperius* and *Irbisia brachycera* (Hemiptera: Miridae), in Utah, USA. Food and Agriculture Organization Plant Protection Bulletin 39:75–81.

Arno, S. F., and R. D. Ottmar. 1994. Reducing hazard for catastrophic fire. Pages 18–19 *in* R. L. Everett, compiler. Restoration of stressed sites, and processes. U.S. Forest Service General Technical Report PNW-GTR-330. Pacific Northwest Research Station, Portland, Oreg.

Asquith, A., and J. D. Lattin. 1991. A review of the introduced Lygaeidae of the Pacific Northwest, including the newly discovered *Plinthisus brevipennis* (Latreille) (Heteroptera). The Pan-Pacific Entomologist 67:259–271.

Aubry, K. B. 1984. The recent history and present distribution of the red fox in Washington. Northwest Science 58:69–79.

Aubry, K. B., and D. B. Houston. 1992. Distribution and status of the fisher (*Martes pennanti*) in Washington. Northwestern Naturalist 73:69–79.

Aubry, K. B., L. L. C. Jones, and P. A. Hall. 1988. Use of woody debris by plethodontid salamanders in Douglas-fir forests in Washington. Pages 32–44 *in* R. C. Szaro, K. E. Severson, and D. R. Patton, technical coordinators. Management of amphibians, reptiles, and small mammals in North America. U.S. Forest Service General Technical Report RM-GTR-166. Rocky Mountain Forest and Range Experiment Station, Fort Collins, Colo.

Bahls, P. 1992. The status of fish populations and management of high mountain lakes in the western United States. Northwest Science 66:183–193.

Banci, V. 1994. Wolverine. Pages 99–127 *in* L. F. Ruggiero, K. B. Aubry, S. W. Buskirk, L. J. Lyon, and W. J. Zielinski, technical editors. The scientific basis for conserving forest carnivores: American marten, fisher, lynx, and wolverine in the western United States. U.S. Forest Service General Technical Report RM-GTR-254. Rocky Mountain Forest and Range Experiment Station, Fort Collins, Colo.

Barbour, M. G., and W. D. Billings, editors. 1988. North American terrestrial vegetation. Cambridge University Press, New York. 434 pp.

Beamesderfer, R. C., and A. A. Nigro, editors. 1993. Status and habitat requirements of the white sturgeon populations in the Columbia River downstream of McNary Dam. 2 volumes. Final report on Project 86-50 to the U.S. Department of Energy, Bonneville Power Administration, Division of Fish and Wildlife, Portland, Oreg.

Behnke, R. J. 1992. Native trout of western North America. American Fisheries Society Monograph 6, Bethesda, Md. 275 pp.

Belnap, J. 1990. Microbiotic crusts: their role in past and present ecosystems. Park Science 10(3):3–4.

Beschta, R. L. 1991. Stream habitat management for fish in the northwestern United States: the role of riparian vegetation. American Fisheries Society Symposium 10:53–58.

Beschta, R. L., R. E. Bilby, G. W. Brown, L. B. Holtby, and T. D. Hofstra. 1987. Stream temperature and aquatic habitat: fisheries and forestry interactions. Pages 191–232 *in* E. O. Salo and T. W. Cundy, editors. Streamside management: forestry and fishery interactions. Institute of Forest Resources Contribution 57. University of Washington, Seattle.

Betts, B. J. 1990. Geographic distribution and habitat preferences of Washington ground squirrels (*Spermophilus washingtoni*). Northwestern Naturalist 71:27–37.

Bilby, R. E., and J. W. Ward. 1991. Characteristics and function of large woody debris in streams draining old-growth, clear-cut, and second-growth forests in southwestern Washington. Canadian Journal of Fisheries and Aquatic Sciences 48:2499–2508.

Bisson, P. A., R. E. Bilby, M. D. Bryant, C. A. Dolloff, G. B. Grette, R. A. House, M. L. Murphy, K. V. Koski, and J. R. Sedell. 1987. Large woody debris in forested streams in the Pacific Northwest: past, present, and future. Pages 143–190 *in* E. O. Salo and T. W. Cundy, editors. Streamside management: forestry and fishery interactions. Institute of Forest Resources Contribution 57. University of Washington, Seattle.

Bisson, P. A., T. P. Quinn, G. H. Reeves, and S. V. Gregory. 1992. Best management practices, cumulative effects, and long-term trends in fish abundance in Pacific Northwest river systems. Pages 189–232 *in* R. J. Naiman, editor. Watershed management: balancing sustainability and environmental change. Springer-Verlag, New York.

Bjornn, T. C., and D. W. Reiser. 1991. Habitat requirements of salmonids in streams. American Fisheries Society Special Publication 19:83–138.

Black, J. C. 1992. Silvicultural approaches to animal damage management in Pacific Northwest forests. U.S. Forest Service General Technical Report PNW-GTR-287. Pacific Northwest Research Station, Portland, Oreg. 422 pp.

Blaustein, A. R., P. D. Hoffman, D. G. Hokit, J. M. Kiesecker, S. C. Walls, and J. B. Hays. 1994a. UV repair and resistance to solar UV-B in amphibian eggs: a link to population declines? Proceedings of the National Academy of Sciences 91:1791–1795.

Blaustein, A. R., and D. B. Wake. 1990. Declining amphibian populations: a global phenomenon? Trends in Ecology and Evolution 5:203.

Blaustein, A. R., D. B. Wake, and W. P. Sousa. 1994b. Amphibian declines: judging stability, persistence, and susceptibility of populations to local and global extinction. Conservation Biology 8:60–71.

Bolsinger, C. L., and K. L. Waddell. 1993. Area of old-growth forests in California, Oregon, and Washington. U.S. Forest Service Research Bulletin PNW-RB-197. Pacific Northwest Research Station, Portland, Oreg. 26 pp.

Botkin, D. B., K. Cummins, T. Dunne, H. Regier, M. Sobel, and L. M. Talbot. 1994. Status and future of salmon of western Oregon and northern California: findings and options—draft. The Center for the Study of Environment Research Report 941005. Santa Barbara, Calif. 400 pp.

Boula, K. M., and P. L. Sharp. 1985. Distribution and abundance of small mammals on native and converted rangelands in southeastern Oregon. Oregon Department of Fish and Wildlife Technical Report 85-5-01. Nongame Wildlife Program, Portland. 31 pp.

Brinkley, D., and R. L. Graham. 1981. Biomass, production, and nutrient cycling of mosses in old-growth Douglas-fir forests. Ecology 62:1387–1389.

Brown, E. R., technical editor. 1985. Management of wildlife and fish habitats in forests of western Oregon and Washington. 2 parts. U.S. Forest Service Publication R6-F&WL-192-1985. Pacific Northwest Region, Portland, Oreg. 332 pp.

Brown, L. R., P. B. Moyle, and R. M. Yasakawa. 1994. Historical decline and current status of coho salmon in California. North American Journal of Fisheries Management 14:237–261.

Bryan, T., and E. D. Forsman. 1987. Distribution, abundance, and habitat of great gray owls in southcentral Oregon. Murrelet 68:45–49.

Bull, E. L. 1987. Ecology of the pileated woodpecker in northeastern Oregon. Journal of Wildlife Management 51:472–481.

Bull, E. L. 1991. Summer roosts and roosting behavior of Vaux's swifts in old-growth forests. Northwestern Naturalist 72:78–82.

Bull, E. L. 1994. Effect of logging on tailed frog populations in northeastern Oregon. Northwest Science 68:117. Abstract.

Bull, E. L., and H. D. Cooper. 1991. Vaux's swift nests in hollow trees. Western Birds 22:85–91.

Bull, E. L., and M. G. Henjum. 1990. Ecology of the great gray owl. U.S. Forest Service General Technical Report PNW-GTR-265. Pacific Northwest Research Station, Portland, Oreg. 39 pp.

Bull, E. L., M. G. Henjum, and R. S. Rohweder. 1988. Nesting and foraging habitat of great gray owls. Journal of Raptor Research 22:107–115.

Bull, E. L., and J. E. Hohmann. 1993. The association between Vaux's swifts and old growth forests in northeastern Oregon. Western Birds 24:38–42.

Bull, E. L., and R. S. Holthausen. 1993. Habitat use and management of pileated woodpeckers in northeastern Oregon. Journal of Wildlife Management 57:335–345.

Bull, E. L., R. S. Holthausen, and M. G. Henjum. 1992. Roost trees used by pileated woodpeckers in northeastern

Oregon. Journal of Wildlife Management 56:786–793.

Bull, E. L., S. R. Peterson, and J. W. Thomas. 1986. Resource partitioning among woodpeckers in northeastern Oregon. U.S. Forest Service Research Note PNW-RN-444. Pacific Northwest Research Station, Portland, Oreg. 19 pp.

Bull, E. L., A. L. Wright, and M. G. Henjum. 1990. Nesting habitat of flammulated owls in Oregon. Journal of Raptor Research 24:52–55.

Bury, R. B. 1988. Habitat relationships and ecological importance of amphibians and reptiles. Pages 61–76 in K. J. Raedeke, editor. Streamside management: riparian wildlife and forestry interactions. Institute of Forest Resources Contribution 59. University of Washington, Seattle.

Bury, R. B. 1994. Vertebrates in the Pacific Northwest: species richness, endemism and dependency on old-growth forests. Pages 392–404 in S. K. Majumdar, F. J. Brenner, J. E. Lovich, J. F. Schalles, and E. W. Miller, editors. Biological diversity: problems and challenges. Pennsylvania Academy of Science, Easton.

Bury, R. B., and P. S. Corn. 1988. Douglas-fir forests in the Oregon and Washington Cascades: relation of the herpetofauna to stand age and moisture. Pages 11–22 in R. C. Szaro, K. E. Severson, and D. R. Patton, technical coordinators. Management of amphibians, reptiles, and small mammals in North America. U.S. Forest Service General Technical Report RM-GTR-166. Rocky Mountain Forest and Range Experiment Station, Fort Collins, Colo.

Bury, R. B., P. S. Corn, K. B. Aubry, F. F. Gilbert, and L. L. C. Jones. 1991a. Aquatic amphibian communities in Oregon and Washington. Pages 353–362 in L. F. Ruggiero, K. B. Aubry, A. B. Carey, and M. H. Huff, technical coordinators. Wildlife and vegetation of unmanaged Douglas-fir forests. U.S. Forest Service General Technical Report PNW-GTR-285. Pacific Northwest Research Station, Portland, Oreg.

Bury, R. B., P. S. Corn, and K. B. Aubry. 1991b. Regional patterns of terrestrial amphibian communities in Oregon and Washington. Pages 341–352 in L. F. Ruggiero, K. B. Aubry, A. B. Carey, and M. H. Huff, technical coordinators. Wildlife and vegetation of unmanaged Douglas-fir forests. U.S. Forest Service General Technical Report PNW-GTR-285. Pacific Northwest Research Station, Portland, Oreg.

Bury, R. B., and J. A. Whelan. 1984. Ecology and management of the bullfrog. U.S. Fish and Wildlife Service Resource Publication 155. Washington, D.C. 23 pp.

Buskirk, S. W., and L. F. Ruggiero. 1994. American marten. Pages 7–37 in L. F. Ruggiero, K. B. Aubry, S. W. Buskirk, L. J. Lyon, and W. J. Zielinski, technical editors. The scientific basis for conserving forest carnivores: American marten, fisher, lynx, and wolverine in the western United States. U.S. Forest Service General Technical Report RM-GTR-254. Rocky Mountain

Forest and Range Experiment Station, Fort Collins, Colo.

California Department of Fish and Game. 1992. Annual report on the status of California state listed threatened and endangered animals and plants. California Department of Fish and Game, Sacramento. 203 pp.

California Department of Fish and Game. 1994. Special animals: August 1994. California Department of Fish and Game, Natural Heritage Division, Natural Diversity Data Base, Sacramento. 28 pp.

California Department of Fish and Game. 1995a. Special plants list: March 1995. California Department of Fish and Game, Natural Heritage Division, Natural Diversity Data Base, Sacramento. 74 pp.

California Department of Fish and Game. 1995b. Endangered and threatened animals of California: April 1995. California Department of Fish and Game, Natural Heritage Division, Natural Diversity Data Base, Sacramento. 14 pp.

Carey, A. B. 1995. Sciurids in Pacific Northwest managed and old-growth forests. Ecological Applications 5:648–661.

Carey, A. B., and M. L. Johnson. 1995. Small mammals in managed, naturally young, and old-growth forests. Ecological Applications 5:336–352.

Carlson, J. Y., C. W. Andrus, and H. A. Froehlich. 1990. Woody debris, channel features, and macroinvertebrates of streams with logged and undisturbed riparian timber in northeastern Oregon, USA. Canadian Journal of Fisheries and Aquatic Sciences 47:1103–1111.

Carter, M. F., and K. Barker. 1993. An interactive database for setting conservation priorities for western Neotropical migrants. Pages 120–144 in D. M. Finch and P. W. Stangel, editors. Status and management of Neotropical migratory birds. U.S. Forest Service General Technical Report RM-GTR-229. Rocky Mountain Forest and Range Experiment Station, Ft. Collins, Colo.

Castellano, M. A., and R. Molina. 1989. Mycorrhizae. Pages 101–167 in T. D. Landis, R. W. Tinus, S. E. McDonald, and J. P. Barnett, editors. The container tree nursery manual. Volume 5. U.S. Department of Agriculture Agricultural Handbook 674. Washington, D.C.

Christy, R. E., and S. D. West. 1993. Biology of bats in Douglas-fir forests. U.S. Forest Service General Technical Report PNW-GTR-308. Pacific Northwest Research Station, Portland, Oreg. 28 pp.

Close, D. A., M. S. Fitzpatrick, H. W. Li, B. Parker, and G. James. 1995. Status report of the Pacific lamprey (Lampetra tridentata). Draft final report. Bonneville Power Administration, Portland, Oreg. 23 pp.

Cobb, N., E. Gruber, E. Heske, A. Flecker, D. Lightfoot, A. Masters, N. McClintock, J. Price, T. Seibert, and M. Smith. 1981. An ecological study of the Alvord Basin sand dunes, southeastern Oregon. Final Technical Report, National Science

Foundation Grant SP7905328. Oregon State University, Corvallis. 142 pp.

Coleman, M. E., A. M. McGie, and D. L. Bottom. 1987. Evaluate causes for the decline of the shortnose and Lost River suckers in Klamath Lake, Oregon. Oregon Department of Fish and Wildlife, Fish Division Annual Progress Report, Portland. 13 pp.

Conservation Data Center. 1994. Rare, threatened and endangered plants and animals of Idaho. 3rd edition. Idaho Department of Fish and Game, Boise. 39 pp.

Corn, P. S. 1994. What we know and don't know about amphibian declines in the West. Pages 59–67 in W. W. Covington and L. F. DeBano, technical coordinators. Sustainable ecological systems: implementing an ecological approach to land management. U.S. Forest Service General Technical Report RM-GTR-247. Rocky Mountain Forest and Range Experiment Station, Ft. Collins, Colo.

Corn, P. S., and R. B. Bury. 1989. Logging in western Oregon: responses of headwater habitats and stream amphibians. Forest Ecology and Management 29:39–57.

Corn, P. S., and R. B. Bury. 1991. Terrestrial amphibian communities in the Oregon Coast Range. Pages 305–317 in L. F. Ruggiero, K. B. Aubry, A. B. Carey, and M. H. Huff, technical coordinators. Wildlife and vegetation of unmanaged Douglas-fir forests. U.S. Forest Service General Technical Report PNW-GTR-285. Pacific Northwest Research Station, Portland, Oreg.

Cowlin, R. W., P. A. Briegleb, and F. L. Moravets. 1942. Forest resources of the ponderosa pine region of Washington and Oregon. U.S. Department of Agriculture Miscellaneous Publication 490. Washington, D.C. 99 pp.

Cross, S. P. 1977. A survey of the bats of Oregon Caves National Monument. National Park Service Report on Contract CX-9000-6-005. Oregon Caves National Monument, Cave Junction. 40 pp.

Cross, S. P. 1988. Riparian systems and small mammals and bats. Pages 93–112 in K. J. Raedeke, editor. Streamside management: riparian wildlife and forestry interactions. Institute of Forest Resources Contribution 59. University of Washington, Seattle.

Cross, S. P. 1991. Censusing a colony of Brazilian free-tailed bats, Tadarida brasiliensis, in northern California. Paper presented to the Oregon Chapter of the Wildlife Society, Bend.

Cross, S. P. 1993. 1992 studies of Townsend's big-eared bat at Salt Caves, Klamath River Canyon, Klamath County, Oregon. Final Report 1991. Bureau of Land Management, Lakeview District, Lakeview, Oreg. 26 pp.

Cross, S. P., and C. Schoen. 1989. Bats at Oregon Caves: 1988 status report. National Park Service Report on Contract CX-9000-6-0051. Oregon Caves National Monument, Cave Junction. 25 pp.

Cross, S. P., and D. Waldien. 1994. Inventory and monitoring of bats in the Medford District of the BLM in 1993. Final Report to the Bureau of Land Management. Medford, Oreg. 45 pp.

Cross, S. P., and D. Waldien. 1995. Survey of bats and their habitats in the Roseburg District of the BLM in 1994. Final Report to the Bureau of Land Management, Roseburg, Oreg. 64 pp.

Cummins, K., D. Botkin, H. Regier, M. Sobel, and L. Talbot. 1994. Status and future of salmon of western Oregon and northern California: management of the riparian zone for the conservation and production of salmon—draft. The Center for the Study of Environment Research Report 941004. Santa Barbara, Calif. 44 pp.

Daubenmire, R. 1970. Steppe vegetation of Washington. Washington Agricultural Experiment Station Technical Bulletin 62. Washington State University, Pullman. 131 pp.

DePuit, E. J., and J. G. Coenenberg. 1979. Methods for establishment of native plant communities on topsoiled coal strip-mine spoils in the northern Great Plains. Reclamation Review 2:75–83.

DeSante, D. F., and T. L. George. 1994. Population trends in the landbirds of western North America. Studies in Avian Biology 15:173–190.

Dimick, R. E., and F. Merryfield. 1945. The fishes of the Willamette River system in relation to pollution. Oregon State College, Engineering Experiment Station Bulletin Series 20:7–55.

Dobkin, D. S. 1994a. Conservation and management of Neotropical migrant landbirds in the northern Rockies and Great Plains. University of Idaho Press, Moscow. 234 pp.

Dobkin, D. S. 1994b. Community composition and habitat affinities of riparian birds on the Sheldon–Hart Mountain Refuges, Nevada and Oregon, 1991–93. U.S. Fish and Wildlife Service Final Report. Lakeview, Oreg. 61 pp. + tables (58 pp.) + figures (39 pp.) + appendixes (51 pp.).

Dobkin, D. S. 1995. Management and conservation of sage grouse, denominable species for the ecological health of shrubsteppe ecosystems. Bureau of Land Management, Portland, Oreg. 26 pp.

Dobkin, D. S., A. C. Rich, J. A. Pretare, and W. H. Pyle. 1995a. Nest site relationships among cavity-nesting birds of riparian and snowpocket aspen woodlands in the northwestern Great Basin. The Condor 97:694–707.

Dobkin, D. S., R. D. Gettinger, and M. G. Gerdes. 1995b. Springtime movements, roost use, and foraging activity of Townsend's big-eared bat (*Plecotus townsendii*) in central Oregon. Great Basin Naturalist 55:315–321.

Dobler, F. C. 1992. The shrub–steppe ecosystem of Washington: a brief summary of knowledge and nongame wildlife conservation needs. Washington Department of Wildlife, Olympia. 17 pp.

Dorband, W. R. 1980. Benthic macroinvertebrate communities in the Lower Snake River Reservoir system. Ph.D. dissertation, University of Idaho, Moscow. 95 pp.

Eddleman, L. E. 1987. Establishment of western juniper in central Oregon. Pages 255–259 in R. L. Everett, compiler. Proceedings—pinyon–juniper conference 1986. U.S. Forest Service General

Technical Report INT-GTR-215. Intermountain Research Station, Ogden, Utah.

Everest, F. H., N. B. Armantrout, S. M. Keller, W. D. Parante, J. R. Sedell, T. E. Nickelson, J. M. Johnston, and G. N. Haugen. 1985. Salmonids. Pages 199–230 in E. R. Brown, technical editor. Management of wildlife and fish habitats in forests of western Oregon and Washington—part 1: chapter narratives. U.S. Forest Service Publication R6-F&WL-192-1985. Pacific Northwest Region, Portland, Oreg.

Everest, F. H., R. L. Beschta, J. C. Scrivener, K. V. Koski, J. R. Sedell, and C. J. Cederholm. 1987. Fine sediment and salmonid production: a paradox. Pages 98–142 in E. O. Salo and T. W. Cundy, editors. Streamside management: forestry and fishery interactions. Institute of Forest Resources Contribution 57. University of Washington, Seattle.

Everett, R., P. Hessburg, M. Jensen, and B. Bormann. 1994. Eastside forest ecosystem health assessment. Volume 1: executive summary. U.S. Forest Service General Technical Report PNW-GTR-317. Pacific Northwest Research Station, Portland, Oreg. 61 pp.

Farr, R. A., and D. L. Ward. 1993. Fishes of the lower Willamette River, near Portland, Oregon. Northwest Science 67:16–22.

Fellers, G. M., and C. A. Drost. 1993. Disappearance of the Cascades frog, *Rana cascadae*, at the southern end of its range, California, U.S.A. Biological Conservation 65:177–181.

Fielding, D. J., and M. A. Brusven. 1993. Grasshopper (Orthoptera: Acrididae) community composition and ecological disturbance on southern Idaho rangeland. Environmental Entomology 22:71–81.

Fielding, D. J., and M. A. Brusven. 1994. Grasshopper community responses to shrub loss, annual grasslands, and crested wheatgrass seedings: management implications. Pages 162–166 in S. B. Monsen and S. G. Kitchen, editors. Proceedings—ecology and management of annual rangelands. U.S. Forest Service General Technical Report INT-GTR-313. Intermountain Research Station, Ogden, Utah.

Fielding, D. J., and M. A. Brusven. 1996. Historical trends in grasshopper populations in southern Idaho. Pages v.2.1–v.2.2 in Grasshopper integrated pest management user handbook, 1996. Section V. Rangeland Management. U.S. Animal and Plant Health Inspection Service Technical Bulletin 1809. Washington, D.C.

Filip, G. M., and C. L. Schmitt. 1990. Rx for *Abies*: silvicultural options for diseased firs in Oregon and Washington. U.S. Forest Service General Technical Report PNW-GTR-252. Pacific Northwest Research Station, Portland, Oreg. 34 pp.

Finch, D. M., and P. W. Stangel, editors. 1992. Status and management of Neotropical migratory birds. U.S. Forest Service General Technical Report RM-GTR-229. Rocky Mountain Forest and Range Experiment Station, Fort Collins, Colo. 422 pp.

Fleischner, T. L. 1994. Ecological costs of livestock grazing in western North America. Conservation Biology 8:629–644.

Forest Ecosystem Management Assessment Team. 1993. Forest ecosystem management: an ecological, economic, and social assessment. U. S. Department of the Interior, U. S. Department of Agriculture, U.S. Department of Commerce, and U.S. Environmental Protection Agency, Washington, D.C. 729 pp.

Forsman, E. D., E. C. Meslow, and H. M. Wight. 1984. Distribution and biology of the spotted owl in Oregon. Wildlife Monographs 87. 64 pp.

Franklin, J. F., K. Cromack, Jr., W. Dennison, A. McKee, C. Maser, J. Sedell, F. Swanson, and G. Juday. 1981. Ecological characteristics of old-growth Douglas-fir forests. U.S. Forest Service General Technical Report PNW-GTR-118. Pacific Northwest Research Station, Portland, Oreg. 48 pp.

Franklin, J. F. and C. T. Dyrness. 1973. Natural vegetation of Oregon and Washington. U.S. Forest Service General Technical Report PNW-GTR-8. Pacific Northwest Research Station, Portland, Oreg. 417 pp.

Franklin, J. F., and R. T. T. Forman. 1987. Creating landscape patterns by forest cutting: ecological consequences and principles. Landscape Ecology 1:5–18.

French, N. R., W. E. Grant, W. Grodzinski, and D. M. Swift. 1976. Small mammal energetics in grassland ecosystems. Ecological Monographs 46:201–220.

Frest, T. C., and E. J. Johannes. 1993. Mollusc species of special concern within the range of the northern spotted owl. Final Report prepared for the Forest Ecosystem Management Assessment Team. U.S. Forest Service, Portland, Oreg. 98 pp.

Frissell, C. A. 1993. Topology of extinction and endangerment of native fishes in the Pacific Northwest and California (USA). Conservation Biology 7:342–354.

Furniss, R. L., and V. M. Carolin. 1977. Western forest insects. U.S. Forest Service Miscellaneous Publication 1339. Washington, D.C. 654 pp.

Gaines, W. L. 1987. Secondary production of benthic insects in three cold-desert streams. U.S. Department of Energy Report on Contract DE-AC06-76RLO. Pacific Northwest Laboratory, Battelle Memorial Institute, Wash. 39 pp.

Gaines, W. L., C. E. Cushing, and S. D. Smith. 1992. Secondary production estimates of benthic insects in three cold desert streams. Great Basin Naturalist 52:11–24.

Gast, W. R., D. W. Scott, C. Schmitt, D. Clemens, S. Howes, C. G. Johnson, R. Mason, F. Mohr, and R. Clapp. 1991. Blue Mountains forest health report: new perspectives in forest health. U.S. Forest Service, Pacific Northwest Region, Malheur, Umatilla, and Wallowa–Whitman National Forests. 326 pp.

Gilbert, F. F., and R. Allwine. 1991a. Terrestrial amphibian communities in the Oregon Cascade Range. Pages 319–326 in L. F. Ruggiero, K. B. Aubry, A. B. Carey, and M. H. Huff, technical coordinators.

Wildlife and vegetation of unmanaged Douglas-fir forests. U.S. Forest Service General Technical Report PNW-GTR-285. Pacific Northwest Research Station, Portland, Oreg.

Gilbert, F. F., and R. Allwine. 1991b. Spring bird communities in the Oregon Cascade Range. Pages 145–158 *in* L. F. Ruggiero, K. B. Aubry, A. B. Carey, and M. H. Huff, technical coordinators. Wildlife and vegetation of unmanaged Douglas-fir forests. U.S. Forest Service General Technical Report PNW-GTR-285. Pacific Northwest Research Station, Portland, Oreg.

Goggans, R. 1985. Habitat use by flammulated owls in northeastern Oregon. M.S. thesis, Oregon State University, Corvallis. 54 pp.

Goggans, R., R. D. Dixon, and L. C. Seminara. 1989. Habitat use by three-toed and black-backed woodpeckers in Deschutes National Forest, Oregon. Oregon Department of Fish and Wildlife Technical Report 87-3-02. Nongame Wildlife Program, Deschutes National Forest, Bend. 43 pp.

Gregory, S. V., F. J. Swanson, W. A. McKee, and K. W. Cummins. 1991. An ecosystem perspective of riparian zones. Bioscience 41:540–551.

Griffith, J., and D. M. Daley. 1984. Re-establishment of Shoshone sculpin (*Cottus greenei*) in the Hagerman Valley, Idaho. Idaho Department of Fish and Game Final Report. Nongame Program, Boise. 12 pp.

Griffiths, R., and T. Beebee. 1992. Decline and fall of the amphibians. New Scientist 134:25–29.

Groves, C. 1987. Distribution of the wolverine (*Gulo gulo*) in Idaho, 1960–1987. Idaho Department of Fish and Game, Natural Heritage Program, Boise. 22 pp.

Groves, C. R., and B. L. Keller. 1983. Ecological characteristics of small mammals on a radioactive waste disposal area in southeastern Idaho. American Midland Naturalist 109:253–265.

Groves, C. R., and C. Peterson. 1992. Distribution and population trends of Idaho amphibians as determined by mail questionnaire. Idaho Department of Fish and Game, Conservation Data Center, Boise. 16 pp.

Groves, C. R., and K. Steenhof. 1988. Responses of small mammals and vegetation to wildfire in shadscale communities of southwestern Idaho. Northwest Science 62:205–210.

Hammond, P. C. 1993. Oregon silverspot butterfly response to habitat management (summary for 1986–1992). U.S. Forest Service, Pacific Northwest Region, Siuslaw National Forest, Corvallis, Oreg. 15 pp.

Harmon, M. E., J. F. Franklin, F. J. Swanson, P. Sollins, S. V. Gregory, J. D. Lattin, N. H. Anderson, S. P. Cline, N. G. Aumen, J. R. Sedell, G. W. Lienkaemper, K. Cromack, Jr., and K. W. Cummins. 1986. Ecology of coarse woody debris in temperate ecosystems. Advances in Ecological Research 15:133–302.

Harris, L. 1984. The fragmented forest: island biogeography theory and the preservation of biotic diversity. The University of Chicago Press, Chicago and London. 211 pp.

Hassemer, P. F. 1993. Salmon spawning ground surveys, 1989–92. Idaho Department of Fish and Game Project F-73-R-15. Boise. 31 pp.

Hassemer, P. F., S. W. Kiefer, and C. E. Petrosky. 1997. Idaho's salmon: can we count every last one? Pages 113–125 *in* D. J. Stouder, P. A. Bisson, and R. J. Naiman, editors. Pacific salmon and their ecosystems: status and future options. Chapman and Hall, New York.

Hawkins, C. P., M. L. Murphy, and N. H. Anderson. 1982. Effects of canopy, substrate composition, and gradient on the structure of macroinvertebrate communities in Cascade Range streams of Oregon. Ecology 63:1840–1856.

Hawkins, C. P., M. L. Murphy, N. H. Anderson, and M. A. Wilzbach. 1983. Density of fish and salamanders in relation to riparian canopy and physical habitat in streams of the northwestern United States. Canadian Journal of Fisheries and Aquatic Sciences 40:1173–1185.

Hawkins, C. P., and J. R. Sedell. 1981. Longitudinal and seasonal changes in functional organization of macroinvertebrate communities in four Oregon streams. Ecology 62:387–397.

Hayes, M. P. 1994. Current status of the spotted frog (*Rana pretiosa*) in western Oregon. Report to the Oregon Department of Fish and Wildlife, Portland. 13 pp.

Hayes, M. P., and M. R. Jennings. 1986. Decline of ranid frog species in western North America: are bullfrogs (*Rana catesbeiana*) responsible? Journal of Herpetology 20:490–509.

Hejl, S. J. 1994. Human-induced changes in bird populations in coniferous forests in western North America during the past 100 years. Studies in Avian Biology 15:232–246.

Henjum, M. G., J. B. Karr, D. L. Bottom, D. A. Perry, J. C. Bednarz, S. G. Wright, S. A. Beckwitt, and E. Beckwitt. 1994. Interim protection for late-successional forests, fisheries, and watersheds: national forests east of the Cascade crest, Oregon and Washington. The Wildlife Society Technical Review 94-2. Bethesda, Md. 245 pp.

Henny, C. J., and J. L. Kaiser. 1996. Osprey population increase along the Willamette River, Oregon, and the role of utility structures, 1976–1993. Pages 97–108 *in* D. M. Bird, D. E. Varland, and J. J. Negro, editors. Raptors in human landscapes: adaptations to built and cultivated environments. Academic Press, London, England.

Herrington, R. E., and J. H. Larsen, Jr. 1985. Current status, habitat requirements and management of the Larch Mountain salamander. Biological Conservation 34:169–179.

Hessburg, P. F., R. G. Mitchell, and G. M. Filip. 1994. Historical and current roles of insects and pathogens in eastern Oregon and Washington forested landscapes. U.S. Forest Service General Technical Report PNW-GTR-327. Pacific Northwest Research Station, Portland, Oreg. 72 pp.

Hicks, B. J., R. L. Beschta, and R. D. Harr. 1991. Long-term changes in streamflow following logging in western Oregon and associated fisheries implications. Water Resources Bulletin 27:217–226.

Higgins, P., S. Dobush, and D. Fuller. 1992. Factors in northern California threatening stocks with extinction. American Fisheries Society, Humboldt Chapter, Arcata, Calif. 25 pp.

Hironaka, M., M. A. Fosberg, and A. H. Winward. 1983. Sagebrush–grass habitat types of southern Idaho. University of Idaho, Forest, Wildlife, and Range Experiment Station Bulletin 35. Moscow. 44 pp.

Holland, D. C. 1991. A synopsis of the ecology and current status of the western pond turtle (*Clemmys marmorata*). Final Report to the U.S. Fish and Wildlife Service. San Simeon, Calif. 180 pp.

Holland, D.C., and R. B. Bury. 1997. *Clemmys marmorata* (Baird and Girard 1852), western pond turtle. *In* A. Rhodin and P. Pritchard, editors. The conservation biology of freshwater turtles. Chelonean Research Monographs 2. In press.

Horning, D. S., Jr., and W. F. Barr. 1970. Insects of Craters of the Moon National Monument, Idaho. University of Idaho, College of Agriculture Miscellaneous Series 8. Idaho Falls. 118 pp.

Howell, P. J., and D. V. Buchanan, editors. 1992. Proceedings of the Gearhart Mountain bull trout workshop. American Fisheries Society, Oregon Chapter, Corvallis. 67 pp.

Hughes, R. M., and J. R. Gammon. 1987. Longitudinal changes in fish assemblages and water quality in the Willamette River, Oregon. Transactions of the American Fisheries Society 116:196–209.

Huntington, C. W., W. Nehlsen, and J. Bowers. 1994. Healthy native stocks of anadromous salmonids in the Pacific Northwest and California. Oregon Trout, Portland. 42 pp.

Idaho Department of Fish and Game. 1994. Idaho's amphibians and reptiles: description, habitat and ecology. Nongame Wildlife Leaflet 7. Idaho Department of Fish and Game, Nongame and Endangered Wildlife Program, Boise. 12 pp.

Ingles, L. G. 1965. Mammals of the Pacific States. Stanford University Press, Stanford, Calif. 506 pp.

Isaacs, F. B., and R. G. Anthony. 1994. Bald eagle nest locations and history of use in Oregon 1971 through 1994. Oregon State University, Oregon Cooperative Wildlife Research Unit, Corvallis. 13 pp.

Jennings, M. R., and M. P. Hayes. 1995. Amphibian and reptile species of special concern in California. Final Report to the California Department of Fish and Game, Sacramento. 315 pp.

Johannessen, C. L., W. A. Davenport, A. Millet, and S. McWilliams. 1971. The vegetation of the Willamette Valley. Annals of the Association of American Geographers 61:286–302.

Johansen, J. R., J. Ashley, and W. R. Rayburn. 1993. Effects of rangefire on the soil algal crusts in semiarid shrub–steppe of the lower Columbia Basin and their subsequent recovery. Great Basin Naturalist 53:73–88.

Johnson, N. K., J. F. Franklin, J. W. Thomas, and J. Gordon. 1991. Alternatives for management of late-successional forests of the Pacific Northwest. A Report to the Agricultural Committee of the Merchant Marine Committee of the U.S. House of Representatives, Washington, D.C. 59 pp.

Joyce, L. A. 1989. An analysis of the range forage situation in the United States: 1989–2040. U.S. Forest Service General Technical Report RM-GTR-180. Rocky Mountain Forest and Range Experiment Station, Fort Collins, Colo. 136 pp.

Kaczynski, V. W., and J. F. Palmisano. 1993. Oregon's wild salmon and trout: a review of the impact of management and environmental factors. Oregon Forest Industries Council, Salem. 328 pp.

Kauffman, J. B., and W. C. Krueger. 1984. Livestock impacts on riparian ecosystems and streamside management implications…a review. Journal of Range Management 37:430–437.

Keister, G. P., Jr., and D. Immell. 1994. Continued investigations of kit fox in southeastern Oregon and evaluation of status. Oregon Department of Fish and Wildlife Technical Report 94-5-01. Wildlife Diversity Program, Portland. 33 pp.

Kellogg, E., editor. 1992. Coastal temperate rain forests: ecological characteristics, status and distribution worldwide. Ecotrust Occasional Paper Series 1. Ecotrust and Conservation International, Portland, Oreg., and Washington, D.C. 64 pp.

Kindschy, R. R. 1994. Pristine vegetation of the Jordan Crater kipukas: 1978–91. Pages 85–88 *in* S. B. Monsen and S. G. Kitchen, compilers. Proceedings—ecology and management of annual rangelands. U.S. Forest Service General Technical Report INT-GTR-313. Intermountain Research Station, Ogden, Utah.

Knick, S. T. 1993. Habitat classification and the ability of habitats to support populations of Townsend's ground squirrels and black-tailed jack rabbits. Pages 237–263 *in* K. Steenhof, editor. Snake River Birds of Prey National Conservation Area research and monitoring annual report 1993. Bureau of Land Management, Boise, Idaho.

Knight, R. L. 1988. Relationships of birds of prey and riparian habitat in the Pacific Northwest: an overview. Pages 79–91 *in* K. J. Raedeke, editor. Streamside management: riparian wildlife and forestry interactions. Institute of Forest Resources Contribution 59. University of Washington, Seattle.

Knopf, F. L., R. R. Johnson, T. Rich, F. B. Samson, and R. C. Szaro. 1988. Conservation of riparian ecosystems in the United States. Wilson Bulletin 100:272–284.

Koehler, D. K., and S. H. Anderson. 1991. Habitat use and food selection of small mammals near a sagebrush/crested wheatgrass interface in southeastern Idaho. Great Basin Naturalist 51:249–255.

Koehler, G. M., and K. B. Aubry. 1994. Lynx. Pages 74–98 *in* L. F. Ruggiero, K. B. Aubry, S. W. Buskirk, L. J. Lyon, and W. J. Zielinski, technical editors. The scientific basis for conserving forest carnivores: American marten, fisher, lynx, and wolverine in the western United States. U.S. Forest Service General Technical Report RM-GTR-254. Rocky Mountain Forest and Range Experiment Station, Fort Collins, Colo. 184 pp.

Konkel, G. W., and J. D. McIntyre. 1987. Trends in spawning populations of Pacific anadromous salmonids. U.S. Fish and Wildlife Service Fish and Wildlife Technical Report 9. Washington, D.C. 25 pp.

Kostow, K. 1997. The status of salmon and steelhead in Oregon. Pages 145–178 *in* D. J. Stouder, P. A. Bisson, and R. J. Naiman, editors. Pacific salmon and their ecosystems: status and future options. Chapman and Hall, New York.

Kuda, D. B., and J. S. Griffith. 1993. Establishment of Shoshone sculpin (*Cottus greenei*) in a spring inhabited by mottled sculpin (*C. bairdi*). Great Basin Naturalist 53:190–193.

Kuda, D., J. S. Griffith, and K. Merkley. 1992. Habitat selection by Shoshone sculpin (*Cottus greenei*) in three springs on the Thousand Springs Preserve, Idaho. Final report. The Nature Conservancy, Idaho Field Office, Sun Valley. 16 pp.

Lattin, J. D. 1990. Arthropod diversity in Northwest old-growth forests. Wings 15(2):7–10.

Lattin, J. D. 1993. Arthropod diversity and conservation in old-growth Northwest forests. American Zoologist 33:578–587.

Lattin, J. D. 1994. Non-indigenous arthropods and research natural areas. Natural Areas Report 6(2):2.

Lattin, J. D., A. Christie, and M. D. Schwartz. 1994. The impact of crested wheatgrasses on native black grass bugs in North America: a case for ecosystem management. Natural Areas Journal 14:136–138.

Laufer, J. R., and P. T. Jenkins. 1989. Historical and present status of the grey wolf in the Cascade Mountains of Washington. Northwest Environmental Journal 5:313–327.

Lawson, P. W. 1993. Cycles in ocean productivity, trends in habitat quality, and the restoration of salmon runs in Oregon. Fisheries 18(8):6–10.

Laycock, W. A. 1991. Stable states and thresholds of range condition on North American rangelands: a viewpoint. Journal of Range Management 44:427–433.

Lehman, R. N., K. Steenhof, M. N. Kochert, and L. B. Carpenter. 1994. Raptor abundance and reproductive success in the Snake River Birds of Prey National Conservation Area—1994. Pages 16–40 *in* K. Steenhof, editor. Snake River Birds of Prey National Conservation Area research and monitoring annual report 1994. Bureau of Land Management and National Biological Survey, Boise, Idaho.

Lehmkuhl, J. F., P. F. Hessburg, R. L. Everett, M. H. Huff, and R. D. Ottmar. 1994. Historical and current forest landscapes of eastern Oregon and Washington. Part 1. Vegetation pattern and insect and disease hazard. U.S. Forest Service General Technical Report PNW-GTR-328. Pacific Northwest Research Station, Portland, Oreg. 88 pp.

Lehmkuhl, J. F., and L. F. Ruggiero. 1991. Forest fragmentation in the Pacific Northwest and its potential effects on wildlife. Pages 35–46 *in* L. F. Ruggiero, K. B. Aubry, and M. H. Huff, editors. Wildlife and vegetation in unmanaged Douglas-fir forests. U.S. Forest Service General Technical Report PNW-GTR-285. Pacific Northwest Research Station, Portland, Oreg.

Lehmkuhl, J. F., L. F. Ruggiero, and P. A. Hall. 1991. Landscape-scale patterns of forest fragmentation and wildlife richness and abundance in the southwestern Washington Cascade Range. Pages 425–442 *in* L. F. Ruggiero, K. B. Aubry, and M. H. Huff, editors. Wildlife and vegetation in unmanaged Douglas-fir forests. U.S. Forest Service General Technical Report PNW-GTR-285. Pacific Northwest Research Station, Portland, Oreg.

Leonard, W. P., H. A. Brown, L. L. C. Jones, K. R. McAllister, and R. M. Storm. 1993. Amphibians of Washington and Oregon. Seattle Audubon Society, Seattle, Wash. 168 pp.

Leonard, W. P., and K. R. McAllister. 1996. Past distribution and current status of the northern leopard frog in Washington. Washington Department of Fish and Wildlife, Olympia. 13 pp. + appendixes.

Lesica, P., B. McCune, S. V. Cooper, and W. S. Hong. 1991. Differences in lichen and bryophyte communities between old-growth and managed second-growth forests in the Swan Valley, Montana. Canadian Journal of Botany 69:1745–1755.

Li, H. W., G. A. Lamberti, T. N. Pearsons, C. K. Tait, J. L. Li, and J. C. Buckhouse. 1994. Cumulative effects of riparian disturbances along High Desert trout streams of the John Day Basin, Oregon. Transactions of the American Fisheries Society 123:627–640.

Li, J., K. Wright, and J. Furnish. 1995. A survey of eastside ecosystem benthic invertebrates. Report submitted to the U.S. Forest Service, Interior Columbia Basin Ecosystem Management Project. 63 pp.

Li, J. L. 1991. Review of DEIS for proposed Elwha and Glines Canyon hydroelectric projects. Report to the Lower Elwha Tribal Council, Port Angeles, Wash.

Li, J. L., and S. V. Gregory. 1993. Issues surrounding the biota of the Tualatin River Basin. Oregon State University, Oregon Water Resources Research Institute, Corvallis. 37 pp.

Liss, W. J., G. L. Larson, E. K. Deimling, R. Gresswell, R. Hoffman, M. Kiss, G. Lomnicky, C. D. McIntire, R. Truitt, and T. Tyler. 1995. Ecological effects of stocked fish on naturally fishless high mountain lakes: North Cascades National Park Service Complex, WA, USA. National Park Service, Draft Final Report, Phase 1 (1989–1992). Washington, D.C. 165 pp.

Lonzarich, M. E. R. 1993. Habitat selection and character analysis of *Cottus marginatus*, the margined sculpin. M.S. thesis, University of Washington, Seattle. 88 pp.

Luecke, C. 1990. Changes in abundance and distribution of benthic macroinvertebrates after introduction of cutthroat trout into a previously fishless lake. Transactions of the American Fisheries Society 119:1010–1021.

Lunde, R. E., and A. S. Harestad. 1986. Activity of little brown bats in coastal forests. Northwest Science 60:206–209.

MacCracken, J. G., and R. M. Hansen. 1987. Coyote feeding strategies in southeastern Idaho: optimal foraging by an opportunistic predator? Journal of Wildlife Management 51:278–285.

Mack, R. N. 1986. Alien plant invasion into the Intermountain West: a case history. Pages 191–213 *in* H. A. Mooney, and J. A. Drake, editors. Ecology of biological invasions of North America and Hawaii. Springer-Verlag, New York.

Mangum, F. M. 1992. Aquatic ecosystem inventory: macroinvertebrate analysis (Mt. Hood National Forest). U.S. Forest Service, National Aquatic Ecosystem Monitoring Center, Logan, Utah.

Mannan, R. W., and E. C. Meslow. 1984. Bird populations and vegetation characteristics in managed old-growth forests, northeastern Oregon. Journal of Wildlife Management 48:1219–1238.

Marcot, B. G., R. S. Holthausen, J. Teply, and W. D. Carrier. 1991. Old-growth inventories: status, definitions, and visions for the future. Pages 47–60 *in* L. F. Ruggiero, K. B. Aubry, and M. H. Huff, editors. Wildlife and vegetation in unmanaged Douglas-fir forests. U.S. Forest Service General Technical Report PNW-GTR-285. Pacific Northwest Research Station, Portland, Oreg.

Markle, D. F., T. N. Pearsons, and D. T. Bills. 1989. Taxonomic status and distributional survey of the Oregon chub. Final report to the Oregon Department of Fish and Wildlife, Corvallis. 29 pp.

Marshall, D. B., M. Chilcote, and H. Weeks. 1996. Species at risk: sensitive, threatened, and endangered vertebrates of Oregon. 2nd edition. Oregon Department of Fish and Wildlife, Portland. [unnumbered]

Maser, C., and Z. Maser. 1988. Interactions among squirrels, mycorrhizal fungi, and coniferous forests in Oregon. Great Basin Naturalist 48:358–369.

Maser, C., B. R. Mate, J. F. Franklin, and C. T. Dyrness. 1981. Natural history of Oregon Coast mammals. U.S. Forest Service General Technical Report PNW-GTR-133. Pacific Northwest Research Station, Portland, Oreg. 496 pp.

Maser, C., and J. M. Trappe, editors. 1984. The seen and unseen world of the fallen tree. U.S. Forest Service General Technical Report PNW-GTR-164. Pacific Northwest Research Station, Portland, Oreg. 56 pp.

Maser, C., J. M. Trappe, and R. A. Nussbaum. 1978. Fungal–small mammal interrelationships with emphasis on Oregon coniferous forests. Ecology 59:799–809.

Maser, Z., C. Maser, and J. M. Trappe. 1985. Food habits of the northern flying squirrel (*Glaucomys sabrinus*) in Oregon. Canadian Journal of Zoology 63:1084–1088.

Mason, R. R., and B. E. Wickman. 1988. The Douglas-fir tussock moth in the interior Pacific Northwest. Chapter 10 *in* A. A. Berryman, editor. Dynamics of forest insect populations. Plenum Press, New York.

Mason, R. R., and B. E. Wickman. 1994. Procedures to reduce landscape hazard from insect outbreaks. Pages 20–21 *in* R. L. Everett, compiler. Restoration of stressed sites and processes. U.S. Forest Service General Technical Report PNW-GTR-330. Pacific Northwest Research Station, Portland, Oreg. 123 pp.

Mattson, W. J., editor. 1977. The role of arthropods in forest ecosystems. Springer-Verlag, New York. 104 pp.

McAllister, K. R., W. P. Leonard, and R. M. Storm. 1993. Spotted frog (*Rana pretiosa*) surveys in the Puget Trough of Washington, 1989–1991. Northwestern Naturalist 74:10–15.

McCorkle, D. V., P. C. Hammond, and G. Pennington. 1980. Ecological investigation report: Oregon silverspot butterfly (*Speyeria zerene hippolyta*). U.S. Forest Service, Pacific Northwest Region, Siuslaw National Forest, Corvallis, Oreg. 117 pp.

McCorquodale, S. M., M. Scott, L. L. Eberhardt, and L. E. Eberhardt. 1988. Dynamics of a colonizing elk population. Journal of Wildlife Management 52:309–313.

McCune, B. 1993. Gradients in epiphyte biomass in three *Pseudotsuga–Tsuga* forests of different ages in western Oregon and Washington. Bryologist 96:405–411.

McCune, B., and R. Rosentreter. 1992. *Texosporium sancti-jacobi*, a rare western North American lichen. Bryologist 95:329–333.

McGarigal, K., and W. C. McComb. 1993. Research problem analysis on biodiversity conservation in western Oregon forests. Report to the National Biological Service, Forest and Rangeland Ecosystem Science Center, Corvallis, Oreg. 174 pp.

McGinnis, S. M. 1984. Freshwater fishes of California. University of California Press, Berkeley. 316 pp.

McHenry, M. 1991. The effects of debris torrents on macroinvertebrate populations in tributaries and side-channels of the Hoh River, Washington. Northwest Indian Fisheries Commission Technical Report. Forks, Wash. 27 pp.

McIntosh, B. A. 1992. Historical changes in anadromous fish habitat in the Upper Grand Ronde, Oregon, 1941–1990. M.S. thesis, Oregon State University, Corvallis. 88 pp.

McIntosh, B. A., J. R. Sedell, J. E. Smith, R. C. Wissmar, S. E. Clarke, G. H. Reeves, and L. A. Brown. 1994. Historical changes in fish habitat for select river basins of eastern Oregon and Washington. Northwest Science 68(Special Issue):268–285.

McPhail, J. D. 1994. Status of the Nooksack dace, *Rhinichthys cataractae* ssp., in Canada. Biodiversity Center and Department of Zoology, University of British Columbia, Vancouver. Unpublished report. [unnumbered]

Medin, D. E., and W. P. Clary. 1990. Bird and small mammal populations in a grazed and ungrazed riparian habitat in Idaho. U.S. Forest Service Research Paper INT-RP-425. Intermountain Research Station, Ogden, Utah. 8 pp.

Meehan, W. R., editor. 1991. Influences of forests and rangeland management on salmonid fishes and their habitats. American Fisheries Society Special Publication 19. 751 pp.

Metter, D. E. 1960. The distribution of amphibians in eastern Washington. M.S. thesis, Idaho State University, Pocatello. 89 pp.

Miller, J. C. 1993. Insect natural history, multi-species interactions and biodiversity in ecosystems. Biodiversity and Conservation 2:233–241.

Miller, R. F., and J. A. Rose. 1995. Western juniper expansion in eastern Oregon. Great Basin Naturalist 55:37–45.

Miller, R. F., and P. E. Wigand. 1994. Holocene changes in semiarid pinyon–juniper woodlands. Bioscience 44:465–474.

Miller, R. R., J. D. Williams, and J. E. Williams. 1989. Extinctions of North American fishes during the past century. Fisheries 14(6):22–38.

Mills, T. J., D. R. McEwan, and M. R. Jennings. 1997. California salmon and steelhead: beyond the crossroads. Pages 91–111 *in* D. J. Stouder, P. A. Bisson, and R. J. Naiman, editors. Pacific salmon and their ecosystems: status and future options. Chapman and Hall, New York.

Minshall, G. W. 1984. Aquatic insect-substratum relationships. Pages 358–400 *in* V. H. Resh and D. M. Rosenburg, editors. The ecology of aquatic insects. Praeger Publishers, New York.

Moldenke, A. R. 1990. One hundred twenty thousand little bugs. Wings 15(2):11–14.

Moldenke, A. R., and J. D. Lattin. 1990. Dispersal characteristics of old-growth soil arthropods: the potential for loss of diversity and biological function. Northwest Environmental Journal 6:408–409.

Mongillo, P. E. 1993. The distribution and status of bull trout/dolly varden in Washington State. Washington Department of Wildlife, Fisheries Management Division, Olympia. 45 pp.

Morrison, P. H. 1988. Old growth in the Pacific Northwest: a status report. The Wilderness Society, Washington, D.C. 46 pp.

Morrison, P. H., and F. Swanson. 1990. Fire history and pattern in a Cascade Range landscape. U.S. Forest Service General Technical Report PNW-GTR-254. Pacific Northwest Research Station, Portland, Oreg. 77 pp.

Moyle, P. B. 1994. The decline of anadromous fishes in California. Conservation Biology 8:869–870.

Moyle, P. B., and J. E. Williams. 1990. Biodiversity loss in the temperate zone: decline of the native fish fauna of California. Conservation Biology 4:275–284.

Moyle, P. B., J. E. Williams, and E. D. Wikramanayake. 1989. Fish species of special concern in California. Final Report to the California Department of Fish and Game, Inland Fisheries Division, Rancho Cordova. 222 pp.

Mullan, J. W., K. R. Williams, G. Rhodus, T. W. Hillman, and J. D. McIntyre. 1992. Production and habitat of salmonids in mid-Columbia River tributary streams. U.S. Fish and Wildlife Service Monograph 1. Washington, D.C. 489 pp.

Murphy, M. L., and J. D. Hall. 1981. Varied effects of clear-cut logging on predators and their habitat in small streams of the Cascade Mountains, Oregon. Canadian Journal of Fisheries and Aquatic Sciences 38:137–145.

Murphy, M. L., C. P. Hawkins, and N. H. Anderson. 1981. Effects of canopy modification and accumulated sediment on stream communities. Transactions of the American Fisheries Society 110:469–478.

Mutch, R. W., S. F. Arno, J. K. Brown, C. E. Carlson, R. D. Ottmar, and J. L. Peterson. 1993. Forest health in the Blue Mountains: a management strategy for fire-adapted ecosystems. U.S. Forest Service General Technical Report PNW-GTR-310. Pacific Northwest Research Station, Portland, Oreg. 14 pp.

Mysak, L. A. 1986. El Niño, interannual variability and fisheries in the Pacific Ocean. Canadian Journal of Fisheries and Aquatic Sciences 43:464–497.

Naiman, R. J., T. J. Beechie, L. E. Benda, D. R. Berg, P. A. Bisson, L. H. Macdonald, M. D. O'Connor, P. L. Olson, and E. A. Steel. 1992. Fundamental elements of ecologically healthy watersheds in the Pacific Northwest coastal ecoregion. Pages 127–188 *in* R. J. Naiman, editor. Watershed management: balancing sustainability and environmental change. Springer-Verlag, New York.

National Marine Fisheries Service 1996. Endangered and threatened species; proposed endangered status for five ESUs of the steelhead and proposed threatened status for five ESUs of steelhead in Washington, Oregon, Idaho, and California. Federal Register 61 (155):41541–41561.

National Marine Fisheries Service. 1997. Endangered and threatened species; revision of candidate species list under the Endangered Species Act. Federal Register 62(134):37560–37563.

National Research Council. 1994. Rangeland health: new methods to classify, inventory, and monitor rangelands. National Academy Press, Washington, D.C. 180 pp.

Nehlsen, W., J. E. Williams, and J. A. Lichatowich. 1991. Pacific salmon at the crossroads: stocks at risk from California, Oregon, Idaho, and Washington. Fisheries 16(2):4–21.

Nickelson, T. E. 1986. Influences of upwelling, ocean temperature, and smolt abundance on marine survival of coho salmon (*Oncorhynchus kisutch*) in Oregon. Canadian Journal of Fisheries and Aquatic Sciences 43:527–535.

Nickelson, T. E., and J. A. Lichatowich. 1983. The influence of the marine environment on the interannual variation in coho salmon abundance: an overview. Pages 24–36 *in* W. G. Pearcy, editor. The influence of ocean conditions on the production of salmonids in the North Pacific: a workshop. Oregon State University Report ORESU-W-83-001. Sea Grant College Program, Corvallis.

Nickelson, T. E., J. W. Nicholas, A. M. McGie, R. B. Lindsay, D. L. Bottom, R. J. Kaiser, and S. E. Jacobs. 1992. Status of anadromous salmonids in Oregon coastal basins. Oregon Department of Fish and Wildlife, Corvallis. 83 pp.

North Cascades Grizzly Bear Steering Committee. 1994. North Cascades grizzly bear recovery chapter—final draft. Washington Department of Wildlife, Olympia. 50 pp.

Northwest Power Planning Council. 1986. Compilation of information on salmon and steelhead losses in the Columbia River basin. Northwest Power Planning Council, Portland, Oreg. 252 pp.

Northwest Power Planning Council. 1994. 1994 Columbia River basin fish and wildlife program. Northwest Power Planning Council, Portland, Oreg. 248 pp.

Noss, R. F., E. T. LaRoe III, and J. M. Scott. 1995. Endangered ecosystems of the United States: a preliminary assessment of loss and degradation. National Biological Service Biological Report 28. Washington, D.C. 58 pp.

Nussbaum, R. A., E. D. Brodie, Jr., and R. M. Storm. 1983. Amphibians and reptiles of the Pacific Northwest. University Press of Idaho, Moscow. 332 pp.

Oester, P. T., S. A. Fitzgerald, W. H. Emmingham, A. Campbell III, and G. M. Filip. 1992. Forest health in eastern Oregon. Oregon State University Extension Service Publication EC-1413. Corvallis. 6 pp.

Ohmart, R. D. 1994. The effects of human-induced changes on the avifauna of western riparian habitats. Studies in Avian Biology 15:273–285.

Old-Growth Definition Task Force. 1986. Interim definitions for old-growth Douglas-fir and mixed-conifer forests in the Pacific Northwest and California. U.S. Forest Service Research Note PNW-RN-447. Pacific Northwest Research Station, Portland, Oreg. 7 pp.

Omernik, J. M. 1987. Ecoregions of the conterminous United States. Annals of the Association of American Geographers 77:118–125.

Omernik, J. M. 1995. Ecoregions: a framework for environmental management. Pages 49–62 *in* W. S. Davis and T. P. Simon, editors. Biological assessment and criteria: tools for water resource planning and decision making. Lewis Publishers, Boca Raton, Fla.

Oregon Natural Heritage Program. 1993. Rare, threatened and endangered plants and animals of Oregon. Oregon Natural Heritage Program, Portland. 79 pp.

Oregon Natural Heritage Program. 1995. Database. Oregon Natural Heritage Program, Portland.

Otte, D. 1981. The North American grasshoppers. Volume 1. Acrididae: Gomphocerinae and Acridinae. Harvard University Press, Cambridge, Mass. 304 pp.

Pacific Fishery Management Council. 1992. Assessment of the status of five stocks of Puget Sound chinook and coho as required under the PFMC definition of overfishing. Summary report prepared by Puget Sound Salmon Stock Review Group. Pacific Fishery Management Council, Portland, Oreg. 113 pp.

Pacific Rivers Council. 1994. Restoration: a blueprint for saving wild fish and watersheds in the Northwest. Pacific Rivers Council, Eugene, Oreg. 30 pp.

Palmisano, J. F., R. N. Ellis, and V. W. Kaczynski. 1993. The impact of environmental and management factors on Washington's wild anadromous salmon and trout. Washington Forest Protection Association and Washington Department of Natural Resources, Olympia. 371 pp.

Parsons, G. L., G. Cassis, A. R. Moldenke, J. D. Lattin, N. H. Anderson, J. C. Miller, P. Hammond, and T. D. Schowalter. 1991. Invertebrates of the H. J. Andrews experimental forest, western Cascade Range, Oregon. Part V: an annotated list of insects and other arthropods. U.S. Forest Service General Technical Report PNW-GTR-290. Pacific Northwest Research Station, Portland, Oreg. 168 pp.

Pellant, M. 1990. The cheatgrass–wildfire cycle—are there any solutions? Pages 11–18 *in* E. D. McArthur, E. M. Romney, S. D. Smith, and P. T. Tueller, compilers. Proceedings—symposium on cheatgrass invasion, shrub die-off, and other aspects of shrub biology and management. U.S. Forest Service General Technical Report INT-GTR-276. Intermountain Forest and Range Experiment Station, Ogden, Utah.

Pellant, M., and C. Hall. 1994. Distribution of two exotic grasses in intermountain rangelands: status in 1992. Pages 109–112 *in* S. B. Monsen and S. G. Kitchen, compilers. Proceedings—ecology and management of annual rangelands. U.S. Forest Service General Technical Report INT-GTR-313. Intermountain Forest and Range Experiment Station, Ogden, Utah.

Pellant, M., and S. B. Monsen. 1993. Rehabilitation on public rangelands in Idaho, USA: a change in emphasis from grass monocultures. Proceedings of the International Grassland Congress 17:778–779.

Perkins, J. M. 1990. Results of population monitoring for the category 2 species *Plecotus townsendii* in Oregon and Washington: 1989–90. Oregon Department of Fish and Wildlife Contract Report 90-9-03. Portland, Oreg. 25 pp.

Perkins, J. M. 1994. Results of summer bat surveys Wallowa Valley Ranger District, Eagle Cap Ranger District, and the HCNRA of the Wallowa–Whitman National Forest, Wallowa County, Oregon. Report to the U.S. Forest Service, Enterprise, Oreg. 75 pp.

Perkins, J. M., and S. P. Cross. 1988. Differential use of some coniferous forest habitats by hoary and silver-haired bats in Oregon. Murrelet 69:21–24.

Perkins, J. M., and C. Levesque. 1987. Distribution, status, and habitat affinities of Townsend's big-eared bat (*Plecotus townsendii*) in Oregon. Oregon Department of Fish and Wildlife Technical Report 86-5-01. Nongame Wildlife Program, Portland. 50 pp.

Perry, D. A. 1988. Landscape pattern and forest pests. Northwest Environmental Journal 4:213–228.

Perry, D. A., R. Molina, and M. P. Amaranthus. 1987. Mycorrhizae, mycorrhizospheres, and reforestation: current knowledge and research needs. Canadian Journal of Forest Research 17:929–940.

Perry, D. A., and G. B. Pitman. 1983. Genetic and environmental influences in host resistance to herbivory: Douglas-fir and the western spruce budworm. Zeitschrift für Angewandte Entomologie 96:217–228.

Peterjohn, B. G. 1994. The North American Breeding Bird Survey. Birding 26:386–398.

Peterjohn, B. G., and J. R. Sauer. 1993. North American Breeding Bird Survey annual summary 1990–1991. Bird Populations 1:1–15.

Peterjohn, B. G., J. R. Sauer, and W. A. Link. 1995. The 1992–1993 summary of the North American Breeding Bird Survey. Bird Populations 2:46–61.

Pickering, D., D. Salzer, and C. Macdonald. 1992. Population dynamics and habitat characteristics of the Oregon silverspot butterfly *Speyeria zerene hippolyta* (Lepidoptera, Nymphalidae). The Nature Conservancy, Portland, Oreg. 100 pp.

Platts, W. S. 1988. Density and biomass of trout and char in western streams. U.S. Forest Service General Technical Report INT-GTR-241. Intermountain Research Station, Ogden, Utah. 17 pp.

Platts, W. S. 1991. Livestock grazing. American Fisheries Society Special Publication 19:103–110.

Platts, W. S., and R. L. Nelson. 1989. Stream canopy and its relationship to salmonid biomass in the Intermountain West. North American Journal of Fisheries Management 9:446–457.

Plotnikoff, R. W. 1992. Timber, fish, and wildlife ecoregion bioassessment pilot project. Washington State Department of Ecology Publication 92-63, TFWWQ11-92-001. Olympia. 57 pp.

Plotnikoff, R. W. 1994. Ambient monitoring instream biological assessment: progress report of 1993 pilot survey—draft. Washington State Department of Ecology, Environmental Investigations and Laboratory Services, Ambient Monitoring Section, Olympia, Wash. 13 pp.

Powell, R. A., and W. J. Zielinski. 1994. Fisher. Pages 38–73 *in* L. F. Ruggiero, K. B. Aubry, S. W. Buskirk, L. J. Lyon, and W. J. Zielinski, technical editors. The scientific basis for conserving forest carnivores: American marten, fisher, lynx, and wolverine in the western United States. U.S. Forest Service General Technical Report RM-GTR-254. Rocky Mountain Forest and Range Experiment Station, Fort Collins, Colo. 184 pp.

Puchy, C. A., and D. B. Marshall. 1993. Oregon wildlife diversity plan. 2nd edition. Oregon Department of Fish and Wildlife, Portland. 511 pp.

Pyke, D. A., and M. M. Borman. 1993. Problem analysis for the vegetation diversity project. Bureau of Land Management Technical Note OR-936-01. Pacific Forest and Basin Rangeland Systems Cooperative Research and Technology Unit, Corvallis. 100 pp.

Ralph, C. J. 1994. Evidence of changes in populations of the marbled murrelet in the Pacific Northwest. Studies in Avian Biology 15:286–292.

Ralph, C. J., G. L. Hunt, Jr., M. G. Raphael, and J. F. Piatt, technical editors. 1995. Ecology and conservation of the marbled murrelet. U.S. Forest Service General Technical Report PSW-GTR-152. Pacific Southwest Research Station, Portland, Oreg.

Ralph, C. J., P. W. C. Paton, and C A. Taylor. 1991. Habitat patterns of breeding birds and small mammals in Douglas-fir/hardwood stands in northwestern California and southwestern Oregon. Pages 379–393 *in* L. F. Ruggiero, K. B. Aubry, A. B. Carey, and M. H. Huff, technical coordinators. Wildlife and vegetation of unmanaged Douglas-fir forests. U.S. Forest Service General Technical Report PNW-GTR-285. Pacific Northwest Research Station, Portland, Oreg.

Raphael, M. G. 1988. Long-term trends in abundance of amphibians, reptiles, and mammals in Douglas-fir forests of northwestern California. Pages 23–31 *in* R. C. Szaro, K. E. Severson, and D. R. Patton, technical coordinators. Management of amphibians, reptiles, and small mammals in North America. U.S. Forest Service General Technical Report RM-GTR-166. Rocky Mountain Forest and Range Experiment Station, Fort Collins, Colo.

Raphael, M. G., K. V. Rosenberg, and B. C. Marcot. 1988. Large-scale changes in bird populations of Douglas-fir forests, northwestern California. Pages 63–83 *in* J. A. Jackson, editor. Bird conservation. University of Wisconsin Press, Madison.

Redente, E. F., T. B. Doerr, C. E. Grygiel, and M. E. Biondini. 1984. Vegetation establishment and succession on disturbed soils in northwest Colorado. Reclamation and Revegetation Research 3:153–165.

Reeves, G. H., F. H. Everest, and J. R. Sedell. 1993. Diversity of juvenile anadromous salmonid assemblages in coastal Oregon basins with different levels of timber harvest. Transactions of the American Fisheries Society 122:309–317.

Reeves, G. H., and J. R. Sedell. 1992. An ecosystem approach to the conservation and management of freshwater habitat for anadromous salmonids in the Pacific Northwest. Transactions of the North American Wildlife and Natural Resources Conference 57:408–415.

Reisenbichler, R. R. 1997. Genetic factors contributing to declines of anadromous salmonids in the Pacific Northwest. Pages 223–244 *in* D. J. Stouder, P. A. Bisson, and R. J. Naiman, editors. Pacific salmon and

their ecosystems: status and future options. Chapman and Hall, New York.

Reynolds, R. T. 1989. Accipiters. Pages 92–110 *in* Proceedings of the western raptor management symposium and workshop, 26–28 October 1987, Boise, Idaho. National Wildlife Federation, Institute for Wildlife Research Scientific and Technical Series 12. Washington, D.C.

Richards, C., and K. L. Bacon. 1994. Influence of fine sedimentation on macroinvertebrate colonization of surface and hyporheic stream substrates. Great Basin Naturalist 54:106–113.

Rieman, B. E., and R. C. Beamesderfer. 1990. White sturgeon in the lower Columbia River: is the stock overexploited? North American Journal of Fisheries Management 10:388–396.

Rieman, B. E., and J. D. McIntyre. 1993. Demographic and habitat requirements for conservation of bull trout. U.S. Forest Service General Technical Report INT-GTR-302. Intermountain Research Station, Ogden, Utah. 38 pp.

Ripple, W. 1994. Historic spatial patterns of old forests in western Oregon. Journal of Forestry 92(11):45–49.

Ripple, W. J., G. A. Bradshaw, and T. A. Spies. 1991. Measuring forest landscape patterns in the Cascade Range of Oregon, USA. Biological Conservation 57:73–88.

Robbins, C. S., D. Bystrak, and P. H. Geissler. 1986. The Breeding Bird Survey: its first fifteen years, 1965–1979. U.S. Fish and Wildlife Service Research Publication 157. Washington, D.C. 196 pp.

Robinson, C. T., and G. W. Minshall. 1991. Biological metric development for the assessment of nonpoint pollution in the Snake River ecoregion of southern Idaho. Idaho Snake River ecoregion final report. Idaho State University, Department of Biological Sciences, Pocatello. 70 pp.

Robinson, S. K., J. A. Gryzbowski, S. I. Rothstein, M. C. Brittingham, L. J. Petit, and F. R. Thompson. 1992. Management implications of cowbird parasitism on Neotropical migrant songbirds. Pages 93–102 *in* D. M. Finch and P. W. Stangel, editors. Status and management of Neotropical migratory birds. U.S. Forest Service General Technical Report RM-GTR-229. Rocky Mountain Forest and Range Experiment Station, Fort Collins, Colo.

Rodrick, E., and R. Milner, technical editors. 1991. Management recommendations for Washington's priority species. Washington Department of Wildlife, Wildlife Management, Fish Management, and Habitat Management Divisions, Olympia. [unnumbered]

Rosenberg, K. V., and M. G. Raphael. 1986. Effects of forest fragmentation on vertebrates in Douglas-fir forests. Pages 263–272 *in* J. Verner, M. L. Morrison, and C. J. Ralph, editors. Wildlife 2000: modeling habitat relationships of terrestrial vertebrates. Proceedings of an international symposium, Fallen Leaf Lake, Calif., 7–11 October, 1984. University of Wisconsin Press, Madison.

Roth, B. 1993. Critical review of terrestrial mollusks associated with late-successional and old-growth forests in the range of the northern spotted owl. Final Report prepared for the Forest Ecosystem Management Assessment Team. U.S. Forest Service, Portland, Oreg. 82 pp.

Ruggiero, L. F., K. B. Aubry, A. B. Carey, and M. H. Huff, technical coordinators. 1991a. Wildlife and vegetation of unmanaged Douglas-fir forests. U.S. Forest Service General Technical Report PNW-GTR-285. Pacific Northwest Research Station, Portland, Oreg. 533 pp.

Ruggiero, L. F., L. L. C. Jones, and K. B. Aubry. 1991b. Plant and animal habitat associations in Douglas-fir forests of the Pacific Northwest: an overview. U.S. Forest Service General Technical Report PNW-GTR-285. Pacific Northwest Research Station, Portland, Oreg. 184 pp.

Salo, E. O., and T. W. Cundy, editors. 1987. Streamside management: forestry and fishery interactions. Institute of Forest Resources Contribution 57. University of Washington, Seattle. 471 pp.

Scheerer, P. D., C. H. Stein, and K. K. Jones. 1994. Oregon chub investigations. Oregon Department of Fish and Wildlife, Fish Division Annual Progress Report. Portland. 37 pp.

Schmitt, C. L., D. G. Goheen, T. F. Gregg, and P. F. Hessburg. 1991. Effects of management activities and stand type on pest-caused losses in true fir and associated species on the Wallowa–Whitman National Forest. U.S. Forest Service Report BMPMZ-01-91. Wallowa–Whitman National Forest, Baker City, Oreg. 78 pp.

Schuman, G. E., F. Rauzi, and D. T. Booth. 1982. Production and competition of crested wheatgrass–native grass mixtures. Agronomy Journal 74:23–26.

Sedell, J. R., P. A. Bisson, F. J. Swanson, and S. V. Gregory. 1988. What we know about large trees that fall into streams and rivers. Pages 47–82 *in* C. Maser, R. F. Tarrant, J. M. Trappe, and J. F. Franklin, technical editors. From the forest to the sea: a story of fallen trees. U.S. Forest Service General Technical Report PNW-GTR-229. Pacific Northwest Research Station, Portland, Oreg.153 pp.

Sedell, J. R., and J. L. Froggatt. 1984. Importance of streamside forests to large rivers: the isolation of the Willamette River, Oregon, USA, from its floodplain by snagging and streamside forest removal. International Vereinigung für Theoretische und Angewandte Limnologie Verhandlungen 22:1828–1834.

Sharpe, P. B., R. L. Schooley, and B. Van Horne. 1994. Behavior of Townsend's ground squirrels in different habitat types. Pages 188–199 *in* K. Steenhof, editor. Snake River Birds of Prey National Conservation Area research and monitoring annual report 1994. Bureau of Land Management and National Biological Survey, Boise, Idaho.

Simpson, J. C., and R. L. Wallace. 1982. Fishes of Idaho. University of Idaho Press, Moscow. 238 pp.

Skinner, M. W., and B. M. Pavlik, editors. 1994. Inventory of rare and endangered vascular plants of California. 5th edition. California Native Plant Society Special Publication 1. Sacramento. 338 pp.

Skovlin, J. M. 1991. Fifty years of research progress: a historical document on the Starkey Experimental Forest and Range. U.S. Forest Service General Technical Report PNW-GTR-266. Pacific Northwest Research Station, Portland, Oreg. 58 pp.

Smith, G. W., and D. R. Johnson. 1985. Demography of a Townsend's ground squirrel population in southwestern Idaho. Ecology 66:171–178.

Society for Range Management. 1989. Assessment of rangeland condition and trend of the United States. Society for Range Management, Public Affairs Committee, Denver, Colo. 12 pp.

Spies, T. A. 1991. Plant species diversity and occurrence in young, mature, and old-growth Douglas-fir stands in western Oregon and Washington. Pages 111–121 *in* L. F. Ruggiero, K. B. Aubry, A. B. Carey, and M. H. Huff, technical coordinators. Wildlife and vegetation of unmanaged Douglas-fir forests. U.S. Forest Service General Technical Report PNW-GTR-285. Pacific Northwest Research Station, Portland, Oreg.

Spies, T. A., and J. F. Franklin. 1988. Old-growth and forest dynamics in the Douglas-fir region of western Oregon and Washington. Natural Areas Journal 8:190–201.

Spies, T. A., and J. F. Franklin. 1991. The structure of natural young, mature, and old-growth Douglas-fir forests in Oregon and Washington. Pages 91–109 *in* L. F. Ruggiero, K. B. Aubry, A. B. Carey, and M. H. Huff, technical coordinators. Wildlife and vegetation of unmanaged Douglas-fir forests. U.S. Forest Service General Technical Report PNW-GTR-285. Pacific Northwest Research Station, Portland, Oreg.

Spies, T. A., W. J. Ripple, and G. A. Bradshaw. 1994. Dynamics and pattern of a managed coniferous forest landscape in Oregon. Ecological Applications 4:555–568.

St. Clair, L. L., and J. R. Johansen. 1993. Introduction to the symposium on soil communities. Great Basin Naturalist 53:1–4.

St. Clair, L. L., J. R. Johansen, and S. R. Rushforth. 1993. Lichens of soil crust communities in the Intermountain area of the western United States. Great Basin Naturalist 53:5–12.

Stebbins, R. C. 1985. Western reptiles and amphibians. 2nd edition. The Eaton Press, Norwalk, Conn. 336 pp.

Stubbs, K., and R. White. 1993. Lost River (*Deltistes luxatus*) and shortnose (*Chasmistes brevirostris*) sucker recovery plan. U.S. Fish and Wildlife Service, Portland, Oreg. 107 pp.

Suring, L. H. 1975. Habitat use and activity patterns of the Columbia white-tailed deer along the lower Columbia River. M.S. thesis, Oregon State University, Corvallis. 59 pp.

Swanson, F. J., S. V. Gregory, J. R. Sedell, and A. G. Campbell. 1982. Land–water interactions: the riparian zone. Pages 267–291 *in* R. L. Edmonds, editor. Analysis of coniferous forest ecosystems in the western United States. Hutchinson Ross, Stroudsburg, Penn.

Swanson, F. J., J. A. Jones, D. O. Wallin, and J. H. Cissel. 1994. Natural variability—implications for ecosystem management. Pages 80–94 *in* M. E. Jensen and P. S. Bourgeron, technical editors. Eastside forest ecosystem health assessment. Volume 2. Ecosystem management: principles and applications. U.S. Forest Service General Technical Report PNW-GTR-318. Pacific Northwest Research Station, Portland, Oreg.

Swengel, A. B. 1995. Fourth of July butterfly count. Pages 171–172 *in* E. T. LaRoe, G. S. Farris, C. E. Puckett, P. D. Doran, and M. J. Mac, editors. Our living resources: a report to the nation on the distribution, abundance, and health of U.S. plants, animals, and ecosystems. U.S. Department of the Interior, National Biological Service, Washington, D.C.

Tait, C. K., J. L. Li, G. A. Lamberti, T. N. Pearsons, and H. W. Li. 1994. Relationships between riparian cover and the community structure of high desert streams. Journal of the North American Benthological Society 13:45–56.

Tanaka, J. A., G. L. Starr, and T. M. Quigley. 1995. Strategies and recommendations for addressing forest health issues in the Blue Mountains of Oregon and Washington. U.S. Forest Service General Technical Report PNW-GTR-350. Pacific Northwest Research Station, Portland, Oreg. 18 pp.

Tesch, S. D. 1995. The Pacific Northwest Region. Pages 499–558 *in* J. W. Barrett, editor. Regional silviculture of the United States. 3rd edition. John Wiley & Sons, New York.

Thies, W. G., and R. N. Sturrock. 1995. Laminated root rot in western North America. U.S. Forest Service General Technical Report PNW-GTR-349. Pacific Northwest Research Station, Portland, Oreg.

Thomas, D. W. 1988. The distribution of bats in different ages of Douglas-fir forests. Journal of Wildlife Management 52:619–626.

Thomas, D. W., and S. D. West. 1991. Forest age associations of bats in the Washington Cascade and Oregon Coast ranges. Pages 295–303 *in* L. F. Ruggiero, K. B. Aubry, A. B. Carey, and M. H. Huff, technical coordinators. Wildlife and vegetation of unmanaged Douglas-fir forests. U.S. Forest Service General Technical Report PNW-GTR-285. Pacific Northwest Research Station, Portland, Oreg.

Thomas, J. W., technical editor. 1979. Wildlife habitats in managed forests of the Blue Mountains of Oregon and Washington. U.S. Forest Service Agriculture Handbook 553. 512 pp.

Thomas, J. W., M. G. Raphael, R. G. Anthony, E. D. Forsman, A. G. Gunderson, R. S. Holthausen, B. G. Marcot, G. H. Reeves, J. R. Sedell, and D. M. Solis. 1993.

Viability assessments and management considerations for species associated with late-successional and old-growth forest of the Pacific Northwest. U.S. Forest Service Research Report of the Scientific Analysis Team. Washington, D.C. 530 pp.

Tirhi, M. J. 1995. Draft Washington state management plan for Columbian sharp-tail grouse (*Tympanuchus phasianellus columbianus*). Washington Department of Fish and Wildlife, Olympia. 94 pp.

Trappe, J. M., and D. L. Luoma. 1992. The ties that bind: fungi in ecosystems. Pages 17–27 *in* G. C. Carroll and D. T. Wicklow, editors. The fungal community: its organization and role in the ecosystem. Marcel Dekker, New York.

U.S. Environmental Protection Agency. 1994. Macroinvertebrate communities—Oregon ecoregion data: Coast Range data set. U.S. Environmental Protection Agency, Corvallis, Oreg.

U.S. Fish and Wildlife Service. 1983. Columbian white-tailed deer recovery plan. U.S. Fish and Wildlife Service, Portland, Oreg. 75 pp.

U.S. Fish and Wildlife Service. 1992. Endangered and threatened species recovery program. U.S. Fish and Wildlife Service Report to Congress. Washington, D.C. 279 pp.

U.S. Fish and Wildlife Service. 1993a. Plant taxa for listing as endangered or threatened species: notice of review. Federal Register 58(188):51143–51190.

U.S. Fish and Wildlife Service. 1993b. Proposed reclassifications: MacFarlane's four-o-clock. Endangered Species Technical Bulletin 18(4):21.

U.S. Fish and Wildlife Service. 1994. Endangered and threatened wildlife and plants; animal candidate review for listing as endangered or threatened; proposed rule. Federal Register 59(219):58981–59028.

U.S. Fish and Wildlife Service. 1995. Endangered and threatened wildlife and plants. 50 CFR 17.11 & 17.12, March 31, 1995. 43 pp.

U.S. Fish and Wildlife Service. 1996a. Endangered and threatened wildlife and plants; listing of the Umpqua River cutthroat trout in Oregon. Federal Register 61(179):48412–48413.

U.S. Fish and Wildlife Service. 1996b. Endangered and threatened wildlife and plants; listing of the central California Coast coho salmon as threatened in California. Federal Register 61(225):59028–59029.

U.S. Fish and Wildlife Service. 1997. Endangered and threatened wildlife and plants; threatened status for the Southern Oregon/Northern California Coast evolutionary significant unit of coho salmon. Federal Register 62(117):33038–33039.

U.S. Forest Service. 1993. Blue Mountains: ecosystem restoration strategy. U.S. Forest Service, Pacific Northwest Region, Portland, Oreg. 12 pp.

U.S. Forest Service and Bureau of Land Management. 1994a. Record of decision for the President's forest plan. U.S. Forest Service and Bureau of Land Management, Washington, D.C. 74 pp.

U.S. Forest Service and Bureau of Land Management. 1994b. Environmental assessment for the implementation of interim strategies for managing anadromous fish-producing watersheds in eastern Oregon and Washington, Idaho, and portions of California. U.S. Forest Service and Bureau of Land Management, Washington, D.C. 68 pp.

U.S. General Accounting Office. 1988. Public rangelands: some riparian areas restored but widespread improvement will be slow: a report to congressional requesters. U.S. General Accounting Office Report GAO/RCED-88-105. Washington, D.C. 85 pp.

Van Horne, B., J. G. Corn, R. L. Schooley, and P. B. Sharpe. 1993a. Some processes influencing reproduction and survival of Townsend's ground squirrels. Pages 184–208 *in* K. Steenhof, editor. Snake River Birds of Prey National Conservation Area research and monitoring annual report 1993. Bureau of Land Management, Boise, Idaho.

Van Horne, B., R. L. Schooley, G. S. Olson, and K. P. Burnham. 1993b. Patterns of density, reproduction, and survival in Townsend's ground squirrels. Pages 158–183 *in* K. Steenhof, editor. Snake River Birds of Prey National Conservation Area research and monitoring annual report 1993. Bureau of Land Management, Boise, Idaho.

Vannote, R. L., G. W. Minshall, K. W. Cummins, J. R. Sedell, and C. E. Kushing. 1980. The river continuum concept. Canadian Journal of Fisheries and Aquatic Sciences 37:370–377.

Vaughan, M. R., and M. R. Pelton. 1995. Black bears in North America. Pages 100–103 *in* E. T. LaRoe, G. S. Farris, C. E. Puckett, P. D. Doran, and M. J. Mac, editors. Our living resources: a report to the nation on the distribution, abundance, and health of U.S. plants, animals, and ecosystems. U.S. Department of the Interior, National Biological Service, Washington, D.C.

Vial, J. L., and L. Saylor. 1993. The status of amphibian populations: a compilation and analysis. Working document 1. Declining Amphibians Task Force, IUCN World Conservation Union and Species Survival Commission. 98 pp.

Wallace, R. L., J. S. Griffith, P. J. Connolly, D. M. Daley, G. B. Beckham. 1982. Distribution, relative abundance, life history and habitat preferences of Shoshone sculpin. Final report to the U.S. Fish and Wildlife Service, Boise, Idaho. 23 pp.

Wallace, R. L., J. S. Griffith, P. J. Connolly, D. M. Daley, G. B. Beckham. 1984. Distribution of the Shoshone sculpin (*Cottus greenei*: Cottiae) in the Hagerman Valley of south central Idaho. Great Basin Naturalist 44:324–326.

Walls, S. C., A. R. Blaustein, and J. J. Beatty. 1992. Amphibian biodiversity of the Pacific Northwest with special reference to old-growth stands. The Northwest Environment Journal 8:53–69.

Waring, R. H., and J. F. Franklin. 1979. Evergreen coniferous forests of the Pacific Northwest. Northwest Science 204:1380–1386.

Washington Department of Fisheries, Washington Department of Wildlife, and Western Washington Treaty Indian Tribes. 1993. 1992 Washington State salmon and steelhead stock inventory. Washington Department of Fisheries, Washington Department of Wildlife, and Western Washington Treaty Indian Tribes, Olympia. 212 pp.

Washington Department of Natural Resources. 1994. Endangered, threatened and sensitive vascular plants of Washington. Washington State Department of Natural Resources, Washington Natural Heritage Program, Olympia. 48 pp.

Washington Department of Wildlife. 1991. 1991 Status report: endangered and threatened. Washington Department of Wildlife, Olympia. 13 pp.

Washington Department of Wildlife. 1992. Bull trout/dolly varden management and recovery plan. Washington Department of Wildlife, Fisheries Management Division, Olympia. Report 92-22. 125 pp.

Washington Department of Wildlife. 1993a. Status of the Oregon silverspot butterfly (*Speyeria zerene hippolyta*) in Washington. Washington Department of Wildlife, Wildlife Management Division, Olympia. 25 pp.

Washington Department of Wildlife. 1993b. Status of the Larch Mountain salamander (*Plethodon larselli*) in Washington. Washington Department of Wildlife, Wildlife Management Division, Olympia. 14 pp.

Washington Department of Wildlife. 1993c. Status of the western pond turtle (*Clemmys marmorata*) in Washington. Washington Department of Wildlife, Wildlife Management Division, Olympia. 32 pp.

Washington Department of Wildlife. 1993d. Status of the North American lynx (*Lynx canadensis*) in Washington. Washington Department of Wildlife, Wildlife Management Division, Olympia. 95 pp.

Washington Department of Wildlife. 1993e. Status of pygmy rabbit (*Brachylagus idahoensis*) in Washington. Washington Department of Wildlife, Wildlife Management Division, Olympia. 23 pp.

Washington Department of Wildlife. 1994. Species of special concern in Washington. Washington Department of Wildlife, Olympia. 39 pp.

Washington Natural Heritage Program. 1995. Washington Natural Heritage Information System database. Washington Department of Natural Resources, Natural Heritage Program, Olympia.

Watts, S. E., and S. T. Knick. 1994. Habitat classification and the ability of habitats to support populations of Townsend's ground squirrels and black-tailed jack rabbits. Pages 212–231 *in* K. Steenhof, editor. Snake River Birds of Prey National Conservation Area research and monitoring annual report 1994. Bureau of Land Management and National Biological Survey, Boise, Idaho.

Welsh, H. H., Jr. 1990. Relictual amphibians and old-growth forests. Conservation Biology 4:309–319.

Welsh, H. H., Jr., and A. J. Lind. 1988. Old growth forests and the distribution of the terrestrial herpetofauna. Pages 439–455 *in* R. C. Szaro, K. E. Severson, and D. R. Patton, technical coordinators. Management of amphibians, reptiles, and small mammals in North America. U.S. Forest Service General Technical Report RM-GTR-166. Rocky Mountain Forest and Range Experiment Station, Fort Collins, Colo.

Wenatchee National Forest. 1993. Gold Creek macroinvertebrate study 1993. U.S. Forest Service, Naches Ranger District, Naches, Wash.

Werschkul, D. F. 1982. Species–habitat relationships in an Oregon cold desert lizard community. Great Basin Naturalist 42:380–384.

West, N. E. 1990. Structure and function of microphytic soil crusts in wildland ecosystems of arid and semi-arid regions. Pages 179–223 *in* M. Begon, A. H. Fitter, and A. McFadyen, editors. Advances in ecological research. Volume 20. Academic Press, London.

West, N. E. 1996. Strategies for maintenance of and repair of biotic community diversity on rangelands. *In* R. C. Szaro and D. W. Johnston, editors. Biodiversity in managed landscapes: theory and practice. Oxford University Press, New York.

West, N. E., and M. A. Hassan. 1985. Recovery of sagebrush–grass vegetation following wildfire. Journal of Range Management 38:131–134.

Whisenant, S. G. 1990. Changing fire frequencies on Idaho's Snake River Plains: ecological and management implications. Pages 4–10 *in* E. D. McArthur, E. M. Romney, S. D. Smith, and P. T. Tueller, editors. Proceedings—symposium on cheatgrass invasion, shrub die-off, and other aspects of shrub biology and management. U.S. Forest Service General Technical Report INT-GTR-276. Intermountain Research Station, Ogden, Utah.

Whitaker, J. O., Jr., C. Maser, and S. P. Cross. 1977. Food habits of bats of western Oregon. Northwest Science 51:46–55.

Whittier, T. R., R. M. Hughes, and D. P. Larsen. 1988. Correspondence between ecoregions and spatial patterns in stream ecosystems in Oregon. Canadian Journal of Fisheries and Aquatic Sciences 45:1264–1278.

Wiederholm, T. 1984. Responses of aquatic insects to environmental pollution. Pages 508–557 *in* V. H. Resh and D. M. Rosenburg, editors. The ecology of aquatic insects. Praeger Publishers, New York.

Wilderness Society. 1993. The living landscape. Volume 2. Pacific salmon and federal lands: a regional analysis. The Wilderness Society, Bolle Center for Forest Ecosystem Management, Washington, D.C. 88 pp.

Williams, J. E., and C. E. Bond. 1983. Status and life history notes on the native fishes of the Alvord Basin. Great Basin Naturalist 43:409–420.

Williams, J. E., J. E. Johnson, D. A. Hendrickson, S. Contreras-Balderas, J. D. Williams, M. Navarro-Mendoza, D. E. McAllister, and J. E. Deacon. 1989. Fishes of North America—endangered, threatened, or of special concern: 1989. Fisheries 14(6):2–20.

Williams, J. E., M. A. Stern, A. V. Munhall, and G. A. Anderson. 1990. Conservation status of threatened fishes in Warner Basin, Oregon. Great Basin Naturalist 50:243–248.

Willis, C. F., D. L. Ward, and A. A. Nigro. 1994. Development of a systemwide program: stepwise implementation of a predation index, predator control fisheries, and evaluation plan in the Columbia River basin. 2 volumes. U.S. Department of Energy, Bonneville Power Administration, Division of Wildlife, Annual Report 1992, Project 90-077. Portland, Oreg. 651 pp.

Willoughby, J., E. Hastey, K. Berg, P. Dittberner, R. Fellows, R. Holmes, J. Knight, B. Radtkey, and R. Rosentreter. 1992. Rare plants and natural communities: a strategy for the future. Fish and Wildlife 2000, National Strategy Plan Series. Bureau of Land Management, Washington, D.C. 60 pp.

Wisseman, R. W. 1992a. Benthic invertebrate biomonitoring: Deschutes National Forest. Aquatic Biology Associates, Corvallis, Oreg.

Wisseman, R. W. 1992b. Benthic invertebrate biomonitoring: Bureau of Land Management, Klamath Falls Resource Area. Aquatic Biology Associates, Corvallis, Oreg.

Wissmar, R. D., J. E. Smith, B. A. McIntosh, H. W. Li, G. H. Reeves, and J. R. Sedell. 1994. Ecological health of river basins in forested regions of eastern Washington and Oregon. U.S. Forest Service General Technical Report PNW-GTR-326. Pacific Northwest Research Station, Portland, Oreg. 65 pp.

Wood, C. A., A. P. Martin, and J. E. Williams. 1993. Bring back the natives: restoring native aquatic species on public lands. Endangered Species Technical Bulletin 18(4):5–10.

Wydoski, R. S., and R. R. Whitney. 1979. Inland fishes of Washington. University of Washington Press, Seattle. 220 pp.

Wygodzinsky, P., and P. Stys. 1970. A new genus of Aenictopecheine bugs from the holarctic. American Museum of Natural History Novitates 2411:1–17.

Wyman, R. L. 1990. What's happening to the amphibians? Conservation Biology 4:350–352.

Yahner, R. H. 1989. Changes in wildlife communities near edges. Conservation Biology 2:333–339.

Yensen, E., D. L. Quinney, K., Johnson, K. Timmerman, and K. Steenhof. 1992. Fire, vegetation changes, and population fluctuations of Townsend's ground squirrels. American Midland Naturalist 128:299–312.

Zarriello, T. J., S. T. Knick, and J. T. Rotenberry. 1994. Producing a burn/disturbance map for the Snake River Birds of Prey National Conservation Area. Pages 333–343 *in* K. Steenhof, editor. Snake River Birds of Prey National Conservation Area research and monitoring annual report 1994. Bureau of Land Management and National Biological Survey, Boise, Idaho.

Zimmerman, G. T., and R. D. Laven. 1984. Ecological interrelationships of dwarf mistletoe and fire in lodgepole pine forests. Pages 123–131 *in* F. G. Hawksworth and R. F. Scharpf, technical coordinators. Biology of dwarf mistletoes: proceedings of the symposium. U.S. Forest Service General Technical Report RM-GTR-111. Rocky Mountain Forest and Range Experiment Station, Fort Collins, Colo. 131 pp.

Complex Interactions of Introduced Trout and Native Biota in High-Elevation Lakes

Anderson, R. S. 1971. Crustacean plankton of 146 alpine and subalpine lakes and ponds in western Canada. Journal of the Fisheries Research Board of Canada 28:311–321.

Anderson, R. S. 1972. Zooplankton composition and change in an alpine lake. Verhandlungen der Internationalen Vereinigung für Theoretische und Angewandte Limnologie 17:264–268.

Anderson, R. S. 1980. Relationships between trout and invertebrate species as predators and the structure of the crustacean and rotiferan plankton in mountain lakes. Pages 635–641 *in* W. C. Kerfoot, editor. Ecology and evolution of zooplankton communities. Special symposium volume 3, American Society of Limnology and Oceanography. University Press of New England, Hanover, N.H.

Bahls, P. 1992. The status of fish populations and management of high mountain lakes in the western United States. Northwest Science 66:183–193.

Bradford, D. F. 1989. Allopatric distribution of native frogs and introduced fishes in high Sierra Nevada lakes of California: implications of the negative impact of fish introductions. Copeia 1989:775–1183.

Bradford, D. F., F. Tabatabai, and D. M. Graber. 1993. Isolation of remaining populations of the native frog, *Rana mucosa*, by introduced fishes in Sequoia and Kings Canyon national parks, California. Conservation Biology 7:882–888.

Liss, W. J., G. L. Larson, E. K. Deimling, R. Gresswell, R. Hoffman, M. Kiss, G. Lomnicky, C. D. McIntire, R. Truitt, and T. Tyler. 1995. Ecological effects of stocked fish on naturally fishless high mountain lakes: North Cascades National Park Service Complex, WA, USA. National Park Service, Draft Final Report, Phase I (1989–1992). Washington, D.C. 165 pp. + appendixes.

Stoddard, J. L. 1987. Microcrustacean communities of high-elevation lakes in the Sierra Nevada, California. Journal of Plankton Research 9:631–650.

Taylor, J. 1983. Orientation and flight behavior of a neotenic salamander (*Ambystoma gracile*) in Oregon. American Midland Naturalist 109:40–49.

Zaret, T. M. 1990. Predation and freshwater communities. Yale University Press, New Haven, Conn. 187 pp.

Spotted Frogs in the Western Pacific Northwest

Baird, S. F., and C. Girard. 1853. Descriptions of new species of reptiles collected by the U.S. Exploring Expedition under the command of Captain Charles Wilkes, U.S.N. First part including the species from the western coast of America. Proceedings of the National Academy of Natural Science, Philadelphia, Penn. 6:174–177.

Bury, R. B., and J. A. Whelan. 1984. Ecology and management of the bullfrog. U.S. Fish and Wildlife Service, Resource Publication 155. 23 pp.

Dickerson, M D. 1906. The frog book: North American toads and frogs with a study of the habits and life histories of those of the northeastern United States. 1969 edition. Dover Publications, New York. 253 pp.

Dumas, P. C. 1966. Studies of the *Rana* species complex in the Pacific Northwest. Copeia 1966(1):60–74.

Green, D. M., T. F. Sharbel, J. Kearsley, and H. Kaiser. 1996. Postglacial range fluctuation, genetic subdivision and speciation in the western North American spotted frog complex, *Rana pretiosa*. Evolution 50:374–390.

Green, D. M., H. Kaiser, T. F. Sharbel, J. Kearsley, and K. R. McAllister. 1997. Cryptic species of spotted frogs, *Rana pretiura* complex, in western North America. Copeia 1997(1):1–8.

Hayes, M. P. 1994. Current status of the spotted frog (*Rana pretiosa*) in western Oregon. Report to the Oregon Department of Fish and Wildlife, Portland. 13 pp.

Hayes, M. P., and M. R. Jennings. 1986. Decline of ranid frog species in western North America: are bullfrogs (*Rana catesbeiana*) responsible? Journal of Herpetology 20:490–509.

Jewett, S. G., Jr. 1936. Notes on amphibians of the Portland, Oregon area. Copeia 1936(1):71–72.

Leonard, W. P., H. A. Brown, L. L. C. Jones, K. R. McAllister, and R. M. Storm. 1993. Amphibians of Washington and Oregon. Seattle Audubon Society, Seattle, Wash. 168 pp.

McAllister, K. R., W. P. Leonard, and R. M. Storm. 1993. Spotted frog (*Rana pretiosa*) surveys in the Puget Trough of Washington, 1989–1991. Northwestern Naturalist 74:10–15.

Nussbaum, R. A., E. D. Brodie, Jr., and R. M. Storm. 1983. Amphibians and reptiles of the Pacific Northwest. University Press of Idaho, Moscow. 332 pp.

Slater, J. R. 1939. Some species of amphibians new to the State of Washington. Department of Biology, College of Puget Sound, Tacoma, Wash. Occasional Papers 1:4.

Storm, R. M. 1966. Endangered plants and animals of Oregon. Part 2. Amphibians and reptiles. Oregon State University, Agricultural Experiment Station. Special Report 206. Corvallis. 13 pp.

Northern Spotted Owl

Burnham, K., D. R. Anderson, and G. C. White. 1994. Estimation of vital rates of the northern spotted owl. Appendix J, Part 1 *in* Final environmental impact statement on management of habitat for late-successional and old-growth forest related species within the range of the northern spotted owl. U.S. Forest Service and Bureau of Land Management, Portland, Oreg.

Carey, A. B., J. A. Reid, and S. P. Horton. 1992a. Spotted owl home range and habitat use in southern Oregon Coast Ranges. Journal of Wildlife Management 54:11–17.

Carey, A. B., S. P. Horton, and B. L. Biswell. 1992b. Northern spotted owls: influences of prey base and landscape character. Ecological Monographs 62:223–250.

Forest Ecosystem Management Assessment Team. 1993. Forest ecosystem management: an ecological, economic, and social assessment. U. S. Department of the Interior, U. S. Department of Agriculture, U. S. Department of Commerce, and U. S. Environmental Protection Agency, Washington, D.C. 729 pp.

Forsman, E. D., E. C. Meslow, and H. M. Wight. 1984. Distribution and biology of the spotted owl in Oregon. Wildlife Monographs 87. 64 pp.

Hamer, T. E., E. D. Forsman, A. D. Fuchs, and M. L. Walters. 1994. Hybridization between barred and spotted owls. Auk 111:487–492.

Lehmkuhl, J. F., and M. G. Raphael. 1993. Habitat pattern around spotted owl locations on the Olympic Peninsula, Washington. Journal of Wildlife Management 57:302–315.

Ripple, W. J., D. H. Johnson, K. T. Hershey, and E. C. Meslow. 1991. Old-growth and mature forests near spotted owl nests in western Oregon. Journal of Wildlife Management 55:316–318.

Taylor, A. L., Jr., and E. D. Forsman. 1976. Recent range extensions of the barred owl in western North America, including first records for Oregon. Condor 78:560–561.

Thomas, J. W., E. D. Forsman, J. B. Lint, E. C. Meslow, B. R. Noon, and J. Verner. 1990. A conservation strategy for the northern spotted owl: a report to the Interagency Scientific Committee to address the conservation of the northern spotted owl. U.S. Forest Service, U.S. Fish and Wildlife Service, and National Park Service, Washington, D.C. 427 pp.

Thomas, J. W., M. G. Raphael, R. G. Anthony, E. D. Forsman, A. G. Gunderson, R. S. Holthausen, B. G. Marcot, G. H. Reeves, J. R. Sedell, and D. M. Solis. 1993. Viability assessments and management considerations for species associated with late-successional and old-growth forest of the Pacific Northwest. Research Report of the Scientific Analysis Team. U.S. Forest Service, Washington, D.C. 530 pp.

U.S. Fish and Wildlife Service. 1992. Recovery plan for the northern spotted owl final draft. 2 volumes. U.S. Fish and Wildlife Service, Portland, Oreg. 662 pp.

Wintering Canada Geese in the Willamette Valley

Campbell, B. 1992. Pacific Flyway Management Plan for dusky Canada geese (revised 1992). Pacific Flyway Study Committee, U.S. Department of the Interior, Fish and Wildlife Service, Portland, Oreg. 28 pp.

Chapman, J. A., C. J. Henny, and H. M. Wight. 1969. The status, population dynamics, and harvest of the dusky Canada goose. Wildlife Monographs 18. 48 pp.

Delacour, J. 1954. The waterfowl of the world. Volume 1. Country Life Ltd., London. 284 pp.

Gullion, G. W. 1951. Birds of the southern Willamette Valley, Oregon. Condor 53:129–149.

Jarvis, R. L., and J. E. Cornely. 1985. Recent changes in wintering populations of Canada geese in western Oregon and southwestern Washington. Pages 517–528 *in* M. W. Weller, editor. Waterfowl in winter. University of Minnesota Press, Minneapolis.

Shepherd, P. E. K. 1965. A preliminary evaluation of earthquake damage to waterfowl habitat in south-central Alaska. Proceedings of the annual conference of the Western Association of State Game and Fish Commissioners 45:76–80.

Simpson, S. G., and R. L. Jarvis. 1979. Comparative ecology of several subspecies of Canada geese during winter in western Oregon. Pages 223–241 *in* R. L. Jarvis and J. C. Bartonek, editors. Management and biology of Pacific Flyway geese. Oregon State University, Corvallis.

Sage Grouse in Oregon

Barnett, J. K. 1993. Diet and nutrition of female sage grouse during the pre-laying period. M.S. thesis, Oregon State University, Corvallis. 46 pp.

Barnett, J. K., and J. A. Crawford. 1994. Pre-laying nutrition of sage grouse hens in Oregon. Journal of Range Management 47:114–118.

Crawford, J. A., and R. S. Lutz. 1985. Sage grouse population trends in Oregon, 1941–1983. Murrelet 66:69–74.

DeLong, A. K. 1994. Relationships between vegetative structure and predation rates of artificial sage grouse nests. M.S. thesis, Oregon State University, Corvallis. 50 pp.

DeLong, A. K., J. A. Crawford, and D. C. DeLong, Jr. 1995. Relationships between vegetational structure and predation of artificial sage grouse nests. Journal of Wildlife Management 59:88–92.

Dobkin, D. S. 1995. Management and conservation of sage grouse, denominative species for the ecological health of shrub-steppe ecosystems. Bureau of Land Management, Portland, Oreg. 26 pp.

Drut, M. S., J. A. Crawford, and M. A. Gregg. 1994a. Brood habitat use by sage grouse in Oregon. Great Basin Naturalist 54:170–176.

Drut, M. S., W. H. Pyle, and J. A. Crawford. 1994b. Technical note: diets and food selection by sage grouse chicks in Oregon. Journal of Range Management 47:90–93.

Gabrielson, I. N., and S. G. Jewett. 1940. Birds of Oregon. Oregon State College, Corvallis. 650 pp.

Gregg, M. A. 1992. Use and selection of nesting habitat by sage grouse in Oregon. M.S. thesis, Oregon State University, Corvallis. 46 pp.

Gregg, M. A., J. A. Crawford, and M. S. Drut. 1993. Summer habitat use and selection by female sage grouse (*Centrocercus urophasianus*) in Oregon. Great Basin Naturalist 53:293–298.

Gregg, M. A., J. A. Crawford, M. S. Drut, and A. K. DeLong. 1994. Vegetational cover and predation of sage grouse nests in Oregon. Journal of Wildlife Management 58:162–166.

Patterson, R. L. 1952. The sage grouse in Wyoming. Sage Book, Denver, Colo. 341 pp.

Ramsey, F. L., M. McCracken, J. A. Crawford, M. S. Drut, and W. J. Ripple. 1994. Habitat association studies of the northern spotted owl, sage grouse, and flammulated owl. Pages 189–209 *in* N. Lange, L. Billard, L. Conquest, L. Ryan, D. Brillinger, and J. Greenhouse, editors. Case studies in biometry. John Wiley & Sons, New York.

Translocated Sea Otter Populations off the Oregon and Washington Coasts

Estes, J. A. 1990. Growth and equilibrium in sea otter populations. Journal of Animal Ecology 59:385–401.

Jameson, R. J., K. W. Kenyon, A. M. Johnson, and H. M. Wight. 1982. History and status of translocated sea otter populations in North America. Wildlife Society Bulletin 10:100–107.

Jameson, R. J., K. W. Kenyon, S. Jeffries, and G. R. VanBlaricom. 1986. Status of a translocated sea otter population and its habitat in Washington. Murrelet 67:84–87.

Jeffries, S., and R. Jameson. 1995. Status of sea otters in Washington State. Seventy-fifth anniversary meeting of the Society for Northwestern Vertebrate Biology, Orcas Island, Washington, 23–25 March 1995. Abstract.

Kenyon, K. W. 1969. The sea otter in the eastern Pacific Ocean. North American Fauna 68. Washington, D.C. 352 pp.

Monson, D. H., and A. R. DeGange. 1995. Reproduction, preweaning survival, and survival of adult sea otters at Kodiak Island, Alaska. Canadian Journal of Zoology 73:1161–1169.

Rathbun, G. B., R. J. Jameson, G. R. VanBlaricom, and R. L. Brownell. 1990. Reintroduction of sea otters to San Nicolas Island, California: preliminary results for the first year. Pages 99–144 in P. J. Bryant and J. Remington, editors. Endangered wildlife and habitats in southern California, memoirs of the Natural History Foundation of Orange County. Volume 3. A collection of occasional papers published by the Natural History Foundation of Orange County.

Riedman, M. L., and J. A. Estes. 1990. The sea otter (*Enhydra lutris*): behavior, ecology, and natural history. U.S. Fish and Wildlife Service Biological Report 90(14). Washington, D.C. 126 pp.

Riedman, M. L., J. A. Estes, M. M. Staedler, A. A. Giles, and D. R. Carlson. 1994. Breeding patterns and reproductive success of California sea otters. Journal of Wildlife Management 58(3):391–399.

Wilson, D. E., M. A. Bogan, R. L. Brownell, Jr., A. M. Burdin, and M. K. Maminov. 1991. Geographic variation in sea otters, *Enhydra lutris*. Journal of Mammalogy 72:22–36.

Roosevelt Elk and Forest Structure in Olympic National Park

Houston, D. B., E. G. Schreiner, B. B. Moorhead, and K. A. Krueger. 1990. Elk in Olympic National Park: will they persist over time? Natural Areas Journal 10:6–11.

Leslie, D. M., E. E. Starkey, and M. Vavra. 1984. Elk and deer diets in old-growth forests in western Washington. Journal of Wildlife Management 48:762–775.

Woodward, A., E. G. Schreiner, and D. B. Houston. 1994. Ungulate forest relationships in Olympic National Park: retrospective exclosure studies. Northwest Science 68:97–110.

Alaska

A laska is the largest state (1,477,270 square kilometers), more than twice as large as Texas and one-fifth the size of the lower 48 contiguous states. Alaska extends more than 20° in latitude, from Point Barrow at 71°23' to Amatignak Island in the Aleutians at 51°20'. It spans 42° in longitude, from 130° at the Portland Canal in the southeast to 172° at Attu Island in the western Aleutians (Fig. 1). The topography, climate, and communities of Alaska are characterized by their great diversity (Selkregg 1974–1976), including Mt. McKinley (Denali), the highest mountain in North America; more glaciers, ice fields, and active volcanoes than in the rest of the United States; several of the largest river systems in the country; a coastline of more than 54,700 kilometers that borders the North Pacific, Bering Sea, and Arctic Ocean; and extensive lowland boreal forest, wetlands, and coastal tundra.

Mean annual temperatures range from -2°C to 2°C along the southern coast, from -6°C to -2°C in the interior, and from -12°C to -10°C in the interior mountains and Arctic. The length of the frost-free period varies from more than 200 days in parts of southeastern Alaska and the Aleutian Islands to 40 days in the Arctic. In summer, the long day length in the interior and the 24 hours of sunlight in the Arctic partially compensate for the short growing season and account for relatively high plant productivity at these high latitudes. Precipitation can exceed 500 centimeters per year in parts of the Alexander Archipelago of southeastern Alaska, but it is generally less than 25 centimeters per year in the Arctic, which is an amount that would result in desert conditions at lower latitudes. Most of the precipitation in winter, however, remains on the land as snow until spring, and low summer evaporation rates and impeded drainage (because of permafrost) allow adequate moisture for plant growth. Permafrost is present throughout most of the Arctic and northwestern Alaska except beneath lakes, rivers, and adjacent riparian zones. South of the Brooks Range in the interior and in southwestern Alaska, permafrost is discontinuous and mostly occurs on north-facing slopes, at higher elevations, and in the lowlands.

During the Ice Ages of the Pleistocene Epoch, when the great continental ice sheets swept down across Canada into the northern tier of states, glaciation in Alaska was extensive only in the southern portion of the state from the Alaska Range, Wrangell and Saint Elias mountains, and Coast Range to the Gulf of Alaska (Hamilton et al. 1986). Portions of the Brooks Range and a few of the higher mountains in interior and western Alaska were also glaciated, but most of interior, western, and arctic Alaska was not glaciated. Because of the great volume of water tied up in the continental ice sheets, sea levels then were as much as 90 meters lower than today, and a continuous ice-free land area connected Alaska with eastern Siberia (that is, the Russian Far East) during the Tertiary and several times during the Quaternary, including the last exposure that occurred between about 30,000 and 11,000 years ago. The land bridge extended from just north of the Aleutian arc northward to the edge of the continental shelf in the Arctic Ocean (Hopkins et al. 1982). Plants, animals, and the first humans entered the New World by this route. This region connecting Asia and America—where a free exchange of species occurred until about 11,000 years ago when meltwaters again raised sea levels—is called Beringia. The area of Alaska and adjacent Asia that was

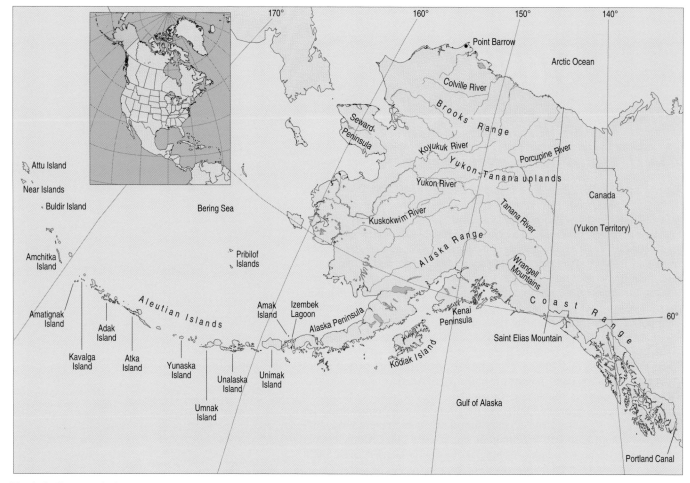

Fig. 1. Outline map of Alaska showing the major rivers and mountain ranges.

not glaciated and adjacent parts of northwestern Canada constituted a vast refugium for plants and animals that were isolated from those south of the great continental ice sheets (Hopkins et al. 1982). As a result of Alaska's ice age history, much of its present flora and fauna are the same as or share genetic affinities with those of northern Asia and therefore provide a unique and diverse component of the total biological diversity of the United States.

The first humans in the Western Hemisphere are believed to have come across Beringia into Alaska 12,000 to 15,000 years ago. They were the Paleo-Indians who spread throughout North and South America and from whom most Native American cultures derived, including the Haida and Tlingit peoples of the southeastern coast of Alaska (Greenberg 1987). Later movements of people are believed to have been responsible for the Athabaskan cultures that are present throughout interior and south-central Alaska and in parts of northwestern Canada. The marine-oriented Eskimos of arctic, western, and southwestern Alaska (represented today by the Inupiat, Yupik, and Koniak cultures) arrived much later, apparently by boat across the Bering Strait. The Aleut culture of

the Aleutian Islands and adjacent Alaska Peninsula has its closest affinity with early Eskimo cultures.

Today, the human population of Alaska is about 600,000. Most people live in and around Anchorage, Fairbanks, Juneau, and smaller southern coastal cities. Alaska Natives constitute about 16% of the population, mostly living in small communities throughout rural Alaska, where they are at least partially dependent on the fishes and wildlife of the land and waters for their subsistence (Selkregg 1974–1976).

Major Biogeographic Regions

Alaska's position between the cold Arctic Ocean and the relatively warm North Pacific Ocean, its extensive coastline and southern islands, its high mountain ranges and associated ice fields, and its large area are responsible for the state's ecological diversity. For this report, Alaska is divided into six major biogeographic regions (Fig. 2). Our classification of Alaska's ecoregions, which varies from classification schemes devised for other purposes (Selkregg 1974–1976; Fig. 3), groups areas of similar climatic and ecological features.

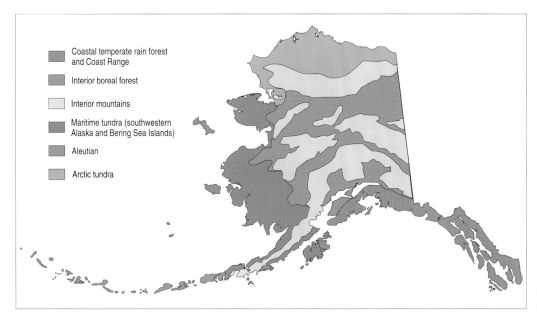

Fig. 2. The major biogeographic regions of Alaska.

Map legend:
- Coastal temperate rain forest and Coast Range
- Interior boreal forest
- Interior mountains
- Maritime tundra (southwestern Alaska and Bering Sea Islands)
- Aleutian
- Arctic tundra

Coastal Temperate Rain Forest and Coast Range

This ecoregion (Fig. 2) is shaped by the moderating maritime influence of the relatively warm ocean currents. It is a continuation of the temperate coniferous rain forest that extends from northern California along the northwestern coast of North America to northern Kodiak Island (Fig. 4). Although the number of tree species and the elevation of the treeline decline with increasing latitude, the extension of temperate rain forest to these high latitudes is globally unique. The abrupt terrain adjacent to the sea culminates in high mountains that capture moisture from the oceanic air as rain and snow. At higher elevations, extensive ice fields feed glaciers that often reach the sea at the heads of fjords in the deeply incised shoreline. Annual precipitation is heavy throughout the region but is highly variable locally because of irregular terrain. In winter, the frequency and accumulation of snow are also variable but increase with increasing latitude and altitude and also from the outer coast and islands to the mainland (Selkregg 1974–1976).

The entire region, except the mountain peaks within the ice fields (known as *nunataks*), was heavily glaciated during the last Ice Age. Ice receded from most of the region about 10,000 years ago, thus most plants and animals that characterize the region arrived from unglaciated areas. Plants dispersed primarily from the south by following a coastal route (Hultén 1937), whereas animals are representative of the fauna of the northern and southern Pleistocene refugia (Klein 1965).

The fauna and flora of the south dominate. The common garter snake, the only reptile in Alaska, has been reported in Revillagigedo Island in southernmost southeastern Alaska. Amphibians are represented by the western toad, three salamander species (the rough-skinned newt, northwestern salamander, and long-toed salamander), and two frog species (the Cascades spotted frog and the wood frog). All but the wood frog, which also occurs throughout the boreal forest, are restricted to southeastern Alaska. No studies have been made of the ecology of these amphibians in the rain forests of southeastern Alaska.

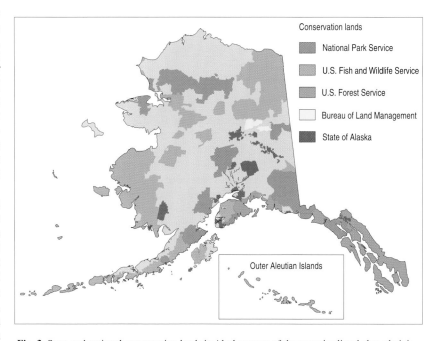

Conservation lands:
- National Park Service
- U.S. Fish and Wildlife Service
- U.S. Forest Service
- Bureau of Land Management
- State of Alaska

Outer Aleutian Islands

Fig. 3. State and national conservation lands in Alaska: some of the agencies listed also administer other land in Alaska. A large portion of the state's land is in protected status: national park lands in Alaska make up 75% of the nation's total, and 90% of national wildlife refuge lands are in Alaska.

Courtesy D. R. Klein, USGS

Fig. 4. The coastal temperate rain forest and Coast Range are a complex interface of mountains and sea.

Vegetation

Lowlands of this region are covered by complex forests of towering trees, by bogs where poor drainage and excessive groundwater preclude tree growth, and by the floodplains of streams and rivers. Lowlands and benches on mountain slopes are dotted with various types of bogs. Many are sphagnum bogs with low ericaceous (heath) shrub vegetation and scattered lodgepole pines (the only pine that occurs in Alaska), some Alaska-cedar, and mountain hemlock. Variability among bogs is great: some are rich in mosses, grasses, and sedges, and some have tussocks and hummocks, whereas others do not (Robuck 1985). Adjacent shallow seas support aquatic marine communities composed mainly of algae but also of surfgrass and eelgrass. In areas where sandy beaches form, the strands are scattered with salt-tolerant plants such as sea-beach sandwort, followed shoreward by beach ryegrass communities with codominants such as hairgrass, beach pea, beachstrawberry, Pacific silverweed, and beach lovage. Most shorelines, however, are narrow, and sea cliffs and rocky shores are common. Between beach and forest lies a dense zone of shrubs where Sitka alder and thorny devilsclub can be common and unforgettable to those who try to pass through.

The structure of the forest varies from south to north in the region. In the south, forests consist primarily of western hemlock and Sitka spruce with fewer western redcedar and Alaska-cedar. Pacific silver fir and Pacific yew reach the northern limits of their distribution in the southernmost portion of the coastal forest. To the north, the proportion of western hemlock increases and mountain hemlock becomes more conspicuous. The understory can be rich in various shrubs and herbaceous species such as grasses, forbs, and large ferns. Lichens and mosses are important on the forest floor and as epiphytes (Alaback 1982).

Species richness of the understory depends in large part on the tree canopy. A moss-dominated, shrub-poor understory is characteristic of young, even-aged stands with closed canopies, particularly in western hemlock forests (Alaback 1982). Frequent shrubs in this kind of understory are Sitka alder, devilsclub, salmonberry, Alaska blueberry and early blueberry, and menziesia.

Original forest stands of this region are ancient, with some trees older than 300 years. Often referred to as old-growth forests, these stands are nevertheless uneven-aged because as older trees die and fall, younger trees become established beneath breaks in the canopy. The resulting layered canopy has openings that let light penetrate to the forest floor, allowing shrubs, ferns, forbs, and tree seedlings to grow.

Local, large-scale natural disturbances are caused by blowdowns and (rarely) by fire. Even-aged forest stands that develop after blowdowns and clear-cutting lack the biological diversity of old-growth forests (Alaback 1982; Schoen et al. 1988). One limitation to diversity in even-aged forests is the closed canopy of second-growth stands, which does not allow sufficient light to reach the forest floor for growth of shrubs and other vascular plants. Also, although forests readily regenerate from seeds in the litter after removal of old-growth forests, species composition of the new growth is usually different from that of the original forests (Alaback 1982; Kirchhoff and Schoen 1987).

The subalpine zone occurs on mountain slopes at elevations of about 400 to 500 meters. Mountain hemlock forests with trees 38–50 centimeters diameter at breast height and a height as tall as 25 meters characteristically define this zone. Heavy snowpack accounts for late-lying snow in the subalpine region throughout the early to middle part of the growing season. In many areas, the subalpine region is a broad belt where small islets of stunted trees are confined to sheltered sites.

The transition to alpine tundra begins with lichen-rich mountain heath and dwarf shrub communities dominated by shrubs and subshrubs of heath and related families. At higher elevations, the landscape is increasingly broken by outcroppings of bedrock, and plant communities are correspondingly fragmented. In stable sites, such as snowbeds and snowflushes, prostrate, mat-forming woody species that occur on ridges are replaced by rich sedge–grass–forb meadows. In open communities these woody species are replaced by a variety of mat and cushion forbs and subshrubs. Ultimately, increasingly severe conditions at higher elevations restrict plants to ledges and crevices. At some localities—such as on Prince of Wales Island—abundant, seasonally dry, calcareous soils are locations of continental species disjunct in the archipelago from the interior (Jaques 1973).

The rare vascular plants of southeastern Alaska were recently surveyed by the Alaska Natural Heritage Program (University of Alaska, Anchorage, unpublished data). The most noteworthy of these are Calder's licorice-root, Alaska mistmaiden, netleaf willow, and cleftleaf groundsel.

Freshwater Fishes

Thirty-five fish species occur in the fresh waters of the coastal temperate rain forest and Coast Range region (Morrow 1980; Table). Species favored by humans are the chinook, chum, coho, pink, and sockeye salmon; the Dolly Varden; the rainbow and cutthroat trout; and the eulachon. The salmon species support a commercial fishery that in 1994 harvested more than 135 million salmon at a value of more than $200 million (Alaska Department of Fish and Game 1995). Salmon also provide subsistence food for Alaskans—in 1993 more than 107,000 salmon were caught for subsistence and personal use (Alaska Department of Fish and Game 1994a). In addition, salmon, Dolly Varden, rainbow trout, and cutthroat trout support a growing recreational fishery that now includes more than 1.5 million angler-days, or about 69% of the total sport effort in the state (Mills 1994).

The region's anadromous and freshwater resident fishes and their eggs provide food for a variety of mammals, birds, and other fishes. Willson and Halupka (1995) list 16 mammal, 31 bird, and 11 fish species in southeastern Alaska that feed on adult salmon and their eggs and young. In addition to the direct benefits these fishes provide to other

Table. Freshwater fishes in Alaska (modified from Morrow 1980).

Species	Region[a]
Lamprey family	
River lamprey	1
Western brook lamprey	1
Arctic lamprey	2, 4, 6
Pacific lamprey	1, 4, 5
Sturgeon family	
Green sturgeon	1, 5
White sturgeon	1
Shad family	
American shad	1
Salmon, trout, char, grayling, whitefish family	
Whitefish subfamily	
Sheefish, inconnu	2, 4
Humpback whitefish	2, 4, 6
Least cisco	2, 4, 6
Arctic cisco	6
Broad whitefish	2, 4, 6
Bering cisco	2, 3, 4, 6
Round whitefish	1, 2, 3, 4, 6
Pygmy whitefish	4
Grayling subfamily	
Arctic grayling	1, 2, 3, 4, 6
Trout, char, salmon subfamily	
Brook trout	1
Lake trout	1, 2, 3, 4, 6
Arctic char	3, 4, 5, 6
Dolly Varden	1, 2, 3, 4, 5, 6
Rainbow trout, steelhead	1, 2, 4
Cutthroat trout	1
Sockeye salmon, kokanee	1, 2, 3, 4, 5
Coho salmon	1, 2, 4, 5
Chinook salmon	1, 2, 4
Chum salmon	1, 2, 4, 5, 6
Pink salmon	1, 4, 5, 6
Smelt family	
Longfin smelt	1
Rainbow smelt	1, 4, 6
Eulachon	1, 5
Pond smelt	1, 4
Pike family	
Northern pike	1, 2, 4, 6
Blackfish family	
Alaska blackfish	2, 4, 6
Minnow family	
Lake chub	2
Sucker family	
Longnose sucker	1, 2, 3, 4, 6
Trout-perches family	
Trout-perch	2, 4
Cod family	
Burbot	1, 2, 3, 4, 6
Arctic cod	4, 6
Saffron cod	1, 4
Stickleback family	
Ninespine stickleback	1, 4, 6
Threespine stickleback	1, 4
Sculpin family	
Pacific staghorn sculpin	1
Fourhorn sculpin	4, 6
Prickly sculpin	1
Coastrange sculpin	1, 5
Slimy sculpin	1, 2, 3, 4, 6
Sharpnose sculpin	1, 5
Surfperch family	
Shiner perch	1
Flounder family	
Arctic flounder	4, 6
Starry flounder	1, 4, 6

[a] Regions: 1 = coastal temperate rain forest and Coast Range; 2 = interior boreal forest; 3 = interior mountains; 4 = maritime tundra of southwest and western Alaska; 5 = Aleutian Islands and adjacent Alaska Peninsula; 6 = arctic tundra.

vertebrates, nutrients from their carcasses he sustain productivity of streamside and lacustrine communities (Ritchey et al. 1975; Kline et al. 1990). The most obvious food chain relationships are the importance of salmon carcasses and spawning eulachons in bald eagle diets (Hansen et al. 1984; Hansen 1987; Armstrong 1995) and of salmon carcasses in brown bear diets (Schoen and Beier 1990). Some systems—such as Anan Creek near Ketchikan, where black bears congregate, and Pack Creek on Admiralty Island, which attracts brown bears—are set up as special areas where the public can view bears (Quinlan et al. 1983). The less obvious food chain relationships involving the region's fishes are the consumption of staghorn sculpins by greater yellowlegs and the consumption of threespine sticklebacks by arc-

tic terns, great blue herons, and river otters (O'Clair et al. 1992).

Stocks of salmon and eulachon attract the largest-known concentrations of bald eagles in North America. The most notable concentrations are those of more than 4,000 eagles attracted by spawning chum salmon in the Chilkat River near Haines and 1,500 eagles drawn by spawning eulachons in the Stikine River (Hughes 1982; Armstrong 1995). Spawning eulachons that congregate during early May each year in the Berners, Lace, and Antler rivers of Berners Bay in southeastern Alaska attract a world-class concentration of more than 50,000 gulls, hundreds of bald eagles, numerous harbor seals, sea lions, and even an occasional humpback whale (Armstrong and Gordon 1995).

Pacific Halibut in Glacier Bay National Park, Alaska

The Pacific halibut is a large (up to 3 meters long) predatory fish in the flatfish family. Glacier Bay National Park is the site of extensive and controversial commercial halibut fisheries that began before the park was established in 1925. These fisheries continue despite prohibitive regulations, including the Wilderness Act and National Park Service regulations. Today, more than 70 commercial boats (1991 and 1992 data) harvest between 136,200 and 181,600 kilograms of Pacific halibut per year within the park.

Commercial halibut fishing in Alaska began as a fishery open to all people and was managed by controlling its duration. The duration of the open fishing season has gradually been shortened, until in 1994 the openings were only a one- to two-day "derby." The large numbers of vessels fishing the waters of Glacier Bay National Park resulted in an additional conflict for the National Park Service, which severely limits the numbers of boats other than commercial fishing vessels permitted to enter Glacier Bay in order to reduce negative impacts on humpback whales and other park resources. In 1995 an individual fishing quota system replaced the derby-style fishery. This system was predicted to result in greater local resource use by fishing vessels, and preliminary data bear out this prediction. Because Glacier Bay is near many fishing communities and its waters are relatively protected, this new quota system may cause the bay to experience increases in fishing activity and more conflicts between visitors and fishing vessels. In addition, unlike the derby-style fishery, the individual fishing quota system permits fishing throughout the summer,

which is when both whale abundance and visitor attendance also peak. The majority of previous studies on Pacific halibut have been directed toward maintaining maximum sustainable yield. In contrast, the National Park Service is directed to manage its resources in such a manner as to maintain their natural state and to provide for visitor enjoyment, which is why Glacier Bay Field Station's research efforts have been directed at such basic ecological questions as diet, home range, site fidelity, habitat selection, distribution patterns, and the relationships between halibut and other species.

To analyze the Pacific halibut's diet, we examined the stomach contents of 947 sport-caught fishes. Content analysis revealed a shift in diet as halibut mature, from small crustaceans consumed by young halibut to fish eaten by large, mature halibut. Two foraging modes by halibuts were revealed by analysis of stomachs with multiple prey items. Individual halibut often exhibited only prey items found during active foraging (for example, large numbers of juvenile crabs) or else only prey items associated with sit-and-wait predation (such as walleye pollock).

Courtesy USGS

Fig. 1. Sonic tag being internally implanted in a Pacific halibut.

Long-distance movements of Pacific halibut have been emphasized in previous studies (Skud 1977; St-Pierre 1984), and most population models developed for this species assume that movement is relatively unrestricted among areas (Quinn et al. 1985). In our own studies of movement patterns, we internally implanted long-life sonic tags in 97 halibut in Glacier Bay and individually wire-tagged more than 1,500 halibut (Fig. 1). Results from the sonic-tagging study indicate that home range patterns shift during the course of halibut development. Juvenile fish move widely but often still remain within the Glacier Bay area, whereas large, sexually mature individuals have much smaller home ranges—often less than 0.5 square kilometers—and many mature fish stay in the same area both during the course of a year and from one year to another with little simultaneous spatial overlap (Fig. 2). Larger fish occasionally alter their pattern of small home ranges to travel more widely before returning to a relatively sedentary pattern. A few larger fish, though, appear never to establish home ranges.

Wire-tagging data corroborate the apparent site fidelity of adult Pacific halibut. Of the halibut originally wire-tagged in Glacier Bay and then recaptured later, more than 95% were recaptured within Glacier Bay, with an additional 3% caught in the adjoining Icy Strait area. Individuals have been recaptured 5 years after tagging within a few hundred meters of their original capture and release location.

Sonic-tracking data also indicate that although some individuals leave during the winter, many appear to remain within Glacier Bay. Researchers have hypothesized that halibut may spawn only in certain areas off the outer coast during the winter (Skud 1977; St-Pierre 1984). The presence of reproductively mature individuals within Glacier Bay during winter may indicate either that Pacific halibut do not spawn every year or that spawning can also occur within the bay.

To study the species' habitat selection and distribution patterns in Glacier Bay, we have sampled Pacific halibut by setting 149 research long-lines with 400 hooks each throughout the bay (Fig. 3). *Long-lining* is a method of fishing that uses a piece of ground line (often kilometers in length) to which short (0.5 to 1 meter long) gangion lines with baited hooks are attached at intervals of 1 to 10 meters. Results of these studies and those from sonic-tracked individuals suggest halibut engage in two broad patterns of habitat choice and dispersion. The first pattern seems to depend on the developmental stage of the fish, with larger individuals distributed relatively uniformly in deeper water and smaller individuals preferring shallow water and areas of steep topographic relief in a much more aggregated distribution pattern. The second observed pattern, which reflects changes in distribution along the length of Glacier Bay's recently deglaciated fjord system, is characterized by decreased halibut abundance with greater proximity to glacial termini. Sampling of water conditions—salinity, temperature, and the amount of silt and phytoplankton (indicated by chlorophyll *a*)—indicates that this pattern is due to oceanographic conditions rather than the habitat's physical or biological successional processes. The presence of tidewater glaciers dramatically affects oceanographic conditions, and Glacier Bay, with its many tidewater glaciers, is a profound example of this. Tremendous amounts of cold, freshwater flow and as much as 6 centimeters of siltation a day come down into the bay from the tidewater glaciers. Comparisons between long-lining data and sonic-tag data indicate that many larger individuals are not caught by long-lines.

As a result of investigations into Pacific halibut ecological relationships, we hypothesize that the way halibut in Glacier Bay hunt depends on their life stage, and that these foraging modes determine their movement patterns, distribution, and catchability. Juveniles probably range widely, actively searching for areas of high prey abundance, where they are easily caught on long-lines. Many adults probably establish small, discrete home ranges where they wait for large fish or invertebrates; rarely does a long-line come close enough for these individuals to be captured.

We plan to continue our studies on long-term site fidelity, to test the hypotheses on foraging modes, and to further our understanding of the distribution and abundance of halibut predators and prey. If future regulatory fishing closures occur, we also hope to use them as experiments to examine the effects of local depletion or broader-scale ecological factors in fished and unfished areas.

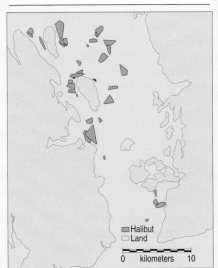

Fig. 2. Home ranges of a few larger halibut in Glacier Bay.

Fig. 3. Research long-lines being retrieved.

Our research results have management implications for the Pacific halibut. As stated, the studies at Glacier Bay indicate that halibut have much smaller home ranges and greater site fidelity than previously thought. These findings, coupled with the potential increase in local commercial fishing due to the individual fishing quota system and a rise in halibut sport fishing in southeast Alaska, indicate a potential for local resource depletion of this species. On the other hand, our hypotheses about differing foraging patterns and catchability in Pacific halibut suggest that this species' behaviors may help protect larger halibut from commercial long-line fishing but not from trawling with nets. This "behavioral refugium" could act to buffer the population from the collapses experienced by other fisheries that have been managed based on maximum sustainable yield models and catch-per-unit effort (Ludwig et al. 1993; Rosenberg et al. 1993).

See end of chapter for references

Authors

Philip N. Hooge
Spencer J. Taggart
U.S. Geological Survey
Biological Resources Division
Alaska Biological Science Center
Glacier Bay Field Station
P.O. Box 140
Gustavus, Alaska 99826

Birds

The coastal temperate rain forest and the Coast Range region is the major migration route of Pacific Flyway birds during their seasonal transits between nesting areas in Alaska and adjacent Siberia and their southern wintering areas. The region also supports a rich diversity of nesting migratory species, and the number of resident species is greater in the coastal region than in the other biogeographic regions of Alaska.

Marbled murrelets nest in old-growth forests in south-central and southeastern Alaska and, less commonly, on the ground in south-central and southwestern Alaska (Quinlan and Hughes 1990). Although data from boat surveys of marbled and Kittlitz's murrelet populations have suggested a total combined population of about 600,000–900,000 murrelets in Alaska (Agler et al. 1995), it is difficult for researchers to determine the actual number of breeding marbled murrelets in the dense coastal forests (Ewins et al. 1993; Piatt and Ford 1993).

Waterfowl are common migrants that breed in wetlands in the coastal temperate rain forest, particularly in major river deltas (for example, the Copper River delta). Several subspecies of geese have limited breeding distributions in this region; for example, the Vancouver Canada goose breeds in southeastern Alaska forests, the dusky Canada goose in the Copper River delta, and the Tule white-fronted goose in the western Cook Inlet. Migrating sandhill cranes from the Pacific Flyway stop in the Copper River delta, at wetlands near Yakutat, and at several points in southeastern Alaska. The southern coastal area is also important for breeding trumpeter swans, whose population has increased since the middle of the century. However, suitable trumpeter swan habitat throughout this region may be saturated (Conant et al. 1993; Kessel and Gibson 1994) or adversely affected by human disturbance. Population sizes of Pacific loons and common loons, which nest in lakes in the southern coastal forest region, seem stable, although increasing human habitation and heavy recreational use of many lakes may potentially disrupt nesting (Ruggles 1994).

Raptors are common in prey-rich coastal areas. As mentioned previously, the largest U.S. populations of bald eagles occur in the coastal forests of southeastern and south-central Alaska (Schempf 1989); the bald eagle population in southeastern Alaska seems to be increasing (Kessel and Gibson 1994). Peale's peregrine falcon is a resident, but insufficient data are available to determine this subspecies' population trends (Ambrose et al. 1988; Schempf 1989).

Southern-coastal Alaska is important to the world's shorebirds because this region contains several large river deltas that serve as stopover sites for millions of migrating shorebirds (for example, the Copper River delta in south-central Alaska and, of somewhat lesser importance, the Stikine River delta in southeastern Alaska; Senner 1979; Bishop and Green 1994). During migration to their breeding grounds in western Alaska, several million western sandpipers and dunlins stop in the Copper River delta (Senner 1979); the western sandpipers make up a substantial proportion of the species' breeding population. Although gulls are common residents in coastal areas, the abundances of some gull populations (for example, glaucous-winged gulls) in Prince William Sound seem to have declined because of closure of some canneries and changes in methods of handling fish waste at other facilities (Ainley et al. 1994). Also, oil that leaked from the *Exxon Valdez* spill narrowly missed contaminating some of the major stopover sites for shorebirds in Prince William Sound.

The coastal temperate rain forest and mountains support abundant populations of resident and migrant passerines, forest-dwelling grouse, and alpine-inhabiting ptarmigan, but few data are available on the population trends of these species. Extensive clear-cutting of old-growth forests throughout the region is causing major habitat alteration for many of these species (Schoen et al. 1988); however, specific details of their habitat requirements are not well understood.

Mammals

The biogeography of mammals is more complex in this region than in the rest of Alaska, a consequence of the region's insularity. Mammals encounter many physical and temporal obstructions to moving, including water barriers, mountain ranges, ice fields, glaciers, and the relatively short geologic time that has passed since ice covered most of the region. More than 30 mammal taxa are endemic to the region, and more than 10 additional taxa are largely confined here (MacDonald and Cook 1994). The many natural barriers to the free movement of mammals into and within the region have acted as filters and affect mammal dispersal in different ways. For example, in the Alexander Archipelago of southeastern Alaska, among the larger mammals, Sitka black-tailed deer are present on all the larger islands, whereas their primary predator, the gray wolf, occurs only on the islands south of Frederick Sound. Bears are also uniquely distributed: brown bears are found only on the Admiralty, Baranof, and Chichagof islands of the northern complex, whereas black bears only occur on the southern island complex, which is also occupied by gray wolves. The distribution of small mammals also varies widely among the islands. Throughout the entire ecoregion, among the smaller mammals, species richness declines from the mainland to the outermost islands, whereas endemism is greatest on the outermost islands and decreases as one moves closer to the mainland (MacDonald and Cook 1994).

Several mammal species that are typical of the temperate forest at lower latitudes reach the northern extensions of their Alaskan ranges in this ecoregion: Sitka black-tailed deer, four bat species (Keen's myotis, long-legged myotis, California myotis, and silver-haired bat), Keen's mice, southern red-backed voles, and most recently, fishers and mountain lions (MacDonald and Cook 1994). Several mainland species have not reached the islands; however, transplantations established mountain goats on Baranof and Revillagigedo islands. From 1916 through 1934, deer were introduced to the islands of Prince William Sound, from which they spread to the adjacent mainland, Kodiak Island, and the Yakutat Bay area, although they no longer are present in the Yakutat Bay area (MacDonald and Cook 1994). Moose were able to gain access to southeastern Alaska only in places where major rivers penetrated the Coast Range in the Stikine, Taku, and Chilkat river valleys; in the early 1900's moose also were able to enter the Yakutat forelands when a glacier that had blocked the Alsek River valley receded. Transplantations also established moose in the Copper River delta area and at Berners Bay north of Juneau (Burris and McKnight 1973).

Before Alaska's statehood in 1959, the Alaska Game Commission, often under the direction of the Territorial Legislature, engaged in wildlife transplantations to establish game species and furbearers and their prey on several of this region's islands where these species had not occurred. Consequently, the muskrat, beaver, marten, mink, snowshoe hare, and red squirrel were introduced to the Kodiak Island group; the marten to Baranof, Chichagof, and Prince of Wales islands in southeastern Alaska; the red squirrel to Baranof and Chichagof islands; the mink to Montague Island; the snowshoe hare to Admiralty, Prince of Wales, and the Shumigin islands; and the hoary marmot to Prince of Wales Island. Most but not all of these transplanted species established populations. Established nonindigenous species include elk on Afognak Island and on a few islands in the central Alexander Archipelago and European rabbits on Middleton Island in the Gulf of Alaska. Raccoons were introduced on the western coast of Prince of Wales Island, but the population subsequently died out (Burris and McKnight 1973).

With some exceptions, mammal populations throughout the ecoregion are at their historical levels and are characterized by natural fluctuations often tied to interannual weather variations. For example, the periodic heavy snow accumulation that occurs throughout the entire deer range causes heavy losses of deer in winter and subsequently depresses deer abundance for several years. In places where gray wolves and black bears are present, predation may further delay recovery. The number of wolves tracks the fluctuations in deer density but with lags of a few years (Klein 1995). Gray wolves have been most abundant in the southern islands of the Alexander Archipelago where deer densities are also high. These wolves are classified as an endemic subspecies.

Although hunters harvest all ungulate species for either subsistence or recreation, annual harvests in the region are constrained by seasons and bag limits. However, because of the vast areas, difficulty of access to many places, and the relatively small number of hunters, the actual kill (as many as 25,000 deer) is considered well below the annual population recruitment, except in areas close to human population centers (Alaska Department of Fish and Game 1994b).

Bears are hunted primarily by resident and nonresident trophy hunters, although some hunting of bears is for subsistence by Alaska Natives. The region's annual bear harvests may exceed 400 brown bears and 900 black bears (Alaska Department of Fish and Game 1994b); the state and federal agencies responsible for bear management consider these harvests sustainable throughout most of the region.

Interior Boreal Forest

High mountains in much of interior Alaska shelter the interior from the moist maritime air that occurs in the south and the cold arctic air characteristic of the north (Fig. 2). The climate is continental, with cold and long winters and short and warm summers. Seasonal climatic changes are rapid. Altitude strongly influences plant growth, the presence and composition of forests, and the presence and extent of permafrost. Fire, mostly caused by lightning, is a natural feature of the ecology of the interior boreal forest ecoregion. The vegetation of the interior boreal forest is a complex array of plant communities shaped by fire, soil temperature, drainage, and exposure (Fig. 5). Permafrost is mostly continuous in the northern portion of this region, except in riverbeds, beneath lakes, and on steep, south-facing bluffs. In the central and southern portions of the region, permafrost is discontinuous, absent on most southern exposures, and irregularly present adjacent to rivers and lakes. In the lowlands of the broad interior valleys, permafrost restricts drainage and accounts for the presence of extensive wetlands that form a complex of marshes, shrub thickets, small ponds, and forested islands (Selkregg 1974–1976).

Vegetation

The conifers of the boreal forest are white spruce on well-drained floodplain soils, on uplands, and on south-facing slopes where seasonal thaw is deep. Trees with diameters of 69–90 centimeters at breast height and heights of 30 meters generally occur on floodplain islands that have not been burned for a long time. Most white spruce stands in floodplains and on uplands consist of smaller, younger trees 12–15 meters tall. Black spruce grows in lowlands and on north-facing slopes where the annual thaw is shallow and permafrost is close to the surface. Black spruce stands are the most widespread of all stand types. The largest black spruce trees reach diameters of 18 centimeters at breast height and heights of 17 meters, but many are no larger than 10 centimeters in diameter at breast height and 9 meters tall.

A broad-leaved deciduous forest of quaking aspen, balsam poplar, and Alaska paper birch is prominent on well-drained uplands. The largest quaking aspens reach diameters of 25–36 centimeters and heights of 18 meters. In contrast, floodplain forests are composed of balsam poplar, white spruce, some paper birch mixed with mountain alder, and several species of willow. Balsam poplars can attain diameters of 40 centimeters and heights of 30 meters. Primary succession on the *point bars* (areas where rivers deposit silt, sand, and gravel on the inside of their curves) of large rivers passes through open-shrub willows and poplars and closed-shrub alder–poplar, two stages of dominant vegetation, before white spruce is well established (Van Cleve et al. 1991).

Fig. 5. The interior boreal forest region is characterized by a diversity of vegetation types, including coniferous and deciduous forests broken by river floodplains and marshy wetlands.

Courtesy D. R. Klein, USGS

Forest types are mixed, with species composition determined by steepness of slopes, *aspects* (the cardinal direction a slope faces), and fire histories. Natural wildfires, which are a critical component of the boreal forest environment, occur about every 50–70 years; stands older than 170 years are rare (Van Cleve et al. 1983). A complex fire history has created a patchwork landscape of stands of different species and ages. Fires occur often enough that the landscape is continually returned to successional forms, and true climax forest is seldom reached on uplands. Fires tend to remove white spruce, which is first replaced by paper birch and quaking aspen; eventually white spruce and then black spruce slowly return to these stands (Van Cleve et al. 1991). Although white spruce is fire-sensitive, the common broad-leaved quaking aspen and paper birch can quickly regenerate by root suckers, leading to pure stands of these species. Because of the ability of black spruce to reproduce vegetatively and because of its retention of seeds in persistent cones, this species is better adapted to withstand fire than white spruce. Postfire successional changes are more marked among understory shrubs and the moss and lichen ground cover than among trees in black spruce stands. The result of this fire history is an array of forest type combinations of white spruce and paper birch or quaking aspen. Extensive, nearly pure stands of paper birch form on floodplains; on warm, dry slopes, stands of quaking aspens persist for long periods without being replaced by spruce.

Forest understory varies greatly with stand density and the amount of moisture on the forest floor. Common tall shrubs found in various mixtures in white spruce forest are green alder and Bebb willow; low shrubs include Labrador tea, alpine blueberry, and especially lingonberry. In mixed stands on floodplains, horsetails carpet the forest floor, with feathermosses and foliose lichens prominent in the moist habitats. Shrubs are less important in quaking aspen forests, but highbush cranberry and prickly rose are common there. Mats of bunchberry, twinflower, and wintergreen are important shrubs, especially in mixed stands of paper birch and quaking aspen; in aspen stands in especially warm and dry settings, large patches of kinnikinnick develop.

Black spruce occurs with tamarack and even with paper birch, which more characteristically occurs on well-drained uplands. This forest type merges with bogs in which black spruce trees are scattered and stunted, forming a landscape called *muskeg*. The muskeg understory is characteristic of bogs—that is, rich in mosses, sedges (including tussock cottongrass), ericaceous shrub and subshrub taxa, and herbs such as roundleaf sundew. Treeless bogs, fens, and other wetlands are also common in this region.

Extensive deposits of sand and sand dunes were formed over some present-day interior boreal forest areas in the late glacial time. Many of these deposits were stabilized by forest cover, but others are exposed along river banks and deltas. The exceptional, extensive, active dune fields of the Great Kobuk Sand Dunes occur on the middle Kobuk River—where the oxytrope *Oxytropis kobukensis* is endemic—and on the Nogahabara Sand Dunes of the Koyukuk River, which is the sole Alaskan locality of *Carex sabulosa* (R. Lipkin, Alaska National Heritage Program, Anchorage, personal communication), a sedge of desert–steppe landscapes in Asia. This species is known from North America only from similar habitats in a few localities in the southwestern Yukon Territory, Canada (Porsild 1966). These unique landscapes and their plant complexes are protected because they are in national parks or national wildlife refuges.

Steppe vegetation, found on steep bluffs along the Yukon, Porcupine, Tanana, Matanuska, and Copper rivers, can be located and defined by its south-facing topographic aspect. The steepest portions of slopes are generally treeless, presumably because of drought and geomorphic instability. Stunted quaking aspens grow in a band on the more level ground that occurs at the crest of the slope between the upper edge of the steppe and the white spruce forest. Warmth-loving, drought-resistant grasses, sedges, forbs, and wormwoods are common steppe species (Murray et al. 1983).

Each steppe site can be thought of as a small island in a sea of forest. The steppe bluffs, especially in the upper Yukon River regions, are characterized by rare plant taxa and a restricted fauna of solitary bees (Armbruster and Guinn 1989). The vascular plants of these steppe bluffs—for example, the disjunct species American alyssum and the wormwood *Artemisia laciniatiformis*—occur only in the subarctic interior of Alaska and in the adjacent Yukon Territory. Other taxa of steppe sites in Alaska and in southwestern Yukon Territory are disjunct from taxa in dry grasslands in the western states, are subspecific variations of species that are more widespread farther to the south, or are narrowly restricted endemic species of eastern Alaska and the adjacent Yukon Territory and have no close relatives in Asia or elsewhere in North America. Researchers are exploring how these isolated plant communities became established on these bluffs and why they remain so restricted. The similarity of the modern bluff flora to pollen remains deposited at the peak of the last glacial maximum has led some researchers to postulate that the steppe vegetation type could be a relict of that time (Hopkins et al. 1982).

Freshwater Fishes

Twenty-two fish species occur in the fresh waters of the interior boreal forest region (Morrow 1980; Table). Humans favor the chinook, chum, and coho salmon; rainbow trout; sheefish; humpback and round whitefish; least cisco; arctic grayling; lake trout; northern pike; and burbot. These species support a growing recreational fishery of almost 500,000 angler-days on the region's fresh waters—about 19% of the total sport fishery in the state (Mills 1994). Commercial fisheries and subsistence and personal-use anglers harvest salmon (particularly chinook, coho, and chum) that are bound for major rivers of the interior boreal forest. Chum salmon are especially important to the Yukon River subsistence fishery, which harvested 188,825 fish in 1993 (Alaska Department of Fish and Game 1994a).

Throughout this region, almost all stocks of freshwater fishes targeted by sport anglers have declined, especially those stocks that are accessible by road. For example, lake trout and burbot stocks have been overharvested in most lakes that are accessible by road (Arvey 1993). Likewise, the density of arctic grayling has declined in the Tanana River drainage, especially in the Chena River, which once supported the largest sport fishery of arctic grayling in North America (Clark 1993). In the Susitna River drainage, the harvest of wild rainbow trout declined by 50% during the last 10 years despite a doubled harvesting effort (Rutz 1992). In recent years, chinook salmon returning to some interior rivers to spawn have failed to reach established population goals (Arvey 1993). Similarly, recent studies of northern pike populations in the Tanana River drainage indicate that exploitation rates of this species are higher than can be sustained by some populations (Arvey 1993). Harvests of whitefish species in the Tanana drainage increased steadily to a peak of 26,810 fish in 1986 but dropped drastically thereafter to a low of 739 in 1991. Fishing closures, reduced bag limits, and catch-and-release provisions were imposed on most of these fisheries in the hope of rebuilding or maintaining fish populations (Mills 1994).

Coho salmon *escapements* (that is, the successful return of fishes to their spawning grounds) have increased in some of the region's waters since 1985 (Bergstrom et al. 1992), probably because of commercial fishing restrictions in the lower Yukon River and because of the species' low natural mortality (Arvey 1993). Burbot populations in the Tanana River also seem healthy and do not seem to be overharvested, although the species is growing more popular with anglers (Arvey 1993; Evenson 1993a,b).

Waters in this region are stocked with fishes that include rainbow trout, arctic grayling, lake trout, arctic char, and coho, chinook, and sockeye salmon (Engel and Vincent-Lang 1992; Arvey 1993). Stocking, which currently provides about 65% of the fish harvested from the Tanana River drainage (Arvey 1993), is designed to ease harvest pressure on wild stocks and to diversify angling opportunities. Only species of direct importance to sport and subsistence fisheries of the region are monitored and then usually only in select systems, thus the status and trends of most other freshwater fish species are unknown.

The interior valleys of this region contain many rivers that originate from different water sources. Arctic grayling, which evolved and adapted to these different rivers, take advantage of the varied freshwater habitats, even though some rivers freeze solid in winter. According to Tack (1980), arctic grayling use glacier-fed systems mainly for overwintering or as migratory routes to other systems, spring-fed systems mainly for feeding but not for spawning or overwintering, bog-fed systems for spawning and feeding but not for overwintering, and large unsilted runoff systems for spawning, feeding, and overwintering. Arctic grayling are able to adapt to such a wide variety of aquatic habitats because they spawn in spring and early summer and because their eggs and larvae develop relatively quickly. This early and rapid development enables the young to leave the streams before the water becomes frozen and uninhabitable in winter.

The adaptations of fish species to different systems or to different parts of the same system have sometimes caused complex migrations to overwintering, spawning, and feeding sites (Armstrong 1986). Spring-fed systems like the Clearwater River delta provide ice-free spawning habitats for coho salmon, which typically spawn late in the year from mid-September to early November (Parker 1991). Such systems also provide ice-free habitats for young coho salmon, which remain as long as 3 years in the Clearwater River delta before going to sea as smolts.

The Alaska blackfish, an unusual fish in this region, occurs only in Alaska and eastern Siberia. Blackfishes are unique because they have a modified esophagus capable of gas absorption, which allows them to breathe air. This ability allows the fishes to live in small, stagnant tundra pools that are almost devoid of oxygen in summer and to survive in moist tundra mosses during extended dry periods until rain again fills the tundra pools (Armstrong

1988). Their great abundance and ease of capture make the Alaska blackfish an excellent subsistence fish for Alaska Natives, especially when other food is in short supply. One native name for the fish—*oonyeeyh*—literally means *to sustain life*.

Birds

Large numbers of breeding waterfowl summer on wetlands of the interior boreal forest, and thousands more pass through this region during migration. The number of trumpeter swans has increased since the early 1980's; similarly, tundra swans have expanded their breeding range into some parts of the boreal forest (Conant et al. 1991, 1993; Wilk 1993; Kessel and Gibson 1994). The interior boreal forest region is important for canvasbacks that winter on the Atlantic coast and for greater white-fronted geese that winter in the central United States.

Trends in sizes of duck populations are variable, however. The abundances of dabbling duck species (such as the northern pintail) and diving duck species (such as the scaups) have increased since 1977, although scoters (also a diving duck) are declining in abundance (Conant and Groves 1994). Wintering populations of waterfowl are small and restricted to a few duck species that inhabit the warmer, open waters used for cooling of utility plants (Christmas Bird Count data) or that inhabit spring-fed areas in rivers that retain some open water in winter (Ritchie and Ambrose 1992).

Although rock ptarmigan, willow ptarmigan, ruffed grouse, spruce grouse, and sharp-tailed grouse are hunted during fall and winter in the interior boreal forest, recent data on population-size trends of these resident species are incomplete. Reports from local residents of the region in 1994, however, indicated that the numbers of ptarmigan were low, whereas grouse numbers were moderate and increasing. Winter counts of ptarmigan and grouse in Fairbanks indicated that the sizes of local populations vary, generally on a several-year cycle that is possibly associated with the snowshoe hare cycle (Christmas Bird Count data) because the species share common food resources and because they are preyed upon by lynx. Thus, when the hare cycle is at a peak, lynx are abundant and prefer to eat hares, thereby reducing predatory pressure on ptarmigan, which then also reach peak levels.

Birds of prey have been closely monitored. Bald eagles that breed along major river systems have maintained relatively stable populations. Along the Tanana River, the population may even be growing (R. J. Ritchie, Alaska Biological Research, Inc., Fairbanks, personal communication). Breeding peregrine falcons on the Yukon River have recovered from their low population sizes during the 1970's and continue to increase in numbers, although populations on the Tanana River have not completely recovered, perhaps because of the growing presence of humans there (Ambrose et al. 1988; White 1994). The American peregrine falcon remains listed as endangered but may soon be moved to threatened status. No data are available on population-size trends of resident northern goshawks, a species recommended for listing in other parts of its range and currently a species of concern (U.S. Fish and Wildlife Service 1996). Population-size trends are also unknown for small birds of prey such as the sharp-shinned hawk, American kestrel, and merlin. Sizes of owl populations seem to fluctuate with prey availability, but trend data are not available.

Population-size trends are largely unknown for shorebirds and gulls, which are present in the interior boreal forest only during summer (May–September). Breeding shorebirds such as the common snipe and the yellowlegs nest in forested bogs, and the spotted sandpiper nests on gravel bars of large rivers. Few data are available on breeding populations of gulls, whose numbers seem to be greater near human settlements.

Passerine populations in the interior boreal forest primarily include migrant breeders and a few residents. Data for evaluating long-term population-size trends for passerines are generally lacking, except in the Fairbanks area. The abundances of Swainson's thrushes, yellow warblers, orange-crowned warblers, and white-crowned sparrows in the Fairbanks area seem to have declined since about 1977 (Kessel and Gibson 1994). Four other species are of special concern because of their declining population trends throughout North America: the olive-sided flycatcher, gray-cheeked thrush, Townsend's warbler, and blackpoll warbler (Alaska Department of Fish and Game 1994c). Researchers believe that the declines of these Neotropical migrants are a consequence of the loss of their wintering habitats in Central America and South America (Greenberg 1992). Wintering populations of resident passerines have been monitored since 1964 in the annual Christmas Bird Count in Fairbanks. The common raven, gray jay, boreal chickadee, black-capped chickadee, and redpolls (common and hoary) are the most common species in winter and show no distinct population-size trends.

Mammals

During the last Ice Age, most of interior Alaska north and west of the Alaska Range and south of the Brooks Range was not glaciated. Interior Alaska and similar, unglaciated areas in the adjacent Yukon Territory and north of the Brooks Range were part of the Beringian Refugium (Hopkins et al. 1982). The unglaciated area included a broad land connection with Asia and adjacent parts of northeastern Siberia that allowed for the presence of many Asian species. The steppelike environment of Beringia supported large numbers of grazing mammals, such as the woolly mammoth, steppe bison, and Lambe's horse, as well as their predators, including the lion, dire wolf, and short-faced bear. Most of these species went extinct when the climate and vegetation changed as glaciers receded about 10,000 years ago (Pielou 1991), with only the badger and black-footed ferret surviving in the Great Plains area of the United States.

As the cold climate ameliorated and the ice sheets melted, winds that had been generated by the adjacent ice sheets subsided, precipitation increased, and the present boreal forest became established throughout the region (Pielou 1991). In interior Alaska, the present boreal forest is an extension of the northern boreal forest that occurs from the Atlantic coast of Canada across the northern portion of the continent into Alaska. With few exceptions, mammals characteristic of this major biome are similar throughout its extent across Canada and the northern tier of states to Alaska (Pielou 1991).

The interior boreal forest is interrupted by several mountain complexes that support typical montane mammal species, some of which (for example, the caribou, grizzly bear, and wolverine) occupy the adjacent forest ecoregion during part of each year. The mammalian faunas of the boreal forest and the alpine habitat of the interior mountain complexes overlap in the transition zones between the two regions, where mammalian biological diversity and productivity of these faunas are consequently particularly high. Denali National Park and Preserve, which is internationally known for the diversity and abundance of its mammalian species, is an example of the ecotonal interface of boreal forest and montane habitats (Selkregg 1974–1976).

The abundances of mammals in the largely intact interior boreal forest community fluctuate primarily in response to variations in snow depth in winter and to plant succession after wildfires. Occasional winters with deep snows are detrimental to moose survival, especially to young-of-the-year animals, but are beneficial to small mammals that live beneath the insulating snow cover (Pruitt 1967). As a result, predators and scavengers are also affected by the weather-induced fluctuations of prey abundances and availability. Most mammal populations in the interior boreal forest region are healthy, and their reproduction rates are at moderately high

levels. The abundance of snowshoe hares, whose population levels fluctuate over a cycle of about 10 years, is increasing slowly from the low of its present cycle. The abundance of the lynx, a specialized predator of the hare, is also low.

The extent of trapping of all furbearers remains low because of low fur prices. Past localized reductions of some furbearers resulted from intensive trapping stimulated by high fur prices.

Interior Mountains

The interior mountain biogeographical region (Fig. 2) is a complex of mountain ranges characterized by extreme physiographic variability. Wide differences in elevation, slope steepness, and exposure exist locally and between major mountain masses. The composition of vegetation throughout the region, although primarily dominated by alpine forms, reflects the variability of the terrain, and at lower elevations includes elements from the boreal forest (Fig. 6). The sizes of plant communities and habitats for animals are small, which means that the home ranges of a lot of animals include many habitat types.

The climate of this region is also variable. The seaward sides of the mountain ranges receive rain or snow from moist oceanic air masses, whereas the interior sides are arid. Plant growth occurs earlier in the year on south-facing slopes and at lower elevations than it does on north-facing slopes and at higher elevations. This variability in the timing of plant growth creates a complex pattern of vegetation in different annual growth stages and of different quality and quantity throughout the year. Consequently, large herbivores in search of food may move extensively over such a landscape.

The treeline varies throughout the region from about 900 meters in the Talkeetna Mountains to about 300 meters in the southern drainages of the Brooks Range, north of the Arctic Circle. Few plants and animals occur above about 2,000 meters in elevation. In this highest region, the high rate of erosion and the glacial recession that characterize the rugged mountains expose mineral soil to pioneering plant species that initiate a succession of plants and animals (Viereck 1979).

Vegetation

Alpine tundra generally occurs on land above 750 meters in elevation. The transition from forest to tundra varies from abrupt, where the terrain is steep, to a gradual shift from closed forest to open woodland and then tundra

Courtesy D. R. Klein, USGS

Fig. 6. The interior mountains support complex and diverse habitats resulting from variation in elevation, substrate, slope, and exposure.

in places where changes in elevation are more gentle. The greatest penetration of forest into tundra occurs in sheltered riparian stands in the narrow headwater valleys. Throughout most of interior Alaska, the predominant tree species at treeline is white spruce (Viereck 1979). Balsam poplars, though, also reach the treeline and even occur as isolated groves well beyond treeline (Viereck and Foote 1972). Vegetation at the transition from forest to tundra is frequently a zone of alder and willow or birch and willow shrubs. In the birch–willow zone, dwarf birch dominates. Outliers of paper birch contain numerous hybrids between tree and shrub birches.

Alpine tundra is a complex of low shrubs, grasses, sedges, and forbs that develops in response to topographic relief, slope, aspect, and surface stability, which in turn determine snow accumulation, soil temperature, depth of thaw, amount of moisture, and soil texture and nutrients. Perhaps the most widespread tundra type in this region is formed around species of mountain avens, plants that occur with sedges or ericaceous prostrate shrubs in extensive closed communities and with lichens and a variety of forbs in open communities.

Uplands with sufficient relief for treeless slopes and summits occur in huge areas of the interior, with some uplands extending beyond this region west to the Bering Sea. The Yukon–Tanana upland extends roughly east–west from the eastern boundary of Alaska and Yukon Territory (Canada) westward to the

Koyukuk River. The upland is a complex of ranges that includes the White Mountains, Ray Mountains, and Kokrines Hills. Isolated mountain groups rise above the upland surface, reaching 1,200 meters or higher. Similarly, the Nulato Hills, a belt of mountains between 800 and 1,200 meters, flank Norton Sound and the Kuskokwim Mountains and extend southwest from the interior to the coast at Bristol Bay. Except for local areas of glaciation, these ancient surfaces remained ice-free throughout the last glacial maximum (Hamilton et al. 1986). The well-formed cirques at the head of many valleys were formed during earlier glaciations. *Solifluction* (that is, soil creep caused by forest action) terraces and lobes are common on many slopes, and stone circles and nets are common on many summits and saddles. (Stone circles and nets are the patterning of the ground surface by frost sorting, which results in stones or coarser materials being moved into lines around finer materials, resulting in circle or net-like patterns.) These geomorphic features contain surface irregularities that greatly enrich the diversity of the resident plant communities; such features exist only in places where cold-climate geomorphic processes have persisted undisturbed for long periods

The bedrock of mountains and uplands is too complex for generalization, but rock types have had an important effect on the region's flora. The Brooks Range and White Mountains of the Yukon–Tanana upland, for example, have limestone backbones and consequently are rich, calcareous habitats. Certain plants in these habitats are absent from the more acid environments of granitic rocks that are widespread in the Chugach Mountains, Wrangell Mountains, and parts of the Alaska Range.

Freshwater Fishes

Several lakes in the interior mountains of Alaska support a small but unique assemblage of freshwater fishes with a known total of ten species (Morrow 1980; Table). Of these, the arctic grayling, lake trout, and burbot are the most sought after by anglers (Mills 1994). About 30,000 angler-days were spent at these interior lakes during 1993 (Mills 1994); this is too many angler-days for waters of such small size with relatively low fish abundances and reflects the excessive fish harvest in some lakes.

Slimy sculpin and round whitefish are extremely important food sources for the predatory lake trout and burbot, whose numbers have declined in lakes with road access (for example, Lake Louise, Lake Paxson, and Summit Lake in the Glennallen area and the Tangle Lakes in the upper Tanana drainage; Lafferty et al. 1992; Szarzi 1992). Lake trout typically occur in deep mountain lakes, where fishes grow slowly; in Alaska lake trout may live longer than 50 years (Burr 1987). These fish usually spawn for the first time when they are 7–14 years old and may not spawn every year. This species' spawning behavior and its allure to anglers make it vulnerable to excessive harvest (Szarzi 1992). The catch and release of lake trout probably increases the mortality of the species because certain types of hooking gear have raised mortalities of lake trout to 71% (Loftus et al. 1988).

Many lakes and streams in the interior mountains freeze severely in winter, often to the bottom. Consequently, habitat becomes extremely limited in winter, and fishes may become concentrated in small areas of rivers and at the bottom of lake basins (Burr 1987). Fishes of this region are in a delicate balance with other species and their habitats. Successfully maintaining these fish populations requires a thorough understanding of their biology.

Birds

Mountain lakes support small numbers of breeding waterfowl (primarily ducks) during summer. Harlequin ducks nest along fast-moving mountain streams and rivers. Although no data on population-size trends are available for this species in the interior mountains, it is a species of concern (U.S. Fish and Wildlife Service 1996), primarily because of concerns over population declines in the coastal regions, particularly Prince William Sound. Pacific loons, common loons, and red-throated loons also breed in mountain lake systems with abundant prey (invertebrates and fishes) for these species.

White-tailed ptarmigan are common local breeders in the Alaska Range and in mountains to the south, and willow and rock ptarmigan are residents of the mountain ranges throughout the interior. The latter two species migrate vertically in winter from higher slopes to lower slopes, where willows are abundant. Little is known about population-size trends in these species.

Golden eagles commonly breed in the interior mountain region, especially on south-facing slopes in the Alaska Range (C. McIntyre, National Park Service, Anchorage, Alaska, personal communication). Gyrfalcons and peregrine falcons nest in some mountains, particularly where suitable cliff-nesting habitats are available. Merlins are also common breeders in the interior mountains (Ritchie and Ambrose 1992).

Upland-nesting shorebirds that are locally common breeders in some interior mountains include surfbirds, American golden-plovers, whimbrels, wandering tattlers, Baird's sandpipers, and least sandpipers. Upland sandpipers are a less common shorebird in the region, but evidence suggests that they breed in small numbers in some interior mountains (Ritchie and Ambrose 1992); this species was classified as a species of ecological concern by the Alaska Natural Heritage Program (West 1991). However, data on population-size trends of all shorebird species are lacking.

Gulls generally are uncommon in the interior mountains, except where substantial wetlands are present that support nesting mew gulls and sometimes nesting herring gulls. Small numbers of long-tailed jaegers also breed in the interior mountains (Gabrielson and Lincoln 1959).

Similarly, limited data are available on the population status or on population-size trends of passerines, which commonly breed in interior mountains in summer (Cooper 1984). Almost all passerines are present only to breed during summer because winter in the mountains is too harsh. American dippers, however, are residents of mountain streams with open water throughout winter. Some species that breed in alpine regions of the interior mountains, such as the snow bunting and Lapland longspur, also breed farther north in the arctic tundra. However, other species (for example, the golden-crowned sparrow, water pipit, and horned lark) breed primarily in the interior mountains. At least two species of Asiatic origin, the northern wheatear and arctic warbler, breed in Alaska's interior mountains.

Mammals

Several mammals of the interior mountain region, including the Dall sheep, collared pika, arctic ground squirrel, and singing vole, do not occur in other states. These species survived the last glaciation in the interior Alaska refugium and are adapted to the short summers and long winters of their mountain habitats. Their ranges do not extend southeastward beyond the interior of northwestern Canada. Dall sheep occur in the Brooks Range mountains of the Arctic National Wildlife Refuge to nearly 70° north, where the sun remains below the horizon for 3 months in the arctic winter. Other common arctic–alpine species in this region include the caribou (Fig. 7), hoary marmot, wolverine, grizzly bear, tundra vole, and brown lemming. The more ubiquitous species that are important components of this mountain ecosystem include the gray wolf, red fox, snowshoe hare, lynx, coyote, moose, northern red-backed vole, ermine, least weasel, and shrews (Hall and Kelson 1959).

An interesting relationship exists between the three canid species of the interior mountains—the coyote, red fox, and gray wolf. The coyote, which is a more recent species in Alaska, suppresses the abundance of the red fox, but wolves in turn suppress coyote populations. In interior Alaska, coyotes are most common close to human settlements and major road and railroad transportation corridors where gray wolves are less frequently present. In areas that are remote from human settlements, wolves are more common and seem to suppress the abundance of coyotes.

The high diversity of ungulate species and the high productivity of other vertebrate wildlife throughout the interior mountains provide a large prey base for gray wolves (Alaska Department of Fish and Game 1994c). The number of wolves is moderate to high in this region, and populations are productive. Control of wolves by the Alaska Department of Fish and Game in 1994 reduced the number of wolves in the northern portions of the central Alaska Range by an estimated 40% to 50% (P. Valkenburg, Alaska Department of Fish and Game, Fairbanks, personal communication). This reduction should be temporary and of little consequence to the long-term status of wolves in the area.

Mammal species and their habitats in the interior mountains biogeographic region have been, for the most part, far removed from the effects of most human development. Widely scattered exceptions are regions with small-scale gold mining, the highways and the Alaska Railroad that transect the Alaska Range, the highways that penetrate the Tanana Hills, and

Courtesy D. R. Klein, USGS

Fig. 7. The interior mountains offer productive habitats for mountain sheep, moose, and caribou, shown here in late winter.

the Dalton Highway that runs through the Brooks Range (Foster 1985). Large representative habitat components of the region are protected in the federal reserve lands of the National Park Service, U.S. Fish and Wildlife Service, and the Bureau of Land Management. Hunting is restricted to traditional subsistence users in some of the national parks in Alaska (Foster 1985).

Most hunting in the interior mountains is for subsistence, sport, and trophies. Regulations that govern hunting on state and federal lands are conservative (Alaska Board of Game 1994; U.S. Fish and Wildlife Service 1994a). For example, most sport and trophy hunters are restricted to hunting only males of mountain sheep and moose, and female bears accompanied by young are protected. Registration, permit hunts, and the length of hunting seasons are regulated to restrict harvests to quotas based on surveys of population status and levels. Because of these hunting restrictions and the favorable condition of habitats in the region, most resident mammal populations are considered healthy and at moderate to high population levels. Exceptions are small caribou herds in the northern foothills and mountains of the Alaska Range that experienced low recruitment for several years. Their estimated population sizes in 1994 were 2,050 in the Denali herd, 4,350 in the Delta herd, 800 in the Chrisana herd, 1,346 in the Tonsona herd, and 750 in the Mentasta herd (Valkenburg, personal communication). Recent low recruitment in the much larger (about 22,000 animals in 1994) Fortymile herd in the Tanana Hills slowed this population's increase from a population low of 6,000 animals in 1977 (Valkenburg, personal communication).

Maritime Tundra of Southwestern and Western Alaska

The maritime tundra that dominates throughout southwestern and western Alaska (Fig. 2) is the product of the cool climate generated by the cold Bering Sea waters. There is a gradient, however, from the more humid and milder conditions that prevail in Bristol Bay and the coastal Alaska Peninsula next to the Aleutian region, to the Seward Peninsula where the adjacent seas may remain ice-covered for 8 months of the year. A gradual transition from maritime to arctic tundra occurs in the region of Kotzebue Sound, and a transition from maritime to alpine tundra occurs in places where mountains extend into the region. The extensive coastal wetlands throughout the region dominate the broad expanse of the ancient delta of the Yukon and Kuskokwim rivers (Selkregg 1974–1976; Fig. 8).

Fig. 8. The wetlands of the Yukon and Kuskokwim River deltas provide some of the most important nesting areas for waterfowl and shorebirds in North America.

Vegetation

The climate and plant life of this region are transitional between the arctic tundra and the subarctic or maritime tundras of the Alaska Peninsula and Aleutian Islands. Furthermore, islands of the Bering Sea are not floristically similar. The vegetation of Saint Matthew and Saint Lawrence islands has affinities with arctic tundra vegetation on the mainland to the north (Young 1971), whereas the vegetation of the Pribilof Islands is clearly distinct and allied more closely with vegetation of the Aleutian Islands (Hultén 1968).

The Asian character of much of Alaska's flora is most evident in the Bering Sea region. This influence becomes less conspicuous

eastward in Canada, and many species common in Alaska do not occur east of the Mackenzie delta (Hultén 1968). The Asian component in Alaska's flora is, of course, a consequence of the state's proximity to Asia (as little as 75 kilometers separate Cape Prince of Wales, Alaska, from Cape Dezhnev, Chukotka) and the geologic history of the region, which included a land bridge between Asia and America.

The distribution of so many vascular plants that are restricted to Alaska and Chukotka prompted Eric Hultén (1937) to propose the concept of Beringia. During glacial maxima when the Bering Land Bridge was exposed, Alaska was essentially the easternmost extension of Asia because ice sheets blocked access to the rest of North America. The emergent Bering Sea floor and the largely unglaciated lands east and west were a glacial refugium. Beringia—the refuge area during glacial maxima when the Bering Land Bridge was exposed—was a route of dispersal for plants. Consequently, it was the place where speciation occurred and where regional endemics developed. Among endemic Beringian taxa are a sagebrush (*Artemisia senjavinensis*), Wright's alkaligrass, Chukchi primrose, and Krange's sorrel (Murray and Lipkin 1987). Two other rather narrowly restricted endemics of Alaska, boreal primrose and Bering douglasia, are mostly confined to the maritime tundra region. In postglacial time, Beringia was an important source of plants for revegetation of the Arctic as ice sheets and glaciers receded and exposed the land to recolonization (Hultén 1937).

The Seward Peninsula is topographically diverse; rugged ranges there have peaks that rise to 1,460 meters from their base at or near sea level. Acidic volcanic rock in some areas and limestone in others support separate and distinctive flora and vegetation. Shrublands of willow, alder, and dwarf birch are well developed on low rolling hills in the southern portions of the uplands. To the north, the same terrain is covered for many kilometers by tussock tundra, which is broken only by low ridges and water tracks. Tussock cotton-grass is the dominant plant on the Seward Peninsula, but in some areas the Bigelow sedge is abundant and also can form tussocks. In spaces between tussocks, the combination of permafrost close to the surface and fine-grained soils accounts for high soil moisture and the frequent appearance of frost scars. Low shrubs such as dwarf birch, Labrador tea, blueberry, and lingonberry are common in tussock–heath phases.

On the Seward Peninsula, lava fields of recent origin provide unusual sites for plants. Groves of balsam poplar and other boreal forbs and ferns, which are common in the boreal forest farther east and south but unusual here, occur in the immediate vicinity of hot springs, presumably because soils are suffused with warm mineral waters. Clusters of pingos and thermokarst lakes (sites of erosion and subsidence by thawing of permafrost) occur in the interior lowlands, which were formed by large rivers, and may also occur in association with isolated groves of balsam poplar where other trees are absent. A Coastal Plain with wet meadows and beaches forms the perimeter of the peninsula, especially in the area of Cape Espenberg.

The central portion of the Bering Sea region is remarkable for its huge extent of low-lying coastal tundra over combined deltas and estuaries of the Yukon and Kuskokwim rivers. The diurnal tidal range in this area is about 2 meters, and the landscape rises so little in elevation from the coast that the influence of tides extends up the rivers to 55 kilometers inland. Storm surges of 1 meter or more in spring and fall allow saline water to penetrate rich wetlands as far as 16 kilometers inland (Kincheloe and Stehn 1991).

Important taxa in the sedge–graminoid meadows where flooding occurs are the Ramenski sedge, loose-flowered alpine sedge, Lyngby sedge, reedgrass, and the forbs silverweed cinquefoil and low chickweed. Pingos 3–30 meters tall in lakes and wet meadows provide a sharp contrast in topography and vegetation to the surrounding terrain (Burns 1964). Bluejoint and the low shrub *Beauverd spiraea* can be codominants. This microrelief provides nesting and denning habitat for several bird and mammal species.

Sandy beaches are common in the maritime region, some of which are associated with dune fields, as on Nunivak Island where dune vegetation is heavily grazed by introduced muskoxen. Mudflats support open communities of halophytic grasses, sedges, and forbs (for example, creeping alkaligrass, Hoppner sedge, seabeach sandwort, and oysterleaf) that are typical of the strand zone in this and other regions to the south and north. The sandy beaches are dominated by beach ryegrass and forbs such as beach pea and seaside ragwort. In places where dunes formed, strong floristic differences exist between plants on prominences and those in depressions, and between plants on dunes and those on backslopes. The taxa are derived from the tundra immediately inland. South of the Yukon–Kuskokwim delta, coastal headlands and uplands of the Kuskokwim Mountains reach the coast. These uplands are covered with tundra and have many of the same characteristics as interior uplands, including topographic, geologic, and floristic similarities.

Freshwater Fishes

Thirty-four fish species occur in the fresh waters of the maritime tundra region (Morrow 1980; Table). The species most used by humans are the sheefish, whitefishes, arctic grayling, arctic char, Dolly Varden, rainbow trout, northern pike, Alaska blackfish, and five salmon species—sockeye, coho, chinook, chum, and pink. All salmon species support a commercial fishery in the region's marine waters that in 1994 yielded more than 39 million fishes at a value of about $152 million (Alaska Department of Fish and Game 1995). Salmon and other freshwater fish species also provide subsistence for the region's residents. In 1993 about 770,000 salmon were taken, making up more than 80% of the total subsistence catch for the state (Lean et al. 1993; Alaska Department of Fish and Game 1994a). In addition, the region supports a growing sport fishery that now consists of more than 150,000 angler-days, or about 6% of the state's total sport fishery effort (Mills 1994).

In the northwestern part of the maritime tundra region, fresh waters are subject to severe freezing in winter. Here, springs are important to the overwinter survival of freshwater fishes (DeCicco 1993a). The region's anadromous and freshwater resident fishes and their eggs provide food for a diversity of mammals, birds, and other fishes. The arctic chars of the Wood River Lakes, for example, depend at times on sockeye salmon smolts for food (McBride 1980). Similarly, in Bristol Bay the eggs of sockeye salmon provide a food source for young rainbow trout, and decaying salmon carcasses provide nutrients for aquatic invertebrates that are, in turn, food for trout. Russell (1977) presented evidence that the condition factor (length–weight relation) of rainbow trout was significantly lower in years of low sockeye salmon escapement than in years of high salmon escapement. Chemical nutrients derived from sockeye salmon carcasses are critical for the productivity of these ecosystems and perhaps even for subsequent generations of sockeye salmon (Kline et al. 1993). The famous opportunities for viewing brown bears at the Brooks and McNeil rivers are only possible because of the seasonal concentrations of bears feeding on spawning salmon (Quinlan et al. 1983). The escapements of salmon to fresh waters of this region have been strong, and most stocks seem to be maintaining their densities. However, in the Norton Sound area stocks of chum salmon declined from 1987 to 1993, which is a cause for concern (Alaska Department of Fish and Game 1994a).

In 1990 the Alaska Board of Fisheries imposed catch-and-release regulations for fly fishing in several waters of this region. Since then, the abundance of rainbow trout in these waters seems to have remained stable (Dunaway 1993). In the Kobuk and Selawik rivers, sheefish are harvested for commercial purposes, subsistence, and sport. These fishes are often harvested throughout the year and are possibly overharvested (Arvey 1993). On the Seward Peninsula, northern pike support a subsistence fishery, but sport fishing of this species has also increased (Burkholder 1993). Arctic grayling populations on the Seward Peninsula may be declining. More restrictive bag limits on arctic graylings were imposed in 1988, and in 1992 the Nome River was closed to fishing of arctic grayling to allow the recovery of this population (DeCicco 1993b).

The status and trends of most other freshwater fish species are not known. Only species of direct importance to commercial, subsistence, and sport fisheries of the region are monitored and then usually only in select systems. The residents of most villages in this region depend on anadromous and freshwater fishes for subsistence: this region accounts for more than 70% of the total subsistence harvest of salmon in Alaska (Alaska Department of Fish and Game 1994a), and more sheefish and Dolly Varden are taken here for subsistence than in any other region.

Runs of sockeye salmon into the Bristol Bay area are the single most valuable stocks of salmon in Alaska. In 1994 the catch of sockeye salmon in Bristol Bay was valued at more than $136 million, or about 30% of the entire value of the harvested salmon in Alaska in that year (Alaska Department of Fish and Game 1995). Sockeye salmon contribute significantly to the biological diversity in this area; the species is probably indirectly responsible for the famous wild rainbow trout stocks that also occur here because young rainbow trout feed heavily on sockeye salmon eggs, which likely improves the growth rate of the rainbow trout. The rainbow trout support an annual multimillion-dollar recreational industry (Dunaway 1993). The region is well known for producing trophy-sized arctic grayling, sheefish, northern pike, rainbow trout, and Dolly Varden. The largest Dolly Varden (nearly 9 kilograms) in the state was caught in this region in the Noatak River (Arvey 1993). The sheefish in this region are as large as 27 kilograms (Alt 1987). Dolly Varden of the Seward Peninsula and northwestern Alaska have migrated between unconnected freshwater drainages via the adjacent sea. Some marked Dolly Varden migrated more than 1,600 kilometers, even from Alaska to Russia (DeCicco 1992).

Birds

The maritime tundra of the Yukon–Kuskokwim River delta of western Alaska is one of the nation's most important nesting areas for geese; it supports large populations of brant, cackling Canada geese, emperor geese, and greater white-fronted geese (Kessel and Gibson 1994). The populations of all of these species underwent significant declines from the mid-1960's through the mid-1980's as a consequence of subsistence hunting in spring on the breeding grounds and, except for emperor geese, of hunting on the wintering grounds. Late snowmelt and heavy predation by foxes during nesting and rearing of young have also contributed to the decline and continued suppression of populations (J. Sedinger, University of Alaska, Fairbanks, personal communication). Recently, populations of Canada geese and white-fronted geese rebounded significantly (Platte and Butler 1993; Kessel and Gibson 1994; Schmutz et al. 1994). Most of the Pacific Flyway brant population stages in the fall on the Izembek Lagoon on the Alaska Peninsula before making a transoceanic migration to wintering grounds in Mexico (Reed et al. 1989). (Staging birds are migratory species that gather in certain areas and usually feed there to gain weight and energy before moving on to the next migratory segment.) Large numbers of ducks, tundra swans, and sandhill cranes also nest on the coastal tundra of western Alaska, particularly on the Yukon–Kuskokwim River delta. The number of spectacled eiders that breed in this area was large but has declined dramatically during the past 20 years (Kertell 1991; Stehn et al. 1993). The spectacled eider has been listed as threatened since 1993, and Steller's eider, which formerly occurred in much smaller numbers, was federally listed as threatened in 1997 (U.S. Fish and Wildlife Service 1996).

Birds of prey are relatively rare in this area, although the *pealei* subspecies of peregrine falcons is common around seabird colonies. Population-size trends of all birds of prey are poorly known but are thought to be stable (Schempf 1989). The number of bald eagles is smaller in western Alaska than in areas to the east (Kessel and Gibson 1994).

The large numbers of shorebirds that breed on coastal maritime tundra in western Alaska include the entire world's population of black turnstones and most of the world's population of bristle-thighed curlews (West 1991; Handel and Gill 1992). The bristle-thighed curlew is classified as a species of concern by the U.S. Fish and Wildlife Service (1996). In spite of the abundance of shorebirds in this region, population-size trends for most of these species are unknown. Gulls and jaegers also breed on coastal tundra but, as for most shorebirds, estimates of their population sizes and trends are unavailable.

Passerines breed on the coastal tundra and on the Bering Sea islands, but species richness on the Bering Sea islands is low. McKay's bunting is endemic on several Bering Sea islands, commonly breeds on Saint Matthew and Hall islands, rarely breeds on Saint Lawrence and the Pribilof islands, and winters on all the islands just listed and in coastal western Alaska (Kessel and Gibson 1978). The species is not endangered, but its limited distribution makes it a species of ecological concern (West 1991). Some species of Asiatic origin (for example, the arctic warbler, red-throated pipit, and yellow wagtail) also breed during summer in western Alaska.

Mammals

The mammalian fauna of this region is composed of shared elements from the boreal forest (the muskrat, northern red-backed vole, tundra vole, and red fox) and from the arctic tundra (the Greenland collared lemming, arctic ground squirrel, and arctic fox). The number of arctic foxes fluctuates widely in response to prey abundance. In turn, predation by arctic foxes on eggs and young of brant in the Yukon–Kuskokwim delta is considered a major factor in the alarming decline of brant in the late 1980's and in the species' subsequent slow recovery (Sedinger, personal communication).

Species that have been absent from much of the area in the recent past include the moose, caribou, snowshoe hare, lynx, beaver, coyote, and gray wolf. Moose, however, became established in recent decades on the Seward Peninsula and are extending their distribution in riparian habitats downstream along the Yukon and Kuskokwim drainages. Human residents in these areas are cooperating with wildlife management authorities by restraining subsistence hunting of this species until viable populations are established. Like moose, beavers and snowshoe hares are also expanding their ranges onto the Seward Peninsula and the Yukon–Kuskokwim delta, apparently in response to the increased growth of willows (Hopkins 1972), a primary winter food for all three species.

Caribou, which disappeared from the Seward Peninsula in the 1800's and have not been present in the Yukon–Kuskokwim delta region for more than a century (Murie 1935), recently began to return to these areas. The number of caribou in the adjacent Western Arctic herd is the highest ever recorded. In recent winters, the Western Arctic herd moved onto the Seward Peninsula, and in 1994 it moved south of the Yukon River (J. Coady, Alaska Department of Fish and Game,

Fairbanks, personal communication). The Mulchatna caribou herd, which is also at an all-time high (180,000 animals) also expanded its distribution in winter into the maritime tundra adjacent to Bristol Bay.

By 1994 many native villages in the maritime tundra region had access to caribou for subsistence hunting for the first time in more than a century. In the past, the low numbers of caribou throughout much of this region were at least partly associated with the expansion of reindeer (domestic caribou) husbandry in western Alaska during the first part of this century (Klein 1980). Reindeer herds expanded throughout this entire region; their abundance reached a peak in the late 1920's but then drastically declined. Reindeer herds have persisted only on the Seward Peninsula and Bering Sea islands. The expanding caribou herd on the Seward Peninsula has created serious problems for the native reindeer herders because their reindeer mingle with the caribou and tend to leave with the caribou when they migrate to their calving grounds in the Arctic. However, there is renewed interest in reindeer husbandry in the region.

Wolves are beginning to return to the Seward Peninsula because of the establishment of moose, an incursion of caribou in winter, and the discontinuance of government-sponsored control of wolves around reindeer herds. As ungulates expand their distributions into other portions of the region, wolves are expected to follow.

The tundra hare is endemic to the region, occurring from the tip of the Alaska Peninsula to the Kotzebue Sound drainages. This species was formerly found in the western Alaskan Arctic (Bee and Hall 1956), but its range decreased southward and its numbers seem to have declined throughout its remaining range. Researchers have not yet identified the causes of the decreased distribution and declining numbers of tundra hare but believe they may be related to vegetational changes caused by the advance of trees and shrubs into the region; this new habitat type is favored by snowshoe hares, whose numbers are increasing (D. R. Klein, U.S. Geological Survey, personal observation).

High densities of minks and river otters in the Yukon–Kuskokwim delta depend on the abundant resident and anadromous fishes for prey (Burns 1964). This rich resource base has also long supported the highest density of indigenous people in all of Alaska.

Muskoxen were introduced to Nunivak Island in 1935 and 1936 to reestablish the species in its historical Alaskan Arctic range. To prevent overgrazing of the limited coastal habitat in winter, muskoxen from Nunivak Island were transplanted elsewhere yearly from

1967 to 1970 and in 1977 and 1981 (Klein 1988). Additional muskox populations derived from these transplanted animals were established on Nelson Island in the Yukon–Kuskokwim delta and on the Seward Peninsula. The populations on Nunivak and Nelson islands have been hunted under a limited permit system for several years. The first hunt of the rapidly expanding population on the Seward Peninsula was scheduled in 1995 (Coady, personal communication), but no hunt had occurred by late 1997. With the growth of the Seward Peninsula muskox population, some reindeer herders there are concerned about potential competition for forage between the muskoxen and reindeer or the displacement of reindeer from their preferred grazing areas by the muskoxen.

The Bering Sea islands of Saint Lawrence, Saint Matthew, Hall, and the Pribilofs were part of Beringia during the last glacial period and support relict populations of small mammals. These include the arctic ground squirrel, northern collared lemming, red-backed voles, tundra vole, and shrews; researchers believe that shrews are an endemic species on Saint Lawrence Island. Other species believed to be endemic in the Bering Sea islands are a shrew on the Saint Paul Islands (Manville and Young 1965), a vole on the Saint Matthew Island group (Rausch and Rausch 1968), and the brown lemming on Saint George Island (Rausch 1953). The only record of a population of polar bears that were resident on land throughout the summer is from the Saint Matthew Island complex, where several hundred polar bears were present until the late 1800's (Elliott 1882), when the population was apparently eliminated by fur trappers. Although polar bears occasionally reach Saint Matthew Island over the pack ice in winter, they have not reestablished themselves there (Rausch and Rausch 1968).

Aleutian Region

The Aleutian Islands and adjacent Alaska Peninsula are the interface between the North Pacific Ocean and the Bering Sea and include the southernmost land area of Alaska (Fig. 2). The Aleutians extend nearly 1,900 kilometers from the tip of the eastern Alaskan Peninsula to the western tip of Attu Island. The island arc is the product of the convergence of the Earth's crustal plates, formed when the massive Pacific plate was forced downward beneath the Bering Sea plate (Gard 1977). This rupturing of the Earth's crust is characterized by extreme tectonic activity, frequent earthquakes, and extensive volcanism. Of the 76 volcanoes throughout the Aleutians, about 40 have been active in the last 250 years.

The Aleutians are geologically young; scientists believe they originated in the early to mid-Tertiary period (Gard 1977). During the last Ice Age, these islands were mostly covered by glacial ice (Hamilton et al. 1986), although extensive lava flows continued building islands throughout the Pleistocene (Péwé et al. 1965).

Persistent cloudy weather, fog, mist, drizzle, and rain borne on powerful driving winds characterize the climate of the Aleutians (Armstrong 1977). Cold ocean currents keep land temperatures consistently cool, even during the warmest summer weather. The mean daily temperature of 3.9°C has an annual range of only ±9.4°C.

The Aleutian Islands are geologically and biogeographically unique in the United States. These islands have acted as a filter for life forms, most of which have island-hopped from the Alaska mainland. As a consequence, the diversity of plant and animal life declines progressively across the nearly 2,000 kilometers from the Alaska Peninsula to the outermost islands. The stormy nature of the North Pacific Ocean and the Bering Sea and the often wide gaps among island groups in the Aleutians have also fostered some endemic plant and animal species.

Although the sea acted as an obstacle to the movement of terrestrial life to the Aleutian Islands, it provided access for sailing ships. The Danish explorer Vitus Bering reached the islands in 1741 during his voyage of discovery from the Siberian coast. More than a century of exploitation of sea otters and other resources of the region rapidly followed, accompanied by major disruption of the lifestyle and culture of the native Aleut people. Animals were also affected: populations of ground-nesting birds were drastically reduced by introductions of the arctic fox to many of the islands by Russians and later by Americans and by the accidental escape of Norway rats from ships to several islands, especially during World War II (Murie 1959).

Vegetation

The Aleutian Islands are an arc of submerged mountains with their visible summits making up the island chain. Some (like Amchitka) have flat summits, whereas others, (like Buldir) have steep summits. The islands are a maritime, treeless zone that is often classified as a distinct subarctic or maritime tundra (Hultén 1968; Fig. 9) because the origin, history, and floristic composition of their vegetation differ from mainland northern tundras. Many islands are bordered by cliffs and bluffs, which can be as high as 60 meters, and by boulder beaches and small dune fields, some of which are on the tops of coastal bluffs. The strand and cliff plants differ little from those in the coastal

Courtesy D. R. Klein, USGS

Fig. 9. Active volcanoes and treeless maritime tundra are characteristic of the Aleutian Island chain.

forest and in the Bering Sea regions. Eelgrass beds in shallow waters of the Alaska Peninsula and the Aleutians account in part for the huge flocks of migratory birds that stop at Izembek Lagoon each year.

Uplands of the Aleutian Islands have deep, peaty soils with thick mats of heath tundra in which crowberry and sedges such as Alaska long-awn sedge are prominent. Rich meadows consist of Pacific reedgrass, sedges, and numerous large forbs such as Aleutian wormwood, umbellifers such as angelica and cow parsnip, and ferns. Trees are absent (except where they were planted in recent times), and even erect shrubs are rare (Shacklette et al. 1969; Amundsen and Clebsch 1971). Snowbed communities are common in sheltered areas at high elevations. The forb and prostrate shrub component of alpine vegetation has many species that are not present farther north or in the interior, including Alaska arnica, Siberian spring beauty, caltha-leaved avens, western buttercup, Kamchatka rhododendron, and club-moss mountain-heather (Hultén 1968).

Although plant cover in the Aleutian Islands is sparse, the mountainous backbone of the islands and the fell-fields on the exposed slopes and ridge crests (even near sea level) provide habitats for some plants that are endemic to the Aleutians: Aleutian draba, Aleutian chickweed, Aleutian wormwood, Aleutian shield-fern, and Aleutian saxifrage (compare Hultén 1968). Aleutian wormwood is known from only two islands, and the Aleutian shield fern is known only from Adak and is federally listed as an endangered species. Personnel at the Alaska Maritime National Wildlife Refuge, which administers the area, are attempting to find additional Aleutian shield fern populations and to protect the species from damage by introduced caribou. Research is under way to develop techniques for propagating this species in the greenhouse. Several other regional plant species that originate in Japan, Korea, and China are found in the western Aleutians but were not part of the Asiatic element that arrived in Alaska via Beringia. These species are found nowhere else in North America (Hultén 1968).

Freshwater Fishes

Eleven fish species occur in the fresh waters of this region (Morrow 1980; Table). The species most used by humans are the Dolly Varden and sockeye, pink, coho, and chum salmon. All communities use the salmon for subsistence when available, and the total annual harvest is usually between 4,000 and 5,000 fish (Alaska Department of Fish and Game 1994a). The 1994 commercial fishery harvested about 18 million salmon at a value of nearly $45 million in the marine waters of the Alaska Peninsula–Aleutian Islands (Alaska Department of Fish and Game 1995). A small amount of sport fishing is done by local residents near villages in the eastern part of the chain, by military personnel and their families adjacent to the bases, and by commercial fishing crews that occasionally come ashore to angle (U.S. Fish and Wildlife Service 1988). The number of streams in which salmon spawn is greater in the Aleutian Islands unit of the Alaska Maritime National Wildlife Refuge than in any other refuge in the United States. In the comprehensive conservation plan for this refuge, 360 such streams are listed (U.S. Fish and Wildlife Service 1988). The status and trends of freshwater fishes in this region are poorly known.

Locally produced salmon, especially their young, may be important forage for the millions of seabirds and numerous marine mammals that frequent the Aleutian Islands. The islands provide nesting habitat for about 10 million seabirds, which all feed heavily on fishes in the marine environment and may eat locally spawned young salmon. In turn, the seabirds enrich the ocean environment with thousands of tons of their droppings, which contain phosphates and nitrates that foster the growth of phytoplanktons (U.S. Fish and Wildlife Service 1990), which then provide food for the zooplankton eaten by the young salmon in the fresh waters of the region.

Birds

Because of the rugged maritime nature of this region, the diversity of breeding waterfowl is limited, though several unique species breed here. The Aleutian Canada goose recovered from near-extirpation caused from predation by foxes introduced on the islands for fur farming (Banks and Springer 1994; Kessel and Gibson 1994). After the foxes were eradicated, the geese were protected on their wintering grounds and were reintroduced to some islands. Now, nesting populations of Aleutian Canada geese are established on three islands, which in 1991 allowed them to be downlisted from endangered to threatened status (Banks and Springer 1994). The emperor goose, which breeds in western Alaska, winters almost exclusively on the rocky coasts of the Aleutian Islands and the Alaska Peninsula (Eisenhauer and Kirkpatrick 1977). This species marginally recovered from a major decline, the causes of which are still unclear (Schmutz et al. 1994), during the 1970's to the mid-1980's. The Steller's eider breeds mainly in Russia, but a small number formerly nested on the maritime tundra of western Alaska. This species winters along the Alaska Peninsula and in the eastern Aleutian Islands (Kertell 1991).

Two raptors are residents in the Aleutian Islands. Approximately 300 pairs of the *pealei* subspecies of the peregrine falcon nest in the Aleutians; this subspecies seems to be maintaining a stable population (Ambrose et al. 1988; Schempf 1989). Although bald eagles have been abundant around military installations on Adak, even nesting there, downscaling of the base may reduce the size of eagle populations if natural prey does not replace human garbage as food. At present, bald eagles nest in the Aleutians only east of the Rat Islands, although in the past they nested on the Rat Islands as well (Kessel and Gibson 1994). The proximity of the Aleutian Islands to Asia allows the occasional occurrence of white-tailed eagles and Steller's sea eagles, particularly on Attu Island.

The isolated nature of the Aleutian Islands gave rise to several endemic subspecies of the rock ptarmigan (Murie 1959). The Evermann's rock ptarmigan, which is restricted to Attu Island, and the Yunaska rock ptarmigan, which is restricted to Yunaska Island, are species of concern (U.S. Fish and Wildlife Service 1996). Few data have been collected on the population size or status of many of the other rock ptarmigan subspecies, but research is ongoing (Holder 1994).

Small numbers of shorebirds breed on some islands but not at the same level as in mainland tundra areas (Kessel and Gibson 1994). Rock sandpipers and black oystercatchers are the most common shorebirds in the Aleutians. Although some Asiatic shorebird species occur in the Aleutians during spring migration, none regularly remain there to breed. Gulls, particularly glaucous-winged gulls and kittiwakes, breed throughout the region. Glaucous-winged gulls probably benefit from commercial fishing offal at sea and in major ports with canneries, such as Dutch Harbor (Kessel and Gibson 1994).

Passerines are resident in the region, although only six species nest west of the eastern Aleutians. The isolated nature of the island chain resulted in the development of several subspecies of resident passerines that are exclusive to one or more islands (Murie 1959). The classic example of this island effect is the largest-bodied subspecies of the song sparrow in North America, which is resident in the central and western Aleutians. One song sparrow subspecies on Amak Island is a species of concern (U.S. Fish and Wildlife Service 1996). Asiatic passerines occur regularly during migration, when they are pushed ashore by winds and storms in the North Pacific or when they stray from their usual migration route. Gibson (1981) and Byrd and Day (1986) suggested that the westernmost islands of the Aleutians are in the regular migration routes of these Palearctic migrants, which accounts for the species' regular occurrence there.

Mammals

The ranges of 19 terrestrial mammals that occur in North America reach the tip of the Alaska Peninsula where the tundra habitat closely resembles that of Unimak Island, which is the closest of the chain to the peninsula. Fourteen of these species, including the brown bear, caribou, and gray wolf, also occur on Unimak Island; only three species occur on Unalaska and only two, the northern collared lemming and red fox, on Umnak Island (Murie 1959). The remaining islands to the west, which extend more than three-quarters of the length of the island chain, are without native terrestrial mammals.

Many mammal species have been introduced to the Aleutian Islands; for example, the arctic fox of the blue phase and the red fox were introduced on most of the larger islands of the outer Aleutians for the harvest of their furs (Murie 1959). The nonindigenous arctic ground squirrel was released on Unalaska Island in the 1890's and on Kavalga Island in 1920 and has become established. European rabbits became established after they were released on Umnak Island and on small islands near Umnak and Unalaska islands (Burris and McKnight 1973). Feral populations of reindeer were established on Umnak and Atka islands after introductions in 1913 and 1914; their numbers fluctuate widely, though in recent decades researchers believed their numbers remain in the range of 1,000–2,000 animals. Reindeer are occasionally hunted for food by local Aleuts.

Caribou were introduced to Adak Island in 1958 and 1959 to provide an emergency source of food as well as recreational hunting for Naval personnel and their dependents at the Adak Naval Air Base (Boone 1993). Management of this population—established jointly by the Alaska Department of Fish and Game, the Navy, and the U.S. Fish and Wildlife Service—has been by harvest quotas for hunters to hold the herd, in the absence of natural predators, to about 200–250 animals. In recent years, however, the population greatly exceeded this number (more than 800 in 1994), and downscaling of the Navy base after the Cold War subsided left too few people to meet the prescribed annual harvest quotas. To prevent overgrazing and consequent erosion from a growing caribou herd and to end the threat to the endangered

Aleutian shield fern found only on Adak Island, the U.S. Fish and Wildlife Service (1994b) proposed the elimination of caribou from the island. Removing the caribou from the island by a method that is publicly acceptable will be difficult, however. The proposal for the removal (U.S. Fish and Wildlife Service 1994b), has been made available for public comment.

Arctic Tundra

The rapid rise in elevation of the Brooks Range mountains, which are part of the interior mountain biogeographic region, account for the abrupt demarcation between the boreal forest and the arctic tundra (Fig. 2; Fig. 10). This is unlike the very gradual transition from boreal forest to tundra in northern Canada. The arctic tundra of Alaska extends northward from the higher and drier ground at its southern limit to the broad and wet Coastal Plain that continues to the Arctic Ocean. The tundra of the Coastal Plain is interspersed with thousands of shallow lakes and is the only arctic biogeographic province in the United States.

In winter, the arctic tundra region experiences strong northeasterly winds that are generated by the arctic high pressure system over the frozen Arctic Ocean (Selkregg 1974–1976). The little snow that falls throughout the long winter is redistributed by winds and accumulates in concavities of the microrelief of the tundra and as drifts on the leeside of stream and river bluffs in the Coastal Plain. In the upland tundra and foothills, riparian shrubs and the leeside of hills accumulate large drifts. The snow-free summer is brief, but the daily 24 hours of sunlight during much of summer accelerate plant growth without the cost of respiration during the darkness of night. The life forms in this

Courtesy D. R. Klein, USGS

Fig. 10. The arctic tundra region extends from the northern foothills of the Brooks Range to the Arctic Ocean. It is the calving grounds and summer habitat for hundreds of thousands of caribou.

region are extremely productive in spite of climatic extremes, the short plant growth season, low mean annual temperature, and cold soils underlain by permafrost. This is particularly true of caribou and most migratory birds that are present in the area only during summer.

The Alaskan Arctic obviously is no longer remote from the influences of the industrialized world to the south. The energy and mineral resources of the Alaskan Arctic are already under intensive exploitation, and scientists are finding that the Arctic basin is a sink for pollutants transported by global air movements from industrial centers to the south (Barrie 1986). These pollutants are entering Arctic marine and terrestrial food chains and are being concentrated in the tissues of animals at upper trophic levels, including the human residents of the Arctic (Muir et al. 1988). In addition, the thinning ozone layer in the atmosphere over the poles increases the ultraviolet radiation that reaches the plants and animals of the Arctic, the consequences of which are not fully known. Global climate change from increasing carbon dioxide and other greenhouse gases in the atmosphere is expected to generate the greatest warming in Arctic regions, especially in the Alaskan Arctic, the ecological consequences of which are also unknown (Committee on Earth and Environmental Sciences 1994).

Vegetation

Arctic tundra lies beyond the latitudinal treeline. The area from the summit of the Brooks Range northward to the Arctic Ocean has been known as the *Arctic Slope* and, since development of oil fields there, also as the *North Slope*. The North Slope consists of mountains, foothills, and the Coastal Plain. The higher mountains were glaciated during the last glacial maximum, with revegetation beginning about 12,000 years ago (Hopkins 1972). Portions of the arctic foothills escaped glacial advances; these areas were not glaciated throughout the 1.5 million years of the Quaternary and therefore served as plant refugia (Hamilton et al. 1986). Likewise, the Coastal Plain was not glaciated, but because marine transgressions during higher sea levels as recently as 0.7–1.9 million years ago reached far inland, the contemporary vegetation postdates that event (Hopkins 1967).

Alaska's environmental and climatic conditions are the result of the state's far northern location. As a direct consequence of the region's geographic position at a high latitude, the Alaska Arctic experiences less intense solar radiation and more pronounced variation in seasons. The region is characterized by a short growing season but long days; continuous permafrost close beneath the land surface; and

low summer temperatures that are exacerbated by frequent coastal fog and drizzle, although the overall annual precipitation of the region is low (compare Murray 1987, 1992).

A vascular plant flora is largely shared by each landform; only a few of about 500 species in the total flora are unique to any one landform. Among endemics, however, a sagebrush, *Artemisia artica* ssp. *comata*, is restricted to the Coastal Plain province, Alaska bluegrass and Drummond's bluebells to the transition from plain to foothills, Muir's fleabane to the foothills, and Kokrine's locoweed to the far western mountains.

Except along the Chukchi Sea coast where cliffs are extensive, coastal communities consist of eroding bluffs, mudflats and sandy beaches, and dune fields at the mouths of major rivers along the Beaufort Sea. Along the Beaufort Sea coast, deposition has formed spits, barrier islands, and lagoons. Plants of the strand include some of the same salt-tolerant taxa of southern regions but differ by the absence of distinctly southern species and the addition of arctic grasses and forbs, such as Anderson alkaligrass, purplish braya, and few-flowered whitlowgrass, which are taxa more common in the high Arctic of Russia, Canada, and Greenland. Although the tidal range is minor, storm tides in fall have carried driftwood as far as 3 meters above sea level and 5 kilometers inland (Reimnitz and Maurer 1979); the position of driftwood indicates how far inland the influence of the sea can penetrate.

Wet meadows are extensive in the Coastal Plain, where continuously distributed permafrost maintains a water table close to the surface. Despite low annual precipitation, lakes and ponds are abundant, and their margins in certain seasons are red with arctic pendantgrass. Wet meadows are dominated by pure and mixed stands of water sedge, cottongrass, and tundra grass. The plants in several localities were comprehensively described during the past two decades, yet the early paper by Britton (1967) remains the best generalization of conditions on the Coastal Plain.

Habitat diversity on the Coastal Plain is primarily a result of the region's raised beaches and drained lakes of thermokarst origin. Beach ridges and some lake margins are dry and exposed and thus are the floristic equivalent of alpine ridges. Exposed lake bottoms offer bare soil for colonization by plants. Networks of high-center and low-center ice-wedge polygons offer sharp, repeated contrasts of well- and poorly drained sites (Fig. 11). Slopes of pingos are also alpinelike. Floodplains of large rivers are a complex of terraces of various ages and a patchwork of sands and gravels in various stages of temporary stabilization.

Outside the reach of the modifying effects of the ocean, rises in temperature and changes in plants are significant. Tussock tundra is absent near the Arctic coast but is the dominant vegetation type in much of the interior Coastal Plain and arctic foothills. Only prostrate shrubs occur near the coast, but the abundance of willows increases inland, especially in riparian settings. Dwarf birch, which is prostrate in places where it is exposed but as tall as 1 meter in sheltered areas, forms thickets on the southern uplands. Balsam poplar stands persist well north of the treeline in the headwaters of several arctic rivers where well-watered gravels (that is, gravels through which groundwater passes) are sheltered by benches and bluffs (Murray 1980; Edwards and Dunwiddie 1985). The tundra on mountain slopes is alpine in many respects and shares the floristic and basic features of the physical environment with the interior uplands. Closed communities are replaced on summits and ridge crests by massive unstable screes and widely scattered plants.

Freshwater Fishes

Twenty-three fish species occur in the fresh waters of this region (Morrow 1980; Table). Species most used by humans are the arctic cisco, least cisco, humpback whitefish, arctic grayling, and Dolly Varden. A commercial fishery on the Colville River takes about 20,000 whitefishes every year (Bergstrom et al. 1992). Whitefishes and the Dolly Varden provide subsistence along the entire North Slope where they may be taken throughout the year for human consumption and for dog food. Craig (1987) reports that Dolly Varden and arctic cisco are the principal anadromous species harvested from the Colville River east to the Canadian border, and pink salmon, chum salmon, broad whitefish, humpback whitefish, round whitefish, and least cisco are the principal species harvested west of the Colville River. Anglers spent 5,600 angler-days in 1993 in the arctic tundra and harvested an estimated 1,632 arctic grayling and 1,092 Dolly Varden (Mills 1994). The abundance of whitefishes in this region provides a forage base for other species such as the Dolly Varden (Arvey 1993).

Angling in this region is relatively light, and most stocks of freshwater fishes are probably stable. Arvey (1991) reviewed the stock status of Dolly Varden on the North Slope and found no evidence of depletion. The status of many of the freshwater fish species in this area seems poorly known, however. Springs in the rivers of the North Slope provide critical habitat for spawning and overwintering of Dolly Varden because the sections of streams without springs usually freeze solid (Craig and McCart 1974;

Courtesy D. R. Klein, USGS

Fig. 11. The permafrost that underlies the Coastal Plain of the Alaskan Arctic and the seasonal frost action at the land's surface have generated the polygonally patterned ground that is common throughout much of the area.

McCart 1980). Fishes that overwinter at sea in this area have evolved antifreeze compounds in their blood—temperatures of seawater may drop to below that at which a fish's blood would otherwise freeze. Because species such as the Dolly Varden and the arctic grayling have not evolved these antifreeze compounds, they must overwinter in fresh water.

Birds

Waterfowl breed across most of the arctic tundra of Alaska, although their total numbers there do not approach those on the maritime tundra of western Alaska. In recent years, tundra swans that breed in the region (and winter on the Atlantic coast) experienced increasing populations, which now seem to have stabilized (Earnst 1991; Stickney et al. 1993, 1994; Brackney and King 1994; Anderson et al. 1995). The population of the greater white-fronted goose, the most common breeding goose, seems to be stable or increasing (Brackney and King 1994). Scattered, small breeding colonies of brant have been monitored closely only in the Colville River delta and in the North Slope oil fields; their numbers seem to be stable (Stickney et al. 1993, 1994; Smith et al. 1994; Anderson et al. 1995). Large numbers of brant from breeding populations in Alaska, western Canada, and Siberia, as well as Canada geese and greater white-fronted geese, also molt in the Teshekpuk Lake area southeast of Barrow (Derksen et al. 1982).

In the United States, snow geese currently nest only in a few small colonies in Arctic Alaska (Johnson 1991; Ritchie, unpublished data). The largest colony (about 400 pairs in 1990) is on Howe Island near Prudhoe Bay (Burgess et al. 1994). Although this colony had been increasing since its inception in the mid-1970's (Johnson 1991; Burgess et al. 1994), its long-term viability is uncertain because of the nearly complete lack of reproduction during 1991–1994 when late springs deterred nesting and when arctic foxes gained access to the colony island and preyed upon the eggs (Burgess and Rose 1993; S. Johnson, LGL Ltd., Sidney, British Columbia, personal communication). In autumn these small breeding populations and a major portion (as many as 325,000 birds) of the western Arctic population of snow geese from Canada stage in the Arctic National Wildlife Refuge in northeastern Alaska (Brackney and Hupp 1993).

The threatened spectacled eider also breeds on the Arctic Coastal Plain, but its current population status is still under investigation; the size of this population may have been declining (Warnock and Troy 1992; Troy Ecological Research Associates 1993a; Anderson and Cooper 1994; Larned and Balogh 1994; Anderson et al. 1995). Oldsquaws and northern pintails are the most abundant ducks in the region, and king eiders are common breeders in this westernmost part of their range (Brackney and King 1994). During droughts in the mid-continental United States, the region serves as habitat for large numbers of nonbreeding northern pintails (Derksen et al. 1979). From Wainright to the Colville River, the North Slope also hosts small numbers of breeding yellow-billed loons.

Five species of raptors regularly breed in the arctic tundra region: peregrine falcon, gyrfalcon, rough-legged hawk, short-eared owl, and snowy owl. The arctic peregrine falcon, which became endangered because of pesticide contamination, has sufficiently recovered to be removed from the endangered species list (U.S. Fish and Wildlife Service 1996). Few data are available on population trends of breeding rough-legged hawks and snowy owls; monitoring of snowy owls is especially difficult because of their annual variability in breeding locations, which seem to coincide with locations of high abundances of lemmings, one of their preferred foods.

Willow and rock ptarmigan breed on the arctic coastal tundra, but their population sizes and trends are unknown. Rock ptarmigan make short migrations to winter in the foothills of the south slopes of the Brooks Range where willows are more abundant (Johnson and Herter 1989). Willow ptarmigan also move inland to the foothills, although some birds move south of the Brooks Range into more forested habitats. During spring, thousands of ptarmigan move north across the foothills to reach their breeding areas on the tundra.

The arctic tundra is one of the most productive and abundant habitats for shorebirds in Alaska and supports a diversity of breeding species (Connors et al. 1979; Page and Gill 1994). Monitoring of shorebirds across the Arctic coast has been sporadic and usually of limited duration. Results from the longest study (more than 10 years), which is still ongoing in the Prudhoe Bay area, indicate that large annual variations in nesting densities and nesting success (because of harsh weather and predation) of most species make determinations of long-term trends difficult (Troy and Wickliffe 1990; Troy Ecological Research Associates 1993b).

Only a few passerine species breed on the arctic tundra; as with shorebirds, their population sizes seem to fluctuate in response to weather conditions and predation levels. The snow bunting may have benefited from the recent increase in artificial structures in their habitat (for example, oil-field buildings and pipelines) by using them as new nest sites. As in western Alaska, some species of Asiatic origin breed on the arctic tundra (for example, bluethroat and yellow wagtail).

Mammals

Mammals of the arctic tundra of Alaska are the caribou, muskox, northern collared lemming, brown lemming, singing vole, arctic ground squirrel, arctic fox, and the barren-ground grizzly, a subspecies of the brown bear. Other mammals in the arctic tundra region—such as the arctic ground squirrel, tundra vole, and the wolverine—are also typical in alpine habitats of the interior mountains, whereas the red fox, wolf, ermine, and least weasel are more widespread in places where prey species are also present. The snowshoe hare and moose are typical boreal forest dwellers that have expanded their distribution into riparian shrub communities along several of the major stream drainages north of the Brooks Range (Bee and Hall 1956; LeResche et al. 1974).

Strategies vary among species for dealing with the long, severe arctic winters and low food availability. Shrews, voles, lemmings, and weasels live in a *subnivian* (beneath the snow) environment where temperatures are mediated by the insulating snow. All of these mammals are active throughout winter. In contrast, arctic ground squirrels and grizzly bears hibernate during most of the winter, foregoing the need to feed. They must, however, accumulate adequate body reserves during the arctic summer to accomplish this; thus, the large body size of the arctic ground squirrel in contrast to that of its conspecifics farther south, seems to be an adaptation to help ensure sufficient fat storage (Barnes 1989).

Muskoxen and caribou are adapted to the climatic extremes of the Arctic and remain active throughout winter, finding plant food by digging through the snow or by seeking areas where winds blow snow from vegetation. Caribou in the large migratory herds cannot remain at high densities in the Arctic in winter when suitable forage biomass is low. Consequently, most caribou migrate long distances through the Brooks Range to winter on the northern fringes of the boreal forest in open woodlands with lichens.

Wolf densities are low in the arctic tundra because of the region's relatively low availability of ungulates and other prey throughout winter. The highest densities of wolves in this region are in the northern foothills of the Brooks Range, where caribou that winter on the North Slope tend to concentrate and where prey such as moose, mountain sheep, and snowshoe hares are available. Wolverines and foxes are wide-ranging during winter and supplement prey with food they cached during summer (Magoun 1985).

Because ecological communities and food chains are less complex and involve fewer mammals in the Arctic than at lower latitudes, such communities are more easily disrupted by human activities. Mammals of the arctic tundra are vulnerable to human activities, but the level of each species' vulnerability depends in part on its body size. The smaller mammals are more secure in the low vegetation of the tundra, which offers them adequate escape cover. Among the large ungulate species, the muskox, which was extirpated from Alaska in the late nineteenth century but has since been reestablished, is particularly vulnerable to uncontrolled hunting and displacement from its habitats by disturbance (Barr 1991). In summer and early winter, muskoxen make heavy use of riparian habitats, also the primary habitats of moose. Humans in the Arctic also favor riparian zones for their use as transportation corridors for road and pipeline construction and associated gravel extraction, or for recreational boating and hiking. In the 1970's humans displaced moose from riparian habitats during the construction of the Trans-Alaska Pipeline and the adjacent Dalton Highway (Klein 1979).

A relationship between the effects of the expanding moose and muskoxen populations and the shrub communities that provide a major food source for these ungulates has not been documented, although riparian shrub growth may possibly be increasing because of climatic change. Even though long-term oscillation between and among these herbivore populations and the extent and biomass of the willows on which they feed is expected, research is required.

The larger carnivores may be directly harmed by human development. The wide-ranging habits of bears, foxes, wolverines, and gray wolves and their propensity for being attracted to any food often bring them in contact with humans in the Arctic. Bears and foxes become nuisances around oil fields, construction camps, and villages where food wastes are available. Bears in these situations lose fear of humans and may damage buildings or threaten humans and are usually shot. Foxes also readily habituate to the availability of human food wastes and to being deliberately fed by construction workers. Consequently, these animals often become tame and, because foxes in the Arctic may become rabid, they then pose a potential threat to humans (Klein 1979).

Increased access to the Alaskan Arctic has become possible by the all-season Dalton Highway, which extends 563 kilometers from the highway system in interior Alaska to Prudhoe Bay. This highway, originally built for commercial use to service the oil fields, was officially opened to the public in 1994. Coupled with the improved designs and reliability of snowmobiles, the Dalton Highway may increase hunting and trapping of carnivores in the Arctic. In the Arctic's open terrain, carnivores can be readily tracked with snowmobiles in late winter when light and snow conditions are favorable. Such ease of hunting may, however, be countered by low fur prices.

The effects of petroleum development on caribou have received major attention in the media. The ecological complexity of this species, which differs greatly from other northern ungulates, makes it difficult to assess the response of caribou to anthropogenic development. Unlike moose, deer, and muskoxen, most caribou are seasonally migratory and thus expand their relationships to food and predators across two or more ecosystems. The winter diet of caribou is dominated by lichens, which unlike vascular plants are unrooted, extremely slow-growing plants that are vulnerable to destruction from overgrazing and fire. Caribou tend to avoid humans in the same manner in which they avoid predators. They give birth in specific, traditionally used areas that favor survival of their young. Caribou are particularly vulnerable to harassment by insects and the parasites associated with insects and in summer require free access to habitat that offers relief from insects (Klein 1991).

Development and operation of the Prudhoe Bay, Kuparuk, and satellite oil fields coincided with a period of increases in the sizes of caribou herds across northern North America, and thus complicated an assessment of the effects of development on the Central Arctic caribou herd that uses the oil-field area. Because other caribou herds were also experiencing population increases, presumably because of favorable environmental conditions, possible detrimental effects of oil-field development may have been overridden. In recent years, reduced reproduction in the Central Arctic herd has been recorded (Klein, unpublished information). A reliable assessment of the effects of this development will require several decades rather than merely a few years. Some anthropogenic developments are obviously detrimental, including the ones that displace caribou from calving and post-calving habitat or obstruct their movements to and from the coastal habitats that offer them relief from insects. Reduced reproduction in caribou seems to reflect the effects of such developments (Cameron et al. 1992, 1995; Cameron 1994).

Caribou herds undergo long-term population fluctuations in response to such environmental constraints as limited forage, extreme weather conditions, predation, and hunting by humans. The estimated size of the Western Arctic herd is close to 500,000 (Valkenburg, personal communication), which greatly exceeds all previous population estimates. A population decline seems imminent in view of previous declines from peak numbers. Although food is expected to limit the herd, assessments of forage resources and the effect of caribou on them have not been made. The Western Arctic caribou herd is extremely important to the subsistence of native peoples in more than 25 villages—more than 5,000 households—in northwestern Alaska (Klein 1989). The Porcupine caribou herd of northeastern Alaska and adjacent Canada, which consisted of an estimated 160,000 animals in 1994 (Valkenburg, personal communication), has been the primary subsistence food source for Gwichin people in Alaska and Canada for millennia. Caribou from this herd are also harvested by the Inupiat and Inuit Eskimos of Alaska and Canada.

Major Issues and Research Needs

Humans have had a much smaller effect on natural communities in Alaska than elsewhere in the United States. Development of land for agriculture and the extent of alterations of land for mining and petroleum development have been minimal. Most human settlement is concentrated along the coast. More than 40% of Alaska is in national conservation lands of the National Park Service, U.S. Fish and Wildlife Service, U.S. Forest Service, and U.S. Bureau of Land Management (Foster 1985). The most extensive alteration of natural communities has been the widespread clear-cutting of old-growth forests in southeastern Alaska.

Coastal Temperate Rain Forest and Coast Range

Extensive clear-cutting on national forest, state, and native-owned lands throughout much of this region is causing a reduction in the biological diversity of the regenerating forests as the uneven-aged old-growth forests are replaced by even-aged second-growth forests. Long-term studies of changes in forest structure after logging and the consequences of these changes for the ecological nature of the region are necessary to formulate effective future land-use policies.

Salmon populations throughout the coastal temperate rain forest and Coast Range are considered healthy. Recent commercial catches of all salmon species in this region have been among the highest on record, and escapements to most spawning streams are considered adequate (Geiger and Savikko 1990; Burger and Wertheimer 1995). However, some rather severe declines in the abundances of steelhead, cutthroat trout, Dolly Varden, and perhaps eulachon stocks have occurred throughout the region, especially in freshwater systems with many recreational anglers. Similarly, steelhead stocks have declined significantly in the Karta River (Harding and Jones 1993), the Situk River (Glynn and Elliott 1993), Peterson Creek (Harding and Jones 1992), and several systems on the Kenai Peninsula (Larson 1990). Harvests of cutthroat trout by anglers have declined throughout southeastern Alaska to about half of what they were 16 years ago; at the same time angling effort has doubled (Mills 1994). The abundance of Dolly Varden also declined in many systems, including in the Juneau area (Armstrong 1979) and on the Kenai Peninsula (Larson 1990). These declines prompted the Alaska Board of Fisheries to reduce bag limits of steelhead, cutthroat trout, and Dolly Varden, to completely close some systems to harvest, and to establish catch-and-release-only fisheries on other streams (Larson 1991; Harding and Jones 1992, 1993). A greater number of anglers are releasing the fish they catch; however, excessive mortality from hooking and releasing fishes remains a concern.

The release of many salmon from the region's hatcheries into the marine environment is not without potential problems. An increased harvest of wild salmon stocks could take place during harvest of returning hatchery fish (Pahlke 1992). Also, the drastic reduction in the body sizes (weight by age group) of salmon (as much as 27%) across the North Pacific may be due to excessive grazing of the ocean by too many hatchery fishes (J. Helle, National Marine Fisheries Service, Juneau, Alaska, personal communication). Releases of hatchery-reared coho and chinook salmon have contributed substantially to sport fisheries in some areas; elsewhere, such enhancement has proved costly and largely unsuccessful (Suchanek and Bingham 1992).

Alaska's anadromous and freshwater resident fishes are managed primarily for human use. Little or no consideration is given to their value and use by other vertebrates or to their value to the natural community as a whole. For example, fishery managers commonly refer to salmon runs as the escapement needed to perpetuate a fishery but rarely consider the value of salmon needed as food for other vertebrates or for the well being of the environment (Willson and Halupka 1995). Notable exceptions are studies that include the importance of salmon to bald eagles (Hansen 1987; Hansen et al. 1984) and to brown bears (Schoen et al. 1994), and studies of the value of salmon carcasses to a stream's ecosystem (Kline et al. 1990, 1993). Investigations of the use of these fishes often focused on the role of other vertebrates as predators of salmon species. In some instances, results of such studies led to misguided predator-control efforts to protect salmon species; these efforts resulted in the killing of many animals, including more than 6 million Dolly Varden between 1921 and 1946 and more than 100,000 bald eagles between 1917 and 1952 to protect Alaska's salmon fisheries (O'Clair et al. 1992). More recently, thousands of arctic chars were captured and held in pens until sockeye salmon smolts had safely passed (McBride 1980).

The abundance of marbled murrelets in Prince William Sound has been declining (Hatch 1993), and data suggest that this trend is also occurring in other parts of the species' range in Alaska (Hatch 1993; Piatt and Ford 1993; West 1993; Ralph 1994). The marbled murrelet in Alaska is a species of concern that could suffer from the loss of nesting habitat resulting from clear-cutting of old-growth forests. Even less is known about trends in the population sizes of Kittlitz's murrelets, which nest on alpine slopes in the coastal mountain ranges. The abundance of this species is also declining (Hatch 1993); it is a species of concern in Alaska. Declining abundances of the dusky Canada goose on the Copper River delta seem to be associated with the 1964 earthquake and its resulting habitat changes that favored predation of eggs and young in nests (West 1993). Habitat loss and fragmentation from logging jeopardize resident northern goshawks on forested islands in southeastern Alaska (West 1993) and have led to the classification of this population as a species of special concern by the state of Alaska.

In winters with deep snows, deer are confined to old-growth forests at low elevations where snow is intercepted by the crowns of large trees. Clear-cutting of these old-growth forests with their high timber volume has greatly reduced the availability of winter refuge areas to deer. New clear-cuts produce an abundance of deer forage but accumulate deep snows that make forage unavailable to deer in winter (Wallmo and Schoen 1980; Schoen et al. 1988). Eventually, the closed canopy of the even-aged second-growth forests does not allow sufficient light to reach the forest floor for forage growth.

In the absence of wolves, the numbers of introduced deer on Kodiak Island and on islands in Prince William Sound have increased to such high densities that deer are altering the vegetation and reducing other animals' food resources. The abundance of plant species preferred by deer, especially shrubs that are heavily browsed in winter, is declining. Many of these plants, such as alpine blueberry, Pacific red elderberry, and Sitka mountain-ash, produce berries and fruit that are important in the diets of brown bears and other mammals and birds.

Increased road access from logging, especially on Chichagof and Prince of Wales islands, is generating concern that easier human access will result in an excessive harvest of brown and black bears on these islands. Wolves may also be threatened by continued clear-cutting on Prince of Wales Island and on other islands in the region. The loss of old-growth forest habitat lowers densities of deer, the primary prey of wolves, and the extensive road systems built during logging opened up much of the region to intensive trapping of wolves by local residents (Person et al. 1996). Consequently, the wolf was considered for listing as a threatened genotype under the Endangered Species Act of 1973. Although the listing was found unwarranted, the species was subsequently (1995) designated as a candidate for federal listing as an endangered species. Following a review period, this proposed listing was also denied in September 1997.

Ecological relationships and population statuses of many mammal species of this region are not understood sufficiently to accurately assess the consequences of extensive clear-cutting on populations. In addition to deer, bears, and wolves, other species that may be affected by logging are the marten, river otter, mink, mountain goat, northern flying squirrel, and bats.

Interior Boreal Forest and Interior Mountains

The role that all species of freshwater fishes play in the natural communities of Alaska is poorly understood. Because of the concerns of funding sources and the need to respond to immediate public interest, research is usually done only on important sport, subsistence, and commercial species of freshwater fishes. Activities that threaten fish populations throughout the interior boreal forest and interior mountain region include proposed timber harvest in the western drainages of the Susitna River; such harvesting may harm the recreational use of the area as well as the fishes and wildlife (Engel and Vincent-Lang 1992). In addition, new public boat-launching areas may increase fishing from boats in areas that are already too congested or may result in further restrictions on the use of motorized boats (Engel and Vincent-Lang 1992). Fish habitat and water quality in the Chuitna River drainage may be reduced because of the development of the Diamond Chuitna coal project. Mine workers at the new site may also increase fishing pressure on all fisheries in the Tyonek–Beluga area (Engel and Vincent-Lang 1992).

Placer mining is also a problem for freshwater fishes in Alaska; it often elevates stream turbidity, which causes gill abrasion and prevents feeding by sight feeders such as arctic grayling. Increased turbidity also limits the growth of aquatic plants and invertebrates that are important for the overall health of natural communities (Weber 1986). Additionally, placer mining can increase the levels of toxic metals such as arsenic and mercury (Alaska Department of Environmental Conservation 1986) and change the physical characteristics of stream channels and riparian habitat (Arvey 1993). In recent years, more than 3,000 new mining claims per year were recorded in the region (Bundtzen et al. 1991).

The number of commercial-use permits on rivers classified for recreation by the Alaska legislature and an increase in the number of wilderness lodges in this region have become issues of environmental concern (DeCicco and Barnes 1992; Engle and Vincent-Lang 1992). These types of activities can increase pressure on certain fish stocks, particularly those in more remote, roadless areas. Vast areas without roads exist, and little is known about the fish populations in many lakes and rivers. For example, more than 40,000 lakes and ponds occur in the area of the Yukon Flats National Wildlife Refuge (U.S. Fish and Wildlife Service 1987), yet little information is available on resident fishes that use these vast wetlands (Arvey 1993).

In Alaska, the boreal forest has been little influenced by human activities, except in habitat adjacent to major settlements where forests were cleared for residences, limited industrial development, and small-scale agriculture. Logging and cutting of firewood have also been localized, primarily in areas with road access.

Proposals for large-scale commercial logging in the Tanana Valley generated concern for the potential effect of clear-cutting of the forests on wildlife habitats and their long-term productivity. Because information about the long-term effects of extensive logging in this region does not exist, research is needed on the ecological features of the interior Alaska boreal forests and the long-term response of the vegetation and animals to clear-cutting there.

In the boreal forest and the interior mountain regions, local issues may influence mammal populations or their habitats. In winters of heavy snows, the number of moose killed when struck by trains on the Alaska Railroad and to a lesser extent by autos on the highways reaches into the hundreds (Alaska Department of Fish and Game 1994b). Although this loss of moose and the associated property damage and injury (or occasional death) of humans are regrettable, the overall effect on moose populations is not long lasting.

The control of wolves by the Alaska Board of Game, supported by the Alaska legislature, met with public opposition from both inside and outside Alaska. The objective of the control was to reduce the number of gray wolves by about 70% in specific areas where management goals are to increase the sustainable annual harvest of moose and caribou by hunters. Effectiveness of the control of the gray wolf was limited because of external influences—the chosen methods of control were restricted as a consequence of public opposition and, in late 1994, the governor of Alaska placed a moratorium on further wolf control until previous control efforts had been evaluated. A wolf bounty bill has been introduced in recent Alaska legislative sessions but has failed to become law. Supporters were expected to propose similar legislation in 1996.

The relationship among the distributions and habitats of small mammals throughout the region is poorly understood. Intensive surveys are required to assess the biological diversity and ecosystem relations of the small mammal fauna.

Maritime Tundra

The maritime tundra region (southwestern Alaska and the Bering Sea Islands) supports populations of freshwater fishes that are so unique that they have intrinsic ecological values that reach beyond Alaska. These populations have not been genetically altered by releases of fishes from hatcheries and represent some of the only truly wild populations left in the world. Any type of proposed development, such as oil exploration, deserves careful consideration of the possible ramification for these fish stocks. Likewise, the growing popularity of recreational fishing in this region must be carefully monitored. Considering the value of many of these freshwater stocks for subsistence use, additional harvests may lead to overexploitation.

Mining is also of environmental concern in the maritime tundra region. The Cominco Alaska Corporation, for example, annually mines more than 2 million tons of zinc–lead–silver concentrate from the Red Dog deposit north of Kotzebue (Bundtzen et al. 1991). A potential threat of this operation is heavy-metal contamination of the Red Dog and Ikalukrok creeks from natural leaching of the ore body as it is stripped for ore production and from discharge of impounded waters from which contaminants are not removed (Arvey 1993). Such contamination may affect both resident and migrating fishes, such as Dolly Varden, in the Wulik River. Regulatory surveillances by the Alaska Department of Environmental Conservation are intended to minimize this hazard (Arvey 1993).

Overall, human effects on natural habitats have been minimal throughout most of the maritime region, although overgrazing was widespread during a population peak of reindeer in the 1920's and after the introduction of reindeer on the Pribilof, Saint Matthew, Nunivak, and Hagemeister islands (Scheffer 1951; Klein 1968; Lay 1994). Lightning-caused tundra fires in the drainages of the Kotzebue Sound in infrequent dry summers destroyed lichens, the primary food for caribou and reindeer on their winter ranges (Fig. 12). Fire also favors shrub growth, which supports an increased number of moose. The numbers of shrubs and trees in the tundra are expected to increase even more because global climate warming from the greenhouse effect is expected to be among the greatest in the Alaskan Arctic and subarctic. This could have a cascading effect throughout the region. For example, during the nineteenth century, when caribou disappeared from the Seward Peninsula and much of the maritime tundra adjacent to the Bering Sea, gray wolves, bears, and scavengers such as wolverines and foxes also declined. In turn, however, the recent and ongoing return of caribou to much of this region also could have a cascading effect on other species that could help restore the former complexity of the region's natural communities and of its biological diversity.

Aleutian Region

Many areas of concern in the Aleutian region will require future consideration and research effort. For example, the short drainage systems of the Aleutians have small runs of salmon and Dolly Varden, making these populations vulnerable to excessive harvest by subsistence, commercial, and sport fisheries. Another concern is that the relationships among seabirds, marine mammals, and salmon in this region are not well understood and should be a focus of study. Except for the effect of the intensive commercial fisheries of the region on seabirds that nest in the Aleutians, the major influence of human activities has been the introduction of non-indigenous mammals, particularly foxes and rats, and perhaps reindeer, caribou, and

Fig. 12. Wildfires contribute to the patterning of vegetation in this drainage to Kotzebue Sound and have both short- and long-term effects on caribou and moose. The burned areas in the photo are the result of a fire on the Selawaik National Wildlife Refuge in the summer of 1988.

Courtesy D.R. Klein, USGS

Authors

David R. Klein*
U.S. Geological Survey
Biological Resources Division
Alaska Cooperative Fish and
Wildlife Research Unit
University of Alaska
Fairbanks, Alaska 99775

David F. Murray
University of Alaska Museum
University of Alaska Fairbanks
Fairbanks, Alaska 99775

Robert H. Armstrong
5870 Thane Road
Juneau, Alaska 99801

Betty A. Anderson
Alaska Biological Research, Inc.
P.O. Box 80410
Fairbanks, Alaska 99708

*Current address:
Institute of Arctic Biology
University of Alaska Fairbanks
Fairbanks, Alaska 99775

domestic livestock (Burris and McKnight 1973). Although foxes were eradicated from several islands where the abundances of nesting birds subsequently increased, they remain a problem on many other islands. Rats are not as widely distributed as foxes and are believed to have fewer effects on nesting birds (Bailey 1993). However, further investigation of the effect of rats on the bird community is needed, and effective eradication techniques have not been developed. Similarly, the effects of reindeer, caribou, and livestock grazing on nesting habitat and nesting success have not been evaluated.

Grazing by fenced and free-ranging sheep, cattle, horses, and pigs has been practiced on Unalaska and Umnak islands since at least the early 1930's. These activities, which have been of limited economic success, have sometimes resulted in the establishment of feral or semiferal populations of livestock. Little is known, however, about the effects of this grazing on the native flora and fauna, although decreased plant species diversity was observed on fenced ranges grazed by sheep on Umnak Island (Klein, personal observation).

The widespread introduction of foxes throughout the Aleutian Islands from 1750 to 1932 for fur production had a devastating effect on the abundant bird life there (Bailey 1993). Ground-nesting waterfowl, shorebirds, ptarmigan, and burrow-nesting seabirds were most detrimentally affected. Only a few hundred breeding pairs of the endangered Aleutian Canada goose survived on three islands where foxes had never been introduced. In 1949, the U.S. Fish and Wildlife Service began eliminating foxes from specific islands by shooting, trapping, biological controls, and approved toxicants. By 1992, the foxes were believed to have been exterminated from 21 islands (Bailey 1993), and the recovery of breeding birds on these islands has been spectacular. The removal of introduced foxes is continuing on the remaining islands and is expected to further restore the biological diversity of the avifauna of this highly productive biogeographical region.

Arctic Tundra

Transportation corridors (such as the Dalton Highway and the Trans-Alaska Pipeline) and the extraction of gravel for roads, pipeline, and construction pads for buildings, drill rigs, and other oil-field facilities on the North Slope have disturbed fish habitat in the arctic tundra (Arvey 1993). Evidence suggests that the construction of Prudhoe Bay's West Dock Causeway has disrupted the migration and recruitment of the arctic cisco in the Colville and Sagavanirktok rivers (Moulton et al. 1986; Gallaway et al. 1987). Continued monitoring of the subsistence fishery for whitefishes may be useful for determining the extent of changes in stock status of species from the construction of the West Dock Causeway. Documentation of oil spills and their adverse effects on fish populations from contamination on the North Slope is lacking (Arvey 1993).

Shorebirds and waterfowl in the oil fields are potentially threatened by their possible access to contaminants. Although contaminants have been found in the tissues of eiders collected in western Alaska (Henny et al. 1995), no evidence of uptake of contaminants by eiders in the oil fields has yet been noted. Close monitoring of contaminant levels in the oil fields and investigation of their possible influence on birds that nest in the tundra are needed.

Generally, mammals of the arctic tundra are at moderate to high population levels. With the exception of the Prudhoe Bay oil-field complex and the associated Trans-Alaska Pipeline and Dalton Highway, loss of habitat from human activities has not been significant. The increasing demand for fossil fuels and other minerals by world markets may, however, bring about a major expansion of petroleum exploration and development, large-scale mining of coal reserves, mineral extraction, and the construction of roads and other infrastructures in the Alaskan Arctic. Integration of these large-scale developments and their associated transportation corridors with natural arctic communities will not be possible without substantial effect on the mammals of the Alaskan Arctic.

Major issues confronting the biological environment of Alaska are few and widely dispersed in contrast to other biogeographic regions. Nevertheless, the consequences of clear-cut logging in the southern coastal areas, oil exploration in the Arctic, small- and large-scale mining activities throughout Alaska, intensive and expanding exploitation of marine fisheries, and anticipated climate change due to the "greenhouse effect" all pose threats to the biodiversity of Alaska's ecosystems and their sustained productivity. Research efforts should be focused on assessing the consequences of these human activities on Alaska's fish and wildlife and their habitats under the wide diversity of conditions that occur throughout the state's biogeographic regions. Other human activities with more localized consequences, such as expanding tourism; increased hunting, fishing, and other recreation; and conflicts between reindeer grazing and wildlife, also merit research efforts to provide a basis for the protection of natural ecosystems and their sustainable management.

Acknowledgments

Constructive comments by L. Adams, D. Dirksen, E. Knudsen, L. A. Viereck, J. Schoen, and W. Seitz are greatly appreciated.

Cited References

Agler, B. A., S. J. Kendall, P. E. Seiser, and D. B. Irons. 1995. Abundance of marbled and Kittlitz's murrelets (*Brachyramphus marmoratus* and *brevirostris*) in south-central and southeast Alaska. Abstract from the 22nd annual meeting of the Pacific Seabird Group, 10–13 January 1995, San Diego, Calif.

Ainley, D. G., W. J. Sydeman, S. A. Hatch, and U. W. Wilson. 1994. Seabird population trends along the west coast of North America: causes and the extent of regional concordance. Pages 119–133 *in* J. R. Jehl, Jr., and N. K. Johnson, editors. A century of avifaunal change in western North America. Cooper Ornithological Society, Studies in Avian Biology 15.

Alaback, P. B. 1982. Dynamics of understory biomass in Sitka spruce–western hemlock forests of southeast Alaska. Ecology 63:1932–1948.

Alaska Board of Game. 1994. Alaska state hunting regulations. (Effective dates 1 July 1994–30 June 1995). Alaska Department of Fish and Game, Juneau. 112 pp.

Alaska Department of Environmental Conservation. 1986. Water quality in Alaska, report 305(b) to the Environmental Protection Agency. Alaska Department of Environmental Conservation, Juneau. 30 pp.

Alaska Department of Fish and Game. 1994a. 1993 Alaska subsistence personal use salmon catches. Alaska Department of Fish and Game, Juneau. 1 p.

Alaska Department of Fish and Game. 1994b. Alaska wildlife harvest summary, 1992–1993. Alaska Department of Fish and Game, Division of Wildlife Conservation, Juneau. 491 pp.

Alaska Department of Fish and Game. 1994c. State of Alaska species of special concern list. Administrative list (1 March 1994) from the commissioner of the Alaska Department of Fish and Game, Juneau. 1 p.

Alaska Department of Fish and Game. 1995. 1994 salmon season, preliminary data. Alaska Department of Fish and Game, Commercial Fisheries Management and Development Division, Juneau. 2 pp.

Alt, K. T. 1987. Review of sheefish (*Stenodus leucichthys*) studies in Alaska. Alaska Department of Fish and Game Fishery Manuscript 3. 69 pp.

Ambrose, R. E., R. J. Ritchie, C. M. White, P. F. Schempf, T. Swem, and R. Dittrick. 1988. Changes in the status of peregrine falcon populations in Alaska. Pages 73–82 *in* T. J. Cade, J. H. Enderson, C. G. Thelander, and C. M. White, editors.

Peregrine falcon populations—their management and recovery. The Peregrine Fund, Inc., Boise, Idaho.

Amundsen, C. C., and E. E. C. Clebsch. 1971. Dynamics of the terrestrial ecosystem vegetation of Amchitka Island, Alaska. BioScience 21:619–623.

Anderson, B. A., and B. A. Cooper. 1994. Distribution and abundance of spectacled eiders in the Kuparuk and Milne Point oil fields, Alaska, 1993. Final report prepared for ARCO Alaska, Inc., and the Kuparuk River Unit, Anchorage, by Alaska Biological Research, Inc., Fairbanks. 71 pp.

Anderson, B. A., A. A. Stickney, R. J. Ritchie, and B. A. Cooper. 1995. Avian studies in the Kuparuk oil field, Alaska, 1994. Final report prepared for ARCO Alaska, Inc., and the Kuparuk River Unit, Anchorage, by Alaska Biological Research, Inc., Fairbanks. 29 pp.

Armbruster, W. S., and D. A. Guinn. 1989. The solitary bee fauna (Hymenoptera: Apoidea) of interior and arctic Alaska: flower associations, habitat use, and phenology. Journal of the Kansas Entomological Society 62:468–483.

Armstrong, R. H. 1977. Weather and climate. Pages 53–58 *in* M. L. Merritt and R. G. Fuller, editors. The environment of Amchitka Island, Alaska. Technical Information Center, Energy Research and Development Administration, Springfield, Va.

Armstrong, R. H. 1979. Where have all the dollies gone? Alaska Fish and Game Trails, March/April:18–19.

Armstrong, R. H. 1986. A review of arctic grayling studies in Alaska, 1952–1982. Biological Papers of the University of Alaska 23. 17 pp.

Armstrong, R. H. 1988. Alaska blackfish. Alaska Department of Fish and Game Wildlife Notebook Series. 2 pp.

Armstrong, R. H. 1995. The importance of fish to bald eagles in southeast Alaska, a review. Proceedings of a bald eagle symposium held in Juneau. In press.

Armstrong, R. H., and R. J. Gordon. 1995. Finding birds in the Juneau area. U.S. Forest Service, Juneau, Alaska. 80 pp.

Arvey, W. D. 1991. Stock status of anadromous Dolly Varden in waters of Alaska's North Slope. Alaska Department of Fish and Game Fishery Manuscript 91-3. 20 pp.

Arvey, W. D. 1993. Annual management report for sport fisheries in the Arctic–Yukon–Kuskokwim region, 1991. Alaska Department of Fish and Game Fishery Management Report 93-1. 119 pp.

Bailey, E. P. 1993. Introduction of foxes to Alaskan islands—history, effects on avifauna, and eradication. U.S. Fish and Wildlife Service Resource Publication 193. 53 pp.

Banks, R. C., and P. F. Springer. 1994. A century of population trends of waterfowl in western North America. Pages 134–146 *in* J. R. Jehl, Jr., and N. K. Johnson, editors. A century of avifaunal change in western North America. Cooper Ornithological Society, Studies in Avian Biology 15.

Barnes, B. M. 1989. Freeze avoidance in a mammal: body temperatures below 0°C in an arctic hibernator. Science 244:1521–1616.

Barr, W. 1991. Back from the brink: the road to muskox conservation in the Northwest Territories. The Arctic Institute of North America Komatik Series 3. 127 pp.

Barrie, L. A. 1986. Arctic air pollution: an overview of current knowledge. Atmosphere and Environment 20:643–663.

Bee, J. W., and E. R. Hall. 1956. Mammals of northern Alaska on the Arctic slope. University of Kansas, Museum of Natural History, Lawrence. 309 pp.

Bergstrom, D. J., C. Blaney, K. C. Shultz, R. R. Holder, G. J. Sandone, D. J. Scheiderhan, L. H. Barton, and D. Mesiar. 1992. Annual management report Yukon area, 1991. Alaska Department of Fish and Game Regional Information Report 3A92-26.

Bishop, M. A., and S. P. Green. 1994. Shorebird migration on the Copper River delta: 1991–1993. Final report to the National Fish and Wildlife Foundation by Copper River Delta Institute, U.S. Forest Service, Cordova, Alaska. 33 pp.

Boone, D. L. 1993. Annual report on caribou management at Adak Island, Alaska 1992–1993. U.S. Fish and Wildlife Service Report AMNWR 93/19, Adak, Alaska. 9 pp.

Brackney, A. W., and J. W. Hupp. 1993. Autumn diet of lesser snow geese staging in northeastern Alaska. Journal of Wildlife Management 57:55–61.

Brackney, A. W., and R. J. King. 1994. Aerial breeding pair survey of the arctic coastal plain of Alaska, 1993. U.S. Fish and Wildlife Service, Migratory Bird Management, Fairbanks, Alaska. 13 pp.

Britton, M. E. 1967. Vegetation of the arctic tundra. Pages 67–130 *in* H. P. Hansen, editor. Arctic biology: 18th biology colloquium (2nd ed.). Oregon State University Press, Corvallis.

Bundtzen, T. K., R. C. Swainbank, J. E. Wood, and A. H. Clough. 1991. Alaska's mineral industry, 1991. Division of

Economic Development, Division of Geology and Geophysical Surveys, and Division of Mining, Special Report 46. Anchorage, Alaska. 89 pp.

Burger, C. V., and A. C. Wertheimer. 1995. Pacific salmon in Alaska. Pages 343–347 *in* E. T. LaRoe, G. S. Farris, C. E. Puckett, P. D. Doran, and M. J. Mac, editors. Our living resources: a report to the nation on the distribution, abundance, and health of U.S. plants, animals, and ecosystems. U.S. Department of the Interior, National Biological Service, Washington, D.C.

Burgess, R. M., and J. R. Rose. 1993. Snow goose. 1992 Endicott Environmental Monitoring Program. Draft report prepared by Science Applications International Corporation, Anchorage, Alaska, for U.S. Army Corps of Engineers, Alaska District, Anchorage. 78 pp.

Burgess, R. M., A. A. Stickney, J. R. Rose, and R. J. Ritchie. 1994. Snow goose. 1990 Endicott Environmental Monitoring Program. Prepared by Science Applications International Corporation, Anchorage, Alaska, for the U.S. Army Corps of Engineers, Alaska District, Anchorage. 183 pp.

Burkholder, A. 1993. Abundance and length–age composition of northern pike near the confluence of the Pilgrim and Kuzitrin rivers, 1992. Alaska Department of Fish and Game Fishery Data Series 93-16. 25 pp.

Burns, J. J. 1964. Pingos in the Yukon–Kuskokwim delta, Alaska: their plant succession and use by mink. Arctic 17:202–210.

Burr, J. M. 1987. Synopsis and bibliography of lake trout (*Salvelinus namaycush*) in Alaska. Alaska Department of Fish and Game Fishery Manuscript 5. 50 pp.

Burris, O. E., and D. E. McKnight. 1973. Game transplants in Alaska. Alaska Department of Fish and Game Wildlife Technical Bulletin 4. 57 pp.

Byrd, G. V., and R. H. Day. 1986. The avifauna of Buldir Island, Aleutian Islands, Alaska. Arctic 39:109–118.

Cameron, R. D. 1994. Distribution and productivity of the central Arctic caribou herd in relation to petroleum development: case history studies with a nutritional perspective. Final report to the Federal Aid in Wildlife Restoration Project W-23-1 through W-24-3, Alaska Department of Fish and Game, Juneau. 35 pp.

Cameron, R. D., R. A. Lenart, D. J. Reed, K. N. Whitten, and W. T. Smith. 1995. Abundance and movements of caribou in the oilfield complex near Prudhoe Bay, Alaska. Rangifer 15:3–7.

Cameron, R. D., D. J. Reed, J. N. Dau, and W. T. Smith. 1992. Redistribution of calving caribou in response to oil field development on the Arctic Slope of Alaska. Arctic 45:338–342.

Clark, R. A. 1993. Stock status and rehabilitation of Chena River arctic grayling during 1991 and 1992. Alaska Department of Fish and Game Fishery Data Series 93-5. 66 pp.

Committee on Earth and Environmental Sciences. 1994. Our changing planet: the FY 1994 U.S. Global Change Research Program. National Science Foundation, Washington, D.C. 84 pp.

Conant, B., and D. J. Groves. 1994. Alaska–Yukon waterfowl breeding population survey, 15 May to 13 June 1994. U.S. Fish and Wildlife Service, Migratory Bird Management, Juneau, Alaska. 25 pp.

Conant, B., J. I. Hodges, D. H. Groves, S. L. Cain, and J. G. King. 1993. An atlas of the distribution of trumpeter swans in Alaska. Volume 1. U.S. Fish and Wildlife Service, Migratory Bird Management, Juneau, Alaska.

Conant, B., J. I. Hodges, and J. G. King. 1991. Continuity and advancement of trumpeter swan *Cygnus buccinator* and tundra swan *Cygnus columbianus* population monitoring in Alaska. Wildfowl Supplement 1:125–136.

Connors, P. G., J. P. Myers, and F. A. Pitelka. 1979. Seasonal habitat use by arctic Alaskan shorebirds. Studies in Avian Biology 2:101–111.

Cooper, B. A. 1984. Seasonal bird use of alpine and subalpine habitats in the upper Susitna River basin, Alaska. M.S. thesis, University of Alaska, Fairbanks. 112 pp.

Craig, P., and P. J. McCart. 1974. Fall spawning and overwintering areas of fish populations along routes of proposed pipeline between Prudhoe Bay and the Mackenzie delta 1972–1973. Chapter 3 *in* P. J. McCart, editor. Fisheries research associated with proposed gas pipeline routes in Alaska, Yukon, and Northwest Territories. Arctic Gas Biological Report Series 15.

Craig, P. C. 1987. Subsistence fisheries at coastal villages in the Alaskan Arctic, 1970–1986. Report prepared for the Minerals Management Service. Alaska Outer Continental Shelf Socioeconomic Studies Program Technical Report 129. 63 pp.

DeCicco, A. L. 1992. Long-distance movements of anadromous Dolly Varden between Alaska and the U.S.S.R. Arctic 45:120–123.

DeCicco, A. L. 1993a. Assessment of Dolly Varden overwintering in selected streams of the Seward Peninsula, Alaska, during 1992. Alaska Department of Fish and Game Fishery Data Series 93-20. 42 pp.

DeCicco, A. L. 1993b. Assessment of selected stocks of arctic grayling in streams of the Seward Peninsula, Alaska, during 1992. Alaska Department of Fish and Game Fishery Data Series 93-36. 46 pp.

DeCicco, A. L., and R. M. Barnes. 1992. Listing of guiding services for recreational fishing in the Arctic–Yukon–Kuskokwim (AYK) region. Alaska Department of Fish and Game Special Publication 92-3. 31 pp.

Derksen, D. V., W. D. Eldridge, and M. W. Weller. 1982. Habitat ecology of Pacific black brant and other geese moulting near Teshekpuk Lake, Alaska. Wildfowl 33:39–57.

Derksen, D. V., M. W. Weller, and W. D. Eldridge. 1979. Distributional ecology of

geese molting near Teshekpuk Lake, National Petroleum Reserve—Alaska. Pages 189–207 *in* R. L. Jarvis and J. C. Bartonek, editors. Management and biology of Pacific Flyway geese. Oregon State University Bookstores, Corvallis.

Dunaway, D. O. 1993. Status of rainbow trout stocks in the Agulowak and Agulukpak rivers of Alaska during 1992. Alaska Department of Fish and Game Fishery Data Series 93-41. 46 pp.

Earnst, S. L. 1991. The third international swan symposium: a synthesis. Wildfowl Supplement 1:7–14.

Edwards, M. E., and P. W. Dunwiddie. 1985. Dendrochronological and palynological observations on *Populus balsamifera* in northern Alaska, U.S.A. Arctic and Alpine Research 17:271–278.

Eisenhauer, D. I., and C. M. Kirkpatrick. 1977. Ecology of the emperor goose in Alaska. Wildlife Monographs 57. 61 pp.

Elliott, H. W. 1882. A monograph of the seal-islands of Alaska. U.S. Commission of Fish and Fisheries Special Bulletin 176. 176 pp.

Engel, L. J., and D. Vincent-Lang. 1992. Area management report for the recreational fisheries of northern Cook Inlet. Alaska Department of Fish and Game Fishery Management Report 92-3. 272 pp.

Evenson, M. J. 1993a. A summary of abundance, catch per unit effort, and mean length estimates of burbot sampled in rivers of interior Alaska, 1986–1992. Alaska Department of Fish and Game Fishery Data Series 93-15. 28 pp.

Evenson, M. J. 1993b. Seasonal movements of radio-implanted burbot in the Tanana River drainage. Alaska Department of Fish and Game Fishery Data Series 93-47. 27 pp.

Ewins, P. J., H. R. Carter, and Y. V. Shibaev. 1993. The status, distribution and ecology of inshore fish-feeding alcids (*Cepphus* guillemots and *Brachyramphus* murrelets) in the North Pacific. Pages 164–175 *in* K. Vermeer, K. T. Briggs, K. H. Morgan, and D. Siegel-Causey, editors. The status, ecology, and conservation of marine birds of the North Pacific. Canadian Wildlife Service Special Publication, Ottawa, Ontario.

Foster, S. 1985. Alaska blue book 1985. Department of Education, Division of State Libraries, Juneau, Alaska. 329 pp.

Gabrielson, I. N., and F. C. Lincoln. 1959. The birds of Alaska. The Stackpole Company, Harrisburg, Pa. 922 pp.

Gallaway, B., W. Gazey, and L. Moulton. 1987. Population trends for arctic cisco in the Colville River of Alaska as reflected by the commercial fishery. Biological Papers, University of Alaska, Fairbanks.

Gard, L. M., Jr. 1977. Geologic history. Pages 13–34 *in* M. L. Merritt and R. G. Fuller, editors. The environment of Amchitka Island, Alaska. Technical Information Center, Energy Research and Development Administration, Springfield, Va.

Geiger, H. J., and H. Savikko. 1990. Preliminary forecasts and projections for

1990. Alaska salmon fisheries. Alaska Department of Fish and Game Regional Information Report 5J90-03. 78 pp.

Gibson, D. D. 1981. Migrant birds at Shemya Island, Aleutian Islands, Alaska. The Condor 83:65–77.

Glynn, B., and S. Elliott. 1993. Situk River steelhead trout counts, 1992. Alaska Department of Fish and Game Fishery Data Series 93-29. 32 pp.

Greenberg, J. H. 1987. Language in the Americas. Stanford University Press, Stanford, Calif. 438 pp.

Greenberg, R. 1992. Forest migrants in non-forest habitats on the Yucatan Peninsula. Pages 273–286 in J. M. Hagan III and D. W. Johnston, editors. Ecology and conservation of Neotropical migrant landbirds. Smithsonian Institution Press, Washington, D.C.

Hall, E. R., and K. R. Kelson. 1959. The mammals of North America. The Ronald Press Company, New York. 1083 pp.

Hamilton, T. D., K. M. Reed, and R. M. Thorson. 1986. Glaciation in Alaska: the geologic record. Alaska Geological Society, Anchorage. 265 pp.

Handel, C. M., and R. E. Gill, Jr. 1992. Breeding distribution of the black turnstone. Wilson Bulletin 104:122–135.

Hansen, A. J. 1987. Regulation of bald eagle reproductive rates in southeast Alaska. Ecology 68:1387–1392.

Hansen, A. J., E. L. Boeker, J. I. Hodges, and D. R. Cline. 1984. Bald eagles of the Chilkat Valley, Alaska: ecology, behavior, and management. Final report of the Chilkat River Cooperative Bald Eagle Study, National Audubon Society, and U.S. Fish and Wildlife Service. National Audubon Society, New York. 27 pp.

Harding, R., and D. Jones. 1992. Peterson Creek and lake system steelhead evaluation, 1991. Alaska Department of Fish and Game Fishery Data Series 92-46. 33 pp.

Harding, R., and D. Jones. 1993. Karta River steelhead: 1992 escapement and creel survey studies. Alaska Department of Fish and Game Fishery Data Series 93-30. 25 pp.

Hatch, S. A. 1993. Population trends of Alaskan seabirds. Pacific Seabird Bulletin 20:3–12.

Henny, C. J., D. D. Rudis, T. J. Roffe, and E. Robinson-Wilson. 1995. Contaminants and sea ducks in Alaska and the circumpolar region. Environmental Health Perspectives 103 (Supplement 4):41–49.

Holder, K. 1994. Evolutionary divergence of rock ptarmigan: an Aleutian solution. Information North 20:1–4.

Hopkins, D. M. 1967. The Bering Land Bridge. Stanford University Press, Stanford, Calif. 495 pp.

Hopkins, D. M. 1972. The paleogeography and climatic history of Beringia during Cenozoic time. Inter-nord 12:121–150.

Hopkins, D. M., J. V. Matthews, Jr., C. E. Schweger, and S. B. Young, editors. 1982. Paleoecology of Beringia. Academic Press, New York. 489 pp.

Hughes, J. 1982. Spring concentration of bald eagles along the Stikine River estuary. Page 82 in W. N. Ladd and P. F. Schempf, editors. Proceedings of the Raptor Management and Biology Symposium. U.S. Fish and Wildlife Service, Alaska Regional Office, Anchorage.

Hultén, E. 1937. Outline of the history of arctic and boreal biota during the Quaternary period. Bokforlags Aktiebolaget Thule, Stockholm. 168 pp. + 43 plates.

Hultén, E. 1968. Flora of Alaska and neighboring territories. Stanford University Press, Stanford, Calif. 1008 pp.

Jaques, D. R. 1973. Reconnaissance botany of alpine ecosystems on Prince of Wales Island, southeast Alaska. M.S. thesis, Oregon State University, Corvallis. 133 pp.

Johnson, S. R. 1991. The status of snow geese in the Sagavanirktok River delta area, Alaska: a 12-year summary report: 1980–1991. Prepared for British Petroleum Exploration (Alaska) Inc., Anchorage, Alaska, by LGL Alaska Research Associates, Inc., Anchorage. 25 pp.

Johnson, S. J., and D. R. Herter. 1989. The birds of the Beaufort Sea. British Petroleum Exploration (Alaska) Inc., Anchorage. 372 pp.

Kertell, K. 1991. Disappearance of the Steller's eider from the Yukon–Kuskokwim delta, Alaska. Arctic 44:177–187.

Kessel, B., and D. D. Gibson. 1978. Status and distribution of Alaska birds. Studies in Avian Biology 1. 100 pp.

Kessel, B., and D. D. Gibson. 1994. A century of avifaunal change in Alaska. Pages 4–13 in J. R. Jehl, Jr., and N. K. Johnson, editors. A century of avifaunal change in western North America. Cooper Ornithological Society Studies in Avian Biology 15. 348 pp.

Kincheloe, K. L., and R. A. Stehn. 1991. Vegetation patterns and environmental gradients in coastal meadows on the Yukon–Kuskokwim delta, Alaska. Canadian Journal of Botany 69:1616–1627.

Kirchhoff, M. D., and J. W. Schoen. 1987. Forest cover and snow: implications for deer in southeastern Alaska. Journal of Wildlife Management 51:28–33.

Klein, D. R. 1965. Postglacial distribution patterns of mammals in the southern coastal regions of Alaska. Arctic 18:7–20.

Klein, D. R. 1968. The introduction, increase, and crash of reindeer on Saint Matthew Island. Journal of Wildlife Management 32:350–367.

Klein, D. R. 1979. The Alaska oil pipeline in retrospect. Transactions of the North American and Natural Resources Conference 44:235–246.

Klein, D. R. 1980. Conflicts between domestic reindeer and their wild counterparts: a review of Eurasian and North American experience. Arctic 33:739–756.

Klein, D. R. 1988. The establishment of muskox populations by translocation. Pages 298–318 in L. Nielsen and R. D. Brown, editors. Translocations of wild animals. Wisconsin Humane Society, Milwaukee, and Caesar Kleberg Wildlife Research Institute, Kingsville, Tex.

Klein, D. R. 1989. Northern subsistence hunting economies. Pages 96–111 in R. J. Hudson, K. R. Drew, and L. M. Baskin, editors. Wildlife production systems. Cambridge University Press, Cambridge, England.

Klein, D. R. 1991. Caribou in the changing North. Applied Animal Behavior Science 29:279–291.

Klein, D. R. 1995. The introduction, increase and demise of wolves on Coronation Island, Alaska. Pages 275–280 in L. N. Carbyn, S. H. Fritts, and D. R. Seip, editors. Ecology and conservation of wolves in a changing world. Canadian Circumpolar Institute Occasional Publication 35.

Kline, T. C., Jr., J. J. Goering, O. A. Mathisen, P. H. Poe, and P. L. Parker. 1990. Recycling of elements transported upstream by runs of Pacific salmon. I. N and C evidence in Sashin Creek, southeastern Alaska. Canadian Journal of Fisheries and Aquatic Sciences 47:136–144.

Kline, T. C., Jr., J. J. Goering, O. A. Mathisen, P. H. Poe, P. L. Parker, and R. S. Scalan. 1993. Recycling of elements transported upstream by runs of Pacific salmon. II. N and C evidence in the Kvichak River watershed, Bristol Bay, southwestern Alaska. Canadian Journal of Fisheries and Aquatic Sciences 50:2350–2365.

Lafferty, R., J. F. Parker, and D. R. Bernard. 1992. Stock assessment and biological characteristics of burbot in lakes of interior Alaska during 1991. Alaska Department of Fish and Game Fishery Data Series 92-20. 71 pp.

Larned, B., and G. Balogh. 1994. Eider breeding population survey, Alaska Arctic Coastal Plain, 1993. Progress report, U.S. Fish and Wildlife Service, Migratory Bird Management, Anchorage, Alaska. 15 pp.

Larson, L. L. 1990. Statistics for selected sport fisheries on the Anchor River, Alaska, during 1989 with emphasis on Dolly Varden char. Alaska Department of Fish and Game Fishery Data Series 90-57. 100 pp.

Larson, L. L. 1991. Statistics for Dolly Varden on the Anchor River, Alaska, during 1990. Alaska Department of Fish and Game Fishery Data Series 91-13. 52 pp.

Lay, J. S. 1994. Researchers probe Hagemeister Island reindeer die-off of 1992. Agroborealis 26:17–19.

Lean, C. F., F. J. Bue, and T. L. Lingnau. 1993. Annual management report 1992: Norton Sound—Port Clarence—Kotzebue. Alaska Department of Fish and Game Regional Information Report 3A93-15. 143 pp.

LeResche, R. E., R. H. Bishop, and J. W. Coady. 1974. Distribution and habitats of moose in Alaska. Le Naturaliste Canadien 101:143–178.

Loftus, A. J., W. W. Taylor, and M. Keller. 1988. An evaluation of lake trout (*Salvelinus namaycush*) hooking mortality in the upper Great Lakes. Canadian Journal of Fisheries and Aquatic Sciences 45:1473–1479.

MacDonald, S. O., and J. A. Cook. 1994. The mammals of southeast Alaska. University of Alaska Museum, Fairbanks. 152 pp.

Magoun, A. J. 1985. Population characteristics, ecology, and management of wolverines in northwestern Alaska. Ph.D. dissertation, University of Alaska, Fairbanks. 197 pp.

Manville, R. H., and S. P. Young. 1965. Distribution of Alaskan mammals. Bureau of Sport Fisheries and Wildlife Circular 211. 74 pp.

McBride, D. N. 1980. Homing of arctic char, *Salvelinus alpinus* (Linnaeus) to feeding and spawning sites in the Wood River lake system, Alaska. Alaska Department of Fish and Game Informational Leaflet 184 iv + 23 pp.

McCart, P. J. 1980. A review of the systematics and ecology of arctic char, *Salvelinus alpinus*, in the western Arctic. Canadian Technical Report of Fisheries and Aquatic Sciences 935. Western Region, Department of Fisheries and Oceans, Winnipeg, Manitoba, Canada.

Mills, M. J. 1994. Harvest, catch, and participation in Alaska sport fisheries during 1993. Alaska Department of Fish and Game Fishery Data Series 94-28, Anchorage.

Morrow, J. E. 1980. The freshwater fishes of Alaska. Alaska Northwest Publishing Company, Anchorage. 248 pp.

Moulton, L., and others. 1986. 1984 central Beaufort Sea fish study. Prudhoe Bay waterflood project. Environmental monitoring program. Volume 2, Chapter 3. LGL, Woodward-Clyde consultants and U.S. Army Corps of Engineers, Anchorage, Alaska. 204 pp.

Muir, D. C. G., R. J. Norstrom, and M. Simon. 1988. Organochlorine contaminants in arctic marine food chains: accumulation of specific polychlorinated biphenyls and chlordane-related compounds. Environmental Science and Technology 22:1071–1079.

Murie, O. J. 1935. Alaska–Yukon caribou. U.S. Bureau of Wildlife and Sport Fisheries. North American Fauna 54. 93 pp.

Murie, O. J. 1959. Fauna of the Aleutian Islands and Alaska Peninsula. U.S. Bureau of Wildlife and Sport Fisheries. North American Fauna 61. 406 pp.

Murray, B. M. 1992. Bryophyte flora of Alaskan steppes. Bryobrothera 1:9–33.

Murray, D. F. 1980. Balsam poplar in arctic Alaska. Canadian Journal of Anthropology 1:29–32.

Murray, D. F. 1987. Breeding systems in the vascular flora of arctic North America. Pages 239–262 *in* K. M. Urbanska, editor.

Differentiation patterns in higher plants. Academic Press, London.

Murray, D. F. 1992. Vascular plant diversity in Alaskan arctic tundra. Northwest Environmental Journal 8:29–52.

Murray, D. F., and R. Lipkin. 1987. Candidate threatened and endangered plants of Alaska. University of Alaska Museum, Fairbanks. 75 pp.

Murray, D. F., B. M. Murray, B. A. Yurtsev, and R. Howenstein. 1983. Biogeographic significance of steppe vegetation in subarctic Alaska. Pages 883–888 *in* Permafrost: fourth international conference proceedings. National Academy Press, Washington, D.C.

O'Clair, R. M., R. H. Armstrong, and R. Carstensen. 1992. The nature of southeast Alaska. Alaska Northwest Books, Seattle, Wash. 254 pp.

Page, G. W., and R. E. Gill, Jr. 1994. Shorebirds in western North America: late 1800's to late 1900's. Pages 147–160 *in* J. R. Jehl, Jr., and N. K. Johnson, editors. A century of avifaunal change in western North America. Studies in Avian Biology 15, Cooper Ornithological Society. 348 pp.

Pahlke, K. A. 1992. Escapements of chinook salmon in southeast Alaska and transboundary rivers in 1991. Alaska Department of Fish and Game, Fishery Data Series 92-32. 55 pp.

Parker, J. F. 1991. Status of coho salmon in the delta Clearwater River of interior Alaska. Alaska Department of Fish and Game, Fishery Manuscript 91-4. 39 pp.

Person, D. K., M. Kirchhoff, V. Van Ballenberge, G. C. Iverson, and E. Grossman. 1996. Alexander Archipelago wolf: a conservation assessment. U.S. Forest Service General Technical Report PNW-GTR-384. 42 pp.

Péwé, T. L., D. M. Hopkins, and J. L. Giddings. 1965. The Quaternary geology and archaeology of Alaska. Pages 355–374 *in* H. E. Wright, Jr., and D. G. Frey, editors. The Quaternary of the United States. Princeton University Press, Princeton, N. J. 922 pp.

Piatt, J. F., and R. G. Ford. 1993. Distribution and abundance of marbled murrelets in Alaska. The Condor 95:662–669.

Pielou, E. C. 1991. After the Ice Age: the return of life to glaciated North America. University of Chicago Press, Chicago. 366 pp.

Platte, R. M., and W. I. Butler, Jr. 1993. Waterbird abundance and distribution on Yukon Delta National Wildlife Refuge, Alaska. U.S. Fish and Wildlife Service Migratory Bird Management Project, Anchorage, Alaska. 11 pp. + figures.

Porsild, A. E. 1966. Contributions to the flora of southwestern Yukon Territory. Natural Museum of Canada, Bulletin 216:1–86.

Pruitt, W. O. 1967. Animals of the North. Harper and Row, New York. 173 pp.

Quinlan, S. E., and J. H. Hughes. 1990. Location and description of a marbled

murrelet tree nest site in Alaska. The Condor 92:1068–1073.

Quinlan, S. E., N. Tankersley, and P. D. Arneson. 1983. A guide to wildlife viewing in Alaska. Alaska Department of Fish and Game, Juneau. 170 pp.

Ralph, C. J. 1994. Evidence of changes in populations of the marbled murrelet in the Pacific Northwest. Pages 286–292 *in* J. R. Jehl, Jr., and N. K. Johnson, editors. A century of avifaunal change in western North America. Studies in Avian Biology 15, Cooper Ornithological Society.

Rausch, C. J. 1953. On the status of some arctic mammals. Arctic 6:91–148.

Rausch, R. L., and V. R. Rausch. 1968. On the biology and systematic position of *Microtus abbreviatus* Miller, a vole endemic to the Saint Matthew Islands, Bering Sea. Sonderdruck aus Zeitschrift für Säugetierkunde 33:65–99.

Reed, A., R. Stehn, and D. Ward. 1989. Autumn use of Izembek Lagoon, Alaska, by brant from different breeding areas. Journal of Wildlife Management 53:720–725.

Reimnitz, E., and D. K. Maurer. 1979. Effects of storm surges on the Beaufort Sea coast, northern Alaska. Arctic 32:329–344.

Ritchey, J. E., M. A. Perkins, and C. R. Goldman. 1975. Effects of kokanee salmon (*Oncorhynchus nerka*) decomposition on the ecology of a subalpine stream. Journal of the Fisheries Research Board of Canada 32:817–820.

Ritchie, R. J., and R. E. Ambrose. 1992. The status of selected birds in east-central Alaska. Canadian Field-Naturalist 106:316–320.

Robuck, O. W. 1985. The common plants of the muskegs of southeast Alaska. U.S. Forest Service Miscellaneous Publication. 131 pp.

Ruggles, A. K. 1994. Habitat selection of loons in southcentral Alaska. Hydrobiologia 279/280:421–430.

Russell, R. 1977. Rainbow trout life history studies in lower Talarik Creek–Kvichak drainage. Alaska Department of Fish and Game, Federal Aid in Fish Restoration Completion Report G-II-E. 48 pp.

Rutz, D. S. 1992. Age and size statistics for rainbow trout collected in the Susitna River drainage during 1991. Alaska Department of Fish and Game, Fishery Data Series 92-26. 26 pp.

Scheffer, V. B. 1951. The rise and fall of a reindeer herd. Scientific Monthly 73:356–362.

Schempf, P. F. 1989. Raptors in Alaska. Pages 144–154 *in* Proceedings of the western raptor management symposium and workshop. National Wildlife Federation, Washington, D.C.

Schmutz, J. A., S. E. Cantor, and M. R. Petersen. 1994. Seasonal and annual survival of emperor geese. Journal of Wildlife Management 58:525–535.

Schoen, J. W., and L. Beier. 1990. Brown bear habitat preferences and brown bear logging and mining relationships in southeast Alaska. Federal Aid in Wildlife

Restoration Project final report W-22-3, Job 4.17R. Alaska Department of Fish and Game, Juneau. 90 pp.

Schoen, J. W., R. W. Flynn, L. H. Suring, K. Titus, and L. R. Beier. 1994. Habitat capability model for brown bear in southeast Alaska. International Conference on Bear Research and Management 9:327–337.

Schoen, J. W., M. D. Kirchhoff, and J. H. Hughes. 1988. Wildlife and old-growth forests in southeastern Alaska. Natural Areas Journal 8:138–145.

Selkregg, L. L. 1974–1976. Alaska regional profiles—Volume 1: Southcentral Alaska. 255 pp.; Volume 2: Arctic. 218 pp.; Volume 3: Southwest. 313 pp.; Volume 4: Southeast. 233 pp.; Volume 5: Northwest. 265 pp.; Volume 6: Yukon. 346 pp. Arctic Information and Data Center, Anchorage, Alaska.

Senner, S. E. 1979. An evaluation of the Copper River delta as critical habitat for migrating shorebirds. Studies in Avian Biology 2:131–145.

Shacklette, H. T., L. W. Erdman, J. R. Keith, W. M. Klein, H. Krog, H. Persson, H. Skuja, and W. A. Weber. 1969. Vegetation of Amchitka Island, Aleutian Islands, Alaska. U.S. Geological Survey Professional Papers 648. 66 pp.

Smith, L. N., L. C. Byrne, C. B. Johnson, and A. A. Stickney. 1994. Wildlife studies on the Colville River delta, Alaska, 1993. Final report prepared for ARCO Alaska, Inc., Anchorage, by Alaska Biological Research, Inc., Fairbanks. 95 pp.

Stehn, R. A., C. P. Dau, B. Conant, and W. I. Butler, Jr. 1993. Decline of spectacled eiders nesting in western Alaska. Arctic 46:264–277.

Stickney, A. A., R. J. Ritchie, B. A. Anderson, and D. A. Flint. 1993. Tundra swan and brant surveys on the Arctic Coastal Plain, Colville River to Sagavanirktok River, 1993. Final report prepared for ARCO Alaska, Inc., Anchorage, by Alaska Biological Research, Inc., Fairbanks. 83 pp.

Stickney, A. A., R. J. Ritchie, B. A. Anderson, and D. A. Flint. 1994. Tundra swan and brant surveys on the Arctic Coastal Plain, Colville River to Sagavanirktok River, 1993. Final report prepared for ARCO Alaska, Inc., Anchorage, by Alaska Biological Research, Inc., Fairbanks. 62 pp.

Suchanek, P. M., and A. E. Bingham. 1992. Harvest estimates for selected enhanced roadside sport fisheries near Juneau, Alaska, during 1991. Alaska Department of Fish and Game Fishery Data Series 92-42. 40 pp.

Szarzi, N. 1992. Evaluation of lake trout stock status and abundance in Paxson Lake and Lake Louise. Alaska Department of Fish and Game Fishery Data Series 92-34. 46 pp.

Tack, S. L. 1980. Migrations and distribution of arctic grayling in interior and arctic Alaska. Alaska Department of Fish and Game Project F-9-12, 21(R-I). Federal Aid in Fish Restoration, annual performance report, 1979–1980. 32 pp.

Troy, D. M., and J. K. Wickliffe. 1990. Trends in bird use of the Point McIntyre reference area 1981–1989. Final report prepared for British Petroleum Exploration (Alaska) Inc., Anchorage, by Troy Ecological Research Associates, Anchorage. 46 pp.

Troy Ecological Research Associates. 1993a. Distribution and abundance of spectacled eiders in the vicinity of Prudhoe Bay, Alaska: 1992 status. Final report prepared for British Petroleum Exploration (Alaska) Inc., Anchorage. 14 pp.

Troy Ecological Research Associates. 1993b. Population dynamics of birds in the Point McIntyre reference area 1981–1992. Final report prepared for British Petroleum Exploration (Alaska) Inc., Anchorage. 30 pp.

U.S. Fish and Wildlife Service. 1987. National wildlife refuges of Alaska. Brochure 1987 795–176. 2 pp.

U.S. Fish and Wildlife Service. 1988. Alaska Maritime National Wildlife Refuge, comprehensive conservation plan and environmental impact statement. U.S. Fish and Wildlife Service, Anchorage. 175 pp.

U.S. Fish and Wildlife Service. 1990. Aleutian Islands unit of the Alaska Maritime National Wildlife Refuge. U.S. Fish and Wildlife Service Brochure 1990-793-161. 1 sheet.

U.S. Fish and Wildlife Service. 1994a. Subsistence management regulations for federal public lands in Alaska. U.S. Fish and Wildlife Service, Office of Subsistence Management, Anchorage, Alaska. 173 pp.

U.S. Fish and Wildlife Service. 1994b. Environmental assessment for removal of introduced caribou from Adak Island, Alaska. U.S. Fish and Wildlife Service, Homer, Alaska. 15 pp.

U.S. Fish and Wildlife Service. 1996. Endangered and threatened species and species of concern in Alaska. U.S. Fish and Wildlife Service, Division of Endangered Species, Anchorage, Alaska. 3 pp.

Van Cleve, K., F. S. Chapin III, C. T. Dyrness, and L. A. Viereck. 1991. Element cycling in taiga forests: state-factor control. BioScience 41:78–88.

Van Cleve, K., C. T. Dyrness, L. A. Viereck, J. Fox, F. S. Chapin III, and W. Oechel. 1983. Taiga ecosystems in interior Alaska. BioScience 33:39–44.

Viereck, L. A. 1979. Characteristics of tree-line plant communities in Alaska. Holarctic Ecology 2:228–238.

Viereck, L. A., and J. M. Foote. 1972. The status of *Populus balsamifera* and *P. trichocarpa* in Alaska. Canadian Field-Naturalist 84:169–173.

Wallmo, O. C., and J. W. Schoen. 1980. Response of deer to secondary forest succession in southeast Alaska. Forestry Science 26:448–462.

Warnock, N. D., and D. M. Troy. 1992. Distribution and abundance of spectacled eiders at Prudhoe Bay, Alaska: 1991. Final report prepared for British

Petroleum Exploration (Alaska) Inc., Anchorage, by Troy Ecological Research Associates, Anchorage. 21 pp.

Weber, P. K. 1986. Downstream effects of placer mining in the Birch Creek basin, Alaska. Alaska Department of Fish and Game Technical Report 86-7. 21 pp.

West, E. W. 1991. An annotated list of vertebrate species of ecological concern in Alaska. Alaska Natural Heritage Program, The Nature Conservancy, Anchorage. 104 pp.

West, E. W. 1993. Rare vertebrate species of the Chugach and Tongass national forests, Alaska. Alaska Natural Heritage Program, The Nature Conservancy, Anchorage, and U.S. Forest Service, Juneau, Alaska. 254 pp.

White, C. M. 1994. Population trends and current status of selected western raptors. Pages 161–172 in J. R. Jehl, Jr., and N. K. Johnson, editors. A century of avifaunal change in western North America. Studies in Avian Biology 15, Cooper Ornithological Society.

Wilk, R. J. 1993. Observations on sympatric tundra, *Cygnus columbianus*, and trumpeter swans, *C. buccinator*, in north-central Alaska, 1989–1991. Canadian Field-Naturalist 107:64–68.

Willson, M. F., and K. C. Halupka. 1995. Anadromous fish as keystone species in vertebrate communities. Conservation Biology 9 (3):489–497.

Young, S. B. 1971. The vascular flora of Saint Lawrence Island with special reference to floristic zonation in the arctic regions. Contributions from the Gray Herbarium of Harvard University 201:11–115.

Pacific Halibut in Glacier Bay National Park, Alaska

Ludwig, D., R. Hilborn, and C. Walters. 1993. Uncertainty, resource exploitation, and conservation: lessons from history. Science 260:17–36.

Quinn, T. J., R. B. Deriso, and S. H. Hoag. 1985. Methods of population assessment of Pacific halibut. International Pacific Halibut Commission Science Report 72. 52 pp.

Rosenberg, A. A., M. J. Fogarty, and M. P. Sissenwine. 1993. Achieving sustainable use of renewable resources. Science 262:828–829.

Skud, B. E. 1977. Drift migration and inter-mingling of Pacific halibut stocks. International Pacific Halibut Commission Science Report 63. 42 pp.

St-Pierre, G. 1984. Spawning locations and season for Pacific halibut. International Pacific Halibut Commission Science Report 70. 46 pp.

Hawaii and the Pacific Islands

Overview of the Pacific

The Pacific is the world's largest ocean—20,000 kilometers across from Singapore to Panama. The Pacific Islands are commonly divided into the three geographic areas of Polynesia, Micronesia, and Melanesia (Fig. 1). These areas together contain 789 habitable islands (Douglas 1969) ranging in size from New Guinea at 800,000 square kilometers to the tiniest coral atolls. Although these geographic divisions were originally based on the appearance and culture of the peoples who lived within their boundaries (Oliver 1989), they are also somewhat meaningful from geological and biogeographical standpoints.

The Melanesian islands (Melanesia), which are close to the Asian and Australian continents, are composed of rock that originated from volcanoes or sediments similar to those found on the continents. These relatively large islands usually have many kinds of animals and plants, but few species are limited to single islands or island groups. Within Melanesia, as one travels greater distances from the Asian mainland and the distance between islands becomes greater, the total number of animal and plant species found on each island decreases, but the number of species found only on each island (endemic species) increases.

Micronesia has a large number of very small islands and a total land area of less than 3,000 square kilometers. Micronesia has two main cultural and environmental divisions, with the volcanic Palau and Mariana islands of western Micronesia differing markedly from the atolls (low-lying, ringlike coral islands) of the Caroline, Marshall, and Gilbert island groups.

Polynesia includes 287 islands and is the largest of the Pacific geographic areas, although its land area of approximately 300,000 square kilometers is much less than that of Melanesia (Bellwood 1979). Polynesia is triangular, with Hawaii, New Zealand, and Easter Island at the apexes. New Zealand (268,570 square kilometers) and the Hawaiian Islands (16,558 square kilometers) are the largest island groups, respectively accounting for 89.5% and 5.5% of the land area of Polynesia. Polynesian islands generally lack sedimentary rocks other than recent alluvial deposits, and they are much smaller and more widely separated than the islands of Melanesia. The islands of Polynesia (except Tonga and New Zealand) are formed of basaltic rocks extruded from ancient volcanoes. The only active volcanoes in Polynesia are in Hawaii and New Zealand. Although the best-known Polynesian islands are high volcanic islands with jagged profiles, deep gorgelike valleys, and cascading waterfalls (for example, Society, Hawaiian, and Samoa islands), Polynesia has many atolls. Animals and plants of Polynesian high islands originated from ancestors that came across thousands of kilometers of empty ocean; the islands generally have few animal and plant species, many of which are found nowhere else.

Many kinds of political systems exist on the Pacific islands. Those islands associated with the U.S. government, in Polynesia and Micronesia, are listed in the Table. We have focused this discussion of the status and trends of nonmarine animals and plants and their ecosystems in the Pacific on the state of Hawaii, which includes all the Hawaiian Islands. To a certain extent, status and trends of Hawaii's biological resources and habitats are similar to those throughout the Pacific region,

Courtesy B. Gagné, Hawaii Department of Land and Natural Resources

☐ Polynesia
☐ Micronesia
☐ Melanesia

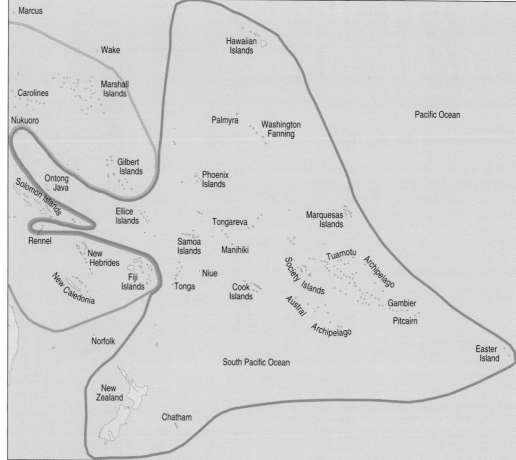

——— Polynesia

——— Micronesia

——— Melanesia

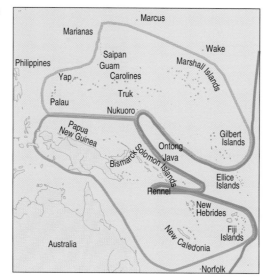

Fig. 1. The three geographic areas of the Pacific, based on ethnicity of the human inhabitants, but also somewhat relevant geologically and biologically (after Oliver 1989).

although important differences do exist. The Hawaiian Islands are larger, have more topographic diversity, and have higher biological uniqueness than most other Pacific islands. Although invasive nonindigenous species are problematic throughout the Pacific, Hawaii is more in the mainstream of commerce, thus more nonindigenous animals and plants are accidentally carried there by planes and ships.

Finally, the Hawaiian Islands have been studied more intensively and received more conservation management attention than most other Pacific islands (except New Zealand), primarily because Hawaii is a U.S. state and because the Hawaiian Islands have a set of unique animals and plants that are renowned for their evolution in geographic isolation.

Overview of the Hawaiian Islands

The Hawaiian archipelago, the most isolated island group of comparable size and topographic diversity on Earth, is about 4,000 kilometers from the nearest continent and 3,200 kilometers from the nearest high-island group (the Marquesas Islands of French Polynesia). The state of Hawaii consists of 132 islands, reefs, and shoals stretching 2,400 kilometers in a northwest–southeast direction between latitudes 28°N and 19°N (Price 1983). The eight major high islands, located at the southeast end of the chain (Fig. 2), make up more than 99% of the total land area. The youngest island, Hawaii, has an area of more than 10,000 square kilometers (63% of the total area of the state) and has elevations of more than 4,000 meters. The island of Hawaii has five volcanoes, two of which are highly active. These islands are part of a longer chain that was produced during at least a 70-million-year period by the northwestward movement of the ocean floor over a hot spot below the Earth's crust (Fig. 3). The oldest rocks of the eight high islands range from 420,000 years old (the island of Hawaii) to about 5 million years old (Kauai). The relatively tiny eroded and submerged islands and reefs extending to the northwest are remnants of high islands that existed millions of years ago (Macdonald et al. 1983).

Rainfall averages from 63 to 75 centimeters per year over the open ocean near Hawaii, yet the islands themselves receive up to 15 times as much rain in some places and less than one-third as much rain in other places. These great differences in rainfall are the result of the moisture-laden trade winds flowing from the northeast over the steep, complicated terrain of the islands (Price 1983). The resulting combinations of temperature and rainfall (and some snow at the highest elevations) account for nearly 95% of the climatic variation in the Earth's tropics (George et al. 1987).

The Hawaiian animals and plants began to evolve as much as 70 million years ago in nearly complete isolation; successful colonization through long-distance transport of species from elsewhere was infrequent. Many groups of organisms common on continents were never able to successfully make the journey to Hawaii. Hawaii lacks any native examples of ants, conifers, or most bird families, for example, and has only one native land-dwelling mammal (a bat). The low number of colonizers

Table. Pacific islands associated with the government of the United States[a] (H. Smith and others, U.S. Fish and Wildlife Service, personal communication).

Island(s)	Area (square kilometers)	Number of islands	1990–1991 population
Polynesia			
Hawaiian Islands			
State of Hawaii	16,558	132	1,108,229
Midway Atoll	5.2	2	453
Samoa Islands, including U.S. Territory of American Samoa	199	7	43,052
Central Pacific islands, including:			
Johnston Atoll	2.8	1	1,325
Palmyra Atoll	10	1	–
Kingman Atoll	1.0	1	–
Howland Island	about 2	1	0[b]
Baker Island	about 2	1	0[b]
Jarvis Island	4.5	1	–
Micronesia			
Mariana Islands			
Commonwealth of the Northern Mariana Islands	759	9	23,494
Territory of Guam	541	1	144,928
Caroline Islands, including Republic of Palau	458	200+	107,662
Federated States of Micronesia, Marshall Islands (the Republic of the Marshall Islands)	180	31	48,091
Wake Atoll	6.5	3	195

[a] Subunits of Pacific Islands Ecoregion of the U.S. Fish and Wildlife Service.
[b] Island uninhabited.
– Data not available

has been partially offset by enrichment of biological diversity through evolution after establishment of these species. The few animals and plants that reached Hawaii over thousands of kilometers of open ocean—on the winds, by floating, or attached to storm-driven birds—arrived in a remarkably diverse potential habitat (Carlquist 1974).

The percentage of endemic species (species found nowhere in the world except Hawaii) is very high. The land-dwelling animals and plants of Hawaii (including liverworts, mosses, fungi,

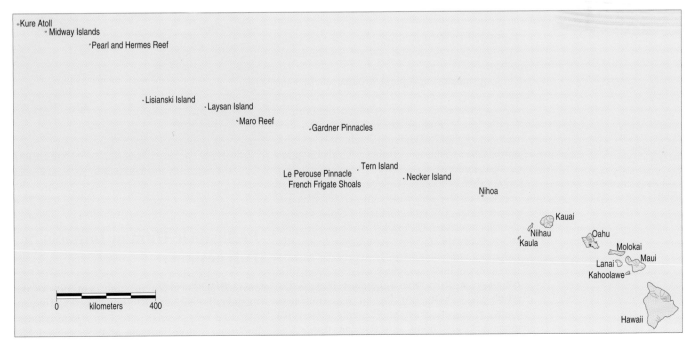

Fig. 2. The Hawaiian Islands.

Fig. 3. Map showing islands produced by a hot spot below the Earth's crust, beginning with the Emperor Seamounts 70 million years ago and actively extending the Hawaiian chain at present. The numbers alongside the islands are ages in millions of years (© University of Hawaii Press).

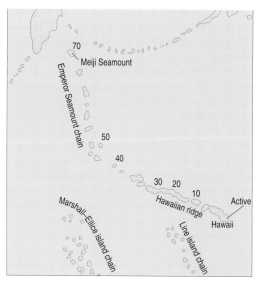

in the family Campanulaceae (including the endemic genera for brighamia, clermontia, cyanea, rollandia, and false lobelia; Wagner et al. 1990).

Among birds, the Hawaiian honeycreepers (Fig. 4), and among insects, the vinegar flies (Kaneshiro 1988), are the best-known examples of adaptive radiation in Hawaiian animals.

The Galapagos Islands have gained considerable fame from Charles Darwin's 1832 visit there and his observations of classic radiations among the island group's animals. These observations led to the birth of Darwin's theory of evolution. The Hawaiian Islands, however, have become an even more important site for modern evolutionary studies because their native animals and plants have evolved much longer in isolation and have a much greater variety of habitats to occupy.

and lichens, in addition to the groups listed in this section) are believed to have evolved from roughly 2,000 ancestors that arrived on the islands by chance. Based on the age of the islands, an average of only one successful immigrant arrived every 35,000 years. Researchers believe that a flora of about 960 species of flowering plants (Wagner et al. 1990) has evolved from about 270 colonizing ancestors (Fosberg 1948); 168 species of ferns and fern allies evolved from about 135 original immigrants (Fosberg 1948). The 6,000–10,000 insects and allied forms native to Hawaii evolved from about 300 to 400 ancestral immigrant species (Hardy 1983; Gagné and Christensen 1985); about 1,200 native land snails evolved from as few as 22–24 long-distance immigrants, probably carried by birds (Zimmerman 1948); and about 115 endemic land birds (including some species only known as fossils) evolved from as few as 20 ancestral immigrants (Olson and James 1982; James 1995).

Some animal and plant groups reached the geologically developing Hawaiian Islands more slowly than others, leaving much opportunity for the early immigrants to evolve into new roles and habitats. Beginning with only a single colonizing species, certain animal or plant groups underwent a sequence of speciation events that produced large numbers of related species that live in a wide range of habitats and play a variety of ecological roles. Only a few of the successful colonizing species are the sources of spectacular evolutionary adaptive radiation. Only about 1 in 10 successful flowering plant colonists was the source of a radiation (Wagner et al. 1990). Examples include the silversword-tarweed group in the sunflower family (comprised of three unique genera for dubautia, silversword, and wilkesia; Robichaux et al. 1990; Baldwin et al. 1991) and the lobelias

Original art © H. D. Pratt, Louisiana State University Museum of Zoology

Fig. 4. The impressive adaptive radiation in Hawaiian honeycreepers.

Prehistoric Human Impacts in Polynesia

Until about 15 years ago, most biologists mistakenly believed that the native Pacific peoples had only a small effect on island habitats and their native animals and plants and that drastic modification of island environments is almost entirely due to more recent effects brought on by European and American cultures. This misinterpretation was partly a result of the cultural influence of the "noble savage" view championed by Rousseau in the eighteenth century, but was even more attributable to a lack of interdisciplinary research. In fact, Polynesians were changing their environments in a major way for thousands of years before their first contacts with Europeans. Polynesians changed native habitats through cutting, burning, and the introduction of nonindigenous plants for agriculture (Kirch 1997). The most convincing evidence for environmental change is from bird bones of species forced to extinction, found in archaeological sites, lava tubes, and sand deposits (Kirch 1982; Olson and James 1982; Steadman 1995).

Prehistoric extinctions of animals and plants that resulted from human activity have been documented throughout the world (Martin 1984). Effects of prehistoric humans in altering island environments may have been more consistently severe because islands are smaller than continents and lack alternative space for species survival. Virtually all habitable islands in the tropical Pacific, except for the Galapagos archipelago, were inhabited at some time in prehistory by humans (Steadman 1995). Although the postglacial fauna of the Galapagos was unaffected until its discovery by Europeans in 1535, major human-caused loss of biodiversity had occurred before European contact almost everywhere else, based on studies by Olson and James in the Hawaiian Islands and by Steadman and his colleagues in the Marquesas Islands, Society Islands, and Cook Islands, and on Henderson Island, Easter Island, Samoa, Tonga, and Polynesian outliers in Melanesia (summarized by Steadman 1995, 1997).

Numbers of bird extinctions caused by humans in Polynesia before European discovery of the islands are far greater than those since then. More than 2,000 bird species, mostly flightless rails (small chickenlike birds), may have been eliminated by prehistoric humans in Polynesia (Steadman 1995).

Seabird populations supplied a major food source for early Polynesians, and their exploitation by native peoples led to significant decreases in the number of seabird species and the size of remaining populations. An analysis of 11,000 bird bones from an archaeological site on Ua Huka, Marquesas Islands, shows that seven species of shearwaters and petrels were abundant and easily obtained from their nesting burrows. Six of the 20 seabird species from the site are no longer found near the island. Today, most of the other 14 seabird species are found only on tiny offshore islets (Steadman 1997).

Humans have caused an estimated 100–1,000-fold loss of nesting seabirds in the Pacific during the past 3,000 years (Steadman 1997). Direct removal of birds and eggs by humans was likely the major factor in this reduction. A second important factor is likely to have been the Polynesian rat, which was introduced throughout Polynesia prehistorically, either intentionally or as a stowaway in voyaging canoes (Steadman 1997). The extremely conservative breeding strategy of seabirds, who normally produce no more than a single young bird per breeding pair per year, makes them quite vulnerable to predation by rats or humans (Atkinson 1985).

The Hawaiian Islands were reached by colonizing Polynesians, probably from the Marquesas Islands, in about the fourth century A.D., somewhat later than most other Polynesian islands. The human population reached 200,000 before contact with Europeans in 1778. Landscapes were modified through the use of shifting cultivation and fire and through the creation of sizeable wetlands for aquaculture. Throughout the Hawaiian Islands, most land below 600 meters elevation with even moderately good soils was cultivated by the Hawaiians in the thirteenth to eighteenth centuries (Kirch 1982). Polynesians introduced pigs, jungle fowl, dogs, Polynesian rats, and various stowaway geckos, skinks, and snails (Kirch 1982), as well as at least 32 plant species, including major food plants and species providing source materials for the manufacture of cloth, rope, and musical instruments (Nagata 1985).

About half of Hawaii's native land birds were extinct before European scientists could observe them. As more extinct birds are discovered in lava tubes and sand dune deposits, the proportion may have to be revised upward (James 1995).

If the known pollen record is a good indicator, plant species may have been more resistant to prehistoric extinction than animals (James 1995). One instance of plant loss was recently shown by a lost pollen type, which was once abundant in Oahu's lowland pollen record but did not match the pollen of any known living plant (Athens et al. 1992). Meanwhile, two plants whose pollen matched the once-abundant type were found on a cliff on the island of Kahoolawe and were described as a new genus and species (Lorence and Wood 1994). The best guess for the unusual match is that the disappearance of the plant from Oahu was due to habitat modification by the early Hawaiians and predation by the introduced Polynesian rat (L. Mehrhoff, U.S. Fish and Wildlife Service, personal communication).

Human Effects on Hawaii after Western Contact

Destruction of native ecosystems and losses of native plants and animals accelerated after the arrival of Captain James Cook's ships in the Hawaiian Islands in 1778. The greatest early effects were from grazing and browsing animals—especially feral cattle, goats, and sheep. These animals continue to affect their surroundings, but feral pigs are now causing the most destruction to remaining native ecosystems. Commercial exploitation of sandalwood and firewood for whaling ships, sugar and pineapple production, logging of koa and ohia, ranching, and real estate development have also progressively destroyed native habitats over the past two centuries (Cuddihy and Stone 1990). Although quite significant direct conflicts exist in Hawaii between proponents of economic development and those supporting preservation of native biota and natural resources, the greater conflicts are indirect, through continued introduction of aggressive animals and plants from elsewhere in the world, both intentionally and inadvertently, by modern transportation and commerce.

Effects of Invasive Animals and Plants

Oceanic islands throughout the world are notoriously vulnerable to biological invasions. Islands experience long periods of evolution in isolation from those forces faced routinely by plants and animals on continents (such as browsing and trampling by herbivorous mammals, predation by ants, virulent disease, and frequent and intense fires). This isolation contributes to the vulnerability of islands to biological invasion (Loope and Mueller-Dombois 1989). Smaller numbers of native species on isolated islands and the intensity of human impacts on small land areas of islands have clearly made the situation worse by increasing most islands' susceptibility to invasion.

Upon the arrival of the Polynesians in the fourth century A.D., the rate of species immigration began to accelerate from 1 every 35,000 years over a 70-million-year period to about 3–4 per century over about 1,400 years (Kirch 1982; Nagata 1985). In modern times, Beardsley (1979) found that on the average, 15–20 species of immigrant insects alone become established in Hawaii each year (Fig. 5). The Hawaiian archipelago has more than 8,000 introduced plant species or cultivars (Yee and Gagné 1992), an average of 40 introductions per year over the past two centuries; 861 (11%) of these are now established and have

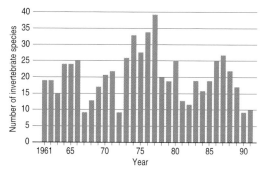

Fig. 5. Number of immigrant invertebrates reported in Hawaii, 1961–1991. Adapted from graphic by The Nature Conservancy of Hawaii and the Natural Resources Defense Council (1992); data were extracted from the Proceedings of the Hawaiian Entomological Society.

reproducing populations (Wagner et al. 1990). Smith (1985) listed 86 invasive nonindigenous plant species present in Hawaii that pose threats to native Hawaiian ecosystems, but this number needs to be revised upward, because new invaders and changing trends of known invaders have become apparent during the past decade.

The following accounts provide sketches of the biology and effects of some of the most destructive invading species in Hawaii. Efforts to control these species for conservation purposes have achieved varying levels of success. In addition, species already present may increase their destructiveness in the future, and new species will invade. Efforts to prevent new invasive species from establishing in Hawaii are extremely important (The Nature Conservancy of Hawaii and the Natural Resources Defense Council 1992; U.S. Congress, Office of Technology Assessment 1993). Active management is needed to protect native Hawaiian ecosystems from being eventually overwhelmed by forces that began to act 1,500 years ago when the Polynesians arrived on Hawaiian shores.

Feral Pigs

Feral pigs are currently the primary modifiers of remaining Hawaiian rain forest and have substantial effects on other ecosystems. Although pigs were brought to the Hawaiian Islands by Polynesians as early as the fourth century A.D., the current severe environmental damage inflicted by pigs apparently began much more recently and seems to have resulted entirely from release of domestic, non-Polynesian genotypes (Diong 1982). Polynesian pigs were much smaller, more docile, and less prone to taking up a feral existence than those introduced in historical times (Tomich 1986). In the early 1900's, the damage caused by feral European pigs in native rain forests was recognized; the Hawaii Territorial Board of Agriculture and Forestry subsequently started a feral pig eradication project that lasted until 1958 and removed 170,000 pigs from forests statewide (Diong 1982).

Perhaps the best example of how recently pigs have damaged native ecosystems comes from the island of Maui. Although goats, cattle, and wild dogs were reported by nineteenth-century explorers of the island, they made no mention of pigs (Diong 1982). Feral pigs were absent in West Maui until someone introduced them in the 1960's; although they have since spread throughout the northeastern portion of the West Maui Mountains, the remainder of high-elevation West Maui has never been degraded by pigs (R. W. Hobdy, Hawaii Department of Land and Natural Resources, personal communication). Feral pigs were first seen in high-elevation East Maui (Haleakala Crater) in the 1930's at Paliku (elevation 1,920 meters). These pigs were probably descended from runaway domestic breeds in lowland areas. Perhaps aided by a seasonally abundant and expanding carbohydrate source—the invading nonindigenous strawberry guava—and by an enhanced protein source from abundant non-indigenous earthworms, truly feral pig populations developed and spread into adjacent pristine forest. By 1945 pigs had moved into the upper Kipahulu Valley of East Maui, though an expedition in that year found no pigs in the valley between 610 and 1,375 meters elevation. By 1967 pig damage could be found for the first time throughout Kipahulu Valley, although damage at that time was still moderate. After 1967 pig densities greatly increased. By 1979–1981, pig densities in Kipahulu Valley ranged from 5 to 31 per square kilometer (Diong 1982). Similar trends in feral pig populations have unfolded across the entire north and northeastern slopes of East Maui during the past 30 years (Hobdy, personal communication).

In addition to eating the foods more common in their diets (starch from tree-fern trunks, strawberry guava fruits, earthworms), pigs selectively seek out certain currently rare plant species for food. Plants with particularly fragile stems and leaves have drastically declined because of such predation by pigs. The ground cover of mosses and small ferns has probably been altered most, based on observations made in the few remaining pig-free areas of the Hawaiian Islands—areas protected by surrounding cliffs such as Lihau on West Maui, Olokui on Molokai, and a few kipukas (areas with soil and vegetation surrounded by recent lava flows) on Hawaii. These mosses and ferns have, for the most part, not been totally eliminated because they survive as epiphytes (plants physically growing on other plants) on tree trunks (especially trunks of native tree-ferns)

and on downed logs. In rain forests of East Maui, native tree-ferns—originally the dominant subcanopy species at elevations up to 1,500 meters—are rapidly being depleted by pig predation. As mature tree-ferns become further reduced in abundance, fewer individuals of rare species will be able to survive as epiphytes.

Opportunistic plant species, often non-indigenous, occupy the habitats remaining after feral pigs have eliminated native species. Seeds of nonindigenous plants are carried on pigs' coats or in their digestive tracts, and they thrive upon germination on the forest floor where pigs have exposed mineral soil. Once these aggressive plant invaders have obtained a new foothold in the forest, they spread opportunistically, aided by pigs and nonindigenous birds. The spread of nonindigenous plants has been better documented for Maui's Kipahulu Valley than for most remote areas in Hawaii (Yoshinaga 1980; Anderson et al. 1992). Removal of feral pigs from the valley in the late 1980's by the National Park Service (Anderson and Stone 1993) has substantially slowed the rate of nonindigenous plant invasion and has already resulted in substantial recovery of the forest understory.

Feral Goats

The negative effect of goats on vegetation is well known worldwide. Damage caused by goats in Hawaii is worse than in other places because Hawaiian plant species evolved in the absence of mammalian grazing and browsing pressure. Goats were introduced to the eight major Hawaiian islands before 1800 and within a few decades had reached remote areas (Cuddihy and Stone 1990). Goats have reduced or eliminated whole populations of native plants, have aided in nonindigenous plant invasion, and have hastened soil erosion (Yocum 1967; Baker and Reeser 1972; Spatz and Mueller-Dombois 1973; Mueller-Dombois and Spatz 1975). Sustained, locally high populations of goats have obliterated even the most unpalatable native plant species. Indeed, some plant species survive only on ledges (Fig. 6) and other sites that are inaccessible to goats, or they are older trees that had reached a size that made them less vulnerable when goats occupied their habitats (Stone and Loope 1987).

Goats were eradicated on Niihau, Lanai (1981), and Kahoolawe (1990). Eradication of feral goats on uninhabited Kahoolawe (a former bombing range, 115 square kilometers) was conducted by the U.S. Navy before the island was returned to the state of Hawaii; unfortunately, most native vegetation had already been destroyed. Feral goats, once abundant in Hawaii's two largest national parks—Hawaii Volcanoes (Fig. 7) and Haleakala—

Fig. 6. *Haleakala schiedia*, endemic to Haleakala Crater, survives only on ledge or cliff faces out of the reach of feral goats.

were virtually eliminated there in the 1970's and 1980's with the aid of fencing (Fig. 8), resulting in marked changes in vegetation dynamics (Mueller-Dombois 1981a; Stone et al. 1992; Tunison et al. 1995) and at least partial recovery of rare plant populations (Fig. 9; Loope and Medeiros 1994a).

Fig. 7. Feral goats in Hawaii Volcanoes National Park before boundary fencing and control implemented in the 1970's.

Fig. 8. Fencing constructed in the 1980's has eliminated goats within Haleakala National Park and allowed partial recovery of native vegetation.

Fig. 9. Reproduction of *Lysimachia kipahuluensis*, an East Maui endemic, in western Kaupo Gap, an area of Haleakala National Park where it was once eliminated (except on ledges and cliff faces) by feral goats.

Cattle

Feral cattle were abundant in the Hawaiian Islands in the 1800's and early 1900's (Tomich 1986). Cattle have contributed substantially to the decline of many plant species (compare Rock 1913). Most cattle grazing is now on private, managed lands, but wild cattle persist locally, numbering in the thousands in the forested areas on the island of Hawaii (P. M. Vitousek, Stanford University, personal communication). Relatively small numbers of feral or domestic cattle can do appreciable damage to native vegetation. Cattle are a continuing significant threat to the Hawaiian flora, especially in certain coastal and lowland leeward habitats.

Rats and Mice

Of four rodent species introduced to the Hawaiian Islands, the arboreal black rat probably has the greatest effect on native fauna and flora (Stone and Loope 1987). Rodents feed on the fleshy fruits and flowers of Hawaiian plants or they girdle and strip tender branches (Cuddihy and Stone 1990). Past and continuing effects of black rats on birds, snails, and insects of Hawaiian rain forests have been enormous (Atkinson 1977; Hadfield 1986; Stone and Loope 1987).

Mosquitoes and Avian Malaria

Avian malaria is believed to be one of the most important factors limiting Hawaiian forest bird populations. The southern house mosquito, a vector of malaria, arrived in Hawaii before the protozoan parasite was established. The parasite was originally carried by nonindigenous game birds brought to the islands in the early 1900's for recreational hunting (van Riper et al. 1986). Van Riper et al. (1986) demonstrated that the highest incidence of malaria is in wet midelevation forests between 900 and 1,500 meters, where vector mosquito populations overlap the ranges of highly susceptible native birds. Current investigations support those observations (Jacobi and Atkinson 1995).

Invasive Plants

Approximately 90 of the roughly 900 naturalized nonindigenous plant species in Hawaii pose significant threats to native ecosystems. Among the most destructive invading plants are various grasses, banana poka (Jacobi and Warshauer 1992; La Rosa 1992), strawberry guava (Jacobi and Warshauer 1992; Loope et al. 1992), firetree (Vitousek and Walker 1989; Whiteaker and Gardner 1992), kahili ginger (Smith 1985; Fig. 10), Australian tree-fern (Medeiros et al. 1992, 1993), and clidemia.

The spread of the invasive plant clidemia represents the worst of plant invaders in Hawaii

Fig. 10. Dense growth of kahili ginger in Kipahulu Valley, Haleakala National Park, Maui.

and the Pacific. This densely branching shrub (to 4 meters tall) is native to the New World tropics (southern Mexico and West Indies to Argentina). Clidemia has become an aggressive invader in parts of India, Southeast Asia, and the Pacific Islands, including the Hawaiian Islands, where it was introduced about 1940. In Hawaii, clidemia forms dense stands in moist to wet environments. The invasion of this plant is particularly severe on Oahu, where it expanded its distribution, mainly by bird dispersal, from less than 100 hectares in 1952 to an estimated 31,000–38,000 hectares in the late 1970's. More recently, the area occupied by clidemia approaches the total habitat available and is estimated at 100,000 hectares (Smith 1992). It has now spread to Hawaii (1972), Molokai (1973), Maui (1976), Kauai (1982), and Lanai (1988). Its elevational range is from just above sea level to about 1,220 meters. The primary mode of inter-island dispersal of clidemia is believed to be as seeds in mud on people's shoes. Because of its status as one of the most ecologically disruptive plant invaders in Hawaii and the Pacific, clidemia has been a focus of biological control efforts, which involve the import of natural enemies (usually insects or diseases) from the native range of the plant. Several insects that eat the leaves of this plant show promise of limiting the further invasion of clidemia on islands other than Oahu (Nakahara et al. 1992; Smith 1992).

Some of the most disruptive weeds are such formidable competitors that they can gradually convert native forest to single-species stands, particularly when invasion is accompanied by disturbances such as a hurricane (Fig. 11) or ohia dieback. Others can alter ecosystem-level processes such as nutrient cycling, sometimes aiding the invasion of other nonindigenous species (Vitousek 1986; Vitousek and Walker 1989; Aplet 1990). Managers of natural areas in Hawaii are actively combating plant invasions (Holt 1992; Tunison 1992a,b). These efforts should be accompanied by a reduction of entry and establishment of weeds and by developing regional approaches to control (Tunison et al. 1992).

The Special Case of Invasive Nonindigenous Grasses

Alteration of the fire cycle in Hawaii by nonindigenous grasses merits special mention. In contrast to the influence of fire in many terrestrial environments of the world, it does not seem to have played an important evolutionary role in most native ecosystems of the Hawaiian Islands, and relatively few Hawaiian endemic plant species possess adaptations to fire (Mueller-Dombois 1981b). Lightning is uncommon on oceanic islands because small island

Courtesy A. Medeiros, USGS

Fig. 11. Downed trunks of ohia, Alakai Swamp, Kauai, September 1983. Severe hurricanes have affected Hawaiian Island forests in recent years only on the island of Kauai. Forest openings caused by the 1982 and 1992 hurricanes on Kauai have greatly accelerated invasion by nonindigenous plant species.

land mass is not conducive to convective buildup of thunderheads. Many native Hawaiian ecosystems may have lacked adequate fuel to carry fires ignited by lightning or vulcanism (Mueller-Dombois 1981b). Fires in modern Hawaii are mostly human-caused, fueled primarily by nonindigenous grasses, and generally highly destructive to native plant species.

The major grasses fueling fires in Hawaii are beardgrass, broomsedge, buffel grass, fountain grass, and molassesgrass. Invasion by these grasses in otherwise undisturbed native ecosystems adds enough fine fuel to previously fire-free sites to carry fire. Most native species are eventually eliminated by fire (if not by the first fire, then by later ones), whereas invasive grasses recover rapidly after fire, increase flammability of the site, and become increasingly dominant after repeated fires (Hughes et al. 1991; D'Antonio and Vitousek 1992).

Incipient or Recent Invaders: Brown Tree Snake, Miconia, and Two-spotted Leafhopper

The following accounts provide sketches of some recent invaders and potential invaders of some areas of the Pacific Islands and the potentially catastrophic consequences they may have. The brown tree snake is native to the Solomon Islands, Papua New Guinea, and northern Australia. This species became established on Guam about the time of World War II and eventually attained population densities of

4,000–12,000 snakes per square kilometer, feeding on birds, rats, shrews, and lizards. Nine of the 11 native bird species present on Guam in 1945 have been eliminated by the snake (Savidge 1987). The high densities of snakes in general on Guam and brown tree snakes' nocturnal habits make them more likely to be undetected stowaways in air and ship cargo.

The brown tree snake is not known to be established in the Hawaiian Islands, but it has been singled out as a serious and imminent potential invader there (U.S. Congress, Office of Technology Assessment 1993). Seven brown tree snakes were found in Hawaii between 1981 and 1994, the most recent in December 1994 in an Oahu military warehouse containing cargo recently arrived from Guam (*Honolulu Advertiser*, 22 December 1994). The snakes would be virtually impossible to eradicate once escaped in Hawaii. They would be able to spread throughout native bird habitat, almost certainly eliminating most bird species, and would cause a host of problems to society through damaging power lines, biting and poisoning small children, and entering homes.

Miconia, a tree native to New World tropical forests at 300–1,800 meters elevation, is now known to be an unusually aggressive invader of moist island habitats. Introduced to Tahiti in 1937, miconia formed dense thickets that by the 1980's had replaced the native forest over most of the island (Fig. 12), with substantial losses of native animal and plant species (Meyer 1996). Because of its attractive purple and green foliage, miconia was introduced as an ornamental species to the Hawaiian Islands in the 1960's. In 1990 its detection on Maui by conservation agencies raised an alarm (Fig. 13). Nearly 20,000 individual miconia trees were removed from private lands by agency staff and volunteers in 1991–1993, and control seemed feasible. In September 1993, however, an aerial survey revealed a previously undetected miconia population on state land. This population, which occurred on more than 100 hectares, covered an area far larger than all previously known populations on Maui. In January 1994 an interagency working group developed and implemented a containment strategy of spraying from a helicopter individual emergent miconia trees with the herbicide Garlon 4. Efforts to mobilize a successful long-term control effort are gaining momentum (Loope and Medeiros 1994b).

The two-spotted leafhopper was first noted in Hawaii on a farm on Oahu in 1987. Since then, this insect has been found in association with many plant species that exhibit such symptoms as chlorosis, leaf distortion, stunting, and death. The most conspicuous effect of this invader has been the phenomenon of uluhe

Fig. 12. Nearly monospecific cover of the Neotropical tree miconia on slopes of Vaihiria Valley, Tahiti, French Polynesia, in June 1994. Invasion of this species has eliminated much of the biodiversity on that island.

Fig. 13. Dense patch of miconia seedlings near Hana, Maui, Hawaii.

dieback, where single-species stands of the fern uluhe (Fig. 14) have died; this phenomenon was first observed on Kauai and in the Manoa and Palolo valleys of Oahu. Preliminary results showed the leafhopper widely distributed on all islands surveyed (Hawaii, Maui, Oahu, and Kauai) and on a range of hosts including 268 plant species (53 natives) representing 78 plant families (Fukada 1994). Uluhe dieback has not been observed on the island of Hawaii. By early 1995 the tree species ohia was also exhibiting severe leafhopper damage on Oahu, especially in mesic habitats (J. Lau, The Nature

Fig. 14. Extensive monospecific stands of uluhe exist in the montane wet zones of all islands. Extensive uluhe dieback caused by the two-spotted leafhopper has occurred since 1990 on Oahu and Kauai and is also present on Maui.

Conservancy of Hawaii, personal communication). Severe effects are not confined to native species; nonindigenous species, such as Brazilian pepper, firetree, and strawberry guava are also heavily damaged in some localities. Thus far, no damage to native communities has occurred at high elevations or in extremely wet areas.

The mechanism of the leafhopper's damage to plant tissue is not completely understood. However, preliminary studies lead investigators to suspect that the insect injects a protein-degrading enzyme to aid its feeding on leaf tissue, and female leafhoppers damage leaf veins (vascular bundles) when they deposit eggs (V. Jones, University of Hawaii, personal communication).

No other invading insect has been documented in the worldwide literature as causing as much damage across as broad a range of vascular plant families as the two-spotted leafhopper. As of early 1995, this invader from China seems to pose a severe and perhaps unmanageable threat to Hawaiian biota. Biological control may hold much promise, however.

Some Factors Influencing Native Ecosystem Development in Hawaii

Ecologists are beginning to understand the long-term changes that occur in Hawaiian forests from the colonization of bare lava or volcanic cinder to the persistence of plants on soils that are millions of years old. This growing understanding is made possible by recent precise dating of lava flows (Lockwood et al. 1988) and research on such areas conducted by an interdisciplinary team.

Changes in Forest Composition and Structure in Relation to Soil Age and Soil Nutrients

Kitayama and Mueller-Dombois (1995) studied plots at 1,200 meters elevation from Hawaii (Big Island) to Kauai with a range in age of 300 years to 4.1 million years. The eight mesic forest sites studied had an annual rainfall of 2,500 millimeters and soil ages of 300, 2,100, 5,000, 20,000, 150,000 years (all Hawaii Island), 410,000 years (East Maui), 1.4 million years (Molokai), and 4.1 million years (Kauai). In addition, the investigators studied eight wet forest sites with annual rainfall of 4,000 millimeters and substrate ages of 400, 1,400, 5,000, 9,000 years (all Hawaii Island), 410,000 years (East Maui), 1.3 million years (West Maui), 1.4

million years (Molokai), and 4.1 million years (Kauai). A single tree species, ohia, dominates the forest canopies at all these sites. The mean height and diameter at breast height of canopy ohia trees increased from lowest values at the youngest site to maximum values at the 2,100-, 5,000-, and 9,000-year-old sites, and successively declined at older sites. The maximum standing biomass (the weight of all plant material in a plot) was found on young soils. The younger soils consistently had a high availability of phosphorus (Crews et al. 1995), and the leaves of trees growing there exhibited high concentrations of phosphorus and nitrogen (Vitousek et al. 1995). Phosphorus, a crucial element, apparently becomes less available to plant roots as Hawaiian soils age and available phosphorus is leached from the soil or is tied up in insoluble forms.

The pattern of replacement of pioneer species by species that thrive in mature forests was similar in both the wet and mesic sites, although replacement occurred more rapidly at the wetter sites, a situation consistent with the concept that soils of wetter forests lose nutrients more rapidly. The greatest numbers of plant species in plots of both mesic and wet sites were at East Maui sites (with a 410,000-year-old substrate, well past maximum phosphorus availability), indicating that species numbers are not determined by soil fertility alone.

Ohia Dieback

An understanding of nutrient availability in soils of different ages is important to our understanding the phenomenon known as ohia dieback. Because ohia dominates 80% of the relatively intact remaining forest in the Hawaiian Islands, ohia dieback is an extremely important phenomenon. Of added significance is that aggressive nonindigenous plants can establish on a large scale during dieback if seed sources are present in the area, so that native forests are eventually replaced by nonindigenous forests.

Extensive stands of ohia died on Hawaii Island in the 1970's. The death of virtually all canopy trees in entire stands affected about half of a 80,000-hectare area of rain forest on the eastern slopes of Mauna Loa and Mauna Kea from the 1950's through the 1970's (Hodges et al. 1986). The phenomenon was initially thought to be caused by a pathogen, but a decade of research led to the conclusion that the trees were dying from some other cause (Mueller-Dombois 1985; Hodges et al. 1986).

Many stands of ohia undergoing dieback are composed of older, perhaps senescent, trees (Mueller-Dombois 1983). Climatic changes that are difficult to notice may have triggered dieback over the large area. Another possibility is that this dominant species faces reduced competition in its island environment and thus invades sites over an extremely broad ecological spectrum. Ohia is not well adapted for some of these sites, such as those having soils with high (toxic) aluminum levels or waterlogged soils. As a consequence, ohia dies periodically over part of its range in response to environmental changes (Mueller-Dombois 1986). No native species is capable of replacing it as a dominant, however, and ohia seedlings usually reinvade vigorously (Jacobi et al. 1983). The cause for stand-level dieback seems to be the even-aged nature of most ohia forest stands. After canopy closure of the first-generation forest, ohia regeneration is confined to small seedlings only, which remain small and are replaced under the canopy without graduating to the sapling life stage. Seedlings and saplings of ohia grow tall and enter the canopy only after the canopy of older ohia trees collapses. Such synchronous stand dieback is to be expected only in forests with a canopy composed of a single species and thus only (at least in the tropics or subtropics) on islands or mountains where biogeographic isolation has strongly limited colonization by other canopy species (Mueller-Dombois 1987).

Loss of Seabirds and Their Guano

The estimated 100–1,000-fold human-caused loss of nesting seabirds in the Pacific also applies to the main Hawaiian Islands (Steadman 1997) and is consistent with the findings of Olson and James in Hawaii (for example, Olson and James 1991). Seabird guano deposits, rich in nitrogen and phosphorus, represent a large amount of potential natural fertilizer transported from sea to land. Seabirds were probably once abundant throughout the slopes of the major Hawaiian islands before predation by humans and introduced predator species, because the few islands in the world that lack rats and other predators have extensive seabird colonies. The guano deposits of seabirds would have enriched soils and likely affected vegetation growing on the soils.

We have good reason to suspect that seabirds nested in large numbers on the forest floor in the pre-Polynesian Hawaiian Islands, based on observations of expeditions to remote forested islands where predators have not been introduced (A. Medeiros, U.S. Geological Survey, personal communication). The Newell's Townsend's shearwater currently nests in wet forests of Kauai. The endangered Hawaiian dark-rumped petrel survives today primarily on the extreme upper cliffs in Haleakala Crater at elevations of 2,500–3,000 meters (Figs. 15 and 16). This bird was once the most abundant seabird in the Hawaiian Islands, present in coastal areas and uplands as documented by bones found in a lava tube in the rain forest of Kipahulu Valley, Maui, at 1,860 meters elevation (James 1995; H. F. James, Smithsonian Institution, personal communication). If seabirds nested throughout Hawaii's forests in pre-Polynesian times, nutrient cycling was markedly altered by their disappearance, possibly influencing ohia dieback and changes in tree species composition of Hawaii's forests.

Status of Hawaiian Ecosystems

Subalpine–Alpine Ecosystems

In the Hawaiian Islands, the mean temperature decreases about half a degree centigrade for every 100-meter increase in elevation. For example, at sea level the mean annual temperature is about 25°C, but it is only 10°C at 3,000 meters. Closed forest extends to only about 2,000 meters. The subalpine and alpine ecosystems—shrublands and open parkland with small trees (mamane parkland on Mauna Kea)—exist above this elevation (the approximate level of the trade-wind inversion, an atmospheric layer that often restricts upward movement of clouds) on volcanoes of Maui and Hawaii islands. Weather becomes increasingly severe at higher elevations and near mountain summits; vegetation there becomes sparser and plants grow lower to the ground. Near-barren eolian habitats on these mountaintops are notable for endemic predator–scavenger insects and spiders that seem to subsist on insects that are blown by wind up to the summits from below (Howarth and Montgomery 1980; Ashlock and Gagné 1983). Many other insects in these eolian habitats—carabid beetles, lacewings (Fig. 17), moths, and true bugs—are flightless, localized, and found only on these summits (Loope and Medeiros 1994c).

Subalpine and alpine communities were probably little affected by the early Hawaiians and even today remain largely undeveloped, except for the presence of astronomical observatories with their associated buildings and roads. Animal species have altered these communities. Goats, sheep, and cattle have damaged the vegetation through grazing. One of the greatest threats to eolian habitats is the Argentine ant, which seems capable of invading areas in which ants were previously absent, even throughout the evolutionary history of flightless species (Cole et al. 1992; Loope and Medeiros 1995).

Rain Forests

Although most Hawaiian rain forests at low elevations were destroyed long ago, rain forests still cover relatively large expanses on the islands of Maui and Hawaii and are also found on the steep windward slopes, ridges, and peaks of Kauai, Oahu, and Molokai (Cuddihy and Stone 1990). Jacobi and Scott (1985) reported the presence of about 140,000 hectares of wet forest dominated by native species on the island of Hawaii. Cuddihy and Stone (1990) estimate a similar area of wet forest for the other four largest islands combined. Most of these forests are dominated by ohia with a closed canopy and a well-developed subcanopy layer of mixed native tree species, shrubs, and tree-ferns. Koa is locally dominant or codominant with ohia.

The long-term prognosis for ecological integrity of these forests is not good, given the recent invasion of feral pigs. Effects of feral pigs on rain forests range from the slow and long-term (gradual loss of native understory species and gradual invasion by aggressive weeds) to the acute and drastic (aggressive tilling of the forest floor, massive erosion of soil and organic matter, or trunks of tree-ferns broken and the starch consumed).

In the past 30 years, gradual, continuous degradation of Hawaii's rain forests by pigs has come to be expected. The best-documented example (but one where the trend toward degradation has been reversed) is Maui's Kipahulu Valley. The National Park Service initially intended to maintain this rain forest ecosystem in a natural, undisturbed state by keeping people out. Addition of the area to the national park system in 1969 coincided with a rapid invasion and expansion of pig populations and weed invasion (Lamoureux and Stemmermann 1976; Anderson et al. 1992). Removal of pigs in the late 1980's (Anderson and Stone 1993), together with fencing, led to partial recovery of forest understory and a noticeable slowing of invasions by weedy aggressive plants. New invasions, formerly occurring throughout extensive pig-disturbed areas, are now largely confined to areas of frequent natural disturbance such as trailsides, stream courses, and landslides. Removal of invasive plants by park managers holds promise for the nearly full restoration of the ecosystem.

Similarly well-documented rain forest degradation by pigs and invasive plants has occurred in the Olao Rain Forest of Hawaii Volcanoes National Park on Hawaii Island. In 1975 this rain forest was virtually free of invasive weeds, although some feral pig activity was detected (J. D. Jacobi and F. R. Warshauer, Hawaii Volcanoes National Park, unpublished report). Two decades later, much of the area had

Courtesy A. Medeiros, USGS

Courtesy J. W. Larson, National Park Service

© S. Montgomery, Montane Matters

Fig. 15. Inner cliffs of Haleakala Crater provide a relatively secure habitat for the endangered Hawaiian dark-rumped petrel or uau. Once common throughout the Hawaiian Islands, this seabird now nests primarily in deep burrows dug in cinder along cliffs on the upper slopes of Haleakala volcano.

Fig. 16. The chick of the rarely seen endangered Hawaiian dark-rumped petrel.

Fig. 17. This strange-looking flightless lacewing had not been collected since 1945, until rediscovered by Steve Montgomery in October 1994. This remarkable species is known from only a small area near the summit of Haleakala volcano at 2,900 meters elevation.

been fenced and feral pigs removed. In places where the pigs had been removed before extensive weed invasion, excellent forest restoration was achieved. However, where nonindigenous plants (kahili ginger, banana poka, strawberry guava, Himalayan raspberry, palmgrass) had already obtained a foothold before pig exclusion, these weedy plants continue to spread (J. T. Tunison, Hawaii Volcanoes National Park, personal communication).

Only about 10% of Hawaii's remaining rain forest presently receives protection from feral pigs. As a result, although high-elevation Hawaiian rain forests are the most unaltered plant and animal communities remaining in

Hawaii, their ecological integrity is being gradually eroded. This process is difficult to reverse after degradation reaches a certain stage. In contrast, in areas where managers of state, federal, and private conservation lands have removed pigs from rain forests before pigs had excessively degraded the area, excellent ecosystem recovery is under way.

Montane Bogs

Hawaiian bogs exist, mostly at higher elevations, as openings in cloud forest on the islands of Hawaii, Maui, Molokai, Oahu, and Kauai. The largest wetland complex in the Hawaiian Islands is on the Alakai Plateau of Kauai, where rainfall amounts are among the highest recorded on Earth. Numerous bogs also exist on Maui at elevations as high as 2,270 meters. Although most Hawaiian bogs are not large, many endemic plant species are largely confined to bog habitats.

Because of their remote locations and extremely wet climates, most Hawaiian montane bogs have until recently been little disturbed. However, feral pigs had finally reached bogs on all islands by the 1980's (Figs. 18 and 19). An example of recent and major progressive pig damage to a formerly pristine area is the montane bogs of East Maui. Feral pigs arrived in undisturbed sedge-dominated bogs in the upper Hana rain forest of Haleakala National Park in the early 1970's. The pigs were apparently attracted to bog habitats by the availability of nonindigenous earthworms, which are probably an important protein source (Loope et al. 1991a). In extreme situations, removal of plant cover in these bogs by pigs can approach 100% (Loope et al. 1991b). In the early 1980's, digging by pigs increased in bogs at 1,650 to 1,660 meters elevation, and nonindigenous plant invasion was under way. A series of sampling plots was established to determine the trends of damage by pigs. In sites dominated by prickly sedge, cover of nonindigenous species increased from 6% in 1982 to 30% in 1988 (Fig. 20), at which time the habitat was fenced for protection from pigs. Cover of native plant species was simultaneously reduced

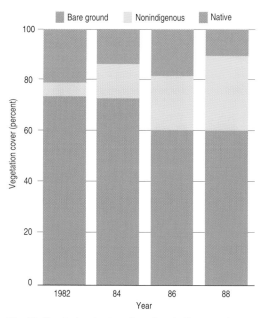

Fig. 20. Graph showing invasion of nonindigenous plants resulting from pig damage to montane bog sites on East Maui, 1982–1988 (from Medeiros et al. 1991).

because of digging by pigs and by invasion of nonindigenous plants; nonindigenous plant species inhibited reproduction of rare endemic plants such as the Maui greensword (Fig. 21; Medeiros et al. 1991).

Fig. 18. Pristine bog vegetation, windward East Maui, at 1,860 meters elevation, before feral pig impacts. Dominant species are Hawaii Island sedge and prickly sedge. Broad-leaved Maui plantain is conspicuous. By 1981, when this bog was fenced, digging by feral pigs had removed 95% of the vegetation cover. By 1984, the dominants had recovered, but plantain was nearly eliminated and remains rare today.

Fig. 19. Feral pig, Hawaii Volcanoes National Park.

Fig. 21. Maui greensword in a montane bog at 1,890 meters elevation, windward East Maui. Feral pigs directly damage greensword plants and their habitat by promoting dispersal and establishment of nonindigenous plant species.

Dryland and Mesic Forests

Lowland dry forests were once considered the richest of all Hawaiian forests in numbers of tree species (Rock 1913), but by the early 1900's they had been largely reduced to small remnants. Direct habitat destruction by humans was the most important factor in the forest's reduction; most leeward forest sites with soil were converted to agriculture by the Hawaiians. In modern times, the effects of browsing animals, invading nonindigenous grasses, and fire have been pervasive. Spectacular undisturbed forests remain in sites such as Auwahi, Maui, and Puuwaawaa, Hawaii, where rich assemblages of tree species persist, but they include only very old individuals, have had no reproduction for many years, and exhibit conditions that seem to make reproduction impossible without heroic restoration efforts. The understory of the remnant native forest at Auwahi, Maui, is covered by a dense mat (about 30–40 centimeters thick) of kikuyu grass, introduced as a pasture grass from East Africa, which smothers any reproduction by native plants (Fig. 22). Dryland forests of Puuwaawaa and other sites on Hawaii Island are being rapidly (over one to two decades) degraded by fires carried by nonindigenous fountain grass from North Africa. The successful perpetuation of dryland forest sites poses one of the greatest challenges to conservation managers (Fig. 23).

Fig. 22. Undisturbed forest in Auwahi, with an understory of kikuyu grass, a nonindigenous species.

Fig. 23. Relatively intact dryland forest of Kanaio Natural Area Reserve, East Maui, elevation 800 meters. Dominant trees are lama and Maui hale pepe.

Intact mesic (moist) forest remnants at higher elevations are commonly dominated by koa. These forests have generally been degraded by browsing goats or cattle, as have forests in Kahikinui, Maui, although excellent potential for recovery exists if these animals are removed. After 20 years of protection, koa parkland in Hawaii Volcanoes National Park has recovered to a great extent from damage by browsing cattle (Tunison et al. 1995).

Subterranean Ecosystems

Unique, interesting, and scientifically important assemblages of endemic, cave-adapted invertebrates (insects and spiders) exist in young lava tubes on at least five of the Hawaiian Islands; about 45 species of obligate cave-dwellers, evolved in most instances from known surface relatives, are now documented (Howarth 1983, 1987). Essential components of the cave ecosystems that are home to these invertebrates include extremely constant conditions of temperature and humidity and the roots of surface plants, which provide a food supply. The primary threats to native cave invertebrates include direct effects of humans (garbage dumping, physical and chemical damage caused by exploring, and root breakage), and destruction of native surface vegetation above the caves by humans, hooved mammals, nonindigenous plants, and fire.

Cave ecosystems are destroyed when the overlying forest that provides the caves their life-giving roots is lost (Howarth 1987); this has happened in several known instances during the past decade (F. G. Howarth, B. P. Bishop Museum, personal communication). On the other hand, many intact cave ecosystems receive protection within national parks.

Freshwater Ecosystems

Except for streams, bodies of fresh water in Hawaii are few and small. Hawaiian streams, islands of freshwater habitat within highly isolated oceanic islands (Fig. 24), are not rich in biological diversity but possess a unique and

Fig. 24. Hawaiian stream habitat.

Courtesy A. Medeiros, USGS

extremely interesting biota. These streams provide examples of how vacant island habitats are colonized from afar. Many members of the stream biota—including all five species of fish (four gobies and an eleotrid), shrimp, and snails—are descendants of marine forms that colonized stream habitats (Ford and Kinzie 1982). In contrast, insects (most notably damselflies, which have undergone spectacular adaptive radiation involving 17 or more endemic species) reached Hawaiian fresh waters from continental freshwater habitats.

The Hawaiian stream macrofauna is still largely intact but is seriously imperiled by habitat destruction (for example, channel alteration), pollution, rampant introduction of nonindigenous species, and withdrawal of water from streams for human use (Maciolek 1977, 1979; Parrish et al. 1978; Devick 1991). The conflict between extensive use of water resources and conservation of Hawaiian freshwater habitats has been severe. In many streams in developed areas, especially on Oahu and Maui, some native species have been extirpated or their populations have been greatly reduced. Careful preservation of relatively few intact stream ecosystems may provide for persistence of most native species, because most freshwater species in Hawaii have naturally extensive distributions within the islands. The recent major reduction in plantation agriculture of sugar and pineapple in much of the state, together with new legislative and administrative mandates for water management, offers opportunities to protect and even restore many damaged or threatened stream habitats. Immediate and vigorous action is required to curtail the alarming increase in accidental, unauthorized introductions of invasive nonindigenous species and to develop control measures for harmful nonindigenous populations.

Only a few Hawaiian streams have protected status. Among the best examples are Pelekunu Stream (within a Nature Conservancy preserve) on Molokai and Palikea Stream in Kipahulu Valley of Haleakala National Park.

Coastal Ecosystems and the Northwestern Hawaiian (Leeward) Islands

A narrow belt of strand vegetation exists around each island. On the main Hawaiian Islands, these communities have been severely damaged, especially by development and grazing mammals. Some largely intact coastal habitats persist, however, the most notable of which is at Moomomi on the north coast of Molokai. Many native coastal plant species generally persist in rocky shoreline habitats. Relatively few

coastal species have been lost to extinction, largely because of steep or rugged coastal habitats that are not easily developed. Restoration of coastal habitats has much potential. Excellent small-scale restoration examples exist at Wailea Point on Maui, Kilauea Point on Kauai, and Kaena Point on Oahu.

The northwestern Hawaiian Islands (politically part of the state of Hawaii except for Midway, which belongs to the U.S. Navy) extend northwest of Kauai for 1,210 kilometers from Nihoa to Kure Atoll. About 40 emergent islands make up a total of about 10.1 square kilometers of dry land. They vary in maximum elevation from 3 to 277 meters. The strand vegetation that covers most of these islands is more intact than the strand of the main islands because these islands receive less disturbance, although past disturbance to individual islands (for example, Laysan and Lisianski, where rabbits were introduced) has been great. Although the islands have only 8 endemic and 42 indigenous plant species, 114 nonindigenous species are naturalized; the nonindigenous species seem largely in balance with the rest of the flora, although close monitoring is warranted (Herbst and Wagner 1992).

Selected Biological Resource Issues of Other Pacific Islands

Pacific islands other than Hawaii are extremely varied and can only be discussed superficially here. The land area of most of these islands is small in relation to Hawaii, but biological diversity is in many instances quite significant. For example, the Republic of Palau has a land area of 458 square kilometers and has approximately 839 species and varieties of higher plants, of which 9% are endemic (Canfield 1981). Tutuila Island (142 square kilometers) of American Samoa supports 308 native flowering plants and 115 native fern species—over half the flora of the Samoan archipelago—and about 30% of its native flowering plants are endemic (Whistler 1993).

Nonindigenous Species

We have already discussed in some detail the effects of invasive nonindigenous plants and animals in Hawaii. Other islands in the Pacific have, in extreme instances, been subjected to even more drastic impacts. The brown tree snake in Guam (Savidge 1987) and miconia in Tahiti (Meyer 1996) should serve as important lessons for Hawaii and other islands. A third example, the effect of the purposely introduced snail *Euglandina* on endemic snails of Moorea, French Polynesia (see box)

The Effect of Introduced Euglandina Snails on Endemic Snails of Moorea, French Polynesia

Critics sometimes point out correctly that conservation biologists have little direct, unambiguous scientific evidence implicating invasive nonindigenous species in the extinction of native species. Indeed, long-term intensive studies of native island species are rare, and most damage done by invasive nonindigenous species is observed sporadically, providing only anecdotal evidence. The extinction of land snails of the genus *Partula* in French Polynesia provides a sad but noteworthy exception to this circumstance. *Partula* snails on the island of Moorea had been studied intensively by biologists in the 1920's through 1930's, and again in the 1960's through 1970's to examine patterns of genetic variation and formation of new species (Johnson et al. 1977). The Euglandina snail (Fig. 1), a predatory species native to Florida and Central America, was introduced purposely by two government agencies in 1977 to one location on the island of Moorea. The species was intended to act as a biological control (for which it proved ineffective) for the giant African snail, which became established on Moorea about 1970. Moorea is a small island with a diameter of about 12 kilometers and a maximum elevation of 1,207 meters. After 1980 the invading snail was advancing its distribution 1.2 kilometers per year. Clarke et al. (1984) provided precise status and trends information and predicted the extinction of all 14 taxa of *Partula* snails on Moorea by 1986–1987 (Fig. 2); populations of these snails were removed from the wild for captive breeding. The last wild population of *Partula* snails was seen in November 1986. An expedition to the highest ridgetops in June–July 1987 failed to find a single individual *Partula* snail (Murray et al. 1988).

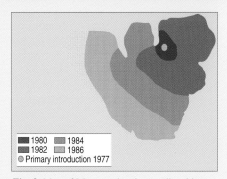

See end of chapter for references

Fig. 1. *Euglandina rosea.*
Courtesy R. Hue, Hawaii Department of Agriculture

■ 1980   ▨ 1984
■ 1982   ▨ 1986
◉ Primary introduction 1977

Fig. 2. Map of Moorea, showing predicted invasion of *Euglandina rosea*, based on information reported in 1983 (Clarke et al. 1984). This spread actually took place and all native *Partula* snails were eliminated by 1987.

Author

Lloyd L. Loope
U.S. Geological Survey
Biological Resources Division
Haleakala National Park Field Station
P. O. Box 369
Makawao, Hawaii 96768

illustrates even more clearly the urgent need for educating government officials to safeguard the biological diversity of the Pacific.

Although these specific examples of nonindigenous species introductions have had extreme effects, many Pacific island ecosystems have suffered significantly less than Hawaii from inadvertent and intentional introductions. Islands not heavily damaged by direct human habitat modification and without introduced hooved mammals such as goats and pigs have been found to be relatively resistant to plant invasion (Merlin and Juvik 1993). In the proposed national park of American Samoa and other cloud forest areas of Samoa, for example, Whistler (1993, 1994) found that feral pigs are largely kept at low numbers by local villagers and thus have not yet caused extensive damage to the forests and, with one exception (clidemia), invasive plants are not a serious problem.

Disruption of Mangrove Communities

Mangroves are tropical trees, classified in several plant families, that can tolerate flooding and high salinity and that usually have aerial roots. Mangrove forests occur in shallow coastal waters in sheltered conditions. No mangrove species are native to the Hawaiian Islands, whereas about 25 species exist elsewhere in the Pacific, and substantial areas of mangrove forest occur in the Mariana Islands and Micronesia (J. C. Ellison, East–West Center, Honolulu, Hawaii, unpublished report). Nine mangrove species exist on Guam, 12 on Palau, and 12 in the Federated States of Micronesia. The 355-square-kilometer island of Pohnpei (Federated States of Micronesia) has 55 square kilometers of mangrove forest.

Mangrove communities are widely recognized as an important biological resource and are traditionally used in the Pacific for fishing and gathering of clams and crabs, as a source of wood for construction, handicrafts, and fuel, and for many other specialized uses (Ellison, unpublished report).

Mangrove communities of small islands throughout the Pacific are rapidly disappearing or undergoing degradation. Creation of the large naval installations around Apra Harbor has destroyed the most extensive mangrove areas formerly existing on Guam; other areas have been lost to urban and aquaculture developments. Mangroves on Guam also suffer from pollutants such as oil spillage and heavy metals. In Palau, mangrove areas have been subjected to extensive clearing for development. Heavy harvesting has removed a large portion of the mangroves on Kosrae (Federated States of Micronesia). Mangroves on Yap (Federated States of Micronesia) were damaged by sedimentation caused by road construction. Degradation of mangroves also occurs from natural causes, especially cyclones. In February 1990, for example, 6 hectares of 16-meter-tall mangrove forest were destroyed by Cyclone Ofa in Tutuila, American Samoa (Ellison, unpublished report).

Destruction of Montane Cloud Forests

Pacific island cloud forests must be considered a high conservation priority because of their importance as traditional sources of food, building materials, medicines, and various other material products for indigenous peoples, as well as for their highly endemic biological diversity (Raynor 1993; Hamilton et al. 1995). Samoa has a large area and rugged topography; it retains more intact native forest than many other Polynesian islands, yet its cloud forests are subjected to increasing inroads by forest industry to support a growing population (Whistler 1993). Montane cloud forest is found on only two high islands in Micronesia, Pohnpei, and Kosrae, where the high-elevation ecosystem has remained largely intact for 2,500 years of human habitation. Population pressure and increasing cultivation of the narcotic crop *sakau* now pose a severe threat to the cloud forests (Raynor 1993).

Subsistence Harvest of Wildlife

The effects of prehistoric human harvest of island birds were enormous (Steadman 1995). Hunting on a small land area continues to threaten the surviving wildlife of many Pacific islands. On the island of Tutuila in American Samoa, an estimated 2,100–4,200 Pacific pigeons, 500–1,000 purple-capped fruit-doves, and 500–1,600 fruit bats were killed by hunters in recent years (Craig et al. 1994a). These removal rates are extremely high in comparison with current population levels of these animals. After a hurricane in 1990, when bats were particularly vulnerable to hunting because of food limitations, more bats were killed (about 3,100) than survived in 1992. In 1987–1992, populations of the two fruit bat species on American Samoa were reduced in size by 80% to 90% (Craig et al. 1994b).

Global Climate Change and Sea-Level Rise

The Intergovernmental Panel on Climate Change estimates that a worldwide sea-level rise of 30–100 centimeters will take place over the next century as a result of melting of glaciers and ice sheets and thermal expansion of ocean water (Markham et al. 1993). Mangrove swamps, particularly those of low islands, are likely to be sensitive to a rise in sea level (Ellison, unpublished report). A sea-level rise poses an obviously severe threat to low-lying Pacific islands. The Republic of the Marshall Islands, for example, has a human population of 48,000 living on a land area of 110 square kilometers and a mean height above sea level of 2.0 meters.

Other Issues

Pacific island ecosystems, because of their very small land size, are highly vulnerable to numerous human activities such as road construction, tourism development, military use, mining, hydropower development, forest fires, and pollution. The rapidly growing human populations of some island nations make such problems more severe. In the Republic of the Marshall Islands, for example, the population growth rate is approximately 4% annually (Thistlethwait and Votaw 1992). The human population of Tutuila of American Samoa, which is only 142 square kilometers in area, has grown 44% in the past 10 years, a rate that, if continued, will result in 100,000 people by the year 2010 (Whistler 1994). Increased tourism and internationalization of trade generate growing transportation to Pacific islands, a trend that poses a severe threat to areas not yet hit by major nonindigenous species invasion (R. A. Holt, The Nature Conservancy, personal communication).

Status and Trends of Animal and Plant Species of Hawaii and Other Pacific Islands

The native Hawaiian biota has suffered substantial extinction and endangerment. This is not surprising in view of the vulnerability of island biota and the drastic changes brought about by humans through habitat modification and species introductions. Much has been lost, with land snails and birds more decimated than other groups, but the Hawaiian Islands still have a rich native biological diversity. Relatively intact ecosystems exist at upper elevations and in the coastal strand ecosystems of the northwest Hawaiian Islands. Biological invasions, both from organisms already present and those that may arrive in the next decades, present the greatest threat to sustained survival of diverse native ecosystems. Active management of biological invasions is essential if most of the remaining diversity is to be saved over the next century.

Birds

The most spectacular land bird assemblage ever found on any remote oceanic archipelago evolved in the Hawaiian Islands (Freed et al. 1987). This status remains arguably true today for the surviving land birds, even though an extremely high percentage of them became extinct through the actions of humans. The precise percentage of this loss is not easy to determine for several reasons: methods differ for delineating the Hawaiian bird fauna (for example, residents versus visitors [Pyle 1995]); numbers depend on whether numbers of species or of subspecies are used; and the recent fossil record is continually expanding and is certainly incomplete. Numbers given here are based on species, include only Hawaiian endemics (based on taxonomy used by Olson and James 1991), and include fossil data (including 22 undescribed species) from Olson and James (1991) and James and Olson (1991).

Of the 76 species of endemic Hawaiian perching birds and songbirds, 31 are known only as fossils. Of the 45 species known historically, 19 are now extinct, 18 are federally listed as endangered (including 9 that have not been seen for several years and are possibly extinct), and only 8 species are not classified as endangered. Of the 39 species of endemic Hawaiian nonpasserine species (including geese, ducks, rails, ibises, and raptors), 31 are known only as fossils. Of the eight species known historically, two are now extinct and six are federally listed as endangered. Surviving endemic Hawaiian land bird species not listed as endangered, all passerines, include six Hawaiian honeycreepers (apapane, anianiau, common amakihi, Kauai creeper, Maui creeper, and iiwi); a thrush (omao); and a monarch (elepaio).

The Hawaiian honeycreepers, with a minimum of 57 species derived from a single finchlike ancestral species (James and Olson 1991), make up by far the largest bird family in Hawaii; their fame in evolutionary biology rivals that of Darwin's finches, a group in the Galapagos archipelago where 14 fairly similar but distinct species are derived from a common ancestor. Whereas all species and most island populations of the Galapagos finches survive today and are subject to little immediate endangerment, the story is quite different for the Hawaiian honeycreepers. At least 29 species of Hawaiian honeycreepers were eliminated prehistorically, and another 8 have been eliminated since Europeans first arrived in 1778.

Nevertheless, the 20 living species (14 of which are federally listed as endangered) still provide the most impressive example of adaptive radiation among birds. Their bills range from stout seed crushers (palila) through parrot shapes (Maui parrotbill), crossed (akepa), and warblerlike forceps (Maui creeper), to decurved probes (iiwi) that in the Hawaiian akialoa span one-third of the bird's overall length (Scott et al. 1988; Fig. 4).

The threats that have reduced the honeycreepers to a fraction of their original diversity and abundance continue largely unabated. Most surviving honeycreeper species inhabit ohia or ohia-koa forests. The factors believed to be most responsible for the continuing decline of honeycreepers and other Hawaiian birds are habitat loss, susceptibility to introduced bird diseases, predation by introduced mammals, competition from introduced birds, and reduction in abundance of arthropod food items (Scott et al. 1986; Jacobi and Atkinson 1995).

The three most common honeycreeper species (apapane, amakihi, and iiwi), which account for 95% of the honeycreeper individuals (Scott et al. 1986), show signs of vulnerability. Even the apapane, with more than 1 million individuals on Hawaii Island alone and densities of more than 1,600 birds per square kilometer, has very marginal populations on Lanai (fewer than 1,000 individuals) and Oahu. The amakihi disappeared from Lanai in the 1970's. The iiwi (Fig. 25), with more than 400,000 individuals in the 1970's (Scott et al. 1986), has disappeared from many midelevation sites over the past 20 years where it was previously common, presumably because of high susceptibility to mosquito-borne malaria (Jacobi and Atkinson 1995).

The poo-uli (Fig. 26) illustrates well the decline toward extinction of the rarest honeycreepers. This bird was not discovered until 1973, by which time it was present within a range of 13 square kilometers on the wet northeast slope of Haleakala volcano (elevation

Courtesy J. Jeffrey, U.S. Fish and Wildlife Service

Fig. 25. Iiwi.

Courtesy P. Baker, USGS

Fig. 26. Poo-uli.

1,500–1,800 meters) on Maui Island (Mountainspring et al. 1990). More recently, poo-uli have been found in the fossil record at 600 meters elevation on the dry, southwestern slope of Haleakala volcano (James et al. 1987). In a 50-hectare study area on upper Haleakala, numbers of poo-uli declined by 90% from 1975 to 1985. During this period, feral pig activity (based on ground disturbance) increased 473% (Mountainspring et al. 1990). Ground disturbance by pigs was found to be 9 to 24 times greater in nearby areas outside poo-uli range than within the range. Disturbance by feral pigs is apparently particularly disruptive to the poo-uli, a species that feeds on invertebrates found in the forest understory (Scott et al. 1988). No poo-uli were detected, in spite of sporadic efforts at locating them, between 1988 and 1993. In surveys throughout the range in 1994–1996, birds have been difficult to locate. As of 1996, the poo-uli population was estimated at 10 birds or less (T. Pratt, U.S. Geological Survey, personal communication).

Of 12 Mariana Islands bird species federally listed as endangered, 10 are endemic to those islands; on Guam, those 10 have been nearly or completely wiped out by the brown tree snake. The other two listed bird species are the Ponape mountain starling and the Ponape greater white-eye (U.S. Fish and Wildlife Service 1994).

The northwestern Hawaiian Islands are renowned for their seabird colonies. Whereas 86% of the seabird species in Hawaii and 48% of the populations exist on main islands, 95% of the breeding pairs nest in the northwestern Hawaiian Islands (Scott et al. 1988). Pyle (1995) states that seabird population estimates

made in the 1960's and 1970's are not comparable for trends analysis because of the varying techniques used. Harrison (1990) discussed the difficulties involved in making representative counts and finds no evidence of long-term trends in species numbers, although some wide fluctuations occurred earlier in this century.

Important seabird populations survive in other areas of the American Pacific, including the Northern Mariana Islands; Howland, Baker, and Jarvis islands; Wake Atoll; Palau; and the Marshall Islands. Johnston Atoll provides the only roosting and breeding habitat for seabirds in 2.1 million square kilometers of ocean (R. Smith and others, U.S. Fish and Wildlife Service, personal communication).

Mammals

As a result of its extreme isolation from other land masses, Hawaii has only one native land mammal, the Hawaiian hoary bat, and one amphibious mammal, the Hawaiian monk seal. Both species are federally listed as endangered.

Two fruit bat species from Guam are federally listed as endangered as a result of habitat destruction and overharvesting by humans for food. Two fruit bat species from American Samoa are in rapid decline (see the section on subsistence harvest of wildlife, this chapter).

Reptiles and Amphibians

Hawaii has no native reptiles or amphibians that spend most of their lives on land. However, five sea turtle species come ashore in the Hawaiian Islands to deposit eggs; none are local endemics but are widespread in the Pacific (see chapter on Marine Resources). Two species, the leatherback and the hawksbill, are federally listed as endangered. The three other species—the loggerhead (see chapter on Caribbean), the green turtle, and the olive ridley—are listed as threatened. A saltwater crocodile found on Palau is federally listed as endangered (U.S. Fish and Wildlife Service 1994).

Fishes

None of the five species of native freshwater fishes (four gobies and an eleotrid), descendants of marine forms that colonized Hawaiian stream habitats, has been federally listed, even though many stream habitats have been drastically altered. The oopu alamoo (Fig. 27) is considered a potential candidate for listing as endangered or threatened, based on its reduced distribution.

Only a few other native fish species occupy freshwater habitat in the American Pacific; none have been federally listed as endangered.

Fig. 27. Oopu alamoo.

Courtesy M. Yamamoto, Hawaii Department of Land and Natural Resources

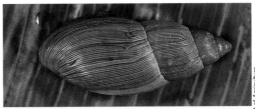

Fig. 29. *Euglandina rosea* (top image is printed actual size).

Land Snails

Approximately 900 species (71%) of about 1,263 historically described species of Hawaiian land snails are extinct (S. Miller, U.S. Fish and Wildlife Service, personal communication). Additional extinct species are abundant in archaeological sites and sediments in limestone sinkholes (Gagné and Christensen 1985). The entire genus *Achatinella* (Fig. 28), with 41 species, was federally listed as endangered in the 1970's; all but 2 species are either extinct or near extinction. The endemic Hawaiian family Amastridae, with more than 330 species, has been virtually eliminated from Hawaii.

No endemic Hawaiian land snail species can be regarded as secure, as a result of predation by rats and by the nonindigenous snail *Euglandina rosea* (Fig. 29), which was introduced in 1958 as a biological control agent to control (unsuccessfully) another introduced snail. Vulnerability of Hawaiian land snails to predation is to a large extent a result of life-history patterns. Hadfield (1986) and coworkers have found that the Hawaiian snails in the genera *Achatinella* and *Partulina* mature slowly (after about 6–7 years) and live to a maximum age of roughly 20 years. All those studied have low reproductive rates (as few as one offspring each year and no more than about seven per year). When a comparison is made with the life-history characteristics of the predatory snail Euglandina—which takes less than a year to mature, produces more than 600 eggs per individual per year, and has a life span of up to 5 years—it is easy to see why the native snails are being driven to extinction by predation (Simon 1987). American Samoa, the Mariana Islands, and Palau have endemic tree snails that are subjected to similar threats as Hawaiian snails but are not yet federally listed.

Arthropods

The number of native Hawaiian arthropods is probably near 10,000, but only about 5,500 of these have been described (including about 5,100 insect species and 150 spiders; Howarth and Mull 1992). The conservation status of these organisms is poorly known compared with that of most other groups, as intensive surveys are almost completely lacking. Only 36 species of arthropods have been recognized as extinct by the U.S. Fish and Wildlife Service; although this is probably a gross underestimate, several of those believed extinct have been rediscovered (Howarth et al. 1995). Because many Hawaiian insects are highly host-specific, many species have likely been and will continue to be lost as their host plants become rare and are destroyed.

The primary threats to the arthropod fauna are habitat and host-plant destruction and introduction of nonindigenous arthropods. About 2,600 insect species have been established in Hawaii through human activities (Howarth et al. 1995). In places where land has been converted to agriculture, few native arthropod species survive; for example, Asquith and Messing (1993) found that less than 10% of the insect fauna of a lowland agricultural area on Kauai consists of native species. Of nonindigenous insects, certain predators, especially ants (Cole et al. 1992; Gillespie and Reimer 1993; Loope and Medeiros 1995) and yellowjackets (Gambino et al. 1987, 1990; Foote and Carson 1995) may have devastating effects on native species. Nonindigenous parasites, especially ones that have broad host ranges, are also believed responsible for many instances of rarity and extinction of arthropods. Such instances are not

Fig. 28. *Achatinella* tree snail, Oahu.

© W.P. Mull, Volcano, Hawaii

Courtesy R. Hue, Hawaii Department of Agriculture

easy to document, but abundant circumstantial evidence implicates even some insect parasites introduced for biological control (Howarth et al. 1995).

Probably the best quantitative information on decline of an insect group in Hawaii comes from the relatively well-studied endemic picture-wing vinegar flies of the Olao Rain Forest of Hawaii Island (Foote and Carson 1995; Figs. 30 and 31). Declines of these flies were attributed in part to declines in host-plant populations, and predation by yellowjackets.

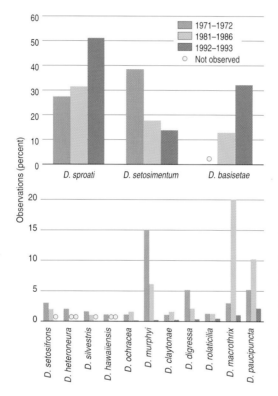

Fig. 30. A comparison of the relative frequencies of 14 endemic species of picture-wing vinegar flies *(Drosophila* species) over three periods of observation from 1971 to 1993 in Olao Rain Forest in Hawaii Volcanoes National Park. The data are expressed as a percentage of total observations within a survey period (Foote and Carson 1995).

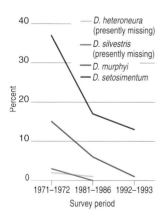

Fig. 31. Long-term trends within one host-specific guild of picture-wing vinegar flies *(Drosophila* species) that breed in rotting bark of native lobelioids (Foote and Carson 1995).

The poorly known conservation status of Hawaiian arthropods (Howarth et al. 1995) and the few detailed studies that show unidirectional declines (Foote and Carson 1995) should not create the assumption that most highly radiated arthropod groups may be doomed to extinction in the near future. Neither can one assume that all highly radiated arthropod groups have already been discovered. One newly explored group exhibiting remarkable adaptive radiation as well as resilience in the face of disturbance and biological invasions is the ecologically important but inconspicuous (nocturnal) long-jawed spiders. Before the 1990's, only nine species of long-jawed spiders had been described in the Hawaiian Islands, based on collections made before 1900. Gillespie and colleagues (1994) recently described 19 new species in this group and collected some 50 additional new types that apparently merit

species status. The Hawaiian long-jawed spiders seem to have radiated from two ancestral species (Gillespie et al. 1994). The Hawaiian tetragnathids span a huge spectrum of colors, shapes, sizes, habitats, and behaviors, far exceeding the limits to form and behavior of this group outside Hawaii. Few of the known Hawaiian species, all of which occupy wet or mesic forest habitats, seem imperiled, and some survive in highly modified habitats (R. G. Gillespie, University of Hawaii, personal communication). These spiders are, however, extremely vulnerable to ant predation (Gillespie and Reimer 1993).

The state of knowledge of Pacific arthropods outside Hawaii is notably inadequate, making even a preliminary assessment of their status very difficult (A. Asquith, U.S. Fish and Wildlife Service, personal communication).

Flowering Plants, Ferns, and Fern Allies

The known Hawaiian vascular flora includes about 1,302 taxa (including species, subspecies, and varieties), of which 1,158 are endemic to Hawaii. About 106 (8%) of these taxa are extinct. An additional 373 taxa (29%) are considered at some risk of becoming extinct in the near future. As of March 1995, Hawaii had 199 plant taxa federally listed as endangered or threatened; this is 38% of the total listed plants nationwide. The Endangered Species Act mandates the U.S. Fish and Wildlife Service to provide for recovery of all listed species, an extremely difficult task because 104 taxa in Hawaii have fewer than 20 known individuals remaining in the wild (Mehrhoff, personal communication).

Some success stories exist for the Hawaiian flora. The most notable may be that of the Haleakala silversword in the Mt. Haleakala crater. Reduced to no more than 4,000 plants in the 1920's and 1930's through human vandalism and goat and cattle browsing, this local endemic has thrived under National Park Service protection and now numbers more than 60,000 (Loope and Medeiros 1995).

Plant endangerment and extinction are also problems throughout the Pacific, but not at the extreme levels found in Hawaii (Mehrhoff, personal communication). Only one plant species, a tree of the Mariana Islands, has been listed as endangered.

Conclusions

"This extraordinary Hawaiian biota would have continued its remarkable adaptive radiation at a rapid rate had man not caused its recent decimation," observes E. C. Zimmerman.

Now a drastic new set of unfavorable conditions faces the delicately adapted biota, and a large fraction of it is doomed to extermination. What the future holds for it we cannot predict, and we shall not know anything like it again. Many of its glorious products, the fruit of ages, have already vanished, and its very mountains are being washed back into the sea from whence they came. (1970)

Much has been written about the tragic destruction and fragmentation of the Hawaiian biota. The extent of the losses (for example, over 70% of endemic land birds and land snail species) is unequaled in any other region of the United States. Hawaii is well known as the extinction capital of the United States, possessing one-third of the species federally listed as endangered. What is not generally appreciated is that much of Hawaii's unique biological heritage remains and can be protected with careful management. Large tracts of near-pristine ecosystems remain at high elevations (Figs. 32 and 33). Even with the high incidence of extinction and endangerment in the Hawaiian Islands, Hawaii has more nonendangered endemic species of vascular plants, birds, and insects

Fig. 33. Cloud forest, East Maui.

than any other state except California. Equally good opportunities exist elsewhere in the Pacific for long-term conservation.

Degradation of Hawaii's ecosystems has been going on for more than 1,500 years, since the first arrival of Polynesians in the fourth century A.D. Nevertheless, prevention of further invasions and management of existing ones require urgent attention. The movement, establishment, and spread of species to new geographic areas have created enough worldwide havoc to support the emerging view of biological invasions as a major component of global environmental change (D'Antonio and Vitousek 1992). Few areas in the world have suffered such negative effects of biological invasions as Hawaii, yet much remains to be lost. The specter of the establishment of some new devastating nonindigenous species such as the brown tree snake is always on the immediate horizon in the Hawaiian and other Pacific islands. Careful management of the flow of species from the continental United States and from foreign countries is crucial to long-term protection of the natural heritage of all Pacific islands.

More than any other invader, the feral pig is ideally suited to degrade the remaining pristine high-elevation rain forest environments. Feral pigs have spread into formerly pristine areas in recent decades, where they have aided in the establishment of invasive plants and invertebrates. How effectively feral pigs are managed

Fig. 32. Kipahulu Valley, in Haleakala National Park, supports extensive, relatively pristine rain forest vegetation.

in the coming decades will probably be the major determinant of how much more of Hawaii's near-pristine environments will be irreparably degraded.

Although the Hawaiian biota is relatively well known compared with biota of other Pacific islands, huge gaps in biological information exist throughout the region. For example, many endemic Hawaiian insect and spider species have yet to be discovered and described, as illustrated by Gillespie's recent work with spiders. Almost nothing is known of the ecology and status of most invertebrate species (Howarth and Ramsey 1991). We also need to better understand the implications of the human-caused massive reduction of seabirds and the associated reduction in nutrient cycling from sea to land. Our lack of biological information is not limited to animals. We have only superficial knowledge of factors limiting recovery of the 300 or more endangered, proposed, and candidate plant species of Hawaii. In spite of these knowledge gaps, lack of research is not the major factor limiting management.

Some conservation biologists argue that too much emphasis in the total conservation effort is placed upon research and management of individual rare species. Ultimately, they argue, almost nothing will be safe without active management, given the progressive pervasiveness of feral pigs into Hawaiian forest and the continuing associated onslaught of nonindigenous plants. A more accurate assessment is that the work with rare species is crucial, but much more must be done to protect additional large ecosystem tracts from pigs and weeds if a substantial part of Hawaii's biological diversity is to be preserved. The efforts under way in Hawaii to protect parks and reserves from invasive species, to prevent new invasive species from establishing, and to develop regional (island-by-island) approaches to invasive species management should be strengthened. Continuing research is needed to understand the biology and effects of invasive species, to provide the tools needed to manage the most destructive invasive species and ensure ecological restoration, and to develop and refine conservation strategies.

The surviving natural heritage of Pacific island ecosystems is a unique biological treasure. Because humans introduced invasive nonindigenous species into these ecosystems, this natural heritage is in serious jeopardy. Continuing ecological research and refinement of management strategies have important roles. However, conservation biologists know enough at present to confidently predict that much can be saved in the long run if the political will exists to implement the needed management. The crucial factor limiting conservation of biological diversity in the Pacific seems not to be a dearth of research but the lack of an adequate base of public understanding and support, at both the state and national levels. Ironically, protection of terrestrial biodiversity in Hawaii and the Pacific may be threatened less by economic conflicts than by apathy of a large segment of the public.

Acknowledgments

I thank the following colleagues for helpful suggestions: R. Gillespie, A. Holt, A. Medeiros, J. Parrish, J. Scott, and P. Vitousek. The following people have made important contributions to the thoughts in this report, although some of them may not entirely endorse the findings: S. Anderson, A. Asquith, C. Atkinson, J. Beardsley, M. Buck, J. Canfield, G. Carr, R. Cole, P. Conant, S. Conant, P. Connally (deceased), C. D'Antonio, C. Diong, P. Dunn, J. Ewel, D. Foote, R. Fosberg (deceased), B. Gagné, W. Gagné (deceased), R. Gillespie, S. Gon, M. Hadfield, D. Herbst, P. Higashino, R. Hobdy, A. Holt, F. Howarth, J. Jacobi, H. James, P. Kirch, C. Lamoureux, J. Lau, J. Leialoha, A. Medeiros, L. Mehrhoff, D. Mueller-Dombois, R. Nagata, L. Nakahara, S. Olson, K. Pang, S. Perlman, E. Petteys, L. Pratt, T. Pratt, D. Reeser, R. Robichaux, M. Scott, P. Schuyler, P. Scowcroft, T. Simons, C. Smith, H. Springer, D. Steadman, L. Stemmermann (deceased), C. Stone, T. Tunison, C. van Riper, P. Vitousek, R. Warshauer, and A. Yoshinaga.

Author

Lloyd L. Loope
U.S. Geological Survey
Biological Resources Division
Haleakala National Park Field Station,
P. O. Box 369
Makawao, Hawaii 96768

Cited References

Anderson, S. J., and C. P. Stone. 1993. Snaring to control feral pigs *Sus scrofa* in a remote Hawaiian rain forest. Biological Conservation 63:195–201.

Anderson, S. J., C. P. Stone, and P. K. Higashino. 1992. Distribution and spread of alien plants in Kipahulu Valley, Haleakala National Park, above 2,300 feet elevation. Pages 300–338 *in* C. P. Stone, C. W. Smith, and J. T. Tunison, editors. Alien plant invasions in native ecosystems of Hawaii: management and research. University of Hawaii Cooperative National Park Resources Studies Unit, Honolulu.

Aplet, G. H. 1990. Alteration of earthworm community biomass by the alien *Myrica faya* in Hawaii. Oecologia 82:414–416.

Ashlock, P. D., and W. C. Gagné. 1983. A remarkable new micropterous *Nysius* species from the eolian zone of Mauna Kea, Hawaii Island (Hemiptera: Heteroptera: Lygaeidae). International Journal of Entomology 25:47–55.

Asquith, A., and R. H. Messing. 1993. Contemporary Hawaiian insect fauna of a lowland agricultural area on Kauai: implications for local and island-wide fruit-fly eradication programs. Pacific Science 47:1–16.

Athens, J. S., J. V. Ward, and S. Wickler. 1992. Late Holocene lowland vegetation, Oahu, Hawaii. New Zealand Journal of Archeology 14:9–34.

Atkinson, I. A. E. 1977. A reassessment of factors, particularly *Rattus* L., that influenced the decline of endemic forest birds in the Hawaiian Islands. Pacific Science 31:109–133.

Atkinson, I. A. E. 1985. The spread of commensal species of *Rattus* to oceanic islands and their effects on island avifaunas. Pages 35–81 *in* P. J. Moors, editor. Conservation of island birds. International Council for Bird Preservation. Technical Publication 3. Cambridge, England.

Baker, J. K., and D. W. Reeser. 1972. Goat management problems in Hawaii Volcanoes National Park: a history, analysis, and management plan. National Park Service Natural Resources Report 2. 22 pp.

Baldwin, B. G., D. W. Kyhos, J. Dvorak, and G. D. Carr. 1991. Chloroplast DNA evidence for a North American origin of the Hawaiian silversword alliance. Proceedings of the National Academy of Sciences (USA) 88:1840–1843.

Beardsley, J. W. 1979. New immigrant insects in Hawaii: 1962 through 1976. Proceedings of the Hawaiian Entomological Society 23:35–44.

Bellwood, P. S. 1979. The oceanic context. Pages 6–26 *in* J. D. Jennings, editor. The prehistory of Polynesia. Harvard University Press, Cambridge, Mass.

Canfield, J. E. 1981. Palau: diversity and status of the native vegetation of a unique Pacific Island ecosystem. Newsletter of the Hawaiian Botanical Society 20:14–21.

Carlquist, S. 1974. Island biology. Columbia University Press, New York. 660 pp.

Cole, F. R., A. C. Medeiros, L. L. Loope, and W. W. Zuehlke. 1992. Effects of the Argentine ant (*Iridomyrmex humilis*) on the arthropod fauna of high-elevation shrubland, Haleakala National Park, Maui, Hawaii. Ecology 7:1313–1322.

Craig, P., T. E. Morrell, and K. So`oto. 1994a. Subsistence harvest of birds, fruit bats, and other game animals in American Samoa, 1990–91. Pacific Science 48:344–352.

Craig, P., P. Trail, and T. E. Morrell. 1994b. The decline of fruit bats in American Samoa due to hurricanes and overhunting. Biological Conservation 69:261–266.

Crews, T. E., K. Kitayama, J. H. Fownes, R. H. Riley, D. A. Herbert, D. Mueller-Dombois, and P. M. Vitousek. 1995. Changes in soil phosphorus fractions and ecosystem dynamics across a long soil chronosequence in Hawaii. Ecology 76(5):1407–1424.

Cuddihy, L. W., and C. P. Stone. 1990. Alteration of native Hawaiian vegetation: effects of humans, their activities and introductions. University of Hawaii Cooperative National Park Resources Studies Unit, Honolulu. 138 pp.

D'Antonio, C. M., and P. M. Vitousek. 1992. Biological invasions by exotic grasses, the grass/fire cycle, and global change. Annual Review of Ecology and Systematics 23:63–87.

Devick, W. S. 1991. Patterns of introductions of aquatic organisms to Hawaiian freshwater habitats. Pages 189–213 *in* New directions in research, management and conservation of Hawaiian freshwater stream ecosystems. Division of Aquatic Resources, Hawaii Department of Land and Natural Resources, Honolulu.

Diong, C. H. 1982. Population biology and management of the feral pig (*Sus scrofa* L.) in Kipahulu Valley, Maui. Ph.D. dissertation, University of Hawaii, Honolulu. 408 pp.

Douglas, G. 1969. Checklist of Pacific oceanic islands. Micronesica 5:327–464.

Foote, D., and H. L. Carson. 1995. *Drosophila* as monitors of change in Hawaiian ecosystems. Pages 368–372 *in* E. T. LaRoe, G. S. Farris, C. E. Puckett, P. D. Doran, and M. J. Mac, editors. Our living resources: a report to the nation on the distribution, abundance, and health of U.S. plants, animals, and ecosystems. U.S. Department of the Interior, National Biological Service, Washington, D.C.

Ford, J. I., and R. A. Kinzie III. 1982. Life crawls upstream. Natural History 91:60–67.

Fosberg, F. R. 1948. Derivation of the flora of the Hawaiian Islands. Pages 107–119 *in* E. C. Zimmerman, editor. Insects of Hawaii 1. University of Hawaii Press, Honolulu.

Freed, L. A., S. Conant, and R. C. Fleischer. 1987. Evolutionary ecology and radiation of Hawaiian passerine birds. Trends in Ecology and Evolution 2:196–203.

Fukada, M. T. 1994. Two-spotted leafhopper continues its invasion of Hawaii. Hawaii's Forests and Wildlife 9:13.

Gagné, W. C., and C. C. Christensen. 1985. Conservation status of native terrestrial invertebrates in Hawaii. Pages 105–126 *in* C. P. Stone and J. M. Scott, editors. Hawaii's terrestrial ecosystems: preservation and management. University of Hawaii Cooperative National Park Resources Studies Unit, Honolulu.

Gambino, P., A. C. Medeiros, and L. L. Loope. 1987. Introduced vespids *Paravespula pensylvanica* prey on Maui's endemic arthropod fauna. Journal of Tropical Ecology 3:169–170.

Gambino, P., A. C. Medeiros, and L. L. Loope. 1990. Invasion and colonization of upper elevations on East Maui (Hawaii, U.S.A.) by the western yellowjacket *Vespula pensylvanica* (Hymenoptera: Vespidae). Annals of the Entomological Society of America 83:1087–1095.

George, T., B. B. Bohlool, and P. W. Singleton. 1987. *Brachyrhizobium japonicum*–environment interactions: nodulation and interstrain competition in soils along an environmental gradient. Applied and Environmental Microbiology 53:1113–1117.

Gillespie, R. G., H. B. Croom, and S. R. Palumbi. 1994. Multiple origins of a spider radiation in Hawaii. Proceedings of the National Academy of Sciences 91:2290–2294.

Gillespie, R. G., and N. Reimer. 1993. The effect of predatory ants (Hymenoptera, Formicidae) on Hawaiian endemic spiders

(Araneae, Tetragnathidae). Pacific Science 47:21–33.

Hadfield, M. G. 1986. Extinction in Hawaiian achetelline snails. Malacologia 27:67–81.

Hamilton, L. S., J. O. Juvik, and F. N. Scatena. 1995. The Puerto Rico tropical cloud forest symposium: introduction and workshop synthesis. Pages 1–16 *in* L. S. Hamilton, J. O. Juvik, and F. N. Scatena, editors. Tropical montane cloud forests: ecological studies 110. Springer-Verlag, New York.

Hardy, D. E. 1983. Insects. Pages 80–82 *in* R. W. Armstrong, editor. Atlas of Hawaii. University of Hawaii Press, Honolulu.

Harrison, C. S. 1990. Seabirds of Hawaii: natural history and conservation. Cornell University Press, Ithaca, N.Y. 249 pp.

Herbst, D. R., and W. L. Wagner. 1992. Alien plants on the northwestern Hawaiian Islands. Pages 189–224 *in* C. P. Stone, C. W. Smith, and J. T. Tunison, editors. Alien plant invasions in native ecosystems of Hawaii: management and research. University of Hawaii Cooperative National Park Resources Studies Unit, Honolulu.

Hodges, C. S., K. T. Adee, J. D. Stein, H. B. Wood, and R. D. Doty. 1986. Decline of ohia (*Metrosideros polymorpha*) in Hawaii: a review. U.S. Forest Service General Technical Report PSW-86. 22 pp.

Holt, R. A. 1992. Control of alien plants on Nature Conservancy preserves. Pages 525–535 *in* C. P. Stone, C. W. Smith, and J. T. Tunison, editors. Alien plant invasions in native ecosystems of Hawaii: management and research. University of Hawaii Cooperative National Park Resources Studies Unit, Honolulu.

Howarth, F. G. 1983. Ecology of cave arthropods. Annual Review of Entomology 28:365–389.

Howarth, F. G. 1987. Evolutionary ecology of eolian and subterranean habitats in Hawaii. Trends in Ecology and Evolution 2:220–223.

Howarth, F. G., and S. L. Montgomery. 1980. Notes on the ecology of the high altitude eolian zone on Mauna Kea. Elepaio 41:21–22.

Howarth, F. G., and W. P. Mull. 1992. Hawaiian insects and their kin. University of Hawaii Press, Honolulu. 160 pp.

Howarth, F. G., G. Nishida, and A. Asquith. 1995. Insects of Hawaii. Pages 365–368 *in* E. T. LaRoe, G. S. Farris, C. E. Puckett, P. D. Doran, and M. J. Mac, editors. Our living resources: a report to the nation on the distribution, abundance, and health of U.S. plants, animals, and ecosystems. U.S. Department of the Interior, National Biological Service, Washington, D.C.

Howarth, F. G, and G. W. Ramsey. 1991. The conservation of island insects and their habitats. Pages 71–107 *in* N. M. Collins and J. A. Thomas, editors. The conservation of insects and their habitats. Academic Press, London.

Hughes, R. F., P. M. Vitousek, and T. Tunison. 1991. Alien grass invasion and fire in the seasonal submontane zone of Hawaii. Ecology 72:743–746.

Jacobi, J. D., and C. T. Atkinson. 1995. Hawaii's endemic birds. Pages 376–380 *in*

E. T. LaRoe, G. S. Farris, C. E. Puckett, P. D. Doran, and M. J. Mac, editors. Our living resources: a report to the nation on the distribution, abundance, and health of U.S. plants, animals, and ecosystems. U.S. Department of the Interior, National Biological Service, Washington, D.C.

Jacobi, J. D., G. Gerrish, and D. Mueller-Dombois. 1983. Ohia dieback in Hawaii: vegetation changes in permanent plots. Pacific Science 37:327–337.

Jacobi, J. D., and J. M. Scott. 1985. An assessment of the current status of native upland habitats and associated endangered species on the island of Hawaii. Pages 3–22 in C. P. Stone and J. M. Scott, editors. Hawaii's terrestrial ecosystems: preservation and management. University of Hawaii Cooperative National Park Resources Studies Unit, Honolulu.

Jacobi, J. D., and F. R. Warshauer. 1992. Distribution of six alien plant species in upland habitats on the island of Hawaii. Pages 155–188 in C. P. Stone, C. W. Smith, and J. T. Tunison, editors. Alien plant invasions in native ecosystems of Hawaii: management and research. University of Hawaii Cooperative National Park Resources Studies Unit, Honolulu.

James, H. F. 1995. Prehistoric change to diversity and ecosystem dynamics on oceanic islands. Pages 87–101 in P. Vitousek, L. Loope, and H. Adsersen, editors. Biological diversity and ecosystem function on islands. Springer-Verlag, Heidelberg, Germany.

James, H. F., and S. L. Olson. 1991. Descriptions of thirty-two new species of birds from the Hawaiian Islands. Part 2. Passeriformes. Ornithological Monographs 46. The American Ornithological Union, Washington, D.C. 88 pp.

James, H. F., T. W. Stafford, Jr., D. W. Steadman, S. L. Olson, P. S. Martin, A. J. T. Jull, and P. C. McCoy. 1987. Radiocarbon dates on bones of extinct birds from Hawaii. Proceedings of the National Academy of Sciences (USA) 84:2350–2354.

Kaneshiro, K. Y. 1988. Speciation in the Hawaiian Drosophila. BioScience 38(4):258–263.

Kirch, P. V. 1982. The impact of prehistoric Polynesians on the Hawaiian ecosystem. Pacific Science 36:1–14.

Kirch, P. V. 1997. Introduction: the environmental history of oceanic islands. Pages 1–21 in P. V. Kirch and T. L. Hunt, editors. Historical ecology in the Pacific Islands: prehistoric environmental and landscape change. Yale University Press, New Haven, Conn.

Kitayama, K., and D. Mueller-Dombois. 1995. Vegetation changes along gradients of long-term soil development in the Hawaiian montane rainforest zone. Vegetatio 120:1–20.

Lamoureux, C. H., and L. Stemmermann. 1976. Kipahulu expedition 1976. Technical Report 11. University of Hawaii Cooperative National Park Resources Studies Unit, Honolulu. 18 pp.

La Rosa, A. M. 1992. The status of banana poka in Hawaii. Pages 271–299 in C. P. Stone, C. W. Smith, and J. T. Tunison, editors. Alien plant invasions in native ecosystems of Hawaii: management and research. University of Hawaii Cooperative National Park Resources Studies Unit, Honolulu.

Lockwood, J. P., P. W. Lipman, L. D. Petersen, and F. R. Warshauer. 1988. Generalized ages of surface lava flows of Mauna Loa Volcano, Hawaii. U.S. Geological Survey Miscellaneous Investigations Series Map I-1908.

Loope, L. L., and A. C. Medeiros. 1994a. Impacts of biological invasions, management needs, and recovery efforts for rare plant species in Haleakala National Park, Maui, Hawaiian Islands. Pages 143–158 in M. Bowles and C. J. Whelan, editors. Restoration of endangered species, Cambridge University Press, Cambridge, England.

Loope, L. L., and A. C. Medeiros. 1994b. Interagency efforts to control spread of a highly invasive tree (Miconia calvescens) on Maui, Hawaii (Abstract). Bulletin of the Ecological Society of America 75:136.

Loope, L. L., and A. C. Medeiros. 1994c. Biotic interactions in Hawaiian high-elevation ecosystems. Pages 337–354 in P. W. Rundel, A. P. Smith, and F. C. Meinzer, editors. Tropical alpine environments: plant form and function. Cambridge University Press, England.

Loope, L. L., and A. C. Medeiros. 1995. Haleakala silversword (Argyroxiphium sandwicense DC. ssp. macrocephalum). Pages 363–364 in E. T. LaRoe, G. S. Farris, C. E. Puckett, P. D. Doran, and M. J. Mac, editors. Our living resources: a report to the nation on the distribution, abundance, and health of U.S. plants, animals, and ecosystems. U.S. Department of the Interior, National Biological Service, Washington, D.C.

Loope, L. L., A. C. Medeiros, and B. H. Gagné. 1991a. Aspects of the history and biology of the montane bogs of Haleakala National Park. Technical Report 76. University of Hawaii Cooperative National Park Resources Studies Unit, Honolulu. 43 pp.

Loope, L. L., A. C. Medeiros, and B. H. Gagné. 1991b. Recovery of vegetation of a montane bog in Haleakala National Park following protection from feral pig rooting. Technical Report 77. University of Hawaii Cooperative National Park Resources Studies Unit, Honolulu. 23 pp.

Loope, L. L., and D. Mueller-Dombois. 1989. Characteristics of invaded islands. Pages 257–280 in H. A. Mooney and others, editors. Ecology of biological invasions: a global synthesis. John Wiley & Sons, Chichester, England.

Loope, L. L., R. J. Nagata, and A. C. Medeiros. 1992. Introduced plants in Haleakala National Park. Pages 551–576 in C. P. Stone, C. W. Smith, and J. T. Tunison, editors. Alien plant invasions in native ecosystems of Hawaii: management and research. University of Hawaii Cooperative National Park Resources Studies Unit, Honolulu.

Lorence, D. H., and K. R. Wood. 1994. Kanaloa, a new genus of Fabaceae (Mimosoideae) from Hawaii. Novon 4:137–145.

Macdonald, G. A., A. T. Abbott, and F. L. Petersen. 1983. Volcanoes in the sea: the geology of Hawaii. University of Hawaii Press, Honolulu. 519 pp.

Maciolek, J. A. 1977. Taxonomic status, biology, and distribution of Hawaiian Lentipes, a diadromous goby. Pacific Science 31:355–362.

Maciolek, J. A. 1979. Hawaiian streams: diversion versus natural quality. Pages 604–606 in Proceedings of the Mitigation Symposium, Fort Collins, Colo., 16–20 July 1979.

Markham, A., N. Dudley, and S. Stolton. 1993. Some like it hot: climate change, biodiversity and the survival of species. World Wildlife Fund International, Gland, Switzerland. 144 pp.

Martin, P. S. 1984. Prehistoric overkill: the global model. Pages 354–403 in P. S. Martin and R. G. Klein, editors. Quaternary extinctions. University of Arizona Press, Tucson.

Medeiros, A. C., L. L. Loope, and S. Anderson. 1993. Differential colonization by epiphytes on native (Cibotium spp.) tree-ferns in a Hawaiian rain forest. Selbyana 14:71–74.

Medeiros, A. C., L. L. Loope, T. Flynn, L. Cuddihy, K. A. Wilson, and S. Anderson. 1992. The naturalization of an Australian tree-fern (Cyathea cooperi) in Hawaiian rain forests. American Fern Journal 82:27–33.

Medeiros, A. C., L. L. Loope, and B. H. Gagné. 1991. Degradation of vegetation in two montane bogs in Haleakala National Park: 1982–1988. Technical Report. University of Hawaii Cooperative National Park Resources Studies Unit, Honolulu. 31 pp.

Merlin, M. D., and J. O. Juvik. 1993. Relationships among native and alien plants on Pacific Islands with and without significant human disturbance. Pages 597–624 in C. P. Stone, C. W. Smith, and J. T. Tunison, editors. Alien plant invasions in native ecosystems of Hawaii: management and research. University of Hawaii Cooperative National Park Resources Studies Unit, Honolulu.

Meyer, J.-Y. 1996. Status of Miconia calvescens (Melastomataceae), a dominant invasive tree in the Society Islands (French Polynesia). Pacific Science 50(1):66–76.

Mountainspring, S., T. L. C. Casey, C. B. Kepler, and J. M. Scott. 1990. Ecology, behavior, and conservation of the poo-uli (Melamprosops phaeosoma). Wilson Bulletin 102:109–122.

Mueller-Dombois, D. 1981a. Vegetation dynamics in a coastal grassland of Hawaii. Vegetatio 46:131–140.

Mueller-Dombois, D. 1981b. Fire in tropical ecosystems. Pages 137–176 in H. A. Mooney, T. M. Bonnicksen, N. L. Christensen, J. E. Lotan, and W. A. Reiners, editors. Fire regimes and ecosystem properties. U.S. Forest Service General Technical Report WO-26. Proceedings of a conference, 11–15 December 1978, Honolulu, Hawaii.

Mueller-Dombois, D. 1983. Canopy dieback and successional processes in Pacific forests. Pacific Science 37:317–325.

Mueller-Dombois, D. 1985. Ohia dieback in Hawaii: 1984 synthesis and evaluation. Pacific Science 39:150–170.

Mueller-Dombois, D. 1986. Perspectives for an etiology of stand-level dieback. Annual Review of Ecology and Systematics 17:221–243.

Mueller-Dombois, D. 1987. Forest dynamics in Hawaii. Trends in Ecology and Evolution 2:216–220.

Mueller-Dombois, D., and G. Spatz. 1975. The influence of feral goats on the lowland vegetation in Hawaii Volcanoes National Park. Phytocoenologica 3:1–29.

Nagata, K. M. 1985. Early plant introductions in Hawaii. Hawaiian Journal of History 19:35–61.

Nakahara, L. M., R. M. Burkhart, and G. Y. Funasaki. 1992. Review and status of biological control of clidemia in Hawaii. Pages 452–465 in C. P. Stone, C. W. Smith, and J. T. Tunison, editors. Alien plant invasions in native ecosystems of Hawaii: management and research. University of Hawaii Cooperative National Park Resources Studies Unit, Honolulu.

Nature Conservancy of Hawaii and the Natural Resources Defense Council. 1992. The alien pest species invasion in Hawaii: background study and recommendations for interagency planning. Honolulu, Hawaii. Nature Conservancy of Hawaii and the Natural Resources Defense Council, Honolulu. 123 pp.

Oliver, D. L. 1989. The Pacific Islands. 3rd edition. University of Hawaii Press, Honolulu. 304 pp.

Olson, S. L., and H. F. James. 1982. Prodromus of the fossil avifauna of the Hawaiian Islands. Smithsonian Contributions in Zoology 365:1–59.

Olson, S. L., and H. F. James. 1991. Descriptions of thirty-two new species of birds from the Hawaiian Islands. Part 1. Non-passeriformes. Ornithological Monographs 45. The American Ornithological Union, Washington, D.C. 88 pp.

Parrish, J. D., J. A. Maciolek, A. S. Timbol, C. B. Hathaway, Jr., and S. E. Norton. 1978. Stream channel modification in Hawaii. Part D: summary report. U.S. Fish and Wildlife Service, Columbia, Mo. 18 pp.

Price, S. 1983. Climate. Pages 53–62 in R. W. Armstrong, editor. Atlas of Hawaii. University Press of Hawaii, Honolulu.

Pyle, R. L. 1995. Birds of Hawaii. Pages 372–376 in E. T. LaRoe, G. S. Farris, C. E. Puckett, P. D. Doran, and M. J. Mac, editors. Our living resources: a report to the nation on the distribution, abundance, and health of U.S. plants, animals, and ecosystems. U.S. Department of the Interior, National Biological Service, Washington, D.C.

Raynor, B. 1993. Montane cloud forests in Micronesia: status and future management. Pages 176–182 in L. S. Hamilton, J. O. Juvik, and F. N. Scatena, editors. Tropical montane cloud forests: proceedings of an international symposium. East–West Center, Honolulu, Hawaii.

Robichaux, R. H., G. D. Carr, M. Liebman, and R. W. Pearcy. 1990. Adaptive radiation of the silversword alliance (Compositae: Madiinae): ecological, morphological, and physiological diversity. Annals of the Missouri Botanical Garden 77:64–72.

Rock, J. F. 1913 [Reprinted 1974, with annotations]. The indigenous trees of the Hawaiian Islands. Pacific Tropical Botanical Garden, Lawai, Kauai, Hawaii, and Charles E. Tuttle Company, Rutland, Vermont, and Tokyo, Japan. 548 pp.

Savidge, J. A. 1987. Extinction of an island forest avifauna by an introduced snake. Ecology 68:660–668.

Scott, J. M., C. B. Kepler, C. van Riper III, and S. I. Fefer. 1988. Conservation of Hawaii's vanishing avifauna. BioScience 38:238–253.

Scott, J. M., S. Mountainspring, F. L. Ramsey, and C. B. Kepler. 1986. Forest bird communities of the Hawaiian Islands: their dynamics, ecology, and conservation. Studies in Avian Biology 9. Cooper Ornithological Society, Calif. 431 pp.

Simon, C. 1987. Hawaiian evolutionary biology: an introduction. Trends in Ecology and Evolution 2:175–178.

Smith, C. W. 1985. Impacts of alien plants on Hawaii's native biota. Pages 180–250 in C. P. Stone and J. M. Scott, editors. Hawaii's terrestrial ecosystems: preservation and management. University of Hawaii Cooperative National Park Resources Studies Unit, Honolulu.

Smith, C. W. 1992. Distribution, status, phenology, rate of spread, and management of clidemia in Hawaii. Pages 241–253 in C. P. Stone, C. W. Smith, and J. T. Tunison, editors. Alien plant invasions in native ecosystems of Hawaii: management and research. University of Hawaii Cooperative National Park Resources Studies Unit, Honolulu.

Spatz, G., and D. Mueller-Dombois. 1973. The influence of feral goats on koa tree reproduction in Hawaii Volcanoes National Park. Ecology 54:870–876.

Steadman, D. W. 1995. Prehistoric extinctions of Pacific Island birds: biodiversity meets zooarcheology. Science 267:1123–1131.

Steadman, D. W. 1997. Extinctions of Polynesian birds: reciprocal impacts of birds and people. Pages 51-79 in P. V. Kirch and T. L. Hunt, editors. Historical ecology in the Pacific Islands: prehistoric environmental and landscape change. Yale University Press, New Haven, Conn.

Stone, C. P., L. W. Cuddihy, and J. T. Tunison. 1992. Responses of Hawaiian ecosystems to removal of feral pigs and goats. Pages 666–704 in C. P. Stone, C. W. Smith, and J. T. Tunison, editors. Alien plant invasions in native ecosystems of Hawaii: management and research. University of Hawaii Cooperative National Park Resources Studies Unit, Honolulu.

Stone, C. P., and L. L. Loope. 1987. Reducing negative effects of introduced animals on native biotas in Hawaii: what is being done, what needs doing, and the role of national parks. Environmental Conservation 14:245–258.

Thistlethwait, R., and G. Votaw. 1992. Environment and development: a Pacific Island perspective. Asian Development

Bank, Manila, Philippines. Technical document for the United Nations Conference on Environment and Development, June 1992.

Tomich, P. Q. 1986. Mammals in Hawaii: a synopsis and notational bibliography. 2nd edition. B. P. Bishop Museum Special Publication 76. Bishop Museum Press, Honolulu, Hawaii. 375 pp.

Tunison, J. T. 1992a. Alien plant control strategies in Hawaii Volcanoes National Park. Pages 485–505 in C. P. Stone, C. W. Smith, and J. T. Tunison, editors. Alien plant invasions in native ecosystems of Hawaii: management and research. University of Hawaii Cooperative National Park Resources Studies Unit, Honolulu.

Tunison, J. T. 1992b. Fountain grass control in Hawaii Volcanoes National Park: management considerations and strategies. Pages 376–393 in C. P. Stone, C. W. Smith, and J. T. Tunison, editors. Alien plant invasions in native ecosystems of Hawaii: management and research. University of Hawaii Cooperative National Park Resources Studies Unit, Honolulu.

Tunison, J. T., W. L. Markiewicz, and A. A. McKinney. 1995. The expansion of koa forest after cattle and goat removal, Hawaii Volcanoes National Park. Technical Report 99. University of Hawaii Cooperative National Park Resources Studies Unit, Honolulu. 32 pp.

Tunison, J. T., C. W. Smith, and C. P. Stone. 1992. Alien plant management in Hawaii: conclusions. Pages 821–833 in C. P. Stone, C. W. Smith, and J. T. Tunison, editors. Alien plant invasions in native ecosystems of Hawaii: management and research. University of Hawaii Cooperative National Park Resources Studies Unit, Honolulu.

U.S. Congress, Office of Technology Assessment. 1993. Harmful non-indigenous species in the United States. OTA-F-565. Washington, D.C. 391 pp.

U.S. Fish and Wildlife Service. 1994. Endangered and threatened wildlife and plants. 50 CFR 17.11 and 17.12. Washington, D.C. 42 pp.

van Riper, C., III, S. G. van Riper, M. L. Goff, and M. Laird. 1986. The epizootiology and ecological significance of malaria in Hawaiian land birds. Ecological Monographs 56:327–344.

Vitousek, P. M. 1986. Biological invasions and ecosystem properties: can species make a difference? Pages 163–176 in H. A. Mooney and J. Drake, editors. The ecology of biological invasions of North America and Hawaii. Springer-Verlag, New York.

Vitousek, P. M., and L. R. Walker. 1989. Biological invasion by *Myrica faya* in Hawaii: plant demography, nitrogen fixation, ecosystem effects. Ecological Monographs 59:247–265.

Vitousek, P. M., G. Gerrish, D. R. Turner, L. R. Walker, and D. Mueller-Dombois. 1995. Litterfall and nutrient cycling in four Hawaiian montane rain forests. Journal of Tropical Ecology 11:189–204.

Wagner, W. L., D. R. Herbst, and S. H. Sohmer. 1990. Manual of the flowering plants of Hawaii. Bishop Museum and University of Hawaii Presses, Honolulu. 1,853 pp.

Whistler, W. A. 1993. The cloud forest of Samoa. Pages 231–236 *in* L. S. Hamilton, J. O. Juvik, and F. N. Scatena, editors. Tropical montane cloud forests: proceedings of an international symposium. East–West Center, Honolulu, Hawaii.

Whistler, W. A. 1994. Botanical inventory of the proposed Tutuila and Ofu units of the National Park of American Samoa. Technical Report 87. University of Hawaii Cooperative National Park Resources Studies Unit, Honolulu. 142 pp.

Whiteaker, L. D., and D. E. Gardner. 1992. Firetree (*Myrica faya*) distribution in Hawaii. Pages 225–240 *in* C. P. Stone, C. W. Smith, and J. T. Tunison, editors. Alien plant invasions in native ecosystems of Hawaii: management and research. University of Hawaii Cooperative National Park Resources Studies Unit, Honolulu.

Yee, R. S. N., and W. C. Gagné. 1992. Activities and needs of the horticulture industry in relation to alien plant problems in Hawaii. Pages 712–725 *in* C. P. Stone, C. W. Smith, and J. T. Tunison, editors. Alien plant invasions in native ecosystems of Hawaii: management and research. University of Hawaii Cooperative National Park Resources Studies Unit, Honolulu.

Yocum, C. F. 1967. Ecology of feral goats in Haleakala National Park, Maui, Hawaii. American Midland Naturalist 77:418–451.

Yoshinaga, A. Y. 1980. Upper Kipahulu Valley weed survey. Technical Report 33. University of Hawaii, Cooperative National Park Resources Studies Unit, Honolulu. 17 pp.

Zimmerman, E. C. 1948. Insects of Hawaii. Volume 1. Introduction. University of Hawaii Press, Honolulu. 206 pp.

Zimmerman, E. C. 1970. Adaptive radiation in Hawaii with special reference to insects. Biotropica 32:32–38.

The Effect of Introduced Euglandina Snails on Endemic Snails of Moorea, French Polynesia

Clarke, B., J. Murray, and M. S. Johnson. 1984. The extinction of endemic species by a program of biological control. Pacific Science 38:97–104.

Johnson, M. S., B. Clarke, and J. Murray. 1977. Genetic variation and reproductive isolation in *Partula*. Evolution 31:116–126.

Murray, J., E. Murray, M. S. Johnson, and B. Clarke. 1988. The extinction of *Partula* on Moorea. Pacific Science 42:150–153.

Marine Resources

There are five large marine regions of the United States: Northeast, Southeast, Alaska, Pacific coast, and western Pacific oceanic (Fig. 1). The wide geographical spread of the United States covers a broad range of physical and oceanographic conditions, from warm tropical to freezing Arctic and coastal to oceanic environments that support a wide variety of marine resources and associated wildlife. The commercial use of these resources provides 5% of the world's fisheries production, making the United States the fifth largest producer of seafood in the world. The resources also have great recreational and aesthetic value.

The stewardship of the nation's living marine resources is largely entrusted to the National Oceanic and Atmospheric Administration. The agency's strategic plan assigns responsibility for research and conservation of the ocean's living marine resources to the National Marine Fisheries Service, with the Secretary of Commerce having the final authority for managing these resources.

The National Marine Fisheries Service carries out its charge under many laws and mandates from the U.S. Congress. Most of its responsibilities arise from six statutes: the Magnuson-Stevens Fishery Conservation and Management Act, which regulates fisheries within the U.S. Exclusive Economic Zone; the Endangered Species Act, which protects threatened or endangered species; the Marine Mammal Protection Act, which regulates the taking or harassment of marine mammals; the Lacey Act, which prohibits fishery transactions that violate state, federal, American tribal, or foreign laws; the Fish and Wildlife Coordination Act, which authorizes the National Marine Fisheries Service to collect fisheries data and to advise other agencies on environmental decisions affecting living marine resources; and the Agricultural Marketing Act, which authorizes a voluntary seafood inspection program. Numerous other statutes and international conventions and treaties also guide federal management of living marine resources.

Northeast Region

Environment and Physical Features

The Northeast marine region encompasses 260,000 square kilometers of the Northeast shelf ecosystem, which includes four major areas: the Gulf of Maine, Georges Bank, southern New England, and the Middle Atlantic Bight (Fig. 2). These separate areas are distinguished largely by bottom topography, productivity, food web dynamics, and water mass characteristics (Sherman et al. 1988).

The Gulf of Maine is a semi-enclosed continental shelf sea area bounded landward by the northeastern United States, including waters west of 66°W longitude between Georges Bank and the entrance to the Bay of Fundy. The gulf is strongly influenced by freshwater inflow. Maine rivers—principally the Androscoggin, Penobscot, Merrimack, and Kennebec—flow into the gulf, forming a plume of relatively brackish stratified water in the western gulf during spring. These local rivers contribute significantly to the upper 40 meters of the water column and, along with the outflow of the Bay of Fundy (the St. John River), maintain the counterclockwise circulation of the gulf, which seems strongest in spring (Fig. 2). The most significant inputs of fresh water originate from the Gulf of St. Lawrence.

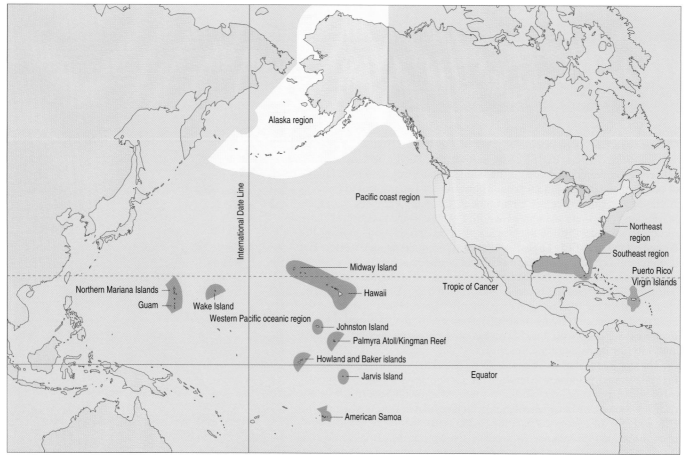

Fig. 1. Geographical coverage of the five major marine regions of the United States, colored to indicate their boundaries within their 200-mile Exclusive Economic Zones.

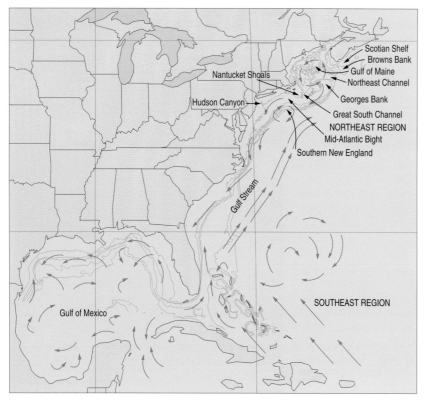

Fig. 2. General surface circulation and major areas of the Atlantic and gulf coasts.

The Gulf of Maine includes more than 59,570 square kilometers of estuarine drainage areas. Estuaries in this region were formed by glaciers that removed soil cover, leaving behind rocky shorelines and steep-sided river channels. These estuaries are smaller on average and generally deeper than those found in other regions. The strong tides and basin geometry in these estuaries result in tidally dominated circulation patterns. Three large human population centers (Boston, Massachusetts; Manchester, New Hampshire; and Portland, Maine) are located in the Gulf of Maine estuarine drainage area.

Georges Bank is generally delineated from the surrounding deeper Gulf of Maine by the area of equal depth (bathymetric contour) of 100 meters. Along the northern flank, sharp depth gradients between the Gulf of Maine and Georges Bank define the bank and generate equally sharp hydrographic and biological fronts. Changes in bottom depths and hydrographic conditions are more gradual along the eastern and southern areas of the bank. To the southwest, Georges Bank is bounded by the Great South Channel, which separates Georges Bank from Nantucket Shoals. The shallowest waters are found on the northwestern part of the bank. Daily and twice-daily tides, interacting with the shallow bottom topography of the bank, maintain exceptionally strong currents that result in a vertically well-mixed water column within the 60-meter isobath throughout the year (Butman 1987; Twichell et al. 1987). This vertical mixing exerts a major influence on the abundance and richness of biological populations on the bank.

The southern New England and Middle Atlantic Bight areas include the shelf area between Cape Hatteras, North Carolina, and the Great South Channel off Cape Cod. The shelves in these areas slope gently offshore and are relatively shallow. Much of the bight from Long Island southward is less than 60 meters deep. Waters in southern New England, with respect to productivity and plankton, represent a transition between the more oceanic characteristics of the Georges Bank area and the dominant estuarine influences on the Middle Atlantic Bight. The water column of the Middle Atlantic Bight is well mixed during winter and strongly stratified during summer, with the exception of shallow coastal areas, which experience episodes of vertical mixing from storms, upwelling, and downwelling.

Fresh water enters the Middle Atlantic Bight principally at the mouth of the Hudson–Raritan, Delaware, and Chesapeake bays. Such local sources of fresh water are responsible for approximately 70% of the large year-to-year variation in salinity in the Middle Atlantic Bight and strongly influence hydrographic conditions (Manning 1991). The general circulation in the Middle Atlantic Bight is characterized by a net flow of shelf water from Georges Bank through the bight in a southwesterly direction toward Cape Hatteras. At the shelf break, a persistent hydrographic front separates the shelf from offshore slope waters. This front undergoes seasonal onshore–offshore movements. Though the front is coherent from Georges Bank to Cape Hatteras, warm core rings drifting southwest between the northern edge of the Gulf Stream current and the continental shelf break, in addition to intrusions of Gulf Stream water along the southern portion of the Middle Atlantic Bight, may entrain and displace significant amounts of shelf water and organisms (Fig. 3).

The combined area of the southern New England–Middle Atlantic Bight accounts for more than 124,320 square kilometers of estuarine drainage, extending from Buzzards Bay in Massachusetts through Chesapeake Bay in Virginia. Chesapeake Bay has the largest total drainage area in the region and is one of the largest estuaries in the world. Almost half of all fresh water entering estuaries in the Northeast region flows into Chesapeake Bay (see chapter on Northeast). The mid-Atlantic coast ranks

third, behind the Gulf of Mexico and South Atlantic regions, in total wetland area contained in estuarine drainages (National Oceanic and Atmospheric Administration 1991a), with forested wetlands being the most common wetland type. The major human population centers in the region—New York, Philadelphia, Baltimore, and Washington, D.C.—make this area the most densely populated coastal region in the nation. Urban land use still lags behind

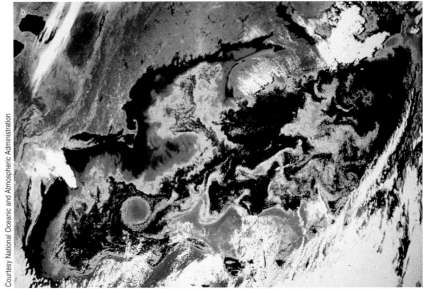

Courtesy National Oceanic and Atmospheric Administration

Fig. 3. a) General surface layer circulation of coastal and offshore surface waters of the Northeast region. b) Color-enhanced satellite image of relative productivity of the Northeast coast and offshore waters based on imagery from Nimbus 7 Coastal Zone Color Scanner. The yellow, red, and brown pigments represent chlorophyll levels, and blue depicts low-productivity Gulf Stream waters and the warm core ring.

forest and agricultural land uses in the region's estuarine drainages; however, compared with other geographical regions, mid-Atlantic estuarine drainages include the greatest percentage of urban land (National Oceanic and Atmospheric Administration 1990a).

Primary Production

Primary production is relatively high throughout the Northeast shelf ecosystem and is exceptional on Georges Bank and in the nearshore waters of the southern New England–Middle Atlantic Bight, where the phytoplankton, influenced by nutrient-enriched estuarine plumes, produce three times the mean productivity of the world's continental shelves (O'Reilly et al. 1987; Fig. 4). Generally, phytoplankton abundance decreases from nearshore to the shelf break (Fig. 5). The highest concentrations of phytoplankton are usually observed near the mouths of the Hudson–Raritan, Delaware, and Chesapeake bays, over the shallow water on Georges Bank, and in a small area along the southeast edge of Nantucket Shoals. Lowest abundances are usually restricted to the most seaward stations sampled along the shelf break and the central deep waters in the Gulf of Maine. When relatively high primary production occurs in the Gulf of Maine, it is in the inshore western portion.

The annual cycle of phytoplankton biomass follows the pattern expected for temperate continental shelf ecosystems, where there is at least a two-fold seasonal variation. The greatest phytoplankton concentrations and largest standing stocks occur during the winter–spring bloom; the lowest biomass is during summer, when vertical density stratification is greatest.

The winter–spring bloom is a major biological event, but its timing, duration, and intensity vary significantly by area. In the southern New England–Middle Atlantic Bight, the bloom progresses from nearshore to the slope, occurring January through March in the nearshore and midshelf and during March in offshore shelf and slope waters. On Georges Bank, the winter–spring bloom peaks in shallow water during

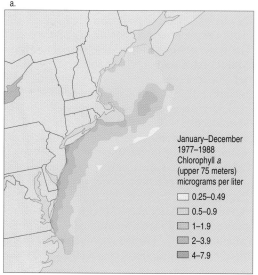

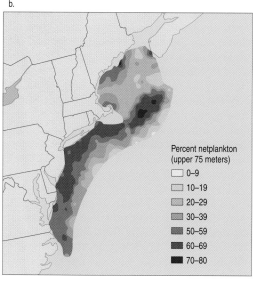

Fig. 5. a) Distribution of Chlorophyll *a* showing primary productivity, and b) distribution of netplankton representing larger phytoplankton (percentage of chlorophyll sample retained in a 20-micron sampling mesh) in the Northeast region, 1977–1988.

Fig. 4. Mean annual primary productivity (grams of organic carbon per square meter per year) of the coastal and offshore waters of the Northeast shelf ecosystem.

March and along the southern flank of the bank during April. The winter–spring blooms appear in February through March in the nearshore western Gulf of Maine and during April throughout the deeper waters of the gulf.

In most regions of the ecosystem, a secondary phytoplankton bloom is evident in the fall. As with the winter–spring bloom, the timing and intensity of the fall bloom vary regionally. In the shallow nearshore Middle Atlantic Bight, fall blooms appear in September through October. In the midshelf and outer-shelf regions and along the Middle Atlantic Bight shelf break, the bloom occurs a month later, in October, and follows the expected delay in destratification of deeper water. In the shallow water on Georges Bank, fall bloom starts in September and continues through November. In the deeper water, blooms appear during October. In some areas of the Gulf of Maine, the fall bloom approaches the magnitude of the winter–spring bloom, but Georges Bank, southern New England, and Middle Atlantic Bight fall blooms are clearly subordinate in magnitude to winter–spring blooms.

There are distinct patterns in the seasonal, regional, and vertical variation in plankton abundance which seem to reflect significant spatial diversity in species composition. Netplankton (larger than 25 microns) tends to dominate nearshore areas of the southern New England–Middle Atlantic Bight and the shallow water on Georges Bank, where chlorophyll concentrations are usually high. Nannoplankton (smaller than 25 microns) dominates deeper waters at the shelf break, as well as deep water in the Gulf of Maine, where chlorophyll levels are usually low (Fig. 3).

Findings support the relatively new idea (Malone et al. 1983) that netplankton prevails during the unstratified period when turbulence, vertical mixing, and nutrient levels are high and light for photosynthesis in the upper mixed layer is variable. Smaller nannoplankton and motile species prevail during density stratification, when turbulence is low and average light for photosynthesis is high in the upper mixed layer.

Planktonic Secondary Production

Secondary production patterns of the Northeast shelf ecosystem are distinguishable in four geographic areas characterized by distinctive water depth, hydrography, food chain relationships, seasonal species succession, and pulses of zooplankton abundance. On Georges Bank, the annual peak is in spring, followed by a sharp decline from spring through summer. In the Gulf of Maine, the annual peak of zooplankton abundance occurs in May and declines

less precipitously in autumn than on Georges Bank. In southern New England, zooplankton abundance is bimodal, with an initial peak in May and a secondary peak in August. In contrast to the other three areas, in the Middle Atlantic Bight zooplankton reaches an annual peak in autumn. The zooplankton of the Northeast region (Fig. 6) includes 394 taxa, with 50 dominant in at least one location. The taxa represented include chaetognaths, barnacle larvae, copepods, water fleas, crab larvae, echinoderm larvae, appendicularians, doliolids, and thaliaceans. Significant differences in patterns of dominance were observed in the four areas, from fewer dominant species in the northern portions of the ecosystem to increased biodiversity and dominance in the southern areas (Sherman et al. 1983).

Courtesy National Marine Fisheries Service

Fig. 6. Zooplankton.

Based on analysis of 9,942 zooplankton volume measurements collected by the National Oceanic and Atmospheric Administration's Marine Resources Monitoring, Assessment, and Prediction program from 1977 to 1986, zooplankton was declining during the 1980's (Sherman et al. 1994). The trend for Georges Bank was more variable than for the other three areas. The persistent downward trend in zooplankton abundance coincided with the increase in abundance of principal open-ocean fishes— herring, mackerel, and butterfish. Zooplankton populations have been sufficiently robust in the Northeast shelf ecosystem, even through a period of decline, to sustain the recovery of both herring and mackerel populations. Sufficient zooplankton will also be needed to support a recovery of the depressed cods and related fishes, together with flounder stocks during their larval stages. This dependence of recovering fish populations on zooplankton represents yet another factor to be considered in the recovery of zooplankton itself. Whether the observed decline in zooplankton reflects a response to increasing predation by the growing populations of pelagic fishes or a biofeedback response to an environmental signal remains an important unanswered question. There are some indications of renewed growth in zooplankton populations. From further analysis of the data through 1992 over Georges Bank, Sherman et al. (1994) found that the zooplankton standing stocks may again be increasing (Fig. 7).

To gain insights into the possibility of an environmentally mediated biofeedback, the zooplankton was examined for shifts in biodiversity over time as indicated by the abundance of five zooplankton species: *Calanus finmarchicus*, *Pseudocalanus* spp., *Centropages typicus*, *Metridia lucens*, and *Centropages hamatus*, for the four Northeast shelf areas from 1977 to 1987. Significant trends for Georges Bank and the Gulf of Maine were found. It appears that

the dominant copepod community was undergoing a shift from *Calanus finmarchicus*, *Pseudocalanus* spp., *Metridia lucens*, and *Centropages typicus* toward an increasing abundance of *C. hamatus* on Georges Bank and in the Gulf of Maine (Sherman et al. 1994). The shift in species abundance may have been environmentally induced, as it is unlikely that the principal pelagic fishes would selectively avoid preying on *C. hamatus*. An ecosystem monitoring program presently under way through the National Marine Fisheries Service's Northeast Fisheries Science Center, in cooperation with the Environmental Protection Agency, is designed to provide more definitive information on factors that cause changes in the structure of the zooplankton community. Major shifts in biodiversity of the zooplankton community could affect the recruitment success of the presently depressed cod, haddock, and flounder populations.

Benthic Resources

Information on the bottom of the Northeast shelf ecosystem is available from qualitative surveys beginning in the 1870's and quantitative sampling in the 1950's–1960's (Wigley and Theroux 1981; Fig. 8) and the 1970's–1980's (Steimle 1990). These surveys found the benthos to consist of a mosaic of assemblages. The 1970's–1980's surveys yielded 1,250 taxa, with polychaete worms most numerous (45% of the total), followed by crustaceans (23%), bivalves (12%), and gastropods (11%). Composition and abundances of dominant species vary temporally and spatially, but subsequent surveys of benthic assemblages have found them to be relatively stable. The only notable exceptions to this stability are in areas where waste disposal has altered the benthos or where shellfishes have been harvested. Larger bottom-dwelling animals like Atlantic rock crabs and lady crabs, the most abundant species, had fairly stable population sizes between 1978 and 1987. Stocks of the horseshoe crab, whose eggs are critically important as food for migrating shorebirds, were stable from the mid-1970's to mid-1980's, with no evidence that harvesting levels were depleting the resource, but continued monitoring was advised for this species (Botton and Ropes 1987).

The larger animals of submarine canyons on the shelf edge and continental slope, such as those off New Jersey and Delaware, have been sampled since the 1970's, with intensive work in the 1980's and early 1990's, in order to estimate the potential effects of oil and gas exploration and of waste disposal at Deepwater Dumpsite 106. Deepwater Dumpsite 106 is located approximately 170 kilometers southeast of New

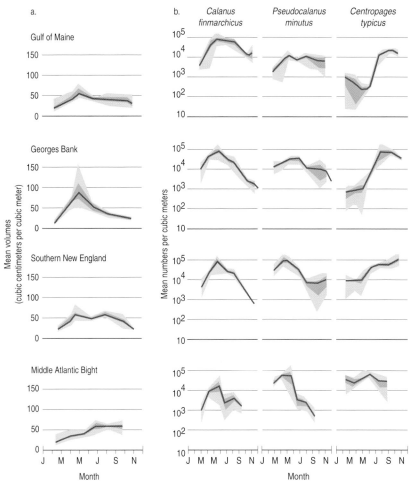

Fig. 7. a) Seasonal changes in the standing stock of zooplankton for four areas of the Northeast shelf ecosystem—the Gulf of Maine, Georges Bank, southern New England, and the Middle Atlantic Bight. Values are based on mean monthly volumes in cubic centimeters per 100 cubic meters of water and include the range and standard error of the mean. b) Seasonal pulses in abundance of the dominant copepods—*Calanus finmarchicus*, *Pseudocalanus minutus*, and *Centropages typicus*—in four areas of the Northeast shelf ecosystem are shown.

York City in depths of 1,700–2,750 meters. The benthic fauna of this area was found to be remarkably naturally diverse, with 798 species (58% new to science) recorded in the pre-oil exploration study.

Surveys of larger bottom-dwelling organisms have been conducted in all the major estuaries of the Northeast shelf ecosystem except Cobscook Bay in northern Maine. Only Chesapeake Bay has a major program for monitoring status and trends of benthic organisms; indicators there signaled recent improvements at some sites and increasing degradation at others. Data from the 1970's indicated that Delaware Bay and Raritan Bay had lower abundances of benthic fauna than did other Middle Atlantic estuaries, but the number of species in these bays was typical for the region. Studies in the early 1990's suggested no major changes had occurred in the Delaware Bay benthic ecosystem since the earlier work, whereas higher abundances and richness were reported for Raritan Bay in the late 1980's (Cerrato et al. 1989). In Chesapeake Bay and western Long Island Sound, chronic hypoxia (defined as less than 2 milligrams dissolved oxygen per liter of water) has altered the benthos and spurred nutrient reduction efforts. The decline of eelgrass in many northeastern estuaries since a 1930's wasting disease affected the plant has probably led to reduced abundance and diversity of benthic organisms in those areas.

Information is becoming available on the secondary production of benthic invertebrates, especially harvestable and forage species and community dominants. Annual productivity estimates of benthic organisms on Georges Bank suggest that production is relatively high there (more than 100 kilocalories per square meter) compared with other offshore and coastal areas in the northwest Atlantic (Steimle 1987).

Fisheries Resources

The Northeast shelf ecosystem contains a myriad of finfish and invertebrate species, some of which exist wholly within the ecosystem and others that are migratory and cross the regional boundary. Briggs (1974) reported about 250 shore fishes in this area, and Grosslein and Azarovitz (1982) indicated that 180 species of fishes and invertebrates were recorded in 1974 in the spring and autumn bottom trawl surveys conducted between Nova Scotia and Cape Hatteras, North Carolina, by the National Marine Fisheries Service's Northeast Fisheries Science Center. This comprehensive survey, which began in 1963, provides an invaluable time-series of relative abundance estimates of all the fish and invertebrate species caught by

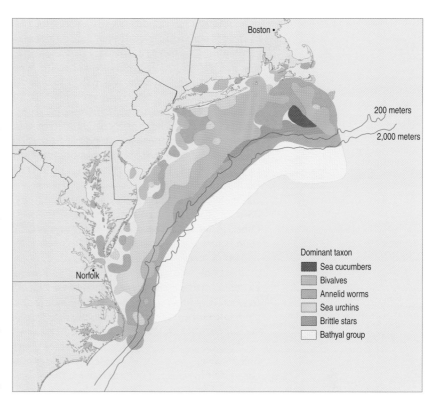

Dominant taxon
- ■ Sea cucumbers
- ▨ Bivalves
- ▨ Annelid worms
- ▨ Sea urchins
- ▨ Brittle stars
- □ Bathyal group

Fig. 8. Geographic distribution of biomass for each dominant benthic species or group in the Middle Atlantic Bight, part of the Northeast shelf ecosystem (from Wigley and Theroux 1981).

this gear. The number of species is not comprehensive, however, because some are not caught in the otter trawls used in those surveys (for example, sharks, large pelagic fishes such as tuna and billfish, small or elusive pelagic fishes, and species residing in water deeper than 365 meters). About 50–75 of the species caught during the surveys are harvested in commercial and recreational fisheries. Although not supporting any fisheries, the remaining species are nevertheless important elements in the complex web of the ecosystem, serving as a food source for those species more directly used by humans.

Recent catches of these economically important fishes, both offshore and nearshore–estuarine, have been reduced because of the overexploitation of many stocks—such as most groundfishes and sea scallops—and the underutilization of others—such as Atlantic mackerel and Atlantic herring. A major shift in dominance has taken place within the finfish community from groundfishes in the 1960's—Atlantic cod, haddock, pollock, white hake, silver hake, and flounders—to predominantly elasmobranchs such as spiny dogfish and skates, and pelagic fishes such as mackerel and herring in recent years (Fig. 9). This shift is the consequence of excessive trawl fishing on groundfish and has led to reduced numbers of mature fishes, diminished spawning success, abnormally few older fishes, and stock depletion.

The demersal groundfish group (about 35 species) has a long-term potential yield of more than 500,000 metric tons; nearly every stock is overexploited. This overall abundance declined substantially in the past 30 years to an all-time low in 1994 (Fig. 10). Initially the decline was the result of heavy distant-water fleet fishing in the 1960's–1970's and, more recently, it is because of excessive domestic fishing pressure. Recent (1994) catches are about 160,000 metric tons per year. The status and trends of the major fish resources in the northeastern region are summarized in Table 1. Haddock, yellowtail flounder, and Atlantic cod stocks on Georges Bank have collapsed—1994 landings were only 23% of long-term potential yield (Fig. 11). Other groundfish species are generally low in abundance, and populations are expected to remain depressed. Consequently, large portions of the bank and southern New England were closed to fishing, effective December 1994, to reduce fishing pressure and avert reducing the abundance of these species to levels that would not support fishing of any kind (Fig. 12).

Abundance of spiny dogfish and skates combined (seven species), expressed as kilograms per tow from research bottom trawl surveys, has increased four- to five-fold since the late 1960's (Fig. 13). The biomass of spiny dogfish alone was estimated to be about 650,000 metric tons in 1994 (Northeast Fisheries Science Center 1994). This increase, coupled with the decline in cods and related fishes and flounders, has resulted in dogfish and skates increasing in

relative abundance from 25% of the total fish biomass on Georges Bank in 1963 to about 75% in recent years. Currently, dogfish and skates are high in abundance and declining slightly.

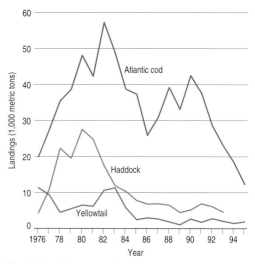

Fig. 11. Total landings of Atlantic cod, haddock, and yellowtail flounder from Georges Bank, 1976–1995.

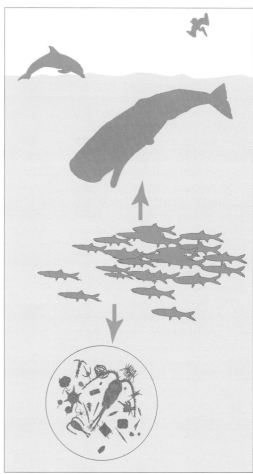

Fig. 12. The removal of hundreds of thousands of metric tons of fishes through fishing activity may result in cascade effects down the marine food chain to phytoplankton and zooplankton, and up the food chain to marine birds and mammals.

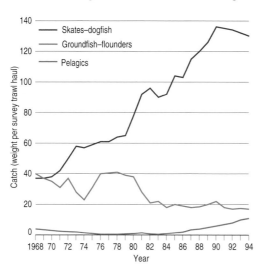

Fig. 9. Trends in indices of abundance of three groups of fishes in the Northeast region, 1968–1994.

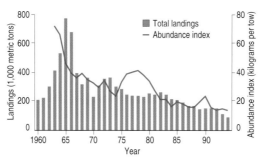

Fig. 10. Landings and abundance index of principal Northeast flounders and groundfishes, 1960–1994 (from National Oceanic and Atmospheric Administration 1996).

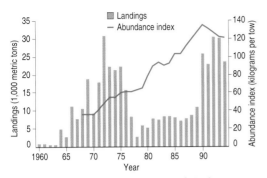

Fig. 13. Landings and abundance index of Northeast skates and dogfish, 1960–1994 (from National Oceanic and Atmospheric Administration 1996).

The status and trends of anadromous species—which breed in rivers—and pelagic species are also summarized in Table 1. The long-term potential yield of anadromous species such as striped bass, river herrings, Atlantic salmon, American shad, and sturgeons is unknown, but current annual landings of less than 4,000 metric tons are far below historical levels. In contrast to most marine fishes, these anadromous species have been affected greatly by nonfishing activities, including the damming of rivers and environmental contamination (see chapters on Water Use and Environmental Contaminants). Atlantic salmon abundance has declined precipitously (Fig. 14), and the National Oceanic and Atmospheric

Administration and the U.S. Fish and Wildlife Service have accepted a State of Maine Conservation Plan to conserve a population complex of Atlantic salmon in seven Maine rivers as an alternative to listing it as threatened under the U.S. Endangered Species Act. The striped bass is another species that was reduced to low numbers in the late 1970's and early 1980's by excessive fishing pressure (Fig. 15). Striped bass populations, however, have responded positively to a series of restrictive management measures imposed in the 1980's; this species' recruitment and abundance are now increasing (National Oceanic and Atmospheric Administration 1993).

The pelagic fishes (Atlantic mackerel, Atlantic menhaden, Atlantic herring, bluefish, and butterfish) have a long-term potential yield of 900,000 metric tons and support a catch of 500,000 metric tons per year. Mackerel and herring, currently underutilized, were heavily fished by distant-water fleets from the mid-1960's to mid-1970's (Fig. 16). Their populations decreased to minimal levels around 1980 but have subsequently increased to present record-high levels, measured in excess of 2 million metric tons each (Northeast Fisheries Science Center 1993). Mackerel and herring

Table 1. Status and trends of major fish resources in the Northeast region.

Fish species	Status and trend
Groundfish/flounders	
American plaice	Average abundance, overutilized, stable
Atlantic cod	Very low abundance, overutilized, depressed
Haddock	Very low abundance, overutilized, depressed
Pollock	Average abundance, fully utilized, stable
Redfish	Low abundance, overutilized, depressed
Red hake	Average-low abundance, underutilized, stable
Silver hake	Low abundance, fully utilized, depressed
Skates and dogfish	High abundance, fully utilized, declining
Summer flounder	Average abundance, overutilized, increasing
Windowpane	Low abundance, fully utilized, declining
Winter flounder	Low abundance, overutilized, depressed
Witch flounder	Low abundance, overutilized, stable
Yellowtail flounder	Very low abundance, overutilized, depressed
Anadromous fishes	
American shad	Low abundance, overutilized, variable
Atlantic salmon	Low abundance, fully utilized, variable
River herring	Low abundance, overutilized, variable
Striped bass	High abundance, fully utilized, increasing
Sturgeons	Low abundance, overutilized, stable
Pelagic fishes[a]	
Atlantic herring	High abundance, underutilized, stable to increasing
Atlantic mackerel	High abundance, underutilized, stable to increasing
Bluefish	Low abundance, overutilized, declining
Butterfish	Average abundance, underutilized, stable
Sharks	Unknown abundance, some species possibly fully or overutilized, unknown trend

[a]For highly migratory pelagic fishes see text discussion of the Southeast region.

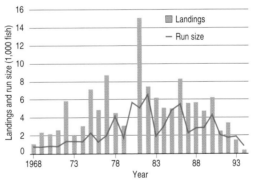

Fig. 14. Spawning run size of Atlantic salmon returning to Maine rivers, and total catch by U.S. anglers and at-sea catch by foreign vessels, 1968–1994 (from National Oceanic and Atmospheric Administration 1996).

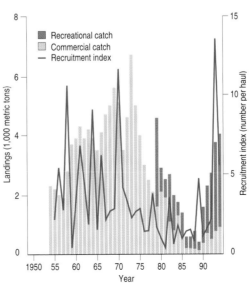

Fig. 15. Striped bass catch and recruitment index of young striped bass in Chesapeake Bay, 1950–1994 (from National Oceanic and Atmospheric Administration 1996).

stocks can support long-term potential yields of 200,000 and 120,000 metric tons, respectively. Bluefish, an important recreational species that supported catches as high as 76,500 metric tons in 1980, has since declined and is now overexploited. Atlantic menhaden, distributed from Canada to Florida, is fully utilized, with a long-term potential yield of 480,000 metric tons per year and recent annual catches of 360,000 metric tons.

The long-term potential yield of highly migratory pelagic fishes in the Atlantic (such as swordfish, bluefin tuna, yellowfin tuna, bigeye tuna, albacore, skipjack tuna, blue marlin, white marlin, sailfish, and spearfish) is about 250,000 metric tons, less than 20,000 metric tons of which is the U.S. share (prorated based on previous catches). Recent catches have averaged about 230,000 metric tons worldwide, with 16,000 metric tons taken by the United States

(Fig. 17). Virtually all the species are either fully or overutilized, with the abundance of several species (for example, swordfish, bluefin tuna, and marlin) below or far below optimal levels.

Little is known about the abundance of large and small coastal and pelagic sharks in the Atlantic, but their long-term potential yield is estimated to be about 10,000 metric tons per year, the level of current catches. Sharks traditionally were fished moderately, but in recent years, large coastal sharks have been particularly subject to increased fishing pressure, resulting in overexploitation (Fig. 18).

Valuable crustacean and molluscan resources occur in offshore and nearshore (estuarine) waters. Offshore species such as American lobster, sea scallop, Atlantic surf-clam, ocean quahog, long-finned squid, short-finned squid, and northern shrimp have a long-term potential yield of 160,000–200,000 metric tons, with smaller recent landings. Their population status and trends are summarized in Table 2. With the exception of short-finned squid, all are either fully utilized or overutilized. Sea scallops in particular (7,400 metric tons landed in the Northeast region in 1993; Fig. 19), and to a lesser extent lobsters in the Gulf of Maine (25,500 metric tons landed in 1993 throughout the Northeast; Fig. 20), are exploited at rates exceeding management definitions of overfishing. Sea scallops, whose reproductive success exhibits high annual variability, are now at a low level of abundance. Green sea urchins (for which we have no estimate of long-term potential yield) have been subjected to ever-increasing fishing pressure since a fishery began in 1987. Landings in 1993 were more than 19,000 metric tons, a three-fold increase from 1990, with a strong likelihood of overexploitation in some areas (for example, coastal Maine).

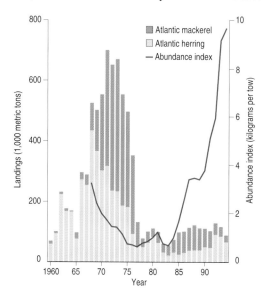

Fig. 16. Landings of Atlantic herring and Atlantic mackerel in the Northeast region and their combined abundance index, 1960–1994 (from National Oceanic and Atmospheric Administration 1996).

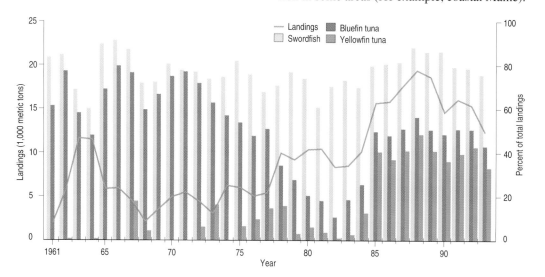

Fig. 17. U.S. landings of tunas, swordfish, marlins, and spearfish from the western North Atlantic Ocean, and percentages of selected species, 1960–1994 (from National Oceanic and Atmospheric Administration 1996).

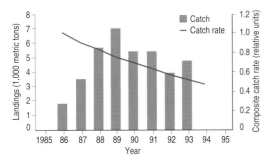

Fig. 18. Landings and relative catch rates of large coastal sharks in the western Atlantic Ocean and Gulf of Mexico by commercial vessels, 1985–1995 (from National Oceanic and Atmospheric Administration 1996).

Various species of crab (for example, Jonah, Atlantic rock, and red crabs), some found both offshore and nearshore, produced about 4,500 metric tons of landings in 1993 and are underutilized.

Nearshore and estuarine species (for example, eastern oysters, northern quahogs, softshell clams, blue mussels, bay scallops, whelks or conchs, and blue crabs) are generally fully or overutilized (Table 2) and have a long-term potential yield of about 100,000 metric tons (meat weight only, except for blue crab). Eastern oyster landings fell sharply in the late

1980's (Fig. 21) and are now, at about 11,000 metric tons per year, far below long-term potential yield. Populations in the principal historical eastern oyster areas, the Chesapeake and Delaware bays, are largely depleted by effects of disease (MSX and dermo); Connecticut now accounts for most eastern oyster production. Bay scallops, with recent landings averaging about 160 metric tons (1990–1992), are overexploited and at reduced levels of abundance. Recent landings of blue mussels (Fig. 22), with an uncertain long-term potential yield, have increased to an average 3,500 metric tons per year. Northern quahog (hardshell clam) landings (Fig. 22) are now fairly steady at about 4,700 metric tons per year and are probably overexploited. Softshell clams (average catches of 2,100 metric tons in 1990–1992; Fig. 22) appear to be fully exploited, with decreased abundance in Maryland and Maine. Blue crab off the Atlantic coast is fully utilized, with commercial landings averaging 44,000 metric tons per year since 1985 (Fig. 21). Their greatest concentration is in Chesapeake Bay, where the recreational harvest has made up approximately 30%–40% of the total catch in recent years.

Habitat/species	Status and trend
Offshore	
American lobster	High abundance, overutilized, variable
Atlantic surfclam	Average abundance, fully utilized, declining
Crabs	High abundance, underutilized, relatively stable
Long-finned squid	Average abundance, fully utilized, variable
Northern shrimp	Average abundance, fully utilized, relatively stable
Ocean quahog	Average abundance, fully utilized, declining
Sea scallop	Low abundance, overutilized, variable
Sea urchins	Average abundance, in danger of being overutilized, unknown trend
Short-finned squid	Average abundance, underutilized, variable
Nearshore and estuarine	
Bay scallop	Low abundance, overutilized, relatively stable
Blue crab	Average abundance, fully utilized, relatively stable
Blue mussel	Average abundance, fully utilized, stable
Eastern oyster	Low to depleted abundance, overutilized, stable
Hardshell clam	Low abundance, overutilized, stable
Softshell clam	Low abundance, overutilized, stable

Table 2. Status and trends of selected crustaceans and mollusks in the Northeast region.

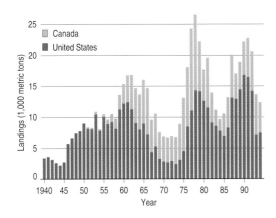

Fig. 19. U.S. and Canadian landings of sea scallops from the mid-Atlantic coast, Gulf of Maine, and Georges Banks, 1940–1994 (from National Oceanic and Atmospheric Administration 1996).

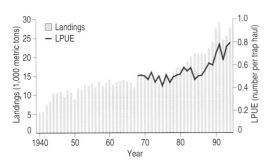

Fig. 20. Landings of American lobsters in U.S. waters, and landings per unit effort (LPUE) in Maine, 1940–1994 (from National Oceanic and Atmospheric Administration 1996).

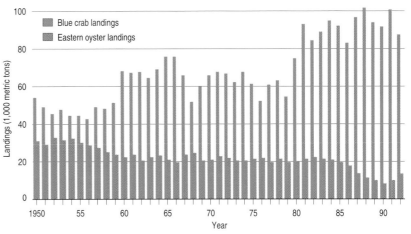

Fig. 21. Commercial landings of blue crabs and eastern oysters (1950–1992) along U.S. Atlantic and Gulf of Mexico coasts (from National Oceanic and Atmospheric Administration 1996).

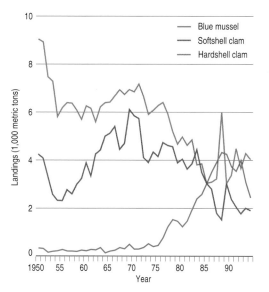

Fig. 22. Commercial landings of hardshell clams, softshell clams, and blue mussels (1950–1994) along U.S. Atlantic and Gulf of Mexico coasts (from National Oceanic and Atmospheric Administration 1996).

Marine Birds

Marine birds, which breed, forage, or migrate through marine environments, consist of several ecological groupings. Some breeding marine birds nest on nearshore beaches, coastal cliffs, or islands near the coast. These species forage in either coastal, nearshore, or pelagic habitats during the breeding season—usually summer and fall—but often migrate to other regions during other seasons. During winter, marine birds that breed farther north may spend the winter living and foraging in more southern environments. For example, some birds that nest in the Canadian arctic and subarctic may be prevalent in the winter in the Northeast. Another group of marine birds found in all U.S. marine systems, sometimes in immense numbers, are birds that breed in tropical environments or in the Southern Hemisphere. Birds of this latter group are almost always found offshore near the continental shelf break and include tropicbirds, shearwaters, albatrosses, petrels, storm-petrels, and skuas.

Buckley and Buckley (1984) reported that at least 20 marine bird species nest regularly in the Northeast region, while 9 others irregularly nest there or in the near vicinity. Schneider and Heinemann (1996) completed a recent overview of predominant marine birds inhabiting the Northeast shelf ecosystem. Gulls, terns, and herons are important species breeding within the ecosystem. Other species of marine birds that do not breed in the ecosystem are nevertheless important and occupy two ecological regimes within the ecosystem, coastal and pelagic. In the coastal zone, plovers, sandpipers, and other shorebirds are important predators of beach and intertidal invertebrates. Other nonbreeding but abundant species are for the most part pelagic, breeding in colonies around the margin of the northwestern Atlantic basin.

In the Gulf of Maine, the dominant bird populations include Leach's storm-petrel, great cormorant, razorbill, and Atlantic puffin. Farther south, terns, gulls, and jaegers are the dominant pelagic birds. Species abundances in the Northeast region are highly seasonal. The most numerous species in early summer is the sooty shearwater, a regular migrant from its Southern Hemisphere breeding areas; in late summer, Wilson's storm-petrel is more abundant. The greater shearwater is abundant in autumn. During winter, the ecosystem is inhabited by southward-migrating flocks of Iceland gulls, glaucous gulls, black-legged kittiwakes, Leach's storm-petrels, dovekies, and razorbills.

Researchers observed the changing population densities of pelagic marine birds during the National Oceanic and Atmospheric Administration's Marine Resources Monitoring, Assessment, and Prediction surveys of the Northeast shelf ecosystem, conducted from 1976 to 1988. Some of these pelagic marine species breed in the area, but others are only winter visitors or passage migrants, thus the trends revealed do not necessarily relate to breeding population sizes. Based on analysis of the surveys, northern fulmar, great black-backed gull, and herring gull populations declined from 1978 to 1980. Other species showing similar decreasing trends in abundance were greater shearwater, sooty shearwater, Cory's shearwater, Wilson's storm-petrel, black-legged kittiwake, and pomarine jaeger. Some of these species feed heavily on trawler-discarded offal at sea. Schneider and Heinemann (1996) hypothesized that as fish stocks decreased in abundance in the early 1980's, the amount of offal discarded at sea also declined, which may have reduced the survival of some marine birds. Another possibility is that the birds moved their foraging activities elsewhere. Significant year-to-year variability was observed among other marine birds, including ring-billed gulls, common terns, northern gannets, laughing gulls, Leach's storm-petrels, red phalaropes, and dovekies. The causes of this variability have not been determined.

Marine Mammals

Since at least the 1500's, humans have interacted with marine mammals in waters off the coast of what is now the United States. Historically, much of this interaction has been the result of the whaling industry; although the last whaling in eastern U.S. waters took place around the 1920's, these interactions have continued. Along the New England coast, nonconsumptive "uses" of whales have replaced commercial whaling—whale watching has emerged as a multimillion dollar industry. It is probable that the economic value of whalewatching today exceeds that of whale harvesting in previous eras. Despite the general change in attitude of the American public to a nonconsumptive interest in marine mammals, as well as the recreational, educational, and research benefits of whalewatching, marine mammals today are still in jeopardy from human interactions, such as entanglement in fishing nets, bycatch (the incidental take of nontarget species), and ship collisions. The larger whales occasionally become entangled in fishing nets and line, which leads to their injury and reduced viability, although the animals often escape or are released. The smaller whales, primarily harbor porpoises and dolphins, are also caught in fishing nets, which is almost always fatal. Large whales are occasionally involved in vessel collisions, events that also may be fatal or leave an animal with debilitating injuries. Indirect effects of human influences are more difficult to assess but include habitat degradation and loss.

Expanding seal populations result in more frequent interactions with coastal fishing activities and aquaculture facilities. Harbor seals and other species are an increasing component of the marine mammal bycatch of fisheries. The additional dimension of indirect fishery interactions is that marine mammals compete with humans (or vice versa) for food (fishes and squid).

The Northeast region supports 34 stocks of at least as many species of marine mammals, for which the National Marine Fisheries Service has management responsibility (Table 3). Twenty-one of these species are classified as *strategic stocks* according to the 1994 amendments to the Marine Mammal Protection Act. Strategic stocks are those that are listed as endangered or threatened under the U.S. Endangered Species Act, are depleted under the Marine Mammal Protection Act, have total annual mortality rates that are not sustainable, or are those for which there are known removals but with uncertain status. Mortality of some strategic stocks exceeds their Potential Biological Removal (PBR), a management term set by 1994 amendments to the Marine Mammal Protection Act to define the removal rate beyond which a marine mammal stock would be impeded from recovery and reaching or maintaining its optimal sustainable population level. Under the Marine Mammal Protection Act, the short-term goal is to reduce incidental takes (mortality and serious injury caused by humans) to at or below PBR level.

In the western North Atlantic, strategic stock species include endangered right whales, humpback whales, fin whales (Figs. 23 and 24), sei

Courtesy Cetacean and Turtle Assessment Program, University of Rhode Island

Fig. 23. The fin whale, the most abundant large cetacean in the western North Atlantic.

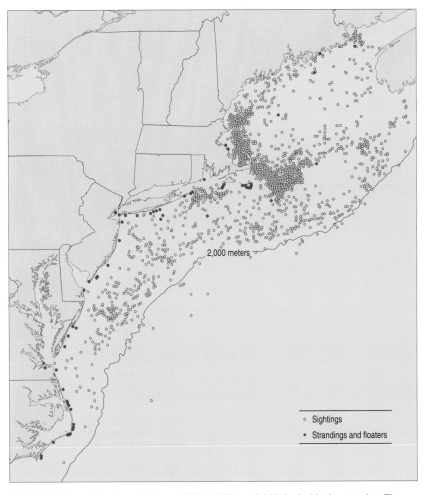

2,000 meters

○ Sightings
• Strandings and floaters

Fig. 24. Sightings of the fin whale from 1966 to 1989 ($n = 3,858$) in the Northeast region. The 2,000-meter depth contour defines the boundary of the survey area. During the peak spring–summer feeding season, there may be between 2,500 and 3,000 fin whales in the Northeast region waters. Modified from Hain et al. (1992). *Strandings* and *floaters* are dead whales found ashore or at sea.

Table 3. Status and trends of marine mammal stocks in waters of U.S. jurisdiction.

Region/species	Area	Minimum population[a]	Status[b]
Alaska			
Seals, sea lions, and walruses			
Bearded seal	Alaska	N/A	N/S
Harbor seal	Bering Sea	12,648	N/S
Harbor seal	Gulf of Alaska	22,427	N/S
Harbor seal	Southeast Alaska	35,226	N/S
Northern fur seal	Eastern Pacific	969,595	D
Ribbon seal	Alaska	N/A	N/S
Ringed seal	Alaska	N/A	N/S
Spotted seal	Alaska	N/A	N/S
Northern sea lion	Eastern U.S. stock	37,166	T
Northern sea lion	Western U.S. stock	42,536	E
Pacific walrus	Alaska	188,316	N/S
Whales, dolphins, and porpoises			
Baird's beaked whale	Alaska	N/A	N/S
Beluga whale	Beaufort Sea	32,453	N/S
Beluga whale	Bristol Bay	1,316	N/S
Beluga whale	Cook Inlet	752	*
Beluga whale	E. Chukchi Sea	3,710	N/S
Beluga whale	E. Bering Sea	6,439	N/S
Bowhead whale	W. Arctic	7,738	E
Cuvier's beaked whale	Alaska	N/A	N/S
Fin whale	Northeast Pacific	N/A	E
Gray whale	E. North Pacific	21,597	N/S
Humpback whale	Central North Pacific	1,407	E
Humpback whale	W. North Pacific	N/A	E
Killer whale	E. North Pacific, northern resident	764	N/S
Killer whale	E. North Pacific, transient	314	N/S
Minke whale	Alaska	N/A	N/S
Northern right whale	North Pacific	N/A	E
Sperm whale	North Pacific	N/A	E
Stejneger's beaked whale	Alaska	N/A	N/S
Pacific white-sided dolphin	North Pacific	486,719	N/S
Dall's porpoise	Alaska	76,874	N/S
Harbor porpoise	Bering Sea	8,549	N/S
Harbor porpoise	Gulf of Alaska	7,085	N/S
Harbor porpoise	Southeast Alaska	8,156	N/S
Sea otters and polar bears			
Sea otter	Alaska	100,000	N/S
Polar bear	Southern Beaufort Sea	1,579	N/S
Polar bear	Chukchi/Bering Sea	N/A	N/S
Atlantic and Gulf of Mexico			
Seals			
Gray seal	N.W. North Atlantic	2,035	N/S
Harbor seal	W. North Atlantic	28,810	N/S
Harp seal	N.W. North Atlantic	N/A	N/S
Hooded seal	N.W. North Atlantic	N/A	N/S
Whales, dolphins, and porpoises			
Blainville's beaked whale	N. Gulf of Mexico	N/A	N/S
Blainville's beaked whale	W. North Atlantic	N/A	N/S
Blue whale	W. North Atlantic	N/A	E
Bryde's whale	N. Gulf of Mexico	17	N/S
Cuvier's beaked whale	N. Gulf of Mexico	20	N/S
Cuvier's beaked whale	W. North Atlantic	895	*
Dwarf sperm whale	N. Gulf of Mexico	N/A	*
Dwarf sperm whale	W. North Atlantic	N/A	*
False killer whale	N. Gulf of Mexico	236	N/S
Fin whale	W. North Atlantic	1,704	E
Gervais' beaked whale	N. Gulf of Mexico	N/A	N/S
Gervais' beaked whale	W. North Atlantic	N/A	N/S
Humpback whale	W. North Atlantic	4,848	E

Region/species	Area	Minimum population[a]	Status[b]
Killer whale	N. Gulf of Mexico	197	N/S
Killer whale	W. North Atlantic	N/A	N/S
Long-finned pilot whale	W. North Atlantic	4,968	*
Melon-headed whale	N. Gulf of Mexico	2,888	N/S
Minke whale	Canadian east coast	2,053	N/S
North Atlantic right whale	W. North Atlantic	295	E
Northern bottlenose whale	W. North Atlantic	N/A	N/S
Pygmy killer whale	N. Gulf of Mexico	285	N/S
Pygmy killer whale	W. North Atlantic	6	N/S
Pygmy sperm whale	N. Gulf of Mexico	N/A	*
Pygmy sperm whale	W. North Atlantic	N/A	*
Sei whale	W. North Atlantic	N/A	E
Short-finned pilot whale	N. Gulf of Mexico	186	*
Short-finned pilot whale	W. North Atlantic	457	*
Sowerby's beaked whale	W. North Atlantic	N/A	N/S
Sperm whale	N. Gulf of Mexico	411	E
Sperm whale	W. North Atlantic	1,617	E
True's beaked whale	W. North Atlantic	N/A	N/S
Atlantic spotted dolphin	N. Gulf of Mexico	2,255	N/S
Atlantic spotted dolphin	W. North Atlantic	1,617	*
Atlantic white-sided dolphin	W. North Atlantic	19,196	N/S
Bottlenose dolphin	E. Gulf of Mexico, coastal	8,963	N/S
Bottlenose dolphin	Gulf of Mexico, bay, sound, estuarine	3,933	*
Bottlenose dolphin	Gulf of Mexico, outer continental shelf	43,233	N/S
Bottlenose dolphin	Gulf of Mexico, shelf edge and slope	4,530	N/S
Bottlenose dolphin	W. North Atlantic, coastal	2,482	D
Bottlenose dolphin	W. North Atlantic, offshore	8,794	N/S
Bottlenose dolphin	N. Gulf of Mexico, coastal	3,518	N/S
Bottlenose dolphin	W. Gulf of Mexico, coastal	2,938	N/S
Clymene dolphin	N. Gulf of Mexico	4,120	N/S
Common dolphin	W. North Atlantic	15,470	*
Fraser's dolphin	N. Gulf of Mexico	66	N/S
Pantropical spotted dolphin	N. Gulf of Mexico	26,510	N/S
Pantropical spotted dolphin	W. North Atlantic	1,617	*
Risso's dolphin	N. Gulf of Mexico	2,199	N/S
Risso's dolphin	W. North Atlantic	11,140	N/S
Rough-toothed dolphin	N. Gulf of Mexico	660	N/S
Spinner dolphin	N. Gulf of Mexico	4,465	N/S
Spinner dolphin	W. North Atlantic	N/A	N/S
Striped dolphin	N. Gulf of Mexico	3,409	N/S
Striped dolphin	W. North Atlantic	18,220	N/S
White-beaked dolphin	W. North Atlantic	N/A	N/S
Harbor porpoise	Gulf of Maine/ Bay of Fundy	48,289	*
Manatees			
West Indian manatee	Antillean	86	E
West Indian manatee	Florida	1,822	E
Pacific and Western Pacific			
Sea otters			
Sea otter	California	2,376	E
Sea otter	Washington	360	N/S
Seals, sea lions, and walruses			
Guadalupe fur seal	Mexico to California	3,028	T

— continued —

Table 3. Continued.

Region/species	Area	Minimum population[a]	Status[b]
Harbor seal	Washington, inland waters	15,349	N/S
Harbor seal	California	27,962	N/S
Harbor seal	Oregon to Washington, coastal	25,665	N/S
Hawaiian monk seal	Hawaii	1,366	E
Northern elephant seal	California, breeding	51,625	N/S
Northern fur seal	San Miguel Island	5,018	N/S
California sea lion	United States	111,339	N/S
Whales, dolphins, and porpoises			
Baird's beaked whale	California to Washington	252	N/S
Blainville's beaked whale	Hawaii	N/A	N/S
Blue whale	California to Mexico	1,463	E
Blue whale	Hawaii	N/A	E
Bryde's whale	E. Tropical Pacific	11,163	N/S
Bryde's whale	Hawaii	N/A	N/S
Cuvier's beaked whale	California to Washington	6,070	N/S
Cuvier's beaked whale	Hawaii	N/A	N/S
Dwarf sperm whale	California to Washington	N/A	N/S
Dwarf sperm whale	Hawaii	N/A	N/S
False killer whale	Hawaii	N/A	N/S
Fin whale	California to Washington	747	E
Fin whale	Hawaii	N/A	E
Humpback whale	California to Mexico	563	E
Killer whale	California to Washington	436	N/S
Killer whale	Eastern N. Pacific, southern resident	96	N/S
Killer whale	Hawaii	N/A	N/S
Melon-headed whale	Hawaii	N/A	N/S
Mesoplodont beaked whales	California to Washington	1,169	*
Minke whale	California to Washington	122	*
Pygmy killer whale	Hawaii	N/A	N/S
Pygmy sperm whale	California to Washington	2,059	N/S
Pygmy sperm whale	Hawaii	N/A	N/S
Sei whale	E. North Pacific	N/A	E
Short-finned pilot whale	California to Washington	741	*
Short-finned pilot whale	Hawaii	N/A	N/S
Sperm whale	California to Washington	896	E
Sperm whale	Hawaii	N/A	E
Bottlenose dolphin	California, coastal	134	N/S

Region/species	Area	Minimum population[a]	Status[b]
Bottlenose dolphin	California to Washington, offshore	1,904	N/S
Bottlenose dolphin	Hawaii	N/A	N/S
Central American spinner dolphin	E. Tropical Pacific	N/A	N/S
Common dolphin	E. Tropical Pacific, central	297,400	N/S
Common dolphin, long-beaked	California	5,504	N/S
Common dolphin	E. Tropical Pacific, northern	353,100	N/S
Common dolphin, short-beaked	California to Washington	309,717	N/S
Common dolphin	E. Tropical Pacific, southern	1,845,600	N/S
Eastern spinner dolphin	E. Tropical Pacific	518,500	D
Northern right-whale dolphin	California to Washington	15,080	N/S
Pacific white-sided dolphin	California to Washington	82,939	N/S
Pantropical spotted dolphin	Hawaii	N/A	N/S
Risso's dolphin	California to Washington	22,388	N/S
Risso's dolphin	Hawaii	N/A	N/S
Rough-toothed dolphin	Hawaii	N/A	N/S
Spinner dolphin	Hawaii	677	N/S
Spotted dolphin	E. Tropical Pacific, coastal	22,500	N/S
Spotted dolphin	E. Tropical Pacific, northeastern	648,900	D
Spotted dolphin	E. Tropical Pacific, west/south	1,145,100	N/S
Striped dolphin	California to Washington	19,248	N/S
Striped dolphin	E. Tropical Pacific	1,745,900	N/S
Striped dolphin	Hawaii	N/A	N/S
Whitebelly spinner dolphin	E. Tropical Pacific	872,000	N/S
Dall's porpoise	California to Washington	34,393	N/S
Harbor porpoise	Central California	3,431	N/S
Harbor porpoise	N. California	7,640	N/S
Harbor porpoise	Oregon to Washington, coast	22,046	N/S
Harbor porpoise	Washington, inland	2,681	N/S

[a]N/A = not available or not determined.
[b]N/S = no special status under the Endangered Species Act/ Marine Mammal Protection Act; D = depleted; T = threatened; E = endangered; and * = Marine Mammal Protection Act strategic stock because human-caused take exceeds potential biological removal.

whales, blue whales, and sperm whales (Fig. 25); the mid-Atlantic coastal bottlenose dolphin, which is listed as depleted under the Marine Mammal Protection Act; and stocks whose estimated mortality exceeds PBR: Cuvier's beaked whale, dwarf sperm whale, long-finned pilot whale, pygmy sperm whale, short-finned pilot whale, Atlantic spotted dolphin, common dolphin, pantropical spotted dolphin, and the Gulf of Maine–Bay of Fundy harbor porpoise. Of these strategic stocks, North Atlantic right whales are believed to be increasing, Western North Atlantic coastal bottlenose dolphins are believed to be stable, and the population trends for the remaining species, such as the Atlantic white-sided dolphin (Fig. 26), are unknown.

In general, the distribution and abundance of marine mammals ebb and flow with the seasons. This phenomenon is almost solely linked

Courtesy Cetacean and Turtle Assessment Program, University of Rhode Island

Fig. 25. Sperm whales in deep water over the continental slope just south of Georges Bank.

Courtesy Cetacean and Turtle Assessment Program, University of Rhode Island

Fig. 26. The white-sided dolphin is the most abundant dolphin in continental shelf waters of the northeastern United States.

to food, because cetacean (whale, dolphin, and porpoise) distributions are far from regular or random but almost always correspond with the location of food sources. Cetaceans generally enter the Northeast region in spring and circulate through the feeding grounds during summer and into fall. During the feeding season, cetaceans in the Northeast region probably intermix with animals of the Canadian Scotian shelf and northward. As the year wanes, whales migrate to wintering grounds in waters of the mid-Atlantic and southeastern United States, offshore deep-ocean areas, and the West Indies. A few individuals of most species, however, occupy the Northeast region throughout the winter.

Although some migration patterns have been described, it is generally observed that whales

appear and disappear and that their whereabouts and habitats, when the animals are absent from their better-known feeding grounds, remain largely unknown. For most species, this is particularly true for the wintering and calving grounds. Where whales are concerned, the waters of the northeastern United States are not a discrete entity but rather part of a continuum that stretches through large sections of the Atlantic Ocean throughout the Northern Hemisphere. The various calving and feeding grounds are not residences but temporary stopping points.

In terms of numbers and ecological effects, marine mammals are significant components of the northeastern marine ecosystem, although many of their populations have been reduced in historical times. The fin whale and humpback whale, for example, are recovering from commercial depletion and presently represent the largest component of the cetacean population in northwest Atlantic waters; their abundance is estimated at 6,000 animals or more during the peak feeding season. The mid-Atlantic coastal and offshore populations of bottlenose dolphin are depleted following a significant die-off in 1987–1988. The abundance of harbor porpoises has been estimated at 47,200 individuals in the Gulf of Maine–Bay of Fundy area; they have been proposed for listing as threatened under the Endangered Species Act because their annual incidental death rate exceeds their birth rate. Populations of harbor seal, gray seal, harp seal, and hooded seal appear to be increasing and expanding their ranges along New England's coasts. The seasonal abundance and population trends of other marine mammals are not well known.

The majority of present research on and management activities for marine mammals is aimed at reducing resource use conflicts and mitigating adverse effects of interactions between humans and these animals. Research on northwest Atlantic marine mammal populations has focused on three primary questions: (1) Have fisheries interactions and other human-related activities directly or indirectly disadvantaged marine mammals or adversely altered their environment? (2) Are the depleted and endangered marine mammals recovering, and have adequate steps been taken to encourage their recovery? and (3) What management actions are and will be necessary in the future to minimize or avoid potential conflicts between marine mammal conservation and human activities, such as commercial fishing and coastal development? For example, in 1994, Cape Cod Bay, Stellwagen Bank, the Great South Channel, and coastal waters of the southeastern United States all were designated under the U.S. Endangered Species Act as critical habitats

North Atlantic Right Whale

The most endangered large whale off the east coast of the United States is the North Atlantic right whale (Figs. 1 and 2). Although a great deal has been learned about this species and such information will aid in its recovery, many questions remain. The population presently numbers about 300, and its growth rate has been estimated at 2.5%, a rate lower than that for right whale populations in the Southern Hemisphere (Knowlton et al. 1994).

North Atlantic right whales move from wintering and calving grounds in coastal

within regions may be more extensive than previously thought. Based on results from satellite-tracked animals, sightings of individuals made perhaps two weeks apart cannot be assumed to indicate a stationary or resident animal. Rather, movement data have shown rather lengthy and somewhat distant excursions within a few weeks within a season (Mate et al. 1992). These findings cast new light on the extent of movements and habitat use by right whales and raise new questions about the purpose and strategies for such excursions.

New England waters are a primary feeding habitat for the North Atlantic right whale, where it appears to feed primarily on calanoid copepods. These dense zooplankton patches are likely a primary component of the spring, summer, and fall right whale habitat (Kenny et al. 1986). During the peak feeding season in Cape Cod and Massachusetts bays, the acceptable surface copepod resource is limited to perhaps 3% of the region. Although feeding in the coastal waters off Massachusetts has been better studied, feeding by right whales has also been observed over Georges Bank, in the Gulf of Maine, in the Bay of Fundy, and over the Scotian Shelf. The characteristics of acceptable prey distribution in these areas are not as well known as those for the area off Massachusetts. New England waters also serve as a nursery for calves and, in some cases, as sites for mating.

Genetic analyses of tissue samples are providing insights to population definition. Schaeff et al. (1993) have suggested that western North Atlantic right whales probably represent a single breeding population that may be based on three female lines of descent. Genetics also suggests that, in addition to nursery areas in New England waters and in the Bay of Fundy, there exists somewhere an additional and undescribed summer nursery area used by approximately one-third of the population. A related question is where individuals other than calving females and a few juveniles overwinter, as one or more major wintering and summering grounds remain to be described for this animal.

The western North Atlantic right whale population in 1992 was estimated, from a census using photoidentification techniques, to be 295 individuals. This population may have been as small as 50 or fewer animals at the turn of the century, suggesting that it is now showing signs of slow recovery. Approximately one-third of all right whale deaths are caused by humans. Further, their small population size and low annual reproductive rate suggest that human sources of mortality may have a greater effect relative to population growth rates than they do for other whales. The principal factors believed to be retarding growth and perhaps population recovery for North Atlantic right whales are ship strikes and net entanglement in fishing gear (Kraus 1990). This population is considered to be small relative to its optimal size, and the species is listed as endangered under the U.S. Endangered Species Act. A recovery plan has been

Fig. 1. The North Atlantic right whale is the most endangered whale off the U.S. Atlantic coast.

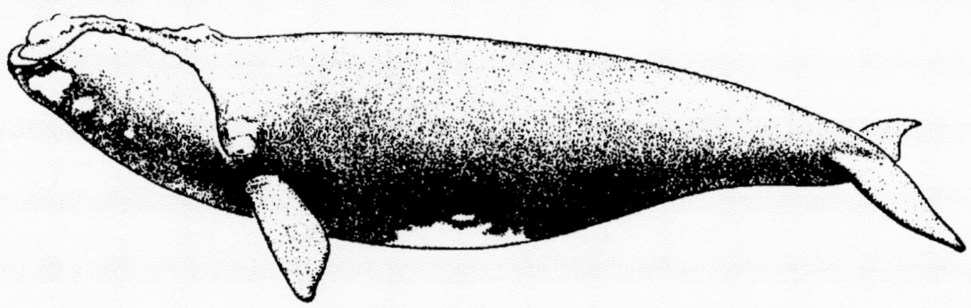

Courtesy S. Landry, New England Aquarium, Boston

waters of the southeastern United States to summer feeding, nursery, and mating grounds in New England waters and in waters northward to the Bay of Fundy and the Scotian Shelf. Long-distance movements as far north as Newfoundland, the Labrador basin, and southeast of Greenland have been reported and suggest an extended range—at least for some individuals—and habitat areas not presently well described. Likewise, while a calving and wintering ground is described for coastal waters of the southeastern United States, 85% of the population is unaccounted for during the winter season, and perhaps 30% is unaccounted for during the summer feeding and nursery season. Sightings from the Gulf of Mexico suggest that the range of this species may be more extensive than previously believed (Blaylock et al. 1995).

Research suggests five major habitats or congregation areas for western North Atlantic right whales: southeastern U.S. coastal waters, the Great South Channel, Cape Cod Bay, the Bay of Fundy, and the Scotian Shelf. However, the results from satellite tracking studies suggest that movements within and between habitats and

Fig. 2. North Atlantic right whale mother and calf, 17 January 1992, just north of Daytona Beach, Florida. Right whales calve in coastal waters during December and January.

Courtesy S. L. Ellis, Sea World airship

published and is in effect, though not completely implemented (National Marine Fisheries Service 1991). The total level of human-caused mortality and serious injury is unknown, but it has exceeded two right whales per year since 1990.

Author

James H. W. Hain
National Marine Fisheries Service
National Oceanic and Atmospheric Administration
Northeast Fisheries Science Center
166 Water Street
Woods Hole, Massachusetts 02543

See end of chapter for references

for the North Atlantic right whale. Amendments to the Marine Mammal Protection Act in 1994 and to the Northeast Multispecies Fishery Management Plan mandate a reduction in harbor porpoise bycatch, and research on gear modifications—including the use of acoustic deterrent devices—is directed at reducing interactions between fisheries and harbor porpoises.

Southeast Region

Environment and Physical Features

The Southeast region, including the islands of Puerto Rico and the U.S. Virgin Islands in the Caribbean, encompasses such diverse marine habitats as extensive coral reefs, salt marshes, tidal flats, seagrass and barrier islands, and nearshore, oceanic, and tropical island areas. The region is composed of three large marine ecosystems: the Gulf of Mexico (Brown et al. 1991), the Southeast shelf (Yoder 1991), and the Caribbean (Richards and Bohnsack 1990). The gulf and Southeast shelf ecosystems share many physical and ecological characteristics. Their fisheries have similar species compositions: both ecosystems have broad continental shelves with many coastal lagoons, estuaries, and marshes, and both systems support reefs with a tropical affinity and a complex multispecies structure. Also, both ecosystems are strongly influenced by Atlantic western boundary current systems (the Loop Current in the Gulf of Mexico and the Gulf Stream in the Southeast shelf), and both receive major freshwater outflows from large drainage systems bordered by extensive estuarine wetlands. Fishes and invertebrates of the insular shelf of the Caribbean ecosystem are similar to those of reefs in the other two systems but are influenced by quite distinct oceanic processes that occur as a result of the Antilles Current, tropical latitudes, and other conditions of the Caribbean.

The Gulf of Mexico ecosystem, "America's inland sea," is fed by warm tropical waters of the Caribbean, which enter through the Yucatan Straits (Fig. 27). This flow develops into the Loop Current and a number of associated eddies in the gulf (Fig. 2). Part of the water entering the

gulf bends to the east, flows through the Straits of Florida, and joins the Florida Current. Some of the water flows farther north into the gulf and then veers to the east to form a clockwise gyre off west Florida. The remaining water turns west, traverses most of the width of the gulf, and contributes to a complex series of anticyclonic warm eddies that eventually exit through the Florida Straits, form the Florida Current, and flow into the Atlantic Ocean. There the water forms the Gulf Stream as it moves northward, carrying its warm waters along the South Atlantic coast (Rabalais and Boesch 1987). The circulation, hydrographic conditions, and productivity of the South Atlantic Bight are influenced by these Gulf Stream waters, which follow the shelf edge from the Straits of Florida to Cape Hatteras.

The Southeast shelf ecosystem is characterized by bays and sounds with extensive coastal marshes. These wetlands form unique habitats that provide important links to production of living marine resources. The Southeast shelf ecosystem accounts for almost 145,040 square kilometers of estuarine drainage, extending from the temperate latitudes of North Carolina to the tropics of southern Florida and the northern Gulf of Mexico (National Oceanic and Atmospheric Administration 1990a). The largest estuary of the Southeast shelf is the combined Albemarle and Pamlico sounds. The Gulf of Mexico ecosystem accounts for more than 248,640 square kilometers of estuarine drainage, extending from south Florida to the Texas–Mexico border. Major features include the Mississippi and Atchafalaya deltas, which form a complex web of estuarine channels and extensive coastal wetlands that are important habitat for many fisheries. This region has the largest estuarine drainage area of all the U.S. regions. The fluvial drainage areas of the Mississippi and Atchafalaya rivers receive water from more than half of the contiguous United States. This drainage system creates tremendous freshwater flow into estuaries, accounting for about half of all the water discharged to U.S. estuaries.

Although wetlands are extensive in the Southeast region, an estimated 1,295 square kilometers of these wetlands were lost each year between the mid-1950's and late 1970's (Day and Craig 1981) because of sea-level rise, land subsidence, and human alterations such as channelization of estuaries, canal dredging through wetlands to accommodate oil and gas production, and impoundments (see chapter on Coastal Louisiana). The dense swamps and marshes of much of the region historically restricted coastal development. However, draining of wetlands uncovered an immensely fertile soil that expanded agricultural activity, particularly in the Mississippi delta and areas near the Florida Everglades. Conflicts over multiple uses of resources abound in this region, where the major population centers are growing rapidly. The Southeast's human population is expected to grow by almost 26% over the next 20 years, a faster rate than most other coastal regions in the country (Culliton et al. 1990).

Primary and Secondary Production

Primary and secondary production are enhanced by upwelling. Upwellings of the tropical and lower temperate waters of the Southeast shelf ecosystem, however, are not as intense and evident as in the higher latitude regions. More important to production in this ecosystem is the role of ichthyoplankton, the component of plankton made up of larval fishes. In the Gulf of Mexico ecosystem, four distinct groups of ichthyoplankton species have been identified: estuarine-dependent, coastal, reef (local or tropical), and oceanic. The general life history pattern of the estuarine-dependent species of the Gulf of Mexico and Southeast shelf ecosystems is characterized by fish that spawn in shallow (less than 200 meters), nearshore waters, with postlarval and juvenile stages inhabiting estuaries. How well the larvae survive—and how effectively they are recruited to estuarine nursery areas—is largely controlled by tides, prevailing winds, and surface currents.

The coastal species are a diverse group of finfish and shellfish that complete all or most of their life cycle in the open ocean or deepwater realms associated with the continental shelf. This group is believed to undergo extreme fluctuations in abundance, probably in response to changes in environmental conditions on the shelf. The locations of principal spawning areas of coastal species and the factors that may determine the species' reproductive success and ultimate recruitment are poorly known. Because spawning usually occurs in shelf waters, major oceanic phenomena, such as gravity-oriented current boundaries or eddies, and riverine fronts may offer key microhabitats for successful reproduction and recruitment of coastal species.

The reef species in the Southeast region spawn virtually continuously, maximizing the potential recruitment of young to a reef. For many reef fishes and invertebrates, spawning is believed to be related to lunar—thus tidal—conditions in a regular fashion that promotes local retention of larvae. Recruitment has been thought to vary little, regulated more by the availability of suitable habitat than by oceanic conditions or number of spawning stock. The oceanic species spend their entire life cycles in the oceanic waters of the Gulf of Mexico and the South Atlantic.

Recent studies indicate that species richness of ichthyoplankton in the Gulf of Mexico ecosystem—in these very different groups—is among the highest of all regions currently reported. A recent gulfwide study by Richards et al. (1993) documented 100 families, including many deepwater species, taken in 87 plankton tows along the boundary of the Loop Current, the major hydrographic feature in the Gulf of Mexico. The boundary was determined by real-time satellite images of the Loop Current.

Benthic Resources

Most studies of the benthic community in the Gulf of Mexico and Southeast shelf ecosystems have been concerned with the characterization of the fauna. Patterns of community composition (Harper and McKinney 1980; Baker et al. 1981) vary by substrate and by factors associated with the depth gradient, such as temperature and temperature variability, freshwater plumes, changes in sediment particle sizes, and the decreasing effects of wind-forced hydraulic factors, including hurricanes (Tenore 1979). The soft-bottom communities are generally dominated by polychaete worms (47%–51%), crustaceans (28%–29%), and mollusks (10%–17%). Hard-bottom communities are characterized by animals and plants living

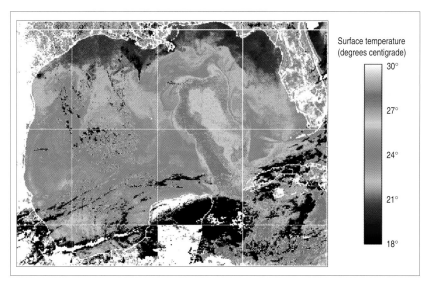

Fig. 27. Color satellite photo showing surface water temperature profiles for the Gulf of Mexico, as detected by a National Oceanic and Atmospheric Administration satellite.

Surface temperature (degrees centigrade)

30°

27°

24°

21°

18°

on the bottom surface, including macroalgae, coralline algae, sponges, soft coral, and often patches of the fungus *Agaricia*. Other areas of soft-bottom sands may be covered with layers of coralline algae nodules or with an algal nodule pavement with *Agaricia* accumulations (Woodward-Clyde Consultants and Continental Shelf Association, Inc., 1983). Characteristically, the benthic fauna of the north inner shelf is an extension of the warm temperate Carolinian province, with divisions at the Rio Grande and east of the Mississippi delta. The southern portion of the South Texas Continental Shelf is inhabited by a more tropical Caribbean fauna. The fauna of the outer shelf of the northwestern gulf has more definite tropical affinities than those of the warm temperate inner shelf.

Overall, the mollusk populations (oysters, clams, conchs, and so forth) in the Southeast region have declined from the levels of 100 years ago. Much of this decline is attributed to disease, overexploitation, and habitat degradation. Habitat degradation is best measured by an index developed by Engle et al. (1994), which is designed for broad spatial generalizations of ecological conditions. Benthic data from selected study sites can be used to estimate the percentage of surface area that exhibits benthic community structures similar to those attributed to known degraded sites. Engle et al. (1994) indicated that 31% of the sediments in the Gulf of Mexico estuaries contained degraded benthic communities. They also noted that 27% of large estuaries, 41% of small estuaries, and 80% of Mississippi River sites exhibited low benthic index scores. These results indicate intense destructive pressures on coastal habitats in the Southeast shelf ecosystem.

Losses of submersed aquatic vegetation, such as reduction of seagrass beds and destruction of coral reefs due to coral bleaching, are occurring throughout the region, resulting in continued loss of primary nursery habitat. Coral reefs of the Gulf of Mexico and Southeast shelf ecosystems are also subjected to harvest of live rock, where part of the reef is taken for the purpose of selling the attached flora and fauna by the aquarium industry. Because of the destructive nature of this industry, it is currently coming under management by the Gulf of Mexico and South Atlantic fishery management councils.

Fisheries Resources

Over the past 40 years, the National Marine Fisheries Service has assembled a vast data base identifying stocks of marine organisms from the Southeast region. Much effort has been directed

toward developing an inventory of species. This work has resulted in an extensive data base of hundreds of new geographical records and an expanding list of previously undescribed species and genera (Donaldson et al. 1996). Additionally, many other state and federal agencies collect and make available resource data from the Southeast region. These data are used in conjunction with the National Marine Fisheries Service data for analysis of management recommendations. Included in this data source are the departments of Interior and Defense, the Environmental Protection Agency, and the several states in the region, as well as other elements within the Department of Commerce, National Oceanic and Atmospheric Administration.

The major fishery resources in the Southeast region are the Southeast and Caribbean invertebrates, gulf menhaden, gulf butterfish, Atlantic shark, Atlantic and Gulf of Mexico coastal pelagic fishes, Atlantic–Gulf of Mexico–Caribbean reef fishes, Southeast drum and croaker, Atlantic highly migratory species, and nearshore fishery resources. Many of these resources are harvested by both the commercial and recreational fisheries. Table 4 provides information on the status of each of these groups.

Southeast and Caribbean Invertebrates

The Southeast and Caribbean invertebrate fisheries are important both recreationally and commercially and include shrimps, Caribbean spiny lobster, stone crab, conches, and corals. Shrimps support one of the most valuable U.S. fisheries, based on revenue accumulated after harvest. Some fisheries, such as those for spiny lobsters and stone crab, have only moderate value on a national basis but are valuable regionally (Southeast Fisheries Science Center 1995).

Penaeid shrimp have been harvested commercially since the late 1800's. Nine species of shrimps contribute to the U.S. shrimp fishery in the Gulf of Mexico and the Atlantic, of which brown shrimp, white shrimp, and pink shrimp make up more than 95% of the commercial harvest. Other species harvested commercially include the royal red shrimp, seabob shrimp, and rock shrimp. These species are generally found in all continental shelf waters in the U.S. Atlantic and the Gulf of Mexico inside 110-meter depths. In the Gulf of Mexico, the largest densities of brown shrimp occur off the Texas–Louisiana coast, the largest concentration of white shrimp occurs off the Louisiana coast, and the greatest densities of pink shrimp occur off the southwestern coast of Florida. Along the south Atlantic, the center of abundance for white shrimp is off the Georgia and South Carolina coasts, while the center of abundance for brown shrimp is off the North Carolina and South Carolina coasts.

Each shrimp species is assessed as two stocks: Gulf of Mexico and U.S. Atlantic. Shrimp in the Southeast are harvested year-round; however, peak seasonal shrimping is species-specific. Shrimp management is under both state and federal control. The commercial fisheries in the Exclusive Economic Zone are managed under federal fishery management plans. Each of the eight states in the Southeast region uses different management measures to control the harvest of shrimp under its jurisdiction. The status and trends of the stocks are summarized in Table 5.

The average annual shrimp catch from the Gulf of Mexico during 1980–1992 was 107,340 metric tons. The largest catch occurred in 1986 (137,949 metric tons) and the smallest in 1983 (86,484 metric tons). On average, brown shrimp account for 57%, white shrimp 31%, and pink shrimp 8% of the total shrimp catch. The other six commercially harvested shrimp species combined account for only 4% of the total. Currently the fishery is overcapitalized, with more effort being expended than is reasonably necessary to harvest the shrimp.

Gulf of Mexico brown shrimp and white shrimp catches have increased significantly over the past 34 years. Pink shrimp catches were level until 1985; all shrimp landings declined in recent seasons. The number of young brown shrimp produced per parent has increased significantly, but white shrimp have not increased. The increase in brown shrimp appears related to marsh alterations. Coastal subsidence and sea-level rise in the northwestern gulf inundate intertidal marshes longer, allowing the shrimp to feed for longer periods within the marsh area. In the gulf, both factors have also expanded estuarine areas, created more marsh edges, and provided more protection from predators. As a result, the nursery function of those marshes has been magnified and brown shrimp production has expanded. However, continued subsidence will lead to marsh deterioration and ultimate loss of supporting wetlands, and current high fishery yields may not be sustainable. Pink shrimp landings reached an all-time low in 1989 and remained lower than the pre-1986 average. Catches began to recover in 1993 and 1994 and exceeded the pre-1986 average in 1995. All commercial shrimp are harvested at maximum levels.

Annual catches of spiny lobsters were stable during the 1980's, running about 2,700 metric tons for the Gulf of Mexico. On Florida's

Table 4. Status and trends of major fish resources in the Southeast region.

Species	Status and trend
Atlantic croaker	Below-average abundance and overutilized, unknown
Butterfish	Above-average abundance and underutilized, stable
Menhaden	
Atlantic	Low abundance and fully utilized, stable
Gulf of Mexico	Average abundance and fully utilized, increasing
Red drum	
Atlantic	Below-average abundance and overutilized, unknown
Gulf of Mexico	Below-average abundance and overutilized, unknown
Atlantic sharks	
Large coastal sharks	Below-average abundance, overutilized, probably declining
Small coastal sharks	Above-average abundance, fully utilized, probably declining
Pelagic sharks	Unknown abundance and trend
Billfishes	
Blue marlin/North Atlantic	Unknown abundance and trend, overexploited
Sailfish/West Atlantic	Unknown abundance and trend, moderately exploited
Swordfish/North Atlantic	Below-average abundance, overexploited, unknown
White marlin/North Atlantic	Unknown abundance and trend, overexploited
Coastal pelagic fishes	
Cero	Unknown status and trend
Cobia	Unknown status and trend
Dolphin	Unknown status and trend
King mackerel/South Atlantic	Average abundance, underexploited, stable
King mackerel/Gulf of Mexico	Low abundance, overexploited, stable
Spanish mackerel/South Atlantic	Average abundance, overexploited, stable
Spanish mackerel/Gulf of Mexico	Low abundance, overexploited, declining
Oceanic pelagic fishes	
Albacore/North Atlantic	Unknown abundance and trend, fully exploited
Bigeye tuna/Atlantic	Unknown abundance and trend, fully exploited
Bluefin tuna/West Atlantic	Below abundance of 1970's, unknown
Skipjack tuna/West Atlantic	Unknown abundance and trend, fully exploited
Yellowfin tuna/West Atlantic	Unknown abundance and trend, approaching full exploitation

Atlantic coast, catches averaged 230 metric tons. This fishery is considered overcapitalized. The recreational fishery is large, but its catch is unknown. Southeastern lobster stock levels remain below the long-term potential yield. Annual catches for Puerto Rico averaged 136 metric tons over the past 23 years. These stocks now appear to be overutilized. U.S. Virgin Islands catches for 1980–1991 were fairly stable and averaged 27 metric tons (Southeast Fisheries Science Center 1995).

Annual catches of stone crab (based on claw weight) varied from 1,200 to 1,400 metric tons in the Gulf of Mexico through the 1980's. Atlantic coast catches averaged around 34

Species	Status and trend
Brown shrimp	
Gulf of Mexico	High abundance, fully exploited, stable
Atlantic	High abundance, fully exploited, stable
Pink shrimp	
Gulf of Mexico	Historical low abundance, fully exploited, depressed
Atlantic	Unknown abundance and trend
Spiny lobster	
Southeastern United States	Below-average abundance, overutilized, trend unknown
Caribbean	Average abundance, stable
Stone crab	Average abundance, stable to declining
White shrimp	
Gulf of Mexico	Average abundance, fully exploited, stable
Atlantic	Average abundance, fully exploited, stable

Table 5. Status and trends of major invertebrate populations in the Southeast region.

Linkages Between Coastal Wetlands and Fishery Resources

Most fishery species within the Southeast shelf ecosystem spend part of their life cycle in estuaries, where there appears to be an important linkage between coastal wetlands and fishery productivity. The Southeast region is characterized by vast expanses of coastal marshland, large beds of seagrasses, and some of the most highly productive fisheries in the country. On a global scale, a positive relationship has long been recognized between the extent of coastal wetlands and fishery landings (Turner 1977). On a smaller scale, investigations of animal distributions within estuaries have documented high densities of juvenile fishes, shrimps, and crabs in seagrass and marsh habitats compared with sites lacking bottom vegetation (Zimmerman and Minello 1984; Hoss and Thayer 1993; Peterson and Turner 1994). These patterns indicate that wetlands provide important nursery functions. Indeed, other research has shown that wetland habitats provide young fishery species with both an abundant source of food to support rapid growth and also protective cover to reduce mortality from predators (Boesch and Turner 1984; Kenworthy et al. 1988; Minello et al. 1989; Minello and Zimmerman 1991).

The linkages between wetlands and fishery productivity, however, can be complex. For example, the importance of marsh availability has only been fully recognized within the last decade. Availability of coastal marshes to fishery organisms is determined by tidal flooding patterns, the amount of marsh–water edge, and the extent of connections between interior marsh and the sea. Within the Southeast, low-elevation marshes in the northern Gulf of Mexico are flooded almost continually during some seasons and are extensively fragmented, providing maximum access for young fishery organisms (Figure). In contrast, marshes along the South Atlantic coast have relatively little marsh–water edge and appear to be infrequently flooded. The density of fishery species using the marsh surface also varies between these areas; densities in the Gulf of Mexico marshes are generally an order of magnitude greater than those on the Atlantic coast (Rozas 1993). Researchers now believe that these differences in wetland availability and use are at least partially responsible for the higher landings of estuarine-dependent species in the Gulf of Mexico compared with the South Atlantic.

One major function of wetlands is to provide food for fishery species, and there is evidence that this function also varies regionally. Historically, salt marshes were thought to contribute mainly to detrital food webs by outwelling plant debris into downstream estuaries (Nixon 1980). Such an indirect use of marsh plant production is consistent with the high elevations and large tidal regimes characteristic of Atlantic coast marshes. In the northern Gulf of Mexico, however, direct use of the marsh surface appears more common and is fostered by low marsh elevations and extensive flooding with small tidal regimes. If organisms have access to the marsh surface, primary producers such as benthic and epiphytic algae, along with abundant small consumers, provide plenty of the high-quality food necessary for young fishery species. Thus, the relative importance of different trophic pathways is probably controlled by wetland availability (McIvor and Rozas 1996).

Overlying and perhaps overshadowing these concepts of relative wetland value are the extensive rates of coastal marsh loss occurring in the Southeast, mainly in the northern Gulf of Mexico. Because of the linkages between wetlands and fishery production, we might expect dramatic declines in estuarine-dependent fisheries as marsh habitats are lost. However, over the last 20 to 30 years, productivity and landings of three dominant fishery species (brown shrimp, white shrimp, and menhaden) in the northern Gulf of Mexico have increased (Klima et al. 1990; Smith 1991). In contrast, production of these species did not increase on the Atlantic coast where wetland loss was low compared with the Gulf of Mexico. We are left with a paradox—increased production of fishery species appears correlated with the degradation of their habitat. The explanation may lie in understanding the process of wetland degradation. Wetland loss in the northern Gulf of Mexico is mainly caused by coastal submergence, canal dredging, levee construction, and erosion (Rozas and Reed 1993; Turner 1997). Concurrently, marsh flooding increases, fragmentation and habitat edge increase, zones of saline and brackish wetland expand, and connections with the sea are shortened. These processes increase the availability and value of the remaining marsh and may be supporting short-term increases in fishery production (Zimmerman et al. 1991; Rozas 1995). If this hypothesis is true, enhanced levels of fishery productivity in the Gulf of Mexico are temporary. Continued wetland loss will overcome any benefits of habitat degradation and result in future declines in fishery production dependent on these coastal wetlands.

See end of chapter for references

Authors

Thomas J. Minello
Roger J. Zimmerman
National Marine Fisheries Service
National Oceanic and
Atmospheric Administration
Fishery Ecology
Galveston Laboratory, 4700 Avenue U
Galveston, Texas 77551

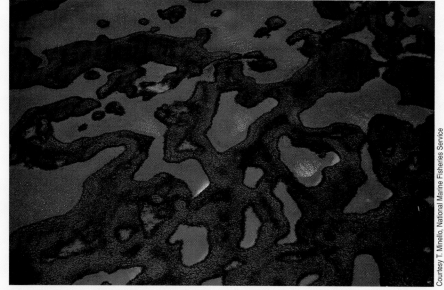

Courtesy T. Minello, National Marine Fisheries Service

Figure. Highly fragmented salt marsh in the northern Gulf of Mexico provides maximum access to the marsh surface for fishery species.

metric tons. Fishing efforts on stone crabs more than doubled during the period from 1979–1980 (295,000 traps set) to 1992–1993 (686,000 traps set) but have been relatively stable in recent years. Thus, more of the catches have been harvested earlier, which has shortened the fishing season. It is unlikely that recent catch levels can be sustained on a long-term basis (Southeast Fisheries Science Center 1995).

Southeast Menhaden and Butterfish

Menhaden are a herringlike species found in coastal and estuarine waters of the U.S. Southeast shelf and Gulf of Mexico ecosystems. The menhaden fishery, one of the larger and older fisheries in the United States, is managed by individual states, with coordination handled by the Atlantic States Marine Fisheries Commission and the Gulf States Marine Fisheries Commission. The Atlantic menhaden population is fully utilized, with a long-term potential yield of 480,000 metric tons per year and recent average catch of 360,000 metric tons (Fig. 28). In the Gulf of Mexico, the fish is also fully utilized, with a long-term potential yield of 660,000 metric tons and recent average catch of 500,000 metric tons (Southeast Fisheries Science Center 1995; Fig. 28).

Abundance of the Atlantic menhaden stock has been low for 30 years and is expected to remain low; however, this fishery is in a rebuilding phase (Table 4). The fishery is supported mainly by young recruit fish. Although Atlantic menhaden may live 10 years, most fish of the catch are 3 years of age or younger because of the biology of the species and the economic dependence of the industry on whatever fish are present, regardless of their age. Thus, overuse because of growth overfishing—or harvest of small fish that could have been significantly larger and heavier if they had survived to an older age—is a prime management concern for this stock. The spawning biomass of the Gulf of Mexico stock has been much higher than the Southeast shelf stock. Menhaden abundance is at average historical levels, having declined by more than 50% from its 1983 peak level. The fishery bottomed out in 1992, and there is now an upswing in recruitment. Gulf menhaden may live for 5 years, but most landed are 1 or 2 years old. Because the gulf stock is short-lived and has a high natural mortality, growth overfishing has not been a management concern.

In 1986, a directed bottom trawl fishery for gulf butterfish began. Currently, only one or two vessels target butterfish, and the fishery is not regulated. Gulf butterfish are underutilized, with a long-term potential yield of 26,500 metric tons and recent average catch of 19,700 metric tons. Butterfish incidentally captured by offshore Gulf of Mexico shrimp fleets have

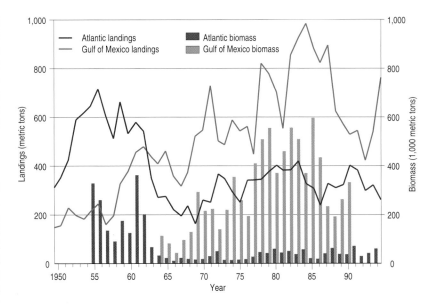

Fig. 28. Landings and spawning biomass trends of menhaden populations in the Southeast region.

composed from 80% to 97% of the annual catch since 1986. The abundance and trends of the stock are not well known.

Atlantic–Gulf of Mexico Sharks

About 350 species of sharks are known worldwide. Of those, 72 species frequent the waters of the U.S. Atlantic, the Gulf of Mexico, Puerto Rico, and the U.S. Virgin Islands. Both recreational and commercial shark fishermen seek coastal sharks along the Atlantic seaboard. Pelagic sharks are targeted by tournament anglers off the mid-Atlantic states and are incidentally caught by swordfish and tuna longliners. Nontournament anglers usually catch small coastal sharks that are generally not targeted by commercial fisheries. The Gulf of Mexico shrimp fishery catches and discards many small coastal sharks. There is also a mobile longline fishery targeting large coastal sharks both in the Atlantic and gulf waters, taking several species that are important to anglers. Many other fisheries catch sharks as bycatch which, depending on the species, are either discarded or marketed.

The long-term potential yield of the large coastal sharks is 3,400 metric tons, and the recent average catch has been 3,800 metric tons. The resource is currently overutilized (Table 4). The small coastal sharks have a long-term potential yield of 3,600 metric tons, and the recent average catch is 3,000 metric tons. This fishery is believed to be fully utilized. The long-term potential yield for the pelagic sharks is currently unknown, but the recent average catch is 2,730 metric tons. The status of use for this group is unknown.

Data are lacking on shark population dynamics. Anecdotal information indicates that the shark populations in the Southeast are declining. Sharks grow and reproduce slowly, thus they are very vulnerable to overfishing.

Recently, a new fishery management plan has been implemented to prevent overfishing, prohibit the practice of finning, discourage the discarding of shark carcasses, rebuild currently overfished stock, and improve data collection (Southeast Fisheries Science Center 1995).

Coastal Pelagic Fishes

Coastal pelagic fishes inhabiting waters off the southeastern United States include king mackerel, Spanish mackerel, cobia, dolphin, and cero. These species range in coastal and continental shelf waters from the northeastern United States through the Gulf of Mexico and the Caribbean Sea and as far south as Brazil. In 1990, more than 90% of coastal pelagic fishes in the Southeast were landed in Florida. Commercial fisheries harvest Spanish mackerel by hook and line. A recreational fishery also exists for Spanish mackerel, accounting for about 25%–42% of the landings of this species. King mackerel are commercially fished from Chesapeake Bay southward. Major production areas include North Carolina and the lower east coast of Florida (Cape Canaveral to Palm Beach), the Florida Keys, and Grand Isle, Louisiana (Southeast Fisheries Science Center 1995).

During 1979–1991, recreational fishermen caught 8,000–17,000 metric tons per year of coastal pelagic fishes, and commercial fishermen caught 5,000–14,000 metric tons per year. King mackerel and Spanish mackerel account for about 95% of all coastal pelagic fishes harvested. The long-term potential yield for this group was estimated to be 22,554 metric tons, and recent average catches were 11,104 metric tons. As a group, coastal pelagic fishes yield only about 56% of long-term potential yield, and certain species are fished near or over maximum production levels (Table 4). Three of the four mackerel stocks are considered overfished. The gulf king mackerel stock is believed to have a large long-term potential yield but is currently severely depleted. Accurate assessment of the status of cobia and dolphin is not yet possible.

Coastal pelagic fishes are jointly managed under the Coastal Migratory Pelagic Resources Fishery Management Plan and the regulations adopted by the South Atlantic and Gulf of Mexico fishery management councils, and are implemented by the National Marine Fisheries Service. Only U.S. fisheries are regulated, although Mexican catches are thought to be large.

Oceanic Pelagic Fishes

The Atlantic oceanic pelagic fishes are wide-ranging and highly migratory. A broad array of species make up this complex, which is harvested by international fishing fleets. The United States is among the major harvesting nations for some of these species (Southeast Fisheries Science Center 1995). The status and trends of these resources are summarized in Table 4.

Northern bluefin tuna is a large migratory pelagic species that is found in the Atlantic and Pacific oceans. In the western Atlantic, bluefin tuna occur from Labrador and Newfoundland south into the Gulf of Mexico, the Caribbean Sea, and off Brazil. The peak yields of bluefin tuna from the western Atlantic (about 8,000–19,000 metric tons) occurred in 1963–1966, when much of the catch was taken by longlines off Brazil. The most recent assessment of west Atlantic bluefin tuna, which was carried out by the International Commission for the Conservation of Atlantic Tunas in 1991 using catch and effort data through 1990, showed that all size classes of fish were substantially below the 1970 levels. The rate of decline slowed with the imposition of catch restrictions in 1982. The long-term potential yield for the stock was estimated at 3,000–13,000 metric tons. Best estimates of current potential yield in the western Atlantic were about 2,000 metric tons. The recent average catch for 1989–1991 was 2,850 metric tons, of which 62% (1,760 metric tons) was taken by U.S. anglers.

Populations of yellowfin tuna are found worldwide in tropical waters. In the Atlantic, their greatest concentrations have been between the equator and the 15° latitudes. However, migrations take place to the north and the south along the North American east coast, and as a result, coastal concentrations of yellowfin tuna are found seasonally off the northeastern United States and Uruguay. In addition, substantial concentrations of yellowfin are found in the Gulf of Mexico. The long-term potential yield was estimated at 33,000 metric tons, with 30% located in the gulf. There is no estimate of the current potential yield; however, the recent average annual catch was 30,000 metric tons for the total stock, with the United States taking 6,200 metric tons. Status of the gulf stock is currently unknown, but a 1991 review suggested that the stock in the Gulf of Mexico may be approaching full exploitation (Southeast Fisheries Science Center 1995).

Bigeye tuna are found in tropical oceans around the world. In the Atlantic they are widely distributed in both tropical and temperate waters, between 45°N and 45°S latitudes. Bigeye tuna catch has increased from levels seen in the early 1960's to a peak of 74,500 metric tons in 1985. Roughly two-thirds of the catch was taken in longline fisheries, with the remaining catch taken by surface gear. Japanese longliners accounted for 40% of total landings in 1991. The long-term potential yield ranged from 61,200 to 74,000 metric tons. There is no estimate of the current potential yield. Recent average catch for the stock is 69,200 metric tons, with the United States taking 782 metric tons. This fishery resource is thought to be fully exploited (Southeast Fisheries Science Center 1995).

Albacore tuna are found in tropical and temperate waters of all oceans and range from 40°S to 50°N latitude. Albacore generally do not grow as large as bluefin, bigeye, or yellowfin tuna of a similar age. The number of North Atlantic albacore catches has declined from 1964 to 1995. Peak catches occurred in 1964, when 64,400 metric tons were landed. The catch was 26,134 metric tons in 1995. Most of the recent catches were taken in surface fisheries, mainly by bait boats and trolled lines, although drift gillnets and pelagic trawls were also used. Before 1987, up to 50% of the catch was by longline gear. In 1993, the International Commission for the Conservation of Atlantic Tunas estimated the long-term potential yield for North Atlantic albacore to range from 37,340 to 68,930 metric tons. There is no estimate of the current potential yield. The recent average catch for the stock was 30,850 metric tons, of which the United States landed 376 metric tons. This fishery is considered to be fully exploited (Southeast Fisheries Science Center 1995).

Skipjack tuna—a relatively small tuna—occurs in tropical and warm temperate seas. The western Atlantic skipjack catch has dramatically increased in recent years from relatively low levels during 1960–1979 to a peak of 40,000 metric tons in 1985. The long-term potential yield for western Atlantic skipjack was estimated at 33,000 metric tons. Its current potential yield is unknown. The recent average catch was 26,900 metric tons, of which the United States harvested 357 metric tons. This resource is considered fully exploited.

At least five species of tunas are included in the "other tuna" category. They include Atlantic bonito, little tunny, frigate tuna, blackfin tuna, and wahoo. Others may also be included but are not discriminated to species level in the international landings statistics. In the Atlantic, the recent average catch of this tuna group (1989–1991) totaled 51,600 metric tons; about 732 metric tons were taken through U.S. fishing efforts (Southeast Fisheries Science Center 1995).

The billfish fishery in the Atlantic Ocean includes swordfish, blue marlin, white marlin, sailfish, and longbill spearfish. The swordfish is the most widely distributed billfish, with the widest water temperature tolerance; they are found in waters with surface temperatures ranging from 5°C to 27°C. Their preferred habitat is near the edge of continental shelves in waters 100–3,000 meters deep, near oceanic frontal zones, and near seamounts and mid-ocean islands. The swordfish fishery is conducted mainly by longline fleets, with the Spanish and U.S. fleets dominating recent catches. These two nations have accounted for about 76% of the total North Atlantic swordfish catch in recent years. The catch and effort data showed a continual increase from 1978, when the United States eased its mercury-content regulation, to a peak catch of nearly 20,000 metric tons in 1987. Since 1987, catches have continued to decline under effort reduction. The species' long-term potential yield is estimated to be between 11,600 and 16,800 metric tons. Its current potential yield ranges from 8,000 to 14,200 metric tons. The recent average catch was 15,300 metric tons, with the United States taking 5,400 metric tons. This fishery is considered overexploited (Southeast Fisheries Science Center 1995).

The other billfish species are taken mostly as incidental catch in the commercial tuna and swordfish fisheries. There is also a growing recreational fishery targeting billfishes by the United States and many other countries in the Caribbean Sea and the eastern Atlantic. The recent development and geographical expansion of longline fisheries in the Gulf of Mexico for tuna and in the Caribbean for swordfish and tunas, as well as geographical expansions of the longline fleet off Africa, are raising concern for the welfare of billfish populations. The growth of these fisheries is expected to result in increased discard mortalities of billfish. Assessment of billfish stocks has generally been hampered by data limitations. Because of domestic concern over the future prospects for billfish resources, U.S. restrictions of billfish landings were implemented in October 1988 under the United States Fishery Management Plan for Atlantic Billfishes. The plan eliminates possession and sale of billfish by commercial fishermen and uses size limits to restrict the allowable catch of billfish by recreational gear (rod and reel).

The long-term potential yields for blue marlin and white marlin were estimated to be 1,718 metric tons and 593 metric tons, respectively. There are no estimates of their current potential yield. Recent average catches were 1,183 metric tons (253 metric tons by the United States) for blue marlin and 253 metric tons (63 metric tons by the United States) for white marlin. Both species are believed to be overexploited. Neither long-term potential yield nor current potential yield is known for sailfish. Its recent average catch was 619 metric tons, with the

United States harvesting 32 metric tons annually (Southeast Fisheries Science Center 1995). The sailfish fishery is thought to be moderately exploited (Table 4).

Reef Fishes

Reef fishes include more than 100 species that prefer coral reefs, artificial structures, or other hard-bottom areas, and tilefishes that prefer muddy bottom areas. They range along the coast to a depth of about 200 meters. Reef fish fisheries compete with fisheries for other reef organisms, including spiny lobsters, conch, stone crab, corals, and live rock and ornamental aquarium species. Nonconsumptive uses of reef resources (such as ecotourism, sport diving, educational programs, and scientific research) are also economically important and can conflict with traditional commercial and recreational fisheries. Although reef fish have been caught for centuries, systematic data collection for this group began only in the late 1970's when recreational fishery surveys began. Data collection remains difficult because of diverse users and scattered landings at many ports. Fishing pressure has increased with growing human populations, greater demands for fishery products, and technological improvements such as the use of longlines, wire fish traps, electronic fish finders, and navigational aids (Southeast Fisheries Science Center 1995).

Reef fisheries vary widely by area. In most cases, the current and long-term potential yields are unknown, though for many species they are probably higher than recent average catches would indicate because the reefs are currently overfished. The reef fish management unit contains about 100 species (excluding those for the marine aquarium trade). In the Southeast region, the unit is managed by the South Atlantic Fishery Management Council, the Gulf of Mexico Fishery Management Council, and the Caribbean Fishery Management Council. The long-term potential yield for this complex fishery has been estimated at 48,054 metric tons, with average recent catches at 35,185 metric tons. While landings and value for individual species are not large, reef fishes as a group produce significant landings and values (Southeast Fisheries Science Center 1995).

Reef fishes are vulnerable to overfishing because—among other factors—they are long-lived, slow growing, easily captured, and have delayed reproduction. More reef fish species are at maximum use or overfished now than in the past, and fishing pressure has shifted to other species in an effort to meet market and ethnic group needs. Most species are probably either fully utilized or overutilized. Red snapper, traditionally the most

important gulf reef fish, is overutilized in part as a result of its incidental catch by the shrimp fishery. Eight of the ten major species in the Atlantic headboat fishery have shown significant size declines since 1972. In the Caribbean, such traditional fishery mainstays as Nassau grouper have practically disappeared, and total landings of species of more recent importance, such as the red hind, have declined since the late 1970's. Landings of amberjacks, lane snapper, vermillion snapper, and similar species have increased as catches of traditional species have declined.

Red snapper less than 1 year old are a major problem as bycatch in the gulf shrimp fishery. Bycatch of other species and undersized fish also pose difficulties on the stocks, even if the bycatch species are returned to the sea. Scientific information for the assessment of most reef fishes is generally lacking or limited. Balancing the interests of the diverse users of reef fishes, especially between commercial and recreational groups, is a difficult management issue.

Southeast Drum and Croaker

Important species in this unit are the weakfish, Atlantic croaker, spot, red drum, black drum, kingfishes (whiting), spotted seatrout, and other seatrouts. Most drum and croaker are harvested in state waters and are therefore under state management. Many of these species are targeted for management under interjurisdictional management plans in both the Gulf of Mexico and the U.S. South Atlantic. In recent years, several states have set regulations favoring recreational use of some species, such as the red drum. However, red drum in the Gulf of Mexico Exclusive Economic Zone continue to be managed under a federal management plan. Many of the species in this complex are affected by the shrimp trawl fishery bycatch, especially weakfish in the U.S. South Atlantic region.

The long-term potential yield of Atlantic croaker is 50,000 metric tons, and the recent average catch is 4,946 metric tons. This fishery is judged to be overutilized because of bycatch from the shrimp fishery. The resource is of below-average abundance, and its trend is unknown (Table 4).

Commercial purse seining of adult red drum in federal waters of the Gulf of Mexico developed rapidly in the middle 1980's as demand grew for the Cajun dish, blackened redfish. Before that, nearly all red drum were harvested in nearshore state waters as juveniles. Commercial landings of all drum peaked in 1956 at more than 32,000 metric tons, more than 20,000 metric tons above the 1953 level. Long-term potential yield of red drum was estimated at 75,815 metric tons, and the recent average catch was 25,689 metric tons for the period 1988–1990. Both the Gulf of Mexico and Southeast shelf red drum stocks are thought to be overutilized. Management plans have been developed in both areas that ban red drum fishing and possession in federal waters until the adult population increases in size. The status of the remaining species in this complex fishery is unknown.

Nearshore Resources

Many U.S. coastal and estuarine species provide important recreational and commercial fisheries that are not federally managed. This diverse unit includes highly prized game fishes. It also includes small fishes used for bait, food, or processing into oil and meal. Valuable invertebrates like blue crabs, northern quahogs, softshell clams, bay scallops, and oysters are also in this group.

Most species in this group live near the shore during much or all of their lives. Some, like the striped bass, are anadromous, moving into fresh water to spawn but spending their adult lives in estuaries or at sea. In contrast, the American eel is catadromous, living much of its life in fresh or brackish water but migrating far offshore to spawn in the Sargasso Sea.

Many of these species are widely distributed. Shads, river herrings, sturgeons, sardines, mullets, Florida pompano, and Atlantic calico scallops are harvested primarily along the middle and southern U.S. Atlantic coast and in the Gulf of Mexico. Many of the game fishes are particularly valuable to the Florida economy, whereas invertebrates, like the blue crab and eastern oysters, support major fisheries from the gulf to Chesapeake Bay. The small bait fishes and food fishes are harvested by both recreational and commercial fisheries that use cast nets, gill nets, seines, dip nets, and pound nets; the southern Florida ballyhoo fishery, for example, supplies bait to the charter boat industry (Southeast Fisheries Science Center 1995).

It is difficult to assess the status of these stocks throughout their ranges because they are under varied management and data collection systems. Comprehensive assessments are scarce. Many of the species in the group are probably overexploited. Stock levels of many are below their historical averages.

Recent catches of nearshore species have been estimated at more than 221,000 metric tons, a conservative figure because data are not available for many species, such as sport catches. Some species, such as tarpon and bonefish, are sought primarily for sport and are usually released alive; consequently, few or no landings data are reported even

though they provide significant local and regional economic benefits.

Sea Turtles

Sea turtles are highly migratory and ply the world's oceans. All marine turtles are listed either as endangered or threatened under the U.S. Endangered Species Act (see box on Sea Turtles in Caribbean Islands chapter). The Kemp's ridley, hawksbill, and leatherback are listed as endangered throughout their ranges. The loggerhead and olive ridley are listed as threatened, as is the green turtle, except its Florida nesting population, which is listed as endangered (Southeast Fisheries Science Center 1995).

Historical data on sea turtle numbers are limited. In addition, the time during which data have been collected is short relative to the long life and low reproductive rate of all sea turtle species. The estimated number of female loggerheads nesting annually in the southeastern United States is about 20,000–28,000. Most nest along Florida's east coast, where nest numbers have been stable for 5 years. Only about 700–800 female Kemp's ridleys nest each year along a limited portion of Mexico's gulf coast, compared to 40,000 females seen nesting on one beach alone on a single day in 1947. This documented decline in the Kemp's ridley is probably indicative of similar population trends for other sea turtles, though the magnitude and period of their respective declines may be different.

Historically, the green turtle has supported commercial harvests along the Florida and Texas coasts. Currently, 400–500 green turtles nest each year along the Florida coast. Recent data suggest that the populations of green turtles and Kemp's ridley turtles in the Gulf of Mexico may be increasing. There are no historical estimates for the numbers of hawksbill or leatherback turtles nesting on U.S. Caribbean beaches. The hawksbill has been heavily exploited, and continued trade of products from this species suggests that further declines are possible. The abundance trend of the leatherback in U.S. waters is unknown.

Sea turtles are threatened by many factors, most related to human actions. A principal concern is incidental capture in commercial fisheries. Trawl, longline, and gill net fisheries pose the greatest threats. Before the implementation of turtle excluder device regulations, the National Academy of Science estimated that a maximum of 44,000 turtles, mostly loggerheads and Kemp's ridleys, were killed annually in the Gulf of Mexico and southeast U.S. Atlantic shrimp fishery. Although use of turtle excluder devices is mandated for the shrimp fishery and most of the summer flounder trawl fishery, recent sea turtle kills indicate that significant mortality still occurs in some areas. Thus, there is a continuing need to refine the turtle excluder device and to continue enforcement of its use. Sea turtles are also taken and killed in pelagic longline fisheries and gill net fisheries, and they are vulnerable to dredging operations and outer continental shelf oil and gas activities. Of particular concern are the gill net fisheries for coastal species, including sharks.

Propeller strikes and vessel collisions also pose significant threats to sea turtles, especially in areas of high human population, where recreational boat traffic is heavy and coastal ports are active. A disease known as green turtle fibropapilloma is affecting a significant number of green turtles, most notably in Florida and Hawaii. The tumors from green turtle fibropapilloma infection, which occur primarily on the skin and eyes, can be fatal. The cause of the disease remains unknown.

Protected nesting beaches of the southeastern United States are essential to the recovery and survival of sea turtles. Many nesting beaches have already been significantly degraded or destroyed. Nesting habitat is threatened by rigid shoreline protection or coastal armoring such as sea walls, groins, revetments, and sandbag installations. Additionally, nesting habitat is negatively affected by beach nourishment projects that are improperly timed to occur during the nesting season or that use poor-quality fill material. Artificial beachfront lighting, increased human activity, and beach vehicles also threaten species recovery. Thus, conservation and long-term protection of marine turtle nesting habitat are high management priorities.

Marine Birds

In the Southeast region, marine birds fall into several groups. First are species that nest and forage on beaches or strands and adjacent islands. Many of these shore and nearshore species are nearly year-round residents. In the winter, the populations of the former group are augmented by large numbers of wintering individuals from more northern ecosystems, some marine and some freshwater or terrestrial. Birds that forage in the pelagic environment comprise species mixtures that originate from many sources. These include birds that nest in more northern areas and may be migrating to or from more southerly environments—a group that includes several gulls, terns, jaegers, and phalaropes; birds such as boobies, noddies, and terns that nest on tropical islands, including Florida's Dry Tortugas; and birds from the Southern Hemisphere or far-offshore islands that follow ocean currents during their nonbreeding seasons, especially tropicbirds, petrels, storm-petrels, and shearwaters. At least 26 marine bird species nest in the Southeast region, including Puerto Rico and the U.S. Virgin Islands (Clapp and Buckley 1984; van Halewyn and Norton 1984; also see chapter on the Caribbean Islands).

Knowledge of marine birds is greatest for coastal and nearshore species, which can be studied from onshore areas, but it is rudimentary for oceanic species. Several studies sponsored by the Minerals Management Service (Clapp et al. 1982a,b; Clapp et al. 1983; Fritts et al. 1983) provide information on the occurrence and distribution of coastal and oceanic birds, their life histories, their risk from oil contamination, and the importance of populations in the southeastern United States relative to their global distribution and abundance. These studies covered the area from North Carolina to the Mexican border of Texas and included transect surveys in selected areas.

Since these studies, more recent work (Patteson and Brinkley 1994; Peake and Elwonger 1996), although less quantitative, has added greatly to our knowledge about the birds that frequent the continental shelf break and the Gulf Stream. Patteson and Brinkley (1994) provide a listing of 63 nearshore and pelagic birds found during dozens of birding trips off the coast of Virginia, Maryland, and North Carolina and also describe the seasonality of movements by different species. On trips to the Gulf Stream off North Carolina, they discovered several pelagic species new to U.S. waters. Rare pelagic birds sighted on their trips included yellow-nosed albatross, black-browed albatross, little shearwater, Audubon's shearwater, Bermuda petrel, Herald petrel, Bulwer's petrel, cape petrel, band-rumped storm-petrel, white-faced storm-petrel, white-tailed tropicbird, red-tailed tropicbird, great skua, Antarctic skua, and South Polar skua.

The Minerals Management Service studies identified 115 species of shorebirds and seabirds: 53 species of ducks, geese and swans; 40 species of loons, grebes, albatrosses, shearwaters, storm-petrels, tropicbirds, frigatebirds, cormorants, boobies, northern gannet, and pelicans; and 22 species of gulls, terns, and skimmers (Clapp et al. 1982a,b; Clapp et al. 1983; Fritts et al. 1983). Some of these bird populations have declined in numbers or have been extirpated locally by pesticides and petroleum. The brown pelican in the northern Gulf of Mexico was in serious decline in the 1940's and 1950's because of weather and pesticide effects. By 1960 only four breeding pairs of brown pelicans remained in the area. Since DDT was banned and individuals were reintroduced from the more abundant Florida populations, the species has made a substantial but incomplete recovery (Clapp et al. 1982a).

Distribution of marine birds varies throughout the Southeast region. In the lower coastal region of Texas the most abundant nearshore species are laughing gull, royal tern, herring gull, and ring-billed gulls. Seven other species of gulls and terns have been observed in low numbers. One of these, Franklin's gull, a wintering species, has been seen only on the lower Texas coast, where gulls and terns account for 91% of all marine birds. Phalaropes, jaegers, and black skimmers are observed only occasionally. Six species of pelicans and their allies have been observed in low numbers. The American white pelican, a wintering species that breeds in the interior far to the north, has been seen in great numbers. Brown pelicans in Texas suffered a great reduction in numbers in the 1950's, and sightings in the early 1980's indicate continued low numbers. In the early studies, only two species of shearwaters were identified off Texas—Cory's shearwater and Audubon's shearwater—but sooty shearwaters, Manx shearwaters, and greater shearwaters have been reported more recently (Peake and Elwonger 1996). Previous sightings of shearwaters off Texas had been infrequent. Storm-petrels had also been seen off Texas on a regular basis, but in low numbers; some species such as band-rumped storm-petrel, Wilson's storm-petrel, and Leach's storm-petrel are now known to occur regularly in certain seasons over deep offshore waters (Peake and Elwonger 1996).

Twenty-five species of birds have been observed off the west-central Louisiana coast. Gulls and terns accounted for 96% of all marine birds seen in that area. The most abundant species are laughing gull, herring gull, royal tern, and ring-billed gulls. Phalaropes, jaegers, and black skimmers have been observed in small numbers. Seven species of pelicans and their allies have been identified. The most abundant species nearshore is the northern gannet, a winter visitor that breeds on rocky cliffs around the North Atlantic. The American white pelican and double-crested cormorant are common in coastal or inland habitats. The white-tailed tropicbird, brown pelican, masked booby, brown booby, and bridled tern are rare. The abundance of northern gannets in west-central Louisiana was reported as greater than expected. Large numbers of magnificent frigatebirds are reported from eastern Louisiana, on the Chandeleur Islands, but none were reported

from the west-central Louisiana area. The large number of birds seen in the west-central Louisiana area may be related to the high productivity of Louisiana coastal waters.

Pelagic bird populations off the Gulf of Mexico coast were little known until the 1990's. Peake and Elwonger (1996) detail up-to-date findings based on trips off Louisiana and Texas, and a recently completed marine mammal study also recorded pelagic birds (Davis and Fargion 1996). Although truly pelagic habitats are limited in U.S. gulf waters, many previously unrecorded or rare pelagic birds are known to be regular inhabitants. Peake and Elwonger (1996) list 43 species of nearshore and pelagic birds, including previously unrecorded rarities such as the white-chinned petrel, four species of boobies, bridled terns, sooty terns, and brown noddies.

Twenty-four species of marine birds were observed off the southwest coast of Florida near Naples (Peake and Elwonger 1996). The number of species observed was similar to that of other areas; however, the number of individuals observed in this vicinity was the second lowest of all four areas in the Minerals Management Service studies (Clapp et al. 1982a,b; Clapp et al. 1983; Fritts et al. 1983). The southwest coast of Florida supported more tern species and larger numbers of terns than the coasts of Texas and Louisiana, and the east coast of Florida. On the west coast of Florida, terns accounted for 67% of all seabirds seen, and black skimmers and black terns were more abundant off Florida's southwest coast than in the other areas studied. Laughing gulls, herring gulls, and ring-billed gulls were seen in low numbers. Pelicans and their allies accounted for only 9% of the seabirds recorded. The most abundant species, in order of decreasing abundance, were the brown pelican, the double-crested cormorant, and the magnificent frigatebird. Frigatebirds were more abundant in this area than in other parts of the Southeast region. Four species of shearwaters, storm-petrels, and their allies occurred in low numbers. The most abundant species was Audubon's shearwater, followed by storm-petrels, Cory's shearwaters, and greater shearwaters. Common loons occurred in large numbers compared to other areas of the Southeast and were observed as far as 114 kilometers from shore.

The abundance of many species of seabirds along the west coast of Florida—such as the common loon, brown pelican, magnificent frigatebird, and several terns—is probably related to the abundance of fish found in the shallow waters of the wide shelf in that area. The occurrence and abundance of sooty terns, magnificent frigatebirds, and brown noddies are probably related to the proximity of the nesting sites located in the Marquesas Keys and Dry Tortugas off Key West, Florida.

The number of species of seabirds observed (25) off the east-central Florida coast is similar in number to other areas, yet the number of individuals there was the lowest of all the areas (Peake and Elwonger 1996). The percentage of gulls and terns in this area was not as high as in other areas, although bridled terns and sooty terns were seen in higher numbers. Seven species of terns were identified. Only three species of gulls were found in this area: herring gulls, laughing gulls, and Bonaparte's gulls. The pelicans and their allies accounted for about 25% of the marine birds observed, and the brown pelican was the most abundant species of this group. The American white pelican was seasonally common in estuarine areas during winter. Shearwaters and storm-petrels and their allies were more numerous and diverse than in the other areas.

The most obvious oceanographic feature that separates the Atlantic coast—from Cape Hatteras south to Florida—from other areas is the presence of the Gulf Stream moving north along the coast. At Cape Hatteras, the Gulf Stream veers eastward into the Atlantic. The higher relative abundance of several pelagic birds in the Southeast, including the shearwaters, the black-capped petrel, the sooty tern, the bridled tern, the jaegers, and the tropicbirds,

may be related to food availability provided by warm-water fauna in the Gulf Stream. Many of the species opportunistically feed along current boundaries and convergences. The black-capped petrel has been seen primarily along the east coast, indicating its near restriction to Atlantic waters. Black-capped petrels are very rare in the Gulf of Mexico (Peake and Elwonger 1996).

Marine birds, like other animal species in the Southeast, are showing signs of stress. Probably the most pressing issue regarding the health of bird populations in the Southeast is the continued rapid development of the area due to increased human populations, resulting in destruction of habitat for both the birds and the organisms supporting their food chain.

Marine Mammals

Blaylock et al. (1995) reported that the Southeast region supports 26 marine mammals that range in distribution from the southeastern Atlantic coast to the Gulf of Mexico (Table 3). The U.S. Fish and Wildlife Service has management authority for two populations of West Indian manatee (Florida and Antillean), and the National Marine Fisheries Service has responsibility for the management of the remaining cetacean (whale and porpoise) and pinniped (seal) populations. These include populations that are

Bottlenose Dolphins

The Gulf of Mexico bays, sounds, and estuaries are home to numerous small populations of bottlenose dolphins. Research relying on identification of individual dolphins indicates that bottlenose dolphins inhabiting many of the bays, sounds, and other estuaries adjacent to the Gulf of Mexico form discrete communities. Although breeding may occur between adjacent communities, the geographic characteristics of these areas suggest that each community exists as a functioning unit of its ecosystem and, under the Marine Mammal Protection Act, must be maintained as such. Therefore, each of those areas forming a contiguous and enclosed or semi-enclosed body of water is considered to contain a distinct bottlenose dolphin population.

Capture-release studies (that is, animals are caught, sampled, and then released) and photoidentification of individual dolphins (using unique marks on each dolphin's dorsal fin) near Sarasota and Tampa bays in

Florida showed that individual dolphins remain in a given area year-round (Wells 1986; Scott et al. 1990). Three distinct dolphin "communities" have been described in the area in and around Sarasota Bay. One community was formed by dolphins residing in the Gulf of Mexico coastal waters, another consisted of the dolphins in Tampa Bay (adjacent to Sarasota Bay), and a third community resided in the shallow waters of Sarasota Bay.

Genetic tests showed significant differences between dolphins from Charlotte Harbour to the south of Tampa and Sarasota bays and those of the Sarasota community and the Tampa Bay community. The tests, however, showed a high degree of interbreeding with other populations, indicating that the Sarasota community was not genetically isolated (Duffield and Wells 1986). The Sarasota community is probably one of a number of communities that make up an extended population, the limits of which are

unknown. The continuous distribution of bottlenose dolphins around the Gulf of Mexico coast theoretically allows genetic exchange between and among adjacent communities; however, the females of the highly structured Sarasota community form a stable, discrete, long-term breeding unit that tends to stay in one area.

Recent photoidentification and radio-tracking studies confirmed that some individual dolphins remain in the same general areas within Matagorda Bay, Texas, throughout the year; thus, the situation there may be similar to that of the Florida west coast (Lynn 1995). Movement of resident bottlenose dolphins in Texas through passes linking bays with the Gulf of Mexico appears to be relatively limited but does occur, suggesting that these populations may not be reproductively isolated from the coastal populations (Lynn 1995). For example, two bottlenose dolphins previously seen in the coastal area of South Padre Island,

Texas, coastal area were seen in Matagorda Bay, 285 kilometers farther north, in May 1992 and May 1993. Preliminary analyses of genetic tests suggest that Matagorda Bay dolphins appear to be a localized population. More than 1,000 individual bottlenose dolphins have been identified in bay and coastal waters near the northeast end of Galveston Island, Texas, but most of these were sighted only once, and only 200 individuals are reported to use the area over the long term (Henningsen 1991; Brager et al. 1994).

Much less is known about the movements of resident bottlenose dolphins in estuaries of the northern Gulf of Mexico. Seasonal differences in bottlenose dolphin abundance in the Mississippi Sound suggest seasonal migration; however, the spatial migration patterns are not known. It appears that some exchange occurs between the Mississippi Sound population and the coastal population in this area (Lohoefener et al. 1990).

Estimates of population sizes are insufficient to determine population trends for these bottlenose dolphins. However, three die-offs occurred in the northern Gulf of Mexico stock between 1990 and 1994, and although these events may have resulted in population declines, the effects of the die-offs on abundance have not yet been determined. It is not possible to accurately partition the mortalities between the bay and coastal stocks. Even though the status of the inland bay and sound stocks of bottlenose

dolphin is unknown, and this species is not listed as threatened or endangered under the Endangered Species Act, the occurrence of three die-offs among bottlenose dolphins along the U.S. Gulf of Mexico coast since 1990 is cause for concern. Available evidence suggests that bottlenose dolphins in the northern and western coastal portion of the U.S. Gulf of Mexico may have experienced a morbillivirus epidemic in 1993 (Lipscomb 1994). Seven of 35 bottlenose dolphins captured alive in 1992 from Matagorda Bay, Texas, tested positive for previous exposure to cetacean morbillivirus, and it is possible that other estuarine resident stocks have been exposed to the morbillivirus as well.

Limited population monitoring surveys in Mississippi Sound indicated a significantly lower average abundance of bottlenose dolphins in the summers between 1985 and 1993. The apparent decline in summer abundance of bottlenose dolphins in Mississippi Sound indicates a possible downward trend in abundance; however, insufficient data are available to conduct a meaningful trend analysis (Peterson and Hubard 1996). The relatively high number of bottlenose dolphin deaths that occurred during the recent die-offs suggests that some of these stocks may be physiologically stressed, possibly from nearshore pollution and chemical contamination or other human-induced causes. Given the levels of human-produced contaminants that flow into the coastal Gulf of

Mexico, there is concern that contaminants are finding their way up the food chain to bottlenose dolphins and perhaps contributing to these die-offs.

See end of chapter for references

Authors

Robert A. Blaylock*
National Marine Fisheries Service
National Oceanic and Atmospheric
Administration
Southeast Fisheries Science Center
75 Virginia Beach Drive
Miami, Florida 33149

Larry J. Hansen
National Marine Fisheries Service
National Oceanic and Atmospheric
Administration
National Ocean Survey
Charleston Laboratory
219 Fort Johnson Road
Charleston, South Carolina 29412

Keith D. Mullin
National Marine Fisheries Service
National Oceanic and Atmospheric
Administration
Mississippi Laboratories
P.O. Drawer 1207
Pascagoula, Mississippi 39568

*Deceased

classified as strategic under the Marine Mammal Protection Act. Strategic stocks are those listed under the U.S. Endangered Species Act as endangered or threatened, or under the Marine Mammal Protection Act as depleted or with estimated mortality exceeding PBR. Stategic stocks have annual mortality that is not sustainable, or they suffer from known fisheries-related mortality, yet their status is unknown. On the southeastern Atlantic coast, these strategic stocks include the Western North Atlantic coastal bottlenose dolphin. In the northern Gulf of Mexico, strategic stocks include the endangered sperm whale, bay and sound stocks of northern bottlenose dolphins, dwarf sperm whales, pygmy sperm whales, short-finned pilot whales, and the Florida and Antillean stocks of endangered West Indian manatees. Mid-Atlantic coastal bottlenose dolphins are believed to be stable, and both stocks of West Indian manatees are believed to be declining. Population trends are unknown for the remaining marine mammals, including the killer whale (Fig. 29) and the striped dolphin (Fig. 30).

Courtesy K. D. Mullin, National Marine Fisheries Service

Fig. 29. Killer whale in the Gulf of Mexico.

Fig. 30. Striped dolphin in the Gulf of Mexico.

Courtesy C. L. Roden, National Marine Fisheries Service

Alaska Region

Environment and Physical Features

Two large marine ecosystems have been identified off Alaska, the eastern Bering Sea–Aleutians and the Gulf of Alaska. The Bering Sea is a semi-enclosed high-latitude sea. Of its total area of 2.3 million square kilometers, 44% is continental shelf, 13% is continental slope, and 43% is deepwater basin.

Its broad continental shelf is biologically one of the most productive areas of the world. A special feature of the Bering Sea is the pack ice that covers most of its eastern and northern continental shelf during winter and spring. The dominant circulation of the water (Fig. 31) begins with the passage of North Pacific water (the Alaskan Stream) through the major passes separating the Aleutian Islands (Favorite et al. 1976) and into the Bering Sea. There is net water transport eastward along the Aleutian chain, and a turn northward at the continental shelf break and at the eastern perimeter of Bristol Bay. Eventually Bering Sea water exits northward through the Bering Straits, or westward and south along the Russian coast, entering the western North Pacific via the Kamchatka Strait. Some resident water joins new North Pacific water entering Near Strait, which sustains a permanent gyre around the deep basin in the central Bering Sea.

The dominant circulation in the Gulf of Alaska (Musgrave et al. 1992) is characterized by the cyclonic flow of the Alaska Gyre. The circulation consists of the eastward-flowing Subarctic Current System at about 50°N latitude and the Alaska Current System along the northern Gulf of Alaska. Large seasonal variations in the wind-stress curl in the Gulf of Alaska affect the meanders of the Alaska Stream and eddies of nearshore areas. It is the

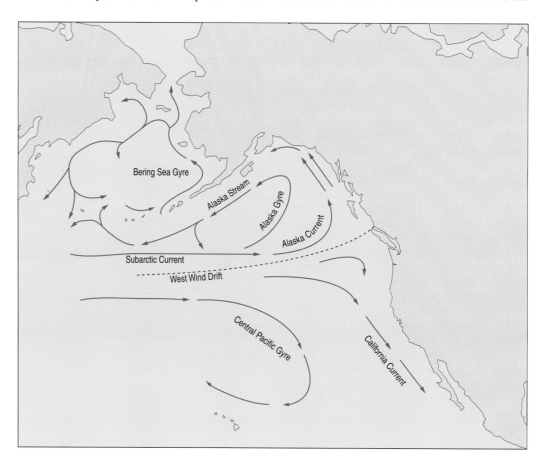

Fig. 31. General surface circulation and major areas of the North Pacific Ocean.

variations in these flows and eddies of the nearshore environment that affect a large part of the biological variability of the region. The Gulf of Alaska has about 160,000 square kilometers of continental shelf, which is less than 25% of the eastern Bering Sea shelf. At the northeast end of the Gulf of Alaska lies Prince William Sound, site of the 1989 *Exxon Valdez* oil spill.

Primary and Secondary Production

Ice-edge dynamics play an extremely important role in boosting Bering Sea production. In late winter and early spring, enough light penetrates the Bering Sea ice cover to promote growth of the algal community living in the lower portion of the sea ice. As radiation increases, blooms begin under the ice and the open water south of the ice edge. The most intense primary production, however, does not occur until the ice edge begins to break up in spring over the wide continental shelf. As the ice separates into many smaller floes, light penetration into the sea increases significantly while the partial ice cover continues to minimize wind mixing. Under these conditions, intense phytoplankton blooms occur. As the ice edge retreats northward, further blooms continue, prolonging the period of high primary production over the Bering Sea shelf. As much as 65% of the annual primary production in the

Bering Sea occurs during April and May (Niebauer 1981). In other areas off the Alaska coast, phytoplankton blooms occur over areas of upwelling. Such areas are widespread, particularly at edges of the various water domains on the shelf and shelfbreak, at the heads of submarine canyons, at the edges of gullies on the continental shelf, in island passes of the Aleutian Islands, and around submerged seamounts.

The organic matter formed by primary production is transferred to secondary zooplankton producers, which graze throughout the water column. The zooplankton grazers are often an assemblage of copepods and krill that form prey–predator links between the primary producers and upper trophic levels (Fig. 32). Shrimplike krill are important in the food web, since they are eaten by almost all other animals in the area.

Benthic Resources

The benthic invertebrate community off Alaska consists of at least 472 species of invertebrates making up the macroinfauna; the major groups include 143 species of polychaete worms, 76 species of amphipods, 76 species of gastropods, and 54 species of bivalves (Stoker 1981). More information is available for the more conspicuous forms of macrobenthos. The smaller organisms of the benthos (microbenthos

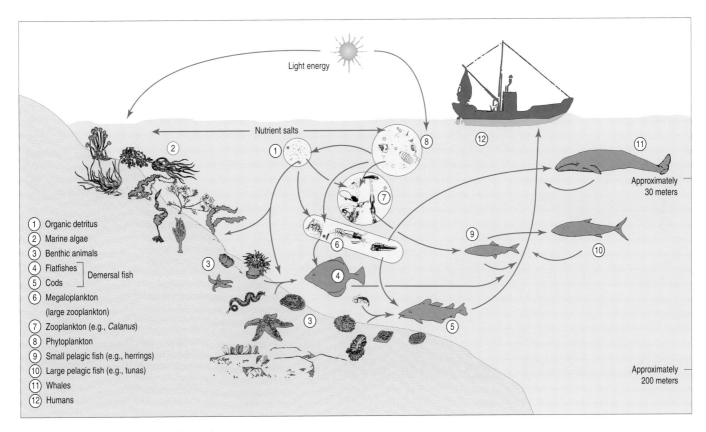

Fig. 32. Food relationships in the marine environment.

and meiobenthos) are poorly known. The microbenthos consists of bacteria and protozoans, which are essential for the mineralization of organic debris that settles to the sea bottom. The meiobenthos consists of small animals such as minute crustaceans, nematode worms, and young stages of benthic invertebrates. Macrobenthos of direct commercial value include king crabs, Tanner crabs, and shrimps.

Three species of king crab are economically important: red, blue, and brown or golden king crab. With minor exceptions, king crab populations off Alaska are low in abundance and are projected to remain low in the near future (Table 6). King crab populations have generally been declining (Fig. 33) because of the occurrence of weak year-classes and increased adult mortality. Increased mortality has resulted from many factors, including natural predation by halibut, Pacific cod, and yellowfin sole; competition for space and food with bottom fishes; handling mortality by the crab-pot fisheries; and disease.

Two species of Tanner crabs predominate, Tanner crab and snow crab. Tanner crabs are much less abundant than snow crabs (Fig. 34) and have been declining. Korean hair crabs are also declining and are substantially less abundant than king crabs and Tanner crabs.

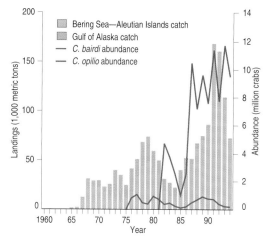

Fig. 34. Abundance and landings of Tanner and snow crabs, 1960–1994.

Dungeness crab is a nearshore species found mainly in the Gulf of Alaska. Its abundance is cyclic and currently low in all areas.

All shrimp populations in Alaskan waters are currently low in abundance and have supported only small localized fisheries since 1973. Pink shrimp constituted a major commercial fishery in the Gulf of Alaska through the late 1970's, yet this and other shrimp species are presently depressed with no apparent signs of increase. The Kodiak stock of weathervane scallops appears low, and stocks in other areas are generally unknown. Several other species of scallops smaller than weathervane scallops may have additional commercial potential, but their abundance is unknown.

The sea snail resource of the eastern Bering Sea is composed of about 15 species broadly distributed over the continental shelf. The Pribilof whelk is the most abundant species, with many of the same genus (lyre whelk and fat whelk) also very common. Other common species include the Oregon triton, as well as several other whelks. The sea snail resource is known to be fairly large and is currently harvested in limited areas around the Pribilof Islands (Pereyra et al. 1976).

Although bivalves are widely distributed on the shelf, they are concentrated in the midshelf region of the Bering Sea. Some species are found in the nearshore surf zones. The Pacific razor clam is found on sand beaches of the Alaska Peninsula. Other clams inhabiting the Alaska Peninsula include the surfclam, the Alaska great-tellin, two species of cockles, and other less frequently taken species. The status and trends of bivalve populations are largely unknown.

Sea cucumbers and sea urchins are recent fisheries in Alaska. Hand-picked by divers in the southeast and Kodiak areas, their abundance and trends are not known. The sea urchin

Table 6. Status and trends of major shellfish resources off the coast of Alaska.

Region/species	Status and trend
Bering Sea	
Blue king crab	Average abundance, fully utilized, stable
Golden king crab	Low abundance, fully utilized, declining
Korean hair crab	Low abundance, fully utilized, stable
Red king crab	Low abundance, overutilized, declining
Shrimps	Low abundance, little utilized, stable
Sea snails	High abundance, underutilized, stable
Snow crabs	Above-average abundance, fully utilized, declining
Tanner crabs	Above-average abundance, fully utilized, declining
Gulf of Alaska	
Dungeness crab	Low abundance, fully utilized, cyclic
Scallops	Low or unknown abundance, fully utilized, unknown trend
Sea urchins	Unknown abundance, underutilized, unknown trend
Shrimps	Low abundance, fully utilized, stable
Tanner crabs	Low to average abundance, fully utilized, stable

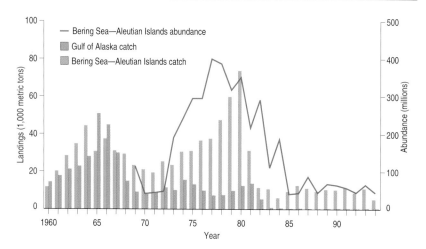

Fig. 33. Alaska king crab landings and abundance, 1960–1994.

fishery is being jeopardized by heavy predation from increased sea otter populations.

Thirty-four species of corals are found in Alaskan waters—21 species of octocorals, 2 species of hexacorals, and 11 species of hydrocorals. Two of the more abundant species are the red tree coral and the sea raspberry coral, which occur in rugged habitat of boulders and bedrock. Southeast Alaska probably has the largest number of coral species in Alaska because of the variety of habitats there, which differ in depth, substrate, temperature, and currents. The status of corals in Alaska is unknown.

Demersal Fish Resources

The total exploitable biomass of Alaska's groundfish resources has been declining slowly since 1995, when it reached 19.6 million metric tons in the Bering Sea–Aleutian region and 4.2 million metric tons in the Gulf of Alaska (National Oceanic and Atmospheric Administration 1996). These are very large biomass levels compared to the standing stocks of other groundfish resources in the world. The long-term potential yield or maximum sustainable yield for the groundfish complex in the Bering Sea–Aleutian Islands region is about 3 million metric tons. The current potential yield is also 3 million metric tons, which reflects the present good health of the resources. Recent average catch has been regulated to be below 2.0 million metric tons, which reflects a conservative management philosophy of limiting catches to protect less abundant species.

All the groundfish populations in the Bering Sea–Aleutian Islands, with one exception, are abundant and in good to excellent condition (Fig. 35; Table 7). The two main walleye pollock stocks are moderately abundant and until 1994 were above the biomass levels that would produce long-term potential yield. However, the

Table 7. Status and trends of major fish resources off the coast of Alaska.

Region/species	Status and trend
Bering Sea and Aleutians groundfish	
Arrowtooth flounder	Very high abundance, underutilized, increasing
Atka mackerel	Average abundance, fully utilized, decreasing
Greenland turbot	Low abundance, fully utilized, stable
Pacific cod	High abundance, fully utilized, stable to increasing
Rock sole	Very high abundance, underutilized, increasing
Rockfishes	Average abundance, fully utilized, stable
Sablefish	Average abundance, fully utilized, decreasing
Walleye pollock	Average abundance, fully utilized, stable to increasing
Yellowfin sole	Very high abundance, underutilized, stable to increasing
Other flatfishes	Very high abundance, underutilized, increasing
All other fishes	Above-average abundance, underutilized, stable
Gulf of Alaska groundfish	
Atka mackerel	Low abundance, fully utilized, unknown trend
Flatfishes	Very high abundance, underutilized, stable to increasing
Pacific cod	High abundance, fully utilized, stable to increasing
Pacific halibut	Above-average abundance, fully utilized, declining
Rockfishes	Below-average abundance, fully utilized, declining
Sablefish	Average abundance, fully utilized, declining slowly
Walleye pollock	Below-average abundance, fully utilized, declining
Salmonids	
Chinook salmon	Below-average abundance, overutilized, stable
Chum salmon	Below-average abundance, fully utilized, stable
Coho salmon	Above-average abundance, fully utilized, stable
Pink salmon	High abundance, fully utilized, stable
Sockeye salmon	High abundance, fully utilized, stable

transboundary eastern Bering Sea pollock stock may be showing stress in 1996 due to heavy harvesting taking place off Cape Navarin, Russia. The Aleutian Basin pollock stock, which spawns in the U.S. Exclusive Economic Zone and migrates through the international zone of the central Bering Sea known as the "Donut Hole," is very low in abundance. All of the flatfish populations but one are abundant and in good to excellent condition. Greenland turbot, a deepwater slope species, is the only depressed flatfish stock. Its abundance is projected to remain low through the 1990's. Pacific cod abundance is near a historical high and is expected to be relatively stable over the next few years. Sablefish or blackcod is low to average in abundance, and its recruitment has been relatively weak. Abundance of rockfish groups, particularly Pacific ocean perch, dropped sharply because of intensive foreign fisheries in the 1960's and remained low into the early 1980's. In recent years, their catch levels have been set well below current potential yield to help rebuild the stocks, and the Pacific ocean perch stocks are now recovering. Little is known about other rockfishes. The Atka mackerel stock that occurs mainly in the Aleutian Islands region is at average abundance and is decreasing.

In the Gulf of Alaska, overall abundance of groundfish has been relatively stable (Fig. 35). The most abundant species in the gulf are flatfishes, particularly arrowtooth flounder, and Pacific cod. Sablefish abundance is at an average level, whereas abundances of pollock and Pacific ocean perch are low. The abundances of

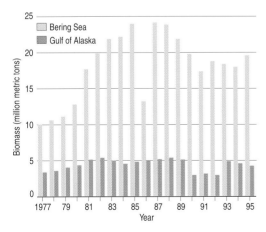

Fig. 35. Biomass trends of groundfish populations in the Bering–Aleutians region and the Gulf of Alaska, 1977–1995.

deepwater flatfish, shallow-water flatfish, flathead sole, demersal shelf rockfish, northern rockfish, pelagic shelf rockfish, other slope rockfish, thornyheads, and Atka mackerel are not well known. The long-term potential yield for Gulf of Alaska groundfish is about 450,000 metric tons. The current potential yield is 535,000 metric tons, well above long-term potential yield because of high flatfish abundance. Recent average catches are less than half of current potential yield because flatfishes, especially Greenland turbot and arrowtooth flounder, are substantially underutilized in the Gulf of Alaska.

Pacific halibut is another groundfish species that has supported an important traditional fishery for the United States and Canada. Its long-term potential yield in U.S. waters is 30,000 metric tons. The 1995 potential yield is 23,000 metric tons, reflecting the relatively good, albeit declining, condition of the stock. The exploitable biomass of Pacific halibut peaked at 270,000 metric tons in 1988–1989. The population began declining at about 5% per year from 1990 to 1993 (Fig. 36a) and accelerated to a 10% loss from 1993 to 1994.

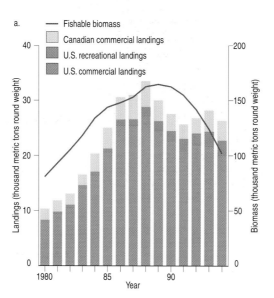

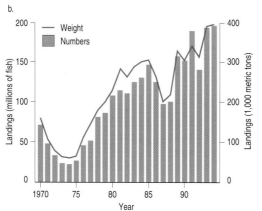

Fig. 36. a) Landings and abundance of Pacific halibut in the northeastern Pacific, 1980–1994, and b) Alaska salmon landings, 1970–1994.

The Fisheries–Oceanography Coordinated Investigations is a National Oceanic and Atmospheric Administration research program seeking to understand recruitment processes of Alaska fishes (Schumacher and Kendall 1995). The scientists in the program study both living organisms and their environment—especially processes within early life history stages—through integrated field, laboratory, and modeling studies. The initial focus of the Fisheries–Oceanography Coordinated Investigations studies was walleye pollock spawning in Shelikof Strait, Gulf of Alaska (Fig. 37). The choice of this population for research was based on the development of a large fishery and the substantial variation in this species' recruitment in the late 1970's and early 1980's. Also, the early life history of this population is quite predictable and is restricted both temporally and spatially. Walleye pollock spawn consistently in a small part of Shelikof Strait in early spring, producing a large patch of eggs and subsequently larvae. In most years, this concentration of larvae drifts to the southwest through the strait during April and May. Large numbers of larvae are often found in eddies, where feeding conditions are favorable. Fisheries–Oceanography Coordinated Investigations found that first-feeding larvae have a higher survival rate during calm sea conditions than in storms, and that in many years recruitment is largely already determined by the end of the larval period.

Pelagic Resources

Along with zooplankton, pelagic fishes and invertebrates form important linkages in the food web of the Alaska marine ecosystems. Two of the biggest groups, squid and forage fish, are particularly important. Squid are eaten by marine mammals, seabirds, and, to a lesser extent, by fish. Squid make up more than 80% of the diets of sperm, bottlenose, and beaked whales and about one-half of the diet of Dall's porpoise. Marine birds are also known to feed heavily on squid. About 5% or less of the diet of most groundfish consists of squid; however, squid play a larger role in the diet of salmon. The biomass of squid is generally very large because of its lower-order position in the food chain.

Livingston (1993) estimated that other groundfish consumed the most walleye pollock in terms of both biomass and numbers, with marine mammal and bird consumption 15–100 times less. Most of the pollock eaten by groundfish are less than 1 year old, putting groundfish in direct competition with marine birds and some marine mammals (for example, northern fur seals, juvenile Northern sea lions, harbor seals, spotted seals, and cetaceans) for this food resource. The fishery, on the other hand, takes pollock 3 years and older and therefore competes less directly with marine birds and some marine mammals (for example, adult Northern sea lions) than with those groundfish predators of adult pollock, such as Pacific halibut and arrowtooth flounder.

The major pelagic forage fishes off Alaska are Pacific herring, capelin, eulachon, rainbow smelt, Pacific sand lance, lanternfishes, and juvenile walleye pollock. Other important forage species, which usually occur either inshore and seasonally or in the northern Bering Sea, include smelts (for example, surf smelt), juvenile Pacific sandfish, Arctic cod, saffron cod, and juvenile Pacific cod. The status and trends of these pelagic forage species are largely unknown.

There is, however, anecdotal information that osmerid abundances, particularly capelin and eulachon, may have declined significantly since the mid-1970's. Major stocks of Pacific herring occur in southeast Alaska, Prince

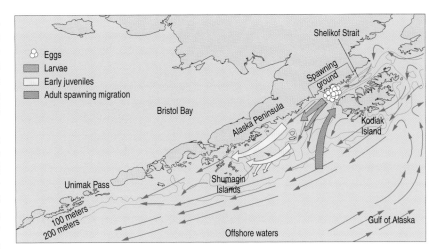

William Sound, and Kodiak Island–Cook Inlet. In the Bering Sea, they occur in northern Bristol Bay and Norton Sound. In the Chukchi Sea, commercial concentrations are located only in Kotzebue Sound. The potential yield of Pacific herring stocks is difficult to determine because stock abundance is highly variable. In the Gulf

Fig. 37. Features of early life history of walleye pollock in Shelikof Strait, Gulf of Alaska.

The *Exxon Valdez* Oil Spill

Shortly after midnight on 24 March 1989, the supertanker *Exxon Valdez* ran aground on a reef in Prince William Sound. The tanker ripped open, resulting in the largest oil spill in U.S. history. Within a few hours, 10.8 million gallons of Alaska North Slope crude oil leaked into one of the most bountiful and diverse ecosystems in the world (Alaska Department of Environmental Conservation 1993). Over the subsequent weeks, storm winds and ocean currents broadcast the oil out of the sound, oiling 2,400 kilometers of beaches from the site of the wreck westward to the Alaska Peninsula (Figure).

Many wildlife populations were exposed to the toxins in the petroleum and to the viscous coating effect of the crude oil. The intertidal communities were the most immediately affected. The largest deposit of oil was stranded in the upper and middle intertidal zones on sheltered rocky shores. In these areas, seaweeds, barnacles, limpets, periwinkles, clams, mussels, amphipods, isopods, and marine worms were killed. The oil that sank into the subtidal bottom affected eelgrass beds, small crustaceans, worms, and clams. Populations of oil-degrading bacteria, which bloomed shortly after the spill, played a major role in cleaning up the nearshore areas. Bird mortalities due to the spill may have totaled as many as half a

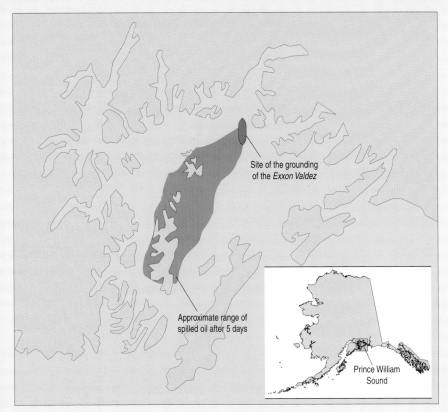

Figure. Prince William Sound, Alaska, site of the 1989 grounding of the *Exxon Valdez*, range of the spilled oil after 5 days (Alaska Department of Environmental Conservation 1993).

million, affecting roughly 90 species of birds (Alaska Department of Environmental Conservation 1993). Species which spend most of their time on the water's surface were most vulnerable, especially seabirds like the common murre. The marbled murrelet was another open-water foraging bird affected by the spill; as many as 12,000 died. More than 150 bald eagles were found dead after the spill. They encountered floating oil while preying on fish and consuming oil-contaminated carcasses. The marine mammal populations in the vicinity of the oil were also affected. The immediate death toll of sea otters was probably 4,500. The oil slicks that spread from the spill blackened many prime haulouts for hundreds of harbor seals just as the pupping season approached. The larger marine mammals, like humpback whales, were less affected. Over a 2-year period, 14 killer whales from a resident pod were missing and presumed dead following the oil spill (Loughlin 1994). Salmon and herring

were the most seriously affected fishes. The causes of deaths for many species will never be known because the corpses sank, were washed out to sea, or were eaten by scavengers that may have then also died from ingesting oil.

The Exxon Corporation mounted an extensive cleanup effort in the spring and summer of 1989, employing thousands of people to wash the beaches with hot and cold water, remove oiled sediments, and apply chemical fertilizers to aid bacterial breakdown of the oil residues. Cleanup crews returned on a smaller scale during the summers of 1990–1992 to remove oiled sediments and keep track of changing conditions on the beaches. A trustee council, representing state and federal resource agencies, was appointed to plan and mobilize a natural resource damage assessment program. With the assistance of the Environmental Protection Agency, the trustee council developed scientific plans to restore the environment. With immediate

cleanup efforts and more than 5 years of natural healing, many of the resources of Prince William Sound are well on their way to recovery or have already recovered. However, some parts of the ecosystem have not recovered and may never be the same. It is still unclear when full recovery will occur and just what it will really mean after such a large-scale catastrophic event.

See end of chapter for references

Author

William P. Hines
National Marine Fisheries Service
National Oceanic and Atmospheric
Administration
Alaska Regional Office
P. O. Box 21668
Juneau, Alaska 99802

of Alaska, Pacific herring stocks are at moderate to high abundance. In the Bering Sea, abundance of the stocks is about average, but the trend is not well defined.

Five species of salmon are dominant off Alaska: sockeye or red salmon, pink salmon, chum or dog salmon, king or chinook salmon, and coho or silver salmon. There are also two species from exclusively Asian river systems, masu salmon and amago salmon, that are found in the North Pacific. Atlantic salmon that are occasionally found in southeast Alaska are believed to be strays from net pens in British Columbia. It is not known whether the species will establish wild runs or to what extent its presence might affect other species.

Salmon are anadromous species that live as adults at sea and return to freshwater streams to spawn. Juvenile salmon migrate out to sea to feed, and upon reaching maturity, return to the freshwater systems from which they originated. The adults deposit and fertilize eggs in gravel beds. After hatching, salmon fry mature for a variable period of time in fresh water and then become smolts and prepare for migration into the marine environment in spring or early summer. In addition to natural populations of salmon, hatcheries contribute fry and smolts to the marine environment. Alaska has 38 hatcheries which released 1.5 billion fish in 1993, and hatcheries in Japan release approximately 2.2 billion fish each year (North Pacific Anadromous Fish Commission 1993). The early survival of these fish is unknown, but hatchery-produced adults do contribute significantly to catches in some areas.

All five species of Pacific salmon in Alaska are fully utilized or overutilized (Table 7). Management of salmon is directed at achieving escapement goals, reducing incidental harvests, rebuilding wild runs, and increasing production through enhancement. The catch trends for salmon have increased significantly over the past 20 years (Fig. 36b). Chinook and chum salmon stocks are generally below the abundance levels that would produce long-term potential yield. Coho salmon stocks are near long-term potential yield levels, and pink salmon and sockeye salmon are above long-term potential yield levels. The salmon stock levels tend to be cyclic and are within normal parameters at present.

Marine Birds

The Bering Sea and Gulf of Alaska support the largest and most diverse assemblages of marine birds in the northern hemisphere (Hatch and Piatt 1995). An overview of general information on Alaska seabirds and related management concerns has been prepared by the U.S. Fish and Wildlife Service (1992). The status and trends of marine bird populations in Alaska from this and other publications are summarized here.

More than 100 million birds of over 100 species depend on Alaska marine ecosystems during some part of their life cycle. These include loons, grebes, and waterfowl, as well as typical marine birds such as shearwaters, gulls, and alcids. At least three-fourths of these species breed in Alaska, and the rest are visitors from a wide variety of locations throughout the Pacific Ocean. Some visitors are transequatorial migrants from as far away as Antarctica. Twelve breeding species each have Alaska populations estimated to exceed one million individuals: northern fulmar, fork-tailed storm-petrel, Leach's storm-petrel, common murre, thick-billed murre, black-legged kittiwake, parakeet auklet, Cassin's auklet, crested auklet, least auklet, horned puffin, and tufted puffin. Two nonbreeding visitor species, sooty shearwater and short-tailed shearwater, also exceed one million individuals when in Alaska (Lensink 1984; Sowls et al. 1978). The Alaska population of many marine bird species represents a significant portion of their total world and North American populations; examples are the emperor goose, the red-faced cormorant, the red-legged kittiwake, the marbled murrelet, and the whiskered auklet.

At sea, marine bird densities are highest over the continental shelf and in the vicinity of breeding colonies (Gould et al. 1982). Important habitats include areas where environmental processes enhance prey abundance and availability, such as ice edges, upwellings, frontal systems, and eddies. Marine birds are secondary and tertiary consumers. Their major food items in Alaska include small fish such as young Arctic cod, capelin, Pacific sand lance, and young walleye pollock and invertebrates such as small squid, small crustaceans, and mollusks. Several species—for example, albatrosses, northern fulmars, and gulls—consume significant quantities of bait, discarded bycatch, and offal at commercial fishing operations. Recently, population changes of several Alaska marine birds have been linked to episodic and decadal changes in prey abundance.

The U.S. Fish and Wildlife Service and the U.S. Geological Survey in Anchorage, Alaska, maintain data bases for monitoring population size and productivity of marine bird colonies. For half of Alaska's marine bird species (for example, storm-petrels, alcids, and jaegers), information on trends is scant or nonexistent. Most populations for which there are data appear to be stable. The major colonies are protected as either national or state wildlife refuges.

Alaska populations of marine birds listed as endangered or threatened are the short-tailed albatross and the spectacled eider. Steller's eider has been proposed for listing as threatened. Declines in populations are thought to be significant in several other marine bird populations, including American goldeneye, several scoters, harlequin duck, red-legged kittiwake, marbled murrelet,

Kittlitz's murrelet, and pigeon guillemot. One bright standout among these troubled populations is a continuing increase in sightings of short-tailed albatrosses. The Aleutian Canada goose, a federally listed endangered species, nests on the Aleutian Islands but is not a seabird in any sense.

Marine Mammals

The Alaska region has 37 stocks of more than 25 species of marine mammals (Table 3). Management of marine mammals is carried out under the Marine Mammal Protection Act of 1972 and the U.S. Endangered Species Act of 1973. Additional management actions are addressed by the Magnuson Fishery Conservation and Management Act of 1976. According to the criteria provided in the 1994 Amendments to the Marine Mammal Protection Act, there are 10 strategic stocks found in waters of the Alaska Region: northern fur seal (listed as depleted under the Marine Mammal Protection Act); the sperm whale, western North Pacific and central North Pacific humpback whales, fin whale, North Pacific right whale, the western U.S. stock of Northern sea lion, and bowhead whale (all listed as endangered under the Endangered Species Act); the eastern U.S. stock of Northern sea lion (listed as threatened under the Endangered Species Act); and the Cook Inlet stock of beluga whale (designated as strategic because total annual removals by humans exceed potential biological removal). Of the 37 Alaskan marine mammal stocks, 9 are believed to be increasing, 7 are believed to be stable, 2 are declining, and the population status of the remaining 19 is unknown.

Some of the most commonly observed species in the eastern North Pacific Ocean, such as the gray whale, Northern sea lion, harbor seal, California sea lion, and sea otter, are normally found close to shore. Other species, like humpback whales, traverse ocean basins while traveling to coastal feeding areas along Alaska's shores. Still others, such as fin whales, usually remain in offshore waters, whereas North Pacific right whales are so rare they are seldom seen. Most of the large whales make long-distance migrations, and other species may move hundreds of kilometers between seasons. In Alaska the annual ingress and egress of sea ice and the annual cycle in the day–night photoperiod influence these seasonal movements. Seasonal migrations from high-latitude summer feeding areas to winter breeding areas in lower latitudes are therefore common among large whales.

Northern Sea Lion

Northern sea lions range along the North Pacific Ocean rim from northern Japan to California, with centers of abundance and distribution in the Gulf of Alaska and the Aleutian Islands, respectively. The species is not known to migrate, but individuals disperse widely outside of the breeding season (late May through early July), potentially intermixing with animals from other areas. Two separate stocks of Northern sea lions are recognized within U.S. waters: an eastern stock, which includes animals east of Cape Suckling, Alaska (that is, southeastern Alaska, British Columbia, Washington, Oregon, and California), and a western stock, which includes animals at and west of Cape Suckling (that is, the western Gulf of Alaska, the Aleutian Islands, and the Bering Sea).

In the 1960's, the worldwide population of Northern sea lions was about 250,000 adult and juvenile animals (nonpups) and was presumed to be healthy and stable. By the mid-1970's, the population in Alaska began to decline at about the same time as declines in other top predators were occurring. Subsequent declines reduced the worldwide population of Northern sea lions to 91,000 by 1989 (Loughlin et al. 1992). The current Northern sea lion population (including British Columbia) is estimated at 80,900 animals, with 43,200 in the western stock and 37,700 in the eastern stock.

The Alaska population of nonpups, which had numbered more than 157,000 animals in the 1970's, declined by 60% to less than 64,000 animals by 1989. Counts of sea lions from the Gulf of Alaska through the Aleutian Islands declined from 67,600 in 1985 to 18,700 in 1994, a decline of 72% in 9 years and more than 80% since the early 1970's. In some areas, such as the eastern Aleutian Islands, the decline was nearly 95%. The only area essentially unaffected by the decline has been southeast Alaska to northern California.

As a result of these population declines, the western stock of Northern sea lions was listed as an endangered species under provisions of the Endangered Species Act in 1990. The eastern stock was reclassified as threatened in 1996. The recovery outlook for the western stock is poor.

See end of chapter for references

Author

Howard W. Braham
National Marine Fisheries Service
National Oceanic and Atmospheric Administration
National Marine Mammal Laboratory
7600 Sand Point Way Northeast
Seattle, Washington 98115

Pacific Coast Region

Environment and Physical Features

The Pacific coast region is adjacent to the three Pacific coast states, California, Oregon, and Washington. The continental shelf is narrow—the 200-meter depth contour typically occurs from 8 to 32 kilometers offshore. The eastward-moving North Pacific Current (also called the West Wind Drift) impinges on North America off the state of Washington and splits to form the northward warm Alaska Current and the southward cold California Current (Fig. 31). The California Current travels the full length of the Pacific coast and turns offshore to flow west across the Pacific off the coast of Baja California. A northward-flowing countercurrent underlies the California Current and surfaces each winter to form the warm, inshore Davidson Current.

Variability in the Pacific coast ocean climate is driven by large-scale atmospheric influences, including the El Niño-Southern Oscillation in the equatorial Pacific and the North Pacific low pressure atmospheric cell known as the Aleutian Low. Every few years the California Current weakens and upwelling is reduced, resulting in a warm condition that has become known as El Niño. The El Niño conditions strongly affect the southern California end of the California Current but may influence ocean conditions as far north as Alaska.

There is also a between-decades pattern of climate variability that is clearest at the southern end of the region. The California Current appears to exhibit warm and cold periods, or *regimes*, each lasting decades and causing large biological shifts in the ecosystem. The most recent temperature shifts were from warm to cold in the early 1940's and back to warm in 1976 (lasting to the present). Samples of fish scales preserved in anaerobic sediments off Santa Barbara, California, provide a 2,000-year record of sardine and anchovy abundances, whose fluctuations show a periodicity of about 60 years. These temperature regimes appear to be associated with long-term variability in the amount of North Pacific Current flow that is directed northward into the Alaska Current as opposed to southward into the California Current.

The predominant ecosystem types on the Pacific coast include nearshore–coastal, continental shelf, and oceanic systems. The nearshore–coastal ecosystem extends from the intertidal zone out to 3 kilometers from shore and includes a diversity of habitats. The rocky intertidal habitat is dominated by macroalgae and sea anemones, barnacles, snails, starfishes, abalones, and mussels. The fish found here include smaller individuals of more offshore taxa like rockfishes and sculpins. Several

Effect of El Niño on the Southern California Bight

El Niño is a warmwater phenomenon in the Pacific Ocean that can have a great effect on the distribution and production of marine mammals and fishes. In the Southern California Bight ecosystem, reduced availability of prey species such as chub mackerel resulted in reduced reproductive success of California sea lions. Along the west coast of North America, the particularly strong 1982–1983 El Niño was associated with unusually warm sea surface temperatures (+4°C off California), high coastal sea levels, and strong northward currents along the California and Oregon coasts in the summer of 1983 (Bailey and Incze 1985; Pearcy and Schoener 1987). These changes altered the distribution of chub mackerel, including the areas where they spawned and the distribution and survival of their eggs and larvae. A 1983–1984 decrease of chub mackerel in the diet of California sea lions in the California Channel Islands was likely caused by the northward displacement of the fish out of the bight in response to ocean warming (Brodeur and Pearcy 1986). Chub mackerel were uncharacteristically abundant off the Washington and Oregon coasts during that period; it was the second most common species collected in purse seines off the Oregon coast in 1983–1984, whereas in most other years it did not rank in the top ten species collected (Pearcy and Schoener 1987).

The 1982–1983 El Niño resulted in decreased prey availability for female sea lions (Fig. 1) in the Southern California Bight. This reduced food availability meant that females had to spend more time and energy foraging for food, resulting in lowered reproductive success and pup growth. Thus, there were large declines in both pup numbers and weights at Channel Island rookeries in 1983 and 1984. Mortality of adults and juveniles was also higher than normal. Although the California sea lion population had increased about 5% annually

Courtesy S. R. Melin, National Marine Fisheries Service

Fig. 1. California sea lions on San Miguel Island, California.

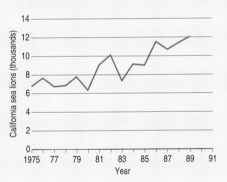

Fig. 2. California sea lion live pup counts, San Miguel Island, California, 1975–1989 (R. L. DeLong and S. R. Melin, National Marine Fisheries Service National Marine Mammal Laboratory, unpublished data).

between 1971 and 1981, their numbers declined 30%–70% at all rookeries in the Channel Islands in 1983 (DeLong et al. 1991). However, births declined the least (a drop of 30%) at San Miguel (Fig. 2), the most northern island, where sea surface temperatures were not as warm.

See end of chapter for references

Author

Howard W. Braham
National Marine Fisheries Service
National Oceanic and Atmospheric
Administration
National Marine Mammal Laboratory
7600 Sand Point Way Northeast
Seattle, Washington 98115

estuarine bays and major embayments are found in the nearshore–coastal ecosystem. The Pacific coast region accounts for almost 98,420 square kilometers of estuarine drainage area. The larger estuaries—Puget Sound, Columbia River, and San Francisco Bay—together account for 53% of the estuarine drainage along the Pacific coast (National Oceanic and Atmospheric Administration 1990a). San Francisco Bay contains more than half of all wetlands in this region, even though it is estimated to have lost almost 95% of its

wetlands since the time of its settlement by humans. The continental shelf ecosystem encompasses the upwelling zone and the inner portions of the California Current, between the nearshore–coastal ecosystem and the offshore oceanic ecosystem. The continental shelf extends to the 200-meter isobath, beyond which lies the continental slope. Associated benthic habitats are highly productive and support extensive fisheries for benthic fishes and invertebrates. Beyond the continental shelf edge (200-meter isobath) is the oceanic ecosystem, where three broad depth zones are recognized: epipelagic (0–200 meters), mesopelagic (200–600 meters), and bathypelagic (600 meters to near bottom). The epipelagic zone is the most productive of the three zones.

Primary and Secondary Production

Like other boundary currents on eastern rims of large oceans, the California Current is an area of high primary and secondary productivity. Nutrients are added to the ecosystem through a combination of coastal upwelling and advection. Initially upwelled water is characteristically low in plankton, but microalgae grow rapidly in the presence of nutrients and can achieve substantial blooms in a few days. Diatoms are the most abundant phytoplankton, but dinoflagellates can be especially important as food for larval fishes or as the source of red tides. Primary production is followed by secondary production of zooplankton, especially copepods and krill. Secondary production is most notable in the core of the California Current, far offshore. Also, krill densities are highest at the northern end of the system, where they are preyed on by marine birds, baleen whales, and coastal migrating fishes such as Pacific whiting, jack mackerel, and salmon. Large gelatinous zooplankton are very abundant on the shelf in late summer, particularly north of Cape Mendocino, although their importance in the ecosystem is not well understood.

The continental shelf ecosystem is trophically based on phytoplankton production responding to nutrients obtained from coastal processes, including upwelling, turbulent mixing, and river discharging. Secondary production of copepods, krill, and other zooplankton provides a rich pelagic environment for schooling fishes (anchovy, sardine), marine birds, and large whales. Other piscivorous fish and marine mammals also occupy higher trophic levels. The mesopelagic and bathypelagic zones are probably the largest habitats in oceanic ecosystems. However, since they depend on settling of production from the surface layer for most of their productivity, they generally contain small, slow-growing nekton species of little commercial importance, such as lanternfishes.

Benthic Resources

Within the nearshore–coastal ecosystem, kelp forests are a prominent feature of much of the rocky coastal areas and the few offshore islands present in the region (Fig. 38). These forests are vertically stratified much as terrestrial forests are and include a canopy of fast-growing giant or bull kelp, an upper understory of larger macrophytes, and a lower understory

of coralline and filamentous algae. Only a few invertebrates—such as limpets, kelp crabs, and sea urchins—are commonly found well off the bottom, although many species use the cover provided by the understory and holdfasts of the kelp. The subtidal sandy-bottom habitats are generally devoid of most attached vegetation but may contain a substantial biomass of detritus that has been deposited from nearshore regions. This detritus supports a high biomass of scavenging epifauna such as starfishes, crabs, and small crustaceans in the inshore area, which progressively gives way, with increasing depth, to suspension-feeding polychaete worms, sea pens, sand dollars, shrimp, and bivalves.

The nearshore–coastal ecosystem is heavily exploited for a variety of shellfish, including crabs, sea urchins, and abalone. The Dungeness crab population is naturally cyclic, so the fishery experiences booms and busts with a 6- to 7-year periodicity. Along the central California coast, foraging by the recovering sea otter population has eliminated most nearshore shellfish of commercial or recreational interest, but elsewhere these resources are subject to intense exploitation. The once-thriving commercial abalone industry has declined severely (see box on California Abalone in California chapter). Sea urchins are harvested for their eggs, mainly for use in Japanese sushi. This industry grew rapidly during the 1980's but has recently declined as the favored red sea urchins have become depleted. Kelp forests briefly

Courtesy National Marine Fisheries Service

Fig. 38. Kelp forest teeming with marine life.

flourished when grazing by urchins was reduced, and the population of smaller purple urchins subsequently grew in the absence of competing red urchins. Markets are now becoming established for purple sea urchins, and the population is expected to decline substantially from its present high level.

Only a few benthic invertebrate resources are harvested as extensively as demersal fishes along the Pacific coast. The most commonly exploited is the Dungeness crab, which ranges throughout the region but occurs in commercial quantities mainly north of Cape Mendocino. This species is found in many bays and shallow coastal water, where the crabs are caught by recreational fishermen, and extends onto the shelf to a depth of about 90 meters, where the crabs are caught in commercial crabbing operations using baited traps. Commercial landings of Dungeness crab are characterized by substantial fluctuations over time and indicate a 10–12-year cycle. A 1991 outbreak of domoic acid produced by toxic dinoflagellates, which commonly affects bivalves and is bioconcentrated by crabs, curtailed the fishery for a season because of perceived health threats to humans. The outbreak was no longer a problem by the following year.

Pink shrimp are abundant on the continental shelf from Washington through northern California but are found deeper (70–200 meters) than Dungeness crabs. Shrimps are important in the demersal food web. They are pursued commercially mainly by trawling with special shrimp nets. The fishery peaked in 1977, declined for a number of years following the 1982–1983 El Niño event, and then was revived in the early 1990's. The population size of pink shrimp appears to be related to levels of summer upwelling and is not as cyclic as that of the Dungeness crab.

Recently, new benthic communities have been discovered that are associated with hydrothermal vents off Washington and Oregon. These are composed of animal life forms that are entirely *chemosynthetic* in origin (that is, they synthesize organic substances using energy solely from chemical reactions rather than from photosynthesis).

Fisheries Resources

Most of the fisheries production in the Pacific coast region is taken from the continental shelf ecosystem. The status and trends of the major pelagic fishes are summarized in Table 8. Sea turtles, including the olive ridley, leatherback, hawksbill, loggerhead, and green turtle, also inhabit the epipelagic zone and undertake extensive migrations.

Table 8. Status and trends of major fish resources off the Pacific coast region.

Species	Status and trend
Pelagic	
Chub mackerel	Average abundance, fully utilized, declining
Jack mackerel	Below-average abundance, underutilized, stable
Northern anchovy	Below-average abundance, fully utilized, declining slowly
Pacific herring	Low abundance, fully utilized, cyclic
Pacific sardine	Abundance rebuilt from depleted levels, fully utilized, increasing
Groundfish	
Dover sole	Above-average abundance, fully utilized, stable
English sole	Average abundance, fully utilized, stable
Pacific whiting	Above-average abundance, fully utilized, stable
Petrale sole	Average abundance, fully utilized, stable
Rockfishes	Average to below-average abundances, fully to overutilized, declining
Sablefish	Average abundance, fully utilized, declining
Salmonids	
Chinook salmon	Many depressed stocks, overutilized, low abundance
Chum salmon	Fairly good condition, overutilized, stable
Coho salmon	Low abundance, overutilized, depressed
Sockeye salmon	Overutilized, and stable at low levels
Cutthroat trout	Low abundance, overutilized, depressed
Pink salmon	Fairly good condition, overutilized, stable
Steelhead	Low abundance, overutilized, depressed

Nearshore Fishes

Minimal fish spawning occurs in kelp beds in nearshore areas; however, juveniles of many fish families such as rockfishes, kelpfishes, and sculpins use kelp beds as nurseries. Many small, cryptic fishes are found within the understory. Fishes occupying the sandy-bottom segments of the nearshore zone include bottom-feeding flatfishes and plankton-feeding surfperch, northern anchovy, and smelt, as well as juvenile salmon migrating through the region.

Continental Shelf Pelagic Fishes

Pacific sardine is the most well known of Pacific coast pelagic fishes. Its fishery was made famous by John Steinbeck's novel *Cannery Row*, a story woven around the sardine fleet in Monterey, California, during the 1930's. The actual fishery underwent a tremendous boom and decline when sardine biomass dropped from 3 million metric tons in the early 1930's to unmeasurably low levels (perhaps a few hundred metric tons) by the 1970's. This decline engendered much debate as to whether fishing or an adverse change in the natural environment was to blame. Discovery and analysis of a 2,000-year paleosedimentary record off Santa Barbara, combined with recent data on status of the stock, have allowed resolution of the debate. Coastal pelagic fishes such as the sardine show natural long-term fluctuations of very large amplitude, presumably in response to decadal-scale warm and cold ocean regimes. During warm periods, the populations can withstand high fishing rates, but during cold periods they decline even in the absence of fishing (that is, sustainable yields are very low, and no sustainable harvest may be possible during coldest conditions). When the ocean entered a cold regime in the 1940's, continued heavy fishing accelerated the otherwise natural decline and drove the sardine population to an extremely low level, hindering recovery once conditions warmed and became favorable again beginning in the late 1970's. Sardines have staged a moderate recovery in the last decade, and the fishery has reached 200,000 metric tons (Alaska Fisheries Science Center 1993).

The northern anchovy is another pelagic fish that has responded to between-decades climate change. Under virtually unfished conditions, the northern anchovy resource grew from tens of thousands of metric tons in the early 1950's to more than 2 million metric tons briefly in the mid-1970's. Reduction fisheries (for production of fish meal) subsequently developed in California and Mexico but ceased operation several years ago. The resource biomass is now about 200,000 metric tons and continues to decline slowly.

Chub mackerel has a history parallel to that of sardines. The chub mackerel supported a large fishery in the 1930's and 1940's and declined to very low levels by the early 1970's. In response to the post-1976 warming, this resource grew rapidly, reaching unprecedented size in the 1980's, but has declined precipitously since then. Because of a combination of population pressure and warm ocean conditions (including El Niño) in recent years, chub mackerel have occasionally migrated as far north as Vancouver Island, British Columbia.

Pacific herring is an important pelagic fish in the Pacific coast region, although it has a more northerly affinity and is most abundant off Canada and Alaska. Small populations are associated with estuaries, and the San Francisco Bay spawning population is at the southern end of the range. Not surprisingly, the San Francisco Bay population has declined in recent years, with warm ocean conditions being compounded by a long drought in California that has reduced freshwater flows thought to be favorable for herring reproduction. The heavy winter rains associated with the 1997 El Niño event have reversed this condition, with unknown long-term consequences.

Continental Shelf and Slope Groundfish

The commercial groundfish fishery presently includes 83 species, 55 of which are rockfishes. Most of the abundant species, with the possible exception of the short-lived shortbelly rockfish, are fully utilized, with harvests near long-term potential yield (Table 8). Several of the rockfish species are abundant enough to support directed commercial trawl and hook-and-line fisheries. Because these species are

often very long-lived (greater than 50 years), and some concentrate in schools that are easily harvested, the potential for overharvesting rockfishes is great. Most species are found near high-relief bottom, but some species (widow rockfish and shortbelly rockfish) are pelagic. The Pacific ocean perch supported a substantial foreign fishery in the 1970's, but its population level is below that which will support long-term maximum potential yield. A fishery for another deepwater rockfish, the thornyhead, has recently developed and is reducing population levels of this species as well.

Of the dominant species, Pacific whiting (also known as hake) contributed 70% of the groundfish biomass in 1992. Pacific whiting are generally found in midwater and migrate annually from offshore spawning grounds off southern California to summer feeding grounds as far north as British Columbia. They migrate back in the fall, following an offshore migration route. Whiting mature at 3 years and undergo large fluctuations in recruitment success. The successful year-classes have fueled a substantial trawl fishery that started as a foreign fishery in the 1960's, switched to a joint-venture fishery in the 1980's, and has been a domestic fishery since 1991. Pacific whiting reached full use in 1989, but potential yield will vary substantially because this species has greater short-term natural fluctuations than most other groundfish species.

Sablefish (or blackcod) are found along the outer shelf and upper slope, with the larger fish found deeper. Their distribution shifts from inshore in summer to offshore in winter. Sablefish can live up to 60 years but mature at 5 years. They are caught by pots and longlines and also in mixed-species trawl fisheries. During the last decade, their stock size appears to have declined.

Dover sole is the most important commercial species of the flatfish group. English and petrale soles have long histories of stable harvests, and the arrowtooth flounder fishery has recently expanded. All flatfish assessments are hampered by limited and inadequate survey data.

Oceanic Pelagic Fishes

The epipelagic zone of the oceanic system is inhabited by fast-growing fishes and squids whose ranges extend well beyond this region in these animals' search for concentrations of prey fishes or zooplankton. Among the most important of these predators are albacore tuna, yellowtail, and swordfish. Other species of lesser commercial importance are pomfret, ocean sunfish, blue sharks, and pelagic squids. The zone is also a rich feeding ground for migrating salmon and steelhead.

Pacific salmon are perhaps the most important living marine resource in the region. Salmon support important traditional, commercial, and recreational fisheries in Washington, Oregon, and California. They have long been an integral part of the Native American culture and heritage in the Pacific Northwest. Pacific salmonids are anadromous. There are seven salmonid species: chinook salmon, coho salmon, sockeye salmon, pink salmon, chum salmon, steelhead, and sea-run cutthroat trout. Salmon runs are inherently highly variable in abundance. Catches have fluctuated widely, and all seven species are overutilized. Environmental conditions tend to play a large role in salmon survival and abundance. For example, an El Niño event of unusual warm ocean and nutrient-poor conditions devastated chinook salmon and coho salmon ocean survival in 1983–1985. Stocks later rebounded but have recently declined, necessitating very restrictive salmon-fishing seasons in 1994. There are many competing user groups vying to catch salmon, and strict limitations are required to protect the stocks. Thus, salmon management issues are concentrated on catch allocations among user groups and adequate protection of migrating adults, spawners, and juvenile salmon during their outbound migration from home streams to the ocean. The status and trends of salmon off the Pacific coast are summarized in Table 8. The history of catches, which largely reflects changes in abundance of the populations, is shown in Fig. 39.

Recent annual commercial salmon catches were 1.8 million sockeye salmon (1990–1993 average), 2.6 million pink salmon (1989–1993 average), and 1.2 million chum salmon (1990–1993 average). Recreational catches of these species, though important, are comparatively small. Most Pacific Northwest stocks of pink salmon and chum salmon appear to be fairly stable and are in good condition. However, some Oregon and Washington chum salmon stocks are of special concern and, in 1991, the Snake River stock of sockeye salmon was listed as endangered under the U.S. Endangered Species Act.

For the period 1990–1992, commercial catches of chinook salmon, both natural and hatchery-produced, averaged 815,000 fish, while recreational catches averaged 354,000 fish. Because of catch restrictions, the landings have been much smaller in recent years. A significant share of the catch now comes from hatchery-produced fish. Some chinook salmon stocks are extremely depressed. The spring–summer and fall chinook salmon runs in the Snake River have been listed as endangered under the U.S. Endangered Species Act; so has the Sacramento River winter run of chinook

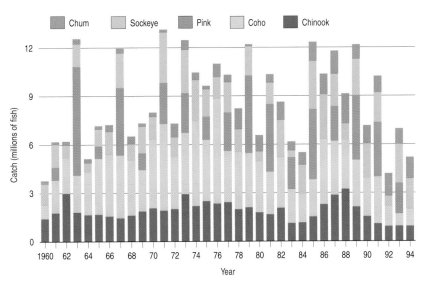

Fig. 39. Trends in salmon catches off the Pacific coast, 1960–1994.

salmon. In recent years, other chinook salmon stocks like the Columbia River summer run, Shasta River run, Skagit River spring run, Stillaguamish River summer–fall run, Snohomish River summer–fall run, Rogue River runs, and Klamath River runs have generally been depressed and have not met escapement goals. The National Marine Fisheries Service is conducting a status review under the U.S. Endangered Species Act of Pacific coast chinook salmon populations.

For 1990–1993, commercial catches of coho salmon averaged 1.1 million fish, while recreational catches averaged 706,575 fish. To an even greater extent than with chinook salmon, hatchery-produced coho salmon have become an increasingly important proportion of the catch, and in some areas account for more than 80% of the catch. Combined coho salmon landings from the ocean fisheries peaked at more than 5 million fish in 1976, then declined rather steeply in recent years to about 1 million fish. This decrease is in part due to a shifting of most of the allowable catch from the ocean fisheries to fisheries within rivers and estuaries. The shift mainly resulted from a 1974 federal judicial decision that entitles tribal fishermen to as much as 50% of the catch of returning salmon that migrate through their traditional tribal fishing areas.

In 1996 the National Marine Fisheries Service listed the central California and southern Oregon/northern California stocks of coho salmon as threatened under the U.S. Endangered Species Act. Also in 1996, stocks of coho salmon from Puget Sound/Strait of Georgia and Lower Columbia River/southwest Washington were placed on the candidate list. The release of hatchery-reared fish, degradation of salmon habitat, and overharvesting are major factors that contributed to the decline of the wild coho salmon populations in the Columbia

River. Oregon coast coho salmon stocks were placed on the candidate list in 1997.

Unlike other salmonids, steelhead and sea-run cutthroat trout do not always die after spawning. A very valuable sport fish because of its size, fighting ability, and flavor, steelhead are harvested commercially only by Native American fishermen. All fisheries occur primarily in rivers and streams. During 1990–1993, a minimum average estimate of 266,000 steelhead (data on California catches are not available) were caught by recreation anglers. Tribal fisheries harvested an estimated 177,425 steelhead during 1987–1988 for commercial, ceremonial, and subsistence purposes. As with other salmonids, steelhead harvests have been augmented by large hatchery releases. Wild populations in many rivers are at very low levels, often restricting sport harvests to only hatchery fish. In response to a May 1992 petition to list Illinois River winter steelhead, the National Marine Fisheries Service initiated a status review for the broader unit that included this population. In August 1997, the National Marine Fisheries Service listed five Pacific coast steelhead populations. Listed as endangered are steelhead in the upper Columbia River and in southern California from the Santa Maria River to just south of Malibu Creek, which is north of Los Angeles; listed as threatened are steelhead in the Snake River Basin (encompassing parts of Idaho, Washington, and Oregon), and along the central California coast and the south-central California coast.

Sea-run cutthroat trout is another popular game fish and is not harvested commercially. The total catch is unknown, but populations appear to be severely depressed in many rivers. The north and south Umpqua River run was petitioned for listing under the U.S. Endangered Species Act, and as a result of its status review, the National Marine Fisheries Service listed Umpqua River cutthroat trout as endangered. The National Marine Fisheries Service is currently conducting a status review under the U.S. Endangered Species Act of all Pacific coast cutthroat trout populations.

Marine Birds

More than 2 million marine birds of 29 species nest along the California, Oregon, and Washington coasts (Carter et al. 1995), including three species listed on the U.S. List of Threatened and Endangered Wildlife: the brown pelican, least tern, and marbled murrelet. The marbled murrelet nests inland in old-growth conifer forests and only forages at sea. Including breeding species, migrants, overwintering birds, and rare vagrants, more than 80 species of marine birds have been found

in Pacific coast nearshore and pelagic waters (Stallcup 1990). For example, many millions of sooty shearwaters pass along the coast during their annual migrations between the Arctic and the southern hemisphere nesting islands. Moreover, the marine bird populations inhabiting the oceanic zone are also highly migratory and move seasonally between northern feeding and breeding grounds and southern overwintering grounds; these species include albatrosses, storm-petrels, shearwaters, phalaropes, and jaegers. Large marine fish-eating birds include loons, murres, cormorants, and gulls and are generally found in nearshore waters.

Breeding marine bird populations along the Pacific coast have declined since European settlement began in the 1700's. Numerous causes for their decline include human occupation or disturbance of nesting sites; introduction of mammalian predators (such as rats or cats) to nesting islands; and most recently, intense commercial use of natural resources, resulting in habitat loss, declines in forage abundance, and direct mortality of birds due to oil spills and entanglement in fishing gear (Jehl 1984). Many of these factors have been or are being resolved through improved management of marine and coastal habitats, fisheries, and resources, but many critical problems remain.

The brown pelican is a good example of an *effect and recovery* situation. In the 1940's and 1950's, a northern breeding colony at Point Lobos, near Monterey, disappeared simultaneously with the end of the sardine fishery in Monterey, suggesting a direct causal relationship. In the late 1960's, pesticide contamination resulted in severe eggshell thinning and nearly complete reproductive failure at the southern California colony at Anacapa Island. Shortly thereafter, the brown pelican was formally declared an endangered species. With reductions in the use of DDT, reproductive success improved, but immigration from Gulf of California colonies may also have contributed to the recovery. In southern California, the northern anchovy is the primary prey of breeding pelicans, and fluctuations in pelican reproductive success have been clearly associated with fluctuations in the abundance of northern anchovies. Thus the potential effect of the fishery on anchovy abundance became an explicit concern to management and led to a management plan for maintaining a large anchovy biomass below which level the commercial fishery would be curtailed. The brown pelican is now being considered for downlisting from its endangered status, based on recent population increases.

The least tern and marbled murrelet are both listed endangered species and are both suffering the effects of habitat loss. The least tern nests in sandy beach areas that are frequently affected by recreational use and by introduced red foxes; however, intense habitat management seems to be stabilizing the population. Like the better-known spotted owl, in the Pacific Northwest the marbled murrelet nests only in old-growth forests, although farther north it also nests in tundra. In the past, the marbled murrelet has experienced significant declines (Marshall 1988), and the health of its remaining populations in the Pacific Northwest is tied to the continuing existence of sufficient coastal old-growth forests (U.S. Fish and Wildlife Service 1995).

Marine Mammals

When Hawaii is considered, the Pacific region is home to 67 stocks of at least 37 species of marine mammals, including 10 stocks of eastern tropical Pacific dolphins (Table 3). Like the birds and fish in the region, the marine mammal populations inhabiting the oceanic zone are also highly migratory and move seasonally between northern feeding–breeding grounds and southern overwintering grounds. Included in this category are blue whales, fin whales, sei whales, dolphins, and fur seals. In the nearshore zone, sea otters are common in kelp beds. In coastal areas, mammalian predators such as migrating California sea lions, gray whales, dolphins, and harbor seals are relatively common (Stallcup 1990). The U.S. Fish and Wildlife Service is responsible for managing two stocks of sea otters (in central California and in Washington), while the National Marine Fisheries Service has management authority for the remaining cetacean and pinniped stocks. According to the criteria provided in the 1994 amendments to the Marine Mammal Protection Act, these include 16 strategic stocks. In the eastern Pacific (that is, the waters of Washington, Oregon, California, and northern Mexico) these include 8 strategic whale species (humpback, blue, fin, sei, short-finned pilot, minke, sperm, mesoplodont beaked) and a threatened seal species (the Guadalupe fur seal). In addition, the northeastern spotted dolphin and eastern spinner dolphin stocks in the eastern tropical Pacific are considered depleted under the Marine Mammal Protection Act. Of the 43 stocks in the Pacific region, 11 are believed to be increasing, 13 are believed to be stable, 3 are declining, and the population trends of 16 stocks are unknown.

Estuarine Resources

San Francisco Bay

The San Francisco Bay is fed by two main riverine systems. The south San Francisco Bay receives very little freshwater input and is poorly flushed, whereas the northern portion is part of an active river outflow system. The Sacramento and San Joaquin rivers form a complicated network of waterways in California's Central Valley, forming the Sacramento delta. After exiting the delta, the system sequentially opens into Suisun Bay, San Pablo Bay, and San Francisco Bay, then flows out through the Golden Gate. The entire San Francisco Bay–Sacramento delta system is subject to high rates of siltation.

Introduced species have continually reshaped the San Francisco Bay ecosystem. The east coast's shipworm arrived with the Gold Rush of 1849 and quickly destroyed the many unprotected wooden structures in the bay. Striped bass and American shad were introduced in the later 1800's and quickly became dominant fishes in the system. More recently, various species of nonindigenous clams have sequentially claimed the benthic habitat, with the Japanese littleneck recently reaching remarkably high densities of 10,000 per square meter. Currently the green crab (a predaceous swimming crab related to the east coast's blue crab but of little human use) is increasing rapidly in abundance and is expected to reshape the system once again (San Francisco Estuary Project 1992).

The Sacramento and associated rivers historically supported several different large runs of chinook salmon, but present salmon production is mostly from the fall run produced by hatcheries. The winter run was recently added to the U.S. List of Endangered and Threatened Wildlife. Diversion of fresh water has posed problems for many species in the system. Large volumes of fresh water are diverted from the Sacramento delta into two large aqueducts serving San Joaquin Valley farmers and the Los Angeles metropolitan area. Reproductive habitat for native fishes and for popular introduced species such as striped bass depends on flows and interfaces of fresh and saline water. Freshwater diversion can greatly modify those patterns, resulting in poor reproductive success for fishes. Also, diversions may physically divert young life stages of several fish species and are a major source of mortality. Restricted freshwater flow also reduces flushing in portions of the bay, allowing contaminants to build up to high levels.

The Columbia River Estuary

The Columbia River is the largest river on the Pacific coast of the United States. Its estuary covers about 48,564 hectares, about three-quarters of its 1870 size. The Columbia River and its tributaries drain an area of 670,000 square kilometers. In late summer, the

Columbia River contributes 90% of the fresh water entering the sea between the Strait of Juan de Fuca and San Francisco Bay, when the river plume extends 300–400 kilometers seaward and is detectable in a 1,000-square kilometer area.

The Columbia River once had the largest chinook salmon and steelhead runs in the world. Chinook salmon originally migrated nearly 1,935 kilometers up the Columbia River to Lake Windermere, Canada, and 967 kilometers up the Snake River to Shoshone Falls near Twin Falls, Idaho. Historical estimates of salmonid runs in the Columbia River range between 10 and 16 million fish. However, access to many spawning areas in the Columbia River basin is presently blocked by hydroelectric development. Fourteen hydroelectric dams have been constructed on the main-stem Columbia River, 7 on the Snake River, and 150 on Columbia River basin tributaries. More than 800 kilometers of the upper Columbia River, excluding tributaries, were blocked by the construction of Grand Coulee Dam in 1941. Only 80 kilometers of the Columbia River above Bonneville Dam are now free-flowing. More than 50% of the Snake River basin is no longer accessible to salmonids. Many salmonid hatcheries were built to compensate for losses in spawning and rearing habitat. Currently 57 hatcheries operate in the basin, placing almost 200 million juvenile salmonids into the Columbia River system each year. Hatchery production now accounts for about three-quarters of all adult salmonids returning to the Columbia River. However, hatcheries have high costs and maintenance requirements, often experience low adult returns, and negatively affect wild salmon stocks. Hatchery salmonids often cause overharvesting of wild salmonid stocks, yet hatcheries do not address many inherent salmonid population problems, such as habitat degradation, overfishing, and challenges faced with upstream and downstream migration.

The various Columbia River estuary habitats—including mudflats, estuarine and freshwater marshes, and major channels—provide food, refuge from predation, a physiological transition area for anadromous species, and nursery areas for a variety of species. Many of these habitats have been reduced in size or highly modified since the early part of this century. The Columbia River estuary has four large embayments, all of which contain expansive intertidal mud flats that harbor abundant benthic invertebrate populations. These benthic invertebrates, particularly amphipods, are primary prey for many fish and wildlife species. Detritus from rivers, estuaries, and the sea is the primary food for most benthic invertebrate species.

The economy of the lower Columbia River was historically directly tied to its salmonid resources, which were central to the Native American way of life. Native Americans still hold treaty fishing rights and continue to harvest salmon for both commercial and ceremonial purposes. Commercial landings of salmon and steelhead were once as high as 14,350 metric tons (2,112,500 fish). By 1993, only 6,360 metric tons or 115,500 salmon were commercially harvested, in part because of very restricted seasons. Some restrictions resulted from the threatened or endangered listings of three Snake River salmon stocks, sockeye salmon, fall chinook salmon, and spring–summer chinook salmon. All salmonid species in the Columbia River basin are fully utilized to overutilized. The stocks are generally at low to depleted levels. Many are threatened or endangered.

Two species of sturgeon live in the Columbia River estuary, white sturgeon and green sturgeon. The overall white sturgeon population is healthy and productive, but populations upstream from Bonneville Dam, which are confined to reservoirs, are less productive and are considered depressed. In 1993, the total Columbia River white sturgeon catch was more than 50,000 fish, with most caught by the sport fishery in the estuary. In recent years, anglers have spent more days fishing for sturgeon than for any other species in the Columbia River estuary. Green sturgeon are much less abundant than white sturgeon. In 1993, 2,200 green sturgeon were caught commercially, and 100 recreationally. The status of green sturgeon in Oregon and Washington is unknown.

American shad were introduced into the Columbia River in 1885 and are now very abundant. The largest run was an estimated 4 million adult shad in 1990. In 1993, the minimum run size was 2.7 million fish. Poor market conditions and the concurrent timing of depressed salmon runs and the shad run in the spring have hindered the commercial harvesting of this species.

Columbia River smelt, or eulachon, spawn in coastal streams and in the lower Columbia River and its tributaries from mid-January through March and sometimes April. Although commercial landings for 1993 were the second lowest on record (227 metric tons), if the landing numbers for the coastal streams and the Columbia River are combined, the total run may have been near normal.

Other Columbia River estuarine fisheries include Dungeness crab, surf perch, and starry flounder. Northern anchovies are caught by seines and sold as bait for the sturgeon and salmonid sport fisheries. Some of the other species that use the Columbia River estuary, particularly as a nursery area, are longfin smelt, Pacific sand lance, Pacific herring, Pacific tomcod, shiner perch, Pacific staghorn sculpin, and

English sole. The status of these populations in the Columbia River is unknown.

Many species of birds use the Columbia River estuary and its two wildlife refuges. Besides many migrating waterfowl, shorebirds, and marine birds, the estuary supports at least 22 nesting pairs of bald eagles. However, bald eagle reproductive success has been poor. Agricultural and industrial contaminants, specifically DDE, PCB's, and TCDD are at levels that adversely affect eggshell thickness and nestling viability. The recently established Caspian tern colony in the Columbia River estuary may now be the largest in the country, with 5,000 pairs in 1994. Because these birds prey on small fishes, there are concerns that terns are consuming many migrating juvenile salmonids. There is also a relatively large cormorant rookery located on the same island as the Caspian tern colony. Gulls nest in two or three locations throughout the estuary.

Waterfowl primarily use the Columbia River estuary during winter. Although relatively abundant, duck populations in the estuary fluctuate widely, depending on breeding success in other areas, and remain below historical levels. The Canada goose population in the Columbia River estuary is healthy and has doubled in the last couple of years.

Many species of marine, terrestrial, and freshwater mammals use the Columbia River estuary. Species such as harbor seals and California sea lions are increasing in abundance, and because they eat salmon, they are causing conflicts with anglers. The once-abundant mink populations have all but disappeared because mink reproduction is extremely sensitive to PCB's, which in the Columbia River are at concentrations that adversely affect reproduction. Other mammal populations, such as nutria and river otter, are healthy and stable. Marine species such as elephant seals, killer whales, gray whales, harbor porpoises, and Northern sea lions are rarely observed in the estuary.

Puget Sound

Puget Sound is located in Washington State and is one of the three major water systems in the Georgia Basin. The Strait of Juan de Fuca connects Puget Sound and its northern neighbor, the Strait of Georgia, to the Pacific Ocean. Puget Sound covers an area of 2,632 square kilometers at mean high tide. Of the eight major habitat types in the sound, kelp beds and eelgrass cover the largest area, almost 1,000 square kilometers. The other predominant habitat types include subaerial and intertidal wetlands (176 square kilometers), mudflats, and sandflats (246 square kilometers). Marked losses of these habitats have occurred during the last

century. At least 76% of the wetlands have been eliminated, mainly within highly urbanized estuaries. Substantial declines of mudflats and sandflats have also occurred in the deltas of these estuaries (Levings and Thom 1994).

Several invertebrate species are commercially and recreationally important in Puget Sound. Among the crustaceans, Dungeness crab and several species of shrimp (for example, sidestripe shrimp and pink shrimp) are the most commonly harvested. The Pacific oyster accounts for 90% of the bivalve harvest. Japanese littlenecks and Pacific geoducks are the primary components of remaining bivalve harvests. Blooms of dinoflagellate species have been linked to the uptake of paralytic shellfish poisoning by various bivalves. Humans who ingest bivalves containing paralytic shellfish poisoning may experience neurological changes and, in a few cases, death. In Puget Sound, illnesses related to paralytic shellfish poisoning were not reported until the 1970's, and since then high toxin levels have been reported in various parts of the sound every year (Taylor and Horner 1994).

Within Puget Sound, salmon are presently the most economically valuable fish resource. In 1994, 45% of the approximately 210 stocks were classified as healthy, 21% as depressed, and 5% as critical by the Washington Department of Fish and Wildlife and the Oregon Department of Fish and Wildlife (1994). Chum salmon and pink salmon had the highest percentages of healthy stocks (70% and 60%, respectively) compared to chinook, coho, and sockeye salmon (35%, 44%, and 0%, respectively).

Historically, the most important commercial marine fish species in Puget Sound have been Pacific herring, Pacific whiting, Pacific cod, rockfish, and English sole. The population sizes of all of these species have declined during the last 10 to 20 years. Stocks of herring reached peak abundance in the mid-1970's, declined to low levels in the 1980's, and have since remained low. Commercial harvests of herring are now used primarily as bait for recreational salmon fishing. Whiting populations declined from a peak in the early 1980's to levels in the 1990's that are below the threshold required for a commercial fishery. Stocks of whiting in the waters north of Puget Sound and the Strait of Georgia are currently at a historical peak. Although cod and walleye pollock in Puget Sound are at their southern limit of abundance, commercial and recreational catches were relatively high during the 1980's. Present populations are severely depressed and the commercial fishery has been closed for several years. In contrast, the abundance of rockfish species has remained at low but constant levels during the last two decades. Populations of English sole are thought to be relatively healthy in the central portions of Puget Sound; however, significant declines have been recorded in populations in localized embayments, such as Bellingham Bay and Discovery Bay.

Approximately 66,000 marine birds breed in or near Puget Sound. About 70% of them breed on Protection Island, located just outside the northern entrance to the sound. The most abundant species are rhinoceros auklet, glaucous-winged gull, pigeon guillemot, cormorants, marbled murrelet, and Canada goose. Examples of less abundant species include common murres and tufted puffins. Populations of rhinoceros auklets and pigeon guillemots appear to be stable, whereas populations of glaucous-winged gulls have increased slightly in recent years, especially in urban areas (Mahaffy et al. 1994). Accurate estimates of current populations of marbled murrelets and Canada geese are not available, but the population of marbled murrelets has been greatly reduced, and this species has been listed as threatened. Thirty years ago, year-round resident Canada geese were rare in Puget Sound, but their population is growing rapidly. The common murre and tufted puffin populations have declined drastically during the last two decades.

The nine primary marine mammal species that are short-term or long-term residents of Puget Sound include (listed in order of abundance): harbor seal, California sea lion, Northern sea lion, northern elephant seal, harbor porpoise, Dall's porpoise, killer whale, gray whale, and minke whale. Harbor seals and California sea lions are year-round residents, and their abundance has increased each year in Puget Sound by 5%–15% at most sites (Calambokidis and Baird 1994; Fig. 40).

Populations of the remaining species are quite small. Northern sea lions and elephant seals are transitory residents, and whereas the Northern sea lion is currently listed as threatened in the United States, the elephant seal is abundant in the eastern North Pacific but has few haul-out areas in Puget Sound. Although harbor porpoises are also abundant in the eastern North Pacific and were common in Puget Sound 50 or more years ago, they are now rarely seen there (Calambokidis and Baird 1994). Low numbers of Dall's porpoise are observed in Puget Sound throughout the year, but little is known about their population size; they are also abundant in the North Pacific. Three pods of resident fish-feeding killer whales, numbering about 100 total, reside just north of the entrance to Puget Sound, and the size of this group is increasing about 2% each year. Minke whales are also primarily observed is this same northern area, but their population size is unknown. Gray whales migrate past the Georgia Basin en route to or from their feeding or breeding grounds; a few of them enter Puget Sound during the spring and summer to feed.

Southern California Bight

The area called the Southern California Bight stretches from Point Conception to the United States–Mexico border (about 1,000 kilometers) and is around 300 kilometers at its widest point. The mainland nearshore habitats of the Southern California Bight are mostly composed of sandy beaches, whereas the roughly equivalent length of coastline on the islands is composed of rocky shores. Most of the sublittoral bottom of the Southern California Bight is sandy with rocky outcrops. Dense kelp forests and other aquatic macrophytes are particularly common around the Channel Islands, providing substantial habitat for the invertebrates, fishes, and marine mammals found there. Many of the islands, because of their relatively undisturbed condition, provide nesting and rearing sites for numerous pelagic birds, although bird production within the Southern California Bight is fairly low. Fish production, however, is high relative to primary production and approaches that of some upwelling areas because of high turnover rates and the low trophic levels of many of the pelagic fishes.

Western Pacific Oceanic Region

The western Pacific region contains a huge number of species and few resource assessment agencies, compared to the continental United States. Great distances between the islands of the region, many of which are uninhabited,

Fig. 40. California sea lions in Puget Sound.

Courtesy P. J. Gearin, National Marine Fisheries Service

make assessment of the region's living resources difficult. In addition, economic conditions of many islands limit the effort that can be expended on resource assessment. For these reasons, there is little information on the status and trends of noncommercial species in the region, with the exception of corals and coral reefs, which have high aesthetic and conservation value. Quantitative data over time are lacking even for commercially important species in this region.

Environment and Physical Features

The United States has jurisdiction over more than 50 Pacific Ocean islands (United Nations Environment Programme/International Union for the Conservation of Nature and Natural Resources 1988; Boehlert 1993). These include the Hawaiian Islands, American Samoa, the Mariana Islands (including Guam), Johnston

Island, Kingman Reef, Palmyra Atoll, and Jarvis, Howland, Baker, and Wake islands (Table 9). Although these islands have only a total of 17,870 square kilometers of emergent land, their isolation enables the United States to claim more than 5.18 million square kilometers of ocean in the 200-mile Exclusive Economic Zone. The U.S. Pacific Islands (Fig. 41) span the tropics and subtropics across the Northern and Southern hemispheres between latitudes of 28°25' North (Kure Atoll) and 14°32' South (Rose Atoll), and across most of the central and western Pacific from the longitude of 155°30' West (Hawaii Island) to 144°38' East (Guam).

The living resources of the Pacific Islands are strongly influenced by dramatic differences in island location, size, elevation above sea level, annual rainfall, and human population (Table 9). The Pacific islands are categorized into two types, high and low islands, with significant differences in their habitats and biota

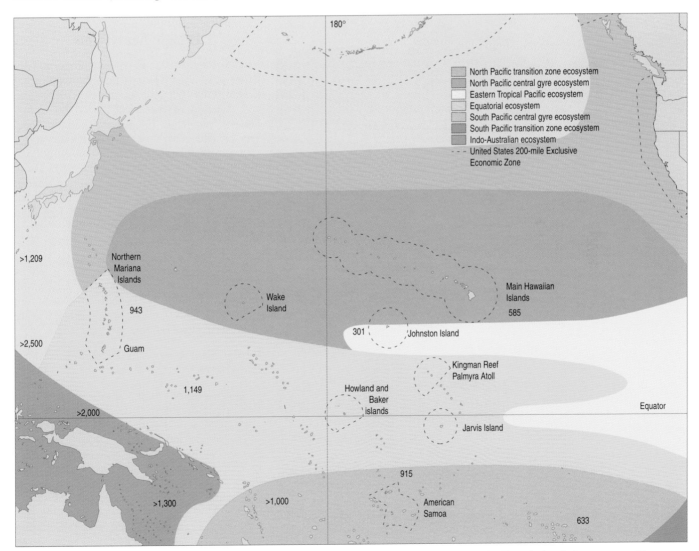

Fig. 41. U.S. island affiliates and the Exclusive Economic Zone in the western Pacific. The areas in color are the Pacific region's pelagic ecosystems. The numbers of island-associated, shallow-water fish species illustrate the west-to-east gradient in species richness in the Pacific. Island sizes are exaggerated for the sake of visibility.

Table 9. Physical characteristics and human populations of U.S. island possessions in the central and western Pacific Ocean.

Island or archipelago	Number of islands	Land area (square kilometers)	Elevation above sea level (meters)	Average precipitation (millimeters per year)[a]	Human population[a]
Main Hawaiian Islands	8 + several islets	1.1–10,414	168–4,205	630–>11,400, depending on area	1,159,600 (1992)
Northwestern Hawaiian Islands	>12	13.5 total	3–277	low, <500	No permanent population
Johnston Atoll	1	2.5	5	663	N/D (military base)
Palmyra Atoll and Kingman Reef	2	6.0 total	3	>5,080	No permanent population
Howland Island	1	1.6	7	low, <500	Uninhabited
Baker Island	1	1.4	8	low, <500	Uninhabited
Jarvis Island	1	4.5	8	low, <500	Uninhabited
American Samoa	7	2.1–135	3–966	<500 (Rose Atoll)–5,080 (Tutuila)	50,900 (1992)
Guam	1	541	405	2,286	133,152 (1990)
Northern Mariana Islands	16	0.8–122	81–965	2,000–2,500	48,216 (1993)
Wake Island	1	7.4	3	N/D	N/D (military base)

[a]N/D = no data available

Fig. 42. Tutuila, American Samoa, is a high island with lush vegetation and estuaries.

Fig. 43. Palmyra Atoll is a low island that is unusual in having abundant rainfall (more that 500 centimeters per year) and lush vegetation.

(Kay 1980). High islands are by definition mountainous (Fig. 42), with numerous rocky headlands and narrow coastal plains fringed by beaches or rocky coastal shelves. They are frequently edged with coral reefs. Subsidence leaves many low islands (Fig. 43) as rings of islets and reefs surrounding a central, marine lagoon (United Nations Environment Programme/International Union for Conservation of Nature and Natural Resources 1988). Examples of these coral atolls are Kingman Reef, French Frigate Shoals (at the Northwestern Hawaiian Islands), and Rose Atoll (American Samoa).

Marine ecosystems within the oceanic Pacific are categorized here as coastal, shelf, slope or seamount, and pelagic. Although the pelagic ecosystems are relatively homogeneous, the coastal ecosystems of the Pacific Islands include numerous habitat types, and the slope ecosystem is intermediate in complexity (Holthus and Maragos 1995).

Pelagic Ecosystems

U.S. Pacific Islands possess proportionally larger pelagic ecosystems than the continental United States because the number, dispersion, and isolation of the islands give the United States a very large Exclusive Economic Zone. The pelagic ecosystems of the Pacific correspond to major biogeographic regions defined by ocean currents (McGowan 1974). These areas are the northern and southern central gyres, the northern and southern currents that border the gyres (in regions called transition zones), the equatorial currents, and a unique area in the eastern tropical Pacific. Most pelagic fishes have ranges that encompass two or more of these areas. Large species such as whales, porpoises (Jefferson et al. 1993), sea turtles (Márquez 1990), billfish (Nakamura 1985), and tunas (Collette and Nauen 1983) are of greatest social importance.

Insular Shelf, Slope, and Seamount Ecosystems

Insular shelves, slopes, and seamounts form the interface between coastal and pelagic ecosystems of the Pacific islands. These deeperwater habitats have unique biotas (Chave and Jones 1991; Rogers 1994). Seamounts are widely scattered throughout the Pacific Ocean, but those near U.S. islands are concentrated in the vicinity of the Marianas and the Hawaiian Islands. Most organisms at seamounts (Fig. 44) have not been threatened by exploitation because of their remoteness and great depth of occurrence (Rogers 1994). However, populations at seamounts where intensive fisheries occur may be easily overexploited

(Grigg 1993; National Oceanic and Atmospheric Administration 1996). Habitat degradation of Pacific island slope and seamount habitats has been minimal.

The oceanic islands of the Pacific lack broad continental shelves. Instead, narrow shelves fringe the larger islands (Brock and Chamberlain 1968). In many areas, these are actually series of terraces with escarpments of 2–35 meters. Biological communities of the shelves, such as the pen shell beds, are poorly studied. Recent surveys have shown that sandy shelf areas are important recruitment and nursery areas for one of the slope snappers, opaka-paka (Parrish 1989). Thus, the importance of this featureless habitat has been underestimated. Island shelves support fisheries for coastal species that extend into deeper waters. Certain migratory species, such as humpback whales, use island slopes or shelves during critical stages of their life histories (Fig. 45). With the exception of larger species taken by fisheries, biological resources of insular shelves are probably not under threat from habitat damage or overexploitation.

Deeper coastal habitats of many high islands consist of escarpments and narrow insular shelves (Brock and Chamberlain 1968). Wave erosion, during geological periods when sea level differed from that of the present, created rocky undersea walls or cliffs 2–35 meters high alternating with sandy terraces. Small atolls may lack shelves or terraces (Colin et al. 1986). The slope ecosystem begins beyond the coastal terraces, or where atoll slopes exceed 30° (Colin et al. 1986; Chave and Jones 1991). High-gradient slopes are rocky, whereas low-gradient areas accumulate sediments. Escarpments exist on slopes as deep as 1,100 meters at some islands, and limestone outcrops may be found to 1,650 meters. Island slopes often descend steeply (from 45° to more than 60°) to the abyssal plain 2,000–4,000 meters below.

Reef-building corals, which thrive above 50 meters, become rare below 100 meters, but many animals usually associated with corals may be found to a depth of 200–300 meters. A deepwater biota begins at 200–500 meters (Chave and Mundy 1994). Below 500 meters, the fauna (there are no plants) gradually becomes abyssal, composed of animals that are often lightly pigmented, with reduced tissue density and eyes that are reduced or absent.

Because islands rise abruptly from the ocean floor, most of their Exclusive Economic Zones are over deepwater. Traveling less than 5 kilometers offshore of Pacific islands usually places one over water more than 2,000 meters deep. Indeed, the greatest ocean depth, 11,022 meters at the Mariana Trench, borders U.S. islands.

Courtesy R. Humphreys, National Marine Fisheries Service

Fig. 44. Pelagic armorhead, sea urchins, and soft corals at about 250 meters depth on the summit of Hancock Seamount, north of Midway Island.

Courtesy L. Consiglieri, National Marine Fisheries Service

Fig. 45. Humpback whale near Kailua-Kona, Hawaii.

Coastal Ecosystems

Coastal ecosystems of the Pacific Islands are composed of several characteristic habitats, most of which are highly structured, topographically complex, and biologically diverse (Holthus and Maragos 1995). Coastal ecosystems of the Pacific Islands are small compared to those of the continental United States, but island coasts have larger ratios of perimeter to land area. Coral reefs, rocky shorelines, seagrass and algal beds, mangrove forests, small estuaries, and terrestrial seabird and turtle rookeries are important habitats for Pacific island marine organisms.

Coastal habitats found on high Pacific islands but not low atolls include estuaries, mangrove forests, rocky reefs with little coral, and brackish shoreline ponds (Kay 1987). Actively volcanic islands such as Hawaii Island

Fig. 46. An ala kumu (also known as the 7–11 crab) in a basalt escarpment reef at 11 meters on Oahu, Hawaii. Although orange cup corals and brightly colored sponges are abundant on this rocky reef, reef-building corals are sparse.

Courtesy B. Mundy, National Marine Fisheries Service

or Aguijan, Farallon de Medinilla, and several smaller Commonwealth of the Northern Mariana Islands have coastal habitats that are rocky, without elaborate coral or mangrove development (United Nations Environment Programme/International Union for the Conservation of Nature and Natural Resources 1988). Rocky reefs and shelves with or without coral make up a larger proportion of coastal habitat in high Pacific islands than is commonly recognized. Freshwater streams and rivers on

high islands provide habitats for amphidromous species, which live in fresh waters through most of their life cycles and breed there, but have planktonic larvae that drift out to sea and return to fresh waters when they are transformed to their juvenile stage (Kinzie 1990). At the Pacific Islands, these are species in families that are usually marine, including fishes (gobies and sleepers), shrimps or prawns, and snails (nerites). Thus, high islands have a greater aquatic biological diversity than low islands.

Shallow coastal habitats common to both island types include algal beds, seagrass beds, sand flats, rocky reefs, and rubble-covered bottom, but the most characteristic habitat of the Pacific Islands is coral (Kay 1987). Coral reefs are formed by the accumulated carbonate structure of ancient reefs capped by growing coral and calcareous algae, typical of low atolls. The majority of coral habitats of high Pacific islands is thus not true coral reef but coral growth on rocky substrates (Fig. 46).

Algal beds are a nearshore habitat often unrecognized as important. The algal habitat in Hawaii is composed of beds or clumps of *Microdictyon* species less than 5 centimeters high and fields containing a variety of genera including *Halimeda, Sargassum, Padina,* and *Dictyopterus* species exceeding 10 centimeters (Parrish and Polovina 1994). This important algal habitat is used by a variety of organisms for food, shelter, and nursery grounds. Calcareous algae are a major source of sand, which in turn forms habitat for numerous other species (Kay 1987; Fig. 47).

The extent and importance of algal habitat are frequently overlooked by resource managers. Adult green turtles, for example, forage primarily on algae and seagrasses (Forsyth and Balazs 1989). In the northwestern Hawaiian Islands, expansive algal areas are thought to provide shelter for newly recruited lobster and fish in depths of about 18–30 meters (Parrish and Polovina 1994; Figs. 48 and 49). This little-studied habitat is extensive on shallow banks throughout the Pacific basin and is vulnerable to the same environmental stress (that is, pollution and sedimentation) as coral reefs. The role of algal beds as habitat for early life stages of Pacific marine organisms should receive high priority for future research.

Seagrass beds are a less widespread coastal habitat in these islands. Unlike temperate seagrasses, most tropical seagrass species are small, with rapid growth rates and the potential to recover rapidly from disturbance (Coles and Leelong 1997). Like algal beds, seagrasses provide shelter for crustaceans and fish, food for sea turtles and other species, stability for bottom sediments, and a means for trapping detritus and nutrients, all of which increase

Fig. 47. Colorful calcareous algae on a patch of dead coral at 3 meters on Palmyra Atoll. The importance of algae in coral habitats and as a separate nonreef habitat in tropical ecosystems is frequently underestimated.

Courtesy B. Mundy, National Marine Fisheries Service

productivity. Seagrasses are most diverse in the western Pacific, with seven species in Guam, but the single Hawaiian species is endemic and known from only a few sites (Coles and Kuo 1995). Little is known about the status of seagrasses in the U.S. Pacific Islands.

Lagoons and estuaries of Pacific tropical islands are usually small (United Nations Environment Programme/International Union for the Conservation of Nature and Natural Resources 1988), relative to North American estuaries. Most lagoons occur in low atolls, and only a small number of species are known to complete their entire life histories within them (Boehlert and Mundy 1993; Leis 1994). Estuaries in the western and southern Pacific are typically fringed with mangrove forests, although the clearing of mangroves has altered estuarine habitats in many areas (Ellison 1997). For example, the 10 hectares of mangrove in Pago Pago Harbor, American Samoa, were cleared for harbor development, as were mangroves in Apra Harbor, Guam, where most of the island's forests grew. Mangrove forests are extremely susceptible to damage by clearing and coastal development and are of special concern for conservation. The Coastal Management Program in American Samoa now provides for mangrove protection and management (Ellison 1997). In contrast, mangroves are an introduced species in Hawaii (Kay 1987). Transported to Hawaii in the early 20th century, they now form altered habitats in restricted locations. Removal of introduced mangroves from Pearl Harbor has been proposed to restore natural estuarine habitat for indigenous birds and fishes (U.S. Army Corps of Engineers 1995).

Few indigenous island species are obligate estuarine organisms, but island estuaries do serve as nurseries or primary adult habitat for some (Kay 1987). No assessments of stock status for estuarine populations in the Pacific Islands are available. In Hawaii, an example of an estuarine organism is the nehu, an endemic anchovy that is the primary bait for the local skipjack tuna fisheries. About 70% of the catch of nehu comes from Kaneohe Bay and Pearl Harbor, Oahu. Nehu numbers fluctuate greatly, but abundance is primarily measured by the bait fish catch, which varies more with socioeconomic factors than the nehu stock size (Clarke 1992). Recent studies suggest an inverse relation between nehu population abundance and salinity, but the influence of environmental factors on the marine resources of most island estuaries is poorly understood.

Estuaries of the Pacific Islands may be exceptionally vulnerable to coastal development because of their small size and number and their proximity to human populations (Maragos 1993). Examples of human

Courtesy B. Mundy, National Marine Fisheries Service

Fig. 48. A slipper lobster at 33 meters on the insular shelf of Oahu, Hawaii.

Courtesy B. Mundy, National Marine Fisheries Service

Fig. 49. An endemic Hawaiian spiny lobster at 8 meters on Oahu, Hawaii.

alterations to island estuarine or lagoon habitats include Pago Pago Harbor and Pala Lagoon, American Samoa. Alteration of Pago Pago Harbor began during 1942–1945 with dredging that continued through 1960 (United Nations Environment Programme/International Union for the Conservation of Nature and Natural Resources 1988). Tuna canneries were built in the northern end of the harbor in 1956. Canneries and other sources of organic pollution added to dredging, oil spills, bilge and ballast water discharges, sewage, and agricultural pollution, and by 1977 about 95% of the corals in the inner harbor were dead. Treatment projects and reduction of direct discharge of wastes into the harbor are now under way. Pala Lagoon is a smaller embayment to the west of Pago Pago Harbor that was once open directly to the sea. About 1960, an airfield constructed on the west side of Pala Lagoon closed most of this opening (Maragos 1993). Resulting habitat damage was manifold: adjacent reefs were dredged and filled, shoreline erosion increased at nearby beach areas, and the water quality was reduced in what had been American Samoa's most important shellfish ground.

The coastal habitats of special concern are terrestrial nesting sites critical to the survival of seabirds and sea turtles (Fig. 50). These animals nest at remote sites that lack indigenous land mammals and are removed from most human activity. They are thus extremely vulnerable to habitat alteration and predation by introduced species. Conservation of marine bird and sea turtle nesting habitat is a high management priority for the region.

Fig. 50. A green turtle swimming at 2 meters beneath the surface over a sand flat on Oahu, Hawaii.

Patterns of Biodiversity

Biodiversity patterns in the oceanic Pacific have been shaped primarily by volcanism, tectonic plate movement, sea-level change, and island subsidence—forces which together create high islands and then transform them to low atolls—as well as by ocean currents, human-caused extinctions, and species introductions (Kay 1980; Springer 1982). Substantial human populations often live on high islands in urban centers that low atolls cannot support. High islands often have substantial rainfall, allowing forests, freshwater streams, and estuaries to exist. Most atolls are dry and lack dense vegetation and estuaries. High and low islands are distributed unevenly across the Pacific, with only low islands present in the central equatorial region. This uneven distribution has greatly influenced the biogeography of the region (Stoddart 1992).

A west-to-east gradient of species richness exists in the Pacific, emanating from a region bounded by the Philippines, Indonesia, and New Guinea (Stoddart 1992). This region contains the greatest number of marine species on Earth, and numbers of species decline with distance away from it (Fig. 41). For example, the Mariana Islands, which are nearest this region, have 943 shore fishes, and Samoa to the southeast has 915. Hawaii, much farther to the northeast, has 585, but Johnston Island, a low atoll far to the southwest of Hawaii, has only 301 (Myers 1989; Kosaki et al. 1991; Randall 1995; Myers and Donaldson 1996).

The Hawaiian Islands are a biodiversity hot spot, despite their relatively low species richness, because a large percentage of Hawaiian species are endemic (Randall 1992). About 18%–25% of Hawaiian shore fishes, mollusks, polychaete worms, seastars, and algae exist only in that archipelago (Kay 1987). This proportion ranks among the highest degree of marine endemism of any of the Earth's islands.

Status of Taxonomy

Much remains to be learned about the taxonomy and biogeography of Pacific island marine organisms (National Research Council 1995). Oceanic organisms are relatively well known, whereas mesopelagic and planktonic species associated with islands or seamounts need more study. Our knowledge of insular coastal biotas is related to the existence of research institutions in some areas (Oahu, Guam) but not in others (Wake, Howland). Hawaii and Guam have been surveyed more than other U.S. islands, whereas the species taxonomy of the small U.S. island possessions in the central equatorial Pacific is very poorly known. The deep-sea biota has received little study except near Hawaii, and the biota below 2,000 meters is almost unknown.

Taxonomic knowledge also varies greatly among systematic groups. Large organisms like mangroves (Ellison 1995), birds (Garnett 1984), and mammals (Jefferson et al. 1993) are well known. Although new species of seagrasses (Coles and Kuo 1995), corals (Veron 1995), and shallow-water fishes (Pyle 1995) continue to be discovered, taxonomic knowledge of these conspicuous groups is essentially at the level of systematic revision and biogeographic description. For example, of 3,392 nonpelagic fish species living above 200 meters, only 442 (13%) were awaiting description as of 1995 (Pyle 1995). The proportion of reef and shore fish families from the region that are in need of taxonomic revision (20%) is high relative to those in North America but is low relative to other taxa in the Pacific Islands.

Algae (Abbott 1995), mollusks (Kay 1995), and echinoderms (Pawson 1995) are somewhat well known, although much remains to be

Courtesy B. Mundy, National Marine Fisheries Service

discovered about cryptic species. Their taxonomy is still at the level of species description, and it is estimated that about 25%–30% of the flora remains unknown. Several reviews of Pacific mollusks exist, but an average of 24 new species a year are still described (Kay 1995). Knowledge of Pacific echinoderms (sea urchins, sea stars, and so forth) is exemplified by a 1986 review of New Caledonian species (Guille et al. 1986), which increased the number of known species by 46%; similar results are expected from other Pacific islands (Pawson 1995).

Poorly known taxa in the central Pacific include sponges (Kelly-Borges and Valentine 1995), cnidarians other than corals, segmented worms (Bailey-Brock 1995), crustaceans (Eldredge 1995), and other arthropods. Numerous species undoubtedly remain undiscovered. Tiny organisms and those in esoteric taxa are almost uninvestigated in the Pacific Islands. These include marine viruses, bacteria, protists, flatworms (Fig. 51), nematodes, groups of marine worms, sipunculids, and several other phyla (for example, rotifers, gastrotrichs, phoronids).

Little is known about the genetic diversity of Pacific island organisms, particularly the ones of less economic value (see, for example, studies reviewed by Palumbi 1994). The larger, socially important pelagic organisms are better studied (Graves and McDowell 1994; Hoelzel 1994), but information is still meager. Evidence for within-species genetic diversification ranges from strong (humpback whales, Baker et al. 1993; green turtles, Bowen et al. 1992) to absent (yellowfin tuna, Scoles and Graves 1993; skipjack tuna, Graves et al. 1984).

Primary Production, Secondary Production, and Benthic Resources

Primary and secondary production are difficult to summarize for the western Pacific region because of the area's extensive geographic range and heterogeneity. Benthic resources of the region are even more difficult to characterize for these reasons and also because of the large number of species present. Our rudimentary taxonomic knowledge about benthic organisms exacerbates this difficulty. A few examples of regional benthic resources are discussed in the sections on coastal ecosystems, western Pacific invertebrates, reef animals, and nearshore resources (coral reefs). There is little information available on ecosystem processes, energy flow, or linkages between primary, secondary and tertiary production in nearshore U.S. Pacific Island ecosystems, with the exception of a study in the northwestern Hawaiian Islands (Polovina 1984).

Courtesy B. Mundy, National Marine Fisheries Service

Primary production in the tropical and subtropical Pacific tends to be oligotrophic compared to that of continental shelves (Mann and Lazier 1991; Olaizola et al. 1993). Oligotrophic systems are characterized by low primary productivity associated with low levels of dissolved nutrients. Reduced primary productivity and consequent low secondary productivity create the clear, blue waters for which the tropics are famous. Two processes that transport nutrients to the ocean surface along continental coasts, estuarine outflow and coastal upwelling, are not dominant features of most of the Pacific Islands. Areas of enhanced productivity exist, however, at regions of equatorial upwelling (Lindley et al. 1995; Mackey et al. 1995) and near islands (Gilmartin and Revelante 1974). Island and seamount topography also act with winds, tides, and prevailing currents to create eddies and other small or mesoscale oceanographic features that locally enhance primary productivity (Boehlert and Genin 1987; Olaizola et al. 1993). These areas of greater productivity support fisheries for tunas, swordfish, marlin, squids, and other species (Boehlert 1993).

Studies of the primary production, secondary production, and benthic communities of the leeward (west) coast of Oahu, Hawaii, were reviewed by Harrison (1987). Although this site cannot be considered representative of average or even typical conditions at other Pacific islands, it does serve as an example of island conditions for comparison to marine regions of the continental United States. The studies showed that Chlorophyll *a* concentrations at west Oahu were low at the surface (average of 0.01 microgram per liter), with a maximum (average of 0.03 microgram per liter) below the mixed layer in a depth range of 50–120 meters. Chlorophyll *a* concentrations were sometimes

Fig. 51. A flatworm on a reef at 14 meters on Oahu, Hawaii, is a species not identified until 1995 despite its abundance and conspicuous appearance. Less is known about the taxonomy of marine life in the Pacific Islands than in other U.S. waters.

increased near shore. Picoplankton (cells smaller than 3 microns) constituted about 65% of the total phytoplankton biomass. High Chlorophyll *a* concentrations just below the mixed layer and a predominance of picoplankton are both characteristics of waters poor in nutrients. The average depth-integrated primary production at leeward Oahu was 13.8 milligrams of organic carbon per square meter per hour, and the average annual production was estimated at 60.4 grams of organic carbon per square meter per year. There was a semiannual cycle, with high primary production in January through May, and lower production from May to December. Secondary production was also typical of oligotrophic waters. Microzooplankton standing stocks were about 166 milligrams of organic carbon per square meter in the upper 200 meters, with a maximum at 75–125 meters as with phytoplankton. Macrozooplankton biomass was variable, as high as 4.72 milligrams of organic carbon per cubic meter in the upper 25 meters at night. Greater concentrations of macrozooplankton occurred nearshore than offshore. Benthic communities at leeward Oahu were dominated by coral reef biota and the fauna of extensive sand flats. Coral communities were rich in macrofauna, whereas sedimented areas were macrofauna-poor and characterized by a lower-diversity fauna. No quantitative estimates were made of percent cover, biomass, or productivity of benthic organisms.

Fisheries Resources

Western Pacific Invertebrates

The only significant lobster fishery in the region exists in Hawaii (Polovina 1993; National Oceanic and Atmospheric Administration 1996). Lobster fisheries at other U.S. islands are small (Pitcher 1993). Recruitment of Hawaiian spiny lobsters and slipper lobsters has been low in recent years (Fig. 52) as a result of prevailing oceanographic factors (Polovina et al. 1994; see box on Hawaiian

Monk Seal). The long-term potential yield of one million individuals in the early 1980's was revised downward to 200,000 in the 1990's (National Oceanic and Atmospheric Administration 1993). Annual catches have recently fluctuated around the long-term potential yield of 200,000, with 424,000 animals harvested in 1992 and 131,000 in 1994 (Haight and DiNardo 1995). The lobster fishery was closed in 1993 to allow stocks to rebuild, opened briefly in 1994, and closed again in 1995. Stocks are thought to be recovering (National Oceanic and Atmospheric Administration 1996), and new management strategies are being developed for this fishery (Wetherall et al. 1995).

Attempts to develop deepwater shrimp fisheries at the Pacific Islands have been unsuccessful. The only sustained island shrimp fishery existed in Hawaii during 1977–1991. Declining catches and high costs made the Hawaiian shrimp fishery economically unfeasible, and the resource is not currently exploited (Polovina 1993). Shrimp fisheries at other U.S. islands have been limited to exploratory efforts (King 1993; Polovina 1993).

Precious corals have been harvested at the northern Hawaiian seamounts and on island slopes in the main and northwestern Hawaiian Islands (Grigg 1993; Fig. 53). No precious coral fisheries are established in other U.S. islands. Sustainable harvest of precious corals depends upon selective fishing techniques, but the prohibitive costs of these techniques has discouraged precious coral harvests in U.S. waters since 1988 (National Oceanic and Atmospheric Administration 1996).

Coral beds harvested by scuba divers or submersibles tend to remain productive, whereas deep coral beds exploited by nonselective tangle-nets or dredges are quickly depleted. Pink coral at seamounts north of Hawaii yielded 70 metric tons to tangle-net fishers in 1979, but by

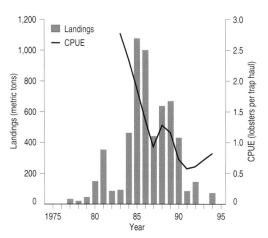

Fig. 52. Hawaiian spiny lobster and slipper lobster landings and catch per unit effort (CPUE) in Hawaiian waters, 1975–1995.

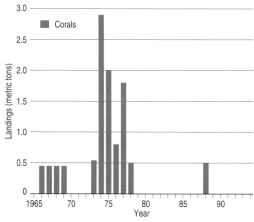

Fig. 53. Landings of precious corals in Hawaiian waters, 1965–1994.

1990 the coral was depleted and the fishery ceased. Deep coral beds on island slopes have also been overfished. In contrast, harvests of black coral and other shallow-dwelling species (in less than 100-meter depths) that are conducted by divers and by submersibles in the main Hawaiian Islands are selective and are now managed by minimum harvest size and quota (Grigg 1993; National Oceanic and Atmospheric Administration 1996). One unexplored question is the effect of precious coral harvest on other organisms, because soft corals serve as habitat for commensal species such as the longnose hawkfish, soft-coral gobies, and some shrimp species (Fig. 54). The extent of such associations is unknown. Coral tangle-netting and dredging obviously are habitat-destructive fishing techniques.

Giant clams are an extreme example of invertebrates threatened by fishery overexploitation (Munro 1993; Fig. 55). Their high monetary value as food and the ease of their harvest have resulted in their overexploitation in many islands. The largest giant clam is locally extirpated in Guam and the Commonwealth of the Northern Mariana Islands, and the small giant clam is rare in Samoa (Lucas 1994) except at Rose Atoll (E. Flint, U.S. Fish and Wildlife Service, Honolulu, personal communication). Giant clams are currently protected by the Convention on International Trade in Endangered Species of Wild Fauna and Flora, but their market value in Taiwan and other Asian nations continues to encourage poaching. Cultured clams currently supply much of the demand, and pilot projects to restock depleted reefs with cultured clams are being evaluated (Lucas 1994).

Fig. 54. A commensal shrimp on wire coral at 12-meter depths on Oahu, Hawaii.

Fig. 55. A small giant clam at Palau.

Western Pacific Sharks

Pelagic sharks form a large part of the bycatch of longline fisheries, though few except the shortfin mako and thresher sharks contribute major revenue to the fisheries (Bonfil 1994). Other sharks taken include the blue shark, oceanic whitetip, and silky shark. Almost 2 million blue sharks were taken by the North Pacific longline fishery in 1988, with another half million in the southern and western Pacific in 1989. The pelagic shark bycatch of disbanded North Pacific driftnet fisheries was also thought to have reached 2 million individuals by 1988. Blue sharks made up most of the driftnet bycatch, although salmon shark, shortfin mako, and cookie-cutter sharks also were caught. Much of the shark bycatch may not survive even if released. More sharks may be finned rather than landed whole; that is, the fins are cut from the sharks and retained for sale later, while the carcasses are discarded. Slow growth, large size at first reproduction, low fecundity, and long gestation periods relative to bony fishes make shark populations vulnerable to over exploitation.

Oceanic Pelagic Fishes

Tunas and billfishes have been considered rather resistant to overexploitation because of their high fecundity, vast ranges, and migratory capabilities. With its nearly cosmopolitan distribution, the swordfish is typical of large pelagic fishes. The rapid growth of a night longline fishery based in Hawaii increased the central-western Pacific swordfish catch from 50 metric tons in 1988 to 4,490 metric tons in 1991 (Skillman et al. 1993) and 6,100 metric tons in 1993 (National Oceanic and Atmospheric Administration 1996; Fig. 56). However, the Hawaii-based fishery declined to 3,100 metric tons in 1994. Total catches of swordfish in the U.S. and foreign Pacific Ocean region were about 29,000 metric tons in 1990–1992, with an estimated guess of long-term potential yield of 25,000 metric tons per year. A recent analysis suggested that Pacific stocks are now stable and

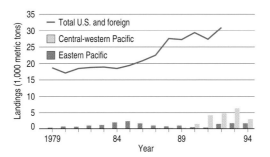

Fig. 56. Landings of swordfish by U.S. and foreign vessels in the Pacific Ocean (total), and U.S. landings in the eastern Pacific and the central-western Pacific, 1979–1994.

underutilized, although catches in years since 1989 have exceeded the estimated long-term potential yield (National Oceanic and Atmospheric Administration 1996).

Marlin, sailfish, and spearfish species are abundant in the open ocean and near island slopes, which arise steeply from the deep sea. The Kona coast of Hawaii is among the world's premier sport-fishing sites for marlin. However, the large size and great mobility of billfish make it exceedingly difficult to assess the status of their stocks (Skillman et al. 1993; National Oceanic and Atmospheric Administration 1996). Population levels are poorly known, and estimates rely on commercial bycatch statistics, which may reflect economic or technical fishery changes as much as biological fluctuations. Indo-Pacific blue marlin stocks are very important to recreational anglers and may be overutilized (National Oceanic and Atmospheric Administration 1996). Catch rates of these stocks in the Hawaii-based longline fishery declined from 115 kilograms per 1,000 hooks set in 1959 to less than 10 kilograms per 1,000 hooks in 1985 (Boggs and Ito 1993). In 1989, rates rose to 50 kilograms per 1,000 hooks, but this recent increase probably reflects altered fishing techniques rather than a stock increase. Limited entry and area closures in the Hawaii-based longline fishery have stabilized or reduced the longline catch of marlin. In Guam, blue marlin catches increased steadily from 15 metric tons in 1980 to 60 metric tons in 1990 (Myers 1993), and the small tournament and artisanal fishery for blue marlin in American Samoa (3–7 metric tons per year) has shown no long-term trend (Craig et al. 1993). Basic biological data on billfish species beyond catch per unit effort are grossly inadequate, making evaluation of stock changes tentative.

The current annual catch of fisheries for all highly migratory Pacific pelagic fishes until 1995 was near the estimated long-term potential yield of 1.8 million metric tons (National Oceanic and Atmospheric Administration

1996). U.S. contributions made up an average annual catch of 264,300 metric tons in 1990–1992. Knowledge of the status of most large pelagic fishes is poor because of these species' wide migratory ranges and the absence of a Pacific-wide international organization to monitor stocks. Tunas account for the greatest portion of the pelagic fisheries catch (Skillman et al. 1993; National Oceanic and Atmospheric Administration 1996; Table 10).

Coastal Pelagic Fishes

Small pelagic fishes that live near islands or seamounts form a trophic connection between coastal, slope, and pelagic ecosystems. Two of these coastal pelagic fishes, the bigeye scad (Fig. 57) and mackerel scad, are a major component of many nearshore Pacific island fisheries (Boehlert 1993; Dalzell 1993; Table 11). Bigeye scad is often the most frequently caught nearshore species in Hawaii and Guam, consistently ranking fifth or sixth among all nearshore commercial species harvested in these areas (see unpublished data reports cited in Hamm [1993], and later reports in the same series).

Courtesy B. Mundy, National Marine Fisheries Service

Fig. 57. Bigeye scad at 2-meter depth off Oahu, Hawaii.

Table 11. Western Pacific nearshore fisheries—productivity in metric tons and status of fisheries.

Area/species	Recent average yield (metric tons)	Fishery utilization level	Stock level relative to long-term potential yield
Western Pacific region			
Bigeye scad[a]	310	Full	Unknown
Mackeral scad[a]	160	Full	Unknown
Other inshore fisheries[a]	700	Full	Unknown
Inshore reef fishes[b]	90	Full	Unknown
Inshore reef fishes[c]	160	Full	Unknown
Total	1,420		

[a]Main Hawaiian Islands (1993–1995 average).
[b]Guam (1993–1995).
[c]American Samoa (1993–1995 average).

Table 10. Status and trends of tuna stocks in the western Pacific.

Species	Status and trend
Albacore	Northern stocks overutilized in 1980–1992, presently recovered but catches increasing; southern stocks uncertain
Bigeye tuna	Estimates vary from under- to fully exploited
Skipjack tuna	Underutilized
Yellowfin tuna	Estimates vary from lightly to fully exploited

Juvenile bigeye scad are targeted as they migrate into nearshore areas, forming an important component of the recreational fishery in Hawaii. The mackerel scad fishery is more localized, largely concentrated off the Kona coast of the island of Hawaii (Smith 1993). Both species undergo cyclic changes in abundance and seasonal migration patterns, resulting in wide variations in annual harvests. In Hawaii, 29% (184 metric tons) and 18% (113 metric tons) of the 1980–1990 mean annual commercial catch consisted of bigeye scad and mackerel scad, respectively (see unpublished data reports cited in Hamm [1993], and later reports in the same series). A small-boat fishing survey conducted on Oahu suggested that 115 metric tons of bigeye scad and 10 metric tons (combined recreational and commercial) of mackerel scad were landed in 1990 (Hamm and Lum 1992). Large seasonal runs of scad accounted for the majority of the annual fish catch in Guam in 1989 (76.5 metric tons; Myers 1993) and Samoa in 1991 (46 metric tons; Craig et al. 1993). There is no indication that bigeye scad are overutilized in Guam, Commonwealth of Northern Mariana Islands, and Samoa, but these fisheries should be closely monitored to better evaluate their status (Table 11). Life-history and population dynamics are not well known for either species, but state biologists conclude that bigeye scad and mackerel scad are fully utilized throughout the main Hawaiian Islands.

Bottom Fish

The pelagic armorhead fishery at the northern Hawaiian and Emperor seamounts exemplifies seamount fisheries vulnerable to overexploitation. This fishery peaked in 1972, with catches of more than 54 metric tons per hour, but dropped to 0.5% of that level by the early 1980's (National Oceanic and Atmospheric Administration 1993; Figs. 58 and 59). A 6-year moratorium on the fishery was declared in U.S. waters in 1986. Improved recruitment of armorhead was observed in 1990–1993, but increased catches resulting from this recruitment were short-lived, and populations of armorhead have not returned to their former levels (National Oceanic and Atmospheric Administration 1996).

Bottom fish taken by slope fisheries include deepwater snappers, groupers, and jacks (Moffitt 1993). About 90% of the commercial bottom fish catch is taken in Hawaii (National Oceanic and Atmospheric Administration 1996; Fig. 60). Most U.S. mid-Pacific bottom fish stocks, including those of the northwestern Hawaiian Islands, have remained stable since 1985 after increasing from low levels in earlier years (Fig. 60). In contrast, bottom fish landings have declined in the main Hawaiian Islands

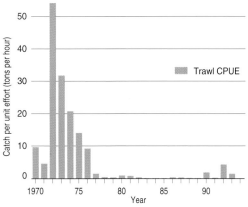

Fig. 58. Catch of pelagic armorheads per unit effort (CPUE) by Japanese trawl vessels on North Pacific seamounts, 1970–1994.

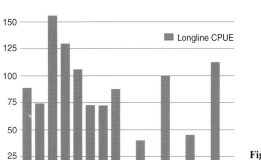

Fig. 59. Catch of pelagic armorhead on Southeast Hancock Seamount—catch per unit effort (CPUE) from bottom longlines of research cruises, 1985–1994.

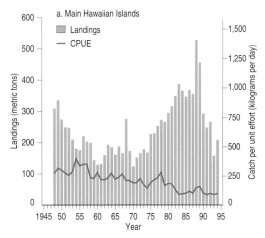

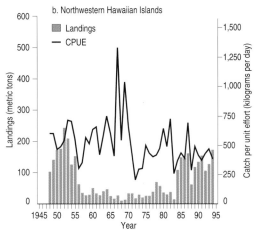

Fig. 60. Bottom fishes landings and catch per unit effort (CPUE) at the main Hawaiian Islands (a), and the northwestern Hawaiian Islands (b), 1948–1994.

since 1989 (Fig. 60), and stocks of four snappers and one grouper in the main Hawaiian Islands are thought to be at only 10%–20% of their original levels (National Oceanic and Atmospheric Administration 1993,1996; Table 12). Little information exists on the biology of slope resources in Guam, Commonwealth of the Northern Mariana Islands, or Samoa, making resource assessment difficult, but concern has been expressed about declining catches of slope fishes in Guam and Samoa (Craig et al. 1993; Myers 1993). Almost nothing is known about deepwater resources at the smaller, uninhabited U.S. islands.

Table 12. Western Pacific bottom fish and armorhead—productivity in metric tons and status of populations.

Species/area	Recent average yield (metric tons)	Current potential yield (metric tons)	Fishery utilization level	Stock level relative to long-term potential yield
Bottomfish				
Main Hawaiian Islands	249	111	Overutilized[a]	Below
Northwestern Hawaiian Islands	184	222	Underutilized	Near
American Samoa	18	34	Underutilized	Near
Guam	24	25	Near	Near
Commonwealth of the Northern Mariana Islands	17	78	Underutilized	Near
Pelagic armorhead	0	0		
Total	492	470		

[a]Overutilized for five species; other species underutilized.

Reef Animals

Monitoring of reef fisheries in the Pacific Islands is especially problematic because much of the harvest effort is by local peoples for subsistence, with more than 80%–90% of the catch on smaller islands utilized directly by extended family groups instead of entering the cash economy. Some measure of this can be obtained from fisheries for octopus, which have been monitored only in recent years. In Hawaii, statewide reports during 1980–1990 revealed an average yearly commercial octopus catch of 2.9 metric tons (Smith 1993), but an island-wide survey of Oahu during 1990–1993 estimated that the catch of octopus on that island alone was 24 metric tons, 10 times greater than the statewide commercial catch. During 1991–1992, an estimated 6–8 metric tons were taken in Kaneohe Bay, Oahu, alone (Everson 1995). This evidence indicates that the recreational or subsistence octopus catch far exceeds the amount taken in commercial fisheries. Few data are available to compare octopus catches in other islands, although a recently established program in American Samoa obtained a preliminary estimate of 6–12 metric tons of benthic octopus per year taken at Tutuila during 1991–1994. Octopus populations are not considered overexploited in any of the islands, however. Little of the recreational or subsistence fishery harvest in most islands is reported in data bases, making evaluation of the status and trends of the affected resources impossible.

Most local reef fisheries use various gear types and seek several species. In more heavily populated areas, high fishing pressure from multigear, multispecies fisheries results in overexploitation of rare species whose harvest would be otherwise unprofitable (Munro and Fakahau 1993). This situation occurs with rare species that are highly susceptible to fishing gear and with the more abundant target species. In such instances, rare species continue to be caught as fishermen harvest the numerous other fish that occur in biologically diverse island habitats. An example may be the emperors, which make up a large portion of the fisheries catch in American Samoa (Craig et al. 1993), Guam (Myers 1993), and the Commonwealth of the Northern Marianas. Potential overexploitation is indicated by the estimated decline of annual emperor catches in Samoa and the Marianas from about 27 metric tons in the mid-1980's to about 12 metric tons in the early 1990's (see unpublished data reports cited in Hamm [1993], and later reports in the same series). Other species susceptible to continued fishery overexploitation in multigear, multispecies fisheries are the larger groupers, and also the giant humphead or Napoleon wrasse, which have become rare or locally extirpated at many populated Pacific islands. The emperors, groupers, and wrasses are protogynous hermaphrodites. They assume a functional female condition first during development before changing to a functional male state. The groupers often aggregate for spawning. These life-history characteristics make them particularly susceptible to overfishing because the fish are often caught before they change sex, thereby reducing the number of males in the population, and because fishers can target spawning aggregations for harvesting (Wright 1993). Both of these harvesting impacts reduce the reproductive success of these fishes to levels below that necessary to sustain the fishes' populations.

Fishing methods that involve habitat damage represent a unique threat to the region's marine resources (Munro et al. 1987; Norse 1993). In the past, extremely damaging practices of fishing with explosives or poisons were prevalent in the Pacific Islands (Alcala and Gomez 1987; Eldredge 1987; United Nations Environment Programme/International Union for the Conservation of Nature and Natural Resources 1988). Although these methods are now almost universally outlawed, poaching remains a problem in remote regions. Less extreme habitat-destructive harvest methods are also of concern. For example, in recent years, feather-duster worms have been harvested for the aquarium industry (Fig. 61). More than 90,000 feather-duster worms were estimated to have been harvested from Kaneohe Bay during 1991 and 1992 (Everson 1995). Feather-duster worms burrow into live coral, rubble, and rock, often necessitating breaking apart the substrate to remove them. The potential effect on Kaneohe's reefs from this growing industry is a point of concern. In answer to this, the State of Hawaii enacted legislation in 1993 which makes it "unlawful to intentionally take, break or damage, with crowbar, chisel or any other implement, any rock or coral to which marine life is attached or affixed."

Nearshore Resources: Coral Reefs

Coral reefs are recognized as among the most productive and biologically diverse marine habitats (Fig. 62). They are also very susceptible to environmental degradation (United Nations Environment Programme/International Union for the Conservation of Nature and Natural Resources 1988; Maragos 1993). Although reefs were once thought to be extremely fragile, it is now believed that their ability to recover from natural disturbance contributes to their high diversity. Even so, human exploitation of and damage to coral habitats add to natural disturbances, slowing natural recovery (Wilkinson 1992). For this reason, corals are now protected from human harvest by the Convention on International Trade in Endangered Species of Wild Fauna and Flora. In general, practices destructive to corals and coral reefs are discouraged or illegal in U.S. Exclusive Economic Zones.

An example of human perturbation, habitat degradation, and recovery of a coral reef system can be found in the changes that occurred in Kaneohe Bay, Oahu, Hawaii (Hunter and Evans 1995). Kaneohe Bay's southern basin once contained some of Hawaii's largest pristine coral reefs. Extensive dredging occurred between 1939 and 1945 during construction of Kaneohe Marine Corps Air Station, which directly damaged reefs, while much of the dredge material was dumped into the bay, often burying coral. In later years, runoff resulting from clearing the surrounding hillsides for housing developments increased sedimentation. Beginning in 1951 and 1963, raw and secondary treated sewage were discharged directly into the bay from two sewage treatment plants. By the mid-1970's, nutrient enrichment from the pollution of Kaneohe Bay's southern basin caused a bloom of bubble algae, which grew in large mats, smothering coral and associated fauna. Sewage was diverted to deepwater outfalls outside the bay in 1977–1978, after the University of Hawaii's Institute of Marine Biology published several reports detailing the problem. Sewage diversion, combined with stricter control over sediment runoff during

Courtesy B. Mundy, National Marine Fisheries Service

Fig. 61. A feather-duster worm at 1-meter depth on a nearshore reef off Oahu, Hawaii.

Courtesy B. Mundy, National Marine Fisheries Service

Fig. 62. Agile damselfish over finger coral at 12-meter depth on Hawaii Island.

construction projects, gradually improved Kaneohe Bay's water quality. Renewed threats to the bay's habitats in the 1990's have come in the form of nonpoint-source pollution and water diversion (Hunter and Evans 1995).

Dredging, coastal development, and nonpoint-source pollution have damaged coral reefs on most U.S. islands (United Nations Environment Programme/International Union for the Conservation of Nature and Natural Resources 1988; Maragos 1993). Dredging and other military activities were extensive in the Pacific during World War II, but recovery of reefs has occurred on the smaller, uninhabited islands. Postwar population growth and urbanization have affected many of the larger islands. Oahu, Hawaii, is the most extreme example. Dredging of harbors and inland construction in the vicinity of Honolulu deposited sediment over reefs, while other reefs were filled to provide new land, including an airport runway. Attempts to stabilize beaches at Waikiki included adding seawalls and groins, which changed beach erosion and deposition patterns. Continued erosion prompted replacement of lost sand with material brought from elsewhere. The eroded sand, including imported material, moved offshore, smothering reefs seaward of Waikiki. Other shorelines on U.S. islands with heavy coastal development include the area from Tumon Bay through Apra Harbor on the west side of Guam, and the northwest shore of Saipan. Nonpoint-source pollution occurs not only in urbanized areas of Pacific Islands but also where extensive erosion occurs from inland land-clearing or where nutrients, herbicides, and pesticides from agricultural activities join freshwater runoff flowing into nearshore habitats.

Sea Turtles

Five sea turtle species occur at U.S. Pacific Islands (Márquez 1990; Eckert 1993), all protected by the Endangered Species Act (National Oceanic and Atmospheric Administration 1996; Table 13). Although there is one record of the threatened olive ridley turtle nesting in Hawaii, most of its reproductive colonies are in continental coastal areas, and olive ridleys are nearly unknown around oceanic islands. The loggerhead turtle has been recorded in the Mariana Islands, Samoa, and Hawaii, but it is uncommon off of Hawaii, being primarily associated with continents and boundary currents rather than islands. Loggerheads do not nest in the U.S. Pacific Islands. Loggerheads are listed as threatened under the U.S. Endangered Species Act

and as vulnerable by the International Union for the Conservation of Nature and Natural Resources. The leatherback is a pelagic sea turtle that probably occurs near all U.S. Pacific Islands—although it nests in none—with records from the Commonwealth of the Northern Mariana Islands, Guam, and Hawaii. Leatherbacks are listed as threatened by the U.S. Endangered Species Act and as endangered by the Convention on International Trade in Endangered Species of Wild Fauna and Flora and the International Union for the Conservation of Nature and Natural Resources. The hawksbill is present at all U.S. islands and nests at many. Hawksbills are listed as endangered by the U.S. Endangered Species Act, the International Union for the Conservation of Nature and Natural Resources, and the Convention on International Trade in Endangered Species of Wild Fauna and Flora. About 150 adult hawksbills were estimated in Hawaii during 1997, with only about 35 nesting females. The number of hawksbills in Samoa and Guam is unknown, but nesting has been observed at Rose Atoll and the Manua Islands. The status of leatherback, hawksbill, and loggerhead sea turtles throughout the Pacific is poorly known, but mortality of leatherbacks from fishery bycatch and continued exploitation of hawksbills for their shells in areas outside of the United States make them a special conservation concern (Eckert 1993; Wetherall et al. 1993; National Oceanic and Atmospheric Administration 1996).

The green turtle is the most abundant sea turtle in the region, probably present at all U.S. islands (Eckert 1993). Green turtles are listed as threatened in most areas by the U.S. Endangered Species Act and as endangered by the International Union for the Conservation of Nature and Natural Resources and the Convention on International Trade in Endangered Species of Wild Fauna and Flora. The historical Pacific green turtle population is estimated to have been about 10,000 (National Oceanic and Atmospheric Administration 1993). In 1997, about 1,000 individuals were estimated to nest in Hawaii, with more than 500 of these at French Frigate Shoals in the northwestern Hawaiian Islands, where more than 90% of their nesting in Hawaii occurs. An additional 100–300 green turtles are estimated to be nesting in American Samoa, primarily at Rose Atoll, but the numbers at Guam were unknown, with only sporadic nesting recorded (National Marine Fisheries Service, unpublished data). The population nesting at French Frigate Shoals is thought to be increasing. In the main Hawaiian Islands green turtles are under threat from lethal fibropapillomas (tumors) of unknown cause (Herbst 1994). In 1991, 31% of

Table 13. Status of sea turtles in U.S. waters.

Area/species	Historical level	Current level[a]	Trend	Status in United States
Atlantic region				
Loggerhead	Unknown	20,000–28,000[b]	Stable[c]	Threatened[d]
Green	Unknown	400–500[b]	Increasing	Threatened, endangered[e]
Kemp's ridley	40,000	700–800[f]	Stable[c]	Endangered
Leatherback	Unknown	Unknown	Unknown	Endangered
Hawksbill	Unknown	Unknown	Declining	Endangered
Pacific region				
Loggerhead	Unknown	Unknown	Stable	Threatened
Green	10,000	1,200[g]	Increasing[h]	Threatened
Olive ridley	Unknown	Unknown	Increasing	Threatened
Leatherback	Unknown	Unknown	Declining	Endangered
Hawksbill	Unknown	>150[i]	Unknown	Endangered

[a]Estimates of sea turtle abundance as indicated by number of female sea turtles nesting on U.S. beaches.
[b]Using 2.5 nests per female.
[c]Stable but critically low.
[d]Listed as threatened under the Endangered Species Act.
[e]Listed as endangered in Florida; threatened in the U.S. Atlantic and Pacific.
[f] Using 1.5 nests per female. Kemp's ridley turtles nest only on one Mexican beach.
[g]Historical level for Hawaii only. Estimated 1995 nesting female population is 1,000 in Hawaii; 100–300 in American Samoa; current level in Guam is unknown.
[h]Trend in Hawaii only, monitored at French Frigate Shoals; however, great concern exists over increasing frequency of fibropapilloma disease in all Hawaiian green turtles.
[i] Estimated total adult population in Hawaii; average number of female hawksbills nesting annually in Hawaii is about 35. Current abundance in Guam and American Samoa is unknown.

Kaneohe Bay's green turtles were afflicted, and in 1989–1990, 77%–85% of the stranded turtles on Maui had fibropapillomas. Research to determine the cause of this disease is a high priority for green turtle conservation because the Hawaiian population was among the few increasing stocks before the disease became prevalent (Fig. 63).

Marine Birds

Pacific island seabird populations are only a remnant of what they were before the arrival of humans and introduced mammalian predators (Steadman 1995). Different seabird species are vulnerable to different introduced predators, depending on their size and nesting habits (Croxall et al. 1984; Vermeer et al. 1993). Tree-nesting species are relatively immune to predation by most nonclimbing animals (Flint 1997); white terns may even be observed nesting in urban areas such as Honolulu.

Small, ground-nesting birds (Fig. 64) are particularly defenseless against the entire range of introduced predators, from rats to dogs. Of 53 extant Pacific island seabird species (Flint 1997), 34 are vulnerable to the rats introduced during the migrations of indigenous peoples and European colonization (Harrison et al. 1984; Flint 1997). On most of the Pacific Islands, small seabird species vulnerable to predation by rats are now extinct (Garnett 1984). Larger ground-nesting birds are able to defend nests and young from rats, and the very largest can sometimes fend off larger predators like domestic cats or mongoose. Cats, introduced by Europeans, have been a major problem for seabirds that can survive rat predation. Recent cat elimination programs have been successful on some islands, giving populations of larger seabirds a chance for survival. Populations of larger seabirds on Howland and Jarvis islands seem to be recovering following this effort by the U.S. Fish and Wildlife Service (Flint 1997). Cats were eradicated from Baker Island in the 1960's, and all seabird populations except the small, burrow-nesting shearwaters, petrels, and storm-petrels have recovered there. Three burrow-nesting species (that is, procellariiform or tubenoses) have been seen nesting at Jarvis, giving hope that all seabird populations will eventually recover there. An experimental rat control program was conducted in 1990 at Rose Atoll, the major seabird nesting site of American Samoa, with apparent success; rats were not found there by a U.S. Fish and Wildlife Service survey in 1993 and in subsequent surveys.

Although relatively immune from predation by introduced species, albatross populations have been diminished by hunting and other factors. Black-footed albatrosses are now restricted

Courtesy B. Mundy, National Marine Fisheries Service

Fig. 63. A green turtle afflicted with a fibropapilloma disease of unknown cause.

Courtesy E. Kridler, U.S. Fish and Wildlife Service

Fig. 64. A Bonin petrel in its nesting burrow in Hawaii.

to breeding colonies in the Northwestern Hawaiian Islands and Japan's Izu and Ryukyu islands (Croxall et al. 1984; Fig. 65). Breeding colonies are extinct at Wake, Johnston, and the northern Marianas. Laysan albatrosses exist in greater abundance (Fig. 66). Albatrosses can constitute a significant bycatch of epipelagic fisheries (McDermond and Morgan 1993). For example, an estimated 17,500 Laysan albatrosses were taken as bycatch of North Pacific driftnet fisheries in 1990, reducing the population's recovery rate by 0.4%–1.6% per year (Gould and Hobbs 1993). The estimated annual bycatch of black-footed albatrosses was 4,400 birds, or 2.2% of the population. The problem of high rates of albatross mortality from the bycatch in driftnets is presumed to be eliminated now that driftnet fisheries are outlawed, but the bycatch of albatross by pelagic longline fisheries remains a conservation concern.

Fig. 65. Black-footed albatross at Tern Island, French Frigate Shoals, Northwest Hawaiian Island National Wildlife Refuge.   Courtesy E. Flint, U.S. Fish and Wildlife Service

Courtesy E. Flint, U.S. Fish and Wildlife Service

Fig. 66. Adult Laysan albatross at Lisianski Island in the Northwest Hawaiian Islands National Wildlife Refuge.

The status and trends of tropical Pacific island seabird populations are best known for the Hawaiian Islands (Table 14). Species nesting in the northwestern Hawaiian Islands have often suffered declines on one or more of the islands, but most generally have stable populations or populations that have not been adequately assessed to determine trends (Harrison et al. 1984; Flint 1997). Several smaller seabird species in Hawaii are a source of special conservation concern: the dark-rumped petrel, Bonin petrel, Newell's Townsend's shearwater, band-rumped storm-petrel, Bulwer's petrel, Tristram's storm-petrel, and Christmas shearwater (Harrison et al. 1984; Flint 1997; U.S. Fish and Wildlife Service Pacific Island Ecoregion, unpublished data).

Less is known about seabird populations on other U.S. Pacific Islands. At least three seabird species once present on Johnston Island no longer nest there, but all 15 species that survived have stable or increasing populations (Harrison et al. 1984; Flint 1997). Rats are established on Palmyra, and dogs have been introduced at various times. Small, ground-nesting shearwaters and petrels were absent in a 1992 survey of Palmyra, perhaps as a result of rat predation (U.S. Fish and Wildlife Service Pacific Islands Ecoregion, unpublished data). Most seabird nesting in the northern Marianas occurs on Naftan Rock, Gaguan, Maug, and Uracus, all wildlife refuges, and on Farallon de Medinilla, a bombing range (United Nations Environment Programme/International Union for the Conservation of Nature and Natural Resources 1988; Reichel 1991). Introduced mammals threaten seabirds at all of these. On Guam, the introduced brown tree snake threatens populations of white terns, brown noddies, and white-tailed tropicbirds (Reichel 1991). This snake has extirpated most of the terrestrial birds in Guam not already driven to

Table 14. Status and trends of selected seabird species in the western Pacific with emphasis on the Hawaiian Islands.

Species	Status and trend
Audubon's shearwater	Status and trends unknown in most of range. Recolonizing Jarvis Island following U.S. Fish and Wildlife Service predator eradication programs
Black-footed albatross	Most nesting occurs in the northwestern Hawaiian Islands with about 67,000 pairs. Populations declining in last decade
Blue-gray ternlet	Population trends unknown but probably stable in areas unaffected by human disturbance or mammalian predators, such as Nihoa and Necker islands in the northwestern Hawaiian Islands
Bonin petrel	Hawaiian populations greatly reduced. Increasing at Laysan and Lisianski islands. Declining at Midway Island, from 500,000 pairs (1984) to 7,500 pairs; successful nesting occurs only where U.S. Fish and Wildlife Service controls rats
Christmas shearwater	About 14,000 individuals in Hawaii. Declined at Laysan and Lisianski islands but populations stable at reduced levels. Declining at Midway Island. Status and trends unknown in other archipelagos
Gray-backed tern	Populations stable at Laysan and Lisianski islands but reduced from historical levels. Declining at Midway and Kure islands because of rat predation
Harcourt's storm-petrel	Status and trends unknown; proposed for listing under the Endangered Species Act. Fledglings found on Kauai Island suggest breeding occurs there, but at risk from mammalian predators
Hawaiian dark-rumped petrel	Endangered; subspecies endemic to Hawaii with most of the approximately 400–600 individuals nesting at Haleakala National Park, Maui. Nesting colonies of unknown sizes recently discovered at high elevations on Kauai. Trends unknown
Laysan albatross	Most nesting occurs in the northwestern Hawaiian Islands with about 616,000 pairs. Populations stable
Newell's Townsend's shearwater	Threatened; Hawaiian endemic. Most nesting colonies are on Kauai Island, which has about 4,000–6,000 individuals. Trends unknown
Short-tailed albatross	Endangered. Number of sightings is increasing
Sooty storm-petrel	Status and trends unknown. Important breeding colony on Nihoa Island in northwestern Hawaiian Islands
Sooty tern	Populations stable in northwestern Hawaiian Islands except declining at Midway and Kure islands. Population trends unknown in main Hawaiian Islands
Wedge-tailed shearwater	Status unknown. Trends variable. Declining where rats occur (for example, on Midway Island). Recolonizing Jarvis Island following predator eradication

extinction by rats and is clearly a major conservation problem (Fritts and Rodda 1995).

Only six of the eight island countries with healthy seabird communities have wildlife reserves, and only three of these, including the United States, have eradication programs for nonindigenous predators (Flint 1997). The small number of reserves and predator control programs gives the most protected seabird nesting islands extreme global importance.

Notable among pristine Pacific Island nesting sites are the northwestern Hawaiian Islands, Howland, Baker, and Jarvis islands, and Rose Atoll.

Marine Mammals

The western Pacific Ocean region supports 26 species of marine mammals (Jefferson et al. 1993; Table 3). Three are recorded from the

Genetic Diversity of Central Pacific Marine Mammals

Recent studies of genetic diversity in marine mammals have investigated patterns of paternity and kinship, the effects of near-extinctions on reductions of genetic variability, migration patterns of populations, and the differentiation of populations applicable to the management of stocks (Hoelzel 1993,1994). To date, paternity and kinship patterns have not been investigated in western Pacific Island marine mammal populations (Hoelzel 1994). Reductions of genetic variability were created by the near-extinctions in the late 19th and early 20th century of the Hawaiian monk seal (Figure), North Pacific right whale, and northern elephant seal, creating historical genetic bottlenecks for these species (Hoelzel 1993; National Oceanic and Atmospheric Administration 1996). Genetic investigation of cetacean migration patterns in the North Pacific established that humpback whales from the western and Hawaiian breeding populations disperse to several discrete feeding grounds around the Pacific (Baker et al. 1993). In contrast, minke whales from distinct breeding populations mingle in common feeding areas (Hoelzel 1994).

Most investigations of genetic diversity in western Pacific cetaceans have concentrated on identifying the degree of divergence of geographic or morphological populations. Divergent genetic populations are considered stocks that can be used to evaluate the status and trends of these species (Dizon et al. 1992). Less is known about this differentiation in the western Pacific than in most other areas. For example, no information exists on genetic variability of 12 species of small cetaceans found in the U.S. Pacific Islands. Even the species identity of the tropical bottlenose whale is unknown (Jefferson et al. 1993).

Genetic investigations have clarified the status of controversial cetacean populations. Genetic divergence of minke whale populations in the North Atlantic, North Pacific, and Antarctic is as great as that found between other whale species, suggesting that these populations deserve recognition as at least subspecies (Wada and Numachi 1991; Dizon et al. 1992). The Bryde's whale, which may be the most common great whale of the tropical Pacific Islands, is differentiated into a larger offshore form and smaller inshore form (Wada and Numachi 1991). Studies of mitochondrial DNA suggest that the large form is a distinct species. Morphologically distinct short- and long-nosed forms of the common dolphin also show a high level of genetic distinctiveness (Rosel et al. 1994).

Many of the smaller cetaceans found near U.S. Pacific Islands have more complex population structures. Killer whales are highly mobile with little morphological variability. Although their stock structure is not well understood, their social behavior and preliminary genetic data suggest considerable genetic variability among pods in a single region (Hoelzel 1993,1994). In contrast, eastern Pacific spinner dolphins have morphologically distinctive forms that are not genetically divergent, compared to a population in the far western Indo-Pacific (Dizon et al. 1991). Genetic intergradation and a zone of hybridization in eastern Pacific spinner dolphin stocks have been suggested as causes of this morphological differentiation, in the absence of genetic divergence (Dizon et al. 1991, 1994).

See end of chapter for references

Author

Bruce C. Mundy
National Marine Fisheries Service
National Oceanic and Atmospheric
Administration
Honolulu Laboratory
2570 Dole Street
Honolulu, Hawaii 96822

Figure. A female Hawaiian monk seal with her 4-week old pup at Laysan Island, northwest Hawaiian Islands.

Courtesy B. Becker, National Marine Fisheries Service

Hawaiian Monk Seal

Restricted almost entirely to the north-western Hawaiian Islands, the Hawaiian monk seal population declined by 50% between the late 1950's and the late 1970's, to 1,406 individuals in 1993 and perhaps 1,300 individuals in 1995 (National Oceanic and Atmospheric Administration 1996; Fig. 1). Progress in managing the recovery of this endemic species varies among the six main breeding islands. The status of the species as a whole is indicated by the annual mean beach counts of seals and the number of pups born (Fig. 2; Table). The total number of births has been highly variable since the early 1980's, increasing between 1985 and 1988, declining 35% in 1990, and increasing during 1991–1992. However, mean counts of seals (excluding pups) at the five main breeding sites have fallen steadily from a peak of 575 in 1986 to 378 in 1993, largely as a result of declines at French Frigate Shoals, the largest rookery (Ragen and Lavigne 1997). In addition, high mortality of females at Lisianski and Laysan islands is due to male mobbing behavior, where multiple males simultaneously attempt to mate with a single female. Efforts were undertaken to reverse this trend through rehabilitation of undernourished juvenile females at French Frigate Shoals and elimination of mobbing-related mortality.

Recent trends in monk seal survival suggest a link between population size and large-scale natural perturbations in ecosystem productivity (Polovina et al. 1994; Fig. 3). Declines in Hawaiian monk seal, seabird (Fig. 4), and lobster reproductive success in the late 1980's appear to have been linked to a large-scale climatic event. From the mid-1970's to late 1980's, the central North Pacific experienced increased vertical mixing, with a deepening of the wind-stirred surface layer into nutrient-rich lower waters and probable increased injection of nutrients into the upper ocean. Resulting increased productivity likely provided a larger food base for fish, lobsters, and eventually seals and seabirds in the region.

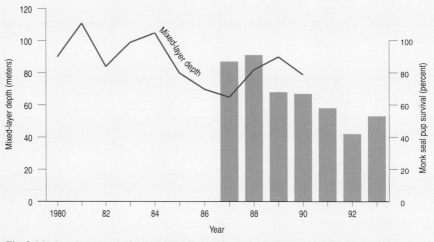

Fig. 1. Hawaiian monk seal and pup at Kure Atoll, Northwest Hawaiian Islands.

Courtesy T. Gerrodette, National Marine Fisheries Service

Table. Status and trends of the Hawaiian monk seal populations at their main breeding islands (Ragen and Lavigne 1997; National Marine Fisheries Service, unpublished data).

Island	Status and trends	Major conservation concerns
Kure	Increasing; births up to 18 in 1997 from 1 in 1986	Past management program for rehabilitating undernourished pups resulted in continued population growth
Midway	Increasing; colony disrupted in 1960's but now recolonized	Reestablishment of this colony because of its significant reproductive potential, disturbance, juvenile survival
Pearl and Hermes	Increasing	None immediate—the only island where numbers are increasing without management intervention
Lisianski	Stable or decreasing	Mortality of females from mobbing by males, entanglement
Laysan	Stable or decreasing	Mortality of females from mobbing by males, juvenile survival
French Frigate Shoals	Decreasing to <55% since 1989	Reduced environmental carrying capacity, male aggression, juvenile survival, shark predation
Total population	About 1,300–1,400 seals, decreasing by about 5% per year	All of above with emphasis on reduced carrying capacity from changing oceanographic conditions and juvenile survival

Fig. 3. Monk seal pup survival (percent) and mixed-layer depth (in meters). Decadal declines in mixed-layer depth may have caused a reduction in primary production and a concomitant subsequent reduction in monk seal prey items that led to a decline in pup survival (Polovina et al. 1994; data from J. Polovina and T. Ragan, National Marine Fisheries Service, Honolulu).

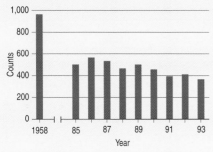

Fig. 2. Hawaiian monk seal mean beach counts.

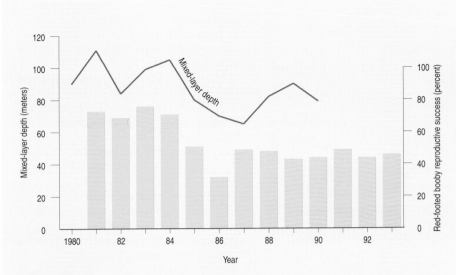

The declines in seabird and monk seal reproduction and survival in the late 1980's and early 1990's may have been caused by a return to less productive oceanographic conditions. The effects of climatic change on the large marine ecosystems of the oceanic Pacific are, however, very poorly understood and thus cannot be definitively linked to changes in animal populations.

See end of chapter for references

Fig. 4. Red-footed booby reproductive success and mixed-layer depth (in meters). Decadal decline in mixed-layer depth may have caused a reduction of primary production and subsequent reduction of prey items for seabirds that in turn led to a decline in seabird reproductive success (Polovina et al. 1994; data from J. Polovina, National Marine Fisheries Service, Honolulu, and E. Flint, U.S. Fish and Wildlife Service, Hawaiian and Pacific Islands National Wildlife Refuge Complex).

Author

Bruce C. Mundy
National Marine Fisheries Service
National Oceanic and Atmospheric Administration
Honolulu Laboratory
2570 Dole Street
Honolulu, Hawaii 96822

region only as strays, and two are restricted to the subarctic and transitional waters of the North Pacific, occurring only within the region at the northernmost Hawaiian Ridge seamounts and islands. Bottlenose dolphin, otherwise circumglobal, are known only from Hawaii on the Pacific tectonic plate; this distribution pattern has been seen for other organisms (Springer 1982) and is probably not an artifact of observation. In contrast, whitebelly spinner dolphin is absent from the central North Pacific, including Hawaii, but probably occurs at the other islands. The tropical bottlenose whale is known only from the equatorial Pacific. The remaining cetaceans probably occur at all U.S. islands, although records are uneven and many have migration patterns that concentrate them in particular regions seasonally (Stone et al. 1997).

There are few assessments of population sizes (Klinowska 1991) or estimates of trends in population changes (National Oceanic and Atmospheric Administration 1993, 1996) for marine mammals at the U.S. Pacific Islands. Regional population information is totally absent for 14 species. Four dolphin species have large, stable populations in the eastern Pacific, but population estimates are lacking at U.S. islands, except for those of spinner dolphins in Hawaii. Estimates of population size for the northern right-whale dolphin vary widely, creating great uncertainty in the assessment of mortality and subsequent recovery from the high-seas drift net fisheries (Hobbs and Jones 1993).

According to the criteria provided in the 1994 amendments to the Marine Mammal Protection Act, there are four strategic marine mammal stocks in Hawaiian waters: blue whales, fin whales, and sperm whales—all of which are listed as endangered under the U.S. Endangered Species Act—and the Hawaiian monk seal, which is both endangered and listed as depleted under the Marine Mammal Protection Act. The Hawaiian Islands are also visited each winter by endangered North Pacific humpback whales, which breed throughout the islands but concentrate near Maui, Kahoolawe, Lanai, and Molokai. In addition, endangered sei whales occur throughout the Pacific Island region. With the exception of the Hawaiian monk seal, whose decline has been discussed in the accompanying highlight box, the population trends of these stocks are unknown (Klinowska 1991; National Oceanic and Atmospheric Administration 1996). Nothing is known of the status of marine mammal stocks at U.S. Pacific Islands other than Hawaii.

Humpback whales in the eastern North Pacific Ocean use the insular shelves of Hawaii during the winter calving season after migrating from the temperate and subarctic waters of northern California and Alaska where they feed (Klinowska 1991; National Oceanic and Atmospheric Administration 1993; Fig. 67). A western population of humpback whales winters as far south as the Commonwealth of the Northern Mariana Islands, although most of this population remains in the Ryukyu and

Courtesy L. Consiglieri, National Marine Fisheries Service

Fig. 67. Humpback whale breeching, western Maui.

Ogasawara islands of southern Japan. Previously, North Pacific humpback whales were estimated to be at 13% of their pre-whaling (around 1850) population of 15,000 (National Oceanic and Atmospheric Administration 1996). Photographic identification of individual whales and other evidence suggest that the total population now exceeds 3,000 (National Oceanic and Atmospheric Administration 1996). There is too little information on humpback whales to judge whether populations are increasing and whether human activities may be affecting their recovery. Some female humpbacks with calves apparently abandoned traditional nearshore nursery habitats near Maui, Hawaii, possibly in response to repeated human contact. Females with calves appear to have returned to these areas after restrictions on "thrill-craft" operations were implemented. In 1993, a National Marine Sanctuary for humpback whales was designated near Maui.

National Status and Trends

Estuarine Resources

Many of the nation's commercial and recreational fisheries depend on estuaries. For example, the species making up the top four fisheries in the Gulf of Mexico (shrimp, menhaden, oyster, and blue crab) use estuaries extensively, and the quantity, quality, and timing of freshwater inflow to these areas can be particularly important to the within-year and between-year success of these fisheries (Stickney 1984; Turek et al. 1987). Several of the most valuable south Atlantic fisheries (for example, shrimp, blue crab, hard clam, and summer flounder) are also estuarine-dependent. In the North Atlantic, soft-shell clam, Atlantic salmon, Atlantic herring, winter flounder, and other fishery species require or prefer estuarine areas at some time of

the year. Although the relative area of estuarine habitat in the Pacific region is small, three of the top five most valuable fisheries (salmon, Dungeness crab, and oyster) are all estuarine-dependent.

Mollusks are of special concern in estuaries because they are sessile. In 1990, more than 69,000 square kilometers of estuarine waters nationwide were classified as shellfish harvest areas (National Oceanic and Atmospheric Administration 1991b). However, many areas were occasionally restricted for harvest because of public health threats from bacterial or viral contamination. Urban stormwater runoff, sewage treatment plant effluent, agricultural runoff, and increased boating activity are the primary causes of harvest restrictions (National Oceanic and Atmospheric Administration 1990a, National Oceanic and Atmospheric Administration, unpublished data).

Humans place a high value on estuarine areas for living, working, and recreating. Estuaries provide cooling waters for industry and energy production and sites for aquaculture, accommodate the needs of large ships and tanker traffic, buffer coastal areas against storm and wave damage, provide wetlands and bottom habitat, supply space for coastal development, and accumulate pollutants from the rivers and streams entering coastal waters. Estuarine areas are among the most densely populated and heavily used areas in the United States and are home to an estimated 45% of the country's human population (Culliton et al. 1990). As human populations grow, demands for increased use of estuarine resources are expected to continue.

Coral Reef Ecosystems

Coral reef ecosystems can be classified in two broad categories: pristine coral reefs and coral reefs at risk. Pristine coral reef ecosystems are those in remote locations with little or no human threat to ecosystem health. By definition (and with minor localized exceptions) the status of these ecosystems is good and the trend in health is steady. Areas under U.S. jurisdiction with pristine coral reef ecosystems include the Flower Garden Banks in the Gulf of Mexico, and, in the Pacific Ocean, the uninhabited northwest Hawaiian Islands, Wake Island, the Northern Mariana Islands (except Saipan), Palmyra Atoll and Kingman Reef, Howland Island, Baker Island, and Jarvis Island.

Coral reef ecosystems at risk are located near human population centers, with some or all of the reefs experiencing local anthropogenic stress. Some important sources of stress include nutrient enrichment from sewage and agriculture, overfishing, and stress from high sedimentation caused by deforestation, agriculture, vessel traffic, and coastal runoff. The status of many coral reefs within these areas is poor, and the trends in their health are declining. However, because of the lack of comprehensive understanding of these ecosystems, it is difficult to make a statement on their status and trends on an islandwide or even reef-tract basis. Coral reef ecosystems at risk within U.S. jurisdiction include the Florida reef tract (Puerto Rico and the U.S. Virgin Islands in the western Atlantic and Caribbean) and, in the Pacific Ocean, the inhabited parts of the main Hawaiian Islands, Johnston Atoll, Guam, Saipan (Northern Mariana Islands), and American Samoa.

Time is running out for many of the coral reef ecosystems that are at risk. In a social climate of increasing human population and declining budgets for environmental issues, effects of stresses to the ecosystems are compounding faster than cases can be made for protective measures. Swift legislative efforts and public works programs to reduce nutrients, sediments, and overfishing may be the only way to save many of these areas for the benefit and enjoyment of present and future generations. Immediate remedies in these areas are therefore of critical importance.

Deep-Sea Resources

Little information is available to assess the status and trends of deep-sea animals. General data suggest that deep-sea animals are long-lived and slow growing, except in the case of animals associated with hydrothermal vents and seeps. In such areas, where temperatures are significantly higher than normal at similar depths, communities of organisms have evolved that are specifically adapted to life based on bacteria that use sulfur from the vents as energy. The animals (for example, tube worms, mussels, clams, crabs, and shrimps) seem to have very high growth rates made possible by high temperatures and locally abundant food. In fact, such growth is a necessity because vents may last only for a few months, during which time animals must colonize the area, mature, and reproduce if the species are to endure. Jannasch (1995) reported that there are still about 50,000 square kilometers of unexplored submarine tectonic spreading zones girdling the globe, where several thousand hydrothermal vent sites are predicted.

Although it is difficult to reach, the deep sea can provide sources for minerals, food, and energy. Exploitable resources of all three are known to exist, although at present the cost of exploiting them is generally greater than their value. Hydrothermal vents are being considered as mineral sources with implications for the unique communities they support. Deep-sea animals seem unlikely to form the basis for viable fisheries, and because of their slow growth rate they are expected to be highly vulnerable to overfishing.

One potential but unexploited resource of the deep sea may be marine microorganisms. Most antibiotics of microbial origin come from one terrestrial taxonomic group, the order Actinomycetales. According to Fenical and Jensen (1993), the rate of discovery of new antibiotics from terrestrial actinomycetes is declining, meaning that new sources must be explored. Although terrestrial organisms exhibit great species diversity, marine organisms represent greater diversity based on evolutionary origin. Only a superficial examination of their potential as sources of bioactive agents has been made. Bacteria and microalgae, including the dinoflagellates that occupy a unique position between the plants and the animals, are showing extraordinary promise. Fenical and Jensen (1993) suggest that the diverse marine microorganisms "must be considered a major, but largely undeveloped (and undescribed), biomedical resource."

Fisheries Resources

Fisheries resources are those taken for their commercial and recreational value. Their status has been comprehensively reported and periodically updated—by species and by major ecological group—by the National Marine Fisheries Service in the federal report *Our Living Oceans* (National Oceanic and Atmospheric Administration 1991c, 1992, 1993, and 1996). According to the 1996 report, the long-term potential yield of all fishery resources fished by the United States within its 200-nautical mile Exclusive Economic Zone was estimated at 8.14 million metric tons (Table 15). This yield represents about 7% of the world's yield potential of 120 million metric tons of marine and freshwater fisheries production.

The total recent average annual catch by the United States during 1992–1994 was 5 million metric tons, or 62% of the long-term potential yield. The additional 38% potential yield was

Region	U.S. long-term potential yield[a]	U.S. recent average yield
Northeast	844,810	449,730
Southeast	1,474,340	1,168,530
Alaska	4,423,670	2,733,300
Pacific coast	1,116,210	462,760
Western Pacific	283,340	242,490
Total	**8,142,370**	**5,056,810**

Table 15. Productivity of fisheries resources of the United States, 1992–1994 (National Oceanic and Atmospheric Administration 1996).

[a]Compiled from Table 2 from *Our Living Oceans*, 1995 (National Oceanic and Atmospheric Administration 1996).

not realized because some of the stocks were underutilized while others have been overutilized and are currently no longer producing at their full long-term potential. By region, the percentage distribution of recent average catches was 9% Northeast, 23% Southeast, 54% Alaska, 9% Pacific coast, and 5% western Pacific.

The Northeast region's finfish and invertebrate resources have a combined long-term potential yield of 844,810 million metric tons, or 10% of the U.S. long-term potential yield. Recent annual catch totaled about 449,730 metric tons, or about 53% of the region's long-term potential yield because of significant overuse of some 18 stocks, principally groundfish species (Northeast Fisheries Science Center 1993). The long-term potential yield for the Southeast region is about 1.47 million metric tons (18% of the U.S. long-term potential yield); recent catches have run about 1.17 million metric tons (National Oceanic and Atmospheric Administration 1996). The Alaska region dominates the fisheries resources that could be obtained in the long term for the United States (long-term potential yield of 4.42 million metric tons). Those resources are generally healthy, with current potential yield only 9% below long-term potential yield. Recent average catches have been steady at about 2.73 million metric tons, or 62% of Alaska's long-term potential yield. The Pacific coast fisheries resources have a combined long-term potential yield of 1.12 million metric tons (14% of the U.S. long-term potential yield). Recent catches were only 462,760 metric tons, or 41% of long-term potential yield, due to low abundance of coastal pelagic fishes. Most of the stocks are fully utilized or overutilized. The western Pacific fisheries resources include highly migratory pelagic fishes that migrate to and from other U.S. regions and to other countries in the Pacific. For the purpose of yield comparison, the highly migratory pelagic fishes are apportioned by the share that would accrue to the United States only. Thus, the long-term potential yield of fisheries resources in the western Pacific totals only 283,340 metric tons, or 3% of U.S. long-term potential yield. If all the available yields from this group are added, long-term potential yield would be almost seven times higher (or 2.04 million metric tons).

A rough estimate of the recreational marine finfish catch for the Atlantic and Gulf of Mexico coasts is about 300 million fish per year. On the Pacific coast, the recreational catch is at least 55 million fish, 25% of which are salmon. All recreational catch estimates are uncertain, since they are determined from random surveys of anglers.

It is difficult to assess the status of all nearshore species around the entire U.S. coast because they come under varied jurisdiction and data collection regimes. No firm estimates exist for long-term potential yield or current potential yield. Management authority is typically a regional, state, or local responsibility because most fisheries occur within the 3-mile interior boundary to the federally controlled Exclusive Economic Zone. Generally, Atlantic oysters, hard clams, softshell clams, bay scallops, and abalones are overutilized, at least in part of their ranges. Fully utilized resources include Pacific shrimps and clams, Dungeness crab, blue crab, and calico scallop. The 1992–1994 averaged catch has been conservatively estimated at 284,000 metric tons.

Marine Mammals

Approximately 163 stocks of at least 62 species of marine mammals are found within waters of U.S. jurisdiction (Table 3). The U.S. Fish and Wildlife Service has authority for stocks of North Pacific walrus, Alaska polar bear, West Indian manatee, and Alaska and California sea otters, and the National Marine Fisheries Service is responsible for the remaining cetaceans and pinnipeds (155 stocks, including 10 eastern tropical Pacific dolphins). The 1994 Amendments to the Marine Mammal Protection Act identified strategic stocks as those that are listed as endangered or threatened under the U.S. Endangered Species Act or that are declining and likely to be listed in the foreseeable future, those designated as depleted under the Marine Mammal Protection Act (that is, below the optimal sustainable population level), and those for which human-caused mortality exceeds the estimated replacement yield. Of the 153 marine mammal stocks managed under Section 117 of the Marine Mammal Protection Act (using 1995 totals), 54 are classified as strategic. These include 2 stocks that are depleted under the Marine Mammal Protection Act, 4 that are listed as threatened and 24 listed as endangered under the Endangered Species Act, 16 for which the total annual mortality exceeds Potential Biological Removal, and 6 stocks for which information on stock status or fisheries-related mortality is uncertain. In addition, 2 of 10 stocks of eastern tropical Pacific dolphins managed under Section 104(h) of the Marine Mammal Protection Act are listed as depleted under the Marine Mammal Protection Act.

Of the total 163 marine mammal stocks in U.S. waters, there is sufficient long-term population information to describe trends for only 55 stocks (33%); the status of the remaining 108 stocks (66%) is unknown. Of those for which we have information, 24 (15%) are known to be increasing, 8 (5%) are declining, and 23 (14%) are believed to be stable. Once implemented by the National Marine Fisheries Service and the U.S. Fish and Wildlife Service, the new management regime established by the 1994 Marine Mammal Protection Act amendments will contribute to the long-term data base needed to detect and evaluate trends for all marine mammal stocks.

National Issues and Threats

Overfishing

Overfishing is recognized as a potential threat to living marine resources. Three historically important groundfish species (cod, haddock, and yellowtail flounder) on Georges Bank off New England are presently among the most overfished stocks in U.S. waters. Catches were projected to be only 23% of their combined long-term potential yield of 98,000 metric tons in 1994 and even lower in 1995. Haddock and yellowtail flounder are classified as collapsed by virtue of their current low abundance (biomass of spawners is about 10% of the mid-1970's levels)—due to prolonged excessive fishing pressure—and by catches less than 25% of long-term potential yield. The cod stock was in imminent danger of collapse in 1994, but drastic management measures reduced the fishing mortality rate by 83% and improved spawning stock biomass by 48% from 1994 to 1996. The cod stock, however, is still considered overexploited and at low population levels.

All five species of Pacific salmon (chinook, coho, sockeye, pink, and chum), which begin their lives in the rivers and streams of Washington, Oregon, and California, are considered overfished. However, the main cause for their decline appears to be related to freshwater habitat alterations, such as water diversion and river–stream blockage by hydroelectric dams, which cause severe restrictions on upstream (adult spawning) and downstream (juvenile migration to the ocean) movements. Intense fishing pressure from competing user groups is an additional problem. Commercial and recreational catches of these five species have fluctuated widely and have averaged in total about 75% of the long-term potential yield (11.8 million fish per year) in the last several years.

In the Gulf of Mexico, king mackerel was severely depleted because of excessively high catches in the late 1970's and early 1980's. The population supports catches about 28% of long-term potential yield. Red snapper, traditionally the most important reef fish in the gulf, is taken mainly as incidental catch in the shrimp fishery and currently supports catches of only about 15% of long-term potential yield. Red drum was subjected to intense fishing in the 1980's in response to consumer demands for the Cajun dish blackened redfish and currently provides catches at about 35% of the long-term potential.

Examples of many other overfished stocks can be found throughout the country. Many are disproportionately affected by fishing because of their low numbers in relation to more abundant target species. In recent years there is growing awareness of responsibility and federal actions needed to mitigate overfished situations. To prevent overfishing, all federal fishery management plans require that numerical overfishing levels be defined and implemented according to sound population dynamics principles derived for each species' biological situation. Restrictions on catches are normally regulated in order to rebuild overfished stocks. Despite a vigilant watch for overfishing, resources do decline—some naturally, some through habitat change, and some through excessive fishing efforts.

Bycatch

The incidental take of nontarget species in fishing operations reflects the situation that species do not live in pure, discrete, exploitable patches but as members of intermixing communities. The groundfish fisheries have notoriously visible bycatch problems. These fisheries, whether conducted with trawl gear, longlines, or pot gear, catch and discard large volumes of animals that are of the wrong size, species, maturity stage, or other distinguishing factor. For instance, as much as 80% of the catch of rock sole in the Bering Sea may be discarded because the individual fish are not the valuable roe-bearing females. According to a comprehensive report on bycatch amounts worldwide, Alverson et al. (1994) estimated that 20%–40% of groundfish caught may be discarded routinely in fisheries of the United States and other parts of the world. Bycatch in these fisheries may be a serious threat to species already low in abundance.

Mitigation of bycatch problems is complex. Although the problem is partly biological and technological, it is also social. Whereas some technological innovations can work to reduce bycatch, such as the use of turtle exclusion devices to minimize effects on sea turtles in shrimp fisheries, most problems of bycatch are not as simple to resolve. In most cases, the catching of highly abundant species will always threaten less abundant species; thus the catch of the desired species may have to be severely restricted. Bycatch levels and control measures have occupied the attention of most fisheries management actions of the eight regional fishery management councils charged with managing the nation's living marine resources. Even when apparent solutions are found, the dynamics and abundances of the species change in space and time, shifting the character of the problems and requiring continual adjustments to carry out their solutions.

Habitat Alterations

Habitat degradation and loss affect mostly inshore and estuarine ecosystems. The primary threats are wetland destruction, alteration of freshwater flows, toxic chemicals, and nutrient overenrichment. Alterations to the freshwater input through damming and diversions of major rivers have affected coastal systems adapted to seasonal discharges of fresh water. Loss of aquatic plant-based habitats (for example, wetlands, eelgrass, and kelp beds) resulting from development, such as for marinas and docking facilities, adversely affects a variety of food webs that are important to adults and juveniles of several marine and anadromous species. Dredging and dredge disposal in estuaries and bays also cause significant habitat destruction. Marine ecosystems are damaged by habitat loss or alterations in rivers, such as effects due to certain forestry, industrial, and agricultural practices (for example, excess sedimentation, hydroelectric dams, dike construction).

Estuaries and coastal systems near urban areas are degraded by runoff from farmlands and by urban development. Much of the contaminant input to waters flowing seaward consists of organic substances having nutritional value for phytoplankton, which form the base of the food chain. Nitrogenous substances, a range of carbohydrates and fats, phosphates, and other nutrients from atmospheric contamination or discharges to rivers in the coastal zone, result in nutrient enrichment. For instance, some of the greatest standing stocks of phytoplankton and highest rates of primary production have been measured in coastal waters of the New York Bight, enriched by ocean dumping and nonpoint sources (O'Reilly et al. 1987).

Coral reefs subject to habitat degradation have exhibited a natural ability to recover from disturbance, but this recovery is hampered by further damage from human exploitation. The Convention on International Trade in Endangered Species of Wild Fauna and Flora protects coral reefs from destruction by human harvests.

Destructive fisheries methods also damage habitat. In the past, extremely damaging practices of fishing with explosives or poisons were prevalent in the Pacific Islands. These are now almost universally outlawed, but poaching by these methods remains a problem in remote regions of the western Pacific Islands. Less extreme habitat-destructive harvest methods are also of concern. Habitat alterations—for example, artificial reefs—can also be purposefully beneficial to living marine resources.

Nonindigenous Species

Marine organisms have been transported from their original ranges to new localities since humans began using boats to cross oceans (Carlton 1987). Many of these introductions have been beneficial to humans as food sources, but introduced organisms can also affect indigenous species, threaten human health, and create financial burdens on human societies (U.S. Congress, Office of Technology Assessment 1993). No nonindigenous marine species have become as notorious or costly as the freshwater zebra mussel, introduced to the Great Lakes from Eurasia via ship ballast water in the 1980's (Nalepa and Schloesser 1993). The increasing alteration of community structure and function in estuaries and other coastal habitats by invasive species has prompted speculation that these may now be among the most threatened ecosystems on earth (Carlton and Geller 1993).

Among marine and estuarine examples, cholera bacteria, found in eastern oysters of Mobile Bay, Alabama (Motes et al. 1994), were probably transported via ballast water in ships arriving from South America (McCarthy and Khambaty 1994). Asian clams introduced to San Francisco Bay filter plankton from water so efficiently that they capture much of the bay's productivity at the expense of its fishes (Nichols et al. 1990). The capacity of the San Francisco Bay drainage to sustain valuable fisheries production may be reduced as a consequence. Many other nonindigenous marine organisms are found in all regions (DeVoe 1992; Rosenfield and Mann 1992). Because the taxonomy and biogeography of many families of organisms are poorly investigated, the number of nonindigenous species in U.S. waters is likely underestimated (Chapman and Carlton 1991).

Many introductions of aquatic species were intended to enhance fisheries productivity or to establish organisms of cultural importance as people immigrated to new regions. For example, Pacific oysters introduced to the U.S. Pacific coast from Japan now account for much of the aquaculture production in that area. The introduction of this species to Chesapeake Bay was proposed as a way to rebuild a declining Eastern oyster fishery. Managers argued that introduced Pacific oysters would compete for food and nursery space with the native oyster, or carry new diseases and predators to the east coast (Mann et al. 1992). The dangers of introductions are now better appreciated, and numerous regulations exist to control them (DeVoe 1992; Rosenfield and Mann 1992; Aquatic Nuisance Species Task Force 1994a,b; also see chapter on Nonindigenous Species). In most cases, the risks from intentional introductions are compared to the benefits before decisions are made. Even so, unforeseen adverse effects from introductions do occur (Aquatic Nuisance Species Task Force 1994a).

Unintentional introductions are increasing even as intentional introductions have declined (Aquatic Nuisance Species Task Force 1994b). Unintentional aquatic introductions include the release of ornamental organisms from the aquarium trade, the accidental release of cultured species, and the release of species into new areas by transport via ships' ballast waters. Release of aquarium species is a problem of increasing magnitude in fresh water; in Hawaii, for example, about 20–25 freshwater ornamental species have become established since 1980 (Devick 1991). Records of nonindigenous marine aquarium organisms from Florida (Courtenay 1995) and Hawaii (Randall 1995) indicate that this is a potential problem, but thus far, aquarium releases have had little effect on marine systems.

The discovery of nonindigenous shrimp species in the waters of South Carolina is an example of the more common problem of introductions from aquaculture sources. Pacific black tiger prawns and Pacific white shrimp, introduced when animals escaped from aquaculture ponds, have both been taken by South Carolina shrimp trawlers in the 1980's (Wenner and Knott 1992). Pacific white shrimp in particular are known to sometimes harbor infectious hypodermal virus, hematopoietic necrosis virus, and other diseases that could devastate native shrimp stocks (Brock 1992). Another mechanism of unintentional introduction is net pen escapes. Atlantic salmon cultured in net pens in the Atlantic Northeast and Pacific Northwest are known to escape and may spawn with natural salmon runs, leading to erosion of local genetic integrity (Wing et al. 1992; Thomson and McKinnell 1995). Double-mesh nets and the use of sterile animals (triploidy) are ways to possibly reduce effects of net pen escapes.

The major source of successful introductions of marine species is presently thought to be ballast water (Carlton and Geller 1993). The huge quantities of ballast water that are transported and exchanged from port to port each year carry planktonic stages of numerous species around the world. Ballast water introductions are suspected of causing the increased incidence of toxic dinoflagellate blooms (red tides) in Australia (Hallegraeff and Bolch 1992), the collapse of fisheries in the Black Sea (Harbison

and Volovik 1993), and the alteration of estuarine community structure in San Francisco Bay (Nichols et al. 1990). Ballast water introductions are clearly a major concern for the future.

Nonindigenous organisms are difficult or impossible to eradicate once they become established (U.S. Congress, Office of Technology Assessment 1993). The effects of introductions are impossible to predict. Slowing the spread of nonindigenous marine species and reversing the effects of introduced species already established are conservation problems of growing importance (see chapter on Nonindigenous Species).

Transboundary Stocks

Many living marine resources straddle the boundaries between states and between countries, complicating their exploitation and management. Effective conservation and management of these resources require coordination, cooperation, and agreement among the respective management entities. Stocks located within the waters of more than one state are, to an increasing extent, managed by interstate compacts. An example of a successful interstate management program for a shared resource is the recovery of Atlantic striped bass (National Oceanic and Atmospheric Administration 1993). After Atlantic striped bass were reduced to very low abundance in the 1970's, a plan was developed by the Atlantic States Marine Fisheries Commission which called for a series of tight restrictions on fishing during 1981–1990, followed by a gradual relaxation of controls as remedial measures began to take effect, and culminating in a declaration in January 1995 of full stock restoration.

Stocks living totally or primarily within federal waters are managed by plans prepared by regional fishery management councils and implemented and enforced by the National Marine Fisheries Service. Stocks whose distribution overlaps the jurisdiction of more than one regional council require the participation of several councils.

Most stocks that extend beyond the U.S. Exclusive Economic Zones are managed wholly or in part under international conventions. In the Atlantic, highly migratory species such as bluefin tuna and swordfish fall under the jurisdiction of the International Commission for the Conservation of Atlantic Tunas. Regulation of these particular species is difficult because international consensus is not always reached on catch levels for these high-value fish, nor are agreed measures always equally enforced by all nations.

Another example in which shared management of transboundary stocks has not worked well is on Georges Bank. A 1984 World Court decision resolving the United States–Canada maritime boundary dispute awarded each country part of the bank. The resulting Hague Line gave each country custody of the fisheries occurring in their respective waters on stocks that were distributed over the entire bank. In spite of the obvious need for joint management of important stocks such as cod, haddock, pollock, yellowtail flounder, and sea scallops, each country has so far exercised separate control of fisheries in its own zone. The lack of joint management has been an important factor in the present severely overfished condition of most of these stocks.

In some cases, foreign fisheries target migrating stocks originating in the United States but found outside the U.S. Exclusive Economic Zone, which can be a threat to U.S. management of those stocks. One such example was the interception of U.S. pollock stocks found in the international zone of the central Bering Sea, known as the Donut Hole area. Several international meetings were organized to solve the problem, and a temporary agreement was reached among the parties to cease fishing for 1992–1995. In 1994, an international treaty was drawn up to replace the temporary agreements and to manage the Aleutian Basin pollock stock.

Salmon in both the Atlantic and the Pacific, which begin life in freshwater rivers and streams and migrate to the open ocean to mature before returning to their home waters to spawn, are subjected to fishing in both domestic and foreign waters. Heavy exploitation in commercial fisheries off Greenland of Atlantic salmon that originate in the United States has recently been reduced through control measures implemented by the North Atlantic Salmon Conservation Organization. On the Pacific coast, some salmon of U.S. origin are intercepted by Canadian fishermen while the fish migrate through Canadian waters, whereas some sockeye and pink salmon originating in the Fraser River of Canada are caught by U.S. fishermen when the fish move through from U.S. waters. The always-contentious allocation of catches from stocks originating in each country to fishermen from the two countries is handled by the United States–Canada Pacific Salmon Commission. In the Pacific Ocean, driftnet fishing for salmon in high seas is now banned under a United Nations General Assembly Resolution and by the North Pacific Anadromous Fish Commission.

Effects of Commercial Fisheries on Marine Birds and Mammals

The ecological effects of commercial fisheries on marine birds and mammals are still largely unknown. Estimates of kills of marine mammals by direct interactions with fishing gear are generally low for most U.S. fisheries; however, there are significant fisheries-related mortalities of some marine mammals. The magnitude of direct kills of marine birds due to interactions with fishing activities is not well known (Johnson et al. 1993).

An important indirect threat to marine birds and mammals by fishing appears to be competition for fish. Marine birds and mammals consume a wide variety of fish species, some of which are commercially important. Many marine mammals in Alaska, particularly seals and sea lions, consume juvenile groundfish, whereas fisheries tend to target adult-sized groundfish. Thus, although direct competition for prey is reduced, commercial fisheries may disrupt prey availability through bycatch of small fish, removal of spawning fish, or general disruption of the food web.

Another ecological consequence of commercial fisheries is the fish-processing waste that can alter the feeding habits of some marine birds and mammals. Gulls, sea lions, bottlenose dolphin, and killer whales feed on fish wastes discharged by processing vessels and plants. Disposal of this waste at sea may create an artificial dependency that is not beneficial for the long-term well-being of the animals. Increases in predator populations (such as large gulls) resulting from this supplemental feeding may be detrimental to populations of their prey, such as other marine birds, and their buildup may also displace other bird species from their nesting areas.

Toxic Wastes and Public Health

Toxic contaminants in estuarine and marine areas originate from a variety of sources, including industrial and sewage treatment plants, urban and agricultural runoff, and air pollution (see chapter on Environmental Contaminants). These contaminants include toxic metals, organic chemicals such as chlorinated hydrocarbons (for example, PCB's and DDT's), and petroleum compounds (for example, polycyclic aromatic hydrocarbons). Toxic effects can range from sublethal changes—such as growth inhibition—to death. Bottom fish species from a number of chemically contaminated areas in the vicinity of such major cities as Boston, New York, Los Angeles, and Seattle have exhibited liver cancers and other lesions that lead to liver cancer (Johnson et al. 1994). Although cancers are the most dramatic of the lesions observed, early degenerative or proliferative conditions are much more prevalent in animals from moderately contaminated sites. Others may suffer various types of reproductive impairment, such as inhibited spawning ability and reduced viability of eggs and larvae.

In addition to demersal animals, certain anadromous fish species, such as salmon, feed on sediment-dwelling organisms as the fish migrate through urban estuaries. During their relatively short residency, the fish can bioaccumulate significant concentrations of toxic chemicals through their diet. Little is known about the effects of contaminants on marine mammals, although there are ongoing studies on the effects of DDT and PCB's on California sea lions.

Chemical contaminants, microbial pathogens, and marine toxins are both real and perceived risks to the health of humans who consume seafood products. In most U.S. coastal areas, the human health risks projected from current levels of environmental contamination in commercial seafood products are either negligible or undetectable. However, commercially and recreationally collected fish and shellfish from some areas of the United States, such as Boston Harbor or San Pedro Bay near Los Angeles, occasionally contain elevated levels of toxic chemicals that could be potentially harmful to human health. A variety of human health effects are also associated with the consumption of seafood containing marine toxins, such as paralytic shellfish poisoning and domoic acid or amnesiac shellfish poisoning. Paralytic shellfish poisoning results from the consumption of mollusks infected with specific strains of dinoflagellates. Domoic acid outbreaks have occurred in a variety of seafood products and have caused huge losses to the seafood industry. For example, in the fall of 1991, an outbreak of domoic acid poisoning in Monterey Bay, California, moved north along the coast and reached Washington and Oregon. Like paralytic shellfish poisoning, the primary route of human exposure to domoic acid is from the consumption of mollusks (for example, razor clams), although the viscera of Dungeness crab is also known to accumulate high levels of the toxin. The mechanism by which clams accumulate domoic acid is not clear, even though the diatom *Pseudonitschia australis* has been implicated in the domoic acid poisoning of marine birds. Infectious microorganisms from agricultural and domestic sources can also contaminate clams, oysters, and other shellfish consumed by humans (Bourne and Chew 1994). This risk to humans has resulted in the closing of vast areas of coastal waters to shellfish growing and harvesting. Other types of poisonings from marine toxins have been reported but are less common in the United States.

Overall, the risks to human health from consumption of most seafood products containing chemical contaminants, microbial pathogens, or marine toxins from U.S. coastal waters are generally low, largely due to present management practices. However, a sufficient likelihood of occasional episodes of environmental contamination exists to warrant implementation of a seafood safety inspection program by the National Marine Fisheries Service, in cooperation with other federal agencies (such as the U.S. Food and Drug Administration) to ensure seafood quality and the detection of low levels of contamination.

Marine Debris

Marine debris pollutes all of the world's oceans, but the problem hardly starts in the ocean itself. Close to 80% of the debris in the ocean is washed, blown, or dumped from shore. Such debris casts a wide net of effects over marine and littoral animals, plants, and perhaps even entire ecosystems. Biological studies tell us that waterborne litter entangles wildlife and masquerades as a food source, smothers beaches and bottom-growing plants, and provides a surface for colonizing small organisms that travel on marine debris to distant shores, perhaps with adverse ecological consequences. Entanglement is the most obvious of all biological effects on living marine resources. Entangled northern fur seals were spotted as early as the 1930's. We get only fleeting glimpses of entangled animals from planes and ships. Many die and sink or are eaten without being detected; others stay submerged and hidden beneath the debris. At least 135 species of marine vertebrates and 8 invertebrates have been reported entangled in marine debris. The list now includes most of the world's sea turtle species, more than 25% of marine mammal species, and more than 15% of marine bird species (Faris and Hart 1995; Fig. 68).

Ingestion of marine debris can also be a serious threat to wildlife. Sea turtles mistake clear plastic bags for jellyfish, one of their favorite meals. Birds mistake plastic pellets for fish eggs. Other times, plastic is accidentally eaten in association with natural food. All are harmful, if not fatal. Ingested debris damages the digestive tract, causes starvation by blocking food, may be toxic, and often kills marine animals.

For centuries, humans, at sea and on land, have used the oceans as dumping grounds for waste. At present there are two global conventions to combat marine debris pollution from sea sources, the London Dumping Convention and the International Convention for the Prevention of Pollution from Ships (MARPOL 73/78). The land-based sources of marine pollution are beginning to be addressed globally in follow-up meetings to the 1992 United Nations

Courtesy J. Henderson, National Marine Fisheries Service

Fig. 68. Hawaiian monk seal killed by entanglement in marine debris.

Conference on Environment and Development. The United States is party to these forums and takes an active role in mitigating marine debris issues. Education and promotion of awareness of the problem and its many solutions seem the best way to combat this widespread pollution.

Taxonomic Knowledge

There is much more to discover and learn about the taxonomy of many marine organisms (Winston 1992). This is due in part to the large number of animal and plant groups that are entirely or primarily marine. For the most part, these groups are composed of small species; the paucity of knowledge about these species is made evident by their lack of common names. On a regional basis, taxonomic knowledge may be ranked as best (first) for the Northeast because many of the nation's oldest and most numerous universities and museums are located there. Second or third in rank might be the Northwest region or perhaps Alaska, with the Southwest region ranked fourth because of elements of a richer subtropical biota found in its southern fringes. The Southeast region ranks fifth despite the long history of academic research in the region because subtropical areas and the tropical Caribbean contain large numbers of marine species that need taxonomic review. Last, with the most poorly known biota, is the western Pacific region; this region contains the largest number of species, covers the largest area, includes several global biogeographic regions, and has the shortest history of taxonomic research.

Taxonomic knowledge decreases with the depth of the ocean being considered (Winston 1992; Norse 1993). The marine biota most accessible to humans is that of the shoreline,

thus it is the best known taxonomically. Sampling becomes a huge problem as one travels deeper into the ocean. Organized sampling of the deep sea did not begin until 1869 and 1872 with the *Porcupine* and *Challenger* expeditions (Mills 1983). Despite numerous oceanographic expeditions throughout this century, less is known about the identities and distributions of organisms the deeper one looks in the sea. For example, almost no biological samples have been taken from below 2,000 meters near any of the U.S. islands in the Pacific.

The ocean contains many more species than previously thought (National Research Council 1995). Entire communities of organisms, many undescribed, were found to live around deep-sea hydrothermal vents, at seepages of natural petroleum from the sea floor, and at areas of deepwater brine seepage. Studies of tropical invertebrates and algae continue to find that 25%–70% of the smallest species are new to science, even in areas like Waikiki, Hawaii, virtually at the front door of local universities and museums. Even well-known organisms—including certain mussels, oysters, shrimps, and crabs of the northeastern United States—have recently been found, with the advent of new genetic techniques, to consist of several physically similar but genetically distinct sibling species. Entire new groups of organisms continue to be found in the sea as exploration widens, including tiny algae (picoplankton) species that contribute to a significant portion of the earth's carbon cycling. The discoveries of previously unknown marine bacteria and the Archaea—organisms as different from bacteria as bacteria are from plants or animals—have altered perceptions of the role of marine microbes in the earth's geochemical cycles.

Ecosystem Biodiversity and Sustainability

Human intervention and ocean warming cause changes in biodiversity and sustainability of biomass yields in marine ecosystems that are important to the economies of U.S. coastal regions. The diversity of ecological processes and the health of coastal ecosystems are being diminished by increased stress from human-induced toxic effluents, habitat degradation, excessive nutrient loadings, harmful algal blooms, emergent diseases, fallout from aerosol contaminants, and losses of living marine resources from pollution effects and overexploitation. A growing awareness that the quality of coastal ecosystems is being adversely affected by multiple forces has accelerated legislative efforts to encourage federal agencies to assess, monitor, and mitigate coastal stressors from an ecosystem perspective. Among recent

Congressional efforts is the enactment of the National Coastal Monitoring Act of 1992. The act provides for assessing the changing states of coastal ecosystem health and reporting the findings to the U.S. Congress as a recurring responsibility of the National Oceanic and Atmospheric Administration and the Environmental Protection Agency (National Coastal Monitoring Act 1992).

Since the enactment of the National Coastal Monitoring Act, greater emphasis has been focused within the National Oceanic and Atmospheric Administration on the assessment and mitigation of coastal stressors on the biodiversity of marine populations within the geographic extent of seven large marine ecosystems. These ecosystems are each characterized by distinct bathymetry, hydrography, productivity, and trophic linkages (American Association for the Advancement of Science 1986, 1989, 1990, 1991, 1993). They are depicted by the National Oceanic and Atmospheric Administration in a Folio Map Series (National Oceanic and Atmospheric Administration 1988) and include the Northeast shelf, Southeast shelf (Ray et al. 1980), Gulf of Mexico (National Oceanic and Atmospheric Administration 1986), California Current, Gulf of Alaska (National Oceanic and Atmospheric Administration 1990b), Eastern Bering Sea (National Oceanic and Atmospheric Administration 1989), and Insular Pacific ecosystems, including the waters around the Hawaiian Islands (Fig. 69). The growing interest in ecosystem-wide assessments has been triggered by the recent evidence of apparent climate-generated ocean warming over the past 40 years that has led to a 70% reduction in zooplankton volume within the California Current ecosystem from the 1950's to the late 1980's and early 1990's (Roemmich and McGowan 1995). This reduction appears to be related to Pacific-wide El Niño-driven shifts in marine oceanographic regimes that in turn have affected fish populations. There have been multiple-year declines and increases in salmon, sardines, and anchovies in the California Current ecosystem (Bakun 1995), in salmon and pollock in the East Bering Sea ecosystem (Bakun 1995; Livingston et al. 1995), and in salmon of the Gulf of Alaska ecosystem (Hare and Francis 1995). In contrast, the persistent decline of the haddock, cod, and flounder stocks of the Northeast shelf ecosystem from 1960 through 1994 is the result of excessive fishing effort rather than any significant oceanographic regime changes or coastal pollution (Northeast Fisheries Science Center 1993; Murawski 1996). Less is known of the causes of the fluctuations in fisheries yields of the Southeast shelf and Gulf of Mexico ecosystems. Reduced oxygen levels caused by

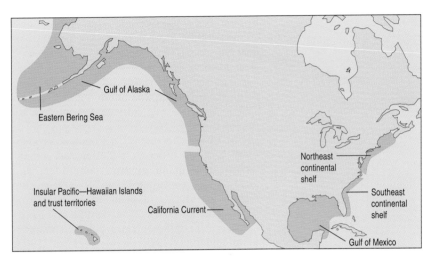

Fig. 69. The seven large marine ecosystems.

excessive nutrient loadings within the northern Gulf of Mexico from the Mississippi River drainage basin have been linked to extensive die-offs of coastal fishes (Brown et al. 1991). Excessive nutrient loadings from river basin drainages have also been observed within the Northeast shelf ecosystem and may be the cause of the growing frequency and extent of harmful algal blooms (Smayda 1991) and the emergence of marine mammal and human pathogens (Epstein 1995).

Adjusting the level of fishing effort to changing ecosystem states through adaptive management is an evolving concept that has led to promising management practices for limiting fishing effort to allow for the recovery of depleted stocks (Collie 1991; Rosenberg et al. 1993). However, more effort is required to improve degraded marine habitats. The growing problem of eutrophication of coastal ecosystems is a trend that must be reversed. Effective strategies must be implemented to reduce excessive nutrient loading in coastal ecosystems, particularly along the Atlantic and gulf coasts. Such strategies should be science-based and implemented at local, state, and regional levels of government to ensure adequate control over both point-source and nonpoint-source coastal eutrophication. This effort will require collaboration between the state and federal agencies responsible for the stewardship of coastal ecosystems, including their habitats and drainage basin effluents, and the health of their marine fish, bird, and mammal populations.

A modularized approach for improving the health of coastal ecosystems has been developed by the National Oceanic and Atmospheric Administration for application within the ecosystems of the United States. This assessment strategy is being implemented in the United States and elsewhere, including the Gulf of Guinea, the Yellow Sea, the Baltic Sea, the Black Sea, and the Benguela Current ecosystems. Five component models developed by the National Oceanic and Atmospheric Administration are being used as measures of changing ecosystem states and guidelines to sustainability. They include indicators of biodiversity, stability, yields, productivity, and resilience (Sherman and Solow 1992; Sherman 1994). The data for deriving the models are obtained from time-series assessments of key ecosystem parameters that include productivity, fish and fisheries, pollution and ecosystem health, socioeconomic conditions, and governance. Sustainable use of renewable marine resources and maintenance of marine biodiversity are possible—according to scientific assessment of biological, social, and economic considerations (Rosenberg et al. 1993; Mangel et al. 1993)—when humans use living components of ecosystems in ways that allow natural processes to replace what is removed. Under these conditions, an ecosystem can renew itself indefinitely, and human use is sustainable.

Empirical and theoretical aspects of biomass, along with biodiversity yield models for large marine ecosystems, have been reviewed by several ecologists. According to Beddington (1986), Daan (1986), Levin (1990), and Mangel (1991), published dynamic models of marine ecosystems offer little guidance on the detailed behavior of communities. However, these authors concur on the need for covering the common ground between observation and theory by monitoring key components of ecosystems over large areas and long periods of time (decades). Understanding the possible mechanisms underlying observed patterns in large marine ecosystems is described by Levin (1990) as: (1) statistical analyses of observed distributional patterns of physical and biological variables; (2) construction of competing models of variability and patchiness based on statistical analyses and natural scales of variability of critical processes; (3) evaluation of competing models through experimental and theoretical studies of component systems; and (4) integration of validated component models to provide predictive models for population dynamics and redistribution. Likens (1992) has developed a conceptual diagram of processes driving changing ecosystem states.

Significant multispecies, ecosystem-oriented assessments in support of the management of marine resources have not been widely practiced until recently. In 1991, the Advisory Committee on Fishery Management of the International Council for the Exploration of the Sea reached an important milestone by including guidance for providing "the advice necessary to maintain viable fisheries within sustainable ecosystems" (International Council for the Exploration of the Sea 1992).

The designation and management of large marine ecosystems involve, at present, an evolving scientific and geopolitical process (Morgan 1988; Alexander 1989; American Association for the Advancement of Science 1993). There has been sufficient progress to make useful comparisons among different processes influencing large-scale changes in the biomass yields of large marine ecosystems (Bax and Laevastu 1990; Bakun 1993). Among the more recent findings is the marked decline in biodiversity among the principal fish stocks of the Black Sea, based, in part, on the introduction of a nonindigenous species of a ctenophore that has undergone a population explosion, causing severe predation on both early life stages of fish and their prey (Mee 1992). The influence of changes in the gene pool of wild stocks from inadvertent releases of cultured stocks is another concern of scientists engaged in large marine ecosystem studies.

The broad-spectrum approach to research and monitoring of large marine ecosystems provides a conceptual framework for collaboration in process-oriented studies conducted by the National Science Foundation–National Oceanic and Atmospheric Administration-sponsored Global Ocean Ecosystems Dynamics program in the United States and the International GLOBEC program. The development of large marine ecosystem assessment strategies is compatible with the proposed Global Ocean Observing System of the Intergovernmental Oceanographic Commission, United Nations Educational, Scientific, and Cultural Organization, including those modules to be focused upon living marine resources and ecosystem health (Intergovernmental Oceanographic Commission 1993).

Effective management strategies for large marine ecosystems will be contingent on identifying the major driving forces behind large-scale changes in biomass. Improving the understanding of the physical factors forcing biological change will enhance management of species that respond to strong environmental signals, thereby enhancing forecasts of El Niño-type events. In other large marine ecosystems, where the prime driving force is predation, options can be explored for implementing adaptive management strategies (Collie 1991). Remedial actions are required to ensure that the pollution of the coastal zones of large marine ecosystems is reduced and does not force changes in biodiversity or resource sustainability in any large marine ecosystem.

Outlook

A range of human activities affects living marine resources, including fishes, marine mammals, and marine birds. Increasingly intensive fishing efforts, in conjunction with uses of more sophisticated fishing gear and electronics, have resulted in gross overfishing of some marine populations. Associated with fishing is bycatch, when nontarget animals are taken in fishing operations. These problems have caused marine resource management to create controlled access to fishing, whereby catch quotas are allocated to individual fishing vessels instead of allowing free and open access. Quota allocation to individual vessels is gaining acceptance.

Another major issue today is mitigation necessary to protect endangered or threatened species. A recent example of this situation is with certain salmon runs in Pacific coast streams. The use of river waters for irrigation, power generation, and domestic consumption by large urban areas has compromised these streams and the survival of the salmon runs. The people and the courts have become more involved with the mitigation of the issues, and some rather extreme options, such as the removal of some Columbia River dams, are under consideration.

Habitat alterations have taken place in rivers and estuaries, as well as in coastal zones, as a result of urbanization. Along with urbanization come alteration of freshwater flows, erosion, introduction of toxic chemicals and other contaminants into the waters, introduction of nonindigenous species, and degradation of the marine habitats essential to the survival of living marine resources. There are numerous demographic trends that suggest these conditions and threats are not likely to change in the immediate future. Some 50% of our population reportedly lives within a 2- or 3-hour drive of major freshwater systems (the Great Lakes) or coastal waters. As the nation grows, there will be further growth in coastal zones. Our desire to live within sight of the ocean has never abated and, increasingly, urban dwelling areas are being developed on or near shorelines. Often these developments alter coastal and marine ecosystems and affect living marine resources.

There are also many natural changes in the dynamics of ecosystems that may exacerbate the aforementioned issues. For instance, attention to global warming over the past two decades is now culminating in data that show widespread climatic effects on living marine resources. Articles in recent issues of major scientific publications such as *Science* indicate a progressive subtle warming of Pacific coastal waters and consequently an extirpation of certain species that would have been found in these waters a half-century ago. Similar trends have been speculated about and, to some degree, measured in shelf waters as well as in the coastal zone of the northwest Atlantic. We need a far better understanding of how human activities, in concert with short- and long-term climatic changes, compromise living marine resources.

Thus far, the use and management of the nation's living marine resources have been approached mainly from a resource-by-resource point of view. As information on multispecies relationships, ecosystem interactions, and environmental influences becomes more readily available, management has moved toward more explicit considerations of effects of human activities on all components of the ecosystem, including marine mammals and marine birds. The goal is to move toward use of living marine resources for their combined ecological, recreational, commercial, and other aesthetic values. These optimal values and uses are no simple matter to define, let alone achieve. This task remains a formidable challenge for science and management.

On the whole, the outlook for the welfare of the nation's living marine resources is guarded, with a need to remain vigilant. The watch is conducted by the National Marine Fisheries Service and regional fishery management councils, for fisheries and marine mammal resources within the U.S. Exclusive Economic Zone; by the Department of the Interior, for marine birds; by the various coastal states, for nearshore resources; and by international commissions, for specific species. The crash of the Northeast groundfish fisheries, the poor welfare of some Pacific coast salmon stocks, and declines in some marine mammal populations are examples of situations that need special attention. Although many other of the nation's marine resources are in good condition, they too must be attentively managed and conserved under a suite of federal laws and international treaties. The science of the marine environment and the state of its resources must be improved, for without reliable scientific knowledge, resource use and management must necessarily be more conservative.

Acknowledgments

We express thanks and appreciation to countless unnamed colleagues for their diligence in collecting the data, performing the research, and conducting the analyses that provide the basis for much of the information presented here. Particular assistance for unpublished data, figures, and critical review were rendered by P. Bowman, Louisiana Department of Wildlife and Fisheries, Baton Rouge, Louisiana; W. F. Gandy, National Marine Fisheries Service, Stennis Space Center, Mississippi; L. L. Massey, National Marine Fisheries Service, Miami, Florida; G. M.

Meaburn, National Marine Fisheries Service, Charleston, South Carolina; J. V. Merriner, National Marine Fisheries Service, Beaufort, North Carolina; J. M. Nance, National Marine Fisheries Service, Galveston, Texas; J. M. Coe, D. P. DeMaster, B. J. Goiney, Jr., A. W. Kendall, Jr., T. R. Loughlin, L. L. Jones, and J. A. Pearce, National Marine Fisheries Service, Seattle, Washington; E. N. Flint (seabirds), U.S. Fish and Wildlife Service Hawaiian and Pacific Islands National Wildlife Refuge Complex, Honolulu, Hawaii; L. G. Eldredge, Pacific Science Association, Bernice P. Bishop Museum, Honolulu, Hawaii; S. Abbott-Stout, G. H. Balazs (sea turtles), C. H. Boggs, R. P. Clarke, E. E. DeMartini, W. R. Haight, D. C. Hamm, D. R. Kobayashi, R. M. Laurs, J. J. Naughton, E. T. Nitta, J. J. Polovina, T. J. Ragen (Hawaiian monk seal), M. P. Seki, and J. A. Wetherall, National Marine Fisheries Service, Honolulu, Hawaii; A. E. Dizon, National Marine Fisheries Service, La Jolla, California; and C. B. Grimes, National Marine Fisheries Service, Santa Cruz, Californa.

Authors

Organizers and principal authors

Loh-Lee Low
National Marine Fisheries Service, NOAA
Alaska Fisheries Science Center
7600 Sand Point Way Northeast
Seattle, Washington 98115

Allen M. Shimada
Steven L. Swartz*
Michael P. Sissenwine**
National Marine Fisheries Service, NOAA
Office of Science and Technology
1315 East West Highway
Silver Spring, Maryland 20910

*Current address:
National Marine Fisheries Service, NOAA
Southeast Fisheries Science Center
75 Virginia Beach Drive
Miami, Florida 33149

**Current address:
National Marine Fisheries Service, NOAA
Northeast Fisheries Science Center
166 Water Street
Woods Hole, Massachusetts 02543

Northeast Region

John B. Pearce
Emory D. Anderson
National Marine Fisheries Service, NOAA
Northeast Fisheries Science Center
166 Water Street
Woods Hole, Massachusetts 02543

Kenneth Sherman
John E. O'Reilly
National Marine Fisheries Service, NOAA
Narragansett Laboratory
28 Tarzwell Drive
Narragansett, Rhode Island 02882

Robert N. Reid
Frank W. Steimle
National Marine Fisheries Service, NOAA
James J. Howard Marine Sciences
Laboratory
75 McGruder Road
Highlands, New Jersey 07732

James H. W. Hain
National Marine Fisheries Service, NOAA
Northeast Fisheries Science Center
166 Water Street
Woods Hole, Massachusetts 02543

Southeast Region

Thomas D. McIlwain
National Marine Fisheries Service, NOAA
Mississippi Laboratories
P.O. Drawer 1207
Pascagoula, Mississippi 39568

Alaska Region

Loh-Lee Low
National Marine Fisheries Service, NOAA
Alaska Fisheries Science Center
7600 Sand Point Way Northeast
Seattle, Washington 98115

Howard W. Braham
National Marine Fisheries Service, NOAA
National Marine Mammal Laboratory
7600 Sand Point Way Northeast
Seattle, Washington 98115

James C. Olsen
National Marine Fisheries Service, NOAA
Auke Bay Laboratory
11305 Glacier Highway
Juneau, Alaska 99801

Patrick J. Gould
U. S. Geological Survey
Alaska Science Center
1011 East Tudor Road
Anchorage, Alaska 99503

Allen M. Shimada
National Marine Fisheries Service, NOAA
Office of Science and Technology
1315 East West Highway
Silver Spring, Maryland 20910

Pacific Coast Region

Bruce B. McCain
National Marine Fisheries Service, NOAA
Hatfield Marine Science Center
2030 S. Marine Science Drive
Newport, Oregon 97365

Alec D. MacCall
National Marine Fisheries Service, NOAA
Tiburon Laboratory
3150 Paradise Drive
Tiburon, California 94920

Robert L. Emmett
National Marine Fisheries Service, NOAA
Hatfield Marine Science Center
2030 S. Marine Science Drive
Newport, Oregon 97365

Richard D. Brodeur
National Marine Fisheries Service, NOAA
Alaska Fisheries Science Center
7600 Sand Point Way Northeast
Seattle, Washington 98115

Western Pacific Oceanic Region

Bruce C. Mundy
Alan R. Everson***
National Marine Fisheries Service, NOAA
Southwest Fisheries Science Center
Honolulu Laboratory
2570 Dole Street
Honolulu, Hawaii 96822

***Current address:
U.S. Army Corps of Engineers
Pacific Ocean Division, Building 230
Fort Shafter, Hawaii 96858

National Status and Trends

Loh-Lee Low
National Marine Fisheries Service, NOAA
Alaska Fisheries Science Center
7600 Sand Point Way Northeast
Seattle, Washington 98115

Howard W. Braham
National Marine Fisheries Service, NOAA
National Marine Mammal Laboratory
7600 Sand Point Way Northeast
Seattle, Washington 98115

James H. W. Hain
National Marine Fisheries Service, NOAA
Northeast Fisheries Science Center
166 Water Street
Woods Hole, Massachusetts 02543

Stephen C. Jameson
Steven H. Jury
Mark E. Monaco
National Ocean Survey, NOAA
Office of Ocean Resources Conservation
and Assessment
1305 East West Highway
Silver Spring, Maryland 20910

Allen M. Shimada
National Marine Fisheries Service, NOAA
Office of Science and Technology
1315 East West Highway
Silver Spring, Maryland 20910

David L. Stein
Office of Oceanic and Atmospheric
Research, NOAA
Program Development and Coordination
1315 East West Highway
Silver Spring, Maryland 20910

Steven L. Swartz*
National Marine Fisheries Service, NOAA
Office of Science and Technology
1315 East West Highway
Silver Spring, Maryland 20910

*Current address:
National Marine Fisheries Service, NOAA
Southeast Fisheries Science Center
75 Virginia Beach Drive
Miami, Florida 33149

National Issues and Threats

Loh-Lee Low
National Marine Fisheries Service, NOAA
Alaska Fisheries Science Center
7600 Sand Point Way Northeast
Seattle, Washington 98115

Bruce C. Mundy
National Marine Fisheries Service, NOAA
Southwest Fisheries Science Center
Honolulu Laboratory
2570 Dole Street
Honolulu, Hawaii 96822

Bruce B. McCain
National Marine Fisheries Service, NOAA
Hatfield Marine Science Center
2030 S. Marine Science Drive
Newport, Oregon 97365

Thomas D. McIlwain
National Marine Fisheries Service, NOAA
Mississippi Laboratories
P.O. Drawer 1207
Pascagoula, Mississippi 39568

Emory D. Anderson
National Marine Fisheries Service, NOAA
Northeast Fisheries Center
166 Water Street
Woods Hole, Massachusetts 02543

Kenneth Sherman
National Marine Fisheries Service, NOAA
Narragansett Laboratory
28 Tarzwell Drive
Narragansett, Rhode Island 02882

Cited References

Abbott, I. A. 1995. The state of systematics of marine algae in tropical island Pacific. Pages 25–38 in J. E. Maragos, M. N. A. Peterson, L. G. Eldredge, J. E. Bardach, and H. F. Takeuchi, editors. Marine and coastal biodiversity in the tropical island Pacific region. Volume 1. Species systematics and information management priorities. Program on Environment, East–West Center, Honolulu, Hawaii.

Alaska Fisheries Science Center. 1993. Status of living marine resources off the Pacific coast of the United States for 1993. National Oceanic and Atmospheric Administration Technical Memorandum NMFS-AFSC-26. 90 pp.

Alcala, A. C., and E. D. Gomez. 1987. Dynamiting coral reefs for fish: a resource-destructive fishing method. Pages 51–60 in B. Salvat, editor. Human effects on coral reefs: facts and recommendations. Museum National d'Histoire Naturelle, École Pratique des Hautes Etudes, Antenna de Tahiti, French Polynesia.

Alexander, L. M. 1989. Large marine ecosystems as global management units. Pages 339–344 in K. Sherman, L. M. Alexander, editors. Biomass yields and geography of large marine ecosystems. American Association for the Advancement of Science Selected Symposium 111. Westview Press, Boulder, Colo.

Alverson, D. L., M. H. Freeberg, J. G. Pope, and S. A. Murawski. 1994. A global assessment of fisheries bycatch and discards. United Nations Food and Agriculture Organization Fisheries Technical Paper 339. Food and Agriculture Organization, Rome. 233 pp.

American Association for the Advancement of Science. 1986. Variability and management of large marine ecosystems. American Association for the Advancement of Science Selected Symposium 99. Westview Press, Boulder, Colo. 319 pp.

American Association for the Advancement of Science. 1989. Biomass yields and geography of large marine ecosystems. American Association for the Advancement of Science Selected Symposium 111. Westview Press, Boulder, Colo. 493 pp.

American Association for the Advancement of Science. 1990. Large marine ecosystems: patterns, processes and yields. American Association for the Advancement of Science Press, Washington, D.C. 242 pp.

American Association for the Advancement of Science. 1991. Food chains, yields, models, and management of large marine ecosystems. Westview Press, Boulder, Colo. 320 pp.

American Association for the Advancement of Science. 1993. Large marine ecosystems: stress, mitigation, and sustainability. American Association for the Advancement of Science Press, Washington, D.C. 376 pp.

Aquatic Nuisance Species Task Force. 1994a. Findings, conclusions, and recommendations of the intentional introductions policy review. Report to Congress. U.S. Fish and Wildlife Service and National Oceanic and Atmospheric Administration, Washington, D.C. 53 pp.

Aquatic Nuisance Species Task Force. 1994b. Proceedings of the conference and workshop: nonindigenous estuarine and marine organisms (NEMO). National Oceanic and Atmospheric Administration Office of the Chief Scientist, Silver Spring, Md. 125 pp.

Bailey-Brock, J. H. 1995. Polychaetes of western Pacific Islands: a review of their systematics and ecology. Pages 121–134 in J. E. Maragos, M. N. A. Peterson, L. G. Eldredge, J. E. Bardach, and H. F. Takeuchi, editors. Marine and coastal biodiversity in the tropical island Pacific region. Volume 1. Species systematics and information management priorities. Program on Environment, East–West Center, Honolulu, Hawaii.

Baker, J. H., K. T. Kimball, W. D. Jobe, J. Janousek, C. L. Howard, and P. R. Chase. 1981. Part 6, Benthic biology. Pages 1–137 in C. A. Bedinger, Jr., editor. Ecological investigations of petroleum production platforms in the central Gulf of Mexico. Pollutant fate and effects studies. Volume 1. Report to Bureau of Land Management. Contract AA551-CT8-17. Southwest Research Institute, San Antonio, Tex.

Baker, C. S., A. Perry, J. L. Bannister, M. T. Weinrich, R. B. Abernathy, J. Calambokidis, J. Lien, R. H. Lambertson, J. Urban, R. O. Vasquez, P. J. Clapham, A. Alling, U. Arnason, S. J. O'Brien, and S. R. Palumbi. 1993. Abundant mitochondrial DNA variation and world-wide population structure in humpback whales. Proceedings National Academy Sciences (USA) 90:8239–8243.

Bakun, A. 1993. The California Current, Benguela Current and southwestern Atlantic Shelf ecosystems: a comparative approach to identifying factors regulating biomass yields. Pages 199–221 in K. Sherman, L. M. Alexander, and B. D. Gold, editors. Large marine ecosystems: stress, mitigation, and sustainability. American Association for the Advancement of Science Press, Washington, D.C.

Bakun, A. 1995. A dynamical scenario for simultaneous "regime-scale" marine population shifts in widely separated LME's of the Pacific. Pages 47–72 in Q. Tang and K. Sherman, editors. The large marine ecosystems of the Pacific Rim: a report of a symposium held in Qingdao, People's Republic of China, 8–11 October 1994. A marine conservation and development report. World Conservation Union (IUCN), Gland, Switzerland.

Bax, N. J., and T. Laevastu. 1990. Biomass potential of large marine ecosystems: a systems approach. Pages 188–205 in K. Sherman, L. M. Alexander, and B. D. Gold, editors. Large marine ecosystems: patterns, processes and yields. American Association for the Advancement of Science Press, Washington, D.C.

Beddington, J. R. 1986. Shifts in resource populations in large marine ecosystems. Pages 9–18 in K. Sherman and L. M. Alexander, editors. Variability and management of large marine ecosystems. American Association for the

Advancement of Science Selected Symposium 99. Westview Press, Boulder, Colo.

Blaylock, R. A., J. W. Hain, L. J. Hansen, D. L. Palka, and G. T. Waring. 1995. U.S. Atlantic and Gulf of Mexico marine mammal stock assessments. National Oceanic and Atmospheric Administration Technical Memorandum NMFS-SEFSC-363. 211 pp.

Boehlert, G., editor. 1993. Fisheries of Hawaii and the U.S.-associated Pacific Islands. Marine Fisheries Review 55:1–138.

Boehlert, G. W., and A. Genin. 1987. A review of the effects of seamounts on biological processes. Pages 319–334 in B. H. Keating, P. Fryer, R. Batiza, and G. W. Boehlert, editors. Seamounts, islands, and atolls. American Geophysical Union. Geophysical Monograph 43.

Boehlert, G. W., and B. C. Mundy. 1993. Ichthyoplankton assemblages at seamounts and oceanic islands. Bulletin of Marine Science 53:336–361.

Boggs, C. H., and R. Y. Ito. 1993. Hawaii's longline fisheries. Pages 69–82 in G. Boehlert, editor. Fisheries of Hawaii and U.S.-associated Pacific Islands. Marine Fisheries Review 55.

Bonfil, R. 1994. Overview of world elasmobranch fisheries. United Nations Food and Agriculture Organization Fisheries Technical Paper 341. Food and Agriculture Organization, Rome. 119 pp.

Botton, M. L., and J. W. Ropes. 1987. The horseshoe crab, *Limulus polyphemus*, fishery and resource in the United States. Marine Fisheries Review 49:57–61.

Bourne, N. F., and K. K. Chew. 1994. The present and future of molluscan shellfish resources in the Strait of Georgia–Puget Sound–Juan de Fuca Strait areas. Canadian Technical Report of Fisheries and Aquatic Sciences 1948:205–217.

Bowen, B. W., A. B. Meylan, J. P. Ross, C. J. Limpus, G. H. Balazs, and J. C. Avise. 1992. Global population structure and natural history of the green turtle (*Chelonia mydas*) in terms of matriarchal phylogeny. Evolution 46:865–881.

Briggs, J. C. 1974. Marine zoogeography. McGraw-Hill Series in Population Biology. McGraw-Hill, New York. 450 pp.

Brock, J. A. 1992. Selected issues concerning obligate pathogens of non-native species of marine shrimp. Pages 165–172 in M. R. DeVoe, editor. Proceedings of the conference and workshop: introductions and transfers of marine species. Achieving a balance between economic development and resource protection. South Carolina Sea Grant Consortium, Charleston.

Brock, V. E., and T. C. Chamberlain. 1968. A geological and ecological reconnaissance off western Oahu, Hawaii, principally by means of the research submarine *Asherah*. Pacific Science 22:373–391.

Brown, B. E., J. A. Browder, J. Powers, and C. D. Goodyear. 1991. Biomass, yield models, and management strategies for the Gulf of Mexico ecosystem. Pages 125–163 in K. Sherman, L. W. Alexander, and B. D. Gold, editors. Food chains, yields, models, and management of large marine ecosystems. Westview Press, Boulder, Colo.

Buckley, P. A., and F. G. Buckley. 1984. Seabirds of the north and middle Atlantic coast of the United States: their status and conservation. Pages 101–133 in J. P. Croxall, P. G. H. Evans, and R. W. Schreiber, editors. Status and conservation of the world's seabirds. International Council for Bird Preservation Technical Publication 2.

Butman, B. 1987. Physical processes causing surficial sediment movement. Pages 147–162 in R. H. Backus and D. W. Bourne, editors. Georges Bank. Massachusetts Institute of Technology Press, Cambridge.

Calambokidis, J., and R. W. Baird. 1994. Status of marine mammals in the Strait of Georgia, Puget Sound, and Juan de Fuca Strait, and potential human effects. Canadian Technical Report of Fisheries and Aquatic Sciences 1948:282–303.

Carlton, J. T. 1987. Patterns of transoceanic marine biological invasions in the Pacific Ocean. Bulletin of Marine Science 41:452–465.

Carlton, J. T., and J. B. Geller. 1993. Ecological roulette: the global transport of nonindigenous marine organisms. Science 261(5117):78–82.

Carter, H. R., D. S. Gilmer, J. E. Takekawa, R. W. Lowe, and U. W. Wilson. 1995. Breeding seabirds in California, Oregon, and Washington. Pages 43–49 in E. T. LaRoe, G. S. Farris, C. E. Puckett, P. D. Doran, and M. J. Mac, editors. Our living resources: a report to the nation on the distribution, abundance, and health of U.S. plants, animals, and ecosystems. U.S. Department of the Interior, National Biological Service, Washington, D.C.

Cerrato, R. M., H. J. Bokuniewicz, and M. H. Wiggins. 1989. A spatial and seasonal study of the benthic fauna of the lower bay of New York Harbor. Marine Science Resources Center, Special Report Number 84, State University of New York, Stony Brook. Unpaginated.

Chapman, J. W., and J. T. Carlton. 1991. A test of criteria for introduced species: the global invasion by the isopod, *Synidotea laevidorsalis* (Miers 1881). Journal of Crustacean Biology 11:386–400.

Chave, E. H., and A. T. Jones. 1991. Deepwater megafauna of the Kohala and Haleakala slopes, Alenuihaha Channel, Hawaii. Deep-Sea Research 38:781–803.

Chave, E. H., and B. C. Mundy. 1994. Deepsea benthic fish of the Hawaiian Archipelago, Cross Seamount, and Johnston Atoll. Pacific Science 48:367–409.

Clapp, R. B., R. C. Banks, D. Morgan-Jacobs, and W. A. Hoffman. 1982a. Marine birds of the southeastern United States and Gulf of Mexico. Part I: Gaviiformes through Pelecaniformes. U.S. Fish and Wildlife Service FWS/OBS-82/01. 637 pp.

Clapp, R. B., and P. A. Buckley. 1984. Status and conservation of seabirds in the southeastern United States. Pages 135–155 in J. P. Croxall, P. G. H. Evans, and R. W. Schreiber, editors. Status and conservation of the world's seabirds. International

Council for Bird Preservation Technical Publication 2.

Clapp, R. B., D. Morgan-Jacobs, and R. C. Banks. 1982b. Marine birds of the southeastern United States and Gulf of Mexico. Part II: Anseriformes. U.S. Fish and Wildlife Service FWS/OBS-82/20. 429 pp.

Clapp, R. B., D. Morgan-Jacobs, and R. C. Banks. 1983. Marine birds of the southeastern United States and Gulf of Mexico. Part III: Charadriiformes. U.S. Fish and Wildlife Service FWS/OBS-83/30. 853 pp.

Clarke, T. A. 1992. Egg abundance and spawning biomass of the Hawaiian anchovy or nehu, *Encrasicholina purpurea*, during 1984–1988 in Kaneohe Bay, Hawaii. Pacific Science 42:325–343.

Coles, R., and J. Kuo. 1995. Seagrasses. Pages 39–57 in J. E. Maragos, M. N. A. Peterson, L. G. Eldredge, J. E. Bardach, and H. F. Takeuchi, editors. Marine and coastal biodiversity in the tropical island Pacific region. Volume 1. Species systematics and information management priorities. Program on Environment, East–West Center, Honolulu, Hawaii.

Coles, R., and W. Leelong. 1997. Seagrasses. *In* L. Eldredge, editor. Proceedings of a workshop on marine/coastal biodiversity in the tropical Pacific region. Volume 2. Population, development and conservation priorities. Program on Environment, East–West Center, Honolulu, Hawaii. In press.

Colin, P. L., D. M. Devaney, L. Hillis-Colinvaux, T. H. Suchanek, and J. T. Harrison III. 1986. Geology and biological zonation of the reef slope, 50–360 meters depth at Enewetak Atoll, Marshall Islands. Bulletin of Marine Science 38:111–128.

Collette, B. B., and C. E. Nauen. 1983. FAO species catalogue. Volume 2. Scombrids of the world. An annotated and illustrated catalogue of tunas, mackerels, bonitos, and related species known to date. United Nations Food and Agriculture Organization Fisheries Synopsis 125. Food and Agriculture Organization, Rome. 137 pp.

Collie, J. S. 1991. Adaptive strategies for management of fisheries resources in large marine ecosystems. Pages 225–242 in K. Sherman, L. M. Alexander, and B. D. Gold, editors. Food chains, yields, models, and management of large marine ecosystems. Westview Press, Boulder, Colo.

Courtenay, W. R., Jr. 1995. Marine fish introductions in southeastern Florida. American Fisheries Society, Introduced Fish Section Newsletter 14(1):2–3.

Craig, P., B. Ponwith, F. Aitaoto, and D. Hamm. 1993. The commercial, subsistence, and recreational fisheries of American Samoa. Pages 69–82 in G. Boehlert, editor. Fisheries of Hawaii and U.S.-associated Pacific Islands. Marine Fisheries Review 55.

Croxall, J. P., P. G. H. Evans, and R. W. Schreiber, editors. 1984. Status and conservation of the world's seabirds. International Council for Bird Preservation Technical Publication 2. 778 pp.

Culliton, T. J., M. A. Warren, T. R. Goodspeed, D. G. Remer, C. M. Blackwell, and J. J. McDonough III. 1990. Fifty years of population change along the nation's coasts. Coastal Trends Series Report 2. National Oceanic and Atmospheric Administration, National Ocean Service, Strategic Assessment Branch, Rockville, Md. 41 pp.

Daan, N. 1986. Results of recent time-series observations for monitoring trends in large marine ecosystems with a focus on the North Sea. Pages 145–174 *in* K. Sherman and L. M. Alexander, editors. Variability and management of large marine ecosystems. American Association for the Advancement of Science Selected Symposium 99. Westview Press, Boulder, Colo.

Dalzell, P. J. 1993. Small pelagic fishes. Pages 97–133 *in* A. Wright and L. Hill, editors. Nearshore marine resources of the South Pacific. Information for fisheries development and management. Forum Fisheries Agency, Honiara, Solomon Islands, and Institute of Pacific Studies, Suva, Fiji.

Davis, R. W., and G. S. Fargion, editors. 1996. Distribution and abundance of cetaceans in the north-central and western Gulf of Mexico: Final report. Vol. II. Technical Report OCS Study MMS 96-0027. Prepared by the Texas Institute of Oceanography and National Marine Fisheries Service. U.S. Department of the Interior, Minerals Management Service, Gulf of Mexico OCS Region, New Orleans, La. 357 pp.

Day, J. W., Jr., and N. J. Craig. 1981. Comparisons of effectiveness of management options for wetlands loss in the coastal zone of Louisiana. Pages 232–239 *in* Proceedings of the conference on coastal erosion and wetland modification in Louisiana: causes, consequences, and options. 5–7 October 1981. Baton Rouge, La.

Devick, W. S. 1991. Patterns of introductions of aquatic organisms to Hawaiian freshwater habitats. Pages 189–213 *in* W. S. Devick, editor. New directions in research, management and conservation of Hawaiian freshwater stream ecosystems. Proceedings of the 1990 symposium on freshwater stream biology and fisheries management. Department of Land and Natural Resources, Honolulu, Hawaii.

DeVoe, M. R., editor. 1992. Proceedings of the conference and workshop: Introductions and transfers of marine species. Achieving a balance between economic development and resource protection. South Carolina Sea Grant Consortium and National Oceanic and Atmospheric Administration, Charleston. 198 pp.

Donaldson, D. M., N. Sanders, Jr., R. Minkler, and P. A. Thompson, editors. 1996. SEMAP environmental and biological atlas of the Gulf of Mexico. Gulf States Marine Fisheries Commission, Ocean Springs, Miss. 284 pp.

Eckert, K. L. 1993. The biology and population status of marine turtles in the North Pacific Ocean. National Oceanic and Atmospheric Administration Technical Memorandum NMFS-SWFSC-186. 156 pp.

Eldredge, L. G. 1987. Poisons for fishing on coral reefs. Pages 61–64 *in* B. Salvat, editor. Human effects on coral reefs: facts and recommendations. Museum National d'Histoire Naturelle, École Pratique des Hautes Etudes, Antenne de Tahiti, Moorea, French Polynesia.

Eldredge, L. G. 1995. Status of crustacean systematics. Pages 161–169 *in* J. E. Maragos, M. N. A. Peterson, L. G. Eldredge, J. E. Bardach, and H. F. Takeuchi, editors. Marine and coastal biodiversity in the tropical island Pacific region. Volume 1. Species systematics and information management priorities. Program on Environment, East–West Center, Honolulu, Hawaii.

Ellison, J. C. 1995. Systematics and distribution of Pacific island mangroves. Pages 59–74 *in* J. E. Maragos, M. N. A. Peterson, L. G. Eldredge, J. E. Bardach, and H. F. Takeuchi, editors. Marine and coastal biodiversity in the tropical island Pacific region. Volume 1. Species systematics and information management priorities. Program on Environment, East–West Center, Honolulu, Hawaii.

Ellison, J. C. 1997. Mangroves. *In* L. Eldredge, editor. Proceedings of a workshop on marine/coastal biodiversity in the tropical Pacific region. Volume 2. Population, development and conservation priorities. Program on Environment, East–West Center, Honolulu, Hawaii. In press.

Engle, V. D., J. K. Summers, and G. R. Gaston. 1994. A benthic index of environmental condition of Gulf of Mexico estuaries. Estuaries 17:372–384.

Epstein, P. R. 1995. Framework for an integrated assessment of health, climate change, and ecosystem vulnerability. *In* M. E. Wilson, R. Levins, and A. Spielman, editors. Disease in evolution: global changes and emergence of infectious diseases. Annals of the New York Academy of Sciences 740:423–435.

Everson, A. E. 1995. Fishery utilization ("creel") survey of Kaneohe Bay, Hawaii. Pages 4–31 *in* Proceedings of the first biennial symposium for the main Hawaiian Island marine resource investigation. State of Hawaii Department of Land and Natural Resources, Division of Aquatic Resources Technical Report 95-01.

Faris, J., and K. Hart. 1995. Sea of debris: a summary of the third international conference of marine debris. Alaska Fisheries Science Center, National Marine Fisheries Service, Seattle, Wash. 54 pp.

Favorite, F., A. J. Dodimead, and K. Nasu. 1976. Oceanography of the subarctic Pacific region, 1960–71. International North Pacific Fisheries Commission, Bulletin 33. Vancouver, B.C. 187 pp.

Fenical, W., and P. R. Jensen. 1993. Marine microorganisms: a new biomedical resource. Pages 419–457 *in* D. H. Attaway and O. R. Zaborsky, editors. Marine biotechnology. I: Pharmaceutical and bioactive natural products. Plenum Publishing Company, New York.

Flint, E. N. 1997. Status of seabird populations and conservation in the tropical island Pacific. *In* L. Eldredge, editor. Proceedings of a workshop on marine/coastal biodiversity in the tropical Pacific region. Volume 2. Population, development and conservation priorities. Program on Environment, East–West Center, Honolulu, Hawaii. In press.

Forsyth, R. G., and G. H. Balazs. 1989. Species profiles: life histories and environmental requirements of coastal vertebrates and invertebrates, Pacific Ocean region. Report 1. Green turtle, *Chelonia mydas*. Department of the Army, U.S. Army Corps of Engineers, Environmental Effect Research Program Technical Report EL-89-10. 20 pp.

Fritts, T. H., A. B. Irvine, R. D. Jennings, L. A. Collum, W. Hoffman, and M. A. McGehee. 1983. Turtles, birds, and mammals in the northern Gulf of Mexico and nearby Atlantic waters. U.S. Fish and Wildlife Service FWS/OBS-82/65. 455 pp.

Fritts, T. H., and G. H. Rodda. 1995. Invasions of the brown tree snake. Pages 454–456 *in* E. T. LaRoe, G. S. Farris, C. E. Puckett, P. D. Doran, and M. J. Mac, editors. Our living resources: a report to the nation on the distribution, abundance, and health of U.S. plants, animals, and ecosystems. U.S. Department of the Interior, National Biological Service, Washington, D.C.

Garnett, M. C. 1984. Conservation of seabirds in the South Pacific region: a review. Pages 547–558 *in* J. P. Croxall, P. G. H. Evans, and R. W. Schreiber, editors. Status and conservation of the world's seabirds. International Council for Bird Preservation Technical Publication 2.

Gilmartin, M., and N. Revelante. 1974. The "island mass" effect on the phytoplankton and primary production of the Hawaiian Islands. Journal of Experimental Marine Biology and Ecology 16:181–204.

Gould, P. J., D. J. Forsell, and C. J. Lensink. 1982. Pelagic distribution and abundance of seabirds in the Gulf of Alaska and eastern Bering Sea. U.S. Fish and Wildlife Service FWS/OBS-82/48. 294 pp.

Gould, P. J., and R. Hobbs. 1993. Population dynamics of the Laysan and other albatrosses in the North Pacific. International North Pacific Fisheries Commission Bulletin 53 (III):485–497.

Graves, J. E., S. D. Ferris, and A. E. Dizon. 1984. High genetic similarity of Atlantic and Pacific skipjack tuna demonstrated with restriction endonuclease analysis of mitochondrial DNA. Marine Biology 79:315–319.

Graves, J. E., and J. R. McDowell. 1994. Genetic analysis of striped marlin (*Tetrapturus audax*) population structure in the Pacific Ocean. Canadian Journal of Fisheries and Aquatic Sciences 51:1762–1768.

Grigg, R. W. 1993. Precious coral fisheries of Hawaii and the U.S. Pacific Islands. Marine Fisheries Review 55:50–60.

Grosslein, M. D., and T. R. Azarovitz. 1982. Fish distribution. MESA New York Bight Atlas Monograph 15, New York Sea Grant Institute, Albany. 182 pp.

Guille, A., P. Laboute, and J. -L. Menon. 1986. Guide des étoiles de mer, oursins et autres échinoderms der lagon de Nouvelle-Calédonie. Editions de l'ORSTOM, Paris: 1-238.

Haight, W. R., and G. T. DiNardo. 1995. Status of lobster stocks in the northwestern Hawaiian Islands, 1994. Southwest Fisheries Science Center, National Marine Fisheries Service, Honolulu, Hawaii. Administrative Report 95-03. 17 pp.

Hain, J. H. W., M. J. Ratnaswamy, R. D. Kenneg, and H. E. Winn. 1992. The fin whale, *Balaenoptera physalus*, in waters of the northeastern United States continental shelf. Report of the International Whaling Commission 42:653-669.

Hallegraeff, G. M., and C. J. Bolch. 1992. Transport of diatom and dinoflagellate resting spores in ship's ballast water: implications for plankton biogeography and aquaculture. Journal of Plankton Research 14:1067–1084.

Hamm, D.C. 1993. The western Pacific fishery information network: a fisheries information system. Pages 102–108 *in* G. Boehlert, editor. Fisheries of Hawaii and U.S.-associated Pacific Islands. Marine Fisheries Review 55.

Hamm, D.C., and H. K. Lum. 1992. Preliminary results of the Hawaii smallboat fisheries survey. Southwest Fisheries Science Center, National Marine Fisheries Service, Administrative Report 92-08. Honolulu, Hawaii. 35 pp.

Harbison, G. R., and S. P. Volovik. 1993. The ctenophore, *Mnemiopsis leidyi*, in the Black Sea: a holoplanktonic organism transported in the ballast water of ships. Pages 25–36 *in* Aquatic Nuisance Species Task Force. Proceedings of the conference and workshop: nonindigenous estuarine and marine organisms (NEMO). National Oceanic and Atmospheric Administration Office of the Chief Scientist, Washington, D.C.

Hare, S. R., and R. C. Francis. 1995. Intervention analysis: a time series method for detecting interdecadal climate regime shifts. *In* Q. Tang and K. Sherman, editors. The large marine ecosystems of the Pacific Rim: a report of a symposium held in Qingdao, People's Republic of China, 8–11 October 1994. A Marine Conservation and Development Report. International Union for the Conservation of Nature and Natural Resources (IUCN), Gland, Switzerland. 168 pp.

Harper, D. E., Jr., and L. D. McKinney. 1980. Benthos. Chapter 5 *in* R. W. Hann, Jr., and R. E. Randall, editors. Evaluation of brine disposal from the Bryan Mound Site of the Strategic Petroleum Reserve Program. Report to the U.S. Department of Energy, contract DE-FC96-79P010114. Texas A&M University Research Foundation, College Station.

Harrison, C. S., M. B. Naughton, and S. I. Fefer. 1984. The status and conservation of seabirds in the Hawaiian Archipelago and Johnston Atoll. Pages 513–526 *in* J. P. Croxall, P. G. H. Evans, and R. W. Schreiber, editors. Status and conservation of the world's seabirds. International Council for Bird Preservation Technical Publication 2.

Harrison, J. T. 1987. The 40 MWe OTEC plant at Kahe Point, Oahu, Hawaii: a case study of potential biological impacts. National Oceanic and Atmospheric Administration Technical Memorandum NMFS-SWFC-68. 105 pp.

Hatch, S. A., and J. F. Piatt. 1995. Seabirds in Alaska. Pages 49–52 *in* E. T. LaRoe, G. S. Farris, C. E. Puckett, P. D. Doran, and M. J. Mac, editors. Our living resources: a report to the nation on the distribution, abundance, and health of U.S. plants, animals, and ecosystems. U.S. Department of the Interior, National Biological Service, Washington, D.C.

Herbst, L. H. 1994. Fibropapillomatosis of marine turtles. Annual Review of Fish Diseases 4:389–425.

Hobbs, R. C., and L. L. Jones. 1993. Impacts of high seas driftnet fisheries on marine mammal populations in the North Pacific. International North Pacific Fisheries Commission Bulletin 53:409–434.

Hoelzel, A. R. 1994. Genetics and ecology of whales and dolphins. Annual Review of Ecology and Systematics 25:377–399.

Holthus, P. F., and J. E. Maragos. 1995. Marine ecosystem classification for the tropical island Pacific. Pages 239–278 *in* J. E. Maragos, M. N. A. Peterson, L. G. Eldredge, J. E. Bardach, and H. F. Takeuchi, editors. Marine and coastal biodiversity in the tropical island Pacific region. Volume 1. Species systematics and information management priorities. Program on Environment, East–West Center, Honolulu, Hawaii.

Hunter, C. L., and C. W. Evans. 1995. Coral reefs in Kaneohe Bay, Hawaii: two centuries of western influence and two decades of data. Bulletin of Marine Science 57:501–515.

Intergovernmental Oceanographic Commission. 1993. Report of the IOC Blue Ribbon Panel for a Global Ocean Observing System (GOOS). The case for GOOS. IOC/INF-915 Corr. Paris. SC-93/WS3.

International Council for the Exploration of the Sea. 1992. Reports of the ICES Advisory Committee on Fishery Management, 1991, part 1. ICES Cooperative Research Report 179. 368 pp.

Jannasch, H. W. 1995. Deep-sea vents as sources of biotechnologically relevant microorganisms. Journal of Marine Biotechnology 3:5–8.

Jefferson, T. A., S. Leatherwood, and M. A. Webber. 1993. FAO species identification guide. Marine mammals of the world. United Nations Food and Agriculture Organization, Rome. 320 pp.

Jehl, J. R., Jr. 1984. Conservation problems of seabirds in Baja California and the Pacific Northwest. Pages 41–48 *in* J. P. Croxall, P. G. H. Evans, and R. W. Schreiber, editors. Status and conservation of the world's seabirds. International Council for Bird Preservation. Technical Publication 2.

Johnson, D. H., T. L. Shaffer, and P. J. Gould. 1993. Incidental catch of marine birds in the north Pacific high seas driftnet fisheries in 1990. Pages 473–483 *in* J. Ito, W. Shaw, and R. L. Burgner, editors. Symposium on biology, distribution and stock assessment of species caught in the high seas driftnet fisheries in the North Pacific Ocean. International North Pacific Fisheries Commission Bulletin 53.

Johnson, L. L., M. S. Myers, D. Goyette, and R. F. Addison. 1994. Toxic chemicals and fish health in Puget Sound and the Strait of Georgia. Canadian Technical Report of Fisheries and Aquatic Sciences 1948:304–327.

Kay, E. A. 1980. Little worlds of the Pacific. An essay on Pacific basin biogeography. University of Hawaii Lyon Arboretum Lecture 9. University of Hawaii Press, Honolulu. 40 pp.

Kay, E. A. 1987. Introduction. Pages 1–9 *in* D. M. Devaney and L. G. Eldredge, editors. Reef and shore fauna of Hawaii. Section 2: Platyhelminthes through Phoronida. Section 3: Sipuncula through Annelida. Bishop Museum Special Publication 64 (2 and 3). Bishop Museum Press, Honolulu, Hawaii.

Kay, E. A. 1995. Pacific island marine mollusks: systematics. Pages 135–159 *in* J. E. Maragos, M. N. A. Peterson, L. G. Eldredge, J. E. Bardach, and H. F. Takeuchi, editors. Marine and coastal biodiversity in the tropical island Pacific region. Volume 1. Species systematics and information management priorities. Program on Environment, East–West Center, Honolulu, Hawaii.

Kelly-Borges, M., and C. Valentine. 1995. The sponges of the tropical island region of Oceania: a taxonomic status review. Pages 83–120 *in* J. E. Maragos, M. N. A. Peterson, L. G. Eldredge, J. E. Bardach, and H. F. Takeuchi, editors. Marine and coastal biodiversity in the tropical island Pacific region. Volume 1. Species systematics and information management priorities. Program on Environment, East–West Center, Honolulu, Hawaii.

King, M. 1993. Deepwater shrimp. Pages 513–538 *in* A. Wright and L. Hill, editors. Nearshore marine resources of the South Pacific. Information for fisheries development and management. Forum Fisheries Agency, Honiara, Solomon Islands, and Institute of Pacific Studies, Suva, Fiji.

Kinzie, R. A., III. 1990. Species profiles: life histories and environmental requirements of coastal vertebrates and invertebrates, Pacific Ocean region; Report 3.

Amphidromous macrofauna of Hawaiian Island streams. U.S. Army Corps of Engineers, Environmental Impact Research Program Technical Report EL-89-10. 28 pp.

Klinowska, M., compiler. 1991. Dolphins, porpoises, and whales of the world. The IUCN Red Data Book. International Union for the Conservation of Nature and Natural Resources (IUCN), Gland, Switzerland. 429 pp.

Kosaki, R. K., R. L. Pyle, J. E. Randall, and D. K. Irons. 1991. New records of fishes from Johnston Atoll, with notes on biogeography. Pacific Science 45:186–203.

Leis, J. M. 1994. Coral Sea atoll lagoons: closed nurseries for the larvae of a few coral reef fishes. Bulletin of Marine Science 54:206–227.

Lensink, C. J. 1984. The status and conservation of seabirds in Alaska. Pages 13–17 *in* J. P. Croxall, P. G. H. Evans, and R. W. Schreiber, editors. Status and conservation of the world's seabirds. International Council for Bird Preservation, Technical Publication 2.

Levin, S. A. 1990. Physical and biological scales, and modeling of predator-prey interactions in large marine ecosystems. Pages 179–187 *in* K. Sherman, L. M. Alexander, and B. D. Gold, editors. Large marine ecosystems: patterns, processes, and yields. American Association for the Advancement of Science Press, Washington, D.C.

Levings, C. D., and R. M. Thom. 1994. Habitat changes in Georgia Basin: implications for resource management and restoration. Canadian Technical Report of Fisheries and Aquatic Sciences 1948:330–351.

Likens, G. E. 1992. The ecosystem approach: its use and abuse. *In* O. Kinne, editor. Excellence in ecology. Volume 3. Ecology Institute, W-2124 Oldendorf/Luhe, Germany. 166 pp.

Lindley, S. T., R. R. Bidigare, and R. T. Barber. 1995. Phytoplankton photosynthesis parameters along 140W in the equatorial Pacific. Deep-Sea Research Part II, 42(2–3):441–463.

Livingston, P. A. 1993. Importance of predation by groundfish, marine mammals and birds on walleye pollock and Pacific herring in the eastern Bering Sea. Marine Ecology Progress Series 102:205–215.

Livingston, P., L. Low, and R. Marasco. 1995. Eastern Bering Sea ecosystem trends. *In* Q. Tang and K. Sherman, editors. The large marine ecosystems of the Pacific Rim: a report of a symposium held in Qingdao, People's Republic of China, 8–10 October 1994. A Marine Conservation and Development Report. International Union for the Conservation of Nature and Natural Resources (IUCN), Gland, Switzerland. 168 pp.

Lucas, J. S. 1994. The biology, exploitation, and mariculture of giant clams (Tridacnidae). Reviews in Fisheries Science 2:181–224.

Mackey, D. J., J. Parslow, H. W. Higgins, F. B. Griffiths, and J. E. O'Sullivan. 1995. Plankton productivity and biomass in the western equatorial Pacific: biological and physical controls. Deep-Sea Research Part II 42(2–3):499–533.

Mahaffy, M. S., D. R. Nysewander, K. Vermeer, T. R. Wahl, and P. E. Whitehead. 1994. Status, trends and potential threats related to birds in the Strait of Georgia, Puget Sound and Juan de Fuca Strait. Canadian Technical Report of Fisheries and Aquatic Sciences 1948:256–281.

Malone, T. C., T. S. Hopkins, P. G. Falkowski, and T. E. Whitledge. 1983. Production and transport of phytoplankton biomass over the continental shelf of the New York Bight. Continental Shelf Research 1:305–337.

Mangel, M. 1991. Empirical and theoretical aspects of fisheries yield models for large marine ecosystems. Pages 243–261 *in* K. Sherman, L. M. Alexander, and B. D. Gold, editors. Food chains, yields, models, and management of large marine ecosystems. Westview Press, Inc., Boulder, Colo.

Mangel, M., R. J. Hofman, E. A. Norse, and J. R. Twiss, Jr. 1993. Sustainability and ecological research. Ecological Applications 3:573–575.

Mann, K. H., and J. R. N. Lazier. 1991. Dynamics of marine ecosystems. Biological-physical interactions in the oceans. Blackwell Scientific Publications, Boston, Mass. 466 pp.

Mann, R., E. M. Burreson, and P. K. Baker. 1992. The decline of the Virginia oyster fishery in Chesapeake Bay: considerations for introduction of a non-endemic species, *Crassostrea gigas* (Thunberg). Pages 107–120 *in* M. R. DeVoe, editor. Proceedings of the conference and workshop: introductions and transfers of marine species. Achieving a balance between economic development and resource protection. South Carolina Sea Grant Consortium, Charleston.

Manning, J. 1991. Middle Atlantic Bight salinity: interannual variability. Continental Shelf Research 11:123–137.

Maragos, J. E. 1993. Impact of coastal construction on coral reefs in the U.S.-affiliated Pacific Islands. Coastal Management 21:235–269.

Márquez, M., R. 1990. FAO species catalogue. Volume 11: Sea turtles of the world. An annotated and illustrated catalogue of sea turtle species known to date. United Nations Food and Agriculture Organization Fisheries Synopsis, 125. Food and Agriculture Organization, Rome. 81 pp.

Marshall, D. B. 1988. Status of the marbled murrelet in North America: with special emphasis on populations in California, Oregon, and Washington. U.S. Fish and Wildlife Service Biological Report 88(30). 19 pp.

McCarthy, S. A., and F. M. Khambaty. 1994. International dissemination of epidemic *Vibrio cholerae* by cargo ship ballast and other nonpotable waters. Applied Environmental Microbiology 60:2597–2601.

McDermond, D. K., and K. H. Morgan. 1993. Status and conservation of North Pacific albatrosses. Pages 70–81 *in* K. Vermeer, K. T. Briggs, K. H. Morgan, and D. Siegel-Causey, editors. The status, ecology, and conservation of marine birds in the North Pacific. Canadian Wildlife Service Special Publication, Ottawa.

McGowan, J. A. 1974. The nature of oceanic ecosystems. Pages 9–28 *in* C. B. Miller, editor. The biology of the oceanic Pacific. Proceedings of the thirty-third annual biology colloquium. Oregon State University Press, Corvallis. 157 pp.

Mee, L. 1992. The Black Sea in crisis: a need for concerted international action. Ambio 21:278–286.

Mills, E. L. 1983. Problems of deep-sea biology: an historical perspective. Pages 1–79 *in* G. T. Rowe, editor. Deep-sea biology. The sea. Volume 8. John Wiley & Sons, New York.

Moffitt, R. B. 1993. Deepwater demersal fish. Pages 73–95 *in* A. Wright and L. Hill, editors. Nearshore marine resources of the South Pacific. Information for fisheries development and management. Forum Fisheries Agency, Honiara, Solomon Islands, and Institute of Pacific Studies, Suva, Fiji.

Morgan, J. R. 1988. Large marine ecosystems: an emerging concept of regional management. Environment 29(10):4–9, 26–34.

Motes, M., A. DePaola, S. Zywno-Van Ginkel, and M. McPhearson. 1994. Occurrence of toxigenic *Vibrio cholerae* 01 in oysters in Mobile Bay, Alabama: an ecological investigation. Journal of Food Protection 57:975–980.

Munro, J. L. 1993. Giant clams. Pages 431–449 *in* A. Wright and L. Hill, editors. Nearshore marine resources of the South Pacific. Information for fisheries development and management. Forum Fisheries Agency, Honiara, Solomon Islands, and Institute of Pacific Studies, Suva, Fiji.

Munro, J. L., and S. T. Fakahau. 1993. Appraisal, assessment and monitoring of small-scale coastal fisheries in the South Pacific region. Pages 15–53 *in* A. Wright and L. Hill, editors. Nearshore marine resources of the South Pacific. Information for fisheries development and management. Forum Fisheries Agency, Honiara, Solomon Islands, and Institute of Pacific Studies, Suva, Fiji.

Munro, J. L., J. D. Parrish, and F. H. Talbot. 1987. The biological effects of intensive fishing upon coral reef communities. Pages 41–49 *in* B. Salvat, editor. Human impacts on coral reefs: facts and recommendations. Museum National d'Histoire Naturelle, Ecole Pratique des Hautes Etudes, Antenne de Tahiti, Moorea, French Polynesia.

Murawski, S. A. 1996. Can we manage our multispecies fisheries? Pages 491–510 *in* K. Sherman, N. A. Jaworski, and T. J. Smayda, editors. The Northeast shelf ecosystem: assessment, sustainability, and management. Blackwell Science, Cambridge, Mass.

Musgrave, D. L., T. J. Weingartner, and T. C. Royer. 1992. Circulation and hydrography in the northwest Gulf of Alaska. Deep-Sea Research 39:1499–1519.

Myers, R. F. 1989. Micronesian reef fishes: a practical guide to the identification of the inshore marine fishes of the tropical central and western Pacific. Coral Graphics, Barrigada, Territory of Guam. 298 pp.

Myers, R. F. 1993. Guam's small-boat-based fisheries. Marine Fisheries Review 55:117–128.

Myers, R. F., and T. J. Donaldson. 1996. New and recent records of fishes from the Mariana Islands. Micronesica 28:207–266.

Nakamura, I. 1985. FAO species catalogue. Volume 5. Billfishes of the world. An annotated and illustrated catalogue of marlins, sailfishes, spearfishes and swordfishes known to date. United Nations Food and Agriculture Organization Species Synopsis 125. Food and Agriculture Organization, Rome. 65 pp.

Nalepa, T. F., and D. Schloesser, editors. 1993. Zebra mussels: biology, impacts, and control. Lewis Publishers, Boca Raton, Fla. 810 pp.

National Coastal Monitoring Act. 1992. One hundred second congress of the United States of America, at the second session. The National Oceanic and Atmospheric Administration Authorization Act of 1992, H.R. 2130, 29 October 1992. Washington, D.C.

National Oceanic and Atmospheric Administration. 1986. Gulf of Mexico coastal and ocean zones strategic assessment: data atlas. U.S. Government Printing Office, Washington, D.C., 163 maps and text.

National Oceanic and Atmospheric Administration. 1988. West coast of North America strategic assessment: data atlas. Marine mammal volume. Rockville, Md. 33 pp.

National Oceanic and Atmospheric Administration. 1989. Bering, Chukchi, and Beaufort seas coastal and ocean zones strategic assessment: data atlas. U.S. Government Printing Office, Washington, D.C. 107 maps and text.

National Oceanic and Atmospheric Administration. 1990a. Estuaries of the United States: vital statistics of a national resource base. National Oceanic and Atmospheric Administration/National Oceanic Survey, Strategic Assessment Branch, Rockville, Md. 79 pp.

National Oceanic and Atmospheric Administration. 1990b. West coast of North America strategic assessment: data atlas. Invertebrate and fish volume. Pre-publication edition. Rockville, Md. 112 pp.

National Oceanic and Atmospheric Administration. 1991a. Coastal wetlands of the United States: an accounting of a valuable national resource. National Oceanic and Atmospheric Administration, Strategic Assessment Branch, Ocean Assessments Division, Rockville, Md. 59 pp.

National Oceanic and Atmospheric Administration. 1991b. The 1990 national shellfish register of classified estuarine waters. National Oceanic and Atmospheric Administration, Strategic Assessment Branch, Ocean Assessments Division, Rockville, Md. 100 pp.

National Oceanic and Atmospheric Administration. 1991c. Our living oceans: report on the status of U.S. living marine resources, 1991. National Marine Fisheries Service, National Oceanic and Atmospheric Administration Technical Memorandum NMFS-F/SPO-1. Silver Spring, Md. 123 pp.

National Oceanic and Atmospheric Administration. 1992. Our living oceans: report on the status of U.S. living marine resources, 1992. National Marine Fisheries Service, National Oceanic and Atmospheric Administration Technical Memorandum NMFS-F/SPO-2. Washington, D.C. 148 pp.

National Oceanic and Atmospheric Administration. 1993. Our living oceans: report on the status of U.S. living marine resources, 1993. National Marine Fisheries Service, National Oceanic and Atmospheric Administration Technical Memorandum NMFS-F/SPO-15. Washington, D.C. 156 pp.

National Oceanic and Atmospheric Administration. 1996. Our living oceans: report on the status of U.S. living marine resources, 1995. National Marine Fisheries Service, National Oceanic and Atmospheric Administration Technical Memorandum NMFS-F/SPO-19. Washington, D.C. 160 pp.

National Research Council. 1995. Understanding marine biodiversity: a research agenda for the nation. National Academy Press, Washington, D.C. 114 pp.

Nichols, F. H., J. K. Thompson, and L. E. Schemel. 1990. Remarkable invasion of San Francisco Bay (California, USA) by the Asian clam *Potamocorbula amurensis*. II. Displacement of a former community. Marine Ecology Progress Series 66:95–101.

Niebauer, H. J. 1981. Recent fluctuations in sea ice distribution in the eastern Bering Sea. Pages 133–140 *in* D. Hood and J. Calder, editors. The eastern Bering Sea shelf: oceanography and resources. Volume 1. Alaska office, Office of Oceanography and Marine Assessment, National Ocean Survey, National Oceanic and Atmospheric Administration and Alaska Outer Continental Shelf Region, Minerals Management Service, Department of the Interior, Washington, D.C.

Norse, E. A. 1993. Global marine biological diversity: a strategy for building conservation into decision making. Island Press, Washington, D.C. 383 pp.

North Pacific Anadromous Fish Commission. 1993. Annual report 1993. North Pacific Anadromous Fish Commission, Vancouver, B.C. 82 pp.

Northeast Fisheries Science Center. 1993. Status of fishery resources off the northeastern United States for 1993. National Oceanic and Atmospheric Administration Technical Memorandum NMFS-F/NEC-101. 140 pp.

Northeast Fisheries Science Center. 1994. Report of the eighteenth Northeast regional stock assessment workshop (18th SAW), Stock Assessment Review Committee (SARC) consensus summary of assessments. NOAA/NMFS/NEFSC Reference Document 94-22. 281 pp.

Olaizola, M., D. A. Ziemann, P. K. Bienfang, W. A. Walsh, and L. D. Conquest. 1993. Eddy-induced oscillations of the pycnocline affect the floristic composition and depth distribution of phytoplankton in the subtropical Pacific. Marine Biology 116:533–542.

O'Reilly, J. E., C. Evans-Zetlin, and D. A. Busch. 1987. Primary production. Pages 220–233 *in* R. H. Backus and D. W. Bourne, editors. Georges Bank. Massachusetts Institute of Technology Press, Cambridge, Mass.

Palumbi, S. R. 1994. Genetic divergence, reproductive isolation, and marine speciation. Annual Review of Ecology and Systematics 25:547–572.

Parrish, F. A. 1989. Identification of habitat of juvenile snappers in Hawaii. U.S. Fishery Bulletin 87:1001–1005.

Parrish, F. A., and J. J. Polovina. 1994. Habitat thresholds and bottlenecks in production of the spiny lobster (*Panulirus marginatus*) in the northwestern Hawaiian Islands. Bulletin of Marine Science 54:151–163.

Patteson, J. B., and E. S. Brinkley. 1994. Birding pelagic frontiers: North Carolina, Virginia, and Maryland. Winging It 6(1):1, 4–8.

Pawson, D. L. 1995. Echinoderms of the tropical island Pacific: status of their systematics and notes on their ecology and biogeography. Pages 171–192 *in* J. E. Maragos, M. N. A. Peterson, L. G. Eldredge, J. E. Bardach, and H. F. Takeuchi, editors. Marine and coastal biodiversity in the tropical island Pacific region. Volume 1. Species systematics and information management priorities. Program on Environment, East–West Center, Honolulu, Hawaii.

Peake, D. E., and M. Elwonger. 1996. A new frontier: pelagic birding in the Gulf of Mexico. Winging It 8(1):1, 4–9.

Pereyra, W. T., J. E. Reeves, and R. G. Bakkala. 1976. Demersal fish and shellfish resources of the eastern Bering Sea in the baseline year 1975. National Marine Fisheries Service, Northwest and Alaska Fisheries Center, Processed Report. Seattle, Wash. 619 pp.

Pitcher, C. R. 1993. Spiny lobster. Pages 539–607 *in* A. Wright and L. Hill, editors. Nearshore marine resources of the South Pacific. Information for fisheries development and management. Forum Fisheries Agency, Honiara, Solomon Islands, and Institute of Pacific Studies, Suva, Fiji.

Polovina, J. J. 1984. ECOPATH—an ecosystem model applied to French Frigate Shoals. Pages 375–398 *in* R. W. Grigg and K. Y. Tanoue, editors. Proceedings of the second symposium on resource investigations in the northwestern Hawaiian Islands. Volume 1. Sea Grant Miscellaneous Report UNIHISEAGRANT-84-01. University of Hawaii, Honolulu.

Polovina, J. J. 1993. The lobster and shrimp fisheries in Hawaii. Marine Fisheries Review 55:28–33.

Polovina, J. J., G. T. Mitchum, N. E. Graham, M. P. Craig, E. E. DeMartini, and E. N. Flint. 1994. Physical and biological consequences of a climate event in the central North Pacific. Fisheries Oceanography 3:15–21.

Pyle, R. L. 1995. Pacific reef and shore fishes. Pages 205–238 *in* J. E. Maragos, M. N. A. Peterson, L. G. Eldredge, J. E. Bardach, and H. F. Takeuchi, editors. Marine and coastal biodiversity in the tropical island Pacific region. Volume 1. Species systematics and information management priorities. Program on Environment, East–West Center, Honolulu, Hawaii.

Rabalais, N. N., and D. F. Boesch. 1987. Dominant features and processes of continental shelf environments of the United States. Pages 71–147 *in* D. F. Boesch and N. N. Rabalais, editors. Long-term environmental effects of offshore oil and gas development. Elsevier Applied Science, London and New York.

Randall, J. E. 1992. Endemism of fishes in Oceania. Pages 55–67 *in* United Nations Environmental Program: coastal resources and systems of the Pacific basin: investigation and steps toward protective management. United Nations Environmental Program Regional Seas Reports and Studies 147.

Randall, J. E. 1995. Zoogeographic analysis of the inshore Hawaiian fish fauna. Pages 193–203 *in* J. E. Maragos, M. N. A. Peterson, L. G. Eldredge, J. E. Bardach, and H. F. Takeuchi, editors. Marine and coastal biodiversity in the tropical island Pacific region. Volume 1. Species systematics and information management priorities. Program on Environment, East–West Center, Honolulu, Hawaii.

Ray, G. C., M. G. McCormick-Ray, J. A. Dobbin, C. N. Ehler, and D. J. Basta. 1980. Eastern United States coastal and ocean zones data atlas. National Oceanic and Atmospheric Administration. Washington, D.C. 127 maps and text.

Reichel, J. D. 1991. Status and conservation of seabirds in the Mariana Islands. International Council for Bird Preservation Technical Publication 11:249–262.

Richards, W. J., and J. A. Bohnsack. 1990. The Caribbean Sea: a large marine ecosystem in crisis. Pages 44–53 *in* K. Sherman, L. M. Alexander, and B. D. Gold, editors. Large marine ecosystems: patterns, processes and yields. American Association for the Advancement of Science Press, Washington, D.C. 242 pp.

Richards, W. J., M. G. McGowan, T. Leming, J. T. Lamkin, and S. Kelley. 1993. Larval fish assemblages at the Loop Current boundary in the Gulf of Mexico. Bulletin of Marine Science 53:475–537.

Roemmich, D., and J. McGowan. 1995. Climatic warming and the decline of zooplankton in the California Current. Science 267:1324–1326.

Rogers, A. D. 1994. The biology of seamounts. Advances in Marine Biology 30:305–350.

Rosenberg, A. A., M. J. Fogarty, M. P. Sissenwine, J. R. Beddington, and J. G. Shepherd. 1993. Achieving sustainable use of renewable resources. Science 262:828–829.

Rosenfield, A., and R. Mann, editors. 1992. Dispersal of living organisms in aquatic systems. Maryland Sea Grant College, University of Maryland, College Park. 471 pp.

San Francisco Estuary Project. 1992. State of the estuary. Association of Bay Area Governments, Oakland, Calif. 269 pp.

Schneider, D.C., and D. Heinemann. 1996. The state of marine bird populations from Cape Hatteras to the Gulf of Maine. Pages 197–216 *in* K. Sherman, N. A. Jaworski, and T. J. Smayda, editors. The Northeast shelf ecosystem: assessment, sustainability, and management. Blackwell Science, Cambridge, Mass.

Schumacher, J. D., and A. W. Kendall, Jr. 1995. An example of fisheries oceanography: walleye pollock in Alaskan waters. Review of Geophysics Supplement 1153-1163.

Scoles, D. R., and J. E. Graves. 1993. Genetic analysis of the population structure of yellowfin tuna, *Thunnus albacares*, from the Pacific Ocean. U.S. Fishery Bulletin 91:690–698.

Sherman, K. 1994. Sustainability, biomass yields, and health of coastal ecosystems: an ecological perspective. Marine Ecology Progress Series 112:277–301.

Sherman, K., J. R. Green, J. R. Goulet, and L. Ejsymont. 1983. Coherence in zooplankton of a large Northwest Atlantic ecosystem. U.S. Fishery Bulletin 81:855–862.

Sherman, K., M. Grosslein, D. Mountain, D. Busch, J. O'Reilly, and R. Theroux. 1988. The continental shelf ecosystem off the northeast coast of the United States. Pages 279–337 *in* H. Postma and J. J. Zijlstra, editors. Continental shelves. Ecosystems of the world, Volume 27. Elsevier Press, New York and Amsterdam.

Sherman, K., and A. R. Solow. 1992. The changing states and health of a large marine ecosystem. International Council for the Exploration of the Sea, Document C. M. 1992/L:38. 31 pp.

Sherman, K., A. Solow, J. Green, and J. Jossi. 1994. Multidecadal stability, resilience, and diversity of the zooplankton in a stressed large marine ecosystem. International Council for the Exploration of the Sea Symposium on Zooplankton Production, Plymouth, U. K.

Skillman, R. A., C. H. Boggs, and S. G. Pooley. 1993. Fishery interaction between the tuna longline and other pelagic fisheries in Hawaii. U.S. Department of Commerce, National Oceanic and Atmospheric Administration Technical Memorandum, NMFS-SWFSC-189. 46 pp.

Smayda, T. 1991. Global epidemic of noxious phytoplankton blooms and food chain consequences in large ecosystems. Pages 275–308 *in* K. Sherman, L. M. Alexander, and B. D. Gold, editors. Food chains, yields, models, and management of large marine ecosystems. Westview Press, Boulder, Colo.

Smith, M. K. 1993. An ecological perspective on inshore fisheries in the main Hawaiian Islands. Pages 34–49 *in* G. Boehlert, editor. Fisheries of Hawaii and U.S.-associated Pacific Islands. Marine Fisheries Review 55.

Southeast Fisheries Science Center. 1995. Status of fishery resources off the southeastern United States for 1993. National Oceanic and Atmospheric Administration Technical Memorandum NMFS-SEFC.

Sowls, A. L., S. A. Hatch, and C. J. Lensink. 1978. Catalog of Alaskan seabird colonies. U.S. Fish and Wildlife Service FWS/OBS-78/78. 32 pp.

Springer, V. G. 1982. Pacific plate biogeography, with special reference to shorefishes. Smithsonian Contributions to Zoology 367:1–182.

Stallcup, R. 1990. Ocean birds of the nearshore Pacific. Point Reyes Bird Observatory, Stinson Beach, Calif. 214 pp.

Steadman, D. W. 1995. Prehistoric extinctions of Pacific Island birds: biodiversity meets zooarchaeology. Science 267:1123–1131.

Steimle, F. 1987. Benthic faunal production. Pages 310–314 *in* R. H. Backus and D. W. Bourne, editors. Georges Bank. Massachusetts Institute of Technology Press, Cambridge, Mass.

Steimle, F. 1990. Benthic macrofauna and habitat monitoring on the continental shelf of the northeastern United States. I. Biomass. National Oceanic and Atmospheric Administration Technical Report NMFS-86. 28 pp.

Stickney, R. F. 1984. Estuarine ecology of the southeastern United States and Gulf of Mexico. Texas A&M University Press, College Station. 310 pp.

Stoddart, D. R. 1992. Biogeography of the tropical Pacific. Pacific Science 46:276–293.

Stoker, S. 1981. Benthic invertebrate macrofauna of the eastern Bering/Chuckchi Continental Shelf. Pages 1069–1990 *in* D. W. Hood and J. A. Calder, editors. The eastern Bering Sea shelf: oceanography and resources. University of Washington Press, Seattle.

Stone, G. S., R. R. Reeves, S. Leatherwood, and L. G. Eldredge. 1997. Marine mammals in the area served by the South Pacific Regional Environment Program (SPREP). *In* L. Eldredge, editor. Proceedings of a workshop on marine/coastal biodiversity in the tropical Pacific region. Volume 2. Population, development and conservation priorities. Program on Environment, East–West Center, Honolulu, Hawaii. In press.

Taylor, F. J. R., and R. A. Horner. 1994. Red tides and other problems with harmful algal blooms in Pacific Northwest coastal waters. Canadian Technical Report Fisheries and Aquatic Science 1948:175–186.

Tenore, K. R. 1979. Macroinfaunal benthos of South Atlantic/Georgia Bight. Pages 281–308 *in* South Atlantic Benchmark Program, Outer Continental Shelf (OCS) Environmental Studies, Volume 3, Results of Georgia Bight of South Atlantic Ocean. Final Report. Contract AA550-CT7-2 to Bureau of Land Management, Texas Instruments, Inc., Dallas, Texas.

Thomson, A. J., and S. McKinnell. 1995. Summary of reported Atlantic salmon (*Salmo salar*) catches and sightings in British Columbia and adjacent waters in 1994. Canadian Manuscript Report Fisheries and Aquatic Science 2304. 30 pp.

Turek, J. G., T. E. Goodger, T. E. Bigford, and J. S. Nichols. 1987. Influence of freshwater inflows on estuarine productivity. National Oceanic and Atmospheric Administration

Technical Memorandum NMFS-F/NEC-46. 26 pp.

Twichell, D., B. Butman, and R. Lewis. 1987. Shallow structure, surficial geology, and the processes currently shaping the bank. Pages 31–37 *in* R. H. Backus and D. W. Bourne, editors. Georges Bank. Massachusetts Institute of Technology Press, Cambridge, Mass.

United Nations Environment Programme/ International Union for the Conservation of Nature and Natural Resources. 1988. Coral reefs of the world. Volume 3. Central and western Pacific. United Nations Environmental Program Regional Seas Directories and Bibliographies. IUCN, Gland, Switzerland, and Cambridge, UK/UNEP, Nairobi, Kenya. 329 pp.

U.S. Army Corps of Engineers. 1995. U.S. Army Corps of Engineers Permit PODCO 95-002, Special Condition 2, issued in September 1995. U.S. Army Corps of Engineers, Honolulu, Hawaii.

U.S. Congress, Office of Technology Assessment. 1993. Harmful non-indigenous species in the United States. OTA-F-565. U.S. Office of Technology Assessment, Washington, D.C. 391 pp.

U.S. Fish and Wildlife Service. 1992. Alaska Seabird Management Plan. U.S. Fish and Wildlife Service, Anchorage, Alaska. 102 pp.

U.S. Fish and Wildlife Service. 1995. Draft marbled murrelet (*Brachyramphus marmoratus*) (Washington, Oregon, and California population) Recovery Plan. Portland, Oreg. 171 pp.

van Halewyn, R., and R. L. Norton. 1984. The status and conservation of seabirds in the Caribbean. Pages 169–222 *in* J. P. Croxall, P. G. H. Evans, and R. W. Schreiber, editors. Status and conservation of the world's seabirds. International Council for Bird Preservation, Technical Publication 2.

Vermeer, K., K. T. Briggs, K. H. Morgan, and D. Seigel-Causey, editors. 1993. The status, ecology, and conservation of marine birds of the North Pacific. Special Publication, Canadian Wildlife Service. 263 pp.

Veron, J. E. N. 1995. Corals of the tropical island Pacific region: biodiversity. Pages 75–82 *in* J. E. Maragos, M. N. A. Peterson, L. G. Eldredge, J. E. Bardach, and H. F. Takeuchi, editors. Marine and coastal biodiversity in the tropical island Pacific region. Volume 1. Species systematics and information management priorities. Program on Environment, East–West Center, Honolulu, Hawaii.

Washington Department of Fish and Wildlife and Oregon Department of Fish and Wildlife. 1994. Status report: Columbia River fish runs and fisheries, 1938–93. Washington Department of Fish and Wildlife, Olympia, and Oregon Department of Fish and Wildlife, Portland. 271 pp.

Wenner, E. L., and D. M. Knott. 1992. Occurrence of Pacific white shrimp, *Penaeus vannamei*, in coastal waters of South Carolina. Pages 173–181 *in* M. R. DeVoe, editor. Proceedings of the conference and workshop: introductions and transfers of marine species. Achieving a balance between economic development and resource protection. South Carolina Sea Grant Consortium.

Wetherall, J. A., G. H. Balazs, R. A. Tokunaga, and M. Y. Y. Yong. 1993. Bycatch of marine turtles in North Pacific high-seas driftnet fisheries and impacts on the stocks. International North Pacific Fisheries Commission Bulletin 53 (III):519–538.

Wetherall, J. A., W. R. Haight, and G. T. DiNardo. 1995. Computation of the preliminary 1995 catch quota for the northwestern Hawaiian Islands lobster fishery. National Marine Fisheries Service, Southwest Fisheries Science Center Administrative Report 95-04. Honolulu, Hawaii. 35 pp.

Wigley, R., and R. Theroux. 1981. Atlantic continental shelf and slope of the United states—macrobenthic invertebrate fauna of the Middle Atlantic Bight region—faunal composition and quantitative distribution. U.S. Geological Survey Professional Paper 529-N. 198 pp.

Wilkinson, C. R. 1992. Coral reefs of the world are facing widespread devastation: can we prevent this through sustainable management practices? Proceedings of the seventh annual international coral reef symposium, Guam. Volume 1:11–21.

Wing, B. L., C. M. Guthrie, and A. J. Gharett. 1992. Atlantic salmon in marine waters of southeastern Alaska. Transactions of the American Fisheries Society 121:814–818.

Winston, J. E. 1992. Systematics and marine conservation. Pages 144–168 *in* N. Eldridge, editor. Systematics, ecology, and the biodiversity crisis. Columbia University Press, New York.

Woodward-Clyde Consultants and Continental Shelf Association, Inc. 1983. Southwest Florida shelf ecosystem study— year 1. Final report. Contract 14-12-0001-29142 to Mineral Management Service, Gulf of Mexico Outer Continental Shelf Region, Metairie, La. 2 volumes. Unpaginated.

Wright, A. 1993. Shallow water reef-associated finfish. Pages 203–284 *in* A. Wright and L. Hill, editors. Nearshore marine resources of the South Pacific. Information for fisheries development and management. Forum Fisheries Agency, Honiara, Solomon Islands, and Institute of Pacific Studies, Suva, Fiji.

Yoder, J. A. 1991. Warm-temperate food chains of the Southeast shelf ecosystem. Pages 49–66 *in* K. Sherman, L. M. Alexander, and B. D. Gold, editors. Food chains, yields, models, and management of large marine ecosystems. Westview Press, Boulder, Colo.

North Atlantic Right Whale

Blaylock, R. A., J. W. Hain, L. J. Hansen, D. L. Palka, and G. T. Waring. 1995. U.S. Atlantic and Gulf of Mexico marine mammal stock assessments. National Oceanic and Atmospheric Administration Technical Memorandum NMFS-SEFSC-363. 211 pp.

Kenny, R. D., A. M. Hyman, R. E. Owen, G. P. Scott, and H. E. Winn. 1986. Estimation of prey densities required by western North Atlantic right whales. Marine Mammal Science 2:1–13.

Knowlton, A. R., J. Sigurjonsson, J. N. Ciano, and S. D. Kraus. 1994. Long distance movements of North Atlantic right whales (*Eubalaena glacialis*). Marine Mammal Science 8:397–405.

Kraus, S. D. 1990. Rates and potential causes of mortality in North Atlantic right whales (*Eubalaena glacialis*). Marine Mammal Science 6:278–291.

Mate, B. R., S. Nieukirk, R. Mescar, and T. Martin. 1992. Application of remote sensing methods for tracking large cetaceans: North Atlantic right whales (*Eubalaena glacialis*). Final report to the U.S. Minerals Management Service, Contract Number 14-12-0001-30411, Reston, Va. 167 pp.

National Marine Fisheries Service. 1991. Recovery plan for the northern right whale (*Eubalaena glacialis*). Prepared by the Right Whale Recovery Team for the National Marine Fisheries Service, Silver Spring, Md. 86 pp.

Schaeff, C. M., S. D. Kraus, M. W. Brown, and B. N. White. 1993. Assessment of the population structure of the western North Atlantic right whales (*Eubalaena glacialis*) based on sighting and mtDNA data. Canadian Journal of Zoology 71:339–345.

Linkages Between Coastal Wetlands and Fishery Resources

Boesch, D. F., and R. E. Turner. 1984. Dependence of fishery species on salt marshes: the role of food and refuge. Estuaries 7:460-468.

Hoss, D. E., and G. W. Thayer. 1993. The importance of habitat to the early life history of estuarine dependent fishes. American Fisheries Society Symposium 14:147-158.

Kenworthy, W. J., G. W. Thayer, and M. S. Fonseca. 1988. The utilization of seagrass meadows by fishery organisms. Pages 548–560 *in* D. D. Hook, et al. editors. Ecology of wetlands. Timber Press, Oregon.

Klima, E. F., J. M. Nance, E. X. Martinez, and T. Leary. 1990. Workshop on definition of shrimp recruitment overfishing. NOAA Technical Memorandum SEFC-NMFS-264.

McIvor, C., and L. P. Rozas. 1996. Direct use of intertidal saltmarsh habitat and linkage with adjacent habitats: a review from the southeastern United States. Pages 311–334 *in* K. F. Nordstrom and C. T. Roman, editors. Estuarine shores: evolution, environments and human alterations. John Wiley & Sons, New York.

Minello, T. J., and R. J. Zimmerman. 1991. The role of estuarine habitats in regulating growth and survival of juvenile penaeid shrimp. Pages 1–16 *in* P. DeLoach, W. J. Dougherty, and M. A. Davidson, editors. Frontiers in shrimp research. Elsevier Scientific, Amsterdam.

Minello, T. J., R. J. Zimmerman, and E. X. Martinez. 1989. Mortality of young brown shrimp *Penaeus aztecus* in estuarine nurseries. Transactions of the American Fisheries Society 118:693–708.

Nixon, S. W. 1980. Between coastal marshes and coastal waters—a review of twenty years of speculation and research on the role of salt marshes in estuarine productivity and water chemistry. Pages 437–524 *in* P. Hamilton, and K. B. Macdonald, editors. Estuarine and wetland processes with emphasis on modeling. Plenum Press, New York.

Peterson, G. W., and R. E. Turner. 1994. The value of salt marsh edge vs. interior as a habitat for fish and decapod crustaceans in a Louisiana tidal marsh. Estuaries 17:235–262.

Rozas, L. P. 1993. Nekton use of salt marshes of the Southeast region of the United States. Pages 528–536 *in* O. Magoon, W.S. Wilson, H. Converse, and L. T. Tobin, editors. Coastal Zone '93, Volume 2. Proceedings of the 8th Symposium on Coastal and Ocean Management. American Society Of Civil Engineers, New York.

Rozas, L. P. 1995. Hydroperiod and its influence on nekton use of the salt marsh: a pulsing ecosystem. Estuaries 18:579–590.

Rozas, L. P., and B. J. Reed. 1993. Nekton use of marsh-surface habitats in Louisiana (USA) deltaic salt marshes undergoing submergence. Marine Ecology Progress Series 96:147–157.

Smith, J. 1991. The Atlantic and Gulf menhaden purse seine fisheries: origins, harvesting technologies, biostatistical monitoring, recent trends in fisheries statistics, and forecasting. Marine Fisheries Review. 53:28–41.

Turner, R. E. 1977. Intertidal vegetation and commercial yields of penaeid shrimp. Transactions of the American Fisheries Society 106:411–416.

Turner, R. E. 1997. Wetland loss in the northern Gulf of Mexico: multiple working hypotheses. Estuaries 20:1–13.

Zimmerman, R. J., and T. J. Minello. 1984. Densities of *Penaeus aztecus*, *P. setiferus* and other natant macrofauna in a Texas salt marsh. Estuaries 7: 421–433.

Zimmerman, R. J., T. J. Minello, E. F. Klima, and J. M. Nance. 1991. Effects of accelerated sea-level rise on coastal secondary production. Pages 110–124 *in* H. S. Bolton, editor. Coastal wetlands. American Society of Civil Engineers, New York.

Bottlenose Dolphins

Brager, S., B. Wursig, A. Acevedo, and T. Henningsen. 1994. Association patterns of bottlenose dolphins (*Tursiops truncatus*) in Galveston Bay, Texas. Journal of Mammalogy 75:431–437.

Duffield, D. A., and R. S. Wells. 1986. Population structure of bottlenose dolphin: genetic studies of bottlenose dolphins along the central west coast of Florida. Contract report to the National Marine Fisheries Service. Southeast Fisheries Science Center. Contract Number 45. WCNF-5-00366. 16 pp.

Henningsen, T. 1991. Zur vertreitung ünd okologie des groben tummlers (*Tursiops*

truncatus*) in Galveston, Texas. Ph.D. thesis, Christian-Albrechts-Universitat, Kiel, Germany. 80 pp.

Lipscomb, T. P. 1994. Morbilliviral disease in an Atlantic bottlenose dolphin (*Tursiops truncatus*) from the Gulf of Mexico. Journal of Wildlife Diseases 30:572–576.

Lohoefener, R., W. Hoggard, K. Mullin, R. Ford, and J. Benigno. 1990. Studies of Mississippi Sound bottlenose dolphins: estimates of bottlenose dolphin density in Mississippi Sound from small boat surveys. Part 2 (of 2) of the Final Report to the U.S. Marine Mammal Commission, MM2910909-2, Washington, D.C.

Lynn, S. K. 1995. Movements, site fidelity, and surfacing patterns of bottlenose dolphins on the central Texas coast. M.S. thesis, Texas A&M University, College Station. 92 pp.

Peterson, J., and C. Hubard. 1996. Abundance and photo-identification studies of bottlenose dolphins in Mississippi Sound. Southeast Fisheries Science Center, Pascagoula, Miss.

Scott, M. D., R. S. Wells, and A. B. Irvine. 1990. A long-term study of bottlenose dolphins on the west coast of Florida. Pages 235–244 *in* S. Leatherwood and R. R. Reeves, editors. The bottlenose dolphin. Academic Press, San Diego, Calif. 635 pp.

Wells, R. S. 1986. Population structure of bottlenose dolphins: behavioral studies along the central west coast of Florida. Contract report to National Marine Fisheries Service, Southeast Fisheries Science Center. Contract Number 45-WCNF-5-00366. 58 pp.

The *Exxon Valdez* Oil Spill

Alaska Department of Environmental Conservation. 1993. The *Exxon Valdez* oil spill. Final report, state of Alaska response, Juneau. 184 pp.

Loughlin, T., editor. 1994. Marine mammals and the *Exxon Valdez*. Academic Press, San Diego, Calif. 395 pp.

Northern Sea Lion

Loughlin, T., A. Perlov, and V. Vladimirov. 1992. Range-wide survey and estimation of total number of Steller sea lions in 1989. Marine Mammal Science 8 (3): 220–239.

Effect of El Niño on the Southern California Bight

Bailey, K. M., and L. S. Incze. 1985. El Niño and the early life history and recruitment of fishes in temperate waters. Pages 143–165 *in* W. S. Wooster and D.L. Fluherty, editors. El Niño north: Niño effects in the eastern subarctic Pacific Ocean. College of Ocean and Fisheries Sciences, Washington Sea Grant Program Publication WSG-WO 85-3. University of Washington, Seattle.

Brodeur, R. D., and W. G. Pearcy. 1986. Distribution and relative abundance of pelagic nonsalmonid nekton off Oregon

and Washington, 1979–84. National Oceanic and Atmospheric Administration Technical Report NMFFS-46-85.

DeLong, R. L., G. A. Antonelis, C. W. Oliver, B. S. Stewart, M. S. Lowry, and P. K. Yochem. 1991. Effects of the 1982–1983 El Niño on several population parameters and diet of California sea lions on the California Channel Islands. Pages 166–172 *in* F. Trillmich and K. A. Ono, editors. Pinnipeds and El Niño. Ecological Studies Volume 88. Springer-Verlag, Berlin.

Pearcy, W. G., and A. Schoener. 1987. Changes in the marine biota coincident with the 1982–1983 El Niño in the northeastern subarctic Pacific Ocean. Journal of Geophysical Research (Oceans) 92:14417–14428.

Genetic Diversity of Central Pacific Marine Mammals

Baker, C. S., A. Perry, J. L. Bannister, M. T. Weinrich, R. B. Abernathy, J. Calambokidis, J. Lien, R. H. Lambertson, J. Urban, R. O. Vasquez, P. J. Clapham, A. Alling, U. Arnason, S. J. O'Brien, and S. R. Palumbi. 1993. Abundant mitochondrial DNA variation and world-wide population structure in humpback whales. Proceedings National Academy Sciences (USA) 90:8239–8243.

Dizon, A. E., C. Lockyer, W. F. Perrin, D. P. DeMaster, and J. Sisson. 1992. Rethinking the stock concept: a phylogeographic approach. Conservation Biology 6:24–36.

Dizon, A. E., W. F. Perrin, and P. A. Akin. 1994. Stocks of dolphins (*Stenella* spp. and *Delphinus delphis*) in the eastern tropical Pacific: a phylogeographic classification. National Oceanic and Atmospheric Administration Technical Memorandum NMFS-119:1–20.

Dizon, A. E., S. O. Southern, and W. F. Perrin. 1991. Molecular analysis of mtDNA types in exploited populations of spinner dolphins (*Stenella longirostris*). Report of the International Whaling Commission (Special Issue 13):183–202.

Hoelzel, A. R. 1993. Genetic ecology of marine mammals. Symposium of the Zoological Society (London) 66:15–32.

Hoelzel, A. R. 1994. Genetics and ecology of whales and dolphins. Annual Review of Ecology and Systematics 25:377–399.

Jefferson, T. A., S. Leatherwood, and M. A. Webber. 1993. FAO species identification guide. Marine mammals of the world. United Nations Food and Agriculture Organization, Rome. 320 pp.

National Oceanic and Atmospheric Administration. 1996. Our living oceans: report on the status of U.S. living marine resources, 1995. National Marine Fisheries Service, National Oceanic and Atmospheric Administration Technical Memorandum NMFS-F/SPO-19. Washington, D.C. 160 pp.

Rosel, P. E., A. E. Dizon, and J. E. Heyning. 1994. Genetic analysis of sympatric morphotypes of common dolphins (genus *Delphinus*). Marine Biology 119:159–167.

Wada, S., and K. Numachi. 1991. Allozyme analyses of genetic differentiation among the populations and species of the *Balaenoptera*. Report of the International Whaling Commission. Special Issue 13:125–154.

Hawaiian Monk Seal

National Oceanic and Atmospheric Administration. 1996. Our living oceans: report on the status of U.S. living marine resources, 1995. National Marine Fisheries Service, National Oceanic and Atmospheric Administration Technical Memorandum NMFS-F/SPO-19. Washington, D.C. 160 pp.

Polovina, J. J., G. T. Mitchum, N. E. Graham, M. P. Craig, E. E. DeMartini, and E. N. Flint. 1994. Physical and biological consequences of a climate event in the central North Pacific. Fisheries Oceanography 3:15–21.

Ragen, T. J., and D. M. Lavigne. 1997. The Hawaiian monk seal. *In* J. Twiss and R. Reeves, editors. Marine mammals. Vol. 2. Smithsonian Institution Press, Washington, D.C. In press.

Courtesy R. Stottlemyer

Appendix, Glossary, and Index

Common and Scientific Name Appendix

The following appendix lists the common and scientific names of microorganisms, fungi, plants, and animals mentioned in the report. The names are listed in the following order: viruses, bacteria, fungi, algae, plants, and animals. Species, or the lowest grouping referred to in the text, are given in alphabetical order by common name, followed by the appropriate scientific name. The intermediate headings are not in a taxonomic classification hierarchy but instead are meant to lead the reader to smaller groupings familiar to most persons.

The major source for verifying and correcting the scientific names of organisms was the Integrated Taxonomic Information System [http://www.itis.usda.gov/itis]. Additional sources are listed below:

American Ornithologists' Union. 1983. Check-list of North American birds, 6th edition. American Ornithologists' Union, Washington, D.C. 877 pp.

American Ornithologists' Union. 1993. Thirty-ninth supplement to the American Ornithologists' Union Check-list of North American Birds. Auk 110:675-682.

American Ornithologists' Union. 1995. Fortieth supplement to the American Ornitologists' Union Check-list of North American Birds. Auk 112:819-830.

American Society of Microbiology. 1989. Bergey's manual of systematic bacteriology. Williams and Wilkins, Baltimore, 4 vols.

Banks, R. C., R. W. McDiarmid, and A. L. Gardner, editors. 1987. Checklist of vertebrates of the United States, the U.S. Territories, and Canada. U.S. Fish and Wildlife Service Resource Publication 166. 79 pp.

Brusca, R. C., and G. J. Brusca. 1990. Invertebrates. Sinauer Associates, Inc., Sunderland, Mass. 922 pp.

Flora of North America Editorial Committee. 1993a. Flora of North America north of Mexico. Volume 1. Introduction. Oxford University Press, New York. 372 pp.

Flora of North America Editorial Committee. 1993b. Flora of North America north of Mexico. Volume 2. Pteridophytes and gymnosperms. Oxford University Press, New York. 372 pp.

Hickman, J. C., editor. 1993. The Jepson manual: higher plants of California. University of California Press, Berkeley and Los Angeles. 1400 pp.

Hitchcock, A. S. 1950. Manual of the grasses of the United States. U.S. Department of Agriculture Miscellaneous Publication 200. 1051 pp.

Kartesz, J.T. 1994. A synonymized checklist of the vascular flora of the United States, Canada, and Greenland. Timber Press, Portland, Ore.

Little, E. L., Jr. 1979. Checklist of United States trees (native and naturalized). U.S. Department of Agriculture, Forest Service, Agriculture Handbook 541. 375 pp.

Little, E. L., Jr., and F. H. Wadsworth. 1964. Common trees of Puerto Rico and the Virgin Islands. U.S. Department of Agriculture, Forest Service, Agriculture Handbook 249. 548 pp.

Little, E. L., Jr., R. O. Woodbury, and F. H. Wadsworth. 1988. Arboles de Puerto Rico y las Islas Virgenes. Volume 2. U.S. Department of Agriculture, Forest Service, Agriculture Handbook 449-S. 1,177 pp.

North American Butterfly Association. 1995. Checklist and English names of North American butterflies. North American Butterfly Association, Morristown, N.J. 43 pp.

Nowak, R. M. and J. L. Paradiso. 1983. Walker's mammals of the world. 4th edition. 2 volumes. Johns Hopkins University Press, Baltimore, Md. 1362 pp.

Parker, S.P., editor. 1982. Synopsis and classification of living organisms. 2 volumes. McGraw-Hill Book Co., New York. 2398 pp.

Pennak, R. W. 1989. Fresh-water invertebrates of the United States. 3rd edition. John Wiley & Sons, Inc., New York. 628 pp.

Robins, C. R., chairman. 1991. Common and scientific names of fishes from the United States and Canada. Fifth edition. American Fisheries Society Special Publication 20. 183 pp.

Turgeon, D. D., chair. 1988. Common and scientific names of aquatic invertebrates from the United States and Canada. Mollusks. American Fisheries Society Special Publication 16. 277 pp., 12 color plates.

U.S. Fish and Wildlife Service. 1994a. Endangered and threatened wildlife and plants. 50 CFR 17.11 & 17.12. U.S. Fish and Wildlife Service, Washington, D.C. 42 pp.

U.S. Fish and Wildlife Service. 1994b. Endangered and threatened wildlife and plants; animal candidate review for listing as endangered or threatened species; proposed rule. Federal Register 59(219):58982-59028.

Werner, F. G., chairman. 1982. Common names of insects and related organisms. Entomological Society of America, College Park, Md. 132 pp.

Compiled by Paul A. Opler
(USGS retired)
Department of Bioagricultural Sciences
Colorado State University
Fort Collins, CO 80523

Viruses

Viruses have no genus and species names because they are not strictly complete self-replicating organisms.

Black band disease
Cetacean morbillivirus
Duck plague
Green turtle papilloma [Causative agent is uncertain]
Hematopoietic necrosis virus
Herpes virus
Infectious hypodermal virus
Rabies [Causative agent is a rhabdovirus]
White band disease

Bacteria

Actinomycete bacteria	Actinomycetales
Avian botulism	*Clostridium botulinum*
Avian cholera	*Pasteurella multocida*
Bacillus "B.T."	*Bacillus thuringiensis* variety *kurstaki*
Bacteria [No common name]	Deinococcaceae
Bacteria [No common name]	*Mycoplasma* species
Bacteria [No common name]	*Spiroplasma* species
Bioluminescing bacteria	[No scientific name]
Brucellosis	*Brucella abortus*
Cholera	*Vibrio cholera*
Coliform bacteria	[No scientific name]
Marine bacteria	Archaea
Withering foot disease	[No scientific name]

Blue-green algae

Blue-green alga [No common name]	*Anabaena* species
Blue-green alga [No common name]	*Anabaenopsis elenkinii*
Blue-green alga [No common name]	*Aphanizomenon* species
Blue-green alga [No common name]	*Chroococcus* species
Blue-green alga [No common name]	*Gloeocapsus* species
Blue-green alga [No common name]	*Gomphospheria* species
Blue-green alga [No common name]	*Lyngbya contorta*
Blue-green alga [No common name]	*Lyngbya* species
Blue-green alga [No common name]	*Merismopedium* species
Blue-green alga [No common name]	*Microcystis aeruginosa*
Blue-green alga [No common name]	*Rivularia* species
Coccoid cyanobacteria	Cyanobacteria

Fungi

Agaricias	*Agaricia* species
Bracket fungi	Family Polyporaceae
Chestnut blight	*Endothia parasitica*
Coral fungus	*Ramaria araiospora*
Cryptobiotic (Microbiotic) crusts	Complexes of fungi and blue-green algae
Dogwood anthracnose	[No scientific name]
Dutch elm disease	*Ophiostoma ulmi*
Elm phloem necrosis	[No scientific name]
Field mushrooms	*Agaricus* species
Foliose lichens	[No scientific name]
Fungus [No common name]	*Entomophaga maimaiga*
Fungus [No common name]	*Piloderma fallax*
Honey mushroom	*Armillaria bulbosa* or *A. gallica*
Laminated root rot	*Phellinus weirii*
Lichens	Several families
Lobaria lichens	*Lobaria* species
Mildews	Order Mucorales
Molds	Division Zygomycota
Mycorrhizal fungi	(Mycorrhizal fungi belong to several fungal divisions)
Root rot	Family Tricholomataceae
Rusts	Order Uredinales
San Joaquin Valley fever (coccidiomycosis)	*Coccidioides immitis* [causative organism]
Smuts	Order Ustilaginales
Stalked puffballs	Family Tulostomataceae
Stalked puffballs	*Tulostoma* species
Toadstools, bracket fungi, mushrooms, puffballs, stinkhorns	Division Basidiomycota
Truffles	Order Tuberales
White pine blister rust	*Cronartius ribicola*
Woven spored lichen	*Texosporium sancti-jacobi*

Algae

Algae and diatoms

Alga [No common name]	*Bostrichia* species
Alga [No common name]	*Ectocarpus* species
Alga [No common name]	*Enteromorpha* species
Alga [No common name]	*Polysiphonia* species
Calcareous algae	[No scientific name]
Macroalgae	[No scientific name]
Alga [No common name]	*Microdictyon*

Algae, autotrophic protists

Diatoms and golden-brown algae

Brown tide	[No scientific name]
Chrysophyte [No common name]	*Cryptomonas* species
Chrysophyte [No common name]	*Ochromonas* species
Diatom [No common name]	*Actinoptychus undulatus*
Diatom [No common name]	*Biddulphia aurita*
Diatom [No common name]	*Chaetoceros breve*
Diatom [No common name]	*Coscinodiscus* species
Diatom [No common name]	*Cyclotella* species
Diatom [No common name]	*Cylindrotheca* species
Diatom [No common name]	*Diploneis elliptica*
Diatom [No common name]	*Gyrosigma* species
Diatom [No common name]	*Navicula* species
Diatom [No common name]	*Nitzschia* species
Diatom [No common name]	*Pseudonitzschia australis*
Diatom [No common name]	*Surirella robusta*
Giant kelp	*Macrocystis pyrifera*
Giant kelp, bull kelp	*Nereocystis leutkesana*
Kelps	*Laminaria* species
Luminescent dinoflagellates	*Gonyaulax* species

Marine alga	*Dictyota*
Microscopic diatoms	Class Bacillariophyceae
MSX and Dermo	[No scientific names]

Green algae

Green alga [No common name]	*Chaetomorpha* species
Green alga [No common name]	*Chara vulgaris*
Green alga [No common name]	*Chlamydomonas* species
Green alga [No common name]	*Chlorella vulgara*
Green alga [No common name]	*Halimeda* species
Green alga [No common name]	*Microdictyon* species
Green alga [No common name]	*Monostroma* species
Green alga [No common name]	*Pediastrum biradiatum*
Green alga [No common name]	*Rhizoclonium* species
Green alga [No common name]	*Scenedsmus quadricauda*
Green alga [No common name]	*Staurastrum americanum*
Green alga [No common name]	*Ulothrix* species
Green alga [No common name]	*Ulva* species

Red algae

Bubble algae	[No scientific name]
Coralline algae	Family Corallinaceae
Filamentous algae	[No scientific name]

Plants

Nonseed plants

Club-mosses

Club-mosses	*Lycopodium* species

Ferns and allies

Aleutian shield-fern	*Polystichum aleuticum*
American hart's-tongue fern	*Asplenium scolopendrium*
Australian tree-fern	*Cyathea cooperi*
Bracken	*Pteridium aquilinum*
Common tree-fern	*Cyathea* species
Elfin tree-fern	*Cyathea dryopteroides*
Fern [No common name]	*Adiantum vivesii*
Fern [No common name]	*Elaphoglossum serpens*
Fern [No common name]	*Polystichum calderonense*
Fern [No common name]	*Tectaria estremerana*
Fern [No common name]	*Thelypteris inaborensis*
Fern [No common name]	*Thelypteris verecunda*
Fern [No common name]	*Thelypteris yaucoensis*
Giant tree-fern	*Cyathea arborea*
Hapuu (tree-ferns)	*Cibotium* species
Helecho gigante espinoso or spiny tree-fern	*Nephelea portoricensis*
Howell's quillwort	*Isotes howellii*
Lady fern	*Athyrium filix-femina*
Marsilea ferns	*Marsilea mucronata*
Moonwort	*Botrychium lunaria*
Northern beech fern	*Phegopteris connectilis*
Royal fern	*Osmunda regalis*
Southern marsh fern	*Thelypteris palustris*
Sword fern	*Polystichum* species
Uluhe	*Dicranopteris linearis*

Horsetails

Horsetails	*Equisetum pratense, E. sylvaticum*

Liverworts, hornworts, and mosses

Feathermosses	[No scientific name]
Liverworts	Class Hepaticae
Mosses	Class Bryopsida
Peat moss	Family Sphagnaceae
Sphagnum moss	*Sphagnum fimbriatum*

Needleleaf plants

Conifers

Alaska-cedar	*Chamaecyparis nootkatensis*
Alligator juniper	*Juniperus deppeana*
Apache pine	*Pinus engelmannii*
Arizona cypress	*Cupressus arizonica*
Arizona pine	*Pinus ponderosa* variety *arizonica*
Atlantic white-cedar	*Chamaecyparis thyoides*
Baldcypress	*Taxodium distichum*
Baldcypresses	*Taxodium* species
Balsam fir	*Abies balsamea*
Bishop pine	*Pinus muricata*
Black spruce	*Picea mariana*
Blue spruce	*Picea pungens*
Border pinyon	*Pinus discolor*
Brewer spruce	*Picea brewerana*
Bristlecone pine	*Pinus aristata*
California red fir	*Abies magnifica*
California white fir	*Abies concolor* variety *lowiana*
Chihuahua pine	*Pinus leiophylla* variety *chihuahuana*
Coast redwood	*Sequoia sempervirens*
Colorado pinyon pine	*Pinus edulis*
Digger pine	*Pinus sabiniana*
Douglas-fir	*Pseudotsuga menziesii*
Eastern hemlock	*Tsuga canadensis*
Eastern redcedar	*Juniperus virginiana*
Eastern white pine	*Pinus strobus*
Engelmann spruce	*Picea engelmannii*
Florida torreya	*Torreya taxifolia*
Florida yew	*Taxus floridana*
Foxtail pine	*Pinus balfouriana*
Fraser fir	*Abies fraseri*
Giant sequoia	*Sequoiadendron giganteum*
Grand fir	*Abies grandis*
Great Basin bristlecone pine	*Pinus longaeva*
Incense-cedar	*Calocedrus decurrens*
Jack pine	*Pinus banksiana*
Jeffrey pine	*Pinus jeffreyi*
Junipers	*Juniperus* species
Knobcone pine	*Pinus attenuata*
Larches	*Larix* species
Limber pine	*Pinus flexilis*
Loblolly pine	*Pinus taeda*
Lodgepole pine	*Pinus contorta*
Longleaf pine	*Pinus palustris*
Long-leafed pines	*Pinus* species
Mexican pinyon	*Pinus cembroides*
Modoc cypress	*Cupressus bakeri*
Monterey cypress	*Cupressus macrocarpa*
Monterey pine	*Pinus radiata*
Mountain hemlock	*Tsuga mertensiana*
Noble fir	*Abies procera*
Northern whitecedar	*Thuja occidentalis*
Pacific silver fir	*Abies amabilis*
Pacific yew	*Taxus brevifolia*
Pines	*Pinus* species
Pitch pine	*Pinus rigida*
Pondcypress	*Taxodium distichum* variety *nutans*
Ponderosa pine	*Pinus ponderosa*
Red pine	*Pinus resinosa*
Red spruce	*Picea rubens*
Rocky Mountain juniper	*Juniperus scopulorum*
Sand pine	*Pinus clausa*
Santa Cruz cypress	*Cupressus goveniana* variety *abramsiana*
Sargent cypress	*Cupressus sargentii*
Short-leaf pine	*Pinus echinata*
Sierra juniper	*Juniperus occidentalis* variety *australis*
Singleleaf pinyon	*Pinus monophylla*
Sitka spruce	*Picea sitchensis*
Southwestern white pine	*Pinus strobiformis*
Spruces	*Picea* species
Subalpine fir	*Abies lasiocarpa*
Sugar pine	*Pinus lambertiana*
Table Mountain pine	*Pinus pungens*
Tamarack	*Larix laricina*
Tecate cypress	*Cupressus guadalupensis* variety *forbesii*
Torrey pine	*Pinus torreyana*
Two-leaf pinyon	*Pinus edulis*
Utah juniper	*Juniperus osteosperma*
Virginia pine	*Pinus virginiana*
Western hemlock	*Tsuga heterophylla*
Western juniper	*Juniperus occidentalis*
Western larch	*Larix occidentalis*
Western redcedar	*Thuja plicata*
White fir	*Abies concolor*
White spruce	*Picea glauca*
Whitebark pine	*Pinus albicaulis*

Joint-firs

Mormon-tea	Family Ephedraceae

Flowering plants

Epiphytes

Cupey	*Clusia rosea*
Bromeliad	Family Bromeliaceae
Orchid [No common name]	*Cranichis ricartii*
Orchid [No common name]	*Epidendrum krugii*
Orchid [No common name]	*Lepanthes eltoroensis*
Pineapple	*Ananas comosus*

Forbs

Acanthus	Family Acanthaceae
Alabama pitcher-plant	*Sarracenia rubra* subspecies *alabamensis*
Alaska arnica	*Arnica unalaschcensis*
Aleutian chickweed	*Cerastium beringinanum* variety *aleuticum*
Aleutian draba	*Draba aleutica*
Aleutian saxifrage	*Saxifraga aleutica*
Aleutian wormwood	*Artemisia aleutica*
Alpine sagebrush	*Artemisia scopulorum*
Altered andesite buckwheat	*Eriogonum robusta*
American alyssum	*Alyssum obovatum*
American chaffseed	*Schwalbea americana*
American sea-rocket	*Cakile edentula*
Angelica	*Angelica lucida*
Applegate's milk-vetch	*Astragalus applegatei*
Arnica	*Arnica unalaschcensis*
Asters	*Aster* species
Beach-bur	*Ambrosia chamissonis*
Beach lovage	*Ligusticum scoticum*
Beach morningglory	*Ipomoea imperati*
Beach pea	*Lathyrus japonicus* subspecies *maritimus*
Beach strawberry	*Fragraria chiloensis*
Bering douglasia	*Douglasia beringensis*
Birdfoot violet	*Viola pedata*
Black-eyed Susan	*Rudbeckia hirta*
Blazing star	*Liatris scariosa*
Blue lupine	*Lupinus perennis*
Boreal primrose	*Primula anvilensis*
Bradshaw's desert-parsley	*Lomatium bradshawii*
Calder's sea-lovage	*Ligusticum calderi*
California bearpaw poppy	*Arctomecon californica*
California buckwheat	*Erigonum fasciculatum*
California goldfield	*Lasthenia californica*
California poppy	*Escholtzia californica*
Caltha-leaved avens	*Geum calthifolium*
Canby's dropwort	*Oxypolis canbyi*
Chukchi primrose	*Primula tschuktschorum*
Clovers	*Trifolium* species
Coastal verbena	*Abronia latifolia*
Common pitcher-plant	*Sarracenia purpurea*
Common ragweed	*Ambrosia artemisifolia*
Common stickyseed	*Blennosperma nanum*
Cow parsnip	*Heracleum maximum*

Crowberry	*Empetrum nigrum*
Crystalline iceplant	*Mesembryanthemum crystallinum*
Daffodil	*Narcissus pseudonarcissus*
Dandelion	*Taraxacum officinale*
Decurrent false aster	*Boltonia decurrens*
Desert-parsleys	*Lomatium* species
Drummond's bluebells	*Mertensia drummondii*
Dwarf crested iris	*Iris cristata*
Dwarf lake iris	*Iris lacustris*
Eastern prairie fringed orchid	*Platanthera leucophaea*
Eschscholtz's buttercup	*Ranunculus eschscholtzii*
Fassett's locoweed	*Oxytropis campestris* variety *chartacea*
Fawn-lilies	*Erythronium* species
Few flowered whitlowgrass	*Draba pauciflora*
Filaree	*Erodium cicutarium*
Fireweed	*Epilobium angustifolium*
Furbish lousewort	*Pedicularis furbishiae*
Garlic mustard	*Alliaria petiolata*
Gladecresses	*Leavenworthia* species
Goldenrods	*Solidago* species
Gosmore	*Hypochaeris radicata*
Grime's vetchling	*Lathyrus grimesii*
Groundsel (see Seaside ragwort)	*Senecio pseudoarnica*
Hairypod cowpea	*Vigna luteola*
Haleakala schiedea	*Schiedea haleakalensis*
Hawksbeards	*Crepis* species
Hedge-hyssop	*Gratiola ebracteata*
Heliotropes	*Heliotropium* species
Hogwallow starfish	*Hesperevax caulescens*
Houghton's goldenrod	*Solidago houghtonii*
Hulten's sea-lovage (see Beach lovage)	
Iceplants	*Mesembryanthemum* species
Jesup's milk-vetch	*Astragalus robbinsii*
Jones' pitcher-plant	*Sarracenia rubra* subspecies *jonesii*
Kearney buckwheat	*Eriogonum nummulare*
Kobuk locoweed	*Oxytropis kobukensis*
Kokrine's locoweed	*Oxytropis kokrinensis*
Krause's sorrel	*Rumex krausei*
Ladies' tresses	*Spiranthes romanzoffiana*
Lakeside daisy	*Hymenoxys herbacea*
Leafy prairie-clover	*Dalea foliosa*
Leedy's roseroot	*Sedum integrifolium* subspecies *leedyi*
Little sagebrush	*Artemisia arbuscula*
Liverleaf wintergreen	*Pyrola asarifolia*
Lobb buckwheat	*Eriogonum lobbii*
Lomatium	*Lomatium* species
Louseworts	*Pedicularis* species
Low chickweed	*Stellaria humifusa*
MacFarlane's four-o-clock	*Mirabilis macfarlanei*
Malheur wire lettuce	*Stephanomeria malheurensis*
Maritime sea-rocket	*Cakile maritima*
Marsh sandwort	*Arenaria paludicola*
McDonald's rock-cress	*Arabis macdonaldiana*
Mead's milkweed	*Asclepias meadii*
Meadowfoam	*Limnanthes* species
Michigan monkeyflower	*Mimulus glabratus* variety *michiganensis*
Microsteris	*Microsteris gracilis*
Milk-vetches	*Astragalus* species
Minnesota dwarf trout lily	*Erythronium propullans*
Monkeyflowers	*Mimulus* species
Moresby's groundsel, cleftleaf groundsel	*Senecio moresbiensis* variety *cymbalarioides*
Mountain avens	*Dryas integrifolia* and *Dryas octopetala*
Mountain-dandelions	*Agoseris* species
Mouse-tail	*Myosurus minimus*
Muir's fleabane	*Erigeron muirii*
Mustards	Family Brassicaceae
Nelson's checker-mallow	*Sidalcea nelsoniana*
Northern wild monkshood	*Aconitum noveboracense*
Oxalis	*Oxalis* species
Oysterleaf	*Mertensia maritima*
Pacific silverweed	*Potentilla egedii*
Paintbrushes	*Castilleja* species
Pearly everlasting	*Anaphalis margaritacea*
Penstemons or beard-tongues	*Penstemon* species
Phloxes	*Phlox* species
Pilewort	*Erechtites hieracifolia*
Pink verbena	*Abronia umbellata*
Pitcher-plant	Family Sarraceniaceae
Pitcher's thistle	*Cirsium pitcheri*
Prairie bush-clover	*Lespedeza leptostachya*
Prairie lupine	*Lupinus* species
Prickly poppies	*Argemone* species
Purple coneflower	*Echinacea purpurea*
Purplish braya	*Braya globella* subspecies *purpurascens*
Pussytoes	*Antennaria* species
Pygmy epilobium	*Epilobium pygmaeum*
Ram's head lady's slipper	*Cypripedium arietinum*
Raspberries	*Rubus* species
Robbins' cinquefoil	*Potentilla robbinsiana*
Roughleaf yellow loosestrife	*Lysimachia asperulifolia*
Roundleaf sundew	*Drosera rotundifolia*
Rugelia, Rugel's ragwort	*Rugelia nudicaulis*
Running buffalo clover	*Trifolium stoloniferum*
Russian thistle	*Salsola kali*
Ruth's golden aster	*Pityopsis ruthii*
Sagebrush	*Artemisia arctica* subspecies *comata*
Sagebrush	*Artemisia laciniata*
Sagebrush	*Artemisia senjavinensis*
Sagebrushes	*Artemisia* species
Salt marsh camphor-weed	*Pluchea camphorata*
Salt marsh bird's-beak	*Cordylanthus maritimus* subspecies *maritimus*
San Clemente Island Indian paintbrush	*Castilleja grisea*
San Clemente Island larkspur	*Delphinium variegatum kinkiense*
Sandplain gerardia	*Agalinis acuta*
Santa Barbara Island liveforever	*Dudleya traskiae*
Schweinitz's sunflower	*Helianthus schweinitzii*
Sea-beach sandwort	*Honckenya peploides*
Sea-oxeye	*Borrichia frutescens*
Seabeach amaranth	*Amaranthus pumilus*
Seaside ragwort	*Senecio pseudoarnica*
Sensitive joint-vetch	*Aeschynomene noveboracense*
Shasta snow-wreath	*Neviusia cliftonii*
Shooting stars	*Dodecatheon* species
Siberian spring beauty	*Claytonia sibirica*
Silver bur ragweed (see Beach-bur)	
Silverweed cinquefoil	*Argentina anserina*
Small whorled pogonia	*Isotria medeoloides*
Smartweeds	*Polygonum* species
Smooth beggartick	*Bidens laevis*
Smooth coneflower	*Echinacea laevigata*
Smooth tidytips	*Layia chrysanthemoides*
Snowy orchid	*Platanthera nivea*
Soybean	*Glycine max*
Spotted knapweed	*Centaurea biebersteinii*
Spreading pogonia	*Cleistes divaricata*
Spring-flowering goldenrod	*Solidago verna*
Steamboat buckwheat	*Eriogonum ovalifolium*
Succulents	Family Crassulaceae
Sulphur Springs buckwheat	*Eriogonum argophyllum* variety *williamsiae*
Sundew	Family Droseraceae
Sunflower	*Helianthus annuus*
Swamp pink	*Helonias bullata*
Sweet pitcher-plant	*Sarracenia rubra*
Tall goldenrod	*Solidago canadensis*
Thistles	*Cirsium* species
Tiehm buckwheat	*Eriogonum tiehmii*
Tricolored monkeyflower	*Mimulus tricolor*
Tuberous grasspink	*Calopogon tuberosus*
Tulips	*Tulipa* species
Tumble mustard	*Sisymbrium altissimum*
Venus flytrap	*Dionaea muscipula*
Violets	*Viola* species
Western buttercup	*Ranunculus occidentalis*
Western lily	*Lilium occidentale*
White bog-orchid	*Platanthera dilatata*

White bursage	*Franseria dumosa*
White clover	*Trifolium repens*
White trout lily	*Erythronium albidum*
Wild buckwheats	*Eriogonum* species
Woolly beachheather	*Hudsonia tomentosa*
Wormwood	*Artemisia caudata*
Yellow pitcher-plant	*Sarracenia flava*
Yellow star-thistle	*Centaurea solstitialis*
[No common name]	*Chamaecrista glandulosa* variety *mirabilis*
[No common name]	*Microsteris* species
[No common name]	*Vernonia proctorii*

Grasses

Alaska bluegrass	*Poa hartzii alaskana*
American beachgrass	*Ammophila breviligulata*
Anderson alkaligrass	*Puccinellia andersonii*
Annual hairgrass	*Deschampsia danthonoides*
Arctic pendantgrass	*Arctophila fulva*
Beach ryegrass	*Elymus arenarius*
Beardgrass	*Schizachyrium condensatum*
Bermuda grass	*Cynodon dactylon*
Big bluestem	*Andropogon gerardii*
Black grama	*Bouteloua eriopoda*
Blue grama	*Bouteloua gracilis*
Blue wild rye	*Elymus glaucus*
Bluejoint	*Calamagrostis canadensis*
Bluestems	*Andropogon* and *Schizachyrium* species
Brome grasses	*Bromus* species
Broomsedge	*Andropogon virginicus*
Buffalo grass	*Buchloe dactyloides*
Buffel grass	*Cenchrus ciliaris*
Bunchgrasses	Not a unified group. Western grasses that grow in clumps rather than as sod.
Calder's licoriceroot	*Ligusticum calderi*
Canes	*Arundinaria* species
Cheatgrass	*Bromus tectorum*
Cleftleaf groundsel	*Senecio cymbalaroides*
Coast sandbur	*Cenchrus incertus*
Coastal dropseed	*Sprorobolus virginicus*
Common reed	*Phragmites australis*
Cordgrasses	*Spartina* species
Corn, maize	*Zea mays*
Creeping alkaligrass	*Puccinellia phryganodes*
Crested wheatgrass	*Agropyron cristatum*
Deergrass	*Muhlenbergia rigens*
Dropseed	*Sporobolus* species
European beachgrass	*Ammophila arenaria*
Fescues	*Festuca* species
Fountain grass	*Pennisetum setaceum*
Foxtail	*Alopecurus saccatus*
Galleta	*Hilaria jamesii*
Giant wild-rye	*Elymus condensatus*
Goatgrasses	*Aegilops* species
Grama grasses	*Bouteloua* species
Hairgrass	*Deschampsia* species
Indian ricegrass	*Oryzopsis hymenoides*
Junegrass	*Koeleria cristata*
Kentucky bluegrass	*Poa pratensis*
Kikuyu grass	*Pennisetum clandestinum*
Lehmann lovegrass	*Eragrostis lehmanniana*
Little bluestem	*Schizachyrium scoparium*
Lovegrasses	*Eragrostis* species
Maidencane	*Panicum hemitomon*
Maize (see Corn)	
Mediterranean grass	*Schismus arabicus* and *S. barbatus*
Medusahead	*Taeniatherum caput-medusae*
Melicgrass	*Melica* species
Molassesgrass	*Melinis minutiflora*
Mountain muhly	*Muhlenbergia montana*
Needle-and-thread	*Stipa comata*
Needlegrasses	*Stipa* species
Oatgrasses	*Danthonia* species
Pacific reedgrass	*Calamagrostis nutkaensis*
Palmgrass	*Setaria palmifolia*

Pelos del Diablo	*Aristida portoricensis*
Purple needlegrass	*Stipa pulchra*
Red brome	*Bromus rubens*
Reedgrass	*Calamagrostis deschampsioides*
Rice	*Oryza sativa*
Ricegrasses	*Oryzopsis* species
Ringgrass	*Muhlenbergia torreyi*
Sacaton	*Sporobolus wrightii*
Salt meadow cordgrass	*Spartina patens*
Saltgrass	*Distichlis spicata*
Sand dropseed	*Sporobolus cryptandrus*
Sand-reeds	*Calamovilfa* species
Sandburs	*Cenchrus* species
Sea oats	*Uniola paniculata*
Seashore paspalum	*Paspalum vaginatum*
Six-weeks fescue	*Festuca octoflora*
Soft chess	*Bromus mollis*
Sorghum	*Sorghum bicolor*
Sugarcane	*Saccharum officinarum*
Switch cane	*Arundinaria tecta*
Switchgrass	*Panicum virgatum*
Three awns	*Aristida* species
Tobosa grass	*Hilaria mutica*
Tufted hairgrass	*Deschampsia caespitosa*
Tundra grass	*Dupontia fisheri*
Walter's millet	*Echinochloa walteri*
Western wheatgrass	*Agropyron smithii*
Wheat	*Triticum aestivum*
Wheatgrasses	*Agropyron* species
Wild oats	*Avena fatua*
Wire-grass (see Bermuda grass)	
Wright's alkaligrass	*Puccinellia wrightii*
Wright's pendantgrass	*Arctophila wrightii*
[No common name]	*Aristida chaseae*

Marine and freshwater plants

Alaska mistmaiden	*Romanzoffia unalaschensis*
Alligatorweed	*Alternanthera philoxeroides*
American bulrush	*Schoenoplectus americanus*
American lotus	*Nelumbo lutea*
American spongeplant	*Limnobium spongia*
American white waterlily	*Nymphaea odorata*
Anglestem primrosewillow	*Ludwigia leptocarpa*
Arrowhead species	*Sagittaria* species
Baby pondweed	*Potamogeton pusillus*
Baldwin's spikerush	*Eleocharis baldwinii*
Big cordgrass	*Spartina cynosuroides*
Bigelow sedge	*Carex bigelowii*
Blue waterhyssop	*Bacopa caroliniana*
Broad-leaved arrowhead	*Sagittaria latifolia*
Bulltongue arrowhead	*Sagittaria lancifolia*
Carolina fanwort	*Cabomba caroliniana*
Carolina mosquitofern	*Azolla caroliniana*
Carolina spiderlily	*Hymenocallis caroliniana*
Cattails	*Typha* species
Coastal waterhyssop	*Bacopa monnieri*
Common cattail	*Typha latifolia*
Common duckweed	*Lemna minor*
Common fogfruit	*Phyla nodiflora*
Common spikerush	*Eleocharis palustris*
Coontail	*Ceratophyllum demersum*
Cotton-grass	*Eriophorum angustifolium*
Creeping primrosewillow	*Ludwigia repens*
Creeping waterprimrose	*Ludwigia peploides*
Curly leaf pondweed	*Potamogeton crispus*
Dwarf spikerush	*Eleocharis parvula*
Eelgrass	*Zostera marina*
Eurasian watermilfoil	*Myriophyllum spicatum*
Flat-stem pondweed	*Potamogeton zosteriformis*
Giant bulrush	*Schoenoplectus californicus*
Greater duckweed	*Lemna* species
Hawaii Island sedge	*Oreobolus furcatus*
Hoppner sedge	*Carex subspathacea*
Horned bladderwort	*Utricularia cornuta*

Hydrilla	*Hydrilla verticillata*
Knieskern's beaked-rush	*Rhynchospora knieskernii*
Largeleaf pennywort	*Hydrocotyle bonariensis*
Leafy threesquare	*Bolboschoenus robustus*
Leafybract dwarf rush	*Juncus capitatus*
Lesser duckweed	*Lemna aequinoctialis*
Lizard's tail	*Saururus cernuus*
Long-awn sedge	*Carex macrochaeta*
Long-leaf pondweed	*Potamogeton nodosus*
Loose-flowered alpine sedge	*Carex rariflora*
Looseflower waterwillow	*Justicia ovata*
Lyngby sedge	*Carex lyngbyei*
Manatee grass	*Syringodium filiforme*
Manyflower marshpennywort	*Hydrocotyle umbellata*
Naiads	*Najas* species
Narrow-leaf pondweed	*Potamogeton foliosus*
Needlegrass rush	*Juncus roemerianus*
Northeastern bulrush	*Scirpus ancistrochaetus*
Parrot feather watermilfoil	*Myriophyllum aquaticum*
Pennywort	*Centella erecta*
Perennial glasswort	*Salicornia virginica*
Pickleweeds	*Salicornia* species
Pondweed	Family Najadaceae
Prickly sedge	*Carex echinata*
Purple loosestrife	*Lythrum salicaria*
Ramenski sedge	*Carex ramenskii*
Richardson pondweed	*Potamogeton richardsonii*
Sago pondweed	*Potamogeton pectinatus*
Saltwort	*Batis maritima*
Sargassoweed	*Sargassum* species
Sawgrass	*Cladium mariscus* subspecies *jamaicense*
Shoal grass	*Halodule wrightii*
Small pondweed	*Potamogeton pusillus*
Small spikerush	*Eleocharis parvula*
Smooth cordgrass	*Spartina alterniflora*
Southern waternymph	*Najas guadalupensis*
Spikerushes	*Eleocharis* species
Surfgrass	*Phyllospadix scouleri*
Toad rush	*Juncus bufonius*
Turtle grass	*Thalassia testudina*
Tussock cotton-grass	*Eriophorum vaginatum*
Twoleaf watermilfoil	*Myriophyllum heterophyllum*
Variable flatsedge	*Cyperus difformis*
Water chestnut	*Trapa natans*
Water howellia	*Howellia aquatilis*
Water hyacinth	*Eichhornia crassipes*
Water sedge	*Carex aquatilis*
Water smartweed	*Polygonum amphibium* variety *stipulaceum*
Water star-grass	*Zosterella dubia*
Watershield	*Brasenia shreberi*
Waterweed	*Elodea canadensis*
Waterweeds	*Elodea* species
Waterweeds	Family Hydrocharitaceae
Widgeongrass	*Ruppia maritima*
Wildcelery	*Vallisneria americana*
Wildrice	*Zizania aquatica*
[No common name]	*Carex sabulosa*
[No common name]	*Dictyopterus* species
[No common name]	*Halimeda* species
[No common name]	*Padina* species

Shrubs

Agaves	*Agave* species
Alaska blueberry	*Vaccinium alaskense*
Alpine blueberry	*Vaccinium* species
Barbwire thistle	*Salsola paulsenii*
Basin big sagebrush	*Artemisia tridentata tridentata*
Bayahonda (see Mesquite, under Trees)	
Beach plum	*Prunus maritima*
Bear oak	*Quercus ilicifolia*
Bearberry (see Kinnikinnick)	
Big sagebrush	*Artemisia tridentata*
Bigleaf sumpweed	*Iva frutescens*
Bitterbrush	*Purshia tridentata*

Black sage	*Salvia mellifera*
Black sagebrush	*Artemisia nova*
Blackbrush	*Coleogyne ramossisima*
Blue blossom	*Ceanothus thyrsiflorus*
Blueberries	*Vaccinium* species
Brighamia	*Brighamia* species
Brittlebush	*Encelia farinosa*
Broom snakeweed	*Gutierrezia sarothrae*
Buckthorn	*Frangula alnus*
Bunchberry	*Cornus canadensis*
Burro-weed	*Ambrosia dumosa*
Bush chinquapin	*Chrysolepis sempervirens*
Buttonbush	*Cephalanthus occidentalis*
California adolphia	*Adolphia californica*
California sagebrush	*Artemisia californica*
California-lilacs, buckbrushes, ceanothuses	*Ceanothus* species
Chamise	*Adenostoma fasciculatum*
Chickasaw plum	*Prunus angustifolia*
Clermontia	*Clermontia* species
Clidemia	*Clidemia hirta*
Club-moss mountain-heather	*Cassiope lycopodioides*
Cotton	*Gossypium hirsutum*
Cranberries	*Vaccinium* species
Creosotebush	*Larrea tridentata*
Currants	*Ribes* species
Cyanea	*Cyanea* species
Deer brush	*Ceanothus integerrimus*
Dubautia	*Dubautia* species
Dwarf birch	*Betula nana*
Dwarf palmetto	*Sabal minor*
Early blueberry	*Vaccinium ovalifolium*
Eastern baccharis	*Baccharis halimifolia*
Emajagua	*Hibiscus tiliaceus*
False azalea	*Menziesia ferruginea*
False lobelia	*Trematolobelia* species
Franklinia	*Franklinia altamaha*
Gooseberries	*Ribes* species
Greasewood	*Sarcobatus vermiculatus*
Green alder	*Alnus crispa*
Green hawthorn	*Crataegus viridis*
Gulf croton	*Croton punctatus*
Hairy ceanothus	*Ceanothus olignathus*
Haleakala silversword	*Argyroxiphium sandwicense* variety *macrocephalum*
Havard oak	*Quercus havardii*
Hawthorns	*Crataegus* species
Heaths	Family Ericaceae
Highbush cranberry	*Viburnum edule*
Higuillo	*Piper aduncum*
Himalayan raspberry	*Rubus ellipticus*
Hoaryleaf ceanothus	*Ceanothus crassifolius*
Huckleberries	*Gaylussacia* species
Huckleberry oak	*Quercus vaccinifolia*
Interior silk-tassel	*Garrya congdonii*
Iodinebush	*Allenrolfea occidentalis*
Kahili ginger	*Hedychium gardnerianum*
Kamchatka rhododendron	*Rhododendron camtschaticum*
Kinnikinnick, bearberry	*Arctostaphylos uva-ursi*
Labrador tea	*Ledum groenlandicum*
Leadplant	*Amorpha canescens*
Leather oak	*Quercus durata*
Lingonberry	*Vaccinium vitis-idaea*
Lobelia	Family Lobeliaceae
Mahogany sumac	*Rhus integrifolia*
Manzanitas	*Arctostaphylos* species
Maui greensword	*Argyroxiphium grayanum*
Maui hala pepe	*Pleomele auwahiensis*
Maui plantain	*Plantago pachyphylla*
Menziesia	*Menziesia ferruginea*
Mexican cliffrose	*Purshia stansburiana*
Miconia	*Miconia calvescens*
Mountain heather	*Cassiope* species
Mountain laurel	*Kalmia latifolia*

Mountain-mahoganies	*Cercocarpus* species
Musk brush	*Ceanothus jepsonii*
Nuttall's goosefoot	*Chenopodium berlandieri* subspecies *nuttalliae*
Nuttall's scrub oak	*Quercus dumosa*
Our Lord's candle	*Yucca whipplei*
Pacific red elderberry	*Sambucus racemosa*
Poison ivy	*Rhus toxicodendron*
Possumhaw	*Ilex decidua*
Prickly pear	*Opuntia rubescens*
Prickly rose	*Rosa acicularis*
Purple sage	*Salvia leucophylla*
Rabbitbrush	*Chrysothamnus* species
Red manjack	*Cordia nitida*
Red shank	*Adenostoma sparsifolium*
Rhododendrons	*Rhododendron* species
Rollandia	*Rollandia* species
Rosemary	*Rosmarina officinalis*
Sacahuista	*Nolina microcarpa*
Sakau	*Piper methysticum*
Salmonberry	*Rubus spectabilis*
Saltbush	*Atriplex leucophylla*
Saltbushes	*Atriplex* species
San Clemente Island broom	*Lotus dendroideus* variety *traskiae*
San Clemente Island bush-mallow	*Malacothamnus clementinus*
Sand-myrtle	*Leiophyllum buxifolium*
Sandsage	*Artemisia filifolia*
Sedge	Family Cyperaceae
Shadscale	*Atriplex confertifolia*
Shaw's agave	*Agave shawii*
Silverswords	*Argyroxiphium* species
Sitka alder	*Alnus sitchensis*
Sitka mountain-ash	*Sorbus sitchensis*
Smooth sumac	*Rhus glabra*
Snakeweeds	*Gutierrezia* species
Sotols	*Dasylirion* species
Spicebush	*Lindera benzoin*
Stevens' spiraea	*Spiraea stevenii*
Strangler fig	*Ficus* species
Sugar bush	*Rhus ovata*
Sumacs	*Rhus* species
Swamp dogwood	*Cornus stricta*
Swamp laurel	*Kalmia microphylla*
Swamp-privet	*Forestiera acuminata*
Tarbush	*Flourensia cernua*
Thorny devilsclub	*Oplopanax horridus*
Tobacco	*Nicotiana* species
Toyon	*Hetermeles arbutifolia*
Twin-flower	*Linnaea borealis*
Vasey's sagebrush	*Artemisia tridentata* subspecies *vaseyana*
Viburnums and haws	*Viburnum* species
Virginia spiraea	*Spiraea virginiana*
Virginia-willow	*Itea virginica*
Water viburnum	*Viburnum obovatum*
Wax myrtle	*Morella cerifera*
White sage	*Salvia apiana*
Whitebark rabbitbrush	*Chrysothamnus viscidiflorus*
Whiteleaf manzanita	*Arctostaphylos viscida*
Wilkesia	*Wilkesia* species
Winterfat	*Halogeton glomeratus*
Wintergreen	*Gaultheria procumbens*
Yuccas	*Yucca* species
[No common name]	*Artemisia laciniatiformis*
[No common name]	*Lyonia truncata* variety *proctorii*

Trees

Alaska paper birch	*Betula neoalaskana*
Alderleaf mountain-mahogany	*Cercocarpus montanus*
Alders	*Alnus* species
Alelí	*Plumeria alba*
Almendra	*Terminalia catappa*
American basswood	*Tilia americana*
American beech	*Fagus grandifolia*

American chestnut	*Castanea dentata*
American elm	*Ulmus americana*
American holly	*Ilex opaca*
American snowbell	*Styrax americanus*
Arizona white oak	*Quercus arizonica*
Ashes	*Fraxinus* species
Australian melaleuca	*Melaleuca quinquenervia*
Australian-pine	*Casuarina equisetifolia*
Ausubo	*Manilkara bidentata*
Balsa	*Ochroma pyramidale*
Balsam poplar	*Populus balsamifera*
Bariaco	*Trichilia triacantha*
Bebb willow	*Salix bebbiana*
Birches	*Betula* species
Bitter-ash	*Rauvolfia nitida*
Black birch	*Betula lenta*
Black cherry	*Prunus serotina*
Black locust	*Robinia pseudacacia*
Black oak	*Quercus velutina*
Black willow	*Salix nigra*
Black-mangrove	*Avicennia germinans*
Blackgum, swamp tupelo	*Nyssa sylvatica* variety *biflora*
Blackjack oak	*Quercus marilandica*
Blue oak	*Quercus douglassii*
Box-briar	*Randia aculeata*
Boxelder	*Acer negundo*
Brazilian pepper	*Schinus terebinthifolius*
Bur oak	*Quercus macrocarpa*
Buttonwood-mangrove	*Conocarpus erectus*
Caimitillo	*Micropholis chrysophylloides*
Caimitillo verde	*Micropholis garciniaefolia*
Calambreña	*Coccoloba venosa*
California bay	*Umbellularia californica*
California black oak	*Quercus kelloggii*
California buckeye	*Aesculus californica*
Canyon live oak	*Quercus chrysolepis*
Capá rosa	*Callicarpa ampla*
Cardon cactus	*Pachycereus pringlei*
Carolina ash	*Fraxinus caroliniana*
Cat-claw acacia	*Acacia greggii*
Catalina cherry	*Prunus ilicifolia* subspecies *lyonii*
Catalina ironwood	*Lyonothamnus floribundus*
Cedar elm	*Ulmus crassifolia*
Chestnut oak	*Quercus prinus*
Chinese chestnut	*Castanea mollissima*
Chinquapin	*Chrysolepis chrysophylla*
Chokecherry	*Prunus virginiana*
Chupacallos	*Pleodendron macranthum*
Coast live oak	*Quercus agrifolia*
Cóbana negra	*Stahlia monosperma*
Coffee	*Coffea arabica*
Common guava	*Psidium guajava*
Common lignumvitae	*Guaiacum officinale*
Common persimmon	*Diospyros virginiana*
Cook's holly	*Ilex cookii*
Corcho bobo	*Pisonia alba*
Cottonwoods and poplars	*Populus* species
Coyote willow	*Salix exigua*
Curlleaf mountain-mahogany	*Cercocarpus ledifolius*
Custard-apple	*Annona reticulata*
Dogwoods	*Cornus* species
Eastern cottonwood (see plains cottonwood)	
Emory oak	*Quercus emoryi*
Engelmann oak	*Quercus engelmannii*
Erubia	*Solanum drymophilum*
False mastic	*Sideroxylon foetidissimum*
Firetree	*Myrica faya*
Florida fiddlewood	*Citharexylum fruticosum*
Flowering dogwood	*Cornus florida*
Fremont cottonwood	*Populus fremontii*
Gambel oak	*Quercus gambelii*
Golden-spined cereus	*Bergerocactus emoryi*
Gray oak	*Quercus grisea*
Green alder	*Alnus viridis* subspecies *crispa*

Green ash	*Fraxinus pennsylvanica*	Red alder	*Alnus rubra*
Guaba	*Inga vera*	Red mangrove	*Rhizophora mangle*
Guama	*Inga fagifolia*	Red maple	*Acer rubrum*
Guayabota de sierra	*Eugenia borinquensis*	River birch	*Betula nigra*
Guayacán blanco	*Guaiacum sanctum*	Roble blanco	*Tabebuia heterophylla*
Gumbo limbo	*Bursera simaruba*	Roble de sierra	*Tabebuia rigida*
Hackberry	*Celtis occidentalis*	Royen's tree cactus	*Cephalocereus royenii*
Hazel alder	*Alnus serrulata*	Russian-olive	*Elaeagnus angustifolia*
Hickories	*Carya* species	Saguaro	*Carnegiea gigantea*
Higo chumbo	*Harrisia portoricensis*	Saltcedar or tamarisk	*Tamarix chinensis*
Higuero de sierra	*Crescentia portoricensis*	Sandalwood	*Santalum* species
Honduras mahogany	*Swietenia macrophylla*	Sarrasuela	*Thouinia portoricensis*
Honey mesquite	*Prosopis glandulosa*	Shortleaf fig	*Ficus citrifolia*
Honeylocust	*Gleditsia triacanthos*	Siberian elm	*Ulmus pumila*
Indio	*Erythroxylon areolatum*	Sierra palm	*Prestoea montana*
Interior live oak	*Quercus wislizeni*	Silk cotton tree	*Ceiba pentandra*
Island live oak	*Quercus tomentella*	Silver buffaloberry	*Shepherdia argentea*
Joshua-tree	*Yucca brevifolia*	Silver maple	*Acer saccharinum*
Jusillo	*Calycogonium squamulosum*	Silverleaf oak	*Quercus hypoleucoides*
Kentucky coffeetree	*Gymnocladus dioicus*	Sitka alder	*Alnus viridis*
Koa	*Acacia koa*	Slippery elm	*Ulmus rubra*
Lama	*Diospyros sandwichensis*	Southern magnolia	*Magnolia grandiflora*
Leadtree	*Leucaena glauca*	Spanish cedar	*Cedrela odorata*
Live oak	*Quercus virginiana*	Spanish lime	*Melicoccus bijugatus*
Madrone	*Arbutus menziesii*	St. Thomas prickly-ash	*Zanthoxylum thomasianum*
Magnolias	*Magnolia* species	Strawberry guava	*Psidium cattleianum*
Mamani	*Sophora chrysophylla*	Sugar maple	*Acer saccharum*
Mangrove	Family Rhizophoraceae	Sugarberry	*Celtis laevigata*
Maple	Family Aceraceae	Swamp white oak	*Quercus bicolor*
Matabuey	*Goetzia elegans*	Sweet acacia	*Acacia farnesiana*
Matchwood	*Didymopanax morototoni*	Sweet gum	*Liquidamber styraciflua*
Mesquite	*Prosopis juliflora*	Sweetbay	*Magnolia virginiana*
Mesquites	*Prosopis* species	Sycamore	*Platanus occidentalis*
Mexican blue oak	*Quercus oblongifolia*	Tachuelo	*Pictetia aculeata*
Motillo	*Sloanea berteriana*	Tamarind	*Tamarindus indica*
Mountain alder	*Alnus incana* subspecies *tenuifolia*	Tamarisks	*Tamarix* species
Mountain immortelle	*Erythrina poeppigiana*	Tambonuco	*Dacryodes excelsa*
Nemocá	*Ocotea spathulata*	Tanoak	*Lithocarpus densiflora*
Netleaf oak	*Quercus rugosa*	Teak	*Tectona grandis*
Netleaf willow	*Salix reticulata* subspecies *glabellicarpa*	Tibet	*Albizia lebbek*
Northern red oak	*Quercus rubra*	Trumpet tree	*Cecropia peltata*
Nuttall oak	*Quercus nuttallii*	Tulipán Africano	*Spathodea campanulata*
Oaks	*Quercus* species	Tupelos	*Nyssa* species
Ocotillo	*Fouquieria splendens*	Turkey oak	*Quercus laevis*
Ohia	*Metrosideros polymorpha*	Uvilla	*Coccoloba diversifolia*
Oreganillo	*Weinmannia pinnata*	Uvillo	*Eugenia haematocarpa*
Oregon oak	*Quercus garryana*	Vahl's boxwood	*Buxus vahlii*
Organpipe cactus	*Lemaireocereus thurberi*	Valley oak	*Quercus lobata*
Overcup oak	*Quercus lyrata*	Velvet ash	*Fraxinus velutina*
Oxhorn bucida	*Bucida buceras*	Vine maple	*Acer circinatum*
Palma de lluvia	*Guassia attenuata*	Water hickory	*Carya aquatica*
Palma de manaca	*Calyptronoma rivalis*	Water oak	*Quercus nigra*
Palmetto	*Sabal minor*	Water tupelo	*Nyssa aquatica*
Palms	Family Palmae	Water-elm	*Planera aquatica*
Palo colorado	*Cyrilla racemiflora*	Waterlocust	*Gleditsia aquatica*
Palo colorado	*Ternstroemia luquillensis*	Wavyleaf oak	*Quercus pauciloba* (hybrid)
Palo de jazmín	*Styrax portoricensis*	West Indian locust	*Hymenaea courbaril*
Palo de nigua	*Cornutia obovata*	Wheeler's peperomia	*Peperomia wheeleri*
Palo de pollo	*Pterocarpus officinalis*	White mulberry	*Morus alba*
Palo de Ramón	*Banara vanderbiltii*	White oak	*Quercus alba*
Palo de rosa	*Ottoschultzia rhodoxylon*	White prickle	*Zanthoxylum martinicense*
Paper birch	*Betula papyrifera*	White-mangrove	*Laguncularia racemosa*
Pecan	*Carya illinoensis*	Wild cherries	*Prunus* species
Pin cherry	*Prunus pensylvanica*	Willow bustic	*Dipholis salicifolia*
Pin oak	*Quercus palustris*	Willow oak	*Quercus phellos*
Plains cottonwood, eastern		Willows	*Salix* species
cottonwood	*Populus deltoides*	Winged elm	*Ulmus alata*
Pomarrosa	*Eugenia jambos*	Yellow birch	*Betula alleghaniensis*
Pond-apple	*Annona glabra*	Yellow paloverde	*Cercidium microphyllum*
Prickly-ash	*Zanthoxylum* species	Yellow-poplar	*Liriodendron tulipifera*
Puerto Rican acrocomia	*Acrocomia media*	[No common name]	*Auerodendron pauciflorum*
Puerto Rican royal palm	*Roystonea borinquena*	[No common name]	*Calyptranthes thomasiana*
Pumpkin ash	*Ulmus profunda*	[No common name]	*Daphnopsis hellerana*
Quaking aspen	*Populus tremuloides*	[No common name]	*Eugenia woodburyana*

[No common name]   *Ilex sintenisii*
[No common name]   *Leptocereus grantianus*
[No common name]   *Mitracarpus maxwelliae*
[No common name]   *Mitracarpus polycladus*
[No common name]   *Myrcia paganii*
[No common name]   *Schoepfia arenaria*
[No common name]   *Ternstroemia subsessilis*

Vines

Banana poka   *Passiflora mollissima*
Beach morningglory   *Ipomoea imperati*
Beans   *Phaseolus* species
Dodder   *Cuscuta* species
Grape   *Vitis* species
Greenbriers   *Smilax* species
Kudzu   *Pueraria lobata*
Peperomia vines   *Peperomia* species
Peppervine   *Ampelopsis arborea*
Pickering's morning-glory   *Stylisma pickeringii*
Sweet pea   *Lathyrus odoratus*
Virginia creeper   *Parthenocissus quinquefolia*

Animals

Protozoans

Dinoflagellates   Class Dinoflagellata
Dinoflagellate [No common name] *Ceratium hircus*
Dinoflagellate [No common name] *Gymnodinium* species
Dinoflagellate [No common name] *Peridinium cinctum*
Dinoflagellate [No common name] *Prorocentrum* species
Malaria   *Plasmodium* species [causative agents]
Protozoan [No common name]   *Perkinsus marinus*
Protozoan encephalitis
  (Toxoplasmosis)   *Toxoplasma gondii*
Schistosome parasite   *Schistosoma mansoni*
Whirling disease   *Myxobolus cerebralis*

Euglenoids

Euglenoid [No common name]   *Trachelomonas hispida*

Sponges

Boring sponge   *Cliona celata*
Calcareous sponges   [No scientific name]
Tubular sponges   [No scientific name]

Jellyfish, sea anemones, and corals

Corals

Black coral   Order Antipatheria
Brain corals   Order Scleractinia (in part)
Elkhorn coral   *Acropora palmata*
Finger coral   *Porites* species
Gorgonian corals   Order Gorgonacea
Hexacorals   Subclass Hexacorallia
Hydrocorals   Class Hydrozoa

Jellyfish

Orange cup coral   *Tubastraea coccinea*
Pink coral   *Corallium* species
Red tree coral   *Primnoa walleyi*
Scleractinian corals   Order Scleractinia
Sea anemones   Order Actinaria
Sea pens and octocorals   Subclass Octocorallia
Sea raspberry coral   *Eunephyta* species
Soft coral   Order Alcyonaceae
Staghorn coral   *Acropora cervicornis*
Star coral   *Monastrea annularis*
Wire coral   *Cirripathes anguinea*

Sea stars, urchins, and sea cucumbers

Brittlestars   Class Ophiuroidea
California sea cucumber   *Parastichopus californicus*
Green sea urchin   *Strongylocentrotus droebachiensis*
Purple sea urchin   *Strongylocentrotus purpuratus*
Red sea urchin   *Strongylocentrotus franciscanus*
Sand dollar   Family Clypeasteridae
Sand dollars and sea urchins   Class Echinoidea
Sea cucumbers   Class Holothuroidea
Sea stars   Class Asteroidea
Sea stars   *Pisaster* species
Sea urchins   *Strongylocentrotus* species

Rotifers and phoronids

Ctenophores   Phylum Ctenophora
Gastrotrichs   Class Gastrotricha
Phoronids   Phylum Phoronida
Rotifers   Class Rotifera

Worms

Acanthocephalan parasite   *Polymorphus* species
Earthworms   Class Oligochaeta
Feather-duster worms   Subclass Sabelliformia
Flatworms   Phylum Platyhelminthes
Flatworms, including freshwater
  microturbellarians   Class Turbellaria
Human blood fluke   *Schistosoma mansoni*
Leeches   Class Hirudinea
Lungworm pneumonia   *Dictyocaulus viviparus*
                                  [Causative agent]
Nematodes   Phylum Nematoda
Nemerteans   Phylum Nematoda
Oligochaetes   Class Oligochaeta
Oligochaete [No common name]   *Corophium* species
Oligochaete [No common name]   *Mediomastus* species
Oligochaete [No common name]   *Streblospio* species
Polychaete worm
  [No common name]   *Hobsonia florida*
Polychaete worms   Class Polychaeta
Polychaetes   Class Polychaeta
Segmented worms   Phylum Annelida
Tube worms   Family Serpulidae

Mollusks

Clams, bivalves

Alaska great-tellin   *Tellina lutea*
Angelwing   *Cyrtopleura costata*
Arctic surfclam   *Mactromeris polynyma*
Asian clam   *Corbicula fluminea*
Asian clams   Family Corbiculidae
Atlantic calico scallop   *Argopecten gibbus*
Atlantic rangia clam   *Rangia cuneata*
Atlantic surfclam   *Spisula solidissima*
Bay scallop   *Argopecten irradians*
Blue mussel   *Mytilus edulis*
Broad cockle   *Serripes laperousi*
Clubshell   *Pleurobema clava*
Cockles   Family Cardiidae
Cracking pearlymussel   *Hemistena lata*
Dwarf wedgemussel   *Alasmidonta heterodon*
Eastern oyster   *Crassostrea virginica*
Elktoe   *Alasmidonta marginata*
Falsemussels   Family Dreissenidae
Fanshell   *Cyprogenia stegaria*
Fat pocketbook   *Potamilus capax*
Fingernailclams   Family Sphaeriidae
Giant clam   *Tridacna gigas*
Giant clams   Family Tridacnidae
Greenland cockle   *Serripes groenlandicus*

Hard clam	*Mercenaria mercenaria*
Hiatellas	Family Hiatellidae
Higgins eye	*Lampsilis higginsi*
Japanese littleneck	*Tapes philippinarum*
Limpets	Family Fissurellidae
Long fingernailclam	*Musculium transversum*
Mangrove oyster	*Crassostrea rhizophorae*
Marshclam	Family Corbiculidae
Mussels	Family Mytilidae
Mussels	Several genera, especially *Mytilus*
Narrowmouth hydrobe	*Texadina spinctosoma*
Neosho mucket	*Lampsilis rafinesqueana*
Northern quahog	*Mercenaria mercenaria*
Northern riffleshell	*Epioblasma torulosa rangiana*
Ocean quahog	*Arctica islandica*
Orange-foot pimpleback	*Plethobasus cooperianus*
Oyster	Family Ostreidae
Oyster piddock	*Diplothyra smithii*
Oysters	*Crassostrea* species
Pacific calico scallop	*Argopecten circularis*
Pacific geoduck	*Panopea abrupta*
Pacific oyster	*Crassostrea gigas*
Pacific razor clam	*Siliqua patula*
Pearlymussels	Family Unionidae
Penshells	Family Pinnidae
Piddocks	Family Pholadidae
Pink mucket	*Lampsilis abrupta*
Purple cat's-paw	*Epioblasma obliquata obliquata*
Razor clams	Family Solenidae
Ring pink	*Obovaria retusa*
Scaleshell	*Leptodea leptodon*
Scallops	Family Pectinidae
Sea scallop	*Placopecten magellanicus*
Small giant clam	*Tridacna maxima*
Snuffbox	*Epioblasma triquetra*
Softshell clam	*Mya arenaria*
Softshell clams	Family Myidae
Spectaclecase	*Cumberlandia monodonta*
Surfclams	Family Mactridae
Tar spinymussel	*Elliptio steinstansana*
Tellins	Family Tellinidae
Tellins	*Tellinia* species
Venus clams	Family Veneridae
Weathervane scallops	*Patinopecten caurinus*
White cat's-paw	*Epioblasma sulcata delicata*
Winged mapleleaf	*Quadrula fragosa*
Zebra mussel	*Dreissena polymorpha*

Snails

Abalones	Family Haliotididae
Abalones	*Haliotis* species
Ambersnails	Family Succineidae
Amistrid tree snails	Family Amistridae
Angular triton	*Cymatium femorale*
Appalachian ambersnail	*Succinea chittenangoensis*
Apple snail	*Pomacea* species
Argus desert snail	*Eremariontoides argus*
Arions	Family Arionidae
Atlantic trumpet triton	*Charonia tritonis variegata*
Banana slug	*Ariolimax dolichophallus*
Banbury Springs limpet	*Lanx* species
Black abalone	*Haliotis cracherondii*
Bliss Rapids snail	*Taylorconcha serpenticola*
Blood fluke planorb	*Biomphalaria glabrata*
Brown gardensnail	*Helix aspersa*
Bruneau Hot Springs snail	*Pyrgulopsis bruneauensis*
Buccinium whelks	*Buccinium* species
Cactus snails	*Xerarionta* species
Cameo helmet	*Cassis madagarcariensis* and *C. m. spinella*
Caribbean helmet	*Cassis tuberosa*
Catalina cactus snail	*Xerarionta kelletti*
Catalina mountain snail	*Radiocentrum avalonense*
Cheat threetooth	*Triodopsis platysayoides*
Chittenango ovate ambersnail (see Appalachian ambersnail)	

Clench's nerite	*Neritina clenchi*
Coachella desert snail	*Eremarionta indioensis*
Columbia pebblesnail	*Fluminicola columbiana*
Conchs	Family Strombidae
Cuban physa	*Physa cubensis*
Dalles sideband snail	*Monadenia fidelis minor*
Decollate snail	*Rumina decollata*
Desert snails	*Eremarionta* species
El Paso shoulderband	*Mohavelix micrometaleus*
Euglandina (see Rosey wolfsnail)	
Fat whelk	*Neptunea ventricosa*
Flame helmet	*Cassis flammea*
Flat abalone	*Haliotis walallensis*
Florida tree snails	*Liguus fasciatus*
Giant African snail	*Achatina fulica*
Green abalone	*Haliotis fulgens*
Guadalupe shelled slug	*Binneya guadalupensis*
Hawaiian tree snails	*Achatinella, Partulina* species
Hawkwing conch	*Strombus raninus*
Hydrobe	Family Hydrobiidae
Island snails	*Micrarionta* species
Karok hesperian	*Vespericola karakorum*
Lyre whelk	*Neptunea lyrata*
Milk conch	*Strombus costatus*
Mimic shoulderband	*Helminthoglypta micrometalleoides*
Mohave shoulderband	*Helminthoglypta greggi*
Morro shoulderband	*Helminthoglypta walkeriana*
Mountain snail	Family Oreohelicidae
Oregon triton	*Fusitriton oregonensis*
Panamint shoulderband	*Helminthoglypta fisheri*
Periwinkles	*Littorina* species
Pink abalone	*Haliotis corrugata*
Pinto abalone	*Haliotis kamtschatkana*
Plain cactus snail	*Xerarionta intercisa*
Polynesian tree snails	*Partula* species
Pribilof whelk	*Neptunea pribiloffensis*
Pyrgs	*Pyrgulopsis* species
Queen conch	*Strombus gigas*
Quilted melania	*Tarebia granifera*
Red abalone	*Haliotis rufescens*
Roostertail conch	*Strombus gallus*
Rosey wolfsnail	*Euglandina rosea*
Sad nerite	*Puperita tristis*
San Miguel shoulderband	*Helminthoglypta ayresiana*
San Nicolas island snail	*Micrarionta feralis*
Santa Barbara shelled slug	*Binneya notabilis*
Sea snails	[No scientific name]
Shasta sideband	*Monadenia troglodytes*
Shelled slug	*Binneya* species
Shipworm	Family Teredinidae
Shoulderband, hemminthoglyptids	Family Helminthoglyptidae
Shoulderbands	*Helminthoglypta* species
Slugs	Several orders and families
Snake River physa	*Physa natricina*
Soledad shoulderband	*Helminthoglypta fontiphila*
South American snail	*Australorbis glabratus*
Springsnails	*Fontelicella* species
Surf shoulderband	*Helminthoglypta fieldi*
Talus snails	*Sonorella* species
Threaded abalone	*Haliotis assimilis*
Trinity bristlesnail	*Monadenia setosa*
Triton	Family Ranellidae
Utah valvata	*Valvata utahensis*
West Indian fighting conch	*Strombus pugilis*
West Indian topsnail	*Cittarium pica*
Whelk	*Mohinia* species
Whelks	Family Buccinidae
White abalone	*Haliotis sorenseni*
Wreathed cactus snail	*Xerarionta redimita*
Zebra nerite	*Puperita pupa*
[No common name]	*Neptunea heros*

Squids and octopuses

Brief squid	*Lolliguncula brevis*

Long-finned squid — *Loligo pealei*
Octopus — *Octopus cyahea*
Short-finned squid — *Illex illecebrosus*
Squid — Family Loliginidae
Squids — Subclass Teuthoidea

Arachnids and related groups

Millipedes — Class Diplopoda
Spiders — Class Araneida
Long-jawed spider — Family Tetragnathidae
Mites — Order Acarina
Pseudoscorpions — Order Pseudoscorpionidae
Sarcoptic mange — [No scientific name]
Scorpions — Order Scorpionidae
Spruce-fir moss spider — *Microhexura montivaga*
Centipedes — Class Chilopoda

Crustaceans

Amphipod [No common name] — *Ampelisca* species
Amphipod [No common name] — *Corophium* species
Amphipod [No common name] — *Gradidierella* species
Amphipods — Order Amphipoda
Barnacles — Subclass Cirripedia
Cladocerans — Order Cladocera
Copepod [No common name] — *Acartia tonsa*
Copepod [No common name] — *Calanus finmarchicus*
Copepod [No common name] — *Centropages hamatus*
Copepod [No common name] — *Centropages typicus*
Copepod [No common name] — *Diaptomus dorsalis*
Copepod [No common name] — *Diaptomus reighardi*
Copepod [No common name] — *Diaptomus siciloides*
Copepod [No common name] — *Metridia lucens*
Copepod [No common name] — *Osphrantieum labronectum*
Copepod [No common name] — *Pseudocalanus* species
Copepods — Order Copepoda
Copepods — *Pseudocalanus* species
Copepods — Subclass Copepoda
Cyclopoid copepods — Family Cyclopidae
Diaptomid copepods — Family Diaptomidae
Luciculous amphipods — [No scientific name]
Ostracods — Subclass Ostracoda
[No common name] — *Diaptomus kenai*
[No common name] — *Diaptomus tyrrelli*

Crabs, crayfishes, lobsters, and shrimps

Ala kumu — *Carpilius maculatus*
American lobster — *Homarus americanus*
Atlantic rock crab — *Cancer irroratus*
Blue crab — *Callinectes sapidus*
Blue king crab — *Paralithodes platypus*
Brine shrimp (see fairy shrimp)
Brown shrimp — *Penaeus aztecus*
California freshwater shrimp — *Syncaris pacifica*
Caribbean spiny lobster — *Panulirus argus*
Cave shrimp — Family Atyidae
Common land crab — [No scientific name]
Crayfishes — *Pacifastacus* species
Crayfishes — *Procambarus* species
Deepwater shrimps — *Heterocarpus* species
Dungeness crab — *Cancer magister*
Fairy shrimp — *Branchinecta* species
Fairy shrimps — Family Branchinectidae
Fairy shrimps and brine shrimps — Order Anostraca
Golden (brown) king crab — *Lithodes aequispina*
Green crab — *Carcinus maenas*
Hawaiian spiny lobster — *Panulirus marginatus*
Horseshoe crab — *Limulus polyphemus*
Isopods and sow bugs — Order Isopoda
Jonah crab — *Cancer borealis*
Kelp crab — *Pugettia producta*
King crabs — *Lithodes* and *Paralithodes* species
Korean hair crab — *Erimacrus isenbeckii*

Krill — *Euphausia* species
Krill — Order Euphausiacea
Krill — *Stylocheiron* species
Krill — *Thysanoessa* species
Lady crab — *Ovalipes ocellatus*
Large native shrimp — [No scientific name]
Lesser blue crab — *Callinectes similis*
Lobster — Infraorders Astacidea and Palinura
Millipedes — Class Diplopoda
Mona cave shrimp — *Typhlatya monae*
Northern shrimp — *Pandalus borealis*
Opossum shrimp — *Mysis relicta*
Pacific black tiger shrimp — *Penaeus monodon*
Pacific white shrimp — *Penaeus vannamei*
Penaeid shrimps — *Penaeus* species
Pink shrimp — *Penaeus duorarum*
Portunid crabs — Family Portunidae
Red crab — *Geryon quinquedens*
Red king crab — *Paralithodes camtschatica*
Riverside fairy shrimp — *Streptocephalus woottoni*
Rock shrimp — *Sicyonia brevirostris*
Royal red shrimp — *Menopenaeus robustus*
Seabob shrimp — *Xiphopenaeus kroyeri*
Shasta crayfish — *Pacifastacus fortis*
Shrimps — Family Penaeidae
Sidestripe shrimp — *Pandalopsis dispar*
Slipper lobster — *Scyllarides squamosus*
Snow crab — *Chionoecetes opilio*
Spiny lobster — *Panulirus argus*
Spiny lobsters — *Panulirus* species
Stone crab — *Menippe mercenaria*
Tanner crabs — *Chionoecetes bairdi, C. opilio*
Water fleas — Order Cladocera
White shrimp — *Penaeus setifer*

Insects

Ants, bees, and wasps

Africanized honey bee — *Apis mellifera adansonii*
Ants — Family Formicidae
Argentine ant — *Iridomyrmex humilis*
Blue orchard bee — *Osmia lignaria*
Boll weevil — *Anthonomus grandis*
Bumble bees — *Bombus* species
Honey bee — *Apis mellifera*
Honey bee and bumble bees — Family Apidae
Imported red fire ant — *Solenopsis invicta*
Leaf-cutting bee — Family Megachilidae
Paper wasps, yellowjackets and hornets — Family Vespidae
Western yellowjacket — *Vespula pensylvanica*
Yellowjackets — *Vespula* species

Aquatic insects

Amargosa naucorid — *Pelocoris shoshone amargosus*
Ash Meadows naucorid — *Ambrysus amargosus*
Burrowing mayflies — *Hexagenia bilineata, Pentagenia vittigera*
Burrowing mayfly — *Tortopus incertus*
Caddisflies — Order Trichoptera
Green-eyed skimmers — Family Corduliidae
Hexagenia burrowing mayflies — *Hexagenia* species
Hydropsychid caddisflies — Family Hydropsychidae
Lake Tahoe benthic stonefly — *Capnia lacustra*
Mayflies — Family Ephemeridae
Mayflies — Order Ephemeroptera
Narrow-winged damselflies — Family Coenagrioniidae
Net-spinning caddisflies — *Hydropsyche* species
Ozark emerald — *Somatochlora ozarkensis*
Ozark snaketail dragonfly — *Ophiogomphus westfalli*
Polymitarcid mayflies — Family Polymitarcidae
Stoneflies — Order Plecoptera
Wahkeena Falls flightless stonefly — *Zapada wahkeena*

Beetles

Alligatorweed flea beetle	*Agasicles hygrophila*
American burying beetle	*Nicrophorus americanus*
Aphodiine scarabs	*Aphodius* species
Bark beetles	Family Scolytidae
Beetle [No common name]	*Dictyopterus* species
Darkling ground beetle	*Eusattus muricatus*
Darkling ground beetles	Family Tenebrionidae
Death Valley agabus diving beetle	*Agabus rumppi*
Delta green ground beetle	*Elaphrus viridis*
Devil's Hole warm spring riffle beetle	*Stenelmis calida calida*
Elm bark beetle	*Scolytus multistriatis*
Giuliani's dune scarab	*Pseudocotalpa giulianii*
Globose dune beetle	*Coelus globosus*
Hungerford's crawling water beetle	*Brychius hungerfordi*
Large aegialian scarab	*Aegalia magnifica*
Moapa warm spring riffle beetle	*Stenelmis calida moapa*
Mountain pine beetle	*Dendroctonus ponderosae*
Nelson's miloderes weevil	*Miloderes rulieni*
Sixbanded longhorn beetle	*Dryobius sexnotatus*
Southern pine beetle	*Dendroctonus frontalus*
Spruce beetle	*Dendroctonus rufipennis*
Travertine band-thigh diving beetle	*Hygrotus fontinalis*
Valley elderberry longhorn beetle	*Desmocerus californicus dimorphus*

Cicadas, leafhoppers, and aphids

Balsam woolly adelgid	*Adelges piceae*
Beech scale insect	*Cryptococcus fagisuga*
Cottonycushion scale	*Icerya purchasi*
Hemlock woolly adelgid	*Adelges tsugae*
Mealybug	*Humococcus ceraricus*
Scale insect	*Acanthococcus arenosus*
Scale insect	*Lecanodiapsis prosopidis*
Scale insect	*Lecanodiapsis rufescens*
Two-spotted leafhopper	*Sophonia rufofascia*

Flies and midges

Blackflies	Family Simuliidae
Brine flies	Family Ephydridae
Delhi Sands flower-loving fly	*Rhaphiomidas terminatus abdominalis*
Midge, chironomids	Family Chironomidae
Mosquitoes	Family Culicidae
Phantom midges	*Chaoborus* species
Picture-wing vinegar flies	Family Drosophilidae
Southern house mosquito	*Culex quinquefasciatus*
Tube-building midges	Family Chironomidae
Vinegar flies	*Drosphila basisetae, claytonae, digressa, hawaiiensis, heteroneura, macrothrix, murphyi, ochracea, paucipuncta, rolaticilia, setosifrons, setosimentum, silvestris, sproati*

Grasshoppers, crickets, and allies

German cockroach	*Blatella germanica*
Idaho pointheaded grasshopper	*Acrolophitus punchellus*
Kelso giant sand treader cricket	*Macrobaenetes kelsonii*
Kelso Jerusalem cricket	*Ammopelmatus kelsonii*
Lake Huron locust	*Trimerotropis huroniana*
Prairie mole cricket	*Gryllotalpa major*
Superb spharagemon grasshopper	*Spharagemon superbum*
Tuna Cave roach	*Aspiduchus cavernicola*

Moths and butterflies

Apache nokomis fritillary	*Speyeria nokomis* subspecies *apacheana*
Arogos skipper	*Atrytone arogos*
Avalon hairstreak	*Strymon avalona*
Bay checkerspot butterfly	*Euphydryas editha bayensis*
Byssus skipper	*Problema byssus*
Common wood nymph	*Cercyonis pegala*
Creole pearly eye	*Enodia creola*
Dakota skipper	*Hesperia dacotae*
Diana fritillary	*Speyeria diana*
Douglas-fir tussock moth	*Orgyia pseudotsugata*
Eastern spruce budworm	*Choristoneura fumiferana*
Edith's checkerspot	*Euphydryas editha*
El Segundo blue butterfly	*Euphilotes enoptes allyni*
Gypsy moth	*Lymantria dispar*
Hessel's hairstreak	*Callophrys hesseli*
Karner blue butterfly	*Lycaeides melissa samuelis*
Kern primrose sphinx moth	*Euproserpinus euterpe*
Lange's metalmark butterfly	*Apodemia mormo langei*
Larch casebearer	*Coleophora laricella*
Lotis blue butterfly	*Lycaeides idas lotis*
Maculated manfreda skipper	*Stallingsia maculosus*
Mission blue butterfly	*Plebejus icarioides missionensis*
Mitchell's satyr	*Neonympha mitchellii*
Myrtle's silverspot butterfly	*Speyeria zerene myrtleae*
Nokomis fritillary	*Speyeria nokomis*
Oregon silverspot butterfly	*Speyeria zerene hippolyta*
Ottoe skipper	*Hesperia ottoe*
Palos Verdes blue butterfly	*Glaucopsyche lygdamus palosverdesensis*
Persius duskywing	*Erynnis persius*
Phlox moth	*Schinia indiana*
Poweshiek skipperling	*Oarisma powesheik*
Rattlesnake-master borer moth	*Papaipema eryngii*
Regal fritillary	*Speyeria idalia*
San Bruno elfin butterfly	*Callophrys mossii bayensis*
Sand Mountain pallid blue	*Euphilotes pallescens* subspecies
Silver-bordered fritillary	*Boloria selene sabulicollis*
Smith's blue butterfly	*Euphilotes enoptes smithi*
Sugarcane borer	*Diatraea saccharalis*
Tawny cresent	*Phyciodes batesi*
Weidemeyer's admiral	*Limenitis weidemeyeri*
Western spruce budworm	*Choristoneura occidentalis*

Other insect groups

Black grass bugs	*Irbisia* and *Labops* species
Flightless lacewing	*Micromus* species
Springtails	Order Collembola
Termites	Order Isoptera

Vertebrate animals

Amphibians

African clawed frog	*Xenopus laevis*
Amargosa toad	*Bufo nelsoni*
American toad	*Bufo americanus*
Appalachian salamander	*Plethodon jordani*
Arboreal salamander	*Aneides lugubris*
Arizona toad	*Bufo microscaphus microscaphus*
Arroyo toad	*Bufo microscaphus californicus*
Black salamander	*Aneides flavipunctatus*
Black toad	*Bufo exsul*
Black-bellied salamander	*Desmognathus quadramaculatus*
Black-bellied slender salamander	*Batrachoseps nigriventris*
Black-spotted newt	*Notophthalmus meridionalis*
Blanchard's cricket frog	*Acris crepitans blanchardi*
Boreal chorus frog	*Pseudacris triseriata maculata*
Breckenridge Mountain slender salamander	*Batrachoseps* undescribed species
Bronze frog	*Rana clamitans clamitans*
Bullfrog	*Rana catesbeiana*
California giant salamander	*Dicamptodon ensatus*
California red-legged frog	*Rana aurora draytonii*
California slender salamander	*Batrachoseps attenuatus*
California tiger salamander	*Ambystoma californiense*
California toad	*Bufo boreas halophilus*
California treefrog	*Hyla cadaverina*
Canadian toad	*Bufo hemiophrys*
Canyon treefrog	*Hyla arenicolor*
Caribbean white-lipped frog	*Leptodactylus albilabris*
Cascade torrent salamander	*Rhyacotriton cascadae*

Cascades chorus frog	*Pseudacris regilla cascadae*
Cascades frog	*Rana cascadae*
Channel Islands slender salamander	*Batrachoseps pacificus*
Chiricahua leopard frog	*Rana chiricahuensis*
Chorus frogs	*Pseudacris* species
Clouded salamander	*Aneides ferreus*
Coast chorus frog	*Pseudacris regilla pacifica*
Coast Range newt	*Taricha torosa torosa*
Columbia spotted frog	*Rana luteiventris*
Columbia torrent salamander	*Rhyacotriton kezeri*
Common coqui	*Eleutherodactylus coqui*
Common mud frog	[No scientific name]
Cope's giant salamander	*Dicamptodon copei*
Couch's spadefoot	*Spea couchii*
Crawfish frog	*Rana areolata*
Cricket coqui	*Eleutherodactylus gryllus*
Cricket frog	*Acris* species
Cuban treefrog	*Osteopilus septentrionalis*
Del Norte salamander	*Plethodon elongatus*
Desert slender salamander	*Batrachoseps aridus*
Dunn's salamander	*Plethodon dunni*
Eastern narrow-mouthed toad	*Gastrophryne carolinensis*
Eastern newt	*Notophthalmus viridescens*
Eastern red-backed salamander	*Plethodon cinereus*
Eastern tiger salamander	*Ambystoma tigrinum tigrinum*
Ensatina	*Ensatina eschscholtzii*
Fairview slender salamander	*Batrachoseps* undescribed species
Flatwoods salamander	*Ambystoma cingulatum*
Foothill yellow-legged frog	*Rana boylii*
Forest coqui	*Eleutherodactylus portoricensis*
Four-toed salamander	*Hemidactylium scutatum*
Fowler's toad	*Bufo woodhousei fowleri*
Gabilan slender salamander	*Batrachoseps* undescribed species
Garden slender salamander	*Batrachoseps major*
Giant salamanders	Family Dicamptodontidae
Golden coqui	*Eleutherodactylus jasperi*
Gray tiger salamander	*Ambystoma tigrinum diabolis*
Gray treefrog	*Hyla versicolor*
Great Basin spadefoot	*Spea intermontana*
Great Plains narrow-mouthed toad	*Gastrophryne* species
Great Plains toad	*Bufo cognatus*
Green frog	*Rana clamitans*
Green toad	*Bufo debilis*
Green treefrog	*Hyla cinerea*
Ground coqui	*Eleutherodactylus richmondi*
Guadalupe slender salamander	*Batrachoseps* undescribed species
Gulf Coast toad	*Bufo valliceps*
Hell Hollow slender salamander	*Batrachoseps* undescribed species
Houston toad	*Bufo houstonensis*
Huachuca tiger salamander	*Ambystoma tigrinum stebbinsi*
Inyo Mountains slender salamander	*Batrachoseps campi*
Jemez Mountains salamander	*Plethodon neomexicanus*
Kern Canyon slender salamander	*Batrachoseps simatus*
Kern Plateau slender salamander	*Batrachoseps* undescribed species
Larch Mountain salamander	*Plethodon larselli*
Large-blotched ensatina	*Ensatina eschscholtzii klauberi*
Leopard frogs	*Rana* species
Lesser siren	*Siren intermedia*
Limestone salamander	*Hydromantes brunus*
Long-toed salamander	*Ambystoma macrodactylum*
Lungless salamanders	Family Plethodontidae
Marine toad	*Bufo marinus*
Monterey ensatina	*Ensatina eschscholtzii eschscholtzii*
Mottled coqui	*Eleutherodactylus eneidae*
Mount Lyell salamander	*Hydromantes platycephalus*
Mountain yellow-legged frog	*Rana muscosa*
Mudpuppy	*Necturus maculosus*
Newts	Family Salamandridae
Northern cricket frog	*Acris crepitans*
Northern leopard frog	*Rana pipiens*
Northern red-legged frog	*Rana aurora aurora*
Northern rough-skinned newt	*Taricha granulosa granulosa*
Northwestern salamander	*Ambystoma gracile*
Olympic torrent salamander	*Rhyacotriton olympicus*

Oregon ensatina	*Ensatina eschscholtzii oregonensis*
Oregon slender salamander	*Batrachoseps wrighti*
Oregon spotted frog	*Rana pretiosa*
Owens Valley web-toed salamander	*Hydromantes* undescribed species
Pacific chorus frog	*Pseudacris regilla*
Pacific giant salamander	*Dicamptodon tenebrosus*
Pacific slender salamander	*Batrachoseps pacificus*
Painted ensatina	*Ensatina eschscholtzii picta*
Pickerel frog	*Rana palustris*
Pig frog	*Rana grylio*
Plains leopard frog	*Rana blairi*
Plains spadefoot	*Spea bombifrons*
Puerto Rican coqui (see Forest coqui)	*Eleutherodactylus portoricensis*
Ramsey Canyon leopard frog	*Rana subaquavocalis*
Red-bellied newt	*Taricha rivularis*
Red-legged frog	*Rana aurora*
Red-spotted toad	*Bufo punctatus*
Relict leopard frog	*Rana onca*
Relictual slender salamander	*Batrachoseps relictus*
Ridge-headed toad	*Peltophryne lemur*
Rio Grande leopard frog	*Rana berlandieri*
Rock coqui	*Eleutherodactylus cooki*
Rough-skinned newt	*Taricha granulosa*
Sacramento Mountain salamander	*Aneides hardii*
San Gabriel slender salamander	*Batrachoseps* undescribed species
Santa Cruz long-toed salamander	*Ambystoma macrodactylum croceum*
Santa Lucia slender salamander	*Batrachoseps* undescribed species
Shasta salamander	*Hydromantes shastae*
Sierra chorus frog	*Pseudacris regilla sierrae*
Sierra Nevada ensatina	*Ensatina eschscholtzii platensis*
Sierra newt	*Taricha torosa sierrae*
Siskiyou Mountains salamander	*Plethodon stormi*
Sonoran Desert toad	*Bufo alvarius*
Southern long-toed salamander	*Ambystoma macrodactylum sigillatum*
Southern toad	*Bufo terrestris*
Southern torrent salamander	*Rhyacotriton variegatus*
Southwestern chorus frog	*Pseudacris regilla hypochondriaca*
Southwestern toad	*Bufo microscaphus*
Spadefoots	Family Pelobatidae
Spotted chorus frog	*Pseudacris clarkii*
Spotted salamander	*Ambystoma maculatum*
Spring peeper	*Pseudacris crucifer*
Squirrel treefrog	*Hyla squirella*
Striped chorus frog	*Pseudacris triseriata*
Striped newt	*Notophthalmus perstriatus*
Tailed frog	*Ascaphus truei*
Tarahumara frog	*Rana tarahumarae*
Tehachapi slender salamander	*Batrachoseps stebbinsi*
Three-toed amphiuma	*Amphiuma tridactylum*
Tiger salamander	*Ambystoma tigrinum*
Toads	Family Bufonidae
Tongueless frogs	Family Pipidae
Torrent salamanders	Family Rhyacotritonidae
Tree-hole coqui	*Eleutherodactylus hedricki*
True frogs and ranid frogs	Family Ranidae
Van Dyke's salamander	*Plethodon vandykei*
Vegas Valley leopard frog	*Rana fisheri*
Virgin Islands coqui	*Eleutherodactylus schwartzi*
Warty coqui	*Eleutherodactylus locustus*
Web-footed coqui	*Eleutherodactylus karlschmidti*
Western narrow-mouthed toad	*Gastrophryne olivacea*
Western red-backed salamander	*Plethodon vehiculum*
Western spadefoot	*Spea hammondii*
Western spotted frog	*Rana pretiosa* unnamed population
Western toad	*Bufo boreas*
Wood frog	*Rana sylvatica*
Woodhouse's toad	*Bufo woodhousii*
Wrinkled coqui	*Eleutherodactylus wightmanae*
Wyoming toad	*Bufo hemiophrys baxteri*
Yavapai leopard frog	*Rana yavapaiensis*
Yellow-blotched ensatina	*Ensatina eschscholtzii croceater*
Yellow-eyed ensatina	*Ensatina eschscholtzii xanthoptica*
Yosemite toad	*Bufo canorus*

Birds

Acorn woodpecker	*Melanerpes formicivorus*
Akepa	*Loxops coccineus*
Albatrosses	Family Diomedeidae
Alcids	Family Alcidae
Alder flycatcher	*Empidonax alnorum*
Aleutian Canada goose	*Branta canadensis leucopareia*
Allen's hummingbird	*Selasphorus sasin*
American avocet	*Recurvirostra americana*
American bittern	*Botaurus lentiginosus*
American black duck	*Anas rubripes*
American coot	*Fulica americana*
American crow	*Corvus brachyrhynchos*
American dipper	*Cinclus mexicanus*
American golden-plover	*Pluvialis dominica*
American goldfinch	*Carduelis tristis*
American kestrel	*Falco sparverius*
American peregrine falcon	*Falco peregrinus anatum*
American redstart	*Setophaga ruticilla*
American robin	*Turdus migratorius*
American white pelican	*Pelecanus erythrorhynchos*
American wigeon	*Anas americana*
American woodcock	*Scolopax minor*
Anhinga	*Anhinga anhinga*
Anianiau	*Hemignathus parvus*
Antarctic skua	*Catharcta maccormicki*
Antillean mango	*Anthracothorax dominicus*
Antillean nighthawk	*Chordeiles gundlachii*
Apapane	*Himatione sanguinea*
Aplomado falcon	*Falco femoralis*
Arctic peregrine falcon	*Falco pereginus tundrius*
Arctic tern	*Sterna paradisaea*
Arctic warbler	*Phylloscopus borealis*
Arizona Bell's vireo	*Vireo bellii arizonae*
Ash-throated flycatcher	*Myiarchus cinerascens*
Atlantic puffin	*Fratercula arctica*
Attwater's greater prairie-chicken	*Tympanuchus cupido attwateri*
Audubon's shearwater	*Puffinus lherminieri*
Auks, guillemots, murres, and puffins	Family Alcidae
Avocets and stilts	Family Recurvirostridae
Babblers	Family Timaliidae
Bachman's sparrow	*Aimophila aestivalis*
Bachman's warbler	*Vermivora bachmanii*
Baird's sandpiper	*Calidris bairdii*
Baird's sparrow	*Ammodramus bairdii*
Bald eagle	*Haliaeetus leucocephalus*
Baltimore oriole	*Icterus galbula*
Band-rumped storm-petrel	*Oceanodroma castro*
Band-tailed pigeon	*Columba fasciata*
Bank swallow	*Riparia riparia*
Barn-owls	Family Tytonidae
Barn swallow	*Hirundo rustica*
Barred owl	*Strix varia*
Bay-breasted warbler	*Dendroica castanea*
Belding's savannah sparrow	*Passerculus sandwichensis beldingi*
Bell's vireo	*Vireo bellii*
Belted kingfisher	*Ceryle alcyon*
Bendire's thrasher	*Toxostoma bendirei*
Bermuda petrel	*Pterodroma chow*
Bewick's wren	*Thryomanes bewickii*
Black-and-white warbler	*Mniotilta varia*
Black-backed woodpecker	*Picoides arcticus*
Black-bellied plover	*Pluvialis squatarola*
Black-bellied whistling-duck	*Dendrocygna autumnalis*
Black-billed cuckoo	*Coccyzus erythropthalmus*
Black-billed magpie	*Pica pica*
Black-browed albatross	*Thalassarche melanophris*
Black-capped chickadee	*Poecile atricapillus*
Black-capped petrel	*Pterodroma hasitata*
Black-chinned sparrow	*Spizella atrogularis*
Black-crowned night-heron	*Nycticorax nycticorax*
Black-footed albatross	*Phoebastria nigripes*
Black guillemot	*Cepphus grylle*
Black-headed grosbeak	*Pheucticus melanocephalus*
Black-legged kittiwake	*Rissa tridactyla*
Black oystercatcher	*Haematopus bachmani*
Black-necked stilt	*Himantopus mexicanus*
Black phoebe	*Sayornis nigricans*
Black rail	*Laterallus jamaicensis*
Black skimmer	*Rynchops niger*
Black tern	*Chlidonias niger*
Black turnstone	*Arenaria melanocephala*
Black-throated blue warbler	*Dendroica caerulescens*
Black-throated sparrow	*Amphispiza bilineata*
Blackbirds and allies	Family Icteridae
Blackpoll warbler	*Dendroica striata*
Blue-gray gnatcatcher	*Polioptila caerulea*
Blue-gray noddy	*Procelsterna cerulea*
Blue-gray ternlet	*Procelsterna cerulea saxatilis*
Blue-winged teal	*Anas discors*
Bluethroat	*Luscinia svecica*
Bobolink	*Dolichonyx oryzivorus*
Bonaparte's gull	*Larus philadelphia*
Bonin petrel	*Pterodroma hypoleuca*
Boobies	*Sula* species
Boreal chickadee	*Poecile hudsonicus*
Boreal owl	*Aegolius funereus*
Brant	*Branta bernicla*
Brewer's blackbird	*Euphagus cyanocephalus*
Brewer's sparrow	*Spizella breweri*
Bridled tern	*Sterna anaethetus*
Bristle-thighed curlew	*Numenius tahitiensis*
Broad-tailed hummingbird	*Selasphorus platycercus*
Broad-winged hawk	*Buteo platypterus*
Brown booby	*Sula leucogaster*
Brown creeper	*Certhia americana*
Brown-headed cowbird	*Molothrus ater*
Brown-headed nuthatch	*Sitta pusilla*
Brown noddy	*Anous stolidus*
Brown pelican	*Pelecanus occidentalis*
Brown thrasher	*Toxostoma rufum*
Buff-breasted flycatcher	*Empidonax fulvifrons*
Buff-breasted sandpiper	*Tryngites subruficollis*
Bufflehead	*Bucephala albeola*
Bullock's oriole	*Icterus bullocki*
Bulwer's petrel	*Bulweria bulwerii*
Burrowing owl	*Athene cunicularia*
Bushtit	*Psaltriparus minimus*
Bushtits and long-tailed tits	Family Aegithalidae
Cackling Canada goose	*Branta canadensis minima*
Cactus wren	*Campylorhynchus brunneicapillus*
California black rail	*Laterallus jamaicensis coturniculus*
California clapper rail	*Rallus longirostris obsoletus*
California condor	*Gymnogyps californianus*
California gnatcatcher	*Polioptila californica*
California least tern	*Sterna antillarum browni*
California towhee	*Pipilo crissalis*
Calliope hummingbird	*Stellula calliope*
Canada goose	*Branta canadensis*
Canada warbler	*Wilsonia canadensis*
Canvasback	*Aythya valisineria*
Cape petrel	*Daption capensis*
Cape Sable seaside sparrow	*Ammodramus maritimus mirabilis*
Caracaras and falcons	Family Falconidae
Cardinaline finches	Family Cardinalidae
Caribbean coot	*Fulica caribaea*
Carolina chickadee	*Poecile carolinensis*
Carolina parakeet	*Conuropsis carolinensis*
Caspian tern	*Sterna caspia*
Cassin's auklet	*Ptychoramphus aleuticus*
Cassin's finch	*Carpodacus cassinii*
Cassin's sparrow	*Aimophila cassinii*
Cattle egret	*Bubulcus ibis*
Cedar waxwing	*Bombycilla cedrorum*
Chestnut-backed chickadee	*Poecile rufescens*
Chestnut-collared longspur	*Calcarius ornatus*
Chickadees and titmice	Family Paridae

Chimney swift	*Chaetura pelagica*
Chipping sparrow	*Spizella passerina*
Christmas shearwater	*Puffinus nativitatis*
Clapper rail	*Rallus longirostris*
Clark's nutcracker	*Nucifraga columbiana*
Clay-colored sparrow	*Spizella pallida*
Cliff swallow	*Petrochelidon pyrrhonota*
Coastal California gnatcatcher	*Polioptila californica californica*
Columbian sharp-tailed grouse	*Tympanuchus phasianellus columbianus*
Common amakihi	*Hemignathus virens*
Common eider	*Somateria mollissima*
Common goldeneye	*Bucephala clangula*
Common grackle	*Quiscalus quiscula*
Common loon	*Gavia immer*
Common merganser	*Mergus merganser*
Common moorhen	*Gallinula chloropus*
Common murre	*Uria aalge*
Common nighthawk	*Chordeiles minor*
Common raven	*Corvus corax*
Common redpoll	*Carduelis flammea*
Common snipe	*Gallinago gallinago*
Common tern	*Sterna hirundo*
Common yellowthroat	*Geothlypis trichas*
Coots, gallinules, and rails	Family Rallidae
Cormorants	Family Phalacrocoracidae
Cory's shearwater	*Calonectris diomedea*
Cowbirds	*Molothrus* species
Cranes	Family Gruidae
Creepers	Family Certhiidae
Crested auklet	*Aethia cristatella*
Crested caracara	*Caracara plancus*
Crows, jays, and magpies	Family Corvidae
Cuban (Lesser Puerto Rican) crow	*Corvus nasicus*
Cuckoos and relatives	Family Cuculidae
Curve-billed thrasher	*Toxostoma curvirostre*
Dark-eyed junco, including slate-colored junco	*Junco hyemalis*
Dark-rumped petrel	*Pterodroma phaeopygia*
Darwin's finches (see Galapagos finches)	
DeBooy's rail	*Nesotrichas debooyi*
Dickcissel	*Spiza americana*
Dippers	Family Cinclidae
Double-crested cormorant	*Phalacrocorax auritus*
Dovekie	*Alle alle*
Doves and pigeons	Family Columbidae
Downy woodpecker	*Picoides pubescens*
Ducks, geese, and swans	Family Anatidae
Dunlin	*Calidris alpina*
Dusky Canada goose	*Branta canadensis occidentalis*
Dusky flycatcher	*Empidonax oberholseri*
Dusky seaside sparrow	*Ammodramus maritimus nigrescens*
Eared grebe	*Podiceps nigricollis*
Eastern Bewick's wren	*Thryomanes bewickii altus*
Eastern bluebird	*Sialia sialis*
Eastern brown pelican	*Pelecanus occidentalis carolinensis*
Eastern kingbird	*Tyrannus tyrannus*
Eastern meadowlark	*Sturnella magna*
Eastern phoebe	*Sayornis phoebe*
Eastern towhee	*Pipilo erythrophthalmus*
Eastern wood-pewee	*Contopus virens*
Eiders	*Polysticta* and *Somateria* species
Elegant trogon	*Trogon elegans*
Elepaio	*Chasiempis sandwichensis*
Elf owl	*Micrathene whitneyi*
Elfin Woods warbler	*Dendroica angelae*
Emberizid finches	Family Emberizidae
Emperor goose	*Chen canagica*
Eskimo curlew	*Numenius borealis*
Estrildid finches	Family Estrildidae
European starling	*Sturnus vulgaris*
Evening grosbeak	*Coccothraustes vespertinus*
Everglade snail kite	*Rostrhamus sociabilis plumbeus*
Evermann's rock ptarmigan	*Laggus mutus evermanni*
Ferruginous hawk	*Buteo regalis*
Ferruginous pygmy-owl	*Glaucidium brasilianum*
Field sparrow	*Spizella pusilla*
Fish crow	*Corvus ossifragus*
Flammulated owl	*Otus flammeolus*
Florida scrub-jay	*Aphelocoma caerulescens*
Fork-tailed storm-petrel	*Oceanodroma furcata*
Forster's tern	*Sterna forsteri*
Franklin's gull	*Larus pipixcan*
Frigatebirds	*Fregata* species
Fulmars, petrels, and shearwaters	Family Procellariidae
Fulvous whistling-duck	*Dendrocygna bicolor*
Gadwall	*Anas strepera*
Galapagos finches	*Camarhynchus, Certhidia, Geospiza, Pinaroloxias* species
Gila woodpecker	*Melanerpes uropygialis*
Gilded flicker	*Colaptes chrysoides*
Glaucous gull	*Larus hyperboreus*
Glaucous-winged gull	*Larus glaucescens*
Glossy ibis	*Plegadis falcinellus*
Golden-crowned kinglet	*Regulus satrapa*
Golden-crowned sparrow	*Zonotrichia atricapilla*
Golden eagle	*Aquila chrysaetos*
Golden-winged warbler	*Vermivora chrysoptera*
Grasshopper sparrow	*Ammodramus savannarum*
Gray-backed tern	*Sterna lunata*
Gray catbird	*Dumetella carolinensis*
Gray-cheeked thrush	*Catharus minimus*
Gray-crowned rosy-finch	*Leucosticte tephrocotis*
Gray flycatcher	*Empidonax wrightii*
Gray jay	*Perisoreus canadensis*
Gray partridge	*Perdix perdix*
Gray vireo	*Vireo vicinior*
Great auk	*Pinguinus impennis*
Great Basin Canada goose	*Branta canadensis moffitti*
Great black-backed gull	*Larus marinus*
Great blue heron	*Ardea herodias*
Great cormorant	*Phalacrocorax carbo*
Great egret	*Ardea alba*
Great gray owl	*Strix nebulosa*
Great horned owl	*Bubo virginianus*
Great skua	*Catharacta skua*
Greater flamingo	*Phoenicopterus ruber*
Greater prairie-chicken	*Tympanuchus cupido*
Greater sandhill crane	*Grus canadensis tabida*
Greater scaup	*Aythya marila*
Greater shearwater	*Puffinus gravis*
Greater white-fronted goose	*Anser albifrons*
Greater yellowlegs	*Tringa melanoleuca*
Grebes	Family Podicipedidae
Green heron	*Butorides virescens*
Green-tailed towhee	*Pipilo chlorurus*
Green-winged teal	*Anas crecca*
Grouse, pheasants, and quails	Family Phasianidae
Gull-billed tern	*Sterna nilotica*
Gulls, terns, and allies	Family Laridae
Gyrfalcon	*Falco rusticolus*
Hairy woodpecker	*Picoides villosus*
Hammond's flycatcher	*Empidonax hammondii*
Harcourt's storm-petrel	*Oceanodroma castro*
Harlequin duck	*Histrionicus histrionicus*
Hawaiian akialoa	*Hemignathus obscurus*
Hawaiian dark-rumped petrel	*Pterodroma phaeopygia sandwichensis*
Hawks, kites, and eagles	Family Accipitridae
Henslow's sparrow	*Ammodramus henslowii*
Herald petrel	*Pterodroma arminjoniana*
Hermit thrush	*Catharus guttatus*
Hermit warbler	*Dendroica occidentalis*
Herons, egrets, and bitterns	Family Ardeidae
Herring gull	*Larus argentatus*
Hill myna	*Gracula religiosa*
Hispaniolan parakeet	*Aratinga chloroptera*
Hoary redpoll	*Carduelis hornemanni*
Hooded merganser	*Lophodytes cucullatus*
Horned grebe	*Podiceps auritus*

Horned lark	*Eremophila alpestris*
Horned puffin	*Fratercula corniculata*
House finch	*Carpodacus mexicanus*
House sparrow	*Passer domesticus*
House wren	*Troglodytes aedon*
Hudsonian godwit	*Limosa haemastica*
Hummingbirds	Family Trochilidae
Ibises and spoonbills	Family Threskiornithidae
Iceland gull	*Larus glaucoides*
Iiwi	*Vestiaria coccinea*
Indigo bunting	*Passerina cyanea*
Interior least tern	*Sterna antillarum athalassos*
Island scrub-jay	*Aphelocoma insularis*
Ivory-billed woodpecker	*Campephilus principalis*
Jaegers	*Stercorarius* species
Jungle fowl	*Gallus gallus*
Kauai creeper	*Oreomystis bairdi*
Key West quail-dove	*Geotrygon chrysias*
Killdeer	*Charadrius vociferus*
King eider	*Somateria spectabilis*
King rail	*Rallus elegans*
Kingfishers	Family Alcedinidae
Kinglets	Family Regulidae
Kirtland's warbler	*Dendroica kirtlandii*
Kittiwakes	*Rissa* species
Kittlitz's murrelet	*Brachyramphus brevirostris*
Ladder-backed woodpecker	*Picoides scalaris*
Lapland longspur	*Calcarius lapponicus*
Lark bunting	*Calamospiza melanocorys*
Lark sparrow	*Chondestes grammacus*
Larks	Family Alaudidae
Laughing gull	*Larus atricilla*
Laysan albatross	*Phoebastria immutabilis*
Lazuli bunting	*Passerina amoena*
Le Conte's sparrow	*Ammodramus leconteii*
Leach's storm-petrel	*Oceanodroma leucorhoa*
Least auklet	*Aethia pusilla*
Least Bell's vireo	*Vireo bellii pusillus*
Least bittern	*Ixobrychus exilis*
Least flycatcher	*Empidonax minimus*
Least sandpiper	*Calidris minutilla*
Least tern	*Sterna antillarum*
Lesser Canada goose	*Branta canadensis parvipes*
Lesser goldfinch	*Carduelus psaltria*
Lesser prairie-chicken	*Tympanuchus pallidicinctus*
Lesser scaup	*Aythya affinis*
Lewis's woodpecker	*Melanerpes lewis*
Light-footed clapper rail	*Rallus longirostris levipes*
Limpkin	*Aramus guarauna*
Little blue heron	*Egretta caerulea*
Little shearwater	*Puffinus assimilis*
Little tern	*Sterna albifrons*
Loggerhead shrike	*Lanius ludovicianus*
Long-billed curlew	*Numenius americanus*
Long-billed dowitcher	*Limnodromus scolopaceus*
Long-tailed jaeger	*Stercorarius longicaudus*
Long-eared owl	*Asio otus*
Loons	Family Gaviidae
Lucy's warbler	*Vermivora luciae*
MacGillivray's warbler	*Oporornis tolmiei*
Magnificent frigatebird	*Fregata magnificens*
Magnolia warbler	*Dendroica magnolia*
Mallard	*Anas platyrhynchos*
Manx shearwater	*Puffinus puffinus*
Marbled godwit	*Limosa fedoa*
Marbled murrelet	*Brachyramphus marmoratus*
Marsh wren	*Cistothorus palustris*
Masked booby	*Sula dactylatra*
Maui creeper	*Paroreomyza montana*
Maui parrotbill	*Pseudonestor xanthophrys*
McCown's longspur	*Calcarius mccownii*
McKay's bunting	*Plectrophenax hyperboreus*
Merlin	*Falco columbarius*
Mew gull	*Larus canus*
Mexican spotted owl	*Strix occidentalis lucida*
Mississippi kite	*Ictinia mississippiensis*
Mississippi sandhill crane	*Grus canadensis pulla*
Mockingbirds and thrashers	Family Mimidae
Monarchs	Family Monarchidae
Monk parakeet	*Myiopsitta monachus*
Mottled duck	*Anas fulvigula*
Mountain bluebird	*Sialia currucoides*
Mountain chickadee	*Poecile gambeli*
Mountain plover	*Charadrius montanus*
Mountain quail	*Oreortyx pictus*
Mourning dove	*Zenaida macroura*
Mynas and starlings	Family Sturnidae
Nashville warbler	*Vermivora ruficapilla*
Nelson's sharp-tailed sparrow	*Ammodramus nelsoni*
New World quail	Family Odontophoridae
New World vultures	Family Cathartidae
Newell's Townsend's shearwater	*Puffinus auricularis newelli*
Nightjars	Family Caprimulgidae
Noddies	*Anous* species
Northern bobwhite	*Colinus virginianus*
Northern cardinal	*Cardinalis cardinalis*
Northern flicker	*Colaptes auratus*
Northern fulmar	*Fulmarus glacialis*
Northern gannet	*Morus bassanus*
Northern goshawk	*Accipiter gentilis*
Northern harrier	*Circus cyaneus*
Northern mockingbird	*Mimus polyglottos*
Northern oriole	*Icterus bullocki* and *I. galbula*
Northern parula	*Parula americana*
Northern pintail	*Anas acuta*
Northern rough-winged swallow	*Stelgidopteryx serripennis*
Northern shoveler	*Anas clypeata*
Northern spotted owl	*Strix occidentalis caurina*
Northern wheatear	*Oenanthe oenanthe*
Northwestern crow	*Corvus caurinus*
Nuthatches	Family Sittidae
Old World finches and allies	Family Fringillidae
Old World flycatchers and allies	Family Sylviidae
Old World sparrows	Family Passeridae
Oldsquaw	*Clangula hyemalis*
Olive-sided flycatcher	*Contopus cooperi*
Omao	*Myadestes obscurus*
Orange-crowned warbler	*Vermivora celata*
Orange-fronted parakeet	*Aratinga canicularis*
Orchard oriole	*Icterus spurius*
Osprey	*Pandion haliaetus*
Oystercatchers	Family Haematopodidae
Pacific golden-plover	*Pluvialis fulva*
Pacific loon	*Gavia pacifica*
Pacific pigeon	*Ducula pacifica*
Pacific-slope flycatcher	*Empidonax difficilis*
Painted bunting	*Passerina ciris*
Palila	*Loxioides bailleui*
Parakeet auklet	*Aethia psittacula*
Passenger pigeon	*Ectopistes migratorius*
Peale's peregrine falcon	*Falco peregrinus pealei*
Pectoral sandpiper	*Calidris melanotos*
Pekin duck (see Mallard)	
Pelicans	Family Pelecanidae
Peregrine falcon	*Falco peregrinus*
Petrels	Family Procellariidae
Phainopepla	*Phainopepla nitens*
Pied-billed grebe	*Podilymbus podiceps*
Pigeon guillemot	*Cepphus columba*
Pileated woodpecker	*Dryocopus pileatus*
Pine grosbeak	*Pinicola enucleator*
Pine siskin	*Carduelis pinus*
Pine warbler	*Dendroica pinus*
Piping plover	*Charadrius melodus*
Pipits and wagtails	Family Motacillidae
Plain pigeon	*Columba inornata*
Plain titmouse	*Baelophus inornatus*
Plovers	Family Charadriidae

Plumbeous solitary vireo	*Vireo solitarius plumbeus*	Savannah sparrow	*Passerculus sandwichensis*
Pomarine jaeger	*Stercorarius pomarinus*	Say's phoebe	*Sayornis saya*
Ponape greater white-eye	*Rukia longirostra*	Scarlet ibis	*Eudocimus ruber*
Ponape mountain starling	*Aplonis pelzelni*	Scaups	*Aythya affinis* and *A. marila*
Poo-uli	*Melamprosops phaeosoma*	Scoters	*Melanitta* species
Prairie falcon	*Falco mexicanus*	Scott's oriole	*Icterus parisorum*
Prairie warbler	*Dendroica discolor*	Sedge wren	*Cistothorus platensis*
Prothonotary warbler	*Protonotaria citrea*	Sharp-shinned hawk	*Accipiter striatus*
Ptarmigans	*Lagopus* species	Sharp-tailed grouse	*Tympanuchus phasianellus*
Puerto Rican barn-owl	*Tyto cavatica*	Shearwaters	*Calonectris* and *Puffinus* species
Puerto Rican broad-winged hawk	*Buteo platypterus brunnescens*	Shiny cowbird	*Molothrus bonariensis*
Puerto Rican nightjar	*Caprimulgus noctitherus*	Short-eared owl	*Asio flammeus*
Puerto Rican parrot	*Amazona vittata*	Short-tailed albatross	*Phoebastria albatrus*
Puerto Rican plain pigeon	*Columba inornata wetmorei*	Short-tailed shearwater	*Puffinus tenuirostris*
Puerto Rican quail-dove	*Geotrygon larva*	Shrikes	Family Laniidae
Puerto Rican screech-owl	*Otus nudipes*	Silky-flycatchers	Family Ptilogonatidae
Puerto Rican sharp-shinned hawk	*Accipiter striatus venator*	Skimmers (see Black skimmer)	
Puerto Rican woodcock	*Scolopax anthonyi*	Skuas	*Catharcta* species
Purple-capped fruit-dove	*Ptilinopus porphyraceus*	Snail kite	*Rostrhamus sociabilis*
Purple finch	*Carpodacus purpureus*	Snow bunting	*Plectrophenax nivalis*
Purple gallinule	*Porphyrula martinica*	Snow goose	*Chen caerulescens*
Purple martin	*Progne subis*	Snowy egret	*Egretta thula*
Purple sandpiper	*Calidris maritima*	Snowy owl	*Nyctea scandiaca*
Razorbill	*Alca torda*	Snowy plover	*Charadrius alexandrinus*
Red-bellied woodpecker	*Melanerpes carolinus*	Solitary vireo	*Vireo solitarius*
Red-breasted merganser	*Mergus serrator*	Song sparrow	*Melospiza melodia*
Red-breasted nuthatch	*Sitta canadensis*	Sooty shearwater	*Puffinus griseus*
Red-cockaded woodpecker	*Picoides borealis*	Sooty storm-petrel	*Oceanodroma tristrami*
Red crossbill	*Loxia curvirostra*	Sooty tern	*Sterna fuscata*
Red-eyed vireo	*Vireo olivaceus*	Sora	*Porzana carolina*
Red-faced cormorant	*Phalacrocorax urile*	South polar skua	*Catharacta maccormicki*
Red-footed booby	*Sula sula*	Southwestern willow flycatcher	*Empidonax traillii extimus*
Red-headed woodpecker	*Melanerpes erythrocephalus*	Spectacled eider	*Somateria fischeri*
Red-legged kittiwake	*Rissa brevirostris*	Spotted dove	*Streptopelia chinensis*
Red-necked grebe	*Podiceps grisegena*	Spotted owl	*Strix occidentalis*
Red-necked phalarope	*Phalaropus lobatus*	Spotted sandpiper	*Actitis macularia*
Red phalarope	*Phalaropus fulicarius*	Spotted towhee	*Pipilo maculatus*
Red-shouldered hawk	*Buteo lineatus*	Sprague's pipit	*Anthus spragueii*
Red siskin	*Carduelis cucullata*	Spruce grouse	*Falcipennis canadensis*
Red-tailed hawk	*Buteo jamaicensis*	St. Croix macaw	*Ara autocthones*
Red-tailed tropicbird	*Phaethon ribricauda*	Steller's eider	*Polysticta stelleri*
Red-throated loon	*Gavia stellata*	Steller's jay	*Cyanocitta stelleri*
Red-throated pipit	*Anthus cervinus*	Steller's sea-eagle	*Haliaeetus pelagicus*
Red-winged blackbird	*Agelaius phoeniceus*	Stilt sandpiper	*Calidris himantopus*
Reddish egret	*Egretta rufescens*	Storks	Family Ciconiidae
Redhead	*Aythya americana*	Storm-petrels	Family Hydrobatidae
Rhinoceros auklet	*Cerorhinca monocerata*	Surfbird	*Aphriza virgata*
Ring-billed gull	*Larus delawarensis*	Swainson's hawk	*Buteo swainsoni*
Ring-necked duck	*Aythya collaris*	Swainson's thrush	*Catharus ustulatus*
Ring-necked pheasant	*Phasianus colchicus*	Swainson's warbler	*Limnothlypis swainsonii*
Rock dove	*Columba livia*	Swallows	Family Hirundinidae
Rock ptarmigan	*Lagopus mutus*	Swamp sparrow	*Melospiza georgiana*
Rock sandpiper	*Calidris ptilocnemis*	Swifts	Family Apodidae
Rock wren	*Salpinctes obsoletus*	Tanagers	Family Thraupidae
Rose-breasted grosbeak	*Pheucticus ludovicianus*	Taverner's Canada goose	*Branta canadensis taverneri*
Roseate tern	*Sterna dougallii*	Thick-billed murre	*Uria lomvia*
Ross's goose	*Chen rossii*	Thick-billed parrot	*Rhynchopsitta pachyrhyncha*
Rough-legged hawk	*Buteo lagopus*	Three-toed woodpecker	*Picoides tridactylus*
Royal tern	*Sterna maxima*	Thrushes	Family Turdidae
Ruby-crowned kinglet	*Regulus calendula*	Townsend's shearwater	*Puffinus auricularis*
Ruby-throated hummingbird	*Archilochus colubris*	Townsend's solitaire	*Myadestes townsendi*
Ruddy duck	*Oxyura jamaicensis jamaicensis*	Townsend's warbler	*Dendroica townsendi*
Ruddy ground-dove	*Columbina talpacoti*	Tree swallow	*Tachycineta bicolor*
Ruffed grouse	*Bonasa umbellus*	Tricolored blackbird	*Agelaius tricolor*
Rufous hummingbird	*Selasphorus rufus*	Tricolored heron	*Egretta tricolor*
Sage grouse	*Centrocercus urophasianus*	Tristram's storm-petrel (see Sooty storm-petrel)	
Sage sparrow	*Amphispiza belli*	Trogons	Family Trogonidae
Sage thrasher	*Oreoscoptes montanus*	Tropicbirds	Family Phaethontidae
San Clemente loggerhead shrike	*Lanius ludovicianus mearnsi*	Troupial	*Icterus icterus*
San Clemente sage sparrow	*Amphispiza belli clementeae*	Trumpeter swan	*Cygnus buccinator*
San Diego cactus wren	*Campylorhynchus brunneicapillus couesi*	Tufted puffin	*Fratercula cirrhata*
Sandhill crane	*Grus canadensis*	Tufted titmouse	*Baelophus bicolor*
Sandpipers and allies	Family Scolopacidae	Tule white-fronted goose	*Anser albifrons gambelli*
Sandwich tern	*Sterna sandvicensis*	Tundra swan	*Cygnus columbianus*

Turkey vulture	*Cathartes aura*
Typical owls	Family Strigidae
Tyrant flycatchers	Family Tyrannidae
Upland sandpiper	*Bartramia longicauda*
Vancouver Canada goose	*Branta canadensis fulva*
Varied bunting	*Passerina versicolor*
Varied thrush	*Ixoreus naevius*
Vaux's swift	*Chaetura vauxi*
Veery	*Catharus fuscescens*
Vermilion flycatcher	*Pyrocephalus rubinus*
Vesper sparrow	*Pooecetes gramineus*
Violet-crowned hummingbird	*Amazilia violiceps*
Violet-green swallow	*Tachycineta thalassina*
Vireos	Family Vireonidae
Virgin Islands screech-owl	*Otus nudipes newtoni*
Virginia rail	*Rallus limicola*
Wandering tattler	*Heteroscelus incanus*
Warbling vireo	*Vireo gilvus*
Water pipit	*Anthus rubescens*
Waxwings	Family Bombycillidae
Wedge-tailed shearwater	*Puffinus pacificus*
West Indian ruddy duck	*Oxyura jamaicensis* subspecies
West Indian whistling-duck	*Dendrocygna arborea*
Western bluebird	*Sialia mexicana*
Western Canada goose	*Branta canadensis moffitti*
Western grebe	*Aechmophorus occidentalis*
Western kingbird	*Tyrannus verticalis*
Western meadowlark	*Sturnella neglecta*
Western sandpiper	*Calidris mauri*
Western scrub-jay	*Aphelocoma californica*
Western snowy plover	*Charadrius alexandrinus nivosus*
Western tanager	*Piranga ludoviciana*
Western wood-pewee	*Contopus sordidulus*
Western yellow-billed cuckoo	*Coccyzus americanus occidentalis*
Whimbrel	*Numenius phaeopus*
Whiskered auklet	*Aethia pygmaea*
White-breasted nuthatch	*Sitta carolinensis*
White-cheeked pintail	*Anas bahamensis*
White-chinned petrel	*Procellaria aequinoctialis*
White-crowned pigeon	*Columba leucocephala*
White-crowned sparrow	*Zonotrichia leucophrys*
White-eyed vireo	*Vireo griseus*
White-faced ibis	*Plegadis chihi*
White-faced storm-petrel	*Pelagodroma marina*
White-headed woodpecker	*Picoides albolarvatus*
White ibis	*Eudocimus albus*
White-necked crow	*Corvus leucognaphalus*
White-rumped sandpiper	*Calidris fuscicollis*
White-tailed eagle	*Haliaeetus albicilla*
White-tailed ptarmigan	*Lagopus leucurus*
White-tailed tropicbird	*Phaethon lepturus*
White tern	*Gygis alba*
White-throated sparrow	*Zonotrichia albicollis*
White-vented myna	*Acridotheres javanicus*
White-winged dove	*Zenaida asiatica*
Whooping crane	*Grus americana*
Wild turkey	*Meleagris gallopavo*
Willet	*Catoptrophorus semipalmatus*
Willow flycatcher	*Empidonax traillii*
Willow ptarmigan	*Lagopus lagopus*
Wilson's phalarope	*Phalaropus tricolor*
Wilson's storm-petrel	*Oceanites oceanicus*
Wilson's warbler	*Wilsonia pusilla*
Winter wren	*Troglodytes troglodytes*
Wood duck	*Aix sponsa*
Wood stork	*Mycteria americana*
Wood thrush	*Hylocichla mustelina*
Wood warblers	Family Parulidae
Woodpeckers and wrynecks	Family Picidae
Wrens	Family Troglodytidae
Wrentit	*Chamaea fasciata*
Yellow-billed cuckoo	*Coccyzus americanus*
Yellow-billed loon	*Gavia adamsii*
Yellow-billed magpie	*Pica nuttalli*

Yellow-breasted chat	*Icteria virens*
Yellow-crowned night-heron	*Nyctanassa violacea*
Yellow-headed blackbird	*Xanthocephalus xanthocephalus*
Yellow-nosed albatross	*Thallasarche chlororhynchos*
Yellow rail	*Coturnicops noveboracensis*
Yellow-rumped warbler	*Dendroica coronata*
Yellow-shafted flicker	*Colaptes auratus auratus*
Yellow-shouldered blackbird	*Agelaius xanthomus*
Yellow-throated vireo	*Vireo flavifrons*
Yellow wagtail	*Motacilla flava*
Yellow warbler	*Dendroica petechia*
Yuma clapper rail	*Rallus longirostris yumanensis*
Yunaska rock ptarmigan	*Lagopus mutus yunaskensis*
Zenaida dove	*Zenaida aurita*

Fishes

Agile damselfish	*Pomacentrus* species
Akule	*Selar crumenophthalmus*
Alabama sturgeon	*Acipenser suttkusi*
Alaska blackfish	*Dallia pectoralis*
Albacore	*Thunnus alalunga*
Alewife	*Alosa pseudoharengus*
Alligator gar	*Lepisosteus spatula*
Alvord chub	*Gila alvordensis*
Alvord cutthroat trout	*Salmo clarki* subspecies
Amago salmon	*Oncorhynchus rhodurus*
Amargosa pupfish	*Cyprinodon nevadensis*
Amberjack (see Greater amberjack)	
American eel	*Anguilla rostrata*
American plaice	*Hippoglossoides platessoides*
American shad	*Alosa sapidissima*
Anchovies	Family Engraulidae
Angelfishes	Family Pomacanthidae
Arctic char	*Salvelinus alpinus*
Arctic cisco	*Coregonus autumnalis*
Arctic cod	*Boreogadus saida*
Arctic grayling	*Thymallus arcticus*
Arctic lamprey	*Lampetra japonica*
Arkansas darter	*Etheostoma cragini*
Arkansas River shiner	*Notropis girardi*
Arkansas River speckled chub	*Macrhybopsis tetranema*
Armorheads	Family Pentacerotidae
Arrowtooth flounder	*Atheresthes stommias*
Arroyo chub	*Gila orcuttii*
Ash Meadows pupfish	*Cyprinodon nevadensis mionectes*
Atka mackerel	*Pleurogrammus monopterygius*
Atlantic bonito	*Sarda sarda*
Atlantic bumper	*Chloroscombrus chrysurus*
Atlantic cod	*Gadus morhua*
Atlantic croaker	*Micropogonias undulatus*
Atlantic herring	*Clupea harengus*
Atlantic mackerel	*Scomber scombrus*
Atlantic menhaden	*Brevoortia tyrannus*
Atlantic salmon	*Salmo salar*
Atlantic sharks	Families Alopiidae, Carcharhinidae, Lamnidae, Sphrynidae, and Squalidae
Atlantic sturgeon	*Acipenser oxyrhynchus*
Atlantic threadfin	*Polydactylus octonemus*
Barracudas	Family Sphyraenidae
Bay anchovy	*Anchoa mitchilli*
Bayou killifish	*Fundulus pulvereus*
Bering cisco	*Coregonus laurettae*
Bering Sea pollock (see Pollock)	
Bigeye marbled sculpin	*Cottus klamathensis macrops*
Bigeye scad	*Selar crumenophthalmus*
Bigeye tuna	*Thunnus obesus*
Bighead searobin	*Prionotus tribulus*
Bigmouth buffalo	*Ictiobus cyprinellus*
Billfishes	Family Istiophoridae
Black crappie	*Pomoxis nigromaculatus*
Black drum	*Pogonias cromis*
Black grouper	*Mycteroperca bonaci*
Blackcod (see Sablefish)	
Blackfin tuna	*Thunnus atlanticus*

Blacknose dace	*Rhinichthys atratulus*
Blindfishes (see Cavefish)	
Bloater	*Coregonus hoyi*
Blue catfish	*Ictalurus furcatus*
Blue chromis	*Chromis cyanea*
Blue chub	*Gila coerulea*
Blue marlin	*Makaira nigricans*
Blue pike	*Stizostedion vitreum glaucum*
Blue shark	*Prionace glauca*
Blue sucker	*Cycleptus elongatus*
Blue tang	*Acanthurus coeruleus*
Blue tilapia	*Tilapia aurea*
Blueback herring	*Alosa aestivalis*
Bluefin tuna	*Thunnus thynnus*
Bluefish	*Pomatomus saltatrix*
Bluegill	*Lepomis macrochirus*
Bluntnose shiner	*Notropis simus*
Bonytail chub	*Gila elegans*
Borax Lake chub	*Gila boraxobius*
Bowfin	*Amia calva*
Boxfishes	Family Ostraciidae
Broad whitefish	*Coregonus nasus*
Brook trout	*Salvelinus fontinalis*
Brown bullhead	*Ameiurus nebulosus*
Brown trout	*Salmo trutta*
Bucktooth parrotfish	*Sparisoma radians*
Buffaloes	*Ictiobus* species
Bull trout	*Salvelinus confluentus*
Bullhead catfishes	Family Ictaluridae
Burbot	*Lota lota*
Butter hamlet	*Hypoplectrus unicolor*
Butterfish	*Peprilus triacanthus*
Butterflyfishes	Family Chaetodontidae
California golden trout	*Oncorhynchus aquabonita* subspecies
California killifish	*Fundulus parvipinnis*
California roach	*Hesperoleucus symmetricus*
Capelin	*Mallotus villosus*
Carps and minnows	Family Cyprinidae
Catlow Valley redband trout	*Oncorhynchus mykiss* subspecies
Catlow Valley tui chub	*Gila bicolor* subspecies
Cavefish	Family Amblyopsidae
Cero	*Scomberomorus regalis*
Chalk bass	*Serranus tortugarum*
Channel catfish	*Ictalurus punctatus*
Chinook salmon	*Oncorhynchus tshawytscha*
Chub mackerel	*Scomber japonicus*
Chum salmon	*Oncorhynchus keta*
Cichlids	Family Cichlidae
Cisco or lake herring	*Coregonus artedii*
Clear Lake splittail	*Pogonichthys ciscoides*
Coastal cutthroat trout	*Oncorhynchus clarki clarki*
Coastrange sculpin	*Cottus aleuticus*
Cobia	*Rachycentron canadum*
Cods	Family Gadidae
Coho salmon	*Oncorhynchus kisutch*
Colorado squawfish	*Ptychocheilus lucius*
Columbia River smelt (see Eulachon)	
Common carp	*Cyprinus carpio*
Cookie-cutter shark	*Isistius brasiliensis*
Cowhead Lake tui chub	*Gila bicolor vaccaceps*
Crappies	*Pomoxis* species
Crevalle jack	*Caranx hippos*
Croakers and drums	Family Sciaenidae
Cui-ui	*Chasmistes cujus*
Cutthroat trout	*Oncorhynchus clarki*
Damselfishes	Family Pomacentridae
Deepwater cisco	*Coregonus johannae*
Deepwater flounder	*Monolene sessilicauda*
Deepwater sculpin	*Myoxocephalus thompsonii*
Delta smelt	*Hypomesus transpacificus*
Demersal shelf rockfish	*Sebastes* species
Devils Hole pupfish	*Cyprinodon diabolis*
Dolly Varden	*Salvelinus malma*
Dolphin	*Coryphaena hippurus*

Dover sole	*Microstomus pacificus*
Eastern mosquitofish	*Gambusia holbrookii*
Elasmobranchs	Class Elasmobranchiomorphi
Eleotrids	Various genera
Emerald shiner	*Notropis atherinoides*
Emperors	*Lethrimus* species
English sole	*Pleuronectes vetulus*
Eulachon	*Thaleichthys pacificus*
Fathead minnow	*Pimephales promelas*
Finfish	Fish of the Class Pisces [as opposed to shellfish]
Flannelmouth sucker	*Catostomus latipinnis*
Flatfish	Order Pleuronectiformes
Flathead catfish	*Pylodictis olivaris*
Flathead chub	*Platygobio gracilis*
Flathead sole	*Hippoglossoides elassodon*
Florida pompano	*Trachinotus carolinus*
Flounders	Order Pleuronectiformes
Foskett speckled dace	*Rhinichthys osculus* subspecies
Foureye butterflyfish	*Chaetodon capistratus*
Fourhorn sculpin	*Myoxocephalus quadricornis*
French grunt	*Haemulon flavolineatum*
Freshwater drum	*Aplodinotus grunniens*
Frigate mackerel (Tuna)	*Auxus thazard*
Fringed filefish	*Monacanthus ciliatus*
Gafftopsail catfish	*Bagre marinus*
Gars	*Lepisosteus* species
Gila topminnow	*Poeciliopsis occidentalis*
Gila trout	*Oncorhynchus gilae*
Gizzard shad	*Dorosoma cepedianum*
Goatfishes	Family Mullidae
Gobies	Family Gobiidae
Golden shiner	*Notropis crysoleucas*
Golden trout	*Oncorhynchus aguabonita*
Goose Lake lamprey	*Lampetra tridentata* subspecies
Goose Lake sucker	*Catostomus occidentalis lacusanserinus*
Goose Lake tui chub	*Gila bicolor thalassina*
Grass carp	*Ctenopharyngodon idella*
Gray angelfish	*Pomacanthus arcuatus*
Gray snapper	*Lutjanus griseus*
Graylings	Subfamily Thymallinae
Great barracuda	*Sphyraena barracuda*
Greater amberjack	*Seriola dumerili*
Green sturgeon	*Acipenser medirostris*
Greenback cutthroat trout	*Oncorhynchus clarki stomias*
Greenland halibut (Turbot)	*Reinhardtius hippoglossoides*
Groupers	*Epinephelus* and *Mycteroperca* species
Grunts	Family Haemulidae
Gulf butterfish	*Peprilus burti*
Gulf killifish	*Fundulus grandis*
Gulf menhaden	*Brevoortia patronus*
Gulf sturgeon	*Acipenser oxyrhynchus desotoi*
Guppy	*Poecilia reticulata*
Haddock	*Melanogrammus aeglefinus*
Hake (see Pacific hake)	
Hardhead	*Mylopharodon conocephalus*
Hardhead catfish	*Arius felis*
Harelip sucker	*Lagochila lacera*
Herrings and shads	Family Clupeidae
Hickory shad	*Alosa mediocris*
Hitch	*Lavinia exilicauda*
Humpback chub	*Gila cypha*
Humpback whitefish	*Coregonus pidschian*
Hutton Spring tui chub	*Gila bicolor* subspecies
Inland silverside	*Menidia beryllina*
Interior redband trout	*Oncorhynchus mykiss gibbsi*
Jack mackerel	*Trachurus symmetricus*
Jacks	*Caranx* species
Jenny Creek sucker	*Catostomus rimiculus* subspecies
Kelpfishes	Family Dactyloscopidae
Kern brook lamprey	*Lampetra hubbsi*
Killifishes	*Fundulus* species
King mackerel	*Scomberomorus cavalla*
Kingfishes	*Menticirrus* species

Klamath largescale sucker	*Catostomus snyderi*
Klamath River lamprey	*Lampetra similis*
Klamath smallscale sucker	*Catostomus rimiculus*
Kokanee (see Sockeye salmon)	
Lahontan cutthroat trout	*Oncorhynchus clarki henshawi*
Lahontan shiner	*Richardsonius egregius*
Lake chub	*Couesius plumbeus*
Lake sturgeon	*Acipenser fulvescens*
Lake trout	*Salvelinus namaycush*
Lake whitefish	*Coregonus clupeaformis*
Lampreys	Family Petromyzontidae
Lane snapper	*Lutjanus synagris*
Lanternfishes	Family Myctophidae
Largemouth bass	*Micropterus salmoides*
Least cisco	*Coregonus sardinella*
Least darter	*Etheostoma microperca*
Leatherjackets	Family Balistidae
Leatherside chub	*Gila copei*
Little Kern golden trout	*Oncorhynchus aquabonita whitei*
Little Kern rainbow trout	*Oncorhynchus mykiss* subspecies
Little tunny	*Euthynnus alletteratus*
Loach minnow	*Rhinichthys cobitis*
Longbill spearfish	*Tetrapturus pfluegeri*
Longfin smelt	*Spirinchus thaleichthys*
Longjaw mudsucker	*Gillichthys mirabilis*
Longnose dace	*Rhinichthys cataractae*
Longnose killifish	*Fundulus similis*
Longnose sucker	*Catostomus catostomus*
Lost River sucker	*Deltistes luxatus*
Lower Columbia River fall	
chinook salmon	*Oncorhynchus tshawytscha* subspecies
Mackerel scad	*Decapterus macarellus*
Mackerels	Family Scombridae
Malheur mottled sculpin	*Cottus bairdii* subspecies
Marbled sculpin	*Cottus klamathensis*
Margined sculpin	*Cottus marginatus*
Marsh killifish	*Fundulus confluentus*
Maryland darter	*Etheostoma sellare*
Masu salmon	*Oncorhynchus masou*
McCloud River redband trout	*Oncorhynchus mykiss* subspecies
Menhaden	*Brevoortia* species
Miller Lake lamprey	*Lampetra minima*
Millicoma dace	*Rhinichthys cataractae* subspecies
Minnows	Family Cyprinidae
Modoc sucker	*Catostomus microps*
Monkey Springs pupfish	*Cyprinodon* species
Mottled sculpin	*Cottus bairdii*
Mountain mullet	*Agonostomus monticola*
Mountain sucker	*Catostomus platyrhynchus*
Mudminnows	Family Umbridae
Mullets	Family Mugilidae
Muskellunge	*Esox masquinongy*
Mutton hamlet	*Epinephelus afer*
Nassau grouper	*Epinephelus striatus*
Nehu	*Encrasicholina purpurea*
Neosho madtom	*Noturus placidus*
Ninespine stickleback	*Pungitius pungitius*
Nooksack dace	*Rhinichthys cataractae* subspecies
Northern anchovy	*Engraulis mordax*
Northern bluefin tuna (see Bluefin tuna)	
Northern kingfish	*Menticirrhus saxatilis*
Northern pike	*Esox lucius*
Northern rockfish	*Sebastes polyspinis*
Northern squawfish	*Ptychocheilus oregonensis*
Ocean sunfish	*Mola mola*
Oceanic whitetip shark	*Carcharhinus longimanus*
Olympic mudminnow	*Novumbra hubbsi*
Oopu alamoo	*Lentipes concolor*
Opakapaka	*Pristipomoides filamentosus*
Opelu	*Decapterus macarellus*
Oregon chub	*Oregonichthys crameri*
Oregon Lakes tui chub	*Gila bicolor oregonensis*
Pacific cod	*Gadus macrocephalus*
Pacific hake	*Merluccius productus*

Pacific halibut	*Hippoglossus stenolepus*
Pacific herring	*Clupea pallasi*
Pacific lamprey	*Lampetra tridentata*
Pacific ocean perch	*Sebastes alutus*
Pacific sandfish	*Trichodon trichodon*
Pacific sardine	*Sardinops sagax*
Pacific staghorn sculpin	*Leptocottus armatus*
Pacific swordfish (see Swordfish)	
Pacific tomcod	*Microgadus proximus*
Pacific whiting (see Pacific hake)	
Paddlefish	*Polyodon spathula*
Pahrump poolfish	*Empetrichthys latos*
Paiute cutthroat trout	*Oncorhynchus clarki* subspecies
Paiute sculpin	*Cottus beldingii*
Pallid sturgeon	*Scaphirhynchus albus*
Pelagic armorhead	*Pentaceros pectoralis*
Pelagic shelf rockfish	*Sebastes* species
Petrale sole	*Eopsetta jordani*
Phantom shiner	*Notropis orca*
Pikes	Family Esocidae
Pink salmon	*Oncorhynchus gorbuscha*
Pit sculpin	*Cottus pitensis*
Pitt-Klamath brook lamprey	*Lampetra lethophaga*
Plains minnow	*Hybognathus placitus*
Plains topminnow	*Fundulus sciadicus*
Pollock	*Pollachius virens*
Pomfrets	Family Bramidae
Pond smelt	*Hypomesus olidus*
Porgies	Family Sparidae
Prickly sculpin	*Cottus asper*
Pupfishes	Family Cyprinodontidae
Pygmy whitefish	*Prosopium coulterii*
Pyramid Lake cutthroat trout	*Oncorhynchus clarki henshawi*
Rainbow smelt	*Osmerus mordax*
Rainbow trout	*Oncorhynchus mykiss*
Razorback sucker	*Xyrauchen texanus*
Red drum	*Sciaenops ocellatus*
Red grouper	*Epinephelus morio*
Red hake	*Urophyssus chuss*
Red snapper	*Lutjanus campechanus*
Redband trout	*Oncorhynchus mykiss* subspecies
Redear sunfish	*Lepomis microlophus*
Redfish	*Sebastes* species
Redhorses	*Moxostoma* species
Reticulate sculpin	*Cottus perplexus*
Riffle sculpin	*Cottus gulosus*
Righteye flounders	Family Pleuronectidae
Rio Grande bluntnose shiner	*Notropis simus simus*
Rio Grande chub	*Gila pandora*
Rio Grande shiner	*Notropis jemezanus*
Rio Grande silvery minnow	*Hybognathus amarus*
River carpsucker	*Carpiodes carpio*
River herrings	*Alosa aestivalis* and *A. pseudoharengus*
River lamprey	*Lampetra ayresii*
Roanoke logperch	*Percina rex*
Rock hind	*Epinephelus adscensionis*
Rock sole	*Pleuronectes bilineatus*
Rockfishes	*Sebastes* species
Rough sculpin	*Cottus asperrimus*
Rough silverside	*Membras martinica*
Round goby	*Neogobius melanostomus*
Round whitefish	*Prosopium cylindraceum*
Roundtail chub	*Gila robusta*
Ruffe	*Gymnocephalus cernuus*
Sablefish	*Anoplopoma fimbria*
Sacramento blackfish	*Orthodon microlepidotus*
Sacramento perch	*Archoplites interruptus*
Sacramento splittail	*Pogonichthys macrolepidotus*
Sacramento squawfish	*Ptychocheilus grandis*
Sacramento sucker	*Catostomus occidentalis*
Saffron cod	*Eliginus gracilis*
Sailfin molly	*Poecilia latipinna*
Sailfish	*Istiophorus platypterus*
Salish sucker	*Catostomus* species

Salmon shark — *Lamna ditropis*
Salmons and trouts — Family Salmonidae
Sand roller — *Percopsis transmontana*
Sand seatrout — *Cynoscion arenarius*
Santa Ana speckled dace — *Rhinichthys osculus* subspecies
Santa Ana sucker — *Catostomus santaanae*
Saratoga Springs pupfish — *Cyprinodon nevadensis nevadensis*
Sardines — Family Clupeidae
Sauger — *Stizostedion canadense*
Scrawled filefish — *Aluturus scriptus*
Sculpins — Family Cottidae
Sea lamprey — *Petromyzon marinus*
Seatrout — *Cynoscion* species
Shads — Family Clupeidae
Shallow-water flatfish — Order Pleuronectiformes
Sheefish (Inconnu) — *Stenodus leucichthys*
Sheepshead — *Archosargus probatocephalus*
Sheepshead minnow — *Cyprinodon variegatus*
Sheldon tui chub — *Gila bicolor eurysoma*
Shiner perch — *Cymatogaster aggregata*
Shortbelly rockfish — *Sebastes jordani*
Shortfin mako — *Isurus oxyrhinhus*
Shorthead sculpin — *Cottus confusus*
Shortnose sucker — *Chasmistes brevirostris*
Shoshone pupfish — *Cyprinodon nevadensis shoshone*
Shoshone sculpin — *Cottus greenei*
Shovelnose sturgeon — *Scaphirhynchus platorynchus*
Sicklefin chub — *Macrhybopsis meeki*
Silky shark — *Carcharhinus falciformis*
Silver hake — *Merluccius bilinearis*
Silver perch — *Bairdiella chrysoura*
Silversides — Family Atherinidae
Skates — Family Rajidae
Skipjack herring — *Alosa chrysochloris*
Skipjack tuna — *Katsuwonus pelamis*
Slender chub — *Erimystax cahni*
Slimy sculpin — *Cottus cognatus*
Smallmouth bass — *Micropterus dolomieu*
Smallmouth buffalo — *Ictiobus bubalus*
Smelts — Family Osmeridae
Snake River chinook salmon — *Oncorhynchus tshawytscha* subspecies
Snake River fine-spotted cutthroat trout — *Oncorhynchhus clarki* subspecies
Snake River sockeye salmon — *Oncorhynchus nerka* subspecies
Sockeye salmon — *Oncorhynchus nerka*
South coast fall chinook salmon — *Oncorhynchus tshawytscha* subspecies
Southeast drum — Family Sciaenidae
Southern flounder — *Paralichthys lethostigma*
Southern steelhead — *Oncorhynchus mykiss* subspecies
Spanish mackerel — *Scomberomorus maculatus*
Spearfish — *Tetrapturus* species
Speckled chub — *Macrhybopsis aestivalis*
Speckled dace — *Rhinichthys osculus*
Spikedace — *Meda fulgida*
Spiny dogfish — *Squalus acanthias*
Spoonhead sculpin — *Cottus ricei*
Spot — *Leiostomus xanthurus*
Spotted seatrout — *Cynoscion nebulosus*
Spotted shiner — *Notropis hudsonius*
Spotted snake eel — *Ophichthus ophis*
Spotted sunfish — *Lepomis punctatus*
Squirrelfishes — *Holocentrus* species
Starry flounder — *Platichthys stellatus*
Steelhead (Sea-run) — *Oncorhynchus mykiss*
Sticklebacks — Family Gasterosteidae
Striped anchovy — *Anchoa hepsetus*
Striped bass — *Morone saxatilis*
Striped mullet — *Mugil cephalus*
Sturgeon chub — *Macrhybopsis gelida*
Sturgeons — Family Acipenseridae
Suckers — Family Catostomidae
Summer flounder — *Paralichthys dentatus*
Sunapee char — *Salvelinus alpinus*
Surf smelt — *Hypomesus pretiosus*

Surfperch — Family Embiotocidae
Swordfish — *Xiphias gladius*
Tahoe sucker — *Catostomus tahoensis*
Tarpon — *Megalops atlanticus*
Tecopa pupfish — *Cyprinodon nevadensis calidae*
Thicktail chub — *Gila crassicauda*
Thornyheads — *Sebastolobus* species
Threespine stickleback — *Gasterosteus aculeatus*
Thresher shark — *Alopias vulpinus*
Tidewater goby — *Eucyclogobius newberryi*
Tidewater silverside — *Menidia peninsulae*
Tilefishes — Family Malacanthidae
Topeka shiner — *Notropis topeka*
Trout–perch — *Percopsis omiscomaycus*
Trout–perches — Family Percopsidae
Tubenose goby — *Proterorhinus marmoratus*
Tui chub — *Gila bicolor*
Tule perch — *Hysterocarpus traskii*
Tunas — *Thunnus* species
Umpqua chub — *Oregonichthys kalawatseti*
Unarmored threespine stickleback — *Gasterosteus aculeatus williamsoni*
Vermilion snapper — *Rhomboplites aurorubens*
Virgin River chub — *Gila semidnuda*
Virgin spinedace — *Lepidomeda mollispinus*
Waccamaw silverside — *Menidia extensa*
Wahoo — *Acanthocybium solandri*
Walleye — *Stizostedion vitreum*
Walleye pollock — *Theragra chalcogramma*
Warm Springs pupfish — *Cyprinodon nevadensis pectoralis*
Warmouth — *Lepomis gulosus*
Warner sucker — *Catostomus warnerensis*
Warner Valley redband trout — *Oncorhynchus mykiss* subspecies
Weakfish — *Cynoscion regalis*
Western brook lamprey — *Lampetra richardsoni*
Western mosquitofish — *Gambusia affinis*
Western silvery minnow — *Hybognathus argyritis*
Westslope cutthroat trout — *Oncorhynchus clarki lewisi*
White bass — *Morone chrysops*
White crappie — *Pomoxis annularis*
White hake — *Urophycis tenuis*
White sturgeon — *Acipenser transmontanus*
Whitefishes — *Coregonus* species
Whiteline topminnow — *Fundulus albolineatus*
Widow rockfish — *Sebastes entomelas*
Windowpane — *Scophthalmus aquosus*
Winter flounder — *Pleuronectes americanus*
Witch flounder — *Glyptocephalus cynoglossus*
Wood River sculpin — *Cottus leiopomus*
Woundfin — *Plagopterus argentissimus*
Wrasses — Family Labridae
Yellow bass — *Morone mississippiensis*
Yellow bullhead — *Ameiurus natalis*
Yellow perch — *Perca flavescens*
Yellowfin sole — *Pleuronectes asper*
Yellowfin tuna — *Thunnus albacares*
Yellowstone cutthroat trout — *Oncorhynchus clarki bouvieri*
Yellowtail — *Seriola lalandi*

Mammals

Abert's squirrel — *Sciurus aberti*
Agile kangaroo rat — *Dipodomys agilis*
Allen's big-eared bat — *Idionycteris phyllotis*
Amargosa vole — *Microtus californicus scirpensis*
Ancient bison — *Bison antiquiuus*
Arctic fox — *Alopex lagopus*
Arctic ground squirrel — *Spermophilus parryii*
Armadillos — Family Dasypodidae
Ash Meadows montane vole — *Microtus montanus nevadensis*
Atlantic spotted dolphin — *Stenella frontalis*
Atlantic white-sided dolphin — *Lagenorhynchus acutus*
Audubon bighorn sheep — *Ovis canadensis auduboni*
Axis deer — *Axis axis*
Badger — *Taxidea taxus*
Baird's beaked whale — *Berarius bairdii*

Baird's shrew	*Sorex bairdii*
Baleen whales	Suborder Mysticeti
Barbary sheep	*Ammotragus lervia*
Barren-ground grizzly bear	*Ursus arctos richardsoni*
Bearded seal	*Erignathus barbatus*
Bears	Family Ursidae
Beaver	*Castor canadensis*
Beluga whale	*Delpinapterus leucas*
Big brown bat	*Eptesicus fuscus*
Big free-tailed bat	*Nyctinomops macrotis*
Bighorn sheep	*Ovis canadensis*
Bison	*Bison bison*
Black bear	*Ursus americanus*
Black-footed ferret	*Mustela nigripes*
Black rat	*Rattus rattus*
Black right whale	*Balaena glacialis*
Black-tailed jack rabbit	*Lepus californicus*
Black-tailed prairie dog	*Cynomys ludovicianus*
Blainville's beaked whale	*Mesoplodon densirostris*
Blue whale	*Balaenoptera musculus*
Bobcat	*Lynx rufus*
Botta's pocket gopher	*Thomomys bottae*
Bottlenose dolphin	*Tursiops truncatus*
Bottlenose whales	*Hyperoodon* species
Bowhead whale	*Balaena mysticetus*
Brazilian free-tailed bat	*Tadarida brasiliensis*
Broad-footed mole	*Scapanus latimanus*
Brown bear	*Ursus arctos*
Brown lemming	*Lemmus sibiricus*
Brush rabbit	*Sylvilagus bachmani*
Bryde's whale	*Balaenoptera edeni*
Burro	*Equus asinus*
Bushy-tailed woodrat	*Neotoma cinerea lucida*
California bighorn sheep	*Ovis canadensis californiana*
California grizzly bear (see Grizzly bear)	
California ground squirrel	*Spermophilus beecheyi*
California leaf-nosed bat	*Macrotus californicus*
California mouse	*Peromyscus californicus*
California myotis	*Myotis californicus*
California pocket mouse	*Chaetodipus californicus*
California sea lion	*Zalophus californianus*
California vole	*Microtus californicus*
California wolverine	*Gulo gulo luteus*
Camels	Family Camelidae
Caribou	*Rangifer tarandus*
Cats	Family Felidae
Cave myotis	*Myotis velifer*
Central American spinner dolphin	(see Spinner dolphin)
Cetaceans	Order Cetacea
Channel Islands gray fox	*Urocyon littoralis*
Clymene dolphin	(see Short-snouted spinner dolphin)
Common dolphin	*Delphinus delphis*
Collared lemmings	*Lemmus* species
Collared pika	*Ochotona collaris*
Colonial pocket gopher	*Geomys pinetis colonus*
Columbian black-tailed deer	*Odocoileus hemionus columbianus*
Columbian white-tailed deer	*Odocoileus virginianus leucurus*
Common squirrel monkey	*Saimiri sciureus*
Coyote	*Canis latrans*
Creeping vole	*Microtus oregoni*
Cumberland Island pocket gopher	*Geomys pinetis cumberlandius*
Cuvier's beaked whale	*Ziphius cavirostris*
Dall sheep	*Ovis dalli*
Dall's porpoise	*Phocoenoides dalli*
Deer mouse	*Peromyscus maniculatus*
Delmarva Peninsula fox squirrel	*Sciurus niger cinereus*
Dense-beaked whale	*Mesoplodon densirostris*
Desert bighorn sheep	*Ovis canadensis nelsoni*
Desert Valley kangaroo mouse	*Microdipodops megacephalus albiventer*
Dire wolf	*Canis dirus*
Dolphins	*Stenella, Steno*, and *Tursiops* species
Douglas' squirrel	*Tamiasciurus douglasii*
Dusky-footed woodrat	*Neotoma fuscipes*
Dwarf sperm whale	*Kogia simus*

Eastern cottontail	*Sylvilagus floridanus*
Eastern cougar	*Puma concolor cougar*
Eastern elk	*Cervus elaphus canadensis*
Eastern mountain lion (see Eastern cougar)	
Eastern small-footed myotis	*Myotis leibii*
Elephant seal (see Northern elephant seal)	
Elk (see Wapiti)	
Ermine	*Mustela erminea*
European rabbit	*Oryctolagus cuniculus*
Fallow deer	*Dama dama*
False killer whale	*Pseudorca crassidens*
Fin whale	*Balaenoptera physalus*
Fish Spring pocket gopher	*Thomomys umbrinus abstrusus*
Fisher	*Martes pennanti*
Fletcher dark kangaroo mouse	*Microdipodops megacephalus nasutus*
Florida black bear	*Ursus americanus floridanus*
Florida mouse	*Podomys floridanus*
Florida panther	*Puma concolor coryi*
Fog shrew	*Sorex sonomae*
Fox squirrel	*Sciurus niger*
Franklin's ground squirrel	*Spermophilus franklinii*
Fraser's dolphin	*Lagenodelphis hosei*
Fresno kangaroo rat	*Dipodomys nitratoides exilis*
Fringed myotis	*Myotis thysanodes*
Gemsbok	*Oryx gazella*
Gervais' beaked whale	*Mesoplodon europaeus*
Giant bison	*Bison* species
Giant kangaroo rat	*Dipodomys ingens*
Giant sea mink	*Mustela macrodon*
Golden-mantled ground squirrel	*Spermophilus lateralus*, including subspecies *bernardinus* and *certus*
Goose-beaked whale	*Ziphius cavirostris*
Gray bat	*Myotis grisescens*
Gray fox	*Urocyon cinereoargenteus*
Gray seal	*Halichoerus grypus*
Gray squirrel	*Sciurus carolinensis*
Gray whale	*Eschrichtius robustus*
Gray wolf	*Canis lupus*
Greater western mastiff-bat	*Eumops perotis californicus*
Greenland collared lemming	*Dicrostonyx groenlandicus*
Grizzly bear	*Ursus arctos horribilis*
Ground sloth	Family Mylodontidae
Guadalupe fur seal	*Arctocephalus townsendi*
Harbor porpoise	*Phocoena phocoena*
Harbor seal	*Phoca vitulina*
Harp seal	*Phoca groenlandica*
Hawaiian hoary bat	*Lasiurus cinereus semotus*
Hawaiian monk seal	*Monachus schauinslandi*
Heermann's kangaroo rat	*Dipodomys heermanni*
Hidden Forest Uinta chipmunk	*Tamias umbrinus nevadensis*
Hoary bat	*Lasiurus cinereus*
Hoary marmot	*Marmota caligata*
Hooded seal	*Cystophora cristata*
Horse	*Equus caballus*
House cat	*Felis catus*
House mouse	*Mus musculus*
Humboldt beaver	*Castor canadensis baileyi*
Humpback whale	*Megaptera novaeangliae*
Ibex	*Capra ibex*
Idaho ground squirrel	*Spermophilus brunneus*
Indian mongoose	*Herpestes javanacus*
Indiana bat	*Myotis sodalis*
Jaguar	*Panthera onca*
Jaguarundi	*Harpailurus yagouaroundi*
Japanese macaque	*Macaca fuscata*
Keen's myotis	*Myotis keenii*
Key deer	*Odocoileus virginianus clavium*
Killer whale	*Orcinus orca*
Kit fox	*Vulpes macrotis*
Lambe's horse	*Equus* species
Least weasel	*Mustela nivalis*
Lemmings	*Lemmus* species
Lesser long-nosed bat	*Leptonycteris curasoae yerbabuenae*
Lion	*Panthera leo*

Little brown bat	*Myotis lucifugus*
Lodgepole chipmunk	*Tamias speciosus*
Long-eared chipmunk	*Eutamias quadrimaculatus*
Long-eared myotis	*Myotis evotis*
Long-finned pilot whale	*Globicephala melas*
Long-legged myotis	*Myotis volans*
Long-snouted spinner dolphin	*Stenella longirostris*
Long-tailed vole	*Microtus longicaudus*, including subspecies *latus*
Long-tailed weasel	*Mustela frenata*
Louisiana black bear	*Ursus americanus luteolus*
Lynx	*Lynx canadensis*
Mammoths	*Mammuthus* species
Margay	*Leopardus wiedii*
Marten	*Martes americana*
Mastodons	*Mammut* species
Meadow vole	*Microtus pennsylvanicus*
Melon-headed whale	*Peponocephala electra*
Merriam's chipmunk	*Eutamias merriami*
Mesoplodont beaked whale	*Mesoplodon* species
Mexican vole	*Microtus mexicanus*
Mink	*Mustela vison*
Minke whale	*Balaenoptera acutorostrata*
Mohave ground squirrel	*Spermophilus mohavensis*
Moose	*Alces alces*
Morro Bay kangaroo rat	*Dipodomys herrmanni morroensis*
Mountain beaver	*Aplodontia rufa*
Mountain goat	*Oreamnos americanus*
Mountain lion	*Puma concolor*
Mountain pocket gopher	*Thomomys monticola*
Mt. Ellen chipmunk	*Tamias umbrinus sedulus*
Mt. Ellen long-tailed vole	*Microtus longicaudus incanus*
Mt. Ellen pocket gopher	*Thomomys bottae dissimilis*
Mt. Lyell shrew	*Sorex lyelli*
Mule deer	*Odocoileus hemionus*
Muskox	*Ovibos moschatus*
Muskrat	*Ondatra zibethicus*
Mustelids	Family Mustelidae
Narrow-faced kangaroo rat	*Dipodomys venustus*
Nelson's antelope squirrel	*Ammospermophilus nelsoni*
New England cottontail	*Sylvilagus transitionalis*
Nine-banded armadillo	*Dasypus novemcinctus*
North American lynx	*Lynx canadensis canadensis*
North American wolverine	*Gulo gulo luscus*
North Atlantic beaked whale	*Mesoplodon bidens*
North Pacific bottle-nosed whale	*Berardius bairdii*
Northern collared lemming	*Lemmus* species
Northern elephant seal	*Mirounga angustirostris*
Northern flying squirrel	*Glaucomys sabrinus*
Northern fur seal	*Callorhinus ursinus*
Northern pygmy mouse	*Baiomys taylori*
Northern red-backed vole	*Clethrionomys rutilus*
Northern right-whale dolphin	*Lissodelphis borealis*
Northern sea lion	*Eumetopias jubatus*
Norway rat	*Rattus norvegicus*
Nutria	*Myocastor coypus*
Ocelot	*Leopardus pardalis*
Oldfield mouse	*Peromyscus polionotus*
Olive-backed pocket mouse	*Perognathus fasciatus*
Ornate shrew	*Sorex ornatus*
Owens Valley vole	*Microtus californicus vallicola*
Pacific fisher	*Martes pennanti pacifica*
Pacific marsh shrew	*Sorex bendirii*
Pacific shrew	*Sorex pacificus*
Pacific walrus	*Odobenus rosmarus*
Pacific white-sided dolphin	*Lagenorhynchus obliquidens*
Pahranagat Valley montane vole	*Microtus montanus fucosus*
Pale Townsend's big-eared bat	*Plecotus townsendii pallescens*
Pallid bat	*Antrozous pallidus*
Palmer's chipmunk	*Tamias palmeri*
Panamint chipmunk	*Tamias panamintinus*, including suspecies *acrus*
Panamint kangaroo rat	*Dipodomys panamintinus*
Pantropical spotted dolphin	*Stenella attenuata*

Peninsular bighorn sheep	*Ovis canadensis cremnobates*
Perdido Key beach mouse	*Peromyscus polionotus trissyllepsis*
Pig	*Sus scrofa*
Pika	*Ochotona princeps*, including subspecies *albata, nevadensis, sheltoni,* and *tutelata*
Pinniped seals, sea lions, and walruses	Families Otariidae, Odobenidae, and Phocidae
Plains bison	*Bison bison bison*
Plains harvest mouse	*Reithrodontomys montanus*
Plains pocket gopher	*Geomys bursarius*
Plains pocket mouse	*Perognathus flavescens*
Polar bear	*Ursus maritimus*
Polynesian rat	*Rattus exulans*
Porcupine	*Erithizon dorsalis*
Porpoises	Family Phocoenidae
Prairie dogs	*Cynomys* species
Prairie vole	*Microtus ochrogaster*
Preble's shrew	*Sorex preblei*
Pribilof Island shrew	*Sorex hydrodromus*
Pronghorn	*Antilocapra americana*
Puerto Rican hutia	*Isolobodon portoricensis*
Pygmy killer whale	*Feresa attenuata*
Pygmy rabbit	*Sylvilagus idahoensis*
Pygmy sperm whale	*Kogia breviceps*
Raccoon	*Procyon lotor*
Red-backed voles	*Clethrionomys* species
Red bat	*Lasiurus borealis*
Red fig-eating bat	*Stenoderma rufum*
Red fox	*Vulpes vulpes*
Red squirrel	*Tamiasciurus hudsonicus*
Red tree vole	*Arborimus longicaudus*
Red wolf	*Canis rufus*
Reindeer (see Caribou)	
Rhesus macaque	*Macaca mulatta*
Rhinoceroses	Family Rhinocerotidae
Ribbon seal	*Phoca fasciata*
Right and bowhead whales	Family Balaenidae
Right whale	*Balaena* species
Ringed seal	*Phoca hispida*
Risso's dolphin	*Grampus griseus*
River otter	*Lontra canadensis*
Roosevelt elk	*Cervus elaphus roosevelti*
Rough-toothed dolphin	*Steno bredanensis*
Round-tailed muskrat	*Neofiber alleni*
Sabertooth cats	*Smilodon* species
Saddle-backed dolphin	*Delphinus delphis*
Salt marsh harvest mouse	*Reithrodontomys raviventris*
Sambar	*Cervus unicolor*
San Antonio pocket gopher	*Thomomys umbrinus curtatus*
San Diego pocket mouse	*Chaetodipus fallax*
San Joaquin kit fox	*Vulpes macrotis muticus*
San Joaquin pocket mouse	*Perognathus inornatus*
Sea lions	*Eumetopias* species
Sea otter	*Enhydra lutris*
Sei whale	*Balaenoptera borealis*
Sherman's pocket gopher	*Geomys pinetus fontanelus*
Short-faced bear	*Arctodus simus*
Short-finned pilot whale	*Globicephala macrorhynchus*
Short-snouted spinner dolphin	*Stenella clymene*
Shrew-mole	*Neurotrichus gibbsii*
Shrews	Family Soricidae
Sierra Nevada red fox	*Vulpes vulpes necator*
Sierra Nevada snowshoe hare	*Lepus americanus tahoensis*
Silver-haired bat	*Lasionycteris noctivagans*
Singing vole	*Microtus miurus*
Sitka black-tailed deer	*Odocoileus hemionus sitchensis*
Snowshoe hare	*Lepus americanus*
Sonoma chipmunk	*Tamias sonomae*
Sonoran pronghorn	*Antilocapra americana sonoriensis*
South African fur seal	*Arctocephalus pusillus*
Southeastern pocket gopher	*Geomys pinetus*
Southern bottle-nosed whale	*Hyperodon planifrons*
Southern red-backed vole	*Clethrionomys gapperi*

Southern sea otter — *Enhydra lutris nereis*
Southwestern otter — *Lontra canadensis sonora*
Sowerby's beaked whale — *Mesoplodon bidens*
Sperm whale — *Physeter catodon*
Spinner dolphin — *Stenella longirostris*
Spotted bat — *Euderma maculatum*
Spotted seal — *Phoca largha*
Spotted skunk — *Spilogale putorius*
St. Lawrence Island shrew — *Sorex cinereus jacksoni*
St. Matthews Island vole — *Microtus abbreviatus*
Stejneger's beaked whale — *Mesoplodon stejnegeri*
Stephens' kangaroo rat — *Dipodomys stephensi*
Steppe bison — *Bison priscus*
Striped dolphin — *Stenella coeruleoalba*
Striped skunk — *Mephitis mephitis*
Swamp rabbit — *Sylvilagus aquaticus*
Swift fox — *Vulpes velox*
Tehachapi white-eared pocket mouse — *Perognathus alticola inexpectatus*
Thirteen-lined ground squirrel — *Spermophilus tridecemlineatus*
Tipton kangaroo rat — *Dipodomys nitratoides nitratoides*
Townsend's big-eared bat — *Plecotus townsendii*
Townsend's ground squirrel — *Spermophilus townsendii*
True's beaked whale — *Mesoplodon mirus*
Tule elk — *Cervus elaphus nannodes*
Tundra hare — *Lepus othus*
Tundra vole — *Microtus oeconomus*
Uinta chipmunk — *Tamias umbrinus*
Vagrant shrew — *Sorex vagrans*
Virginia opossum — *Didelphis virginiana*
Voles — *Microtus* species
Walrus — *Odobenus rosmarus*
Wapiti, elk — *Cervus elaphus*
Washington ground squirrel — *Spermophilus washingtoni*
Weasels — *Mustela* species
West Indian manatee — *Trichechus manatus*
Western jumping mouse — *Zapus princeps*, including subspecies *curtatus*
Western mastiff-bat — *Eumops perotis*
Western pipistrelle — *Pipistrellus hesperus*
Western red-backed vole — *Clethrionomys californicus*
Western red bat — *Lasiurus blossevillii*
Western small-footed myotis — *Myotis ciliolabrum*
White-beaked dolphin — *Lagenorhynchus albirostris*
White-footed vole — *Arborimus albipes*
White-sided jack rabbit — *Lepus callotis*
White-tailed deer — *Odocoileus virginianus*
White-tailed jack rabbit — *Lepus townsendii*
White whale — *Delphinapterus leucas*
Whitebelly spinner dolphin (see Fraser's dolphin)
Wild boar (see Pig)
Wolverine — *Gulo gulo*
Woodland bison — *Bison bison athabascae*
Woodland caribou — *Rangifer tarandus caribou*
Woodrats — *Neotoma* species
Woolly mammoth — *Mammuthus primigenius*
Yellow-eared pocket mouse — *Perognathus xanthanotus*
Yellow-bellied marmot — *Marmota flaviventris*, including subspecies *fortirostris* and *parvula*
Yellow-pine chipmunk — *Tamias amoenus monoensis*
Yuma myotis — *Myotis yumanensis*
Yuma puma — *Puma concolor browni*

Reptiles

Alameda striped racer — *Masticophis lateralis euryxanthus*
Alligator snapping turtle — *Macroclemys temminckii*
American alligator — *Alligator mississippiensis*
American crocodile — *Crocodylus acutus*
Banded rock rattlesnake — *Crotalus lepidus willardi*
Barefoot gecko — *Coleonyx switaki*
Big Bend slider — *Trachemys gaigeae*
Blanding's turtle — *Emydoidea blandingii*
Blue-tailed mole skink — *Eumeces egregius lividus*
Blunt-nosed leopard lizard — *Gambelia silus*
Bog turtle — *Clemmys muhlenbergii*

Broad-banded water snake — *Nerodia* species
Broad-headed skink — *Eumeces laticeps*
Brown tree snake — *Boiga irregularis*
Bunch grass lizard — *Sceloporus scalaris*
California black-headed snake — *Tantilla planiceps*
California king snake — *Lampropeltis getulus californiae*
California legless lizard — *Anniella nigra*
California mountain kingsnake — *Lampropeltis zonata*
Canyon spotted whiptail — *Cnemidophorus burti*
Checkered garter snake — *Thamnophis marcianus*
Chuckwalla — *Sauromalus obesus*
Coachella Valley fringe-toed lizard — *Uma inornata*
Coachwhip — *Masticophis flagellum*
Collared lizard — *Crotaphytus collaris*
Common caiman — *Caiman crocodilus*
Common garter snake — *Thamnophis sirtalis*
Common kingsnake — *Lampropeltis getula*
Common map turtle — *Graptemys geographica*
Common musk turtle — *Stenotherus odoratus*
Common slider — *Trachemys scripta*
Common snapping turtle — *Chelydra serpentina*
Corn snake — *Elaphe guttata*
Couch's garter snake — *Thamnophis couchii*
Cuban ground iguana — *Cyclura nubila*
Culebra Island giant anole — *Anolis roosevelti*
Dekay's brown snake — *Storeria dekayi*
Desert collared lizard — *Crotaphytus insularis*
Desert horned lizard — *Phrynosoma platyrhinos*
Desert iguana — *Dipsosaurus dorsalis*
Desert tortoise — *Gopherus agassizii*
Diamondback terrapin — *Malaclemys terrapin*
Diamondback water snake — *Nerodia rhombifera*
Eastern fence lizard — *Sceloporus undulatus*
Eastern indigo snake — *Drymarchon corais couperi*
European pond turtle — *Emys orbicularis*
Flattened musk turtle — *Kinosternon depressus*
Florida pine snake — *Pituophis melanoleucus mugitus*
Florida sand skink — *Neoseps reynoldsi*
Giant garter snake — *Thamnophis gigas*
Gila monster — *Heloderma suspectum*
Gilbert's skink — *Eumeces gilberti*
Glossy snake — *Arizona elegans*
Gopher tortoise — *Gopherus polyphemus*
Granite night lizard — *Xantusia henshawi*
Great Plains rat snake — *Elaphe obsoleta obsoleta*
Great Plains skink — *Eumeces obsoletus*
Green anole — *Anolis carolinensis*
Green iguana — *Iguana iguana*
Green turtle — *Chelonia mydas*
Green water snake — *Nerodia cyclopion*
Ground snake — *Sonora semiannulata*
Gulf salt marsh snake — *Nerodia fasciata clarki*
Hawksbill — *Eretmochelys imbricata*
Horned ground iguana — *Cyclura cornuta*
Indigo snake — *Drymarchon corais*
Island night lizard — *Xantusia riversiana*
Kemp's ridley — *Lepidochelys kempii*
Leatherback — *Dermochelys coriacea*
Leopard lizards — *Gambelia* species
Lesser earless lizard — *Holbrookia maculata*
Lined snake — *Tropidoclonion lineatum*
Little striped whiptail — *Cnemidophorus inornatus*
Loggerhead — *Caretta caretta*
Long-nosed snake — *Rhinocheilus lecontei*
Long-nosed leopard lizard — *Gambelia wislizenii*
Map turtles — *Graptemys* species
Massasauga — *Sistrurus catenatus*
Mexican garter snake — *Thamnophis eques*
Milk snake — *Lampropeltis triangulum*
Mississippi mud turtle — *Kinosternon subrubrum hippocrepis*
Mojave Desert sidewinder — *Crotalus cerastes cerastes*
Mojave fringe-toed lizard — *Uma scoparia*
Mona boa — *Epicrates monensis*
Mona ground iguana — *Cyclura stejnegeri*

Monito dwarf gecko	*Sphaerodactylus micropithecus*
Musk turtles	*Kinosternon* species
Narrow-headed garter snake	*Thamnophis rufipunctatus*
New Mexican ridge-nosed rattlesnake	*Crotalus willardi obscurus*
Night snake	*Hypsiglena torquata*
Northern water snake	*Nerodia sipedon*
Northwestern garter snake	*Thamnophis ordinoides*
Olive ridley	*Lepidochelys olivacea*
Orange-throated whiptail	*Cnemidophorus hyperythrus*
Oregon garter snake	*Thamnophis hydrophilus*
Ornate box turtle	*Terrapene ornata*
Painted turtle	*Chrysemys picta*
Panamint alligator lizard	*Elgaria panamintina*
Panamint rattlesanke	*Crotalus mitchellii*
Pine snake	*Pituophis melanoleucus*
Plain-bellied water snake	*Nerodia erythrogaster*
Plains black-headed snake	*Tantilla nigriceps*
Plains garter snake	*Thamnophis radix*
Plateau striped whiptail	*Cnemidophorus velox*
Plymouth red-bellied turtle	*Pseudemys rubriventris bangsi*
Prairie kingsnake	*Lampropeltis calligaster*
Prairie skink	*Eumeces septentrionalis*
Puerto Rican boa	*Epicrates inornatus*
Puerto Rican racer	*Alsophis portoricensis*
Queen snake	*Regina septemvittata*
Racer	*Coluber constrictor*
Red-bellied snake	*Storeria occipitomaculata*
Red-bellied turtle	*Pseudemys rubriventris*
Red-eared turtle	*Chrysemys scripta elegans*
Red-footed tortoise	*Geochelone carbonaria*
Reticulate collared lizard	*Crotaphytus reticulatus*
Ridge-nosed rattlesnake	*Crotalus willardi*
Ringed map turtle	*Graptemys oculifera*
Ring-necked snake	*Diadophis punctatus*
Rock rattlesnake	*Crotalus lepidus*
Rosy boa	*Lichanura trivirgata*
Rubber boa	*Charina bottae*
Saltwater crocodile	*Crocodylus porosus*
San Francisco garter snake	*Thamnophis sirtalis tetrataenia*

Santa Cruz Island gopher snake	*Pituophis melanoleucus pumilis*
Sea turtles	Family Cheloniidae
Sharp-tailed snake	*Contia tenuis*
Side-blotched lizard	*Uta stansburiana*
Sierra garter snake	*Thamnophis couchii*
Six-lined racerunner	*Cnemidophorus sexlineatus*
Skinks	Family Scincidae
Slender glass lizard	*Ophisaurus attenuatus*
Smooth green snake	*Liochlorophis vernalis*
Smooth softshell turtle	*Trionyx muticus*
Snapping turtles	Family Chelydridae
Softshell turtles	Family Trionychidae
Sonoran mud turtle	*Kinosternon sonoriense*
Southern rubber boa	*Charina bottae umbratica*
Speckled kingsnake	*Lampropeltis getulus holbrooki*
Spiny softshell	*Trionyx spinifer*
Spot-tailed earless lizard	*Holbrookia lacerata*
Spotted turtle	*Clemmys guttata*
St. Croix ground lizard	*Ameiva polops*
St. Croix racer	*Alsophis sancticrucis*
St. Croix tree snake	*Epicrates* species
Striped racer	*Masticophis lateralis*
Striped whipsnake	*Masticophis taeniatus*
Switak's banded gecko	*Coleonyx switaki* subspecies
Texas horned lizard	*Phrynosoma cornutum*
Texas slender blind snake	*Leptotyphlops dulcis*
Timber rattlesnake	*Crotalus horridus*
Tortoises	Family Testudinidae
Twin-spotted rattlesnake	*Crotalus pricei*
Virgin Islands tree boa	*Epicrates monensis granti*
Western box turtle	*Terrapene ornata*
Western cottonmouth	*Agkistrodon piscivorus leucostoma*
Western hog-nosed snake	*Heterodon nasicus*
Western pond turtle	*Clemmys marmorata*
Western rattlesnake	*Crotalus viridis*
Western ribbon snake	*Thamnophis proximus*
Wood turtle	*Clemmys insculpta*
Yellow mud turtle	*Kinosternon flavescens*
[No common name]	*Cyclura portoricensis*

Glossary

The definitions below are taken from the following references: *The Concise Oxford Dictionary of Ecology*, edited by M. Allaby, published by Oxford University Press, Oxford, England, 1994; *The Dictionary of Ecology and Environmental Science*, edited by H. W. Art, published by Henry Holt and Company, 1993; *A Dictionary of Ecology, Evolution, and Systematics*, edited by R. J. Lincoln, G. A. Boxshall, and P. F. Clark, published by Cambridge University Press, Cambridge, England, 1982; *Glossary of Oceanography and the Related Geosciences with References*, by Steven K. Baum, Texas Center for Climate Studies, Texas A&M University; *Marine Biology: Function, Biodiversity, Ecology* by Jeffrey Levinton, published by Oxford University Press, New York; and *Webster's Third New International Dictionary*, published by Merriam-Webster, Springfield, Massachusetts, 1961, 1993.

Abyssal. Pertaining to zones of great depth in the oceans or lakes into which light does not penetrate; occasionally restricted to depths below 2,000 meters but more usually used for depths between 4,000 and 6,000 meters.

Accretion. Deposition of material by sedimentation which increases land area.

Achene. A small, usually single-seeded, dry fruit which remains closed at maturity. The achene is the simplest of any fruit.

Active layer. A seasonally thawed surface layer of soil in arctic or alpine regions that lies above permanently frozen ground and is between a few centimeters and about 3 meters thick.

Adaptive radiation. The evolutionary diversification of a taxon into a number of different forms, usually as a result of encounters with new resources or habitats. Adaptive radiation generally occurs over a relatively short period of time.

Adventive plant. A species of plant that is not native and has been introduced into the area but has not become permanently established.

Aeolian. Pertaining to the action or effect of the wind; wind-borne.

Aerenchyma. Spongy, modified cork tissue of many aquatic plants that facilitates gaseous exchange and maintains buoyancy.

Afforestation. The establishment of forest by natural succession or by the planting of trees on land where they did not grow formerly.

Albedo. A measure of surface reflectivity, usually expressed as a percentage, such as the proportion of solar radiation that is reflected back into space from the Earth, clouds, and atmosphere without heating the receiving surface. Studying a planet's albedo can help determine the composition of its surface.

Alcids. Any of the Alcidae family (Order Charadriiformes) of marine birds having a stout bill, short wings and tail, webbed feet, a large head and heavy body, and thick, compact plumage. Confined to the northern parts of the Northern Hemisphere, alcids include auks, guillemots, murres, and puffins.

Alevin. A young fish, particularly a young salmon that is still attached to the yolk sac.

Algae. The common name for the relatively simple type of unicellular or multicellular plant which is never differentiated into root, stem, and leaves, contains chlorophyll *a* as its photosynthetic pigment, has no true vascular system, and has no sterile layer of cells surrounding its reproductive organs. Found in most habitats on Earth, though the majority occur in freshwater or marine environments.

Alliance. A group of related botanical or zoological families, especially a group of plants intermediate between a class and an order.

Allogenic. Resulting from factors acting from outside a system or material transported into an area from outside which alters the system's habitat.

Alluvial. Of or relating to river and stream deposits.

Alluvial soil. Soil formed in material deposited by the action of running water, such as a floodplain or delta.

Alpine tundra. A treeless region above the treeline of high mountains, characterized by cold winters and short, cool summers and having permafrost below a surface layer that may melt in summer.

Alvar. A plant community dominated by mosses and herbs, occurring on shallow, alkaline limestone soils.

Amphidromous. Referring to the migratory behavior of fishes moving from fresh water to the sea and vice versa, not for breeding purposes but occurring regularly at some stage of the life cycle (such as feeding or overwintering).

Amphipod. Any of a large order of small, usually aquatic crustaceans with a laterally compressed body, for example, beach fleas.

Anadromous. Referring to the life cycle of fishes, such as salmon, in which adults travel upriver from the sea to breed, usually returning to the area where they were born.

Anaerobic. Referring to an environment in which oxygen is absent, or to a process which occurs only in the absence of oxygen, or to an organism which lives, is active, or occurs in the absence of oxygen, such as some yeasts or bacteria.

Annelids. Any of a phylum (Annelida) of usually elongated, segmented coelomate invertebrates, such as earthworms, various marine worms, and leeches.

Anoxic. Greatly deficient in oxygen; oxygenless.

Anthropogenic. Of, relating to, or resulting from the influence of humans on nature.

Anticyclonic. Referring to an area or system of high atmospheric pressure having a characteristic pattern of air circulation which usually induces settled weather conditions. Light winds flow clockwise in the northern hemisphere and counterclockwise in the southern hemisphere.

Appendicularia. A genus of small, free-swimming, pelagic tunicates shaped somewhat like a tadpole and remarkable for their resemblance to larvae of other tunicates.

Arboreal. Resembling a tree, or inhabiting or frequenting trees.

Archaebacteria. A taxonomic kingdom of bacteria, including sulphur-dependent bacteria, methane-producing bacteria, and halophilic bacteria.

Aromatic hydrocarbon. One of a group of hydrocarbon compounds containing one or more six-carbon rings characteristic of the benzene series. They are called aromatic because many of the earlier ones discovered, such as turpentine and wintergreen oil, have strong odors; many odorless aromatic compounds are now known.

Arroyo. A watercourse (such as a creek) or a water-carved gully or channel in an arid region.

Arthropod. Invertebrate animals with a segmented body and jointed appendages, for example, spiders, bees, and crabs.

Association. A stable grouping of two or more plant species that characterize or dominate a type of biotic community.

Avens. Any of a genus of perennial herbs of the rose family with white, purple, or yellow flowers.

Avian. Of, relating to, or derived from birds.

Avifauna. The birds of a specific region or period.

Barrens. A level area with poor, usually sandy or serpentine soils that is sparsely forested or unable to support normal vegetative cover and that generally has a low level of productivity. Plants growing in barrens are usually much smaller and stunted in comparison to individuals grown on more fertile soils. Barrens are frequently dominated by specialized groups of endemic plants.

Bathymetry. The measurement of the depth of the ocean floor from the water surface; the oceanic equivalent of topography.

Bathypelagic. Of, relating to, or living in the depths of the ocean, especially in the area between about 600 and 3,000 meters deep. The number of species and populations is relatively low in the bathypelagic zone, where no light source exists other than bioluminescence, temperature is uniformly low, and pressures are great.

Beach face. A strip of land that fronts a beach.

Benthic. Occurring at the bottom of a body of water, for example, a seabed, riverbed, or lake bottom.

Benthos. In freshwater and marine ecosystems, the collection of organisms both attached to or resting on the bottom sediments and burrowed into the sediments. In terms of size, benthos are generally divided into three categories: *meiobenthos*, the organisms that pass through a 0.5 millimeter sieve; *macrobenthos*, those that are caught by grabs or dredges but retained on the 0.5 millimeter sieve, and *epibenthos*, those organisms than live on rather than in the seabed.

Bight. A large indentation in a coastline or continental shelf margin forming an open bay.

Bioaccumulation. (Also called *biomagnification*.) The process by which chemical contaminants become more concentrated in the tissues of organisms as they pass higher up the food chain. Heavy metals and pesticides such as DDT are stored in the fatty tissues of animals and are passed along to predators of those animals. The resulting concentrations eventually reach harmful levels in predators at the top of the food chain.

Biodiversity. See *biological diversity*

Biogeochemical. Related to the partitioning and cycling of chemical elements and compounds between living organisms and nonliving components of the environment.

Biogeographical region. Any geographical region characterized by distinctive flora or fauna (such as a *biome* or a *province*).

Biogeography. The science that deals with the geographical distribution of animals and plants.

Biological diversity. (Often called *biodiversity*.) Used to describe species richness, ecosystem complexity, and genetic variation.

Biomass. The total mass of all living organisms or of a particular set of organisms in an ecosystem or at a trophic level in a food chain; usually expressed as a dry weight or as the carbon, nitrogen, or caloric content per unit area.

Biome. A major regional ecological community characterized by distinctive life forms and principal plant or animal species, such as a tropical rain forest, a tundra, a grassland, or a desert.

Biota. The plants and animals of a specific region or period, or the total aggregation of organisms in the biosphere.

Bivalve. A mollusk whose body is enclosed by two hinged valves or shells.

Blowdown. An extensive toppling of trees by wind within a relatively small area which significantly alters the small-scale climate within the ecosystem.

Boreal forest. The circumpolar, subarctic forest of high northern latitudes that is dominated by conifers. The boreal forest stretches across North America, Europe, and northern Asia (regions characterized by short summers and long, cold winters). It is found south of the tundra in the Northern Hemisphere and often contains peaty or swampy areas.

Brachyuran. Of or belonging to the Brachyura, a group of crustaceans with a greatly reduced abdomen which is more or less folded against the ventral surface of the thorax, such as the typical crabs.

Brackish. Water that is saline but not as salty as seawater.

Braided channel. A stream consisting of a network of interlacing small channels separated by bars, which may be vegetated and stable or barren and unstable.

Breeding Bird Survey. The North American Breeding Bird Survey (BBS) was begun in 1966 to collect standardized data on bird populations along more than 3,400 survey routes across the continental United States and southern Canada. A cooperative effort, the BBS has been used to document distributions and establish continental, regional, and local population trends for more than 250 species.

Brood parasitism. (Also called *nest parasitism* or *breeding parasitism*.) The laying of eggs by one bird species in the nest of another bird species and the subsequent brooding of the egg and raising of the young by the parasitized host, usually to the detriment of the host's young.

Bunchgrass. Any of several grasses, especially of the western United States, that grow in tufts rather than forming turf, for example, the genus *Andropogon*.

Bycatch. Nontarget organisms that are caught in fishing or other harvest operations and are usually discarded.

Calcareous. Consisting of or containing calcium carbonate; a soil rich in calcium salts, derived from limestone or chalk. Also, an organism which has an affinity for such an alkaline or basic soil.

Carrying capacity. The maximum population of a given organism that a particular environment or habitat can sustain; implies continuing yield without environmental damage; often denoted as *K*.

Catadromous. An organism which lives in fresh water and goes to the sea to spawn, such as some eels.

Catchment. The area drained by a river or body of water.

Cation. An ion or group of ions having a positive charge and characteristically moving toward the negative electrode in electrolysis.

Cay. A low island or reef of sand or coral.

Cetacean. Any of an order of aquatic, mostly marine mammals that include the whales, dolphins, porpoises, and related forms.

Chaetognaths. A group of small, active, transparent marine worms of uncertain systemic position with horizontal lateral and caudal fins and a row of moveable, curved spines around the mouth, for example, arrowworms.

Channelization. The straightening of rivers or streams by means of an artificial channel.

Chaparral. A vegetation type dominated by shrubs and small trees, especially evergreen trees with thick, small leaves.

Charismatic megafauna. Large vertebrate animals that evoke sentimental support from the general public, for example, deer, sea turtles, and wolves.

Chenier. A large, ridge-shaped deposit of sandy material built up by wave action in a marshy area. Chenier deposits are common on the Gulf of Mexico coast of North America.

Chironomids. Any of a family (Chironomidae) of midges that lack piercing mouthparts.

Chlorofluorocarbons. (Also called greenhouse gases or CFC's.) A group of gaseous compounds that contain carbon, chlorine, fluorine, and sometimes hydrogen and are used as refrigerants, cleaning solvents, and aerosol propellants and in the manufacture of plastic foams. They are suspected of being a major cause of stratospheric ozone depletion as well as of absorbing long-wave electromagnetic radiation.

Chlorophyll *a*. Chlorophyll, the green photosynthetic pigment found chiefly in chloroplasts of plants, occurs in variants of *a, b, c,* and *d.* Chlorophyll *a* is a waxy, blue-black microcrystalline, $C_{55}H_{72}MgN_4O_5$, with a characteristic blue-green alcohol solution.

Cirque. A steep hollow, often containing a small body of water, found at the upper end of a mountain valley.

CITES. Convention on International Trade in Endangered Species of Wild Fauna and Flora, an agreement between 103 nations to restrict international commerce involving endangered and threatened species of animals and plants, such as tropical birds, rhinoceros horns, orchids, and ivory.

Cladocerans. Any of an order (Cladocera) of minute, freshwater brachiopod crustaceans, including the water fleas.

Climax. The final stage of succession in an ecosystem. Also, a community that reached a steady state under a particular set of environmental conditions.

Clonal dispersal. The separation of parts of a modular organism, or the growth of such parts away from each other without actually detaching.

Clone. To reproduce individual organisms asexually, as in plant propagation through budding or layering. Also, an organism or group of organisms (or group of cells) so produced.

Close-crowned. Descriptive of crowded forests where closely spaced trees have tops that touch or overlap.

Cloud forest. A wet, tropical forest, often near peaks of coastal mountains and at an altitude usually between 1,000 and 2,500 meters, that is characterized by a profusion of epiphytes and the presence of clouds even in the dry season.

Cluster analysis. A method grouping those variables within a set of variables that are highly correlated and excluding from clusters those that are negatively correlated or uncorrelated. Used in numerical taxonomy as a procedure for arranging taxonomic units into homogenous clusters based on their mutual similarities.

Commensal. Referring to the relationship between two kinds of organisms in which one obtains food or other benefits from the other without damaging or benefitting it. Also, an organism which lives in this way.

Community. Any grouping of populations of different organisms that live together in a particular environment.

Conspecific. Of or relating to the same species.

Continental shelf. The shallow, gradually sloping seabed around a continental margin, not usually deeper than 200 meters and formed by submergence of part of a continent.

Copepods. Any of a large subclass (Copepoda) of usually minute freshwater and marine crustaceans that form an important element of the plankton in the marine environment and in some fresh waters.

Corridor. A more or less continuous connection between land masses or habitats; a migration route that allows more or less uninhibited migration of most of the animals of one faunal region to another. In terms of conservation biology, a connection between habitat fragments in a fragmented landscape.

Crevasse. A breach in a levee along the bank of a river through which floodwater may flow and produce sheetlike deposits of gravel or sandy sediment; or, a large, open fissure forming in a glacier as it moves and is deformed.

Crown fires. Fires that spread from tree crown to tree crown, usually indicative of particularly hot fires in dry conditions.

Crustacean. Any of a large class (Crustacea) of mostly aquatic mandibulate arthropods that have a chitinous or calcareous and chitinous exoskeleton, a pair of often modified appendages on each segment, and two pairs of antennae; includes lobsters, shrimps, crabs, wood lice, water fleas, and barnacles.

Ctenophore. Any of a phylum (Ctenophora) of marine animals superficially resembling jellyfishes but having biradial symmetry and swimming by means of eight meridional bands of transverse ciliated plates; also called *comb jellies.*

Cultivar. A variety of a plant produced and maintained by horticultural techniques and not normally found in wild populations.

Cyanobacteria. A large and varied group of bacteria which possess chlorophyll *a* and which carry out photosynthesis in the presence of light and air, producing oxygen. They were formerly regarded as algae and were called "blue-green" algae. The group is very old and is believed to have been the first oxygen-producing organisms on Earth.

Cyclonic. Referring to a region of low atmospheric sea level pressure; or, the wind system around such a low pressure center that has a clockwise rotation in the Northern Hemisphere and a counterclockwise rotation in the Southern Hemisphere.

Debouch. To emerge or issue; often used in reference to rivers or streams.

Debris torrent. A flood of debris (branches, shrubs, rocks, mud, and so forth) and water rushing down a stream channel, caused by excessive rainfall or snow melt. Debris torrents have a significant scouring effect on the stream ecosystem.

Deciduous. Plants having structures that are shed at regular intervals or at a given stage in development, such as trees that shed their leaves seasonally.

Degradation. The breaking down of a substance into smaller or simpler parts, usually by erosion.

Degree-days. Units used in the measurement of the duration of a life cycle or a particular growth phase of an organism; calculated as the product of time and temperature averaged over a specified interval.

Delta. An alluvial deposit at the mouth of a river or tidal inlet. Deltas occur when a sediment-laden current enters an open body of water, at which point there is a reduction in the velocity of the current, resulting in rapid deposition of the sediment, as at the mouth of a river where the river discharges into the sea or a lake.

Delta plain. A nearly horizontal portion of delta that, during low tide or other regression of water, is largely exposed to the atmosphere.

Demersal. Living at or near the sea floor but having the capacity for active swimming.

Dendritic. Branched, like a tree.

Dendrochronology. The science of dating events and variations in the environment by the comparative study of annual growth rings of trees.

Detritus. Debris or waste material, usually organic, such as dead or partially decayed plants and animals, often important as a source of nutrients; or, small particles of minerals from weathered rock, such as sand or silt.

Diapir. An anticlinal fold in which a mobile core, such as salt or gypsum, has broken through brittle overlying rocks.

Diel. A 24-hour period, usually encompassing 1 day and 1 night.

Dinoflagellates. Any of an order (Dinoflagellata) of chiefly marine, planktonic, usually solitary phytoflagellates (which have many characteristics in common with algae) that includes luminescent forms, forms important in marine food chains, and forms causing red tides.

Disjunct. Distinctly separate; a discontinuous range in which one or more populations are separated from other potentially interbreeding populations by a sufficient distance to preclude gene flow between them.

Distributary. A river branch flowing away from the main stream.

Disturbance. An event or change in the environment that alters the composition and successional status of a biological community and may deflect succession onto a new trajectory, such as a forest fire or hurricane, glaciation, agriculture, and urbanization.

Diurnal. Occurring or active only in daylight.

Diversity. The variety of species in a sample, community, or area.

Doliolids. Any of a small family of oceanic tunicates.

Dominance. The extent to which a given species or individual influences community composition or form because of its size, abundance, or coverage.

Downwelling. The downward movement of surface waters caused by the convergence of different water masses or where surface waters flow toward the coast.

Drawdown. A lowering of the water level in a reservoir or other body of water.

Echinoderms. Any of a phylum (Echinodermata) of radially symmetrical coelomate marine animals including the starfishes, sea urchins, and related forms.

Ecological succession. The chronological sequence of vegetation and associated animals in an area; or, continuous colonization, extinction, and replacement of species' populations at a particular site, due either to environmental changes or to the intrinsic properties of the plants and animals.

Ecoregion. See *biogeographical region*.

Ecosystem. A community of organisms and their physical environment that interact as an ecological unit.

Ecotone. The boundary or transitional zone between adjacent communities containing the characteristic species of each, such as the edge of a woodland next to a field or lawn.

Ectotherm. A cold-blooded animal, one having a body temperature determined primarily by the temperature of its surrounding environment. Terrestrial reptiles are ectotherms.

Edaphic. Pertaining to soil or to the physical, chemical, and biological properties of the soil or substratum which influence associated biota, such as pH and organic matter content.

Edge effect. The tendency for a transitional zone between communities (an *ecotone*) to contain a greater variety of species and more dense populations of species than either community surrounding it.

El Niño. (Also called *El Niño–Southern Oscillation Event*, or *ENSO.*) A warmwater current which periodically flows southward along the coast of Ecuador, associated with the Southern Oscillation in the atmosphere, and which affects climate throughout the Pacific region. Approximately once every seven years in late December, prevailing trade winds weaken and the equatorial countercurrent strengthens. Warm surface waters, normally driven westward by the wind to form a deep layer off Indonesia, flow eastward to overlie the cold waters of the Peru Current. The Southern Oscillation is a fluctuation of the intertropical atmospheric circulation in which air moves between the southeastern Pacific subtropical high and the Indonesian equatorial low, driven by the temperature difference between the two areas.

Elasmobranchs. Any of a subclass (Elasmobranchii) of cartilaginous fishes that have five to seven lateral gill openings on each side, comprising sharks, rays, skates, and extinct related fishes.

Electrophoresis. A technique for separating mixtures of organic molecules based on their different rates of travel in electric fields.

Emergent. An aquatic plant having most of its vegetative parts above water. Also, a tree which reaches or exceeds the level of the surrounding canopy.

Emersed. Rising above the surface of the water.

Empirical. Originating in or based upon observation or experience; capable of being verified or disproved by observation or experiment.

Encinal. Referring to live oaks; any of several American evergreen oaks, such as the Coast live oak (*Quercus agrifolia*) of coastal California and *Quercus virginiana* of northern Baja California, Mexico, and the southeastern United States.

Endangered Species Listing. See *List of Endangered or Threatened Species.*

Endemic. Belonging or native to a particular people or geographic region; a genetically unique life form.

Endotherm. A warm-blooded animal, one that maintains a body temperature largely independent of the temperature of the environment. Mammals are endothermic.

Epipelagic. The oceanic zone extending from the surface to about 200 meters, where enough light penetrates to allow photosynthesis.

Epiphyte. A plant that uses another plant (usually a tree) for support or anchorage but not for water or nutrients.

Epizootic. An outbreak of disease (an epidemic) in nonhuman animals, or pertaining to such an outbreak.

Ericaceous. Of, relating to, or being a heath or of the heath family of plants, which are mostly shrubby, dicotyledonous, and often evergreen plants that thrive on open, barren soil that is usually acidic and poorly drained.

Escapement. The number of fish that are permitted to survive and spawn (as by adjustment of fishing season or by provision of fishways).

Estuary. A semi-enclosed coastal body of water which has a free connection with the open sea and where fresh water derived from land drainage (usually mouths of rivers) is mixed with seawater; often subject to tidal action and cyclic fluctuations in salinity.

Euryhaline. Able to live in waters with a wide range of salinity.

Eustatic. Relating to or characterized by worldwide change in sea level such as that caused by tectonic movements or by the growth or decay of glaciers.

Eutrophication. The process by which a body of water acquires a high concentration of nutrients, especially phosphates and nitrates, which typically promote excessive growths of algae. As the algae die and decompose, the amount of available oxygen in the water is depleted, in turn causing the death of other organisms, such as fishes. Normally, eutrophication is a natural, slow-aging process for a body of water, but human activity can greatly speed up the process.

Evapotranspiration. Loss of water from the soil both by evaporation and by transpiration from plants.

Extinction. The dying out of a species, or the condition of having no remaining living members; also, the process of bringing about such a condition.

Extirpation. The loss or removal of a species from one or more specific areas but not from all areas.

Fauna. The animal life of a region or geological period.

Fecundity. The potential reproductive capacity of an organism or population.

Fen. A marshy, low-lying wetland covered by shallow, usually stagnant, and often alkaline water that originates from groundwater sources.

Feral. Of or relating to plants or animals which have escaped from domestication and to their descendants.

Ferrous. Of or containing iron.

Fetch. The distance along open water or land over which the wind blows; the distance traversed by waves without obstruction.

Flora. Plant or bacterial life forms of a region or geological period.

Floristic. Relating to all of the plant species of a geographic or ecological area or region.

Fluvial. Pertaining to rivers or streams and their action.

Fold. A curve or bend in a stratum of rock.

Forb. An herbaceous plant which is not a grass.

Fragmentation. See *habitat fragmentation*.

Fringing reef. A coral reef that forms near the shoreline.

Gallery forest. A narrow strip of forest along the margins of a river in an otherwise unwooded landscape.

Gamete. A mature male or female germ cell possessing a haploid chromosome set and capable of fusing with a gamete of the opposite sex to produce a fertilized egg.

GAP analysis. The process of identifying and classifying components of biological diversity to determine which components already occur in protected areas and which are not present or are underrepresented in protected areas.

Gastropod. Any of a large class (Gastropoda) of mollusks, usually with a univalve shell or no shell and a distinct head bearing sensory organs, such as snails and slugs.

Geographic Information System (GIS). A combination of computer hardware and software that allows storage and manipulation of information suitable for mapping. A GIS software system synthesizes geographic position data and other data (such as the type of bottom sediment) in order to create a map. Data on processes (current speed, for example) can be incorporated to make a geographic model of flow, for example.

Geomorphology. The study of landforms on a planet's surface and of the processes that have fashioned them.

Geotropic. Referring to the involuntary response of a plant or one of its parts to gravity. Geotropism may be a positive or negative response: primary taproots show positive geotropism; vertical primary shoots show negative geotropism.

Glade. An open space in the forest.

Gorgonian. Any of an order (Gorgonacea) of colonial anthozoans (corals), usually with a horny and branching radial skeleton.

Gradient. A rate of change of a variable with distance; a regularly increasing or decreasing change in a factor, such as ambient temperature; a character gradient.

Graminoid. Grasses and grasslike plants, such as sedges.

Gravid. Carrying eggs or young; pregnant.

Greenhouse effect. Heating of the Earth's atmosphere that is loosely analogous to the glass of a greenhouse letting light in but not letting heat out. Radiation from the sun easily enters the atmosphere as light waves, heating the Earth's surface and causing it to emit infrared radiation. Gases such as water vapor, carbon dioxide, methane, and chlorofluorocarbons absorb infrared radiation, preventing its energy from leaving the Earth.

Gregarious. Tending to form into groups which possess a social organization, such as schools of fish, herds of mammals, flocks of birds.

Groin. A breakwater structure extending seaward at a right angle to the shoreline, designed to inhibit the longshore drift of sediments.

Groundfish. A bottom-dwelling fish, especially one of commercial importance such as cod, haddock, pollack, or flounder.

Guild. A group of species having similar ecological resource requirements and foraging strategies and therefore having similar roles in the community.

Gymnosperm. A plant, such as a cycad or a conifer, whose seeds are not enclosed in an ovary (fruit).

Gyre. A circular or spiral system of movement, especially a giant circular oceanic surface current.

Habitat. The place, including physical and biotic conditions, where a plant or an animal usually occurs.

Habitat fragmentation. The breaking up of a habitat into unconnected patches interspersed with other habitat which may not be inhabitable by species occupying the habitat that was broken up. The breaking up is usually by human action, as, for example, the clearing of forest or grassland for agriculture, residential development, or overland electrical lines.

Habitat sharing. A situation in which species occupy the same habitat without competition, either through requiring different resources or being present at different times.

Halophytic. Referring to a plant that can tolerate or thrive in alkaline soil rich in sodium or calcium salts; tolerant of saline conditions.

Hard mast. Fruit of hardwood trees such as beech and oaks.

Hardwood hammock. A somewhat elevated area with hardwood trees and deeper soils, often surrounded by pine forest on shallower soils. Most often found in the Southeast.

Heavy metals. A metallic element of high specific gravity, such as antimony, bismuth, cadmium, copper, gold, lead, mercury, nickel, silver, tin, and zinc. These metals, which are toxic even in low concentrations, persist in the environment and can accumulate to levels that stunt plant growth and interfere with animal life.

Hectare (ha). A metric unit of measure for area, equal to 2.47 acres.

Hermaphroditic. An individual that possesses both male and female sex organs.

Herptiles. Reptiles and amphibians collectively.

Heterogeneous. Consisting of diverse or dissimilar parts; having nonuniform structure or composition.

Heterotrophic. An organism that is unable to manufacture its own food from simple chemical compounds and therefore consumes other organisms, living or dead, as its main source of carbon.

Heterozygous. Having two different alleles at a particular gene locus on a chromosome pair. Provides a measure of genetic variation either in a population or in an individual.

Holocene. The present, post-Pleistocene geologic epoch of the Quaternary period, including the last 10,000 years; the recent or Post-glacial period.

Hybridization. Any crossing of individuals of different genetic composition, often belonging to separate species, resulting in hybrid offspring.

Hydric. Characterized by, relating to, or requiring an abundance of moisture.

Hydrocarbon. A naturally occurring organic compound that contains carbon and hydrogen; may be gaseous, solid, or liquid, for example, natural gas, bitumens, and petroleum.

Hydrographic. Relating to the characteristic features of bodies of water, such as depth and flow.

Hydrological cycle. The movement of water from the sea through the air to the land and back to the sea.

Hydrology. The study of the movement of water from the sea through the air to the land and back to the sea; the properties, distribution, and circulation of water on or below the Earth's surface and in the atmosphere.

Hydromorphic. Descriptive of an intrazonal soil formed under waterlogged or poorly drained conditions.

Hydroperiod. The duration and frequency of flooding.

Hypoxic. Deficient in oxygen.

Igneous rock. Rock formed by solidification of molten magma.

Impoundment. A natural or artificial body of water held back by a dam.

Indicator species. An organism whose presence or state of health is used to identify a specific type of biotic community or as a measure of ecological conditions or changes occurring in the environment.

Indigenous. A species that occurs naturally in an area; native.

Infauna. Benthic organisms that dig into the sea bed or construct tubes or burrows. They are most common in the subtidal and deeper zones.

Insolation. Incoming solar radiation. Also, a measurement of the amount of this solar energy falling on a surface perpendicular to the sun's rays, of a specified size and over a specified period of time.

Interglacial. A warm period between glacial epochs.

Intermediate host. The host occupied by juvenile stages of a parasite prior to the definitive host and in which asexual reproduction often occurs.

Intertidal. Relating to the littoral zone above the low-tide mark.

Invertebrate. An animal without a backbone, such as snails, worms, and insects.

Invertivore. An animal or plant that eats invertebrate animals.

Ion. An atom or group of atoms that carries a positive or negative electric charge as a result of having lost or gained one or more electrons.

Isobath. A line on a map or chart that connects all points having the same depth below the surface of a body of water; also, having constant depth.

Isopod. Any of a large order of sessile-eyed crustaceans with the body composed of seven free thoracic segments, each bearing a similar pair of legs.

Isotherm. A line on a map or chart of the Earth's surface connecting points having the same temperature at a given time or the same mean temperature for a given period.

Karst. A limestone landscape that is characterized by sinks, underground streams, and caverns.

Keystone species. Organisms that play dominant roles in an ecosystem and affect many other organisms. The removal of a keystone predator from an ecosystem causes a reduction of the species diversity among its former prey.

Krill. Planktonic crustaceans and larvae that constitute the primary food of baleen whales.

Krummholz. A discontinuous belt of stunted forest or scrub typical of windswept alpine regions close to treeline; a wind-deformed tree at high elevations.

Lacustrine. Pertaining to or living in lakes or ponds.

Lagoon. A shallow waterbody that is near or connected to a larger body of water.

Larvae. The wingless and often wormlike hatchlings of insects; also, the early form of an animal (such as a frog or sea urchin) which at birth or hatching is fundamentally unlike its parent and must metamorphose before assuming adult characteristics.

Leachate. The solution that is formed when water percolates through a permeable medium. A leachate may contain toxic material or bacteria.

Leaching. The removal of readily soluble components, such as chlorides, sulfates, organic matter, and carbonates, from soil by percolating water. The remaining upper layer of leached soil becomes increasingly acidic and deficient in plant nutrients.

Lentic. Related to still waters such as ponds, lakes, or swamps.

Levee. A raised embankment along the edge of a river channel. Natural levees result from periodic overbank flooding, when coarser sediment is immediately deposited because of a reduction in river velocity. Levees are often constructed by humans living in low-lying areas as protection against flooding.

Lichen. A composite organism consisting of a fungus and an alga or cyanobacterium living in symbiotic association. Lichens may be crustlike, scaly or leafy, or shrubby in form and are classified on the basis of the fungal partner. Many lichens are extremely sensitive to atmospheric pollution and have been used as pollution indicators.

Life history. The significant features of the life cycle through which an organism passes, with particular reference to strategies influencing survival and reproduction.

Life zone. See *biome.*

Limnic. Pertaining to lakes or to other bodies of standing fresh water; often used with reference only to the open water of a lake away from the bottom; limnetic.

List of Endangered or Threatened Species. A listing of animals and plants administratively determined to meet legal criteria for protection under provisions of the U.S. Endangered Species Act.

Littoral zone. The biogeographic zone in a body of fresh water where light penetration is sufficient for the growth of plants; the intertidal zone of the seashore.

Loess. Unconsolidated sediment deposited by wind. Loess is usually composed of unstratified fine sand or silt.

Long-line fishing. A method of fishing which utilizes a piece of ground line (often kilometers in length) to which short (0.5 to 1.0 meter) ganglion lines are attached at intervals of 1 to 10 meters, with a baited hook at the end of each ganglion line.

Lotic. Relating to or living in moving water, such as a river or stream.

Macrofauna. Animals large enough to be seen with the naked eye.

Marsh. An ecosystem of more or less continuously waterlogged soil dominated by emersed herbaceous plants but without a surface accumulation of peat. A marsh differs from a swamp in that it is dominated by rushes, reeds, cattails, and sedges, with few if any woody plants, and differs from a bog in having soil rather than peat as its base.

Maximum Sustainable Yield. The maximum yield or crop which may be harvested year after year without damage to the system, or the theoretical point at which the size of a population is such as to produce a maximum rate of increase. The concept has been applied widely to commercial fisheries, forming the basis for models that predict stocking density required to maintain optimum fish production and the harvest methods and food supply required to maintain production at that level.

Megafauna. The largest size category of animals in a community.

Meiofauna. That part of the microfauna which inhabits algae, rock fissures, and superficial layers of the muddy sea bottom. They are smaller than 1 millimeter but larger than 0.1 millimeter.

Meristems. The undifferentiated, growing parts of plants, consisting of groups of cells capable of actively dividing.

Meroplankton. Temporary zooplankton, such as the larval stages of some organisms (fishes and crabs, for example).

Mesic. Neither wet (*hydric*) nor dry (*xeric*); intermediate in moisture, without extremes.

Mesopelagic. The ocean zone from 200 to 1,000 meters deep, where little light penetrates and the temperature gradient is even and gradual with little seasonal variation. This zone contains an oxygen minimum layer and usually has the maximum concentrations of the nutrients nitrate and phosphate. It overlies the *bathypelagic* zone and is overlain by the *epipelagic* zone.

Metabolite. A product of metabolism or a substance that is essential to the metabolism of an organism or to a metabolic process.

Metamorphic rock. Preexisting rock that is restructured by high temperature and pressure.

Metapopulation. A group of populations, usually of the same species, which exist at the same time but in different places.

Microclimate. The climate that prevails in a small area, usually in the layer near the ground.

Microfauna. The smallest animals in a community, not visible to the naked eye.

Microturbellarians. Any of a class of mostly aquatic and free-living flatworms (Turbellaria), for example, planarians.

Midden. A heap of refuse. Also, a pile of seeds or of various items that were gathered by a rodent, for example, by a squirrel or packrat.

Miocene. A geologic epoch within the Tertiary period (about 26 to 5 million years B.C.).

Mollusk. An organism in the phylum Mollusca (for example, snails, clams, or squids), whose soft, unsegmented body parts are frequently enclosed in a shell.

Montane. Of, relating to, growing in, or being the biogeographical zone of relatively moist, cool upland slopes below the timberline, often dominated by large coniferous trees.

Moraine. An accumulation of boulders, stones, or other debris carried and deposited by a glacier.

Morphology. The form and structure of organisms.

Mosaic. Heterogeneous ecological conditions on a landscape, usually produced by the variable, patchy effects of disturbances; a patchwork of vegetation communities within a landscape as determined by environmental conditions.

Mustelid. One of a large, widely distributed family of small, lithe, carnivorous mammals, including weasels, otters, skunks, wolverines, and minks.

Mutagen. Any agent that produces a mutation or enhances the rate of mutation in an organism, for example, x-rays, gamma rays, and certain chemicals.

Mutualism. An interaction between members of two species which benefits both; in strict terms, obligatory mutualism, in which neither species can survive under natural conditions without the other.

Mycorrhizae. The mutually beneficient association between a fungus and the roots of a plant; a mycorrhizal root takes up nutrients more efficiently than an uninfected root. Some plants seem to be incapable of normal development in the absence of their mycorrhizal fungi.

Nannoplankton. Minute planktonic organisms with a body diameter between 0.2 and 20 micrometers.

Nekton. (Also spelled *necton*.) Free-swimming organisms in aquatic ecosystems; unlike plankton, they are able to navigate at will (such as fishes, amphibians, and large swimming insects).

Nematode. Any of a phylum (Nematoda or Nemata) of elongated cylindrical worms parasitic in animals or plants, or free-living in soil or water.

Nemerteans. Any of a phylum (Nemertea) of often vividly colored marine worms, most of which burrow in the mud or sand along seacoasts; often called ribbon worms.

Neotenic. Referring to an organism which has attained sexual maturity while retaining juvenile characteristics.

Neotropical migrant. A bird that nests in temperate regions and migrates to the Neotropical faunal region, which includes the West Indies, Mexico, Central America, and that part of South America within the tropics.

Neritic. Of, relating to, or inhabiting the shallow water, or nearshore marine zone extending from the low-tide level to a depth of 200 meters. The neritic zone is populated by benthic organisms because of the penetration of sunlight to these shallow depths.

Netplankton. Plankton larger than 25 micrometers in diameter.

Nitrogen fixation. The process of converting inorganic, atmospheric nitrogen into an organic form of nitrogen, ammonia. This process can be carried out by lightning, by photochemical fixation in the atmosphere, or by the action of microorganisms. Also, the chemical processes used in the manufacture of fertilizers.

Nival. Of, relating to, or growing under or in snow.

Nonindigeneous. (Also called *exotic*, *nonnative*, *introduced*, and *alien*.) A plant or animal that is not native to the area in which it occurs; it was either purposely or accidentally introduced.

Nonpoint. Not from a single, well-defined site. Nonpoint sources are pollution-producing entities not tied to a specific origin, such as an individual smokestack. Nonpoint sources of water pollution include runoff which washes pollutants from roads into storm sewers and bodies of water or agricultural chemicals from lawns, fields, and golf courses.

Nunatak. An exposed hill or mountain completely surrounded by glacial ice.

Obligate. Essential, necessary; unable to exist in any other state, mode, or relationship; restricted to one particularly characteristic mode of life. An obligate predator lives off of only one species or specific group of prey; an obligate parasite is capable of living naturally only as a parasite and only on its single host; an obligate halophyte is a plant that requires salty soils in order to thrive; an obligate relationship is a relationship between organisms in which neither can exist without the other.

Old-growth. Referring to an ecosystem or community, particularly a forest, which has not experienced intense or widespread disturbance for a long time relative to the lifespans of the dominant species and which has entered a late successional stage; usually associated with high diversity of species, specialization, and structural complexity.

Oligochaetes. Any of a class or order (Oligochaeta) of hermaphroditic terrestrial or aquatic annelid worms that lack a specialized head.

Oligotrophic. Waters or soils that are poor in nutrients and have low primary productivity.

Ontogenetic. Relative to the course of growth and development of an individual organism.

Osmerid. A member of the family of fishes (Osmeridae) to which the true smelts belong; smelts and smeltlike fishes.

Ovigerous. Carrying eggs, or modified for carrying eggs.

Oxidation. A reaction in which atoms or molecules gain oxygen or lose hydrogen or electrons.

Pair bonding. The forming of a pair for breeding.

Paleoecology. The application of ecological concepts to fossil communities.

Palustrine. Pertaining to wet or marshy habitats.

Palynology. The study of living and fossil pollen and spores.

Parasite. An organism that is intimately associated with and metabolically dependent on another living organism (the host) for completion of its life cycle, and which is typically detrimental to the host.

Passerine. Of or relating to the largest order (Passeriformes) of birds, which includes more than half of all living birds and consists primarily of perching songbirds, whose young are hatched in an immature and helpless condition.

Patch dynamics. The idea that communities are a mosaic of different areas (patches) within which nonbiological disturbances (such as climate) and biological interactions proceed.

Pathogen. A specific causative agent of a disease, such as a bacterium or a virus.

Patterned ground. An assemblage of small, geometric features (circles, polygons, nets, steps) at the surface of nonconsolidated, weathered rock, resulting from disturbance by frost action such as cracking, heaving, and mass movement.

Pelagic. Referring to or occurring in the open sea.

Percent cover. In descriptions of plant communities, the proportion of ground, expressed as a percentage, that is occupied by the perpendicular projection down onto it of the aerial parts of individuals of the species under consideration.

Perennial. A plant that normally lives for more than two seasons and which, often after an initial period, produces flowers annually.

Permafrost. A permanently frozen layer of soil at variable depth below the surface in frigid regions of a planet. It may be discontinuous, that is, it may be interspersed with areas that are free of permafrost.

pH. A measure of acidity and alkalinity of a solution, taken by measuring the relative concentration of hydrogen ions in the solution.

Phenology. The study of the relationship between climate and the timing of periodic natural phenomena such as migration of birds, bud bursting, or flowering of plants.

Phenotype. The observable manifestation of a specific genetic makeup; those observable properties of structure and function of an organism as modified by genetic structure in conjunction with the environment.

Piscivores. Fish-eaters; those organisms that subsist exclusively or primarily on fish.

Photic zone. The surface zone of the sea or a lake having sufficient light penetration for photosynthesis.

Photoperiod. The length of time an organism is daily exposed to light, especially with regard to how that exposure affects growth and development.

Phreatophyte. A plant that absorbs groundwater from the permanent watertable.

Phylogenetic. Pertaining to the evolutionary history of a group or lineage, or the evolutionary relationships within and between taxonomic levels; the relationships of groups of organisms as reflected by their evolutionary history.

Physiognomy. The physical features of something. For example, the physiognomy of a landscape includes its topography and vegetation.

Physiographic province. A region of the landscape with distinctive geographical features.

Physiography. Landform; physical geography.

Phytoplankton. One of two groups into which *plankton* are divided, the other being *zooplankton*. Phytoplankton comprise all the freely floating photosynthetic forms in the oceans.

Pingo. A low hill or mound forced up by hydrostatic pressure in an area underlain by permafrost and consisting of an outer layer of soil covering a core of solid ice. Pingos range from 2 to 50 meters in height.

Pinniped. Any of a suborder of aquatic carnivorous mammals with all four limbs modified into flippers; includes seals, sea lions, and walruses.

Pioneer. The first species or community to colonize or recolonize a barren or disturbed area, thereby commencing a new biological succession.

Placer mining. The removal of ore from placers, which are glacial or alluvial deposits of sand or gravel containing valuable minerals.

Plankton. One of three major ecological groups into which marine organisms are divided, the other two being the *nekton* and the *benthos*. Plankton are small aquatic organisms (animals and plants) that, generally having no locomotive organs, drift with the currents. The animals in this category include protozoans, small crustaceans, and the larval stages of larger organisms, while plant forms are mainly diatoms.

Playa. A nearly level area at the bottom of an undrained desert basin, sometimes temporarily covered with water during wet periods. Playas are barren and usually saline.

Pleistocene. The earlier epoch of the Quaternary period or the corresponding system of rocks; 1.6 million–10,000 years ago; the "Ice Age."

Plutonic. Of or relating to conditions of rock formation from magma within the crust of the Earth.

Pluvial. Characterized by abundant rain.

Pocosin. A swamp or marsh in an upland coastal region. The term is chiefly used in the South Atlantic states of Virginia, Maryland, Delaware, and the Carolinas.

Polychaetes. Any of a class (Polychaeta) of chiefly marine annelid worms (such as clam worms), usually with paired segmental appendages, separate sexes, and a free-swimming trochophore larva.

Polychlorinated biphenyls (PCB's). A group of toxic, carcinogenic organic compounds containing more than one chlorine atom. PCB's were used in the manufacture of plastics and as insulating fluids in electrical transformers and capacitors. They behave much like DDT in the environment in that they are very stable compounds and are also fat-soluble; therefore, they accumulate in ever-higher concentrations as they move up the food chain. The use of PCB's was banned in the United States in 1979.

Population. A group of organisms, all of the same species, which occupies a particular area. Also, the total number of individuals of a species within an ecosystem, or of any group of similar individuals.

Primary producer. An organism capable of using the energy derived from light or a chemical substance in order to manufacture energy-rich organic compounds, mainly green plants.

Primary productivity. The rate at which biomass is produced by organisms which synthesize complex organic substances from simple inorganic substrates, such as in photosynthesis and chemosynthesis.

Primary production. The biomass produced through photosynthesis and chemosynthesis in a community or group of communities.

Progradation. The outward building of a sedimentary deposit, such as the seaward advance of a delta or shoreline, or the outbuilding of an alluvial fan.

Province. An area of land, less extensive than a region, having a characteristic plant and animal population.

Purse seine. A large seine net designed to be set by two boats around a school of fish and so arranged that after the ends have been brought together, the bottom can be closed.

Pyroclastic. Formed by or involving fragmentation as a result of volcanic or igneous action.

Quaternary. The period of geological time, a sub-era of the Cenozoic, which covers the last 1.6 million years; comprises the Pleistocene ("Ice Age") and the Holocene epochs to the present and is noted for numerous major ice sheet advances in the northern hemisphere.

Radiation. In ecology, the spread of a group of organisms into new habitats.

Reagent. A compound involved in a chemical or biochemical reaction, especially one used in chemical analysis to produce a characteristic reaction in order to determine the presence of another compound.

Recovery plan. A plan that lists the actions that must be taken and the objectives that must be reached before an organism is no longer endangered or threatened and may be removed from the list of endangered and threatened species.

Recruitment. The influx of new members into a population by reproduction or immigration.

Reduction. A chemical reaction in which atoms or molecules either lose oxygen or gain hydrogen or electrons.

Refugium. An isolated area where extensive changes, typically due to changing climate (such as glaciation) but also due to large-scale disturbances such as those caused by humans, have not occurred and where plants and animals typical of a region may survive. Such a refuge is a center of *relict* forms from which dispersion and speciation may take place after environmental readjustment.

Regime. A regular pattern of occurrence or action.

Region. An area of the Earth having a distinctive plant or animal life.

Relict. Persistent remnants of a formerly widespread species in certain isolated areas.

Remigrant. An individual migrant that returns to its previous or former location.

Remote sensing. Methods for gathering data on a large or landscape scale which do not involve on-the-ground measurement, especially satellite photographs and aerial photographs; often used in conjunction with Geographic Information Systems.

Resource partitioning. Division of some resource or resources among two or more co-occurring species; for example, eating slightly different foods.

Revetment. A facing made of supporting material, such as masonry or concrete, used to support an embankment.

Rhyolite. An acidic, volcanic rock that is the lava form of granite.

Riparian. Relating to, living, or located on the bank of a natural watercourse (such as a river) or sometimes of a lake or a tidewater.

Riprap. A general term for large, blocky stones that are artificially placed to stabilize and prevent erosion along a riverbank or shoreline.

Rookery. A breeding or nesting place for some gregarious mammals and birds.

Rotifers. Any of a class (Rotifera of the phylum Aschelminthes) of minute, usually microscopic but many-celled, chiefly freshwater aquatic invertebrates having the anterior end modified into a retractile disk bearing circles of strong cilia that often give the appearance of rapidly revolving wheels. Rotifers mostly live in freshwater environments and eat a variety of bacteria and planktonic species.

Runoff. Precipitation on land that runs off to a body of water.

Salinity. A measure of the total concentration of dissolved salts in water. The salinity of ocean water is in the range 33–38 parts per thousand.

Salmonid. Any of a family of elongate bony fishes (such as salmon or trout) that have the last three vertebrae upturned.

Savanna. (Also spelled *savannah.*) A grassland–woodland mosaic vegetation type found in tropical and subtropical regions with long dry periods and receiving more rainfall than desert areas but not enough to support complete forest cover. A savanna is characterized by scattered trees or scattered clumps of trees and drought-resistant grasses. Fire often plays an important role in maintaining the vegetation.

Seamount. An underwater mountain (usually a submarine volcanic mountain peak) rising from the ocean floor whose summit is below the water's surface.

Secondary production. The biomass production resulting from the assimilation of organic matter produced by a primary consumer; production by organisms (mainly animals) which consume primary producers (mainly plants).

Secondary productivity. The rate of biomass production resulting from the assimilation of organic matter produced by a primary consumer; production by organisms (mainly animals) which consume primary producers (mainly plants).

Sediment. Materials that sink to the bottom of a body of water or materials that are deposited by wind, water, or glaciers.

Seleniferous. Referring to an ore containing selenium, or referring to a plant that absorbs selenium from the soil and concentrates the selenium within its tissues.

Senescence. The aging process in mature individuals; or, the period near the end of an organism's life cycle; in deciduous plants, the process that occurs before the shedding of leaves.

Seral. Relating to a phase in the sequential development of ecological communities formed in ecological succession in a particular habitat and leading to a particular climax association; intermediate communities in an ecological succession.

Serotinous. Late in developing or blooming.

Serotinous cones. Pine cones that remain on the tree for many years and are tightly closed until stimulated by the heat of a forest fire to open and release seeds.

Serpentine. A mineral rock consisting essentially of a hydrous magnesium silicate (chrysolite and antigorite) and usually having a dull green color and often a mottled appearance; or, the usually infertile, excessively well-drained soil derived from serpentine.

Sessile. Permanently attached to a substrate or established; not free to move about. Also, attached without a stalk.

Short-stopping. The process of creating habitat improvements to hold geese or ducks throughout the winter in an area that was historically used only as a migratory stopover point en route to wintering grounds farther south. Examples of short-stopping methods include measures that keep water bodies open (that is, unfrozen) and available to birds all winter, or providing food items on which birds can forage throughout the winter in areas where little or no food was formerly available.

Silviculture. The management of forests or woodlands for the production of timber and other wood products; growing trees as a crop.

Sink. A sinkhole; or, an area with a demand for metabolic substances. For example, growing meristems are sinks for energy compounds from photosynthesis, mitochondria are oxygen sinks, and tropical rainforests or deep oceans may act as carbon sinks, absorbing carbon dioxide from the atmosphere.

Sinkhole. A hollow place or depression in which water collects and goes underground, generally occurring in limestone regions and formed by solution or by collapse of a cavern roof.

Slough. A swamp, marsh, or muddy backwater.

Smolt. The stage in the life of salmon and similar fishes in which the subadult individuals acquire a silvery color and migrate down the river to begin their adult lives in the open sea.

Snag. A standing dead tree or stump that provides habitat for a broad range of wildlife, from beetle larvae (and the birds that feed upon them) to dens for raccoons. Or, a tree or branch embedded in a river or lake.

Solifluction. The slow creeping of fragmented material such as soil down a slope.

Spawn. The eggs of certain aquatic organisms; also, the act of producing such eggs or egg masses.

Species. A group of organisms formally recognized as distinct from other groups; the taxon rank in the hierarchy of biological classification below genus; the basic unit of biological classification, defined by the reproductive isolation of the group from all other groups of organisms.

Species diversity. See *diversity*.

Species richness. The absolute number of species in an assemblage or community.

Staging area. A traditional area, usually a lake, where birds that migrate in flocks rest and feed either immediately before or during migration. Many flocks may be gathered in such an area.

Standard error. In statistics, the standard deviation of the sampling distribution of a statistic; an estimate of the range by which the means of a number of sets of data deviate about the mean of those means.

Standing stock. Biomass; the total mass of organisms comprising all or part of a population or other specified group or within a given area; measured as volume, mass, or energy.

Steppe. Specifically, the temperate, semiarid areas of treeless grassland in the midlatitudes of Europe and Asia; more generally, any such grassland.

Stochastic. Random.

Subaerial. Occurring immediately above the surface of the ground.

Subalpine. The zone just below treeline on temperate mountains, usually dominated by a coniferous forest ecologically similar to boreal forest. The elevation of this zone increases with a decrease in latitude.

Sublittoral zone. The deeper zone of a lake below the limit of rooted vegetation; the marine zone extending from the lower margin of the intertidal (littoral) to the outer edge of the continental shelf at a depth of about 200 meters; sometimes used for the zone between low tide and the greatest depth to which photosynthetic plants can grow.

Submersed. Pertaining to a plant or plant structure growing entirely underwater.

Subnivian. Beneath the snow cover; specifically, the interface between snow and the surface of the ground where small mammals are active in winter.

Subsidence. The process of sinking or settling of a land surface or a crustal elevation because of natural or artificial causes.

Subspecies. A race of a species that is granted a taxonomic name; rules for designating subspecies are subjective, but subspecies are generally geographically distinct and form populations (not merely morphs) which differ to some degree from other geographic populations of the species.

Substrate. The surface or medium that serves as a base for something.

Subterranean. Under the surface of the Earth.

Subtidal. Applied to that portion of a tidal-flat environment which lies below the level of mean low water for spring tides. Normally it is covered by water at all states of the tide. Often used as a general descriptive term for a subaqueous but shallow marine depositional environment.

Subtropical. The latitudinal zone between 23.5° and 34.0° in either hemisphere, bordering the tropical zone. Also can refer to vegetation, organisms, or weather typical of subtropical habitats.

Succession. See *ecological succession*.

Succulent. A plant that has a specialized fleshy tissue in roots, stems, or leaves for the conservation of water. Most succulents are xerophytes, plants preferring dry climates, such as cactus or aloe, but some are halophytes, adapted for living in salty soils where water retention is a problem.

Suspended sediment. Sediment suspended in a fluid by the upward components of turbulent currents, moving ice, or wind.

Sustainability. Economic development that takes full account of the environmental consequences of economic activity and is based on the use of resources that can be replaced or renewed and therefore are not depleted.

Swale. A low tract of land, especially when moist or marshy.

Sympatric. Referring to populations, species, or taxa occurring together in the same geographical area; they may occupy the same habitat or different habitats within the same geographical area.

Synergistic. Of or pertaining to the cooperative action of two or more agencies such that the total is greater than the sum of the component actions; combined action or operation.

Syntopic. Relating to or displaying conditions as they exist simultaneously over a broad area, as of the atmosphere or weather.

Tailings. The fine-grained waste materials from an ore-processing operation.

Taxon (taxa). Any organism or group of organisms of the same taxonomic rank; for example, members of an order, family, genus, or species.

Tectonic movement. The formation of faults and folds on the crust of a planet.

Temporal niche. The functional position of an organism in its environment as determined by the periods of time during which it occurs and is active there.

Teratogen. A substance which interferes with the normal development of a fetus or embryo.

Tertiary. The first period of the Cenozoic Era which began about 65 million years and lasted to 1.6 million years before the present, marked by formation of high mountains, the dominance of mammals on land, and angiosperms superseding gymnosperms as dominant plants.

Thaliaceans. Any of a small order of tunicates consisting of various aberrant, free-swimming pelagic forms, including those of the genera *Salpa* and *Doliolum*.

Thermokarst. A landscape characterized by shallow pits and depressions caused by selective thawing of ground ice, or permafrost.

Topography. The natural and constructed relief of an area.

Transect. A line or narrow belt used in ecological surveys to provide a means of measuring and representing graphically the distributions of organisms across a given area.

Transpiration. The loss of water vapor from a plant to the outside atmosphere, mainly through the stomata of leaves and the lenticels of stems.

Treeline. The upper limits of tree growth in mountains or at high latitudes.

Triploid. A polyploid having three sets of homologous chromosomes.

Trophic. Pertaining to nutrition or to a position in a food web, food chain, or food pyramid.

Tropical. Referring to the zone between the Tropic of Cancer (23°27'N) and the Tropic of Capricorn (23°27'S); characterized by a climate with high temperatures, humidity, and rainfall. Also can refer to vegetation, organisms, or weather typical of tropical conditions.

Tundra. A level or rolling treeless plain in the arctic or subarctic regions; the soil is black and mucky, the subsoil is permanently frozen, and the vegetation is dominated by mosses, lichens, herbs, and dwarf shrubs. A similar environment occurs in mountainous areas above the timberline.

Tunicate. Any of a subphylum (Urochordata or Tunicata) of marine chordate animals that have a thick secreted covering layer, a greatly reduced nervous system, a heart able to reverse the direction of blood flow, and a notochord in the larval stage.

Turbid. Having sediment or foreign particles stirred up or suspended; muddy.

Tussock. A compact tuft of grass or sedges, or an area of raised solid ground, which is held together by roots of low vegetation, found in a wetland or tundra.

Ultisol. A member of a soil characterized by acidic, highly weathered layers with accumulations of silicate clays in subsurface layers; usually forms in tropical and subtropical climates.

Ultraviolet radiation. Electromagnetic radiation at wavelengths between 10 and 400 nanometers lying just beyond the high-energy (violet) end of the visible-light band of the solar spectrum; about 5% of the radiation the Earth receives from the sun, much of which is absorbed by the atmosphere. UV-A = 315 to 400 nanometers wavelength; UV-B = 280 to 315 nanometers; UV-C = 100 to 280 nanometers. UV-B is the part of the spectrum that causes sunburn and has been linked to skin cancer.

Umbellifer. A plant bearing flat-topped or rounded flower clusters in which the individual flower stalks arise from about the same point, such as the geranium, milkweed, onion, and chive.

Understory. The vegetation layer between the overstory or canopy and the groundcover of a forest community, usually formed by shade-tolerant species or young individuals of emergent species. May also refer to the groundcover if no tree or shrub layer is present.

Ungulate. Any four-footed, hoofed, grazing mammal (such as a ruminant, swine, camel, hippopotamus, horse, tapir, rhinoceros, elephant, or hyrax) that is adapted for running but is not necessarily related to other ungulates.

Upwelling. The upward movement of cold, nutrient-rich water from ocean depths, produced by wind or diverging currents.

Vascular plant. A plant with a specialized conducting system (for the transport of water and nutrients) that includes xylem and phloem; includes familiar higher plants such as trees, shrubs, and grasses.

Vegetative reproduction. (Also called *vegetative propagation*.) A reproductive process that is asexual and so does not involve a recombination of genetic material. It involves unspecialized plant parts which may become reproductive structures (such as roots, stems, or leaves). Compared with sexual reproduction, it represents a savings of material and energy for the plant. It is especially common among grasses.

Vertebrate. An animal with a backbone; includes mammals, birds, reptiles, amphibians, and fishes.

Volant. Flying or capable of flying.

Watershed. An area or a region that is bordered by a divide and from which water drains to a particular watercourse or body of water.

Wetland. A general term applied to land areas which are seasonally or permanently waterlogged, including lakes, rivers, estuaries, and freshwater marshes; an area of low-lying land submerged or inundated periodically by fresh or saline water.

Woodland. A vegetation community that includes widely spaced large trees. The tree crowns are typically more spreading in form than those of forest trees and do not form a closed canopy. Grass, heath, or scrub may develop between the trees.

Xeric. Dry; tolerating or adapted to dry conditions.

Year-class. Fish of a given species spawned or hatched in a given year; for example, a three-year-old fish caught in 1998 would be a member of the 1995 year-class.

Zoeae. The free-swimming, planktonic larval forms of many decapod crustaceans (especially crabs) that have a relatively large cephalothorax, conspicuous eyes, and fringed antennae and mouthparts.

Zooplankton. See *plankton*.

f = figure
t = table

western Pacific region, 830, 838–39
sea urchins, 207, 612, 785t, 808–9, 808t, 816
  purple, 817
  red, 816–17
seaweeds, 207
secondary production
  Alaska, 807
  defined, 907
  Pacific coast region, 816
  Southeast region, 793
  western Pacific oceanic region, 831–32
secondary productivity, defined, 907
second-growth forests, Northeast, 185
sedge-graminoid meadows, 725
sedges, 476, 477, 549, 710, 717, 729
  Alaska long-awn, 729
  Hawaii Island, 760f
  Hoppner, 725
  loose-flowered alpine, 725
  Lyngby, 725
  prickly, 760f
  Ramenski, 725
sediment-dwelling species, cancers in, 850–51
sediments
  aquatic plant loss and, 74
  coral reef effects, 323, 324
  defined, 907
  delta lobe subsidence rates and, 393
  grain size, coastal Louisiana, 403, 416
  habitat modification and, 75
  hurricane damage to wetlands, 21
  Lower Mississippi River, 396
  Mississippi River, 356
  Missouri River, 443
  Rio Grande, 562
  river channel effects, 74
  Southeast, 274
  suspended, 909
  water storage capacity and, 74
seine nets, purse, 906
seleniferous, defined, 907
selenium, 138, 143, 152
  poisoning, 143
Selway-Bitterroot Wilderness, 483
semidesert grasslands, Southwest, 555–59
sequoia, giant, 602
Sequoia National Park, 485
seral, defined, 907
serotinous, defined, 907
serotinous cones, defined, 907–8
serpentine, defined, 908
serpentine chaparral, 607–8
serpentine soils, California, 599, 600, 605
sessile, defined, 908
settlement patterns, 39, 40f
sewage
  eutrophication and, 50
  recovery from pollution caused by, 364
Seward Peninsula, 724, 726, 727
sex determination, temperature-dependent, 102
sex-selective harvest, 172
shadowscale scrub, 510
shads, 801
  American, 204–5, 620, 783, 821, 822
  Atlantic, 783t
  gizzard, 242, 369
shadscale zone, 510
sharks, 783t, 802
  Atlantic, 795t, 797–98
  blue, 818, 833
  cookie-cutter, 833
  Gulf of Mexico, 797–98
  oceanic whitetip, 833
  pelagic, 784, 797–98
  salmon, 833
  shortfin mako, 833
  silky, 833
  western Pacific, 833
shearwaters, 628t, 751, 786, 802, 804, 812, 820
  Audubon's, 802, 803, 840t
  Christmas, 840
  Cory's, 786, 803
  greater, 786, 803
  little, 802

Manx, 803
  Newell's Townsend's, 758, 840
  short-tailed, 813
  sooty, 786, 803, 813, 820
  wedge-tailed, 840f
sheefish, 711t, 718, 725, 726
sheep, 611, 740
  Audubon bighorn, 460
  Barbary, 580, 634
  bighorn, 265, 477, 495, 496f, 579, 635
  California bighorn, 525, 525t, 633t
  Dall, 723
  desert bighorn, 579
  mountain, 735
  peninsular bighorn, 633t
sheep grazing, *See also* grazing
  Southwest, 546
sheepshead, 419, 420t
Sheldon National Wildlife Refuge, 522
shellfish, *See also specific species*
  amnesiac shellfish poisoning, 851
  Pacific coast region, 816
  paralytic shellfish poisoning, 823, 851
shield-ferns, Aleutian, 729
shieldwort, arrowleaf, 384
shiners
  Arkansas River, 449
  emerald, 239, 242
  Lahontan, 662t
  phantom, 567
  Rio Grande, 567
  Rio Grande bluntnose, 567
  spottail, 242
  Topeka, 449
shipworm, 821
shooting stars, 514
shorebirds, 786
  Alaska, 722, 734, 740
  Southeast, 802–3
short-grass prairies, 438, 439t, 440
  leafhoppers, 448
  plant assemblages, 445
  Rocky Mountains, 476
short-stopping, 423
  defined, 908
Shoshoni, 478
shoulderbands, 617
  El Paso, 618
  mimic, 618
  Mojave, 618
  Morro, 618, 619t
  Panamint, 618
  San Miguel, 618
  Soledad, 618
  surf, 618
shovelers, 424t
  northern, 142, 425t, 456–57
shrew-mole, 683t
shrews, 248, 304, 579, 723, 728, 734
  Baird's, 683t
  fog, 683t
  Mt. Lyell, 633t
  ornate, 633t
  Pacific, 633t, 683t
  Pacific marsh, 683t
  Preble's, 525t
shrikes
  loggerhead, 189t, 523, 524, 678t, 679f
  San Clemente loggerhead, 626t
shrimp, 406, 420t, 762, 794, 808, 833, 844, 852
  brown, 357, 418, 419f, 420t, 794–95, 795t, 796
  by-catch, 418f, 419
  California freshwater, 619t
  estuaries and, 796
  fairy, 510
  freshwater, 331
  Hawaii, 832
  Mona Cave, 331
  mysid, 221
  nonindigenous, 849
  northern, 784, 785t
  opossum, 122
  Pacific, 846
  Pacific white, 849

Author Index

Alderman, John M., 307
Allen, Craig D., 552, 578, 582
Anderson, Betty A., 740
Anderson, Emory D., 856, 857
Armstrong, Robert H., 740
Austin, Jane E., 457

Ball, Lianne C., 534
Barnett-Lawrence, Matthew, 307
Belnap, Jayne, 559, 583
Bishop, Kathleen A., 534
Blaylock, Robert A., 805
Bogan, Michael A., 563, 580, 581, 582
Boulon, Ralf, 337
Braham, Howard W., 814, 815, 856
Britten, Hugh B., 534
Brodeur, Richard D., 856
Brussard, Peter F., 534
Bury, R. Bruce, 690

Carpenter, Stephen R., 56
Castellano, Michael A., 691
Chambers, Steven M., 618
Chapman, Duane C., 272
Charlet, David A., 534
Clugston, David A., 195, 200
Cole, Rebecca A., 613
Coleman, James M., 430
Collopy, Michael W., 690
Crawford, Dale, 72
Crawford, John A., 680
Creekmore, Lynn H., 613
Crisafulli, Charles M., 26
Cross, Stephen P., 691

Davis, Gary E., 615
Dobkin, David S., 534, 691
Droege, Sam, 188

Edsall, Thomas A., 250
Emmett, Robert L., 856
Euliss, Ned, 451
Evans, Michael, 337
Everson, Alan R., 856

Fagre, Daniel B., 107
Farley, Greg H., 583
Fetterolf, Carlos, 239
Finley, William L., 677
Fleishman, Erica, 534
Fleury, Scott A., 534
Forsman, Eric D., 673
Fremling, Calvin R., 378
Friend, Milton, 460

Garrison, Virginia Haney, 327
Gibbons, J. Whitfield, 307
Gibson, Thomas C., 307
Gilmer, David S., 637
Gould, Patrick J., 856
Graber, David M., 637
Graham, Tim, 637
Greenwood, Raymond J., 463
Gosselink, James G., 430
Guntenspergen, Glenn, 22
Gustafson, Eric J., 56

Hagar, Joan, 691
Hain, James H. W., 792, 856
Hansen, Larry J., 805
Hawkins, Charles P., 26
Henny, Charles J., 677
Henry, Mary G., 146
Herrmann, Raymond, 83
Hill, Jennifer A., 212
Hillis-Starr, Zandy-Marie, 337
Hines, William P., 812
Hooge, Philip N., 714
Houston, Douglas B., 688
Huckaby, Laurie S., 637

Igl, Lawrence, D., 455
Jameson, Ronald J., 686
Jameson, Stephen C., 856
Jenni, Tom, 534
Jennings, Mark R., 638
Johnson, Catherine M., 208
Johnson, Douglas H., 455
Jury, Steven H., 856

Kendall, Katherine C., 485, 486
Kennedy, Tom B., 534
Kenow, Kevin P., 378
Kirsch, Eileen M., 378
Klein, David R., 740
Knopf, Fritz L., 466
Korschgen, Carl E., 378
Krohn, William B., 208

Lannoo, Michael J., 451
Larson, Diane L., 451
Larson, Gary L., 660
Lattin, John D., 691
Lee, David S., 307
Li, Judith, 691
Liss, William J., 660
Longcore, Jerry R., 200
Loope, Lloyd L., 763, 770
Loope, Walter L., 229
Lovich, Jeff, 531
Low, Loh-Lee, 856, 857

Mac, Michael J., 148
MacCall, Alec D., 856
Marlow, Ron, 534
Marnell, Leo, 491
Martin, Karl J., 691
McAllister, Kelly R., 669
McAuley, Daniel, 195
McCain, Bruce B., 856, 857
McComb, William C., 691
McEachern, Kathryn, 229, 638
McIlwain, Thomas D., 856, 857
Meffe, Gary K., 121, 124, 127
Mehlhop, Patricia, 564, 583
Mertz, Tawna, 206
Meteyer, Carol U., 613
Miller, Jeffrey C., 691
Minello, Thomas J., 796
Molina, Randy, 691
Monaco, Mark E., 856
Moyle, Peter B., 638
Muldavin, Esteban H., 583
Mullen, Christine O., 534
Mullin, Keith D., 805
Mundy, Bruce C., 841, 843, 856, 857
Murray, David F., 740
Mushet, David M., 451

Naiman, Robert J., 56
Naughton, Maura B., 677
Nudds, Thomas D., 175

Olsen, James C., 856
Onuf, Chris P., 272
Opler, Paul A., 637
O'Reilly, John E., 856
Ostlie, Wayne R., 466

Papoulias, Diana M., 272
Peacock, Mary M., 534
Pearce, John B., 856
Pearson, Scott M., 56
Pelton, Michael R., 307
Penrose, David, 307
Perkins, J. Mark, 691
Pickett, T. A. Steward, 32
Platania, Steven P., 568, 583
Porter, William F., 212
Powell, Abby, 631
Prusso, Don, 534
Pyke, David A., 691